SURVEY SAMPLING THEORY AND APPLICATIONS

SURVEY SAMPLING THEORY AND APPLICATIONS

RAGHUNATH ARNAB

University of Botswana, Botswana
and
University of Kwazulu-Natal, South Africa

ELSEVIER

ACADEMIC PRESS
An imprint of Elsevier
elsevier.com

Academic Press is an imprint of Elsevier
125 London Wall, London EC2Y 5AS, United Kingdom
525 B Street, Suite 1800, San Diego, CA 92101-4495, United States
50 Hampshire Street, 5th Floor, Cambridge, MA 02139, United States
The Boulevard, Langford Lane, Kidlington, Oxford OX5 1GB, United Kingdom

Library of Congress Cataloging-in-Publication Data
A catalog record for this book is available from the Library of Congress

British Library Cataloguing-in-Publication Data
A catalogue record for this book is available from the British Library

ISBN: 978-0-12-811848-1

Publisher: Jonathan Simpson
Acquisition Editor: Glyn Jones
Editorial Project Manager: Ana Claudia A. Garcia
Production Project Manager: Poulouse Joseph
Designer: Mark Rogers

Dedicated to the memory of my brother in law

Late Sunil Kumar Biswas

CONTENTS

PREFACE

This proposed book provides a chronological development of survey sampling theory and applications from the basic level of concepts, theories, principles, and their practical applications to the very advanced level. The book covers a wide spectrum of topics on the subject. Some of the topics discussed here are not available in other text books. Theories are illustrated with appropriate theoretical and numerical examples for further clarity. This book will be useful for the graduate students and researchers in the field of survey sampling. It will also serve practitioners engaged in surveys because it contains almost every aspect of survey sampling.

DESCRIPTIONS OF CHAPTERS

The book comprises 26 chapters. The first 15 chapters are devoted to the basic concepts of survey sampling, which may be considered as a text for graduate students. Theories of each of the chapters are developed whenever possible in a unified setup—that can be generalized for wider classes of estimators and sampling designs. The remaining chapters 16–26 consist of advanced materials useful for researchers and practitioners engaged in the field of survey sampling.

Chapter 1 introduces terminologies and basic concepts such as sampling designs, inclusion probabilities, and sampling schemes. It also contains equivalency of sampling designs and sampling schemes—Hanurav's algorithm, sampling from finite and various types of infinite populations.

Chapter 2 is devoted to inferential problems for finite population sampling, e.g., various classes of unbiased estimators, uniformly minimum variance of unbiased estimation, nonexistence theorem, admissibility, sufficiency, and Rao-Blackwellization technique.

Chapters 3–5 comprise details of simple random sampling, and systematic and unequal probability sampling.

Chapter 6 introduces superpopulation model, model-based inference, and model/design-based (model-assisted) inferences; optimal sampling strategies for various superpopulation models, e.g., product-measure, equicorrelated, transformation, exchangeable, and random permutation models; and robustness of various sampling designs, Bayesian inferences and comparisons under various superpopulation models, and comparisons of various sampling strategies under superpopulation models.

Chapters 7–9 discuss stratified sampling, ratio method, regression, product, and calibrated method–based estimation in detail. The expressions of the bias and mean square error of the proposed estimators are derived under various sampling designs.

Chapter 10 deals with two-phase sampling where data collected in the first phase sample are used at the stages of estimation, selection of sample, and stratification, along with their combinations.

Chapter 11 provides repetitive sampling under various sampling schemes such as simple random, probability proportional to size with replacement, and Rao–Hartley–Cochran sampling schemes, which are not available in other text books.

Chapters 12 and 13 provide various aspects of cluster and multistage sampling designs such as general method of estimation of the population total, mean, and proportion and methods of estimation of their variances.

Chapter 14 presents unbiased estimation of mean square errors of homogeneous unbiased estimators based on various sampling designs and conditions of nonnegativity of the proposed mean-squared estimators.

Chapter 15 discusses various aspects of nonsampling errors and methods of controlling such errors, e.g., poststratification, use of response probabilities, various types of imputations, measurement errors, and interpenetrating subsamples.

Chapter 16 gives a comprehensive review of randomized response techniques for qualitative and quantitative characteristics and unified theory of estimation of population characteristics, e.g., mean and proportions. Methods of variance estimation are also discussed in detail. Various methods of optional randomized response techniques and measure of protection of privacy are also discussed. Optimal sampling strategies under various superpopulation models are also established.

Chapter 17 introduces methods of estimation of population characteristics for domain (larger areas) and small areas. Various methods of small area estimation have been proposed. This includes symptomatic accounting techniques and direct, synthetic, and composite methods. Methods of borrowing strength, use of various superpopulation models, empirical best linear unbiased prediction (EBLUP), empirical Bayes (EB), and hierarchical Bayes (HB) approach have also been explained.

Chapter 18 gives various methods of estimation of variance/mean square errors of estimators originated from complex survey design. Methods of linearization, jackknife, balanced repeated replication, and bootstrap methods are discussed for various sampling designs. The method of generalized variance functions is also included.

Chapters 19 and 20 describe various adjustments that are needed for the traditional chi-square test statistics for categorical data and regression analysis when data are obtained from complex survey designs.

Chapter 21 introduces the methods of ranked set sampling for estimating finite population characteristics based on SRSWR and SRSWOR, judgment ranking, ranking based on concomitant variables, moments of judgment order statistics, size-based probability of selections, etc.

Chapter 22 introduces concepts of estimating functions and estimating equations, optimal estimating function, estimating function for survey populations, and interval estimation, among others.

Chapter 23 gives different methods of estimating distribution function from finite population. The design-based, model-based, model-assisted, nonparametric regression method and calibration methods are also introduced. The methods of estimation of quantiles and medians are treated as a special case.

Chapter 24 gives various methods of controlled sampling such as experimental design and application of linear and nonlinear programming. The methods nearest proportional to size and coordination of samples over time are also discussed.

Chapter 25 introduces concepts of empirical likelihood in survey sampling. The concept of pseudo—empirical likelihood and model-calibrated pseudo—empirical likelihood and their applications are also introduced. Empirical likelihood methods for estimation of confidence interval are also given.

Chapter 26 comprises the different methods of collection of data for rare and mobile populations. The methods include methods of screening, disproportionate sampling, multiplicity or network sampling, multiframe sampling, snowball sampling, location sampling, adapted sampling, and capture—recapture methods.

Overall, the book addresses itself to a wide spectrum of survey sampling theory and applications. The book will be useful for graduate students, researchers, and practitioners in the field of survey sampling theory and applications.

ACKNOWLEDGMENTS

I wish to acknowledge my brother Mr. Biswarup Arnab and sisters Mrs. Gayatri Biswas and Mrs. Putul Ghosh for their help, encouragement, and support in building my academic carrier. I wish to thank my wife Mrs. Rita Arnab for her moral support. My gratitude also goes to my children Bubai, Buima, and Kintoshi for proofreading my work.

I would like to thank the colleagues of the Department of Statistics and the Faculty of the Social Sciences for their fruitful input, support, and inspiration to complete this book project. In addition, my sincere thanks to Dr. Glyn Jones, Mr. Poulouse Joseph and Ana Claudia A. Garcia of the production team of Elsevier publishing company for publishing the book.

CHAPTER 1

Preliminaries and Basics of Probability Sampling

1.1 INTRODUCTION

Various government organizations, researchers, sociologist, and businesses often conduct surveys to get answers to certain specific questions, which cannot be obtained merely through laboratory experiments or simply using economic, mathematical, or statistical formulation. For example, the knowledge of the proportion of unemployed people, those below poverty line, and the extent of child labor in a certain locality is very important for the formulation of a proper economic planning. To get the answers to such questions, we conduct surveys on sections of people of the locality very often. Surveys should be conducted in such a way that the results of the surveys can be interpreted objectively in terms of probability. Drawing inference about aggregate (population) on the basis of a sample, a part of the populations, is a natural instinct of human beings. Surveys should be conducted in such a way that the inference relating to the population should have some valid statistical background. To achieve valid statistical inferences, one needs to select samples using some suitable sampling procedure. The collected data should be analyzed appropriately. In this book, we have discussed various methods of sample selection procedures, data collection, and methods of data analysis and their applications under various circumstances. The statistical theories behind such procedures have also been studied in great detail.

In this chapter we introduce some of the basic definitions and terminologies in survey sampling such as population, unit, sample, sampling designs, and sampling schemes. Various methods of sample selection as well as Hanurav's algorithm which gives the correspondence between a sampling design and a sampling scheme have also been discussed.

1.2 DEFINITIONS AND TERMINOLOGIES

1.2.1 Population and Unit

A population is an aggregate or collection of elements or objects in a certain region at a particular point in time and is often a subject of study. Each

Survey Sampling Theory and Applications
ISBN 978-0-12-811848-1
http://dx.doi.org/10.1016/B978-0-12-811848-1.00001-7

element of the population is called a unit. Suppose we want to study the prevalence of HIV in the province of KwaZulu-Natal in 2016, the collection of all individuals, i.e., male or female and child or adult, residing in KwaZulu-Natal will be termed as population and each individual will be called a unit. Suppose we consider air pollution in a certain region. In this case, the air under consideration constitutes the population, but we cannot divide it into identifiable parts or elements. This type of population is called a continuous population.

1.2.2 Finite and Infinite Populations

A finite population is a collection of a finite number of identifiable units. The total number of elements will be denoted by N and refers to the size of the population. The students in a class, tigers in a game park, and households in a certain locality are examples of finite population as the units are identifiable and finite in number. Bacteria in a test tube, however, are identifiable, but they are very large in number. In this case $N \to \infty$, and hence it is considered an infinite population. The size of the population may be known or unknown before a survey. Sometimes, surveys are conducted to determine the unknown population size N, such as the total number of illegal immigrants or certain kinds of animals in a game park.

1.2.3 Sampling Frame

It is a list of all the units of a population with proper identification. The list is the basic material for conducting a survey. So, the sampling frame must be complete, up-to-date and free from duplication or omission of units. We denote a list of finite population or sampling frame as

$$U = (u_1, \ldots, u_i, \ldots, u_N)$$

where $u_i (i = 1, \ldots, N)$ is the ith unit of the population U. For simplicity we will denote the population U as

$$U = \{1, \ldots, i, \ldots, N\} \tag{1.2.1}$$

1.2.4 Parameter and Parameter Space

For a given population U, we may be interested in studying certain characteristics of it. Such characteristics are known as study variables. When considering a population of students in a certain class, we may be interested to know the age, height, racial group, economic condition, marks on different subjects, and so forth. Each of the variables under study is called a study variable, and it will be denoted by y. Let y_i be the value of a study variable y for the ith unit of the population U, which is generally not known before the survey. The N-dimension vector $\mathbf{y} = (y_1,\ldots, y_i,\ldots, y_N)$ is known as a parameter of the population U with respect to the characteristic y. The set of all-possible values of the vector $\mathbf{y}$ is the N-dimensional Euclidean space $R^N = (-\infty < y_1 < \infty,\ldots,-\infty < y_i < \infty,\ldots,-\infty < y_N < \infty)$ and it is known as a parameter space. In most of the cases we are not interested in knowing the parameter $\mathbf{y}$ but in a certain parametric function of $\mathbf{y}$ such as, $Y = \sum_{i=1}^{N} y_i =$ population total, $\overline{Y} = \frac{1}{N}\sum_{i=1}^{N} y_i =$ population mean, $S_y^2 = \frac{1}{N-1}\sum_{i=1}^{N}\left(y_i - \overline{Y}\right)^2 =$ population variance, $C_y = S_y/\overline{Y}$ $=$ population coefficient of variation, and so forth.

1.2.5 Complete Enumeration and Sample Survey

To know the value of a parameter or parametric function for a certain study variable y, we can follow two routes. The first route is to survey all the elements of the population and get all the values of y_i's, $i = 1,\ldots, N$. The second route is to select only a part of the population, which is termed as a sample. Then survey all the selected units in the sample and obtain the y-values from the selected units. From the y-values obtained in the sample, we predict (estimate) the population parameter under consideration. The first route is known as a complete enumeration or census, whereas the second route is called a sample survey.

1.2.6 Sampling and Nonsampling Errors

Obviously, using the complete enumeration method, we get the correct value of the parameter, provided all the y-values of the population obtained are correct. This would mean that there is no nonresponse, i.e., a response from each unit is obtained, and there is no measurement error in measuring y-values. However, in practice, at least for a large-scale survey, nonresponse is unavoidable, and y-values are also subject to error because the respondents report untrue values, especially when y-values relate to confidential characteristics such as income and age. The error in a survey, which is originated from nonresponse or incorrect measurement of y-values, is termed as the nonsampling error. The nonsampling errors increase with the sample size.

From a sample survey, we cannot get the true value of the parameter because we surveyed only a sample, which is just a part of the population. The error committed by making inference by surveying a part of the population is known as the sampling error. In complete enumeration, sampling error is absent, but it is subjected more to nonsampling error than sample surveys. When the population is large, complete enumeration is not possible as it is very expensive, time-consuming, and requires many trained investigators. The advantages of sample surveys over complete enumeration were advocated by Mahalanobis (1946), Cochran (1977), and Murthy (1977), to name a few.

1.2.7 Sample

A sample is an ordered sequence of elements from a population U, and it will be denoted by $s = (i_1, \ldots, i_j, \ldots, i_{n_s})$, where $i_j \in U$. The units in s need not be distinct and they may be repeated. The number of units in s, including repetition, is called the size of the sample s and will be denoted by n_s. The number of the distinct units in s is known as effective sample size and will be denoted by $\nu(s)$.

Example 1.2.1

Let $U = (1, 2, 3, 4)$ be a population of size 4, then $s = (1, 1, 2)$ is a sample of size $n_s = 3$ and effective sample size $\nu(s) = 2$.

1.2.8 Probability and Purposive Sampling

In probability sampling, a sample is selected according to a certain rule or method (known as sampling design), where each sample has a definite preassigned probability of selection. In purposive sampling or subjective sampling, the selection of sample is subjective; it totally depends on the choice of the sampler. Thus probability sampling reduces to purposive sampling when the probability of selection of a particular sample is assigned to 1.

1.3 SAMPLING DESIGN AND INCLUSION PROBABILITIES

1.3.1 Sampling Design

Let $\mathfrak{I}$ be the collection of all possible samples s. A sampling design p is a function defined on $\mathfrak{I}$, which satisfies the following conditions: (i) $p(s) \geq 0 \ \forall \ s \in \mathfrak{I}$ and (ii) $\sum_{s \in \mathfrak{I}} p(s) = 1$.

Example 1.3.1

Consider a finite population $U = (1, 2, 3, 4)$. Let $s_1 = (1, 1, 2)$, $s_2 = (1, 2, 2)$, $s_3 = (3, 2)$, and $s_4 = (4)$ be the possible samples and their respective probabilities are $p(s_1) = 0.25$, $p(s_2) = 0.30$, $p(s_3) = 0.20$, and $p(s_4) = 0.25$.

Here $\mathfrak{I} = (s_1, s_2, s_3, s_4)$ and p is a sampling design selecting the sample s_j with probability $p(s_j)$ for $j = 1, 2, 3, 4$.

1.3.2 Inclusion Probabilities

The inclusion probability of the unit i is the probability of inclusion of the unit i in any sample with respect to the sampling design p and will be denoted by π_i. Thus,

$$\pi_i = \Pr ob(i \in s) = \sum_{s \supset i} p(s) = \sum_{s \in \mathfrak{I}} I_{si} p(s)$$

where $I_{si} = 1$ if $i \in s$ and $I_{si} = 0$ if $i \notin s$ and $s \supset i$ denotes the sum over the samples containing the ith unit. Similarly, inclusion probability for the ith and jth unit $(i \neq j)$ is denoted by

$$\pi_{ij} = \sum_{s \supset i,j} p(s) = \sum_{s \in \mathfrak{I}} I_{si} I_{sj} p(s) = \pi_{ji}$$

The inclusion probabilities π_i and π_{ij} are called first- and second-order inclusion probabilities, respectively. The higher order inclusion probabilities are defined similarly. For the sake of convenience, we write $\pi_{ii} = \pi_i$.

1.3.3 Consistency Conditions of Inclusion Probabilities

The consistency conditions of the inclusion probabilities obtained by Godambe (1955) and Hanurav (1966) are given in the following theorem:

Theorem 1.3.1

(i) Godambe (1955): $\sum_{i=1}^{N} \pi_i = \nu$

(ii) Hanurav (1966): $\sum_{i \neq}^{N} \sum_{j=1}^{N} \pi_{ij} = V_p(\nu(s)) + \nu(\nu - 1)$

where $\nu = E_p(\nu(s))$ = expected effective sample size for the design $p = \sum_{s \in \mathfrak{I}} \nu(s) p(s)$, and $V_p(\cdot)$ is the variance with respect to the design p.

Proof

(i) $$\sum_{i=1}^{N} \pi_i = \sum_{i=1}^{N} \left\{ \sum_{s \in \mathfrak{I}} I_{si} p(s) \right\}$$

$$= \sum_{s \in \mathfrak{I}} p(s) \sum_{i=1}^{N} I_{si}$$

Now noting $\sum_{i=1}^{N} I_{si} = \nu(s)$ = number of distinct units in s, we find

$$\sum_{i=1}^{N} \pi_i = \sum_{s \in \mathfrak{I}} p(s) \nu(s)$$

$$= \nu$$

$$\text{(ii)} \quad \sum_{i\neq}^{N}\sum_{j=1}^{N}\pi_{ij} = \sum_{i=1}^{N}\left\{\sum_{j(\neq i)}^{N}\sum_{s\in\mathfrak{S}} I_{si}I_{sj}p(s)\right\}$$
$$= \sum_{s\in\mathfrak{S}} p(s)\sum_{i=1}^{N} I_{si}\sum_{j(\neq i)}^{N} I_{sj}$$
$$= \sum_{s\in\mathfrak{S}} p(s)\sum_{i=1}^{N} I_{si}(\nu(s) - I_{si})$$
$$= \sum_{s\in\mathfrak{S}} p(s)\left\{(\nu(s))^2 - \nu(s)\right\}$$
$$= E_p(\nu(s))^2 - E_p(\nu(s))$$
$$= V_p(\nu(s)) + \nu(\nu - 1)$$

1.3.4 Fixed Effective Size Design

The number of distinct units in a sample s is known as the effective sample size and is denoted by $\nu(s)$. A sampling design for which all the samples with positive probability have exactly n distinct units, i.e., $P(\nu(s) = n) = 1$ is known as a fixed effective size (n) sampling design (FESD(n)). The class of fixed effective size sampling designs will be denoted by $\mathcal{P}_n$.

1.3.5 Fixed Sample Size Design

A sampling design p is said to be a fixed sample size (FSS) design if $p\{n_s = n\} = 1$, i.e., sample size n_s is fixed as n for each of the samples s with $p(s) > 0, s\in\mathfrak{S}$.

Corollary 1.3.1 (Yates and Grundy, 1953)

For a fixed effective size (ν), sampling design p, $V_p(\nu(s)) = 0$ and in this case, Theorem 1.3.1 yields

$$\text{(i)} \quad \sum_{i=1}^{N}\pi_i = \nu \quad \text{and} \quad \text{(ii)} \quad \sum_{i\neq}^{N}\sum_{j}^{N}\pi_{ij} = \nu(\nu - 1) \tag{1.3.1}$$

Corollary 1.3.2

For a fixed effective size (ν) design

$$\sum_{j(\neq i)}^{N}\pi_{ij} = (\nu - 1)\pi_i \tag{1.3.2}$$

Proof

$$\sum_{j(\neq i)}^{N} \pi_{ij} = \sum_{j(\neq i)}^{N} \sum_{s\in \mathfrak{S}} I_{si} I_{sj} p(s) = \sum_{s\in \mathfrak{S}} I_{si} \sum_{j\neq i}^{N} I_{sj} p(s)$$
$$= \sum_{s\in \mathfrak{S}} I_{si}(\nu - I_{si}) p(s) = (\nu - 1)\pi_i$$

Example 1.3.2

Consider Example 1.3.1. Here the first-order inclusion probabilities for the units 1, 2, 3 and 4 are $\pi_1 = p(s_1) + p(s_2) = 0.55$, $\pi_2 = p(s_1) + p(s_2) + p(s_3) = 0.75$, $\pi_3 = p(s_3) = 0.20$, and $\pi_4 = p(s_4) = 0.25$, respectively. The second-order probabilities are $\pi_{12} = p(s_1) + p(s_2) = 0.55$, $\pi_{13} = \pi_{14} = 0$, $\pi_{23} = p(s_3) = 0.20$, and $\pi_{24} = \pi_{34} = 0$. The expectation and variance of the effective sample size are obtained as follows:

s	s_1	s_2	s_3	s_4	Total
$p(s)$	0.25	0.30	0.20	0.25	1.00
$\nu(s)$	2	2	2	1	
$\nu(s)p(s)$	0.50	0.60	0.40	0.25	1.75
$\nu(s)^2 p(s)$	1.0	1.20	0.80	0.25	3.25

(i) $E_p(\nu(s)) = \nu = 1.75$, (ii) $V_p(\nu(s)) = 3.25 - (1.75)^2 = 0.1875$. Here we can easily verify the consistency conditions $\sum_{i=1}^{4} \pi_i = 1.75 = \nu$ and $\sum_{i\neq}^{4} \sum_{j=1}^{4} \pi_{ij} = 1.50 = \nu(\nu - 1) + V_p(\nu(s))$.

1.4 METHODS OF SELECTION OF SAMPLE

We can use the following two methods of selection of sample.

1.4.1 Cumulative Total Method

Here we label all possible samples of $\mathfrak{S}$ as $s_1,\ldots, s_i,\ldots, s_M$, where M = total number of samples in $\mathfrak{S}$. Then we calculate the cumulative total $T_i = p(s_1) + \cdots + p(s_i)$ for $i = 1,\ldots, M$ and select a random sample R (say) from a uniform population with range (0, 1). This can be done by choosing a five-digit random number and placing a decimal preceding it. The sample s_k is selected if $T_{k-1} < R \leq T_k$, for $k = 1,\ldots, M$ with $T_0 = 0$.

Example 1.4.1

Let $U = (1, 2, 3, 4)$; $s_1 = (1, 1, 2)$, $s_2 = (1, 2, 2)$, $s_3 = (3, 2)$, $s_4 = (4)$; $p(s_1) = 0.25$, $p(s_2) = 0.30$, $p(s_3) = 0.20$, and $p(s_4) = 0.25$.

s	s_1	s_2	s_3	s_4
$p(s)$	0.25	0.30	0.20	0.25
T_k	0.25	0.55	0.75	1

Let a random sample $R = 0.34802$ be selected from a uniform population with range (0, 1). The sample s_2 is selected as $T_1 = 0.25 < R = 0.34802 \leq T_2 = 0.55$.

The cumulative total method mentioned above, however, cannot be used in practice because here we have to list all the possible samples having positive probabilities. For example, suppose we need to select a sample of size 15 from a population size $R = 30$ following a sampling design, where all possible samples of size $n = 15$ have positive probabilities, we need to list $M = \binom{30}{15}$ possible samples, which is obviously a huge number.

1.4.2 Sampling Scheme

In a sampling scheme, we select units one by one from the population by using a preassigned set of probabilities of selection of a unit in a particular draw. For a fixed sample of size n (FSS(n)) design, we select the ith unit at kth draw with probability $p_i(k)$ for $k = 1,\ldots, n$; $i = 1,\ldots, N$. $p_i(k)$'s are subject to

$$0 \leq p_i(k) \leq 1 \text{ and } \sum_{i=1}^{N} p_i(k) = 1; k = 1, \ldots, n, i = 1, \ldots, N \qquad (1.4.1)$$

There are various sampling schemes available in literature. We have given some of FSS designs, which are commonly used in practice below.

1.4.3 With and Without Replacement Sampling

In a with replacement (WR) sampling scheme, a unit may occur more than once in a sample with positive probability, whereas in a without replacement (WOR) sampling scheme, all the units of the sample are distinct, i.e., no unit is repeated in a sample with positive probability.

1.4.4 Simple Random Sampling With Replacement

In a simple random sampling WR (SRSWR) sampling scheme, $p_i(k) = 1/N$ for $k = 1, \ldots, n$. So, for an SRSWR, the probability of selection of a unit at any draw is the same and is equal to $1/N$. Hence the probability of selecting i_1 at the first draw, i_2 at the second draw, and i_n at the nth draw is

$$p(i_1, i_2, \ldots, i_n) = 1/N^n \ \text{ for } \ 1 \le i_1, i_2, \ldots, i_n \le N \tag{1.4.2}$$

1.4.5 Simple Random Sampling Without Replacement

In a simple random sampling WOR (SRSWOR),

$$p_i(k) = \begin{cases} \dfrac{1}{N-(k-1)} & \text{if } i\text{th unit is not selected in first } k-1 \text{ draws,} \\ & k = 1, \ldots, n \\ 0 & \text{if the } i\text{th unit is selected in first } k-1 \text{ draws} \end{cases} \tag{1.4.3}$$

So, under SRSWOR, the probabilities of selecting units i_1 at the first draw, $i_2 (i_2 \neq i_1)$ at the second draw, and $i_n (i_n \neq i_{n-1} \neq \cdots \neq i_1)$ at the nth draw are $1/N$, $1/(N-1)$, and $1/(N-n+1)$, respectively. So the probability of selection of such a sample $(i_1, i_2, \ldots, i_n)$ is

$$p(i_1, i_2, \ldots, i_n) = 1/\{N(N-1)\cdots(N-n+1)\} \ \text{ for } \ 1 \le i_1 \neq i_2 \neq \cdots \neq i_n \le N \tag{1.4.4}$$

1.4.6 Probability Proportional to Size With Replacement Sampling

For a probability proportional to size WR (PPSWR) sampling scheme, the probability of selecting the ith unit at any draw is $p_i \left(p_i \ge 0, \sum_{i=1}^{N} p_i = 1 \right)$, which is called the normed size measure for the ith unit. So for a PPSWR sampling scheme, $p_i(k) = p_i$ for $k = 1, \ldots, n$; $i = 1, \ldots, N$. Hence the probability of selecting i_1 at the first draw, i_2 at the second draw, and i_n at the nth draw under a PPSWR sampling scheme is

$$p(i_1, i_2, \ldots, i_n) = p_{i_1} p_{i_2} \ldots p_{i_n} \ \text{ for } \ 1 \le i_1, i_2, \ldots, i_n \le N \tag{1.4.5}$$

Clearly the PPSWR sampling scheme reduces to SRSWR sampling scheme if $p_i = 1/N$ for $i = 1,\ldots, N$.

1.4.7 Probability Proportional to Size Without Replacement Sampling

In probability proportional to size WOR (PPSWOR) sampling scheme, probability of selection of i_1 at the first draw is $p_{i_1}(1) = p_{i_1}$. Probability of selecting i_2 at the second draw is $p_{i_2}(2) = \dfrac{p_{i_2}}{1 - p_{i_1}}$ if the unit $i_1 (i_2 \neq i_1)$ is selected at the first draw and $p_{i_2}(2) = 0$ when the unit i_2 is selected at the first draw, i.e., $i_2 = i_1$. In general, the probability of selection of i_k at the kth draw is $p_{i_k}(k) = \dfrac{p_{i_k}}{1 - p_{i_1} - p_{i_2} - \cdots - p_{i_{k-1}}}$, if the units $i_1, i_2,\ldots, i_{k-1}$ are selected in any of the first $k - 1$ draws and $p_{i_k}(k) = 0$ if the unit i_k is selected in any of the first $k - 1$ draws for $k = 2,\ldots, n$; $i = 1,\ldots, N$. So, for a PPSWOR sampling scheme, the probability of selecting i_1 at the first draw, i_2 at the second draw, and i_n at the nth draw is

$$p(i_1, \ldots, i_n) = p_{i_1}\frac{p_{i_2}}{1 - p_{i_1}} \cdots \frac{p_{i_k}}{1 - p_{i_1} - \cdots - p_{i_{k-1}}} \cdots \frac{p_{i_n}}{1 - p_{i_1} - \cdots - p_{i_{n-1}}} \quad \text{for} \quad 1 \leq i_1 \neq i_2 \neq \cdots \neq i_n \leq N \tag{1.4.6}$$

It should be noted that PPSWOR reduces to SRSWOR sampling scheme if $p_i = 1/N$ for $i = 1,\ldots, N$.

1.4.8 Lahiri–Midzuno–Sen Sampling Scheme

In Lahiri (1951)–Midzuno (1952)–Sen (1953) (LMS) sampling scheme, at the first draw ith unit is selected with a normed size measure p_i, after which the remaining $n - 1$ units are selected by the SRSWOR method from units not selected at the first draw, i.e., where $p_{i_1}(1) = p_{i_1}$ and $p_{i_j}(k) = \dfrac{1}{(N-1)\cdots(N-k+1)}$ for $k = 2,\ldots, n$ if the unit i_j is not selected in earlier $k - 1$ draws, otherwise $p_{i_j}(k) = 0$. Thus the probability of selecting i_1 at the first draw, i_2 at the second draw, and i_n at the nth draw under the LMS sampling scheme is

$$p(i_1, i_2, \ldots, i_n) = p_{i_1} \cdot \frac{1}{N-1} \cdots \frac{1}{N-n+1} \quad \text{for } 1 \leq i_1 \neq i_2 \neq \cdots \neq i_n \leq N \tag{1.4.7}$$

The LMS sampling scheme reduces to SRSWOR sampling scheme if $p_i = 1/N$ for every $i = 1,\ldots, N$.

1.5 HANURAV'S ALGORITHM

Hanurav (1966) established a correspondence between a sampling design and a sampling scheme. He proved that any sampling scheme results in a sampling design. Similarly, for a given sampling design, one can construct at least one sampling scheme, which can implement the sampling design. In fact, Hanurav proposed the most general sampling scheme, known as Hanurav's algorithm, using which one can derive various types of sampling schemes or sampling designs. Henceforth, we will not differentiate between the terms "sampling design" and "sampling scheme".

Let n_0 denote the maximum sample size that might be required from a sampling scheme. Then, Hanurav's (1966) algorithm is defined as follows:

$$\mathcal{A} = \mathcal{A}\{q_1(i); q_2(s); q_3(s, i)\} \tag{1.5.1}$$

where

(i) $0 \leq q_1(i) \leq 1, \quad \sum_{i=1}^{N} q_1(i) = 1$ for $i = 1,\ldots, N$

(ii) $0 \leq q_2(s) \leq 1$ for any sample $s \in \mathfrak{J}$, where $\mathfrak{J}$ be the set of all possible samples.

(iii) $q_3(s, i)$ is defined when $q_2(s) > 0$ and subject to $0 \leq q_3(s, i) \leq 1$, $\sum_{i=1}^{N} q_3(s, i) = 1$ for $i = 1,\ldots, N$

Samples are selected using the following steps:

Step 1: At the first draw a unit i_1 is selected with probability $q_1(i_1)$; $i_1 = 1,\ldots, N$

Step 2: In this step, we decide whether the sampling procedure will be terminated or continued. Let $s_{(1)} = i_1$ be the unit selected in the first draw. A Bernoulli trial is performed with success probability $q_2(s_{(1)})$. If the trial results in a failure, the sampling procedure is terminated and the selected sample is $s_{(1)} = i_1$. On the other hand, if the trial results in a success, we go to step 3.

Step 3: In this step, a second unit i_2 is selected with probability $q_3(s_{(1)}, i_2)$ and we denote $s_{(2)} = (i_1, i_2)$. After selection of the sample $s_{(2)}$, we go back to step 2 and perform a Bernoulli trial with success probability $q_2(s_{(2)})$. If the trial results in a failure, then the sample procedure is terminated and the selected sample is $s_{(2)}$. Otherwise, another unit i_3 is selected with probability $q_3(s_{(2)}, i_3)$, and we denote $s_{(3)} = (i_1, i_2, i_3)$ as the selected sample. This procedure is continued until the sampling procedure is terminated. The sampling procedure is terminated with probability 1 after a selection of a sample of size n_0, as we assign $q_2\left(s_{(n_0)}\right) = 0$.

The probability of selection of a sample $s_{(n)} = (i_1,..., i_n)$ using the algorithm $\mathscr{A}$ is

$$p\left(s_{(n)}\right) = q_1\left(i_1\right)\cdot q_2\left(i_1\right)\cdot q_3\left(i_1, i_2\right)\cdot q_2\left(i_1, i_2\right)\cdot\cdots\cdot q_3\left(i_1, i_2 ..., i_n\right)\cdot \\ \left\{1 - q_2\left(i_1, i_2 ..., i_n\right)\right\}$$

Corollary 1.5.1

Hanurav's (1966) algorithm reduces to an FSS (n) sampling scheme, if

$$q_2\left(i_1\right) = q_2\left(i_1, i_2\right) = \cdots = q_2\left(i_1, i_2 ..., i_{n-1}\right) = 1 \text{ and } q_2\left(i_1, i_2 ..., i_n\right) \\ = 0 \text{ for } i_1, i_2, ..., i_n = 1, 2, ..., N.$$

The following examples show that (i) SRSWR, (ii) SRSWOR, (iii) PPSWR, (iv) PPSWOR, and (v) LMS sampling schemes are particular cases of Hanurav's algorithm.

Example 1.5.1

SRSWR of size n:

Here we choose (i) $q_1(i_1) = 1/N$, (ii) $q_2(i_1) = \cdots = q_2(i_1, i_2,..., i_{n-1}) = 1$ and $q_2(i_1, i_2,..., i_n) = 0$ for $i_1, i_2,..., i_n = 1, 2,..., N$, and (iii) $q_3(s, i) = 1/N$ for $i = 1,..., N$.

Example 1.5.2

SRSWOR of size n:

Here we choose (i) $q_1(i_1) = 1/N$, (ii) $q_2(i_1) = \cdots = q_2(i_1, i_2,..., i_{n-1}) = 1$ and $q_2(i_1, i_2,..., i_n) = 0$ for $i_1, i_2,..., i_n = 1, 2,..., N$, and (iii) $q_3(s, i) = 1/(N - k)$ for $i = 1,..., N$ if $s = (i_1,..., i_k)$ does not contain the unit i otherwise $q_3(s, i) = 0$.

Example 1.5.3

PPSWR of size n:

Here we choose (i) $q_1(i_1) = p_{i_1}$, (ii) $q_2(i_1) = \cdots = q_2(i_1, i_2,..., i_{n-1}) = 1$ and $q_2(i_1, i_2,..., i_n) = 0$ for $i_1, i_2,..., i_n = 1, 2,..., N$, and (iii) $q_3(s, i) = p_i$ for $i = 1,..., N$.

Example 1.5.4

PPSWOR of size n:

Here we choose (i) $q_1(i_1) = p_{i_1}$, (ii) $q_2(i_1) = \cdots = q_2(i_1, i_2, \ldots, i_{n-1}) = 1$ and $q_2(i_1, i_2, \ldots, i_n) = 0$ for $i_1, i_2, \ldots, i_n = 1, 2, \ldots, N$, and (iii) $q_3(s, i) = p_i/(1 - p_{i_1} - p_{i_2} - \cdots - p_{i_{k-1}})$ for $i = 1, \ldots, N$, if $s = (i_1, \ldots, i_{k-1})$ does not contain the unit i and $i_1 \neq \cdots \neq i_{k-1}$; $q_3(s, i) = 0$ if s contains i.

Example 1.5.5

LMS of size n:

Here we choose (i) $q_1(i_1) = p_{i_1}$, (ii) $q_2(i_1) = \cdots = q_2(i_1, i_2, \ldots, i_{n-1}) = 1$ and $q_2(i_1, i_2, \ldots, i_n) = 0$ for $i_1, i_2, \ldots, i_n = 1, 2, \ldots, N$, and (iii) $q_3(s, i) = 1/(N - k)$ for $i = 1, \ldots, N$, if $s = (i_1, \ldots, i_{k-1})$ does not contain the unit i and $i_1 \neq \cdots \neq i_{k-1}$; $q_3(s, i) = 0$, if s contains i.

A correspondence between a sampling design and a sampling scheme is given in the following theorem:

Theorem 1.5.1

(i) Sampling according to the algorithm $\mathscr{A} = \mathscr{A}\{q_1(i); q_2(s); q_3(s, i)\}$ results in a sampling design.

(ii) For a given sampling design p, there exists an algorithm $\mathscr{A}$, which results in the design p.

Proof

(i) Here we have to show $\sum_{s \in \mathfrak{J}} p(s) = 1$.

Let $\mathfrak{J}_k$ = collection of all samples whose size is k. Then,

$$\mathfrak{J} = \bigcup_{k=1}^{n_0} \mathfrak{J}_k \text{ and } \sum_{s \in \mathfrak{J}} p(s) = \sum_{k=1}^{n_0} \sum_{s \in \mathfrak{J}_k} p(s). \tag{1.5.2}$$

Now,

$$\sum_{s \in \mathfrak{J}_1} p(s) = \sum_{i_1=1}^{N} p(i_1) = \sum_{i_1=1}^{N} q_1(i_1)\{1 - q_2(i_1)\} = 1 - \sum_{i_1=1}^{N} q_1(i_1) q_2(i_1) \tag{1.5.3}$$

$$\begin{aligned}\sum_{s \in \mathfrak{J}_2} p(s) &= \sum_{i_1=1}^{N} \sum_{i_2=1}^{N} p(i_1, i_2) = \sum_{i_1=1}^{N} \sum_{i_2=1}^{N} q_1(i_1) q_2(i_1) q_3(i_1, i_2)\{1 - q_2(i_1, i_2)\} \\ &= \sum_{i_1=1}^{N} q_1(i_1) q_2(i_1) - \sum_{i_1=1}^{N} \sum_{i_2=1}^{N} q_1(i_1) q_2(i_1) q_3(i_1, i_2) q_2(i_1, i_2)\end{aligned} \tag{1.5.4}$$

$$\sum_{s\in \mathfrak{S}_{n_0-1}} p(s)=\sum_{i_1=1}^{N}\sum_{i_2=1}^{N}\cdots\sum_{i_{n_0-1}=1}^{N} p(i_1,i_2,\ldots,i_{n_0-1})$$
$$=\sum_{i_1=1}^{N}\sum_{i_2=1}^{N}\cdots\sum_{i_{n_0-1}=1}^{N} q_1(i_1)q_2(i_1)q_3(i_1,i_2)\cdots q_2(i_1,i_2,\ldots,i_{n_0-2})$$
$$q_3(i_1,i_2,\ldots,i_{n_0-1})\{1-q_2(i_1,i_2,\ldots,i_{n_0-1})\}$$
$$=\sum_{i_1=1}^{N}\sum_{i_2=1}^{N}\cdots\sum_{i_{n_0-2}=1}^{N} q_1(i_1)q_2(i_1)q_3(i_1,i_2)\cdots q_2(i_1,i_2,\ldots,i_{n_0-2})$$
$$-\sum_{i_1=1}^{N}\sum_{i_2=1}^{N}\cdots\sum_{i_{n_0-1}=1}^{N} q_1(i_1)q_2(i_1)q_3(i_1,i_2)\cdots q_2(i_1,i_2,\ldots,i_{n_0-2})$$
$$q_3(i_1,i_2,\ldots,i_{n_0-1})q_2(i_1,i_2,\ldots,i_{n_0-1}) \tag{1.5.5}$$

$$\sum_{s\in \mathfrak{S}_{n_0}} p(s)=\sum_{i_1=1}^{N}\sum_{i_2=1}^{N}\cdots\sum_{i_{n_0}=1}^{N} p(i_1,i_2,\ldots,i_{n_0})$$
$$=\sum_{i_1=1}^{N}\sum_{i_2=1}^{N}\cdots\sum_{i_{n_0}=1}^{N} q_1(i_1)q_2(i_1)q_3(i_1,i_2)\cdots q_2(i_1,i_2,\ldots,i_{n_0-1})$$
$$q_3(i_1,i_2,\ldots,i_{n_0})\{1-q_2(i_1,i_2,\ldots,i_{n_0})\}$$
$$=\sum_{i_1=1}^{N}\sum_{i_2=1}^{N}\cdots\sum_{i_{n_0}=1}^{N} q_1(i_1)q_2(i_1)q_3(i_1,i_2)\cdots q_2(i_1,i_2,\ldots,i_{n_0-1}) \tag{1.5.6}$$

(noting $q_2(i_1,i_2,..,i_{n_0})=0$ since n_0 is the maximum sample size)

Finally, from (1.5.2) to (1.5.6), we get

$$\sum_{s\in\mathfrak{S}} p(s)=\sum_{k=1}^{n_0}\sum_{s\in\mathfrak{S}_k} p(s)=1$$

(ii) Here we have given a sampling design p where $\mathfrak{S}$ = all possible samples and $p(s)$ is the probability of selection of a sample s $(\in\mathfrak{S})$. Here we are to find q_1, q_2, and q_3 so that sampling according to the algorithm $\mathcal{A}(q_1,q_2,q_3)$ implements the design p.

Let $\mathfrak{S}_{(i)}=\{s|i_1=i\}$ = collection of samples, whose first element is i; $\mathfrak{S}_{(i,j)}=\{s|i_1=i,i_2=j\}$ = collection of samples, whose first element is i and the second element is j; $\mathfrak{S}_{(j_1,\ldots,j_k)}$'s are similarly defined.

Let $\beta(i_1, i_2, \ldots, i_n)$ = probability of selection of the sample $(i_1, i_2, \ldots, i_n)$ $= p(i_1, i_2, \ldots, i_n)$, where the unit i_1 is selected at the first draw, i_2 at the second draw, and i_n at the nth draw.

$\alpha(j_1) = \sum_{s \in \mathfrak{S}_{(j_1)}} p(s)$, $\alpha(j_1, \ldots, j_k) = \sum_{s \in \mathfrak{S}_{(j_1, \ldots, j_k)}} p(s)$ are defined similarly.

Here we check $\mathfrak{S} = \bigcup_{i=1}^{N} \mathfrak{S}_{(i)}$, $\mathfrak{S}_{(i)} = \bigcup_{j=1}^{N} \mathfrak{S}_{(i,j)} \cup (i)$, $\sum_{i=1}^{N} \alpha(i) = 1$, $\sum_{j=1}^{N} \alpha(i,j) + \beta(i) = \alpha(i)$, etc. Now following Hanurav (1966), we define

$$q_1(i_1) = \alpha(i_1)$$

$$q_2(i_1, i_2, \ldots, i_k) = \begin{cases} 1 - \dfrac{\beta(i_1, i_2, \ldots, i_k)}{\alpha(i_1, i_2, \ldots, i_k)} \\ \quad \text{if} \quad \alpha(i_1, i_2, \ldots, i_k) > 0 \\ \quad 0 \quad \text{otherwise} \end{cases}$$

$$q_3\{(i_1, i_2, \ldots, i_k), i_{k+1}\} = \begin{cases} \dfrac{\alpha(i_1, i_2, \ldots, i_{k+1})}{\alpha(i_1, i_2, \ldots, i_k) - \beta(i_1, i_2, \ldots, i_k)} \\ \quad \text{if} \quad q_2(i_1, i_2, \ldots, i_k) > 0 \\ \quad 0 \quad \text{otherwise} \end{cases}$$

So, the probability of drawing a sample $(i_1, i_2, \ldots, i_n)$ by the algorithm $\mathscr{A}$ is

$$\begin{aligned} p(i_1, i_2, \ldots, i_n) &= q_1(i_1) \times q_2(i_1) \times q_3(i_1, i_2) \times q_2(i_1, i_2) \times \cdots \times q_3(i_1, i_2, \ldots, i_n) \\ &\quad \times \{1 - q_2(i_1, i_2, \ldots, i_n)\} \\ &= \alpha(i_1) \times \left(1 - \frac{\beta(i_1)}{\alpha(i_1)}\right) \times \cdots \times \frac{\alpha(i_1, i_2, \ldots, i_n)}{\alpha(i_1, i_2, \ldots, i_{n-1}) - \beta(i_1, i_2, \ldots, i_{n-1})} \\ &\quad \times \frac{\beta(i_1, i_2, \ldots, i_n)}{\alpha(i_1, i_2, \ldots, i_n)} = \beta(i_1, i_2, \ldots, i_n) \end{aligned}$$

Example 1.5.6

Let us consider the sampling design where the population $U = \{1, 2, 3\}$; $\mathfrak{S}$ consists of the samples $s_1 = (1, 1)$, $s_2 = (3)$, and $s_3 = (2, 3)$ with respective probabilities $p(s_1) = 0.2$ and $p(s_2) = p(s_3) = 0.4$.

Here, $n_0 = 2 =$ maximum sample size. $\mathfrak{I}_1 = \{s_1\}\mathfrak{I}_2 = \{s_3\}$, $\mathfrak{I}_3 = \{s_2\}$; $\mathfrak{I}_{11} = \{s_1\}$ and $\mathfrak{I}_{1j} = \phi =$ (null set) for j = 2, 3; $\mathfrak{I}_{23} = \{s_3\}$, $\mathfrak{I}_{2j} = \phi$ for $j = 1, 2$ and $\mathfrak{I}_{3j} = \phi$ for $j = 1, 2, 3$; $\beta(3) = p(s_2) = 0.4$; $\beta(j) = 0$ for $j = 1, 2$; $\beta(1, 1) = p(s_1) = 0.2$, $\beta(2, 3) = p(s_3) = 0.4$ and other values of $\beta(i, j)$'s are equal to zero.

$\alpha(1) = p(\mathfrak{I}_1) = 0.2$, $\alpha(2) = p(\mathfrak{I}_2) = 0.4$, $\alpha(3) = p(\mathfrak{I}_3) = 0.4$; $\alpha(1, 1) = p(s_1) = 0.2$, $\alpha(1, j) = p(\phi) = 0$ for j = 2, 3; $\alpha(2, 3) = p\{s_3\} = 0.4$ and $\alpha(2, j) = p(\phi) = 0$ for $j = 1, 2$; $\alpha(3, j) = p(\phi) = 0$ for $j = 1, 2, 3$.

$q_1(1) = \alpha(1) = 0.2$, $q_1(2) = \alpha(2) = 0.4$, $q_1(3) = \alpha(3) = 0.4$;

$q_2(3) = 1 - \beta(3)/\alpha(3) = 1 - 1 = 0$, $q_2(j) = 1 - \beta(j)/\alpha(j) = 1$ for $j = 1, 2$, $q_2(1, 1) = 1 - \beta(1, 1)/\alpha(1, 1) = 0$, $q_2(1, j) = 0$ for $j = 2, 3$ since in this case $\alpha(1, j) = 0$; $q_2(2, 3) = 1 - \beta(2, 3)/\alpha(2, 3) = 0$ and $q_2(2, j) = 1 - \beta(2, j)/\alpha(2, j) = 0$ for $j \neq 3$ as $\alpha(2, j) = 0$; $q_2(3, j) = 1 - \beta(3, j)/\alpha(3, j) = 0$ as $\alpha(3, j) = 0$ for $j = 1, 2, 3$;

$q_3(1, 1) = \alpha(1, 1)/\{\alpha(1) - \beta(1)\} = 1$, $q_3(1, j) = 0$ for $j = 2, 3$ since $\alpha(1, j) = 0$; $q_3(2, 3) = \alpha(2, 3)/\{\alpha(2) - \beta(2)\} = 1$, $q_3(2, j) = 0$ for $j \neq 3$ as $\alpha(2, j) = 0$ for $j \neq 3$ and $q_3(3, j) = 0$ as $\alpha(3, j) = 0$ for j = 1, 2, 3.

Now using $q_1(\cdot)$, $q_2(\cdot)$, and $q_3(\cdot)$ we can check,

$$\begin{aligned} p(s_1) = p\{(1, 1)\} &= q_1(1) \times q_2(1) \times q_3(1, 1) \times \{1 - q_2(1, 1)\} \\ &= 0.2 \times 1 \times 1 \times (1 - 0) = 0.2 \end{aligned}$$

$$\begin{aligned} p(s_2) = p\{(3)\} &= q_1(3) \times \{1 - q_2(3)\} \\ &= 0.4 \times (1 - 0) = 0.4 \end{aligned}$$

$$\begin{aligned} p(s_3) = p\{(2, 3)\} &= q_1(2) \times q_2(2) \times q_3(2, 3) \times \{1 - q_2(2, 3)\} \\ &= 0.4 \times 1 \times 1 \times (1 - 0) = 0.4 \end{aligned}$$

1.6 ORDERED AND UNORDERED SAMPLE

A sample is said to be an ordered sample if it retains information about which draw selects which unit. So, from an ordered sample, we know the number of times a particular unit is selected in a sample and also in which draw it was selected. If we pick up the set of distinct units from an ordered sample and arrange them in ascending order of their labels, then the resulting sample is known as an unordered sample. Thus an unordered sample does not retain information about which draw a particular unit was selected and its multiplicity. Let $s = (i_1, \ldots, i_k, \ldots, i_{n_s})$ be an ordered sample of size n_s, where the unit i_k is selected at the kth draw. Let $\widetilde{s} = (j_1, \ldots, j_k, \ldots, j_{\nu(s)})$ be the set of distinct units in s of size $\nu(s)$ with $j_1 < \cdots < j_l < \cdots < j_{\nu(s)}$, then $\widetilde{s}$ is an unordered sample obtained from s.

Example 1.6.1

Suppose from a population $U = (1, 2, 3, 4, 5)$, a sample of three units is selected as follows; On the first draw the unit 5, second draw unit 2, and at the third draw the unit 5 is selected. Then the sample $s = (5, 2, 5)$ is an ordered sample as we know, from the sample, that the unit 5 is selected twice, once in the first draw and again in the third draw whereas the unit 2 is selected in the second draw. Now, selecting the distinct units of the sample s and arranging in ascending order, we get an unordered sample $\tilde{s} = (2, 5)$.

1.7 DATA

After selection of a sample s, we collect information on one or more characters of interest from the selected units in the sample s. Consider the simplest situation where a single character y is of interest, and y_i is the value of the character obtained from the ith unit. The information related to the units selected in a sample and their y-values obtained from the survey are known as data and will be denoted by d. Thus data corresponding to an ordered sample $s = (i_1,\dots, i_k,\dots, i_{n_s})$ will be denoted by

$$\begin{aligned} d(s) &= \left((i_1, y_{i_1}), \dots, (i_k, y_{i_k}), \dots, \left(i_{n_s}, y_{i_{n_s}}\right)\right) \\ &= ((j, y_j), j \in s). \end{aligned}$$

The data $d(s)$ based on the ordered sample are known as ordered data.

Similarly, the data obtained from the unordered sample $\tilde{s} = \left(j_1, \dots, j_l, \dots, j_{\nu_s}\right)$ are known as unordered data and are denoted by

$$\begin{aligned} d(\tilde{s}) &= \left(\left(j_1, y_{j_1}\right), \dots, \left(j_k, y_{j_k}\right), \dots, \left(j_{\nu_s}, y_{j_{\nu(s)}}\right)\right) \\ &= \left((j, y_j), j \in \tilde{s}\right). \end{aligned}$$

If the label part of the data, i.e., information of the selected units in the sample, is deleted from the data, then the resulting data are called unlabeled data. Thus unlabeled data from an ordered and unordered sample may be denoted by $\left(y_{i_1}, \dots, y_{i_k}, \dots, y_{i_{n_s}}\right)$ and $\left(y_{j_1}, \dots, y_{j_k}, \dots, y_{j_{\nu(s)}}\right)$, respectively.

1.7.1 Sample Space

The sample space corresponding to a sampling design p is defined as the collection of all possible data that could be obtained from the design. The sample space corresponding to an ordered and unordered design will be denoted by $\mathcal{X}_o = (d(s), s \in \mathfrak{S})$ and $\mathcal{X} = (d(\tilde{s}), \tilde{s} \in \mathfrak{S})$, respectively.

1.8 SAMPLING FROM HYPOTHETICAL POPULATIONS

Let X be a random variable with a distribution function $F(x) = P(X \leq x)$. To draw a sample from this population, we use the property that $F(x)$ follows uniform distribution over (0, 1). Let R be a random sample from a uniform distribution. Then $x = F^{-1}(R)$ is a random sample from a population, whose distribution function is $F(x)$.

1.8.1 Sampling From a Uniform Population

Here we select a five-digit random number (selection of more digits gives better accuracy) from a random number table and then place a decimal point preceding the digits. The resulting number is a sample from the uniform distribution over (0, 1). For example, if the selected five-digit random number is 56342 the selected sample from a uniform population (0, 1) is $R = 0.56342$.

1.8.2 Sampling From a Normal Population

Suppose we want to select a random sample from a normal population with mean $\mu = 50$ and variance $\sigma^2 = 25$. We first select a five-digit random number 89743 and put a decimal place preceding it. The resulting number $R = 0.89743$ is a random sample from a uniform distribution (0, 1). A random sample x from a normal population $N(\mu, \sigma)$ with mean $\mu(=50)$ and variance $\sigma^2(=25)$ is obtained from the equation,

$$P(X \leq x) = 0.89743 \text{ i.e., } P\left(\frac{X-\mu}{\sigma} = z \leq \frac{x-50}{5}\right) = 0.89743.$$

Noting that $\frac{X-\mu}{\sigma} = z$ is a standard normal variable, we find from the normal deviate table, $\frac{x-50}{5} \cong 1.27$. Hence $x = 56.35$ is a random sample from $N(50, 5)$.

1.8.3 Sampling From a Binomial Population

Suppose we want to select a sample from a binomial population with $n = 5$ and $p = 0.342$. Let X be a Bernoulli variable with a success probability $p = 0.342$. Then, $P\{X = 1\} = p$ and $P\{X = 0\} = 1 - p$. We first select five independent random samples $R_1 = 0.302$, $R_2 = 0.987$, $R_3 = 0.098$, $R_4 = 0.352$, and $R_5 = 0.004$ from a uniform distribution over (0, 1) using Section 1.8.1. From the random samples R_i, select a random sample from Bernoulli population X_i, which is equal to 1 (success) if $R_i \leq p(=0.342)$

and $X_i = 0$ (failure) if $R_i > p$. Then $Y = X_1 + X_2 + X_3 + X_4 + X_5 = 1 + 0 + 1 + 0 + 1 = 3$ is a random sample from the Binomial population with $n = 5$ and $p = 0.342$.

1.9 EXERCISES

1.9.1 Define the following terms giving suitable examples: (i) population, (ii) sampling frame, (iii) sample, (iv) sampling scheme, (v) sampling design, and (vi) effective sample size.

1.9.2 **(a)** Define inclusion probabilities of the first two orders. Compute inclusion probabilities of the first two orders of the following sampling designs: (i) SRSWR, (ii) SRSWOR, and (iii) PPSWR.

(b) Find (i) expectation and (ii) variance of the number of distinct units in a sample of size 5, selected from a population of size 10, by the SRSWR method.

1.9.3 Let the expected effective sample size of a sampling design be $\nu = E(\nu(s)) = [\nu] + \theta$, where $[\nu]$ is the integer part of ν. Then show that

(i) $\theta(1-\theta) \leq Var(\nu(s)) \leq (N-\nu)(\nu-1)$ and (ii) $\nu(\nu-1) + \theta(1-\theta) \leq \sum_{i \neq} \sum_{j} \pi_{ij} \leq N(\nu-1)$ (Hanurav, 1966).

1.9.4 **(a)** Let π_i^* and π_{ij}^* be the exclusion (noninclusion) probabilities for the ith, and ith and jth $(i \neq j)$ units, then show that $\pi_i^* \pi_j^* - \pi_{ij}^* = \pi_i \pi_j - \pi_{ij}$ (Lanke, 1975a,b).

(b) Show that the first two order exclusion probabilities of units in SRSWOR sampling of size n selected from a population of size N are $(N-n)/N$ and $(N-n)(N-n-1)/\{N(N-1)\}$, respectively.

1.9.5 Let π_{ijk} be the inclusion probability of the unit i, j and $k (i \neq j \neq k)$ for a fixed effective size ν design. Show that (i) $\sum_{k(\neq i,j)} \pi_{ijk} = (\nu-2)\pi_{ij}$ and (ii) $\sum_{i \neq} \sum_{j \neq} \sum_{k} \pi_{ijk} = \nu(\nu-1)(\nu-2)$

1.9.6 Consider the sampling design where $U = (1, 2, 3, 4)$ and

Sample (s)	(1, 2, 1)	(1, 4, 3)	(3)	(4)	(3, 2, 1)	(2, 2)
Probability $p(s)$	0.2	0.1	0.4	0.1	0.1	0.1

(a) Calculate (i) inclusion probabilities of first two orders, (ii) $E(\nu(s))$, and (iii) $Var(\nu(s))$

(b) Select a sample using the cumulative total method

1.9.7 Use Hanurav's algorithm to select a sample using the following sampling designs:

(a) $U = (1, 2, 3, 4, 5, 6)$

Sample (s)	(1, 4)	(2, 5)	(3, 6)
Probability $p(s)$	1/3	1/3	1/3

(b) $U = (1, 2, 3, 4)$

Sample (s)	(1, 2)	(1, 3)	(1, 4)	(2, 3)	(2, 4)	(3, 4)
Probability $p(s)$	1/6	1/6	1/6	1/6	1/6	1/6

(c) $U = (1, 2, 3, 4, 5)$

Sample (s)	(1, 2, 2)	(1, 3, 4)	(4)	(1, 5, 3)	(1, 1, 4)	(3, 4)
Probability $p(s)$	0.20	0.10	0.25	0.25	0.10	0.10

1.9.8 Using a random number table, select a sample of size 5 from the following populations:

(i) Uniform distribution over (0, 1)
(ii) Uniform distribution over (10, 100)
(iii) Bernoulli population with parameter $p = 0.1234$.
(iv) Binomial distribution with parameters $n = 8$ and $p = 0.673$.
(v) Hypergeometric distribution with $N_1 = 10$, $N_2 = 15$, and $n = 8$.
(vi) Poisson distribution with parameter $\lambda = 4$.
(vii) Normal population with mean $\mu = 50$ and standard deviation $\sigma = 5$.
(viii) Chi-square distribution with degrees of freedom 10.
(ix) Bivariate normal distribution with parameters $\mu_1 = 50$, $\mu_2 = 60, \sigma_1^2 = 16, \sigma_2^2 = 25$, and $\rho = 0.8$.
(x) Cauchy population $f(x|\theta) = 1/[\pi\{1 + (x - \theta)^2\}]$; $\theta = 5$, $-\infty < x < \infty$

1.9.9 The following table gives a list of households in 10 localities.

Localities	1	2	3	4	5	6	7	8	9	10
Number of households	5	15	20	10	15	25	15	10	10	15

Select 15 households at random by (i) SRSWR and (ii) SRSWOR methods.

1.9.10 Select five points at random in a (i) circle of radius 5 cm, and (ii) square of sides 5 cm.

1.9.11 The following table gives the number of students in different sections and grades. Select a sample of size 5 by (i) SRSWR and (ii) SRSWOR methods.

Grades	Section 1	Section 2	Section 3	Section 4
A	10		5	5
B	12	30	10	15
C	40	25	20	
D	5	25	30	20
E		20	5	10

CHAPTER 2

Unified Sampling Theory: Design-Based Inference

2.1 INTRODUCTION

In this chapter we consider the inferential aspects of sampling from a finite population under a fixed population setup, where each of the unit is associated with a fixed unknown real number. A sample is selected from the population using some man-made randomization procedure called sampling design. The design-based inference is based on all possible samples that might be selected according to the sampling design. Expectation is the average of all possible samples. Different types of linear unbiased estimators have been proposed and conditions of unbiasedness of the estimators have been derived. The nonexistence theorems proposed by Godambe (1955), Hanurav (1966), and Basu (1971) have been discussed. Concepts of admissibility and sufficient statistics in finite population sampling have been introduced as well.

2.2 DEFINITIONS AND TERMINOLOGIES

2.2.1 Noninformative and Adaptive (Sequential) Sampling Designs

A sampling design p is said to be noninformative if the selection probability $p(s)$ of a sample s does not depend on the value of the study variable y. In adaptive or sequential sampling procedures, the selection probability $p(s)$ may depend on the values of the variable of interest y for the units selected in the sample s.

2.2.2 Estimator and Estimate

After selection of a sample s using a suitable sampling design p, information on the study variable y is collected from each of the units selected in the sample. Here we assume that all units in the sample have responded and there is no measurement error in measuring a response, i.e., the true value

Survey Sampling Theory and Applications
ISBN 978-0-12-811848-1
http://dx.doi.org/10.1016/B978-0-12-811848-1.00002-9

y_i, of the study variable y is obtained from each of the ith unit ($i \in s$). The information gathered from the selected units in the sample and their values y_i's is known as data, and it will be denoted by $d = ((i,y_i), i \in s)$. The collection of all possible values of d is known as the sample space, and it will be denoted by $\mathscr{X}$. A real valued function $T(s,\mathbf{y}) = T(d)$ of d on the sample space $\mathscr{X}$ is known as a statistic. When the statistic $T(s,\mathbf{y})$ is used as a guess value of a certain parametric function $\theta = \theta(\mathbf{y})$ of interest (such as the population mean, total, median etc.), we call $T(s,\mathbf{y})$ as an estimator of the parameter θ. Obviously, an estimator is a random variable whose value depends on the sample selected (i.e., data). The numerical value of an estimator for a given data is called an estimate.

2.2.3 Unbiased Estimator

An estimator $T = T(s,\mathbf{y})$ is said to be design unbiased (p-unbiased or unbiased) for estimating a population parameter θ if and only if

$$E_p(T) = \sum_{s \in \mathfrak{I}} T(s,\mathbf{y})p(s) = \theta \;\; \forall \;\; \mathbf{y} \in R^N \tag{2.2.1}$$

where, E_p denotes the expectation with respect to the sampling design p, $p(s)$ is the probability of the selection of the sample s according to design p, $\mathfrak{I}$ is the collection of all possible samples, and R^N is the N-dimensional Euclidean space. The class of all unbiased estimators of θ satisfying (2.2.1) will be denoted by C_θ.

An estimator, which is not unbiased, is called a biased (or design biased) estimator. The amount of bias of an estimator T is defined as

$$B(T) = E_p(T) - \theta = \sum_{s \in \mathfrak{I}} T(s,\mathbf{y})p(s) - \theta \tag{2.2.2}$$

2.2.4 Mean Square Error and Variance

The mean square error of an estimator T is denoted by

$$M(T) = E_p(T - \theta)^2 = \sum_{s \in \mathfrak{I}} [T(s,\mathbf{y}) - \theta]^2 \tag{2.2.3}$$

The mean square error measures the closeness of an estimator T around the true value θ.

The variance of an estimator T with respect to the sampling design p is denoted by

$$V_p(T) = E_p[T - E_p(T)]^2 = \sum_{s \in \mathfrak{I}} [T(s,\mathbf{y}) - E_p(T)]^2 p(s) \tag{2.2.4}$$

It can be easily checked that

$$M(T) = V_p(T) + [B(T)]^2 \tag{2.2.5}$$

For an unbiased estimator $B(T) = 0$ and hence the mean square is equal to its variance.

2.2.5 Uniformly Minimum Variance Unbiased Estimator

Let T_1 and $T_2(\neq T_1)$ be two unbiased estimators that belong to a certain class of unbiased estimators C_θ. The estimator T_1 is said to be better than T_2 if:

(i) $V_p(T_1) \leq V_p(T_2) \quad \forall \quad \mathbf{y} \in R^N$
and
(ii) the strict inequality $V_p(T_1) < V_p(T_2)$ holds for at least one $\mathbf{y} \in R^N$.

In case at least one of the estimators T_1 and T_2 is biased, T_1 is said to be better than T_2 if:

(i) $M(T_1) \leq M(T_2) \ \forall \ \mathbf{y} \in R^N$
and
(ii) the strict inequality $M(T_1) < M(T_2)$ holds for at least one $\mathbf{y} \in R^N$.

An estimator T_0 belonging to the class of unbiased estimators C_θ is called the uniformly minimum variance unbiased estimator (UMVUE) for estimating a parametric function θ if T_0 is better than any other unbiased estimators belonging to the class C_θ, i.e., any $\widetilde{T}(\neq T_0) \in C_\theta$ satisfies

(i) $V_p(T_0) \leq V_p(\widetilde{T}) \ \forall \ \mathbf{y} \in R^N$
and (2.2.6)
(ii) $V_p(T_0) < V_p(\widetilde{T})$ for at least one $\mathbf{y} \in R^N$

2.3 LINEAR UNBIASED ESTIMATORS

In case θ is a linear function of $\mathbf{y}$, such as population total Y or mean $\overline{Y}$, we very often use a linear estimator for Y as follows:

$$t^* = t^*(s, \mathbf{y}) = a_s + \sum_{i \in s} b_{si} y_i \tag{2.3.1}$$

where, a_s, a known constant, depends on the selected sample s but is independent of the units selected in the sample and their y-values. b_{si}'s are known constants free from y_i's, $i \in s$, but may be dependent on the selected sample s and units $i(\in s)$. $\sum_{i \in s}$ denotes the sum over distinct units in s.

In case a_s in (2.3.1) is equal to zero, then t^* reduces to a linear homogeneous unbiased estimator for Y and it is given by

$$t = t(s, \mathbf{y}) = \sum_{i \in s} b_{si} y_i \tag{2.3.2}$$

The different choices of the constants a_s and b_{si}'s yield different estimators. Our objective is to choose certain specific estimators, which must possess certain desirable properties.

2.3.1 Conditions of Unbiasedness

The estimator t^* in (2.3.1) will be unbiased for the population total Y if and only if

$$E_p(t^*) = Y \quad \forall \quad \mathbf{y} \in R^N \tag{2.3.3}$$

Now noting

$$\begin{aligned}
E_p(t^*) &= \sum_{s \in \mathscr{S}} t^*(s, \mathbf{y}) p(s) \\
&= \sum_{s \in \mathscr{S}} a_s p(s) + \sum_{s \in \mathscr{S}} \left(\sum_{i \in s} b_{si} y_i \right) p(s) \\
&= \sum_{s \in \mathscr{S}} a_s p(s) + \sum_{s \in \mathscr{S}} \left(\sum_{i=1}^{N} I_{si} b_{si} y_i \right) p(s) \\
&\quad (\text{writing } I_{si} = 1 \text{ if } i \in s \text{ and } I_{si} = 0 \text{ if } i \notin s) \\
&= \sum_{s \in \mathscr{S}} a_s p(s) + \sum_{i=1}^{N} y_i \left(\sum_{s \in \mathscr{S}} I_{si} b_{si} p(s) \right) \\
&= \alpha_0 + \sum_{i=1}^{N} \alpha_i y_i
\end{aligned} \tag{2.3.4}$$

where, $\alpha_0 = \sum_{s \in \mathscr{S}} a_s p(s)$ and $\alpha_i = \sum_{s \in \mathscr{S}} I_{si} b_{si} p(s) = \sum_{s \supset i} b_{si} p(s)$.

From Eqs. (2.3.3) and (2.3.4), we note that t^* is unbiased for Y if and only if

$$\alpha_0 + \sum_{i=1}^{N} \alpha_i y_i = Y \ \forall \ \mathbf{y} \in R^N \tag{2.3.5}$$

Now, putting $\mathbf{y} = \mathbf{y}^{(0)} = (0,\ldots,0,\ldots,0)$, all coordinates of $\mathbf{y}$ are zero and $\mathbf{y} = \mathbf{y}^{(i)} = (0,\ldots,y_i,\ldots,0)$ whose ith coordinate y_i is nonzero and the remaining coordinates are zero, in (2.3.5) the unbiasedness condition (2.3.5) reduces to

$$\alpha_0 = \sum_{s\in\mathcal{S}} a_s p(s) = 0 \quad \text{and} \quad \alpha_i = \sum_{s\supset i} b_{si} p(s) = 1 \;\; \forall \;\; i = 1,\ldots,N \qquad (2.3.6)$$

Substituting $a_s = 0$ in (2.3.5), we find the condition of unbiasedness of a linear homogeneous estimator $t = \sum_{i\in s} b_{si} y_i$ for the total Y as

$$\alpha_i = \sum_{s\supset i} b_{si} p(s) = 1 \quad \forall \quad i = 1,\ldots,N \qquad (2.3.7)$$

If $\pi_i > 0$, then $b_{si} = 1/\pi_i$ meets the unbiased condition (2.3.7). On the other hand, if $\pi_i = \sum_{s\supset i} p(s) = \sum_{s\in\mathcal{S}} I_{si} p(s) = 0$, then $I_{si} = 0$ for s with $p(s) > 0$ and hence we cannot find b_{si}'s ($\neq 0$) that satisfies the unbiasedness condition (2.3.7). This leads to the following theorem attributed to Godambe (1955).

Theorem 2.3.1

The necessary and sufficient condition for existence of a linear unbiased estimator t of the population total Y is that the inclusion probability π_i should be positive for all $i = 1,\ldots,N$.

2.3.2 Horvitz–Thompson Estimator

Consider the linear homogeneous unbiased estimator $t = \sum_{i\in s} b_{si} y_i$ for the total Y. If we put $b_{si} = c_i$ in the expression of t, then the unbiasedness condition (2.3.7) yields $c_i = 1/\pi_i$. In this case the estimator t reduces to

$$\widehat{Y}_{ht} = \sum_{i\in s} y_i/\pi_i \qquad (2.3.8)$$

The estimator $\widehat{Y}_{ht}$ is known as Horvitz–Thompson (1952) estimator for the population total Y.

For an simple random sampling without replacement (SRSWOR), sampling design of size n, $\pi_i = n/N$ and the Horvitz–Thompson estimator (HTE) reduces to

$$\widehat{Y} = N\,\overline{y}(s) \;\; \text{with} \;\; \overline{y}(s) = \sum_{i\in s} y_i/n \qquad (2.3.9)$$

2.3.3 Hansen–Hurwitz Estimator

If we take $b_{si} = k\, n_i(s)$, with k as a constant and $n_i(s)$ = number of times ith unit is repeated in s, then the unbiasedness condition (2.3.7) reduces to

$$k\sum_{s\supset i} n_i(s)p(s) = k\sum_{s\in \mathfrak{I}} n_i(s)p(s) = k\, E_p\{n_i(s)\} = 1$$

and the estimator t reduces to

$$t = \sum_{i\in s} n_i(s)y_i/E_p(n_i(s)) \qquad (2.3.10)$$

In particular, for a probability proportional to size with replacement (PPSWR) sampling design with normed size measure p_i for the ith unit, $E_p(n_i(s)) = np_i$ and (2.3.10) reduces to Hansen–Hurwitz (1943) estimator

$$\widehat{Y}_{hh} = \frac{1}{n}\sum_{i\in s} n_i(s)\frac{y_i}{p_i} \qquad (2.3.11)$$

The PPSWR sampling reduces to simple random sampling with replacement (SRSWR), if $p_i = 1/N\ \forall i = 1,\ldots,N$ and in this case we get

$$t = \frac{N}{n}\overline{y}_n$$

where, $\overline{y}_n$ is the sample mean of all the n units including repetition.

2.3.4 Unbiased Ratio Estimator

Let us choose $b_{si} = c_i/p(s)$. In this case the unbiasedness condition (2.3.7) reduces to $c_i = 1/\beta_i$, where $\beta_i = \sum_{s\supset i} = \sum_{s\in\mathfrak{I}} I_{si}$ = total number of times ith unit appears in all possible samples with $p(s) > 0$ and the estimator (2.3.2) reduces to

$$t = \sum_{i\in s}\frac{1}{\beta_i}\frac{y_i}{p(s)}$$

In case $\mathfrak{I}$ consists of all possible $\binom{N}{n}$ samples each of n distinct units with positive probabilities, then $\beta_i = \binom{N-1}{n-1} = M_1$ (say) and the expression of t becomes

$$t = \frac{1}{M_1}\sum_{i\in s}\frac{y_i}{p(s)} \qquad (2.3.12)$$

For the Lahiri–Midzuno–Sen (LMS) sampling scheme, $p(s) = x_s/(M_1 X)$, where $x_s = \sum_{i\in s} x_i$, $X = \sum_{i\in U} x_i$ and $x_i(>0)$ is a known positive number (measure of size) for the ith unit, the estimator (2.3.12) reduces to the unbiased ratio estimator for population total Y proposed by LMS (1951, 1952, 1953) and it is given by

$$\widehat{Y}_{lms} = (y_s/x_s)X \tag{2.3.13}$$

where $y_s = \sum_{i\in s} y_i$

2.3.5 Difference and Generalized Difference Estimator

Let $t\ (s, y) = \sum_{i\in s} b_{si}y_i$ be a linear homogeneous unbiased estimator of the total Y, x_i the known value of a certain character x of the ith unit, and $X = \sum_{i=1}^{N} x_i$. Then the linear estimator

$$t_D = \sum_{i\in s} b_{si}y_i - \beta\left(\sum_{i\in s} b_{si}x_i - X\right) \tag{2.3.14}$$

is unbiased for the total Y for any known value β. The estimator (2.3.14) is known as a difference estimator. In particular when $b_{si} = 1/\pi_i$ and $\beta = 1$, (2.3.14) takes the following elegant form

$$t_D = \sum_{i\in s} y_i/\pi_i - \left(\sum_{i\in s} x_i/\pi_i - X\right) \tag{2.3.15}$$

The estimator t_D is known as generalized difference estimator.

2.4 PROPERTIES OF THE HORVITZ–THOMPSON ESTIMATOR

The HTE of the finite population total Y is

$$\widehat{Y}_{ht} = \sum_{i\in s} y_i/\pi_i$$

where, $\sum_{i\in s}$ denotes sum over distinct units in s and $\pi_i > 0\ \forall\ i = 1,\ldots,N$.

The HTE is widely used in survey sampling theory because it possesses some important properties. The properties are presented in this section.

Theorem 2.4.1

Let $I_{si} = 1$ for $i \in s$ and $I_{si} = 0$ for $i \notin s$. Then,
(i) $E_p(I_{si}) = \pi_i$, (ii) $V_p(I_{si}) = \pi_i(1 - \pi_i)$, and (iii) $C_p(I_{si}, I_{sj}) = -\Delta_{ij}$ for $i \neq j$. where C_p is the covariance operator with respect to the sampling design p and $\Delta_{ij} = \pi_i\pi_j - \pi_{ij}$.

Proof

(i) $$E_p(I_{si}) = \sum_s I_{si}p(s) = \pi_i$$

(ii) $$V_p(I_{si}) = E_p(I_{si})^2 - \pi_i^2 = E_p(I_{si}) - \pi_i^2 = \pi_i(1-\pi_i)$$

and

(iii) $$C_p(I_{si}, I_{sj}) = E_p(I_{si}I_{sj}) - E_p(I_{si})E_p(I_{sj}) = \sum_s I_{si}I_{si}p(s) - \pi_i\pi_j = \pi_{ij} - \pi_i\pi_j = -\Delta_{ij}$$

Theorem 2.4.2

(i) $\widehat{Y}_{ht}$ is unbiased for Y i.e., $E_p\left(\widehat{Y}_{ht}\right) = Y$

(ii) The variance of $\widehat{Y}_{ht}$ is

$$V_p\left(\widehat{Y}_{ht}\right) = \sum_{i=1}^{N} y_i^2\left(\frac{1}{\pi_i} - 1\right) + \sum_{i\neq}^{N}\sum_{j=1}^{N} y_i y_j\left(\frac{\pi_{ij}}{\pi_i\pi_j} - 1\right) = V_{ht}$$

(iii) An unbiased estimator of $V_p\left(\widehat{Y}_{ht}\right)$ is

$$\widehat{V}_{ht} = \sum_{i\in s} y_i^2\frac{1-\pi_i}{\pi_i^2} + \sum_{i\neq}\sum_{j\in s} y_i y_j\frac{\pi_{ij} - \pi_i\pi_j}{\pi_{ij}}$$

for $\pi_{ij} > 0$ for $i \neq j = 1,\ldots,N$.

Proof

(i) $$E_p\left(\widehat{Y}_{ht}\right) = E_p\left(\sum_{i\in s}\frac{y_i}{\pi_i}\right) = E_p\left(\sum_{i=1}^{N}\frac{y_i}{\pi_i}I_{si}\right) = \sum_{i=1}^{N}\frac{y_i}{\pi_i}E_p(I_{si}) = Y \text{ (since } E_p(I_{si}) = \pi_i)$$

(ii) $$V_p\left(\widehat{Y}_{ht}\right) = V_p\left(\sum_{i=1}^{N}\frac{y_i}{\pi_i}I_{si}\right)$$

$$= \sum_{i=1}^{N}\frac{y_i^2}{\pi_i^2}\{V_p(I_{si})\} + \sum_{i\neq}^{N}\sum_{j=1}^{N}\frac{y_i}{\pi_i}\frac{y_j}{\pi_j}\{C_p(I_{si}, I_{sj})\}$$

Now noting $V_p(I_{si}) = \pi_i(1-\pi_i)$ and $C_p(I_{si},I_{sj}) = \pi_{ij} - \pi_i\pi_j$, we find $V_p\left(\widehat{Y}_{ht}\right) = V_{ht}$.

(iii) Writing, $\widehat{V}_{ht} = \sum_{i=1}^{N}\frac{y_i^2}{\pi_i^2}(1-\pi_i)I_{si} + \sum_{i\neq}^{N}\sum_{j=1}^{N}\frac{y_i}{\pi_i}\frac{y_j}{\pi_j}\frac{\pi_{ij}-\pi_i\pi_j}{\pi_{ij}}I_{si}I_{sj}$, we get

$$E_p\left(\widehat{V}_{ht}\right) = \sum_{i=1}^{N}\frac{y_i^2}{\pi_i^2}(1-\pi_i)E_p(I_{si}) + \sum_{i\neq}^{N}\sum_{j=1}^{N}\frac{y_i}{\pi_i}\frac{y_j}{\pi_j}\frac{\pi_{ij}-\pi_i\pi_j}{\pi_{ij}}E_p(I_{si}I_{sj})$$

$$= V_{ht}\ \left(\text{since } E_p(I_{si}) = \pi_i \text{ and } E_p(I_{si}I_{sj}) = \pi_{ij}\right)$$

Theorem 2.4.3

For a fixed effective size (ν) sampling design,

(i) $$V_p\left(\widehat{Y}_{ht}\right) = V_{ht} = \frac{1}{2}\sum_{i\neq}^{N}\sum_{j=1}^{N}\Delta_{ij}\left(\frac{y_i}{\pi_i}-\frac{y_j}{\pi_j}\right)^2 = V_{YG}$$

(ii) An unbiased estimator of $V_p\left(\widehat{Y}_{ht}\right)$ is given by

$$\widehat{V}_{YG} = \frac{1}{2}\sum_{i\neq}\sum_{j\in s}\frac{\Delta_{ij}}{\pi_{ij}}\left(\frac{y_i}{\pi_i}-\frac{y_j}{\pi_j}\right) \quad \text{for } \pi_{ij} > 0 \text{ for } i\neq j = 1,\ldots,N$$

Proof

(i) $$V_{YG} = \frac{1}{2}\sum_{i\neq}^{N}\sum_{j=1}^{N}\Delta_{ij}\left(\frac{y_i}{\pi_i}-\frac{y_j}{\pi_j}\right)^2$$

$$= \frac{1}{2}\sum_{i\neq}^{N}\sum_{j=1}^{N}\Delta_{ij}\left(\frac{y_i^2}{\pi_i^2}+\frac{y_j^2}{\pi_j^2}-2\frac{y_i}{\pi_i}\frac{y_j}{\pi_j}\right)$$

$$= \sum_{i=1}^{N}\frac{y_i^2}{\pi_i^2}\sum_{j(\neq i)=1}^{N}\Delta_{ij} - \sum_{i\neq}^{N}\sum_{j=1}^{N}\Delta_{ij}\frac{y_i}{\pi_i}\frac{y_j}{\pi_j}$$

$$= V_{ht}$$

since

$$\sum_{j(\neq i)=1}^{N} \Delta_{ij} = \pi_i \sum_{j(\neq i)=1}^{N} \pi_j - \sum_{j(\neq i)=1}^{N} \pi_{ij}$$

$$= \pi_i(\nu - \pi_i) - (\nu - 1)\pi_i$$

$$= \pi_i(1 - \pi_i) \text{ (using corollary 1.3.2).}$$

(ii) $$E_p\left(\widehat{V}_{YG}\right) = \frac{1}{2}E_p\left[\sum_{i\neq}\sum_{j\in s} \frac{\Delta_{ij}}{\pi_{ij}}\left(\frac{y_i}{\pi_i} - \frac{y_j}{\pi_j}\right)^2\right]$$

$$= \frac{1}{2}\sum_{i\neq}^{N}\sum_{j=1}^{N} \frac{\Delta_{ij}}{\pi_{ij}}\left(\frac{y_i}{\pi_i} - \frac{y_j}{\pi_j}\right)^2 E_p(I_{si}I_{sj})$$

$$= V_{YG}$$

Remark 2.4.1

The variance estimator $\widehat{V}_{ht}$ was proposed by Horvitz and Thompson (1952) and is applicable for any sampling design with $\pi_{ij} > 0$ for $i \neq j = 1,\ldots,N$. The variance estimator $\widehat{V}_{YG}$ was proposed by Yates and Grundy (1953) and is known as the Yates–Grundy variance estimator. $\widehat{V}_{YG}$ is unbiased for only a fixed effective size sampling design. Both the estimators $\widehat{V}_{ht}$ and $\widehat{V}_{YG}$ suffer from the drawback that they can be negative. No simple sufficient condition of nonnegativity is available for $\widehat{V}_{ht}$. The Yates–Grundy variance estimator $\widehat{V}_{YG}$ is nonnegative for a sampling design, for which $\Delta_{ij} \geq 0$. Various sampling designs have been proposed, for which $\Delta_{ij} \geq 0$ for $i \neq j = 1,\ldots,N$. As far as efficiency is concerned, $\widehat{V}_{ht}$ is known to be admissible but $\widehat{V}_{YG}$ is inadmissible. Details are given in Section 2.6.

2.5 NONEXISTENCE THEOREMS

We will call the collection of the estimators $t = \sum_{i\in s} b_{si}y_i$, whose coefficients b_{si}'s satisfy the unbiasedness condition (2.3.7) as the class of linear homogeneous unbiased estimator C_{lh}. The class of the linear unbiased estimators C_l comprises estimators (2.3.1) and is subject to (2.3.6). The class of all possible unbiased estimators, which includes linear, linear homogeneous, and nonlinear estimators, will be denoted by C_u and clearly, $C_u \supset C_l \supset C_{lh}$.

In Section 2.2, it is shown that for a given sampling design p we can construct numerous linear unbiased estimators for a finite population total Y. Therefore, we need to select the best estimator in the sense of having uniformly minimum variance. Godambe (1955) proved that in the class of linear homogeneous unbiased estimators C_{lh}, the UMVUE does not exist, i.e., none of the estimators can be termed as the best. Hanurav (1966) modified Godambe's result and showed the existence of the UMVUE for a unicluster sampling design (defined in the section next). Basu (1971) generalized the result further and proved the nonexistence of the UMVUE in the class of all unbiased estimators C_u. Godambe (1955) showed that the HTE is the UMVUE in the reduced subspace R_0 of the parameter space R^N, where $R_0 = \bigcup_{i=1}^{N} \mathbf{y}^{(i)}$ and $\mathbf{y}^{(i)}$ = vector $\mathbf{y}$, whose ith coordinate y_i is nonzero and the remaining coordinates are zeros.

2.5.1 Unicluster Sampling Design

A design $\widetilde{p}$ is called as a unicluster design if any two samples $s, s^* \in \mathfrak{I}$ with $\widetilde{p}(s), \widetilde{p}(s^*) > 0$ imply either (i) $s \cap s^* = \phi$ (null set) or (ii) the samples s and s^* are equivalent ($s \sim s^*$) in the sense that they contain the same set of distinct units.

Example 2.5.1

Let $U = (1,2,3,4,5,6)$ be a finite population, and $\mathfrak{I}$ consists of the samples $s_1 = (1,2)$, $s_2 = (3,4)$, $s_3 = (5,6)$ with probabilities $p(s_1) = 0.4$ and $p(s_2) = p(s_3) = 0.3$. Then $\widetilde{p}$ is a unicluster sampling design.

2.5.2 Class of Linear Homogeneous Unbiased Estimators

Theorem 2.5.1

In the class of linear homogeneous unbiased estimators C_{lh}, the UMVUE of a finite population total Y based on a noncensus sampling design, p with $\pi_i > 0 \ \forall \ i = 1,\ldots,N$ does not exist if the sampling design p is non-unicluster. However, the UMVUE does exist if p is a unicluster design.

Proof

The class C_{lh} consists of the estimators $t(s, \mathbf{y}) = \sum_{i \in s} b_{si} y_i$, where the constants b_{si}'s satisfy the unbiasedness condition (2.3.7) viz.

$$\sum_{s \supset i} b_{si} p(s) = 1 \quad \text{for every } i = 1, \cdots, N$$

Here we want to find the constants, b_{si}'s, that minimize

$$V_p(t) = \sum_{s \in \mathfrak{I}} \left(\sum_{i \in s} b_{si} y_i \right)^2 p(s) - Y^2$$

subject to the unbiasedness condition (2.3.7).

For minimization, we consider

$$\Psi = \sum_{s \in \mathfrak{I}} \left(\sum_{i \in s} b_{si} y_i \right)^2 p(s) - Y^2 - \sum_{i=1}^{N} \lambda_i \left\{ \sum_{s \supset i} b_{si} p(s) - 1 \right\} \tag{2.5.1}$$

with λ_i's as undetermined Lagrange multipliers.

Differentiating Ψ with respect to b_{si} and equating to zero, we get

$$\frac{\partial \Psi}{\partial b_{si}} = 2y_i \left(\sum_{i \in s} b_{si} y_i \right) p(s) - \lambda_i p(s) = 0 \tag{2.5.2}$$

Eq. (2.5.2) should be valid for $\forall\ \mathbf{y} \in R^N$. In particular for $\mathbf{y} = \mathbf{y}^{(i)} = (0,\ldots,0,y_i,0,\ldots,0)$ with $y_i \neq 0$, Eq. (2.5.2) yields the optimum value of b_{si} as

$$b_{si} = \lambda_i / \left(2y_i^2\right) = b_{0i} \text{ (say)} \tag{2.5.3}$$

The unbiasedness condition (2.3.7) and (2.5.3) yield

$$b_{si} = b_{0i} = 1/\pi_i \tag{2.5.4}$$

Thus the UMVUE should necessarily be the HTE viz. $\sum_{i \in s} y_i/\pi_i$ and should satisfy Eq. (2.5.2), i.e.,

$$\sum_{i \in s} y_i/\pi_i = \frac{\lambda_i}{2y_i} \quad \text{for } \forall\ i \in s, y_i \neq 0 \tag{2.5.5}$$

For a noncensus non-unicluster sampling design, we always find two different samples, s_1 and s_2, which have a common unit, i.e., $s_1 \cap s_2 \neq \phi$ for which $p(s_1) > 0$ and $p(s_2) > 0$. In this case Eq. (2.5.5) indicates that

$$\sum_{i \in s_1} y_i/\pi_i = \sum_{i \in s_2} y_i/\pi_i \quad \forall\ \mathbf{y} \in R^N, \tag{2.5.6}$$

which is clearly impossible. Hence the UMVUE does not exist for a noncensus non-unicluster sampling design.

For a non-census unicluster sampling design any two samples s_1 and s_2 with $p(s_1), p(s_2) > 0$ implies $s_1 \cap s_2 = \phi$. In this case, the condition of unbiasedness (2.3.7) yields $b_{si} = 1/\pi_i = 1/p(s)$ for $i \in s$. Hence the

Horvitz–Thompson estimator is the unique linear homogeneous unbiased estimator for the total Y. Therefore, it is trivially the MVUE in C_{lh}.

2.5.3 Optimality of the Horvitz–Thompson Estimator

The following theorem shows the optimality the HTE in the restricted parametric space $R_0 = \bigcup_{i=1}^{n} \mathbf{y}^{(i)} (\subset R^N)$.

Theorem 2.5.2

For any sampling design p, with $\pi_i > 0 \ \forall \ i = 1,\ldots,N$, the HTE $\widehat{Y}_{ht} = \sum_{i \in s} y_i/\pi_i$ is the UMVUE in C_{lh}, the class of linear homogeneous unbiased estimators of the population total Y when $\mathbf{y} \in R_0$.

Proof

$$\begin{aligned} V_p[t(s,y)] &= \sum_{s \in \mathfrak{S}} \left(\sum_{i \in s} b_{si} y_i \right)^2 p(s) - Y^2 \\ &= \sum_{s \in \mathfrak{S}} \left(\sum_{i \in s} b_{si}^2 y_i^2 + \sum_{i \neq} \sum_{j \in s} b_{si} b_{sj} y_i y_j \right) p(s) - Y^2 \\ &= \sum_{i=1}^{N} y_i^2 \left(\sum_{s \supset i} b_{si}^2 p(s) - 1 \right) + \sum_{i \neq}^{N} \sum_{j}^{N} y_i y_j \left(\sum_{s \supset i,j} b_{si} b_{sj} p(s) - 1 \right) \end{aligned} \tag{2.5.7}$$

Substituting $\mathbf{y} = \mathbf{y}^{(j)} = (0,\ldots,0,y_j,0,\ldots,0)$ in (2.5.7), we get

$$V_p\left[t\left(s, \mathbf{y}^{(j)}\right)\right] = y_j^2 \left(\sum_{s \supset j} b_{sj}^2 p(s) - 1 \right)$$

$$\geq y_j^2 \left[\frac{\left(\sum_{s \supset j} b_{sj} p(s) \right)^2}{\sum_{s \supset j} p(s)} - 1 \right] \text{(using Cauchy inequality)}$$

The unbiasedness condition $\sum_{s \supset j} b_{si} p(s) = 1$ yields

$$V_p\left[t\left(s, \mathbf{y}^{(j)}\right)\right] \geq y_j^2 \left(\frac{1}{\pi_j} - 1 \right) \tag{2.5.8}$$

The equality in (2.5.8) holds if and only if $b_{sj} = 1/\pi_j$. Hence,

$$V_p[t(s, y)] \geq V_p\left(\sum_{i \in s} \frac{y_i}{\pi_i}\right) \text{ for } \forall \ \mathbf{y} \in R_0 \qquad (2.5.9)$$

The inequality in (2.5.9) is strict if and only if $t(s, \mathbf{y}) \neq \widehat{Y}_{ht}$. Hence for the restricted parametric space R_0, $\widehat{Y}_{ht} = \sum_{i \in s} \frac{y_i}{\pi_i}$ is the UMVUE in the class C_{lh}.

2.5.4 Class of All Unbiased Estimators

Let $T(s,\mathbf{y})$ be an unbiased estimator for an arbitrary parametric function $\theta = \theta(\mathbf{y})$. The value of $T(s,\mathbf{y})$ depends on the values of y_i's belonging to the sample s but is independent of y_i's, which do not belong to s. The value of $\theta = \theta(\mathbf{y})$ depends on all the values of $y_i, i = 1,\ldots N$. Let C_θ be the class of all unbiased estimators of θ. Basu (1971) proved the nonexistence of a UMVUE of $\theta(\mathbf{y})$ in the class C_θ of all unbiased estimators. The theorem is described as follows:

Theorem 2.5.3

For a noncensus design, there does not exist the UMVUE of $\theta = \theta(\mathbf{y})$ in the class of all unbiased estimators C_θ.

Proof

If possible, let $T_0(s,\mathbf{y})(\in C_\theta)$ be the UMVUE of the population parameter $\theta = \theta(\mathbf{y})$. Since the design p is noncensus and the value of $T_0(s,\mathbf{y})$ depends on y_i's for $i \in s$ but not on the values of y_i's for $i \notin s$, we can find a known vector $\mathbf{y}^{(a)} = (a_1,\ldots,a_i,\ldots,a_N)$ for which $T_0(s,\mathbf{y}^{(a)}) \neq \theta(\mathbf{y}^{(a)})$ with $p(s) > 0$. Consider the following estimator

$$T^*(s, \mathbf{y}) = T_0(s, \mathbf{y}) - T_0\left(s, \mathbf{y}^{(a)}\right) + \theta\left(\mathbf{y}^{(a)}\right)$$

$T^*(s,\mathbf{y})$ is unbiased for $\theta(\mathbf{y})$ because

$$E_p[T^*(s, \mathbf{y})] = \theta(\mathbf{y}) - \theta\left(\mathbf{y}^{(a)}\right) + \theta\left(\mathbf{y}^{(a)}\right) = \theta(\mathbf{y}).$$

Since $T_0(s,\mathbf{y})$ is assumed to be the UMUVE for $\theta(\mathbf{y})$, we must have

$$V_p[T_0(s, \mathbf{y})] \leq V_p[T^*(s, \mathbf{y})] \quad \forall \ \mathbf{y} \in R^N \qquad (2.5.10)$$

Now for $\mathbf{y} = \mathbf{y}^{(a)}$, $V_p[T^*(s,\mathbf{y})] = V_p\left[T^*\left(s,\mathbf{y}^{(a)}\right)\right] = V_p\left[\theta\left(\mathbf{y}^{(a)}\right)\right] = 0$ while $V_p[T_0(s,\mathbf{y}^{(a)})] > 0$ since we assumed $T_0\left(s, \mathbf{y}^{(a)}\right) \neq \theta(\mathbf{y}^{(a)})$ with $p(s) > 0$. Hence the inequality (2.5.10) is violated at $\mathbf{y} = \mathbf{y}^{(a)}$ and the nonexistence of the UMVUE for $\theta(\mathbf{y})$ is proved.

In particular, for $\theta(\mathbf{y}) = Y$, Theorem 2.5.3 reduces to following theorem.

Theorem 2.5.4

For a noncensus design, p, the UMVUE of Y in the class of unbiased estimators C_u does not exist.

2.5.5 Class of Linear Unbiased Estimators

Now writing $\theta(\mathbf{y}) = Y$, $T_0(s, \mathbf{y}) = t_0{}^*(s, \mathbf{y}) = a_{s0} + \sum_{i \in s} b_{si0} y_i$ and

$$T^*(s, \mathbf{y}) = t_0{}^*(s, \mathbf{y}) - t_0{}^*\left(s, \mathbf{y}^{(a)}\right) + \sum_{i=1}^{N} a_i = \sum_{i \in s} b_{si0}(y_i - a_i) + \sum_{i=1}^{N} a_i$$

in Theorem 2.5.3, we derive the following theorem.

Theorem 2.5.5

For a noncensus design, p, the UMVUE of the population total, Y, in the class of linear unbiased estimators C_l does not exist.

2.6 ADMISSIBLE ESTIMATORS

We have seen in Section 2.5 that in almost all practical situations, the UMVUE for a finite population total does not exist. The criterion of admissibility is used to guard against the selection of a bad estimator.

An estimator T is said to be admissible in the class C of estimators for a given sampling design p if there does not exist any other estimator in the class C better than T. In other words, there does not exist an alternative estimator $T^*(\neq T) \in C$, for which following inequalities hold.

$$\begin{aligned} &\text{(i)}\ V_p(T^*) \le V_p(T) \quad \forall\, T^*(\neq T) \in C \text{ and } \mathbf{y} \in R^N \\ &\text{and}\quad \text{(ii)}\ V_p(T^*) < V_p(T) \text{ for at least one } \mathbf{y} \in R^N \end{aligned} \tag{2.6.1}$$

Theorem 2.6.1

In the class of linear homogeneous unbiased estimators C_{lh}, the HTE $\widehat{Y}_{ht}$ based on a sampling design p with $\pi_i > 0\ \forall\ i = 1,\ldots,N$ is admissible for estimating the population total Y.

Proof

The proof is immediate from Theorem 2.5.2. Since $\widehat{Y}_{ht}$ is the UMVUE when $\mathbf{y} \in R_0$, we cannot find an estimator $\forall\, T^*\left(\neq \widehat{Y}_{ht}\right) \in C_{lh}$ for which (2.6.1) holds.

The Theorem 2.6.1 of admissibility of the HTE $\widehat{Y}_{ht}$ in the class C_{lh} was proved by Godambe (1960). Godambe and Joshi (1965) proved the admissibility of $\widehat{Y}_{ht}$ in the class of all unbiased estimators C_u, and it is given in Theorem 2.6.2.

Theorem 2.6.2

For a given sampling design p with $\pi_i > 0 \ \forall \ i = 1,\ldots,N$, the HTE $\widehat{Y}_{ht}$ is admissible in the class C_u of all unbiased estimator for estimating the total Y.

Proof

Suppose $\widehat{Y}_{ht}$ is not admissible in C_u, then there exists an estimator $t(s,\mathbf{y})(\neq \widehat{Y}_{ht}) \in C_u$, for which the following inequalities hold.

$$\begin{aligned} &\text{(i)} \ V_p[t(s,\mathbf{y})] \leq V_p\left(\widehat{Y}_{ht}\right) \ \forall \ \mathbf{y} \in R^N \\ &\qquad\qquad \text{and} \\ &\text{(ii)} \ V_p[t(s,\mathbf{y})] < V_p\left(\widehat{Y}_{ht}\right) \ \text{at least one} \ \mathbf{y} \in R^N \end{aligned} \tag{2.6.2}$$

The estimator $t(s,\mathbf{y})$ can be written as

$$t(s,\mathbf{y}) = \widehat{Y}_{ht} + h(s,\mathbf{y}) \tag{2.6.3}$$

$$\text{i.e., } h(s,\mathbf{y}) = t(s,\mathbf{y}) - \widehat{Y}_{ht}$$

Since $t(s,\mathbf{y})$ and $\widehat{Y}_{ht}$ are unbiased for the total Y, we must have

$$E_p[h(s,\mathbf{y})] = \sum_{s\in \mathfrak{S}} h(s,\mathbf{y})p(s) = 0 \tag{2.6.4}$$

Furthermore, (2.6.2) yields

$$V_p[h(s,y)] + 2C_p\left[\widehat{Y}_{ht}, h(s,y)\right] \leq 0$$

i.e.,

$$\sum_{s\in \mathfrak{S}} \{h(s,y)\}^2 p(s) + 2\sum_{s\in \mathfrak{S}} h(s,y)\left(\sum_{i\in s}\frac{y_i}{\pi_i}\right)p(s) \leq 0 \quad \forall \ \mathbf{y} \in R^N \tag{2.6.5}$$

Let us define $\mathbf{y}_{(j)}$ = collection of all vectors $\mathbf{y} = (y_1,\ldots,y_k,\ldots,y_N)$ having exactly j (=0,1,…,N) nonzero coordinates and $\mathfrak{S}_{(j)}(\subset \mathfrak{S})$ collection of samples consisting of units with nonzero $\mathbf{y}$-values for exactly j units and $\mathfrak{S}$ is the collections of all possible samples with positive selection probabilities. Clearly, $\mathbf{y}_{(j)} \cap \mathbf{y}_{(k)} = \phi$ for $j \neq k = 0,1,\ldots,N$ and $\bigcup_{j=1}^{N} \mathbf{y}_{(j)} = R^N$.

To prove this the theorem, we prove the following lemma by Godambe and Joshi (1965).

Lemma 2.6.1

$h(s,\mathbf{y}) = 0$ for $\forall\ s \in \mathfrak{S}$ and $\forall\ \mathbf{y} \in \mathbf{y}_{(j)}$ implies $h(s,\mathbf{y}) = 0$ for $\forall s \in \mathfrak{S}$ and $\forall\ \mathbf{y} \in \mathbf{y}_{(j+1)}$, $j = 0, 1, .., N-1$.

Proof

Let $h(s,\mathbf{y}) = 0$ for $\forall s \in \mathfrak{S}$ and $\forall\ \mathbf{y} \in \mathbf{y}_{(j)}$, then for any $\forall\ \mathbf{y} \in \mathbf{y}_{(j+1)}$, Eqs. (2.6.4) and (2.6.5) yield

$$\sum_{i=0}^{j+1} \sum_{s \in \mathfrak{S}_{(i)}} h(s,y)p(s) = 0 \tag{2.6.6}$$

and

$$\sum_{i=0}^{j+1} \sum_{s \in \mathfrak{S}_{(i)}} \{h(s,y)\}^2 p(s) + 2\sum_{i=0}^{j+1} \sum_{s \in \mathfrak{S}_{(i)}} h(s,y)\left(\sum_{i \in s} \frac{y_i}{\pi_i}\right) p(s) \le 0 \tag{2.6.7}$$

Now $h(s,\mathbf{y}) = 0$ for $\forall s \in \mathfrak{S}$ and $\forall\ \mathbf{y} \in \mathbf{y}_{(j)}$ implies $h(s,\mathbf{y}) = 0$ for $\forall s \in \mathfrak{S}_{(i)}, i = 0, 1, \ldots, j$. Hence (2.6.6) and (2.6.7) implies

$$\sum_{s \in \mathfrak{S}_{(j+1)}} h(s,\mathbf{y})p(s) = 0 \tag{2.6.8}$$

and

$$\sum_{s \in \mathfrak{S}_{(j+1)}} \{h(s,\mathbf{y})\}^2 p(s) + 2\sum_{s \in \mathfrak{S}_{(j+1)}} h(s,\mathbf{y})\left(\sum_{i \in s} \frac{y_i}{\pi_i}\right) p(s) \le 0 \tag{2.6.9}$$

The magnitude of $\sum_{i \in s} \frac{y_i}{\pi_i}$ is the same as $\sum_{i=1}^{N} \frac{y_i}{\pi_i}$ for every $s \in \mathfrak{S}_{(j+1)}$ because, when $\mathbf{y} \in \mathbf{y}_{(j+1)}$, each of the samples belonging to $\mathfrak{S}_{(j+1)}$ contains all the $j+1$ nonzero y-values that belong to $\mathbf{y}_{(j+1)}$. Thus (2.6.9) yields

$$\sum_{s \in \mathfrak{S}_{(j+1)}} \{h(s,\mathbf{y})\}^2 p(s) + \left(\sum_{i=1}^{N} \frac{y_i}{\pi_i}\right) \sum_{s \in \mathfrak{S}_{(j+1)}} h(s,\mathbf{y})p(s) \le 0\ \ \forall\ \ \mathbf{y} \in \mathbf{y}_{(j+1)}$$

Finally using (2.6.8), we get

$$h(s,\mathbf{y}) = 0\ \ \forall s \in \mathfrak{S}_{(j+1)}\ \text{ and }\ \forall\ \ \mathbf{y} \in \mathbf{y}_{(j+1)} \tag{2.6.10}$$

Since $\mathfrak{S} = \bigcup_{i=1}^{j+1} \mathfrak{S}_{(i)}$, (2.6.10) implies

$$h(s,\mathbf{y}) = 0\ \ \forall\ \ s \in \mathfrak{S}\ \text{ and }\ \forall\ \ \mathbf{y} \in \mathbf{y}_{(j+1)} \tag{2.6.11}$$

Hence the lemma is proved.

For $\mathbf{y} \in \mathbf{y}_{(0)}$, $\sum_{i \in s} \frac{y_i}{\pi_i} = 0$, hence using (2.6.5), we find $h(s,\mathbf{y}) = 0\ \forall\ s \in \mathfrak{S}$ with $p(s) > 0$. Now from the Lemma 2.6.1, we see that

$$h(s, \mathbf{y}) = 0\ \forall\ s \in \mathfrak{S} \text{ and } \forall\ \mathbf{y} \in R^N \tag{2.6.12}$$

Finally Eqs. (2.6.3) and (2.6.12) yield $t(s, y) = \widehat{Y}_{ht}$. Hence there does not exist an estimator, $t(s,\mathbf{y})(\neq \widehat{Y}_{ht}) \in C_u$, for which (2.6.2) holds. Thus $\widehat{Y}_{ht}$ is an admissible in the class of unbiased estimators C_u.

Remark 2.6.1

The admissible estimator is not unique. In fact we can construct a number of admissible estimators in a given class of estimators. It follows from the following corollary derived by Cassel et al. (1977).

Corollary 2.6.1

For any sampling design p with $\pi_i > 0\ \forall\ i = 1,\ldots,N$, the generalized difference estimator

$$t_D = \sum_{i \in s} \frac{y_i - a_i}{\pi_i} + A,$$

with a_i's are known constants and $A = \sum_{i=1}^{N} a_i$ is admissible in the class C_u.

Proof

From Theorem 2.6.2, we note that $\sum_{i \in s} \frac{y_i - a_i}{\pi_i}$ is admissible in the class of all unbiased estimators of $Y - A$. Hence $\sum_{i \in s} \frac{y_i - a_i}{\pi_i} + A$ is admissible in the class of all unbiased estimators C_u for estimating Y.

Remark 2.6.2

Godambe and Joshi (1965) showed that Theorem 2.6.2 is also valid for the subspace $-a < y_i < b$ for $\forall\ i = 1,\ldots,N$ with $a,b \geq 0$ of the parameter space R^N.

Remark 2.6.3

Godambe and Joshi (1965) also proved that the variance estimator $\widehat{V}_{ht}$ is admissible in the entire class of unbiased estimators of V_{ht}.

Remark 2.6.4

It is important to note that if an estimator is admissible in a class C, then it is also admissible in a subclass of $C^*(\subset C)$ containing the estimator. Hence, Theorem 2.6.1 can be proved as a corollary of Theorem 2.6.2.

For further results of admissibility, readers are referred to Joshi (1965a,b), Cassel et al. (1977), Patel and Dharmadikari (1978), and Sengupta (1980), among others.

2.7 SUFFICIENCY IN FINITE POPULATION

An estimator $e(s,\mathbf{y})$ is said to be inadmissible in a class of estimators C if there exists an estimator $e^*(s,y)(\in C)$ better than $e(s,y)$. Hence it is natural to question how an inadmissible estimator could be improved. The method of improvement of an inadmissible estimator with the aid of sufficient statistics is known as Rao–Blackwellization. The concept of sufficient statistics in survey sampling was introduced by Basu (1958), while the concepts of linear sufficiency, distribution-free sufficient statistics, and Bayesian sufficiency were also introduced by Godambe (1966, 1968). Details have been given by Cassel et al. (1977), Chaudhuri and Stenger (1992), and Thompson and Seber (1996).

Let $s = (i_1, \ldots, i_k, \ldots, i_{n_s})$ be an ordered sample of size n_s selected from a population U with probability $p(s)$ using a sampling design p, where the unit i_k is selected at the kth draw. After selection of sample s, the responses $y_{i_1}, \ldots, y_{i_{n_s}}$ were obtained from sampled units $i_1, \ldots, i_{n_s}$, respectively. The ordered data based on the ordered sample s are denoted by $d = \left\{\left(i_1, y_{i_1}\right), \ldots, \left(i_k, y_{i_{n_s}}\right)\right\} = \left(i_k, y_{i_k}; i_k \in s\right)$.

Let $\widetilde{s} = \left(j_1, j_2, \ldots, j_{\nu_s}\right)$ with $j_1 < j_2 < \ldots < j_{\nu_s}$ be the unordered sample obtained from s by taking distinct units of s and arranging them in ascending order of their labels. Let us denote the operator r, which transforms the ordered sample s to the unordered sample $\widetilde{s}$, i.e., $r(s) = \widetilde{s}$. The unordered data are denoted by $\widetilde{d} = \left\{\left(j_1, y_{j_1}\right), \ldots, \left(j_{\nu_s}, y_{j_{\nu_s}}\right)\right\} = \left(j, y_j; j \in \widetilde{s}\right)$. We define the operator R to get unordered data $\widetilde{d}$ from ordered data d, i.e., $R(d) = \widetilde{d}$.

Example 2.7.1

Let $U = (1,2,3,4,5)$, $\mathbf{y} = (10,15,15,20,10)$, and $s = (5,2,5)$. Here $y_1 = 10$, $y_2 = 15$, $y_3 = 15$, $y_4 = 20$, and $y_5 = 10$; $r(s) = \widetilde{s} = (2,5)$, $d = \{(5,10),(2,15),(5,10)\}$ and $R(d) = \widetilde{d} = \{(2,15),(5,10)\}$.

2.7.1 Sufficiency and Likelihood

Suppose that prior to the survey the surveyor had no idea about the values of the parameter $\mathbf{y} = (y_1,\ldots,y_N)$ and hence $\Omega_y = R^N$, the N-dimensional Euclidean space was considered the parametric space. After surveying the sample s, the surveyor collects the ordered data $d = \left(i_k, y_{i_k}; i_k \in s\right)$. The data d is said to be consistent with parameter vector $\mathbf{y}_0 = (y_{10}, \ldots, y_{i0}, \ldots, y_{N0})$ if $y_{i_k} = y_{i_k 0}$ for $i_k \in s$. After collection the data d, the surveyor knows that the values of y_i's belong to s and hence the parametric space Ω_y reduces to Ω_{yd} $(\subset \Omega_y)$, where Ω_{yd} consists of the vectors $\mathbf{y}$ with $y_j = y_{j0}$ for $j \in s$.

The unordered data $\tilde{d} = (i_{j_k}, y_{j_k}; j_k \in \tilde{s})$ obtained from d will be consistent with $\mathbf{y}_0$ if $y_{jk} = y_{jk0}$ for $j_k \in \tilde{s}$, i.e., $y_{j_1} = y_{j_1 0}, \ldots, y_{j_{\nu_s}} = y_{j_{\nu_s} 0}$. Here also, given the unordered data $\tilde{d}$, the parametric space Ω_y reduces to $\Omega_{y\tilde{d}}$, which consists of the vectors $\mathbf{y}$ with $y_j = y_{j0}$ for $j \in \tilde{s}$. Clearly, $\Omega_{yd} = \Omega_{y\tilde{d}}$ as s and $\tilde{s}$ consist of the same set of distinct units.

Example 2.7.2

Consider the population $U = (1,2,3,4)$ of 4 units from which an ordered sample $s = (1,2,2)$ is selected. Let the y-values of the units selected in the sample s be $y_1 = 50$ and $y_2 = 100$. In this case, $d = \{(1,50),(2,100),(2,100)\}$; $\Omega_y = (-\infty < y_1 < \infty, -\infty < y_2 < \infty, -\infty < y_3 < \infty, -\infty < y_4 < \infty) = R^4$, $\tilde{d} = \{(1,50),(2,100)\}$, $\Omega_{yd} = \Omega_{y\tilde{d}} = (50, 100, -\infty < y_3 < \infty, -\infty < y_4 < \infty)$. Here both d and $\tilde{d}$ are consistent with parameter $\mathbf{y} = (50,100,500,600)$ but inconsistent with $\mathbf{y} = (100,100,500,600)$. Data ($D$), a random variable, depends on the selection of the sample and realization of the parametric vector $\mathbf{y}$. Given data $D = d$, the likelihood function of the parameter $\mathbf{y}$ was obtained by Godambe (1966) as

$$
\begin{aligned}
L(\mathbf{y}|d) &= P\{D = d\} \\
&= \begin{cases} p(s) & \text{for } \mathbf{y} \in \Omega_{yd} \\ 0 & \text{for } \mathbf{y} \notin \Omega_{yd} \end{cases} \qquad (2.7.1) \\
&= p(s)\ I_d(\mathbf{y}) \ \text{ for } \forall\ \mathbf{y} \in \Omega_y
\end{aligned}
$$

where, $I_d(\mathbf{y})$ is an indicator variable defined as

$$
I_d(\mathbf{y}) = \begin{cases} 1 & \text{for } \mathbf{y} \in \Omega_{yd} \\ 0 & \text{for } \mathbf{y} \notin \Omega_{yd} \end{cases} \qquad (2.7.2)
$$

The likelihood function (2.7.2) is flat (constant), equal to $p(s)$ for $\mathbf{y} \in \Omega_{yd}$, and zero outside Ω_{yd}. Hence no unique maximum likelihood of $\mathbf{y}$ exists, and the likelihood function is noninformative.

Theorem 2.7.1

For a noninformative sampling design p, the unordered data $\tilde{d} = R(d)$ derived from an ordered data d is a sufficient statistic for $\mathbf{y}$.

Proof

Suppose an ordered sample s is selected with probability $p(s)$ using a sampling design p and $d = \left(i_k, y_{i_k}; i_k \in s\right)$ be the ordered data based on s. Here we denote D as a random variable, which takes the value d with probability

$$P\{D = d\} = p(s) \ I_d(\mathbf{y}) \text{ for } \forall \ \mathbf{y} \in \Omega_y$$

Similarly, if we denote $\tilde{D}$ as a random variable of getting an unordered data $\tilde{d}$, then

$$\begin{aligned} P\left(\tilde{D} = \tilde{d}\right) &= p(\tilde{s}) = \sum_{s|r(s)=\tilde{s}} p(s) \text{ for } \mathbf{y} \in \Omega_{y\tilde{d}} \text{ and} \\ P\left(\tilde{D} = \tilde{d}\right) &= 0 \text{ when } \mathbf{y} \notin \Omega_{y\tilde{d}} \end{aligned} \tag{2.7.3}$$

where, $\sum_{s|r(s)=\tilde{s}}$ denotes the sum over the ordered samples s, which produces the unordered sample $\tilde{s}$.

Furthermore, Eq. (2.7.3) can be written as

$$P\left\{\tilde{D} = \tilde{d}\right\} = \left\{\sum_{s|r(s)=\tilde{s}} p(s)\right\} I_{\tilde{d}}(y) \text{ for } \mathbf{y} \in \Omega_y$$

The conditional distribution of obtaining ordered data d_0 (based on an ordered sample s_0), given unordered data $\tilde{d}_1$, is given by

$$P\left(D = d_0 \middle| \tilde{D} = \tilde{d}_1\right) = \frac{P\left(D = d_0 \cap \tilde{D} = \tilde{d}_1\right)}{P\left(\tilde{D} = \tilde{d}_1\right)} \tag{2.7.4}$$

Now there are two cases.

Case I: $P(\tilde{D} = \tilde{d}_1) > 0$ and $R(d_0) = \tilde{d}_1$, here $I_{d_0}(\mathbf{y}) = I_{\tilde{d}_1}(\mathbf{y})$ and

$$P\left(D = d_0 \middle| \tilde{D} = \tilde{d}_1\right) = \frac{P(D = d_0)}{P\left(\tilde{D} = \tilde{d}_1\right)} = \frac{ps_0}{\left\{\sum_{s|r(s)=\tilde{s}_1} p(s)\right\}} \tag{2.7.5}$$

Case II: $P\left(\widetilde{D}=\widetilde{d}_1\right)>0$ and $R(d_0)\neq\widetilde{d}_1$, here $P\left(D=d_0\cap\widetilde{D}=\widetilde{d}_1\right)=0$ and hence

$$P\left(D=d_0\middle|\widetilde{D}=\widetilde{d}_1\right)=0 \qquad (2.7.6)$$

From (2.7.5) and (2.7.6), we note that $P\left(D=d_0\middle|\widetilde{D}=\widetilde{d}_1\right)$ is independent of parameter $\mathbf{y}$ as $p(s)$ is noninformative (does not involve $\mathbf{y}$). Hence, the unordered data $\widetilde{d}=R(d)$ is a sufficient statistic for $\mathbf{y}$.

2.7.2 Minimal Sufficient Statistic

A statistic $f(d)$, a function of data d, partitions the sample space of d. Let $\mathscr{P}_f$ be the partition associated with f. The statistic $f^*(d)$ is called a minimal sufficient statistic if and only if for any statistic $f(\neq f^*)$, each partition set of $\mathscr{P}_f$ is a subset of a partition set $\mathscr{P}_{f^*}$ induced by f^*. In other words, every set of $\mathscr{P}_{f^*}$ can be expressed as the union of the sets of $\mathscr{P}_f$.

Theorem 2.7.2

For a noninformative sampling design p, the unordered data $\widetilde{d}=R(d)$ derived from an ordered data d is a minimal sufficient statistic for $\mathbf{y}$.

Proof

Let s_1 and s_2 be two samples with $p(s_1)>0$ and $p(s_2)>0$ yielding data d_1 and d_2, respectively. Following Thompson and Seber (1996), we note that $\widetilde{d}$ is a minimal sufficient for $\mathbf{y}$ if for any two data points d_1 and d_2, the following holds:

$$\widetilde{d}_1=R(d_1)=R(d_2)=\widetilde{d}_2\Leftrightarrow P(D=d_1)=kP(D=d_2)\,\forall\ \mathbf{y}\in\Omega_y \qquad (2.7.7)$$

where k is a constant independent of $\mathbf{y}$.

Let $\widetilde{d}_1=\widetilde{d}_2$, then $I_{\widetilde{d}_1}(\mathbf{y})=I_{d_1}(\mathbf{y})=I_{\widetilde{d}_2}(\mathbf{y})=I_{d_2}(\mathbf{y})$ and

$$P(D=d_1)=p(s_1)I_{d_1}(\mathbf{y})=\frac{p(s_1)}{p(s_2)}p(s_2)I_{d_2}(\mathbf{y})=kP(D=d_2)$$

where $k=\dfrac{p(s_1)}{p(s_2)}$ is independent of $\mathbf{y}$.

Similarly, $P(D=d_1)=k\,P(D=d_2)\,\forall\ \mathbf{y}\in\Omega_y$ implies

$$p(s_1)I_{d_1}(\mathbf{y})=k\ p(s_2)I_{d_2}(\mathbf{y})\ \ \forall\ \ \mathbf{y}\in\Omega_y \qquad (2.7.8)$$

Since, $p(s_1), p(s_2)>0$ and $I_{d_1}(\mathbf{y})$ and $I_{d_2}(\mathbf{y})$ can take only two values 0 or 1, Eq. (2.7.8) implies $I_{d_1}(\mathbf{y})=I_{d_2}(\mathbf{y})$, i.e., $\widetilde{d}_1=R(d_1)=R(d_2)=\widetilde{d}_2$. Hence $\widetilde{d}$ is a minimal sufficient statistic for $\mathbf{y}$.

2.7.3 Rao–Blackwellization

Let an ordered sample s be selected from a population with a probability $p(s)$ and d be the corresponding ordered data. Let $t(d)$ be an estimator (not necessarily unbiased) for a parametric function $\theta(\mathbf{y}) = \theta$ and $\tilde{t}\left(\tilde{d}\right) = E_p\left[t(d)\middle|\tilde{d}\right]$, where $\tilde{d}$ be the unordered data obtained from d. Then we have the following theorem.

Theorem 2.7.3

(i) $E_p\{t(d)\} = E_p\left\{\tilde{t}\left(\tilde{d}\right)\right\}$

and

(ii) $MSE\{t(d)\} = E_p[t(d) - \theta]^2 \geq MSE\left\{\tilde{t}\left(\tilde{d}\right)\right\} = E_p\left[\tilde{t}\left(\tilde{d}\right) - \theta\right]^2$

Proof

(i) $E_p[t(d)] = E_p\left[E_p\left\{t(d)\middle|\tilde{d}\right\}\right] = E_p\left[\tilde{t}\left(\tilde{d}\right)\right]$

(ii)
$$\begin{aligned} MSE[t(d)] &= E_p[t(d) - \theta]^2 \\ &= E_p\left[\left\{\tilde{t}\left(\tilde{d}\right) - \theta\right\} + \left\{t(d) - \tilde{t}\left(\tilde{d}\right)\right\}\right]^2 \\ &= E_p\left\{\tilde{t}\left(\tilde{d}\right) - \theta\right\}^2 + E_p\left\{t(d) - \tilde{t}\left(\tilde{d}\right)\right\}^2 \end{aligned}$$

Now noting $E_p\left[\left\{\tilde{t}\left(\tilde{d}\right) - \theta\right\}\left\{t(d) \; - \tilde{t}\left(\tilde{d}\right)\right\}\right] = E_p\left[\left\{\tilde{t}\left(\tilde{d}\right) - \theta\right\}\right.$ $\left.E_p\left(\left\{t(d) - \tilde{t}\left(\tilde{d}\right)\right\}\middle|\tilde{d}\right)\right] = 0$, we find

$$\begin{aligned} MSE\{t(d)\} &= MSE\left\{\tilde{t}\left(\tilde{d}\right)\right\} + E_p\left\{t(d) - \tilde{t}\left(\tilde{d}\right)\right\}^2 \\ &\geq MSE\left\{\tilde{t}\left(\tilde{d}\right)\right\} \end{aligned}$$

Corollary 2.7.1

In case $t(d)$ is unbiased for θ, we find that,

(i) $\tilde{t}\left(\tilde{d}\right)$ is also unbiased for θ.

and

(ii) $V_p\left\{\tilde{t}\left(\tilde{d}\right)\right\} \leq V_p\{t(d)\}$.

Corollary 2.7.2

An estimator based on an ordered data can be improved by applying Rao–Blackwellization technique. Hence we conclude that an estimator based on ordered data is inadmissible.

Remark 2.7.1

Theorem 2.7.3 is greatly used for improvement of the conventional ordered estimators of various sampling designs such as SRSWR, PPSWR, and probability proportional to size without replacement (PPSWOR) sampling schemes.

Remark 2.7.2

It is natural to question as to whether two different unbiased estimators based on the same unordered data (sufficient statistic) for a particular population parameter, θ exists. The answer is "yes" because, although the unordered data $\widetilde{d}$ is a sufficient statistic, it is not a complete sufficient statistic (vide Theorem 2.7.4 below). Hence one can construct an infinite number of unbiased estimators as a function of the sufficient statistic for estimating a population parameter θ. But none of them attains the minimum variance for all possible values of θ, as we have shown earlier that the UMVUE does not exist.

Theorem 2.7.4 (Cassel et al., 1977)

The statistic $\widetilde{d}$ is not a complete sufficient statistic.

Proof

By definition, the statistic $\widetilde{d}$ will be a complete sufficient statistic for the parameter $\mathbf{y}$ if for any function $H(\widetilde{d})$ the following condition holds:

$$E\left\{H\left(\widetilde{d}\right)\right\} = 0 \;\; \forall \;\; \mathbf{y} \in \Omega_y \Rightarrow P\{H(\widetilde{d}) = 0\} = 1 \;\; \forall \;\; \mathbf{y} \in \Omega_y \tag{2.7.9}$$

To prove the theorem, consider a finite population $U = \{1,2,3,4\}$ of known size $N = 4$. Let $s_1 = (1,1,2)$, $s_2 = (1,2,2)$, $s_3 = (3,3,4)$ and $s_4 = (4)$ be all the possible samples with respective selection probabilities $p(s_1) = 0.10$, $p(s_2) = 0.20$, $p(s_3) = 0.30$ and $p(s_4) = 0.40$. Here the order samples s_1,s_2,s_3, and s_4 generates the unordered samples $\widetilde{s}_1 = (1,2)$, $\widetilde{s}_2 = (3,4)$, and $\widetilde{s}_3 = (4)$ with probabilities $p(\widetilde{s}_1) = 0.30$, $p(\widetilde{s}_2) = 0.30$, and $p(\widetilde{s}_3) = 0.40$. The inclusion probabilities of the four units are $\pi_1 = \pi_2 = 0.30$, $\pi_3 = 0.30$, and $\pi_4 = 0.70$. Consider the function

$$H\left(\widetilde{d}\right) = \sum_{i \in \widetilde{s}} \frac{1}{\pi_i} - N = \sum_{i \in \widetilde{s}} \frac{1}{\pi_i} - 4$$

Here $E\left\{H\left(\widetilde{d}\right)\right\} = 0 \;\; \forall \;\; \mathbf{y} \in \Omega_y$. But, $H\left(\widetilde{d}_1\right) = \dfrac{1}{\pi_1} + \dfrac{1}{\pi_2} - 4 = 2.667$, $H\left(\widetilde{d}_2\right) = \dfrac{1}{\pi_3} + \dfrac{1}{\pi_4} - 4 = 0.762$, and $H\left(\widetilde{d}_3\right) = \dfrac{1}{0.70} - 4 = -2.571$ are all nonzero. Hence $\widetilde{d}$ is not a complete sufficient statistic.

2.8 SAMPLING STRATEGIES

So far, we have compared performances of several estimators for a given sampling design. It is found that UMVUE exists only for a unicluster sampling design. Among the sampling designs that are used in practice, only the linear systematic sampling scheme (described in Chapter 4) is a unicluster sampling design. Other sampling designs do not belong to this category. So, it is a natural to search for a combination of a sampling design and an estimator, which may yield better estimators than an alternative combination of a sampling design and an estimator. A combination of an estimator and a sampling design is known as a sampling strategy. We will denote a sampling strategy by the pair $h = (p,t)$.

2.8.1 Unbiased Strategy

The strategy $h = (p,t)$ is to be p-unbiased for the parameter θ if the associated estimator t is p-unbiased, i.e., $E_p(t) = \theta \; \forall \; \mathbf{y} \in R^N$.

2.8.2 Uniformly Minimum Variance Unbiased Strategy

An unbiased strategy $h_1 = (p_1,t_1)$ is said to be better than another unbiased strategy $h_2 = (p_2,t_2)$ for estimating θ if

$$V_{p_1}(t_1) \leq V_{p_2}(t_2) \text{ for } \forall \; \mathbf{y} \in R^N$$

and

$$V_{p_1}(t_1) < V_{p_2}(t_2) \text{ for at least one } \mathbf{y} \in R^N.$$

A strategy $h_0 = (p_0,t_0)$ belonging to the class of unbiased strategies $\mathcal{H}$ is said to be the uniformly minimum variance unbiased strategy in $\mathcal{H}$ if

$$V_{p_0}(t_0) \leq V_p(t) \text{ for } h = (p, t)(\neq h_0) \in \mathcal{H} \text{ and } \forall \; \mathbf{y} \in R^N$$

Cassel et al. (1977) showed that there does not exist a UMV strategy in the class $\mathcal{H}_u(n)$ of unbiased fixed effective size n sampling strategies. The class $\mathcal{H}_u(n)$ consists of strategies $h = (p,t)$, where $p \in P_n$ and $t \in C_u$, P_n is the class of fixed effective size n sampling design, and C_u is the class of unbiased estimators.

2.8.3 Admissible Strategies

A strategy h_0 belonging to the class of strategies $\mathcal{H}$ is said to be admissible in $\mathcal{H}$ if there does not exist a strategy in $\mathcal{H}$, which is better than h_0.

Admissibility of strategies was considered by Joshi (1966), Godambe (1969), Ramakrishnan (1975), and Ericson (1970) among others. Good details have been given by Cassel et al. (1977). The concept of hyper-admissibility was introduced by Hanurav (1966). It was found that HTE is the unique hyperadmissible estimator in a certain class of estimators. The results of the comparisons of strategies and properties of admissibility have not been presented here. However, the comparisons of some strategies under super population models have been discussed in Chapter 6.

2.8.4 Minimax Strategy

A strategy $h_0 = (p_0, t_0)$ belonging to the class $\mathcal{H}$ of strategies is said to be a minimax strategy in $\mathcal{H}$ for estimating parameter θ if and only if

$$\underset{h \in \mathcal{H}}{Inf}\left[\underset{\mathbf{y} \in \Omega_y}{S_{up}}\ E_p(t-\theta)^2\right] = \underset{\mathbf{y} \in \Omega_y}{S_{up}}\ E_{p_0}(t_0-\theta)^2$$

The minimax strategies for estimating the finite population mean were studied by several authors including Aggarwal (1959, 1966), Godambe (1960), Godambe and Joshi (1965), and Chaudhuri (1969) and they derived minimax property of the sample mean in the restricted parameter space. Details are given by Cassel et al. (1977).

2.9 DISCUSSIONS

In a fixed population setup, the population vector **y** is considered as fixed, i.e., each unit is associated with a real number. The vector **y** is considered unknown and is called the parameter. Our object is to estimate certain parametric function $\theta = \theta(\mathbf{y})$ by selecting a sample by using a suitable sampling design. After collecting the sample, information of the study variable y is collected from the selected sample. The sample and the observed y-values constitute the data d. An estimator is a function of data d. The estimator is a random variable since its value varies from one data to another. The expectation is the average value of the estimator computed over all possible samples that might be realized by selection through a sampling design. The design-based inference comprises all possible samples that can be selected from a sampling design with positive probabilities. Given a sampling design, an unbiased estimator can be chosen in various ways. Godambe (1955) proved that for a given sampling design, UMVUE doest not exist in the class of linear homogeneous unbiased estimators. Hanurav (1966) showed that the UMVUE exists only in unicluster sampling design, which includes systematic sampling design. Basu (1971) proved the nonexistence of UMVUE in the class of all unbiased estimators. Because of the nonexistence of the UMVUE, alternative criteria for admissibility were proposed to guard against selection of unsuitable estimators. It was found that the HTE is admissible for both the classes of linear unbiased and all unbiased estimators. Numerous admissible estimators for estimating a parametric function exist. The concept of sufficiency in finite population sampling was introduced by Basu (1958). Basu showed that unordered data constitute a sufficient statistic. An estimator based on ordered data is inadmissible. An inadmissible estimator can be improved by using a sufficient statistic. This technique is known as Rao–Blackwellization. The unordered data are a sufficient statistic but not a complete sufficient statistic. So, we can construct numerous unbiased estimators for a particular parameter θ, which are function of a sufficient statistic. But, none of them is the UMVUE. The combination of a sampling design and an estimator is known as a sampling strategy. The best sampling strategy does not exist in a finite population setup. However, several admissible strategies for estimating a parameter θ exist.

2.10 EXERCISES

2.10.1 Consider the marks of four students

Students	1	2	3	4
Marks	80	52	30	70

Select a sample using the following design:

Sample	(1, 2, 1)	(1, 3, 4)	(1, 3)	(2, 4)	(1, 2, 3)	(2, 3, 4)
Probability	0.2	0.1	0.4	0.1	0.1	0.1

(i) Compute the HTE of the total marks of the four students.
(ii) Obtain the variance of the HTE.
(iii) Give an unbiased estimate of the variance of the HTE.

2.10.2 The following data relate consumption of food in six households in a certain locality.

Households	1	2	3	4	5	6
Expenditure on food ($)	1500	2000	3000	2500	1500	3000

Select a sample of three households using the following design.

Sample	(1,2,3)	(1,2,4)	(1,2,5)	(1,2,6)	(1,3,4)	(1,3,5)	(1,3,6)	(1,4,5)	(1,4,6)	(1,5,6)
Probability	0.05	0.05	0.05	0.04	0.04	0.04	0.06	0.06	0.06	0.04
Sample	(2,3,4)	(2,3,5)	(2,3,6)	(2,4,5)	(2,4,6)	(2,5,6)	(3,4,5)	(3,4,6)	(3,5,6)	(4,5,6)
Probability	0.05	0.05	0.05	0.03	0.03	0.07	0.07	0.05	0.05	0.06

(i) Estimate the average consumption of food by using the HTE.

(ii) Obtain the standard error of the HTE using both the HTE and Yates–Grundy's variance estimators.

2.10.3 Show that for an SRSWOR sampling design of size n

(i) HTE for the population total is $\widehat{Y}_{ht} = N\,\overline{y}(s)$, where $\overline{y}(s) = \sum_{i\in s} y_i/n$.

(ii) $V_{ht} = V_{YG} = N^2(1/n - 1/N)S_y^2$,

where $S_y^2 = \sum_{i=1}^{N}\left(y_i - \overline{Y}\right)^2/(N-1)$

(iii) $\widehat{V}_{ht} = \widehat{V}_{YG} = N^2(1/n - 1/N)s_y^2$,

where $s_y^2 = \sum_{i\in s}\left(y_i - \overline{y}(s)\right)^2/(n-1)$

2.10.4

(a) Prove that an unbiased estimator of the population variance $S_y^2 = \sum_{i=1}^{N}\left(y_i - \overline{Y}\right)^2/(N-1)$ is available if and only if $\pi_{ij} > 0\ \forall\ i \neq j = 1,\ldots,N$.

(b) Show that $\widehat{S}_y^2 = a\sum_{i\in s} y_i^2/\pi_i + b\sum\sum_{i\neq j\in s} y_i y_j/\pi_{ij}$ is an unbiased estimator for S_y^2, where $\pi_{ij} > 0\ \forall\ i \neq j$, $a = 1/N$, and $b = -1/N(N-1)$.

(c) Show that $\widehat{S}_y^2$ is admissible in the class of quadratic unbiased estimators of S_y^2.

2.10.5 Show that the estimator $e(s, y) = \sum_{i\in s} c_i y_i$ with $c_i \geq 1, \sum_{i\in s} 1/c_i = 1$

based on a fixed effective sampling design is admissible in the class of linear estimators (non necessarily unbiased) of Y (Godambe and Joshi, 1965).

2.10.6 Show that for a given sampling design (i) the sample mean based of distinct units and (ii) ratio estimator $\widehat{Y}_R = \dfrac{\sum_{i\in s} x_i}{\sum_{i\in s} y_i}\overline{X}$ (where $\overline{X}$ is known population mean of the charter x) are admissible in the class of all estimators for estimating the mean $\overline{Y}$ (Joshi, 1965a,b).

CHAPTER 3

Simple Random Sampling

3.1 INTRODUCTION

Simple random sampling is the simplest, popular, and commonly used probability sampling. In this chapter we will consider simple random sampling with (SRSWR) and without replacement (SRSWOR) procedures. Various results on simple random sampling have been presented in detail. Methods of estimating population mean, proportion, and their standard errors (SEs) have been proposed. Estimators of the domain mean, proportions, and their variances have been derived. Methods of determining confidence intervals for mean and proportions have also been proposed. Use of the Rao–Blackwell technique for improvement of estimators based on SRSWR procedure has been presented. The method of inverse sampling for estimating population proportion has also been discussed.

3.2 SIMPLE RANDOM SAMPLING WITHOUT REPLACEMENT

3.2.1 Sampling Scheme

On the first draw a unit is selected from the population of N units at random (with probability $1/N$). On the second draw, another unit is selected at random from the remaining $N-1$ units, which were not selected in the first draw. In general, at the rth draw, a unit is selected at random with probability $p_i(r)=\dfrac{1}{N-(r-1)}$ from $N-(r-1)$ units, which were not selected in the earlier $r-1$ draws where $r=1,\ldots,n$ and n is the required sample size. So, for an SRSWOR sampling scheme, the probability of selection of an ordered sample $s_o=(i_1\rightarrow i_2,\ldots,\rightarrow i_n)$ is $p(s_o)=\dfrac{1}{N}\cdot\dfrac{1}{N-1}\cdots\dfrac{1}{N-n+1}$, where the unit i_k is selected at the kth draw. Hence the probability of selection of an unordered sample $s=(j_1,\ldots,j_k,\ldots,j_n)$ obtained from s_o by arranging their label in ascending order as $j_1<\cdots<j_k\cdots<j_n$ is given by

$$p(s)=n!\cdot\frac{1}{N}\cdot\frac{1}{N-1}\cdots\frac{1}{N-n+1}=1\Big/\binom{N}{n}$$

Survey Sampling Theory and Applications
ISBN 978-0-12-811848-1
http://dx.doi.org/10.1016/B978-0-12-811848-1.00003-0

Therefore, the total number of distinct unordered samples of size n is $\binom{N}{n}$.

Theorem 3.2.1
The unconditional probability of selection of a unit at any draw is $1/N$.

Proof
The probability of selection of an ordered sample $s_o = (i_1 \rightarrow i_2, \ldots, \rightarrow i_n)$ is $p(s_o) = 1 \Big/ \left[n! \binom{N}{n}\right]$.

Hence the unconditional probability of selection of the ith unit at kth draw $= p(s_o) \times$ number of ways we can permute $(n-1)$ integers $(i_1, \ldots, i_{k-1}, i_{k+1}, \ldots, i_n)$ taking from the $(N-1)$ integers $1, \ldots, i-1, i+1, \ldots, N = \binom{N-1}{n-1}(n-1)!p(s_o) = \dfrac{1}{N}$.

3.2.2 Estimation of Population Mean and Variance

Theorem 3.2.2
Let $\overline{y}(s) = \sum_{i \in s} y_i/n$ be the sample mean based on an unordered sample s. Then,
(i) $\overline{y}(s)$ is an unbiased estimator for the population mean $\overline{Y}$
(ii) Variance of $\overline{y}(s)$ is

$$V(\overline{y}(s)) = \left(\frac{1}{n} - \frac{1}{N}\right) S_y^2 = \frac{1}{n}(1 - f_n) S_y^2$$

where $S_y^2 = \sum_{i=1}^{N} (y_i - \overline{Y})^2/(N-1)$ and $f_n = n/N$ is known as a sampling fraction.
(iii) An unbiased estimator of $V(\overline{y}(s))$ is

$$\widehat{V}(\overline{y}(s)) = \frac{1}{n}(1 - f_n) s_y^2$$

where $s_y^2 = \sum_{i \in s} (y_i - \overline{y}(s))^2/(n-1)$ is the sample variance.

Proof
(i) Let $\mathfrak{J}$ be the set of all possible $\binom{N}{n}$ unordered samples of size n and

$$I_{si} = \begin{cases} 1 \text{ if } i \in s \\ 0 \text{ if } i \notin s \end{cases}.$$

Then, $\overline{y}(s) = \dfrac{1}{n}\sum_{i=1}^{N} I_{si} y_i$

and

$$E(\overline{y}(s)) = \sum_{s\in \mathfrak{S}} \overline{y}(s)p(s) = \frac{1}{n}\sum_{s\in \mathfrak{S}}\sum_{i=1}^{N} y_i I_{si} \Bigg/ \binom{N}{n}$$

$$= \frac{1}{n}\sum_{i=1}^{N} y_i \sum_{s\in \mathfrak{S}} I_{si} \Bigg/ \binom{N}{n}$$

Now noting $\sum_{s\in \mathfrak{S}} I_{si}$ = total number of unordered samples containing the unit $i = \sum_{s\supset i} = \binom{N-1}{n-1}$,

$$\text{we get } E(\overline{y}(s)) = \frac{1}{n}\sum_{i=1}^{N} y_i \binom{N-1}{n-1} \Bigg/ \binom{N}{n}$$

$$= \frac{1}{N}\sum_{i=1}^{N} y_i = \overline{Y}.$$

(ii) $V(\overline{y}(s)) = E(\overline{y}(s))^2 - \overline{Y}^2$

$$= \frac{1}{n^2} E\left(\sum_{i=1}^{N} I_{si} y_i^2 + \sum_{i\neq j}^{N}\sum_{=1}^{N} I_{si} I_{sj} y_i y_j\right) - \overline{Y}^2$$

$$= \frac{1}{n^2}\sum_{s\in \mathfrak{S}} \frac{1}{\binom{N}{n}}\left(\sum_{i=1}^{N} I_{si} y_i^2 + \sum_{i\neq j}^{N}\sum_{=1}^{N} I_{si} I_{sj} y_i y_j\right) - \overline{Y}^2$$

$$= \frac{1}{n^2\binom{N}{n}}\left(\sum_{i=1}^{N} y_i^2 \sum_{s} I_{si} + \sum_{i\neq}^{N}\sum_{j=1}^{N} y_i y_j \sum_{s} I_{si} I_{sj}\right) - \overline{Y}^2$$

Noting $\sum_{s} I_{si} I_{sj} = \sum_{s\supset i,j}$ = number of unordered samples containing the units i and j $(i \neq j) = \binom{N-2}{n-2}$, we can verify part (ii) of the theorem.

(iii) $$(n-1)E\left(s_y^2\right) = E\left(\sum_{i\in s} y_i^2\right) - nE(\overline{y}(s))^2$$
$$= \sum_{s\in\mathfrak{I}}\sum_{i=1}^{N} I_{si}y_i^2 \Big/ \binom{N}{n} - n\left(V(\overline{y}(s)) + \overline{Y}^2\right)$$

Now noting

$$\sum_{s\in\mathfrak{I}}\sum_{i=1}^{N} I_{si}y_i^2 \Big/ \binom{N}{n} = \sum_{i=1}^{N} y_i^2 \binom{N-1}{n-1} \Big/ \binom{N}{n} = \frac{n}{N}\sum_{i=1}^{N} y_i^2,$$ we

find $E\left(s_y^2\right) = S_y^2$, i.e., s_y^2 is an unbiased estimator of S_y^2. Hence, part (iii) of the theorem is proved.

Theorem 3.2.3

The first two order inclusion probabilities π_i and $\pi_{ij}(i \neq j)$ for an SRSWOR design with sample size n are n/N and $n(n-1)/\{N(N-1)\}$, respectively.

Proof

$$\pi_i = \sum_{s\in\mathfrak{I}} I_{si}p(s)$$
$$= \sum_{s\in\mathfrak{I}} I_{si} \Big/ \binom{N}{n}$$
$$= \binom{N-1}{n-1} \Big/ \binom{N}{n} = n/N$$

and

$$\pi_{ij} = \sum_{s\in\mathfrak{I}} I_{si}I_{sj}p(s)$$
$$= \sum_{s\in\mathfrak{I}} I_{si}I_{sj} \Big/ \binom{N}{n}$$
$$= \binom{N-2}{n-2} \Big/ \binom{N}{n} = \{n(n-1)/N(N-1)\}$$

The Horvitz–Thomson estimator for the population total Y for SRSWOR sampling design is given by

$$\widehat{Y}_{ht} = \sum_{i\in s} y_i/\pi_i = N\,\overline{y}(s)$$

Using Theorem 3.2.3, we can prove Theorem 3.2.2 in the following alternative way:

(i) $$E(\overline{y}(s)) = \frac{1}{N}E\left(\widehat{Y}_{ht}\right) = \frac{Y}{N} = \overline{Y}$$

(ii) $$V[\overline{y}(s)] = \frac{1}{N^2}V\left(\widehat{Y}_{ht}\right) = \frac{1}{N^2}\left[\frac{1}{2}\sum_{i\neq}^{N}\sum_{j}^{N}(\pi_i\pi_j - \pi_{ij})\left(\frac{y_i}{\pi_i} - \frac{y_j}{\pi_j}\right)^2\right] = \left(\frac{1}{n} - \frac{1}{N}\right)S_y^2$$

and

(iii) An unbiased estimator of $V(\overline{y}(s))$ is

$$\widehat{V}(\overline{y}(s)) = \frac{1}{N^2}\widehat{V}_{YG} = \frac{1}{N^2}\left[\frac{1}{2}\sum_{i\neq}\sum_{j\in s}\frac{(\pi_i\pi_j - \pi_{ij})}{\pi_{ij}}\left(\frac{y_i}{\pi_i} - \frac{y_j}{\pi j}\right)^2\right] = \left(\frac{1}{n} - \frac{1}{N}\right)s_y^2$$

Remark 3.2.1

Since $\widehat{Y}_{ht} = \sum_{i\in s} y_i/\pi_i = N\,\overline{y}(s)$ is admissible for $Y = N\overline{Y}$, the sample mean $\overline{y}(s)$ is also an admissible estimator of the population mean $\overline{Y}$.

Theorem 3.2.4

Let s be a sample of size n selected from a population of size N by SRSWOR method, s_1 a subsample of size $m\ (\leq n)$ selected from s by SRSWOR method, and $\overline{y}(s_1) = \sum_{i\in s_1} y_i/m$. Then,

(i) $E(\overline{y}(s_1)) = \overline{Y}$

(ii) $V(\overline{y}(s_1)) = \left(\frac{1}{m} - \frac{1}{N}\right)S_y^2$

(iii) An unbiased estimator of $V(\overline{y}(s_1))$ is

$$\widehat{V}(\overline{y}(s)) = \left(\frac{1}{m} - \frac{1}{N}\right)s_{1y}^2, \text{ where } s_{1y}^2 = \sum_{i \in s_1}(y_i - \overline{y}(s_1))^2/(m-1)$$

(iv) $Cov(\overline{y}(s_1), \overline{y}(s)) = V(\overline{y}(s)) = \left(\frac{1}{n} - \frac{1}{N}\right)S_y^2$

where $Cov[\overline{y}(s_1), \overline{y}(s)] =$ covariance between $\overline{y}(s_1)$ and $\overline{y}(s)$.

Proof

(i) $E(\overline{y}(s_1)) = E[E(\overline{y}(s_1)|s)] = E(\overline{y}(s)) = \overline{Y}$

(ii)
$$\begin{aligned} V(\overline{y}(s_1)) &= E[V(\overline{y}(s_1)|s)] + V[E(\overline{y}(s_1)|s)] \\ &= \left(\frac{1}{m} - \frac{1}{n}\right)E\left(s_y^2\right) + V(\overline{y}(s)) \\ &= \left(\frac{1}{m} - \frac{1}{n}\right)S_y^2 + \left(\frac{1}{n} - \frac{1}{N}\right)S_y^2 \\ &= \left(\frac{1}{m} - \frac{1}{N}\right)S_y^2 \end{aligned}$$

(iii)
$$\begin{aligned} Cov(\overline{y}(s_1), \overline{y}(s)) &= E(\overline{y}(s_1)\overline{y}(s)) - \{E(\overline{y}(s_1))\}\{E(\overline{y}(s))\} \\ &= E[E(\overline{y}(s_1)|s)\overline{y}(s)] - \overline{Y}^2 \\ &= E(\overline{y}(s))^2 - \overline{Y}^2 \\ &= V(\overline{y}(s)) \\ &= \left(\frac{1}{n} - \frac{1}{N}\right)S_y^2 \end{aligned}$$

Theorem 3.2.5

Let s be a sample of size n selected from a population U of size N by the SRSWOR method, s^* be a sample of size m selected from $U - s$ (the set of units not belonging to s) by the SRSWOR method, and $\overline{y}(s^*) = \sum_{i \in s^*} y_i/m$.

Then,

(i) $E(\overline{y}(s^*)) = \overline{Y}$

(ii) $V(\overline{y}(s^*)) = \left(\frac{1}{m} - \frac{1}{N}\right)S_y^2$

(iii) An unbiased estimator of $V(\overline{y}(s^*))$ is

$$\widehat{V}(\overline{y}(s^*)) = \left(\frac{1}{m} - \frac{1}{N}\right)s_y^{*2}$$

(iv) $Cov(\overline{y}(s^*), \overline{y}(s)) = -\dfrac{S_y^2}{N}$

where $s_y^{*2} = \sum_{i \in s^*} \left(y_i - \overline{y}(s^*)\right)^2/(m-1)$.

Proof

(i) $E(\overline{y}(s^*)) = E[E(\overline{y}(s^*)|s)]$

$$= E\left(\sum_{i \in U-s} y_i/(N-n)\right)$$

$$= E(Y - n\overline{y}(s))/N - n$$

$$= \overline{Y}$$

(ii) $$V(\overline{y}(s^*)) = E[V(\overline{y}(s^*)|s)] + V[E(\overline{y}(s^*)|s)] \tag{3.2.1}$$

Now $V(\overline{y}(s^*)|s) = \left(\dfrac{1}{m} - \dfrac{1}{N-n}\right)S_{U-s,y}^2$

where $S_{U-s,y}^2 = \dfrac{1}{N-n-1}\left[\sum_{i \in U-s} y_i^2 - \dfrac{1}{(N-n)}\left(\sum_{i \in U-s} y_i\right)^2\right]$.

$$E\left(S_{U-s,y}^2\right) = \frac{1}{N-n-1}E\left[\left(\sum_{i \in U} y_i^2 - \sum_{i \in s} y_i^2\right) - \frac{1}{(N-n)}\left(N\overline{Y} - n\overline{y}(s)\right)^2\right]$$

$$= \frac{1}{N-n-1}\left[\left(\sum_{i \in U} y_i^2 - \frac{n}{N}\sum_{i \in U} y_i^2\right)\right.$$

$$\left. - \frac{1}{(N-n)}E\left\{\left(N\overline{Y}\right)^2 + n^2\left(\overline{y}(s)\right)^2 - 2\,n\,N\,\overline{Y}\,\overline{y}(s)\right\}\right]$$

$$= \frac{1}{N-n-1}\left[\left(\frac{N-n}{N}\sum_{i \in U} y_i^2\right)\right.$$

$$\left. - \frac{1}{(N-n)}\left\{(N-n)^2\,\overline{Y}^2 + n^2 V(\overline{y}(s))\right\}\right] = S_y^2$$

Hence,

$$E[V(\overline{y}(s^*)|s)] = \left(\frac{1}{m} - \frac{1}{N-n}\right)S_y^2 \tag{3.2.2}$$

Furthermore,

$$V[E(\overline{y}(s^*)|s)] = V\left[\frac{N\overline{Y} - n\overline{y}(s)}{N-n}\right] = \left(\frac{n}{N-n}\right)^2 V(\overline{y}(s))$$
$$= \left(\frac{1}{N-n} - \frac{1}{N}\right) S_y^2 \qquad (3.2.3)$$

Finally substituting (3.2.2) and (3.2.3) in (3.2.1), we can verify part (ii) of the theorem.

(iii) $E\left(s_y^{*2}\right) = E\left[E\left(s_y^{*2}|s\right)\right]$

$$= \frac{1}{m-1} E\left[\sum_{i\in s^*} y_i^2 - m(\overline{y}(s^*))^2\right]$$

$$= \frac{1}{m-1}\left[E\left\{E\left(\sum_{i\in s^*} y_i^2|s\right)\right\} - m\left\{V(\overline{y}(s^*) + \overline{Y}^2\right\}\right]$$

$$= \frac{1}{m-1}\left[E\left(\frac{m}{N-n}\sum_{i\in U-s} y_i^2\right) - m\left\{\left(\frac{1}{m} - \frac{1}{N}\right)S_y^2 + \overline{Y}^2\right\}\right]$$

$$= \frac{1}{m-1}\left[\frac{m}{N-n}\left\{\sum_{i\in U} y_i^2 - E\left(\sum_{i\in s} y_i^2\right)\right\} - m\left\{\left(\frac{1}{m} - \frac{1}{N}\right)S_y^2 + \overline{Y}^2\right\}\right]$$

$$= \frac{1}{m-1}\left[\frac{m}{N-n}\left(\sum_{i\in U} y_i^2 - \frac{n}{N}\sum_{i\in U} y_i^2\right) - m\left\{\left(\frac{1}{m} - \frac{1}{N}\right)S_y^2 + \overline{Y}^2\right\}\right]$$

$$= S_y^2 \qquad (3.2.4)$$

Using (3.2.4) we verify $E\left[\widehat{V}(\overline{y}(s^*))\right] = V[\overline{y}(s^*)]$

(iv) $Cov[\overline{y}(s^*), \overline{y}(s)] = E[\overline{y}(s^*), \overline{y}(s)] - [E\{\overline{y}(s^*)\}][E\{\overline{y}(s)\}]$

$$= E[E\{\overline{y}(s^*)|s\}\overline{y}(s)] - Y^2$$

$$= \frac{1}{N-n}[E\{N\overline{Y} - n\overline{y}(s)\}\overline{y}(s)] - \overline{Y}^2$$

$$= \frac{1}{N-n}\left[N\overline{Y}^2 - nV\{\overline{y}(s)\} - n\overline{Y}^2\right] - \overline{Y}^2$$

$$= -S_y^2\big/N$$

3.2.3 Estimation of Population Covariance

Let y_i and x_i be the value of the variables y and x for the ith unit of the population U. Then the finite population covariance between x and y is defined as

$$S_{xy} = \sum_{i=1}^{N} (y_i - \overline{Y})(x_i - \overline{X})/(N-1) \text{ where } \overline{X} = \sum_{i=1}^{N} x_i/N.$$

Let a sample s of size n be selected by SRSWOR. Let $\overline{x}(s)$ and $\overline{y}(s)$ be the sample mean of x and y, respectively. The sample covariance between x and y is defined as

$$s_{xy} = \sum_{i\in s} \{x_i - \overline{x}(s)\}\{y_i - \overline{y}(s)\}/(n-1).$$

Theorem 3.2.6

$$E(s_{xy}) = S_{xy}$$

Proof

$$\begin{aligned}
E(s_{xy}) &= E\left[\sum_{i\in s} x_i y_i - n\overline{x}(s)\overline{y}(s)\right] \Big/ (n-1) \\
&= E\left[\frac{1}{n}\sum_{i\in s} x_i y_i - \frac{1}{n(n-1)}\sum_{i\neq}\sum_{j\in s} x_i y_j\right] \\
&= \left[\frac{1}{n}\sum_{i\in U} x_i y_i \sum_{s\supset i} - \frac{1}{n(n-1)}\sum_{i\neq}\sum_{j\in U} x_i y_j \sum_{s\supset i,j}\right] \Big/ \binom{N}{n} \\
&= \frac{1}{N}\sum_{i\in U} x_i y_i - \frac{1}{N(N-1)}\sum_{i\neq}\sum_{j\in U} x_i y_j \\
&= S_{xy}
\end{aligned}$$

Theorem 3.2.7

(i) The covariance between $\overline{y}(s)$ and $\overline{x}(s)$ is

$$\begin{aligned}
Cov\{\overline{x}(s), \overline{y}(s)\} &= \left(\frac{1}{n} - \frac{1}{N}\right) S_{xy} \\
&= \left(\frac{1}{n} - \frac{1}{N}\right) \rho S_x S_y
\end{aligned}$$

(ii) An unbiased estimator of $Cov\{\overline{x}(s), \overline{y}(s)\}$ is

$$\widehat{Cov}\{\overline{x}(s), \overline{y}(s)\} = \left(\frac{1}{n} - \frac{1}{N}\right) s_{xy}$$
$$= \left(\frac{1}{n} - \frac{1}{N}\right) \widehat{\rho} s_x s_y$$

where

$\rho =$ population correlation coefficient between x and $y = S_{xy}/(S_x S_y)$

and

$\widehat{\rho} =$ sample correlation coefficient between x and $y = s_{xy}/(s_x s_y)$

Proof

(i) $$Cov\{\overline{x}(s), \overline{y}(s)\} = E\{\overline{x}(s)\overline{y}(s)\} - [E\{\overline{x}(s)\}][E\{\overline{y}(s)\}]$$
$$= E\left(\sum_{i \in s} x_i\right)\left(\sum_{i \in s} y_i\right) \Big/ n^2 - \overline{X}\,\overline{Y}$$
$$= E\left(\sum_{i \in s} x_i y_i + \sum_{i \neq} \sum_{j \in s} x_i y_j\right) \Big/ n^2 - \overline{X}\,\overline{Y}$$
$$= \frac{1}{\binom{N}{n}} \sum_{s} \left(\sum_{i \in s} x_i y_i + \sum_{i \neq} \sum_{j \in s} x_i y_j\right) \Big/ n^2 - \overline{X}\,\overline{Y}$$
$$= \left(\sum_{i=1}^{N} x_i y_j \sum_{s \supset i} + \sum_{i \neq}^{N} \sum_{j=1}^{N} x_i y_j \sum_{s \supset i,j}\right) \Big/ \left\{n^2 \binom{N}{n}\right\} - \overline{X}\,\overline{Y}$$
$$= \left\{\sum_{i=1}^{N} x_i y_i \binom{N-1}{n-1} + \sum_{i \neq}^{N} \sum_{j=1}^{N} x_i y_j \right.$$
$$\left. \times \binom{N-2}{n-2}\right\} \Big/ \left\{n^2 \binom{N}{n}\right\} - \overline{X}\,\overline{Y}$$
$$= \frac{1}{Nn}\left(\sum_{i=1}^{N} x_i y_i + \frac{n-1}{N-1} \sum_{i \neq}^{N} \sum_{j=1}^{N} x_i y_j\right) - \overline{X}\,\overline{Y}$$
$$= \left(\frac{1}{n} - \frac{1}{N}\right) S_{xy}$$

(ii) The result follows from Theorem 3.2.6.

An Alternative Proof of the Theorem 3.2.7

(i) Writing $k_i = y_i - x_i$ for every $i \in U$ and $\overline{k}(s) = \sum_{i \in s} k_i/n = \overline{y}(s) - \overline{x}(s)$, we get

$$V\left\{\overline{k}(s)\right\} = V\{\overline{y}(s)\} + V\{\overline{x}(s)\} - 2Cov\{\overline{y}(s), \overline{x}(s)\} \text{ and hence}$$

$$Cov\{\overline{y}(s), \overline{x}(s)\} = \left[V\{\overline{y}(s)\} + V\{\overline{x}(s)\} - V\left\{\overline{k}(s)\right\}\right]/2 \tag{3.2.5}$$

Now using Theorem 3.2.2, we get

$$V\left\{\overline{k}(s)\right\} = \left(\frac{1}{n} - \frac{1}{N}\right) S_k^2$$

where

$$S_k^2 = \sum_{i=1}^{N} \left(k_i - \overline{K}\right)^2/(N-1) = S_y^2 + S_x^2 - 2S_{xy}, \overline{K} = \overline{Y} - \overline{X} \text{ and}$$

$$S_x^2 = \sum_{i=1}^{N} \left(x_i - \overline{X}\right)^2/(N-1) \tag{3.2.6}$$

Finally, substituting $V\{\overline{x}(s)\} = \left(\frac{1}{n} - \frac{1}{N}\right) S_x^2$, $V\{\overline{y}(s)\} = \left(\frac{1}{n} - \frac{1}{N}\right) S_y^2$ and $S_k^2 = S_y^2 + S_x^2 - 2S_{xy}$ in (3.2.5), we can verify the theorem.

(ii) Let $s_k^2 = \frac{1}{n-1} \sum_{i \in s} \left\{k_i - \overline{k}(s)\right\}^2 = s_x^2 + s_y^2 - 2s_{xy}$, where $s_x^2 = \frac{1}{n-1} \sum_{i \in s} \{x_i - \overline{x}(s)\}^2$. Theorem 3.2.2 yields

$$E\left(s_k^2\right) = S_k^2$$

i.e., $E\left(s_x^2\right) + E\left(s_y^2\right) - 2E\left(s_{xy}\right) = S_x^2 + S_y^2 - 2S_{xy}$;

i.e., $E\left(s_{xy}\right) = S_{xy}$

3.2.4 Estimation of Population Proportion

Sometimes we need to estimate the proportion of the population that possesses a certain attribute A, such as smoking, drug addiction, or unemployment. In such a situation, we take $y_i = 1$ if the ith unit belongs to the group A and $y_i = 0$ otherwise (if the ith unit does not belong to the group A). So in this case, $Y = N_A$ = total number of units possessing the attribute A and $\overline{Y} = N_A/N = \pi_A$ = proportion of units in the population belonging to the group A; $\overline{y}(s) = n_A/n = \widehat{\pi}_A$ = proportion of units in the sample that belong to the group A and n_A is the total number of units in the sample that belong to the group A. Now noting

(i) $S_y^2 = \frac{1}{(N-1)}\sum_{i=1}^{N}\left(y_i - \overline{Y}\right)^2 = \frac{1}{(N-1)}\left(\sum_{i=1}^{N} y_i - N\overline{Y}^2\right)$

$= \frac{N\{\pi_A(1-\pi_A)\}}{N-1}$ (since $y_i = 0$ or 1),

(ii) $s_y^2 = \sum_{i \in s}\{y_i - \overline{y}(s)\}^2/(n-1) = \frac{n}{n-1}\widehat{\pi}_A(1-\widehat{\pi}_A)$,

and using Theorem 3.2.2, we have the following.

Theorem 3.2.8

For an SRSWOR sampling of size n,

(i) $\widehat{\pi}_A$ is an unbiased estimator for the population proportion π_A.

(ii) Variance of $\widehat{\pi}_A$ is

$$V(\widehat{\pi}_A) = \frac{N-n}{n(N-1)}\pi_A(1-\pi_A)$$

and

(iii) An unbiased estimator of $V(\widehat{\pi}_A)$ is

$$\widehat{V}(\widehat{\pi}_A) = \frac{N-n}{N(n-1)}\widehat{\pi}_A(1-\widehat{\pi}_A).$$

3.2.5 Estimation of Domain Mean and Total

Suppose we want to estimate $Y_D = \sum_{i \in D} y_i$, the domain total of the y-values of a section D of the population U. Let N_D be the population domain size = total number of units in the domain D, then $\overline{Y}_D = Y_D/N_D$ = population domain mean of y and $S_{yD}^2 = \sum_{i \in D}\left(y_i - \overline{Y}_D\right)^2/(N_D - 1)$ = population domain variance of y. Let $s_D(\subset s)$ be the set of units in the sample s, which belong to D, and $n_D(\leq n)$ be the total numbers of units in s_D, $\overline{y}(s_D) = \sum_{i \in s_D} y_i/n_D$ = sample domain mean and $s_{yD}^2 = \sum_{i \in s_D}\{y_i - \overline{y}(s_D)\}^2/(n_D - 1)$ be the sample domain variance.

Let us define $z_i = d_i y_i$, where $d_i = 1$ if the ith unit belongs to D and $d_i = 0$ otherwise. Then, $Z = \sum_{i=1}^{N} z_i = Y_D$ = population domain total, $\overline{Y}_D = Y_D/N_D$ = population domain mean, $\pi_D = N_D/N$ = population domain proportion, and $\overline{Z} = Z/N = \pi_D\,\overline{Y}_D$; the sample domain mean $= \overline{y}(s_D) = \sum_{i\in s_D} y_i/n_D$ and the sample domain proportion $=$ $\widehat{\pi}_D = n_D/n$. Furthermore,

$$S_z^2 = \sum_{i\in U} \left(z_i - \overline{Z}\right)^2/(N-1)$$

$$= \left[\sum_{i\in D} y_i^2 - N\left(\pi_D \overline{Y}_D\right)^2\right] \Big/ (N-1)$$

$$= (N_D - 1)S_{yD}^2/(N-1) + N_D\overline{Y}_D^2(1-\pi_D)/(N-1)$$

$$= \left[(N\pi_D - 1)S_{yD}^2 + N\pi_D(1-\pi_D)\overline{Y}_D^2\right] \Big/ (N-1),$$

$$\overline{z}(s) = \sum_{i\in s} z_i/n = \sum_{i\in s_D} y_i/n = \widehat{\pi}_D\overline{y}(s_D) \text{ and } s_z^2 = \sum_{i\in s} \left[z_i - \overline{z}(s)\right]^2/(n-1)$$

$$= (n\widehat{\pi}_D - 1)s_{yD}^2 \Big/ (n-1) + n\widehat{\pi}_D(1-\widehat{\pi}_D)\{\overline{y}(s_D)\}^2/(n-1)$$

Now using Theorem 3.2.2, we derive the following results:

Theorem 3.2.9

(i) $\widehat{Y}_D = N\,\overline{z}(s) = N\widehat{\pi}_D\overline{y}(s_D)$ is an unbiased estimator of the domain total Y_D

(ii) The variance of $\widehat{Y}_D$ is

$$V\left[\widehat{Y}_D\right] = N^2\left(\frac{1}{n} - \frac{1}{N}\right)S_z^2$$

$$= N^2\left(\frac{1}{n} - \frac{1}{N}\right)\left[\frac{(N\pi_D - 1)}{N-1}S_{yD}^2 + \frac{N\pi_D(1-\pi_D)}{N-1}\overline{Y}_D^2\right]$$

(iii) An unbiased estimator of $V\left[\widehat{Y}_D\right]$ is

$$\widehat{V}\left[\widehat{Y}_D\right] = N^2\left(\frac{1}{n} - \frac{1}{N}\right)s_z^2$$

$$= N^2\left(\frac{1}{n} - \frac{1}{N}\right)\left[\frac{(n\widehat{\pi}_D - 1)}{n-1}s_{yD}^2 + \frac{n\widehat{\pi}_D(1-\widehat{\pi}_D)}{n-1}\{\overline{y}(s_D)\}^2\right]$$

Remark 3.2.2

In case the domain size N_D is known, one can easily get an unbiased estimator for the domain mean $\overline{Y}_D$ as $\widehat{\overline{Y}}_D = \widehat{Y}_D/N_D = \dfrac{\widehat{\pi}_D}{\pi_D}\overline{y}(s_D)$. But in most situations, N_D is unknown, and we can use the sample mean $\overline{y}(s_D)$ as an unbiased estimator of $\overline{Y}_D$. The following theorem gives the detailed properties of $\overline{y}(s_D)$.

Theorem 3.2.10

(i) $E[\overline{y}(s_D)] = \overline{Y}_D$

(ii) $V[\overline{y}(s_D)] = \left[E\left(\frac{1}{n_D}\right) - \frac{1}{N_D}\right] S_{yD}^2$

(iii) $\widehat{V}[\overline{y}(s_D)] = \left(\frac{1}{n_D} - \frac{1}{N_D}\right) s_{yD}^2$

where $V[\overline{y}(s_D)]$ and $\widehat{V}[\overline{y}(s_D)]$ are respectively the variance of $\overline{y}(s_D)$ and an unbiased estimator of $V[\overline{y}(s_D)]$.

Proof

For fixed n_D, we regard $\overline{y}(s_D)$ as a random sample of the entire domain D selected by SRSWOR method. Hence

$$E\{\overline{y}(s_D)|n_D\} = \overline{Y}_D, V\{\overline{y}(s_D)|n_D\} = \left(\frac{1}{n_D} - \frac{1}{N_D}\right) S_{yD}^2 \text{ and } E\left\{s_{yD}^2 \middle| n_D\right\} = s_{yD}^2 \tag{3.2.7}$$

Now using (3.2.7), we can find

(i) $E[\overline{y}(s_D)] = E[\overline{y}(s_D)|n_D] = \overline{Y}_D$

(ii)
$$V[\overline{y}(s_D)] = E[V\{\overline{y}(s_D)|n_D\}] + V[E\{\overline{y}(s_D)|n_D\}]$$
$$= \left[E\left(\frac{1}{n_D}\right) - \frac{1}{N_D}\right] S_{yD}^2 + V\left(\overline{Y}_D\right)$$
$$= \left[E\left(\frac{1}{n_D}\right) - \frac{1}{N_D}\right] S_{yD}^2$$

(iii)
$$E\left[\widehat{V}\{\overline{y}(s_D)\}\right] = E\left[E\left(\widehat{V}\{\overline{y}(s_D)|n_D\}\right)\right]$$
$$= \left[E\left(\frac{1}{n_D}\right) - \frac{1}{N_D}\right] S_{yD}^2$$
$$= V[\overline{y}(s_D)]$$

Example 3.2.1

Table 3.2.1 gives the height, weight, and gender of 30 students who participated in a sport.

Table 3.2.1

Serial no. of students	Height (cm) x	Weight (kg) y	Gender Male = 1 Female = 0	Serial no. of students	Height (cm) x	Weight (kg) y	Gender Male = 1 Female = 0
1	155	60	0	16	165	70	1
2	160	70	1	17	140	50	0
3	170	72	1	18	165	70	1
4	150	56	0	19	168	72	1
5	170	70	1	20	160	65	1
6	175	75	0	21	158	58	0
7	160	70	1	22	160	60	0
8	160	70	0	23	155	50	1
9	155	55	0	24	170	75	1
10	150	60	1	25	145	78	1
11	158	65	0	26	156	65	1
12	170	80	1	27	156	62	0
13	165	75	1	28	175	42	1
14	165	70	1	29	170	75	1
15	160	72	1	30	150	60	0

Select a sample of size 8 by the SRSWOR method. From the selected sample, (i) estimate the mean height and weight of students (male and female combined) and obtain the variances of the estimator used. Estimate the SEs of the estimators. (ii) Estimate the proportion of the male and female students with their SEs. (iii) Estimate the mean weight of the male and female students and estimate their SEs. (iv) Give an unbiased estimator of the covariance of height and weight of the students (male and female combined).

To select 30 students from a list of 60 students, we consider three sets of two-digit random numbers. The first set consists of random numbers 01−30; the second set 31−60, and the third set 61−90, respectively. We associate random numbers 01, 02,..., 30 of the first set to the student serial numbers 1, 2,..., 30, respectively. Random numbers 31, 32,..., 60 of the second set are associated with the units 1, 2,..., 30, respectively. Similarly, random numbers 61, 62,..., 90 of the third set are associated with the units 1, 2,..., 30, respectively. From the page of a random number table, we select the random numbers row wise as follows:

Random number selected	45	45	27	54	46	66	60	24	67	74
Unit selected	15	—	27	24	16	6	30	—	7	14

Here the symbol "—" indicates selection of no unit because the unit 15 and 24 are already selected in earlier draws.

So, the selected ordered sample is $s_o = (15, 27, 24, 16, 6, 30, 7, 14)$. After arranging the units in ascending order of label, the unordered sample obtained is $s = (6, 7, 14, 15, 16, 24, 27, 30)$. The data obtained from the selected sample s is as follows:

Serial no. of students	Height (cm) x	Weight (kg) y	Gender Male = 1 Female = 0
6	175	75	0
7	160	70	1
14	165	70	1
15	160	72	1
16	165	70	1
24	170	75	1
27	156	62	0
30	150	60	0

(i) Estimated mean height and weight (male and female combined) are given by $\overline{x}_s = 162.625$ cm and $\overline{y}_s = 69.25$ kg, respectively. The variance of $\overline{x}_s$ is $V(\overline{x}_s) = (1/n - 1/N)S_x^2 = (1/8 - 1/30) \times 73.981 = 6.781\ \text{cm}^2$. The variance of $\overline{y}_s$ is $V(\overline{y}_s) = (1/n - 1/N)S_y^2 = (1/8 - 1/30) \times 83.374$ $= 7.642\ \text{kg}^2$

The estimated SEs for $\overline{x}_s$ and $\overline{y}_s$ are $se(\overline{x}_s) = \sqrt{\widehat{V}(\overline{x}_s)} = \sqrt{(1/n - 1/N)s_x^2} = \sqrt{(1/8 - 1/30) \times 62.267} = 2.389$ cm and $se(\overline{y}_s) = \sqrt{\widehat{V}(\overline{y}_s)} = \sqrt{(1/n - 1/N)s_y^2} = \sqrt{(1/8 - 1/30) \times 30.5} = 1.672$ kg, respectively.

(ii) The estimated proportions of male and female students are given by $\widehat{\pi}_m = 5/8 = 0.625$ and $\widehat{\pi}_f = 3/8 = 0.375$. Estimated SEs of $\widehat{\pi}_m$ and $\widehat{\pi}_f$ are given respectively $Se(\widehat{\pi}_m) = \sqrt{\frac{N - n}{N(n - 1)}\widehat{\pi}_m(1 - \widehat{\pi}_m)}$ $= \sqrt{\frac{30 - 8}{30 \times 7} \times 0.625 \times (1 - 0.625)} = 0.156$ and $Se(\widehat{\pi}_f) = \sqrt{\frac{N - n}{N(n - 1)}\widehat{\pi}_f(1 - \widehat{\pi}_f)} = \sqrt{\frac{30 - 8}{30 \times 7} \times 0.375 \times (1 - 0.375)} = 0.156$

(iii) Selected male and female students are given by $s_1 = (7, 14, 15, 16, 24)$ and $s_2 = (6, 27, 30)$, respectively. Estimated mean weights of the male and female students are given by $\overline{y}_{s_1} = 71.4$ kg and $\overline{y}_{s_2} = 65.7$ kg. The sample variances of the weights of male and female students are $s_{1y}^2 = 4.8$ and $s_{2y}^2 = 66.33$. Estimated SEs for the male and females are $Se(\overline{y}_{s_1}) = \sqrt{\widehat{V}(\overline{y}_{s_1})} = \sqrt{(1/5 - 1/8)s_{1y}^2} = \sqrt{(1/5 - 1/8) \times 4.8} = 0.6$ kg and $Se(\overline{y}_{s_2})$ $= \sqrt{\widehat{V}(\overline{y}_{s_2})} = \sqrt{(1/3 - 1/8)s_{2y}^2} = \sqrt{(1/3 - 1/8) \times 66.33} = 3.717$ kg.

(iv) Unbiased estimate of the covariance of the height and weight of the students (male and female combined) = sample covariance = $s_{xy} =$ $\sum_{i \in s}\{x_i - \overline{x}(s)\}\{y_i - \overline{y}(s)\}/(n - 1) = \left(\sum_{i \in s} x_i y_i - 8 \times 162.625 \times 69.25\right)/7$ $= 38.964$.

3.3 SIMPLE RANDOM SAMPLING WITH REPLACEMENT

3.3.1 Sampling Scheme

On the first draw, a unit is selected from a population of N units at random, i.e., with probability $1/N$. For drawing the second unit, the unit selected on the first draw is returned to the population and a unit is selected at random again from the entire population with probability $1/N$. So the unit selected on the first draw may appear again in the second draw. The procedure is continued till a sample of n units is selected. So, in an SRSWR, the probability of selection of the ith unit at the kth draw is $p_i(k) = 1/N$; $i = 1,\ldots, N$, $k = 1,\ldots, n$ and the probability of selection of an ordered sample $s_o = (i_1 \rightarrow i_2,\cdots, \rightarrow i_n)$ is $p(s_o) = 1/N^n$.

3.3.2 Estimation of the Population Mean and Variance

Let y_i be the value of the variable under study y for the ith ($i = 1, 2,\ldots, N$) unit of the population and $y_{(r)}$ the value of the variable y for the unit that is selected at the rth ($r = 1, 2,\ldots, n$) draw. Clearly, $y_{(r)} = y_j$ if the rth draw produces the jth unit, $j = 1,\ldots, N$. Let $\overline{y}(s_o) = \sum_{r=1}^{n} y_{(r)}/n$ and $\widehat{\sigma}_y^2 = \sum_{r=1}^{n} \{y_{(r)} - \overline{y}(s_o)\}^2/(n-1)$ denote the sample mean and sample variance, respectively, based on the ordered sample s_o of size n units with repetition. Then we have the following theorem:

Theorem 3.3.1

(i) $\overline{y}(s_o)$ is an unbiased estimator for the population mean $\overline{Y}$.

(ii) The variance of $\overline{y}(s_o)$ is

$$V[\overline{y}(s_o)] = \sigma_y^2/n$$

(iii) An unbiased estimator of $V[\overline{y}(s_o)]$ is

$$\widehat{V}[\overline{y}(s_o)] = \widehat{\sigma}_y^2/n$$

where $\sigma_y^2 = \sum_{i=1}^{N} \left(y_i - \overline{Y}\right)^2/N$

Proof

(i) $E[\overline{y}(s_o)] = \sum_{r=1}^{n} E\left(y_{(r)}\right)/n = \overline{Y}$ since, $E\left(y_{(r)}\right) = \sum_{r=1}^{N} y_i/N = \overline{Y}$.

(ii) $$V[\overline{y}(s_o)] = V\left(\sum_{r=1}^{n} y_{(r)}/n\right)$$

$$= \left[\sum_{r=1}^{n} V\left(y_{(r)}\right) + \sum_{r\neq}^{n}\sum_{r'}^{n} Cov\left(y_{(r)}, y_{(r')}\right)\right]\Big/ n^2$$

$$= \sigma_y^2\Big/n$$

since, $V\left(y_{(r)}\right) = E\left(y_{(r)} - \overline{Y}\right)^2 = \sum_{i=1}^{N}\left(y_i - \overline{Y}\right)^2/N = \sigma_y^2$ and $Cov\left(y_{(r)},\right.$ $\left.y_{(r')}\right) = 0$ for $r \neq r'$ as the draws are independent.

$$\text{(iii)}\quad E\left(\widehat{\sigma}_y^2/n\right) = \frac{1}{n(n-1)}\left[\sum_{r=1}^{n} E\left(y_{(r)}^2\right) - nE\{\overline{y}(s_o)\}^2\right]$$

$$= \frac{1}{n(n-1)}\left[n\left(\sigma_y^2 + \overline{Y}^2\right) - n\left(\frac{\sigma_y^2}{n} + \overline{Y}^2\right)\right]$$

$$= V[\overline{y}(s_o)]$$

3.3.3 Estimation of Population Proportion

Let $y_i = 1$ if the ith unit belongs to the group A, and let $y_i = 0$ if the ith unit does not belong to the group A. In this case $Y = N_A$ = total number of units that possess the attribute A and $\overline{Y} = N_A/N = \pi_A$ = proportion of units in the population belonging to the group A; $\overline{y}(s_o) = n_A/n = \widehat{\pi}_A$ = proportion of units in the sample s_o belonging to A, where n_A is the total number of units in the sample that fall in group A. Now noting that

(i) $\sigma_y^2 = \sum_{i=1}^{N}\left(y_i - \overline{Y}\right)^2/N = \sum_{i=1}^{N} y_i/N - \overline{Y}^2 = \pi_A(1-\pi_A)$ since $y_i = 0$ or 1;

and

(ii) $\widehat{\sigma}_y^2 = \sum_{r=1}^{n}\{y_{(r)} - \overline{y}(s_o)\}^2/(n-1) = \dfrac{n}{n-1}\widehat{\pi}_A(1-\widehat{\pi}_A)$

we have the following theorem:

Theorem 3.3.2

(i) $\widehat{\pi}_A$ is an unbiased estimator for the population proportion π_A.

(ii) Variance of $\widehat{\pi}_A$ is $V(\widehat{\pi}_A) = \dfrac{\pi_A(1-\pi_A)}{n}$

(iii) An unbiased estimator of $V(\widehat{\pi}_A)$ is $\widehat{V}[\widehat{\pi}_A] = \dfrac{\widehat{\pi}_A(1-\widehat{\pi}_A)}{n-1}$

Remark 3.3.1

It is important to note that Theorems 3.3.1 and 3.3.2 can be obtained from Theorems 3.2.2 and 3.2.8 when N is sufficiently large compared to n so that the finite population correction term $f_n = n/N$ is ignored.

Example 3.3.1

From the list of 30 students given in the Example 3.2.1, select a sample of size 8 by the SRSWR method. From the selected sample, (i) estimate the mean height and weight of students (male and female combined) and obtain the variances of the estimator used. Estimate the SEs of the estimators. (ii) Estimate the proportion of the male and female students with their SEs.

Here we associate serial number of students to the two-digit random numbers 01–90 as in Example 3.2.1, and we select the random numbers as follows:

Random number selected	45	45	27	54	46	66	60	24
Unit selected	15	15	27	24	16	6	30	24

Here the selected SRSWR sample is $s_o = (15, 15, 27, 24, 16, 6, 30, 24)$ and the data corresponding to the sample s_o are given below:

Serial no. of students	Height (cm)	Weight (kg)	Gender Male = 1 Female = 0
15	160	72	1
15	160	72	1
27	156	62	0
24	170	75	1
16	165	70	1
6	175	75	0
30	150	60	0
24	170	75	1

(i) Estimated mean height and weight of the 30 students are given by $\overline{x}_s = 163.25$ cm and $\overline{y}_s = 70.125$ kg. The variance of $\overline{x}_s$ is $\sigma_x^2/n = (N-1)S_x^2/(nN) = 29 \times 73.981/(30 \times 8) = 8.939\ \text{cm}^2$. The variance of $\overline{y}_s$ is $V(\overline{y}_s) = \sigma_y^2/n = (N-1)S_y^2/(nN) = 29 \times 83.374/(30 \times 8) = 10.074\ \text{kg}^2$. The estimated SEs for $\overline{x}_s$ and $\overline{y}_s$ are, respectively, $se(\overline{x}_s) = \sqrt{\widehat{V}(\overline{x}_s)} = \sqrt{s_x^2/n} = \sqrt{68.785/8} = 2.932$ cm and $se(\overline{y}_s) = \sqrt{\widehat{V}(\overline{y}_s)} = \sqrt{s_y^2/n} = \sqrt{35.267/8} = 2.099$ kg.

(ii) The estimated proportions of male and female students are $\widehat{\pi}_m = 5/8 = 0.625$ and $\widehat{\pi}_f = 3/8 = 0.375$. Estimated SEs of $\widehat{\pi}_m$ and $\widehat{\pi}_f$ are, respectively, $Se(\widehat{\pi}_m) = \sqrt{\dfrac{\widehat{\pi}_m(1-\widehat{\pi}_m)}{n-1}} = \sqrt{\dfrac{0.625(1-0.625)}{7}} = 0.183$

and $Se(\widehat{\pi}_f) = \sqrt{\dfrac{\widehat{\pi}_f(1-\widehat{\pi}_f)}{n-1}} = \sqrt{\dfrac{0.375(1-0.375)}{7}} = 0.183$

3.3.4 Rao–Blackwellization

It follows from Section 2.7.3 that the estimator $\overline{y}(s_o)$ is inadmissible because it is based on ordered data, which may consist of repletion of units, hence $\overline{y}(s_o)$ is not a function of a sufficient statistic. Let $s = (j_1,\ldots, j_\nu)$ denote the unordered sample obtained by taking distinct ν units $j_1,\ldots, j_\nu$ with $j_1<\cdots<j_\nu$. The unordered sample s is a sufficient statistic. Hence, we can improve the inefficient estimator $\overline{y}(s_o)$ by applying the Rao–Blackwellization technique. The improved estimator is given by $E[\overline{y}(s_o)|s] = \overline{y}_s = \sum\limits_{i\in s} y_i/\nu =$ the sample mean based on the distinct units. The details have been given in the following theorems.

Theorem 3.3.3

Let $\overline{y}_s = \sum\limits_{i\in s} y_i/\nu = \sum_{k=1}^{\nu} y_{j_k}/\nu$ be the sample mean based on the distinct units of s_o. Then,

(i) $E[\overline{y}_s] = E[\overline{y}(s_o)] = \overline{Y}$

(ii) $V[\overline{y}_s] \leq V[\overline{y}(s_o)]$

Proof

Let $n_i(s_o)$ denote the number of times the ith unit appears in s_o. Then writing,

$$\begin{aligned}\overline{y}(s_o) &= \sum_{r=1}^{n} y_{(r)}/n\\ &= \sum_{i=1}^{N} n_i(s_o)y_i/n\\ &= \sum_{k=1}^{\nu} n_{j_k}(s_o)y_{j_k}/n\end{aligned}$$

where $n_{j_1}(s_o), \ldots, n_{j_\nu}(s_o)$ denote the number of times the distinct units $j_1,\ldots, j_\nu$ appear in s_o.

Now noting $E\left(n_{j_k}(s_o)|s\right) = n/\nu$ we get

(i) $E[\overline{y}(s_o|s)] = \sum\limits_{k=1}^{\nu} E\left(n_{j_k}(s_o)|s\right)y_{j_k}/n = \overline{y}_s$

and

(ii) $V[\overline{y}(s_o)] = V[E\{\overline{y}(s_o)|s\}] + E[V\{\overline{y}(s_o)|s\}] \geq V[E\{\overline{y}(s_o)|s\}] = V[\overline{y}_s]$

Theorem 3.3.4

Let s be an unordered sample derived by taking distinct units from an ordered sample s_o and $\overline{y}_s = \sum_{i\in s} y_i/\nu$, where ν is the number of distinct units of s_o. Then,

(i) Variance of $\overline{y}_s$ is

$$V(\overline{y}_s) = \left[E\left(\frac{1}{\nu}\right) - \frac{1}{N}\right] S_y^2$$

(ii) An unbiased estimator for $V(\overline{y}_s)$ is

$$\widehat{V}(\overline{y}_s) = \left(\frac{1}{\nu} - \frac{1}{N}\right) s_\nu^2$$

where $s_\nu^2 = \frac{1}{\nu - 1} \sum_{i\in s} (y_i - \overline{y}_s)^2$ is the sample variance of the set of distinct units of s_o.

Proof

Let $E(\overline{y}_s|\nu)$ and $V(\overline{y}_s|\nu)$ be the conditional expectation and conditional variance of $\overline{y}(s)$ over the variation of all possible unordered samples of size ν. Then, given ν, s is an SRSWOR sample of size ν from the entire population U. Hence

$$E(\overline{y}_s|\nu) = \overline{Y},\ V(\overline{y}_s|\nu) = \left(\frac{1}{\nu} - \frac{1}{N}\right) S_y^2 \text{ and } E(s_\nu^2|\nu) = S_y^2 \tag{3.3.1}$$

Using (3.3.1), we get

$$\begin{aligned} E(\overline{y}_s) &= E\{E(\overline{y}_s|\nu)\} = \overline{Y}, \\ V(\overline{y}_s) &= E\{V(\overline{y}_s|\nu)\} + V\{E(\overline{y}_s|\nu)\} \\ &= E\left(\frac{1}{\nu} - \frac{1}{N}\right) S_y^2 + V(\overline{Y}) \\ &= \left[E\left(\frac{1}{\nu}\right) - \frac{1}{N}\right] S_y^2 \end{aligned}$$

and

$$\begin{aligned} E\left[\widehat{V}(\overline{y}_s)\right] &= E\left[\left(\frac{1}{\nu} - \frac{1}{N}\right) E(s_\nu^2|\nu)\right] \\ &= \left[E\left(\frac{1}{\nu}\right) - \frac{1}{N}\right] S_y^2. \end{aligned}$$

Remark 3.3.2

(i) The exact probability distribution of ν was derived by Feller (1957) as

$$Prob\{\nu = j\} = \frac{1}{N^n}\binom{N}{j}\sum_{r=0}^{j}(-1)^r\binom{j}{r}(j-r)^n; j = 1, \ldots, \min(n, N) \tag{3.3.2}$$

(ii) The expression of $E\left(\frac{1}{\nu}\right)$ was derived by Pathak (1961) as

$$E\left(\frac{1}{\nu}\right) = \frac{1^{n-1} + 2^{n-1} + \cdots + N^{n-1}}{N^n} \tag{3.3.3}$$

(iii) Substituting (3.3.3) in Theorem 3.3.4, we get

$$V(\overline{y}_s) = \frac{1^{n-1} + 2^{n-1} + \cdots + (N-1)^{n-1}}{N^n} S_y^2$$

Neglecting terms of degree greater than $(1/N)^2$, Murthy (1967) obtained an approximate expression of $V(\overline{y}_s)$ as

$$V(\overline{y}_s) = \left(\frac{1}{n} - \frac{1}{2N} + \frac{n-1}{12N^2}\right) S_y^2$$

Theorem 3.3.5

Let ν_{s_o} be the number of distinct units in an ordered sample s_o of size n selected by the SRSWR method. Then,

(i) $E(\nu_{s_o}) = N[1 - \alpha]$ and

(ii) The variance of ν_{s_o} is

$$V(\nu_{s_o}) = N(N-1)[1 - 2\alpha + \beta] - \{N(1-\alpha)\}\{N(1-\alpha) - 1\}$$

where $\alpha = \left(1 - \frac{1}{N}\right)^n$ and $\beta = \left(1 - \frac{2}{N}\right)^n$.

Proof

The inclusion probability of the ith unit $= \pi_i =$ probability of selection of the ith unit in any of the n draws $= 1 -$ probability that the ith unit will not be selected from any of the n draws $= 1 - \left(1 - \frac{1}{N}\right)^n = 1 - \alpha$.

Inclusion probability of the ith and jth $(i \neq j)$ unit $= \pi_{ij} =$ probability of selection of both the ith and jth units in n draws $= 1$—at least one of the units i and j will not be selected in n draws

$$= 1 - \left\{ \left(1 - \frac{1}{N}\right)^n + \left(1 - \frac{1}{N}\right)^n - \left(1 - \frac{2}{N}\right)^n \right\} = 1 - 2\alpha + \beta.$$

Now using Theorem 1.3.1, we get $E(\nu_{s_o}) = \sum_{i=1}^{N} \pi_i = N(1 - \alpha)$ and

$$V(\nu_{s_o}) = \sum_{i \neq}^{N} \sum_{j=1}^{N} \pi_{ij} - E(\nu_{s_o})\{E(\nu_{s_o} - 1)\}$$

$$= N(N-1)(1 - 2\alpha + \beta) - N(1-\alpha)\{N(1-\alpha) - 1\}.$$

3.4 INTERVAL ESTIMATION

In this section, we will present the formulae for determining confidence intervals for the population mean $\overline{Y}$ and population proportion π_A based on SRSWOR and SRSWR sampling schemes. In general, we cannot determine exact confidence intervals as no distribution of the parameter vector $\mathbf{y}$ is assumed. However, approximate confidence intervals are determined assuming that the central limit theorem holds when the sample size is large. The validity of the normality assumption and requirement of the sample size are given in detail by Cochran (1977) and Sukhatme et al. (1984). For smaller sample sizes, we use t distribution.

3.4.1 Confidence Intervals for Mean and Proportion

3.4.1.1 Large Sample Size

In case the sample size n is large ($n \geq 35$) and the population variance S_y^2 or σ_y^2 is known, $(1 - \alpha)100\%$ confidence interval for $\overline{Y}$ is given by

$$\overline{y}(s) \pm z_{\alpha/2}\sqrt{V[\overline{y}(s)]} = \overline{y}(s) \pm z_{\alpha/2} S_y \sqrt{\frac{N-n}{Nn}} \quad \text{(for SRSWOR)}$$

and

$$\overline{y}(s_o) \pm z_{\alpha/2}\sqrt{V[\overline{y}(s_o)]} = \overline{y}(s_o) \pm z_{\alpha/2} \frac{\sigma_y}{\sqrt{n}} \quad \text{(for SRSWR)}$$

where $z_{\alpha/2}$ is the upper $\alpha/2$ point of standard normal distribution, S_y and σ_y denote the positive square root of S_y^2 and σ_y^2, respectively.

In most situations S_y^2 or σ_y^2 is unknown. In this case S_y^2 and σ_y^2 are replaced by their unbiased estimators, and we set $(1 - \alpha)100\%$ interval for the population mean $\overline{Y}$ as

$$\overline{y}(s) \pm z_{\alpha/2}\sqrt{\widehat{V}[\overline{y}(s)]} = \overline{y}(s) \pm z_{\alpha/2} s_y \sqrt{\frac{N-n}{Nn}} \quad \text{(for SRSWOR)}$$

and

$$\bar{y}(s_o) \pm z_{\alpha/2}\sqrt{\widehat{V}[\bar{y}(s_o)]} = \bar{y}(s_o) \pm z_{\alpha/2}\frac{s_y}{\sqrt{n}} \quad \text{(for SRSWR)}$$

where s_y is the positive square root of s_y^2.

For a large n we set $(1-\alpha)100\%$ interval for the population proportion π_A as

$$\widehat{\pi}_A \pm z_{\alpha/2}\sqrt{\frac{N-n}{N}\cdot\frac{\widehat{\pi}_A(1-\widehat{\pi}_A)}{n-1}} \quad \text{(for SRSWOR)}$$

and

$$\widehat{\pi}_A \pm z_{\alpha/2}\sqrt{\frac{\widehat{\pi}_A(1-\widehat{\pi}_A)}{n-1}} \quad \text{(for SRSWR)}$$

3.4.1.2 Small Sample Size

For small sample size $n < 35$, $z_{\alpha/2}$ is replaced by $t_{\alpha/2,n-1}$ in all the formulae aforementioned, where $t_{\alpha/2,n-1}$ is the upper $(\alpha/2)100\%$ point of t distribution with degrees of freedom (df) $n-1$.

Example 3.4.1

For Example 3.2.1, $N=30$, $n=8$, and the sample s are selected by the SRSWOR method. Here the sample size $n=8$ is small. Hence, a 95% confidence interval for the mean height of the students is $\bar{x}(s) \pm t_{0.025,7}s_x\sqrt{\frac{N-n}{Nn}} =$ 162.625 cm $\pm$ 2.365 $\times$ 2.389 cm $=$ (156.975 cm, 168.275 cm). Similarly for Example 3.2.1, a 90% confidence interval for the proportion of male students would be $\widehat{\pi}_m \pm t_{0.05,7}\sqrt{\frac{N-n}{N}\frac{\widehat{\pi}_m(1-\widehat{\pi}_m)}{n-1}} = 0.625 \pm 1.895 \times 0.156 =$ $(0.329, 0.920)$.

Example 3.4.2

From a population of 10,000 adult males of a certain community, a sample of 500 males were selected by SRSWOR method, and 125 males were found to be HIV positive. Find a 95% confidence interval for the proportion of HIV-positive persons (π_A) in the community.

Here both the sample and population size are very large, and so we use the following formula for the confidence interval for π_A.

$$\widehat{\pi}_A \pm z_{\alpha/2}\sqrt{\frac{\widehat{\pi}_A(1-\widehat{\pi}_A)}{n}} = 0.25 \pm 1.96\sqrt{\frac{0.25(1-0.25)}{500}}$$
$$= (0.212, 0.288)$$

3.5 DETERMINATION OF SAMPLE SIZE

In any large- or small-scale survey, one has to determine the sample size, which will fulfill the objective of the survey. To get a representative sample, it is expected that the sample size should increase with the population size. In general, the sampling error decreases with the increase of sample size. However, the cost of a survey increases with the increase of sample size. The cost also increases with the increase of the size of the questionnaire because the investigator requires more time to complete a long questionnaire. The size of the sample is determined, considering available cost, time, and the degree of precision needed for estimation of parameters under consideration. In this section we will consider how one can determine an appropriate sample size for SRSWR and SRSWOR sampling designs.

3.5.1 Consideration of the Cost of a Survey

Suppose the cost of a survey is expressed by a simple cost function $C = C_o + c\,n$ where C_o is the overhead cost of the survey, which is fixed for a survey, i.e., independent of the sample size, and c is the average cost for surveying a unit. Hence for a fixed available cost $C = C^*$ of a survey, one should select sample size $n = (C^* - C_o)/c$.

3.5.2 Consideration of the Efficiency of Estimators

3.5.2.1 Given Variance

The magnitude of the sampling error of an estimator is determined by its variance. Hence, one can determine the sample size that yields some specific value of variance, V_0, for example. To estimate the population mean $\overline{Y}$, if one uses the sample mean as an estimator and expects that the sample mean will have a specific value of its variance V_0, then n can be derived from the relation

$$V\{\overline{y}(s)\} = \left(\frac{1}{n} - \frac{1}{N}\right)S_y^2 = V_0 \text{ i.e., } n = \left(\frac{V_0}{S_y^2} + \frac{1}{N}\right)^{-1} \text{ (for SRSWOR)}$$

and

$$V\{\overline{y}(s_o)\} = \frac{1}{n}\sigma_y^2 = V_0 \text{ i.e., } n = \frac{\sigma_y^2}{V_0} \quad \text{(for SRSWR)}$$

To determine the value of n, one needs to know the value of S_y^2 $\left(\text{or } \sigma_y^2 = \frac{N-1}{N}S_y^2\right)$, which is generally unknown. Hence, it is customary to use a value of S_y^2 either from the past survey (or experience) or replace S_y^2 by its estimate s_y^2 determined from a pilot survey.

3.5.2.2 Given Coefficient of Variation

The coefficient of variation (CV) of a population characteristic y is defined by $C_y = \sigma_y/\overline{Y}$. It is found that the value of the CV of a population characteristic is much more stable (not subject to as much change over time) than the population variance or mean. Generally, the CV of a population is available from a past survey. In case it is not known, a reliable estimator of CV may be obtained from a pilot survey. The CV of an estimator t is defined by $C_t = \sqrt{Var(t)}/E(t)$. In case C_y is known, we can find the sample size n, which yields a specific value of CV of the estimator t, $C_t = C_{to}$ (say) as follows:

For SRSWOR with $t = \overline{y}(s)$, $C_{\overline{y}(s)} = C_{to}$ yields

$$C_{\overline{y}(s)} = \frac{\sqrt{\frac{N-n}{Nn}S_y^2}}{\overline{Y}} = \frac{\sqrt{\frac{N-n}{n(N-1)}\sigma_y^2}}{\overline{Y}} = \sqrt{\frac{N-n}{n(N-1)}}C_y = C_{to}, \text{ which}$$

in turn gives

$$n = \frac{N\{C_y\}^2}{(N-1)C_{to}^2 + \{C_y\}^2}$$

For SRSWR with $t = \overline{y}(s_o)$, $C_{\overline{y}(s_o)} = C_o$ yields the required sample size

$$n = \left(\frac{C_y}{C_{to}}\right)^2$$

3.5.2.3 Given Margin of Permissible Error

Here, sample size is determined by assigning the probability to a certain level, $(1 - \alpha)$ (say) for the maximum permissible error (difference between the estimated and the true value of the parameter) to a certain value d. For instance, let d be the permissible error, maximum acceptable difference between the estimator t, and the population mean $\overline{Y}$. The sample size is determined from the relation

$$Prob\left\{\left|t-\overline{Y}\right| \le d\right\} = 1-\alpha \tag{3.5.1}$$

The above equation is equivalent to

$$Prob\left\{\frac{\left|t-\overline{Y}\right|}{\sqrt{V(t)}} \le \frac{d}{\sqrt{V(t)}}\right\} = 1-\alpha$$

Assuming the sample size to be so large, enabling $z = \dfrac{\left|t-\overline{Y}\right|}{\sqrt{V(t)}}$ distributed $N(0,1)$, one can determine the value of n using the relation

$$\frac{d}{\sqrt{V(t)}} = z_{\alpha/2} \tag{3.5.2}$$

For an SRSWOR design with $t = \overline{y}(s)$, Eq. (3.5.1) yields

$$n = \left[\frac{1}{N} + \left(\frac{d}{S_y z_{\alpha/2}}\right)^2\right]^{-1} = \left[\frac{1}{N} + \frac{N-1}{N}\left(\frac{k}{C_y z_{\alpha/2}}\right)^2\right]^{-1} \tag{3.5.3}$$

where $k = d/\overline{Y}$

For an SRSWR design with $t = \overline{y}(s_o)$, Eq. (3.5.1) yields

$$n = \left(\frac{\sigma_y\, z_{\alpha/2}}{d}\right)^2 = \left(\frac{C_y\, z_{\alpha/2}}{k}\right)^2 \tag{3.5.4}$$

3.5.3 Use of Chebyshev Inequality

Here we determine sample size by keeping permissible error to a certain level with probability exceeding a certain preassigned value $(1-\alpha)$. Let t be an unbiased estimator for $\overline{Y}$. Using the Chebyshev Inequality, we have

$$Prob\left[\left|t-\overline{Y}\right| \le d\right] \ge 1 - \frac{V(t)}{d^2} \tag{3.5.5}$$

The sample size n is determined from the relation $1 - \dfrac{V(t)}{d^2} = 1-\alpha$ which is equivalent to

$$\frac{V(t)}{d^2} = \alpha \tag{3.5.6}$$

For an SRSWOR design with $t = \overline{y}(s)$, Eq. (3.5.5) yields

$$n = \left[\frac{1}{N} + \alpha \frac{N-1}{N}\left(\frac{d}{\sigma_y}\right)^2\right]^{-1} \qquad (3.5.7)$$

$$= N\left[1 + \gamma^2 \alpha (N-1)\right]^{-1}$$

where $d = \gamma \ \sigma_y$

For an SRSWR design with $t = \overline{y}\,(s_o)$, Eq. (3.5.6) yields

$$n = \frac{\sigma_y^2}{\alpha d^2} = \frac{1}{\alpha \gamma^2} \qquad (3.5.8)$$

Substituting $\alpha = 0.05$ and $\gamma = 1$ in (3.5.8), we get $n = 20$, i.e., selection of $n = 20$ ensures

$$Prob\left[\left|\overline{y}(s_o) - \overline{Y}\right| \leq \sigma_y\right] \geq 0.95$$

Remark 3.5.1

Determination of sample size using Chebyshev Inequality does not require the assumption of the normality of estimator t used for estimating the parameter $\overline{Y}$.

Example 3.5.1

It is decided to estimate the mean consumption of food per household in a region of 1000 households. A pilot study reveals that the average consumption of food per family is \$2500 with a SE \$1250. Determine the minimum sample size required to ensure that the permissible error is 5% of the true value with 99% confidence coefficient under (i) SRSWOR and (ii) SRSWR procedures.

Here $N = 1000$, $\overline{Y} = \$2500$, $S_y = \$1250$, and $d = 0.05\,\overline{Y}$.

(i) For SRSWOR, using (3.5.3) we get the required sample size as

$$n = \left[\frac{1}{N} + \left(\frac{d}{S_y z_{\alpha/2}}\right)^2\right]^{-1} = \left[\frac{1}{1000} + \left(\frac{0.05 \times 2500}{1250 \times z_{.005}}\right)^2\right]^{-1}$$

$$= \left[\frac{1}{1000} + \left(\frac{0.05 \times 2500}{1250 \times 2.576}\right)^2\right]^{-1} = 399$$

(ii) For SRSWR, the formula (3.5.4) yields

$$n = \left(\frac{\sigma_y z_{\alpha/2}}{d}\right)^2 = \left(\frac{1250 \times 2.576}{0.05 \times 2.500}\right)^2 = 664$$

Here we assumed $\sigma_y = S_y$ as $N(=1000)$ is large.

Example 3.5.2

It is known from the past survey that the HIV infection, STD, and TB rates in a locality were 25%, 5%, and 15%, respectively. To find if there has been any improvement of health-care system to reduce the infection rates of HIV, STD, and TB, an SRSWR survey is proposed in a community consisting of 16,000 people. Determine the minimum sample size required, which ensures a permissible error of 10% of the true value with 95% confidence for measuring each of the infections.

Let π_x, π_y, and π_z be the population infection rates for HIV, STD, and TB, respectively and let $\widehat{\pi}_x$, $\widehat{\pi}_y$ and $\widehat{\pi}_z$ be the sample proportions.

Here we find n for which

$$Prob\{|\widehat{\pi}_t - \pi_t| \le d\} = 1 - \alpha \text{ for } t = x, y \text{ and } z \qquad (3.5.9)$$

where $d = \pi_t \times 0.1$ and $1 - \alpha = 0.95$.

Assuming the population size $N(=16{,}000)$ is large, Eq. (3.5.9) yields

$$Prob\left\{ |z| \le \frac{d}{\sqrt{V(\widehat{\pi}_t)}} \right\} = 1 - \alpha \qquad (3.5.10)$$

where $z = (\widehat{\pi}_t - \pi_t)/\sqrt{V(\widehat{\pi}_t)}$ and $V(\widehat{\pi}_t) = \pi_t(1 - \pi_t)/n$

Now equating $\dfrac{d}{\sqrt{V(\widehat{\pi}_t)}} = z_{\alpha/2}$, we get

$$n = \left[\sqrt{\pi_t(1 - \pi_t)} z_{\alpha/2}/d\right]^2 = \left[\sqrt{\pi_t((1 - \pi_t)} \times 1.96/d\right]^2$$

Taking π_t as the previous infection rate, we determine the sample sizes as follows:

Infection	π_t	d	**Required sample size n**
HIV	0.25	0.025	1152.48
STD	0.05	0.005	7299.04
TB	0.15	0.015	2176.90

Considering the table above, we have to take the maximum sample size, which is 7299 to meet the requirement.

3.6 INVERSE SAMPLING

In estimating π, the proportion of units of a population possessing a rare characteristic A such as color blindness or having suffered from the hepatitis viral infection, the conventional sample proportion may not serve as an efficient estimator because very often the sample proportion becomes zero and hence its variance becomes very large. Haldane (1946) proposed an inverse sampling scheme, which is more effective and excludes zero as an estimator of π. In inverse sampling, units are selected one by one by the SRSWOR or SRSWR method until a specified number m of the units is selected from the rare population A.

3.6.1 Simple Random Sampling Without Replacement

Let us suppose that a population consists of N units of which $N_A(=N\pi)$ units possess certain rare characteristics A and that the remaining $N_B = N - N_A$ do not possess this characteristic. Let X be the number of units required to be drawn to get m units that possess characteristic A. Here X is a random variable whose probability distribution depends on m and N_A. The probability distribution of X is given by

$$P(X = x) = f(x|N_A, m)$$

$= $ Probability of getting $(m-1)$ units bearing characteristic A in the first $x-1$ draws and at the xth draw one unit is selected from the group A.

$$= \frac{\binom{N_A}{m-1}\binom{N_B}{(x-1)-(m-1)}}{\binom{N}{x-1}} \frac{N_A-(m-1)}{N-(x-1)};$$

$$x \geq m, m+1, \ldots$$

Theorem 3.6.1

(i) An unbiased estimator of π is

$$\widehat{\pi} = \frac{m-1}{x-1}$$

(ii) An unbiased estimator for the variance of $\widehat{\pi}$ is

$$\widehat{V}(\widehat{\pi}) = \frac{\widehat{\pi}(1-\widehat{\pi})}{x-2}\left(1 - \frac{x-1}{N}\right)$$

Proof

(i) $E(\widehat{\pi}) = \sum_{x\geq m} \frac{m-1}{x-1} f(x|N_A, m)$

$$= \sum_{x\geq m} \frac{m-1}{x-1} \frac{\binom{N_A}{m-1}\binom{N_B}{(x-1)-(m-1)}}{\binom{N}{x-1}} \frac{N_A-(m-1)}{N-(x-1)}$$

$$= \frac{N_A}{N} \sum_{x\geq m} \frac{\binom{N_A-1}{m-2}\binom{N_B}{(x-2)-(m-2)}}{\binom{N-1}{x-2}}$$

$$\times \frac{(N_A-1)-(m-2)}{(N-1)-(x-2)}$$

$$= \pi \sum_{z\geq m-1} \frac{\binom{N_A-1}{m-2}\binom{N_B}{(z-1)-(m-2)}}{\binom{N-1}{z-1}}$$

$$\times \frac{(N_A-1)-(m-2)}{(N-1)-(z-1)}$$

$$= \pi \sum_{z\geq m-1} f(z|N_A-1, m-1)$$

$$= \pi \tag{3.6.1}$$

(ii) Following (3.6.1) we get

$$E\left(\frac{m-1}{x-1}\frac{m-2}{x-2}\right) = \frac{N_A}{N}\frac{N_A-1}{N-1} \sum_{z\geq m-2} f(z|N_A-2, m-2)$$

$$= \pi\frac{\pi N-1}{N-1}$$

$$\text{i.e.,} \quad \frac{(N-1)}{N} E\left\{\frac{m-1}{x-1}\left(\frac{m-2}{x-2}+\frac{1}{N-1}\right)\right\} = \pi^2 \tag{3.6.2}$$

Finally using (3.6.2), an unbiased estimator for the variance of $\widehat{\pi}$ is obtained as

$$\begin{aligned}
\widehat{V}(\widehat{\pi}) &= \widehat{\pi}^2 - \text{unbiased estimator of } \pi^2 \\
&= \widehat{\pi}^2 - \frac{(N-1)}{N}\frac{m-1}{x-1}\left(\frac{m-2}{x-2}+\frac{1}{N-1}\right) \\
&= \widehat{\pi}\left[\widehat{\pi} - \frac{N-1}{N}\frac{\widehat{\pi}(x-1)-1}{x-2} - \frac{1}{N}\right] \\
&= \frac{\widehat{\pi}(1-\widehat{\pi})}{x-2}\left(1-\frac{x-1}{N}\right)
\end{aligned}$$

3.6.2 Simple Random Sampling With Replacement

In inverse sampling with the SRSWR method, we select units one by one with replacement until m units possessing characteristic A are selected. Let X be the required number of draws. Then the probability distribution of X follows negative binomial distribution with parameter π and is given by

$$P(X=x) = f(x|\pi,m) = \binom{x-1}{m-1}\pi^x(1-\pi)^{x-m}; x = m, m+1, \ldots$$

Here we get

$$\begin{aligned}
E\left(\frac{m-1}{x-1}\right) &= \sum_{x\geq m}\frac{m-1}{x-1}f(x|\pi,m) \\
&= \sum_{x\geq m}\frac{m-1}{x-1}\binom{x-1}{m-1}\pi^x(1-\pi)^{x-m} \\
&= \pi\sum_{x\geq m}\binom{x-2}{m-2}\pi^{x-1}(1-\pi)^{(x-1)-(m-1)} \qquad (3.6.3) \\
&= \pi\sum_{z\geq m-1}\binom{z-1}{m-1-1}\pi^z(1-\pi)^{z-(m-1)} \\
&= \pi\sum_{z\geq m-1}f(z|\pi,m-1) \\
&= \pi
\end{aligned}$$

Similarly, we get

$$E\left(\frac{m-1}{x-1}\frac{m-2}{x-2}\right)=\pi^2 \tag{3.6.4}$$

The expressions (3.6.3) and (3.6.4) yield

Theorem 3.6.2

(i) An unbiased estimator of π is

$$\widehat{\pi}=\frac{m-1}{x-1}$$

(ii) An unbiased estimator for the variance of $\widehat{\pi}$ is

$$\begin{aligned}\widehat{V}(\widehat{\pi})&=\frac{m-1}{x-1}\left(\frac{m-1}{x-1}-\frac{m-2}{x-2}\right)\\&=\frac{\widehat{\pi}(1-\widehat{\pi})}{x-2}\end{aligned}$$

3.7 EXERCISES

3.7.1 The following table gives the weight of six students in a first-year class

Serial no of students	1	2	3	4	5	6
Weight (in kg)	55	60	75	77	60	55

(i) Write down all possible samples of size 3, which can be selected by the SRSWOR method.

(ii) Compute the sample means of all the possible samples and construct the frequency distribution of the sample mean.

(iii) From the frequency distribution of the sample mean, verify that the sample mean is an unbiased estimator of the population mean and the sample variance follows the formula given in Theorem 3.2.2.

(iv) Compute the sample variances for all the samples, make a frequency distribution of sample variance, and verify that the sample variance is an unbiased estimator of the population variance.

3.7.2 The ages of five school-going children are given in the following table.

Serial no of children	1	2	3	4	5
Age in years	5	6	5	4	6

(i) Write down all possible samples of size 2 that can be selected by SRSWR method.
(ii) Compute the frequency distribution of the sample means $\overline{y}(s_o)$ and sample variances.
(iii) Show that the sample mean and sample variances are unbiased estimators of the population mean and population variance σ_y^2 respectively.
(iii) Find the frequency distribution of sample means of distinct units $\overline{y}(s)$ and compute the variance of $\overline{y}(s)$.
(iv) Find the frequency distribution of the distinct units $\nu(s)$ and compute (i) $E\{\nu(s)\}$, (ii) $V\{\nu(s)\}$, and (iii) $E\{1/\nu(s)\}$.

3.7.3 The following table gives the marks of Mathematics and Statistics of 30 students by gender in the first-year class of the University of Botswana.

Serial no of students	Marks on Mathematics	Marks on Statistics	Gender	Serial no of students	Marks on Mathematics	Marks on Statistics	Gender
1	50	53	1	16	36	60	1
2	90	81	1	17	71	55	0
3	61	85	1	18	72	57	1
4	38	68	0	19	35	82	0
5	45	73	0	20	62	62	1
6	67	62	1	21	81	20	1
7	40	34	1	22	74	55	0
8	50	49	1	23	88	32	0
9	89	59	1	24	55	59	1
10	77	84	1	25	69	61	1
11	67	23	0	26	32	66	1
12	82	73	0	27	37	74	1
13	48	61	1	28	66	92	1
14	71	37	0	29	34	68	0
15	42	35	1	30	35	71	1

0, female; *1*, male.

(i) Select a sample of 10 students by SRSWR method and give unbiased estimates of the average marks in Mathematics and Statistics and also estimate their variances.
(ii) Compute an unbiased estimate of the difference between the average marks in Mathematics and Statistics with its SE.
(iii) Give an estimate of the population CV of Mathematics marks.
(iv) Estimate the population proportion of female students and estimate its SE. Compute a 95% confidence interval for the proportion of male students in the class.
(v) Give unbiased estimates of the average marks for Mathematics and Statistics of the male students and give unbiased estimates of their variances.
(vi) Find a 90% confidence interval for the difference between the average marks for Mathematics and Statistics of the entire class.

3.7.4 From a list of 100 households, 10 households were selected by the SRSWOR method and the following information was obtained.

Household	Head of household	Household income (in $)	Household size	Education	Expenditure on transport (in $)
1	M	2000	4	Primary	200
2	M	3000	2	Primary	250
3	F	4500	5	Middle	600
4	M	8000	3	High	500
5	F	2000	2	Primary	100
6	F	5000	4	Middle	150
7	M	7500	5	High	300
8	M	4000	3	High	250
9	F	5000	4	Middle	200
10	F	6000	2	High	300

M, Male, *F*, Female.

(i) Estimate the average (a) household income and (b) expenditure on transport and calculate unbiased estimates of the variances of the estimators used.
(ii) Estimate the proportion of people having primary, middle and high education. Give unbiased estimates for the variance—covariance matrix of the proposed estimators.
(iii) Estimate 95% confidence intervals for the population proportion of male-headed and female-headed households.

(iv) Give unbiased estimates of the total household income for the income groups 0–4000, 4001–6000, and 6000, respectively. Also give unbiased estimates of the average income for these income groups when the proportions of persons belonging to these income groups are known, which are 0.4, 0.3, and 0.3, respectively.

(v) Estimate the average income of the female-headed household for the income group 0–5000. Is your estimator biased or unbiased? Explain why.

(vi) Estimate the population correlation coefficient between the household income and household size? Is the estimator unbiased?

3.7.5 From the student population given in 3.7.3, select a sample of size 5 by the SRSWOR method and estimate the following:

(i) The average marks for Mathematics and Statistics and compute their SEs.

(ii) Estimate the proportion of male students and obtain a 95% confidence interval for the proportion.

(iii) Estimate the difference between the average marks for Mathematics and Statistics of the female students and obtain a 90% confidence interval for the difference of the mean marks.

3.7.6 Suppose samples s^* and s of sizes m and n are selected from the entire population U of size N and from $U - s^*$ of size $N - m$ by SRSWOR method. Find the optimum value of w for which the variance of the weighted estimator $\overline{y}_w = w\overline{y}_{s^*} + (1 - w)\overline{y}_s$ of the population mean attains a minimum value.

3.7.7 A sample of 70 migrant workers is selected from a population of 500 migrant workers and they are classified into the following age groups:

Age group	<15	16–50	51–65	>65
No of workers	15	20	30	5
Average daily wages in $	30	50	55	40

(i) Estimate the average daily wage for all the migrant workers (μ) and give the SE of your estimator. Obtain a 95% confidence interval for μ.

(ii) Estimate the proportion of workers in each age group and estimate the variance and covariance matrix of the estimators.

(iii) Estimate the proportion of workers of age less than or equal to 50. Find a 90% confidence interval of this proportion.

3.7.8 Let s be a sample of size n selected from a population U by the SRSWR method and s_m be the set of distinct units in s. Suppose a sample s_u of size u is selected from $U - s_m$ by the SRSWOR method and $\overline{y}_u$ is the sample mean of s_u. Show that $\overline{y}_u$ is an unbiased estimator of the mean of the population U. Derive the variance of $\overline{y}_u$.

3.7.9 Let s and s_m be the same as in 3.7.8 but s'_u be a subsample of size u selected from s_m by the SRSWOR method and $\overline{y}'_u$ be the sample mean of s'_u. Show that $\overline{y}'_u$ is an unbiased estimator of the population mean. Derive the variance of $\overline{y}'_u$.

3.7.10 A survey is conducted to find the prevalence of HIV/AIDS infection in a certain community consisting of 150 residents. Given that an estimate of the prevalence obtained from the pilot study is 25%, determine the sample size required so that the permissible error is 1% of the true value with 95% confidence coefficient if the sample is selected by (i) SRSWOR and (ii) SRSWR methods.

3.7.11 It is known from the past survey that the CV of household income is 15%. Determine the required sample size so that the CV of the estimate of the household mean income is less than 10% when the sample is selected from a population of 500 households by the (i) SRSWOR and (ii) SRSWR methods.

3.7.12 Estimate the minimum sample size required to estimate the population mean household expenditure on food to ensure that the error of an estimator is less than 1% of the true value of the population mean with probability 0.99, when the sample is selected from a population of 500 by the (i) SRSWR and (ii) SRSWOR methods, given that the sample mean and the sample standard deviation of the expenditure on food were obtained as \$4000 and \$625, respectively, from the pilot survey.

CHAPTER 4

Systematic Sampling

4.1 INTRODUCTION

In systematic sampling, the first unit is selected by using a random number, and the remaining units are selected by a predetermined pattern. Systematic sampling is very convenient for selection of a sample and hence it is used extensively in most practical situations. In this chapter we will discuss various types of systematic sampling procedures and methods of estimation of the population mean and its variances for such procedures. The efficiency of systematic sampling will also be compared with simple random and stratified sampling.

4.2 LINEAR SYSTEMATIC SAMPLING

Suppose we are to select a sample of size n from a population U of size N by a linear systematic sampling procedure. We consider the following two situations.

4.2.1 Linear Systematic Sampling With $N/n = k$ an Integer

We first select one unit, r, at random from 1 to k, then every kth unit. The initial unit r selected is called "random start" and k is called "sampling interval." Thus with the selection of a random start r (=1,..., k), a systematic sample $s_r = \{r, r+k, r+2k, \ldots, r+(n-1)k\}$ of size n is selected. The probability of selection of each of the k possible systematic samples is equal to $1/k$. The systematic samples partition the population U into mutually exclusive and exhaustive samples, i.e., $\bigcup_{r=1}^{k} s_r = U$ and $s_i \cap s_j = \phi$ for $i \neq j = 1, \ldots, k$. Since a unit can occur in one and only one sample, the inclusion probability of ith unit is $\pi_i = 1/k$ for $\forall i \in U$ and the inclusion probability for the ith and j ($\neq i$)th unit is $\pi_{ij} = 1/k$ if ith and jth unit belong to the same sample, and zero if the units do not belong to the same systematic sample. Thus for $N = 12$, and $n = 4$, we get $k = 12/4 = 3$ and we have k (=3) possible systematic samples given in Table 4.2.1.

Here inclusion probabilities $\pi_i = 1/3$ for $i = 1, \ldots, 12$. Since the units 1 and 4, 1 and 7, and 1 and 10 belong to the same sample s_1, we get $\pi_{1,4} = \pi_{1,7} = \pi_{1,10} = 1/3$. On the other hand, the units 1 and 5, 1 and 6, and 1 and 8 do not belong to the same sample, hence $\pi_{1,5} = \pi_{1,6} = \pi_{1,8} = 0$.

Survey Sampling Theory and Applications
ISBN 978-0-12-811848-1
http://dx.doi.org/10.1016/B978-0-12-811848-1.00004-2

Table 4.2.1 LSS with *N/n* is integer

Random start (*r*)	**1**	**2**	**3**
Sample (s_r)	$s_1 = (1, 4, 7, 10)$	$s_2 = (2, 5, 8, 11)$	$s_3 = (3, 6, 9, 12)$
Probability	1/3	1/3	1/3

4.2.2 Linear Systematic Sampling With $N/n = k$ Not an Integer

In case N/n is not an integer, k is chosen as an integer close to N/n. In this situation we cannot always realize the fixed sample size n. For example, if $N = nk + t$ with $0 < t < k$, a systematic sample corresponding to random start r less than or equal to t produces sample of size $(n + 1)$, whereas a random start r greater than t produces a sample of size n. Since the probability of selection of each sample is $1/k$, the expected sample size is $(n+1)\frac{t}{k} + n\frac{k-t}{k} = \frac{N}{k}$. If $N = (n-1)k + t$ with $0 < t < k$, we get t systematic samples each of size n and $(k - t)$ samples each of size $(n - 1)$. For more clarity, let us consider Table 4.2.2.

Since the selection of a systematic sample is very easy and ensures even distribution of units over the population, it is used in most practical situations. Suppose an auditor decides to check 20 vouchers out of 100 vouchers. Here $N = 100$, $n = 20$, and $k = 100/20 = 5$. So, the auditor should select first a voucher at random from 1 to 5 (=k), then every fifth voucher. Systematic sampling can be used for selection of every kth row of plants in an agricultural experiment, checking of driver's license for every kth car on the road, reexamining every kth answer scripts, and so on.

Table 4.2.2 LSS with *N/n* not an integer

N	***n***	***k***	**Possible systematic samples**
13	4	3	$s_1 = (1, 4, 7, 10, 13)$, $s_2 = (2, 5, 8, 11)$, $s_3 = (3, 6, 9, 12)$
14	3	5	$s_1 = (1, 6, 11)$, $s_2 = (2, 7, 12)$, $s_3 = (3, 8, 13)$, $s_4 = (4, 9, 14)$, $s_5 = (5, 10)$

4.2.3 Estimation of the Population Mean and Its Variance

It has been discussed in Chapter 2 that the systematic sampling design is a unicluster sampling design. For a unicluster sampling design the Horvitz–Thompson estimator $\widehat{Y}_{ht} = \sum_{i \in s} \frac{y_i}{\pi_i}$ is the only unbiased estimator of the population total Y in the class of linear unbiased estimators and hence it is trivially the best (MVUE) unbiased estimator in that class. Since for systematic sampling scheme $\pi_i = 1/k$, the Horvitz–Thompson estimator, $\widehat{Y}_{ht} = k \sum_{i \in s_r} y_i$ based on the systematic sample s_r, is unbiased for the population total. Hence we have the following theorem.

Theorem 4.2.1

For a systematic sampling, $\widehat{\overline{Y}}_{\mathrm{LSS}} = \dfrac{\hat{Y}_{ht}}{N} = \dfrac{k}{N}\sum_{i\in s_r} y_i$ is an unbiased estimator for the population mean $\overline{Y}$.

Corollary 4.2.1

In particular, when $N/n = k$ is an integer, $\widehat{\overline{Y}}_{\mathrm{LSS}}$ reduces to $\dfrac{\hat{Y}_{ht}}{N} = \dfrac{1}{n}\sum_{i\in s_r} y_i = \overline{y}(s_r) =$ sample mean, which is an unbiased estimator for the population mean.

Theorem 4.2.2

For $N/n = k$ an integer, the variance of $\overline{y}(s_r)$ is given by

$$V_{sy} = V[\overline{y}(s_r)] = \frac{1}{N}(\mathrm{TSS} - \mathrm{WSS}) = \frac{\sigma_y^2}{n}[1 + (n-1)\rho]$$

where $\mathrm{TSS} = \sum_{i=1}^{N}\left(y_i - \overline{Y}\right)^2 =$ total sum of square,

$\mathrm{WSS} = \sum_{r=1}^{k}\sum_{j\in s_r}\left\{y_j - \overline{y}(s_r)\right\}^2 =$ sum of square within the systematic samples, $\rho = \dfrac{\sum_{r=1}^{k}\sum_{i\neq}\sum_{j\in s_r}(y_i - \overline{Y})(y_j - \overline{Y})}{kn(n-1)\sigma_y^2} =$ intraclass correlation between the units of the same systematic sample, and $\sigma_y^2 = \frac{1}{N}\sum_{i=1}^{N}\left(y_i - \overline{Y}\right)^2$.

Proof

$$\begin{aligned} V_{sy} &= V[\overline{y}(s_r)] \\ &= E\left[\overline{y}(s_r) - \overline{Y}\right]^2 \\ &= \frac{1}{k}\sum_{r=1}^{k}\left(\overline{y}(s_r) - \overline{Y}\right)^2 \end{aligned}$$

Now

$$\begin{aligned} \mathrm{TSS} &= \sum_{i=1}^{N}\left(y_i - \overline{Y}\right)^2 \\ &= \sum_{r=1}^{k}\sum_{j\in s_r}\left(y_j - \overline{Y}\right)^2 \\ &= \sum_{r=1}^{k}\sum_{j\in s_r}\left\{y_j - \overline{y}(s_r) + \overline{y}(s_r) - \overline{Y}\right\}^2 \\ &= \sum_{r=1}^{k}\sum_{j\in s_r}\left\{y_j - \overline{y}(s_r)\right\}^2 + n\sum_{r=1}^{k}\left\{\overline{y}(s_r) - \overline{Y}\right\}^2 \\ &= SSW + NV[\overline{y}(s_r)] \end{aligned}$$

Hence

$$V_{sy} = (\text{TSS} - \text{WSS})/N \tag{4.2.1}$$

Now

$$\begin{aligned}
V_{sy} &= \frac{1}{k}\sum_{r=1}^{k}\left[\overline{y}(s_r) - \overline{Y}\right]^2 \\
&= \frac{1}{k}\sum_{r=1}^{k}\left[\frac{1}{n}\sum_{i\in s_r}(y_i - \overline{Y})\right]^2 \\
&= \frac{1}{kn^2}\left[\sum_{r=1}^{k}\sum_{i\in s_r}(y_i - \overline{Y})^2 + \sum_{r=1}^{k}\sum_{i\neq}\sum_{j\in s_r}(y_i - \overline{Y})\left(y_j - \overline{Y}\right)\right] \\
&= \frac{\sigma_y^2}{n}[1 + (n-1)\rho]
\end{aligned} \tag{4.2.2}$$

4.2.4 Nonexistence of Unbiased Variance Estimator

The variance of $\overline{y}(s_r)$ can be written as

$$\begin{aligned}
V_{sy} &= E\left[\overline{y}(s_r) - \overline{Y}\right]^2 \\
&= \sum_{r=1}^{k}\{\overline{y}(s_r)\}^2/k - \overline{Y}^2 \\
&= \sum_{r=1}^{k}\left(\sum_{i\in s_r} y_i\right)^2 \Big/ (kn^2) - \left\{\sum_{r=1}^{k}\left(\sum_{i\in s_r} y_i\right)\right\}^2 \Big/ N^2 \\
&= \frac{1}{N}\left(\frac{1}{n} - \frac{1}{N}\right)\sum_{r=1}^{k}\left\{\sum_{i\in s_r} y_i^2 + \sum_{i\neq}\sum_{j\in s_r} y_i y_j\right\} \\
&\quad - \frac{1}{N^2}\sum_{r\neq}^{k}\sum_{r'=1}^{k}\left(\sum_{j\in s_r} y_i\right)\left(\sum_{j\in s_{r'}} y_j\right)
\end{aligned}$$

Here we note that the quantity $\frac{1}{N}\left(\frac{1}{n} - \frac{1}{N}\right)\sum_{r=1}^{k}\left\{\sum_{i\in s_r} y_i^2 + \sum_{i\neq}\sum_{j\in s_r} y_i y_j\right\}$ can be estimated unbiasedly by $k\frac{1}{N}\left(\frac{1}{n} - \frac{1}{N}\right)\left\{\sum_{i\in s_r} y_i^2 + \sum_{i\neq}\sum_{j\in s_r} y_i\, y_j\right\}$ since the inclusion probability for the ith unit is $1/k$ for $i = 1,\ldots, N$ and the inclusion probability of the ith and jth $(i \neq j)$ unit belonging to the same systematic sample is also $1/k$. However, $\sum_{r \neq}^{k} \sum_{r'=1}^{k} \left(\sum_{i \in s_r} \sum_{j \in s_{r'}} y_i y_j \right)$ cannot be estimated unbiasedly because the inclusion probability for the pair of units i, j $(i \neq j)$ that do not belong to the same systematic sample is zero. Hence we cannot get an unbiased estimator of V_{sy} from a single systematic sample.

4.3 EFFICIENCY OF SYSTEMATIC SAMPLING

From Eqs. (4.2.1) and (4.2.2), we note that systematic sampling becomes efficient if WSS, the variation within the systematic sample, is large. In this case, the intraclass correlation ρ becomes negative. So, labeling the units plays an important role in ensuring the efficiency of the estimator. Thus systematic sampling will be efficient if one labels the units in such a way that units within the systematic samples become as heterogeneous as possible, allowing intraclass correlation to take a high negative value. This can be achieved if one labels the units so that the y-values increase with the label, i.e., $y_i \leq y_j$ for $i < j$. This type of labeling, however, is not possible as y-values are unknown before conducting a survey. However, if auxiliary information, x (such as area under cultivation), is known before the survey, and it is approximately proportional to y, one should label units in ascending values of x to employ a systematic sampling scheme effectively.

4.3.1 Comparison With Simple Random Sampling

In case k is an integer, the variances of the sample means based on the simple random sampling with (SRSWR) and without (SRSWOR) replacement methods of size n are given by $V_{wr} = \frac{1}{n}\sigma_y^2$ and $V_{wor} = \frac{(N-n)}{Nn}S_y^2 = \frac{(N-n)}{(N-1)n}\sigma_y^2$, respectively. From Eq. (4.2.2), we note that the sample mean based on a systematic sampling procedure will be more, equal, or less efficient than the sample mean based on SSRWR, if the intraclass correlation ρ becomes negative, equal to $-1/(n-1)$, or larger than $-1/(n-1)$, respectively. Similarly, systematic sampling mean is more, equal, or less efficient than the sample mean based on SRSWOR if $\rho <, =$ or $> -1/(N-1)$, respectively.

4.3.2 Comparison With Stratified Sampling

Let us write $y_{jr} = y_{(j-1)k+r}$; $j = 1,\ldots, n$; $r = 1,\ldots, k$ as the value of y for the jth unit of the rth systematic sample and arrange y_{jr}'s in an $n \times k$ table as follows:

	Systematic sample					
Stratum	**1**		***r***		***k***	**Stratum mean**
1	y_{11}	·	y_{1r}	·	y_{1k}	$\overline{y}_{1\cdot}$
	·	·	·	·	·	
j	y_{j1}	·	y_{jr}	·	y_{jk}	$\overline{y}_{j\cdot}$
	·	·	·	·	·	
n	y_{n1}	·	y_{nr}	·	y_{nk}	$\overline{y}_{n\cdot}$
Systematic sample mean	$\overline{y}_1$	·	$\overline{y}_r$	·	$\overline{y}_k$	

The systematic sample mean corresponding to the random start r is given by $\overline{y}(s_r) = \overline{y}_r = \frac{1}{n}\sum_{j=1}^{n} y_{jr}$. Now let us assume that y_{jr} is a random sample of size 1 selected from the jth strata consisting of k units $y_{j1},\ldots, y_{jr},\ldots, y_{jk}$. Under this assumption, the variance of $\overline{y}_r$ is given by

$$
\begin{aligned}
V_{st} &= V\left(\frac{1}{n}\sum_{j=1}^{n} y_{jr}\right)\\
&= \frac{1}{n^2}\sum_{j=1}^{n} V(y_{jr})\\
&= \frac{1}{n^2 k}\sum_{j=1}^{n}\sum_{r=1}^{k}\left(y_{jr} - \overline{y}_{j\cdot}\right)^2\\
&= \frac{\sigma^2_{wst}}{n}
\end{aligned}
\qquad (4.3.1)
$$

where

$$
\sigma^2_{wst} = \frac{\sum_{r=1}^{k}\sum_{j=1}^{n}\left(y_{jr} - \overline{y}_{j\cdot}\right)^2}{N(= nk)} \qquad (4.3.2)
$$

The variance of the systematic sample mean is

$$
\begin{aligned}
V_{sy} &= V[\overline{y}(s_r)] \\
&= \frac{1}{k}\sum_{r=1}^{k}\left(\overline{y}_r - \overline{y}\right)^2 \\
&= \frac{1}{k}\sum_{r=1}^{k}\left\{\frac{1}{n}\sum_{j=1}^{n}\left(y_{jr} - \overline{y}_{j\cdot}\right)\right\}^2 \\
&= \frac{1}{n^2k}\left[\sum_{r=1}^{k}\sum_{j=1}^{n}\left(y_{jr} - \overline{y}_{j\cdot}\right)^2 + \sum_{r=1}^{k}\left\{\sum_{j\neq}^{n}\sum_{j'=1}^{n}\left(y_{jr} - \overline{y}_{j\cdot}\right)\left(y_{j'r} - \overline{y}_{j'\cdot}\right)\right\}\right] \\
&= \left[1 + (n-1)\rho_{st}\right]\frac{\sigma^2_{wst}}{n}
\end{aligned}
\tag{4.3.3}
$$

where

$$
\rho_{st} = \frac{\sum_{r=1}^{k}\left\{\sum_{j\neq}^{n}\sum_{j'=1}^{n}\left(y_{jr} - \overline{y}_{\cdot j}\right)\left(y_{j'r} - \overline{y}_{\cdot j'}\right)\right\}}{kn(n-1)\ \sigma^2_{wst}}
$$

denotes the intraclass correlation coefficient between the units of the same systematic sample with deviation measured from stratum means. It should be noted that in ρ (defined in Theorem 4.2.2) the deviations are taken from the overall population mean.

Eqs. (4.3.2) and (4.3.3) indicate that systematic sampling is more efficient than stratified sampling if ρ_{st} is negative. The reverse is true when ρ_{st} is positive. Both are equally efficient if ρ_{st} is zero.

4.3.3 Random Arrangement of Units

If the labels of the units are attached at random, then systematic sampling becomes identical to SRSWOR sampling since, in this case, probability of selecting n distinct units is $1\Big/\binom{N}{n}$. Madaw and Madaw (1944) showed that if the labels of the units are attached at random, then the average variance of a systematic sample mean is exactly equal to the average variance of the sample mean based on SRSWOR sampling.

4.3.4 Population With Linear Trend

It is mentioned earlier that the systematic sampling procedure becomes efficient if y_i's increase or decrease with the label i ($\in U$). Here we will consider an artificial population with a linear trend viz. $y_i = \alpha + \beta i$ for $i = 1,\ldots, N$. In this situation

$$\overline{Y} = \frac{1}{N}\sum_{i=1}^{N}(\alpha+\beta i) = \alpha + \frac{N+1}{2}\beta$$

$$\sigma_y^2 = \frac{1}{N}\sum_{i=1}^{N}\left(y_i-\overline{Y}\right)^2 = \frac{\beta^2}{N}\sum_{i=1}^{N}\left(i-\frac{N+1}{2}\right)^2 = \beta^2\frac{(N^2-1)}{12}$$

Let $\overline{y}(s_r)$ be the sample mean based on a linear systematic sample selected with a random start r and sampling interval $k = N/n$ is an integer. Then we have

$$\overline{y}(s_r) = \frac{1}{n}\sum_{j=1}^{n} y_{r+(j-1)k} = \frac{1}{n}\sum_{j=1}^{n}[\alpha+\beta\{r+(j-1)k\}] = \alpha + \left\{r + k\frac{(n-1)}{2}\right\}\beta$$

The variance of $\overline{y}(s_r)$ is

$$V_{sy} = \frac{1}{k}\sum_{r=1}^{k}\left[\overline{y}(s_r)-\overline{Y}\right]^2 = \frac{\beta^2}{k}\sum_{r=1}^{k}\left(r-\frac{k+1}{2}\right)^2 = \beta^2\frac{(k^2-1)}{12} \tag{4.3.4}$$

If we treat s_r as a stratified sample (see Chapter 7), the variance of $\overline{y}(s_r)$ becomes

$$
\begin{aligned}
V_{st} &= \frac{1}{n^2}\sum_{j=1}^{n}\left(1-\frac{1}{k}\right)\frac{1}{k-1}\sum_{r=1}^{k}\left(y_{jr}-\bar{y}_{j\cdot}\right)^2 \\
&= \frac{\beta^2}{n^2}n\left(\frac{1}{k}-1\right)\frac{1}{k-1}\frac{k(k^2-1)}{12} \qquad (4.3.5) \\
&= \beta^2\frac{k^2-1}{12n}
\end{aligned}
$$

The variances of the sample means based on SRSWOR and SRSWR sampling for this population are, respectively, given by

$$
\begin{aligned}
V_{swor} &= \frac{N-n}{n(N-1)}\sigma_y^2 = \beta^2\frac{(N-n)(N+1)}{12n} \qquad (4.3.6) \\
V_{swr} &= \frac{\sigma_y^2}{n} = \beta^2\frac{(N-1)(N+1)}{12n}
\end{aligned}
$$

From Eqs. (4.3.4)−(4.3.6), we note that the variance of stratified sampling is $1/n$ times of the variance of the systematic sampling scheme, and the systematic sampling scheme has a variance approximately $1/n$ times of that of SRSWOR sampling. Thus we have the following theorem.

Theorem 4.3.1

For a population with $y_i = \alpha + \beta i$ for $i = 1,\ldots, N$

(i) $V_{st} \leq V_{sy} \leq V_{swor} \leq V_{swr}$

(ii) If the finite population correction term ($=n/N$) is ignored,

$$V_{st} : V_{sys} : V_{swor} : V_{swr} = \frac{1}{n^2} : \frac{1}{n} : 1 : 1$$

4.3.4.1 End Corrections

From Theorem 4.3.1, we note that systematic sampling from a population with a linear trend yields a reduction in variance of a magnitude approximately $1/n$ over the SRSWOR sampling scheme. Yates (1948) showed for a population with linear trend, the variance of the systematic sample mean becomes exactly zero if we take a weighted systematic sample mean

$$\begin{aligned}\overline{y}_w(s_r) &= \sum_{j=1}^{n} w_j y_{r+(j-1)k} \\ &= \overline{Y} \\ &= \alpha + \beta\frac{N+1}{2}\end{aligned} \tag{4.3.7}$$

with weights $w_1 = \frac{1}{n} + \frac{2r - k - 1}{2(n - 1)k}$, $w_n = \frac{1}{n} - \frac{2r - k - 1}{2(n - 1)k}$, and $w_j = \frac{1}{n}$ for $j = 2, ..., n - 1$. The changes of weights for the first and last term are known as end corrections.

4.3.4.2 Balanced Systematic Sampling

Balanced systematic sampling was introduced by Sethi (1965) and Murthy (1967). In this sampling scheme the units of a sample are selected from the hypothetical population with a linear trend in such a way that the sample mean becomes exactly equal to the population mean $\overline{Y}$.

In the balanced sampling scheme, we assume that the sample size n is even and k $(=N/n)$ is an integer. Here the population is divided into $n/2$ groups of $2k$ consecutive units each. Here we select a random number r from 1 to k. Corresponding to the random start r, the selected balanced systematic sample is

$$\{r + 2jk, 2(j + 1)k - (r - 1)\},\ j = 0, 1, ..., \frac{n}{2} - 1$$

The sample mean for the balanced systematic sample is

$$\overline{y}_s = \frac{1}{n}\sum_{j=0}^{n/2-1}\left\{y_{r+2jk} + y_{2(j+1)k-(r-1)}\right\} \tag{4.3.8}$$

In presence of linear trend $y_i = \alpha + \beta i$, $\overline{y}_s$ becomes $\alpha + (N + 1)\beta/2$, a constant. Hence in the presence of a linear trend, $\overline{y}_s$ is unbiased with a variance zero.

Singh et al. (1968) proposed an alternative method where pairs of units equidistant from the end points are selected in the sample. Thus when n is even, the random start r $(=1,..., k)$ selects the sample

$$\{r + jk, (N - jk) - (r - 1)\},\ j = 0, 1, ..., \frac{n}{2} - 1$$

The sample mean of Singh's balanced sample is

$$\begin{aligned}\bar{y}_s &= \frac{1}{n}\sum_{j=0}^{n/2-1}\left\{y_{r+jk} + y_{N-jk-(r-1)}\right\}\\ &= \frac{1}{n}\sum_{j=0}^{n/2-1}\{2\alpha+\beta(N+1)\}\\ &= \alpha+(N+1)\beta/2\end{aligned}$$

This method also eliminates the effect of linear trend because the sample mean is equal to the population mean $\alpha + (N+1)\beta/2$.

4.3.5 Population With Periodic Variation

Consider a hypothetical population that is strictly periodic with a period λ, i.e., $y_i = y_{i+\lambda}$. If we select a systematic sample with a sampling interval that is equal to the period $k=\lambda$ or its multiple, then all y_j values belonging to the same systematic samples will be equal and hence the variance of the systematic sample mean will be large. On the other hand, if we choose $k=\lambda/2$ or an odd multiple of it, then the systematic sample mean will be equal to the population mean $\overline{Y}$, and in this situation, the variance of the estimator is exactly zero. In practice, a population that is strictly periodic in nature may not be available, but approximately periodic populations are not uncommon. Examples are number of accidents, telephone calls during the 24 h in a day, and sales over 7 days in department stores. In this situation systematic sampling with an appropriate sampling interval produces an efficient estimator for the population mean.

4.3.6 Autocorrelated Population

In some population units, closer units are well related with respect to the study variable y. For example, the number of illegal immigrants in the villages decreases with the distance of the border from the neighboring country. In this situation, to accommodate relationship of the y-values, some statistical model (superpopulation model) is generally assumed. Details of the superpopulation model are given in Chapter 6. Cochran (1946) assumed the following autocorrelation model to study the performance of systematic sampling with other sampling designs:

$$E_m(y_i)=\mu,\ E_m(y_i-\mu)^2=\sigma^2,\ E_m(y_i-\mu)(y_{i+u}-\mu)=\rho_u\sigma^2 \qquad (4.3.9)$$

where E_m denotes expectation over superpopulation model (4.3.9) and ρ_u, the autocorrelation of order (lag) u, satisfies $\rho_u \geq \rho_v \geq 0$ for $u < v$.

Let $E_m(V_{sy})$, $E_m(V_{st})$, and $E_m(V_{ran})$ denote average variance of the sample mean of the systematic sample, stratified sample, and simple random sample without replacement, respectively. Cochran (1946) derived the following theorems:

Theorem 4.3.3

Under superpopulation model (4.3.9) with $\rho_u \geq \rho_v \geq 0$, $E_m(V_{st}) \leq E_m(V_{ran})$

Proof

We can write $\sum_{i=1}^{N}\left(y_i - \overline{Y}\right)^2$ as

$$\sum_{i=1}^{N}\left(y_i - \overline{Y}\right)^2 = \frac{1}{N}\sum_{i<}^{N}\sum_{j=1}^{N}\left(y_i - y_j\right)^2 \tag{4.3.10}$$

Taking expectation both sides with respect to the model (4.3.9), we get

$$\begin{aligned}
E_m\sum_{i=1}^{N}\left(y_i - \overline{Y}\right)^2 &= \frac{1}{N}\sum_{i<}^{N}\sum_{j=1}^{N}E_m\left(y_i - y_j\right)^2 \\
&= \frac{2\sigma^2}{N}\sum_{i<}^{N}\sum_{j=1}^{N}\left(1 - \rho_{j-i}\right) \\
&= \frac{2\sigma^2}{N}\left[\frac{N(N-1)}{2} - \sum_{i=1}^{N-1}\sum_{j=i+1}^{N}\rho_{j-i}\right] \\
&= \frac{2\sigma^2}{N}\left[\frac{N(N-1)}{2} - \sum_{i=1}^{N-1}\left(\rho_1 + \rho_2 + \cdots + \rho_{N-i}\right)\right] \\
&= \frac{2\sigma^2}{N}\left[\frac{N(N-1)}{2} - \sum_{j=1}^{N-1}(N-j)\rho_j\right] \\
&= \sigma^2(N-1)\left[1 - \frac{2}{N(N-1)}\sum_{j=1}^{N-1}(N-j)\rho_j\right]
\end{aligned} \tag{4.3.11}$$

Eq. (4.3.11) yields

$$E_m(V_{ran}) = \frac{N-n}{Nn}\frac{1}{N-1}E_m\sum_{i=1}^{N}\left(y_i-\overline{Y}\right)^2$$

$$= \sigma^2\frac{k-1}{nk}\left[1-\frac{2}{N(N-1)}\sum_{j=1}^{N-1}(N-j)\rho_j\right] \quad (4.3.12)$$

$$\text{(noting } N = nk\text{)}$$

Let us assume that the systematic sample of size n is a stratified sample where one unit is selected from each of the stratum. Then the variance of the stratified sample mean is obtained from Eq. (4.3.1) as

$$V_{st} = \frac{1}{n^2k}\sum_{j=1}^{n}\sum_{r=1}^{k}\left(y_{jr}-\overline{y}_{j\cdot}\right)^2 \quad (4.3.13)$$

Now using (4.3.11) we get

$$E_m\sum_{r=1}^{k}\left(y_{rj}-\overline{y}_{j\cdot}\right)^2 = \sigma^2(k-1)\left[1-\frac{2}{k(k-1)}\sum_{j=1}^{k-1}(k-j)\rho_j\right] \quad (4.3.14)$$

Now substituting (4.3.14) in (4.3.13), we have

$$E_m(V_{st}) = \sigma^2\frac{(k-1)}{nk}\left[1-\frac{2}{k(k-1)}\sum_{j=1}^{k-1}(k-j)\rho_j\right] \quad (4.3.15)$$

Eqs. (4.3.12) and (4.3.15) yield

$$E_m(V_{ran}) - E_m(V_{st}) = 2\sigma^2\frac{k-1}{nk}[L(N)-L(k)] \quad (4.3.16)$$

where

$$L(t) = 1-\frac{2}{t(t-1)}\sum_{j=1}^{t-1}(t-j)\rho_j$$

$$= 1-\frac{2}{t(t-1)}\sum_{j=1}^{t}(t-j)\rho_j \quad (4.3.17)$$

Now

$$L(t)-L(t+1) = -\frac{2}{t(t^2-1)}\sum_{j=1}^{t}a_j\rho_j \quad (4.3.18)$$

where $a_j = (t+1-2j)$.

The expression $\sum_{j=1}^{t} a_j\rho_j$ can be written as

$$\sum_{j=1}^{t} a_j\rho_j = a_1(\rho_1 - \rho_2) + (a_1 + a_2)(\rho_2 - \rho_3) + \cdots + (a_1 + \cdots + a_i)$$
$$(\rho_i - \rho_{i+1}) + \cdots + (a_1 + \cdots + a_{t-1})(\rho_{t-1} - \rho_t) + (a_1 + \cdots + a_t)\rho_t$$

Now noting (i) $\rho_j - \rho_{j+1} \geq 0$, (ii) $a_1 + \ldots + a_i = i(t-i) \geq 0$ for $t \geq i$, and (iii) $a_1 + \ldots + a_t = 0$, we find

$$L(t) - L(t+1) \leq 0 \tag{4.3.19}$$

Finally the theorem follows from (4.3.16) and (4.3.19).

The expected variance of the systematic sample mean under the superpopulation model (4.3.9) was derived by Cochran (1946) as

$$E_m(V_{sy}) = \frac{\sigma^2}{n}\left(1 - \frac{1}{k}\right)\left\{1 - \frac{2}{N(k-1)}\sum_{j=1}^{N-1}(N-j)\rho_j + \frac{2k}{n(k-1)}\sum_{j=1}^{n-1}(n-j)\rho_{kj}\right\} \tag{4.3.20}$$

Eqs. (4.3.15), (4.3.16), and (4.3.20) do not yield any general comparison of efficiency of the systematic sampling with random sampling and stratified sampling. However, Cochran (1946) reported for the model (4.3.9) that (i) systematic sample is more efficient than stratified sampling if $\rho_j = 0$ for $j > (k-1)$ while (ii) systematic sampling is less efficient than simple random sampling if $\rho_1 = \rho_2 = \ldots = \rho_k$ and $\rho_j = 0$ for $j > k$.

Cochran (1946) derived the following important results relating to the performances of systematic, stratified, and random sampling.

Theorem 4.3.4

Under superpopulation model (4.3.9) with (i) $\rho_i \geq \rho_{i+1} \geq 0$, for $i = 1, 2, \ldots, (N-1)$, and (ii) $\delta_i^2 = \rho_{i+1} - 2\rho_i + \rho_{i-1} \geq 0$, for $i = 2, 3, \ldots, N-2$,

$$E_m(V_{sy}) \leq E_m(V_{st}) \leq E_m(V_{ran})$$

for any sample size n.

Further, $E_m(V_{sy}) < E_m(V_{st})$ unless $\delta_i^2 = 0, i = 2, 3, \ldots, (N-2)$.

Further additional results are given by Quenouille (1949), Gautschi (1957), Jowett (1952), Hájek (1959), and Murthy and Rao (1988), among others.

4.4 LINEAR SYSTEMATIC SAMPLING USING FRACTIONAL INTERVAL

We have seen that linear systematic sampling cannot always produce a sample of the desired size n when N/n is not an integer. The above difficulty can be removed if we take $k = N/n$ without rounding it off. The method of selection is as follows. Firstly, select a random sample r, from a uniform distribution $(0, k)$. Then the random start r selects the ith unit in the sample if $i - 1 < r + jk \leq i$ for $j = 0, 1, \ldots, n - 1$. For instance, let $N = 13$, $n = 4$, and $k = 13/4 = 3.25$. Suppose $r = 2.15$ is a random sample selected from a uniform distribution (0, 3.25). Then the selected systematic sample includes the units 3, 6, 9, and 12. The first-order inclusion probability for each of the unit is equal to $1/k$ and hence an unbiased estimator of the population mean $\overline{Y}$ is

$$\widehat{\overline{Y}}_{ss} = \frac{k}{N} \sum_{i \in s} y_i = \overline{y}_s \tag{4.4.1}$$

4.5 CIRCULAR SYSTEMATIC SAMPLING

In a circular systematic sampling (CSS) scheme, initially a unit r (random start) is selected at random from all the N units of the population, then every kth unit is selected in a circular manner until the desired sample size n is obtained. Here k is taken as an integer nearest to N/n. Thus corresponding to a random start r, a circular systematic sample selects units

$$r + j\,k \quad \text{if } r + j\;k \leq N$$

and

$$r + j\,k - N \quad \text{if } r + j\;k > N$$
$$\text{for } j = 0, \ldots, n - 1.$$

The CSS method was introduced by Lahiri (1954), and it was extensively used by the Indian National Sample Survey Organization for large-scale sample surveys. Here we will consider the following cases.

4.5.1 Circular Systematic Sampling With $k = N/n$ as an Integer

In this situation we get N systematic samples, but all samples are not distinct. Each sample occurs exactly n times in different orders. To illustrate this phenomenon, let us consider the following example.

Example 4.5.1
Consider a population of size $N = 6$, the target sample size $n = 3$ and $k = 6/3 = 2$.

Random start	Systematic samples
1	1, 3, 5
2	2, 4, 6
3	3, 5, 1
4	4, 6, 2
5	5, 1, 3
6	6, 2, 4

Here we note that the random starts 1, 3, and 5 produce the same systematic sample but in different orders and they are (1, 3, 5), (3, 5, 1), and (5, 1, 3). Here also the sample mean $\overline{y}_s$ is an unbiased estimator of the population mean.

4.5.2 Circular Systematic Sampling With N/n is Not an Integer

In this case, if one chooses a value of k greater than N/n, then a circular systematic sample may contain a unit more than once, which is seen in the following example. In this case, the sample mean based on the distinct units and all the units including repetitions are unbiased for the population mean.

Example 4.5.2
Let $N = 30$ and the target sample size $n = 12$. Here $N/n = 2.5$, so both the integers 2 and 3 are nearest to 2.5. If we choose $k = 3$, then the random start 15 will select the following systematic sample:

$$s = (15, 18, 21, 24, 27, 30, 3, 6, 9, 12, 15, 18)$$

In the above systematic sample the units 15 and 18 occur twice.

To avoid the above difficulty, Sudakar (1978) pointed out that if k is taken as $\left[\frac{N}{n}\right]$, the largest integer less than or equal to N/n, then none of the units in the same systematic sample can occur more than once, i.e., all the units are distinct. In this case, the sample mean is an unbiased estimator of the population mean.

In Example 4.5.2, if we take $k = 2$, then the random start 15 would select the following systematic sample where no unit is repeated twice.

$$s = (15, 17, 19, 21, 23, 25, 27, 29, 1, 3, 5, 7)$$

The choice of the sampling interval k depends on the sample size n. Sometimes due to budgetary considerations and inadequate knowledge of sampling cost, the optimal sample size cannot be determined in advance. Hence it is of interest to know for a given population of size N, if there exists a sampling plan that can select a circular systematic sample of any size from 1 to N. Sudakar (1978) established a necessary and sufficient condition on N and k, which results in all n ($\leq N$) units of a sample being distinct. The condition is stated in the following theorem.

Theorem 4.5.1

Let N be the population size, k ($<N$) be an integer chosen as the sampling fraction, and the circular systematic sample selected with the random start r be denoted as

$$s(r, n) = (i_1, i_2, \ldots, i_n)$$

where

$$i_j = r + (j - 1)k \mod(N) \text{ where } j = 1, \ldots, n.$$

A necessary and sufficient condition for the units of the systematic samples s (r, n) to be distinct for all r, $n \leq N$ is that N and k are relatively coprimes.

Proof

Here both the necessary and sufficient conditions are proved by contradiction as follows:

(i) Sufficiency: Suppose N and k are relatively coprimes and there exist r and n for which two elements of s (r, n) are equal. Without loss of generality let us suppose that $i_1 = r$ and i_{j+1} are equal. Now $i_1 = r = i_{j+1} = (r + jk)$ mod (N) implies jk is a multiple of N; $j < n \leq N$. This contradicts that k and N are coprimes.

(ii) Necessary *j*th ($i \neq j$): Suppose for all r, $n \leq N$, all elements of s (r, n) are distinct and N and k are not coprimes. Let the greatest common factor of N and k be α so that $k = \beta\alpha$, $N = \gamma\alpha$, and β and γ are integers smaller than N. Then

$$\begin{aligned} i_{\gamma+1} &= (r + \gamma k)\text{mod}(N) \\ &= (r + \gamma\beta\alpha)\text{mod}(N) \\ &= (r + N\beta)\text{mod}(N) \\ &= r = i_1 \end{aligned}$$

This contradicts assumptions that all elements of s (r, n) are distinct.

Remark 4.5.1

The variance expressions for the estimator of the population mean $\overline{Y}$ based on fractional interval and CSS are not straightforward because all the possible samples may not be disjoint (see Särndal et al., 1992). The variance of $\widehat{\overline{Y}}_{ss}$ cannot be estimated from a single systematic sample as the second-order inclusion probabilities of some of the units are zero.

4.6 VARIANCE ESTIMATION

There are two types of variance estimators. One based on a single systematic sample of size n and the other based on several systematic samples (t) each of size m ($=n/t$). The variance estimators based on a single systematic sample produce biased estimators. Although the variance estimators based on several samples produce unbiased estimators of the variance, the convenience and efficiency of the systematic sampling are lost. The technique of variance estimation based on several samples was introduced by Mahalanobis (1946) and it is known as interpenetrating subsamples. Details of the variance estimations are given by Cochran (1946), Yates (1949), Koop (1971), Wolter (1984), Murthy and Rao (1988), and Bellhouse (1988), among others. Here we shall assume that the population size N is an integer multiple of the sample size n, i.e., $N = nk$ where k is the sampling interval.

4.6.1 Single Systematic Sample

Here the variance estimators are obtained under some assumptions on the population.

4.6.1.1 Random Arrangements of Units

Systematic sampling can be treated as SRSWOR sampling if the arrangement of the units in the population is random. So, treating the selected systematic sample s as an SRSWOR sample of size n, the variance of the systematic sample mean V_{sy} can be estimated as

$$\widehat{v}_1 = \frac{(1-f)}{n} s_y^2 \tag{4.6.1}$$

where $f = n/N$ and $s_y^2 = \sum_{i \in s} \{y_i - \overline{y}(s)\}^2/(n-1)$.

The estimator $\widehat{v}_1$ performs well if the arrangement of the units is random, i.e., if no trend or periodicity in the arrangements of the units persists. However, if the arrangement of the units is such that the systematic sampling is more efficient than SRSWOR sampling, then $\widehat{v}_1$ certainly overestimates the variance. On the other hand, $\widehat{v}_1$ underestimates the variance if in reality systematic sampling is less efficient than SRSWOR.

4.6.1.2 Stratified Sampling With One Unit Per Stratum

Systematic sampling stratifies the population into k strata each of size n. Let n be even and equal to $2q$. Now consider the units $(r, r+k)$ as a stratified sample of size 2 selected by the SRSWOR method from a stratum comprising the first $2k$ units viz. $1, \ldots, 2k$; $(r+2k, r+3k)$ as a stratified sample of size 2 from a stratum comprising the next $2k$ viz. $2k+1, \ldots, 4k$; in general, units $\{r+2(j-1)k, r+(2j-1)k\}$ as a stratified sample of size 2 from the jth stratum comprising units $2(j-1)k+1, \ldots, 2jk$; $j = 1, \ldots, q$.

An approximate estimate of V_{sy} is obtained as

$$\widehat{v}_2 = \frac{(1-f)}{n^2} \sum_{j=1}^{n/2} \delta_j^2 \tag{4.6.2}$$

where $\delta_j = y_{r+(2j-1)k} - y_{r+2(j-1)k}$.

4.6.1.3 Presence of Linear Trend

Let us construct the following difference table from the selected systematic sample s_r:

s_r	Y	Δ	Δ^2
r	y_r		
$r+k$	y_{r+k}	$y_{r+k} - y_r = \Delta(1)$	
$r+2k$	y_{r+2k}	$y_{r+2k} - y_k = \Delta(2)$	$\Delta(2) - \Delta(1) = \Delta^2(1)$
.	.	.	
.	.	.	
$r+(n-1)k$	$y_{r+(n-1)k}$	$y_{r+(n-1)k} - y_{r+(n-2)k}$ $= \Delta(n-1)$	$\Delta(n-1) - \Delta(n-2) =$ $\Delta^2(n-2)$

The variance of the systematic sample mean V_{sy} is estimated by using the following formulae:

$$\widehat{v}_3 = \frac{(1-f)}{n}\frac{1}{2(n-1)}\sum_{j=1}^{n-1}\{\Delta(j)\}^2 \tag{4.6.3}$$

and

$$\widehat{v}_4 = \frac{(1-f)}{n}\frac{1}{6(n-2)}\sum_{j=1}^{n-2}\{\Delta^2(j)\}^2 \tag{4.6.4}$$

The estimators $\widehat{v}_3$ and $\widehat{v}_4$ perform well if the ordering of units produces a linear trend. Otherwise, they underestimate variance. Several variance estimators based on higher-order differences are also proposed, e.g., Yates (1949).

4.6.1.4 Presence of Autocorrelation Between Successive Units

Cochran (1946) provided an alternative variance estimator involving $\widehat{\rho}_k$ as an estimate of ρ_k, intraclass correlation of order k $(=N/n)$. Cochran's variance estimator is given by

$$\widehat{v}_5 = \begin{cases} \dfrac{(1-f)s_y^2}{n}\left[1+2\left\{\dfrac{1}{\mathrm{In}(\widehat{\rho}_k)}+\dfrac{1}{(\widehat{\rho}_k)^{-1}-1}\right\}\right] & \text{if } \widehat{\rho}_k > 0 \\ \dfrac{(1-f)s_y^2}{n} & \text{if } \widehat{\rho}_k \le 0 \end{cases} \tag{4.6.5}$$

where $\widehat{\rho}_k = \left[\dfrac{1}{(n-1)}\sum_{j=0}^{n-2}\left\{y_{r+(j+1)k} - \overline{y}(s)\right\}\left\{y_{r+jk} - \overline{y}(s)\right\}\right]\Big/ s_y^2.$

The proposed variance estimators mentioned earlier are based on underlying certain superpopulation model. The variance estimators may not perform well if the underlying model fails. Wolter (1984) studied in detail the properties of bias, mean-square error, and confidence interval estimation theoretically and empirically.

4.6.1.5 Splitting of a Systematic Sample

Koop (1971) divided the systematic sample into two samples each of size $n/2$ (assuming integer) and proposed the following variance estimator

$$\widehat{v}_6 = \frac{1}{4}\left(\overline{y}_A - \overline{y}_B\right)^2 \tag{4.6.6}$$

where $\overline{y}_A$ and $\overline{y}_B$ are the sample means based on odd and even labels of the systematic sample s. Koop (1971) derived the expression of bias relative to its variance in terms of the intraclass correlation coefficient.

Now noting $\overline{y}_s = \frac{1}{2}\left(\overline{y}_A + \overline{y}_B\right)$, we find

$$V\left(\overline{y}_s\right) = \frac{1}{2}(1+\rho_0)\sigma_0^2$$
$$= \frac{(1+\rho_0)}{(1-\rho_0)}E(\widehat{v}_6)$$

where $\sigma_0^2 = V\left(\overline{y}_A\right) = V\left(\overline{y}_B\right)$ and ρ_0 is the correlation coefficient between the subsample means. In case correlation coefficient ρ_0 is known from the past survey, an almost unbiased estimator of $V\left(\overline{y}_s\right)$ is given by

$$\widehat{v}_7 = \widehat{V}[\overline{y}(s)]$$
$$= \frac{1+\rho_0}{1-\rho_0}\widehat{v}_6 \tag{4.6.7}$$

4.6.2 Several Systematic Samples

Here instead of taking a single systematic sample of size n, several (l) independent systematic samples are selected. The sample sizes for the systematic samples need not be the same. Let t_i be an unbiased estimator of $\overline{Y}$ based on the ith (=1,..., l) systematic sample of size $m_i\left(\sum_{i=1}^{l} m_i = n\right)$. Then we have the following theorem.

Theorem 4.6.1

(i) $\overline{t} = \frac{1}{l}\sum_{i=1}^{l} t_i$ is an unbiased estimator of $\overline{Y}$

(ii) An unbiased estimator of the variance of $\overline{t}$ is $\widehat{v}_8 = \frac{1}{l(l-1)}\sum_{i=1}^{l}(t_i - \overline{t})^2$

Proof

(i) $E(\overline{t}) = \frac{1}{l}\sum_{i=1}^{l} E(t_i) = \overline{Y}$

(ii) Variance of $\overline{t} = V(\overline{t}) = \frac{1}{l^2}\sum_{i=1}^{l} V(t_i)$ and

$$E(\widehat{v}_8) = \frac{1}{l(l-1)}E\left(\sum_{i=1}^{l} t_i^2 - l\,\overline{t}^2\right)$$
$$= \frac{1}{l(l-1)}\left[\sum_{i=1}^{l}\left\{V(t_i) + \overline{Y}^2\right\} - l\left\{V(\overline{t}) + \overline{Y}^2\right\}\right]$$
$$= V(\overline{t})$$

For independent interpenetrating sampling, a unit may be repeated more than once. To avoid repletion of units, we first enumerate all possible $N/m = t$ (assuming integer) systematic samples of size m corresponding to t random starts. From the t random starts, a sample of l random starts is selected by SRSWOR and l systematic samples each of size m are selected using the chosen l random starts.

Let $\overline{y}(s_j)$ be the sample mean for the jth systematic sample and

$$\overline{y}(s) = \sum_{j=1}^{l} \overline{y}(s_j)/l \tag{4.6.8}$$

be the pooled sample mean.

The combined mean $\overline{y}(s)$ is unbiased for $\overline{Y}$ and an unbiased estimator of $V\{\overline{y}(s)\}$ is given by

$$\widehat{v}_9 = \frac{(1-g)}{l(l-1)} \sum_{j=1}^{l} \left\{\overline{y}(s_j) - \overline{y}(s)\right\}^2 \tag{4.6.9}$$

where $g = l/t$.

As an illustration, we give the following example.

Example 4.6.1

The following list gives claims against 30 vouchers.

Voucher number	Amount in $	Voucher number	Amount in $	Voucher number	Amount in $
1	128	11	453	21	303
2	499	12	395	22	112
3	196	13	334	23	256
4	142	14	400	24	256
5	158	15	342	25	229
6	126	16	317	26	225
7	101	17	316	27	225
8	282	18	210	28	488
9	368	19	335	29	117
10	473	20	286	30	329

(i) Select a sample of size six vouchers by systematic sampling procedures. Estimate the mean amount of the vouchers and estimate its variance by using the different procedures described in Section 4.6.1.

(ii) Select three independent samples of sizes 4, 6, and 7 by using the CSS procedure and obtain an estimate of average claim per voucher from each of the samples separately. Obtain an estimator for the mean amount of a voucher based on the three samples combined and find an unbiased estimator of the standard error of the proposed combined estimator.

Solution:

(i) Here $N = 30$ and $n = 6$, so $k = 30/6 = 5$. Let the random start 3 ($=r$) be selected from 1 to 5 ($=k$). The random start select sample $s = (3, 8, 13, 18, 23, 28)$. So, the estimated mean amount of voucher is

$$\bar{y}(s) = \frac{y_3 + y_8 + y_{13} + y_{18} + y_{23} + y_{28}}{6}$$

$$= \frac{196 + 282 + 334 + 210 + 256 + 488}{6}$$

$$= \$294.33$$

The estimated variances based on the Eq. (4.6.1) to (4.6.6) are computed as follows:

$$\widehat{v}_1 = \frac{(1-f)}{n} s_y^2$$

$$= \frac{(1 - 6/30)}{6} \times 11496.67$$

$$= 1532.889$$

$$\widehat{v}_2 = \frac{(1-f)}{n^2} \sum_{j=1}^{n/2} \delta_j^2$$

$$= \frac{(1 - 6/30)}{6^2} \times 76596$$

$$= 1702.133$$

$$\widehat{v}_3 = \frac{(1-f)}{n} \frac{1}{2(n-1)} \sum_{j=1}^{n-1} \{\Delta(j)\}^2$$

$$= \frac{(1 - 6/30)}{6} \frac{1}{2(6-1)} \times 81416$$

$$= 1085.54$$

$$\widehat{v}_4 = \frac{(1-f)}{n} \frac{1}{6(n-2)} \sum_{j=1}^{n-2} \{\Delta^2(j)\}^2$$

$$= \frac{(1 - 6/30)}{6} \frac{1}{6 \times (6-2)} \times 95628$$

$$= 192.2$$

Here correlation coefficient $\widehat{\rho}_5$ between the pairs of observation (282, 196), (334, 282), (210, 334), (256, 210), and (448, 256) is −0.125 (negative). Hence the estimate variance

$$\widehat{v}_5 = \frac{(1-f)}{n} s_y^2$$
$$= 1532.889$$

Sample mean based on odd and even labels are $\overline{y}_A = 262$ and $\overline{y}_B = 326.66$, respectively. Hence

$$\widehat{v}_6 = \frac{1}{4}\left(\overline{y}_A - \overline{y}_B\right)^2$$
$$= 1045.44$$

(ii) For sample size $n = 4$, $N/n = 7.5$, is not an integer. So, we take $k = (30/4) = 7$ and select one unit at random from 1 to 30. Let the selected unit (random start) be 5. Then the selected sample is $s_1 = \{5, 12, 19, 26\}$. The estimator for the mean $\overline{Y}$ is $t_1 = (158 + 395 + 335 + 225)/4 = \$278.25$. For sample size $n = 6$, $k = N/n = 30/6 = 5$ an integer. Here also we select a random start from 1 to 30 and let it be 10. The selected CSS with the random start 10 is $s_2 = \{10, 15, 20, 25, 30, 5\}$; the estimator for $\overline{Y}$ is $t_2 = (473 + 342 + 286 + 229 + 329 + 158)/6 = \$302.83$. Similarly for the sample size $n = 7$, we choose $k = (30/7) = 4$. Let the selected random start from 1 to 30 be 25, then the selected CSS is $s_3 = \{25, 29, 3, 7, 11, 15, 19\}$ and estimator for $\overline{Y}$ is $t_3 = (229 + 117 + 196 + 101 + 453 + 342 + 335)/7 = \$253.29$. The combined estimator of $\overline{Y}$ is $\overline{t} = (t_1 + t_2 + t_3)/3 = \$278.12$ and an estimate of the standard error of t is

$$\sqrt{\widehat{V}(\overline{t})} = \sqrt{\frac{1}{3 \times 2}\sum_{i=1}^{3}\left(t_i - \overline{t}\right)^2} = \$14.30$$

4.7 TWO-DIMENSIONAL SYSTEMATIC SAMPLING

Linear systematic sampling can be generalized to two-dimensional systematic sampling. Consider a population of size $N = ab$ arranged in $a = ml$ rows and $b = tk$ columns. Suppose we want to take a systematic sample of size $n = mt$ units from N units. The procedure is as follows: Instead of selecting single random number as used in a one-dimensional systematic sample, we select a pair of random numbers (r, r'). The random number r is selected from 1 to l, and r' is selected from 1 to k. (r, r') may be regarded as a two-dimensional random start. The first coordinate (r) represents the position of the row and the second (r') the column of the initially chosen unit. Then the two-dimensional systematic sample consists of units with coordinates (x, y), where $x = r + (i - 1)l$ and $y = r' + (j - 1)k$; $i = 1, \ldots, m$; $j = 1, \ldots, t$. Methods of estimation of population mean and its variance can be obtained from the results derived for one-dimensional systematic sampling. Interested readers are referred to Quenouille (1949), Das (1951), Koop (1976), and Bellhouse (1977, 1981).

Example 4.7.1

Consider 180 units arranged into $a = 12$ rows and $b = 15$ columns. We want to select a sample of 20 units. Here we may take $m = 4$, $l = 3$ and $t = 5$, $k = 3$. Suppose a random number $r = 2$ is selected from 1 to l (=3) and another number $r' = 3$ is selected from 1 to t (=5) so that the selected random start is (2, 3). Then our selected two-dimensional systematic sample is as follows:

s = (2,3), (2,6), (2,9), (2,12), (2,15)
(5,3), (5,6), (5,9), (5,12), (5,15)
(8,3), (8,6), (8,9), (8,12), (8,15)
(11,3), (11,6), (11,9), (11,12), (11,15)

	Columns														
Rows	**1**	**2**	**3**	**4**	**5**	**6**	**7**	**8**	**9**	**10**	**11**	**12**	**13**	**14**	**15**
1															
2			√			√			√			√			√
3															
4															
5			√			√			√			√			√
6															
7															
8			√			√			√			√			√
9															
10															
11			√			√			√			√			√
12															

4.8 EXERCISES

4.8.1 Define the systematic sampling procedure and explain how it differs from random sampling. State some of the merits and demerits of systematic sampling with suitable examples.

4.8.2 Explain the LSS. Suggest an unbiased estimator for the population mean. Obtain the expression of its variance. Compute the first- and second-order inclusion probabilities when N/n is an integer. Explain why it is not possible to get an unbiased estimator of the variance based on a single systematic sample.

4.8.3 Compute the inclusion probabilities of the first two orders for an LSS when N/n is an integer. Explain the difficulty you may face for using LSS when N/n is not an integer.

4.8.4 Describe the CSS and explain how you will choose the sampling interval k when N/n is not an integer. Show that the first-order inclusion probabilities of LSS and CSS are the same when N/n is an integer.

4.8.5 The following data relate to the heights of 30 plants for an agricultural experiment.

Serial number of plant	Height in cm	Serial number of plant	Height in cm	Serial number of plant	Height in cm
1	28	11	53	21	33
2	99	12	95	22	12
3	96	13	34	23	56
4	42	14	40	24	56
5	58	15	42	25	29
6	26	16	17	26	25
7	11	17	16	27	25
8	82	18	10	28	88
9	68	19	35	29	17
10	73	20	86	30	29

(a) Write all possible LSSs of size 5 and compute the variance of the estimator of the population mean. Compare the performance of systematic sampling and SRSWOR sampling based on the same sample size 5.

(b) Select an LSS of size 8 by using a fractional interval and find an unbiased estimate of the population mean height of the 30 plants.

(c) Select a circular systematic sample of size 10 and estimate average height of the plants.

4.8.6 Consider the data given in Example 4.6.1. (a) Select three independent systematic samples of sizes 6, 7, and 8 using LSS using fractional intervals. Using the selected samples find an unbiased estimate of the population mean and its standard errors. (b) Select two independent samples of sizes 6 and 8 by CSS and compute a 90% confidence interval of the population mean height.

CHAPTER 5

Unequal Probability Sampling

5.1 INTRODUCTION

In an unequal (or varying) probability sampling (VPS) design, all possible samples need not have the same selection probabilities. In this design, samples are generally selected by using an auxiliary variable, x, which is also known as a measure of size variable or simply as a measure of size. The values of the auxiliary variable are assumed to be positive and known before the survey. It is found that if the auxiliary variable x is well related to the study variable y, then one can use auxiliary information in an appropriate manner for selection of a sample to get an efficient estimator for the population parameter of interest. For instance, to estimate the total number of inpatients treated in hospitals, one may use the number of beds in a hospital as measure of size variable. The hospitals that have a large number of beds are expected to have treated a higher number of inpatients than those with a fewer number of beds. Hence the hospitals having a large number of beds should be given a higher chance of selection in a sample than those with fewer beds to ensure an efficient estimator for the total number of inpatients. In estimating the total production of a crop, the farm size may be used as an auxiliary variable because the production of a crop in a farm is expected to be proportional to the size of the farm. In this chapter we consider various types of VPS schemes such as probability proportional to size with and without replacement sampling schemes, inclusion probability proportional to size sampling schemes, Lahiri−Midzuno−Sen (LMS), and Rao−Hartley−Cochran (RHC) sampling schemes. Methods of unbiased estimation of population characteristics, expressions of the variances of the proposed estimators, and unbiased estimators of the variances for various sampling schemes are also studied in detail.

We will denote x_i as the value of the auxiliary variable x for the ith unit, which is assumed to be known and strictly positive for every $i \in U$, and $X = \sum_{i=1}^{N} x_i$ be their total. The normed size measure for the ith unit is defined by $p_i = x_i/X(>0)$ and it is subject to $\sum_{i=1}^{N} p_i = 1$.

Survey Sampling Theory and Applications
ISBN 978-0-12-811848-1
http://dx.doi.org/10.1016/B978-0-12-811848-1.00005-4

5.2 PROBABILITY PROPORTIONAL TO SIZE WITH REPLACEMENT SAMPLING SCHEME

In the probability proportional to size with replacement (PPSWR) sampling scheme, units are selected independently in each draw so that the probability of selection of the ith unit in any draw is p_i. The probability of selection of an ordered sample $s_o = (i_1, \rightarrow, i_2, \rightarrow, \cdots, \rightarrow i_n) = (i_1, i_2, \cdots, i_n)$ where the unit i_r is selected at the rth draw, $r = 1, \ldots, n$ is $p(s_o) = p_{i_1} p_{i_2} \ldots p_{i_n}$. Here it is important to note that a unit may be selected more than once in a sample. We use cumulative total method and Lahiri's (1951) method for the selection of PPSWR samples. The methods are described as follows.

5.2.1 Cumulative Total Method

Here we first choose a constant k to make $kx_i = X_i$, an integer for each $i = 1, \ldots, N$. We associate random numbers $1(=T_0)$ to $X_1(=T_1)$ for the first unit. For the second unit, random numbers $X_1 + 1$ to $X_1 + X_2(=T_2)$ and in general for the ith unit, random numbers $T_{i-1} + 1$ to T_i are associated where $T_i = X_1 + \cdots + X_i$ for $i = 1, \ldots, N$. For the selection of a sample, we select a number R at random from 1 to $T_n = kX$. The random number R selects the jth unit if $T_{j-1} < R \leq T_j$. Clearly, the probability of selection of the jth unit is $(T_j - T_{j-1})/T_n = X_j/T_n = p_j$. The procedure is repeated n times to select a sample of size n units.

Example 5.2.1

Select a sample of four households from a list of nine by the PPSWR method, using household size as an auxiliary variable (Table 5.2.1).

Here we choose $k = 10$, so that $X_i = 10x_i$ becomes an integer for $i = 1, \ldots, 9$. The values of X_i's and T_i's are given in Table 5.2.2.

From the random number table we select three-digit random numbers between 001 and 285 as given in Table 5.2.3.

Table 5.2.1

Serial number household (i)	1	2	3	4	5	6	7	8	9
Household size (x_i)	2	3.5	2	4	2.5	5	2	5	2.5

Table 5.2.2

i	1	2	3	4	5	6	7	8	9
X_i	20	35	20	40	25	50	20	50	25
T_i	20	55	75	115	140	190	210	260	285

Table 5.2.3

Random number selected	465	238	198	431	215	023
Unit selected	-	8	7	-	8	2

"-" indicates that the units are not selected because corresponding chosen random number exceeds 285.

Hence the selected ordered sample by PPSWR method is $s_o = (8, 7, 8, 2)$ where the unit 8 has been selected twice.

5.2.2 Lahiri's Method

The computation of cumulative totals becomes tedious if the population size N is large. Lahiri (1951) has proposed an alternative method where one need not compute cumulative totals. The method is described as follows.

Select two random numbers, one from 1 to N say "j" and another say "R" from 1 to M where $M \geq \underset{1 \leq i \leq N}{Max} \{x_i\}$. The unit "$j$" is selected in the sample if $R \leq x_j$. If $R > x_j$, no unit is selected and the procedure is repeated till a unit is selected. For selection of a sample size n, the method is repeated n times.

Theorem 5.2.1

Probability of selection of the ith unit for Lahiri's method is p_i for $i = 1, \ldots, N$.

Proof

$$\begin{aligned}\text{Let } a_i &= \text{Probability that the first pair of random numbers selects unit } i \\ &= \text{Probability of selection of random number } i(1 \leq i \leq N) \\ &\quad \text{and a random number } R \text{ less than or equal to } x_i = x_i/(NM)\end{aligned}$$

and

$$\begin{aligned}b &= \text{Probability that the first pair of random numbers fails to select} \\ &\quad \text{any unit} \\ &= \sum_{i=1}^{N} \frac{1}{N} \frac{M - x_i}{M} = 1 - \frac{X}{NM}\end{aligned}$$

Hence the probability of selection of ith unit using this method

$$\begin{aligned}&= \text{Prob (First pair of random numbers selects } i\text{th unit)} + \text{Prob} \\ &\quad \text{(First pair of random numbers fail to select any unit)} \\ &\quad \times \text{Prob (Second pair of random number selects } i\text{th unit)} + \cdots \\ &= a_i + ba_i + b^2 a_i + \cdots = a_i(1 - b)^{-1} = x_i/X = p_i\end{aligned}$$

Example 5.2.2

Select a sample of four households using Lahiri's method for the data given in Table 5.2.1. Here the population size is $N = 9$ and we may take $M = 6 \geq \underset{1 \leq i \leq 9}{Max}\{x_i\}$

Table 5.2.4

Random number between 1 and 9 (j)	4	5	3	2	5
Random number between 1 and 6 (R)	6	2	2	3	2
x_j	4	2.5	2	3.5	2.5
Unit selected	–	5	3	2	5

So, the selected ordered sample by Lahiri's method is $s_o = (5, 3, 2, 5)$.

5.2.3 Hansen–Hurwitz Estimator and its Variance

Let $y_{(r)}$ and $x_{(r)}$, respectively, be the value of the study (y) and auxiliary variable (x) of the unit that are selected at the rth draw, $r = 1,\ldots, n$ and $p_{(r)} = x_{(r)}/X$. Then $y_{(r)}/p_{(r)} = y_i/p_i$ if the rth draw produces the ith unit with probability p_i; $i = 1,\ldots, N$; $r = 1,\ldots, n$. The Hansen–Hurwitz (1943) estimator for the population total Y is defined as

$$\widehat{Y}_{hh} = \frac{1}{n}\sum_{r=1}^{n}\frac{y_{(r)}}{p_{(r)}} \tag{5.2.1}$$

Theorem 5.2.2

(i) $\widehat{Y}_{hh}$ is an unbiased estimator of the population total Y

(ii) Variance of $\widehat{Y}_{hh}$ is $V\left(\widehat{Y}_{hh}\right) = \dfrac{V_{pps}}{n}$

where

$$V_{pps} = \sum_{i=1}^{N} p_i\left(\frac{y_i}{p_i} - Y\right)^2 = \frac{1}{2}\sum_{i\neq}^{N}\sum_{j=1}^{N} p_i p_j\left(\frac{y_i}{p_i} - \frac{y_j}{p_j}\right)^2$$

(iii) An unbiased estimator of $V\left(\widehat{Y}_{hh}\right)$ is

$$\widehat{V}\left(\widehat{Y}_{hh}\right) = \frac{1}{n(n-1)}\sum_{r=1}^{n}\left(\frac{y_{(r)}}{p_{(r)}} - \widehat{Y}_{pps}\right)^2$$

Proof

The expectation and variance of $\dfrac{y_{(r)}}{p_{(r)}}$ are given by

$$E\left(\frac{y_{(r)}}{p_{(r)}}\right) = \sum_{i=1}^{N} \frac{y_i}{p_i} p_i = Y \tag{5.2.2}$$

and

$$\begin{aligned} V\left(\frac{y_{(r)}}{p_{(r)}}\right) &= E\left(\frac{y_{(r)}}{p_{(r)}} - Y\right)^2 \\ &= \sum_{i=1}^{N} p_i \left(\frac{y_i}{p_i} - Y\right)^2 \qquad (5.2.3) \\ &= V_{pps}. \end{aligned}$$

Furthermore,

$$\begin{aligned} \frac{1}{2}\sum_{i\neq}^{N}\sum_{j=1}^{N} p_i p_j \left(\frac{y_i}{p_i} - \frac{y_j}{p_j}\right)^2 &= \sum_{i\neq}^{N}\sum_{j=1}^{N} p_j \frac{y_i^2}{p_i} - \sum_{i\neq}^{N}\sum_{j=1}^{N} y_i y_j \\ &= \sum_{i=1}^{N} \frac{y_i^2}{p_i}(1 - p_i) - \left(Y^2 - \sum_{i=1}^{N} y_i^2\right) \qquad (5.2.4) \\ &= V_{pps}. \end{aligned}$$

Now using Eqs. (5.2.2–5.2.4), we get

(i) $E\left(\widehat{Y}_{hh}\right) = \dfrac{1}{n}\sum\limits_{r=1}^{n} E\left(\dfrac{y_{(r)}}{p_{(r)}}\right) = Y$

(ii) $$\begin{aligned} V\left(\widehat{Y}_{hh}\right) &= \frac{1}{n^2} V\left(\sum_{r=1}^{n} \frac{y_{(r)}}{p_{(r)}}\right) \\ &= \frac{1}{n^2}\left[\sum_{n=1}^{n} V\left(\frac{y_{(r)}}{p_{(r)}}\right) + \sum_{r\neq}^{n}\sum_{t=1}^{n} Cov\left(\frac{y_{(r)}}{p_{(r)}}, \frac{y_{(t)}}{p_{(t)}}\right)\right] \\ &= \frac{V_{pps}}{n} \end{aligned}$$

(because $V\left(\frac{y_{(r)}}{p_{(r)}}\right) = V_{pps}$ and $Cov\left(\frac{y_{(r)}}{p_{(r)}}, \frac{y_{(t)}}{p_{(t)}}\right) = 0$ for $r \neq t$ as the draws are independent)

and

$$\text{(iii)}\quad E\left[\widehat{V}\left(\widehat{Y}_{hh}\right)\right] = \frac{1}{n(n-1)} E\left[\sum_{r=1}^{n}\left(\frac{y_{(r)}}{p_{(r)}}\right)^2 - n\left(\widehat{Y}_{pps}\right)^2\right]$$

$$= \frac{1}{n(n-1)}\left[\sum_{r=1}^{n} E\left(\frac{y_{(r)}}{p_{(r)}}\right)^2 - nE\left(\widehat{Y}_{pps}\right)^2\right]$$

$$= \frac{1}{n(n-1)}\left[n\left(V_{pps} + Y^2\right) - n\left(V\left(\widehat{Y}_{pps}\right) + Y^2\right)\right]$$

$$= \frac{V_{pps}}{n}$$

$$= V\left(\widehat{Y}_{hh}\right)$$

Remark 5.2.1

The variance of $\widehat{Y}_{hh}$ is zero when y_i is exactly proportional to x_i for every $i \in U$. This condition of the exact proportionality may not be realized in practice, but situations of approximate proportionality are not rare, such as the yield of a crop is approximately proportional to an area under cultivation and the income of a person is approximately proportional to his/her tax return. Hence, the Hansen−Hurwitz (HH) estimator becomes efficient when the study variable y is approximately proportional to the auxiliary variable x.

Remark 5.2.2

The estimator $\widehat{Y}_{hh}$ can be written as

$$\widehat{Y}_{hh} = \sum_{i=1}^{N} n_i(s_o)\frac{y_i}{p_i} \tag{5.2.5}$$

where $n_i(s_o)$ denotes the number of times the ith unit occurs in the ordered sample s_o. Now noting $n_i(s_o)$ follows a multinomial distribution with $E[n_i(s_o)] = np_i$, $V[n_i(s_o)] = np_i(1-p_i)$ and $Cov[n_i(s_o), n_j(s_o)] = -np_ip_j$ for $i \neq j$, one can prove the Theorem 5.2.2 in an alternative way.

5.2.4 Rao-Blackwellization

The HH estimator $\widehat{Y}_{hh}$ defined earlier is an ordered estimator because the estimator depends on the multiplicity of the units in the sample s_o. Hence $\widehat{Y}_{hh}$ is inadmissible as explained in Section 2.7. Let s be the unordered sample obtained by taking distinct units and arranging their labels in ascending orders from the ordered sample s_o, then by applying the Rao–Blackwell theorem (Theorem 2.7.3), one can find an improved estimator of the population total Y as follows:

$$t^* = E\left(\widehat{Y}_{hh}|s\right) = E\left[\frac{1}{n}\sum_{r=1}^{n}\frac{y_{(r)}}{p_{(r)}}\bigg|s\right] \tag{5.2.6}$$

But the estimator t^* cannot be used in practice because the expression (5.2.6) does not provide any elegant form. Let us consider the following example.

Example 5.2.3

Let $s_{o1} = (i, i, j)$ be an ordered sample of size $n = 3$ selected by the PPSWR method with $i < j$. The HH estimator based on s_{o1} is given by $\widehat{Y}_{hh}(s_{o1}) = \frac{1}{3}\left(2\frac{y_i}{p_i} + \frac{y_j}{p_j}\right)$. The ordered sample s_{o1} yields the unordered sample $s = (i, j)$. The unordered sample s would have been realized from any of the following ordered samples $s_{o1} = (i, i, j)$, $s_{o2} = (i, j, i)$, $s_{o3} = (j, i, i)$, $s_{o4} = (i, j, j)$, $s_{o5} = (j, i, j)$, and $s_{o6} = (j, j, i)$. Now noting that $\widehat{Y}_{hh}(s_{ok}) = \frac{1}{3}\left(2\frac{y_i}{p_i} + \frac{y_j}{p_j}\right)$ for $k = 1, 2, 3$; $\widehat{Y}_{hh}(s_{ok}) = \frac{1}{3}\left(\frac{y_i}{p_i} + 2\frac{y_j}{p_j}\right)$ for $k = 4, 5, 6$; $p(s_{o1}) = p(s_{o2}) = p(s_{o3}) = p_i^2 p_j$; and $p(s_{o4}) = p(s_{o5}) = p(s_{o6}) = p_i p_j^2$, we get the following improved unordered estimator:

$$\begin{aligned} t^* &= E\left(\widehat{Y}_{hh}|s\right) \\ &= \sum_{k=1}^{6}\widehat{Y}_{hh}(s_{ok})p(s_{ok}) \Big/ \sum_{k=1}^{6} p(s_{ok}) \\ &= \frac{1}{3}\left(\frac{y_i}{p_i} + \frac{y_j}{p_j} + \frac{y_i + y_j}{p_i + p_j}\right) \end{aligned} \tag{5.2.7}$$

Example 5.2.4
The Horvitz–Thompson estimator

$$\widehat{Y}_{ht} = \sum_{i \in s} \frac{y_i}{\pi_i} \tag{5.2.8}$$

with $\pi_i = 1 - (1 - p_i)^n$ as the inclusion probability of the ith unit is an unordered estimator and unbiased for the population total Y. However, this estimator is quite different from Eq. (5.2.7).

Remark 5.2.3
It is very difficult to make a theoretical comparison among unordered estimators in general. However, if y_i is proportional to x_i, the unordered estimator (5.2.7) becomes constant and its variance becomes zero. The Horvitz–Thompson estimator does not possess such properties. Hence the unordered estimator (5.2.7) is expected to fare better than $\widehat{Y}_{ht}$ when y_i is approximately proportional to x_i. Furthermore, Chaudhuri and Arnab (1978) have shown that the estimator $\widehat{Y}_{ht}$ is not consistent for the population total Y because the variance of $\widehat{Y}_{ht}$ does not necessarily decrease with the increase of the sample size n.

Remark 5.2.4
Although estimators (5.2.7) and (5.2.8) based on unordered data are more efficient than $\widehat{Y}_{hh}$, the estimators are seldom used in practice because of their complexity. Conversely, the inadmissible HH estimator $\widehat{Y}_{hh}$ is used extensively in practice mainly because of its simple expression and it possesses an elegant expression of its variance and unbiased estimator of its variance.

5.3 PROBABILITY PROPORTIONAL TO SIZE WITHOUT REPLACEMENT SAMPLING SCHEME

In a probability proportional to size without replacement (PPSWOR) sampling scheme, on the first draw, a unit i_1 (say) is selected with probability $p_{i_1}(1)$, which is proportional to its measure of size x_{i_1}, i.e., $p_{i_1}(1) = x_{i_1}/X = p_{i_1}$. On the second draw, another unit $i_2(\neq i_1)$ is selected from the remaining $N - 1$ units, which were not selected in the first draw with probability $p_{i_2|i_1}(2)$, proportional to its measure of size x_{i_2}. Hence the probability of selection of the unit i_2 in the second draw given that the unit i_1 is selected in the first draw is $p_{i_2|i_1}(2) = x_{i_2}/(X - x_{i_1}) = p_{i_2}/(1 - p_{i_1})$. In general, at the $r(=1,\ldots, n)$th draw, the probability of selection of unit i_r given that the units $i_1,\ldots, i_{r-1}$ are selected in earlier $r-1$ draws is $p_{i_r|i_1\ldots i_{r-1}}(r) = p_{i_r}/(1 - p_{i_1} - \ldots - p_{i_{r-1}})$. So, for the PPSWOR sampling, all units are distinct and the probability of selecting an ordered sample $s_o = (i_1,\ldots, i_n)$ where the unit i_r is selected at the rth draw is

$$p(i_1, \ldots, i_n) = p_{i_1} \times \frac{p_{i_2}}{1 - p_{i_1}} \times \cdots \times \frac{p_{i_r}}{1 - p_{i_1} - \cdots - p_{i_{r-1}}} \times \cdots \times \frac{p_{i_n}}{1 - p_{i_1} - \cdots - p_{i_{n-1}}} \tag{5.3.1}$$

For all practical purposes, PPSWOR sample can be selected in the same manner as selection of a PPSWR sample by the cumulative method or Lahiri's (1951) method, keeping in mind that if a unit is selected once, it cannot be selected again. In other words, the PPSWR method is used until a set of n (desired sample size) distinct units are selected.

Example 5.3.1

Consider the data given in Example 5.2.1. Let us select a sample of size $n = 4$ units by the PPSWOR method using x_i's as the size measure for the ith unit. Here, we also have to start with Table 5.2.2, which was constructed in Example 5.2.1.

From the random number table we selected three-digit random numbers from 001 to 285 as given in Table 5.3.1:

Table 5.3.1

Random number selected	165	240	211	215	023	076
Unit selected	6	8	—	—	2	4

Here we note that we cannot select any unit corresponding to the random number 211 and 215 because the unit 8 has already been selected in the second draw. So the selected PPSWOR sample of size four is $s_o = (6, 8, 2, 4)$.

5.3.1 Raj's Estimator and its Variance

Let us now consider the following ordered estimators for the total Y based on the ordered sample $s_o = (i_1,\ldots, i_n)$.

$$t(1) = \frac{y_{i_1}}{p_{i_1}},\ t(2) = y_{i_1} + \frac{y_{i_2}}{p_{i_2}}(1 - p_{i_1})$$

and in general

$$t(r) = y_{i_1} + \cdots + y_{i_{r-1}} + \frac{y_{i_r}}{p_{i_r}}(1 - p_{i_1} - \cdots - p_{i_{r-1}}) \text{ for } r = 2, \ldots, n \quad (5.3.2)$$

The properties of ordered estimators are stated in the following theorem.

Theorem 5.3.1

(i) $E[t(r)] = Y$

(ii) $Var[t(r)] \leq Var[t(r-1)]$ for $r = 2, \ldots, n$

(iii) $Cov[t(r), t(k)] = 0$ for $r \neq k$

Proof

Let $E_{i_1,\ldots,i_k}$ and $V_{i_1,\ldots,i_k}$ denote unconditional expectation and variance for the selection of units $i_1, \ldots, i_k$. The conditional expectation and variance given that the units $i_1, \ldots, i_k$ are selected in earlier draws are denoted by $E(\bullet|i_1, \ldots, i_k)$ and $V(\bullet|i_1, \ldots, i_k)$, respectively.

$$\text{(i) } E[t(r)] = E_{i_1,\ldots,i_{r-1}}[E(t(r)|i_1, \ldots, i_{r-1})] \quad (5.3.3)$$

Now

$$\begin{aligned} E(t(r)|i_1, \ldots, i_{r-1}) &= y_{i_1} + \cdots + y_{i_{r-1}} + E\left(\frac{y_{i_r}}{p_{i_r}}(1 - p_{i_1} - \cdots - p_{i_{r-1}})|i_1, \ldots, i_{r-1}\right) \\ &= y_{i_1} + \cdots + y_{i_{r-1}} + \sum_{i_r \in U/i_1,\ldots,i_{r-1}} \frac{y_{i_r}}{p_{i_r|i_1,\ldots,i_{r-1}(r)}} \cdot p_{i_r|i_1,\ldots,i_{r-1}} \\ &= Y \end{aligned} \quad (5.3.4)$$

where $U/i_1, \ldots, i_{r-1}$ = the set of $N - (r-1)$ units that were not selected in the first $r - 1$ draws and $p_{i_r|i_1,\ldots,i_{r-1}} = p_{i_r}/(1 - p_{i_1} - \ldots - p_{i_{r-1}})$.

Now substituting Eq. (5.3.4) in Eq. (5.3.3), we find $E[t(r)] = Y$.

$$\begin{aligned} \text{(ii) } Var[t(1)] &= Var\left(\frac{y_{i_1}}{p_{i_1}}\right) \\ &= \frac{1}{2}\sum_{i\neq}^{N}\sum_{j=1}^{N} p_i p_j \left(\frac{y_i}{p_i} - \frac{y_j}{p_j}\right)^2 \\ &= V_{pps} \text{ (Using Eq.5.2.4)} \end{aligned}$$

$$Var[t(r)] = E_{i_1,\ldots,i_{r-1}}[V(t(r)|i_1, \ldots, i_{r-1})] + V_{i_1,\ldots,i_{r-1}}[E(t(r)|i_1, \ldots, i_{r-1})] \tag{5.3.5}$$

Now using Eq. (5.3.4), we get

$$V_{i_1,\ldots,i_{r-1}}[E\{t(r)|i_1, \ldots, i_{r-1}\}] = V_{i_1,\ldots,i_{r-1}}(Y) = 0 \tag{5.3.6}$$

and

$$\begin{aligned} V\{t(r)|i_1, \ldots, i_{r-1}\} &= V\left(\frac{y_{i_r}}{p_{i_r}(r)}|i_1, \ldots, i_{r-1}\right) \\ &= \frac{1}{2}\sum_{i \neq}\sum_{j \in U/i_1,\ldots,i_{r-1}} p_{i|i_1,\ldots,i_{r-1}}p_{j|i_1,\ldots,i_{r-1}}\left(\frac{y_i}{p_{i|i_1,\ldots,i_{r-1}}} - \frac{y_j}{p_{j|i_1,\ldots,i_{r-1}}}\right)^2 \\ &= \frac{1}{2}\sum_{i \neq}\sum_{j \in U/i_1,\ldots,i_{r-1}} p_i p_j\left(\frac{y_i}{p_i} - \frac{y_j}{p_j}\right)^2 \end{aligned} \tag{5.3.7}$$

Substituting Eqs. (5.3.6) and (5.3.7) in Eq. (5.3.5), we get

$$\begin{aligned} Var[t(r)] &= E_{i_1\cdots i_{r-1}}\left[\frac{1}{2}\sum_{i \neq}\sum_{j \in U/i_1,\ldots,i_{r-1}} p_i p_j\left(\frac{y_i}{p_i} - \frac{y_j}{p_j}\right)^2\right] \\ &\leq E_{i_1\cdots i_{r-1}}\left[\frac{1}{2}\sum_{i \neq}\sum_{j \in U/i_1,\ldots,i_{r-2}} p_i p_j\left(\frac{y_i}{p_i} - \frac{y_j}{p_j}\right)^2\right] \\ &= E_{i_1\cdots i_{r-2}}\left[\frac{1}{2}\sum_{i \neq}\sum_{j \in U/i_1,\ldots,i_{r-2}} p_i p_j\left(\frac{y_i}{p_i} - \frac{y_j}{p_j}\right)^2\right] \\ &= Var[t(r-1)] \end{aligned}$$

(iii)

$$\begin{aligned} Cov[t(r), t(k)] &= E[t(r)t(k)] - [E\{t(r)\}][E\{t(k)\}] \\ &= E[t(r)t(k)] - Y^2 \end{aligned} \tag{5.3.8}$$

$$\begin{aligned} \text{Let } r < k, \text{then } E[t(r)t(k)] &= E[t(r)E\{t(k)|i_1, \ldots, i_r\}] \\ &= Y\,E[t(r)] \text{ (Using Eq.5.3.4)} \\ &= Y^2 \end{aligned} \tag{5.3.9}$$

Inserting Eq. (5.3.9) in Eq. (5.3.8), we get $Cov[t(r), t(k)] = 0$.

Raj's (1956) estimator for PPSWOR sampling is defined as

$$\widehat{Y}_{RA} = \frac{1}{n}\sum_{i=1}^{n} t(r) \tag{5.3.10}$$

The properties of Raj's estimator are given in the following theorem.

Theorem 5.3.2

(i) $\widehat{Y}_{RA}$ is unbiased for the total Y

(ii) Variance of $\widehat{Y}_{RA}$ is

$$V\left(\widehat{Y}_{RA}\right) \leq \frac{V_{pps}}{n} = V\left(\widehat{Y}_{hh}\right)$$

(iii) An unbiased estimator of $V\left(\widehat{Y}_{RA}\right)$ is

$$\widehat{V}\left(\widehat{Y}_{RA}\right) = \frac{1}{n(n-1)}\sum_{r=1}^{n}\left(t(r) - \widehat{Y}_{RA}\right)^2$$

Proof

(i) $$E\left(\widehat{Y}_{RA}\right) = \frac{1}{n}\sum_{r=1}^{n} E[t(r)] = Y$$

(ii) $$V\left(\widehat{Y}_{RA}\right) = \frac{1}{n^2}\sum_{r=1}^{n} V[t(r)] \leq \frac{V_{pps}}{n} = V\left(\widehat{Y}_{hh}\right)$$

(using the Theorem 5.3.1)

(iii) $$E\left[\widehat{V}\left(\widehat{Y}_{RA}\right)\right] = \frac{1}{n(n-1)}\left[\sum_{r=1}^{n} E\{t(r)\}^2 - nE\left(\widehat{Y}_{RA}\right)^2\right]$$

$$= \frac{1}{n(n-1)}\left[\sum_{r=1}^{n}\left(V(t(r)) + Y^2\right) - n\left\{V\left(\widehat{Y}_{RA}\right) + Y^2\right\}\right]$$

$$= V\left(\widehat{Y}_{RA}\right)$$

Remark 5.3.1

Theorem 5.3.2 indicates that for a given sample size, n, the Raj's estimator based on PPSWOR sampling is more efficient than the HH estimator based on PPSWR sampling of the same sample size.

Corollary 5.3.1

Let an ordered sample $s_o = (i, j)$ of size $n = 2$ be selected by the PPSWOR method, then

(i) $\widehat{Y}_{RA}(s_o) = \dfrac{1}{2}\left[\dfrac{y_i}{p_i}(1+p_i) + \dfrac{y_j}{p_j}(1-p_i)\right]$

$$\text{(ii) } V\left(\widehat{Y}_{RA}(s_o)\right) = \frac{1}{4}\sum_{i\neq}\sum_{j\in U}\left[\frac{y_i}{p_i}(1+p_i) + \frac{y_j}{p_j}(1-p_i)\right]^2 \frac{p_i p_j}{1-p_i} - Y^2$$

$$= \frac{1}{2}\left(1 - \frac{1}{2}\sum_{i\in U} p_i^2\right)\left\{\sum_{i\in U} p_i\left(\frac{y_i}{p_i} - Y\right)^2\right\}$$

$$- \frac{1}{4}\sum_{i\in U} p_i^2\left(\frac{y_i}{p_i} - Y\right)^2$$

$$\text{(iii) } \widehat{V}\left(\widehat{Y}_{RA}(s_o)\right) = \frac{(1-p_i)^2}{4}\left(\frac{y_i}{p_i} - \frac{y_j}{p_j}\right)^2$$

5.3.2 Rao-Blackwellization

It should be noted that Raj's (1956) estimator $\widehat{Y}_{RA}$ is an ordered estimator because it depends on the order of selection of units in the ordered sample. Raj's estimator based on the ordered sample $s_o = (i, j)$ is $\widehat{Y}_{RA}(s_o) = \dfrac{1}{2}\left[\dfrac{y_i}{p_i}(1+p_i) + \dfrac{y_j}{p_j}(1-p_i)\right]$, which is quite different from Raj's estimator $\widehat{Y}_{RA}\left(s_o^*\right) = \dfrac{1}{2}\left[\dfrac{y_j}{p_j}(1+p_j) + \dfrac{y_i}{p_i}(1-p_j)\right]$, which is based on the ordered sample $s_o^* = (j, i)$. Murthy (1957) improved Raj's estimator by using Rao-Blackwellization as follows.

5.3.2.1 Murthy's Estimator

Murthy (1957) symmetrized Raj's estimator by taking the weighted average of Raj's estimators using weights proportional to the probability of selection of the ordered samples. So, Murthy's estimator derived from the ordered samples $s_o = (i, j)$ or $s_o^* = (j, i)$ is given by

$$\widehat{Y}_M = \frac{\widehat{Y}_{RA}(s_o)p(s_o) + \widehat{Y}_{RA}\left(s_o^*\right)p\left(s_o^*\right)}{p(s_o) + p\left(s_o^*\right)}$$

$$= \frac{\dfrac{y_i}{p_i}(1-p_j) + \dfrac{y_j}{p_j}(1-p_i)}{(2-p_i-p_j)} \tag{5.3.11}$$

Let s be the unordered sample obtained from the ordered sample s_o and $\overset{*}{s}_o$, then we can write Eq. (5.3.11) as follows:

$$
\begin{aligned}
\widehat{Y}_M &= \frac{\widehat{Y}_{RA}(s_o)p(s_o) + \widehat{Y}_{RA}(\overset{*}{s}_o)p(\overset{*}{s}_o)}{p(s)} \\
&= \frac{1}{2}\frac{1}{p(s)}\left[\left\{\frac{y_i}{p_i}(1+p_i) + \frac{y_j}{p_j}(1-p_i)\right\}\frac{p_ip_j}{1-p_i}\right. \\
&= \frac{1}{p(s)}\left(y_i\frac{p_j}{1-p_i} + y_j\frac{p_i}{1-p_j}\right) \\
&= \frac{1}{p(s)}\left(y_ip(s|i) + y_jp(s|i)\right)
\end{aligned}
\tag{5.3.12}
$$

where $p(s|k)$ denotes the conditional probability of obtaining the unordered sample s given that the unit $k(=i, j)$ is selected at the first draw.

Let $s = (j_1,\ldots, j_n)$ be an unordered sample of size n with labels $j_1 <,\ldots, < j_n$ obtained by the PPSWOR method of sampling, then Murthy's (1957) estimator can be written as

$$
\widehat{Y}_M = \frac{1}{p(s)}\sum_{i\in s} y_ip(s|i) \tag{5.3.13}
$$

where $p(s|i)$ = conditional probability of obtaining the unordered sample s when the ith unit is selected at the first draw.

The conditional probabilities satisfy the following relationships:

(i) $\sum_{i\in s} p_ip(s|i) = \sum_i I_{si}p_ip(s|i) = p(s)$,

(ii) $\sum_{s\supset i} p(s|i) = \sum_s I_{si}p(s|i) = 1 \;\text{ for }\; i = 1, \ldots, N$

and

$$
\text{(iii)} \sum_{s\supset i,j} p(s|i,j) = \sum_s I_{si}I_{sj}p(s|i,j) = 1 \tag{5.3.14}
$$

where $I_{si} = 1$ if $i \in s$, $I_{si} = 0$ if $i \notin s$ and $p(s|i, j)$ = conditional probability of s given that the units i and j have been selected in the first two draws.

Theorem 5.3.3

(i) $\widehat{Y}_M = \frac{1}{p(s)}\sum_{i\in s} y_i p(s|i)$ is unbiased of Y

(ii) Variance of $\widehat{Y}_M$ is

$$\begin{aligned} V_M &= \sum_{i=1}^{N}\beta_{ii}y_i^2 - \sum_{i\neq}^{N}\sum_{j=1}^{N}\beta_{ij}y_iy_j \\ &= \frac{1}{2}\sum_{i\neq}^{N}\sum_{j=1}^{N}\beta_{ij}p_ip_j\left(\frac{y_i}{p_i}-\frac{y_j}{p_j}\right)^2 \end{aligned} \tag{5.3.15}$$

(iii) An unbiased estimator of V_M is

$$\widehat{V}_M = \frac{1}{2\{p(s)\}^2}\sum_{i\neq}\sum_{j\in s}\{p(s|i,j)p(s) - p(s|i)p(s|j)\}p_ip_j\left(\frac{y_i}{p_i}-\frac{y_j}{p_j}\right)^2$$

where $\beta_{ii} = \sum_{s\supset i}\frac{\{p(s|i)\}^2}{p(s)} - 1$ and $\beta_{ij} = 1 - \sum_{s\supset i,j}\frac{p(s|i)p(s|j)}{p(s)}$.

Proof

(i)
$$\begin{aligned} E(\widehat{Y}_M) &= \sum_s\left\{\frac{1}{p(s)}\sum_{i\in s}y_ip(s|i)\right\}p(s) \\ &= \sum_{i=1}^{N}y_i\sum_{s\supset i}p(s|i) \\ &= Y \text{ (using (ii) of Eq. 5.3.14)} \end{aligned}$$

(ii)
$$\begin{aligned} V_M = V(\widehat{Y}_M) &= E\left[\frac{1}{(p(s))^2}\left(\sum_{i\in s}y_i^2(p(s|i))^2 + \sum_{i}\sum_{\neq j\in s}y_iy_jp(s|i)p(s|j)\right)\right] - Y^2 \\ &= \sum_{i=1}^{N}y_i^2\left\{\sum_{s\supset i}\frac{(p(s|i))^2}{p(s)} - 1\right\} + \sum_{i\neq}^{N}\sum_{j=1}^{N}y_iy_j\left\{\sum_{s\supset i,j}\frac{p(s|i)p(s|j)}{p(s)} - 1\right\} \\ &= \sum_{i=1}^{N}\beta_{ii}y_i^2 - \sum_{i\neq}^{N}\sum_{j=1}^{N}\beta_{ij}y_iy_j \\ &= V_M \end{aligned}$$

Now

$$\frac{1}{2}\sum_{i\neq}^{N}\sum_{j=1}^{N}\beta_{ij}p_ip_j\left(\frac{y_i}{p_i}-\frac{y_j}{p_j}\right)^2=\sum_{i=1}^{N}\frac{y_i^2}{p_i}\sum_{j(\neq i)=1}^{N}p_j\beta_{ij}-\sum_{i\neq}^{N}\sum_{j=1}^{N}\beta_{ij}\,y_iy_j \tag{5.3.16}$$

and

$$\sum_{j(\neq i)=1}^{N}p_j\beta_{ij}=1-p_i-\sum_{j(\neq i)=1}^{N}p_j\sum_{s}I_{si}I_{sj}p(s|i)p(s|j)/p(s)$$

$$=1-p_i-\sum_{s}I_{si}p(s|i)\sum_{j(\neq i)=1}^{N}I_{sj}p_jp(s|j)/p(s)$$

$$=1-p_i-\sum_{s}I_{si}p(s|i)\{p(s)-I_{si}p_ip(s|i)\}/p(s) \tag{5.3.17}$$

[noting (i) of Eq. 5.3.14]

$$=p_i\left[\sum_{s\supset i}\frac{(p(s|i))^2}{p(s)}-1\right] \quad \text{[using (ii) of Eq. 5.3.14]}$$

$$=p_i\beta_{ii}$$

From Eqs. (5.3.16) and (5.3.17) we get

$$V_M=\frac{1}{2}\sum_{i\neq}^{N}\sum_{j=1}^{N}\beta_{ij}p_ip_j\left(\frac{y_i}{p_i}-\frac{y_j}{p_j}\right)^2$$

(iii) $$E\left(\widehat{V}_M\right)=\frac{1}{2}\sum_{s}\frac{1}{p(s)}\sum_{i\neq}^{N}\sum_{j=1}^{N}I_{si}I_{sj}\{p(s|i,j)p(s)-p(s|i)p(s|j)\}p_ip_j\left(\frac{y_i}{p_i}-\frac{y_j}{p_j}\right)^2$$

$$=\frac{1}{2}\sum_{i\neq}^{N}\sum_{j=1}^{N}p_ip_j\left(\frac{y_i}{p_i}-\frac{y_j}{p_j}\right)^2\sum_{s}\frac{1}{p(s)}I_{si}I_{sj}\{p(s|i,j)p(s)-p(s|i)p(s|j)\}$$

Now noting $\sum_{s}\frac{1}{p(s)}I_{si}I_{sj}\{p(s|i,j)p(s)-p(s|i)p(s|j)\}=\beta_{ij}$ [from (iii) of Eq. 5.3.14], we find

$$E\left(\widehat{V}_M\right)=V_M$$

Remark 5.3.2

Pathak and Shukla (1966) proved that the variance estimator $\widehat{V}_M$ is nonnegative because $p(s|i,j)p(s) - p(s|i)p(s|j) \geq 0$ for $\forall i \neq j$.

Corollary 5.3.2

The expression of the Murthy's estimator $\widehat{Y}_M$, variance of $\widehat{Y}_M$, and an unbiased estimator of the variance of $\widehat{Y}_M$ based on an unordered sample $s = (i, j)$ with $i < j$ are, respectively, given as follows:

(i) $$\widehat{Y}_M = \frac{\frac{y_i}{p_i}(1 - p_j) + \frac{y_j}{p_j}(1 - p_i)}{2 - p_i - p_j}$$

(ii) $$V_M = \frac{1}{2} \sum_{i \neq} \sum_{j \in U} \frac{p_i p_j (1 - p_i - p_j)}{2 - p_i - p_j} \left(\frac{y_i}{p_i} - \frac{y_j}{p_j} \right)^2$$

(iii) $$\widehat{V}_M = \frac{(1 - p_i)(1 - p_j)(1 - p_i - p_j)}{(2 - p_i - p_j)^2} \left(\frac{y_i}{p_i} - \frac{y_j}{p_j} \right)^2$$

Example 5.3.2

The number of publications and length of service of 10 academics in a University are given in Table 5.3.2.

Select a sample of size 3 academics by the PPSWOR method, using length of service as a measure of size variable. Estimate the average number of publications per academic by Raj's and Murthy's estimators. Give unbiased estimates of the variances of each of the estimators.

Table 5.3.2

Academics	1	2	3	4	5	6	7	8	9	10
Years of service (x)	10	5	15	10	25	30	10	10	20	15
No. of publications (y)	20	15	20	10	30	25	10	10	40	20

Here we make Table 5.3.3 for selection of sample.

From a random number table we select a five-digit random number 91050. This random number selects unit 10. The next five-digit random number 17898 selects the unit 3. The next five-digit number 19070 selects the unit 3 again. This selection is discarded because unit 3 has already been selected. The next five-digit number 99225 is also discarded because this number selects the unit 10, which has already been selected. The next five-digit number 32589 selects unit 5. Hence, the selected ordered sample of size 3 by PPSWOR method is $s_o = (10, 3, 5)$. Raj's estimator for total

Table 5.3.3

Unit (i)	x_i	y_i	$p_i = x_i/X$	$T_i =$ (cumulative p_i) $\times$ 100,000
1	10	20	0.06667	06,667
2	5	15	0.03333	10,000
3	15	20	0.10000	20,000
4	10	10	0.06667	26,667
5	25	30	0.16667	43,334
6	30	25	0.20000	63,334
7	10	10	0.06667	70,001
8	10	10	0.06667	76,668
9	20	40	0.13333	90,001
10	15	20	0.10000	100,000
Total	150 = X	—	1	

Y based on $s_o = (10,\ 3,\ 5)$ is $\widehat{Y}_{RA} = \bar{t} = \frac{1}{3}(t_1 + t_2 + t_3) = 194.66$ where $t_1 = \dfrac{y_{10}}{p_{10}} = 20/0.1 = 200$; $t_2 = \dfrac{y_3}{p_3}(1 - p_{10}) + y_{10} = (20/0.1) \times 0.9 + 20 = 200$; and $t_3 = \dfrac{y_5}{p_5} \times (1 - p_{10} - p_3) + y_{10} + y_3 = (30/0.16667) \times (1 - 0.1 - 0.1) + 20 + 20 = 183.99$ Raj's estimator for the population mean $\overline{Y}$ is $\widehat{\overline{Y}}_{RA} = \widehat{Y}_{RA}/10 = 19.466$. An unbiased estimator for variance of $\widehat{\overline{Y}}_{RA}$ is $\widehat{V}\left(\widehat{\overline{Y}}_{RA}\right) = \dfrac{1}{10^2}\dfrac{1}{3 \times 2}\sum_i (t_i - \bar{t})^2 = 0.284.$

Arranging the labels in ascending order of the ordered sample $s_o = (10,\ 3,\ 5)$, we get the unordered sample as $s = (3,\ 5,\ 10)$. This unordered sample would have been realized from any of the following ordered samples $s_{o1} = (3,\ 5,\ 10)$, $s_{o2} = (3,\ 10,\ 5)$, $s_{o3} = (5,\ 3,\ 10)$, $s_{o4} = (5,\ 10,\ 3)$, $s_{o5} = (10,\ 3,\ 5)$, and $s_{o6} = (10,\ 5,\ 3)$. The selection probabilities and Raj's estimator $\widehat{\overline{Y}}_{RA}$ based on the ordered samples are given in Table 5.3.4.

The conditional probabilities $p(s|i)$ are computed using the following formula:

$$p(s|3) = \frac{p_5}{1 - p_3}\frac{p_{10}}{1 - p_3 - p_5} + \frac{p_{10}}{1 - p_3}\frac{p_5}{1 - p_3 - p_{10}};$$

$$p(s|5) = \frac{p_3}{1 - p_5}\frac{p_{10}}{1 - p_5 - p_3} + \frac{p_{10}}{1 - p_5}\frac{p_3}{1 - p_5 - p_{10}} \quad \text{and}$$

$$p(s|10) = \frac{p_3}{1 - p_{10}}\frac{p_5}{1 - p_{10} - p_3} + \frac{p_5}{1 - p_{10}}\frac{p_3}{1 - p_{10} - p_5}$$

Table 5.3.4

Ordered sample (s_o)	$p(s_o)$	$\widehat{\overline{Y}}_{RA}(s_o)$	$p(s_o) \times \widehat{\overline{Y}}_{RA}(s_o)$
3, 5, 10	0.002525	19.28876	0.048710
3, 10, 5	0.002315	19.46657	0.045062
5, 3, 10	0.002727	19.11095	0.052122
5, 10, 3	0.002727	19.11095	0.052122
10, 3, 5	0.002315	19.46657	0.045062
10, 5, 3	0.002525	19.28876	0.048710
Total	$p(s) = 0.015135$		0.291789

Conditional probabilities and the corresponding y_i values are given in Table 5.3.5:

Table 5.3.5

i	$p(s\|i)$	y_i	$y_i p(s\|i)$
3	0.048402	20	0.968035
5	0.032728	30	0.981827
10	0.048402	20	0.968035
Total	—	—	2.917897

Murthy's estimator for the population mean $\overline{Y}$ is

$$\widehat{\overline{Y}}_M = \frac{1}{Np(s)} \sum_{i \in s} y_i p(s|i) = 2.917897/(10 \times 0.015135) = 19.278.$$

(Here we can check that $\sum p(s_o) \times \widehat{\overline{Y}}_{RA}(s_o)/p(s) = 19.278.$)

An unbiased estimate of $V\left(\widehat{\overline{Y}}_M\right)$ is given by

$$\widehat{V}\left(\widehat{\overline{Y}}_M\right) = \frac{1}{100} \frac{1}{2\{p(s)\}^2} \sum_{i \neq} \sum_{j \in s} \{p(s|i,j)p(s) - p(s|i)P(s|j)\} p_i p_j \left(\frac{y_i}{p_i} - \frac{y_j}{p_j}\right)^2$$

$$= \frac{1}{100} \frac{1}{\{p(s)\}^2} \left[\{p(s|3,5)p(s) - p(s|3)P(s|5)\} p_3 p_5 \left(\frac{y_3}{p_3} - \frac{y_5}{p_5}\right)^2 \right.$$

$$+ \{p(s|3,10)p(s) - p(s|3)P(s|10)\} p_3 p_{10} \left(\frac{y_3}{p_3} - \frac{y_{10}}{p_{10}}\right)^2$$

$$\left. + \{p(s|5,10)p(s) - p(s|5)P(s|10)\} p_3 p_5 \left(\frac{y_5}{p_5} - \frac{y_{10}}{p_{10}}\right)^2 \right]$$

Now noting $p(s|3,\ 5) = p(s|5,\ 3) = p_{10}/(1 - p_3 - p_5) = 0.136364$; $p(s|3,\ 10) = p(s|10,\ 3) = p_5/(1 - p_3 - p_{10}) = 0.208338$; and $p(s|5,\ 10) = p(s|10,\ 5) = p_3/(1 - p_5 - p_{10}) = 0.136364$, we prepare Table 5.3.6.

Table 5.3.6

i, j	$p(s\|i, j)$	$\delta_{ij} = p(s\|i, j)p(s) - p(s\|i)p(s\|j)$	$\left(\frac{y_i}{p_i} - \frac{y_j}{p_j}\right)^2$	$\delta_{ij}p_ip_j\left(\frac{y_i}{p_i} - \frac{y_j}{p_j}\right)^2$
3, 5	0.136364	0.00048	400.14401	0.003199698
3, 10	0.208338	0.00081	0	0
5, 10	0.136364	0.00048	400.14401	0.003199698
Total				0.006399306

$$\widehat{V}\left(\widehat{Y}_M\right) = 0.006399396/\{10 \times (0.015135)\}^2 = 0.2793.$$

5.4 INCLUSION PROBABILITY PROPORTIONAL TO MEASURE OF SIZE SAMPLING SCHEME

The Horvitz–Thompson estimator $\widehat{Y}_{ht} = \sum_{i \in s} \frac{y_i}{\pi_i}$, based on a fixed sample size design, becomes constant if y_i's are proportional to the inclusion probabilities π_i's, and in this case the variance of $\widehat{Y}_{ht}$ becomes zero. Because the values of y_i's are unknown before the survey, one cannot construct a sampling design whose inclusion probabilities are proportional to the y_i values. In this situation, if an auxiliary variable x is available whose values are known, positive, and approximately proportional to the study variable y, then the variance of $\widehat{Y}_{ht}$ is expected to be small for a sampling design whose inclusion probabilities are proportional to the measure of size x variable, i.e., $\pi_i = nx_i/X = np_i\,(p_i = x_i/X)$ where n is the sample size. A sampling design is said to be inclusion probability proportional to measure of size (IPPS or πps) sampling scheme if $\pi_i = np_i$ for every $i \in U$. For an IPPS sampling design we must have

(i) $\pi_i = np_i \leq 1$, i.e., $p_i \leq 1/n$

Apart from the condition (i), it is desirable to stipulate the following conditions:

(ii) $\pi_{ij} > 0$ for $i \neq j \in U$

and

(iii) $\pi_i\pi_j - \pi_{ij} \geq 0$ for $i \neq j \in U$

The condition (ii) is required for unbiased estimation of variance while the condition (iii) is a requirement for obtaining nonnegative Yates-Grundy's (1953) variance estimator. To control the variance of the Horvitz–Thompson estimator, Hanurav (1966) set an additional restriction

(iv) $\pi_i\pi_j/\pi_{ij}$ should not be close to zero.

The IPPS sampling designs have been constructed by various authors viz. Narain (1951), Brewer (1963a,b), Fellegi (1963), Rao (1965), Durbin (1967), Hanurav (1966), and Sampford (1967), among others. A detailed review is given by Brewer and Hanif (1983). In this section, we will describe a few of them in detail. In general, IPPS sampling schemes for $n > 2$ are very complex. The expressions for π_{ij}'s are also not simple and hence the expressions of variances are very complex in general.

5.4.1 Inclusion Probability Proportional to Measure of Size Sampling With $n = 2$

5.4.1.1 Brewer's Sampling Scheme

In Brewer's (1963a,b) method, at the first draw, the ith unit is selected with probability

$$p_i(1) = \frac{2p_i(1-p_i)}{A(1-2p_i)}$$

where

$$\begin{aligned} A &= \sum_{i \in U} \frac{2p_i(1-p_i)}{1-2p_i} \\ &= \sum_{i \in U} \frac{p_i(1+1-2p_i)}{1-2p_i} \\ &= 1 + \sum_{i \in U} \frac{p_i}{1-2p_i} \end{aligned} \quad (5.4.1)$$

The conditional probability of selecting the ith unit in the second draw when the jth unit is selected in the first draw is

$$p_{i|j}(2) = \begin{cases} p_i/(1-p_j) & \text{for} \quad i \neq j \\ 0 & \text{for} \quad i = j \end{cases}$$

So, for this method, the inclusion probability of the ith unit is

$$\begin{aligned} \pi_i &= p_i(1) + \sum_{j \neq i} p_j(1) p_{i|j}(2) \\ &= \frac{2p_i}{A}\left(\frac{1-p_i}{1-2p_i} + \sum_{j \neq i} \frac{p_j}{1-2p_j}\right) \\ &= 2p_i. \end{aligned}$$

Inclusion probability for the ith and jth unit ($i \neq j$) is

$$\begin{aligned} \pi_{ij} &= p_i(1) p_{j|i}(2) + p_j(1) p_{i|j}(2) \\ &= \frac{2p_i p_j}{A}\left(\frac{1}{1-2p_i} + \frac{1}{1-2p_j}\right) \end{aligned} \quad (5.4.2)$$

and the difference

$$\begin{aligned}\pi_i\pi_j - \pi_{ij} &= \frac{2p_ip_j}{A}\left\{2A - \left(\frac{1}{1-2p_i} + \frac{1}{1-2p_j}\right)\right\}\\ &= \frac{4p_ip_j}{A}\sum_{k\neq(i,j)}\frac{p_k}{1-2p_k}\\ &> 0\end{aligned}$$

5.4.1.2 Durbin's Sampling Scheme

In Durbin's (1967) sampling scheme, the probability of selection of the ith unit at the first draw is $p_i(1) = p_i$, $i \in U$ and the conditional probability of selection of the ith unit in the second draw given that the jth unit selected at the first draw is

$$p_{i|j}(2) = \begin{cases} p_i\left(\dfrac{1}{1-2p_i} + \dfrac{1}{1-2p_j}\right)\Big/A & \text{for} \quad i\neq j\\ 0 & \text{for} \quad i = j\end{cases}$$

where A is given in Eq. (5.4.1) and clearly $\sum_{i\in U} p_{i|j}(2) = 1$.

The inclusion probability for the ith unit is

$$\begin{aligned}\pi_i &= p_i(1) + \sum_{j\neq i} p_j(1)p_{i|j}(2)\\ &= p_i + \sum_{j\neq i} p_jp_i\left(\frac{1}{1-2p_i} + \frac{1}{1-2p_j}\right)\Big/A\\ &= 2p_i\end{aligned}$$

Inclusion probability for the ith and jth unit ($i \neq j$) is

$$\begin{aligned}\pi_{ij} &= p_i(1)p_{j|i}(2) + p_j(1)p_{i|j}(2)\\ &= 2p_ip_j\left(\frac{1}{1-2p_i} + \frac{1}{1-2p_j}\right)\Big/A \end{aligned} \tag{5.4.3}$$

Remark 5.4.1

Brewer (1963a,b) and Durbin's (1967) schemes are identical in the sense of having the same expressions of the second order inclusion probabilities. In the same sense, Rao's (1965) sampling scheme is also similar to them.

Furthermore, all these three sampling schemes satisfy $\pi_{ij} > 0$ and $\pi_i\pi_j - \pi_{ij} \geq 0$ for $i \neq j \in U$.

5.4.1.3 Hanurav's Sampling Scheme

Hanurav's (1967) IPPS sampling scheme is defined as follows. Select two units with replacement with probability $p_i(1) = p_i$ attached to the ith unit. If the two units are distinct, accept them as a sample. Otherwise cancel this selection and select two fresh units with replacement with probability $p_i(2) = p_i^2 \Big/ \left(\sum_1^N p_k^2\right) = \{p_i(1)\}^2 \Big/ \sum_1^N \{p_k(1)\}^2$ for the ith unit. If the two units are different, select them as a sample, otherwise select two fresh units with replacement with probability $p_i(3) = p_i^4 \Big/ \left(\sum_1^N p_k^4\right) = \{p_i(2)\}^2 \Big/ \sum_1^N \{p_k(2)\}^2$ for the ith unit. If the units are distinct accept them as a sample, otherwise continue the procedure till the two units are distinct.

The inclusion probability of ith unit $\pi_i =$ probability of selection of the ith unit in first attempt + probability of selection of the ith unit in second attempt + probability of selection of the ith unit in third attempt + $\cdots$

$$
\begin{aligned}
&= \left[\sum_{j\neq i}\{p_i(1)p_j(1) + p_j(1)p_i(1)\}\right] + \left[\left(\sum_{j=1}^N \{p_{j(1)}\}^2\right)\sum_{j\neq i}\{p_i(2)p_j(2) + p_j(2)p_i(2)\}\right] \\
&\quad + \left[\left(\sum_{j=1}^N \{p_{j(1)}\}^2\right)\left(\sum_{j=1}^N \{p_{j(2)}\}^2\right)\sum_{j\neq i}\{p_i(3)p_j(3) + p_j(3)p_i(3)\}\right] + \cdots \\
&= [2p_i(1)\{1 - p_i(1)\}] + 2\left[\left(\sum_{j=1}^N \{p_{j(1)}\}^2\right)\left(p_i(2)\{1 - p_i(2)\}\right)\right] \\
&\quad + 2\left[\left(\sum_{j=1}^N \{p_{j(1)}\}^2\right)\left(\sum_{j=1}^N \{p_{j(2)}\}^2\right)\right](p_i(3)\{1 - p_i(3)\}) + \cdots \\
&= 2[p_i(1) - \{p_i(1)\}^2] + 2\left[\{p_i(1)\}^2 - \left[\sum_{j=1}^N \{p_{j(1)}\}^2\right]\{p_i(2)\}^2\right] \\
&\quad + 2\left[\left(\sum_{j=1}^N \{p_{j(1)}\}^2\right)\left(\left(\sum_{j=1}^N \{p_j(2)\}^2\right)p_i(3)\right)\right] \\
&\quad - 2\left[\left(\sum_{j=1}^N \{p_{j(1)}\}^2\right)\left(\sum_{j=1}^N \{p_{j(2)}\}^2\right)\{p_i(3)\}^2\right] + \cdots \\
&= 2p_i(1) \\
&= 2p_i
\end{aligned}
$$

The inclusion probability of ith and jth unit $(i \neq j)$

$$= \pi_{ij}$$
$$= 2p_i(1)p_j(1) + \left[\sum_{k=1}^{N}\{p_k(1)\}^2\right]2p_i(2)p_j(2) + \left[\sum_{k=1}^{N}\{p_k(1)\}^2\right]$$
$$\times\left[\sum_{k=1}^{N}\{p_k(2)\}^2\right]2p_i(3)p_j(3) + \cdots$$

It can be shown that for this sampling scheme $\pi_i\pi_j - \pi_{ij} \geq 0$.

5.4.2 Inclusion Probability Proportional to Measure of Size Sampling with $n > 2$

5.4.2.1 Lahiri–Midzuno–Sen Sampling Design

The simplest IPPS sampling scheme for an arbitrary sample size was proposed independently by Lahiri (1951), Midzuno (1952), and Sen (1953) but is applicable under the restrictive condition $\frac{n-1}{n(N-1)} < p_i < \frac{1}{n}$. For the LMS sampling design, at the first draw, the ith unit is selected with probability θ_i (>0, to be determined and subject to $\sum_{i \in U}\theta_i = 1$), called the revised normed size measure. Then the remaining $n - 1$ units are selected from those $N - 1$ units, which are not selected in the first draw by the SRSWOR method. Here, θ_i's are chosen to make the inclusion probability

$$\begin{aligned}\pi_i &= \theta_i + \sum_{j(\neq i)}\theta_j\frac{n-1}{N-1}\\ &= \frac{N-n}{N-1}\theta_i + \frac{n-1}{N-1}\end{aligned} \tag{5.4.4}$$

equals to np_i.

Now equating $\pi_i = np_i$, we get from Eq. (5.4.4)

$$\theta_i = \frac{(N-1)n}{N-n}p_i - \frac{n-1}{N-n} \tag{5.4.5}$$

Now putting a restriction on θ_i to be positive, we get a restriction on p_i as

$$p_i > \frac{n-1}{n(N-1)}$$

Thus the LMS–IPPS sampling scheme is applicable, if and only if,

$$\frac{n-1}{n(N-1)} < p_i < \frac{1}{n} \tag{5.4.6}$$

The inclusion probability of the ith and jth unit ($i \neq j$) for this sampling scheme is π_{ij} = probability of selection of the ith unit at the first draw and the jth unit in any of the remaining $n-1$ draws + probability of selection of the jth unit at the first draw and the ith unit in any of the remaining $n-1$ draws + probability that none of the ith and jth units were selected in the first draw but are selected in the remaining $n-1$ draws

$$= \theta_i \frac{n-1}{N-1} + \theta_j \frac{n-1}{N-1} + (1-\theta_i-\theta_j)\frac{(n-1)(n-2)}{(N-1)(N-2)}$$

$$= \frac{(n-1)(N-n)}{(N-1)(N-2)}(\theta_i+\theta_j) + \frac{(n-1)(n-2)}{(N-1)(N-2)} \tag{5.4.7}$$

$$= \frac{n(n-1)}{(N-2)}\left[p_i + p_j - \frac{1}{(N-1)}\right] \quad \text{(using Eq. 5.4.5)}$$

It can be easily checked that for this sampling scheme $\pi_i\pi_j - \pi_{ij}$ is positive.

Remark 5.4.2

For the sample size $n=2$, Rao (1963) proved that the Horvitz–Thompson estimator $\widehat{Y}_{ht} = \sum_{i\in s}\frac{y_i}{\pi_i} = \frac{1}{n}\sum_{i\in s}\frac{y_i}{p_i}$ based on the LMS–IPPS sampling scheme possesses lower variance than that of the HH estimator based on the PPSWR sampling scheme of the same sample size. Asok (1974) and Asok and Sukhatme (1978) proved that the Rao's result is valid for $n \geq 2$, i.e.,

$$V(\widehat{Y}_{ht}) \leq V(\widehat{Y}_{hh}) = \sum_{i\in U} p_i\left(\frac{y_i}{p_i} - Y\right)^2 \Big/ n. \tag{5.4.8}$$

5.4.2.2 Probability Proportionate to Size Systematic Sampling Scheme

Madaw (1949) and Goodman and Kish (1950) proposed the PPS systematic sampling procedure, which ensures that $\pi_i = np_i$. The proposed IPPS sampling scheme is very simple to execute and is applicable to any value of n as long as $\pi_i = np_j \leq 1$. But the main drawback of this procedure is that the expressions for the second-order inclusion probabilities are highly complex. The PPS systematic sampling procedure is described as follows.

Let $T_i = n\sum_{j=1}^{i} p_j$ for $i = 1,\ldots, N$ and $T_0 = 0$. Select a random sample (called random start) d from a uniform distribution with the range (0, 1). This random start d selects a sample with those units whose index, "j," satisfies $T_{j-1} \le d + k < T_j$ for $k = 0, 1,\ldots, n-1$.

It can easily be noted that each value of k results in the selection of only one unit because $np_i \le 1$. For this sampling scheme, the inclusion probability of the ith unit is $\pi_i = T_i - T_{i-1} = np_i$. Assuming that the labels of the units are attached at random, Hartley and Rao (1962) gave the following approximate expression of π_{ij} to the order $O(N^{-4})$ when p_i is of $O(N^{-1})$, n is relatively small to N, and N is moderately large.

$$\pi_{ij} = n(n-1)p_ip_j\left[\begin{array}{l} 1 + \left\{(p_i + p_j) - \sum_j p_j^2\right\} + \left\{2\left(p_i^2 + p_j^2\right) - 2\sum_j p_j^3\right. \\ \left. + 2p_ip_j - 3(p_i + p_j)\sum_j p_j^2 + 3\left(\sum_j p_j^2\right)^2\right\} \end{array}\right] \tag{5.4.9}$$

An approximate expression the variance of the $\widehat{Y}_{ht}$ was obtained by Harley and Rao (1962) correct to $O(N^{-2})$ as follows:

$$\begin{aligned} V\left(\widehat{Y}_{ht}\right) &= \frac{1}{n}\left[\sum_i p_i z_i^2 - (n-1)\sum_i p_i^2 z_i^2\right] \\ &\quad - \frac{(n-1)}{n}\left[2\sum_i p_i^3 z_i^2 - \left(\sum_i p_i^2\right)\left(\sum_i p_i^2 z_i^2\right) - 2\left(\sum_i p_i^2 z_i\right)^2\right] \\ &= V_{GK} \end{aligned} \tag{5.4.10}$$

where $z_i = \dfrac{y_i}{p_i} - Y$.

The approximate expression for V_{GK} correct to $O(N^{-1})$ is given by

$$\begin{aligned} V_{GK} &= \frac{1}{n}\sum_{i\in U} p_i\left(\frac{y_i}{p_i} - Y\right)^2 - \frac{n-1}{n}\sum_{i\in U} p_i^2\left(\frac{y_i}{p_i} - Y\right)^2 \\ &< \frac{1}{n}\sum_{i\in U} p_i\left(\frac{y_i}{p_i} - Y\right)^2 = V\left(\widehat{Y}_{hh}\right) \end{aligned} \tag{5.4.11}$$

The expression (5.4.11) indicates that the variance of the Horvitz–Thompson estimator based on the PPS systematic sampling design provides a smaller variance than that of the HH estimator based on PPWR sampling of the same size.

An approximate expression for an unbiased estimator of V_{GK} correct to $O(N)$ was given by Hartley and Rao (1962) as follows:

$$\widehat{V}_{GK} = \left(\frac{1}{n^2(n-1)}\right)\frac{1}{2}\left[\sum_{i\neq}\sum_{j\in s}\left\{1-n(p_i+p_j)+n\left(\sum_{i=1}^{N}p_i^2\right)\right\}\left(\frac{y_i}{p_i}-\frac{y_j}{p_j}\right)^2\right] \tag{5.4.12}$$

Example 5.4.1

Consider a population of size $N = 10$ units from which an IPPS sample of size $n = 4$ is to be selected. The measures of size (x_j) are given in Table 5.4.1.

Table 5.4.1

Unit	1	2	3	4	5	6	7	8	9	10
Measure of size (x_j)	10	25	24	15	12	18	35	32	18	11
np_j	0.20	0.50	0.48	0.30	0.24	0.36	0.70	0.64	0.36	0.22
T_j	0.20	0.70	1.18	1.48	1.72	2.08	2.78	3.42	3.78	4.00

Let a random start $d = 0.382$ be selected from a uniform distribution (0, 1). For $k = 0$, unit 2 is selected because $T_1 \leq d = 0.382 < T_2$; similarly, corresponding to $k = 1$, 2, and 3 units 4, 7, and 8 are selected, respectively, because $T_3 \leq d + 1 = 1.382 < T_4$, $T_6 \leq d + 2 = 2.382 < T_7$, and $T_7 \leq d + 3 = 3.382 < T_8$.

5.4.2.3 Sampford's Sampling Scheme

In Sampford's (1967) IPPS sampling scheme, on the first draw the ith ($i = 1,\ldots, N$) unit is selected with probability p_i, i.e., $p_i(1) = p_i$. Then the remaining $(n - 1)$ units are drawn with replacement from the entire population with probability proportional to $\lambda_i = p_i/(1 - np_i)$, i.e., the probability of selecting the ith unit at the kth draw is $p_i(k) = \lambda_i/\sum_{j=1}^{N}\lambda_j$ $k = 2,\ldots, n$; $i = 1,\ldots N$. The selected units are accepted as a sample if all the n units happened to be different, otherwise the entire selection is discarded, and this process is repeated unless a set of n distinct units is obtained.

Sampford (1967) has shown that the inclusion probability for the selection of ith unit is $\pi_i = np_i$ and $\pi_i\pi_j - \pi_{ij} \geq 0$. The expression for the second-order inclusion probabilities is not simple. However, approximate expression of π_{ij} correct to $O(N^{-4})$, derived by Asok and Sukhatme (1976), is given for $n \geq 3$ as follows:

$$\pi_{ij} = n(n-1)p_i p_j \left[\begin{array}{l} 1 + \left(p_i + p_j - \sum_j p_j^2 \right) \\ + \left(2\left(p_i^2 + p_j^2\right) - 2\sum_j p_j^3 - (n-2)p_i p_j \right. \\ \left. + (n-3)(p_i + p_j)\sum_j p_j^2 - (n-3)\left(\sum_j p_j^2\right)^2 \right) \end{array} \right] \tag{5.4.13}$$

For Sampford's sampling, the variance of the Horvitz–Thompson estimator correct to $O(N^{-2})$ is given by Asok and Sukhatme (1976) as

$$\begin{aligned} V(\widehat{Y}_{ht}) &= \frac{1}{n}\left(\sum_i p_i z_i^2 - (n-1)\sum_i p_i^2 z_i^2 \right. \\ &\quad - \frac{(n-1)}{n}\left(2\sum_i p_i^3 z_i^2 - \left(\sum_i p_i^2\right)\left(\sum_i p_i^2 z_i^2\right) \right. \\ &\quad \left.\left. + (n-2)\left(\sum_i p_i^2 z_i\right)^2 \right) \right) \\ &= V_{SAM} \end{aligned}$$

where z_i is given in Eq. (5.4.10).

5.4.2.3.1 Comparison of Efficiency

Asok and Sukhatme (1976) showed that the variances of the Horvitz–Thompson estimator for a finite population total Y based on the PPS systematic sampling scheme and Sampford's (1967) sampling scheme correct to $O(N^{-1})$ are exactly equal to

$$V_{GK} = V_{SAM} = \frac{1}{n}\left[\sum_i p_i z_i^2 - (n-1)\sum_i p_i^2 z_i^2 \right] \tag{5.4.14}$$

The above expression (5.4.14) indicates that the Horvitz–Thompson estimator based on the PPS systematic as well as Sampford procedures possesses a uniformly smaller variance than that of the HH estimator based on a PPSWR sampling design of the same sample size n. Furthermore, when the variance is considered to $O(N^{-2})$, the Horvitz-Thompson estimator based on Sampford sampling has a uniformly smaller variance than that of the PPS systematic sampling procedure. Their difference

$$V_{GK} - V_{SAM} = (n-1)\left(\sum_i p_i^2 z_i\right)^2$$

is nonnegative and increases with the sample size n.

Remark 5.4.3

For $n = 2$, Sampford (1967), Durbin (1967), and Brewer's (1963a,b) sampling designs are identical in the sense that the probability of getting an unordered sample $s = (i, j)$, $i < j$ is the same and is equal to $2p_ip_j\left(\frac{1}{1-2p_i}+\frac{1}{1-2p_j}\right)/A$ where A is given in Eq. (5.4.1).

Example 5.4.2

Table 5.4.2 relates to the annual income tax paid by 10 employees of a certain university. Select a sample of size 4 by Sampford's IPPS sampling procedure using the amount of tax paid as measure of size variable (x).

In Sampford sampling, the first unit is selected with probability proportional to a measure of size x. So, we select the first unit by the cumulative total method and prepare Table 5.4.3.

Here we select a six-digit random number between 000001 and 360000. From the random number table we select random number 029092. This random number selects unit 1. The remaining three units needed to be selected with replacement from the units 1 to 10 with probability $q_i = \dfrac{p_i/(1-4p_i)}{\sum_{i=1}^{10} p_i/(1-4p_i)}$ attached to the ith unit, $i = 1,\ldots, 10$. The values of q_i's and cumulative totals Q_i are given in Table 5.4.4.

Now we select three four-digit random numbers from a random number table, and putting a decimal point to the left of each number, we get the numbers as follows: 0.9356, 0.1892, 0.4598. Now looking at the cumulative total Q_i values, we select units 9, 3, and 6. Because the selected units (1, 9, 3, 6) are all distinct, we accept $s = (1, 3, 6, 9)$ as a sample according to Sampford's IPPS sampling scheme. It should be noted that if all the units were not distinct, we would have to repeat the procedure until all the four selected units were distinct.

Table 5.4.2

Serial no. of employees (i)	1	2	3	4	5	6	7	8	9	10
Tax paid in dollars (x_i)	30,000	35,000	45,000	25,000	20,000	60,000	40,000	50,000	30,000	25,000

Table 5.4.3

Serial no. of employees (i)	1	2	3	4	5	6	7	8	9	10
Tax paid in dollars (x_i)	30,000	35,000	45,000	25,000	20,000	60,000	40,000	50,000	30,000	25,000
Cumulative total T_i	30,000	65,000	110,000	135,000	155,000	215,000	255,000	305,000	335,000	360,000

Table 5.4.4

Unit	1	2	3	4	5	6	7	8	9	10
$\frac{p_i}{(1-4p_i)}$	0.125	0.1591	0.25	0.0962	0.0714	0.5	0.2	0.3125	0.125	0.0962
q_i	0.0646	0.0822	0.1292	0.0497	0.0369	0.2583	0.1033	0.1615	0.0646	0.0497
Q_i	0.0646	0.1468	0.276	0.3257	0.3626	0.6209	0.7242	0.8857	0.9503	1

5.4.2.4 Poisson (or Bernoulli) Sampling

In the Poisson or Bernoulli sampling scheme, the units are selected by performing N (population size) Bernoulli trials independently. The ith ($i = 1,\ldots, N$) unit is selected by performing a Bernoulli trial with a success probability π_i. If the trial produces a success, the ith unit is selected in the sample, otherwise it is not included in the sample. Clearly, in a Poisson sampling scheme, the sample size is not fixed. It is a random variable. But the expected sample size $\sum_{i=1}^{N} \pi_i$ becomes fixed as n if $\pi_i = np_i$. For the Poisson sampling scheme, the inclusion probability for the ith unit is π_i and inclusion probability of a pair of units i, $j(i \neq j)$ is $\pi_{ij} = \pi_i\pi_j$ (because the draws are independent). Hence for a Poisson sampling $\pi_i\pi_j - \pi_{ij} = 0$. The expression for the variance of $\widehat{Y}_{ht} = \sum_{i\in s} y_i/\pi_i$ and its unbiased estimator are very simple and they are given, respectively, as follows:

$$V\left(\widehat{Y}_{ht}\right) = \sum_{i=1}^{N} y_i^2\left(\frac{1}{\pi_i} - 1\right) \tag{5.4.15}$$

and

$$\widehat{V}\left(\widehat{Y}_{ht}\right) = \sum_{i\in s} \frac{y_i^2}{\pi_i}\left(\frac{1}{\pi_i} - 1\right) \tag{5.4.16}$$

5.4.2.5 Use of Combinatorics

Let $\mathfrak{S} = (s_1,\ldots, s_j,\ldots, s_b)$ be the collection of b possible samples each of effective size n and $p(s_j)$ be the probability of selection of the sample s_j with $p(s_j) \geq 0$, $\sum_{j=1}^{b} p(s_j) = 1$. Then the inclusion probability of the ith unit is $\pi_i = \sum_{j=1}^{b} n_{ij}p(s_j)$, where $n_{ij} = 1$, if the sample s_j contains the ith unit and $n_{ij} = 0$ otherwise; $i = 1,\ldots, N$; $j = 1,\ldots, b$. So, we can write the inclusion probability matrix as

$$\mathbf{\Pi} = \mathbf{N}\cdot\mathbf{P} \tag{5.4.17}$$

where

$$\mathbf{\Pi} = \begin{pmatrix} \pi_1 \\ \bullet \\ \pi_i \\ \bullet \\ \pi_N \end{pmatrix}, \quad \mathbf{N} = \begin{pmatrix} n_{11} & \bullet & n_{1j} & \bullet & n_{1b} \\ \bullet & \bullet & \bullet & \bullet & \bullet \\ n_{i1} & \bullet & n_{ij} & \bullet & n_{ib} \\ \bullet & \bullet & \bullet & \bullet & \bullet \\ n_{N1} & \bullet & n_{Nj} & \bullet & n_{Nb} \end{pmatrix} \text{ and } \mathbf{P} = \begin{pmatrix} p(s_1) \\ \bullet \\ p(s_j) \\ \bullet \\ p(s_b) \end{pmatrix}$$

Thus **P** will be an IPPS sampling design if a nonnegative solution of **P** for Eq. (5.4.17) exists for $\pi_i(=np_i)$ for $i = 1,\ldots, N$. The solution of Eq. (5.4.17) was derived by Gupta et al. (1982) using the properties of a balanced incomplete block design (BIBD). If a unit is considered as a treatment and a sample as a block, then the matrix **N** may be considered as an incidence matrix of an incomplete block design with parameter $v =$ number of treatments $=$ total numbers of units $= N$, $b =$ total number of blocks of a design $=$ total number of samples, and $k =$ total number of treatments in a block $=$ sample size $= n$. Furthermore, the incidence matrix **N** corresponds to a BIBD if every treatment (unit) is exactly repeated in r blocks (samples) and any two treatments (units) occur together in λ blocks (samples). The parameters of a BIBD design are denoted by v, b, r, k, and λ. The parameters satisfy (i) $bk = vr$ and (ii) $\lambda(v-1) = r(k-1)$ [vide Raghavarao, 1971]. Here, we will suppose that the b samples (blocks) are so chosen that **N** corresponds to an incidence matrix of a BIBD.

Theorem 5.4.1

Let the samples s_j's are selected with probability

$$p(s_j) = v\left(r\sum_{i\in s_j} p_i - \lambda\right)\Big/\{b(r-\lambda)\} \quad \text{for} \quad j = 1,\ldots b$$

with $\sum_{i\in s_j} p_i \geq \lambda/r = (n-1)/(N-1)$.

Then, (i) $\pi_i = np_i$

$$\text{(ii)}\,\pi_{ij} = \frac{v}{b(r-\lambda)}\left[r\left\{\lambda(p_i+p_j) + \sum_{l(\neq i,j)=1}^{N} p_l\,\mu_{ijl}\right\} - \lambda^2\right] \tag{5.4.18}$$

where μ_{ijl} is the number of times the ith, jth, and lth $(i \neq j \neq l)$ treatments (units) occur together in the same block (sample).

Proof

(i)
$$\pi_i = \sum_{j=1}^{b} n_{ij}p(s_j) = v\sum_{j=1}^{b} n_{ij}\left(r\sum_{t=1}^{N} n_{tj}p_t - \lambda\right)\Big/\{b(r-\lambda)\}$$

$$= v\left[r\left(p_i\sum_{j=1}^{b} n_{ij} + \sum_{t(\neq i)=1}^{N} p_t\sum_{j=1}^{b} n_{ij}n_{tj}\right) - \lambda\sum_{j=1}^{b} n_{ij}\right]\Big/[b(r-\lambda)]$$

$$= v\left[r\left(r\,p_i + \lambda\sum_{t(\neq i)=1}^{N} p_t\right) - \lambda r\right]\Big/[b(r-\lambda)]$$

$$= vrp_i/b = np_i \;(\text{since for a BIDB, } bk = vr \text{ and } k = n)$$

$$\text{(ii)}\quad \pi_{ij} = \sum_{t=1}^{b} n_{it}\, n_{jt} p(s_t) = v \sum_{t=1}^{b} n_{it}\, n_{jt} \left(r \sum_{l=1}^{N} n_{lt} p_l - \lambda \right) \Big/ [b(r-\lambda)]$$

$$= v \left[\sum_{t=1}^{b} n_{it}\, n_{jt}\, r \left(\left(n_{it} p_i + n_{jt} p_j \right) + \sum_{l(\neq i,j)=1}^{N} n_{lt} p_l \right) - \lambda^2 \right] \Big/ [b(r-\lambda)]$$

$$= v \left[r \left(\lambda \left(p_i + p_j \right) + \sum_{l(\neq i,j)=1}^{N} \mu_{ijl} p_l \right) - \lambda^2 \right] \Big/ [b(r-\lambda)]$$

The previous expression of π_{ij} is complex and one is not sure about the sign of $\pi_i\pi_j - \pi_{ij}$. However, instead of considering BIBD, if we consider doubly balanced incomplete block design (DBIBD) introduced by Calvin (1954), where every triplet of treatment (units) appears together in the same block (sample), an equal number of times (μ, say), we may construct an IPPS sampling design with π_{ij} identical to the LMS−IPPS sampling scheme satisfying $\pi_i\pi_j - \pi_{ij} \geq 0\ \forall i \neq j$. For a DBIBD, $\mu(v-2) = \lambda(k-2)$ [vide Hedayat and Kageyama, 1980].

Theorem 5.4.2

For a DBIBD, the expression (5.4.18) reduces to $\pi_{ij} = \dfrac{n(n-1)}{N-2} \times \left(p_i + p_j\right) - \dfrac{n(n-1)}{(N-1)(N-2)}$.

Proof

From Eq. (5.4.18), we get

$$\pi_{ij} = \frac{v}{b(r-\lambda)} \left[r \left(\lambda(p_i + p_j) + \mu \sum_{k(\neq i,j)=1}^{N} p_k \right) - \lambda^2 \right]$$

$$= \frac{v}{b(r-\lambda)} \left[r \left((\lambda - \mu)(p_i + p_j) + \mu \right) - \lambda^2 \right]$$

Now using the relations (i) $bk = vr$, (ii) $\lambda(v-1) = r(k-1)$, and (iii) $\mu(v-2) = \lambda(k-2)$, we get

$$\pi_{ij} = \frac{k(k-1)}{(v-2)}(p_i + p_j) - \frac{k(k-1)}{(v-1)(v-2)}$$
$$= \frac{n(n-1)}{(N-2)}(p_i + p_j) - \frac{n(n-1)}{(N-1)(N-2)}$$
$$(\text{noting } k = n \text{ and } v = N)$$

Remark 5.4.4

Gupta et al. (1982) IPPS design requires $\sum_{i \in s_j} p_i \geq \lambda / r = (n-1)/(N-1)$, which is less restrictive than that of the LMS–IPPS sampling scheme, which requires $p_i \geq (n-1)/\{n(N-1)\}$ for every $i = 1,\ldots, N$.

Remark 5.4.5

The construction of sampling designs realizing preassigned sets of values of inclusion probabilities of the first and second order were provided by Sinha (1973). Arnab and Roy (1990) have given methods of construction of such sampling designs by using of BIB designs.

5.4.2.6 The Nearest Proportional to Size Sampling

Suppose a sampling design p_0 is preferable because of practical consideration, such as cost. But on the other hand, an IPPS sampling design p^* with inclusion probability $\pi_i = np_i$ is desirable for theoretical considerations, such as efficiency. Gabler (1987a,b) provided a method of constructing of a p^* sampling design, which is nearest to the sampling design p_0 in the sense of minimizing the distance

$$D(p_0, p^*) = \sum_{s \in \mathfrak{I}_0} \frac{\{p^*(s) - p_0(s)\}^2}{p_0(s)}$$

where $\mathfrak{I}_0$ is the support of the sampling design p_0, which is the collection of all possible samples s of n distinct units with $p_0(s) > 0$, $\sum_{s \in \mathfrak{I}_0} p_0(s) = 1$. Here we assume that $\mathfrak{I}^*$, the support of the design p^* based on n distinct units, is a subset of $\mathfrak{I}_0$.

Now minimizing distance $D(p_0, p^*)$ subject to $\sum_{s \supset i} p^*(s) = \pi_i = np_i$, we get

$$p^*(s) = p_0(s) \sum_{i \in s} \lambda_i \tag{5.4.19}$$

where λ_i's are the solution of the equation

$$\mathbf{\Pi}_0\boldsymbol{\lambda} = \boldsymbol{\pi} \tag{5.4.20}$$

with

$$\mathbf{\Pi}_0 = \begin{pmatrix} \pi_1^0 & \pi_{12}^0 & \ldots & \pi_{1N}^0 \\ \pi_{12}^0 & \pi_2^0 & \ldots & \pi_{2N}^0 \\ \ldots & \ldots & \ldots & .. \\ \pi_{N1}^0 & \pi_{N2}^0 & \ldots & \pi_N^0 \end{pmatrix}, \quad \boldsymbol{\pi} = \begin{pmatrix} \pi_1 \\ \pi_2 \\ \ldots \\ \pi_N \end{pmatrix}, \quad \boldsymbol{\lambda} = \begin{pmatrix} \lambda_1 \\ \lambda_2 \\ \ldots \\ \lambda_N \end{pmatrix},$$

$$\pi_i^0 = \sum_{s \supset i} p_0(s) \text{ and } \pi_{ij}^0 = \sum_{s \supset i,j} p_0(s) \text{ for } i \neq j$$

Eq. (5.4.19) provides the solution of sampling design p^* if $\sum_{i \in s} \lambda_i \geq 0$. Clearly, $p^*(s) \geq 0$ whenever all λ_i's are positive.

Example 5.4.3

Let p_0 be an SRSWOR design of size n with $p_0(s) = 1 \Big/ \binom{N}{n}$.

Here $\pi_i^0 = n\big/N = \alpha$, $\pi_{ij}^0 = n(n-1)\big/\{N(N-1)\} = \beta$,

$$\mathbf{\Pi}_0 = \begin{pmatrix} \alpha & \beta & \ldots & \beta \\ \beta & \alpha & \ldots & \beta \\ \ldots & \ldots & \ldots & \ldots \\ \beta & \beta & \ldots & \alpha \end{pmatrix} \text{ and } \mathbf{\Pi}_0^{-1} = \begin{pmatrix} a & b & \ldots & b \\ b & a & \ldots & b \\ \ldots & \ldots & \ldots & \ldots \\ b & b & \ldots & a \end{pmatrix}$$

with $a = \dfrac{N}{n} - (n-1)b$ and $b = -\dfrac{(n-1)N}{n^2(N-n)}$.

From Eq. (5.4.20), we get $\lambda_i = \dfrac{N(N-1)}{n(N-n)}\pi_i - \dfrac{N(n-1)}{n(N-n)}$ for $i = 1,\ldots, N$. Finally Eq. (5.4.19) yields

$$\begin{aligned} p^*(s) &= \left[\frac{N-1}{N-n}\sum_{i \in s}\pi_i - \frac{n(n-1)}{N-n}\right] \Big/ \binom{N-1}{n-1} \\ &= \left[\frac{N-1}{N-n} n \sum_{i \in s} p_i - \frac{n(n-1)}{N-n}\right] \Big/ \binom{N-1}{n-1} \end{aligned} \tag{5.4.21}$$

Eq. (5.4.21) is the selection probability of the LMS–IPPS sampling procedure. Hence the LMS–IPPS sampling scheme is the nearest to the SRSWOR sampling scheme of size n.

Remark 5.4.6

Arnab (2004a,b) provided the conditions of existence of the solution to Eq. (5.4.21).

5.5 PROBABILITY PROPORTIONAL TO AGGREGATE SIZE WITHOUT REPLACEMENT

Lahiri–Midzuno–Sen (1951, 1952, 1953) considered the probability proportional to aggregate size (PPAS) sampling scheme. Here, at the first draw, the ith unit is selected with probability p_i and the remaining $n-1$ units are selected by the SRSWOR method from those units that were not selected in the first draw. So, the probability of selection of an unordered sample, $s=(i_1,\ldots,i_k,\ldots,i_n)$ with $i_1<\cdots<i_k<\cdots<i_n$ is

$$p(s)=\sum_{i\in s}p_i\bigg/\binom{N-1}{n-1}=x_s/(M_1X)$$

where $x_s=\sum_{i\in s}x_i$, $X=\sum_{i=1}^{N}x_i$ and $M_1=\binom{N-1}{n-1}$.

Hence, the probability of selection of an unordered sample is proportional to the aggregate measure of size x_s.

The inclusion probabilities for the ith, and ith and jth units $(i\neq j)$ are, respectively, given as follows:

$\pi_i=$ probability of selection of the ith unit at the first draw + probability that the ith unit is not selected at the first draw and it is selected in any of the remaining $(n-1)$ draws

$$=p_i+(1-p_i)\binom{N-1}{n-1}\bigg/\binom{N}{n}$$

$$=\frac{N-n}{N}p_i+\frac{n}{N}$$

and

$\pi_{ij}=$ probability of selection of the ith unit at the first draw and the jth unit in any of the remaining $(n-1)$ draws + probability of selection of the jth unit at the first draw and the ith unit in any of the remaining $(n-1)$ draws + probability that none of the ith and jth units are selected at the first draw and they are selected in the remaining $(n-1)$ draws

$$=p_i\frac{n}{N}+p_j\frac{n}{N}+(1-p_i-p_j)\frac{n(n-1)}{N(N-1)}$$

$$=\frac{n(N-n)}{N(N-1)}(p_i+p_j)+\frac{n(n-1)}{N(N-1)}$$

Theorem 5.5.1

(i) $\widehat{Y}_{lms} = \dfrac{y_s}{x_s} X$ is unbiased for the total Y

(ii) The variance of $\widehat{Y}_{lms}$ is

$$V_{lms} = V\left(\widehat{Y}_{lms}\right) = \sum_{i=1}^{N} y_i^2 (T_i - 1) + \sum_{i\neq}^{N} \sum_{j=1}^{N} y_i y_j (T_{ij} - 1)$$

where $y_s = \sum_{i\in s} y_i, \quad T_i = \dfrac{X}{M_1} \sum_s I_{si}/x_s = \dfrac{1}{M_1^2} \sum_s I_{si}/p(s),$

$T_{ij} = \dfrac{X}{M_1} \sum_s I_{si} I_{sj}/x_s = \dfrac{1}{M_1^2} \sum_s I_{si} I_{sj}/p(s),\ I_{si} = 1 \text{ if } i \in s, \text{ and } I_{si} = 0 \text{ if } i \notin s.$

(iii) $V\left(\widehat{Y}_{lms}\right)$ can be unbiasedly estimated by any of the following:

(a) $\widehat{V}_{lms} = \sum_{i\in s} y_i^2 (T_i - 1)/\pi_i + \sum_{i\neq} \sum_{j\in s} y_i y_j (T_{ij} - 1)/\pi_{ij}$

or

(b) $\widehat{V}^{*}_{lms} = \left(\widehat{Y}_{lms}\right)^2 - \dfrac{X}{x_s}\left[\sum_{i\in s} y_i^2 + \dfrac{(N-1)}{(n-1)} \sum_{i\neq} \sum_{j\in s} y_i y_j\right]$

Proof

(i)
$$E(\widehat{Y}_{lms}) = E\left(\frac{y_s}{x_s} X\right)$$
$$= E\left(\frac{y_s}{M_1 p(s)}\right)$$
$$= \sum_s \frac{y_s}{M_1}$$
$$= \frac{1}{M_1} \sum_s \sum_{i=1}^{N} I_{si} y_i$$
$$= \frac{1}{M_1} \sum_{i=1}^{N} y_i \sum_s I_{si}$$
$$= Y$$

(ii) $V\left(\widehat{Y}_{lms}\right) = E\left(\widehat{Y}_{lms}\right)^2 - Y^2$

$$= E\left[\frac{1}{M_1}\frac{1}{p(s)}\sum_{i=1}^{N} I_{si}y_i\right]^2 - Y^2$$

$$= \sum_s \frac{1}{(M_1)^2}\frac{1}{p(s)}\left(\sum_{i=1}^{N} I_{si}y_i^2 + \sum_{i\neq}\sum_{j=1}^{N} I_{si}I_{sj}y_iy_j\right) - Y^2$$

$$= \sum_{i=1}^{N} y_i^2\left\{\frac{1}{(M_1)^2}\sum_s I_{si}\frac{1}{p(s)} - 1\right\}$$

$$+ \sum_{i\neq}^{N}\sum_{j=1}^{N} y_iy_j\left\{\frac{1}{(M_1)^2}\sum_s I_{si}I_{sj}\frac{1}{p(s)} - 1\right\}$$

$$= \sum_{i=1}^{N} y_i^2(T_i - 1) + \sum_{i\neq}^{N}\sum_{j=1}^{N} y_iy_j(T_{ij} - 1)$$

(iii) (a) $E\left(\widehat{V}_{lms}\right) = E\left[\sum_{i\in s} y_i^2(T_i - 1)/\pi_i + \sum_{i\neq}\sum_{j\in s} y_iy_j(T_{ij} - 1)/\pi_{ij}\right]$

$$= \sum_{i=1}^{N} y_i^2(T_i - 1)E(I_{si})/\pi_i + \sum_{i\neq}^{N}\sum_{j=1}^{N} y_iy_j(T_{ij} - 1)E(I_{si}I_{sj})/\pi_{ij}$$

$$= V\left(\widehat{Y}_{lms}\right) \text{ (since } E(I_{si}) = \pi_i \text{ and } E(I_{si}I_{sj}) = \pi_{ij})$$

(b) $E\left(\widehat{V}^*_{lms}\right) = \left(V\left(\widehat{Y}_{lms}\right) + Y^2\right) - \sum_s\left\{\frac{X}{x_s}\left(\sum_{i\in s} y_i^2 + \frac{(N-1)}{(n-1)}\sum_{i\neq}\sum_{j\in s} y_iy_j\right)\right\}p(s)$

$$= \left(V\left(\widehat{Y}_{lms}\right) + Y^2\right) - \frac{1}{M_1}\sum_s\left\{\sum_{i\in s} y_i^2 + \frac{(N-1)}{(n-1)}\sum_{i\neq}\sum_{j\in s} y_iy_j\right\}$$

$$= \left(V\left(\widehat{Y}_{lms}\right) + Y^2\right) - \frac{1}{M_1}\left\{\sum_{i=1}^{N} y_i^2\sum_{s\supset i} + \frac{(N-1)}{(n-1)}\sum_{i\neq}^{N}\sum_{j\in s}^{N} y_iy_j\sum_{s\supset i,j}\right\}$$

$$= V\left(\widehat{Y}_{lms}\right)$$

Remark 5.5.1

The variance estimator $\widehat{V}_{lms}$ was proposed by Rao T.J. (1967). Although the estimator $\widehat{V}_{lms}$ seems to be elegant, it is seldom used in practice because

T_i and T_{ij}'s are very tedious to compute. Conversely, the estimator $\widehat{V}^{*}_{lms}$ is used very often in practice because it is very easy to compute. However, both the estimators can take negative values. Chaudhuri and Arnab (1981) studied nonnegativity properties of the unbiased estimators of $\widehat{V}_{lms}$ in detail.

5.6 RAO–HARTLEY–COCHRAN SAMPLING SCHEME

In the Rao–Hartley–Cochran (1962) sampling scheme, the population is divided at random into n disjoint groups $G_1, \ldots, G_n$. The number of units that belong to the jth group G_j is N_j, a preassigned number with $N = \sum_{j=1}^{n} N_j$. From each of the groups, one unit is selected independently with its probability proportional to its measure of size viz. if the unit i_j belongs to the jth group G_j, it is selected with probability $q_{i_j} = x_{i_j} / \sum_{k \in G_j} x_k = p_{i_j}/P_j$, where $p_i = x_i/X$ and $P_j = \sum_{k \in G_j} p_k =$ sum of p_k's for the group G_j.

Suppose that the units $i_1, \ldots, i_j, \ldots, i_n$ are selected from the groups $G_1, \ldots, G_j, \ldots, G_n$, respectively. Then, we have the following theorem.

Theorem 5.6.1

(i) $\widehat{Y}_{rhc} = \sum_{j=1}^{n} \frac{y_{i_j}}{p_{i_j}} P_j$ is an unbiased estimator for the population total Y

(ii) The variance of $\widehat{Y}_{rhc}$ is

$$V_{rhc} = V\left(\widehat{Y}_{rhc}\right) = \frac{\sum_{j=1}^{n} N_j^2 - N}{N(N-1)} \sum_{i=1}^{N} p_i \left(\frac{y_i}{p_i} - Y\right)^2$$

(iii) An unbiased estimator for $V\left(\widehat{Y}_{rhc}\right)$ is

$$\widehat{V}\left(\widehat{Y}_{rhc}\right) = \frac{\sum_{j=1}^{n} N_j^2 - N}{N^2 - \sum_{j=1}^{n} N_j^2} \sum_{j=1}^{n} P_j \left(\frac{y_{i_j}}{p_{i_j}} - \widehat{Y}_{rhc}\right)^2$$

Proof

Let E_G and V_G denote the unconditional expectation and the variance over G, respectively. The conditional expectation and variance for a given $G = (G_1, \ldots, G_j, \ldots, G_n)$ are denoted by $E(\cdot|G)$ and $V(\cdot|G)$ respectively. Then

(i) $$E\left(\widehat{Y}_{rhc}\right) = E_G\left\{\sum_{j=1}^{n} E\left(\frac{y_{i_j}}{p_{i_j}} P_j \middle| G\right)\right\} = E_G\left(\sum_{j=1}^{n} Y_j\right) = Y$$

where $Y_j = \sum_{k \in G_j} y_k$

$$\text{(ii)}\ V\left(\widehat{Y}_{rhc}\right) = E_G\left\{\sum_{j=1}^{n} V\left(\frac{y_{i_j}}{p_{i_j}}P_j\middle| G\right)\right\} + V_G\left\{E\left(\sum_{j=1}^{n}\frac{y_{i_j}}{p_{i_j}}P_j\middle| G\right)\right\} \quad (5.6.1)$$

Now

$$V_G\left\{E\left(\sum_{j=1}^{n}\frac{y_{i_j}}{p_{i_j}}P_j\middle| G\right)\right\} = 0 \quad \left(\text{since } E\left(\sum_{j=1}^{n}\frac{y_{i_j}}{p_{i_j}}P_j\middle| G\right) = Y\right) \quad (5.6.2)$$

and

$$E_G\left\{\sum_{j=1}^{n} V\left(\frac{y_{i_j}}{p_{i_j}}P_j\middle| G\right)\right\} = E_G\left[\sum_{j=1}^{n}\left\{\sum_{k\in G_j}\frac{y_k^2}{p_k}P_j - \left(\sum_{k\in G_j} y_k\right)^2\right\}\right]$$

$$= E_G\left[\sum_{j=1}^{n}\left\{\left(\sum_{k\in G_j} y_k^2 + \sum_{k\neq}\sum_{t\in G_j} p_t\frac{y_k^2}{p_k}\right)\right.\right.$$

$$\left.\left. - \left(\sum_{k\in G_j} y_k^2 + \sum_{k\neq}\sum_{t\in G_j} y_k y_t\right)\right\}\right]$$

$$= \sum_{j=1}^{n} E_G\left\{\sum_{k\neq}\sum_{t\in G_j}\left(p_t\frac{y_k^2}{p_k} - y_k y_t\right)\right\}$$

Now noting that G_j is a random sample of size N_j selected from the population of N units by SRSWOR we get

$$E_G\sum_{k\neq}\sum_{t\in G_j}\left(p_t\frac{y_k^2}{p_k} - y_k y_t\right) = \frac{N_j(N_j-1)}{N(N-1)}\sum_{t\neq}^{N}\sum_{k=1}^{N}\left(p_t\frac{y_k^2}{p_k} - y_k y_t\right)$$

$$= \frac{N_j(N_j-1)}{N(N-1)}\sum_{i=1}^{N} p_i\left(\frac{y_i}{p_i} - Y\right)^2$$

and

$$E_G\left\{\sum_{j=1}^{n} V\left(\frac{y_{i_j}}{p_{i_j}}P_j\middle| G\right)\right\} = \frac{\sum_{j=1}^{n} N_j(N_j-1)}{N(N-1)}\sum_{i=1}^{N} p_i\left(\frac{y_i}{p_i} - Y\right)^2 \quad (5.6.3)$$

Finally, substituting Eqs. (5.6.2) and (5.6.3) in Eq. (5.6.1), we prove (ii) of the theorem.

$$\text{(iii) } E\left[\sum_{j=1}^{n} P_j\left(\frac{y_{i_j}}{p_{i_j}} - \widehat{Y}_{rhc}\right)^2\right] = E\left(\sum_{j=1}^{n} P_j\frac{y_{i_j}^2}{p_{i_j}^2}\right) - E\left(\widehat{Y}_{rhc}^2\right)$$
$$= E\left(\sum_{j=1}^{n} P_j\frac{y_{i_j}^2}{p_{i_j}^2}\right) - V\left(\widehat{Y}_{rhc}\right) - Y^2 \tag{5.6.4}$$

Now

$$E\left(\sum_{j=1}^{n} P_j\frac{y_{i_j}^2}{p_{i_j}^2}\right) = E_G\left[\sum_{j=1}^{n} E\left(P_j\frac{y_{i_j}^2}{p_{i_j}^2}\middle| G\right)\right]$$
$$= \sum_{j=1}^{n} E_G\left(\sum_{i\in G_j}\frac{y_i^2}{p_i}\right)$$
$$= \sum_{j=1}^{n}\frac{N_j}{N}\left(\sum_{i=1}^{N}\frac{y_i^2}{p_i}\right) \tag{5.6.5}$$
$$= \sum_{i=1}^{N}\frac{y_i^2}{p_i}$$

Substituting Eq. (5.6.5) in Eq. (5.6.4), we get

$$E\left[\sum_{j=1}^{n} P_j\left(\frac{y_{i_j}}{p_{i_j}} - \widehat{Y}_{rhc}\right)^2\right] = \sum_{i=1}^{N} p_i\left(\frac{y_i}{p_i} - Y\right)^2 - V\left(\widehat{Y}_{rhc}\right)$$
$$= \left(\frac{N(N-1)}{\sum_j N_j^2 - N} - 1\right) V\left(\widehat{Y}_{rhc}\right) \tag{5.6.6}$$

From Eq. (5.6.6), we get $E\left[\widehat{V}\left(\widehat{Y}_{rhc}\right)\right] = V\left(\widehat{Y}_{rhc}\right)$.

Remark 5.6.1

$\sum_{j=1}^{n} N_j^2$ attains a minimum when $N_j = N/n$. Hence if N/n is an integer, we should choose $N_j = N/n$ for every $j = 1,\ldots, n$ to minimize $V\left(\widehat{Y}_{rhc}\right)$ for a given value of n. In this case

$$V\left(\widehat{Y}_{rhc}\right) = \frac{N-n}{n(N-1)} \sum_{i=1}^{N} p_i \left(\frac{y_i}{p_i} - Y\right)^2 \tag{5.6.7}$$

and its unbiased estimator becomes

$$\widehat{V}\left(\widehat{Y}_{rhc}\right) = \frac{N-n}{N(n-1)} \sum_{j=1}^{n} P_j \left(\frac{y_{i_j}}{p_{i_j}} - \widehat{Y}_{rhc}\right)^2 \tag{5.6.8}$$

In case N/n is not an integer and $N = kn + r$ where k is an integer and $1 \le r < n$, we choose $N_1 = N_2 = \cdots = N_r = k+1$ and $N_{r+1} = N_{r+2} = \cdots = N_n = k$ to minimize $V\left(\widehat{Y}_{rhc}\right)$.

Remark 5.6.2

The estimator $\widehat{Y}_{rhc}$ based on an RHC sampling scheme with N/n as an integer is more efficient than the HH estimator, $\widehat{Y}_{hh}$, based on a PPSWR sampling scheme because $V\left(\widehat{Y}_{rhc}\right) \le V\left(\widehat{Y}_{hh}\right)$. The relative efficiency of $\widehat{Y}_{rhc}$ with respect to $\widehat{Y}_{hh}$ is $R = \dfrac{V\left(\hat{Y}_{hh}\right)}{V\left(\hat{Y}_{rhc}\right)} 100 = \dfrac{N-1}{N-n} 100\%$, which increases with the sample size n.

Remark 5.6.3

Although the estimator $\widehat{Y}_{rhc}$ is simple and possesses elegant expressions for its variance and unbiased estimator of the variance, it suffers from a serious drawback that the estimator $\widehat{Y}_{rhc}$ is inadmissible. This is because $\widehat{Y}_{rhc}$ is an ordered estimator because it depends on the formation of the initial group $G = (G_1, \ldots, G_n)$. The technique of Rao-Blackwellization does not yield any elegant expression of $\widehat{Y}_{rhc}$. For example, let $N = 6$ and $n = 2$, then we can select a sample $s = (1, 2)$ by forming the following six distinct set of groups, each with probability 1/6.

Set 1: $G_1 = (1, 3, 4)$, $G_2 = (2, 5, 6)$; Set 2: $G_1 = (1, 3, 5)$, $G_2 = (2, 4, 6)$;
Set 3: $G_1 = (1, 3, 6)$, $G_2 = (2, 4, 5)$; Set 4: $G_1 = (1, 4, 5)$, $G_2 = (2, 3, 6)$;
Set 5: $G_1 = (1, 4, 6)$, $G_2 = (2, 3, 5)$; Set 6: $G_1 = (1, 5, 6)$, $G_2 = (2, 3, 4)$.

The RHC estimator based on the sets and the probabilities of selection of the respective samples are given in the following table with $P_{ijk} = p_i + p_j + p_k$:

Set (j)	1	2	3	4
$\widehat{Y}_{rhc}(j)$	$\frac{y_1}{p_1}P_{134} + \frac{y_2}{p_2}P_{256}$	$\frac{y_1}{p_1}P_{135} + \frac{y_2}{p_2}P_{246}$	$\frac{y_1}{p_1}P_{136} + \frac{y_2}{p_2}P_{245}$	$\frac{y_1}{p_1}P_{145} + \frac{y_2}{p_2}P_{236}$
Probability $P(j)$	$\frac{1}{6}\frac{p_1 p_2}{P_{134}P_{256}}$	$\frac{1}{6}\frac{p_1 p_2}{P_{135}P_{246}}$	$\frac{1}{6}\frac{p_1 p_2}{P_{136}P_{245}}$	$\frac{1}{6}\frac{p_1 p_2}{P_{145}P_{236}}$

Set (j)	5	6
$\widehat{Y}_{rhc}(j)$	$\frac{y_1}{p_1}P_{146} + \frac{y_2}{p_2}P_{235}$	$\frac{y_1}{p_1}P_{156} + \frac{y_2}{p_2}P_{234}$
Probability $P(j)$	$\frac{1}{6}\frac{p_1 p_2}{P_{146}P_{235}}$	$\frac{1}{6}\frac{p_1 p_2}{P_{156}P_{234}}$

The Rao-Blackwellization of $\widehat{Y}_{rhc}(j)$ gives a uniformly better estimator

$$\widehat{Y}^*_{rhc} = \frac{\sum_j \widehat{Y}(j)P(j)}{\sum_j P(j)}$$

which is obviously not a simple expression.

Example 5.6.1

Table 5.6.1 gives the yield of fish from nine ponds along with their areas. Select a sample of three ponds by the (i) RHC and (ii) PPAS method of sampling taking the area of the pond as measure of size. Estimate the average yield of fish and 90% confidence interval of the average yield for each of the samples separately.

(i) RHC method: For selection of a sample of size $n = 3$, we divide the population of size $N = 9$ at random into three groups each of size 3. For making groups, we select six one-digit random numbers from 1 to 9

Table 5.6.1

Ponds	1	2	3	4	5	6	7	8	9
Area (x, in acre)	1	3	1.5	2.5	1.5	2	2	1.5	2
Yield (y, in kg)	200	500	250	600	300	600	700	250	300

without replacement. The selected random numbers are 6, 1, 9, 3, 5, and 4. So, we allocate units 6, 1, and 9 to group G_1; units 3, 5, and 4 for G_2, and the remaining units 2, 7, and 8 to the group G_3. Now from each of the groups, we need to select one unit by the PPSWR method. For the selection of the sample, we prepare Table 5.6.2:

Table 5.6.2

G_1			G_2			G_3		
Unit	x	Cumulative total × 10	Unit	x	Cumulative total × 10	Unit	x	Cumulative total × 10
6	2	20	3	1.5	15	2	3	30
1	1	30	5	1.5	30	7	2	50
9	2	50	4	2.5	55	8	1.5	65

For selection of a unit from group G_1, we select one two-digit random number from 01 to 50; for G_2, one two-digit random number from 01 to 55; and for group G_3, one two-digit random number from 01 to 65. The selected random numbers are 19, 42, and 56. The random numbers select unit 6 from G_1, 4 from G_2, and 8 from G_3.

So, the selected sample is $s = (i_1 = 6,\ i_2 = 4,\ i_3 = 8)$. The estimator for the population mean $\overline{Y}$ is $\widehat{\overline{Y}}_{rhc} = \dfrac{\hat{Y}_{rhc}}{N} = \dfrac{1}{N}\sum_{j=1}^{n}\dfrac{y_{i_j}}{p_{i_j}}P_j = \dfrac{1}{9}\left[\dfrac{600}{2}\times 5 + \dfrac{600}{2.5}\times\right.$

$$\left. 5.5 + \frac{250}{1.5}\times 6.5\right] = 3903.333/9 = 433.704 \text{ kg}$$

Estimated standard error of $\widehat{\overline{Y}}_{rhc} = SE\left(\widehat{\overline{Y}}_{rhc}\right) = \sqrt{\widehat{V}\left(\widehat{\overline{Y}}_{rhc}\right)}$

$$= \frac{1}{N}\sqrt{\frac{N-n}{N(n-1)}\sum_{j=1}^{n}P_j\left(\frac{y_{i_j}}{p_{i_j}} - \widehat{Y}_{rhc}\right)^2}$$

$$= \frac{1}{9}\sqrt{\frac{9-3}{9\times 2}\left\{\frac{5}{17}\left(\frac{600}{2}\times 17 - 3903.333\right)^2 + \frac{5.5}{17}\left(\frac{600}{2.5}\times 17 - 3903.333\right)^2 + \frac{6.5}{17}\left(\frac{250}{1.5}\times 17 - 3903.333\right)^2\right\}}$$

$= 59.8091$ kg

90% confidence interval for $\overline{Y}$ is

$$\widehat{\overline{Y}}_{rhc} \pm t_{.05,2}SE\left(\widehat{\overline{Y}}_{rhc}\right) = 433.703 \pm 2.92 \times 81.653$$

$$= (259.081 \text{ kg}, 608.325 \text{ kg}).$$

(ii) LMS method: Here the first unit should be selected with the probability proportional to the measure of size. For this selection, we prepare Table 5.6.3:

Table 5.6.3

Units	1	2	3	4	5	6	7	8	9
Measure of size (x)	1	3	1.5	2.5	1.5	2	2	1.5	2
Cumulative total (T) × 10	10	40	55	80	95	115	135	150	170

Here we select one three-digit random number from 001 to 170. The selected random number is 154 and the corresponding unit selected is 9. Now we need to select the remaining two units from 1 to 8 by SRSWOR. For this, we select two one-digit random numbers from 1 to 8. The selected random numbers are 5 and 2. So, the selected sample is $s = (2, 5, 9)$. The estimator for the population mean $\overline{Y}$ is

$$\widehat{\overline{Y}}_{lms} = \widehat{Y}_{lms}/N = \frac{y_s}{x_s}X/N = \left(\frac{500+300+300}{3+1.5+2}\times 17\right)\Big/9 = 2876.92/9$$
$$= 319.65 \text{ kg}$$

Estimated standard error of $\widehat{\overline{Y}}_{lms} = SE\left(\widehat{\overline{Y}}_{lms}\right) = \sqrt{\widehat{V}^{*}_{lms}}\Big/N.$
Now

$$\widehat{V}^{*}_{lms} = \left(\widehat{Y}_{lms}\right)^2 - \frac{X}{x_s}\left\{\sum_{i\in s} y_i^2 + \frac{(N-1)}{(n-1)}\sum_{i\neq}\sum_{j\in s} y_i y_j\right\}$$
$$= (2876.92)^2 - \frac{17}{6.5}(3550000)$$
$$= -1007929$$

Because the estimated variance $\widehat{V}^{*}_{lms} = -1007929$ is negative, we cannot find the confidence interval using $\widehat{V}^{*}_{lms}$ from this selected sample.

5.7 COMPARISON OF UNEQUAL (VARYING) PROBABILITY SAMPLING DESIGNS

VPS provides an efficient estimator for the population total Y if the measure of size variable x is well related to the study variable y. Various VPS designs are available in the literature. The most popular is the PPSWR sampling design. The HH estimator for the population total based on PPSWR sampling becomes efficient if the study variable is proportional to the measure of size variable. However, this estimator is inadmissible because it is based on the repetition of units in the samples. One can improve the HH estimator by employing the Rao-Blackwellization technique. The improved estimator, however, cannot be used in practice because of its complexity. In addition to this difficulty, it does not possess any simple expression of its unbiased variance estimator. Raj's estimator based on PPSWOR sampling is more efficient than the HH estimator based on PPSWR sampling of the same size. Murthy's estimator is obtained by applying Rao-Blackwellization techniques on Raj's estimator. Although Murthy's estimator is more efficient than Raj's estimator, it is seldom used in practice because the estimator itself and also its unbiased variance estimators are very tedious to compute. HTE based on an IPPS sampling design becomes very efficient if the study variable is proportional to the inclusion probability. The construction of an IPPS sampling design is not simple when sample size exceeds 2. The LMS—IPPS sampling scheme requires $p_i > (n-1)/\{n(N-1)\}$, which is seldom satisfied in practice. The PPS systematic sampling scheme is very easy to apply for any sample size, but the unbiased variance estimator is difficult to compute because of the complexity of the second-order inclusion probabilities. Sampford's (1967) IPPS sampling scheme is slightly more difficult to execute than the PPS systematic sampling scheme. The unbiased estimators of the variances of the estimators of the population total are equally complex. The variances of the HTE based on both the PPS systematic and Sampford sampling schemes are smaller than that of the HH estimator based on a PPSWR sampling scheme of the same sample size. The variance of the HTE based on Sampord's sampling is smaller than that of the PPS systematic sampling scheme. The LMS—PPAS sampling scheme is very easy to execute and possesses a smaller variance than that of the HH estimator of the same size, but it again suffers from the drawback that it does not always possess a nonnegative unbiased variance estimator. The RHC sample scheme is easy to execute and possesses elegant expressions for the population total, variance and nonnegative unbiased variance estimator, but it also suffers from the drawback of being inadmissible.

5.8 EXERCISES

5.8.1 Let a sample of size n be selected from a finite population of N units by the PPSWR method of sampling with p_i as the normed size measure for the ith unit. Find the expected number of distinct units in a sample and its variance. Obtain the expression of the Horvitz-Thompson estimator for the population total.

5.8.2 Show that the variance of the Hansen–Hurwitz estimator $\widehat{Y}_{hh}$ can be written as follows:

$$V\left(\widehat{Y}_{hh}\right) = \frac{1}{n}\sum_{i=1}^{N} p_i\left(\frac{y_i}{p_i} - Y\right)^2 = \frac{1}{2n}\sum_{i\neq}^{N}\sum_{j=1}^{N} p_i p_j\left(\frac{y_i}{p_i} - \frac{y_j}{p_j}\right)^2$$

5.8.3 Let $s = (i, j)$, $i < j$ be the set of distinct units in a sample of size 3 obtained by the PPSWR sampling method. List all the possible ordered samples s_o of size 3 that produces the unordered sample $s = (i, j)$. Obtain the expression of the unordered estimator that could be obtained by applying the Rao-Blackwellization technique on an HH estimator. Suggest an unbiased estimator of the variance of the unordered estimator.

5.8.4 (a) Let a sample size n be selected by the PPSWR sampling scheme. Derive the bias and mean square error of the sample mean, $\overline{y}_n$, based on n units with repetition, as an estimator for the population mean (Rao, 1966a,b).

(b) Let a sample of size n be selected by the RHC method with normed size measure p_i attached to the ith unit. Derive the expression of bias and mean square error of the estimator $T = N\sum_{i\in s} y_i P_i$ as an estimator of the population total Y where P_i is the sum of the p_j values for the group containing the ith unit (Rao, 1966a,b).

5.8.5 Let s be the unordered sample of size n obtained from ordered samples $s_1, \ldots, s_M$, each of size n and selected with probability $p(s_1), \ldots, p(s_M)$, respectively. Let $\widehat{\theta}(s_1), \ldots, \widehat{\theta}(s_M)$ be the unbiased estimators of θ based on $s_1, \ldots, s_M$, respectively. Then show that

(i) $\widehat{\theta} = \sum_{j=1}^{M} \widehat{\theta}(s_j)p(s_j)/p(s)$ is unbiased for θ where $p(s) = \sum_{j=1}^{M} p(s_j)$

(ii) $V\left(\widehat{\theta}\right) \leq V\left\{\widehat{\theta}(s_j)\right\}$

(iii) $V\left\{\widehat{\theta}(s_j)\right\} - V\left(\widehat{\theta}\right) = \sum_{s}\sum_{i}\left\{\widehat{\theta}(s_i) - \widehat{\theta}\right\}^2 p(s_i)$

(iv) An unbiased estimator of $V\left(\widehat{\theta}\right)$ is

$$\widehat{V}\left(\widehat{\theta}\right)=\widehat{\theta}^{2}-\left[\sum_{i}\left\{\widehat{\theta}\left(s_{j}\right)\right\}^{2}-\widehat{V}\left\{\widehat{\theta}\left(s_{j}\right)\right\}\right]p(s_j)/p(s)$$

where $\widehat{V}\left\{\widehat{\theta}\left(s_{j}\right)\right\}$ is an unbiased estimator for $V\left\{\widehat{\theta}\left(s_{j}\right)\right\}$ (Murthy, 1957).

5.8.6 Determine the expressions of the first two order inclusion probabilities of the PPSWOR sampling design of size $n = 2$. Derive the expressions of $\widehat{Y}_{ht}$ (Horvitz–Thompson estimator) and $\widehat{Y}_{RA}$ (the Raj's estimator) of the population total Y. Derive the expressions of variances of $\widehat{Y}_{RA}$ and $\widehat{Y}_{ht}$.

5.8.7 Let an ordered sample $s = (i,\ j)$ of size 2 be selected by the PPSWOR method.

(i) Show that the Raj's estimator and Murthy's estimator are

$$\widehat{Y}_{RA}=\frac{1}{2}\left\{(1+p_i)\frac{y_i}{p_i}+\left(1-p_i\right)\frac{y_j}{p_j}\right\} \text{ and}$$

$$\widehat{Y}_{M}=\frac{1}{2-p_i-p_j}\left\{\left(1-p_j\right)\frac{y_i}{p_i}+(1-p_i)\frac{y_j}{p_j}\right\}, \text{ respectively.}$$

(ii) Unbiased estimators of the variances of $\widehat{Y}_{RA}$ and $\widehat{Y}_{M}$ are, respectively $\widehat{V}\left(\widehat{Y}_{RA}\right)=\frac{1}{4}(1-p_i)^2\left(\frac{y_i}{p_i}-\frac{y_j}{p_j}\right)^2$ and $\widehat{V}\left(\widehat{Y}_{M}\right)=$

$$\frac{(1-p_i)\left(1-p_j\right)\left(1-p_i-p_j\right)}{\left(2-p_i-p_j\right)^2}\left(\frac{y_i}{p_i}-\frac{y_j}{p_j}\right)^2.$$

5.8.8 (a) Describe the IPPS sampling scheme and explain where it is useful.

(b) Consider selecting a sample of size 2 with replacement where the probability of selection of the ith unit at the first draw is $p_i(1) = p_i$ and at the second draw is $p_i(2) \propto p_i/(1 - 2p_i)$ for $i = 1,\ldots, N$. If the selected units are distinct, they are accepted as a sample. If they are the same, the entire selection is discarded and two units are selected again with the earlier method. The process is continued until two distinct units are selected. Show that for this sampling scheme (i) $\pi_i = 2p_i$ and (ii) $\pi_i\pi_j - \pi_{ij} \geq 0$ (Rao, 1965).

5.8.9 Let a sample s of size 2 be selected from a population U of size N following a without replacement scheme $d(2)$ of size 2 whose first- and second-order inclusion probabilities $\pi_i(2)$ and $\pi_{ij}(2)$ satisfy $\pi_i(2)\pi_j(2) - \pi_{ij}(2) \geq 0$ for $\forall i \neq j$. Let $d(n)$ be a sampling design of size n whose first two units are selected using $d(2)$ and the remaining $n-2$ units are selected by SRSWOR from the remaining $N-2$ units. Express the first- and second-order inclusion probabilities $\pi_i(n)$ and $\pi_{ij}(n)$ of $d(n)$ in terms of $\pi_i(2)$ and $\pi_{ij}(2)$. Show that $\pi_i(n)\pi_j(n) - \pi_{ij}(n) \geq 0$ (Seth, 1966).

5.8.10 (a) Describe the PPAS sampling scheme. Derive the expression of π_i and π_{ij}'s and show that $\pi_i\pi_j - \pi_{ij} \geq 0$ for $\forall i \neq j$.

(b) Derive the expression of bias and mean square error of the sample mean $\overline{y}(s) = \sum_{i \in s} y_i/n$ as an estimator of the population total mean $\overline{Y}$ when the sample is selected by the PPAS sampling scheme (Rao, 1966).

5.8.11 Table 5.8.1 gives the floor area and rent of 20 flats in Durban. Select a sample of five flats by probability proportional to area of the flat with replacement using (i) the cumulative total method and (ii) Lahiri's method. Estimate the average rent of the flat and its standard error from each of the selected samples.

Table 5.8.1

Serial no. of flat	1	2	3	4	5	6	7	8	9	10
Area (in m^2)	50	70	90	50	60	80	75	40	80	70
Rent (in \$)	3000	4000	5000	2500	3000	4500	4000	2500	5000	4000

Serial no. of the flat	11	12	13	14	15	16	17	18	19	20
Area (in m^2)	60	40	80	50	60	80	75	50	75	80
Rent (in \$)	3000	3000	4000	2000	3500	4000	4500	2000	5500	6000

5.8.12 Consider the data given in Exercise 5.8.11. Select a sample of size two flats by the PPSWOR method using the area of the flat as the measure of size variable. Estimate the average rent by using (i) Raj's estimator, (ii) Murthy's estimator, and (iii) Horvitz–Thompson estimator. Compute standard errors of each of the estimators and comment.

5.8.13 Table 5.8.2 gives the number of cows and production of milk of 10 farms.

Table 5.8.2

Serial no. of farm	1	2	3	4	5	6	7	8	9	10
No. of cows	10	20	8	10	15	20	15	5	25	12
Milk (in L)	100	240	40	150	200	250	175	75	300	130

Select a sample of size 5 by the PPSWOR method using the number of cows as the measure of size variable. Estimate the average yield of milk per farm using (a) Raj's and (b) Murthy's estimator. Estimate 95% confidence intervals for the average yield milk per farm based on the above estimators.

5.8.14 Consider the data given in Exercise 5.8.13. Select a sample of size 2 farms by using (i) Brewer, (ii) Durbin, and (iii) Hanurav's IPPS sampling design, using the number of cows as a measure of size variable. Estimate the average production of milk per farm based on the selected sample. Compare the efficiencies of your estimators.

5.8.15 Consider the data given in Exercise 5.8.13. Select a sample of five farms using the PPS systematic sampling procedure using the number of cows as the size variable. Estimate the average production of milk per farm and estimate its variance.

5.8.16 The weekly wages and expenditure on beer of 15 people are given in Table 5.8.3.

Table 5.8.3

Persons	1	2	3	4	5	6	7	8	9	10
Weekly wages in $	200	250	300	400	450	500	450	400	300	250
Expenditure on beer in $	50	20	30	40	25	40	30	30	25	30

Persons	11	12	13	14	15
Weekly wages in $	200	250	300	400	450
Expenditure on beer in $	50	20	30	40	25

Select samples of four people by the (i) PPAS and (ii) RHC methods using weekly wages as the size variable. Estimate average expenditure on beer and 90% confidence of the average expenditure from each of the samples separately.

CHAPTER 6

Inference Under Superpopulation Model

6.1 INTRODUCTION

In the previous chapters, we considered a fixed population setup, where the vector of variable under study $\mathbf{y} = (y_1,\ldots, y_N)$ is a fixed point in the N-dimensional Euclidean space R^N, and y_i is the unknown value of the study variable y associated with the unit $i(1,\ldots, N)$. Our problem was to find point estimates and confidence intervals of certain real-valued parametric functions of $\mathbf{y}$. Unlike the previous sections, in this section, we assume that the population vector $\mathbf{y}$ is a realization of a random vector $\mathbf{Y} = (Y_1,\ldots, Y_N)$ and its distribution is denoted by ξ. Here the population mean $\overline{Y} = \sum_{i=1}^{N} Y_i/N$ is a random variable. For example, if we consider a fixed set of households in a certain locality as a finite population, and the annual income of households as a study variable, then the ith household income Y_i is a random variable, which changes with time. The change of values of $\mathbf{Y}$ overtime may be characterized by a certain probability distribution ξ. The distribution ξ may be indexed by some parameter θ, which is generally unknown, but it may belong to a certain known parameter space Ω_θ. Such a distribution ξ is called a superpopulation model or simply a model. In most situations, the distribution ξ is related to an auxiliary variable $\mathbf{x} = (x_1,\ldots, x_N)$, whose values are fixed, assumed to be known and positive. We will denote E_ξ, V_ξ and C_ξ as the operators for expectation, variance, and covariance with respect to the model ξ. Our objective is to estimate some parametric function $\phi(\mathbf{y})$, where $\mathbf{y}$ is a particular realization of the random vector $\mathbf{Y} = (Y_1,\ldots, Y_i,\ldots, Y_N)$. For notational simplicity, we write $\mathbf{y}$ as a random vector $\mathbf{Y}$ and y_i for Y_i; $i = 1,\ldots, N$.

In this section we will discuss the concepts of design-based, model-based (dependent), and model design—based inference. Optimal sampling strategies (a combination of a sampling design and an estimator) in the presence of various superpopulation models will be established. The optimal strategy based on a model-based inference is highly dependent on the choice of the

Survey Sampling Theory and Applications
ISBN 978-0-12-811848-1
http://dx.doi.org/10.1016/B978-0-12-811848-1.00006-6

model and the strategy becomes inefficient if the chosen model is inappropriate. The concept of robustness of sampling designs and estimators has been discussed to guard against the use of inappropriate models. The concepts of Bayesian inferences will be introduced, and it is shown that the sampling design does not play any role in making inference about the parametric function $\phi(\mathbf{y})$. Finally, various sampling strategies, which are commonly used in practice, are compared under different superpopulation models.

6.2 DEFINITIONS

6.2.1 Sampling Strategy

A sampling strategy is a combination of sampling design p and an estimator t based on a sample selected using the design p. Let $\mathcal{P}$ be a class of sampling designs and C be a class of estimators. Then a sampling strategy is denoted by $h = (p,\ t)$ with $p \in \mathcal{P}, t \in C$.

6.2.2 Noninformative Sampling Design

A sampling design is said to be noninformative if the selection of a sample does not depend on the study variable y_i's, i.e., the sampling design is nonsequential. For a noninformative sampling design, the operators E_ξ and E_p are commutative in the sense $E_\xi E_p(\cdot) = E_p E_\xi(\cdot)$. In our present discussion, we will consider the noninformative sampling design only.

6.2.3 Design-Unbiased (or *p*-Unbiased) Estimator

An estimator t is said to be design-unbiased or p-unbiased or simply unbiased for the total Y if and only if $E_p(t) = Y$ for all $\mathbf{y} \in R^N$. The class of p-unbiased estimators (including linear and nonlinear) will be denoted by C_{pu}.

6.2.4 Model-Unbiased (or ξ-Unbiased) Estimator

An estimator t is said to be a model unbiased or ξ-unbiased estimator for Y if and only if $E_\xi(t) = E_\xi(Y)\ \ \forall \theta \in \Omega_\theta$. The class of ξ-unbiased estimators will be defined by $C_{\xi u}$.

6.2.5 Model Design—Unbiased (or *p*ξ-Unbiased) Estimator

An estimator is said to be a model design—unbiased or $p\xi$-unbiased estimator for the population total Y, if and only if $E_\xi E_p(t) = E_p E_\xi(t) = E_\xi(Y)\ \ \forall \theta \in \Omega_\theta$. The class of $p\xi$-unbiased estimators will be denoted by $C_{p\xi}$. If an estimator is p-unbiased or ξ-unbiased, then it is necessarily

$p\xi$-unbiased, but the converse is not true. So, the class $C_{p\xi}$ contains both the classes C_{pu} and $C_{\xi u}$. An estimator can be p-unbiased but may not be ξ-unbiased and vice versa.

Example 6.2.1

The Horvitz–Thompson estimator (HTE), $\widehat{Y}_{ht} = \sum_{i \in s} \frac{y_i}{\pi_i}$, is p-unbiased for Y but not ξ-unbiased for Y in general. But if π_i is proportional to $E_\xi(y_i)$ and p is a fixed effective sample of size n design (FESD(n)), then $\widehat{Y}_{ht}$ becomes ξ-unbiased for Y. The sample mean based on an simple random sampling without replacement (SRSWOR) method is p-unbiased but not ξ-unbiased for the population mean $\overline{Y}$ in general. But the HTE, based on any sampling design, is $p\xi$-unbiased for the total Y. The ratio estimator $\widehat{Y}_R = \frac{\overline{y}_s}{\overline{x}_s} X$ with $\overline{y}_s = \sum_{i \in s} y_i/n$, $\overline{x}_s = \sum_{i \in s} x_i/n$ and $X = \sum_{i \in U} x_i$ is ξ-unbiased for Y, when $E_\xi(y_i) \propto x_i$. The ratio estimator is, however, not p-unbiased in general.

6.2.6 Design-Based Inference

In a design-based inference, which is also called the probability sampling approach, we assume that the population vector $\mathbf{y}$ is fixed. From the population, a sample s of n units is selected by following a sampling design p, a man-made randomization procedure. The information concerning the study variable y is obtained from units selected in the sample s and a link between the observed values of y_i's, $i \in s$ and unobserved values y_i's, $i \notin s$ is established through the sampling design. Here, expectation of an estimator is the average value over all possible hypothetical samples.

6.2.7 Model-Based Inference

In a model-based inference, a sample is selected by using a sampling design p and the link between the observed y_i's, $i \in s$, and unobserved y_i's and $i \notin s$ is obtained through a superpopulation model ξ. The assumed model distribution ξ provides valid inferences with respect to the particular selected sample, irrespective of the chosen sampling design.

6.2.8 Model-Assisted Inference

The model-assisted or model-dependent inference is a hybrid of design-based and model-based inferences. Here, inference is based on the chosen sampling design p and a superpopulation model ξ. Model-assisted inference provides valid inferences under the assumed superpopulation model and also protects against model misspecification by providing a valid repeated sampling inference (Rao, 1994).

6.2.9 Optimal Estimator

An estimator $t^0 (\in C)$ is said to be an optimal estimator (or simply optimal) in a certain class of estimators C for estimating Y under a given superpopulation model ξ and a sampling design p if

$$E_\xi E_p (t^0 - Y)^2 \leq E_\xi E_p (t - Y)^2 \;\; \forall t (\neq t^0) \in C, \quad \theta \in \Omega_\theta \tag{6.2.1}$$

and the inequality in Eq. (6.2.1) is strict at least for one $\theta \in \Omega_\theta$.

6.2.10 Optimal Strategy

Let $\mathcal{H}$ be a class of strategies. Then, the strategy $h_1 = (p_1, t_1)$ is said to be better than $h_2 = (p_2, t_2)$ for estimating Y if

$$E_\xi E_{p_1} (t_1 - Y)^2 \leq E_\xi E_{p_2} (t_2 - Y)^2, \quad \forall \theta \in \Omega_\theta$$

and

$$E_\xi E_{p_1} (t_1 - Y)^2 < E_\xi E_{p_2} (t_2 - Y)^2 \text{ for at least one value of } \theta \in \Omega_\theta.$$

The strategy $h^0 = (p^0, t^0)$ $(\in \mathcal{H})$ is said to be optimal in the class of strategies $\mathcal{H}$ if

$$E_\xi E_{p_0} (t^0 - Y)^2 \leq E_\xi E_p (t - Y)^2 \;\; \forall h (\neq h^0) \in \mathcal{H}, \quad \theta \in \Omega_\theta \tag{6.2.2}$$

and the inequality in Eq. (6.2.2) is strict for some $\theta \in \Omega_\theta$.

6.3 MODEL-ASSISTED INFERENCE

In this section we will establish optimal strategies in the class of design-unbiased and model design—unbiased estimators. Here we will call an estimator t a predictor when it is used to estimate a random variable. In the presence of a superpopulation model ξ the population total, Y, is a random variable and hence an estimator, t, of Y will be called a predictor of Y.

6.3.1 Optimal Design-Unbiased Predictors

6.3.1.1 Product Measure Model

Model $M1$: y_i's are independently distributed with mean, $E_\xi(y_i) = \mu_i (-\infty < \mu_i < \infty)$ and variance $V_\xi(y_i) = \sigma_i^2$ for $i = 1, \ldots, N$. Here μ_i's are assumed to be known and σ_i's are positive but unknown. Under the model $M1$, we have the following theorem (Godambe and Joshi, 1965).

Theorem 6.3.1

Under the model $M1$, for a given noninformative sampling design p with inclusion probability $\pi_i > 0 \ \forall i = 1,\ldots, N$, the expected variance of a predictor t belonging to the class of p-unbiased predictors C_{pu} of the population total of Y satisfies the following inequality.

$$E_\xi[V_p(t)] \geq \sum_{i=1}^{N} \sigma_i^2\left(\frac{1}{\pi_i} - 1\right) \quad \forall t \in C_{pu} \tag{6.3.1}$$

Proof

$$E_\xi V_p(t) = E_\xi[E_p(t^2) - Y^2]$$

$$= E_p E_\xi(t^2) - V_\xi(Y) - [E_\xi(Y)]^2$$

(since for a noninformative p, $E_\xi E_p(t) = E_p E_\xi(t)$)

$$= E_p[V_\xi(t)] + E_p[E_\xi(t)]^2 - V_\xi(Y) - [E_\xi(Y)]^2$$

$$= E_p[V_\xi(t)] + E_p[E_\xi(t) - E_\xi(Y)]^2 - V_\xi(Y) \tag{6.3.2}$$

(noting $E_p E_\xi(t) = E_\xi E_p(t) = E_\xi(Y)$)

Let $t(s, y)$ be the value of the predictor t based on the sample s, selected with a probability $p(s)$.

Then we can write

$$t(s, y) = \widehat{Y}_{ht} + h(s, y) \tag{6.3.3}$$

where $\widehat{Y}_{ht} = \sum_{i=1}^{N} \frac{y_i}{\pi_i} I_{si}$, is the Horvitz–Thompson (HT) predictor, $I_{si} = 1$ if $i \in s$ and $I_{si} = 0$ if $i \notin s$, and $h(s, y)$ is a function of y_i's for $i \in s$ only.

Since $t(s, y)$ is p-unbiased of Y, we must have

$$E_p[t(s, y)] = \sum_s t(s, y)p(s)$$

$$= E_p\left(\widehat{Y}_{ht}\right) + \sum_s h(s, y)p(s)$$

$$= Y$$

i.e., $\sum_s h(s, y)p(s) = 0$

Let $\sum_{si}$ be the sum over those samples s that do not contain the unit i.

Then,

$$\sum_s h(s, y)p(s) = \sum_{s \supset i} h(s, y)p(s) + \sum_{si} h(s, y)p(s) = 0$$

implies

$$\sum_{s \supset i} h(s, y)p(s) = -\sum_{si} h(s, y)p(s) = 0 \tag{6.3.4}$$

Furthermore,

$$E_p V_\xi(t) = \sum_s \left[V_\xi\left(\widehat{Y}_{ht}\right) + V_\xi\{h(s, y)\} + 2C_\xi\left\{\widehat{Y}_{ht}(s), h(s, y)\right\}\right] p(s) \tag{6.3.5}$$

Now

$$\sum_s V_\xi\left(\widehat{Y}_{ht}\right)p(s) = \sum_s \left(\sum_{i=1}^{N} \frac{\sigma_i^2}{\pi_i^2} I_{si} p(s) \right) = \sum_{i=1}^{N} \frac{\sigma_i^2}{\pi_i} \tag{6.3.6}$$

and

$$\begin{aligned}
\sum_s C_\xi\left\{\widehat{Y}_{ht}, h(s, y)\right\}p(s) &= \sum_s C_\xi\left\{\widehat{Y}_{ht}, h(s, y)\right\}p(s) \\
&= \sum_s p(s) E_\xi\left\{ \sum_{i=1}^{N} I_{si} \frac{(y_i - \mu_i)}{\pi_i} h(s, y) \right\} \\
&= \sum_{i=1}^{N} E_\xi\left\{ \frac{(y_i - \mu_i)}{\pi_i} \sum_s I_{si} h(s, y) p(s) \right\} \\
&= \sum_{i=1}^{N} E_\xi\left\{ \frac{(y_i - \mu_i)}{\pi_i} \sum_{s \supset i} h(s, y) p(s) \right\} \\
&= -\sum_{i=1}^{N} E_\xi\left\{ \frac{(y_i - \mu_i)}{\pi_i} \sum_{si} h(s, y) p(s) \right\} \\
&= -\sum_{i=1}^{N} E_\xi\left\{ \frac{(y_i - \mu_i)}{\pi_i} E_\xi \sum_{si} h(s, y) p(s) \right\} \\
&= 0
\end{aligned} \tag{6.3.7}$$

$$\left(\text{since } \sum_{si} h(s, y)p(s) \text{ is independent of } y_i \text{ for } i \in s \right)$$

Eqs. (6.3.2) and (6.3.5)−(6.3.7) yield the required result

$$E_\xi[V_p(t)] = \sum_{i=1}^{N} \frac{\sigma_i^2}{\pi_i} + E_p[V_\xi\{h(s,y)\}] + E_p[E_\xi(t) - E_\xi(Y)]^2 - V_\xi(Y)$$

$$\geq \sum_{i=1}^{N} \sigma_i^2 \left(\frac{1}{\pi_i} - 1\right)$$

We note that $E_\xi V_p(t)$ attains the lower bound Eq. (6.3.1) when

$$\text{(i) } E_p[V_\xi\{h(s,y)\}] = 0 \text{ and (ii) } E_\xi(t) - E_\xi(Y) = 0 \tag{6.3.8}$$

The condition (ii) of Eq. (6.3.8) implies

$$E_\xi\{h(s,y)\} = \sum_{i=1}^{N} \mu_i - \sum_{i \in s} \frac{\mu_i}{\pi_i}$$

The condition (i) of Eq. (6.3.8) implies

$$h(s,y) = E_\xi\{h(s,y)\} = \sum_{i=1}^{N} \mu_i - \sum_{i \in s} \frac{\mu_i}{\pi_i}$$

Hence the optimum predictor attaining the lower bound Eq. (6.3.1) is given by

$$t = t_0 = \sum_{i=1}^{N} \frac{y_i - \mu_i}{\pi_i} I_{si} + \sum_{i=1}^{N} \mu_i \tag{6.3.9}$$

The predictor t_0 is known as the generalized difference predictor, which is usable when μ_i's are known for every $i = 1,\ldots, N$. Hence we have the following theorem (vide Chaudhuri and Stenger, 1992).

Theorem 6.3.2

Under the model $M1$, with $\pi_i > 0\ \forall i = 1,\ldots, N$

$$E_\xi[V_p(t)] \geq \sum_{i=1}^{N} \sigma_i^2 \left(\frac{1}{\pi_i} - 1\right) = E_\xi V_p(t_0) \quad \forall t \in C_{pu}$$

Remark 6.3.1

Theorem 6.3.2 was derived by Godambe (1955), when t belongs to C_{lhp}, the class of linear homogeneous unbiased predictors of Y, which is a subclass of unbiased predictors C_{pu}.

Minimizing $\sum_{i=1}^{N}\sigma_i^2\left(\frac{1}{\pi_i}-1\right)$, the right hand side of Eq. (6.3.1) with respect to π_i keeping $\sum_{i=1}^{n}\pi_i = n$, the optimum value of π_i is obtained as $\pi_i = \pi_{i0} = np_i(\sigma)$, where $p_i(\sigma) = \sigma_i / \sum_{i=1}^{N}\sigma_i$. The minimum value of $\sum_{i=1}^{N}\sigma_i^2\left(\frac{1}{\pi_i}-1\right)$ with $\pi_i = \pi_{i0} = np_i(\sigma)$ is obtained as

$$\frac{\left(\sum_{i=1}^{N}\sigma_i\right)^2}{n} - \sum_{i=1}^{N}\sigma_i^2 \tag{6.3.10}$$

Let $p(\pi, \sigma, \mu)$ be an FESD(n) with (i) $\pi_i = np_i(\sigma)$ and (ii) $\sum_{i=1}^{N}\frac{\mu_i}{np_i(\sigma)}I_{si} = \sum_{i=1}^{N}\mu_i$ for every s with $p(s) > 0$. Then the expected variance of the HT predictor $\widehat{Y}_{ht}$ based on the sampling design $p(\pi, \sigma, \mu)$ attains the lower bound Eq. (6.3.10). The findings of Godambe and Joshi (1965) aforementioned have been summarized in the following theorem:

Theorem 6.3.3

Let $\mathcal{P}_n$ be the class of FESD(n). Then under the model $M1$, the strategy $\left(p(\pi,\sigma,\mu), \widehat{Y}_{ht}\right)$ is the optimal in the class of strategies $\mathcal{H} = (p, t)$ with $p \in \mathcal{P}_n, t \in C_{pu}$, i.e.,

$$E_\xi[V_p(t)] \geq \frac{\left(\sum_{i=1}^{N}\sigma_i\right)^2}{n} - \sum_{i=1}^{N}\sigma_i^2 = E_\xi V_{p(\pi,\sigma,\mu)}\left(\widehat{Y}_{ht}\right) \quad \forall p \in \mathcal{P}_n, t \in C_{pu}$$

Remark 6.3.2

Hájek (1959) derived Theorem 6.3.3, when t belongs to C_{lhp}, the class of linear homogeneous p-unbiased predictors.

Consider the model $M1$ with $\mu_i = \beta x_i$, where β is an unknown real number and x_i is the value of the auxiliary variable x for the ith unit, which is known and positive for every $i = 1,\ldots, N$. Let $p(\pi, x)$ be an FESD(n) with $\pi_i = np_i$, $p_i = x_i/X$, and $X = \sum_{i=1}^{N}x_i$. The design $p(\pi, x)$ is termed an inclusion probability proportional to size (IPPS) design in Chapter 5. For a $p(\pi, x)$ design, t_0 reduces to $\widehat{Y}_{ht} = \sum_{i=1}^{N}\frac{y_i}{np_i}I_{si}$ and we have the following theorem.

Theorem 6.3.4

Under the model $M1$ with $\mu_i = \beta x_i$

$$E_\xi\left[V_{p(\pi,x)}(t)\right] \geq \sum_{i=1}^{N} \sigma_i^2\left(\frac{1}{np_i} - 1\right) = E_\xi V_{p(\pi,x)}\left(\widehat{Y}_{ht}\right) \quad \forall t \in C_{pu}$$

Theorem 6.3.4 indicates that the strategy $h^* = \left(p(\pi, x), \widehat{Y}_{ht}\right)$ is optimal in the class of strategies $\mathcal{H}^* = (p(\pi, x), t), \;\; t \in C_{pu}$.

Furthermore, for the model $M1$ with $\mu_i = \beta x_i$ and $\sigma_i^2 = \sigma^2 x_i^2 (\sigma > 0)$, $p(\pi, \sigma, \mu)$ reduces to a $p(\pi, x)$ design and we get the following result.

Theorem 6.3.5

Under the model $M1$, with $\mu_i = \beta x_i$ and $\sigma_i^2 = \sigma^2 x_i^2$, $h^* = \left(p(\pi, x), \widehat{Y}_{ht}\right)$ is the optimal in the class of strategies $\mathcal{H}_1 = (p, t)$ with $p \in \mathcal{P}_n, t \in C_{pu}$, i.e.,

$$E_\xi[V_p(t)] \geq \sigma^2\left(\frac{X^2}{n} - \sum_{i=1}^{N} x_i^2\right) = E_\xi\left[V_{p(\pi,x)}\left(\widehat{Y}_{ht}\right)\right] \quad \forall p \in \mathcal{P}_n, t \in C_{pu}$$

Finally, when $x_i = 1$ for $i = 1,\ldots, N$, we get $\mu_i = \beta, \sigma_i^2 = \sigma^2$ and $p(\pi, x)$ reduces to a sampling design p_0, whose inclusion probability is $\pi_i = n/N$ for $\forall i = 1,\ldots, N$. The sampling design p_0 includes SRSWOR. In this case, $\widehat{Y}_{ht}$ reduces to the expansion predictor $N\overline{y}_s$. Hence, we have the following result as a corollary of the Theorem 6.3.5 above.

Theorem 6.3.6

Under the model $M1$, with $\mu_i = \beta$ and $\sigma_i^2 = \sigma^2$, $h_0 = \left(p_0, N\overline{y}_s\right)$ is the optimal in the class of strategies $\mathcal{H}_1 = (p, t)$ with $p \in \mathcal{P}_n, t \in C_{pu}$, i.e.,

$$E_\xi[V_p(t)] \geq \sigma^2 N\left(\frac{N}{n} - 1\right) = N^2 E_\xi\left[V_{p_0}\left(\overline{y}_s\right)\right] \quad \forall p \in \mathcal{P}_n, t \in C_p$$

Remark 6.3.3

Rao T.J. (1971) considered the model $M1g$, which is the model $M1$ with $\mu_i = \beta x_i$ and $\sigma_i^2 = \sigma^2 x_i^g$, where g is positive. He defined a $g - \pi ps$ sampling design, where (i) $\pi_i = \pi_{i0} = \nu_0\, x_i^{g/2} \Big/ \sum_{i=1}^{N} x_i^{g/2}$, (ii) $\frac{1}{\nu_0}\left(\sum_{i \in s} \frac{x_i}{x_i^{g/2}}\right)\left(\sum_{i \in U} x_i^{g/2}\right) = X$, and ν_0 is the expected effective sample size of the design. Rao T.J. (1971) showed that, under the model $M1g$, the HT predictor $\widehat{Y}_{ht} = \sum_{i=1}^{N} \frac{y_i}{\pi_{i0}} I_{si}$ with $g - \pi ps$ design is the optimal in the class of strategies $\mathcal{H}_g = (p, t)$ with $p \in \mathcal{P}_{\nu_0}, t \in C_{pu}$, where $\mathcal{P}_{\nu_0}$ is the class of sampling designs with ν_0 as the expected effective sample size.

6.3.1.2 Equicorrelation Model

The equicorrelation model is defined as follows:

$$\begin{aligned}&\text{Model } M2: E_\xi(y_i) = \mu_i(-\propto < \mu_i < \propto),\ V_\xi(y_i) = \sigma_i^2(\sigma_i > 0) \text{ and}\\ &\quad C_\xi(y_i, y_j) = \rho\sigma_i\sigma_j(-1/(N-1) < \rho < 1)\\ &\text{for } i \neq j; i, j = 1, \ldots, N.\end{aligned} \tag{6.3.11}$$

Let C_{lp} be the class of linear p-unbiased predictors of Y consisting of the predictors of the form

$$t = a_s + \sum_{i \in s} b_{si} y_i$$

where a_s and b_{si}'s are constants free from y_i's and satisfy the following p-unbiasedness conditions:

$$\text{(i)} \sum_s a_s p(s) = 0 \quad \text{and} \quad \text{(ii)} \sum_{s \supset i} b_{si} p(s) = 1 \ \text{ for } i = 1, \ldots, N \tag{6.3.12}$$

From Eq. (6.3.2), it follows

$$\begin{aligned}E_\xi[V_p(t)] &= E_p[V_\xi(t)] + E_p[E_\xi(t) - E_\xi(Y)]^2 - V_\xi(Y)\\ &\geq E_p[V_\xi(t)] - V_\xi(Y)\end{aligned} \tag{6.3.13}$$

Now,

$$\begin{aligned}E_p[V_\xi(t)] &= E_p\left[\sum_{i \in s} b_{si}^2\sigma_i^2 + \rho \sum_{i \neq} \sum_{j \in s} b_{si} b_{sj} \sigma_i \sigma_j\right]\\ &= \sum_s p(s) \sum_{i \in s} (b_{si}\sigma_i)^2 - (1-\rho)A\end{aligned} \tag{6.3.14}$$

where $A = \sum_s p(s) \left\{ \left(\sum_{i \in s} b_{si}\sigma_i \right)^2 - \sum_{i \in s} b_{si}^2 \sigma_i^2 \right\}$.

Let us maximize A subject to the following condition:

$$\sum_{i=1}^{N} \sigma_i \sum_{s \supset i} b_{si} p(s) = \sum_{i=1}^{N} \sigma_i \tag{6.3.15}$$

Here we note that condition (ii) of Eq. (6.3.12) implies Eq. (6.3.15), but the converse is not true. To maximize A subject to Eq. (6.3.15), consider the following ϕ with λ as a Lagrange multiplier

$$\phi = \sum_{s} p(s) \left\{ \left(\sum_{i \in s} b_{si} \sigma_i \right)^2 - \sum_{i \in s} b_{si}^2 \sigma_i^2 \right\} - 2\lambda \left\{ \sum_{i=1}^{N} \sigma_i \sum_{s \supset i} b_{si} p(s) - \sum_{i=1}^{N} \sigma_i \right\}$$

$\dfrac{\partial \phi}{\partial b_{si}} = 0$ implies

$$\sum_{i \in s} \sigma_i b_{si} - \sigma_i b_{si} = \lambda \tag{6.3.16}$$

Summing Eq. (6.3.16) over $i \in s$ and assuming p is a fixed effective sample size n design, i.e., $p \in \mathcal{P}_n$, we get

$$\sum_{i \in s} \sigma_i b_{si} = n\lambda/(n-1) \tag{6.3.17}$$

Multiplying both sides of Eq. (6.3.17) by $p(s)$ and then summing over all possible samples s yields

$$\sum_{s} p(s) \left(\sum_{i \in s} \sigma_i b_{si} \right) = n\lambda/(n-1)$$

$$\text{i.e.,} \quad \lambda = \frac{(n-1)}{n} \sum_{i=1}^{N} \sigma_i \tag{6.3.18}$$

Eqs. (6.3.16)–(6.3.18) yield the optimum value of b_{si}'s, which maximize A subject to Eq. (6.3.15) as

$$b_{si} = b_{si0} = \frac{1}{np_i(\sigma)} \quad \text{with} \quad p_i(\sigma) = \sigma_i \Big/ \sum_{i=1}^{N} \sigma_i \tag{6.3.19}$$

The maximum value of A with $b_{si} = b_{si0} = \dfrac{1}{np_i(\sigma)}$ comes out as

$$A_0 = \frac{n-1}{n} \left(\sum_{i=1}^{N} \sigma_i \right)^2 \tag{6.3.20}$$

Eqs. (6.3.14) and (6.3.20) yield

$$\begin{aligned} E_p[V_\xi(t)] &\geq \sum_s p(s)\left(\sum_{i\in s}\sigma_i b_{si}\right)^2 - (1-\rho)A_0 \\ &\geq \left\{\sum_s p(s)\sum_{i\in s} b_{si}\sigma_i\right\}^2 \Big/ \sum_s p(s) - (1-\rho)A_0 \\ &= \left(\sum_{i=1}^N \sigma_i\right)^2 - (1-\rho)A_0 \end{aligned} \tag{6.3.21}$$

The aforementioned derivations can be stated in the following theorem as follows.

Theorem 6.3.7

Under model $M2$, $E_\xi[V_p(t)] \geq (1-\rho)\left[\dfrac{\left(\sum_{i=1}^N \sigma_i\right)^2}{n} - \sum_{i=1}^N \sigma_i^2\right]$ for $\forall p \in \mathcal{P}_n$, $t \in C_{lp}$

Equality in Theorem 6.3.7 holds if (i) $A = A_0$ and (ii) equality in Eq. (6.3.13) holds, i.e., $b_{si} = b_{si0} = \dfrac{1}{np_i(\sigma)}$ and $a_s = a_{s0} = \sum_{i=1}^N \mu_i - \sum_{i\in s}\dfrac{\mu_i}{np_i(\sigma)}$.

Hence the equality in Theorem 6.3.7 holds if

$$t = t^* = \sum_{i\in s}\frac{y_i - \mu_i}{np_i(\sigma)} + \sum_{i=1}^N \mu_i \tag{6.3.22}$$

Furthermore, the predictor t^* becomes design-unbiased if and only if the inclusion probability of the ith unit $\pi_i = np_i(\sigma)$ $\forall i = 1,\ldots, N$. We will call an FESD(n) with inclusion probability $\pi_i = np_i(\sigma)$ for $\forall i = 1,\ldots, N$ as $p(\pi, \sigma)$ design. Hence we have the following theorem.

Theorem 6.3.8

Under the model $M2$, the strategy $h_1^* = \left(t^*, p(\pi,\sigma)\right)$ is the optimum in the class of strategies $\mathcal{H}_2^* = (t, p)$, $t \in C_{lp}$, $p \in \mathcal{P}_n$ in the sense that

$$\begin{aligned} E_\xi[V_p(t)] &\geq (1-\rho)\left[\frac{\left(\sum_{i=1}^N \sigma_i\right)^2}{n} - \sum_{i=1}^N \sigma_i^2\right] \\ &= E_\xi\left[V_{p(\pi,\sigma)}\left(t^*\right)\right] \quad \forall t \in C_{lp}, p \in \mathcal{P}_n \end{aligned}$$

where $t^* = \sum_{i\in s}\dfrac{y_i - \mu_i}{np_i(\sigma)} + \sum_{i=1}^N \mu_i$.

Remark 6.3.4

The optimal predictor t^* given in Theorems 6.3.7 and 6.3.8 cannot be used in practice unless μ_i and σ_i's are known. Arnab (1986) showed that for any sampling design p (not necessarily an FESD(n)), the following inequality holds when t belongs to C_{llp}.

$$E_\xi[V_p(t)] \geq (1-\rho)\left[\frac{\left(\sum_{i=1}^{N}\sigma_i\right)^2}{n} - \sum_{i=1}^{N}\sigma_i^2\right] \quad \forall t \in C_{llp}$$

6.3.1.3 Transformation Model

We define model $M3$ as the model $M2$ with $\mu_i = a_i + \beta x_i$ and $\sigma_i^2 = \sigma^2 x_i^2$, where a_i's are known constants, x_i's are known auxiliary variables, β, $\sigma(>0)$, and ρ are unknown. The model $M3$ is called the transformation model because the transformed variables $z_i = \frac{y_i - a_i}{x_i}$'s have the same means, variances, and covariances. Under the model $M3$, t^* reduces to $t_1 = \sum_{i\in s}\frac{y_i - a_i}{nx_i}X + \sum_{i=1}^{N} a_i$ and hence we get the optimum strategy under model $M3$ using Theorem 6.3.8 as follows.

Theorem 6.3.9

Under the model $M3$, the strategy $h_1 = (t_1, p(\pi, x))$ is the optimum in the class of strategies $\mathcal{H}_1^* = (t, p),\ t \in C_{lp},\ p \in \mathcal{P}_n$ in the sense

$$E_\xi[V_p(t)] \geq (1-\rho)\sigma^2\left(\frac{X^2}{n} - \sum_{i=1}^{N} x_i^2\right) = E_\xi\left[V_{p(\pi,x)}(t_1)\right] \quad \forall t \in C_{lp}, p \in \mathcal{P}_n$$

where $p(\pi, x)$ is an FESD(n) with inclusion probability $\pi_i = np_i$.

Corollary 6.3.1

If we put $a_i = 0$ in the model $M3$, t_1 reduces to the HT predictor $\widehat{Y}_{ht} = \sum_{i\in s}\frac{y_i}{np_i}$, and we find that $\widehat{Y}_{ht}$ based on a $p(\pi, x)$ sampling design provides the optimum strategy.

Corollary 6.3.2

If we put $a_i = 0$ and $x_i = 1$ in the model $M3$, t_1 reduces to $\frac{N}{n}\sum_{i\in s} y_i = N\overline{y}_s$, where $\overline{y}_s = \sum_{i\in s} y_i/n =$ the sample mean. In this case, $p(\pi, x)$ sampling design reduces to p_0, which includes SRSWOR. Hence the expansion estimator $N\overline{y}_s$ based on an SRSWOR becomes the optimal strategy.

Corollary 6.3.3

If we put $x_i = 1$ in the model $M3$, then t_1 reduces to the difference estimator $t_D = N\{\bar{y}_s - (\bar{a}_s - \bar{a})\}$, where $\bar{a}_s = \sum_{i\in s} a_i/n$ and $\bar{a} = \sum_{i=1}^{N} a_i/N$. So in this situation, the difference estimator t_D based on SRSWOR sampling design becomes the optimum strategy.

Remark 6.3.5

Mukherjee and Sengupta (1989) considered the optimality of predictors under a more general model viz. $E_\xi(y_i) = \mu_i(-\infty < \mu_i < \infty)$, $V_\xi(y_i) = \sigma_{ii} = \sigma_i^2(\sigma_i > 0)$, and $C_\xi(y_i, y_j) = \sigma_{ij}\ i \neq j;\ i, j = 1,\ldots, N$. They derived the following result (proof is omitted here):

$$E_\xi[V_p(t)] \geq \mathbf{I}'\boldsymbol{\varphi}^{-1}\mathbf{I} - \mathbf{I}'\boldsymbol{\Sigma}\mathbf{I} \ \text{ for } \ p \in \mathcal{P}_n \ \text{ and } \ t \in C_{lp}$$

where $\boldsymbol{\Sigma} = (\sigma_{ij})$, $\mathbf{I}' = (1, \ldots, N)$, $\boldsymbol{\varphi} = (\phi_{ij})$, $\phi_{ij} = \sum_{s\supset i,j} v^{ij}(s)p(s)$, $(v^{ij}(s)) = \boldsymbol{\Sigma}_s^{-1}$, and $\boldsymbol{\Sigma}_s$ is the $n \times n$ submatrix of $\boldsymbol{\Sigma}$ obtained by considering n units in s. They also derived the optimum estimator attaining the lower bound.

6.3.2 Optimal Model Design—Unbiased Prediction

The class of linear model design ($p\xi$)-unbiased predictors for the total Y will be denoted by $C_{lp\xi}$. The class $C_{lp\xi}$ consists of the estimators of the form $t = a_s + \sum_{i\in s} b_{si}y_i$ satisfying the $p\xi$-unbiasedness condition

$$E_pE_\xi(t) = E_\xi E_p(t) = E_\xi(Y) \tag{6.3.23}$$

In this section, the optimal predictor for the total Y will be considered under the following superpopulation model:

$$\begin{aligned}\text{Model } M4\text{: } E_\xi(y_i) = \beta\, x_i,\ V_\xi(y_i) = \sigma^2 x_i^2 \ \text{ and}\\ C_\xi(y_i, y_j) = \rho\sigma^2 x_i x_j \ \text{ for } \ i \neq j\end{aligned} \tag{6.3.24}$$

Here we define $t_0^*(\in C_{lp\xi})$ as optimal in the class $C_{lp\xi}$ for estimating βX if

$$E_pE_\xi\left(t_0^* - \beta X\right)^2 \leq E_pE_\xi(t - \beta X)^2 \ \ \forall t \neq t_0^* \in C_{lp\xi}.$$

The $p\xi$-unbiasedness conditions of the estimator t under the model $M4$ yield

$$\text{(i) } \sum_s a_s p(s) = 0 \ \text{ and } \ \text{(ii) } \sum_s p(s)\sum_{i\in s} b_{si}x_i = X \tag{6.3.25}$$

Now,

$$E_pE_\xi(t-\beta X)^2 = E_pE_\xi(t^2) - \beta^2X^2 = A_1 + A_2 - \beta^2X^2 \qquad (6.3.26)$$

where

$$\begin{aligned} A_1 = E_p[E_\xi(t)]^2 &= \sum_s p(s)\left(a_s + \beta\sum_{i\in s} b_{si}x_i\right)^2 \\ &\geq \left[\sum_s p(s)a_s\left(a_s + \beta\sum_{i\in s} b_{si}x_i\right)\right]^2 \Big/ \sum_s p(s) \qquad (6.3.27) \\ &= \beta^2X^2 \left(\text{noting } \sum_s p(s) = 1\right) \end{aligned}$$

and

$$\begin{aligned} A_2 &= E_p[V_\xi(t)] \\ &= \sigma^2\sum_s p(s)\left[\left(\sum_{i\in s} b_{si}x_i\right)^2 - (1-\rho)\left\{\left(\sum_{i\in s} b_{si}x_i\right)^2 - \sum_{i\in s} b_{si}^2x_i^2\right\}\right] \end{aligned} \qquad (6.3.28)$$

The expression A_2 given in Eq. (6.3.28) can be obtained from the expression Eq. (6.3.14) by writing $\sigma_i^2 = \sigma^2x_i^2$. Hence the optimum value of b_{si} that minimizes A_2 subject to Eq. (6.3.22) is obtained from Eq. (6.3.19) by writing $\sigma_i^2 = \sigma^2x_i^2$. The optimum value of b_{si} therefore comes out as

$$b_{i0} = 1/(np_i) \qquad (6.3.29)$$

It can be easily checked that the equality of Eq. (6.3.27) holds when $a_s = 0$ and $b_{si} = b_{i0} = 1/(np_i)$. From the aforementioned findings we get the following theorem derived by Cassel et al. (1976).

Theorem 6.3.10

Let p be any FESD(n) with $\pi_i > 0$ for $i = 1,\ldots, N$. Then, under model $M4$, $t_0^* = \frac{1}{n}\sum_{i\in s}\frac{y_i}{p_i}$ with $p_i = x_i/X$ is the optimum in the class of linear $p\xi$-unbiased estimators ($C_{lp\xi}$) for βX, i.e.,

$$E_pE_\xi(t-\beta X)^2 \geq \sigma^2X^2\{1+(n-1)\rho\}/n = E_pE_\xi\left(t_0^* - \beta X\right)^2 \quad \forall\, t \in C_{lp\xi}$$

Corollary 6.3.4

For a $p(\pi, x)$ sampling design, the HT estimator $\widehat{Y}_{ht} = \sum_{i\in s} y_i/\pi_i$ is the optimum in $C_{lp\xi}$, the class of linear $p\xi$-unbiased estimators for βX.

Corollary 6.3.5

Putting $a_i = 0$ and $x_i = 1$ $\forall i$ in the model $M4$, we get $E_\xi(y_i) = \beta$, $V_\xi(y_i) = \sigma^2$, and $C_\xi(y_i, y_j) = \rho\sigma^2$ for $i \neq j$. In this situation, the optimum $p\xi$-unbiased estimator t_0^* reduces to $N\overline{y}_s$ and hence $\overline{y}_s$, the sample mean is the optimal in $C_{lp\xi}(\beta)$, the class of linear $p\xi$-unbiased estimators of β, i.e.,

$$E_p E_\xi\left(\widehat{\beta} - \beta\right)^2 \geq \sigma^2\{1 + (n-1)\rho\}/n = E_p E_\xi\left(\overline{y}_s - \beta\right)^2 \quad \forall \widehat{\beta} \in C_{lp\xi}(\beta)$$

6.3.3 Exchangeable Model

The random variables $y_1,\ldots, y_i,\ldots, y_N$ follow exchangeable distribution (model), if the joint distribution of $y_{j_1},\ldots, y_{j_i},\ldots, y_{j_N}$ is the same for all $N!$ permutations of the labels $(j_1 \neq \cdots \neq j_i \neq \cdots \neq j_N) \subset (1,\ldots, j,\ldots, N)$. In other words the joint distribution of $\mathbf{y} = (y_1,\ldots, y_i,\ldots, y_N)$ is symmetric with respect to its coordinates. Hence if $\mathbf{y}$ follows exchangeable distribution, then y_i's must have the same means, variances, and covariances. The notion of exchangeability was introduced by de Finetti (1937) and used extensively in survey sampling by many, including Ericson (1969a,b). We will denote the exchangeable model by $MEX(y)$ when $\mathbf{y}$ follows an exchangeable distribution. Let us suppose that y_i's follow the transformation model $M3$ viz. $E_\xi(y_i) = \mu_i = a_i + \beta x_i$, $V_\xi(y_i) = \sigma_i^2 = \sigma^2 x_i^2$, and $C_\xi(y_i, y_j) = \rho\sigma^2 x_i x_j$, then the transformed variables $z_i = \dfrac{y_i - a_i}{x_i}\overline{X}$'s with $\overline{X} = X/N$, follow the exchangeable model with $E_\xi(z_i) = \beta\overline{X}$, $V_\xi(y_i) = \sigma^2\overline{X}^2$, and $C_\xi(y_i, y_j) = \rho\sigma^2\overline{X}^2$. The following theorem states that $\dfrac{1}{n}\sum_{i\in s} z_i + \overline{a} = \sum_{i\in s}\dfrac{y_i - a_i}{nx_i}\overline{X} + \overline{a} = \overline{t}_0$ with $\overline{a} = \sum_{i\in U} a_i/N$ is optimum for estimating $E_\xi\left(\dfrac{1}{N}\sum_{i\in U} z_i\right) + \overline{a} = \beta\overline{X} + \overline{a} = \overline{\Upsilon}$ (say) in $\overline{C}_{p\xi}$, the class of $p\xi$-unbiased estimator (linear or nonlinear) of $\overline{\Upsilon}$. The proof of the theorem is omitted but it is obtainable from Cassel et al. (1977).

Theorem 6.3.11
Let $\mathcal{P}_n$ be any FESD(n) design with $\pi_i > 0$ for $i = 1,\ldots, N$. Then, under the exchangeable model $MEX(z)$, any $p\xi$-unbiased estimator $\bar{t}$ satisfies

$$E_\xi E_p\left(\bar{t} - \overline{Y}\right)^2 \geq \frac{\{1 + (n-1)\rho\}\sigma^2}{n}\overline{X}^2$$
$$= E_\xi E_p\left(\bar{t}_0 - \overline{Y}\right)^2 \text{ for any } \bar{t} \in \overline{C}_{p\xi} \text{ and } p \in \mathcal{P}_n.$$

The class $\overline{C}_{p\xi}$ of $p\xi$-unbiased estimators of $\overline{Y}$ contains $\overline{C}_{pu}$, the class of p-unbiased predictors of $\overline{Y}$. Furthermore, the predictor $\bar{t}_0 = \overline{X}\sum_{i\in s}\frac{y_i - a_i}{nx_i} + \bar{a}$ is a p-unbiased predictor for $\overline{Y}$ when p is a $p(\pi, x)$ design. Hence, from Theorem 6.3.11, we get the following theorem.

Theorem 6.3.12
Let $\mathcal{P}_n$ be an FESD(n) with $\pi_i > 0$ for $i = 1,\ldots, N$. Then, under the exchangeable model $EX(z)$, the strategy $\bar{h}_0 = (\bar{t}_0, p(\pi, x))$ is optimal in the class of strategies $\overline{H}^* = (p, \bar{t})$, $p \in \mathcal{P}_n, \bar{t} \in \overline{C}_{pu}$ and we have

$$E_\xi E_p\left(\bar{t} - \overline{Y}\right)^2 \geq \frac{\{1 + (n-1)\rho\}\sigma^2}{n}\overline{X}^2$$
$$= E_\xi E_{p(\pi,x)}\left(\bar{t}_0 - \overline{Y}\right)^2 \text{ for any } \bar{t} \in \overline{C}_{pu} \text{ and } p \in \mathcal{P}_n$$

6.3.4 Random Permutation Model

Let us suppose that we know the marks of four students in Statistics as $50 = a_1$, $80 = a_2$, $30 = a_3$, and $90 = a_4$, but we do not know the students' identification. Let the marks of ith student be y_i. If the vector (y_1, y_2, y_3, y_4) can take any of the $4! = 24$ permutations of the number (a_1, a_2, a_3, a_4) with equal probabilities, then the resulting model is known as random permutation model (MRP). For this MRP, Prob $(y_1 = a_2, y_2 = a_3, y_3 = a_1, y_4 = a_4) = 1/24$ and $E_\xi(y_i) = \bar{a} = \sum_{i=1}^{4} a_i/4 = 62.5$, $V_\xi(y_i) = \frac{1}{4}\sum_{i=1}^{4}(a_i - \bar{a})^2 = 568.75$, and $C_\xi(y_i y_j) = \frac{1}{12}\sum_{i\neq}^{4}\sum_{j=1}^{4} a_i a_j - \bar{a}^2 = -V_\xi(y_i)/3 = -189.583$. For any pair $i, j (i \neq j)$ the correlation between y_i and y_j is $\rho = -1/3$. In general for an MRP, the population vector $\mathbf{y} = (y_1,\ldots, y_i,\ldots, y_N)$ of N units can take any one of the $N!$ permutations of the elements of the vector $\mathbf{a} = (a_1,\ldots, a_i,\ldots, a_N)$ with equal probability, i.e., Prob

$\left(y_1 = a_{i_1}, \ldots, y_j = a_{i_j}, \ldots, y_N = a_{i_N}\right) = 1/N!$ for $i_1 \neq \cdots i_j \neq \cdots \neq i_N \in (1, 2, \ldots, N)$. So, for the *MRP* we have

$$E_\xi(y_i) = \bar{a} = \sum_{i=1}^{N} a_i/N,\ V_\xi(y_i) = \sum_{i=1}^{N} (a_i - \bar{a})^2/N = \sigma_a^2 \text{ and}$$
$$C_\xi(y_i, y_j) = -\sigma_a^2/(N-1) \tag{6.3.30}$$

Because MRP is a special case of the exchangeable model, by substituting $\beta = 0$, $x_i = 1$, $a_i = \bar{a}$, and $\overline{Y} = \bar{a}$ in Theorem 6.3.11, we get the following theorem.

Theorem 6.3.13

Let $\mathcal{P}_n$ be any FESD(n) with $\pi_i > 0$ for $i = 1, \ldots, N$. Then, under *MRP*

$$E_\xi E_p(\bar{t} - \bar{a})^2 \geq \frac{N-n}{(N-1)n}\sigma_a^2 = E_\xi E_{p_0}\left(\bar{y}_s - \bar{a}\right)^2 \text{ for any } \bar{t} \in \overline{C}_{p\xi} \text{ and } p \in \mathcal{P}_n$$

where $\bar{y}_s = \sum_{i \in s} y_i/n$ and p_0 is defined in Corollary 6.3.2.

The results of optimality in the MRPs were also provided by Kempthorne (1969), Rao C.R. (1971), and Rao and Bellhouse (1978) under various contexts.

6.4 MODEL-BASED INFERENCE

Let s be a sample selected using a sampling design p yielding the survey data $d = (i,\ y_i,\ i \in s)$. Our objective is to estimate the population total $Y = \sum_{i \in s} y_i + \sum_{i \notin s} y_i$ on the basis of the data d. Here $\sum_{i \in s} y_i = Y_s$, the sum of the distinct units in s, can be obtained from the data d. $\sum_{i \notin s} y_i = Y_{\bar{s}}$, the sum of the distinct units in $\bar{s} = U - s$, cannot be obtained from the data d. Hence we need to predict the unknown part $Y_{\bar{s}}$ through the observed data $d = (i,\ y_i,\ i \in s)$. In a model-based inference (which is also known as the prediction approach) we establish a link between y_i, $i \in s$ and y_i, $i \notin s$ by using an appropriate superpopulation model ξ, whereas in design-based inference, sampling design establishes such a link.

Let $t = t(s, y)$ be a predictor of the total Y. We can write

$$t = Y_s + U_s \tag{6.4.1}$$

where $U_s = U(s, y) = t - Y_s$ is the predictor of the unobserved total $Y_{\bar{s}}$.

The predictor U_s depends on the y_i's for $i \in s$ but is independent of y_i, $i \notin s$. Let t be a ξ-unbiased predictor of Y, i.e., $E_\xi(t) = E_\xi(Y)$, then U_s is a ξ-unbiased predictor of $Y_{\bar{s}}$ in the sense $E_\xi(U_s) = E_\xi(Y_{\bar{s}})$. The predictor U_s is called an implied predictor of t. Our objective is to find a predictor t belonging to $C_{\xi u}$, the class of ξ-unbiased predictors of Y, for which the expected mean-square error of t attains a minimum. Thus the predictor $t_1^0(\in C_{\xi u})$ is optimal in $C_{\xi u}$ for predicting Y if

$$\begin{aligned} M_{\xi p}\left(t_1^0\right) &= E_\xi E_p\left(t_1^0 - Y\right)^2 \le E_\xi E_p(t - Y)^2 \\ &= M_{\xi p}(t) \quad \forall t\left(\neq t_1^0\right) \in C_{\xi u} \text{ and } \theta \in \Omega_\theta \end{aligned} \tag{6.4.2}$$

and the inequality is strict in Eq. (6.4.2) for at least one $\theta \in \Omega_\theta$.

Here we will assume p is noninformative so that the commutativity of the operators E_ξ and E_p are valid. We will now prove the following theorem attributed to Royall (1970).

Theorem 6.4.1

Let U_s and U'_s be the implied ξ-unbiased predictors of t and t', respectively. If ξ is a product measure (y_i's are independent for $i = 1,\ldots, N$), then the inequality

$$V_\xi(U_s) \le V_\xi(U'_s) \tag{6.4.3}$$

for every s with $p(s) > 0$ implies

$$M_{\xi,p}(t) \le M_{\xi,p}(t') \tag{6.4.4}$$

If the inequality is strict for some s with $p(s) > 0$ in Eq. (6.4.3), then the latter inequality in Eq. (6.4.4) is also strict.

Proof

$$\begin{aligned} M_{\xi,p}(t) &= E_\xi E_p(t - Y)^2 \\ &= E_\xi E_p\left(\sum_{i\in s} y_i + U_s - Y\right)^2 \\ &= E_p E_\xi\left(U_s - \sum_{i\notin s} y_i\right)^2 \\ &= \sum_s p(s) E_\xi\left(U_s - \sum_{i\notin s} y_i\right)^2 \\ &= \sum_s p(s)\left\{V_\xi(U_s) + V_\xi\left(\sum_{i\notin s} y_i\right) - 2C_\xi\left(\sum_{i\notin s} y_i, U_s\right)\right\} \end{aligned}$$

Now noting $C_\xi\left(\sum_{i\notin s} y_i, U_s\right) = 0$ as U_s is a function of y_i's for $i \in s$ and y_i's are independent, we find

$$M_{\xi,p}(t) = \sum_s p(s)\left\{V_\xi(U_s) + V_\xi\left(\sum_{i\notin s} y_i\right)\right\}$$

Hence $V_\xi(U_s) \le V_\xi(U'_s)$ for every s with $p(s) > 0$ implies $M_{\xi,p}(t) \le M_{\xi,p}(t')$. Furthermore, if there exists one $s = s_0$ with $p(s_0) > 0$, for which $V_\xi(U_s) < V_\xi(U'_s)$ in addition to Eq. (6.4.3), we get $M_{\xi,p}(t) < M_{\xi,p}(t')$.

Corollary 6.4.1
Let t belong to $C_{l\xi}$, the class of linear ξ-unbiased predictors of Y. Then the implied estimator U_s is also linear and in that case $C_\xi\left(\sum_{i\notin s} y_i, U_s\right) = 0$, whenever y_i's are uncorrelated.

6.4.1 Optimal Model–Unbiased Prediction

6.4.1.1 Product Measure Model

Model $M_1^*: E_\xi(y_i) = \beta x_i,\ V_\xi(y_i) = \sigma^2 v(x_i)$ and $C_\xi(y_i, y_j) = 0$ for $i \ne j$

(6.4.5)

where $v(x_i)$ is a known function of x_i.

The class $C_{l\xi}$ consists of estimators of the form

$$t = a_s + \sum_{i\in s} b_{si} y_i = Y_s + U_s \tag{6.4.6}$$

where $U_s = a_s + \sum_{i\in s}(b_{si} - 1) y_i$.

Under the model M_1^*, $E_\xi(t) = E_\xi(Y) = \beta X$ implies $E_\xi(U_s) = \beta X_{\bar{s}}$, where $X_{\bar{s}} = \sum_{i\notin s} x_i$. Now using the generalized least square theory, we find that the best linear ξ-unbiased predictor (ξ-BLUP) for $\beta X_{\bar{s}}$ is $\widehat{\beta}_{blue} X_{\bar{s}}$, where

$$\widehat{\beta}_{blue} = \sum_{i\in s} \frac{y_i x_i}{v(x_i)} \left\{\sum_{i\in s} x_i^2 / v(x_i)\right\}^{-1} \tag{6.4.7}$$

is the ξ-BLUP of β based on the sample s. Hence for a given s, $V_\xi(t)$ attains a minimum when

$$t = T_0 = Y_s + \widehat{\beta}_{blue} X_{\bar{s}} \tag{6.4.8}$$

Hence we have the following theorem.

Theorem 6.4.2

Under model M_1^*, for a given sampling design p

$$M_{p\xi}(t) = E_\xi E_p(t - Y)^2 \geq E_\xi E_p(T_0 - Y)^2 = M_{p\xi}(T_0) \text{ for } t \in C_{l\xi}$$

where $T_0 = \sum_{i \in s} y_i + \widehat{\beta}_{blue} X_{\bar{s}}$ with $\widehat{\beta}_{blue} = \sum_{i \in s} \frac{y_i x_i}{\nu(x_i)} \left\{ \sum_{i \in s} x_i^2 / \nu(x_i) \right\}^{-1}$.

It follows from Theorem 6.4.2 that the optimum estimators belonging to the class $C_{l\xi}$ for $\nu(x_i)$ equal to 1, x and x^2 become, respectively, as follows

$$T_{00} = Y_s + \frac{\sum_{i \in s} x_i y_i}{\sum_{i \in s} x_i^2} X_{\bar{s}}$$

$$T_{01} = Y_s + \frac{\sum_{i \in s} y_i}{\sum_{i \in s} x_i} X_{\bar{s}} = \frac{Y_s}{X_s} X = \widehat{Y}_R$$

$$T_{02} = Y_s + \left(\frac{1}{\nu_s} \sum_{i \in s} \frac{y_i}{x_i} \right) X_{\bar{s}}$$

where ν_s is the number of distinct units in s.

The estimator $T_{01} = \widehat{Y}_R$ is the well-known ratio estimator. It is interesting to note that when $\nu(x_i) = x_i^2$, the optimum p-unbiased predictor is the HT predictor $\widehat{Y}_{ht} = \frac{1}{n} \sum_{i \in s} \frac{y_i}{x_i}$ based on a $p(\pi, x)$ sampling design whereas the optimum ξ-unbiased predictor of Y is T_{02} as described earlier. In practice $\nu(x_i)$ is not known in advance and hence it is interesting to compare the performance of $\widehat{Y}_{ht}$ with T_{02} under various forms of the functions $\nu(x)$. The following theorem (Royall, 1970) compares performance of T_{02} and $\widehat{Y}_{ht}$.

Theorem 6.4.3

Let $\mathcal{P}_n$ be the class of FESD(n) and $\nu(x_i)/x_i^2$ be a nonincreasing function of x_i with $np_i \leq 1$ for $i = 1, \ldots, N$, then under the model M_1^*

$$E_p E_\xi (T_{02} - Y)^2 \leq E_p E_\xi \left(\widehat{Y}_{ht} - Y \right)^2 \quad \forall p \in \mathcal{P}_n$$

Proof

$$E_\xi\left(\widehat{Y}_{ht}-Y\right)^2-E_\xi\left(T_{02}-Y\right)^2=E_\xi\left\{\sum_{i\in s}\left(\frac{X}{nx_i}-1\right)y_i-Y_{\bar{s}}\right\}^2$$

$$-E_\xi\left\{\left(\frac{1}{n}\sum_{i\in s}\frac{y_i}{x_i}\right)X_{\bar{s}}-Y_{\bar{s}}\right\}^2$$

$$=\frac{\sigma^2}{n^2}X_{\bar{s}}^2\sum_{i\in s}\left\{\left(\frac{X-nx_i}{X_{\bar{s}}}\right)^2-1\right\}\frac{v(x_i)}{x_i^2}$$

$$=\frac{\sigma^2}{n^2}X_{\bar{s}}^2\sum_{i\in s}c_id_i$$

where $c_i=\left(\dfrac{X-nx_i}{X_{\bar{s}}}\right)^2-1$ and $d_i=\dfrac{v(x_i)}{x_i^2}$

Let us consider the sample s contains labels 1, 2,..., n with $x_1\geq x_2\geq\ldots\geq x_n$, then $d_1\leq d_2\leq\ldots\leq d_n$ and $c_1\leq c_2\leq\ldots\leq c_n$ with $\sum_{i=1}^n c_i\geq 0$. Let k be the greatest integer i for which $c_i\leq 0$, then

$$\sum_{i=1}^n c_id_i=\sum_{i=1}^k c_id_i+\sum_{i=k+1}^n c_id_i$$

$$\geq d_k\sum_{i=1}^k c_i+d_{k+1}\sum_{i=k+1}^n c_i$$

$$=d_k\sum_{i=1}^n c_i+\left(d_{k+1}-d_k\right)\sum_{i=k+1}^n c_i$$

$$\geq 0$$

Hence the proof.

6.4.1.1.1 Optimal Strategy and Purposive Sampling Design

Let Ψ_n be the collection of all possible $\binom{N}{n}$ samples of size n and $\mathcal{P}_n$ be the class of FESD(n). From Theorem 6.4.1, we note that under model M_1^*, $E_\xi(t-Y)^2$ attains a minimum value when $T_0=\sum_{i\in s}y_i+\widehat{\beta}_{blue}X_{\bar{s}}$ with

$$\widehat{\beta}_{blue}=\sum_{i\in s}\frac{y_ix_i}{v(x_i)}\left\{\sum_{i\in s}x_i^2/v(x_i)\right\}^{-1}.$$

Hence

$$\begin{aligned} M_{p\xi}(t) &= \sum_{s\in\Psi_n} p(s)E_\xi(t-Y)^2 \\ &\geq \sum_{s\in\Psi_n} p(s)E_\xi(T_0-Y)^2 \\ &= \sigma^2\sum_{s\in\Psi_n} p(s)\left\{\sum_{i\in\bar{s}} v(x_i) + X_{\bar{s}}^2\Big/\sum_{i\in s}\frac{x_i^2}{v(x_i)}\right\} \end{aligned} \tag{6.4.9}$$

So, the optimum sampling strategy should be a combination of the estimator T_0 and a purposive sampling design, which consists of only one sample $s_{\max}(\in\Psi_n)$ with probability 1, for which $V_\xi(T_0)$ attains a minimum. Now if $v(x_i)$ is nondecreasing and $v(x_i)/x_i^2$ is nonincreasing in x_i, then the right hand side of Eq. (6.4.9) is minimized for the sample $s_{\max}$ consisting of those n units having the largest x-values.

Let $\max\limits_{s\in\Psi_n}\sum\limits_{i\in s} x_i = \sum\limits_{i\in s_{\max}} x_i = X_{s_{\max}}$ and $p_{n,\max}$ be a purposive sampling design that contains only the sample $s_{\max}$ with probability 1, i.e.,

$$p_{n,\max}(s_{\max}) = 1 \tag{6.4.10}$$

Then we have the following theorem (Royall, 1970).

Theorem 6.4.4

If $v(x_i)$ is nondecreasing and $v(x_i)/x_i^2$ is nonincreasing in x_i, then under the model M_1^*, the strategy $h_{\max} = (T_{\max}, p_{\max})$ with $T_{\max} = \sum\limits_{i\in s_{\max}} y_i + \widehat{\beta}_{blue}X_{\bar{s}_{\max}}$ is the optimal in the class of strategies $\mathcal{H} = (t,p)$, $t\in C_{l\xi}$, $p\in\mathcal{P}_n$ in the sense that

$$M_{p\xi}(t) \geq M_{p_{\max}\xi}(T_{\max}) = \sigma^2\left\{\sum_{i\in\bar{s}_{\max}} v(x_i) + X_{\bar{s}_{\max}}^2\Big/\sum_{i\in s_{\max}}\frac{x_i^2}{v(x_i)}\right\}$$

Corollary 6.4.2

Consider the model $\widetilde{M}1$, which is the model M_1^* with $v(x_i) = x_i$ and $p\in\mathcal{P}_n$. Under the model $\widetilde{M}1$ $\widehat{\beta}_{blue} = \overline{y}_s/\overline{x}_s$, where $\overline{x}_s = \sum\limits_{i\in s} x_i/n$, $\overline{y}_s = \sum\limits_{i\in s} y_i/n$, the optimal predictor T_0 reduces to the well-known ratio estimator $\widehat{Y}_R = X(Y_s/X_s)$. The mean-square error of $\widehat{Y}_R$ under $\widetilde{M}1$ is given by

$$M_{p\xi}\left(\widehat{Y}_R\right) = \sigma^2 X \sum_s \left(\frac{X}{X_s} - 1\right) p(s) \tag{6.4.11}$$

Here also Eq. (6.4.11) is minimized for the purposive sampling design $p_{\max}$, which selects the sample $s_{\max}$ consisting of n units having the largest x-values with $p(s_{\max}) = 1$. Then $M_{\xi p_{\max}}\left(\widehat{Y}_R\right)$ attains the minimum value $\sigma^2 X\left(\dfrac{X}{X_{s_{\max}}} - 1\right)$. Hence, $h_{0\max} = \left(\widehat{Y}_{R\max}, p_{\max}\right)$ becomes the optimal strategy in the class of strategies $\mathcal{H} = (t, p)$ with $t \in C_{l\xi}$, $p \in \mathcal{P}_n$, and

$$\widehat{Y}_{R\max} = \frac{Y_{s_{\max}}}{X_{s_{\max}}} X.$$

6.4.1.2 Transformation Model

For the transformation model

Model $M3$: $E_\xi(y_i) = a_i + \beta x_i$, $V_\xi(y_i) = \sigma^2 x_i^2$ and $C_m(y_i, y_j) = \rho\sigma^2 x_i x_j; i \neq j$

the linear predictor $t = a_s + \sum_{i\in s} b_{si} y_i = a_s + Y_s + \sum_{i\in s} d_{si} y_i$ with $d_{si} = b_{si} - 1$ will be ξ-unbiased for the total Y if

$$\text{(i) } a_s + \sum_{i\in s} d_{si} a_i = A_{\bar{s}} \text{ and (ii) } \sum_{i\in s} d_{si} x_i = X_{\bar{s}} \text{ where } A_{\bar{s}} = \sum_{i\in \bar{s}} a_i \tag{6.4.12}$$

The expected mean-square error of t is given by

$$\begin{aligned} M_{p\xi}(t) &= E_p E_\xi (t - Y)^2 = E_p E_\xi \left(a_s + \sum_{i\in s} d_{si} y_i - Y_{\bar{s}}\right)^2 \\ &= E_p V_\xi \left(a_s + \sum_{i\in s} d_{si} y_i - Y_{\bar{s}}\right) \end{aligned} \tag{6.4.13}$$

Now,

$$V_{\xi}\left(a_s + \sum_{i\in s} d_{si}y_i - Y_{\bar{s}}\right) = \sigma^2\left[\left(\sum_{i\in s} d_{si}^2 x_i^2 + \rho \sum_{i\neq}\sum_{j\in s} d_{si}d_{sj}x_i x_j\right)\right.$$

$$\left. + \sum_{i\in \bar{s}} x_i^2 + \rho \sum_{i\neq}\sum_{j\in \bar{s}} x_i x_j - 2\rho X_{\bar{s}} \sum_{j\in \bar{s}} d_{sj}x_j\right]$$

$$= \sigma^2\left[\left\{(1-\rho)\sum_{i\in s} d_{si}^2 x_i^2 + \rho\left(\sum_{j\in s} d_{si}x_i\right)^2\right\}\right.$$

$$\left. + (1-\rho)\sum_{i\in \bar{s}} x_i^2 + \rho X_{\bar{s}}^2 - 2\rho X_{\bar{s}} \sum_{j\in s} d_{si}x_i\right]$$

$$\geq \sigma^2(1-\rho)\left\{\frac{1}{\nu_s}X_{\bar{s}}^2 + \sum_{i\in \bar{s}} x_i^2\right\} \tag{6.4.14}$$

(using the unbiasedness condition (ii) of Eq. (6.4.12) and noting $\sum_{i\in s} d_{si}^2 x_i^2 \geq \left(\sum_{j\in s} d_{si}x_i\right)^2 \Big/ \nu_s = X_{\bar{s}}^2/\nu_s$)

The lower bound Eq. (6.4.14) is attained by a predictor t with $d_{si}x_i = X_{\bar{s}}/\nu_s$ and $a_s = A_{\bar{s}} - \dfrac{X_{\bar{s}}}{\nu_s}\sum_{i\in s}\dfrac{a_i}{x_i}$. So, the optimum predictor attaining the lower bound is given by

$$T_0^* = \sum_{i\in s} y_i + \frac{X_{\bar{s}}}{\nu_s}\sum_{i\in s}\frac{y_i - a_i}{x_i} + A_{\bar{s}} \tag{6.4.15}$$

Hence we have the following theorem.

Theorem 6.4.5

For a given sampling design p and under the transformation model $M3$, the best linear ξ-unbiased predictor Y is

$$T_0^* = \sum_{i\in s} y_i + \frac{X_{\bar{s}}}{\nu_s}\sum_{i\in s}\frac{y_i - a_i}{x_i} + A_{\bar{s}} \text{ with } A_{\bar{s}} = \sum_{i\in \bar{s}} a_i.$$

The predictor T_0^* satisfies

$$E_pE_\xi(t-Y)^2 \geq \sigma^2(1-\rho)\sum_s\left(\frac{1}{\nu_s}X_{\bar{s}}^2+\sum_{i\in\bar{s}}x_i^2\right)p(s) = E_pE_\xi\left(T_0^*-Y\right)^2 \text{ for } t\in C_{l\xi}$$

Remark 6.4.1

For an FESD(n), the optimum linear ξ-unbiased predictor $T_0^* = \sum_{i\in s} y_i + \frac{X_{\bar{s}}}{n}\sum_{i\in s}\frac{y_i - a_i}{x_i} + A_{\bar{s}}$ is quite different from the optimum p-unbiased predictor $t_1 = \frac{X}{n}\sum_{i\in s}\frac{y_i - a_i}{x_i} + A$, where $A = \sum_{i\in U} a_i$. In particular, when $a_i = a =$ constant for $i = 1,\ldots, N$, the predictor $T_0^* = \sum_{i\in s} y_i + \frac{X_{\bar{s}}}{n}\sum_{i\in s}\frac{y_i}{x_i}$ is different from the optimum p-unbiased predictor $\widehat{Y}_{ht} = \frac{X}{n}\sum_{i\in s}\frac{y_i}{x_i}$.

For $x_i = 1$ and $i = 1,\ldots, N$, the predictor T_0^* reduces to the well-known difference estimator

$$t_D = N\left[\bar{y}_s - (\bar{a}_s - \bar{a})\right].$$

Furthermore, if $x_i = 1$ and $a_i = a$ for $i = 1,\ldots, N$, the predictor T_0^* reduces to $N\bar{y}_s$.

6.4.1.3 Multiple Regression Model

Suppose that the study variable y is related to the k-auxiliary variables $x_1,\ldots, x_i,\ldots, x_k$ through the following model,

$$\text{Model } M_{reg}: E_\xi(y_i) = \beta_0x_i(0) + \beta_1x_i(1) + \cdots + \beta_jx_i(j) + \cdots + \beta_kx_i(k);$$
$$V_\xi(y_i) = \sigma^2\nu_i \text{ and } C_\xi(y_i, y_j) = 0 \text{ for } i \neq j \tag{6.4.16}$$

where $x_i(0) = 1$ and $x_i(j)$ is the value of the jth auxiliary variable for the ith unit, which is assumed to be known for $j = 1,\ldots, k$; $i = 1,\ldots, N$; $\beta_0,\ldots, \beta_j,\ldots, \beta_k$ and σ^2 are unknown model parameters, and ν_i is a known function of $x_i(0),\ldots, x_i(j),\ldots, x_i(k)$.

Without loss of generality, let us assume that the sample s contains ν_s distinct units with labels $1,\ldots, \nu_s$ and $\bar{s}$ contains units with labels $\nu_s + 1,\ldots, N$.

Let $t = Y_s + U_s$ be a linear ξ-unbiased predictor of Y, where U_s is the linear ξ-unbiased predictor of $Y_{\bar{s}}$. Since t is ξ-unbiased,

$$E_\xi(U_s) = \beta_0X_{\bar{s}}(0) + \cdots + \beta_jX_{\bar{s}}(j) + \cdots + \beta_kX_{\bar{s}}(k)$$

where $X_{\bar{s}}(0) = N - \nu_s$ and $X_{\bar{s}}(j) = \sum_{i \in \bar{s}} x_i(j)$ for $j = 2, \ldots, k$. Using the generalized Gauss–Markov theorem, we note that ξ-BLUP of $Y_{\tilde{s}}$ is

$$U_s(blue) = \widehat{\beta}_0 X_{\bar{s}}(0) + \cdots + \widehat{\beta}_j X_{\bar{s}}(j) + \cdots + \widehat{\beta}_k X_{\bar{s}}(k)$$

where $\widehat{\beta}_0, \ldots, \widehat{\beta}_j, \ldots, \widehat{\beta}_k$ are the least square estimators of $\beta_0, \ldots, \beta_j$ and β_k, respectively. The estimates $\widehat{\beta}_j$'s are given by

$$\widehat{\boldsymbol{\beta}}_{\mathbf{s}} = \left(\mathbf{X}_{\mathbf{s}}^{\mathbf{T}} \mathbf{V}_{\mathbf{s}}^{-1} \mathbf{X}_{\mathbf{s}}\right)^{-1} \mathbf{X}_{\mathbf{s}}^{\mathbf{T}} \mathbf{V}_{\mathbf{s}}^{-1} \mathbf{y}_{\mathbf{s}} \tag{6.4.17}$$

where

$$\mathbf{y}_{\mathbf{s}} = \begin{bmatrix} y_0 \\ \cdot \\ y_i \\ \cdot \\ y_{\nu_s} \end{bmatrix}, \mathbf{X}_{\mathbf{s}} = \begin{bmatrix} x_1(0) & .. & x_1(j) & .. & x_1(k) \\ .. & .. & .. & .. & .. \\ x_i(0) & .. & x_i(j) & .. & x_i(k) \\ .. & .. & .. & .. & .. \\ x_{\nu_s}(0) & .. & x_{\nu_s}(j) & .. & x_{\nu_s}(k) \end{bmatrix},$$

$$\widehat{\boldsymbol{\beta}}_{\mathbf{s}} = \begin{bmatrix} \widehat{\boldsymbol{\beta}}_0 \\ \cdot \\ \widehat{\boldsymbol{\beta}}_{\mathbf{j}} \\ \cdot \\ \widehat{\boldsymbol{\beta}}_{\mathbf{k}} \end{bmatrix} \text{ and } \mathbf{V}_{\mathbf{s}} = \begin{bmatrix} \mathrm{v}_1 & .. & 0 & .. & 0 \\ .. & .. & .. & .. & .. \\ 0 & .. & \mathrm{v}_i & .. & 0 \\ .. & .. & .. & .. & .. \\ 0 & .. & 0 & .. & \mathrm{v}_{\nu_s} \end{bmatrix}$$

So for a given s, the best linear ξ-unbiased predictor of Y is

$$T_{BLUE} = Y_s + \widehat{\boldsymbol{\beta}}_{\mathbf{s}}^{\mathbf{T}} \mathbf{X}_{\bar{\mathbf{s}}}$$

where $\mathbf{X}_{\bar{\mathbf{s}}} = \left[X_{\bar{s}}(0), \ldots, X_{\bar{s}}(j), \ldots, X_{\bar{s}}(k)\right]^{\mathbf{T}}$ and $\widehat{\boldsymbol{\beta}}_{\mathbf{s}}$ is given in Eq. (6.4.17).

The expected mean-square error of T_{BLUE} is given by

$$\begin{aligned} E_p E_\xi (T_{BLUE} - Y)^2 &= E_p\left[V_\xi(Y_{\bar{s}}) + V_\xi\left(\widehat{\boldsymbol{\beta}}_{\mathbf{s}}^{\mathbf{T}} \mathbf{X}_{\bar{\mathbf{s}}}\right)\right] \\ &= \sigma^2 \sum_s \left\{ \sum_{i \in \bar{s}} \nu_i + \mathbf{X}_{\bar{\mathbf{s}}}^{\mathbf{T}} \left(\mathbf{X}_{\mathbf{s}}^{\mathbf{T}} \mathbf{V}_{\mathbf{s}}^{-1} \mathbf{X}_{\mathbf{s}}\right)^{-1} \mathbf{X}_{\bar{\mathbf{s}}} \right\} p(s) \end{aligned}$$

The aforementioned findings have been summarized in the following theorem.

Theorem 6.4.6

Under model M_{reg} and a given sampling design p

$$E_\xi E_p(t-Y)^2 \geq E_\xi E_p(T_{BLUE}-Y)^2$$
$$= \sigma^2 \sum_s \left\{ \sum_{i\in \bar{s}} v_i + \mathbf{X}_{\bar{\mathbf{s}}}^{\mathbf{T}} \left(\mathbf{X}_{\mathbf{s}}^{\mathbf{T}} \mathbf{V}_{\mathbf{s}}^{-1} X_s\right)^{-1} \mathbf{X}_{\bar{\mathbf{s}}} \right\} p(s) \quad \text{for any } t \in C_{\xi l}$$

where $\quad T_{BLUE} = Y_s + \widehat{\boldsymbol{\beta}}_{\mathbf{s}}^{\mathbf{T}} \mathbf{X}_{\bar{\mathbf{s}}}, \qquad \widehat{\boldsymbol{\beta}}_{\mathbf{s}} = (\mathbf{X}_{\mathbf{s}}^{\mathbf{T}} \mathbf{V}_{\mathbf{s}}^{-1} \mathbf{X}_{\mathbf{s}})^{-1} \mathbf{X}_{\mathbf{s}}^{\mathbf{T}} \mathbf{V}_{\mathbf{s}}^{-1} \mathbf{y}_{\mathbf{s}},$
$\mathbf{X}_{\bar{\mathbf{s}}} = [X_{\bar{s}}(0), \ldots, X_{\bar{s}}(j), \ldots, X_{\bar{s}}(k)]^{\mathbf{T}}$ and $X_{\bar{s}}(j) = \sum_{i\in\bar{s}} x_i(j)$.

Corollary 6.4.3

For $k=1$ and $v_i = 1$, M_{reg} reduces to $E_\xi(y_i) = \beta_0 + \beta_1 x_i(1)$; $V_\xi(y_i) - \sigma^2$ and $C_\xi(y_i, y_j) = 0$ for $i \neq j$. Now writing x_i for $x_i(1)$ and $\overline{x}_s = \sum_{i\in s} x_i / v_s$, the least square estimates of β_0 and β_1 become $\widehat{\beta}_0 = \overline{y}_s - \widehat{\beta}_1 \overline{x}_s$ and $\widehat{\beta}_1 = \sum_{i\in s} (y_i - \overline{y}_s)(x_i - \overline{x}_s) / \sum_{i\in s} (x_i - \overline{x}_s)^2$, respectively. In this case, the optimal ξ-unbiased estimator T_{BLUE} reduces to the well-known regression estimator

$$t_{reg} = N\left\{\overline{y}_s - \widehat{\beta}_1\left(\overline{x}_s - \overline{X}\right)\right\} \text{ with } \overline{X} = \sum_{i\in U} x_i/N \qquad (6.4.18)$$

and

$$E_\xi\left(t_{reg}-Y\right)^2 = \sigma^2 \left\{ \frac{N(N-v_s)}{v_s} + \frac{N^2\left(\overline{X}-\overline{x}_s\right)^2}{\sum_{i\in s}\left(x_i - \overline{x}_s\right)^2} \right\} \qquad (6.4.19)$$

For an FESD(n) design $E_\xi(t_{reg} - Y)^2$ reduces to $\sigma^2 \left\{ \frac{N(N-n)}{n} + \frac{N^2\left(\overline{X}-\overline{x}_s\right)^2}{\sum_{i\in s}\left(x_i - \overline{x}_s\right)^2} \right\}$. The optimal sampling strategy in this case is again a purposive sampling, which selects a sample with probability one consisting of those units for which either (i) $\overline{x}_s = \overline{X}$ or (ii) $\frac{\left(\overline{X}-\overline{x}_s\right)^2}{\sum_{i\in s}\left(x_i - \overline{x}_s\right)^2}$ is minimum.

6.5 ROBUSTNESS OF DESIGNS AND PREDICTORS

In the previous sections, we have derived optimal sampling strategies in presence of various superpopulation models. For example, $\widehat{Y}_{ht}$, based on a $p(\pi, x)$ sampling design, was found to be optimum for estimating a finite population total under the superpopulation model $M1$ with $\mu_i = \beta x_i$ and $\sigma_i^2 = \sigma^2 x_i^2$. The optimum $p(\pi, x)$ design is seldom used in practice because it is difficult to construct when the sample size n is greater than 2. Apart from this, unbiased estimation of the variance of $\widehat{Y}_{ht}$ becomes tedious because the second-order inclusion probabilities are difficult to compute. It should be noted that although $\widehat{Y}_{hh}$, the Hansen–Hurwitz (HH) estimator based on probability proportional to size with replacement (PPSWR) is inadmissible, it is extensively used in practice because of its elegant and simple expression of its unbiased variance estimator. In most situations, one is not sure about the validity of a particular superpopulation model. Hence the estimator $\widehat{Y}_{ht}$ based on $p(\pi, x)$ will no longer be optimal if the assumed model $M1$ fails to hold in practice. It is very difficult to choose an appropriate model in a particular situation. Furthermore, in multicharacter surveys, we collect information about more than one character at a time. The $p(\pi, x)$ sampling design provides efficient estimators for estimation of totals of those characters, which are approximately proportional to the auxiliary variable x, but the $p(\pi, x)$ sampling design yields inefficient estimators of the totals for those characters, which are not well related to the auxiliary variable x, as reported by Rao (1966a,b). We call an estimator "robust" if it is still optimum in wider classes of designs and/or models. In this section, we will consider the robustness of predictors for mis-specification of the model only.

6.5.1 Robustness of Predictors

The ratio predictor $\widehat{Y}_R = X\sum_{i\in s} y_i \Big/ \sum_{i\in s} x_i$ is the ξ-BLUP, when the following model $\widetilde{M}1$ is true.

$$\begin{aligned}&\widetilde{M}1: y_i = \beta x_i + \epsilon_i; E_\xi(\epsilon_i) = 0,\ V_\xi(\epsilon_i) = \sigma^2 x_i \ \text{ and}\\ &C_\xi(\epsilon_i, \epsilon_j) = 0 \ \text{ for } \ i \neq j\end{aligned} \tag{6.5.1}$$

Suppose that the model $\widetilde{M}1$ was not true but an alternative model $\widetilde{M}2$, which is similar to $\widetilde{M}1$ in all respects except $V_\xi(\epsilon_i) = \sigma^2 x_i^2$, is true. Then the predictor $\widehat{Y}_R$ is no longer the ξ-BLUP for Y under the model $\widetilde{M}2$.

But $\widehat{Y}_R$ retains its ξ-unbiasedness property with respect to $\widetilde{M}2$ as $E_\xi\left(\widehat{Y}_R\right) = Y$. So, we say that the predictor $\widehat{Y}_R$ is robust with respect to ξ-bias. But as far as expected mean-square error is concerned, the optimum estimator $Y_s + \dfrac{X_{\bar{s}}}{\nu_s}\sum_{i\in s}\dfrac{y_i}{x_i}$ has a much lower expected mean-square error under the model $\widetilde{M}2$ than that of $\widehat{Y}_R$. Hence we say that $\widehat{Y}_R$ is not robust with respect to the expected mean-square error.

6.5.2 Balanced Sampling Design

Let us consider the performance of $\widehat{Y}_R$ when the following linear model $\widetilde{M}3$ is true.

$$\widetilde{M}3: y_i = \alpha + \beta x_i + \in_i \text{ with } E_\xi(\in_i) = 0, V_\xi(\in_i) = \sigma^2 \text{ and}$$
$$C_\xi(\in_i, \in_j) = 0 \text{ for } i \neq j$$

Under the model $\widetilde{M}3$ the ξ-bias of $\widehat{Y}_R$ is

$$B_\xi\left(\widehat{Y}_R\right) = N\alpha\left(\overline{X}/\overline{x}_s - 1\right) \text{ where } \overline{x}_s = \sum_{i\in s} x_i/\nu_s \tag{6.5.2}$$

The ξ-bias $B_\xi\left(\widehat{Y}_R\right)$ is nonzero if $\alpha \neq 0$. However, if the selected sample s is such that $\overline{x}_s = \overline{X}$, then for the given sample s, the estimator $\widehat{Y}_R$ is ξ-unbiased.

A sample s for which $\overline{x}_s = \overline{X}$ is known as a balanced sample. A sampling design that selects a balanced sample with probability 1 is called a balanced sampling design or simply a balanced design, and it will be denoted by p_B.

Under a balanced sampling design, the ratio estimator $\widehat{Y}_R$ reduces to $T_{BL} = N\overline{y}_s$, where $\overline{y}_s = \sum_{i\in s} y_i/\nu_s$. Hence the estimator T_{BL} based on a balanced sample is robust with respect to ξ-bias when the model $\widetilde{M}3$ is true. For an FESD(n), the expected mean square of $T_{BL} = N\overline{y}_s$ based on a balanced design under the model $\widetilde{M}1$ is

$$M_\xi(T_{BL}) = N^2\sigma^2(1-f)\overline{X}/n$$

The expected mean square for the optimum estimator $\widehat{Y}_{R\max}$ under the model $\widetilde{M}1$, based on the optimum design $p_{\max}$ is obtained from Eq. (6.4.11) as

$$M_\xi\left(\widehat{Y}_{R\max}\right) = N^2\sigma^2\frac{(1-f)\overline{X}}{n}\frac{\overline{x}_{\bar{s}_{\max}}}{\overline{x}_{s_{\max}}} \leq M_\xi(T_{BL}) \tag{6.5.3}$$

where $\overline{x}_{s_{\max}} = \sum_{i\in s_{\max}} x_i/n$ and $\overline{x}_{\bar{s}_{\max}} = (X - n\overline{x}_{s_{\max}})/(N-n)$.

Although a balanced design provides a ξ-unbiased estimator, one should accept lower efficiency if a balance strategy $h_b = \left(p_B, N\overline{y}_s\right)$ instead of the optimum strategy $h_{\max} = \left(p_{\max}, \widehat{Y}_{\max}\right)$ is used. Obviously, the balanced design cannot be constructed unless a sample s exists with $\overline{x}_s = \overline{X}$. In this case, one may choose an approximate balance sample for which $\overline{x}_s$ is approximately equal to $\overline{X}$. The simple random sampling with replacement (SRSWR) and SRSWOR provide approximately balanced samples when the sample size n is large.

6.5.3 Polynomial Regression Model

Following Royall and Herson (1973), we denote the polynomial regression model of order J as follows:

$$\xi[\delta_0, \ldots, \delta_j, \ldots, \delta_J : v(x)]: E_\xi(y_i) = \sum_{j=0}^{J} \delta_j \beta_j x_i^j,\ V_\xi(y_i) = \sigma^2 v(x_i) \text{ and }$$
$$C_\xi(y_i, y_j) = 0 \text{ for } i \neq j \tag{6.5.4}$$

where $\delta_j = 0$ if the term x_i^j is absent in the model and $\delta_j = 1$ if x_i^j is present in the model Eq. (6.5.4).

For example, $\xi[1,\ 0,\ 1 : x]$ represents the model $E_\xi(y_i) = \beta_0 + \beta_2 x_i^2$, $V_\xi(y_i) = \sigma^2 x_i$ and $C_\xi(y_i,\ y_j) = 0$ for $i \neq j$.

Let $T = Y_s + U_s$ be a linear ξ-unbiased estimator of Y. Then $E_\xi(U_s) = E_\xi(Y_{\bar{s}}) = \sum_{j=0}^{J} \delta_j \beta_j X_{\bar{s}}^{(j)}$, where $X_{\bar{s}}^{(j)} = \sum_{i \in \bar{s}} x_i^j$. Using the generalized Gauss–Markov theorem, we note that ξ-BLUP of $E_\xi(Y_{\bar{s}})$ is $\sum_{j=0}^{J} \delta_j \widehat{\beta}_j X_{\bar{s}}^{(j)}$, where $\widehat{\beta}_j$ is the weighted least square predictor of β_j. Hence, we have the following theorem by Royall and Herson (1973).

Theorem 6.5.1

Under the model $\xi[\delta_0, \ldots, \delta_j, \ldots, \delta_J : v(x)]$, the ξ-BLUP for the total Y is

$$\widehat{T} = Y_s + \sum_{j=0}^{J} \delta_j \widehat{\beta}_j X_{\bar{s}}^{(j)}$$

where $\widehat{\beta}_j$ is the weighted least square estimate of β_j and $X_{\bar{s}}^{(j)} = \sum_{i \in \bar{s}} x_i^j$ for $j = 0, \ldots, J$.

The ξ-bias of the ratio estimator $\widehat{Y}_R = \dfrac{\sum_{i\in s} y_i}{\sum_{i\in s} x_i} X$ under the model $\xi[\delta_0,\ldots,\delta_j,\ldots,\delta_J : v(x)]$ is

$$B_\xi\left(\widehat{Y}_R\right) = E_\xi\left(\widehat{Y}_R - Y\right) = N\overline{X}\sum_{j=0}^{J}\delta_j\beta_j\left(\frac{\overline{x}_s^{(j)}}{\overline{x}_s} - \frac{\overline{X}^{(j)}}{\overline{X}}\right) \tag{6.5.5}$$

where $\overline{x}_s^{(j)} = \sum_{i\in s} x_i^j \Big/ v_s$ and $\overline{X}^{(j)} = \sum_{i\in U} x_i^j \Big/ N$.

From Eq. (6.5.5), we note that the bias of $\widehat{Y}_R$ is independent of the regression coefficient β_1 and the bias is of order N, the population size. Furthermore, the ξ-bias of $\widehat{Y}_R$ is zero under the model $\xi[\delta_0,\ldots,\delta_j,\ldots,\delta_J : v(x)]$ if and only if

$$\frac{\overline{x}_s^{(j)}}{\overline{x}_s} = \frac{\overline{X}^{(j)}}{\overline{X}} \tag{6.5.6}$$

for all $j = 1,\ldots,J$ with $\delta_j = 1$.

For example, $\widehat{Y}_R$ is ξ-unbiased for the model $\xi[1,\ 1 : v(x)]$

$$\text{if}\quad \frac{\overline{x}_s^{(0)}}{\overline{x}_s} = \frac{\overline{X}^{(0)}}{\overline{X}}\quad \text{i.e.,}\quad \overline{x}_s = \overline{X}.$$

Similarly, $\widehat{Y}_R$ is ξ-unbiased under the model $\xi[1,\ 1,\ 1 : v(x)]$ if and only if

$$\overline{x}_s = \overline{X}\ \text{and}\ \overline{x}_s^{(2)} = \overline{X}^{(2)}.$$

6.5.4 Balanced Sample of Order *k*

A sample s is said to be a balanced sample of order k if

$$\overline{x}_s^{(j)} = \overline{X}^{(j)}\ \text{for}\ j = 1,\ldots,k(\leq J) \tag{6.5.7}$$

A sampling design that selects a balanced sample of order k with probability 1 is called a balanced sampling design of order k or simply a balanced design of order k. Construction of a balanced sample of order k is certainly difficult when k is large (say 5 or more). We will denote a balanced sampling design of order k by p_{Bk}. However, SRSWOR and SRSWR provide an approximate balanced sample when the sample size n is large.

6.5.5 Optimality of Balanced Sampling

Consider an FESD(n) with model $\xi[1,\ 1 : x]$. Under this model the ξ-BLUP is given by

$$T = Y_s + (N - n)\widehat{\beta}_0 + \widehat{\beta}_1 X_{\bar{s}} \tag{6.5.8}$$

where

$$\widehat{\beta}_0 = \left(n\overline{x}_s \sum_{i\in s}\frac{y_i}{x_i} - n^2\,\overline{y}_s\right)\Big/ D,\ \widehat{\beta}_1 = \left(Y_s\sum_{i\in s}\frac{1}{x_i} - n\sum_{i\in s}\frac{y_i}{x_i}\right)\Big/ D \text{ and}$$

$$D = n^2\left\{\overline{x}_s\left(\frac{1}{n}\sum_{i\in s}\frac{1}{x_i}\right) - 1\right\}$$

For a balanced sampling design $\overline{x}_s = \overline{X}$ and hence the ξ-BLUP reduces to the expansion estimator $N\overline{y}_s$. Royall and Herson (1973) proved that ξ-BLUP for the polynomial regression model reduces to the expansion estimator when a balanced sampling design is used.

Theorem 6.5.2

Let p_{BJ} be a balanced sampling design of order J with fixed effective sample size n. Then,

$$\widehat{T}[1, \delta_1, \delta_2, \ldots, \delta_J : 1] = \widehat{T}[\delta_0, 1, \delta_2, \ldots, \delta_J : x] = \cdots =$$
$$\widehat{T}[\delta_0, \delta_1, \delta_2, \ldots, 1 : x^J] = N\overline{y}_s$$

where $\delta_j = 0$ or 1 for $j = 1,\ldots, J$ and $\widehat{T}[\delta_0, \delta_1, \ldots, \delta_{j-1}, 1, \delta_{j+1}, \ldots, \delta_J : x^j]$ is the ξ-BLUP for Y under the model $[\delta_0,\ \delta_1,\ldots,\ \delta_{j-1},\ 1,\ \delta_{j+1},\ldots,\ \delta_J : x^j]$.

Proof

Using Theorem 6.5.1, we find

$$\widehat{T}\left[\delta_0, \delta_1, \ldots, \delta_{j-1}, 1, \delta_{j+1}, \ldots, \delta_J : x^j\right] = Y_s + \sum_{j=0}^{J}\delta_j\widehat{\beta}_j X_{\bar{s}}^{(j)} \tag{6.5.9}$$

The coefficients $\widehat{\beta}_j$ of Eq. (6.5.9) are obtained by minimizing

$$\sum_{i\in s}\frac{1}{x_i^j}\left(y_i - \sum_{r=0}^{J}\delta_r\beta_r x_i^r\right)^2$$

with respect to $\beta_0,\ldots,\ \beta_r,\ldots,\ \beta_J$.

In particular, $\frac{\partial}{\partial \beta_j}\sum_{i\in s}\frac{1}{x_i^j}\left(y_i-\sum_{r=0}^{J}\delta_j\beta_j x_i^r\right)^2=0$ yields

$$\sum_{i\in s} y_i=\sum_{r=0}^{J}\delta_r\widehat{\beta}_r\sum_{i\in s}x_i^r \tag{6.5.10}$$

For a balanced sampling design $\sum_{i\in s}x_i^r=\frac{n}{N-n}\sum_{i\in \bar{s}}x_i^r=\frac{n}{N-n}X_{\bar{s}}^{(r)}$, Eq (6.5.10) reduces to

$$\sum_{i\in s} y_i=\frac{n}{N-n}\sum_{r=0}^{J}\delta_r\widehat{\beta}_r X_{\bar{s}}^{(r)} \tag{6.5.11}$$

Now substituting Eq. (6.5.11) in Eq. (6.5.9), we get

$$\widehat{T}\left[\delta_0,\delta_1,\ldots,\delta_{j-1},1,\delta_{j+1},\ldots,\delta_J : x^j\right]=N\overline{y}_s.$$

6.6 BAYESIAN INFERENCE

Let s be an unordered sample selected from a finite population U using a noninformative sampling design p. Let the unordered sample s contains ν_s distinct units with labels $i_1,\ldots,i_k,\ldots,i_{\nu_s}$, where $i_1<\ldots<i_k<\ldots<i_{\nu_s}$ and $\bar{s}$, the complementary of the sample s, contains $N-\nu_s$ distinct units with labels $i_{\nu_s+1},\ldots,i_N$, where $i_{\nu_s+1}<\ldots<i_N$. Let the unordered data be denoted by $d=\{(i,\ y_i);\ i\in s\}$. In Bayesian inference, we assume that the population vector $\mathbf{y}=(y_1,\ldots,\ y_i,\ldots,\ y_N)$ has a prior distribution $h(\mathbf{y})$, which may be regarded as the statistician's belief or ignorance about parameter $\mathbf{y}$. Here we assume that the parametric space of $\mathbf{y}$ is the N-dimensional Euclidean space $R^N=\Omega_N=\{-\propto<y_1<\propto,\ldots,-\propto<y_N<\propto\}$. After observing the data d, the parametric space will be reduced to Ω_d, where $i_1,\ldots,i_k$ and i_{ν_s}th coordinates are respectively fixed as $y_{i_1},\ldots,y_{i_k}$ and $y_{i_{\nu_s}}$ and the rest $N-\nu_s$ coordinates can take any value between $-\propto$ and $\propto$. We technically say Ω_d is consistent with data d (details given in Section 2.7). The Bayesian approach in finite population sampling was mainly popularized by Godambe (1968), Basu (1969), Ericson (1969a,b), and Cassel et al. (1977). The likelihood of $\mathbf{y}$ given the data d is

$$L(\mathbf{y}|d)=\begin{cases} p(s) & \text{for}\quad \mathbf{y}\in\Omega_d\\ 0 & \text{for}\quad \mathbf{y}\notin\Omega_d\end{cases} \tag{6.6.1}$$

The posterior distribution of $\mathbf{y}$ taking the prior of $\mathbf{y}$ as $h(\mathbf{y})$ is given by

$$h(\mathbf{y}|d) \propto \begin{cases} p(s)h(\mathbf{y}) & \text{for} \quad \mathbf{y} \in \Omega_d \\ 0 & \text{for} \quad \mathbf{y} \notin \Omega_d \end{cases} \tag{6.6.2}$$

Writing the marginal distribution of $y^{(s)} = (y_i, i \in s) = h(\mathbf{y}^{(s)}) = \int_{-\infty}^{\infty} \dots \int_{-\infty}^{\infty} h(\mathbf{y}) dy_{\nu_{s+1}} \dots dy_N$, the constant of proportionality in Eq. (6.6.2) is obtained as $\{p(s) \cdot h(\mathbf{y}^{(s)})\}^{-1}$. Hence the posterior distribution of $\mathbf{y}$ taking the prior of $\mathbf{y}$ as $h(\mathbf{y})$ is given by

$$h(\mathbf{y}|d) = \begin{cases} h(\mathbf{y})/h(\mathbf{y}^{(s)}) & \text{for} \quad \mathbf{y} \in \Omega_d \\ 0 & \text{for} \quad \mathbf{y} \notin \Omega_d \end{cases} \tag{6.6.3}$$

From the expression (6.6.3) above, we note that the posterior distribution of $\mathbf{y}$, given the data d, is independent of the sampling design. Thus Bayesian inference about any function of $\mathbf{y}$ should depend only on the data but not the sampling design, through which the data were collected. We summarize the aforementioned discussions in the following theorem.

Theorem 6.6.1

Let an unordered data d be selected following a sampling design p and let the parameter vector $\mathbf{y}$ has a prior distribution $h(\mathbf{y})$. Then the posterior distribution of $\mathbf{y}$, given the data d, is independent of the sampling design.

Corollary 6.6.1

Let $y_1, \dots, y_N$'s be mutually independent and the density of y_i be $h_i(y_i)$. Then the joint density of $\mathbf{y}$ is $\prod_{i=1}^{N} h_i(y_i)$ and the posterior distribution of $\mathbf{y}$ given d is

$$h(\mathbf{y}|d) = \begin{cases} \prod_{i \notin s}^{N} h_i(y_i) & \text{for} \quad \mathbf{y} \in \Omega_d \\ 0 & \text{for} \quad \mathbf{y} \notin \Omega_d \end{cases}$$

Corollary 6.6.2

In case the prior distribution of $\mathbf{y}$ depends on an unknown parameter vector θ, we write the density of $\mathbf{y}$ as $h(\mathbf{y}|\theta)$. Let the prior distribution of θ be given by $F(\theta)$, then the posterior density of $\mathbf{y}$, given the data d, is

$$h(\mathbf{y}|d) = \begin{cases} \dfrac{\int_{\Theta} h(\mathbf{y}|\theta)dF(\theta)}{\int_{\Theta} h\left(y^{(s)}|\theta\right)dF(\theta)} & \text{for} \quad \mathbf{y} \in \Omega_d \\ 0 & \text{for} \quad \mathbf{y} \notin \Omega_d \end{cases}$$

where $\boldsymbol{\Theta}$ is the parameter space of θ.

6.6.1 Bayes Estimator

An estimator T^* in the class C of estimators is said to be the Bayes estimator for a population total Y under a given sampling design p and prior distribution $h(\mathbf{y})$ if for every other estimator T in C satisfies

$$\int_{\Omega_N}\left[\sum_s p(s)\left(T^* - Y\right)^2\right]h(\mathbf{y})d\mathbf{y} \le \int_{\Omega_N}\left[\sum_s p(s)(T - Y)^2\right]h(\mathbf{y})d\mathbf{y}$$

Following Godambe (1969), the Bayes estimator is obtained as

$$\begin{aligned} T^* &= \sum_{i\in s} y_i + E_{\mathbf{y}}\left(\sum_{i\notin s} y_i|d\right) \\ &= N\left[f_s\overline{y}_s + (1 - f_s)\psi(d)\right] \end{aligned} \tag{6.6.4}$$

where $f_s = \nu_s/N$, $\psi(d) = E_{\mathbf{y}}\left(\overline{y}_{\bar{s}}|d\right)$ and $\overline{y}_{\bar{s}} = \sum_{i\in\bar{s}} y/(N - n_s)$.

Remark 6.6.1

In case $h(\mathbf{y}|\theta)$, the distribution of $\mathbf{y}$, depends on some known parameter vector θ, then $\psi(d)$ involves θ and we write $\psi(d, \theta)$ in place of $\psi(d)$. In particular, if y_i's are independently and identically distributed with common mean $\mu(\theta)$, then $\psi(d, \theta) = \mu(\theta)$ and the Bayes estimator T^* reduces to

$$T^* = N\left[f_s\overline{y}_s + (1 - f_s)\mu(\theta)\right] \tag{6.6.5}$$

Remark 6.6.2

In case $h(\mathbf{y}|\theta)$ involves an unknown parameter θ and θ has a prior distribution $F(\theta|\gamma)$, where γ is a known parameter vector, then the Bayes estimator T^* may be written as

$$T^* = N\left[f_s\bar{y}_s + (1 - f_s)\psi(d, \gamma)\right] \tag{6.6.6}$$

In particular, if y_i's are independently and identically distributed with common mean $\mu(\theta)$, then the Bayes estimator T^* becomes

$$T^* = N\left[f_s\bar{y}_s + (1 - f_s)E_\theta\{\mu(\theta|d, \gamma)\}\right] \tag{6.6.7}$$

where $E_\theta\{\mu(\theta|d, \gamma)\}$ is the conditional expectation $\mu(\theta)$, given the data d.

Example 6.6.1 (Ericson, 1969a,b)

Let y_i's $(i = 1, \ldots, N)$ be independently normally distributed random variables with unknown mean θ and known variance σ^2. Suppose further that θ is normally distributed with the known mean θ_0 and known variance σ_0^2. Here the parameter θ is normally distributed with the known parameter vector $\psi = (\mu_0, \sigma_0)$. It is well known that the posterior distribution of θ given the data $d = \{(i, y_i);\ i \in s\}$ is normally distributed with mean $E_\theta(\theta|d, \gamma) = w_0\bar{y}_s + (1 - w_0)\mu_0$ and variance $\left(\nu_s/\sigma^2 + 1/\sigma_0^2\right)^{-1}$, where $w_0 = \left(\nu_s/\sigma^2\right)\left(\nu_s/\sigma^2 + 1/\sigma_0^2\right)^{-1}$. Hence using Eq. (6.6.7), the Bayes estimator for the population total is obtained as

$$T^* = N\left[f_s\bar{y}_s + (1 - f_s)\left\{w_0\bar{y}_s + (1 - w_0)\mu_0\right\}\right].$$

Example 6.6.2

Suppose that a sample s of size n_s is selected from a finite population of size N. Let n_A and N_A be the number of units that belong to the certain category A (HIV+, say) in the sample and population, respectively. Here $y_i = 1$ if $i \in A$ and $y_i = 0$ if $i \notin A$. Let us assume that y_i's are independent Bernoulli variables with parameter π, and the prior distribution of π has the form $p(\pi) \propto \pi^{\alpha-1}(1 - \pi)^{\beta-1}$ with $\alpha, \beta > 0$. Here, n_A follows the binomial distribution $B(n_s, \pi)$ and the posterior distribution of π, given n_A, has the form $p(\pi|n_A) \propto \pi^{\alpha+n_A-1}(1 - \pi)^{\beta+n_s-n_A-1}$, a beta distribution with parameter $(\alpha + n_A,\ \beta + n_s - n_A)$. The Bayes estimator of the population proportion $\pi_A = N_A/N$ is given by

$$\begin{aligned}\widehat{\pi}_A &= \left[f_sP_A + (1 - f_s)E(\pi|n_A)\right] \\ &= \left[f_sP_A + (1 - f_s)\frac{\alpha + n_sP_A}{\alpha + \beta + n_s}\right]\end{aligned}$$

where $P_A = n_A/n_s =$ proportion of HIV+ persons in the sample.

6.7 COMPARISON OF STRATEGIES UNDER SUPERPOPULATION MODELS

In this section, we will compare the relative efficiencies of a few well-known sampling strategies under the superpopulation model

$$M_1^*: E_\xi(y_i) = \beta x_i,\ V_\xi(y_i) = \sigma^2 v(x_i) \text{ and } C_\xi(y_i, y_j) = 0 \text{ for } i \neq j = 1, \ldots, N \tag{6.7.1}$$

(i) HH strategy (H_1): HH estimator $\widehat{Y}_{hh} = \frac{1}{n}\sum_{r=1}^{n}\frac{y(r)}{p(r)}$ (defined in Section 5.2.3) based on PPSWR sampling design of size.

(ii) HT strategy (H_2): HT estimator $\widehat{Y}_{ht} = \sum_{i \subset s}\frac{y_i}{\pi_i}$ based on an IPPS sampling design of fixed effective size n.

(iii) Rao–Hartley–Cochran (RHC) strategy (H_3): RHC estimator $\widehat{Y}_{rhc} = \sum_{i \in s}\frac{y_i}{p_i}P_i$ based on RHC sampling scheme (defined in Section 5.6).

(iv) Lahiri–Midzuno–Sen (LMS) strategy (H_4): Unbiased ratio estimator $\widehat{Y}_{lms} = \frac{y_s}{x_s}X$ based on probability proportional to aggregate size (PPAS) sampling scheme (defined in Section 5.5).

(v) Ratio estimator strategy (H_5): Ratio estimator $\widehat{Y}_R = \frac{y_s}{x_s}X$ based on SRSWOR sampling scheme.

(vi) Ratio estimator strategy (H_6): Ratio estimator $\widehat{Y}_R^* = \frac{y_s}{x_s}X$ based on SRSWR sampling scheme.

The expected variances of the design-unbiased estimators $\widehat{Y}_{hh}$, $\widehat{Y}_{ht}$, $\widehat{Y}_{rhc}$, $\widehat{Y}_{lms}$ and approximate mean-square errors of the approximate design unbiased estimators $\widehat{Y}_R$ and $\widehat{Y}_R^*$ are given under the model M_1^* as follows:

$$\begin{aligned}
E_\xi\left[V\left(\widehat{Y}_{hh}\right)\right] &= \frac{1}{n}E_\xi\left(\sum_{i \in U}\frac{y_i^2}{p_i} - Y^2\right) \\
&= \frac{\sigma^2}{n}\sum_{i \in U} v(x_i)\left(\frac{1}{p_i} - 1\right) \\
&= \frac{\sigma^2}{n}\sum_{i \in U}\frac{v(x_i)}{x_i}(X - x_i) \\
&= Q_1 \text{(say)}
\end{aligned}$$

$$E_\xi\left[V\left(\widehat{Y}_{ht}\right)\right] = E_\xi\left[\sum_{i\in U} y_i^2\left(\frac{1}{\pi_i}-1\right) + \sum_{i\neq}\sum_{j\in U} y_i y_j\left(\frac{\pi_{ij}}{\pi_i\pi_j}-1\right)\right]$$

$$= \sum_{i\in U}\left\{\sigma^2\nu(x_i)+\beta^2x_i^2\right\}\left(\frac{1}{\pi_i}-1\right) + \beta^2\sum_{i\neq}\sum_{j\in U} x_i x_j\left(\frac{\pi_{ij}}{\pi_i\pi_j}-1\right)$$

For a $p(\pi, x)$ sampling $\pi_i = np_i = nx_i/X$ and we get

$$E_\xi\left[V\left(\widehat{Y}_{ht}\right)\right] = \sigma^2\sum_{i\in U}\nu(x_i)\left(\frac{1}{np_i}-1\right)$$

$$= \sigma^2\sum_{i\in U}\frac{\nu(x_i)}{x_i}\left(\frac{X}{n}-x_i\right)$$

$$= Q_2(\text{say})$$

$$E_\xi\left[V\left(\widehat{Y}_{rhc}\right)\right] = \frac{N-n}{n(N-1)}E_\xi\left(\sum_{i\in U}\frac{y_i^2}{p_i}-Y^2\right)$$

$$= \sigma^2\frac{N-n}{n(N-1)}\sum_{i\in U}\nu(x_i)\left(\frac{1}{p_i}-1\right)$$

$$= \sigma^2\frac{N-n}{n(N-1)}\sum_{i\in U}\frac{\nu(x_i)}{x_i}(X-x_i)$$

$$= Q_3\ (\text{say})$$

$$E_\xi\left[V\left(\widehat{Y}_{lms}\right)\right] = E_\xi\left(\sum_{i\in U} T_i y_i^2 + \sum_{i\neq}\sum_{j\in U} T_{ij}y_i y_j - Y^2\right)\ (\text{using Theorem 5.5.1})$$

where $T_i = X\sum_{s\supset i}\frac{1}{x_s}\bigg/\binom{N-1}{n-1}$ and $T_{ij} = X\sum_{s\supset i,j}\frac{1}{x_s}\bigg/\binom{N-1}{n-1}$.

Hence,

$$E_\xi\left[V\left(\widehat{Y}_{lms}\right)\right] = \sigma^2\sum_{i\in U}(T_i-1)\nu(x_i) + \beta^2\left(\sum_{i\in U}T_i x_i^2 + \sum_{i\neq}\sum_{j\in U}T_{ij}x_i x_j - X^2\right)$$

Finally, noting $\sum_{i\in U} T_i x_i^2 + \sum\sum_{i\neq j\in U} T_{ij} x_i x_j = X^2$ (Rao, T.J., 1967a,b), we get

$$E_\xi\left\{V\left(\widehat{Y}_{lms}\right)\right\} = \sigma^2 \sum_{i\in U} \frac{\nu(x_i)}{x_i}\left(T_i x_i - x_i\right)$$

$$= Q_4 \text{ (say)}$$

An approximate mean-square error of $\widehat{Y}_R$ is given by

$$MSE\left(\widehat{Y}_R\right) = \frac{N-n}{Nn}\frac{1}{N-1}\sum_{i\in U}\left(y_i - Rx_i\right)^2 \text{ where}$$

$$R = Y/X \text{ (vide Eq. 8.4.7)}$$

and

$$E_\xi\left[MSE\left(\widehat{Y}_R\right)\right] = \frac{N-n}{Nn}\frac{1}{N-1}E_\xi\sum_{i\in U}\left(y_i - Rx_i\right)^2$$

$$= Q_5 \text{ (say)}$$

The expected approximate mean-square error of $\widehat{Y}_R^*$ is

$$E_\xi\left[MSE\left(\widehat{Y}_R^*\right)\right] = \frac{1}{n}\frac{1}{N}E_\xi\sum_{i\in U}\left(y_i - Rx_i\right)^2$$

$$= Q_6 \text{ (say)}$$

6.7.1 Hansen–Hurwitz Strategy With Others

The strategy H_i is said to be better than that of H_j under the model M_1^* if $Q_i < Q_j$ for $i \neq j$. It is well known that H_1, the HH strategy, is inferior to RHC strategy H_3 since $V\left(\widehat{Y}_{rhc}\right) < V\left(\widehat{Y}_{hh}\right)$. The strategy H_1 is also inferior to the HT strategy H_2 under the superpopulation model M_1^* since $Q_1 > Q_2$. However, no exact comparisons between the strategies H_1 and H_6 or H_3 and H_5 are available under the model M_1^*. However, if $\nu(x_i) = x_i^g$, following Cassel et al. (1977), we note that for a large N and n, the strategy H_1 is approximately superior, equal, or inferior to the alternative H_6 if $g > 1$, $=$ or <1, respectively.

6.7.2 Horvitz–Thompson and Rao–Hartley–Cochran Strategy

Following Hanurav (1967), we get

$$Q_2 - Q_3 = \sigma^2 \sum_{i \in U} \frac{\nu(x_i)}{x_i} \left(\frac{X}{n} - x_i - \frac{N-n}{n(N-1)} X + \frac{N-n}{n(N-1)} x_i \right)$$

$$= -\frac{N(n-1)}{n(N-1)} \sigma^2 \sum_{i \in U} \frac{\nu(x_i)}{x_i} (x_i - \overline{X})$$

$$= -\frac{N^2(n-1)}{n(N-1)} \sigma^2 Cov\left(\frac{\nu(x_i)}{x_i}, x_i \right)$$

Thus $Q_2 - Q_3$ is $<$, $=$ or >0 if $\nu(x_i)/x_i$ is increasing, constant, or decreasing function of x_i, respectively.

6.7.3 Horvitz–Thompson and Lahiri–Midzuno–Sen Strategy

Following Rao T.J. (1967a,b), we find

$$Q_2 - Q_4 = \sigma^2 \sum_{i \in U} \frac{\nu(x_i)}{x_i} \left(\frac{X}{n} - x_i - T_i x_i + x_i \right)$$

$$= -\sigma^2 N Cov\left(\frac{\nu(x_i)}{x_i}, T_i x_i \right)$$

$$\left(\text{since } \sum_{i \in U} T_i x_i = NX/n \right)$$

Thus $Q_2 - Q_4$ is $<$, $=$ or >0 if $\nu(x_i)/x_i$ is increasing, constant, or decreasing function of x_i respectively, because $T_i x_i$ is an increasing function of x_i as observed by Rao T.J. (1967a,b).

6.7.4 Rao–Hartley–Cochran and Lahiri–Midzuno–Sen Strategy

Following Chaudhuri and Arnab (1979a,b), we note that

$$Q_3 - Q_4 = \sigma^2 \sum_{i \in U} \frac{\nu(x_i)}{x_i} \left(\frac{N-n}{n(N-1)} X - \frac{N-n}{n(N-1)} x_i - T_i x_i + x_i \right)$$

$$= \sigma^2 \sum_{i \in U} \frac{\nu(x_i)}{x_i} z_i$$

where $z_i = \dfrac{N-n}{n(N-1)} X - T_i x_i + \dfrac{N(n-1)}{n(N-1)} x_i.$

Noting $\sum_{i\in U} z_i = \frac{(N-n)N}{n(N-1)}X - \frac{N}{n}X + \frac{N(n-1)}{n(N-1)}X = 0$ as $\sum_{i\in U} T_i x_i = NX/n$, we can write

$$Q_3 - Q_4 = \sigma^2 NCov\left(\frac{v(x_i)}{x_i}, z_i\right) \tag{6.7.2}$$

Furthermore, noting $T_i = X\sum_i \frac{1}{x_i + x_{i_2} + \ldots + x_{i_n}} \Big/ \binom{N-1}{n-1}$, where $\sum_i$ is the sum over distinct $n-1$ integers $i_2, i_3, \ldots, i_n$ $(i_2 \neq i_3 \neq \ldots \neq i_n)$ from $1, 2, \ldots, N(\neq i)$

and

$$T_i + x_i\frac{\partial T_i}{\partial x_i} = \left[X\sum_i \frac{1}{x_i + x_{i_2} + \cdots + x_{i_n}} + x_i\left\{\sum_i \frac{1}{x_i + x_{i_2} + \cdots + x_{i_n}} - X\sum_i \frac{1}{(x_i + x_{i_2} + \cdots + x_{i_n})^2}\right\}\right] \Big/ \binom{N-1}{n-1}$$

$$= \sum_i \frac{x_i(x_i + x_{i_2} + \cdots + x_{i_n}) + X(x_{i_2} + \cdots + x_{i_n})}{(x_i + x_{i_2} + \cdots + x_{i_n})^2} \Big/ \binom{N-1}{n-1}$$

$$\geq \sum_i \frac{x_i(x_i + x_{i_2} + \cdots + x_{i_n}) + (x_i + x_{i_2} + \cdots + x_{i_n})(x_{i_2} + \cdots + x_{i_n})}{(x_i + x_{i_2} + \cdots + x_{i_n})^2} \Big/ \binom{N-1}{n-1}$$

$$= \sum_i 1 \Big/ \binom{N-1}{n-1} = 1$$

We conclude that z_i is a decreasing function of x_i as $\frac{\partial z_i}{\partial x_i} = 1 - \left(T_i + x_i\frac{\partial T_i}{\partial x_i}\right) \leq 0$.

Hence, $Q_3 - Q_4$ is <, = or >0 if $v(x_i)/x_i$ is increasing, constant, or decreasing function of x_i, respectively. These findings lead to the following theorem (Chaudhuri and Arnab, 1979a,b).

Theorem 6.7.1

Under the superpopulation model M_1^*,

$$E_\xi\left[V\left(\widehat{Y}_{ht}\right)\right] < E_\xi\left[V\left(\widehat{Y}_{rhc}\right)\right] < E_\xi\left[V\left(\widehat{Y}_{lms}\right)\right]$$
if $v(x_i)/x_i$ is an increasing function of x_i

$$E_\xi\left[V\left(\widehat{Y}_{ht}\right)\right] = E_\xi\left[V\left(\widehat{Y}_{rhc}\right)\right] = E_\xi\left[V\left(\widehat{Y}_{lms}\right)\right]$$
if $v(x_i)/x_i$ is a constant

$$E_\xi\left[V\left(\widehat{Y}_{ht}\right)\right] > E_\xi\left[V\left(\widehat{Y}_{rhc}\right)\right] > E_\xi\left[V\left(\widehat{Y}_{lms}\right)\right]$$
if $v(x_i)/x_i$ is a decreasing function of x_i

Corollary 6.7.1

In particular if $v(x_i) = \sigma^2 x_i^g$, the theorem above reduces to

$$\begin{aligned}
E_\xi\left[V\left(\widehat{Y}_{ht}\right)\right] &< E_\xi\left[V\left(\widehat{Y}_{rhc}\right)\right] < E_\xi\left[V\left(\widehat{Y}_{lms}\right)\right] \text{ if } g > 1\\
E_\xi\left[V\left(\widehat{Y}_{ht}\right)\right] &= E_\xi\left[V\left(\widehat{Y}_{rhc}\right)\right] = E_\xi\left[V\left(\widehat{Y}_{lms}\right)\right] \text{ if } g = 1\\
E_\xi\left[V\left(\widehat{Y}_{ht}\right)\right] &> E_\xi\left[V\left(\widehat{Y}_{rhc}\right)\right] > E_\xi\left[V\left(\widehat{Y}_{lms}\right)\right] \text{ if } g < 1
\end{aligned}$$

Remark 6.7.1

Rao (1966a) proved that the inequality $E_\xi\left\{V\left(\widehat{Y}_{rhc}\right)\right\} - E_\xi\left\{V\left(\widehat{Y}_{lms}\right)\right\}$ <, = or >1 above, for values of g >, = or <1, respectively, when both the population and sample sizes N and n are large. But the results by Chaudhuri and Arnab (1979a,b) above are exact.

Remark 6.7.2

Under the models M_1^* with $v(x_i) = x_i^g$, Rao (1966a) showed that for a large n $E_\xi MSE\left(\widehat{Y}_R\right) \approx E_\xi MSE\left(\widehat{Y}_{lms}\right)$. Hence using the Corollary 6.7.1, we note that for $g > 1$, both the HT and RHC strategies are more efficient than the ratio strategies based on SRSWOR; whereas for $g < 1$, the ratio strategy fares better than the HT and RHC Strategies. For $g = 1$, all three strategies are equally efficient. Cassel et al. (1977) cited that H_3 is approximately superior, equal, and inferior to H_6 according to $g \geq$ or <1. It is well known that H_5 and H_6 are approximately equal for large N.

6.8 DISCUSSIONS

In a design-based inference or a fixed population setup, the vector of the variable under study **y** is a fixed point in the N-dimensional parametric space R^N. Here, a sample is selected by using a sampling design and expectation of an estimator is the average value of all possible samples that can be selected through the sampling design. It was found that the minimum variance unbiased estimator for estimating a finite population total or mean under a given sampling design p does not exist, and hence one cannot find the best strategy.

In the superpopulation approach, which is also known as model-based or prediction approach, we assume that the vector of the study variable **y** is a realization of a random vector **Y**. The distribution of the random vector **Y** is denoted here by ξ and it is called the superpopulation model or simply model. The superpopulation model may depend on a set of known auxiliary variables involving unknown model parameters. The model-based approach yields the minimum variance unbiased (optimal) estimator in presence of a given sampling design, and it also yields the best sampling strategy in certain classes of strategies. The best strategy very often involves a purposive sampling design. The main demerit of the model-based approach is that its performance is highly dependent on the assumed model. The performance of the optimum strategy becomes poor if the assumed model fails to hold in reality. We call an estimator "robust" if it remains optimum, not only when the assumed model is true, but also in a wider class of models. A balanced sampling design retains optimality of the estimators in wider class of models. It is very difficult to find a balanced design that can control both the bias and the mean-square error together.

In a Bayesian model, the random vector **Y** is assumed to be distributed with one or more unknown parameters and the parameters have some known prior distribution. In Bayesian inference, the posterior distribution of the vector **Y**, given the data, is independent of the sampling design, and hence here the sampling design p has no role in making the inference of any parametric function of **Y**.

The model-design based inference or model-assisted inference is a hybrid of the design-based and model-based inference. Here also the optimum strategies for estimating certain parametric functions such as total or mean are available. The model-assisted strategies are found to be robust in the sense that it provides valid inferences in the event of failure of the assumed model.

Attempts are made to compare performances of the existing popular strategies under various superpopulation models. A good review is given by Cassel et al. (1977). Details of the comparisons of the HH, HT, LMS, and RHC strategies are given under model M_1^*. It is found that each of the strategies HT, LMS, and RHC perform better that than the HH strategy. The HT strategy is the best if $\nu(x)$ is an increasing function of x whereas LMS is the best if $\nu(x)$ is the decreasing function of x. Strategies HT, LMS, and RHC are equally efficient if $\nu(x)$ is proportional to x.

6.9 EXERCISES

6.9.1 Explain the concept of design-based, model-based, and model-assisted inference with reference to a finite population sampling. Show that the design and model unbiasedness imply model design unbiasedness but the converse may not be true.

6.9.2 Find optimum ξ-unbiased linear predictors for the population mean under a given sampling design and the following models, where α, β, γ and σ are unknown but ρ is known.

(a) $E_\xi(y_i) = \alpha + \beta i$, $V_\xi(y_i) = \sigma^2 i$ and $C_\xi(y_i, y_j) = 0$ for $i \neq j$; $i, j = 1,\ldots, N$

(b) $E_\xi(y_i) = \alpha + \beta i + \gamma i^2$, $V_\xi(y_i) = \sigma^2$ and $C_\xi(y_i, y_j) = 0$ for $i \neq j$; $i, j = 1,\ldots, N$

(c) $E_\xi(y_i) = \alpha + \beta x_i$, $V_\xi(y_i) = \sigma^2 x_i^2$ and $C_\xi(y_i, y_j) = \rho\sigma^2 x_i x_j$ for $i \neq j; i, j = 1, \ldots, N$

6.9.3 State the optimality of the HT estimators under different superpopulation models. Derive optimality of the expansion estimator based on an SRSWOR sampling design as a special case.

6.9.4 Consider the superpopulation model $E_\xi(y_i) = \mu$, $V_\xi(y_i) = \sigma^2$ and $C_\xi(y_i, y_j) = 0$.

(i) Derive the optimal estimators for the population mean for a given sampling design p in the class of (a) p-unbiased, (b) ξ-unbiased, and (c) $p\xi$-unbiased predictors.

(ii) Derive optimal sampling strategies under p-unbiased, ξ-unbiased, and $p\xi$-unbiased class predictors, respectively.

6.9.5 Let $y_1,\ldots, y_n$ be independently distributed random variables with $E(y_i) = \mu(\theta)$ and $V(y_i) < \infty$ for $i = 1,\ldots, n$ and let the posterior expectation of $\mu(\theta)$ be given $\mathbf{y} = (y_1,\ldots, y_n)$ is $E\{\mu(\theta)|\mathbf{y}\} = \alpha\overline{y} + \beta$, where α and β are independent of $\mathbf{y}$. Then show that $E\{\mu(\theta)|\mathbf{y}\} = \dfrac{\overline{y}V\{\mu(\theta)\} + mE_\theta V(\overline{y}|\theta)}{V\{\mu(\theta)\} + E_\theta V(\overline{y}|\theta)}$, where $\overline{y} = \sum_{i=1}^{n} y_i/n$ (Ericson, 1969a,b).

6.9.6 Let a finite population $y_1,\ldots, y_n$ be viewed as a sample from a normal superpopulation with unknown mean θ and unknown variance $1/h$. Suppose further that the prior distribution of θ and h have joint density $f(\theta, h) \propto h^{-1}$, $h > 0, -\propto < \theta < \propto$. Obtain the Bayes estimator of the population mean $\overline{Y}$ based on a sample s of n distinct units (Ericson, 1969a,b).

6.9.7 Consider the superpopulation model: $E_\xi(y_i) = \alpha_i + \beta x_i$, $V_\xi(y_i) = \sigma^2 f_i$, $C_\xi(y_i, y_j) = \rho\sigma^2(f_i f_j)^{1/2}$, $i \neq j$; $i, j = 1,\ldots, N$ where α_i, β, x_i, and f_i are known and ρ and σ^2 are unknown. Let C_{lu} be a linear unbiased estimator of the population total Y and $\mathcal{P}_n$ be FESD(n). Then show that

(i) $E_p E_\xi(T - Y)^2 \geq (1-\rho)\sigma^2\left\{\left(\sum_{i\in U} f_i^{1/2}\right)^2 \Big/ n - \sum_{i\in U} f_i\right\}$ for any $T \in C_{lu}$ and $p \in \mathcal{P}_n$

(ii) $E_{p_0} E_\xi(T_0 - Y)^2 = (1-\rho)\sigma^2\left\{\left(\sum_{i\in U} f_i^{1/2}\right)^2 \Big/ n - \sum_{i\in U} f_i\right\}$

where $T_0 = \sum_{i\in s}(y_i - \alpha_i - \beta x_i)/\pi_{i0} + \sum_{i\in U}(\alpha_i + \beta x_i)$

and p_0 is an FESD(n) with inclusion probability $\pi_i = \pi_{i0} = n f_i^{1/2} \Big/ \sum_{i\in U} f_i^{1/2}$ (Tam, 1984).

6.9.8 Let $\widehat{T}_g$ be the ξ-BLUE for the total Y for a given sampling design under the model

$$E_\xi(y_i) = \beta x_i,\ V_\xi(y_i) = \sigma^2 v(x_i) \text{ and } C_\xi(y_i, y_j) = 0$$
$$\text{for } i \neq j;\ i, j = 1, \ldots, N.$$

Then, for $0 \leq g \leq h$, show that (a) $MSE\left(\widehat{T}_g\right) \leq MSE\left(\widehat{T}_h\right)$ if $v(x_i)/x_i^g$ is nonincreasing and (b) $MSE\left(\widehat{T}_g\right) \leq MSE\left(\widehat{T}_h\right)$, if $v(x_i)/x_i^g$ is nondecreasing (Royall, 1970).

CHAPTER 7

Stratified Sampling

7.1 INTRODUCTION

The efficiency of an estimator can be increased by increasing the sample size. It is not wise to select a sample from a large population at a time because it is seldom representative in the sense that the sample units may be distributed unevenly over the population. However, the efficiency can also be increased greatly by dividing the population into homogeneous groups or layers (strata) and then selecting samples from each of the groups separately. Stratified sampling is commonly used in large-scale surveys. In this chapter we will discuss in detail various aspects of stratified sampling.

7.2 DEFINITION OF STRATIFIED SAMPLING

In stratified sampling, the entire population U of N units is divided into a number (K) of mutually exclusive and exhaustive groups, which are called strata. The ith stratum U_i consists of N_i units with $U = \bigcup_{i=1}^{K} U_i$ and $\sum_{i=1}^{K} N_i = N$. From each of the strata, samples of suitable sizes n_i's are selected independently by some suitable sampling design. The sampling procedure for the different strata need not be the same and may depend on the available information about the respective stratum. For example, suppose we want to select a sample of 50,000 households from South Africa to estimate the total number of unemployed persons in South Africa. In this situation, one may divide South Africa into a few mutually exclusive and exhaustive zones such as provinces (strata), and from each of the provinces, samples of households may be selected independently. Suppose further that the number of unemployed persons (x) in the households of some of the provinces is available from the past surveys. Then, using the x values as an auxiliary variable, one can select samples by using the inclusion probability proportional to size (IPPS) sampling procedure for those strata, where x values are available. Thus for stratified sampling, one should carefully consider the problem of forming strata and its number, sampling procedures for different strata, and allocation of sample sizes to the respective stratum.

Survey Sampling Theory and Applications
ISBN 978-0-12-811848-1
http://dx.doi.org/10.1016/B978-0-12-811848-1.00007-8

7.3 ADVANTAGES OF STRATIFIED SAMPLING

Stratified sampling is used in most large-scale surveys because of its various advantages, some of which are described below:

(i) Estimation of subpopulations: In cases where the estimates of the population characteristics are needed not only for the entire population but also for its different subpopulations, one should treat such subpopulations as strata. For example, in a national unemployment survey, the government may be interested in estimating unemployment figures for the entire country as well as at provincial levels. In this case, each province can be taken as a stratum.

(ii) Administrative convenience: The agency conducting the survey may stratify the population such that the survey can be supervised in an efficient manner, e.g., the agency can appoint separate supervisors to conduct survey for each of the strata separately.

(iii) Representativeness of sample: In stratified sampling, formation of strata and allocation of samples to different strata may de done in such a way that the sample can represent the population with respect to the characteristics under study. For instance, if we want to select a sample of students from a school, which represents the different races of South Africa, a simple random sampling without replacement (SRSWOR) sample from the entire school may not be representative. In this situation, a stratified sampling using different racial groups as strata is expected to provide a more representative sample than an SRSWOR sample from the entire school.

(iv) Efficiency: Stratification may increase efficiency of the estimates by forming strata in such a way that each stratum becomes homogeneous with respect to the characteristic under study. Suitable sampling schemes to the respective strata may increase efficiencies of the estimators.

(v) Improved quality of data: Improved quality of data may be obtained by employing different types of investigators to different strata. For example, investigators knowing local languages may be deployed to the rural areas, whereas in urban areas investigators knowing English may be more advantageous.

7.4 ESTIMATION PROCEDURE

Consider a population U consisting of N units, which has been divided into K strata and the ith stratum U_i consists of N_i units and $\sum_{i=1}^{K} N_i = N$. Suppose from the ith stratum a sample s_i of size n_i is selected by some suitable sampling procedure. Samples from each of the stratum are selected independently. Here we will use the following notations:

y_{ij} = value of the character y under study for the jth unit of the ith stratum, $j = 1,\ldots, N_i$, $i = 1,\ldots, K$; $Y_i = \sum_{j=1}^{N_i} y_{ij}$ = ith stratum total; $\overline{Y}_i = Y_i/N_i$ = ith stratum mean, $Y = \sum_{i=1}^{K} Y_i$ = population total; $\overline{Y} = \sum_{i=1}^{K}\sum_{j=1}^{N_i} y_{ij}/N = \sum_{i=1}^{K} W_i\overline{Y}_i$ = population mean, $W_i = N_i/N$ = ith stratum weight; $S_{yi}^2 = \sum_{j=1}^{N_i}\left(y_{ij} - \overline{Y}_i\right)^2/(N_i - 1)$ = ith stratum variance, $S_y^2 = \sum_{i=1}^{K}\sum_{j=1}^{N_i}\left(y_{ij} - \overline{Y}\right)^2/(N - 1)$ = population variance

	Stratum				
	1	...	i	...	K
	y_{11}	...	y_{i1}	...	y_{K1}
	...	...	...	...	...
	y_{1j}	...	y_{ij}	...	y_{Kj}
	...	...	...	...	...
	y_{1N_1}	...	y_{iN_i}	...	y_{KN_K}
Stratum size	N_1	...	N_i	...	N_K
Stratum weight	W_1	...	W_i	...	W_K
Stratum total	Y_1	...	Y_i	...	Y_K
Stratum mean	$\overline{Y}_1$	...	$\overline{Y}_i$	...	$\overline{Y}_K$
Stratum variance	S_{y1}^2	...	S_{yi}^2	...	S_{yK}^2

7.4.1 Estimation of Population Mean

Let the sample s_i of size n_i be selected from the stratum U_i with probability $p(s_i)$ by using a sampling design $p^{(i)}$. Let $\pi_{j|i}(>0)$ and $\pi_{jk|i}(>0)$ be the inclusion probabilities for the jth unit and jth and kth $(j \neq k)$ for the ith stratum, respectively. An unbiased estimator for the ith stratum mean $\overline{Y}_i$ based on the selected sample s_i is given by

$$\widehat{\overline{Y}}_i = \sum_{j \in s_i} b_j(s_i) y_{ij} \tag{7.4.1}$$

where $b_j(s_i)$'s are constants free from y_{ij}'s and satisfy the unbiasedness condition

$$\sum_{s_i \supset j} b_j(s_i) p(s_i) = 1/N_i \text{ for } \forall j \in U_i; \quad i = 1, \ldots, N.$$

Theorem 7.4.1

(i) $\widehat{\overline{Y}}_{st} = \sum_{i=1}^{K} W_i \widehat{\overline{Y}}_i$ is an unbiased estimator for mean $\overline{Y}$

(ii) $V\left(\widehat{\overline{Y}}_{st}\right) = \sum_{i=1}^{K} W_i^2 Q_i$,

where

$Q_i = V\left(\widehat{\overline{Y}}_i\right) = \sum_{j \in U_i} \alpha_{j|i} y_{ij}^2 + \sum_{j \neq}\sum_{k \in U_i} \alpha_{jk|i} y_{ij} y_{ik}$,

$\alpha_{j|i} = \sum_{s_i \supset j} b_j^2(s_i) p(s_i) - 1/N_i^2$ and $\alpha_{jk|i} = \sum_{s_i \supset j,k} b_j(s_i) b_k(s_i) p(s_i) - 1/N_i^2$

(iii) An unbiased estimator of $V\left(\widehat{\overline{Y}}\right)$ is

$$\widehat{V}\left(\widehat{\overline{Y}}_{st}\right) = \sum_{i=1}^{K} W_i^2 \widehat{Q}_i,$$

where $\widehat{Q}_i = \sum_{j \in s_i} \frac{\alpha_{j|i}}{\pi_{j|i}} y_{ij}^2 + \sum_{j \neq}\sum_{k \in s_i} \frac{\alpha_{jk|i}}{\pi_{jk|i}} y_{ij} y_{ik}$

Proof

(i) $E\left(\widehat{\overline{Y}}_{st}\right) = \sum_{i=1}^{K} W_i E\left(\widehat{\overline{Y}}_i\right) = \sum_{i=1}^{K} W_i Y_i = \overline{Y}$

(ii) Since samples are selected independently from each stratum, we get

$$V\left(\widehat{\overline{Y}}_{st}\right) = \sum_{i=1}^{K} W_i^2 \, V\left(\widehat{\overline{Y}}_i\right),$$

$$\text{Now } V\left(\widehat{\overline{Y}}_i\right) = \sum_{s_i}\left(\sum_{j \in s_i} b_j(s_i) y_{ij}\right)^2 p(s_i) - \overline{Y}_i^2$$

$$= \sum_{j \in U_i} y_{ij}^2 \left(\sum_{s_i \supset j} b_j^2(s_i) p(s_i)\right)$$

$$+ \sum_{j \neq}\sum_{k \in U_i} y_{ij} y_{ik}\left(\sum_{s_i \supset j,k} b_j(s_i) b_k(s_i) p(s_i)\right) - \overline{Y}_i^2 = Q_i$$

$$\text{(iii)} \quad E\left[\widehat{V}\left(\widehat{\overline{Y}}_{st}\right)\right] = \sum_{i=1}^{K} W_i^2 E\left(\widehat{Q}_i\right)$$

$$= \sum_{i=1}^{K} W_i^2 Q_i = V\left(\widehat{\overline{Y}}_{st}\right)$$

7.4.1.1 Arbitrary Fixed Sample Size Design

Let s_i's be selected by some fixed effective size sampling design of size $n_i(FESD(n_i))$ and $b_j(s_i) = \frac{1}{N_i \pi_{j|i}}$. Then, $\widehat{\overline{Y}}_i = \sum_{j\in s_i} \frac{y_{ij}}{N_i \pi_{j|i}}$ and $\overline{Y}_{st} = \sum_{i=1}^k W_i \sum_{j\in s_i} \frac{y_{ij}}{N_i \pi_{j|i}}$

$$Q_i = V\left(\widehat{\overline{Y}}_i\right) = \frac{V\left(\widehat{Y}_i\right)}{N_i^2} = \frac{1}{N_i^2}\left[\sum_{j<}\sum_{k\in U_i} \Delta_{jk|i}\left(\frac{y_{ij}}{\pi_{j|i}} - \frac{y_{ik}}{\pi_{k|i}}\right)^2\right]$$

where $\Delta_{jk|i} = \pi_{j|i}\pi_{k|i} - \pi_{jk|i}$.

$$V\left(\widehat{\overline{Y}}_{st}\right) = \frac{1}{N^2}\sum_{i=1}^{k}\sum_{j<}\sum_{k\in U_i} \Delta_{jk}\left(\frac{y_{ij}}{\pi_{j|i}} - \frac{y_{ik}}{\pi_{k|i}}\right)^2$$

Since Q_i can be estimated unbiasedly by any of the following

$$\widehat{Q}_i(1) = \frac{1}{N_i^2}\left[\sum_{j\in s_i}\frac{1}{\pi_{j|i}}\left(\frac{1}{\pi_{j|i}} - 1\right)y_{ij}^2 + \sum_{j\neq}\sum_{k\in s_i}\frac{1}{\pi_{jk|i}}\left(\frac{\pi_{jk|i}}{\pi_{j|i}\pi_{k|i}} - 1\right)y_{ij}y_{ik}\right] \text{ and}$$

$$\widehat{Q}_i(2) = \frac{1}{N_i^2}\sum_{j<}\sum_{k\in s_i}\frac{\Delta_{jk|i}}{\pi_{jk|i}}\left(\frac{y_{ij}}{\pi_{j|i}} - \frac{y_{ik}}{\pi_{k|i}}\right)^2,$$

the variance $V\left(\widehat{\overline{Y}}_{st}\right)$ can be estimated unbiasedly by any of the following estimators:

$$\sum_{i=1}^{K} W_i^2\widehat{Q}_i(1) \text{ and } \sum_{i=1}^{K} W_i^2\widehat{Q}_i(2).$$

7.4.1.2 Simple Random Sampling Without Replacement

For SRSWOR, $\pi_{j|i} = n_i/N_i$ and $\pi_{jk|i} = n_i(n_i - 1)/\{N_i(N_i - 1)\}$. Substituting the values of $\pi_{j|i}$ and $\pi_{jk|i}$ in Section 7.4.1.1, we get

$$\widehat{\overline{Y}}_i = \overline{y}(s_i) = \sum_{j\in s_i} y_{ij}/n_i = \text{sample mean for the } i\text{th stratum},$$

$$\widehat{\overline{Y}}_{st} = \sum_{i=1}^{K} W_i\, \overline{y}(s_i),$$

$$Q_i = V(\overline{y}(s_i)) = (1 - f_i)S_{yi}^2\Big/n_i,\ f_i = n_i/N_i,\ \widehat{Q}_i(1) = \widehat{Q}_i(2) = \widehat{V}(\overline{y}(s_i))$$

$$= (1 - f_i)s_{yi}^2\Big/n_i,\ V\left(\widehat{\overline{Y}}_{st}\right) = \sum_{i=1}^{K} W_i^2(1 - f_i)S_{yi}^2\Big/n_i \text{ and}$$

$$\widehat{V}\left(\widehat{\overline{Y}}_{st}\right) = \sum_{i=1}^{K} W_i^2(1 - f_i)s_{yi}^2\Big/n_i$$

where
$S_{yi}^2 = \sum_{j \in U_i} \left(y_{ij} - \overline{Y}_i\right)^2 \Big/ (N_i - 1)$ and $s_{yi}^2 = \sum_{j \in s_i} \left(y_{ij} - \overline{y}(s_i)\right)^2 \Big/ (n_i - 1)$.

7.4.1.3 Probability Proportional to Size With Replacement Sampling

In case, s_i's are selected by probability proportional to size with replacement (PPSWR) method of sampling using $p_{j|i}\left(\sum_{j \in U_i} p_{j|i} = 1\right)$ as the normed size measure for the jth unit of the ith stratum, we may take $b_j(s_i) = n_j(s_i)/(N_i p_{j|i})$, where $n_j(s_i)$ = number of times jth unit appears in the sample s_i. In this situation we have

$$\widehat{\overline{Y}}_i = \frac{1}{n_i N_i} \sum_j n_j(s_i) \frac{y_{ij}}{p_{j|i}},\ Q_i = V\left(\widehat{\overline{Y}}_i\right) = V_{pps|i} \Big/ \left(N_i^2 n_i\right) \text{ and}$$

$$\widehat{Q}_i = \frac{1}{N_i^2\, n_i(n_i - 1)} \sum_j n_j(s_i) \left(\frac{y_{ij}}{p_{j|i}} - N_i \widehat{\overline{Y}}_i\right)^2$$

where $V_{pps|i} = \sum_{j \in U_i} p_{j|i} \left(\frac{y_{ij}}{p_{j|i}} - Y_i\right)^2$. The expressions for $\widehat{\overline{Y}}_{st}$, $V\left(\widehat{\overline{Y}}_{st}\right)$, and $\widehat{V}\left(\widehat{\overline{Y}}_{st}\right)$ come out respectively as $\widehat{\overline{Y}}_{st} = \frac{1}{N} \sum_{i=1}^{K} \left(\frac{1}{n_i} \sum_j n_j(s_i) \frac{y_{ij}}{p_{j|i}}\right)$,

$$V\left(\widehat{\overline{Y}}_{st}\right) = \frac{1}{N^2} \sum_{i=1}^{K} \frac{V_{pps}(i)}{n_i}, \text{ and}$$

$$\widehat{V}\left(\widehat{\overline{Y}}_{st}\right) = \frac{1}{N^2} \sum_{i=1}^{K} \frac{1}{n_i(n_i - 1)} \sum_j n_j(s_i) \left(\frac{y_{ij}}{p_{j|i}} - N_i \widehat{\overline{Y}}_i\right)^2.$$

7.4.1.4 Simple Random Sampling With Replacement

The PPSWR sampling reduces to simple random sampling with replacement (SRSWR) sampling when $p_{j|i} = 1/N_i\ \forall\ j \in U_i$. So, substituting $p_{j|i} = 1/N_i$ in Section 7.4.1.3, we get $\widehat{\overline{Y}}_i = \sum_j n_j(s_i) y_{ij} / n_i = \widetilde{\overline{y}}_i$ = sample mean based on n_i units with repetition, $Q_i = \sigma_{yi}^2 \big/ n_i$, $\widehat{Q}_i = \widehat{\sigma}_{yi}^2 \big/ n_i$,

$$\widehat{\overline{Y}}_{st} = \sum_{i=1}^{K} W_i \widetilde{\overline{y}}_i,\quad V\left(\widehat{\overline{Y}}_{st}\right) = \sum_{i=1}^{K} W_i^2 \frac{\sigma_{yi}^2}{n_i}, \text{ and } \widehat{V}\left(\widehat{\overline{Y}}_{st}\right) = \sum_{i=1}^{K} W_i^2 \widehat{\sigma}_{yi}^2 \big/ n_i$$

where $\sigma_{yi}^2 = \dfrac{N_i - 1}{N_i} S_{yi}^2$ and $\widehat{\sigma}_{yi}^2 = \widetilde{s}_{yi}^2 = \dfrac{\sum_j n_j(s_i)\left(y_{ij} - \widetilde{\overline{y}}_i\right)^2}{n_i - 1}$.

7.4.2 Estimation of Population Proportion

Let π_A denote the proportion of the individuals that possess a certain attribute or group A such as HIV +ve. In this situation we write $y_{ij}=1$ if jth individual of the ith stratum possesses attribute A and $y_{ij}=0$ otherwise. Let the total number of individuals belonging to A in the ith stratum and the entire population be denoted, respectively, by $N_{iA}=\sum_{j=1}^{N_i} y_{ij}$ and $N_A=\sum_{i=1}^{K}\sum_{j=1}^{N_i} y_{ij}$. The proportion of individuals that possess the attribute A in the ith stratum and the entire population are respectively denoted by $\pi_{iA}=N_{iA}/N_i$ and $\pi_A=N_A/N=\sum_{i=1}^{K} N_{iA}/N=\sum_{i=1}^{K} W_i\,\pi_{iA}$.

7.4.2.1 Simple Random Sampling Without Replacement

Suppose samples s_i's of sizes n_i's are selected independently from each of the strata by the SRSWOR method and let n_{iA} be the number of individuals belong to A. Then from Section 7.4.1.2 we get the following theorem.

Theorem 7.4.2

(i) $\widehat{\pi}_A=\sum_{i=1}^{K} W_i\,\widehat{\pi}_{iA}$ is an unbiased estimator for mean π_A where. $\widehat{\pi}_{iA}=n_{iA}/n$

(ii) $V(\widehat{\pi}_A)=\sum_{i=1}^{K} W_i^2(1-f_i)\dfrac{N_i}{N_i-1}\dfrac{\pi_{iA}(1-\pi_{iA})}{n_i}$ with $f_i=n_i/n$

(iii) An unbiased estimator of $V(\widehat{\pi}_A)$ is

$$\widehat{V}(\widehat{\pi}_A)=\sum_{i=1}^{K} W_i^2(1-f_i)\frac{\widehat{\pi}_{iA}(1-\widehat{\pi}_{iA})}{n_i-1}$$

Proof

The theorem can be proved easily from Section 7.4.1.2 by substituting $y_{ij}=1$ for $i\in A$ and $y_{ij}=0$ for $i\notin A$ that is by writing.

$$\overline{y}(s_i)=\widehat{\pi}_{iA},\ S_{yi}^2=\frac{N_i\pi_{iA}(1-\pi_{iA})}{(N_i-1)}\ \text{and}\ s_{yi}^2=\frac{\sum_{j\in s_i}\left(y_{ij}-\overline{y}(s_i)\right)^2}{(n_i-1)}=\frac{n_i\,\widehat{\pi}_{iA}(1-\widehat{\pi}_{iA})}{(n_i-1)}.$$

7.4.2.2 Simple Random Sampling With Replacement

In case samples s_i's are selected by SRSWR method, we get the following theorem from Section 7.4.1.4 by substituting $y_{ij}=1$ for $i\in A$ and $y_{ij}=0$ for $i\notin A$.

Theorem 7.4.3

(i) $\widehat{\pi}_A = \sum_{i=1}^{K} W_i\, \widehat{\pi}_{iA}$ is an unbiased estimator for the mean π_A where $\widehat{\pi}_{iA} = n_{iA}/n$

(ii) $V(\widehat{\pi}_A) = \sum_{i=1}^{K} W_i^2 \dfrac{\pi_{iA}(1 - \pi_{iA})}{n_i}$

(iii) An unbiased estimator of $V(\widehat{\pi}_A)$ is

$$\widehat{V}(\widehat{\pi}_A) = \sum_{i=1}^{K} W_i^2 \frac{\widehat{\pi}_{iA}(1 - \widehat{\pi}_{iA})}{n_i - 1}$$

7.4.3 Interval Estimation

If the sample size in each stratum is large or the number of strata is large, we can find $(1 - \alpha) \times 100\%$ confidence intervals for the population mean $\overline{Y}$ and proportion π_A by using the central limit theorem as follows:

$$(1 - \alpha) \times 100\% \text{ confidence interval of } \overline{Y}\text{: } \widehat{\overline{Y}}_{st} \pm z_{\alpha/2}\sqrt{\widehat{V}\left(\widehat{\overline{Y}}_{st}\right)} \quad (7.4.2)$$

$$(1 - \alpha) \times 100\% \text{ confidence interval of } \pi_A\text{: } \widehat{\pi}_A \pm z_{\alpha/2}\sqrt{\widehat{V}(\widehat{\pi}_A)} \quad (7.4.3)$$

where $z_{\alpha/2}$ is the upper $\alpha/2\,\%$ point of a standard normal distribution.

In case the total number of strata and sample sizes from each of the strata is small, we find confidence intervals by replacing $z_{\alpha/2}$ in the formulae above by $t_{\alpha/2,n-K}$, which is the upper $\alpha/2 \times 100\%$ point of t distribution with $n - K$ degrees of freedom.

7.5 ALLOCATION OF SAMPLE SIZE

In this section, we will consider the method of allocation of total sample size n to different strata, i.e., the number of units that should be taken from the different strata. The optimum allocation of sample sizes to different strata are determined either by minimizing the cost of the survey, keeping the efficiency (variance) of the estimator to a certain level or by maximizing the efficiency of the survey (minimizing variance) keeping the cost of the survey to a certain level. Here we will consider the following simple cost function used by Cochran (1977).

$$C = c_0 + \sum_{i=1}^{K} n_i c_i \quad (7.5.1)$$

where C = total cost of the survey, c_0 = overall cost, which is fixed for the survey and does not depend on the sample size, and c_i = cost of surveying each unit of the ith stratum.

Let us assume the situation where the $V\left(\widehat{\overline{Y}}_{st}\right)$ can be expressed in the following form

$$V\left(\widehat{\overline{Y}}_{st}\right) = \sum_{i=1}^{K} A_i/n_i + B \quad (7.5.2)$$

where A_i and B are independent of the sample sizes n_i

7.5.1 Optimum Allocation for Fixed Cost

Here we minimize the variance $V\left(\widehat{\overline{Y}}_{st}\right) = \sum_{i=1}^{K} A_i/n_i + B$ subject to the condition that the total cost of the survey is kept fixed as C_0, i.e.,

$$C_0 = c_0 + \sum_{i=1}^{K} n_i c_i \quad (7.5.3)$$

For this minimization problem, consider

$$\Phi = \sum_{i=1}^{K} A_i/n_i + B - \lambda\left(C_0 - c_0 - \sum_{i=1}^{K} n_i c_i\right) \quad (7.5.4)$$

with λ as a Lagrange multiplier.

Now $\frac{\partial \Phi}{\partial n_i} = 0$ implies

$$n_i = \frac{\sqrt{A_i}}{\sqrt{\lambda}\sqrt{c_i}} \quad (7.5.5)$$

On substituting Eq. (7.5.5) in Eq. (7.5.3), we get

$$\frac{1}{\sqrt{\lambda}} = \frac{C_0 - c_0}{\sum_{i=1}^{K} \sqrt{A_i c_i}} \quad (7.5.6)$$

Substituting Eq. (7.5.6) in Eq. (7.5.5), the optimum value of n_i for a given cost $C = C_0$ comes out as

$$opt(n_i|C_0) = n_i(C_0) = \frac{C_0 - c_0}{\left(\sum_{i=1}^{K} \sqrt{A_i c_i}\right)} \sqrt{\frac{A_i}{c_i}} \quad (7.5.7)$$

Finally substituting $n_i = n_i(C_0)$ in Eq. (7.5.2), we get the optimum value of $V\left(\widehat{\overline{Y}}_{st}\right)$ as

$$V_{opt|C_0} = \frac{\left(\sum_{i=1}^{K} \sqrt{A_i c_i}\right)^2}{C_0 - c_0} + B \quad (7.5.8)$$

Hence we get the following theorem:

Theorem 7.5.1
If the total cost of the survey C given in Eq. (7.5.1) is fixed as C_0, then the optimum values of n_i's that minimize the variance $V\left(\widehat{\overline{Y}}_{st}\right)$ given in Eq. (7.5.2) and the minimum value of $V\left(\widehat{\overline{Y}}_{st}\right)$ with the optimum n_i's are given by $n_i(C_0) = \dfrac{C_0 - c_0}{\sum_{i=1}^{K}\sqrt{A_i c_i}}\sqrt{\dfrac{A_i}{c_i}}$ and $V_{opt|C_0} = \dfrac{\left(\sum_{i=1}^{K}\sqrt{A_i c_i}\right)^2}{C_0 - c_0} + B$, respectively.

7.5.2 Optimum Allocation for Fixed Variance

Here the optimum sample size is obtained by minimizing $C = c_0 + \sum_{i=1}^{K} n_i c_i$, the total cost of the survey keeping the variance $V\left(\widehat{\overline{Y}}_{st}\right)$ fixed at V_0. Then, following Section 7.5.1, the optimum value of n_i for a given value of $V = V_0$ is obtained as

$$opt(n_i|V_0) = n_i(V_0) = \frac{\sum_{i=1}^{K}\sqrt{A_i c_i}}{V_0 - B}\sqrt{\frac{A_i}{c_i}} \tag{7.5.9}$$

Finally, putting $n_i = n_i(V_0)$ in Eq. (7.5.2), we get the expression of the optimum cost for the given variance V_0 as

$$C_{opt|V_0} = c_0 + \frac{\left(\sum_{i=1}^{K}\sqrt{c_i A_i}\right)^2}{V_0 - B} \tag{7.5.10}$$

The results above are summarized as follows:

Theorem 7.5.2
If the variance $V\left(\widehat{\overline{Y}}_{st}\right)$ is kept as V_0, then the optimum values of n_i's that minimize the cost C, given in Eq. (7.5.1) and the minimum value of C with the optimum n_i's, are given by $n_i(V_0) = \dfrac{\sum_{i=1}^{K}\sqrt{A_i c_i}}{V_0 - B}\sqrt{\dfrac{A_i}{c_i}}$ and $C_{opt|V_0} = c_0 + \dfrac{\left(\sum_{i=1}^{K}\sqrt{c_i A_i}\right)^2}{V_0 - B}$, respectively.

7.5.3 Simple Random Sampling Without Replacement

For SRSWOR sampling, $A_i = W_i^2 S_{yi}^2$ and $B = -\sum_{i=1}^{K} W_i S_{yi}^2 / N$. Hence using Theorems 7.5.1 and 7.5.2, we get

$$n_i(C_0) = \frac{C_0 - c_0}{\left(\sum_{i=1}^{K} W_i S_{yi} \sqrt{c_i}\right)} \frac{W_i S_{yi}}{\sqrt{c_i}},$$

$$V_{opt|C_0} = \frac{\left(\sum_{i=1}^{K} W_i S_{yi} \sqrt{c_i}\right)^2}{C_0 - c_0} - \frac{\sum_{i=1}^{K} W_i S_{yi}^2}{N}$$

$$n_i(V_0) = \frac{\left(\sum_{i=1}^{K} W_i S_{yi} \sqrt{c_i}\right)}{V_0 + \frac{1}{N} \sum_{i=1}^{K} W_i S_{yi}^2} \cdot \frac{W_i S_{yi}}{\sqrt{c_i}} \text{ and}$$

$$C_{opt|V_0} = c_0 + \frac{\left(\sum_{i=1}^{K} W_i S_{yi} \sqrt{c_i}\right)^2}{V_0 + \sum_{i=1}^{K} W_i S_{yi}^2 / N}$$

7.5.4 Simple Random Sampling With Replacement

For SRSWR sampling $V\left(\widehat{\overline{Y}}_{st}\right) = A_i = W_i^2 \sigma_{yi}^2$ and $B = 0$. Hence Theorems 7.5.1 and 7.5.2 yield

$$n_i(C_0) = \frac{C_0 - c_0}{\left(\sum_{i=1}^{K} W_i \sigma_{yi} \sqrt{c_i}\right)} \frac{W_i \sigma_{yi}}{\sqrt{c_i}}, \quad V_{opt|C_0} = \frac{\left(\sum_{i=1}^{K} W_i \sigma_{yi} \sqrt{c_i}\right)^2}{C_0 - c_0},$$

$$n_i(V_0) = \frac{\left(\sum_{i=1}^{K} W_i \sigma_{yi} \sqrt{c_i}\right)}{V_0} \frac{W_i \sigma_{yi}}{\sqrt{c_i}} \text{ and } C_{opt|V_0} = c_0 + \frac{\left(\sum_{i=1}^{K} W_i \sigma_{yi} \sqrt{c_i}\right)^2}{V_0}.$$

7.5.5 Probability Proportional to Size With Replacement Sampling

For PPSWR sampling $V\left(\widehat{\overline{Y}}_{st}\right)=\frac{1}{N^2}\sum_{i=1}^{K}\frac{V_{pps|i}}{n_i}$. In this case $A_i=V_{pps|i}/N^2$ and $B=0$. Hence,

$$n_i(C_0)=\frac{C_0-c_0}{\left(\sum_{i=1}^{K}\sqrt{V_{pps|i}\,c_i}\right)}\sqrt{\frac{V_{pps|i}}{c_i}},\quad V_{opt|C_0}=\frac{\left(\sum_{i=1}^{K}\sqrt{V_{pps|i}\,c_i}\right)^2}{N^2(C_0-c_0)},$$

$$n_i(V_0)=\frac{\sum_{i=1}^{K}\sqrt{V_{pps|i}\,c_i}}{N^2V_0}\sqrt{\frac{V_{pps|i}}{c_i}}\text{ and }C_{opt|V_0}=c_0+\frac{\left(\sum_{i=1}^{K}\sqrt{c_i\,V_{pps|i}}\right)^2}{N^2V_0}.$$

Remark 7.5.1

For both the SRSWOR and SRSWR sampling schemes, the optimum sample sizes are directly proportional to the stratum size and stratum standard deviation but inversely proportional to the square root of the stratum cost per unit of survey. Hence we can propose a rule of thumb for allocation of sample sizes as follows: Take more samples from those strata whose sizes and variances are large and select small sample sizes from the strata where the cost of surveying units is high.

7.5.6 Neyman Optimum Allocation

Consider the situation where the cost of surveying the units of all the strata are the same, i.e., $c_i=c$ for $i=1,\ldots,K$. In this situation, if the total cost of a survey C is kept fixed as C_0, then the total sample size $\sum_{i=1}^{K}n_i$ becomes fixed and is equal to $\frac{C_0-c_0}{c}=n$. It can be checked that under SRSWOR and SRSWR sampling, the optimum value of n_i with $c_i=c$ is obtained as

$$n_i(C_0)=n\frac{W_i\ S_{yi}}{\left(\sum_{i=1}^{K}W_iS_{yi}\right)}=n\frac{N_iS_{yi}}{\left(\sum_{i=1}^{K}N_iS_{yi}\right)}\qquad(7.5.11)$$

The allocation (7.5.11) is the well known Neyman (1934) allocation, which is usable when both the stratum sizes N_i's and stratum variances S_{yi}^2 are known.

7.5.7 Proportional Allocation

If the cost of surveying units in all the strata and the strata variances are the same, i.e., $c_i = c$ and $S_{yi} = S_y$ for $i = 1,2,\ldots, K$, then Eq. (7.5.11) reduces to

$$n_i(C_0) = nN_i/N = nW_i \tag{7.5.12}$$

The allocation (7.5.12) is known as proportional allocation since the sample sizes in the respective stratum are taken proportional to the stratum sizes when the total sample size is fixed as n. Proportional allocation was introduced by Bowley (1926).

7.6 COMPARISON BETWEEN STRATIFIED AND UNSTRATIFIED SAMPLING

In this section, we will consider the performance of stratified and unstratified sampling under various sampling designs that are commonly used in practice.

7.6.1 Simple Random Sampling Without Replacement

Let an unstratified sample of size n be selected from the entire population U by the SRSWOR method and $\overline{y}_s$ be the sample mean. Then,

$$V(\overline{y}_s) = (1-f)\frac{S_y^2}{n} = V_{ran}(\text{say}) \tag{7.6.1}$$

where

$$f = n/N \text{ and } S_y^2 = \frac{1}{(N-1)}\sum_{i=1}^{K}\sum_{j\in U_i}\left(y_{ij}-\overline{Y}\right)^2 = \frac{1}{N-1}\sum_{i=1}^{K}(N_i-1)S_{yi}^2$$
$$+\frac{1}{N-1}\sum_{i=1}^{K}N_i\left(\overline{Y}_i-\overline{Y}\right)^2$$

Now assuming N_i's are so large that we can take $1/N_i \cong 0$ for $i = 1,\ldots, K$ and hence also $1/N \cong 0$, we can approximate V_{ran} as

$$V_{ran} = \left(\frac{1}{n}-\frac{1}{N}\right)S_y^2 = \left(\frac{1}{n}-\frac{1}{N}\right)\left[\sum_{i=1}^{K}W_iS_{yi}^2+\sum_{i=1}^{K}W_i\left(\overline{Y}_i-\overline{Y}\right)^2\right] \tag{7.6.2}$$

Let a sample s_i of size n_i be selected by an SRSWOR method from the ith stratum keeping $\sum_{i=1}^{K} n_i = n$ as fixed, then the variance of $\widehat{\overline{Y}}_{st} = \sum_{i=1}^{K} W_i\overline{y}(s_i)$ becomes

$$V\left(\widehat{\overline{Y}}_{st}\right) = \sum_{i=1}^{K}W_i^2(1-f_i)S_{yi}^2\Big/n_i \tag{7.6.3}$$

Under Neyman optimum allocation, $n_i = n_{i0} = n\dfrac{W_i\ S_{yi}}{\left(\sum_{i=1}^{K} W_i S_{yi}\right)}$ and

the expression of $V\left(\widehat{\overline{Y}}_{st}\right)$ in Eq. (7.6.3) reduces to

$$V_{opt} = \frac{\left(\sum_{i=1}^{K} W_i\ S_{yi}\right)^2}{n} - \frac{1}{N}\sum_{i=1}^{K} W_i\ S_{yi}^2 \tag{7.6.4}$$

For proportional allocation, $n_i = nN_i/N = nW_i$, the expression (7.6.3) reduces to

$$V_{prop} = \left(\frac{1}{n} - \frac{1}{N}\right)\sum_{i=1}^{K} W_i\ S_{yi}^2 \tag{7.6.5}$$

Eqs. (7.6.4) and (7.6.5) yield

$$V_{prop} - V_{opt} = \frac{1}{n}\sum_{i=1}^{K} W_i\left[S_{yi} - \left(\sum_{i=1}^{K} W_i S_{yi}\right)\right]^2 \geq 0 \tag{7.6.6}$$

From Eqs. (7.6.2) and (7.6.5), we get

$$V_{ran} - V_{prop} = \left(\frac{1}{n} - \frac{1}{N}\right)\sum_{i=1}^{K} W_i\left(\overline{Y}_i - \overline{Y}\right)^2 \geq 0 \tag{7.6.7}$$

Finally using Eqs. (7.6.6) and (7.6.7), the following theorem is obtained.

Theorem 7.6.1

If the sample size n is fixed and $1/N_i \cong 0$ for $i = 1,\ldots, K$, then

$$V_{ran} \geq V_{prop} \geq V_{opt}$$

7.6.2 Probability Proportional to Size With Replacement Sampling

Let $x_{ij}(>0)$ be the measure of size for the jth unit of the ith stratum and $p_{ij} = x_{ij}/X$ be the corresponding normed size measure with $X = \sum_{i=1}^{K} X_i$ and $X_i = \sum_{j=1}^{N_i} x_{ij}$. Suppose an unstratified sample of size n is selected from the population by PPSWR sampling scheme using p_{ij} as normed size measure for the jth unit of the ith stratum. Then the Hansen–Hurwitz (HH) estimator for population mean $\overline{Y}$ is given by

$$\widehat{\overline{Y}}_{hh} = \frac{1}{nN}\sum_{r=1}^{n}\frac{y(r)}{p(r)}$$

where $y(r) = y_{ij}$ and $p(r) = p_{ij}$ if the rth draw produces the jth unit of the ith stratum.

The variance of the HH estimator $\widehat{\overline{Y}}_{hh}$ is given by

$$V\left(\widehat{\overline{Y}}_{hh}\right) = \frac{1}{N^2 n}\left(\sum_{i=1}^{K}\sum_{j=1}^{N_i}\frac{y_{ij}^2}{p_{ij}} - N^2\overline{Y}^2\right) \tag{7.6.8}$$

Consider a stratified sampling where sample s_i of size n_i is selected from the ith stratum for $i = 1,\ldots, K$ by PPSWR method using normed size measure $p_{ij}^* = p_{ij}/P_i$ for the jth unit of the ith stratum, where $P_i = X_i/X$; $i = 1,2,\ldots, K$, $j = 1,\ldots, N_i$. Then the HH estimator for the mean $\overline{Y}$ based on the stratified sample is given by

$$\widehat{\overline{Y}}_{hh}^{st} = \frac{1}{N}\sum_{i=1}^{K}\widehat{Y}_i$$

where $\widehat{Y}_i = \frac{1}{n_i}\sum_{r=1}^{n_i}\frac{y(r,i)}{p(r,i)}$ is an unbiased estimator for the total Y_i, $y(r,i) = y_{ij}$ and $p(r,i) = p_{ij}^*$ if the rth draw produces the jth unit of the ith stratum with probability p_{ij}^*.

The variance of $\widehat{\overline{Y}}_{hh}^{st}$ is given by

$$V\left(\widehat{\overline{Y}}_{hh}^{st}\right) = \frac{1}{N^2}\sum_{i=1}^{K}\frac{1}{n_i}\left(\sum_{j=1}^{N_i}\frac{y_{ij}^2}{p_{ij}^*} - Y_i^2\right) \tag{7.6.9}$$

Consider Arnab allocation (1991), where n_i's are allocated with proportional X_i's, the ith stratum total of the auxiliary variable, i.e., $n_i = nX_i/X = nP_i$. In this case, Eq. (7.6.9) reduces to

$$\begin{aligned} V\left(\widehat{\overline{Y}}_{hh}^{st}\right) &= \frac{1}{N^2}\left[\sum_{i=1}^{K}\frac{1}{nP_i}\left(\sum_{j=1}^{N_i}\frac{y_{ij}^2}{p_{ij}^*} - Y_i^2\right)\right] \\ &= \frac{1}{N^2}\left[\sum_{i=1}^{K}\frac{1}{n}\left(\sum_{j=1}^{N_i}\frac{y_{ij}^2}{p_{ij}} - \frac{Y_i^2}{P_i}\right)\right] \end{aligned} \tag{7.6.10}$$

Eqs. (7.6.8) and (7.6.10) yield

$$\begin{aligned} V\left(\widehat{Y}_{hh}\right) - V\left(\widehat{Y}^{st}_{hh}\right) &= \frac{1}{N^2 n}\left(\sum_{i=1}^{K}\frac{Y_i^2}{P_i} - Y^2\right) \\ &= \frac{1}{N^2 n}\sum_{i=1}^{K} P_i\left(\frac{Y_i}{P_i} - Y\right)^2 \qquad (7.6.11) \\ &\geq 0 \end{aligned}$$

Eq. (7.6.11) derived by Arnab (1991) leads to the following theorem:

Theorem 7.6.2

For a given sample size n, the stratified PPSWR sampling is more efficient than unstratified PPSWR sampling if sample sizes are allocated to each of the strata proportional to the aggregate of the measure of size and in this case

$$V\left(\widehat{Y}_{hh}\right) - V\left(\widehat{Y}^{st}_{hh}\right) = \frac{1}{N^2 n}\sum_{i=1}^{K} P_i\left(\frac{Y_i}{P_i} - Y\right)^2 \geq 0$$

7.6.3 Inclusion Probability Proportional to Size Sampling Scheme

Consider an unstratified sample s of size n selected from the entire population U using a suitable IPPS sampling design with inclusion probability $\pi^u_{j|i} = n\, x_{ij}/X$ attached to the jth unit of the ith stratum. In this case the Horvitz–Thompson estimator (HTE) for the population total Y is given by

$$\begin{aligned} \widehat{Y}^u_{ht} &= \sum_i \sum_{j\in s} \frac{y_{ij}}{\pi^u_{j|i}} \\ &= \frac{X}{n}\sum_i \sum_{j\in s}\frac{y_{ij}}{x_{ij}} \end{aligned}$$

Suppose from the ith ($i = 1,\ldots, K$) stratum a sample s_i of size $n_i\left(\sum_{i=1}^{K} n_i = n\right)$ is selected with inclusion probability $\pi_{j|i} = n_i x_{ij}/X_i$ for the jth unit of the ith stratum. Then the HTE for Y based on the stratified sample is given by

$$\begin{aligned} \widehat{Y}^{st}_{ht} &= \sum_i \sum_{j\in s_i} \frac{y_{ij}}{\pi_{j|i}} \\ &= \sum_i X_i \sum_{j\in s_i} \frac{y_{ij}}{n_i x_{ij}} \end{aligned}$$

Let us suppose that the study variable y be related to the auxiliary variable x through the following superpopulation model ξ

$$E_\xi(y_{ij}) = \beta x_{ij},\ V_\xi(y_{ij}) = \sigma^2 x_{ij}^2 \text{ and } C_\xi(y_{ij}, y_{kl}) = 0 \ \text{ for}(i,j) \neq (k,l) \quad (7.6.12)$$

where E_ξ, V_ξ, and C_ξ denote the operator for expectation, variance, and covariance with respect to the model ξ, respectively.

The expected variances of $\widehat{Y}_{ht}^{u}$ and $\widehat{Y}_{ht}^{st}$ under the superpopulation model ξ are given, respectively, by

$$\begin{aligned} E_\xi V\left(\widehat{Y}_{ht}^{u}\right) &= E_\xi E\left(\widehat{Y}_{ht}^{u} - Y\right)^2 \\ &= \sigma^2 \sum_{i=1}^{K} \sum_{j=1}^{N_i} x_{ij}^2 \left(\frac{1}{\pi_{j|i}^{u}} - 1\right) \\ &= \sigma^2 \left(\frac{X^2}{n} - \sum_{i=1}^{K} \sum_{j=1}^{N_i} x_{ij}^2\right) \end{aligned}$$

and

$$\begin{aligned} E_\xi V\left(\widehat{Y}_{ht}^{st}\right) = E_\xi E\left(\widehat{Y}_{ht}^{st} - Y\right)^2 &= \sigma^2 \sum_{i=1}^{K} \sum_{j=1}^{N_i} x_{ij}^2 \left(\frac{1}{\pi_{j|i}} - 1\right) \\ &= \sigma^2 \left(\sum_{i=1}^{K} \frac{X_i^2}{n_i} - \sum_{i=1}^{K} \sum_{j=1}^{N_i} x_{ij}^2\right) \end{aligned}$$

Now noting,

(i) $\displaystyle \sum_{i=1}^{K} \frac{X_i^2}{n_i} = \sum_{i=1}^{K} n_i \left(\frac{X_i}{n_i}\right)^2 \geq \frac{\left(\sum_{i=1}^{K} X_i\right)^2}{\sum_{i=1}^{K} n_i} = \frac{X^2}{n}$

and

(ii) $\displaystyle \sum_{i=1}^{K} \frac{X_i^2}{n_i} = \frac{X^2}{n}$ if $n_i = nX_i/X$,

we find $E_\xi V\left(\widehat{Y}_{ht}^{st}\right) \geq E_\xi V\left(\widehat{Y}_{ht}^{u}\right)$.

Furthermore, $E_\xi V\left(\widehat{Y}_{ht}^{st}\right) = E_\xi V\left(\widehat{Y}_{ht}^{u}\right)$ if $n_i = nX_i/X$. Hence we have the following theorem.

Theorem 7.6.3

Under the superpopulation model (7.6.12), the unstratified IPPS sampling design is more efficient than the stratified IPPS sampling design based on the same sample size. Both of them are equally efficient if $n_i = nX_i/X$ for $i = 1,\ldots, K$.

Comparisons between the stratified and unstratified sampling were studied in details by Hanurav (1965), Ramachandran and Rao, T.J. (1974), Cassel et al. (1977), and Rao, T.J. (1968, 1977a,b, 1983), among others.

Example 7.6.1

4000 households of a city were stratified into three strata: high-, middle-, and low-income groups. A stratified sample of 400 households was selected from the entire population by using the SRSWOR method. The following table gives the sample sizes, sample mean, and sample variances of the household income and the number of earning members of the sampled households.

(i) Estimate the mean household income of the city and 95% confidence interval of the mean income.

(ii) Estimate the proportion of earning members of the city and 95% confidence interval of this proportion.

Income Group (i)	Number of households N_i	Sampled households n_i	Number of earning members in the sample z_i	Sample mean income (\$) $\bar{y}(s_i)$	Sample standard deviation of income (\$) s_{yi}
High	500	50	10	25,000	300
Middle	1500	150	20	15,000	200
Low	2000	200	20	2500	400
Total	4000	400	50	—	

We first calculate the following:

Strata	W_i	$W_i\bar{y}(s_i)$	$W_i^2(1-f_i)s_{yi}^2/n_i$	$\widehat{\pi}_{iA}$	$W_i\widehat{\pi}_{iA}$	$\frac{W_i^2(1-f_i)\widehat{\pi}_{iA}(1-\widehat{\pi}_{iA})}{n_i-1}$
High	0.125	3125	25.3125	0.2000	0.025	0.00004
Middle	0.375	5625	33.7500	0.1333	0.050	0.0001
Low	0.500	1250	180.0000	0.1000	0.050	0.0001
Total	1	10,000	239.0625	—	0.125	0.00024

(i) Here the estimated mean household income. $\widehat{\bar{Y}}_{st} = \sum_{i=1}^{4} W_i\bar{y}(s_i) = \$10,000.$

Estimated standard error of $\widehat{\overline{Y}}_{st}$ is $\sqrt{\widehat{V}\left(\widehat{\overline{Y}}_{st}\right)} = \sqrt{\sum W_i^2(1-f_i)s_{yi}^2/n_i}$ $= \$15.461$.

95% confidence interval of the mean income is $10{,}000 \pm 1.96 \times 15.461 = (\$\ 9969.695, \$\ 10{,}030.30)$

(ii) Estimated population proportion of earning member is $\widehat{\pi}_A = \sum_{i=1}^{4} W_i \widehat{\pi}_{iA} = 0.125$.

Estimated standard error of $\widehat{\pi}_A$ is $\sqrt{\sum W_i^2(1-f_i)\widehat{\pi}_{iA}(1-\widehat{\pi}_{iiA})/(n_i-1)}$ $= 0.0156$

95% confidence interval of π_A is $0.125 \pm 1.96 \times 0.0156 = (0.094, 0.156)$.

Example 7.6.2

A survey was conducted to determine the average farm size cultivating maize in the KwaZulu- Natal (KZN) province of South Africa. The farmers are classified into five strata according to their farm sizes. The total cost of the survey was limited to \$ 50, 000. The fixed cost of the survey was budgeted as \$ 10,000 to cover expenses for administration and report writing. The variable costs for collection of data per farm, including transport and filling schedule, are given in the following table along with other information such as the number of farms, average farm sizes, and SD of the farm sizes for respective stratum.

(a) Determine the sample sizes that should be selected from each stratum under (i) proportional, (ii) Neyman allocation, and (iii) optimum allocations.

(b) Compute the relative efficiencies of the Neyman and optimum allocation with respect to the proportional allocation for estimating the population mean when samples from each stratum are selected by SRSWOR method.

Stratum (i)	Number of farms (N_i)	Mean area under maize (acre) ($\overline{Y}_i$)	SD of area under maize (S_i)	Cost of collection of data per farm (c_i)
1	500	80	5.25	100
2	400	100	10.75	100
3	300	120	15.50	150
4	200	150	20.50	175
5	100	200	25.00	200

Here the cost function is

$$C = C_o + \sum_{i=1}^{5} c_i n_i \tag{7.6.13}$$

where $C = 50{,}000$ and $C_o = 10{,}000$.

(a) (i) Proportional allocation:

Under proportional allocation $n_i = nN_i/N$, where $N = \sum_{i=1}^{5} N_i$ and n should be obtained from the equation $C - C_o = n\sum_{i=1}^{5} c_i N_i/N$, i.e., $n = \dfrac{(C - C_o)}{\sum_{i=1}^{5} c_i N_i/N}$ and hence n_i under proportional allocation is

$$n_{i0}(p) = \frac{(C - C_o)}{\sum_{i=1}^{5} c_i N_i} N_i$$

(ii) Neyman allocation: Under this allocation $n_i = nN_iS_i/\sum_{i=1}^{5} N_iS_i$ and Eq. (7.6.13) yields $C - C_o = n\sum_{i=1}^{5} c_i N_i S_i/\sum_{i=1}^{5} N_iS_i$, i.e., $n = \dfrac{(C - C_o)\left(\sum_{i=1}^{5} N_iS_i\right)}{\sum_{i=1}^{5} c_i N_i S_i}$ and hence n_i under Neyman allocation is

$$n_{i0}(ney) = \frac{(C - C_o)N_iS_i}{\sum_{i=1}^{5} c_i N_i S_i}$$

(iii) Under optimum allocation

$$n_{i0}(opt) = \frac{(C - C_o)N_iS_i/\sqrt{c_i}}{\sum_{i=1}^{5} \sqrt{c_i} N_i S_i}$$

The values of n_i's under proportional, Neyman and optimum allocation are obtained as follows:

i	N_i	S_i	c_i	N_ic_i	$n_{i0}(p) = \frac{(C-C_o)}{\sum_{i=1}^{5} c_iN_i}N_i$	$N_iS_ic_i$	$n_{i0}(ney) = \frac{(C-C_o)N_iS_i}{\sum_{i=1}^{5} c_iN_iS_i}$	$N_iS_i\sqrt{c_i}$	$n_{i0}(opt) = \frac{(C-C_o)N_iS_i/\sqrt{c_i}}{\sum_{i=1}^{5} \sqrt{c_i}N_iS_i}$
1	500	5.25	100	50,000	105	262,500	40	26,250	49
2	400	10.8	100	40,000	84	430,000	66	43,000	80
3	300	15.5	150	45,000	63	697,500	71	56,950.64	70
4	200	20.5	175	35,000	42	717,500	63	54,237.9	57
5	100	25	200	20,000	21	500,000	38	35,355.34	33
	1500			190,000	315	2,607,500	278	215,793.9	289

(b) The unbiased estimator for the population mean $\overline{Y}$ is $\widehat{\overline{Y}}_{st} = \sum_{i=1}^{5} N_i \overline{y}(s_i)/N$. The variance of $\widehat{\overline{Y}}_{st}$ under proportional, Neyman, and optimum allocations are respectively given by

$$V_{prop} = \sum_{i=1}^{5} \left(\frac{N_i}{N}\right)^2 \left(\frac{1}{n_{i0}(p)} - \frac{1}{N_i}\right) S_i^2 = 0.4659,$$

$$V_{ney} = \sum_{i=1}^{5} \left(\frac{N_i}{N}\right)^2 \left(\frac{1}{n_{i0}(ney)} - \frac{1}{N_i}\right) S_i^2 = 0.4043 \text{ and}$$

$$V_{opt} = \sum_{i=1}^{5} \left(\frac{N_i}{N}\right)^2 \left(\frac{1}{n_{i0}(opt)} - \frac{1}{N_i}\right) S_i^2 = 0.394.$$

The relative efficiency of Neyman and optimum allocations are $\frac{V_{prop}}{V_{ney}} \times 100 = 115.23\%$ and $\frac{V_{prop}}{V_{opt}} \times 100 = 118.26\%$, respectively.

7.7 CONSTRUCTION OF STRATA

It is obvious that the strata should be constructed in such a way that each stratum becomes homogeneous with respect to the character under study so that strata means differ significantly from each other, but at the same time variation within the strata becomes as small as possible. To make the strata homogeneous, one needs to have some prior knowledge of the population such as past survey data or auxiliary information related to the study variable. As far as the number of strata is concerned, more is better because the efficiency of stratified sampling increases as the number of strata increases. According to Dalenius (1953) conjecture $V_{k+1}\left(\widehat{\overline{Y}}_{st}\right)$, the variance of the estimator of the population mean based on $k+1$ strata is approximately equal to $\left(\frac{k}{k+1}\right)^2 V_k\left(\widehat{\overline{Y}}_{st}\right)$. Dalenius (1953) conjecture was supported by Cochran (1961) through empirical studies. Hence the gain in efficiency of increasing the number of strata becomes marginal after a certain stage. Various researchers have paid attention to construction of strata and details have been given by Sukhatme et al. (1984), Murthy (1977), and Cochran (1977), among others. Here we will highlight the following methods of optimum points of stratification when there is plan to divide a population into a predetermined number of strata K.

7.7.1 Optimum Points of Stratification

Suppose that we want to stratify a population into K strata, where K is prespecified. Let y_0 and y_K be the maximum and minimum value of y, respectively. Let the interval (y_0, y_K) be divided at the points $y_1, \ldots, y_{K-1}$ $(y_1 < \cdots < y_{K-1})$ to form K strata. The problem is to determine the points of division so that the variance of the estimator of the population mean attains a minimum. Clearly the optimum points of division depend on the sampling design and the estimators used. For simplicity, let us assume (i) each stratum size is large and (ii) the study variable has a continuous probability density function $f(y)$ so that we can approximate

$$W_i = \int_{y_{i-1}}^{y_i} f(y)dy, \overline{Y}_i = \mu_i = \frac{1}{W_i}\int_{y_{i-1}}^{y_i} yf(y)dy,$$

$$S_{yi}^2 = \sigma_i^2 = \frac{1}{W_i}\int_{y_{i-1}}^{y_i} (y-\mu_i)^2 f(y)dy, i = 1, \ldots, k.$$

The expressions of $V\left(\widehat{\overline{Y}}_{st}\right)$ under proportional and optimum allocation for SRSWR sampling are given respectively as follows:

$$V_{prop} = \frac{\sum_{i=1}^K W_i S_{yi}^2}{n} = \frac{\sum_{i=1}^K W_i\sigma_i^2}{n} \text{ and } V_{opt} = \frac{\left(\sum_{i=1}^K W_i S_{yi}\right)^2}{n}$$
$$= \frac{\left(\sum_{i=1}^K W_i\sigma_i\right)^2}{n}$$

7.7.1.1 Proportional Allocation

Under proportional allocation

$$V_{prop} = \frac{\sum_{i=1}^K W_i\sigma_i^2}{n}$$
$$= \frac{\sum_{i=1}^K \int_{y_{i-1}}^{y_i} (y-\mu_i)^2 f(y)dy}{n}$$
$$= \frac{\sum_{i=1}^K \int_{y_{i-1}}^{y_i} y^2 f(y)dy - \sum_{i=1}^k W_i\mu_i^2}{n}$$
$$= \frac{1}{n}\int_{y_0}^{y_K} y^2 f(y)dy - \frac{1}{n}\sum_{i=1}^k \frac{1}{W_i}\left(\int_{y_{i-1}}^{y_i} yf(y)dy\right)^2$$

Now

$$\frac{\partial V_{prop}}{\partial y_i} = -\frac{1}{n}\frac{\partial}{\partial y_i}\sum_{i=1}^{k}\frac{1}{W_i}\left(\int_{y_{i-1}}^{y_i} yf(y)dy\right)^2$$

$$= -\frac{1}{n}\left[\frac{\partial}{\partial y_i}\frac{1}{W_i}\left(\int_{y_{i-1}}^{y_i} yf(y)dy\right)^2 + \frac{\partial}{\partial y_i}\frac{1}{W_{i+1}}\left(\int_{y_i}^{y_{i+1}} yf(y)dy\right)^2\right]$$

(since the other terms do not involve y_i)

$$= -\frac{1}{n}\left[-\left(\int_{y_{i-1}}^{y_i} yf(y)dy\right)^2\left((W_i)^{-2}\frac{\partial W_i}{\partial y_i}\right) + (W_i)^{-1}\right.$$

$$\left.\times\left(2\int_{y_{i-1}}^{y_i} yf(y)dy\right)\left(\frac{\partial}{\partial y_i}\int_{y_{i-1}}^{y_i} yf(y)dy\right)\right]$$

$$-\frac{1}{n}\left[\left(\int_{y_i}^{y_{i+1}} yf(y)dy\right)^2\left((W_{i+1})^{-2}\frac{\partial W_{i+1}}{\partial y_i}\right) + (W_{i+1})^{-1}\right.$$

$$\left.\times\left(2\int_{y_i}^{y_{i+1}} yf(y)dy\right)\left(\frac{\partial}{\partial y_i}\int_{y_i}^{y_{i+1}} yf(y)dy\right)\right]$$

$$= -\frac{1}{n}\left[-\left(\mu_i^2\frac{\partial W_i}{\partial y_i}\right) + 2\mu_i\frac{\partial}{\partial y_i}\left(\int_{y_{i-1}}^{y_i} yf(y)dy\right)\right.$$

$$\left.-\left(\mu_{i+1}^2\frac{\partial W_{i+1}}{\partial y_i}\right) + 2\mu_{i+1}\frac{\partial}{\partial y_i}\left(\int_{y_i}^{y_{i+1}} yf(y)dy\right)\right]$$

Furthermore, noting

$$\frac{\partial W_i}{\partial y_i}=f(y_i),\frac{\partial W_{i+1}}{\partial y_i}=-f(y_i),\frac{\partial}{\partial y_i}\left\{\int\limits_{y_{i-1}}^{y_i} yf(y)dy\right\}$$

$$=y_if(y_i),\frac{\partial}{\partial y_i}\left\{\int\limits_{y_i}^{y_{i+1}} yf(y)dy\right\}=-y_if(y_i)$$

and equating $\frac{\partial V_{prop}}{\partial y_i}=0$, the optimum points of stratification are obtained as

$$y_i=\frac{\mu_i+\mu_{i+1}}{2}\text{ for } i=1,\ldots,K-1 \tag{7.7.1}$$

Since μ_i's are unknown, y_i's can be determined by using an iterative procedure when $f(y)$ is known.

7.7.1.2 Optimum Allocation

The optimum points of stratification y_i's are obtained by solving the equation

$$\frac{\partial V_{opt}}{\partial y_i}=2\frac{\left(\sum\limits_{i=1}^{K} W_i\sigma_i\right)}{n}\left\{\frac{\partial(W_i\sigma_i)}{\partial y_i}+\frac{\partial(W_{i+1}\sigma_{i+1})}{\partial y_i}\right\}=0 \tag{7.7.2}$$

Now

$$\frac{\partial(W_i\sigma_i)}{\partial y_i}=\sigma_i\frac{\partial W_i}{\partial y_i}+W_i\frac{\partial\sigma_i}{\partial y_i} \tag{7.7.3}$$

Further differentiating both sides of $W_i\sigma_i^2=\int_{y_{i-1}}^{y_i}(y-\mu_i)^2f(y)dy$ with respect to y_i, we get

$$f(y_i)\sigma_i^2+2W_i\sigma_i\frac{\partial\sigma_i}{\partial y_i}=(y_i-\mu_i)^2f(y_i) \tag{7.7.4}$$

Eqs. (7.7.3) and (7.7.4) yield

$$\frac{\partial(W_i\sigma_i)}{\partial y_i}=\frac{f(y_i)}{2}\left[\sigma_i+\frac{(y_i-\mu_i)^2}{\sigma_i}\right] \tag{7.7.5}$$

Similarly we get

$$\frac{\partial(W_{i+1}\sigma_{i+1})}{\partial y_i} = -\frac{f(y_i)}{2}\left[\sigma_{i+1} + \frac{(y_i - \mu_{i+1})^2}{\sigma_{i+1}}\right] \tag{7.7.6}$$

Finally putting Eqs. (7.7.5) and (7.7.6) in Eq. (7.7.2) and setting $\frac{\partial V_{opt}}{\partial y_i} = 0$, the optimum points of stratifications y_i's are obtained from the following equation

$$\sigma_i + \frac{(y_i - \mu_i)^2}{\sigma_i} = \sigma_{i+1} + \frac{(y_i - \mu_{i+1})^2}{\sigma_{i+1}} \quad \text{for } i = 1, \ldots, K-1 \tag{7.7.7}$$

here also y_i's cannot be solved easily from the equation above because μ_i's and σ_i's are unknown. However, if $f(y)$ is known, y_i's may be obtained by using an iterative method when the number of strata is small.

7.7.2 Dalenius and Hodges's Approximation

Dalenius and Hodges (1959) have given a simple formula to determine the optimum points of stratification under the following assumptions:

(i) The stratum weight W_i is approximated as

$$W_i = \int_{y_{i-1}}^{y_i} f(y)dy \cong f(y_i)\int_{y_{i-1}}^{y_i} dy = f(y_i)(y_i - y_{i-1}) \tag{7.7.8}$$

(ii) The distribution of y is uniform for all the strata so that

$$\sigma_i^2 = \frac{1}{W_i}\int_{y_{i-1}}^{y_{i1}} (y - \mu_i)^2 f(y)dy = \frac{(y_i - y_{i-1})^2}{12}$$

So, under the approximations (i) and (ii) of uniform distribution over the strata, the expression of $V_{opt} = \frac{\left(\sum_{i=1}^{K} W_i\sigma_i\right)^2}{n}$ reduces to $\frac{\left(\sum_{i=1}^{K} F_i^2\right)^2}{12n}$, where $F_i = (y_i - y_{i-1})\sqrt{f(y_i)}$

Now

$$V_{opt} = \frac{\left(\sum_{i=1}^{K} F_i^2\right)^2}{12n} \geq \frac{\left(\sum_{i=1}^{K} F_i\right)^4}{12nK^2} \tag{7.7.9}$$

Thus V_{opt} attains a minimum when

$$F_i = \frac{1}{K}\sum_{i=1}^{K}(y_i - y_{i-1})\sqrt{f(y_i)} \cong \frac{1}{K}\sum_{i=1}^{K}\int_{y_{i-1}}^{y_i}\sqrt{f(y)}dy = \frac{1}{K}\int_{y_0}^{y_K}\sqrt{f(y)}dy$$

i.e., when

$$F_i = \int_{y_{i-1}}^{y_i}\sqrt{f(y)}dy = \frac{1}{K}\int_{y_0}^{y_K}\sqrt{f(y)}dy = \overline{F}(\text{say}) \tag{7.7.10}$$

Thus the Dalenius and Hodges (1959) rule is to find the points of stratification, which yield equal cumulative values of the square root of the frequency function $f(y)$.

Example 7.7.1

The following table gives the frequency distribution of daily wages of 820 workers in a certain factory.

Wages ($)	Frequency (f)
Less than 50	80
51−100	120
101−150	250
151−200	120
201−300	100
301−450	70
451−600	50
Above 601	30

It is decided to divide the workers in five strata. Use the Dalenius and Hodges rule to find the optimal strata.

Here we first prepare the cumulative frequency table of $\sqrt{f}$ as follows:

Wages ($)	Frequency (f)	$\sqrt{f}$	Cumulative $\sqrt{f}$
Less than 50	80	8.94	8.94
51−100	120	10.95	19.89
101−150	250	15.81	35.7
151−200	120	10.95	46.65
201−300	100	10.00	56.65
301−450	70	8.37	65.02
451−600	50	7.07	72.09
Above 600	30	5.48	77.57

The computed value of $\overline{F} = 77.57/5 = 15.51$, $2\overline{F} = 31.03$, $3\overline{F} = 46.54$, $4\overline{F} = 62.06$ and $5\overline{F} = 77.57$. Thus the five strata may be formed as follows:

Strata	Wages ($)	Size
I	0–100	200
II	101–150	250
III	151–200	120
IV	201–450	170
V	Above 450	80

7.7.3 Other Methods

Dalenius and Gurney (1951) suggested that strata should be formed so that $W_i\, \sigma_i$ should be constant, whereas Mahalanobis (1952) recommended that $W_i\, \mu_i$ should be constant. Cochran (1961) suggested that the coefficient of variations for all the strata should be the same i.e.,

$$\frac{\sigma_1}{\mu_1} = \cdots = \frac{\sigma_i}{\mu_i} = \cdots = \frac{\sigma_K}{\mu_K} \tag{7.7.11}$$

Furthermore, if the distribution of y within each stratum is uniform so that

$$\mu_i = \frac{y_{i-1} + y_i}{2} \text{ and } \sigma_i = \frac{y_i - y_{i-1}}{\sqrt{12}}$$

Eq. (7.7.11) leads to

$$\frac{y_i - y_{i-1}}{y_i + y_{i-1}} = \frac{y_{i+1} - y_i}{y_{i+1} + y_i} \tag{7.7.12}$$

Gunning and Horgan (2004) derived the following recurrence relation using Eq. (7.7.12)

$$\begin{aligned} y_i^2 &= y_{i+1}\ y_{i-1} \\ \text{i.e., } y_i &= ar^i \text{ for } i = 0, \ldots, K \end{aligned} \tag{7.7.13}$$

Now putting $i = 0$ and $i = K$, we get $a = y_0$ and $r = (y_K/y_0)^{\frac{1}{K}}$. Substituting $a = y_0$ and $r = (y_K/y_0)^{\frac{1}{K}}$ in Eq. (7.7.13), we get

$$y_i = y_0 \left(\frac{y_K}{y_0}\right)^{\frac{i}{K}} \tag{7.7.14}$$

Thus for $K = 4$, $y_0 = 5$, and $y_4 = 50{,}000$, Gunning and Horgan (2004) obtained the optimum points of stratification as $y_1 = 50$, $y_2 = 500$, and $y_3 = 5000$. They showed that the proposed method performs reasonably well for positively skewed populations.

7.8 ESTIMATION OF GAIN DUE TO STRATIFICATION

Suppose from the ith stratum a sample s_i of size n_i is selected by some suitable sampling procedure with $\pi_{l|i}$ and $\pi_{lq|i}$ as the inclusion probabilities for the lth unit and lth and qth unit of the ith stratum l, $q = 1,\ldots, N_i$, $i = 1,\ldots, K$, $l \neq q$. The HTE of $\overline{Y}$ based on the stratified sample is given by

$$\widehat{\overline{Y}}_{ht}^{st} = \frac{1}{N}\sum_{i=1}^{K}\left(\sum_{l\in s_i}\frac{y_{il}}{\pi_{l|i}}\right) \tag{7.8.1}$$

The variance of $\widehat{\overline{Y}}_{ht}^{st}$ and its unbiased estimates are respectively given by

$$V\left(\widehat{\overline{Y}}_{ht}^{st}\right) = \frac{1}{2N^2}\sum_{i=1}^{K}\sum_{l\neq}^{N_i}\sum_{q=1}^{N_i}\left(\pi_{l|i}\pi_{q|i} - \pi_{lq|i}\right)\left(\frac{y_{il}}{\pi_{l|i}} - \frac{y_{iq}}{\pi_{q|i}}\right)^2 \tag{7.8.2}$$

and

$$\widehat{V}\left(\widehat{\overline{Y}}_{ht}^{st}\right) = \frac{1}{2N^2}\sum_{i=1}^{K}\sum_{l\neq}^{N_i}\sum_{q\in s_i}^{N_i}\left(\frac{\pi_{l|i}\pi_{q|i} - \pi_{lq|i}}{\pi_{lq|i}}\right)\left(\frac{y_{il}}{\pi_{l|i}} - \frac{y_{iq}}{\pi_{q|i}}\right)^2 \tag{7.8.3}$$

Suppose an unstratified sample s of size n is selected from the entire population using some sampling design with $\widetilde{\pi}_{l|i}$ and $\widetilde{\pi}_{lq|i,j}$ (>0) as the inclusion probabilities for the lth unit of the ith stratum, and lth unit of the ith stratum and qth unit of the jth stratum, respectively; $l = 1,\ldots, N_i$; $q = 1,\ldots, N_j$; $i, j = 1,\ldots, K$. In this case the expressions for the HTE for $\overline{Y}$ and its variance are respectively given as follows:

$$\widehat{\overline{Y}}_{ht}^{u} = \frac{1}{N}\sum_{i=1}^{K}\sum_{j=1}^{N_i}\frac{y_{ij}}{\widetilde{\pi}_{j|i}}I_{sj|i} \tag{7.8.4}$$

where $I_{sj|i} = 1$, if jth unit of the ith stratum is selected in the sample s and $I_{sj|i} = 0$ otherwise.

$$V\left(\widehat{\overline{Y}}_{ht}^{u}\right) = \frac{1}{N^2}\left[\sum_{i=1}^{K}\left\{\sum_{l=1}^{N_i}\frac{1}{\widetilde{\pi}_{l|i}}y_{il}^2 + \sum_{l'=1}^{N_i}\sum_{l=1}^{N_i}\left(\frac{\widetilde{\pi}_{ll'|ii}}{\widetilde{\pi}_{l|i}\widetilde{\pi}_{l'|i}}\right)y_{il}y_{il'}\right\} + \sum_{i\neq}^{K}\sum_{j=1}^{K}\left\{\sum_{l=1}^{N_i}\sum_{q=1}^{N_j}\left(\frac{\widetilde{\pi}_{lq|ij}}{\widetilde{\pi}_{l|i}\widetilde{\pi}_{q|j}}\right)y_{il}y_{jq}\right\} - Y^2\right] \tag{7.8.5}$$

The gain in efficiency of using stratified sampling over unstratified sampling is given by

$$G = V\left(\widehat{\overline{Y}}_{ht}^{u}\right) - V\left(\widehat{\overline{Y}}_{ht}^{st}\right) \tag{7.8.6}$$

here we will find the formula for estimating G through a sample $s^*(=s_1 \cup s_2 \cup \cdots \cup s_K)$ selected by the stratified sampling procedure.

The quantities $\sum_{i=1}^{K}\sum_{l=1}^{N_i}\frac{1}{\pi_{l|i}}y_{il}^2$ and Y^2 can by estimated unbiasedly by

$$\sum_{i=1}^{K}\sum_{l\in s_i}\frac{1}{\pi_{l|i}\widetilde{\pi}_{l|i}}y_{il}^2 \text{ and } N^2\left[\left(\widehat{\overline{Y}}_{ht}^{st}\right)^2 - \widehat{V}\left(\widehat{\overline{Y}}_{ht}^{st}\right)\right] \tag{7.8.7}$$

respectively, where $\widehat{V}\left(\widehat{\overline{Y}}_{ht}^{st}\right)$ is an unbiased estimator of $V\left(\widehat{\overline{Y}}_{ht}^{st}\right)$.

Similarly, $\sum_{k=1}^{K}\sum_{l'=1}^{N_i}\sum_{l=1}^{N_i}\frac{\widetilde{\pi}_{ll'|ii}}{\widetilde{\pi}_{l|i}\widetilde{\pi}_{l'|i}}y_{il}y_{il'}$ and $\sum_{i\neq}^{K}\sum_{j=1}^{N}\left\{\sum_{l=1}^{N_i}\sum_{q=1}^{N_j}\frac{\widetilde{\pi}_{lq|ij}}{\widetilde{\pi}_{l|i}\widetilde{\pi}_{q|j}}y_{il}y_{jq}\right\}$ can be estimated unbiasedly by

$$\sum_{k=1}^{K}\sum_{l'\neq}\sum_{l\in s_i}\frac{\widetilde{\pi}_{ll'|ii}}{\widetilde{\pi}_{l|i}\pi_{l'|i}\pi_{ll'|i}}y_{il}y_{il'} \text{ and } \sum_{i\neq}^{K}\sum_{j=1}^{K}\left\{\sum_{l\in s_i}\sum_{q\in s_j}\frac{\widetilde{\pi}_{lq|ij}}{\widetilde{\pi}_{l|i}\widetilde{\pi}_{q|j}\pi_{l|i}\pi_{q|j}}y_{il}y_{jq}\right\} \tag{7.8.8}$$

The expressions Eqs. (7.8.7) and (7.8.8) yield an unbiased estimator for $V\left(\widehat{\overline{Y}}_{ht}^{u}\right)$ as

$$\widehat{V}\left(\widehat{\overline{Y}}_{ht}^{u}\right) = \frac{1}{N^2}\left[\begin{array}{l}\sum_{i=1}^{K}\left(\sum_{l\in s_i}\frac{1}{\widetilde{\pi}_{l|i}\pi_{l|i}}y_{il}^2 + \sum_{l\neq}\sum_{q\in s_i}\frac{\widetilde{\pi}_{lq|ii}}{\widetilde{\pi}_{l|i}\widetilde{\pi}_{q|i}\pi_{lq|i}}y_{il}y_{iq}\right)\\ +\sum_{i\neq}^{K}\sum_{j=1}^{K}\left(\sum_{l\in s_i}\sum_{q\in s_j}\frac{\widetilde{\pi}_{lq|ij}}{\pi_{l|i}\widetilde{\pi}_{q|j}\pi_{l|i}\pi_{q|j}}y_{il}y_{jq}\right)\end{array}\right] - \left[\left(\widehat{\overline{Y}}_{ht}^{st}\right)^2 - \widehat{V}\left(\widehat{\overline{Y}}_{ht}^{st}\right)\right] \tag{7.8.9}$$

Finally using Eqs. (7.8.6) and (7.8.9) an unbiased estimator for the gain in efficiency of stratified sampling over the unstratified is obtained as

$$\widehat{G} = \widehat{V}\left(\widehat{\overline{Y}}_{ht}^{u}\right) - \widehat{V}\left(\widehat{\overline{Y}}_{ht}^{st}\right)$$

$$= \frac{1}{N^2}\left[\begin{array}{l}\sum_{i=1}^{K}\left(\sum_{l\in s_i}\frac{1}{\widetilde{\pi}_{l|i}\pi_{l|i}}y_{il}^2 + \sum_{l\neq}\sum_{q\in s_i}\frac{\widetilde{\pi}_{lq|ii}}{\widetilde{\pi}_{l|i}\widetilde{\pi}_{q|i}\pi_{lq|i}}y_{il}y_{iq}\right)\\ +\sum_{i\neq}^{K}\sum_{j=1}^{K}\left(\sum_{l\in s_i}\sum_{q\in s_j}\frac{\widetilde{\pi}_{lq|i,j}}{\widetilde{\pi}_{l|i}\widetilde{\pi}_{q|j}\pi_{l|i}\pi_{q|j}}y_{il}y_{jq}\right)\end{array}\right] - \left(\widehat{\overline{Y}}_{ht}^{st}\right)^2 \tag{7.8.10}$$

7.8.1 Simple Random Sampling Without Replacement

In case both the stratified and unstratified samples are selected by SRSWOR method, we get

$$\pi_{l|i} = n_i/N_i,\ \pi_{lq|i} = n_i(n_i-1)/N_i(N_i-1);\ \widetilde{\pi}_{l|i} = n/N,$$

$$\widetilde{\pi}_{lq|i,j} = n(n-1)/N(N-1);\ \widehat{\overline{Y}}_{ht}^{st} = \sum_{i=1}^{K} W_i\overline{y}(s_i),$$

$$\overline{y}(s_i) = \frac{1}{n_i}\sum_{j\in s_i} y_{ij},\ \widehat{\overline{Y}}_{ht}^{u} = \frac{1}{n}\sum_{i}\sum_{j\in s} y_{ij} = \overline{y}(s),$$

$$V\left(\widehat{\overline{Y}}_{ht}^{st}\right) = \sum_{i=1}^{K} W_i^2(1-f_i)S_{yi}^2\Big/n_i,$$

$$\widehat{V}\left(\widehat{\overline{Y}}_{ht}^{st}\right) = \sum_{i=1}^{K} W_i^2(1-f_i)s_{yi}^2\Big/n_i,\ V\left(\widehat{\overline{Y}}_{ht}^{u}\right) = \frac{(1-f)}{n}S_y^2, f_i = n_i/N_i,$$

$$f = n/N \text{ and } (N-1)S_y^2 = \sum_{i=1}^{K}\sum_{j=1}^{N_i}\left(y_{ij} - \overline{Y}\right)^2.$$

Now $V\left(\widehat{\overline{Y}}_{ht}^{u}\right) = \frac{(1-f)}{n}\frac{1}{N-1}\sum_{i=1}^{K}\sum_{j=1}^{N_i}\left(y_{ij} - \overline{Y}\right)^2$ can be unbiasedly estimated through stratified sample by

$$\widehat{V}\left(\widehat{\overline{Y}}_{ht}^{u}\right) = \frac{(1-f)}{n}\frac{1}{N-1}\left[\sum_{i=1}^{K}\frac{N_i}{n_i}\sum_{j\in s_i}y_{ij}^2 - N\left\{\left(\widehat{\overline{Y}}_{ht}^{st}\right)^2 - \widehat{V}\left(\widehat{\overline{Y}}_{ht}^{st}\right)\right\}\right].$$

After simplification we get

$$\widehat{V}\left(\widehat{\overline{Y}}^{u}_{ht}\right) = \left(\frac{1}{n} - \frac{1}{N}\right)\sum_{i=1}^{K} W_i s_{yi}^2 + \frac{N-n}{n(N-1)} \sum_{i=1}^{K}\left[W_i\left(\overline{y}(s_i) - \widehat{\overline{Y}}^{st}_{ht}\right)^2 - W_i(1-W_i)s_{yi}^2\Big/n_i\right] \tag{7.8.11}$$

So, the estimated gain in efficiency is

$$\widehat{G} = \left(\frac{1}{n} - \frac{1}{N}\right)\sum_{i=1}^{K} W_i s_{yi}^2 + \frac{N-n}{n(N-1)}\sum_{i=1}^{K}\left[W_i\left(\overline{y}(s_i) - \widehat{\overline{Y}}^{st}_{ht}\right)^2 - W_i(1-W_i)s_{yi}^2\Big/n_i\right] - \sum_{i=1}^{K} W_i^2\left(\frac{1}{n_i} - \frac{1}{N_i}\right)s_{yi}^2 \tag{7.8.12}$$

7.8.2 Probability Proportional to Size With Replacement Sampling

Suppose an unstratified sample of size n is selected by PPSWR method using normed size measure p_{ij} for the jth unit of the ith stratum. Then the HH estimator of the population mean $\overline{Y}$ is given by

$$\widehat{\overline{Y}}^{u}_{hh} = \frac{1}{nN}\sum_{r=1}^{n}\frac{y(r)}{p(r)} \tag{7.8.13}$$

where $y(r) = y_{ij}$ and $p(r) = p_{ij}$ if the rth draw selects the jth unit of the ith stratum.

The variance of $\widehat{\overline{Y}}^{u}_{hh}$ is given by

$$V\left(\widehat{\overline{Y}}^{u}_{hh}\right) = \frac{1}{N^2 n}\left(\sum_{i=1}^{K}\sum_{j=1}^{N_i}\frac{y_{ij}^2}{p_{ij}} - N^2\overline{Y}^2\right) \tag{7.8.14}$$

Let the HH estimator based on a stratified sample of size n be

$$\widehat{\overline{Y}}^{st}_{hh} = \frac{1}{N}\sum_{i=1}^{K}\frac{1}{n_i}\sum_{r=1}^{n_i}\frac{y(r,i)}{p(r,i)}$$

where $y(r,i) = y_{ij}$, $p(r,i) = p_{ij}/P_i = p^*_{ij}$ if the rth draw selects the jth unit of the ith stratum and $P_i = \sum_{j=1}^{N_i} p_{ij}$.

Unbiased estimators of $\frac{1}{N^2 n}\sum_{i=1}^{K}\sum_{j=1}^{N_i}\frac{y_{ij}^2}{p_{ij}} = \frac{1}{N^2 n}\sum_{i=1}^{K}\frac{1}{P_i}\sum_{j=1}^{N_i}\frac{y_{ij}^2}{p^*_{ij}}$ and $\frac{\overline{Y}^2}{n}$ based on the stratified sample are given by $\frac{1}{N^2 n}\sum_{i=1}^{K}\frac{1}{n_i P_i}\sum_{r=1}^{n_i}\left(\frac{y(r,i)}{p(r,i)}\right)^2$ and

$\dfrac{\left(\widehat{\overline{Y}}_{hh}^{st}\right)^2 - \widehat{V}\left(\widehat{\overline{Y}}_{hh}^{st}\right)}{n}$, respectively, where $\widehat{V}\left(\widehat{\overline{Y}}_{hh}^{st}\right) = \dfrac{1}{N^2}\sum_{i=1}^{K}\dfrac{1}{n_i(n_i-1)}\sum_{r=1}^{n_i}\left(\dfrac{y(r,i)}{p(r,i)} - \widehat{Y}_i\right)^2$ is an unbiased estimator of $V\left(\widehat{\overline{Y}}_{hh}^{st}\right)$ and $\widehat{Y}_i = \dfrac{1}{n_i}\sum_{r=1}^{n_i}\dfrac{y(r,i)}{p(r,i)}$. Hence an unbiased estimator for $V\left(\widehat{\overline{Y}}_{hh}^{u}\right)$ based on the stratified sample is given by

$$\widehat{V}\left(\widehat{\overline{Y}}_{hh}^{u}\right) = \frac{1}{N^2 n}\sum_{i=1}^{K}\frac{1}{n_i P_i}\sum_{r=1}^{n_i}\left(\frac{y(r,i)}{p(r,i)}\right)^2 - \frac{1}{n}\left[\left(\widehat{\overline{Y}}_{hh}^{st}\right)^2 - \widehat{V}\left(\widehat{\overline{Y}}_{hh}^{st}\right)\right]$$

The expression of the estimated gain in efficiency of stratified sampling over unstratified sampling under PPSWR sampling design is

$$\begin{aligned}\widehat{G} &= \widehat{V}\left(\widehat{\overline{Y}}_{hh}^{u}\right) - \widehat{V}\left(\widehat{\overline{Y}}_{hh}^{st}\right)\\ &= \frac{1}{N^2 n}\left[\sum_{i=1}^{K}\frac{1}{n_i P_i}\sum_{r=1}^{n_i}\left(\frac{y(r,i)}{p(r,i)}\right)^2 - N^2\left(\widehat{\overline{Y}}_{hh}^{st}\right)^2\right] - \left(1-\frac{1}{n}\right)\widehat{V}\left(\widehat{\overline{Y}}_{hh}^{st}\right)\end{aligned} \tag{7.8.15}$$

Example 7.8.1

The garment factories of the KZN province of South Africa are classified into three strata. To estimate the average earnings of the garments, samples of factories are selected by SRSWOR method from each of the strata. Estimate the average earnings of garments per factory and compute its standard error. Estimate the gain in efficiency of stratified sampling over simple random sampling of the same sample size.

Factories	Number of factories (N_i)	Sample size (n_i)	Earnings (000 \$) (y_{ij})
Small	50	5	150, 100, 75, 125,175
Medium	100	10	750, 800, 950, 1000, 750, 750, 1200, 1500, 1200, 1500
Large	75	8	3000, 4000, 5000, 4500

To estimate the gain in efficiency, we prepare the following table.

Factories	N_i	n_i	Sample mean (000 \$) $\overline{y}_i$	Sample SD s_{yi} (000 \$)	$W_i s_{yi}^2$	$W_i(\overline{y}_i - \overline{y}_{st})^2$	$W_i(1-W_i)\frac{s_{yi}^2}{n_i}$
Small	50	5	125	39.52	347.22	672,800	54.01
Medium	100	10	1040	297.02	39,209.88	302,500	2178.32
Large	75	8	4125	583.91	243,055.55	1,702,533.333	20,254.63
Total	225	23	–	–	282,612.65	2,677,833.333	22,486.97

$$\text{Estimated mean income per factory} = \overline{y}_{st} = \sum_{i=1}^{3} N_i \overline{y}_i / N = \$\ 1,865,000$$

$$\text{Estimated variance of stratified sampling} = \widehat{V}\left(\overline{y}_{st}\right) = \sum_{i=1}^{3} W_i^2 \left(\frac{1}{n_i} - \frac{1}{N_i}\right)$$

$s_{yi}^2 = 10,629.35$

Estimated variance of the sample mean based on the stratified sampling is

$$\begin{aligned}\widehat{V}\left(\overline{y}_{wor}\right) &= \left(\frac{1}{n} - \frac{1}{N}\right) \sum_i W_i s_{yi}^2 \\ &\quad + \frac{N-n}{(N-1)n} \left\{ \sum_i W_i \left(\overline{y}_i - \overline{y}_{st}\right)^2 - \sum_i W_i (1 - W_i) s_{yi}^2 / n_i \right\} \\ &= (1/23 - 1/225) \times 282,612.65 + \frac{225 - 23}{(224)23} \\ &\quad \times (2,677,833.33 - 22,486.97) \\ &= 115,142.5\end{aligned}$$

So, the percentage gain in efficiency of stratified sampling $= \left\{ \widehat{V}\left(\overline{y}_{wor}\right) / \widehat{V}\left(\overline{y}_{st}\right) - 1 \right\} \times 100 = 983.25\%$

Example 7.8.2

To estimate the total production of apples (in 00 kg) in a region, the farms were classified into four strata. From each of the strata, samples of farms were selected by PPSWR method using farm size (in acres) as measure of size variable. The details are given as follows:

Types of farm	Number of farms (N_i)	Sample size	Average farm size (in acre) $\overline{X}_i$	Production (y) and farm size (x)				
Small	30	4	3	x	2	4	2	4
				y	25	30	20	40
Medium	50	8	4	x	2	5	5	4
				y	30	40	60	30
				x	4	4	5	4
				y	50	50	60	40
Large	20	4	5	x	5	4	5	4
				y	60	80	60	50

(i) Estimate the total production of apples and estimate the standard error of the estimator used.

(ii) Estimate the gain in efficiency of stratified sampling over unstratified sampling that might be selected by the PPSWR method of the same sample size of 16.

The HH estimate for the total production of apple Y based on the stratified sample is given by

$$\widehat{Y}_{hh}^{st} = \sum_{i=1}^{3} \widehat{Y}_i = 900 + 2237.5 + 1412.5 = 4550$$

where $\widehat{Y}_i = \frac{1}{n_i} \sum_{j=1}^{n_i} y_{ij}/p_{ij}^*,\ p_{ij}^* = x_{ij}/X_i$ and $X_i = N_i \overline{X}_i$

Estimated standard error of $\widehat{Y}_{hh}^{st} = \sqrt{\sum_{i=1}^{3} \widehat{V}\left(\widehat{Y}_i\right)}$ with

$$\widehat{V}\left(\widehat{Y}_i\right) = \frac{1}{n_i(n_i-1)} \sum_{j=1}^{n_i} \left(\frac{y_{ij}}{p_{ij}^*} - \widehat{Y}_i\right)^2$$

We find $\widehat{V}\left(\widehat{Y}_1\right) = 8437.50$, $\widehat{V}\left(\widehat{Y}_2\right) = 31,763.39$ and $\widehat{V}\left(\widehat{Y}_3\right) = 38,489.58$

Hence standard error of $\widehat{Y}_{hh}^{st}$ is $\sqrt{\sum_{i=1}^{3} \widehat{V}\left(\widehat{Y}_i\right)} = \sqrt{78,690.47} = 280.52$

The HH estimator based on an unstratified PPSWR sample of size n is given by

$$\widehat{Y}_{hh}^{u} = \frac{1}{n} \sum_{i=1}^{3} \sum_{j=1}^{N_i} y_{ij}/p_{ij} \text{ with } p_{ij} = x_{ij}/X \text{ and } X = \sum_{i=1}^{3} X_i = 390.$$

The variance of $\widehat{Y}_{hh}^{u}$ is $V\left(\widehat{Y}_{hh}^{u}\right) = \frac{1}{n}\left(\sum_{i=1}^{3}\sum_{j=1}^{N_i} y_{ij}^2 \big/ p_{ij} - Y^2\right) = \frac{1}{n}\left(\sum_{i=1}^{3}\frac{X}{X_i}\sum_{j=1}^{N_i} y_{ij}^2 \big/ p_{ij}^{*} - Y\right)^2$

An unbiased estimate of $\sum_{i=1}^{3}\frac{X}{X_i}\sum_{j=1}^{N_i} y_{ij}^2 \big/ p_{ij}^{*}$ is $\sum_{i=1}^{3}\frac{X}{X_i}\left(\frac{1}{n_i}\sum_{j=1}^{n_i} y_{ij}^2 \big/ \left(p_{ij}^{*}\right)\right) =$ 3,619,687.5 +10,196,062.5 + 8,231,438 = 22,047,188 and an unbiased estimate of Y^2 is $\left(\widehat{Y}_{hh}^{st}\right)^2 - \widehat{V}\left(\widehat{Y}_{hh}^{st}\right) = (4550)^2 - 78,690.47 = 20,623,809.53$. Hence an unbiased estimate of $V\left(\widehat{Y}_{hh}^{u}\right)$ based on the stratified sample is $\widehat{V}\left(\widehat{Y}_{hh}^{u}\right)$ (22,047,188−20,623,809.53)/ 16 = 88,961.15. An estimate of gain in efficiency is $\left(\frac{\hat{V}\left(\hat{Y}_{hh}^{u}\right)}{\hat{V}\left(\hat{Y}_{hh}^{st}\right)} - 1\right) \times 100\% = 13.05\%$.

7.9 POSTSTRATIFICATION

The principle of stratification is to divide the entire population into a number of homogenous strata and then select samples from the each of the strata by some suitable sampling procedure. Sometimes, stratification is not possible before selection of sample for various reasons, e.g., (i) variability of the character under study is not known, (ii) in a multicharacter survey, information about more than one character is studied at a time. In this case, stratification with respect to one character may make the strata homogeneous with respect to that character only, but it makes them more heterogeneous with respect to the other character, and (iii) stratum sizes and sampling frame for each stratum may not be available. For example, in an HIV prevalence study, reliable data on the number of people in different age groups may be available from the past census or surveys, but it is difficult to get a list of people belonging to different age groups. In this situation, one can select a sample from the entire population first and then stratify the sample, looking at the values of the study variable under consideration. The technique of stratifying the population after the selection of the sample is known as poststratification.

Let a sample s of size n be selected from a finite population of size N by SRSWOR method, and let the sample be poststratified into K strata. Let n_i be the number of units falling into the ith stratum and $\overline{y}_i$ be the sample mean, $i = 1,\ldots, K$. Here n_i's are random variables subject to $\sum_{i=1}^{K} n_i = n$. We assume that n is so large that $P(n_i = 0) \cong 0$ for every i.

Theorem 7.9.1

(i) $\overline{y}_{pst} = \sum_{i=1}^{K} W_i \overline{y}_i$ is unbiased for the population mean $\overline{Y}$ when stratum weights W_i's are known.

(ii)
$$V\left(\overline{y}_{pst}\right) = \sum_{i=1}^{K} W_i^2 \left[E\left(\frac{1}{n_i}\right) - 1\right] S_{yi}^2$$

$$\cong \frac{N-n}{Nn} \sum_{i=1}^{K} W_i S_{yi}^2 + \frac{1}{n^2} \sum_{i=1}^{K} (1 - W_i) S_{yi}^2$$

(iii) An unbiased estimator of $V\left(\overline{y}_{pst}\right)$ is

$$\widehat{V}\left(\overline{y}_{pst}\right) = \sum_{i=1}^{K} W_i^2 \left(\frac{1}{n_i} - \frac{1}{N_i}\right) s_{yi}^2$$

where s_{yi}^2 is the sample variance for the ith stratum.

Proof

(i) $E\left(\overline{y}_{pst}\right) = E\left\{\sum_{i=1}^{K} W_i E\left(\overline{y}_i \middle| n_i\right)\right\} = E\left(\sum_{i=1}^{K} W_i \overline{Y}_i\right) = \overline{Y}$

(ii) $V\left(\overline{y}_{pst}\right) = E\left\{V\left(\overline{y}_{pst} \middle| n_i\right)\right\} + V\left\{E\left(\overline{y}_{pst} \middle| n_i\right)\right\}$

$$= E\left\{\sum_{i=1}^{K} W_i^2 \left(\frac{1}{n_i} - \frac{1}{N_i}\right) S_{yi}^2\right\} + V\left(\sum_{i=1}^{K} W_i \overline{Y}_i\right)$$

$$= \sum_{i=1}^{K} W_i^2 \left[E\left(\frac{1}{n_i}\right) - \frac{1}{N_i}\right] S_{yi}^2$$

Now writing $\epsilon_i = \dfrac{n_i - \gamma_i}{\gamma_i}$ with $\gamma_i = E(n_i)$, we approximate

$$E\left(\frac{1}{n_i}\right) = \frac{1}{\gamma_i} E(1 + \epsilon_i)^{-1} \cong \frac{1}{\gamma_i} E\left(1 - \epsilon_i + \epsilon_i^2\right) = \frac{1}{\gamma_i}\left(1 + \frac{V(n_i)}{\gamma_i^2}\right) \tag{7.9.1}$$

Furthermore, noting that n_i follows a hypergeometric distribution with $\gamma_i = E(n_i) = nW_i$ and $V(n_i) = nW_i(1 - W_i)\dfrac{N - n}{N - 1} \cong nW_i(1 - W_i)$ (ignoring f.p.c), we get

$$E\left(\frac{1}{n_i}\right) \cong \frac{1}{nW_i} + \frac{1 - W_i}{(nW_i)^2} \tag{7.9.2}$$

The approximation (7.9.2) was derived by Stephan (1945). Substituting Eq. (7.9.2) in Theorem 7.9.1, we find an approximate expression of $V\left(\overline{y}_{pst}\right)$ as follows:

$$V\left(\overline{y}_{pst}\right) = \frac{N - n}{Nn}\sum_{i=1}^{K} W_i S_{yi}^2 + \frac{1}{n^2}\sum_{i=1}^{K}(1 - W_i) S_{yi}^2 \tag{7.9.3}$$

(iii) $$E\left[\widehat{V}\left(\overline{y}_{pst}\right)\right] = E\left[\sum_{i=1}^{K} W_i^2\left(\frac{1}{n_i} - \frac{1}{N_i}\right)E\left(s_{yi}^2 \middle| n_i\right)\right]$$

$$= \sum_{i=1}^{K} W_i^2\left[E\left(\frac{1}{n_i}\right) - \frac{1}{N_i}\right]S_{yi}^2$$

The first term in Eq. (7.9.3) is the variance of the stratified sample mean $\overline{y}_{st}$ under proportional allocation. The second term is a positive quantity, which may be defined as the cost of poststratification. Thus post-stratification is always less efficient than stratified sampling with proportional allocation. However, for large n, the second term is expected to be much smaller compared to the first one. Hence, poststratified sampling is almost as efficient as stratified sampling if the sample size is very large.

7.10 EXERCISES

7.10.1 Consider a population of $N = nk$ units where the variable under study y possesses a linear trend $y_j = \alpha + \beta j$, $i = 1,\ldots, N$. The population is stratified into n strata of k units each. The ith ($=1,\ldots, k$) stratum consists of units with labels $(i - 1)k + 1,\ldots, ik$. Let $\overline{y}_{st}$, $\overline{y}_{wor}$, and $\overline{y}_{sys}$ be the sample means each of size n, selected by stratified simple random sampling with one unit per stratum, SRSWOR sampling, and systematic sampling from the entire population, respectively. Prove that $V\left(\overline{y}_{st}\right) \le V\left(\overline{y}_{sys}\right) \le V\left(\overline{y}_{wor}\right)$.

7.10.2 Suppose an infinite population $f(y) = \theta e^{-y\theta}$; $y > 0$ (θ is a known positive constant) is divided into two strata by taking a point y_0 in the range of the study variable y. Samples of sizes n_1 and $n_2(=n - n_1)$ are selected from stratum one and stratum two, respectively, by SRSWR method. Find the optimum point of stratification, which minimizes the variance of the estimator of the population mean $\overline{Y}$ under (i) proportional and (ii) optimum allocation, when the sample size n is fixed. Compute the relative efficiency of the optimal allocation over the proportional allocation.

7.10.3 Consider a population consisting of two strata with sizes N_1 and N_2. Let V_{prop} and V_{opt} denote the variances of the estimator of the population mean for the proportional and optimum allocation under the cost function $C = c_0 + c_1 n_1 + c_2 n_2$, respectively, based on SRSWR samples from each of the stratum. Show that when the total cost of the survey C is fixed, strata variances are equal

$$\frac{V_{prop}}{V_{opt}} = \frac{N(N_1 c_1 + N_2 c_2)}{\left(N_1\sqrt{c_1} + N_2\sqrt{c_2}\right)^2}$$

7.10.4 A population is stratified into K strata. From the $i(=1,\ldots, K)$th stratum a sample of n_i units is selected from the N_i units by SRSWOR method and let $\overline{y}_i$ be the sample mean. Show that the optimum n_i that minimizes the variance of $T = \sum_{i=1}^{K} l_i \overline{y}_i$ for a given cost C_0 under the cost function $C = c_0 + \sum_{i=1}^{K} c_i n_i$ is $n_i = \frac{(C_0 - c_0)}{\left(\sum_{i=1}^{k} |l_i| S_i \sqrt{c_i}\right)} \frac{|l_i| S_i}{\sqrt{c_i}}$. Hence, show that the optimum values of n_i and n_j that minimizes the variance of $\overline{y}_i - \overline{y}_j$ are given by $n_i = \frac{(C_0 - c_0)}{\left(\sum_{i=1}^{k} S_i\sqrt{c_i}\right)} \frac{S_i}{\sqrt{c_i}}$ and $n_j = \frac{(C_0 - c_0)}{\left(\sum_{i=1}^{k} S_j\sqrt{c_j}\right)} \frac{S_j}{\sqrt{c_j}}$, respectively.

7.10.5 The following data relate to the household size, monthly income, and ownership of the house of 50 households in South Africa.

Serial number of HH	HH size	HH income ($)	Ownership of house	Serial number of HH	HH size	HH income ($)	Ownership of house
1	1	18,562	0	26	5	11,526	1
2	1	10,864	1	27	5	10,778	0
3	4	8456	1	28	5	17,843	1
4	5	11,990	0	29	1	17,183	0
5	4	16,449	1	30	5	6543	0
6	4	13,563	1	31	5	10,009	0
7	3	2151	0	32	1	19,592	0
8	1	20,786	0	33	4	20,473	1
9	4	16,738	0	34	1	6662	1
10	4	18,003	0	35	1	7052	0
11	5	22,073	0	36	1	16,071	0
12	1	8104	0	37	4	14,222	0
13	3	1565	0	38	2	24,092	0
14	4	20,049	1	39	5	12,970	1
15	5	10,770	0	40	1	23,993	1
16	1	14,678	1	41	1	21,026	0
17	1	5674	1	42	1	17,053	0
18	1	19,169	0	43	4	11,736	1
19	1	24,088	0	44	3	5816	0
20	2	8938	0	45	2	2167	0
21	3	19,090	1	46	3	23,211	1
22	4	13,141	1	47	2	22,293	1
23	3	23,849	1	48	5	9189	0
24	1	9728	1	49	1	5443	1
25	1	12,605	1	50	4	23,163	1

Stratify the 50 households with their income (a) less than \$10000 (stratum 1), (b) \$10001 to \$15000 (stratum 2), and (c) above \$15000 (stratum 3).

(i) From each of the stratum select a sample of five households by SRSWOR method. Estimate the average income of the 50 households. Find the standard error of your estimator. Compare the relative efficiency of your estimator with an unstratified sample mean based on an SRSWOR of the same sample size of 15.

(ii) Select samples of sizes 5 from stratum 1, 6 from stratum 2, and 5 from stratum 3 by using SRSWR, SRSWOR, and PPSWR methods, respectively, taking household size as a measure of size variable. From the selected samples, find an unbiased estimate of the proportion of persons owning a house. Estimate the standard error of the estimate and obtain a 90% confidence interval for the proportion of households owning a house.

(iii) Give the allocation of a total of 16 households using proportional and Neyman allocation.

7.10.6 The following sample of 23 factories selected by SRSWOR method from 50 small, 75 medium, and 25 large factories in a certain region. Information regarding number of workers, output, and membership of a medical aid is also listed.

Small factories				Medium factories			
Serial number of factories	Number of workers	Daily output	Medical aid facility	Serial number of factories	Number of workers	Daily output	Medical aid facility
1	21	885	1	1	29	1021	1
2	20	981	0	2	48	1184	0
3	18	762	1	3	40	1198	0
4	22	539	1	4	50	1112	1
5	14	777	0	5	47	1404	0
6	13	648	0	6	33	1175	0
7	17	972	1	7	50	1286	0
8	16	919	0	8	48	1319	1
				9	45	1358	1
				10	50	18,668	1

Large factories

Serial number of factories	Number of workers	Daily output	Medical Aid facility
1	70	1791	1
2	58	2455	1
3	89	2659	1
4	79	2342	0
5	74	1743	1

(i) Estimate the average daily output of the 150 factories of the region and find a 90% confidence interval of the average output.

(ii) Estimate the proportion of factories having a medical aid facility. Find a 95% confidence interval of this proportion.

(iii) Estimate the average number of workers per factory and obtain its standard error. Estimate the gain in efficiency of stratified sampling over unstratified sampling for estimating the average number of workers.

(iv) Estimate the average daily output of the factories having a medical aid facility when it is known that 40% of the factories provide medical facilities. Estimate the standard error of your estimator.

7.10.7 To estimate the total production of maize in a certain region, the farms were classified in to four strata. It is decided that a total of 50 farms will be selected using (i) proportional, (ii) Neyman, (iii) allocation proportional to the average size of farm, and (iv) optimal allocation. Compare the relative efficiencies of the allocations for estimating the total output if SRSWOR method is used from each of the strata using the following past survey data:

Strata	Number of farms	Average size of the farm (in acre)	Standard deviation of maize production (in 000 kg)	Cost of surveying a factory (in $)
I	50	10	10	50
II	75	15	20	60
III	100	20	20	75
IV	25	25	25	100

7.10.8 The following table gives a summary result of an agricultural survey for estimating the production of milk using SRSWR method of sampling from the each type of farms.

Types of farm	Number of farms	Sampled farms	Production of milk per sampled farm (00 in liter)	Sample standard deviation of production
Small	50	10	125	30
Medium	100	25	200	30
Large	175	30	200	50

(i) Estimate total production of milk and its standard error.

(ii) Estimate the gain in efficiency of the stratified sampling with respect to unstratified sampling based on SRSWR method of the same sample size.

(iii) Compare relative efficiencies of estimating the total production of milk of the present allocation with (i) proportion and (ii) Neyman's allocation keeping the total sample size fixed.

7.10.9 To estimate the average marks of a STA112 course, which comprises of four groups of students in the University of Botswana, samples from each of the groups were selected by PPSWR method using the CA marks as a measure of size variable. Details are given in the following table:

(i) Estimate the average examination marks and find a 90% confidence interval of the average marks.

Groups	Number of students	Sampled students	Average CA	Marks on CA (*x*) and Exam (*y*) obtained from the students selected in the sample										
A	75	10	60	*x*	50	60	60	80	80	25	30	80	70	40
				y	55	50	70	90	95	30	50	70	75	20
B	50	5	75	*x*	30	70	60	45	30					
				y	40	75	55	60	50					
C	75	8	60	*x*	90	25	40	80	35	65	60	20		
				y	98	30	50	70	50	50	80	30		
D	75	5	65	*x*	50	20	10	60	70					
				y	60	25	40	70	90					

(ii) Estimate the gain in efficiency of the stratified sampling over unstratified PPSWR sampling from the entire population with the same sample size.

(iii) Based on the selected sample, compare efficiencies of the present allocation with (a) proportional allocation and (b) allocation proportional to the average CA marks for estimating the average examination marks.

7.10.10 A sample of 200 students was selected from a class of 800 students by SRSWOR method, and their marks in the STAT 121 course were recorded. The students were then classified according to their grades. The following table gives the sample means and standard deviations along with the grades of 800 students obtained from the university record.

		Sampled students		
Grade	**Number of students**	**Number**	**Average marks**	**Standard deviation**
A	80	20	82	10
B	120	35	72	15
C	350	100	65	12
D	150	30	52	10
E	100	15	30	20

(i) Estimate the average marks of the STAT 121 course.

(ii) Give an appropriate estimate of the standard error of your estimator used in "(i)."

7.10.11 A simple random sample of 500 households was selected from an urban community of 5000 households. The following table shows a summary of the survey along with the distribution of the age of the community from the past census report.

		Sampled people			
Age group	**Number of people (obtained from census)**	**Number**	**Average income ($)**	**Standard deviation ($)**	**HIV prevalence rate**
Below 18	1000	80	500	100	0.10
19–25	1500	150	2500	50	0.25
26–45	1500	170	8000	40	0.40
46–60	700	80	12,000	60	0.30
Above 60	300	20	1500	80	0.10

(i) Estimate the mean income of the community and also obtain the 90% confidence interval of the mean income.

(ii) Estimate the HIV prevalence rate for the entire community and its standard error.

7.10.12 The following table gives the income distribution of 1500 households of a certain district. Make five strata using (i) Dalenius and Hodges and (ii) Gunning and Horgan methods.

Household income (US$)	Below 1000	1001 to 2000	2001 to 3000	3001 to 4000	4001 to 5000	5001 to 7000	Above 7000
Frequency	200	350	400	250	150	100	50

CHAPTER 8

Ratio Method of Estimation

8.1 INTRODUCTION

In Chapter 5, it has been shown that when the auxiliary information has a good relation with the study variable, one can use the auxiliary information to construct various sampling deigns such as probability proportional to size with replacement (PPSWR), probability proportional to size without replacement (PPSWOR), inclusion probability proportion to size (IPPS), and Rao-Hartley-Cochran (RHC) to yield efficient estimators for population parameters (e.g., total, mean, and proportions) under some specific circumstances. Furthermore, we have also mentioned that the auxiliary information can be used gainfully in constructing strata for stratified sampling. The auxiliary information can also be used at the stages of estimation to improve the efficiency of the conventional estimator. In this section we consider the ratio method of estimation to improve the efficiency of the conventional estimator by utilizing auxiliary information x, whose values need not be known for all units, but the total X is assumed to be known. For instance, in estimating the total production of a crop, one can use the area under cultivation (x) of the farm growing a particular crop as a size variable because production is strictly related to the area under cultivation. Here, the values of the area under cultivation for all the farms need not be known but the total area under cultivation for the crop (X) should be known.

8.2 RATIO ESTIMATOR FOR POPULATION RATIO

We are very often interested in estimating the ratio R of two characters such as per capita income, proportion of unemployed persons, average cost of treatment per indoor patients, or average cost of educating a person. The ratio estimator of the ratio $R = Y/X$ of the population totals Y and X of the characters y and x is defined as

$$\widehat{R} = \widehat{Y}/\widehat{X} \tag{8.2.1}$$

where $\widehat{X}$ and $\widehat{Y}$ are unbiased estimators of the population totals X and Y, respectively.

Survey Sampling Theory and Applications
ISBN 978-0-12-811848-1
http://dx.doi.org/10.1016/B978-0-12-811848-1.00008-X

8.2.1 Exact Expression of Bias and Mean-Square Error

The estimator $\widehat{R}$ is not unbiased for R in general since $E\left(\widehat{R}\right) \neq E\left(\widehat{Y}\right)/E\left(\widehat{X}\right)$. The bias of $\widehat{R}$ is

$$
\begin{aligned}
B\left(\widehat{R}\right) &= E\left(\widehat{R}\right) - R \\
&= -\frac{1}{X}\left\{Y - XE\left(\widehat{R}\right)\right\} \\
&= -\frac{1}{X}\left\{E\left(\widehat{R}\,\widehat{X}\right) - E\left(\widehat{X}\right)E\left(\widehat{R}\right)\right\} \\
&= -\frac{1}{X}\,Cov\left(\widehat{R}, \widehat{X}\right) \\
&= -\rho_{\hat{R},\hat{X}}\frac{\sqrt{V\left(\widehat{R}\right)}\sqrt{V\left(\widehat{X}\right)}}{X}
\end{aligned} \tag{8.2.2}
$$

where $\rho_{\hat{R},\hat{X}}$ is the correlation coefficient between $\widehat{R}$ and $\widehat{X}$.

Hence, the bias $B\left(\widehat{R}\right)$ is zero if the correlation between $\widehat{R}$ and $\widehat{X}$ is zero. In particular, if y is approximately proportional to x, then $\widehat{R}$ is approximately constant and the bias $B\left(\widehat{R}\right)$ is negligible. Furthermore, since $\left|\rho_{\hat{R},\hat{X}}\right| \leq 1$, we have the following upper bound of the absolute relative bias of $\widehat{R}$ derived by Hartley and Ross (1954).

$$
ARB\left(\widehat{R}\right) = \left|\frac{B\left(\widehat{R}\right)}{R}\right| \leq \frac{\sqrt{V\left(\widehat{R}\right)}}{|R|}\frac{\sqrt{V\left(\widehat{X}\right)}}{|X|} \tag{8.2.3}
$$

$$
\text{i.e., } \frac{\left|B\left(\widehat{R}\right)\right|}{\sqrt{V\left(\widehat{R}\right)}} \leq \left|C_{\hat{X}}\right| \tag{8.2.4}
$$

where $C_{\hat{X}}$ is the coefficient of variation (CV) of $\widehat{X}$.

Eq. (8.2.4) indicates that if $C_{\hat{X}}$, is less than 0.1, the effect of bias is not an appreciable disadvantage in estimating R (Cochran, 1977).

Let a sample s of size n be selected from a population U of size N with probability $p(s)$ and $\widehat{X}_s = \sum_{i \in s} b_{si}x_i$ and $\widehat{Y}_s = \sum_{i \in s} b_{si}y_i$ be the linear unbiased estimators for X and Y, respectively. The constants b_{si}'s are free from y_i's and satisfy the unbiasedness condition $\sum_{s \supset i} b_{si}p(s) = 1\ \forall i = 1, \ldots, N$. Consider the following estimator of R based on an arbitrary sampling design p

$$
\widehat{R} = \frac{\sum_{i \in s} b_{si}y_i}{\sum_{i \in s} b_{si}x_i} \tag{8.2.5}
$$

Then, following Rao T.J. (1967a,b), we get the bias of $\widehat{R}$ as

$$B(\widehat{R}) = \sum_{s}\left(\frac{\sum_{i\in s} b_{si}y_i}{\sum_{i\in s} b_{si}x_i}\right)p(s) - R$$

$$= \frac{1}{X}\left[\sum_{s}\left(\sum_{i\in s} w_{si}y_i\right)p(s) - Y\right]$$

$$\left[\text{where } w_{si} = b_{si}X\Big/\left(\sum_{i\in s} b_{si}x_i\right)\right] \quad (8.2.6)$$

$$= \frac{1}{X}\left[\sum_{i\in U} y_i\left\{\sum_{s\supset i} w_{si}p(s) - 1\right\}\right]$$

$$= \frac{1}{X}\sum_{i\in U} y_i(\alpha_i - 1)$$

where $\alpha_i = \sum_{s\supset i} w_{si}p(s)$.

The mean-square error (MSE) of $\widehat{R}$ is given by

$$M(\widehat{R}) = E(\widehat{R} - R)^2 = \sum_{s}\left[\left(\frac{\sum_{i\in s} b_{si}y_i}{\sum_{i\in s} b_{si}x_i}\right)^2 - 2R\left(\frac{\sum_{i\in s} b_{si}y_i}{\sum_{i\in s} b_{si}x_i}\right) + R^2\right]$$

$$= \frac{1}{X^2}\left[\sum_{s}\left\{\left(\sum_{i\in s} w_{si}y_i\right)^2 - 2Y\left(\sum_{i\in s} w_{si}y_i\right)\right\}p(s) + Y^2\right]$$

$$= \frac{1}{X^2}\left[\sum_{i\in U}\beta_i y_i^2 + \sum_{i\neq}\sum_{j\in U}\beta_{ij}y_iy_j - 2\left(\sum_{i\in U}\alpha_i y_i\right)Y + Y^2\right]$$

$$\left(\text{where } \beta_i = \sum_{s\supset i} w_{si}^2 p(s) \text{ and } \beta_{ij} = \sum_{s\supset i,j} w_{si}w_{sj}p(s)\right)$$

$$= \frac{1}{X^2}\left[\sum_{i\in U}\gamma_i y_i^2 + \sum_{i\neq}\sum_{j\in U}\gamma_{ij}y_iy_j\right] \quad (8.2.7)$$

where $\gamma_i = \beta_i - 2\alpha_i + 1$, $\gamma_{ij} = \beta_{ij} - (\alpha_i + \alpha_j) + 1$.

From Eqs. (8.2.6) and (8.2.7), we set unbiased estimators for $B(\widehat{R})$ and $M(\widehat{R})$, when X is known, as follows:

$$\widehat{B}(R) = \frac{1}{X}\sum_{i\in s}\frac{(\alpha_i - 1)}{\pi_i}y_i \text{ and } \widehat{M}(\widehat{R}) = \frac{1}{X^2}\left(\sum_{i\in s}\frac{\gamma_i}{\pi_i}y_i^2 + \sum_{i\neq}\sum_{j\in s}\frac{\gamma_{ij}}{\pi_{ij}}y_iy_j\right)$$

where π_i and $\pi_{ij}(>0)$ are the inclusion probabilities of the ith unit and ith and jth unit, $i \neq j$.

In case X is not known, we may replace X by its unbiased estimate $\widehat{X}$. The results above are summarized in the following theorem:

Theorem 8.2.1

Let $\widehat{R} = \dfrac{\sum_{i\in s} b_{si}y_i}{\sum_{i\in s} b_{si}x_i}$ be an estimator of R based on a sampling design with inclusion probabilities π_i and $\pi_{ij}(>0)$ for the ith and ith and jth with b_{si}'s are suitably chosen weights. Then,

(i) $B(\widehat{R}) = \dfrac{1}{X}\sum_{i\in U} y_i(\alpha_i - 1)$

(ii) $M(\widehat{R}) = \left(\sum_{i\in U}\gamma_i y_i^2 + \sum_{i\neq}\sum_{j\in U}\gamma_{ij}y_iy_j\right)\Big/X^2$

(iii) $\widehat{B}(R) = \dfrac{1}{X}\sum_{i\in s}\dfrac{(\alpha_i - 1)}{\pi_i}y_i$

and

(iv) $\widehat{M}(\widehat{R}) = \dfrac{1}{X^2}\left(\sum_{i\in s}\dfrac{\gamma_i}{\pi_i}y_i^2 + \sum_{i\neq}\sum_{j\in s}\dfrac{\gamma_{ij}}{\pi_{ij}}y_iy_j\right)$

where $\alpha_i = \sum_{s\supset i} w_{si}p(s), \beta_i = \sum_{s\supset i} w_{si}^2p(s), \beta_{ij} = \sum_{s\supset i,j} w_{si}w_{sj}p(s), \gamma_i = \beta_i - 2\alpha_i + 1, \gamma_{ij} = \beta_{ij} - (\alpha_i + \alpha_j) + 1$ and $p(s)$ is the probability of selection of s.

Remark 8.2.1

Although the expressions of bias, MSE, and their estimators, look elegant in Theorem 8.2.1, they cannot be used in practice because the expressions α_i, β_i, β_{ij} are difficult to compute even in simple random sampling without replacement (SRSWOR) sampling as pointed out by Sukhatme et al. (1984). For example, consider SRSWOR, where $p(s) = 1\Big/\binom{N}{n}$, $b_{si} = N/n$, $w_{si} = b_{si}X\Big/\left(\sum_{i\in s} b_{si}x_i\right) = X/X_s$ and $X_s = \sum_{i\in s} x_i$.

Here we get

$$B(\widehat{R}) = \sum_{i\in U} y_i(\alpha_{i}\cdot - 1),\ \widehat{B}(\widehat{R}) = \frac{1}{X}\sum_{i\in s} y_i(\alpha_{i}\cdot - 1)/\pi_i,$$

$$M(\widehat{R}) = \frac{1}{X^2}\left(\sum_{i\in U}\gamma_{i}\cdot y_i^2 + \sum_{i\neq}\sum_{j\in U}\gamma_{ij}\cdot y_i y_j\right) \text{and}$$

$$\widehat{M}(\widehat{R}) = \frac{1}{X^2}\left(\sum_{i\in s}\gamma_{i}\cdot y_i^2/\pi_i + \sum_{i\neq}\sum_{j\in s}\gamma_{ij}\cdot y_i y_j/\pi_{ij}\right)$$

where

$$\alpha_{i}\cdot = \sum_{s\supset i}\frac{X}{X_s}\bigg/\binom{N}{n},\ \gamma_{i}\cdot = X^2\sum_{s\supset i}\frac{1}{X_s^2} - 2X\sum_{s\supset i}\frac{1}{X_s} + 1 \text{ and}$$

$$\gamma_{ij}\cdot = X^2\sum_{s\supset i,j}\frac{1}{X_s^2} - X\left(\sum_{s\supset i}\frac{1}{X_s} + \sum_{s\supset j}\frac{1}{X_s}\right) + 1$$

8.2.2 Approximate Expression of Bias and Mean-Square Errors

Following Murthy (1977), writing

$$\delta_y = (\widehat{Y} - Y)/Y \text{ and } \delta_x = (\widehat{X} - X)/X \tag{8.2.8}$$

we get $\widehat{R} = R(1+\delta_y)(1+\delta_x)^{-1}$.

Now assuming $|\delta_x| < 1\ \forall\ s$ with $p(s) > 0$, we can write $\widehat{R}$ as

$$\widehat{R} = R(1+\delta_y)\{1 - \delta_x + (\delta_x)^2 - \cdots\}$$

$$= R[1 + \{\delta_y - \delta_x\} + \{(\delta_x)^2 - \delta_y\delta_x\} + \cdots] \tag{8.2.9}$$

$$\text{i.e. } \widehat{R} - R = R[\{\delta_y - \delta_x\} + \{(\delta_x)^2 - \delta_y\delta_x\} + \cdots]$$

For a large sample size n, $\eta_{ij} = E\left[(\delta_y)^i(\delta_x)^j\right]$ is expected to be small for $i+j > 2$ (vide Murthy, 1977). Hence, neglecting η_{ij} for $i+j>2$, we get the following expressions of the bias of $\widehat{R}$ up to the first order of approximation as follows:

$$\text{Bias of } \widehat{R} = B(\widehat{R}) \cong R\left[E\{\delta_y - \delta_x\} + E\{(\delta_x)^2 - \delta_y\delta_x\}\right]$$

$$= R\left[V(\widehat{X})/X^2 - Cov(\widehat{X}, \widehat{Y})/(XY)\right]$$

$$(\text{since } E(\delta_x) = E(\delta_y) = 0) \tag{8.2.10}$$

$$= \left[RV(\widehat{X}) - Cov(\widehat{X}, \widehat{Y})\right]/X^2$$

$$= R\left[C^2_{\hat{X}} - \rho_{\hat{X},\hat{Y}} C_{\hat{X}} C_{\hat{Y}}\right]$$

where $C_{\hat{X}} = \sqrt{V(\widehat{X})/X}$ and $C_{\hat{Y}} = \sqrt{V(\widehat{Y})/Y}$ are the CVs of $\widehat{X}$ and $\widehat{Y}$, respectively and $\rho_{\hat{X},\hat{Y}}$ is the correlation coefficient between $\widehat{X}$ and $\widehat{Y}$.

$$\text{Hence,} \quad B(\widehat{R}) = -\left[Cov(\widehat{X}, \widehat{D})\right]/X^2 \tag{8.2.11}$$

(where $\widehat{D} = \widehat{Y} - R\widehat{X} = \sum_{i\in s} b_{si}d_i$ and $d_i = y_i - R\,x_i$)

$$= -\left[\sum_{i\in U} x_i d_i \theta_i + \sum_{i\neq}\sum_{j\in U} x_i d_j \theta_{ij}\right]\Big/ X^2 \tag{8.2.12}$$

where $\theta_i = \sum_{s\supset i} b^2_{si}p(s)$ and $\theta_{ij} = \sum_{s\supset i,j} b_{si}b_{sj}p(s)$.

Remark 8.2.2

If $d_i = y_i - R\ x_i = 0$ for every $i = 1, \ldots, N$, the regression of y on x passes through the origin and the bias $B(\widehat{R})$ is exactly equal to zero.

An approximate unbiased estimator for bias $B(\widehat{R})$ is given by

$$\widehat{B}(\widehat{R}) \cong \left[\widehat{R}\,\widehat{V}(\widehat{X}) - \widehat{C}ov(\widehat{X}, \widehat{Y})\right]\Big/\widehat{X}^2$$

$$= -\left[\sum_{i\in s} x_i \widehat{d}_i \theta_i \big/ \pi_i + \sum_{i\neq}\sum_{j\in s} x_i \widehat{d}_j \theta_{ij} \big/ \pi_{ij}\right]\Big/ \widehat{X}^2 \tag{8.2.13}$$

where $\widehat{V}(\widehat{X})$ = unbiased estimator of $V(\widehat{X})$, $\widehat{C}ov(\widehat{X}, \widehat{Y})$ = unbiased estimator of $Cov(\widehat{X}, \widehat{Y})$ and $\widehat{d}_i = y_i - \widehat{R}\ x_i$.

From the expression (8.2.9), the expression of MSE of $\widehat{R}$ is obtained as

$$M(\widehat{R}) = E(\widehat{R} - R)^2 = R^2 E\left[(\delta_y - \delta_x) + \{(\delta_x)^2 - \delta_y\delta_x\} + \ldots\right]^2$$

If we neglect the terms $\eta_{ij} = E\left(\delta_x^i \delta_y^j\right)$ for $i + j > 2$, an approximate expression of $M(\widehat{R})$ up to the first order of approximation is obtained as

$$M(\widehat{R}) \cong R^2\left[\frac{V(\widehat{Y})}{Y^2} - 2\frac{Cov(\widehat{Y},\widehat{X})}{YX} + \frac{V(\widehat{X})}{X^2}\right] \tag{8.2.14}$$

$$= R^2\left[C_{\hat{Y}}^2 - 2\rho_{\hat{X},\hat{Y}} C_{\hat{X}} C_{\hat{Y}} + C_{\hat{X}}^2\right]$$

$$= \left[V(\widehat{Y}) - 2R\,Cov(\widehat{Y},\widehat{X}) + R^2\, V(\widehat{X})\right]/X^2 \tag{8.2.15}$$

$$= V(\widehat{D})/X^2 \tag{8.2.16}$$

$$= \left[\sum_{i\in U} d_i^2\theta_i + \sum_{i\neq}\sum_{j\in U} d_i d_j \theta_{ij}\right]\Big/ X^2 \tag{8.2.17}$$

(noting $D = \sum_{i\in U} d_i = 0$).

An approximate unbiased estimator for the mean square $M(\widehat{R})$ is

$$\widehat{M}(\widehat{R}) \cong \left[\widehat{V}(\widehat{Y}) - 2\widehat{R}\,\widehat{C}ov(\widehat{Y},\widehat{X}) + \widehat{R}^2\,\widehat{V}(\widehat{X})\right]\Big/\widehat{X}^2 \tag{8.2.18}$$

(where $\widehat{V}(\widehat{Y})$ is an unbiased estimator of $V(\widehat{Y})$)

$$= \left[\sum_{i\in s} \widehat{d}_i^{\,2}\theta_i\big/\pi_i + \sum_{i\neq}\sum_{j\in s} \widehat{d}_i\widehat{d}_j\theta_{ij}\big/\pi_{ij}\right]\Big/\widehat{X}^2 \tag{8.2.19}$$

All the estimators of biases and MSEs presented here are consistent for large samples. The results derived earlier have been summarized in the following theorem.

Theorem 8.2.2

Approximate expressions of bias and MSE of $\widehat{R}$ up to the first order of approximation and their approximate unbiased estimates are as follows:

(i) $B(\widehat{R}) \cong \left[R V(\widehat{X}) - Cov(\widehat{X},\widehat{Y})\right]/X^2$

$$= -\left[Cov(\widehat{X},\widehat{D})\right]/X^2$$

$$= -\left(\sum_{i\in U} x_i d_i \theta_i + \sum_{i\neq}\sum_{j\in U} x_i d_j \theta_{ij}\right)\Big/ X^2$$

(ii) $M(\widehat{R}) \cong \left[V((\widehat{Y}) - 2RCov(\widehat{X}, \widehat{Y}) + R^2 V(\widehat{X})\right]/X^2$

$$= V(\widehat{D})/X^2$$

$$= \left(\sum_{i\in U} d_i^2\theta_i + \sum_{i\neq}\sum_{j\in U} d_i d_j \theta_{ij}\right) \Big/ X^2$$

$$= R^2\left(C_{\hat{Y}}^2 - 2\rho_{\hat{X},\hat{Y}} C_{\hat{X}} C_{\hat{Y}} + C_{\hat{X}}^2\right)$$

(iii) $\widehat{B}(\widehat{R}) = \left[\widehat{R}\,\widehat{V}(\widehat{X}) - \widehat{C}ov(\widehat{X}, \widehat{Y})\right] \Big/ \widehat{X}^2$

$$= -\left(\sum_{i\in s} x_i \widehat{d}_i \theta_i \big/ \pi_i + \sum_{i\neq}\sum_{j\in s} x_i \widehat{d}_j \theta_{ij} \big/ \pi_{ij}\right) \Big/ \widehat{X}^2$$

and

(iv) $\widehat{M}(\widehat{R}) = \left[\widehat{V}(\widehat{Y}) - 2\widehat{R}\,\widehat{C}ov(\widehat{Y}, \widehat{X}) + \widehat{R}^2\,\widehat{V}(\widehat{X})\right] \Big/ \widehat{X}^2$

$$= \left[\sum_{i\in s} \widehat{d}_i^2 \theta_i \big/ \pi_i + \sum_{i\neq}\sum_{j\in s} \widehat{d}_i \widehat{d}_j \theta_{ij} \big/ \pi_{ij}\right] \Big/ \widehat{X}^2$$

where $\widehat{V}(\widehat{Y})$, $\widehat{V}(\widehat{X})$, and $\widehat{C}ov(\widehat{Y}, \widehat{X})$ are unbiased estimators of $V(\widehat{Y})$, $V(\widehat{X})$, and $Cov(\widehat{Y}, \widehat{X})$; $\theta_i = \sum_{s\supset i} b_{si}^2 p(s)$, $\theta_{ij} = \sum_{s\supset i,j} b_{si} b_{sj}\, p(s)$, $d_i = y_i - R\ x_i$, and $\widehat{d}_i = y_i - \widehat{R}\ x_i$.

8.3 RATIO ESTIMATOR FOR POPULATION TOTAL

The ratio estimator for the population total Y is defined as

$$\widehat{Y}_R = \frac{\widehat{Y}}{\widehat{X}} X = \widehat{R} X \tag{8.3.1}$$

where X is the total of the auxiliary variable, which is known.

Clearly, the ratio estimator $\widehat{Y}_R$ is a biased estimator for the total Y in general. The bias and MSE of $\widehat{Y}_R$ are obtained from the relations $B(\widehat{Y}_R) = XB(\widehat{R})$ and $M(\widehat{Y}_R) = X^2 M(\widehat{R})$, respectively. Approximate expressions of the bias and the MSE of the ratio estimator are directly obtained from Theorem 8.2.2 and have been given in the following theorem.

Theorem 8.3.1

The expressions of bias and MSE of $\widehat{Y}_R$ up to the first order of approximation and their approximate unbiased estimators are given by:

(i) $$B(\widehat{Y}_R) \cong \left[RV(\widehat{X}) - Cov(\widehat{X}, \widehat{Y})\right]/X$$
$$= -\left[Cov(\widehat{X}, \widehat{D})\right]/X$$
$$= -\left(\sum_{i\in U} x_i d_i \theta_i + \sum_{i\neq}\sum_{j\in U} x_i d_j \theta_{ij}\right)\bigg/ X$$

(ii) $$M(\widehat{Y}_R) \cong \left[V(\widehat{Y}) - 2RCov(\widehat{X}, \widehat{Y}) + R^2 V(\widehat{X})\right]$$
$$= V(\widehat{D})$$
$$= -\left(\sum_{i\in U} d_i^2 \theta_i + \sum_{i\neq}\sum_{j\in U} d_i d_j \theta_{ij}\right)$$
$$= Y^2\left(C_{\hat{Y}}^2 - 2\rho_{\hat{X},\hat{Y}} C_{\hat{X}} C_{\hat{Y}} + C_{\hat{X}}^2\right)$$

(iii) $$\widehat{B}(\widehat{Y}_R) = \left[\widehat{R}\widehat{V}(\widehat{X}) - C\widehat{ov}(\widehat{X}, \widehat{Y})\right]/\widehat{X}$$
$$= -\left(\sum_{i\in s} x_i \widehat{d}_i \theta_i \big/ \pi_i + \sum_{i\neq}\sum_{j\in s} x_i \widehat{d}_j \theta_{ij} \big/ \pi_{ij}\right)\bigg/ \widehat{X}$$

and

(iv) $$\widehat{M}(\widehat{Y}_R) = \left[\widehat{V}(\widehat{Y}) - 2\widehat{R}C\widehat{ov}(\widehat{Y}, \widehat{X}) + \widehat{R}^2 \widehat{V}(\widehat{X})\right]$$
$$= \left(\sum_{i\in s} \widehat{d}_i^2 \theta_i \big/ \pi_i + \sum_{i\neq}\sum_{j\in s} \widehat{d}_i \widehat{d}_j \theta_{ij} \big/ \pi_{ij}\right)$$

8.3.1 Efficiency of the Ratio Estimator

The MSE of $\widehat{Y}_R$ becomes zero if $d_i = y_i - Rx_i = 0$ for every $i = 1,\ldots, N$. Hence the ratio estimator becomes optimal if the study variable y is exactly proportional to the auxiliary variable. However, in practice, it is not possible to have d_i exactly equal to zero for every $i = 1,\ldots, N$. But it is quite possible that study variable y is approximately proportional to the auxiliary variable x. In such cases, the use of auxiliary information x makes the ratio estimator highly efficient by reducing the magnitude of the mean square substantially.

The ratio estimator $\widehat{Y}_R$ becomes more efficient than the conventional estimator $\widehat{Y}$ if

$$V(\widehat{Y}) - M(\widehat{Y}_R) \geq 0$$
$$\text{i.e., } 2RCov(\widehat{Y}, \widehat{X}) - R^2 V(\widehat{X}) \geq 0 \tag{8.3.2}$$

$$\text{i.e., } \rho_{\hat{X}\hat{Y}} > \frac{1}{2}\left(\sqrt{V(\widehat{X})}\Big/X\right)\Big/\left(\sqrt{V(\widehat{Y})}\Big/Y\right)$$
$$= \frac{1}{2}(C_{\hat{X}}/C_{\hat{Y}}) \quad \text{when } R > 0 \tag{8.3.3}$$

$$\text{or } \rho_{\hat{X}\hat{Y}} < -\frac{1}{2}\left(\sqrt{V(\widehat{X})}\Big/X\right)\Big/\left(\sqrt{V(\widehat{Y})}\Big/Y\right)$$
$$-\frac{1}{2}|C_{\hat{X}}/C_{\hat{Y}}| \quad \text{when } R < 0 \tag{8.3.4}$$

8.3.2 Optimality of the Ratio Estimator

It is pointed out in Theorem 6.4.2 that the ratio estimator $T_{01} = \widehat{Y}_R$ is the optimal in the class of linear ξ-unbiased predictors $C_{l\xi}$ of the population Y, under the superpopulation model $E_\xi(y_i) = \beta x_i$, $V_\xi(y_i) = \sigma^2 x_i$, and $C_\xi(y_i, y_j) = 0$ for $i \neq j$, i.e., the regression of y on x passes through the origin and the variance of y_i is proportional to x_i.

8.4 BIASES AND MEAN-SQUARE ERRORS FOR SPECIFIC SAMPLING DESIGNS

In this section, the expressions of biases, MSEs, and their approximate unbiased estimators for a few well-known sampling designs, which are commonly used in practice, are given. The conditions of superiority of the ratio estimator over the conventional estimators are also given. These results are derived as special cases of the results presented in Sections 8.2 and 8.3.

8.4.1 Fixed Effective Sample Size (n) Design

Let $b_{si} = y_i/\pi_i$, then we get

$$\widehat{Y} = \sum_{i\in s} y_i/\pi_i,\ \widehat{X} = \sum_{i\in s} x_i/\pi_i,\ \widehat{R} = \frac{\sum_{i\in s} y_i/\pi_i}{\sum_{i\in s} x_i/\pi_i},\ \widehat{Y}_R = \frac{\sum_{i\in s} y_i/\pi_i}{\sum_{i\in s} x_i/\pi_i} X$$

Now noting $\theta_i = \sum_{s \supset i} b_{si}^2 p(s) = 1/\pi_i$ and $\theta_{ij} = \sum_{s \supset i,j} b_{si} b_{sj} p(s)$ $= \pi_{ij}/(\pi_i \pi_j)$, the approximate expressions of the bias, MSE, and their unbiased estimators are obtained as follows:

$$B\left(\widehat{Y}_R\right) = -\left[\sum_i x_i d_i/\pi_i + \sum_{i \neq} \sum_j x_i d_j \pi_{ij}/\pi_i \pi_j\right] \Big/ X \quad (8.4.1)$$

$$\begin{aligned} M\left(\widehat{Y}_R\right) &= \left(\sum_{i \in U} \frac{d_i^2}{\pi_i} + \sum_{i \neq} \sum_{j \in U} d_i d_j \frac{\pi_{ij}}{\pi_i \pi_j}\right) \\ &= \frac{1}{2} \sum_{i \neq} \sum_{j \in U} (\pi_i \pi_j - \pi_{ij}) \left(\frac{d_i}{\pi_i} - \frac{d_j}{\pi_j}\right)^2 \end{aligned} \quad (8.4.2)$$

$$\widehat{B}\left(\widehat{Y}_R\right) = -\left[\sum_{i \in s} x_i \widehat{d}_i \Big/ \pi_i^2 + \sum_{i \neq} \sum_{j \in s} x_i \widehat{d}_j \Big/ (\pi_i \cdot \pi_j)\right] \Big/ \widehat{X} \quad (8.4.3)$$

and

$$\widehat{M}\left(\widehat{Y}_R\right) = \frac{1}{2} \sum_{i \neq} \sum_{j \in s} \frac{(\pi_i \pi_j - \pi_{ij})}{\pi_{ij}} \cdot \left(\frac{\widehat{d}_i}{\pi_i} - \frac{\widehat{d}_j}{\pi_j}\right)^2 \quad (8.4.4)$$

8.4.2 Simple Random Sampling Without Replacement

For SRSWOR $\pi_i = n/N$ and $\pi_{ij} = n(n-1)/\{N(N-1)\}$; hence, for the choice of $b_{si} = 1/\pi_i = N/n$, we get $\widehat{Y} = N\overline{y}_s$, $\widehat{X} = N\overline{x}_s$, $\widehat{R} = \frac{\overline{y}_s}{\overline{x}_s}$, $\widehat{Y}_R = \frac{\overline{y}_s}{\overline{x}_s} X$, where $\overline{y}_s = \sum_{i \in s} y_i/n$ and $\overline{x}_s = \sum_{i \in s} x_i/n$.

Theorem 8.3.1 yields

$$\begin{aligned} B\left(\widehat{Y}_R\right) &= \left[RV\left(\widehat{X}\right) - Cov\left(\widehat{X}, \widehat{Y}\right)\right]/X \\ &= N^2 \left(\frac{1}{n} - \frac{1}{N}\right) \left(RS_x^2 - \rho S_x S_y\right)/X \end{aligned} \quad (8.4.5)$$

$$M\left(\widehat{Y}_R\right) = N^2 \left(\frac{1}{n} - \frac{1}{N}\right) \left(S_y^2 - 2R\rho S_x S_y + R^2 S_x^2\right) \quad (8.4.6)$$

$$= N^2 \left(\frac{1}{n} - \frac{1}{N}\right) S_d^2 \quad (8.4.7)$$

where $(N-1)S_y^2 = \sum_{i \in U} \left(y_i - \overline{Y}\right)^2$, $(N-1)S_x^2 = \sum_{i \in U} \left(x_i - \overline{X}\right)^2$,

$(N-1)S_d^2 = \sum_{i\in U} d_i^2$, $\overline{X} = X/N$, $\overline{Y} = Y/N$, $\rho = S_{xy}/(S_x S_y)$ = population correlation coefficient between x and y, and
$(N-1)S_{xy} = \sum_{i\in U}\left(y_i - \overline{Y}\right)\left(x_i - \overline{X}\right)$.

Approximate unbiased estimators for $B\left(\widehat{Y}_R\right)$ and $M\left(\widehat{Y}_R\right)$ are also obtained from Theorem 8.3.1 as follows:

$$\widehat{B}\left(\widehat{Y}_R\right) = N^2\left(\frac{1}{n} - \frac{1}{N}\right)\left(\widehat{R}s_x^2 - s_{xy}\right)/\widehat{X}$$

$$\widehat{M}\left(\widehat{Y}_R\right) = N^2\left(\frac{1}{n} - \frac{1}{N}\right)\left(s_y^2 - 2\widehat{R}s_{xy} + \widehat{R}^2 s_x^2\right) \qquad (8.4.8)$$

$$= N^2\left(\frac{1}{n} - \frac{1}{N}\right)s_{\hat{d}}^2$$

where $(n-1)s_y^2 = \sum_{i\in s}\left(y_i - \overline{y}_s\right)^2$, $(n-1)s_x^2 = \sum_{i\in s}\left(x_i - \overline{x}_s\right)^2$, $(n-1)s_{\hat{d}}^2 = \sum_{i\in s}\widehat{d}_i^{\,2}$, and $(n-1)s_{xy} = \sum_{i\in s}\left(y_i - \overline{y}_s\right)\left(x_i - \overline{x}_s\right)$.

Finally, using Eqs. (8.3.3) and (8.3.4), we get conditions of superiority of the ratio estimator $\widehat{Y}_R = \frac{\overline{y}_s}{\overline{x}_s}X$ over the conventional estimator $N\overline{y}_s$ for the total Y as

$$\rho > \frac{1}{2}\left(C_x/C_y\right) \text{ with } R > 0 \text{ and } \rho < -\frac{1}{2}\left|C_x/C_y\right| \text{ with } R < 0 \qquad (8.4.9)$$

where C_x and C_y denote the CVs of x and y, respectively. In the case where the CVs of x and y are equal, i.e., $C_x = C_y$, the ratio estimator fares better if $\rho > 0.5$ with $R > 0$ and $\rho < -0.5$ with $R < 0$.

8.4.3 Probability Proportional to Size With Replacement

Let the sample s be selected by PPSWR sampling scheme using $z_i(>0)$ as the known size measure for the ith unit and $p_i = z_i/Z$ $\left(Z = \sum_{i\in U} z_i\right)$ as its normed size measure. In this case,

$$\widehat{Y} = \sum_i \frac{n_i(s)y_i}{np_i},\ \widehat{X} = \sum_i \frac{n_i(s)x_i}{np_i},\ \widehat{R} = \frac{\sum_i n_i(s)y_i/p_i}{\sum_i n_i(s)x_i/p_i};$$

$$\widehat{Y}_R = \left(\frac{\sum_i n_i(s)y_i/p_i}{\sum_i n_i(s)x_i/p_i}\right)X$$

where $n_i(s)$ denotes the number of times ith unit appears in the sample s.

Now writing $\sum_{i \in U} p_i\left(\frac{y_i}{p_i} - Y\right)^2 = V_{pyy}$, $\sum_{i \in U} p_i\left(\frac{x_i}{p_i} - X\right)^2 = V_{pxx}$, $\sum_{i \in U} p_i\left(\frac{x_i}{p_i} - X\right)\left(\frac{y_i}{p_i} - Y\right) = V_{pxy}$, and $\sum_{i \in U} \frac{d_i^2}{p_i} = V_{pdd}$, we get

$$B(\widehat{Y}_R) = \left[RV(\widehat{X}) - Cov(\widehat{X}, \widehat{Y})\right]/X = (RV_{pxx} - V_{pxy})/(nX) \quad (8.4.10)$$

$$M(\widehat{Y}_R) = (V_{pyy} - 2RV_{pxy} + R^2 V_{pxx})/n \quad (8.4.11)$$

$$= V_{pdd}/n \quad (8.4.12)$$

Approximate unbiased estimators of $B(\widehat{Y}_R)$ and $M(\widehat{Y}_R)$ come out as follows:

$$\widehat{B}(\widehat{Y}_R) = (\widehat{R}\,\widehat{V}_{pxx} - \widehat{V}_{pxy})/(n\widehat{X}) \quad (8.4.13)$$

$$\widehat{M}(\widehat{Y}_R) = (\widehat{V}_{pyy} - 2\widehat{R}\,\widehat{V}_{pxy} + \widehat{R}^2\,\widehat{V}_{pxx})/n \quad (8.4.14)$$

$$= \widehat{V}_{p\hat{d}\hat{d}}/n \quad (8.4.15)$$

where $\widehat{V}_{pxx} = \sum_i n_i(s)\left(\frac{x_i}{p_i} - \widehat{X}\right)^2 \Big/ (n-1)$,

$\widehat{V}_{pyy} = \sum_i n_i(s)\left(\frac{y_i}{p_i} - \widehat{Y}\right)^2 \Big/ (n-1)$,

$\widehat{V}_{pxy} = \sum_i n_i(s)\left(\frac{x_i}{p_i} - \widehat{X}\right)\left(\frac{y_i}{p_i} - \widehat{Y}\right) \Big/ (n-1)$,

$\widehat{V}_{p\hat{d}\hat{d}} = \sum_i n_i(s)\left(\frac{\hat{d}_i}{p_i} - \widehat{D}\right)^2 \Big/ (n-1)$, $\widehat{D} = \sum_i \frac{n_i(s)\hat{d}_i}{np_i}$, and $\hat{d}_i = y_i - \widehat{R}x_i$.

8.4.4 Simple Random Sampling With Replacement

PPSWR reduces to simple random sampling with replacement, when $p_i = 1/N$ for every $i = 1, \ldots, N$. So, putting $p_i = 1/N$ in Section 8.4.3, we get the following:

$\widehat{Y} = N\overline{y}_n$, $\widehat{X} = N\overline{x}_n$, $\widehat{R} = \overline{y}_n/\overline{x}_n$, $\widehat{Y}_R = \widehat{Y}_R = (\overline{y}_n/\overline{x}_n)X$, $V(\widehat{Y}) = N(N-1)S_y^2\big/n$, $V(\widehat{X}) = N(N-1)S_x^2/n$, and $Cov(\widehat{Y}, \widehat{X}) = N(N-1)$ S_{xy}/n, where $\overline{x}_n$ and $\overline{y}_n$ denote the sample means of x and y based on all n units including repetition.

$$B(\widehat{Y}_R) = \left[RV(\widehat{X}) - Cov(\widehat{X}, \widehat{Y})\right]/X$$
$$= (N-1)\left(RS_x^2 - \rho S_x S_y\right)/\left(n\overline{X}\right) \quad (8.4.16)$$

$$M(\widehat{Y}_R) = N(N-1)\left(S_y^2 - 2R\rho S_x S_y + R^2 S_x^2\right)/n \quad (8.4.17)$$

$$= N(N-1)S_d^2/n \quad (8.4.18)$$

Approximate unbiased estimator of $B(\widehat{Y}_R)$ and $M(\widehat{Y}_R)$ are given as follows:

$$\widehat{B}(\widehat{Y}_R) = (N-1)\left[\widehat{R}s_x^2 - \widehat{\rho}s_x s_y\right]/(n\,\overline{x}_s) \quad (8.4.19)$$

$$\widehat{M}(\widehat{Y}_R) = N(N-1)\left[s_y^2 - 2\widehat{R}s_{xy} + \widehat{R}^2 s_x^2\right]/n \quad (8.4.20)$$

$$= N(N-1)s_{\widehat{d}}^2/n \quad (8.4.21)$$

where $(n-1)s_x^2 = \sum_i n_i(s)\{x_i - \overline{x}(s)\}^2$, $(n-1)s_y^2 = \sum_i n_i(s)\{y_i - \overline{y}(s)\}^2$, $(n-1)s_{\widehat{d}}^2 = \sum_i n_{is}\widehat{d}_i^{\,2}$, $(n-1)s_{xy} = \sum_i n_i(s)\{x_i - \overline{x}(s)\}\{y_i - \overline{y}(s)\}$, and $\widehat{\rho} = s_{xy}/(s_x\, s_y)$.

Remark 8.4.1

The MSE of $\widehat{Y}_R$ can be estimated by using the standard Jackknife procedure. Details have been given in Chapter 18. The estimators $\widehat{M}_1(\widehat{Y}_R) = X^2\left(\frac{1}{n} - \frac{1}{N}\right)\left(\frac{\overline{X}}{\overline{x}}\right)^2 s_{dy}^2$ and $\widehat{M}_2(\widehat{Y}_R) = X^2\left(\frac{1}{n} - \frac{1}{N}\right)\left(\frac{\overline{X}}{\overline{x}}\right)s_{dy}^2$ were proposed by Rao (1969) and Wu (1982), respectively. Various authors including Rao and Rao (1971), Rao (1981), and Kreweski and Chakrabarti (1981) proposed alternative variance estimators and studied their performances using superpopulation models. None of the estimators was found to be the best in all the situations.

8.5 INTERVAL ESTIMATION

If the sample size $n(>35)$ is large, the distribution of $\widehat{Y}_R$ is asymptotically normal, and the confidence interval of Y is obtained by applying the central limit theorem. Hence, $(1-\alpha)\times 100\%$ confidence interval for the total Y is given by

$$\widehat{Y}_R \pm z_{\alpha/2}\sqrt{\widehat{M}\left(\widehat{Y}_R\right)} \tag{8.5.1}$$

where $z_{\alpha/2}$ is the upper $100\alpha/2$ point of a standard normal distribution.

For SRSWOR the confidence interval for Y and R are obtained, respectively, using the following formulae:

$$N\left[\overline{y}_s \pm z_{\alpha/2}\sqrt{\frac{(1-f)}{n}s_{\hat{d}}^2}\right] \tag{8.5.2}$$

and

$$\widehat{R} \pm z_{\alpha/2}\sqrt{\frac{(1-f)}{n}s_{\hat{d}}^2\Big/(\overline{x}_s)^2} \tag{8.5.3}$$

If the sample size n is small, we cannot approximate the distribution of $\widehat{Y}_R$ as normal in general. In this situation, if we assume that $(x_1, y_1),\ldots, (x_n, y_n)$ are random samples from a bivariate normal distribution, then

$$t_{n-1} = \frac{\left(\overline{y}_s - R\overline{x}_s\right)}{\sqrt{(1-f)\left(s_y^2 - 2Rs_{xy} + R^2 s_x^2\right)\Big/n}}$$

follows t distribution with $n-1$ degrees of freedom.

The $(1-\alpha)\times 100\%$ confidence interval of R is (R_1, R_2), where R_1 and R_2 are the roots of the quadratic equation

$$\left(\overline{y}_s - R\overline{x}_s\right)^2 - t_{\alpha/2,n-1}^2(1-f)\left(s_y^2 - 2Rs_{xy} + R^2 s_x^2\right)\Big/n = 0 \tag{8.5.4}$$

and $t_{\alpha/2,n-1}$ is the upper $100\alpha/2$ point of t distribution with $n-1$ degrees of freedom.

Example 8.5.1

A sample of 15 orchards was selected from 125 orchards of a certain locality by SRSWOR method.

Table 8.5.1 shows the number of plants and production of apples in the selected orchards.

(i) Estimate the average yield of apple per plant and its standard error.

(ii) Estimate the mean yield of apple per orchard and obtain 90% confidence interval of the mean yield when it is known that the average number of plants per orchard is 40 kg. Estimate the relative efficiency of the ratio estimator with respect to the sample mean.

Table 8.5.1 Production of apples in selected orchards

Orchards	Number of plants	Production of apples (kg)
1	25	740
2	30	1000
3	50	1450
4	15	400
5	30	950
6	60	1700
7	30	1000
8	40	1260
9	45	1300
10	20	650
11	40	1250
12	30	850
13	50	1400
14	40	1250
15	20	550

Let x_i and y_i, respectively, be the number of the apple plants (x) and production of apples (y) for the ith orchard, $i = 1,\ldots, 15$. Here, $n = 15$, $N = 125$, $\overline{X} = 40$, $\overline{x}(s) = 35$, and $\overline{y}(s) = 1050$.

Hence, estimated yield per plant $= \widehat{R} = \overline{y}(s)/\overline{x}(s) = 30$ kg. Estimated MSE of $\widehat{R} = \widehat{M}\left(\widehat{R}\right) = (1/n - 1/N)s_{\hat{d}}^2 \Big/ \overline{X}^2 = 0.173$, since $s_{\hat{d}}^2 = \dfrac{1}{n-1}\sum_{i \in s}\left(y_i - \widehat{R}x_i\right)^2 = 4728.571.$

Estimated standard error of $\widehat{R} = SE\left(\widehat{R}\right) = \sqrt{\widehat{M}\left(\widehat{R}\right)} = 0.416.$

The ratio estimate of the mean yield $\left(\overline{Y}\right)$ of apple per orchard is $\widehat{\overline{Y}}_R = 30\overline{X} = 1200$. Estimated standard error of $\widehat{\overline{Y}}_R = SE\left(\widehat{\overline{Y}}_R\right) = \overline{X}\sqrt{\widehat{M}\left(\widehat{R}\right)} = 16.655.$

Assuming that sample size is large, a 90% confidence interval for the mean yield is $\widehat{\overline{Y}}_R \pm t_{0.05,14}\overline{X}\sqrt{(1-f)s_{\hat{d}}^2 \Big/ n} = 1200 \pm 1.761 \times 16.655 =$ $(1170.670, 1229.331)$.

The estimated efficiency of the ratio estimator $\widehat{\overline{Y}}_R = \dfrac{\hat{V}\{\overline{y}(s)\}}{\hat{M}\left(\widehat{\overline{Y}}_R\right)}100 = \dfrac{s_y^2}{s_{\hat{d}}^2}100 = 133942.9 \times 100/4728.571 = 2832.6\%.$

8.6 UNBIASED RATIO, ALMOST UNBIASED RATIO, AND UNBIASED RATIO–TYPE ESTIMATORS

8.6.1 Unbiased Ratio Estimator

It is shown in Section 5.5 that the estimator $\widehat{Y}_{lms} = \dfrac{\sum_{i\in s} y_i}{\sum_{i\in s} x_i} X$ is unbiased for the total Y under the Lahiri-Midzuno-Sen (1951, 1952, 1953) sampling scheme. The estimator $\widehat{Y}_{lms}$ is known as the unbiased ratio estimator. The details properties of $\widehat{Y}_{lms}$ have been discussed in Section 5.5.

8.6.2 Almost Unbiased Ratio Estimator

Let k-independent samples, each of size $m = n/k$, be selected using the same sampling design. Let $\widehat{Y}_i$ and $\widehat{X}_i$ be the unbiased estimators of Y and X, respectively, based on the ith sample s_i. Let the ratio estimator computed from the sample s_i be

$$\widehat{Y}_R(i) = \frac{\widehat{Y}_i}{\widehat{X}_i} X$$

From the samples s_i's, we can find two estimators of Y as follows:

$$\widehat{Y}_{\overline{R}} = \frac{1}{k}\sum_{i=1}^{K} \widehat{Y}_R(i) \text{ and } \widehat{Y}_R(c) = \frac{\left(\sum_{i=1}^{k} \widehat{Y}_i/k\right)}{\left(\sum_{i=1}^{k} \widehat{X}_i/k\right)} X$$

Since all the samples s_i's are of the same size m and are selected by the same sampling design, we have approximate expressions of biases of $\widehat{Y}_R(i)$, $\widehat{Y}_{\overline{R}}$, and $\widehat{Y}_R(c)$ as follows:

$$B\left(\widehat{Y}_R(i)\right) = \left[RV\left(\widehat{X}_i\right) - Cov\left(\widehat{X}_i, \widehat{Y}_i\right)\right]/X = b(\text{say})\text{for } \forall\, i = 1, \ldots, k,$$
$$B\left(\widehat{Y}_{\overline{R}}\right) = b,$$

and

$$\begin{aligned} B\left(\widehat{Y}_R(c)\right) &= \left[RV\left(\frac{1}{k}\sum_{i=1}^{k}\widehat{X}_i\right) - Cov\left(\frac{1}{k}\sum_{i=1}^{k}\widehat{X}_i, \frac{1}{k}\sum_{i=1}^{k}\widehat{Y}_i\right)\right]\Big/ X \\ &= \left[\sum_{i=1}^{k}\left\{RV\left(\widehat{X}_i\right) - Cov\left(\widehat{X}_i, \widehat{Y}_i\right)\right\}\right]\Big/\left(k^2X\right) \\ &= b/k \end{aligned} \tag{8.6.1}$$

Eq. (8.6.1) yields $E\left(\widehat{Y}_{\overline{R}} - \widehat{Y}_R(c)\right) = B\left(\widehat{Y}_{\overline{R}}\right) - B\left(\widehat{Y}_R(c)\right) = b\frac{(k-1)}{k}$. Hence an unbiased estimator of b is

$$\widehat{b} = \frac{k}{k-1}\left(\widehat{Y}_{\overline{R}} - \widehat{Y}_R(c)\right) \tag{8.6.2}$$

Finally, subtracting an estimate of the amount of bias from $\widehat{Y}_R(c)$, we get an almost unbiased ratio estimator of Y as

$$\begin{aligned}\widehat{Y}_{RAU} &= \widehat{Y}_R(c) - \widehat{B}\left(\widehat{Y}_R(c)\right)\\ &= \widehat{Y}_R(c) - \frac{1}{k-1}\left(\widehat{Y}_{\overline{R}} - \widehat{Y}_R(c)\right)\\ &= \frac{k\widehat{Y}_R(c) - \widehat{Y}_{\overline{R}}}{k-1}\end{aligned} \tag{8.6.3}$$

Under SRSWOR sampling, $\widehat{Y}_{\overline{R}} = \frac{1}{k}\sum_{i=1}^{k}\frac{\overline{y}_i}{\overline{x}_i}X$ and

$\widehat{Y}_R(c) = \frac{\sum_{i=1}^{k}\overline{y}_i}{\sum_{i=1}^{k}\overline{x}_i}X = \frac{\overline{y}}{\overline{x}}X$, where $\overline{y}_i$ and $\overline{x}_i$, respectively, denote sample means of y and x for the ith group, whereas $\overline{y}$ and $\overline{x}$, respectively, denote the sample means for y and x based on all the sampled units. It can be easily checked that under SRSWOR, the estimator (8.6.3) reduces to Murthy-Nanjamma (1957) estimator

$$\widehat{Y}_{MN} = \frac{k}{k-1}\frac{\overline{y}}{\overline{x}}X - \frac{1}{k-1}\left(\frac{1}{k}\sum_{i=1}^{k}\frac{\overline{y}_i}{\overline{x}_i}\right)X \tag{8.6.4}$$

8.6.3 Unbiased Ratio–Type Estimators

Let a sample s of size n be selected from a population with probability $p(s)$ and let $\widehat{Y} = \sum_{i\in s} b_{si}y_i$ be an unbiased estimator of Y. The constants b_{si}'s satisfy the unbiasedness condition $\sum_{s\supset i} b_{si}p(s) = 1$. Let $r_i = y_i/x_i$ and $\widehat{Y}_r = X\sum_{i\in s} b_{si}r_i/N$. Then,

$$E\left(\widehat{Y}_r\right) = \frac{X}{N}\sum_{i} r_i \sum_{s\supset i} b_{si}p(s) = \frac{X}{N}\sum_{i\in U} r_i \text{ as } \sum_{s\supset i} b_{si}p(s) = 1$$

and the bias of $\widehat{Y}_r$ is

$$B(\widehat{Y}_r) = E(\widehat{Y}_r) - Y = -\sum_{i \in U} r_i(x_i - \overline{X})$$
$$= -\left[(N-1)\sum_{i \in U} r_i x_i - \sum_{i \neq} \sum_{j \in U} r_i x_j\right] \Big/ N \quad (8.6.5)$$

An exact unbiased estimator of $B(\widehat{Y}_r)$ is

$$\widehat{B}(\widehat{Y}_r) = -\left[(N-1)\sum_{i \in s} r_i x_i/\pi_i - \sum_{i \neq} \sum_{j \in s} r_i x_j/\pi_{ij}\right] \Big/ N$$

Hence the estimator

$$\widehat{Y}_r - \widehat{B}(\widehat{Y}_r) = \widehat{Y}_r + \left[(N-1)\sum_{i \in s} r_i x_i/\pi_i - \sum_{i \neq} \sum_{j \in s} r_i x_j/\pi_{ij}\right] \Big/ N$$
$$= \widehat{Y}_{RU} \quad (8.6.6)$$

is exactly unbiased for the total Y. The estimator $\widehat{Y}_{RU}$ is known as an unbiased ratio-type estimator.

8.6.4 Hartley–Ross Estimator

For SRSWOR sampling, $\pi_i = n/N, \pi_{ij} = n(n-1)/\{N(N-1)\}$. If we choose $b_{si} = 1/\pi_i = N/n$, then Eq. (8.6.6) reduces to

$$\widehat{Y}_{RU} = \widehat{Y}_{HR} = \overline{r}_s X + \frac{n(N-1)}{n-1}\left(\overline{y}_s - \overline{r}_s \overline{x}_s\right) \quad (8.6.7)$$

where $\overline{r}_s = \sum_{i \in s} r_i/n$.

The estimator $\widehat{Y}_{HR}$ is the well-known Hartley–Ross (1954) estimator. Several authors proposed alternative ratio-type unbiased and almost unbiased estimators, e.g., Goodman and Hartley (1958), Murthy and Nanjamma (1957), De Pascal (1961), Beale (1962), Tin (1965), Rao T.J. (1966, 1967a,b), among others. As far as efficiency (MSE) is concerned, none of the proposed alternative estimators are unconditionally superior to the conventional ratio estimator $\widehat{Y}_R$. The general method of bias reduction using Jackknife technique has been given in Chapter 18.

8.7 RATIO ESTIMATOR FOR STRATIFIED SAMPLING

Consider a population stratified into K strata and that X_i and Y_i are, respectively, the total of the auxiliary variable x and the study variable y for the ith stratum $i = 1, \ldots, K$. From the ith stratum of size N_i, a sample s_i of size n_i is selected by some suitable sampling design. Let $\widehat{X}_i$ and $\widehat{Y}_i$ denote the estimators of X_i and Y_i based on the sample s_i. In case X_i is known, the ratio estimator for the ith stratum total Y_i is given by

$$\widehat{Y}_{iR} = \frac{\widehat{Y}_i}{\widehat{X}_i} X_i \tag{8.7.1}$$

Hence, ratio estimator for the population total Y is given by

$$\widehat{Y}_{SR} = \sum_{i=1}^{K} \widehat{Y}_{iR} = \sum_{i=1}^{K} \frac{\widehat{Y}_i}{\widehat{X}_i} X_i \tag{8.7.2}$$

The ratio estimator (8.7.2) is termed as "separate ratio estimator."

Since $\widehat{Y}_{st} = \sum_{i=1}^{K} \widehat{Y}_i$ and $\widehat{X}_{st} = \sum_{i=1}^{K} \widehat{X}_i$ are unbiased for the population totals Y and X, we define an alternative ratio estimator, which is called a "combined ratio estimator" for the total Y as

$$\widehat{Y}_{CR} = \frac{\widehat{Y}_{st}}{\widehat{X}_{st}} X = \widehat{R}_C X \tag{8.7.3}$$

where $\widehat{R}_C = \dfrac{\widehat{Y}_{st}}{\widehat{X}_{st}}$.

8.7.1 Separate Ratio Estimator

Now noting the bias of $\widehat{Y}_{SR} = B(\widehat{Y}_{SR}) = \sum_{i=1}^{K} B(\widehat{Y}_{iR})$ and the MSE of $\widehat{Y}_{SR} = M(\widehat{Y}_{SR}) = \sum_{i=1}^{K} M(\widehat{Y}_{iR})$, we get the following result straight from Theorem 8.3.1 as follows:

Theorem 8.7.1

Approximate expressions of bias, MSE of $\widehat{Y}_{SR}$ (up to the first order of approximation) and their approximate unbiased estimators are given by:

(i) $B(\widehat{Y}_{SR}) \cong \sum_{i=1}^{K} \left(R_i V(\widehat{X}_i) - Cov(\widehat{X}_i, \widehat{Y}_i) \right) / X_i$

(ii) $M(\widehat{Y}_{SR}) \cong \sum_{i=1}^{K} \left(V(\widehat{Y}_i) - 2R_i Cov(\widehat{Y}_i, \widehat{X}_i) + R_i^2 V(\widehat{X}_i) \right)$

(iii) $\widehat{B}(\widehat{Y}_{SR}) \cong \sum_{i=1}^{K} \left(\widehat{R}_i \widehat{V}(\widehat{X}_i) - \widehat{C}ov(\widehat{X}_i, \widehat{Y}_i) \right) / \widehat{X}_i$

(iv) $\widehat{M}(\widehat{Y}_{SR}) \cong \sum_{i=1}^{K} \left(\widehat{V}(\widehat{Y}_i) - 2\widehat{R}_i Cov(\widehat{Y}_i, \widehat{X}_i) + R_i^2 V(\widehat{X}_i) \right)$

where $R_i = Y_i / X_i$ and $\widehat{Cov}(\widehat{X}_i, \widehat{Y}_i)$ is an unbiased estimator of $Cov(\widehat{X}_i, \widehat{Y}_i)$.

8.7.2 Combined Ratio Estimator

Theorem 8.3.1 yields approximate expressions of bias and MSE of $\widehat{Y}_{CR}$ as well as their approximate unbiased estimators as follows:

Theorem 8.7.2

$$B(\widehat{Y}_{CR}) \cong \left(RV(\widehat{X}_{st}) - Cov(\widehat{X}_{st}, \widehat{Y}_{st})\right)/X$$

$$= \sum_{i=1}^{K}\left(RV(\widehat{X}_i) - Cov(\widehat{X}_i, \widehat{Y}_i)\right)/X$$

$$M(\widehat{Y}_{CR}) \cong V(\widehat{Y}_{st}) - 2RCov(\widehat{X}_{st}, \widehat{Y}_{st}) + R^2V(\widehat{X}_{st})$$

$$= \sum_{i=1}^{K}\left(V(\widehat{Y}_i) - 2RCov(\widehat{X}_i, \widehat{Y}_i) + R^2V(\widehat{X}_i)\right)$$

$$\widehat{B}(\widehat{Y}_{CR}) = \sum_{i=1}^{K}\left(\widehat{R}_C\widehat{V}(\widehat{X}_i) - \widehat{C}ov(\widehat{X}_i, \widehat{Y}_i)\right)/\widehat{X},$$

and

$$\widehat{M}(\widehat{Y}_{CR}) = \sum_{i=1}^{K}\left(\widehat{V}(\widehat{Y}_i) - 2\widehat{R}_C\widehat{C}ov(\widehat{X}_i, \widehat{Y}_i) + \widehat{R}_C^2\widehat{V}(\widehat{X}_i)\right).$$

8.7.3 Comparison Between the Separate and Combined Ratio Estimators

(i) To use a separate ratio estimator, one needs to know the values of each stratum's total X_i, whereas for the combined ratio estimator, only the knowledge of the population total X is required.

(ii) With regard to bias, the combined ratio estimator should be preferred because the separate ratio estimator is likely to have more bias than the combined one because X_i's are much smaller than X. Further, if $R_i = R$ $\forall i = 1,\ldots, N$ and each of the stratum biases is of the same sign, then the difference of the absolute biases is

$$\left|B(\widehat{Y}_{SR})\right| - \left|B(\widehat{Y}_{CR})\right| = \sum_{i=1}^{K}\left|RV(\widehat{X}_i) - Cov(\widehat{X}_i, \widehat{Y}_i)\right|\left(\frac{1}{X_i} - \frac{1}{X}\right) \geq 0.$$

(iii) For comparing efficiency, we consider the difference

$$M(\widehat{Y}_{CR}) - M(\widehat{Y}_{SR}) \cong \sum_{i=1}^{K}(R_i - R)^2V(\widehat{X}_i)$$

$$+2\sum_{i=1}^{K}(R_i - R)Cov(\widehat{X}_i, \widehat{Y}_i - R\widehat{X}_i) \qquad (8.7.4)$$

The second term of Eq. (8.7.4) is expected to be small compared to the first term if the regression of y on x in each stratum is linear and passes through the origin. The difference $M(\widehat{Y}_{CR}) - M(\widehat{Y}_{SR})$ is likely to be positive unless R_i's are not equal to R, for every $\forall i = 1,\ldots, N$. Hence, a separate ratio estimator is expected to be more efficient than the combined one unless regression of y on x passes through the origin for every stratum.

(iv) The expressions of bias and MSE of $\widehat{Y}_{SR}$ are valid if n_i's are large for all of the strata, which may not hold in reality. But for $\widehat{Y}_{CR}$, the expressions of bias and MSE are valid because the total sample size is expected to be large in all practical situations. In conclusion, one should use a separate ratio estimator if the sample sizes for every stratum is large. The combined ratio estimator is preferable if R_i's are the same for all the strata.

Example 8.7.1

A factory has three categories of staff: skilled, semiskilled, and unskilled of numbering 50, 75, and 100, respectively. Samples of sizes 10, 12, and 15 are selected from each of the categories by SRSWOR method and information on daily wages and years of service was collected (Table 8.7.1). Estimate the average wages of the factory by using (i) separate and (ii) combined ratio estimators, when the average years of service of the three categories are known to be 15, 8, and 3 years, respectively. Estimate the gain in efficiencies of the separate ratio estimator with respect to the combined ratio estimator.

Table 8.7.1 Wages and years of service of factory workers

Skilled		Semiskilled		Unskilled	
Years of service (years)	**Wages ($)**	**Year of service (years)**	**Wages ($)**	**Year of service (years)**	**Wages ($)**
10	500	10	450	2	100
20	600	6	250	4	200
15	550	12	500	2	150
25	750	5	350	4	200
30	800	5	400	6	300
8	450	12	500	4	250
5	500	10	450	2	150
15	650	8	450	5	350
20	700	10	400	5	350
15	600	10	550	2	150
		12	450	4	300
		8	350	2	250
				5	300
				4	300
				6	550

Here we take years of service (x) as an auxiliary variable and daily wage as a study variable (y). From the table above, we compute the following:

Strata	N_i	$\overline{X}_i$	n_i	$\overline{x}(s_i)$	$\overline{y}(s_i)$	$\widehat{R}_i$	$N_i\overline{x}(s_i)$	$N_i\overline{y}(s_i)$	$\widehat{Y}_{iR}$	$s^2_{\hat{d}_i}$	$\dfrac{W_i^2(1-f_i)s^2_{\hat{d}_i}}{n_i}$	$s^2_{\hat{d}_i^*}$	$\dfrac{W_i^2(1-f_i)s^2_{\hat{d}_i^*}}{n_i}$
Skilled	50	15	10	16.3	610	37.423	815	30,500	28,067.25	34,316.266	135.570	67,343.138	266.046
Semiskilled	75	8	12	9	425	47.222	675	31,875	28,333.20	7,425.645	57.755	7,438.108	57.851
Unskilled	100	3	15	3.8	260	68.421	380	26,000	20,526.30	4,360.902	48.813	4,728.685	52.930
Total	225						1,870	88,375	76,926.75		242.138		376.827

Separate ratio estimate for the population mean $\overline{Y}$ is $\widehat{\overline{Y}}_{SR} = \sum_{i=1}^{K} \widehat{Y}_{iR}/N = 76,926.75/225 = \$341.90.$

Estimated MSE of $\widehat{\overline{Y}}_{SR}$ is

$$\widehat{M}\left(\widehat{\overline{Y}}_{SR}\right) \sum_{i=1}^{K}\left(\widehat{V}\left(\widehat{Y}_i\right) - 2\widehat{R}_i \widehat{Cov}\left(\widehat{Y}_i, \widehat{X}_i\right) + R_i^2 \widehat{V}\left(\widehat{X}_i\right)\right)/N^2$$

$$= \sum_{i=1}^{K} W_i^2(1-f_i)s_{\widehat{d}_i}^2 \Big/ n_i = 242.138$$

where $W_i = N_i/N$.

The combined ratio estimator for $\overline{Y}$ is $\widehat{\overline{Y}}_{CR} = \widehat{Y}_{CR}/N = \dfrac{\hat{Y}_{st}}{\hat{X}_{st}}\overline{X}$

$= \dfrac{88,375}{1870} \times \dfrac{1650}{225} = \$346.57.$

Estimated MSE of $\widehat{\overline{Y}}_{CR}$ is

$$\widehat{M}\left(\widehat{\overline{Y}}_{CR}\right) = \sum_{i=1}^{K}\left(\widehat{V}\left(\widehat{Y}_i\right) - 2\widehat{R}_C Cov\left(\widehat{X}_i, \widehat{Y}_i\right) + \widehat{R}_C V\widehat{X}_i\right)/N^2$$

$$= \sum_{i=1}^{K} W_i^2(1-f_i)s_{\widehat{d}_i^*}^2 \Big/ n_i$$

$$\left(\text{Where } s_{\widehat{d}_i^*}^2 = \sum_{j\in s_i}\left(\widehat{d}_{ij}^* - \overline{\widehat{d}}^*\right)^2 \Big/ (n_i - 1), \widehat{d}_{ij}^* = y_{ij} - \widehat{R}_C x_{ij}, \widehat{R}_C\right.$$

$$\left. = \left(\widehat{Y}_{st}/\widehat{X}_{st}\right) = 47.259 \text{ and } \overline{\widehat{d}}^* = \frac{1}{n_i}\sum_{j\in s_i}\widehat{d}_{ij}^*\right)$$

$$= 376.827$$

Gain in efficiency of the separate ratio estimator over the combined ratio estimator is $\left(\dfrac{\hat{M}\left(\hat{\overline{Y}}_{CR}\right)}{\hat{M}\left(\hat{\overline{Y}}_{SR}\right)} - 1\right) \times 100 = (376.827/242.138 - 1)$ $\times 100 = 55.62\%$

8.8 RATIO ESTIMATOR FOR SEVERAL AUXILIARY VARIABLES

Let us consider the situation where the study variable y is approximately proportional to each of the p auxiliary variables $x_1,\ldots, x_p$. The ratio estimator for the total Y based on the ith auxiliary variable x_i when its total X_i is known is given by

$$\widehat{Y}_{R_i} = \frac{\widehat{Y}}{\widehat{X}_i} X_i \quad \text{for } i = 1, \ldots, p \tag{8.8.1}$$

Olkin (1958) proposed the following composite estimator for Y based on the auxiliary variables $x_1,\ldots, x_p$ as

$$\widehat{Y}_{RO} = \sum_{i=1}^{p} w_i \widehat{Y}_{R_i} = \sum_{i=1}^{p} w_i \left(\frac{\widehat{Y}}{\widehat{X}_i} X_i \right) \tag{8.8.2}$$

where w_i's are suitably chosen weights to minimize the MSE of $\widehat{Y}_{RO}$ subject to $\sum_{i=1}^{p} w_i = 1$.

For large sample size n, the estimator $\widehat{Y}_{RO}$ is approximately unbiased for Y. Hence the MSE of $\widehat{Y}_{RO}$ is approximately equal to its variance and we have

$$\begin{aligned} M\left(\widehat{Y}_{RO}\right) \cong V\left(\widehat{Y}_{RO}\right) &= \sum_{i=1}^{p} w_i^2 V\left(\widehat{Y}_{R_i}\right) + \sum_{i \neq}^{p} \sum_{j=1}^{p} w_i w_j Cov\left(\widehat{Y}_{R_i}, \widehat{Y}_{R_j}\right) \\ &= \sum_{i=1}^{p} w_i^2 C_{ii} + \sum_{i \neq}^{p} \sum_{j=1}^{p} w_i w_j C_{ij} \end{aligned} \tag{8.8.3}$$

where $C_{ii} = V\left(\widehat{Y}_{R_i}\right)$ and $C_{ij} = Cov\left(\widehat{Y}_{R_i}, \widehat{Y}_{R_j}\right)$.

Now minimizing $M\left(\widehat{Y}_{RO}\right)$ subject to $\sum_{i=1}^{p} w_i = 1$, the optimum value of w_i is obtained as

$$w_{i0} = \Delta_i / \Delta \tag{8.8.4}$$

where $\boldsymbol{\Delta}_i$ is the sum of the elements of the ith row of the matrix of $\mathbf{C}^{-1}$, $\mathbf{C}$ is the $p \times p$ matrix with ith and jth element as C_{ij} and $\boldsymbol{\Delta}$ is the sum of all the elements of $\mathbf{C}^{-1}$. The value of $M\left(\widehat{Y}_{RO}\right)$ with $w = w_{i0} = \Delta_i/\Delta$ is denoted by $M_{\min}\left(\widehat{Y}_{RO}\right)$ and is equal to $1/\boldsymbol{\Delta}$.

For example, for $p = 2$, we have $C = \begin{pmatrix} C_{11} & C_{12} \\ C_{12} & C_{22} \end{pmatrix}$,

$$\mathbf{C}^{-1} = \frac{1}{\left(C_{11}C_{22} - C_{12}^2\right)}\begin{pmatrix} C_{22} & -C_{12} \\ -C_{12} & C_{11} \end{pmatrix}, \ \Delta_1 = \frac{C_{22} - C_{12}}{\left(C_{11}C_{22} - C_{12}^2\right)},$$

$$\Delta_2 = \frac{C_{11} - C_{12}}{\left(C_{11}C_{22} - C_{12}^2\right)}, \text{ and } \Delta = \frac{C_{11} + C_{22} - 2C_{12}}{\left(C_{12}C_{22} - C_{12}^2\right)}. \text{ Hence,}$$

$$w_{10} = \frac{C_{22} - C_{12}}{\left(C_{11} + C_{22} - 2C_{12}\right)}, \ w_{20} = \frac{C_{11} - C_{12}}{\left(C_{11} + C_{22} - 2C_{12}\right)}, \text{ and}$$

$$M_{\min} = \Delta^{-1} = \frac{C_{11}C_{22} - C_{12}^2}{C_{11} + C_{22} - 2C_{12}} \tag{8.8.5}$$

Since, C_{ij}'s are unknown, the weights w_{10} and w_{20} are estimated by replacing C_{ij} by their suitable estimates.

8.8.1 Simple Random Sampling Without Replacement

Let a sample s of size n be selected by the SRSWOR method and $\bar{x}_1, \bar{x}_2$, and $\bar{y}$ be the sample means of the auxiliary variables x_1, x_2, and the study variable y, respectively. Then,

$$\widehat{Y}_{R_1} = \frac{\bar{y}}{\bar{x}_1}X_1, \ \widehat{Y}_{R_2} = \frac{\bar{y}}{\bar{x}_2}X_2, \ \widehat{Y}_{RO} = w_1\widehat{Y}_{R_1} + w_2\widehat{Y}_{R_2},$$

$$C_{11} = V\left(\widehat{Y}_{R_1}\right) \cong \frac{(1-f)}{n}Y^2\left(C_y^2 - 2\rho_1 C_{x_1}C_y + C_{x_1}^2\right)$$

$$C_{22} = V\left(\widehat{Y}_{R_2}\right) \cong \frac{(1-f)}{n}Y^2\left(C_y^2 - 2\rho_2 C_{x_2}C_y + C_{x_2}^2\right),$$

and

$$C_{12} \cong \frac{(1-f)}{n}Y^2\left(C_y^2 - \rho_1 C_{x_1}C_y - \rho_2 C_{x_2}C_y + \rho^* C_{x_1}C_{x_2}\right)$$

where C_y and C_{xi} are CVs of y and x_i, ρ_i is the correlation coefficient between x_i and y for $i = 1, 2$, and ρ^* = correlation coefficient between x_1 and x_2.

In particular if $C_{x1} = C_{x2} = C_x$ and $\rho_1 = \rho_2 = \rho$, we get $C_{11} = C_{22}$ and

$$\begin{aligned} M_{\min} &= \frac{C_{11} + C_{12}}{2} \\ &= \frac{(1-f)}{n}Y^2\left(C_y^2 - 2\rho C_x C_y + \frac{(1+\rho^*)}{2}C_x^2\right) \qquad (8.8.6) \\ &= V\left(N\bar{y}_s\right) - \frac{(1-f)}{n}Y^2\left(2\rho C_x C_y - \frac{(1+\rho^*)}{2}C_x^2\right) \end{aligned}$$

The ratio estimator $\widehat{Y}_{RO}$ will be more efficient than the conventional estimator $\widehat{Y} = N\overline{y}_s$.

If $M_{\min} < V\left(N\overline{y}_s\right)$

$$\text{i.e., if} \quad \left(\frac{\rho}{1+\rho^*}\right)\frac{C_y}{C_x} > \frac{1}{4} \tag{8.8.7}$$

8.9 EXERCISES

8.9.1 Describe the ratio method of estimation. Derive approximate expressions of bias and MSEs under SRSWOR sampling. Derive the conditions of superiority of the ratio estimator over the sample mean for estimating the population mean.

8.9.2 Consider the following estimators with α as a constant.

(i) $\widehat{Y}_1 = \alpha\widehat{Y}$ (Searls, 1964)

(ii) $\widehat{Y}_2 = \widehat{Y}\left(\dfrac{X}{\widehat{X}}\right)^{\alpha}$ (Srivastava, 1967)

(iii) $\widehat{Y}_3 = \alpha\widehat{Y} + (1-\alpha)\widehat{Y}\left(\dfrac{X}{\widehat{X}}\right)$ (Chakrabarty, 1968)

(iv) $\widehat{Y}_4 = \alpha\widehat{Y}\left(\dfrac{X}{\widehat{X}}\right)$

For each of the estimators above, determine (a) approximate expression of bias, (b) MSE, (c) optimum value of α, which minimizes the MSE, and (d) expressions of biases and mean squares with the optimum value of α under SRSWOR sampling.

8.9.3 Let s be an SRSWOR sample of size n and $t = \overline{y}(s)\dfrac{\overline{x}(\overline{s})}{\overline{X}}$ be an estimator of $\overline{Y}$ with $\overline{x}(\overline{s}) = \sum_{i \notin s} x_i/(N-n)$. Then show that

(i) the bias of $t = S_{xy}/\left(N\overline{X}\right)$,

(ii) $MSE(t) = (1-f)\left(S_y^2 - 2\rho h R S_x S_y + h^2R^2S_x^2\right)/n$, where $h = n/(N-n)$ and $1/N^2 \cong 0$, and (iii) the estimator t is more efficient than the convention ratio estimator $\widehat{\overline{Y}}_R = \overline{y}(s)\dfrac{\overline{x}(s)}{\overline{X}}$ if $n < N/2$ and $0 < \rho\dfrac{cv(y)}{cv(x)} < N\Big/\{2(N-n)\}$ (Srivenkataramana, 1980).

8.9.4 To determine the access to treatment of HIV/AIDS patients in a certain locality, a sample of 15 clinics from a total of 50 clinics was selected by the SRSWOR method.

Serial number of clinics	Number of HIV/AIDS patients	Total number of patients
1	105	1000
2	30	400
3	65	500
4	120	1500
5	75	1000
6	90	850
7	55	600
8	90	1200
9	45	500
10	35	300
11	50	750
12	80	1000
13	120	1000
14	90	750
15	60	500

(i) Estimate the proportion of HIV/AIDS patients treated and its standard error.

(ii) Using the ratio method, estimate the total number of HIV/AIDS patients in the locality when the average number of patients treated per clinic is known to be 700. Obtain a 90% confidence interval of the total number of HIV/AIDS patients treated in the locality. Estimate the efficiency of the ratio estimator with the sample mean as an estimator.

8.9.5 A simple random sample of 600 households in a region from a total of 12,000 households was selected, and the information of household size (x) and household income (y) was collected. The following is the summarization of data: $\overline{x} = 5.3, \overline{y} = 5000.00, s_x^2 = 0.567, s_y^2 = 150.506$, and $r_{xy} = 0.75$. Estimate the average income of the households and its 95% confidence interval when the average family size is known to be 5, from the past survey.

8.9.6 A sample of 10 households from a population of 50 households is selected by the PPSWR method of sampling using household size as a size variable. The information of household income and household expenditure on food was obtained and has been presented in the following table:

Sampled households	1	2	3	4	5	6	7	8	9	10
Size	5	4	3	2	4	5	6	4	4	3
Expenditure on food ($)	2000	2500	1500	1000	2500	3000	4000	2000	1500	1200
Income ($)	5500	4500	5000	2500	7500	6000	8500	6000	2500	5000

From the past survey, it is known that the average family size and average income of the population are 4.2 and $5000, respectively. Use this information to estimate the following:
(i) Average expenditure on food by the ratio method of estimation.
(ii) Population proportion of expenditure on food.
(iii) Standard error of the estimators used in (i) and (ii).

8.9.7 The academics of an university are classified into three categories. From each of the categories, samples are selected by the SRSWOR method, and information on age and earnings is given in the following table.

			Sample				
Categories	Number of employees	Average age	Size	Mean ages (years)	Mean income ($)	SD of ages (years)	SD of income ($)
Professor	50	55	10	60	22,000	15	200
Senior Lecturer	100	40	15	45	15,000	25	325
Lecturer	200	35	30	35	10,000	10	250

(i) Estimate the average income of the academics of the university by the separate and combined ratio methods of estimation.
(ii) Estimate the relative efficiency of the separate ratio estimator with respect to the combined ratio estimator and comment on your findings.

CHAPTER 9

Regression, Product, and Calibrated Methods of Estimation

9.1 INTRODUCTION

We have seen in Chapter 8 that the ratio estimator is optimal if the regression of y on x is linear and passing through the origin. The ratio estimator may not be efficient if the regression line does not pass through the origin. In this situation, one may use auxiliary information effectively through the regression estimator to improve precision of the conventional estimator. Like the ratio estimator, the regression estimator is not unbiased for the population mean or total. In this section we will study the difference, regression, product, calibrated, and related estimators along with their properties in detail.

9.2 DIFFERENCE ESTIMATOR

If the population total X of the auxiliary variable x is known, then the estimator

$$\widehat{Y}_{\lambda} = \widehat{Y} - \lambda\left(\widehat{X} - X\right) \qquad (9.2.1)$$

is unbiased for the population total Y for any known value of λ. The estimator (Eq. 9.2.1) is known as a difference estimator. To use the difference estimator one needs to find the optimum value of λ, which minimizes the variance of $\widehat{Y}_{\lambda}$.

Now minimizing the variance of $\widehat{Y}_{\lambda}$ with respect to λ, the optimum value of λ comes out as

$$\lambda_{opt} = \frac{Cov\left(\widehat{X}, \widehat{Y}\right)}{V\left(\widehat{X}\right)} = \beta \quad \text{(say)} \qquad (9.2.2)$$

Putting $\lambda = \beta$ in Eq. (9.2.1), we get the optimum estimator

$$\widehat{Y}_{\beta} = \widehat{Y} - \beta\left(\widehat{X} - X\right) \qquad (9.2.3)$$

Survey Sampling Theory and Applications
ISBN 978-0-12-811848-1
http://dx.doi.org/10.1016/B978-0-12-811848-1.00009-1

The variance of the optimum estimator $\widehat{Y}_\beta$ is given by

$$V\left(\widehat{Y}_\beta\right) = \left(1 - \rho^2_{\hat{X},\hat{Y}}\right) V\left(\widehat{Y}\right) \tag{9.2.4}$$

where $\rho_{\hat{X},\hat{Y}}$ is the correlation between $\widehat{X}$ and $\widehat{Y}$.

Because $\left|\rho_{\hat{X},\hat{Y}}\right| \leq 1$, $\widehat{Y}_\beta$ possesses smaller variance than the convention estimator $\widehat{Y}$. The estimator $\widehat{Y}_\beta$ is seldom used in practice because the value of β is unknown in most practical situations. Let β_0 be a guess value of β obtained from a past experiment. Then substituting $\beta = \beta_0$ in Eq. (9.2.3), one gets the estimator

$$\widehat{Y}_{\beta_0} = \widehat{Y} - \beta_0\left(\widehat{X} - X\right) = \widehat{Y}_\beta + (\beta - \beta_0)\left(\widehat{X} - X\right) \tag{9.2.5}$$

The loss of efficiency of using $\widehat{Y}_{\beta_0}$ over the optimum estimator $\widehat{Y}_\beta$ comes out as

$$\begin{aligned}\frac{V\left(\widehat{Y}_{\beta_0}\right)}{V\left(\widehat{Y}_\beta\right)} - 1 &= \frac{(\beta - \beta_0)^2 V\left(\widehat{X}\right)}{V\left(\widehat{Y}_\beta\right)} \\ &= \frac{(\beta - \beta_0)^2 V\left(\widehat{X}\right)}{\left(1 - \rho^2_{\hat{X},\hat{Y}}\right) V\left(\widehat{Y}\right)} \\ &= \left(1 - \frac{\beta_0}{\beta}\right)^2 \frac{\rho^2_{\hat{X},\hat{Y}}}{1 - \rho^2_{\hat{X},\hat{Y}}}\end{aligned} \tag{9.2.6}$$

To keep the loss of efficiency to a certain level γ (say), we must have

$$\left(1 - \frac{\beta_0}{\beta}\right)^2 < \gamma \frac{1 - \rho^2_{\hat{X},\hat{Y}}}{\rho^2_{\hat{X},\hat{Y}}} \tag{9.2.7}$$

For simple random sampling without replacement (SRSWOR) and simple random sampling with replacement (SRSWR), Eq. (9.2.7) reduces to

$$\left(1 - \frac{\beta_0}{\beta}\right)^2 < \gamma \frac{1 - \rho^2}{\rho^2} \tag{9.2.8}$$

where ρ is the correlation coefficient between x and y.

In particular if $\gamma = 0.25$, $\rho = 0.5$, Eq. (9.2.8) reduces to $0.134 < \dfrac{\beta_0}{\beta} < 1.866$.

For the value of $\lambda = 1$, the estimator $\widehat{Y}_\lambda$ reduces to a very simple estimator viz.

$$\widehat{Y}_1 = \widehat{Y} - \left(\widehat{X} - X\right) \tag{9.2.9}$$

This estimator $\widehat{Y}_1$ is unbiased for Y and becomes more efficient than the conventional estimator $\widehat{Y}$ if $\rho_{\hat{X},\hat{Y}} > \frac{1}{2}\sqrt{\frac{V(\widehat{Y})}{V(\widehat{X})}}$. In particular if the variances of $\widehat{X}$ and $\widehat{Y}$ are equal, then the difference estimator given in Eq. (9.2.9) performs better than the conventional estimator $\widehat{Y}$ if $\rho_{\hat{X},\hat{Y}} > 1/2$. For SRSWOR and SRSWR sampling, $\widehat{X} = N\overline{x}_s$ and $\widehat{Y} = N\overline{y}_s$. Here the condition of superiority of the difference estimator over the conventional estimator reduces to $\rho > 1/2$ and it is realized in most practical situations.

9.3 REGRESSION ESTIMATOR

The value of β in Eq. (9.2.2) is generally unknown and it is estimated by

$$\widehat{\beta} = \frac{C\widehat{ov}\left(\widehat{X}, \widehat{Y}\right)}{\widehat{V}\left(\widehat{X}\right)} \tag{9.3.1}$$

where $C\widehat{ov}\left(\widehat{X}, \widehat{Y}\right)$ and $\widehat{V}\left(\widehat{X}\right)$ are unbiased for $Cov\left(\widehat{X}, \widehat{Y}\right)$ and $V\left(\widehat{X}\right)$, respectively.

Hence, replacing β by its estimator $\widehat{\beta}$ given in Eq. (9.3.1), we get the well known linear regression estimator (or simply regression estimator) for the total Y as

$$\widehat{Y}_{reg} = \widehat{Y} - \widehat{\beta}\left(\widehat{X} - X\right) \tag{9.3.2}$$

Like the ratio estimator, the regression estimator is not unbiased for Y, but it is more efficient than the conventional estimator $\widehat{Y}$.

9.3.1 Exact Expression of Bias

The bias of the regression estimator $\widehat{Y}_{reg}$ is given by

$$\begin{aligned} B\left(\widehat{Y}_{reg}\right) &= E\left(\widehat{Y}_{reg} - Y\right) \\ &= -E\left\{\widehat{\beta}\left(\widehat{X} - X\right)\right\} \\ &= -Cov\left(\widehat{\beta}, \widehat{X}\right) \end{aligned} \tag{9.3.3}$$

Because in most situations, $Cov\left(\widehat{\beta}, \widehat{X}\right)$ decreases as the sample size increases, the regression estimator is approximately unbiased when the sample size is large.

9.3.2 Approximate Expression of Bias

Now writing $\varepsilon_{xy} = \dfrac{C\widehat{o}v\left(\widehat{X},\widehat{Y}\right)}{Cov\left(\widehat{X},\widehat{Y}\right)} - 1$ and $\varepsilon_{xx} = \dfrac{\widehat{V}\left(\widehat{X}\right)}{V\left(\widehat{X}\right)} - 1$, we get

$$\begin{aligned}\widehat{\beta} &= \frac{(1+\varepsilon_{xy})Cov\left(\widehat{X},\widehat{Y}\right)}{(1+\varepsilon_{xx})V\left(\widehat{X}\right)} \\ &= \beta(1+\varepsilon_{xy})(1+\varepsilon_{xx})^{-1} \\ &= \beta(1+\varepsilon_{xy})\left(1-\varepsilon_{xx}+\varepsilon_{xx}^2-\cdots\right)\end{aligned} \tag{9.3.4}$$

The expression of bias of the regression estimator comes out as

$$\begin{aligned}B\left(\widehat{Y}_{reg}\right) &= E\left(\widehat{Y}_{reg} - Y\right) \\ &= -\beta E\left[(1+\varepsilon_{xy})\left(1-\varepsilon_{xx}+\varepsilon_{xx}^2-\cdots\right)\left(\widehat{X}-X\right)\right] \\ &= -\beta E\left[\left(\varepsilon_{xy}-\varepsilon_{xx}+\varepsilon_{xx}^2-\varepsilon_{xy}\varepsilon_{xx}+\cdots\right)\left(\widehat{X}-X\right)\right]\end{aligned}$$

Now neglecting the terms containing $E\left[\varepsilon_{xx}^i\varepsilon_{xy}^j\left(\widehat{X}-X\right)\right]$ for $i+j>1$ (which is expected to be small), we get an expression of bias up to the first order of approximation as

$$\begin{aligned}B\left(\widehat{Y}_{reg}\right) &\cong \beta E\left[(\varepsilon_{xx}-\varepsilon_{xy})\left(\widehat{X}-X\right)\right] \\ &= \beta\left[\frac{Cov\left\{\widehat{V}\left(\widehat{X}\right),\widehat{X}\right\}}{V\left(\widehat{X}\right)} - \frac{Cov\left\{C\widehat{o}v\left(\widehat{X},\widehat{Y}\right),\widehat{X}\right\}}{Cov\left(\widehat{X},\widehat{Y}\right)}\right]\end{aligned} \tag{9.3.5}$$

9.3.2.1 Bias Under Simple Random Sampling Without Replacement

For an SRSWOR, $\widehat{X} = N\overline{x}_s$, $V\left(\widehat{X}\right) = N^2\left(\dfrac{1}{n}-\dfrac{1}{N}\right)S_x^2$, $\widehat{V}\left(\widehat{X}\right) = N^2\left(\dfrac{1}{n}-\dfrac{1}{N}\right)s_x^2$, $Cov\left(\widehat{X},\widehat{Y}\right) = N^2\left(\dfrac{1}{n}-\dfrac{1}{N}\right)S_{xy}$, and $C\widehat{o}v\left(\widehat{X},\widehat{Y}\right) = N^2\left(\dfrac{1}{n}-\dfrac{1}{N}\right)s_{xy}$. Hence the approximate bias of the regression estimator is obtained using Eq. (9.3.5) as follows:

$$\begin{aligned}B\left(\widehat{Y}_{reg}\right) &\cong \beta\left[\frac{Cov\left\{N^2\left(\frac{1}{n}-\frac{1}{N}\right)s_x^2, N\overline{x}_s\right\}}{\left\{N^2\left(\frac{1}{n}-\frac{1}{N}\right)S_x^2\right\}} - \frac{Cov\left\{N^2\left(\frac{1}{n}-\frac{1}{N}\right)s_{xy}, N\overline{x}_s\right\}}{\left\{N^2\left(\frac{1}{n}-\frac{1}{N}\right)S_{xy}\right\}}\right] \\ &= N\beta\left[\frac{Cov\left(s_x^2,\overline{x}_s\right)}{S_x^2} - \frac{Cov\left(s_{xy},\overline{x}_s\right)}{S_{xy}}\right]\end{aligned}$$

Now using the following results of Sukhatme (1944) (detailed derivation is given in Appendix 9A), we find

$$Cov\left(s_x^2, \overline{x}_s\right) = \frac{N(N-n)}{(N-1)(N-2)}\frac{\mu_{30}}{n} \text{ and}$$
$$Cov\left(s_{xy}, \overline{x}_s\right) = \frac{N(N-n)}{(N-1)(N-2)}\frac{\mu_{21}}{n}$$

where $\mu_{ij} = \frac{1}{N}\sum_{i\in U}\left(x_i - \overline{X}\right)^i\left(y_i - \overline{Y}\right)^j$.

The expression for the bias (Eq. 9.3.5) comes out as

$$B\left(\widehat{Y}_{reg}\right) = \frac{N(N-n)}{(N-2)}\frac{\beta}{n}\left(\frac{\mu_{30}}{\mu_{20}} - \frac{\mu_{21}}{\mu_{11}}\right)$$

In case x and y follow bivariate normal distribution, we have $\mu_{30} = \mu_{21} = 0$ and the bias of the regression estimator $B\left(\widehat{Y}_{reg}\right)$ reduces to zero.

9.3.3 Approximate Expression of the Mean Square Error

The mean square error (MSE) of the regression estimator $\widehat{Y}_{reg}$ is given by

$$\begin{aligned}
M\left(\widehat{Y}_{reg}\right) &= E\left(\widehat{Y}_{reg} - Y\right)^2 \\
&= E\left[\left(\widehat{Y} - Y\right) - \widehat{\beta}\left(\widehat{X} - X\right)\right]^2 \\
&= E\left[\left(\widehat{Y} - Y\right) - \beta(1+\varepsilon_{xy})\left(1 - \varepsilon_{xx} + \varepsilon_{xx}^2 - \cdots\right)\left(\widehat{X} - X\right)\right]^2 \\
&= E\left[\left\{\left(\widehat{Y} - Y\right) - \beta\left(\widehat{X} - X\right)\right\} - \beta\left(\widehat{X} - X\right)\right. \\
&\qquad \left.\times\left\{\varepsilon_{xy} - \varepsilon_{xx} + \varepsilon_{xx}^2 - \varepsilon_{xy}\varepsilon_{xx} + \cdots\right\}\right]^2
\end{aligned}$$

Neglecting terms $E\left(\widehat{X} - X\right)^2\varepsilon_{xy}^i\varepsilon_{xx}^j$ for $i + j \geq 2$, the expression of MSE up to the first order of approximation is given by

$$\begin{aligned}
M\left(\widehat{Y}_{reg}\right) &\cong E\left[\left(\widehat{Y} - Y\right) - \beta\left(\widehat{X} - X\right)\right]^2 \\
&= V\left(\widehat{Y}\right) + \beta^2 V\left(\widehat{X}\right) - 2\beta Cov\left(\widehat{X}, \widehat{Y}\right)
\end{aligned} \tag{9.3.6}$$

$$= \left(1 - \rho_{\hat{X},\hat{Y}}^2\right)V\left(\widehat{Y}\right) \tag{9.3.7}$$

Let $\widehat{X} = \sum_{i\in s} b_{si}x_i$ and $\widehat{Y} = \sum_{i\in s} b_{si}y_i$ be linear homogeneous unbiased estimators of the population totals X and Y, respectively, based on a sample

s of size n selected with probability $p(s)$. The constants b_{si}'s satisfy the unbiasedness condition $\sum_{s \supset i} b_{si} p(s) = 1$. Then we can write Eq. (9.3.6) as

$$\begin{aligned} M\left(\widehat{Y}_{reg}\right) &\cong E\left[\sum_{i \in s} b_{si}(y_i - \beta x_i) - (Y - \beta X)\right]^2 \\ &= V\left[\sum_{i \in s} b_{si}(y_i - \beta x_i)\right] \\ &= V\left(\sum_{i \in s} b_{si} E_i\right) \\ &= \sum_i (\alpha_i - 1)E_i^2 + \sum_{i \neq} \sum_j (\alpha_{ij} - 1)E_i E_j \\ &= V\left(\widehat{E}\right) \end{aligned} \tag{9.3.8}$$

where $E_i = y_i - \beta x_i$, $\widehat{E} = \sum_{i \in s} b_{si} E_i$, $\alpha_i = \sum_{s \supset i} b_{si}^2 p(s)$, and $\alpha_{ij} = \sum_{s \supset i,j} b_{si} b_{sj} p(s)$.

From Eq. (9.3.8), we set an approximate unbiased estimator for $M\left(\widehat{Y}_{reg}\right)$ as

$$\begin{aligned} \widehat{M}\left(\widehat{Y}_{reg}\right) &= \sum_{i \in s} (\alpha_i - 1)\widehat{E}_i^2 / \pi_i + \sum_{i \neq} \sum_j (\alpha_{ij} - 1)\widehat{E}_i \widehat{E}_j / \pi_{ij} \\ &= \widehat{V}\left(\widehat{E}\right) \end{aligned} \tag{9.3.9}$$

where $\widehat{E}_i = y_i - \widehat{\beta} x_i$, π_i = inclusion probability of ith unit, π_{ij} (>0) inclusion probability of the ith and $j(\neq i)$th unit and $\widehat{\beta}$ as in Eq. (9.3.1).

9.3.4 Mean Square Errors for Some Sampling Designs

9.3.4.1 Arbitrary Fixed Effective Sample Size Design

Consider a fixed effective sample size n design with $b_{si} = 1/\pi_i$. In this case

$$\widehat{X} = \sum_{i \in s} x_i/\pi_i,\ \widehat{Y} = \sum_{i \in s} y_i/\pi_i,\ V\left(\widehat{X}\right) = \frac{1}{2} \sum_{i \neq} \sum_{j \in U} \Delta_{ij} \left(\frac{x_i}{\pi_i} - \frac{x_i}{\pi_j}\right)^2,$$

$$Cov\left(\widehat{X}, \widehat{Y}\right) = \frac{1}{2} \sum_{i \neq} \sum_{j \in U} \Delta_{ij} \left(\frac{x_i}{\pi_i} - \frac{x_j}{\pi_j}\right)\left(\frac{y_i}{\pi_i} - \frac{y_j}{\pi_j}\right),$$

$$\beta = \sum_{i \neq} \sum_{j \in U} \Delta_{ij} \left(\frac{x_i}{\pi_i} - \frac{x_j}{\pi_j}\right)\left(\frac{y_i}{\pi_i} - \frac{y_j}{\pi_j}\right) \Big/ \sum_{i \neq} \sum_{j \in U} \Delta_{ij} \left(\frac{x_i}{\pi_i} - \frac{x_j}{\pi_j}\right)^2$$

$$\widehat{V}\left(\widehat{X}\right) = \frac{1}{2}\sum_{i\neq}\sum_{j\in s}\frac{\Delta_{ij}}{\pi_{ij}}\left(\frac{x_i}{\pi_i} - \frac{x_j}{\pi_j}\right)^2,$$

$$C\widehat{o}v\left(\widehat{X}, \widehat{Y}\right) = \frac{1}{2}\sum_{i\neq}\sum_{j\in s}\frac{\Delta_{ij}}{\pi_{ij}}\left(\frac{x_i}{\pi_i} - \frac{x_j}{\pi_j}\right)\left(\frac{y_i}{\pi_i} - \frac{y_i}{\pi_j}\right)$$

$$\widehat{\beta} = \sum_{i\neq}\sum_{j\in s}\frac{\Delta_{ij}}{\pi_{ij}}\left(\frac{x_i}{\pi_i} - \frac{x_j}{\pi_j}\right)\left(\frac{y_i}{\pi_i} - \frac{y_j}{\pi_j}\right)\Big/\sum_{i\neq}\sum_{j\in s}\frac{\Delta_{ij}}{\pi_{ij}}\left(\frac{x_i}{\pi_i} - \frac{x_j}{\pi j}\right)^2$$

where $\Delta_{ij} = \pi_i\pi_j - \pi_{ij}$.

The expression for the regression estimator $\widehat{Y}_{reg}$, its approximate MSE, and an approximate unbiased estimator of the MSE, respectively, come out as follows:

$$\widehat{Y}_{reg} = \sum_{i\in s}\frac{y_i}{\pi_i} - \widehat{\beta}\left(\sum_{i\in s}\frac{x_i}{\pi_i} - X\right) \tag{9.3.10}$$

$$M\left(\widehat{Y}_{reg}\right) = \frac{1}{2}\sum_{i\neq}\sum_{j\in U}\Delta_{ij}\left(\frac{E_i}{\pi_i} - \frac{E_j}{\pi_j}\right)^2 \tag{9.3.11}$$

$$\widehat{M}\left(\widehat{Y}_{reg}\right) = \frac{1}{2}\sum_{i\neq}\sum_{j\in s}\frac{\Delta_{ij}}{\pi_{ij}}\left(\frac{\widehat{E}_i}{\pi_i} - \frac{\widehat{E}_j}{\pi_j}\right)^2 \tag{9.3.12}$$

9.3.4.2 Simple Random Sampling Without Replacement

For an SRSWOR design $\pi_i = \dfrac{n}{N}$ and $\pi_{ij} = \dfrac{n(n-1)}{N(N-1)}$. In this case $\widehat{X} = N\overline{x}_s$, $\widehat{Y} = N\overline{y}_s$, $V\left(\widehat{X}\right) = N^2\left(\frac{1}{n} - \frac{1}{N}\right)S_x^2$, $\widehat{V}\left(\widehat{X}\right) = N^2\left(\frac{1}{n} - \frac{1}{N}\right)s_x^2$, $V\left(\widehat{Y}\right) = N^2\left(\frac{1}{n} - \frac{1}{N}\right)S_y^2$, $\widehat{V}\left(\widehat{Y}\right) = N^2\left(\frac{1}{n} - \frac{1}{N}\right)s_y^2$, $Cov\left(\widehat{X}, \widehat{Y}\right) = N^2\left(\frac{1}{n} - \frac{1}{N}\right)S_{xy}$, $C\widehat{o}v\left(\widehat{X}, \widehat{Y}\right) = N^2\left(\frac{1}{n} - \frac{1}{N}\right)s_{xy}$, $\beta = \dfrac{S_{xy}}{S_x^2}$, $\widehat{\beta} = \dfrac{s_{xy}}{s_x^2}$, and $\rho_{\widehat{X},\widehat{Y}} = \rho = \dfrac{S_{xy}}{S_xS_y}$, where $\overline{x}_s = \sum_{i\in s} x_i/n$, $\overline{y}_s = \sum_{i\in s} y_i/n$, $S_x^2 = \sum_{i\in U}\left(x_i - \overline{X}\right)^2/(N-1)$, $s_x^2 = \sum_{i\in s}(x_i - \overline{x}_i)^2/(n-1)$, and $S_y^2 = \sum_{i\in U}\left(y_i - \overline{Y}\right)^2/(N-1)$.

So, under SRSWOR

$$\widehat{Y}_{reg} = N\left[\overline{y}_s - \widehat{\beta}\left(\overline{x}_s - \overline{X}\right)\right] \tag{9.3.13}$$

$$M\left(\widehat{Y}_{reg}\right) = N^2\left(\frac{1}{n} - \frac{1}{N}\right)S_E^2 = N^2\left(\frac{1}{n} - \frac{1}{N}\right)(1-\rho^2)S_y^2 \tag{9.3.14}$$

$$\widehat{M}\left(\widehat{Y}_{reg}\right)=N^2\left(\frac{1}{n}-\frac{1}{N}\right)s_{\hat{E}}^2=N^2\left(\frac{1}{n}-\frac{1}{N}\right)(1-\widehat{\rho}^2)s_y^2 \tag{9.3.15}$$

where

$$S_E^2=\sum_{i\in U}\left(E_i-\overline{E}\right)^2/(N-1),\ s_{\hat{E}}^2=\sum_{i\in U}\left(\widehat{E}_i-\overline{\widehat{E}}_s\right)^2/(n-1),$$

$$\overline{E}=\sum_{i\in U}E_i/N,\ \overline{\widehat{E}}_s=\sum_{i\in s}\widehat{E}_i/n \text{ and } \widehat{\rho}=s_{xy}/\left(s_x s_y\right).$$

9.3.5 Efficiency of Regression Estimator

From Eq. (9.3.7) we note that the regression estimator $\widehat{Y}_{reg}$ is more efficient than the conventional estimator $\widehat{Y}$ because $V\left(\widehat{Y}\right)-M\left(\widehat{Y}_{reg}\right)=\rho_{\hat{X}\hat{Y}}^2V\left(\widehat{Y}\right)\geq 0$.

9.3.5.1 Comparison With the Ratio Estimator

Theorem 9.3.1

The regression estimator is more efficient than the ratio estimator.

Proof

The difference between the approximate expressions of the MSEs of the ratio and regression estimators obtained from the Theorem 8.3.1 and Eq. (9.3.7) is

$$\begin{aligned}M\left(\widehat{Y}_R\right)-M\left(\widehat{Y}_{reg}\right)&=V\left(\widehat{Y}\right)+R^2V\left(\widehat{X}\right)-2\rho_{\hat{X}\hat{Y}}R\sqrt{V\left(\widehat{Y}\right)V\left(\widehat{X}\right)}\\&\quad-\left(1-\rho_{\hat{X}\hat{Y}}^2\right)V\left(\widehat{Y}\right)\\&=\left[R\sqrt{V\left(\widehat{X}\right)}-\rho_{\hat{X}\hat{Y}}\sqrt{V\left(\widehat{Y}\right)}\right]^2\\&\geq 0\end{aligned}$$

9.3.6 Optimality of the Regression Estimator

It is pointed out in Corollary 6.4.3 that the regression estimator $\widehat{Y}_{reg}=N\{\overline{y}_s-\widehat{\beta}\left(\overline{x}_s-\overline{X}\right)\}$ is the optimal in the class of linear ξ-unbiased predictors $C_{l\xi}$ of the population Y under the superpopulation model $E_\xi(y_i)=\beta_0+\beta x_i$; $V_\xi(y_i)=\sigma^2$ and $C_\xi(y_i,\ y_j)=0$ for $i\neq j$.

Example 9.3.1

From the data given in Example 8.5.1, estimate total production of apples (Y) by the regression method of estimation. Estimate the standard error of the regression estimator. Compare the relative efficiencies of the regression estimator with the ratio estimator and the expansion estimator for estimating the total Y.

From the given data, we get $N = 125$, $n = 15$, $\overline{X} = 40$, $\overline{x}(s) = 35$, $\overline{y}(s) = 1050$, $s_x^2 = 164.286$, $s_y^2 = 133942.9$, and $s_{xy} - 4617.857$. The sample regression coefficient $\widehat{\beta} = \frac{s_{xy}}{s_x^2} = 28.1087$ and $\widehat{\rho} =$ sample correlation coefficient $= r = 0.984$.

The regression estimator for the population total Y is

$$\widehat{Y}_{reg} = N\left[\overline{y}_s - \widehat{\beta}\left(\overline{x}_s - \overline{X}\right)\right] = 125 \times \{1050 - 28.1087 \times (35 - 40)\}$$
$$= 148,817.9$$

An estimate of the MSE of the regression estimator is

$$\begin{aligned}\widehat{M}\left(\widehat{Y}_{reg}\right) &= N^2(1 - f)(1 - \widehat{\rho}^2)s_y^2\Big/n \\ &= (125)^2(1 - 15/125)(1 - 0.9844^2) \times 13394.9/15 \\ &= 3975,839.8\end{aligned}$$

Estimated standard error of $\widehat{Y}_{reg}$ is $S\widehat{E}\left(\widehat{Y}_{reg}\right) = \sqrt{\widehat{M}\left(\widehat{Y}_{reg}\right)} = 1948.29$.

From the Example 8.5.1, we get $\widehat{Y}_R = N\widehat{R}\overline{X} = 125 \times 30 \times 40 = 150,000$ and the estimated MSE is $\widehat{M}\left(\widehat{Y}_R\right) = N^2\widehat{M}\left(\widehat{\overline{Y}}_R\right) = 125^2 \times 277.389 = 4334203.13$.

An estimate of Y based on the expansion estimator is $\widehat{Y} = N\overline{y}_s = 125 \times 1050 = 131,250$.

Estimated variance of $\widehat{Y}$ is $\widehat{V}\left(\widehat{Y}\right) = N^2(1 - f)s_y^2/n = 125^2(1 - 15/125) \times 133,942.9/15 = 122,780,952.4$.

Estimated efficiency of the regression estimator compared to the ratio estimator is

$$\left\{\widehat{M}\left(\widehat{Y}_R\right)\Big/\widehat{M}\left(\widehat{Y}_{reg}\right)\right\} \times 100 = 114.18\%$$

Estimated efficiency of the regression estimator compared to the expansion estimator is

$$\left\{V\left(\widehat{Y}\right)\Big/\widehat{M}\left(\widehat{Y}_{reg}\right)\right\} \times 100 = 3234.61\%$$

9.3.7 Unbiased Regression Estimator

Singh and Srivastava (1980) proposed a sampling scheme for which the regression estimator (Eq. 9.3.13) is unbiased for the total Y. The sampling scheme is described as follows.

9.3.7.1 Singh and Srivastava Sampling Scheme

Under the Singh and Srivastava (SS) sampling scheme, we first select two units i and $j (i \neq j)$ with probability proportional to $(x_i - x_j)^2$, then select $(n-2)$ units from the remaining $(N-2)$ units by SRSWOR method. So, for the SS sampling scheme, probability of selection a sample s of n units is

$$p(s) \propto \sum_{i \neq} \sum_{j \in s} (x_i - x_j)^2 \tag{9.3.16}$$

$$\text{i.e.}\quad p(s) = \sum_{i \neq} \sum_{j \in s} (x_i - x_j)^2 \Big/ \sum_{s} \sum_{i \neq} \sum_{j \in s} (x_i - x_j)^2$$

Now noting $\sum_{i \neq} \sum_{j \in s} (x_i - x_j)^2 = 2n(n-1)s_x^2$ and $\sum_{s} \sum_{i \neq} \sum_{j \in s} (x_i - x_j)^2$ $= 2N(N-1)S_x^2 \binom{N-2}{n-2}$, we find

$$p(s) = s_x^2 \Big/ \left\{ \binom{N}{n} S_x^2 \right\} \tag{9.3.17}$$

Now noting $\widehat{\beta} = \left(s_{xy}/s_x^2\right)$, we find for the SS sampling scheme

$$E\left(\widehat{Y}_{reg}\right) = N \sum_{s} \left\{ \overline{y}_s - \left(s_{xy}/s_x^2\right)\left(\overline{x}_s - \overline{X}\right) \right\} s_x^2 \Big/ \left\{ S_x^2 \binom{N}{n} \right\}$$

$$= \frac{N}{S_x^2} \sum_{s} \left\{ \overline{Y} s_x^2 + \left(\overline{y}_s - \overline{Y}\right) s_x^2 - \left(s_{xy}\right)\left(\overline{x}_s - \overline{X}\right) \right\} \Big/ \binom{N}{n}$$

$$= N\left[\overline{Y} E\left(s_x^2 \middle| \text{SRSWOR}\right) + Cov\left(\overline{y}_s, s_x^2 \middle| \text{SRSWOR}\right) - Cov\left(s_{xy}, \overline{x}_s \middle| \text{SRSWOR}\right) \right] / S_x^2$$

where

$$E\left(s_x^2 \middle| \text{SRSWOR}\right) = S_x^2, \tag{9.3.18}$$

$$Cov\left(\overline{y}_s, s_x^2 \middle| \text{SRSWOR}\right) = \frac{1}{n} \frac{N(N-n)}{(N-1)(N-2)} \mu_{21} \tag{9.3.19}$$

$$Cov\left(s_{xy}, \overline{x}_s \middle| \text{SRSWOR}\right) = \frac{1}{n} \frac{N(N-n)}{(N-1)(N-2)} \mu_{21} \tag{9.3.20}$$

and $\mu_{21} = \frac{1}{N} \sum\limits_{i \in U} \left(x_i - \overline{X}\right)^2 \left(y_i - \overline{Y}\right)$.

The results (Eqs. 9.3.19 and 9.3.20) were obtained by Sukhatme (1944) (details given in Appendix 9A). From Eqs. (9.3.18–9.3.20), we note that $\widehat{Y}_{reg}$ is an unbiased estimator of the total Y.

The expression of the exact variance of $\widehat{Y}_{reg}$ for this sampling scheme is very complex. However, an approximate expression of variance and its approximate unbiased estimator are stated in the following theorem without derivation.

Theorem 9.3.2

Under the SS sampling scheme

(i) $\widehat{Y}_{reg} = N\left\{\overline{y}_s - \left(s_{xy}/s_x^2\right)\left(\overline{x}_s - \overline{X}\right)\right\}$ is unbiased for the Y

(ii) $$V\left(\widehat{Y}_{reg}\right) \cong N^2\left[\left(1-\rho^2\right)\left(\frac{1}{n} + \frac{1}{n^2}\right)\mu_{02} + \frac{1}{n^2 \mu_{20}}\left(\frac{2\mu_{11}\mu_{31}}{\mu_{20}} - \frac{\mu_{11}^2 \mu_{40}}{\mu_{20}^2} - \mu_{22}\right)\right]$$

(iii) $$\widehat{V}\left(\widehat{Y}_{reg}\right) \cong Y_{reg}^2 - \frac{NS_x^2}{ns_x^2}\left(\sum_{i \in s} y_i^2 + \frac{N-1}{n-1} \sum_{i \neq j} \sum_{\in s} y_i y_j\right)$$

where $\mu_{jk} = \frac{1}{N} \sum\limits_{i \in U} \left(x_i - \overline{X}\right)^j \left(y_i - \overline{Y}\right)^k$

9.3.8 Stratified Regression Estimator

Here we suppose that a population is stratified into K strata. Let y_{ij} and x_{ij} be the value of the study (y) and auxiliary (x) variables for the jth unit of the ith stratum U_i, $i = 1, \ldots, K$. From each of the strata, samples are selected independently. Let the sample s_i of size n_i be selected from the ith stratum of size N_i by some sampling procedure. Let $\widehat{Y}_i$ and $\widehat{X}_i$ be the unbiased estimators of the totals Y_i and X_i of the study and auxiliary variables of the ith stratum, respectively. Then we may construct the following two types of regression estimators, viz. separate and combined regression estimators for the total Y analogous to separate and combined ratio estimators.

9.3.8.1 Separate Regression Estimator

Here we assume that the linear regressions of y and x for the different strata are different and that the population totals of the auxiliary information X_i's of all the strata are known. Under this situation, we construct a regression estimator of the total Y_i as

$$\widehat{Y}_{i(reg)} = \widehat{Y}_i - \widehat{\beta}_i\left(\widehat{X}_i - X_i\right) \tag{9.3.21}$$

where $\widehat{\beta}_i = C\widehat{o}v\left(\widehat{Y}_i, \widehat{X}_i\right)/\widehat{V}\left(\widehat{X}_i\right)$ is an estimate of the population regression coefficient $\beta_i = Cov\left(\widehat{Y}_i, \widehat{X}_i\right)/V\left(\widehat{X}_i\right)$ of the ith stratum.

The separate regression estimators for the population total Y is given by

$$\widehat{Y}_{reg}(s) = \sum_{i=1}^{K} \widehat{Y}_{i(reg)} \tag{9.3.22}$$

The bias of $\widehat{Y}_{reg}(s)$ is

$$B(s) = \sum_{i=1}^{K} B\left(\widehat{Y}_{i(reg)}\right)$$

where $B\left(\widehat{Y}_{i(reg)}\right)$ is the bias of $\widehat{Y}_{i(reg)}$.

Eqs. (9.3.3) and (9.3.5) yield

$$\begin{aligned} B(s) &= -\sum_{i=1}^{K} Cov\left(\widehat{\beta}_i, \widehat{X}_i\right) \\ &\cong \sum_{i=1}^{K} \beta_i \left[\frac{Cov\left\{\widehat{V}\left(\widehat{X}_i\right), \widehat{X}_i\right\}}{V\left(\widehat{X}_i\right)} - \frac{Cov\left\{\widehat{C}ov\left(\widehat{X}_i, \widehat{Y}_i\right), \widehat{X}_i\right\}}{Cov\left(\widehat{X}_i, \widehat{Y}_i\right)}\right] \end{aligned} \tag{9.3.23}$$

An approximate expression for the MSE of $\widehat{Y}_{reg}(s)$ is obtained from Eqs. (9.3.6) and (9.3.7) as

$$M\left[\widehat{Y}_{reg}(s)\right] \cong \sum_{i=1}^{K} V\left\{\left(\widehat{Y}_i\right) - 2\beta_i Cov\left(\widehat{X}_i, \widehat{Y}_i\right) + \beta_i^2 V(X_i)\right\} \tag{9.3.24}$$

$$= \sum_{i=1}^{K}\left(1 - \rho^2_{\hat{X}_i, \hat{Y}_i}\right) V\left(\widehat{Y}_i\right) \tag{9.3.25}$$

where $\rho_{\hat{X}_i, \hat{Y}_i}$ is the correlation between $\widehat{X}_i$ and $\widehat{Y}_i$.

In case samples are selected from each of the strata by SRSWOR, we have

$$\widehat{Y}_{reg}(s) = \sum_{i=1}^{K} N_i\left\{\overline{y}_i - \widehat{\beta}_i\left(\overline{x}_i - \overline{X}_i\right)\right\} \tag{9.3.26}$$

$$B(s) = \sum_{i=1}^{K} \frac{N_i(N_i - n_i)}{(N_i - 2)} \frac{\beta_i}{n_i}\left(\frac{\mu_{30}^{(i)}}{\mu_{20}^{(i)}} - \frac{\mu_{21}^{(i)}}{\mu_{11}^{(i)}}\right) \tag{9.3.27}$$

$$M\left[\widehat{Y}_{reg}(s)\right] \cong \sum_{i=1}^{K} N_i^2(1-f_i)\left(S_{yi}^2 - 2\beta_i S_{xyi} + \beta_i^2 S_{xi}^2\right)\Big/ n_i$$
$$= \sum_{i=1}^{K} N_i^2(1-f_i) S_{qi}^2 \Big/ n_i \tag{9.3.28}$$
$$= \sum_{i=1}^{K} N_i^2(1-f_i)\left(1-\rho_i^2\right) S_{yi}^2 \Big/ n_i$$

$$\widehat{M}\left[\widehat{Y}_{reg}(s)\right] \cong \sum_{i=1}^{K} N_i^2(1-f_i) s_{\hat{q}i}^2 \Big/ n_i$$
$$= \sum_{i=1}^{K} N_i^2(1-f_i)\left(1-\widehat{\rho}_i^2\right) s_{yi}^2 \Big/ n_i \tag{9.3.29}$$

where $f_i = n_i/N_i$, $\overline{y}_i = \sum_{j\in s_i} y_{ij}/n_i$, $\overline{x}_i = \sum_{j\in s_i} x_{ij}/n_i$,

$\widehat{\beta}_i = \sum_{j\in s_i}(x_{ij}-\overline{x}_i)(y_{ij}-\overline{y}_i)\Big/\sum_{j\in s_i}(x_{ij}-\overline{x}_i)^2, S_{yi}^2 = \sum_{j\in U_i}\left(y_{ij}-\overline{Y}_i\right)^2\Big/(N_i-1)$,

$S_{xi}^2 = \sum_{j\in U_i}\left(x_{ij}-\overline{X}_i\right)^2\Big/(N_i-1)$, $S_{xyi} = \sum_{j\in U_i}\left(y_{ij}-\overline{Y}_i\right)\left(x_{ij}-\overline{X}_i\right)\Big/(N_i-1)$,

$\overline{X}_i = \sum_{j\in U_i} x_{ij}\Big/N_i$, $\overline{Y}_i = \sum_{j\in U_i} y_{ij}\Big/N_i$, $\beta_i = S_{xyi}^2\Big/S_{xi}^2$, $\rho_i = S_{xyi}\Big/(S_{xi}S_{yi})$,

$s_{xi}^2 = \sum_{j\in s_i}\left(x_{ij}-\overline{x}_i\right)^2\Big/(n_i-1)$, $s_{yi}^2 = \sum_{j\in s_i}\left(y_{ij}-\overline{y}_i\right)^2\Big/(n_i-1)$,

$s_{xyi} = \sum_{j\in s_i}(x_{ij}-\overline{x}_i)\left(y_{ij}-\overline{y}_i\right)\Big/(n_i-1)$, $\overline{x}_i = \sum_{j\in s_i} x_{ij}\Big/n_i$ $\overline{y}_i = \sum_{j\in s_i} y_{ij}\Big/n_i$,

$\widehat{\rho}_i = s_{xyi}\big/(s_{xi}s_{yi})$, $S_{qi}^2 = \sum_{j\in U_i}\left(q_{ij}-\overline{Q}_i\right)^2\Big/(N_i-1)$, $q_{ij} = y_{ij}-\beta_i x_{ij}$,

$\overline{Q}_i = \sum_{j\in U_i} q_{ij}\Big/N_i$, $s_{\hat{q}i}^2 = \sum_{j\in s_i}\left(\widehat{q}_{ij}-\overline{q}_i\right)^2\Big/(n_i-1)$, $\widehat{q}_{ij} = y_{ij}-\widehat{\beta}_i x_{ij}$,

$\overline{q}_i = \sum_{j\in s_i}\widehat{q}_{ij}\Big/n_i$, and $\mu_{kl}^{(i)} = \sum_{j\in U_i}\left(x_{ij}-\overline{X}_i\right)^k\left(y_{ij}-\overline{Y}_i\right)^l\Big/N_i$.

9.3.8.2 Combined Regression Estimator

Here we assume that a single regression line may be fitted over the entire population. The combined regression estimator for the total Y is given by

$$\widehat{Y}_{reg}(c) = \widehat{Y}_{st} - \widehat{\beta}_c\left(\widehat{X}_{st} - X\right) \tag{9.3.30}$$

where $\widehat{X}_{st} = \sum_{i=1}^{K}\widehat{X}_i$, $\widehat{Y}_{st} = \sum_{i=1}^{K}\widehat{Y}_i$ and $\widehat{\beta}_c = \widehat{Cov}\left(\widehat{X}_{st}, \widehat{Y}_{st}\right)\big/\widehat{V}\left(\widehat{X}_{st}\right)$ = an estimate of the combined regression coefficient $\beta_c = Cov\left(\widehat{X}_{st}, \widehat{Y}_{st}\right)\big/ V\left(\widehat{X}_{st}\right)$.

The bias of $\widehat{Y}_{reg}(c)$ is obtained by using Eqs. (9.3.3) and (9.3.5) as

$$B(c) = -Cov\left(\widehat{\beta}_c, \widehat{X}_{st}\right)$$
$$\cong \beta_c\left[\frac{Cov\{\widehat{V}(\widehat{X}_{st}), \widehat{X}_{st}\}}{V(\widehat{X}_{st})} - \frac{Cov\left\{\widehat{C}ov(\widehat{X}_{st}, \widehat{Y}_{st}), \widehat{X}_{st}\right\}}{Cov(\widehat{X}_{st}, \widehat{Y}_{st})}\right] \quad (9.3.31)$$

An approximate MSE of $\widehat{Y}_{reg}(c)$ upto first order of approximation is obtained from Eq. (9.3.6) as

$$M\left[\widehat{Y}_{reg}(c)\right] \cong V\left(\widehat{Y}_{st} - \beta_c\widehat{X}_{st}\right)$$
$$= \sum_{i=1}^{K}\left\{V\left(\widehat{Y}_i\right) - 2\beta_c \text{cov}\left(\widehat{X}_i, \widehat{Y}_i\right) + \beta_c^2 V(X_i)\right\} \quad (9.3.32)$$
$$= \sum_{i=1}^{K} V\left(\widehat{Q}_{ci}\right)$$

where $\widehat{Q}_{ci} = \widehat{Y}_i - \beta_c\widehat{X}_i$.

An approximate unbiased estimator of $M\left[\widehat{Y}_{reg}(c)\right]$ is

$$\widehat{M}\left[\widehat{Y}_{reg}(c)\right] = \sum_{i=1}^{K}\left\{\widehat{V}\left(\widehat{Y}_i\right) - 2\widehat{\beta}_c\widehat{C}ov\left(\widehat{X}_i, \widehat{Y}_i\right) + \widehat{\beta}_c^2\,\widehat{V}(X_i)\right\} \quad (9.3.33)$$

For SRSWOR sampling, $\widehat{Y}_{reg}(c)$ reduces to

$$\widehat{Y}_{reg}(c) = \left(\sum_{i=1}^{K} N_i\overline{y}_i\right) - \widehat{\beta}_c\left(\sum_{i=1}^{K} N_i\left(\overline{x}_i - \overline{X}\right)\right) \quad (9.3.34)$$

where $\widehat{\beta}_c = \dfrac{\sum_{i=1}^{K} N_i^2(1-f_i)s_{xyi}/n_i}{\sum_{i=1}^{K} N_i^2(1-f_i)s_{xi}^2/n_i}$.

The expressions for MSE and its approximate unbiased estimator of $\widehat{Y}_{reg}(c)$ under SRSWOR are, respectively, given by

$$M\left[\widehat{Y}_{reg}(c)\right] \cong \sum_{i=1}^{K} N_i^2(1-f_i)\left\{S_{yi}^2 - 2\beta_c S_{xyi} + \beta_c^2 S_{xi}^2\right\}\Big/ n_i$$
$$= \sum_{i=1}^{K} N_i^2(1-f_i)S_{gi}^2\Big/ n_i \quad (9.3.35)$$

and

$$\widehat{M}\left[\widehat{Y}_{reg}(c)\right] \cong \sum_{i=1}^{K} N_i^2(1-f_i)\left\{s_{yi}^2 - 2\widehat{\beta}_c s_{xyi} + \widehat{\beta}_c^2 s_{xi}^2\right\}\Big/ n_i$$
$$= \sum_{i=1}^{K} N_i^2(1-f_i)s_{\widehat{g}i}^2\Big/ n_i \quad (9.3.36)$$

where $S_{gi}^2 = \sum_{j \in U_i} \left(g_{ij} - \overline{G}_i\right)^2 / (N_i - 1)$, $g_{ij} = y_{ij} - \beta_c x_{ij}$, $\overline{G}_i = \sum_{j \in U_i} g_{ij} / N_i$,

$s_{gi}^2 = \sum_{j \in s_i} \left(\widehat{g}_{ij} - \overline{g}_i\right)^2 / (n_i - 1)$, $\widehat{g}_{ij} = y_{ij} - \widehat{\beta}_c x_{ij}$, and $\overline{g}_i = \sum_{j \in s_i} \widehat{g}_{ij} / n_i$.

9.3.8.3 Comparison Between Combined and Separate Regression Estimators

Approximate expressions of biases and MSEs for separate and combined regression estimators are obtained assuming sample sizes are large for each of the strata. The bias of a regression estimator decreases with the increase in sample size. The total sample size n may be large, but the sample sizes n_i's for each of the strata may not be large. Hence the bias of $\widehat{Y}_{i(reg)}$ may not be negligible compared to its standard error if the sample size n_i is small. Hence the amount of bias of the separate regression estimator is expected to be much larger than the combined regression estimator.

It is obvious that the separate regression estimator $\widehat{Y}_{reg}(s)$ is more efficient than the combined regression estimator $\widehat{Y}_{reg}(c)$ because

$$M\left[\widehat{Y}_{reg}(c)\right] - M\left[\widehat{Y}_{reg}(s)\right] \cong \sum_{i=1}^{K} \left\{ V\left(\widehat{Y}_i - \beta_c \widehat{X}_i\right) - V\left(\widehat{Y}_i - \beta_i \widehat{X}_i\right) \right\} \geq 0$$

as $V\left(\widehat{Y}_i - \beta_c \widehat{X}_i\right)$ attains a minimum when $\beta_c = \beta_i$.

In practice, the values of β_i's are unknown and they are estimated from the samples. In case the sample sizes are not large enough for each of the strata, the estimate $\widehat{\beta}_i$ becomes unreliable and hence the separate regression estimator becomes inefficient. But, if the regression coefficients are approximately equal, then $\widehat{\beta}_c$, the estimator of β_c becomes more efficient than each of the separate estimators $\widehat{\beta}_i$ because the former is computed using a relatively large sample size. In this situation combined regression estimator certainly becomes more efficient than the separate regression estimators. As a rule of thumb, a combined regression estimator should be used in case regression lines for each of the strata are the same. Otherwise, separate regression estimator should be used.

Example 9.3.2

From the data given in Example 8.7.1, estimate the average wage of the factory workers by using (i) separate and (ii) combined regression estimators when average years of services of the three categories are known to be 15, 8, and 3 years, respectively. Estimate the relative efficiency of the separate regression estimator with respect to the combined regression estimator.

Here we prepare the following tables:

Strata	N_i	$\overline{X}_i$	$N_i\overline{X}_i$	n_i	$\overline{x}(s_i)$	$\overline{y}(s_i)$	$N_i\overline{x}(s_i)$	$N_i\overline{y}(s_i)$	$s_{x_i}^2$	$s_{y_i}^2$	s_{xy_i}	$\widehat{\beta}_i$
Skilled	50	15	750	10	16.3	610	815	30,500	59.122	13,222.220	824.444	13.944
Semiskilled	75	8	600	12	9	425	675	31,875	6.727	6,590.909	150.000	22.297
Unskilled	100	3	300	15	3.8	260	380	26,000	2.171	12,571.430	134.285	61.842
Total	225		1650				1,870	88,375				

$\widehat{Y}_{i(reg)}$	$\widehat{\rho}_i = r_i$	$\dfrac{N_i^2(1-f_i)(1-r_i^2)s_{yi}^2}{n_i}$	$\dfrac{N_i^2(1-f_i)s_{xyi}}{n_i}$	$\dfrac{N_i^2(1-f_i)s_{xi}^2}{n_i}$	$\dfrac{N_i^2(1-f_i)\left\{s_{yi}^2 - 2\widehat{\beta}_c s_{xyi} + \widehat{\beta}_c^2 s_{xi}^2\right\}}{n_i}$
29,593.591	0.932	345,110.568	164,888.889	11,824.440	660,175.087
30,202.702	0.712	1,278,236.333	59,062.500	2,648.864	1,305,202.464
21,052.631	0.813	2,417,919.799	76,095.238	1,230.476	4,665,162.568
80,848.925		4,041,266.701	300,046.627	15,703.780	6,630,540.119

Separate regression estimate for the average wage $\overline{Y}$ is

$$\widehat{\overline{Y}}_{reg}(s) = \widehat{Y}_{reg}(s)/N = \sum_{i=1}^{K} \widehat{Y}_{i(reg)}/N = 80848.925/225 = 359.329$$

$$\text{Estimated MSE of } \overline{Y}_{reg}(s) = \widehat{M}\{\widehat{Y}_{reg}(s)\}/N^2$$
$$= \left\{\sum_{i=1}^{K} N_i^2(1-f_i)(1-\widehat{\rho}_i^2)s_{yi}^2/n_i\right\}\Big/N^2$$
$$= 4041266.701/(225)^2 = 79.827$$

An estimate of the combined regression β_c is

$$\widehat{\beta}_c = \frac{\sum_{i=1}^{3} N_i^2(1-f_i)s_{xyi}/n_i}{\sum_{i=1}^{3} N_i^2(1-f_i)s_{xi}^2/n_i} = 300046.627/15703.780 = 19.107$$

Combined regression estimate of the average wage $\overline{Y}$ is

$$\widehat{\overline{Y}}_{reg}(c) = \widehat{Y}_{reg}(c)/N = \left(\sum_{i=1}^{K} N_i\overline{y}_i/N\right) - \widehat{\beta}_c\left(\sum_{i=1}^{K} N_i\overline{x}_i/N - \overline{X}\right)$$
$$= 88375/225 - 19.107 \times (1870/225 - 1650/225) = 374.096$$

Estimated MSE of $\widehat{M}\left[\widehat{\overline{Y}}_{reg}(c)\right]$ is

$$\widehat{M}\left[\widehat{\overline{Y}}_{reg}(c)\right] = \left[\sum_{i=1}^{K} N_i^2(1-f_i)\left(s_{yi}^2 - 2\widehat{\beta}_c s_{xyi} + \widehat{\beta}_c^2 s_{xi}^2\right)/n_i\right]\Big/N^2$$
$$= 6630540.119/(225)^2 = 130.974$$

Estimated relative efficiency of separate regression estimator with respect to the combined regression estimator for estimating $\overline{Y} = \widehat{M}\left[\widehat{\overline{Y}}_{reg}(c)\right] \times 100\Big/\widehat{M}\left[\widehat{\overline{Y}}_{reg}(s)\right] = (130.974/79.827) \times 100 = 164.071$ percent.

9.3.9 Regression Estimator for Several Auxiliary Variables

Suppose that we have p auxiliary variables $x_1,\ldots, x_p$, each of which is well related to the study variable y. Let x_{ij} be the value of ith unit of the jth auxiliary variable x_j. In this case, we may set the following two types of regression estimators for the total Y when the totals of each of the auxiliary variables are known.

9.3.9.1 Multivariate Regression Estimator

$$\widehat{Y}_{reg}(1)=\widehat{Y}-\lambda_1\left(\widehat{X}_1-X_1\right)-\cdots-\lambda_j\left(\widehat{X}_j-X_j\right)-\cdots-\lambda_p\left(\widehat{X}_p-X_p\right) \tag{9.3.37}$$

where $\widehat{X}_j=\sum_{i\in s} b_{si}x_{ji}$ is an unbiased estimator of the total X_j and $\lambda_1,\ldots,\lambda_p$ are known constants. The optimum values of $\lambda_1,\ldots,\lambda_p$ are obtained by minimizing the variance of $\widehat{Y}_{reg}(1)$. The variance of $\widehat{Y}_{reg}(1)$ is

$$V\left\{\widehat{Y}_{reg}(1)\right\}=V\left(\widehat{Y}\right)+\sum_{i=1}^{p}\lambda_i^2V\left(\widehat{X}_i\right)+\sum_{i\neq}^{p}\sum_{j=p}^{p}\lambda_i\lambda_jCov\left(\widehat{X}_i,\widehat{X}_j\right)$$
$$-2\sum_{i=1}^{p}\lambda_iCov\left(\widehat{Y},\widehat{X}_i\right) \tag{9.3.38}$$

Differentiating $V\left\{\widehat{Y}_{reg}(1)\right\}$ with respect to λ_j and equating to zero for $j=1,\ldots,p$, the optimum values of $\lambda_j=\beta_j$ (say) are obtained as

$$\begin{pmatrix}\beta_1\\ \cdots\\ \beta_j\\ \cdots\\ \beta_p\end{pmatrix}=\begin{pmatrix}V\left(\widehat{X}_1\right) & \cdots & Cov\left(\widehat{X}_1,\widehat{X}_j\right) & \cdots & Cov\left(\widehat{X}_1,\widehat{X}_p\right)\\ \cdots & \cdots & \cdots & \cdots & \cdots\\ Cov\left(\widehat{X}_j,\widehat{X}_1\right) & \cdots & V\left(\widehat{X}_j\right) & \cdots & Cov\left(\widehat{X}_j,\widehat{X}_p\right)\\ \cdots & \cdots & \cdots & \cdots & \cdots\\ Cov\left(\widehat{X}_p,\widehat{X}_1\right) & \cdots & Cov\left(\widehat{X}_p,\widehat{X}_j\right) & \cdots & V\left(\widehat{X}_p\right)\end{pmatrix}^{-1}$$
$$\times\begin{pmatrix}Cov\left(\widehat{Y},\widehat{X}_1\right)\\ \cdots\\ Cov\left(\widehat{Y},\widehat{X}_j\right)\\ \cdots\\ Cov\left(\widehat{Y},\widehat{X}_p\right)\end{pmatrix} \tag{9.3.39}$$

The solution (Eq. 9.3.39) indicates that the optimum value β_j is the partial regression coefficient of $\widehat{Y}$ on $\widehat{X}_j$.

In particular for SRSWOR sampling design $V\left(\widehat{X}_j\right)=N^2(1-f)S_{x_j}^2/n$, $Cov\left(\widehat{X}_j,\widehat{X}_k\right)=N^2(1-f)S_{x_j,x_k}^2/n$ and $Cov\left(\widehat{Y},\widehat{X}_k\right)=N^2(1-f)S_{y,x_k}^2/n$ where $S_{x_j}^2=\sum_{i\in U}\left(x_{ji}-\overline{x}_j\right)^2/(N-1)$,

$$S^2_{x_j,x_k} = \sum_{i\in U}\left(x_{ji} - \overline{x}_j\right)\left(x_{ki} - \overline{x}_k\right)/(N-1),$$
$$S_{y,x_k} = \sum_{i\in U}\left(y_i - \overline{y}\right)\left(x_{ki} - \overline{x}_k\right)/(N-1), \text{ and } \overline{x}_j = \sum_{i\in U} x_{ji}/N.$$

The value of $\widehat{Y}_{reg}(1)$ with the optimum value of $\lambda_i = \beta_i$ comes out as

$$V_{\min}\left(\widehat{Y}_{reg}(1)\right) = N^2(1-f)\left(1 - R^2_{y|x_1,\ldots,x_k}\right)/n \tag{9.3.40}$$

where $R^2_{y|x_1,\ldots,x_k}$ = multiple correlation between y and $x_1,\ldots, x_p$.

The optimum values β_j are generally unknown. Hence regression estimator of Y is obtained by replacing β_j by its suitable estimate $\widehat{\beta}_j$ and it is given as

$$\widehat{Y}_{reg}(1) = \widehat{Y} - \widehat{\beta}_1\left(\widehat{X}_1 - X_1\right) - \cdots - \widehat{\beta}_j\left(\widehat{X}_j - X_j\right) - \cdots - \widehat{\beta}_p\left(\widehat{X}_p - X_p\right)$$

9.3.9.2 Two Auxiliary Variables

For $p = 2$, Eq. (9.3.39) reduces to

$$\begin{pmatrix}\beta_1\\ \beta_2\end{pmatrix} = \begin{pmatrix} S^2_{x_1} & \rho_{x_1,x_2}S_{x_1}S_{x_2}\\ \rho_{x_1,x_2}S_{x_1}S_{x_2} & S^2_{x_2}\end{pmatrix}^{-1}\begin{pmatrix}\rho_{yx_1}S_{x_1}S_y\\ \rho_{yx_2}S_{x_2}S_y\end{pmatrix}$$

$$= \begin{pmatrix} \dfrac{\rho_{yx_1} - \rho_{x_1x_2}\rho_{yx_2}}{1-\rho^2_{x_1,x_2}}\dfrac{S_y}{S_{x_1}}\\[2ex] \dfrac{\rho_{yx_2} - \rho_{x_1x_2}\rho_{yx_1}}{1-\rho^2_{x_1,x_2}}\dfrac{S_y}{S_{x_2}}\end{pmatrix}$$

where ρ_{uv} is the correlation coefficient between u and v.

Hence the variance of $\widehat{Y}_{reg}(1) = \widehat{Y} - \lambda_1\left(\widehat{X}_1 - X_1\right) - \lambda_2\left(\widehat{X}_2 - X_2\right)$ attains a minimum when $\lambda_1 = \beta_1$ and $\lambda_2 = \beta_2$. The corresponding minimum variance is

$$V_{\min}\left\{\widehat{Y}_{reg}(1)\right\} = N^2(1-f)\left(1 - R^2_{y|x_1,x_2}\right)S^2_y/n$$

where $R^2_{y|x_1,x_2} = \dfrac{\rho^2_{yx_1} + \rho^2_{yx_2} - 2\rho_{yx_1}\rho_{yx_2}\rho_{x_1x_2}}{1-\rho^2_{x_1x_2}}$

9.3.9.3 Raj's Regression Estimator

Raj (1965a,b) computed p difference estimators based on each of the auxiliary variables and then obtained a combined regression estimator by taking the weighted average of these separate estimators. Raj's estimator is given by

$$\widehat{Y}_{reg}(R) = \sum_{j=1}^{p} w_j\widehat{Y}_{reg}(j) \tag{9.3.41}$$

where $\widehat{Y}_{reg}(j) = \widehat{Y} - \lambda_j^*(\widehat{X}_j - X_j)$, λ_j^* is a suitably chosen constant, and $\sum_{j=1}^{K} w_j = 1$.

The optimum value of λ_j^* that minimizes the variance of $\widehat{Y}_{reg}(j)$ is $\beta_j^* = \rho_{yx_j} S_y \big/ S_{x_j}$ and the corresponding minimum variance of $\widehat{Y}_{reg}(j)$ is $N^2(1-f)\left(1-\rho_{yx_j}^2\right)S_y^2\big/n$. The optimum values of w_i's that minimize the variance of $\widehat{Y}_{reg}(R)$ are obtained following Section 8.8. It is obvious that the minimum variance of Eq. (9.3.41) is higher than that of Eq. (9.3.40). Hence Raj's estimator is less efficient than the estimator given in Eq. (9.3.40). For $p = 2$, details are given by Sukhatme et al. (1984).

9.4 PRODUCT METHOD OF ESTIMATION

The ratio method of estimation is more efficient than the conventional estimator $\widehat{Y}$ when the study variable y has a high positive correlation with the auxiliary variable x and $R = Y/X$ is positive. Goodman (1960) and Murthy (1964) proposed the product estimator that is used when the study and the auxiliary variables are negatively related. The product estimator is given by

$$\widehat{Y}_p = \frac{\widehat{Y}\widehat{X}}{X} \tag{9.4.1}$$

The product estimator $\widehat{Y}_p$ is not unbiased. The bias is negligible for a large sample size. The expressions of bias and MSE of the product estimator are given as follows.

9.4.1 Bias of the Product Estimator

Writing $\varepsilon_x = \dfrac{\widehat{X}}{X} - 1$ and $\varepsilon_y = \dfrac{\widehat{Y}}{Y} - 1$, we get the bias of the product estimator as

$$\begin{aligned} B(\widehat{Y}_p) &= E\left[\frac{XY(1+\varepsilon_x)(1+\varepsilon_y)}{X} - Y\right] \\ &= YE(\varepsilon_x\varepsilon_y) \qquad (9.4.2) \\ &= \frac{Cov(\widehat{X}, \widehat{Y})}{X} \end{aligned}$$

An estimator of the bias $B(\widehat{Y}_p)$ is given by

$$\widehat{B}(\widehat{Y}_p) = \frac{C\widehat{o}v(\widehat{X}, \widehat{Y})}{\widehat{X}} \tag{9.4.3}$$

9.4.2 Mean Square Error of the Product Estimator

The MSE of the product estimator is given by

$$\begin{aligned} M\left(\widehat{Y}_p\right) &= E\left[\frac{XY(1+\varepsilon_x)(1+\varepsilon_y)}{X} - Y\right]^2 \\ &= Y^2 E(\varepsilon_x + \varepsilon_y + \varepsilon_x\varepsilon_y)^2 \end{aligned} \tag{9.4.4}$$

Neglecting terms $E\left(\varepsilon_x^i \varepsilon_y^j\right)$ with $i+j>2$ in Eq. (9.4.4), we get the expression of the MSE up to the first order of approximation as

$$M\left(\widehat{Y}_p\right) = V\left(\widehat{Y}\right) + R^2 V\left(\widehat{X}\right) + 2RCov\left(\widehat{X}, \widehat{Y}\right) \tag{9.4.5}$$

An approximate unbiased estimator of $M\left(\widehat{Y}_p\right)$ is

$$\widehat{M}\left(\widehat{Y}_p\right) = \widehat{V}\left(\widehat{Y}\right) + \widehat{R}^2 \widehat{V}\left(\widehat{X}\right) + 2\widehat{R}C\widehat{ov}\left(\widehat{X}, \widehat{Y}\right) \tag{9.4.6}$$

9.4.3 Comparison With the Conventional Estimator

The product estimator $\widehat{Y}_p$ becomes superior to the conventional estimator $\widehat{Y}$ if

$$V\left(\widehat{Y}\right) - M\left(\widehat{Y}_p\right) \geq 0$$

$$\text{i.e. if } \rho_{\hat{X},\hat{Y}} \leq -\frac{1}{2}\frac{C_{\hat{X}}}{C_{\hat{Y}}} \quad \text{if } R > 0$$

and

$$\rho_{\hat{X},\hat{Y}} \geq \frac{1}{2}\left|\frac{C_{\hat{X}}}{C_{\hat{Y}}}\right| \quad \text{if } R < 0$$

where $C_{\hat{X}} = \sqrt{V\left(\widehat{X}\right)}/X$ and $C_{\hat{Y}} = \sqrt{V\left(\widehat{Y}\right)}/Y$ are the coefficient of variation of $\widehat{X}$ and $\widehat{Y}$, respectively.

9.4.4 Product Estimator for a Few Sampling Designs

9.4.4.1 Fixed Effective Sample Size Design

Consider a fixed effective sample size n design and $b_{si} = 1/\pi_i$. In this case

$$\widehat{X} = \sum_{i\in s} x_i/\pi_i,\ \widehat{Y} = \sum_{i\in s} y_i/\pi_i,\ \widehat{Y}_p = \frac{\left(\sum_{i\in s} x_i/\pi_i\right)\left(\sum_{i\in s} y_i/\pi_i\right)}{X}$$

$$B\left(\widehat{Y}_p\right) = \frac{1}{X}\frac{1}{2}\sum_{i\neq}\sum_{j\in U}\Delta_{ij}\left(\frac{x_i}{\pi_i} - \frac{x_j}{\pi_j}\right)\left(\frac{y_i}{\pi_i} - \frac{y_j}{\pi_j}\right),$$

$$\widehat{B}\left(\widehat{Y}_p\right) = \frac{1}{\widehat{X}}\frac{1}{2}\sum_{i\neq}\sum_{j\in s}\frac{\Delta_{ij}}{\pi_{ij}}\left(\frac{x_i}{\pi_i} - \frac{x_j}{\pi_j}\right)\left(\frac{y_i}{\pi_i} - \frac{y_j}{\pi_j}\right)$$

$$M\left(\widehat{Y}_p\right)=\frac{1}{2}\sum_{i\neq}\sum_{j\in U}\Delta_{ij}\left(\frac{F_i}{\pi_i}-\frac{F_j}{\pi_j}\right)^2,\ \widehat{M}\left(\widehat{Y}_p\right)=\frac{1}{2}\sum_{i\neq}\sum_{j\in s}\frac{\Delta_{ij}}{\pi_{ij}}\left(\frac{\widehat{F}_i}{\pi_i}-\frac{\widehat{F}_j}{\pi_j}\right)^2$$

where $F_i = y_i + Rx_i$ and $\widehat{F}_i = y_i + \widehat{R}x_i$.

9.4.4.2 Simple Random Sampling Without Replacement

For SRSWOR, $\widehat{Y}_p = \dfrac{N\overline{y}_s\overline{x}_s}{\overline{X}}$, $B\left(\widehat{Y}_p\right) = \dfrac{N(1-f)}{n}\dfrac{S_{xy}}{\overline{X}}$,

$\widehat{B}\left(\widehat{Y}_p\right) = \dfrac{N(1-f)}{n}\dfrac{s_{xy}}{\overline{x}_s}$, $M\left(\widehat{Y}_p\right) = N^2\dfrac{(1-f)}{n}\left(S_y^2 + R^2S_x^2 + 2RS_{xy}\right)$, and

$\widehat{M}\left(\widehat{Y}_p\right) = N^2\dfrac{(1-f)}{n}\left(s_y^2 + \widehat{R}^2 s_x^2 + 2\widehat{R}S_{xy}\right)$.

9.4.5 Unbiased Product Type Estimators

Chaudhuri and Arnab (1982a,b) proposed the unbiased product type estimator and it is constructed as follows:

Let $\widehat{X} = \sum_{i\in s} b_{si}x_i$ and $\widehat{Y} = \sum_{i\in s} b_{si}y_i$ be the unbiased estimators of the population totals X and Y as defined in Section 9.3.3. The sample s of size n is divided at random into k groups so that the ith group s_i consists of n_i units $\left(\sum_{i=1}^{k} n_i = n\right)$. Let the product estimator of Y based on the sample s_j is denoted by

$$\widehat{Y}_p(j) = \widehat{X}_j\widehat{Y}_j/X \tag{9.4.7}$$

where $\widehat{Y}_j = n\sum_{i\in s_j}\xi_i/n_j$, $\widehat{X}_j = n\sum_{i\in s_j}\eta_i/n_j$, $\xi_i = b_{si}y_i$, and $\eta_i = b_{si}x_i$.

The proposed modified product estimator is defined as follows:

$$\widehat{Y}_p(m) = \sum_{i=1}^{k} w_i\widehat{Y}_p(j) + \left(1 - \sum_{i=1}^{k} w_i\right)\widehat{Y}_p \tag{9.4.8}$$

where w_i's are suitably chosen weights.

Denoting $\varepsilon_{jx} = \left\{\widehat{X}_j - X\right\}/X$ and $\varepsilon_{jy} = \left(\widehat{Y}_j - Y\right)/Y$, we have the following theorem due to Chaudhri and Arnab (1982a,b).

Theorem 9.4.1

The modified product estimator $\widehat{Y}_p(m)$ is unbiased for Y and has the same efficiency as $\widehat{Y}_p$ upto the first order of approximation if

$$(i)\sum_{i=1}^{k} c_iw_i + \left(1 - \sum_{i=1}^{k} w_i\right) = 0 \text{ and } (ii)\ w_i = \frac{n_i}{n}\left(\sum_{i=1}^{k} w_i\right)$$

where $c_i = E(\varepsilon_{ix}\varepsilon_{iy})/E(\varepsilon_x\varepsilon_y)$.

Proof

$$E\{\widehat{Y}_p(m)\} = \sum_{i=1}^{k} w_i E\{\widehat{Y}_p(j)\} + \left(1 - \sum_{i=1}^{k} w_i\right) E(\widehat{Y}_p)$$
$$= Y\left\{\sum_{i=1}^{k} w_i E(1 + \varepsilon_{ix} + \varepsilon_{iy} + \varepsilon_{ix}\varepsilon_{iy}) + \left(1 - \sum_{i=1}^{k} w_i\right) E(1 + \varepsilon_x + \varepsilon_y + \varepsilon_x\varepsilon_y)\right\}$$
$$= Y\left[\left\{\sum_{i=1}^{k} w_i c_i + \left(1 - \sum_{i=1}^{k} w_i\right)\right\} E(\varepsilon_x\varepsilon_y) + 1\right] \tag{9.4.9}$$

From Eq. (9.4.9), we note that $E\{\widehat{Y}_p(m)\}$ is unbiased if and only if

$$\sum_{i=1}^{k} c_i w_i + \left(1 - \sum_{i=1}^{k} w_i\right) = 0 \tag{9.4.10}$$

The MSE of $\widehat{Y}_p(m)$ is

$$M\{\widehat{Y}_p(m)\} = E\{\widehat{Y}_p(m) - Y\}^2$$
$$= Y^2\left[\sum_{i=1}^{k} w_i(1 + \varepsilon_{ix} + \varepsilon_{iy} + \varepsilon_{ix}\varepsilon_{iy}) + \left(1 - \sum_{i=1}^{k} w_i\right)(1 + \varepsilon_x + \varepsilon_y + \varepsilon_x\varepsilon_y) - 1\right]^2$$

Now neglecting terms with $E\left(\varepsilon_x^r \varepsilon_y^l\right)$ for $r + l > 2$, we get upto the first order of approximation:

$$M\{\widehat{Y}_p(m)\} \cong Y^2 E\left[\sum_{i=1}^{k} w_i(\varepsilon_{ix} + \varepsilon_{iy}) + \left(1 - \sum_{i=1}^{k} w_i\right)(\varepsilon_x + \varepsilon_y)\right]^2$$
$$= Y^2 E\left[E\left\{\sum_{i=1}^{k} w_i(\varepsilon_{ix} + \varepsilon_{iy}) + \left(1 - \sum_{i=1}^{k} w_i\right)(\varepsilon_x + \varepsilon_y)\right\}^2 \Big| s\right]$$
$$= Y^2 E\left[V\left\{\sum_{i=1}^{k} w_i(\varepsilon_{ix} + \varepsilon_{iy}) + \left(1 - \sum_{i=1}^{k} w_i\right)(\varepsilon_x + \varepsilon_y)|s\right\}\right]$$
$$+ Y^2 E\left[E\left\{\sum_{i=1}^{k} w_i(\varepsilon_{ix} + \varepsilon_{iy}) + \left(1 - \sum_{i=1}^{k} w_i\right)(\varepsilon_x + \varepsilon_y)|s\right\}\right]^2 \tag{9.4.11}$$

Because the first term of Eq. (9.4.11) is nonnegative and $E\left\{\sum_{i=1}^{k} w_i(\varepsilon_{ix}+\varepsilon_{iy})+\left(1-\sum_{i=1}^{k} w_i\right)(\varepsilon_x+\varepsilon_y)|s\right\}=\varepsilon_x+\varepsilon_y$, we find

$$M\left\{\widehat{Y}_p(m)\right\} \geq Y^2 E(\varepsilon_x+\varepsilon_y)^2 = M\left(\widehat{Y}_p\right) \tag{9.4.12}$$

Equality in Eq. (9.4.12) holds if and only if

$$\sum_{i=1}^{k} w_i(\varepsilon_{ix}+\varepsilon_{iy})+\left(1-\sum_{i=1}^{k} w_i\right)(\varepsilon_x+\varepsilon_y)$$
$$= E\left\{\sum_{i=1}^{k} w_i(\varepsilon_{ix}+\varepsilon_{iy})+\left(1-\sum_{i=1}^{k} w_i\right)(\varepsilon_x+\varepsilon_y)|s\right\}=\varepsilon_x+\varepsilon_y \tag{9.4.13}$$

for every s with $p(s) > 0$.

Now noting $\varepsilon_x+\varepsilon_y=\sum_{i=1}^{k}\frac{n_i}{n}(\varepsilon_{ix}+\varepsilon_{iy})$, the above condition (Eq. 9.4.13) reduces to

$$\sum_{i=1}^{k} w_i(\varepsilon_{ix}+\varepsilon_{iy})=\left(\sum_{i=1}^{k} w_i\right)\sum_{i=1}^{k}\frac{n_i}{n}(\varepsilon_{ix}+\varepsilon_{iy}) \tag{9.4.14}$$

The condition (Eq. 9.4.14) is realized if we choose

$$w_i=\frac{n_i}{n}\left(\sum_{i=1}^{k} w_i\right) \quad \text{for } i=1,\ldots,k \tag{9.4.15}$$

The theorem follows from Eqs. (9.4.10) and (9.4.15).

9.4.5.1 *Simple Random Sampling Without Replacement*

Let $b_{si}=N/n$, then $\widehat{X}=N\overline{x}(s)$, $\widehat{Y}=N\overline{y}(s)$, $\widehat{X}_j=N\sum_{i\in s_j} x_i/n_j$, $\widehat{Y}_j=N\sum_{i\in s_j} y_i/n_j$, and $c_j=1+\left(\frac{1}{n_j}-\frac{1}{n}\right)\left(\frac{1}{n}-\frac{1}{N}\right)^{-1}$. In this case Eq. (9.4.10) reduces to

$$\left(\frac{1}{n}-\frac{1}{N}\right)^{-1}\sum_{j=1}^{k} w_j\left(\frac{1}{n_j}-\frac{1}{n}\right)+1=0 \tag{9.4.16}$$

In particular if we choose $n_j=n/k$, Eqs. (9.4.15) and (9.4.16) yield a very simple solution of w_j as

$$w_j = -\frac{1}{k(k-1)}\frac{N-n}{N} \quad \text{for } j = 1, \ldots, k \tag{9.4.17}$$

Shukla (1976) derived Eq. (9.4.17) for $k = 2$.

9.5 COMPARISON BETWEEN THE RATIO, REGRESSION, PRODUCT, AND CONVENTIONAL ESTIMATORS

The regression estimator is always superior to the ratio, product, and the conventional estimator provided the estimate $\widehat{\beta}$ becomes very close to the true value β. In practice, the reliability of the estimator of the regression coefficient β is in question, especially if the sample size is small. Hence, in this situation, the regression estimator may not be efficient. Noting the difference $M(\widehat{Y}_p) - M(\widehat{Y}_R) = 4RCov(\widehat{X}, \widehat{Y})$, we conclude (i) for $R > 0$, the ratio estimator is superior to the product estimator if $\rho_{\hat{X},\hat{Y}} > 0$ and the product estimator is superior to the ratio estimator if $\rho_{\hat{X},\hat{Y}} < 0$; (ii) for $R < 0$, the ratio estimator is superior to the product estimator if $\rho_{\hat{X},\hat{Y}} < 0$ but for $\rho_{\hat{X},\hat{Y}} > 0$, the product estimator is superior. In case $\left|\rho_{\hat{X},\hat{Y}}\right| < \frac{1}{2}\frac{C_{\hat{X}}}{C_{\hat{Y}}}$, the conventional estimator $\widehat{Y}$ is always better than the product and ratio estimator. In particular, if $C_{\hat{X}} = C_{\hat{Y}}$, then the conventional estimator becomes more efficient than both the ratio and product estimators when $\left|\rho_{\hat{X},\hat{Y}}\right| < 1/2$.

9.6 DUAL TO RATIO ESTIMATOR

Srivenkataramana (1980) proposed the following product type estimator, known as a dual to ratio estimator

$$\widehat{Y}_{dR} = \frac{\widehat{Y}\widehat{X}_{\bar{s}}}{X} \tag{9.6.1}$$

where $X_{\bar{s}} = \sum_{i \in \bar{s}} \frac{x_i}{1 - \pi_i}$ is an estimator for the total X based on the set $\bar{s}(= U - s)$ of $N - n$ units, which are not selected in the sample s. So the estimator $\widehat{Y}_{dR}$ is usable when the values of the auxiliary variable x are known at unit level. For SRSWOR, $\pi_i = n/N$ and the estimator $\widehat{Y}_{dR}$ reduces to

$$\widehat{Y}_{dR} = \frac{\widehat{Y}(X - n\overline{x}_s)N}{(N-n)X} = \frac{N\widehat{Y} - n\widehat{Y}_p}{N-n} \quad (9.6.2)$$

where $\widehat{Y}_p$ is the product estimator defined in Eq. (9.4.1).

9.6.1 Bias of the Dual Estimator

Now writing

$$\widehat{Y}_{dR} - Y = \frac{N\left(\widehat{Y} - Y\right) - n\left(\widehat{Y}_p - Y\right)}{N-n} \quad (9.6.3)$$

we find the bias of $\widehat{Y}_{dR}$ as

$$B\left(\widehat{Y}_{dR}\right) = -\frac{n}{N-n}B\left(\widehat{Y}_p\right) \quad (9.6.4)$$

Furthermore, from Section 9.4.4.2 we note

$$B\left(\widehat{Y}_{dR}\right) = -\frac{S_{xy}}{\overline{X}} \quad (9.6.5)$$

The expression (Eq. 9.6.4) indicates that the absolute bias of $\widehat{Y}_{dR}$ is less than the absolute bias of the product estimator $\widehat{Y}_p$ if the sampling fraction $f = n/N < 1/2$. The expression (Eq. 9.6.5) indicates that $\widehat{Y}_{dR}$ is an inconsistent estimator for Y because the bias does not decrease with the increase in the sample size.

9.6.2 Mean Square Error of the Dual Estimator

Writing $\varepsilon_x = \frac{\widehat{X} - X}{X}$ and $\varepsilon_y = \frac{\widehat{Y} - Y}{Y}$, we find

$$\widehat{Y}_{dR} - Y = Y\frac{(N-n)\varepsilon_y - n\varepsilon_x - n\varepsilon_x\varepsilon_y}{N-n} \quad (9.6.6)$$

The expression of the MSE of $\widehat{Y}_{dR}$ upto the first-order approximation comes out as

$$M\left(\widehat{Y}_{dR}\right) = E\left(\widehat{Y}_{dR} - Y\right)^2 = N^2(1-f)\left(S_y^2 + g^2R^2S_x^2 - 2gRS_{xy}\right)/n \quad (9.6.7)$$

where $g = n/(N-n)$.

9.6.3 Comparison With Other Estimators

9.6.3.1 Conventional Estimator

The $\widehat{Y}_{dR}$ becomes more efficient than the conventional estimator $\widehat{Y} = N\overline{y}(s)$ if

$$\text{(i)}\ \rho_{xy} \geq \frac{g}{2}\frac{C_x}{C_y} \quad \text{with } R > 0, \text{ or (ii) } \rho_{xy} \leq -\frac{g}{2}\left|\frac{C_x}{C_y}\right| \quad \text{with } R < 0.$$

9.6.3.2 Ratio Estimator

$$M\left(\widehat{Y}_{dR}\right) - M\left(\widehat{Y}_{R}\right) = (g-1)\left[(g+1)R^2S_x^2 - 2R\rho_{xy}S_xS_y\right] \tag{9.6.8}$$

The estimator $\widehat{Y}_{dR}$ will be more efficient than the ratio estimator $\widehat{Y}_R$ if the difference (Eq. 9.6.8) is negative. Here we consider the following situations:

(i) $g-1>0$, i.e., $f>1/2$, in this case the difference (Eq. 9.6.8) is negative if either $\rho_{xy} > \frac{g+1}{2}\frac{C_x}{C_y}$ with $R>0$ or $\rho_{xy} < -\frac{g+1}{2}\left|\frac{C_x}{C_y}\right|$ with $R<0$. The situation $f>1/2$ rarely happens in practice.

(ii) $g-1<0$, i.e., $f<1/2$, in this case (Eq. 9.6.8) is negative if either $\rho_{xy} < \frac{g+1}{2}\frac{C_x}{C_y}$ with $R>0$ or $\rho_{xy} > -\frac{g+1}{2}\left|\frac{C_x}{C_y}\right|$ with $R<0$. The situation $f<1/2$ most likely happens in practice.

9.6.3.3 Product Estimator

From Eq. (9.6.7) and Section 9.4.4.2, we note that

$$M\left(\widehat{Y}_{dR}\right) - M\left(\widehat{Y}_{P}\right) = (g+1)\left[(g-1)R^2S_x^2 - 2R\rho_{xy}S_xS_y\right] \tag{9.6.9}$$

The estimator $\widehat{Y}_{dR}$ will be more efficient than $\widehat{Y}_P$ if the difference (Eq. 9.6.9) is negative, i.e., if (i) $\rho_{xy} > \frac{g-1}{2}\frac{C_x}{C_y}$ with $R>0$ or (ii) if $\rho_{xy} < -\frac{g-1}{2}\left|\frac{C_x}{C_y}\right|$ with $R<0$.

9.7 CALIBRATION ESTIMATORS

The calibration estimator was proposed by Deville and Särndal (1992). In calibration estimators, the weights of the conventional unbiased estimators are modified by using auxiliary information. The weights of the conventional estimator are calibrated in such a way that the average distances between the calibrated and original weights attain a minimum according to a certain distance measure and at the same time, the calibrated weights satisfy some constraints that are known as calibration equations. The calibrated estimators become asymptotically design consistent and more efficient than the corresponding conventional estimator.

Suppose our objective is to estimate the population total Y on the basis of a sample s selected with probability $p(s)$. Let the conventional estimator for Y be the HTE

$$\widehat{Y}_{ht} = \sum_{i \in s} d_i y_i \tag{9.7.1}$$

where $d_i = 1/\pi_i$ and π_i is the inclusion probability for the ith unit, assumed to be positive for every $i \in U$.

In case X, the population total of the auxiliary variable x is known, we use this known total X to form a calibrated estimator

$$\widehat{Y}_c = \sum_{i \in s} w_i y_i \tag{9.7.2}$$

where w_i's are the weights that satisfy the following calibration constraint or calibration equation.

$$\sum_{i \in s} w_i x_i = X \tag{9.7.3}$$

Here it is important to note that in deriving the constraints (Eq. 9.7.3), no model between x and y has been assumed. To determine the modified weights w_i's, we define the distance function $D(w, d)$, which satisfies the following conditions.

For every fixed $d > 0$, (i) $D(w, d) \geq 0$, differentiable with respect to w; (ii) strictly convex on an interval containing d, and such that $D(d, d) = 0$; and (iii) $D'(w, d) = \dfrac{\partial D(w, d)}{\partial w}$ is continuous.

Deville and Särndal (1992) proposed the following distance functions: (i) $D(w, d) = (w_i - d_i)^2/(d_i)$, (ii) $D(w, d) = w_i \log(w_i/d_i) - w_i + d_i$, (iii) $D(w, d) = 2\left(\sqrt{w_i} - \sqrt{d_i}\right)^2$, (iv) $-d_i \log(w_i/d_i) + w_i - d_i$, and (v) $D(w, d) = (w_i - d_i)^2/(2w_i)$

The most popular is the chi-square type distance function

$$D(w_i, d_i) = \frac{(w_i - d_i)^2}{d_i} \tag{9.7.4}$$

Here we minimize the average chi-square distance

$$\sum_{i \in s} \frac{(w_i - d_i)^2}{d_i q_i} \text{ subject to } \sum_{i \in s} w_i x_i = X \text{ for every } s \tag{9.7.5}$$

where $1/q_i$ is a known positive weight.

For minimization consider

$$\phi = \frac{1}{2} \sum_{i \in s} \frac{(w_i - d_i)^2}{d_i q_i} - \lambda \left(\sum_{i \in s} w_i x_i - X \right) \tag{9.7.6}$$

with λ as an undetermined Lagrange multiplier.

$$\begin{aligned} &\frac{\partial \phi}{\partial w_i} = 0 \text{ implies} \\ &w_i = d_i + \lambda d_i q_i x_i \end{aligned} \tag{9.7.7}$$

On eliminating w_i from Eqs. (9.7.3) and (9.7.7), we get

$$\lambda = \frac{\left(X - \sum_{i \in s} d_i x_i\right)}{\sum_{i \in s} d_i q_i x_i^2} \tag{9.7.8}$$

Eqs. (9.7.7) and (9.7.8) yield

$$w_i = d_i + d_i q_i x_i \frac{\left(X - \sum_{i \in s} d_i x_i\right)}{\sum_{i \in s} d_i q_i x_i^2} \tag{9.7.9}$$

Finally substituting Eq. (9.7.9) in Eq. (9.7.2), we get the calibrated estimator as

$$\begin{aligned} \widehat{Y}_c &= \sum_{i \in s} d_i y_i - B_s \left(\sum_{i \in s} d_i x_i - X \right) \\ &= \widehat{Y}_{ht} - B_s \left(\widehat{X}_{ht} - X \right) \end{aligned} \tag{9.7.10}$$

where

$$\widehat{X}_{ht} = \sum_{i \in s} d_i x_i \text{ and } B_s = \frac{\sum_{i \in s} d_i q_i x_i y_i}{\sum_{i \in s} d_i q_i x_i^2} \tag{9.7.11}$$

The resulting calibrated estimator (Eq. 9.7.10) is known as generalized regression estimator (Cassel et al., 1976 and Särndal, 1982).

Remark 9.7.1

The calibrated weights w_i may take negative values. Distance functions (ii), (iii), (iv) and (v) guarantee positive weights. Details of properties are given by Deville and Särndal (1992), Singh and Mohl (1996), and Anderson and Thorburn (2005), among others.

Example 9.7.1: Ratio Estimator

If we choose $q_i = 1/x_i$, then we get $B_s = \dfrac{\sum_{i \in s} d_i y_i}{\sum_{i \in s} d_i x_i}$ and the estimator $\widehat{Y}_c$ in Eq. (9.7.10) reduces to

$$\widehat{Y}_c = \frac{\widehat{Y}_{ht}}{\widehat{X}_{ht}} X = \widehat{Y}_R \tag{9.7.12}$$

the well known ratio estimator.

Example 9.7.2

If we choice of $q_i = 1$ and $d_i = N/n$, then the estimator (Eq. 9.7.10) reduces to

$$\widehat{Y}_c = N\left[\overline{y}(s) - \frac{\sum_{i \in s} y_i x_i}{\sum_{i \in s} x_i^2}\left(\overline{x}(s) - \overline{X}\right)\right] \tag{9.7.13}$$

The estimator (Eq. 9.7.13) is quite different from the regression estimator given in Eq. (9.7.12).

9.7.1 Efficiency of Calibrated Estimator

The calibrated estimator $\widehat{Y}_c$ is asymptotically design unbiased. The asymptotic variance (AV) of $\widehat{Y}_c$ given by Deville and Särndal (1992) is

$$AV\left(\widehat{Y}_c\right) = \sum_{i \neq} \sum_{j \in U} \Delta_{ij}\left(d_i E_i - d_j E_j\right)^2 \tag{9.7.14}$$

where $\Delta_{ij} = \pi_i \pi_j - \pi_{ij}$, $E_i = y_i - B x_i$, and $B = \sum_{i \in U} x_i y_i q_i \big/ \sum_{i \in U} x_i^2 q_i$.

From the expression (Eq. 9.7.14), it is evident that the estimator $\widehat{Y}_c$ is more efficient than $\widehat{Y}_{ht}$ in most situations since $AV(\widehat{Y}_c)$ is expected to be smaller than $V(\widehat{Y}_{ht})$.

A design-consistent estimator of $AV(\widehat{Y}_c)$ is given by

$$\widehat{V}(\widehat{Y}_c) = \sum_{i\neq}\sum_{j\in s}\frac{\Delta_{ij}}{\pi_{ij}}\left(w_i\widehat{E}_i - w_j\widehat{E}_j\right)^2 \tag{9.7.15}$$

where $\widehat{E}_i = y_i - \widehat{B}_s x_i$ and $\widehat{B}_s = \left(\sum_{i\in s} w_i q_i x_i y_i\right) \Big/ \left(\sum_{i\in s} w_i q_i x_i^2\right)$.

9.7.2 Calibration Estimator for Several Auxiliary Variables

Suppose we have p auxiliary variables $x_1,\ldots, x_p$ with known totals $X_1,\ldots, X_p$, respectively. In this situation the weights w_i are obtained by minimizing

$$\sum_{i\in s}\frac{(w_i - d_i)^2}{d_i q_i}$$

Subject to

$$\sum_{i\in s} w_i x_{1i} = X_1, \ldots, \sum_{i\in s} w_i x_{ji} = X_j, \ldots, \sum_{i\in s} w_i x_{pi} = X_p$$

where x_{ji} be the value of the ith unit of the jth auxiliary variable x_j, $j = 1,\ldots, p$.

For this minimization problem, we consider

$$\phi = \frac{1}{2}\sum_{i\in s}\frac{(w_i - d_i)^2}{d_i q_i} - \lambda_1\left(\sum_{i\in s} w_i x_{1i} - X_1\right) - \cdots - \lambda_j\left(\sum_{i\in s} w_i x_{ji} - X_j\right) - \cdots - \lambda_p\left(\sum_{i\in s} w_i x_{pi} - X_p\right)$$

with $\lambda_1,\ldots, \lambda_p$ as Lagrange multipliers.

$\dfrac{\partial\phi}{\partial w_i} = 0$ implies

$$w_i = d_i + \lambda_1 d_i q_i x_{1i} + \cdots + \lambda_j d_i q_i x_{ji} + \cdots + \lambda_p d_i q_i x_{pi} \quad \text{for } i = 1, \ldots, n \tag{9.7.16}$$

On multiplying both sides of Eq. (9.7.16) by x_{ki} then summing over $i \in s$ and writing $\sum_{i \in s} w_i x_{ki} = X_k$, for $k = 1, \ldots, p$, we get estimated value of $\boldsymbol{\lambda}' = (\lambda_1, \ldots, \lambda_p)$ as

$$\widehat{\boldsymbol{\lambda}} = \mathbf{A}_s^{-1}\mathbf{T}_s \tag{9.7.17}$$

where

$$\mathbf{T}_s = \begin{pmatrix} X_1 - \sum_{i \in s} d_i x_{1i} \\ \cdots \\ X_j - \sum_{i \in s} d_i x_{ji} \\ \cdots \\ X_p - \sum_{i \in s} d_i x_{pi} \end{pmatrix},$$

$$\mathbf{A}_s = \begin{pmatrix} \sum_{i \in s} d_i q_i x_{1i}^2 & \cdots & \sum_{i \in s} d_i q_i x_{ji} x_{1i} & \cdots & \sum_{i \in s} d_i q_i x_{pi} x_{1i} \\ \cdots & \cdots & \cdots & \cdots & \cdots \\ \sum_{i \in s} d_i q_i x_{1i} x_{ji} & \cdots & \sum_{i \in s} d_i q_i x_{ji}^2 & \cdots & \sum_{i \in s} d_i q_i x_{pi} x_{ji} \\ \cdots & \cdots & \cdots & \cdots & \cdots \\ \sum_{i \in s} d_i q_i x_{1i} x_{pi} & \cdots & \sum_{i \in s} d_i q_i x_{pi} x_{ji} & \cdots & \sum_{i \in s} d_i q_i x_{pi}^2 \end{pmatrix} \text{ and}$$

$$\widehat{\boldsymbol{\lambda}} = \begin{pmatrix} \widehat{\lambda}_1 \\ \cdots \\ \widehat{\lambda}_j \\ \cdots \\ \widehat{\lambda}_p \end{pmatrix}$$

Eq. (9.7.17) yields

$$\begin{aligned} \widehat{Y}_c &= \sum_{i \in s} w_i y_i \\ &= \widehat{Y}_{ht} + \widehat{\lambda}_1 \sum_{i \in s} d_i q_i x_{1i} y_i + \cdots + \widehat{\lambda}_j \sum_{i \in s} d_i q_i x_{ji} y_i + \cdots + \widehat{\lambda}_p \sum_{i \in s} d_i q_i x_{pi} \\ &= \widehat{Y}_{ht} - B_{1s}\left(\widehat{X}_{ht1} - X_1\right) - \cdots - B_{js}\left(\widehat{X}_{htj} - X_j\right) - \cdots B_{ps}\left(\widehat{X}_{htp} - X_p\right) \end{aligned} \tag{9.7.18}$$

where $\widehat{X}_{htj} = \sum_{i \in s} d_i x_{ji}$, $\mathbf{B}'_{\mathbf{s}}\mathbf{A}_s = \left(\sum_{i \in s} d_i q_i x_{1i} y_i, \ldots, \sum_{i \in s} d_i q_i x_{ji} y_i, \ldots, \sum_{i \in s} d_i q_i x_{pi} y_i\right)$, and $\mathbf{B}'_{\mathbf{s}} = (B_{1s}, \ldots, B_{js}, \ldots, B_{ps})$.

Following Deville and Särndal (1992), the a symptotic variance (AV) of $\widehat{Y}_c$ and its approximate unbiased estimators are obtained as

$$AV(\widehat{Y}_c) = \sum_{i \neq} \sum_{j \in U} \Delta_{ij} \left(d_i E_i^* - d_j E_j^*\right)^2 \tag{9.7.19}$$

and

$$\widehat{V}(\widehat{Y}_c) = \sum_{i \neq} \sum_{j \in s} \frac{\Delta_{ij}}{\pi_{ij}} \left(d_i \widehat{E}_i^* - d_j \widehat{E}_j^*\right)^2 \tag{9.7.20}$$

where $E_i^* = y_i - B_1 x_{1i} - \ldots - B_j x_{ji} - \ldots - B_p x_{pi}$, $\widehat{E}_i^* = y_i - B_{1s} x_{1i} - \ldots - B_{js} x_{ji} - \ldots - B_{ps} x_{pi}$, and

$$\begin{pmatrix} B_1 \\ \ldots \\ B_j \\ \ldots \\ B_p \end{pmatrix} = \begin{pmatrix} \sum_{i \in U} q_i x_{1i}^2 & \ldots & \sum_{i \in U} q_i x_{ji} x_{1i} & \ldots & \sum_{i \in U} q_i x_{pi} x_{1i} \\ \ldots & \ldots & \ldots & \ldots & \ldots \\ \sum_{i \in U} q_i x_{1i} x_{ji} & \ldots & \sum_{i \in U} q_i x_{ji}^2 & \ldots & \sum_{i \in U} q_i x_{pi} x_{ji} \\ \ldots & \ldots & \ldots & \ldots & \ldots \\ \sum_{i \in U} q_i x_{1i} x_{pi} & \ldots & \sum_{i \in U} q_i x_{pi} x_{ji} & \ldots & \sum_{i \in U} q_i x_{pi}^2 \end{pmatrix}^{-1} \begin{pmatrix} \sum_{i \in U} q_i x_{1i} y_i \\ \ldots \\ \sum_{i \in U} q_i x_{ji} y_i \\ \ldots \\ \sum_{i \in U} q_i x_{pi} y_i \end{pmatrix}$$

The variance estimator (Eq. 9.7.20) provides strictly design-consistent variance estimator. When design-based as well as model-based properties are considered, the following alternative variance estimator is proposed.

$$\widehat{V}_w(\widehat{Y}_c) = \sum_{i \neq} \sum_{j \in s} \frac{\Delta_{ij}}{\pi_{ij}} \left(w_i \widehat{E}_i^* - w_j \widehat{E}_j^*\right)^2$$

9.8 EXERCISES

9.8.1 Describe the regression methods of estimation. Derive approximate expressions of bias and mean square errors of the regression estimator of the population mean. Show that the regression estimator is more efficient than the ratio estimator for estimating the population mean.

9.8.2 Consider the sampling scheme where the first unit is selected with probability proportional to $\left(x_i - \overline{X}\right)^2$ and the rest $n - 1$ units are selected from the remaining $(N-1)$ units by SRSWOR method. Show that for this sampling scheme

(i) probability of selection of an unordered samples of n units is

$$p(s) = \frac{\sum_{i\in s}(x_i-\overline{X})^2}{\binom{N-1}{n-1}\sum_{i\in U}(x_i-\overline{X})^2} \text{ and}$$

(ii) $\frac{n(N-1)}{N(n-1)}\left[\overline{y}(s)-\widehat{\beta}^*\{\overline{x}(s)-\overline{X}\}\right]$ is an unbiased estimator $\overline{Y}$

where $\widehat{\beta}^* = \frac{\sum_{i\in s} y_i(x_i-\overline{X})}{\sum_{i\in s}(x_i-\overline{X})^2}$ (Singh and Srivastava, 1980).

9.8.3 Let s be sample selected from a population by SRSWR method. Show that the regression estimator is more efficient than the ratio estimator under the following superpopulation model where x is used as an auxiliary variable.

$$E_m(y|x_i) = \alpha + \beta x_i,\ V_m(y_i|x_i) = \sigma^2 x_i^g \text{ and } Cov(y_i, y_j|x_i, x_j) = 0 \quad \text{for } i \neq j$$

with σ^2, $g > 0$ (Raj, 1958).

9.8.4 To determine the average height of 50 plants in a certain field, a sample of 10 plants was selected by SRSWOR method. The following table gives the exact height and eye estimate heights.

Serial number of plants	Eye estimate height (in m)	Exact height (in m)
1	15	16.8
2	10	12.2
3	20	18.5
4	8	9.0
5	5	4.8
6	10	10.5
7	15	16.2
8	10	10.0
9	15	16.3
10	14	12.2
11	12	12.0
12	8	9.5
13	12	12.8
14	9	8.5
15	6	5.8

Given that the average eye estimate of heights for the 50 plants is 16.5 m.

(i) Estimate the average height of the plant by ratio, regression, and product method of estimation.

(ii) Estimate the biases of the estimators used in (i).

(iii) Estimate the relative efficiencies of the above estimators with respect to the sample mean.

9.8.5 A sample of 10 households is selected from a population of 50 households by PPSWR method of sampling using household size as a size variable. The information of household income, household expenditure, and amount of income tax paid were obtained and presented in the following table.

Sampled households	1	2	3	4	5	6	7	8	9	10
Size	5	4	3	2	4	5	6	4	4	3
Expenditure on food ($)	2000	2500	1500	1000	2500	3000	4000	2000	1500	1200
Income ($)	5500	4500	5000	2500	7500	6000	8500	6000	2500	5000
Income tax paid ($)	1000	850	950	400	1400	1100	1450	1050	400	950

From the past survey it is known that the average family size and average income of the population are 4.2 and $5000, respectively. The total tax collected from the 50 households was $45,000. Use this information estimate the following:

(i) Average expenditure on food by using the regression estimator taking tax paid and income as auxiliary variables.

(ii) Give two separate estimates of the average expenditure of food by using the regression method using income and income tax paid as auxiliary variables. Give a pooled estimate of these two regression estimates.

(iii) Estimate relative efficiency of the regression estimators used in "i" with respect to the pooled estimator used in "ii."

9.8.6 Using the data given in Exercise 9.8.5: (i) Calibrate the Hansen–Hurwitz estimator for the population mean consumption on food using total tax collection as $45,000. Estimate the standard error of the calibrated estimator you used. (ii) Obtain the calibrated estimator for the average consumption on food using average income and total tax collection, which were \$5000 and \$45,000, respectively. (iii) Compare precision of the calibrate estimator used in "i" and "ii."

9.8.7 To estimate the average production of milk per farm, the farms were classified into three strata: small, medium, and big with respect to cattle population. From each of the strata, samples were selected by SRSWOR method and information on production of milk and the number of cattle was collected.

			Sample				
Strata	Number of farms	Average number of cattle	Size	Mean no. of cattle	Mean yield of milk (kg)	SD of no. of cattle	SD of yield of milk (kg)
Small	550	10	50	12	22,000	15	200
Medium	400	25	30	28	15,000	25	325
Large	250	45	20	50	10,000	10	250

(i) Estimate the average production of milk per farm using (a) separate and (b) combined regression methods taking cattle population as an auxiliary variable.

(ii) Estimate the relative efficiency of the separate regression estimation with respect to the combined regression estimator and comment on your findings.

APPENDIX 9A

$$Cov\left(\overline{y}_s, s_x^2 \middle| \text{SRSWOR}\right) = Cov\left\{s_{xy}, \left(\overline{x}_s - \overline{X}\right)\right\} = \frac{1}{n}\frac{N(N-n)}{(N-1)(N-2)}\mu_{21}$$

Proof:

$$Cov\left(\overline{y}_s, s_x^2 \middle| \text{SRSWOR}\right)$$

$$= E\left[\overline{y}_s - \overline{Y}, s_x^2\right]$$
$$= A_0 \sum_s \left[\sum_{i \in s} h_i \left\{\sum_{i \in s} z_i^2 - \left(\sum_{i \in s} z_i\right)^2 \Big/ n\right\}\right], z_i = x_i - \overline{X}$$

$$\left(\text{where } A_0 = \left[n(n-1)\binom{N}{n}\right]^{-1}\right)$$

$$= \frac{A_0}{n}\left[\sum_s \left\{(n-1)\sum_{i\in s} h_i \sum_{i\in s} z_i^2 - \left(\sum_{i\in s} h_i\right)\left(\sum_{i\neq}\sum_{j\in s} z_i z_j\right)\right\}\right]$$

$$= \frac{A_0}{n}\left[\sum_s (n-1)\left\{\sum_{i\in s} h_i z_i^2 + \sum_{i\neq}\sum_{j\in s} h_i z_j^2\right\} - \left\{2\sum_{i\neq}\sum_{j\in s} h_i z_i z_j + \sum_{i\neq}\sum_{j\neq}\sum_{k\in s} h_i z_j z_k\right\}\right]$$

$$= \frac{A_0}{n}\left[(n-1)\left\{M_1 \sum_{i\in U} h_i z_i^2 + M_2 \sum_{i\neq}\sum_{j\in U} h_i z_j^2\right\} - \left\{2M_2 \sum_{i\neq}\sum_{j\in U} h_i z_i z_j + M_3 \sum_{i\neq}\sum_{j\neq}\sum_{k\in U} h_i z_j z_k\right\}\right];\ M_i = \binom{N-i}{n-i}$$

$$= \frac{A_0}{n}\left[(n-1)(M_1 - M_2)\sum_{i\in U} h_i z_i^2 + 2\left\{M_2 \sum_{i\in U} h_i z_i^2 - M_3 \sum_{i\in U} z_i^2 h_i\right\}\right]$$

$$= \frac{A_0}{n}\left[\left\{(n-1)(M_1 - M_2) + 2(M_2 - M_3)\right\}\right]\sum_{i\in U} h_i z_i^2$$

$$= \frac{A_0}{n}\left[\left\{(n-1)M_1(1 - M_2/M_1) + 2M_2(1 - M_3/M_2)\right\}\right]\sum_{i\in U} h_i z_i^2$$

$$= \frac{A_0}{n} M_1 \left[(n-1)\frac{N-n}{N-1} + 2\frac{(n-1)}{N-1}\frac{N-n}{N-2}\right]\sum_{i\in U} h_i z_i^2$$

$$= \frac{1}{n^2}\frac{n}{N}\frac{(N-n)}{N-1}\frac{N}{N-2}\sum_{i\in U} h_i z_i^2$$

$$= \frac{1}{n}\frac{(N-n)}{N-1}\frac{1}{N-2}\sum_{i\in U} h_i z_i^2$$

$$= \frac{1}{n}\frac{(N-n)}{N-1}\frac{1}{N-2}\sum_{i\in U}(y_i - \overline{Y})(x_i - \overline{X})^2$$

$$= \frac{1}{n}\frac{N(N-n)}{(N-1)(N-2)}\mu_{21} \tag{A1}$$

Similarly,

$$
\begin{aligned}
&Cov\left\{s_{xy}, \left(\overline{x}_s - \overline{X}\right)\right\} \\
&= E\left[\left\{\frac{1}{n-1}\sum_{i\in s}\left(y_i - \overline{Y}\right)\left(x_i - \overline{x}_s\right)\right\}\left\{\frac{1}{n}\sum_{i\in s}\left(x_i - \overline{X}\right)\right\}\right] \\
&= E\left[\frac{1}{(n-1)n}\sum_{i\in s} h_i\left(z_i - \overline{z}_s\right)\sum_{i\in s} z_i\right] \\
&= \frac{1}{(n-1)n^2 M_o}\sum_{s}\left[\left\{(n-1)\sum_{i\in s} h_i z_i - \sum_{i\neq}\sum_{j\in s} h_i z_j\right\}\sum_{i\in s} z_i\right] \\
&= \frac{1}{(n-1)n^2 M_o}\left[(n-1)(M_1 - M_2) + 2(M_2 - M_3)\right]\sum_{i\in U} h_i z_i^2 \\
&= \frac{1}{n}\frac{N(N-n)}{(N-1)(N-2)}\mu_{21} \qquad \text{(A2)}
\end{aligned}
$$

From (A1) and (A2), we verify the result.

CHAPTER 10

Two-Phase Sampling

10.1 INTRODUCTION

It has been explained in Chapters 5, 8, and 9 how one can use auxiliary information at the stages of estimation of parameters, selection of samples, stratification, and their combinations to increase the precision of the estimators. To use auxiliary information for estimation of the population characteristics through ratio and regression methods, one needs to know only the population total of the auxiliary variable. For using the auxiliary variable as a measure of size for the selection of sample, one needs to know the values of the auxiliary variable at unit level. For the purpose of stratification using an auxiliary variable, one needs to know the values of the auxiliary variable for a large number of units so that after formation of the strata, each unit of the population can be assigned to one of the strata. In some situations the auxiliary information is not available in advance but can be made available at little extra cost. For example, in household income and expenditure surveys, one can easily collect information regarding the household size at the time of listing the households by putting little or no additional cost.

Two-phase sampling, also known as double sampling, was introduced by Neyman (1938). In two-phase sampling, samples are selected in phases. In the first phase, a relatively large sample s' of size n' is selected by some sampling design p', and only information on the auxiliary variable is collected. During the second phase, a sample s of size n is selected either from s', the sample selected in the first phase, or from the entire population by using a suitable sampling design p, and information regarding the study and auxiliary variable is collected. Evidently, two-phase sampling is useful if the auxiliary information x is relatively easy and cheaper to collect than the study variable y as well as if the strength of the relationship between the variables x and y is high. If the sample is selected in more than two phases, the resulting sampling design is called a multiphase sampling. In this chapter we will consider the use of the first-phase sample at the stages of estimation of the parameter, stratification, selection of sample, and their combinations in the second phase.

Survey Sampling Theory and Applications
ISBN 978-0-12-811848-1
http://dx.doi.org/10.1016/B978-0-12-811848-1.00010-8

10.2 TWO-PHASE SAMPLING FOR ESTIMATION

In the first phase, a sample s' of size n' be selected from a population with probability $p(s')$ using a sampling design p and the information of the auxiliary variable x is obtained only. Let

$$t(s',x) = \sum_{i \in s'} b_{s'i} x_i \tag{10.2.1}$$

be an unbiased estimator for the population total $X\left(=\sum_{i=1}^{N} x_i\right)$ of the auxiliary variable x based on the first-phase sample s' and $b_{s'i}$'s are known constants satisfying the unbiasedness condition

$$\sum_{s' \supset i} b_{s'i} p'(s') = 1 \quad \forall i = 1, \ldots, N \tag{10.2.2}$$

In the second phase, a subsample s of size n is selected from s' and an unbiased estimator of the total $Y\left(=\sum_{i=1}^{N} y_i\right)$ is obtained as

$$t(s,y) = \sum_{i \in s} b_{si} y_i \tag{10.2.3}$$

Here the weights b_{si}'s are suitably chosen to satisfy

$$E\{t(s,y)|s'\} = \sum_{i \in s'} b_{s'i} y_i \tag{10.2.4}$$

The condition (Eq. 10.2.4) requires

$$\sum_{s \supset i} b_{si} p(s) = b_{s'i} \quad \forall i \in s' \tag{10.2.5}$$

10.2.1 Difference Method of Estimation

A difference estimator for the total Y is given by

$$\widehat{Y}_{dt} = t(s,y) - \lambda\{t(s,x) - t(s',x)\} \tag{10.2.6}$$

where $t(s,x) = \sum_{i \in s} b_{si} x_i$ and λ is a suitably chosen constant.

Clearly, $\widehat{Y}_{dt}$ is unbiased for Y for any chosen value of λ because $E\{t(s,y)\} = E[E\{t(s,y)|s'\}] = E\{t(s',y)\} = Y$ and $E\{t(s,x)\} = E\{t(s',x)\} = X$.

The variance of $\widehat{Y}_{dt}$ is given by

$$V\left(\widehat{Y}_{dt}\right) = V\left\{E\left(\widehat{Y}_{dt}|s'\right)\right\} + E\left\{V\left(\widehat{Y}_{dt}|s'\right)\right\}$$
$$= V\{t(s', y)\} + E[V\{t(s, d_\lambda)|s'\}] \tag{10.2.7}$$

where $t(s', y) = \sum_{i\in s'} b_{s'i} y_i$, $t(s, d_\lambda) = \sum_{i\in s} b_{si} d_{\lambda i}$, and $d_{\lambda i} = y_i - \lambda x_i$.

An unbiased estimator of $V\left(\widehat{Y}_{dt}\right)$ is given by

$$\widehat{V}\left(\widehat{Y}_{dt}\right) = \widehat{V}\{t(s', y)\} + \widehat{V}\{t(s, d_\lambda)|s'\} \tag{10.2.8}$$

where $\widehat{V}\{t(s', y)\}$ and $\widehat{V}\{t(s, d_\lambda)|s'\}$ are unbiased estimators of $V\{t(s', y)\}$ and $V\{t(s, d_\lambda)|s'\}$, respectively.

10.2.1.1 Arbitrary Sampling Design

Let the initial sample s' be selected by a varying probability sampling scheme with $\pi_i(>0)$ and $\pi_{ij}(>0)$ as the inclusion probabilities of the ith, and ith and jth units, respectively. The subsample s is selected from s' by simple random sampling without replacement (SRSWOR) method. In this case, if we choose, $t(s', x) = \sum_{i\in s'} \frac{x_i}{\pi_i}$ and $t(s, y) = \frac{n'}{n} \sum_{i\in s} \frac{y_i}{\pi_i}$, we get

$$\widehat{Y}_{dt} = \frac{n'}{n} \sum_{i\in s} \frac{y_i}{\pi_i} - \lambda\left(\frac{n'}{n} \sum_{i\in s} \frac{x_i}{\pi_i} - \sum_{i\in s'} \frac{x_i}{\pi_i}\right) \tag{10.2.9}$$

$$V\left\{E\left(\widehat{Y}_{dt}|s'\right)\right\} = V\{t(s', y)\}$$
$$= V\left(\sum_{i\in s'} \frac{y_i}{\pi_i}\right) \tag{10.2.10}$$
$$= \frac{1}{2} \sum_{i\neq} \sum_{j\in U} (\pi_i\pi_j - \pi_{ij})\left(\frac{y_i}{\pi_i} - \frac{y_j}{\pi_j}\right)^2$$

and

$$E[V\{t(s, d_\lambda)|s'\}] = n' \frac{(n' - n)}{n(n' - 1)} E\left[\sum_{i\in s'} \frac{d_{\lambda i}^2}{\pi_i^2} - n'\left(\frac{1}{n'} \sum_{i\in s'} \frac{d_{\lambda i}}{\pi_i}\right)^2\right]$$

$$= n' \frac{(n' - n)}{n(n' - 1)} \left[\left(\sum_{i\in U} \frac{d_{\lambda i}^2}{\pi_i} - \frac{D_\lambda^2}{n'}\right) - \frac{1}{n'} \frac{1}{2} \sum_{i\neq} \sum_{j\in U} (\pi_i\pi_j - \pi_{ij})\left(\frac{d_{\lambda i}}{\pi_i} - \frac{d_{\lambda j}}{\pi_j}\right)^2\right] \tag{10.2.11}$$

where $d_{\lambda i} = y_i - \lambda_i x_i$ and $D_\lambda = Y - \lambda X$.

Unbiased estimators of $V\{t(s',y)\}$ and $V(t(s,d_\lambda)|s')$ are, respectively, given by

$$\widehat{V}\{t(s',y)\} = \frac{n'}{n}\frac{1}{2}\sum_{i\neq}\sum_{j\in s}\left(\frac{\pi_i\pi_j - \pi_{ij}}{\pi_{ij}}\right)\left(\frac{y_i}{\pi_i} - \frac{y_j}{\pi_j}\right)^2 \tag{10.2.12}$$

and

$$\widehat{V}\{(t(s,d_\lambda)|s')\} = n'\frac{(n'-n)}{n(n-1)}\left[\sum_{i\in s}\frac{d_{\lambda i}^2}{\pi_i^2} - \frac{1}{n}\left(\sum_{i\in s}\frac{d_{\lambda i}}{\pi_i}\right)^2\right] \tag{10.2.13}$$

The expressions from Eq. (10.2.10) to Eq. (10.2.13) yield the variance of $\widehat{Y}_{dt}$ and its unbiased estimators as follows:

$$V\left(\widehat{Y}_{dt}\right) = \frac{1}{2}\sum_{i\neq}\sum_{j\in U}(\pi_i\pi_j - \pi_{ij})\left(\frac{y_i}{\pi_i} - \frac{y_j}{\pi_j}\right)^2 + n'\frac{(n'-n)}{n(n'-1)}$$
$$\times\left[\left(\sum_{i\in U}\frac{d_{\lambda i}^2}{\pi_i} - \frac{D_\lambda^2}{n'}\right) - \frac{1}{n'}\left\{\frac{1}{2}\sum_{i\neq}\sum_{j\in U}(\pi_i\pi_j - \pi_{ij})\left(\frac{d_{\lambda i}}{\pi_i} - \frac{d_{\lambda j}}{\pi_j}\right)^2\right\}\right] \tag{10.2.14}$$

and

$$\widehat{V}\left(\widehat{Y}_{dt}\right) = \frac{n'}{n}\frac{1}{2}\sum_{i\neq}\sum_{j\in s}\left(\frac{\pi_i\pi_j - \pi_{ij}}{\pi_{ij}}\right)\left(\frac{y_i}{\pi_i} - \frac{y_j}{\pi_j}\right)^2$$
$$+ n'\frac{(n'-n)}{n(n-1)}\sum_{i\in s}\left(\frac{d_{\lambda i}}{\pi_i} - \frac{1}{n}\sum_{i\in s}\frac{d_{\lambda i}}{\pi_i}\right)^2 \tag{10.2.15}$$

10.2.1.2 Simple Random Sampling Without Replacement

In case, both the samples s' and s are selected by SRSWOR method, we get $\pi_i = n'/N$ and $\pi_{ij} = n'(n'-1)/\{N(N-1)\}$. In this situation the expressions (Eqs. 10.2.9, 10.2.14, and 10.2.15) reduce to

$$\widehat{Y}_{dt} = N\left[\overline{y}(s) - \lambda(\overline{x}(s) - \overline{x}(s'))\right] \tag{10.2.16}$$

$$V\left(\widehat{Y}_{dt}\right) = N^2\left[\left(\frac{N-n'}{Nn'}\right)S_y^2 + \frac{(n'-n)}{nn'}S_{d\lambda}^2\right] \tag{10.2.17}$$

and

$$V\left(\widehat{Y}_{dt}\right) = N^2\left[\left(\frac{N-n'}{Nn'}\right)s_y^2 + \frac{(n'-n)}{nn'}s_{d\lambda}^2\right] \tag{10.2.18}$$

where $\overline{x}(s') = \sum_{i\in s'} x_i/n'$, $\overline{x}(s) = \sum_{i\in s} x_i/n$, $\overline{y}(s) = \sum_{i\in s} y_i/n$,

$S_y^2 = \sum_{i\in U} \left(y_i - \overline{Y}\right)^2/(N-1)$, $S_{d\lambda}^2 = \sum_{i\in U} \left(d_{\lambda i} - \overline{D}_\lambda\right)^2/(N-1)$,

$s_y^2 = \sum_{i\in s} \left(y_i - \overline{y}_s\right)^2/(n-1)$, $s_{d\lambda}^2 = \sum_{i\in s} \left(d_{\lambda i} - \overline{d}_{\lambda s}\right)^2/(n-1)$,

$\overline{Y} = Y/N$, $\overline{D}_\lambda = D_\lambda/N$, and $\overline{d}_{\lambda s} = \sum_{i\in s} d_{\lambda i}/n$.

10.2.1.3 Efficiency Under Simple Random Sampling Without Replacement

Let us consider the following simple cost function

$$C = c'n' + cn \tag{10.2.19}$$

where C is the total cost of the survey excluding the fixed cost, c' and c denote, respectively, the cost for collecting information on x and y from a unit.

The cost c is expected to be much higher than c'. If information regarding the study variable y is collected only through direct survey (without using two-phase sampling) keeping C as C_0, then the sample size for the direct survey would be

$$n_0 = C_0/c = (c'/c)n' + n \tag{10.2.20}$$

The variance of the sample mean $\overline{y}_{dr} = \sum_{i\in s_0} y_i/n_0$ based on a sample s_0 of size n_0 selected by SRSWOR method is given by

$$V\left(\overline{y}_{dr}\right) = \left(\frac{1}{n_0} - \frac{1}{N}\right)S_y^2$$

Two-phase sampling will be more precise than direct sampling if

$$V\left(\widehat{Y}_{dt}\right) < V\left(N\overline{y}_{dr}\right)$$

i.e., if $N^2\left[\left(\frac{N-n'}{Nn'}\right)S_y^2 + \frac{(n'-n)}{nn'}\left(S_y^2 + \lambda^2 S_x^2 - 2\lambda\rho S_x S_y\right)\right] < N^2\left(\frac{1}{n_0} - \frac{1}{N}\right)S_y^2$

i.e., if $\rho > \frac{1}{2}\left[h + \left\{h\left(1 - \frac{n}{n'}\right)\left(1 + \frac{nc}{n'c'}\right)\right\}^{-1}\right]$

(10.2.21)

where ρ is the correlation coefficient between x and y and $h = \lambda S_x/S_y$.

10.2.1.4 Probability Proportional to Size With Replacement Sampling

Let the initial sample s' be selected by probability proportional to size with replacement (PPSWR) method using z_i as a measure of size for the ith unit and the second phase sample s be selected from s' by SRSWOR method, treating all the units in s' are distinct. In this case

$$\widehat{Y}_{dt} = \frac{1}{n}\sum_{i\in s}\frac{y_i}{q_i} - \lambda\left(\frac{1}{n}\sum_{i\in s}\frac{x_i}{q_i} - \frac{1}{n'}\sum_{i\in s'}\frac{y_i}{q_i}\right) \tag{10.2.22}$$

where $q_i = z_i/Z$ and $Z = \sum_{i\in U} z_i$ and $\sum_{i\in s}$ denotes the sum over the units in s with repetition. Clearly, $\widehat{Y}_{dt}$ in Eq. (10.2.22) is unbiased for Y and

$$\begin{aligned} V\left\{E\left(\widehat{Y}_{dt}|s'\right)\right\} &= V\left(\frac{1}{n'}\sum_{i\in s'}\frac{y_i}{q_i}\right) \\ &= \frac{1}{n'}\sum_{i\in U} q_i\left(\frac{y_i}{q_i} - Y\right)^2; \end{aligned} \tag{10.2.23}$$

$$\begin{aligned} E\left\{V\left(\widehat{Y}_{dt}|s'\right)\right\} &= \left(\frac{1}{n} - \frac{1}{n'}\right)\frac{1}{n'-1}E\sum_{i\in s'}\left\{\frac{d_{\lambda i}}{q_i} - \frac{1}{n'}\left(\sum_{i\in s'}\frac{d_{\lambda i}}{q_i}\right)\right\}^2 \\ &= \left(\frac{1}{n} - \frac{1}{n'}\right)\sum_{i\in U} q_i\left(\frac{d_{\lambda i}}{q_i} - D_\lambda\right)^2 \end{aligned} \tag{10.2.24}$$

Eqs. (10.2.23) and (10.2.24) yield

$$V\left(\widehat{Y}_{dt}\right) = \frac{1}{n'}\sum_{i\in U} q_i\left(\frac{y_i}{q_i} - Y\right)^2 + \left(\frac{1}{n} - \frac{1}{n'}\right)\sum_{i\in U} q_i\left(\frac{d_{\lambda i}}{q_i} - D_\lambda\right)^2 \tag{10.2.25}$$

Unbiased estimators of $V\left\{E\left(\widehat{Y}_{dt}|s'\right)\right\}$ and $E\left\{V\left(\widehat{Y}_{dt}|s'\right)\right\}$ are given, respectively, by

$$\widehat{V}\left\{E\left(\widehat{Y}_{dt}|s'\right)\right\} = \frac{1}{n'(n-1)}\sum_{i\in s}\left(\frac{y_i}{q_i} - \frac{1}{n}\sum_{i\in s}\frac{y_i}{q_i}\right)^2$$

and

$$\widehat{V}\left(\widehat{Y}_{dt}|s'\right) = \left(\frac{1}{n} - \frac{1}{n'}\right)\frac{1}{n-1}\sum_{i\in s}\left(\frac{d_{\lambda i}}{q_i} - \frac{1}{n}\sum_{i\in s}\frac{d_{\lambda i}}{q_i}\right)^2$$

Hence an unbiased estimator for $V(\widehat{Y}_{dt})$ is

$$\widehat{V}(\widehat{Y}_{dt}) = \frac{1}{n'(n-1)}\sum_{i\in s}\left(\frac{y_i}{q_i} - \frac{1}{n}\sum_{i\in s}\frac{y_i}{q_i}\right)^2 + \left(\frac{1}{n} - \frac{1}{n'}\right)\frac{1}{n-1}\sum_{i\in s}\left(\frac{d_{\lambda i}}{q_i} - \frac{1}{n}\sum_{i\in s}\frac{d_{\lambda i}}{q_i}\right)^2 \qquad (10.2.26)$$

Under the cost function, (Eq. 10.2.19) two-phase sampling will be more precise than the single phase using the same cost if

$$\frac{1}{n'}\sum_{i\in U} q_i\left(\frac{y_i}{q_i} - Y\right)^2 + \left(\frac{1}{n} - \frac{1}{n'}\right)\sum_{i\in U} q_i\left(\frac{d_{\lambda i}}{q_i} - D_\lambda\right)^2 < \frac{1}{n_0}\sum_{i\in U} q_i\left(\frac{y_i}{q_i} - Y\right)^2$$

$$\text{i.e., if } \delta > \frac{1}{2}\left[h^* + \left\{h^*\left(1 - \frac{n}{n'}\right)\left(1 + \frac{nc}{n'c'}\right)\right\}^{-1}\right] \qquad (10.2.27)$$

where

$$\delta = \frac{\sum_{i\in U} q_i\left(\frac{y_i}{q_i} - Y\right)\left(\frac{x_i}{q_i} - X\right)}{\sqrt{\sum_{i\in U} q_i\left(\frac{y_i}{q_i} - Y\right)^2}\sqrt{\sum_{i\in U} q_i\left(\frac{x_i}{q_i} - X\right)^2}}$$

and

$$h^* = \lambda\sqrt{\sum_{i\in U} q_i\left(\frac{x_i}{q_i} - X\right)^2} \Big/ \sqrt{\sum_{i\in U} q_i\left(\frac{y_i}{q_i} - Y\right)^2}$$

.

10.2.2 Ratio Method of Estimation

The ratio estimator for the population total in two-phase sampling is given by

$$\widehat{Y}_{Rt} = \frac{t(s,y)}{t(s,x)}t(s',x) \qquad (10.2.28)$$

Like the ordinary ratio estimator $\widehat{Y}_R$ described in Eq. (8.3.1), $\widehat{Y}_{Rt}$ is a biased estimator for Y.

Now writing,

$$Cov\left\{\frac{t(s,y)}{t(s,x)}, t(s,x)|s'\right\} = E(t(s,y)|s') - \left[E\left\{\frac{t(s,y)}{t(s,x)}|s'\right\}\right]E(t(s,x)|s')$$
$$= E(t(s,y)|s') - \left[E\left\{\frac{t(s,y)}{t(s,x)}|s'\right\}\right]t(s',x) \tag{10.2.29}$$

and taking expectation both sides of Eq. (10.2.29) we get

$$E\left[Cov\left\{\frac{t(s,y)}{t(s,x)}, t(s,x)|s'\right\}\right] = Y - E\left[E\left\{\frac{t(s,y)}{t(s,x)}t(s',x)|s'\right\}\right] \tag{10.2.30}$$
$$= Y - E\left(\widehat{Y}_{Rt}\right)$$

Thus we have

Theorem 10.2.1

The bias of $\widehat{Y}_{Rt}$ under two-phase sampling is

$$B\left(\widehat{Y}_{Rt}\right) = -E\left[Cov\left\{\frac{t(s,y)}{t(s,x)}, t(s,x)|s'\right\}\right]$$

The above expression of bias is exact. However, it is not useful as it does not provide any elegant expression for any specific sampling design.

10.2.2.1 Approximate Expression of Bias

Let us define

$$e_x = \{t(s,x) - X\}/X, e_y = \{t(s,y) - Y\}/Y, \text{ and } e'_x = \{t(s',x) - X\}/X \tag{10.2.31}$$

Then,

$$\widehat{Y}_{Rt} = Y\frac{(1+e_y)}{(1+e_x)}\left(1+e'_x\right)$$
$$= Y(1+e_y)\left(1-e_x+e_x^2-\cdots\right)\left(1+e'_x\right) \tag{10.2.32}$$
$$= Y\left(1+e_y-e_x+e'_x+e_x^2-e_xe_y+e'_xe_y-e_xe'_x+\cdots\right)$$

The expression of bias of $\widehat{Y}_{Rt}$ is given by

$$\begin{aligned} B(\widehat{Y}_{Rt}) &= E(\widehat{Y}_{Rt} - Y) \\ &= YE(e_y - e_x + e'_x + e_x^2 - e_x e_y + e'_x e_y - e_x e'_x + \cdots) \end{aligned} \tag{10.2.33}$$

On neglecting the terms $E\left(e_x^i e_y^j e_x'^k\right)$ with $i+j+k>2$, assuming it small for large sample size n, an approximate expression of bias of $\widehat{Y}_{Rt}$ up to the first order of approximation comes out as

$$\begin{aligned} B(\widehat{Y}_{Rt}) &= E(\widehat{Y}_{Rt} - Y) \\ &= YE(e_y - e_x + e'_x + e_x^2 - e_x e_y + e'_x e_y - e_x e'_x) \\ &= Y\left[E(e_x^2 - e_x e_y) + E\{e'_x E(e_y|s')\} - e'_x E(e_x|s')\right] \\ &= Y\left[E(e_x^2 - e_x e_y) + E\left(e'_x e'_y - e_x'^2\right)\right] \\ &= Y\left[\frac{V\{t(s,x)\}}{X^2} - \frac{Cov\{t(s,x),t(s,y)\}}{XY} + \frac{Cov\{t(s',y),t(s',x)\}}{XY} - \frac{V\{t(s',x)\}}{X^2}\right] \\ &= Y\left[\frac{V\{t(s,x)\} - V\{t(s',x)\}}{X^2} - \frac{Cov\{t(s,x),t(s,y)\} - Cov\{t(s',y),t(s',x)\}}{XY}\right] \end{aligned} \tag{10.2.34}$$

Now substituting

$$\begin{aligned} V\{t(s,x)\} &= E[V\{t(s,x)|s'\}] + V[E\{t(s,x)|s'\}] \\ &= E[V\{t(s,x)|s'\}] + V\{t(s',x)\} \end{aligned}$$

and

$$\begin{aligned} Cov\{t(s,x),t(s,y)\} &= E[Cov\{t(s,x),t(s,y)|s'\}] + Cov[E\{t(s,x)|s'\},E\{t(s,y)|s'\}] \\ &= E[Cov\{t(s,x),t(s,y)|s'\}] + Cov\{t(s',x),t(s',y)\} \end{aligned}$$

in Eq. (10.2.34), the expression of $B(\widehat{Y}_{Rt})$ reduces to

$$B(\widehat{Y}_{Rt}) = Y\left[\frac{E[V\{t(s,x)|s'\}]}{X^2} - \frac{E[Cov\{t(s,x),t(s,y)|s'\}]}{XY}\right] \tag{10.2.35}$$

10.2.2.2 Approximate Expression of Mean Square Error

The mean square error of $\widehat{Y}_{Rt}$ is given by

$$M\left(\widehat{Y}_{Rt}\right) = E\left(\widehat{Y}_{Rt} - Y\right)^2$$
$$= Y^2 E\left(e_y - e_x + e'_x + e_x^2 - e_x e_y + e'_x e_y - e_x e'_x + \cdots\right)^2$$

Neglecting the terms $E\left(e_x^i e_y^j e_x'^k\right)$ with $i+j+k>2$, an approximate expression of $M\left(\widehat{Y}_{Rt}\right)$ up to the first order of approximation is obtained as

$$M\left(\widehat{Y}_{Rt}\right) = Y^2 E\left(e_y^2 + e_x^2 + e_x'^2 - 2e_y e_x - 2e_x e_x'^2 + 2e_y e_x'^2\right)$$
$$= Y^2 E\left[\left(e_y^2 + e_x^2 - 2e_y e_x\right) + \left\{e_x'^2 - 2E(e_x|s')e_x'^2 + 2E(e_y|s')e_x'^2\right\}\right]$$
$$= Y^2 E\left[\left(e_y^2 + e_x^2 - 2e_y e_x\right) - \left(e_x'^2 - 2e'_y e'_x\right)\right]$$
$$= \left[V\{t(s,y)\} + R^2 V\{t(s,x)\} - 2R\,Cov\{t(s,y), t(s,x)\}\right]$$
$$- \left[R^2 V\{t(s',x)\} - 2R\,Cov\{t(s',y), t(s',x)\}\right]$$
$$= V\left[t(s,y) - R\ t(s,x)\right] - V\left[E\{t(s,y) - R\ t(s,x)|s'\right] + V\{t(s',y)\}$$
$$= V\{t(s',y)\} + E\left[V\{t(s,y) - R\ t(s,x)|s'\}\right]$$
$$= V\{t(s',y)\} + E\left[V\{t(s,d)|s'\}\right] \tag{10.2.36}$$

where $t(s,d) = \sum_{i \in s} b_{si} d_i$ and $d_i = y_i - Rx_i$.

An approximate expression for an unbiased estimator of $M\left(\widehat{Y}_{Rt}\right)$ is given by

$$\widehat{M}\left(\widehat{Y}_{Rt}\right) = \widehat{V}\{t(s',y)\} + \widehat{V}\left\{t\left(s,\widehat{d}\right)|s'\right\} \tag{10.2.37}$$

where $\widehat{V}\{t(s',y)\}$ is an unbiased estimator of $V\{t(s',y)\}$ and $\widehat{V}\left\{t\left(s,\widehat{d}\right)|s'\right\}$ is an approximate unbiased estimator of $V\{t(s,d)|s'\}$, which is obtained by replacing d_i with its estimate $\widehat{d}_i = y_i - \widehat{R}x_i$ in the expression of $\widehat{V}\{t(s,d)|s'\}$, an unbiased estimator of $V\{t(s,d)|s'\}$.

Eqs. (10.2.35)–(10.2.37) yield

Theorem 10.2.2

Approximate expressions for

(i) Bias of $\widehat{Y}_{Rt}$ is

$$B(\widehat{Y}_{Rt}) = Y\left[\frac{E[V\{t(s,x)|s'\}]}{X^2} - \frac{E[Cov\{t(s,x),t(s,y)|s'\}]}{XY}\right]$$

(ii) Mean square error of $\widehat{Y}_{Rt}$ is

$$M(\widehat{Y}_{Rt}) = V\{t(s',y)\} + E[V\{t(s,d)|s'\}]$$

(iii) Estimator of $M(\widehat{Y}_{Rt})$ is

$$\widehat{M}(\widehat{Y}_{Rt}) = \widehat{V}\{t(s',y)\} + \widehat{V}\left\{t\left(s,\widehat{d}\right)|s'\right\}$$

10.2.2.3 Simple Random Sampling Without Replacement

Consider the situation when both the samples s and s' are selected by SRSWOR method of sampling. In this case $t(s,x) = N\ \overline{x}(s)$, $t(s,y) = N\ \overline{y}(s)$, $t(s',x) = N\ \overline{x}(s')$, $t(s',y) = N\ \overline{y}(s')$, $t(s,d) = N\{\overline{y}(s) - R\,\overline{x}(s)\}$, $E[V\{N\ \overline{x}(s)|s'\}] = N^2\left(\frac{1}{n} - \frac{1}{n'}\right)S_x^2$, $V\{N\,\overline{y}(s')\} = N^2\left(\frac{1}{n'} - \frac{1}{N}\right)S_y^2$, $E[Cov\{N\overline{x}(s), N\overline{y}(s)|s'\}] = N^2\left(\frac{1}{n} - \frac{1}{n'}\right)S_{xy}$ with $S_{xy} = \sum_{i\in U}(x_i - \overline{X})(y_i - \overline{Y})/(N-1)$, and $E[V(N\{\overline{y}(s) - R\,\overline{x}(s)\}|s')] = N^2\left(\frac{1}{n} - \frac{1}{n'}\right)\left(S_y^2 + R^2S_x^2 - 2RS_{xy}\right)$.

The expressions of $\widehat{Y}_{Rt}$, $B(\widehat{Y}_{Rt})$, $\widehat{B}(\widehat{Y}_{Rt})$, $M(\widehat{Y}_{Rt})$, and $\widehat{M}(\widehat{Y}_{Rt})$ appear as follows:

$$\widehat{Y}_{Rt} = N\frac{\overline{y}(s)}{\overline{x}(s)}\,\overline{x}(s')$$

$$B(\widehat{Y}_{Rt}) = N\left(\frac{1}{n} - \frac{1}{n'}\right)\overline{Y}\left(\frac{S_x^2}{\overline{X}^2} - \frac{S_{xy}}{\overline{X}\,\overline{Y}}\right)$$

$$= N\left(\frac{1}{n} - \frac{1}{n'}\right)\overline{Y}\left(C_x^2 - \rho\ C_xC_y\right)$$

$$\widehat{B}(\widehat{Y}_{Rt}) = N\left(\frac{1}{n} - \frac{1}{n'}\right)\overline{y}(s)\left(\frac{s_x^2}{\{\overline{x}(s)\}^2} - \frac{s_{xy}}{\overline{x}(s)\overline{y}(s)}\right)$$

$$M\left(\widehat{Y}_{Rt}\right) = N^2\left[\left(\frac{1}{n'} - \frac{1}{N}\right)S_y^2 + \left(\frac{1}{n} - \frac{1}{n'}\right)\left(S_y^2 + R^2 S_x^2 - 2RS_{xy}\right)\right]$$

$$\widehat{M}\left(\widehat{Y}_{Rt}\right) = N^2\left[\left(\frac{1}{n'} - \frac{1}{N}\right)s_y^2 + \left(\frac{1}{n} - \frac{1}{n'}\right)\left(s_y^2 + \widehat{R}^2 s_x^2 - 2\widehat{R}s_{xy}\right)\right]$$

where $s_x^2 = \dfrac{\sum_{i\in s}\{x_i - \overline{x}(s)\}^2}{(n-1)}$, $s_y^2 = \dfrac{\sum_{i\in s}\{y_i - \overline{y}(s)\}^2}{n-1}$,

$s_{xy} = \dfrac{\sum_{i\in s}\{x_i - \overline{x}(s)\}\{y_i - \overline{y}(s)\}}{n-1}$; C_x and C_y are the coefficient of variation of x and y, respectively.

If the total cost of the survey is fixed, the two-phase sampling will be more precise than the single phase sampling under the cost function (Eq. 10.2.19) if

$$N^2\left[\left(\frac{1}{n'} - \frac{1}{N}\right)S_y^2 + \left(\frac{1}{n} - \frac{1}{n'}\right)\left(S_y^2 + R^2 S_x^2 - 2RS_{xy}\right)\right] < N^2\left(\frac{1}{n_0} - \frac{1}{N}\right)S_y^2$$

$$\text{i.e., if } \rho > \frac{1}{2}\left[\overline{h} + \left\{\overline{h}\left(1 - \frac{n}{n'}\right)\left(1 + \frac{nc}{n'c'}\right)\right\}^{-1}\right] \tag{10.2.38}$$

where $\overline{h} = C_x/C_y$.

10.2.2.4 Optimal Allocation Under Simple Random Sampling Without Replacement

Here we will find the optimum values of n and n', which minimize $M\left(\widehat{Y}_{Rt}\right)$ when total cost of the survey $C = c'n' + cn$ is fixed to a certain level C_0. For this minimization problem, consider

$$\phi = M\left(\widehat{Y}_{Rt}\right) - \mu(C_0 - cn - c'n')$$

where μ is an undetermined Lagrange multiplier.

For large N, we can write

$$\phi = N^2\left[\frac{1}{n'}S_y^2 + \left(\frac{1}{n} - \frac{1}{n'}\right)S_d^2\right] - \mu(C_0 - cn - c'n')$$

where $S_d^2 = S_y^2 + R^2 S_x^2 - 2RS_{xy}$.

Now $\dfrac{\partial\phi}{\partial n} = 0$ and $\dfrac{\partial\phi}{\partial n'} = 0$ yield

$$n = NS_d/\sqrt{c\mu} \text{ and } n' = N\sqrt{S_y^2 - S_d^2}\Big/\sqrt{c'\mu} \tag{10.2.39}$$

Substituting n and n' from Eq. (10.2.39) in the constraint $C_0 = c'n' + cn$, the optimum values of n and n' are derived as

$$n_{opt} = \frac{C_0}{S_d\sqrt{c} + \sqrt{\left(S_y^2 - S_d^2\right)c'}} \frac{S_d}{\sqrt{c}} \text{ and}$$
$$n'_{opt} = \frac{C_0}{S_d\sqrt{c} + \sqrt{\left(S_y^2 - S_d^2\right)c'}} \frac{\sqrt{S_y^2 - S_d^2}}{\sqrt{c'}} \tag{10.2.40}$$

Placing the optimum values of n and n' in the expression of $M\left(\widehat{Y}_{Rt}\right)$ and taking $1/N \cong 0$, we arrive at the minimum value of $M\left(\widehat{Y}_{Rt}\right)$ for large N as

$$M_{\min}\left(\widehat{Y}_{Rt}\right) = \frac{N^2\left[S_d\sqrt{c} + \sqrt{\left(S_y^2 - S_d^2\right)c'}\right]^2}{C_0} \tag{10.2.41}$$

10.2.3 Regression Method of Estimation

The regression estimator for the population total for two-phase sampling is

$$\widehat{Y}_{regt} = t(s,y) - B(s)[t(s,x) - t(s',x)] \tag{10.2.42}$$

where $B(s) = \dfrac{\widehat{Cov}\{t(s,y), t(s,x)|s'\}}{\widehat{V}\{t(s,x)|s'\}}$, $\widehat{Cov}\{t(s,y), t(s,x)|s'\}$ is an unbiased estimator of $Cov\{t(s,y), t(s,x) - t(s',x)\} = E[Cov\{t(s,y), t(s,x)|s'\}]$, and $\widehat{V}\{t(s,x)|s'\}$ is an unbiased estimator of $V\{t(s,x) - t(s',x)\} = E[V\{t(s,x)|s'\}]$.

10.2.3.1 Approximate Expressions of Bias and Mean Square Errors

Let $e_B = \dfrac{B(s) - B}{B}$ with $B = \dfrac{Cov\{t(s,y), t(s,x) - t(s',x)\}}{V\{t(s,x) - t(s',x)\}}$.

The bias of $\widehat{Y}_{regt}$ is

$$\begin{aligned} B\left(\widehat{Y}_{regt}\right) &= E\left(\widehat{Y}_{regt}\right) - Y \\ &= E\left[Y(1+e_y) - B(1+e_B)X\left(e_x - e'_x\right)\right] - Y \\ &= -B\ X\ E\left\{(e_B)\left(e_x - e'_x\right)\right\} \\ &= -E\{B(s), t(s,x) - t(s',x)\} \\ &= -E\left[Cov\{B(s), t(s,x)|s'\}\right] \end{aligned} \tag{10.2.43}$$

The mean square error of $\widehat{Y}_{regt}$ is given by

$$M\left(\widehat{Y}_{regt}\right) = E\left(\widehat{Y}_{regt} - Y\right)^2$$
$$= E\left[Y(1 + e_y) - BX(1 + e_B)\left(e_x - e'_x\right) - Y\right]^2$$

Now neglecting terms $E\left\{e_B^i e_x^j e_x'^k e_y^l\right\}$ for $i + j + k + l > 2$, an approximate expression of $M\left(\widehat{Y}_{regt}\right)$ is obtained as

$$M\left(\widehat{Y}_{regt}\right) \cong V[t(s,y) - B\{t(s,x) - t(s',x)\}]$$
$$= V\{t(s',y)\} + E[V\{t(s,\in)|s'\}] \tag{10.2.44}$$

where $t(s,\in) = \sum_{i\in s} b_{si}E_i$ and $E_i = y_i - B\,x_i$.

An approximate unbiased estimator of $M\left(\widehat{Y}_{regt}\right)$ is

$$\widehat{M}\left(\widehat{Y}_{regt}\right) = \widehat{V}(t(s',y)) + \widehat{V}(t(s,\widehat{\in})|s') \tag{10.2.45}$$

where $\widehat{V}(t(s',y))$ is an unbiased estimator of $\widehat{V}(t(s,y))$ and $\widehat{V}(t(s,\widehat{\in})|s')$ is an approximate unbiased estimator of $V(t(s,\in)|s')$. The estimator $\widehat{V}(t(s,\widehat{\in})|s')$ is obtained by replacing E_i by its estimate $\widehat{E}_i = y_i - B(s)x_i$ in the expression of $\widehat{V}(t(s,\in)|s')$, an unbiased estimator of $V(t(s,\in)|s')$.

10.2.3.2 Arbitrary Sampling Design

Let the initial sample s' and s be selected as in Section 10.2.1.1 and

$$t(s',y) = \sum_{i\in s'} \frac{y_i}{\pi_i},\ t(s',x) = \sum_{i\in s'} \frac{x_i}{\pi_i} \text{ and } t(s,y) = \frac{n'}{n}\sum_{i\in s} \frac{y_i}{\pi_i}.$$

In this case

$$\widehat{Y}_{regt} = \frac{n'}{n}\sum_{i\in s} \frac{y_i}{\pi_i} - B(s)\left(\frac{n'}{n}\sum_{i\in s} \frac{x_i}{\pi_i} - \sum_{i\in s'} \frac{x_i}{\pi_i}\right) \tag{10.2.46}$$

where

$$B(s) = \frac{\widehat{Cov}\left(\sum_{i\in s}\frac{y_i}{\pi_i}, \sum_{i\in s}\frac{x_i}{\pi_i}|s'\right)}{\widehat{V}\left(\sum_{i\in s}\frac{y_i}{\pi_i}|s'\right)} = \frac{\sum_{i\in s}\left(\frac{x_i}{\pi_i} - \frac{1}{n}\sum_{i\in s}\frac{x_i}{\pi_i}\right)\sum_{i\in s}\left(\frac{y_i}{\pi_i} - \frac{1}{n}\sum_{i\in s}\frac{y_i}{\pi_i}\right)}{\sum_{i\in s}\left(\frac{x_i}{\pi_i} - \frac{1}{n}\sum_{i\in s}\frac{x_i}{\pi_i}\right)^2}.$$

The expression of bias of $\widehat{Y}_{regt}$ is very complicated and hence it is omitted here. On the other hand, the expression of $M\left(\widehat{Y}_{regt}\right)$ is quite elegant and is given as follows:

$$
\begin{aligned}
M\left(\widehat{Y}_{regt}\right) &\cong V\left(\sum_{i\in s'}\frac{y_i}{\pi_i}\right) + E\left\{V\left(\frac{n'}{n}\sum_{i\in s}\frac{E_i}{\pi_i}|s'\right)\right\}\\
&= \frac{1}{2}\sum_{i\neq}\sum_{j\in U}(\pi_i\pi_j - \pi_{ij})\left(\frac{y_i}{\pi_i} - \frac{y_j}{\pi_j}\right)^2\\
&\quad + \frac{n'(n'-n)}{n}\frac{1}{n'-1}E\left[\sum_{i\in s'}\left\{\frac{E_i}{\pi_i} - \frac{1}{n'}\sum_{i\in s'}\frac{E_i}{\pi_i}\right\}^2\right]\\
&= \frac{1}{2}\sum_{i\neq}\sum_{j\in U}(\pi_i\pi_j - \pi_{ij})\left(\frac{y_i}{\pi_i} - \frac{y_j}{\pi_j}\right)^2 + \frac{n'(n'-n)}{n}\frac{1}{n'-1}\\
&\quad \times \left[E\left(\sum_{i\in s'}\frac{E_i^2}{\pi_i^2}\right) - \left\{V\left(\sum_{i\in s'}\frac{E_i}{\pi_i}\right)^2 + \left(\sum_{i\in U}E_i\right)^2\right\}\Big/ n'\right]\\
&= \frac{1}{2}\sum_{i\neq}\sum_{j\in U}(\pi_i\pi_j - \pi_{ij})\left(\frac{y_i}{\pi_i} - \frac{y_j}{\pi_j}\right)^2 + n'\frac{(n'-n)}{n(n'-1)}\\
&\quad \times \left[\left\{\sum_{i\in U}\frac{E_i^2}{\pi_i} - \left(\sum_{i\in U}E_i\right)^2\Big/ n\right\} - \frac{1}{n'}\sum_{i\neq}\sum_{j\in U}(\pi_i\pi_j - \pi_{ij})\left(\frac{E_i}{\pi_i} - \frac{E_j}{\pi_j}\right)^2\right]
\end{aligned}
\tag{10.2.47}
$$

where $E_i = y_i - Bx_i$ and $B = \dfrac{E\left\{Cov\left(\sum_{i\in s}\frac{y_i}{\pi_i}, \sum_{i\in s}\frac{x_i}{\pi_i}\right)|s'\right\}}{E\left\{V\left(\sum_{i\in s}\frac{x_i}{\pi_i}\right)|s'\right\}}$.

An approximate unbiased estimator of $M\left(\widehat{Y}_{regt}\right)$ is given by

$$
\begin{aligned}
\widehat{M}\left(\widehat{Y}_{regt}\right) &\cong \frac{n'(n'-1)}{n(n-1)}\frac{1}{2}\sum_{i\neq}\sum_{j\in s}\left(\frac{\pi_i\pi_j - \pi_{ij}}{\pi_{ij}}\right)\left(\frac{y_i}{\pi_i} - \frac{y_j}{\pi_j}\right)^2\\
&\quad + n'\frac{(n'-n)}{n(n-1)}\left\{\sum_{i\in s}\frac{\widehat{E}_i^2}{\pi_i^2} - \frac{1}{n}\left(\sum_{i\in s}\frac{\widehat{E}_i}{\pi_i}\right)^2\right\}
\end{aligned}
\tag{10.2.48}
$$

with $\widehat{E}_i = y_i - B(s)\,x_i$.

10.2.3.3 Simple Random Sampling Without Replacement

In case samples s' and s are selected by SRSWOR method we have

$$t(s',y) = N\frac{1}{n'}\sum_{i\in s'} y_i = N\,\overline{y}(s'),\ t(s',x) = N\frac{1}{n'}\sum_{i\in s'} x_i = N\overline{x}(s'),$$

$$t(s,y) = n'\frac{1}{n}\sum_{i\in s} y_i = n'\ \overline{y}(s),$$

$$B(s) = \frac{\widehat{C}ov\{\overline{y}(s),\overline{x}(s)|s'\}}{\widehat{V}\{\overline{x}(s)|s'\}} = \frac{s_{xy}}{s_x^2} \text{ and}$$

$$\widehat{Y}_{regt} = N\left[\overline{y}(s) - \frac{s_{xy}}{s_x^2}(\overline{x}(s) - \overline{x}(s'))\right] \tag{10.2.49}$$

An approximate bias of $\widehat{Y}_{regt}$ is obtained from Eq. (10.2.43) as

$$B\left(\widehat{Y}_{regt}\right) = -NE\left[Cov\left\{\frac{s_{xy}}{s_x^2},(\overline{x}(s) - \overline{x}(s'))|s\right\}\right]$$

$$= -NE\left[\frac{s_{xy}}{s_x^2}(\overline{x}(s) - \overline{x}(s'))\right]$$

Now writing $\varepsilon_{xy} = \dfrac{s_{xy} - S_{xy}}{S_{xy}}$, $\varepsilon_{xx} = \dfrac{s_x^2 - S_x^2}{S_x^2}$, and $B = \beta = \dfrac{S_{xy}}{S_x^2}$, we get

$$B\left(\widehat{Y}_{regt}\right) = -N\beta E\left[(1+\varepsilon_{xy})(1+\varepsilon_{xx})^{-1}(\overline{x}(s) - \overline{x}(s'))\right]$$

$$\cong -\ N\beta E\left[(\varepsilon_{xy} - \varepsilon_{xx})(\overline{x}(s) - \overline{x}(s'))\right] \tag{10.2.50}$$

$$= -N\beta\left[\frac{E\left[Cov\{s_{xy},\overline{x}(s)\}|s'\right]}{S_{xy}} - \frac{E\left[\{s_x^2,\overline{x}(s)\}|s'\right]}{S_x^2}\right]$$

Now using Sukhatme et al. (1984), we find

$$E\left[Cov\{s_{xy},\overline{x}(s)\}|s'\right] = \frac{n'(n'-n)}{(n'-1)(n'-2)n}E\left[\frac{1}{n'}\sum_{i\in s'}\{x_i - \overline{x}(s')\}^2\{y_i - \overline{y}(s')\}\right]$$

$$= \frac{n'(n'-n)}{(n'-1)(n'-2)n}\mu_{21} \tag{10.2.51}$$

and

$$
\begin{aligned}
E\left[\left\{s_x^2, \overline{x}(s)\right\}|s'\right] &= \frac{n'(n'-n)}{(n'-1)(n'-2)n} E\left[\frac{1}{n'}\sum_{i\in s'}\left\{x_i - \overline{x}(s')\right\}^3\right] \\
&= \frac{n'(n'-n)}{(n'-1)(n'-2)n}\mu_{30}
\end{aligned}
\tag{10.2.52}
$$

where $\mu_{ij} = \sum_{i\in U}\left(x_i - \overline{X}\right)^i\left(y_i - \overline{Y}\right)^j\Big/N$

Now substituting Eqs. (10.2.51) and (10.2.52) in Eq. (10.2.50), we find

$$
B\left(\widehat{Y}_{regt}\right) \cong -\beta\frac{Nn'(n'-n)}{(n'-1)(n'-2)n}\left(\frac{\mu_{21}}{S_{xy}} - \frac{\mu_{30}}{S_x^2}\right)
$$

If n' is large compared to n, the bias $B\left(\widehat{Y}_{regt}\right)$ can be approximated as

$$
B\left(\widehat{Y}_{regt}\right) \cong -N\beta\left(\frac{1}{n} - \frac{1}{n'}\right)\left(\frac{\mu_{21}}{S_{xy}} - \frac{\mu_{30}}{S_x^2}\right)
\tag{10.2.53}
$$

Following Eqs. (10.2.44) and (10.2.45), approximate expressions for the mean square error of $\widehat{Y}_{regt}$ and its unbiased estimator are obtained as

$$
\begin{aligned}
M\left(\widehat{Y}_{regt}\right) &\cong N^2\left[\left(\frac{1}{n'} - \frac{1}{N}\right)S_y^2 + \left(\frac{1}{n} - \frac{1}{n'}\right)S_E^2\right] \\
&\cong N^2\left[\left(\frac{1}{n'} - \frac{1}{N}\right) + \left(\frac{1}{n} - \frac{1}{n'}\right)\left(1-\rho^2\right)\right]S_y^2
\end{aligned}
\tag{10.2.54}
$$

and

$$
\begin{aligned}
\widehat{M}\left(\widehat{Y}_{regt}\right) &\cong N^2\left[\left(\frac{1}{n'} - \frac{1}{N}\right)s_y^2 + \left(\frac{1}{n} - \frac{1}{n'}\right)s_{\widehat{E}}^2\right] \\
&= N^2\left[\left(\frac{1}{n'} - \frac{1}{N}\right) + \left(\frac{1}{n} - \frac{1}{n'}\right)\left(1-r^2\right)\right]s_y^2
\end{aligned}
\tag{10.2.55}
$$

where $S_E^2 = \sum_{i\in U}\left(E_i - \overline{E}\right)^2/(N-1)$, $s_{\widehat{E}}^2 = \sum_{i\in s}\left(\widehat{E}_i - \overline{\widehat{E}}\right)^2/(n-1)$, $\overline{E} = \sum_{i\in U}E_i/N$, $\overline{\widehat{E}} = \sum_{i\in s}\widehat{E}_i/n$, $\widehat{E}_i = y_i - \widehat{\beta}x_i$, and $r = s_{xy}/(s_x s_y) =$ sample correlation coefficient.

10.2.3.4 Optimum Allocation

For large N, the expression (Eq. 10.2.54) reduces to

$$M\left(\widehat{Y}_{regt}\right) = N^2\left[\frac{(1-\rho^2)}{n} + \frac{\rho^2}{n'}\right]S_y^2 \qquad (10.2.56)$$

The optimum values of n and n' that minimize (Eq. 10.2.56) for a given cost $C = C_0 = cn + c'n'$ are obtained by minimizing $N^2\left[\frac{(1-\rho^2)}{n} + \frac{\rho^2}{n'}\right]S_y^2$, subject to $C_0 = cn + c'n'$. For minimization consider the function

$$\phi = N^2\left[\frac{(1-\rho^2)}{n} + \frac{\rho^2}{n'}\right]S_y^2 - \mu(C_0 - cn - c'n')$$

where μ is an undetermined Lagrange multiplier.

$\frac{\partial \phi}{\partial n} = 0$ and $\frac{\partial \phi}{\partial n'} = 0$ yield the optimum values of n and n' are as follows:

$$n_{opt} = \frac{C_0}{\left(\sqrt{(1-\rho^2)}\sqrt{c} + \rho\sqrt{c'}\right)}\frac{\sqrt{1-\rho^2}}{\sqrt{c}}$$

and

$$n'_{opt} = \frac{C_0}{\left(\sqrt{(1-\rho^2)}\sqrt{c} + \rho\sqrt{c'}\right)}\frac{\rho}{\sqrt{c'}}$$

The value of $M\left(\widehat{Y}_{regt}\right)$ with the optimum values of n and n' is given by

$$M_{\min} = N^2\frac{\left(\sqrt{(1-\rho^2)}\sqrt{c} + \rho\sqrt{c'}\right)^2}{C_0}S_y^2 \qquad (10.2.57)$$

For the fixed cost C_0, the value of $n_0 = C_0/c$. Hence, two-phase sampling with optimum allocation will be more precise than the single-phase sampling with the same cost for a large N

$$\begin{aligned}
&\text{if } M_{\min} < N^2\frac{1}{n_0}S_y^2\\
&\text{i.e. if } \left(\sqrt{(1-\rho^2)}\sqrt{c} + \rho\sqrt{c'}\right)^2 < c\\
&\text{i.e. if } 2\sqrt{1-\rho^2}\sqrt{cc'} < \rho^2(c-c')\\
&\text{i.e. if } \rho^2 > \frac{4cc'}{(c+c')^2}
\end{aligned} \qquad (10.2.58)$$

Example 10.2.1

A sample of 50 plants was selected from a garden of 500 plants by SRSWOR method, and eye estimates of the heights of the selected plants were noted. From the selected sample of 50 plants, a subsample of 10 plants was selected by SRSWOR method and the heights of the selected plants were measured accurately. The data are presented as follows. Estimate the average height of the plants by using (i) ratio and (ii) regression method of estimation using the information that the sample mean height of eye estimation of the 50 plants was 10.5 m. Compare standard errors of the estimators used. (iii) Given that the cost of eye estimation and actual estimation of measuring height are \$10 and \$50, respectively, estimate the optimum proportions of actual measurement that should be made for the ratio and regression estimators.

Eye estimates and actual heights of plants.

Plants	Eye estimate of height (in m) (x)	Actual height (in m) (y)
1	15	13.85
2	10	9.75
3	8	10
4	8.5	10.5
5	12	14
6	6	7
7	6.5	8
8	15	14
9	10	12
10	8	7.5

Here $N = 500$, $n' = 50$, $n = 10$, $\overline{x}(s') = 10.5$, $\overline{x}(s) = 9.9$, $\overline{y}(s) = 10.66$, $s_x^2 = 10.267$, $s_y^2 = 7.337$, $s_{xy} = 7.907$, $\widehat{R} = 10.66/9.9 = 1.077$, $s_d^2 = 2.213$, and sample correlation coefficient $= r = 0.911$.

(i) Ratio estimation:

Estimated mean height of the plants under two-phase sampling scheme is

$$\widehat{\overline{Y}}_{Rt} = \frac{\overline{y}(s)}{\overline{x}(s)}\,\overline{x}(s') = (10.66/9.9) \times 10.5 = 11.306 \text{ m}$$

$$\text{Estimated MSE of } \widehat{\overline{Y}}_R = \widehat{M}\left(\widehat{\overline{Y}}_{Rt}\right) = \left(\frac{1}{n'} - \frac{1}{N}\right)s_y^2 + \left(\frac{1}{n} - \frac{1}{n'}\right)s_d^2$$

$$= \left[\left(\frac{1}{50} - \frac{1}{500}\right) \times 7.336 + \left(\frac{1}{10} - \frac{1}{50}\right) \times 2.213\right]$$

$$= 0.3091$$

$$\text{Estimated standard error of } \widehat{\overline{Y}}_{Rt} = SE\left(\widehat{\overline{Y}}_{Rt}\right) = \sqrt{\widehat{M}\left(\widehat{\overline{Y}}_{Rt}\right)} = 0.556 \text{ m.}$$

(ii) Regression Estimation:

Estimated mean height of plants under two-phase sampling is

$$\widehat{\overline{Y}}_{regt} = \overline{y}(s) - \frac{s_{xy}}{s_x^2}\{\overline{x}(s) - \overline{x}(s')\}$$

$$= 10.66 - (7.116/10.266) \times (9.9 - 10.5) = 11.122 \text{ m}$$

$$\text{Estimated MSE of } \widehat{\overline{Y}}_{regt} = \widehat{M}\left(\widehat{\overline{Y}}_{regt}\right) = \left[\left(\frac{1}{n'} - \frac{1}{N}\right) + \left(\frac{1}{n} - \frac{1}{n'}\right)(1 - r^2)\right]s_y^2$$

$$= \left[\left(\frac{1}{50} - \frac{1}{500}\right) + \left(\frac{1}{10} - \frac{1}{50}\right)(1 - 0.911^2)\right] \times 7.336 = 0.232$$

$$\text{Estimated standard error of } \widehat{\overline{Y}}_{regt} = SE\left(\widehat{\overline{Y}}_{regt}\right) = \sqrt{\widehat{M}\left(\widehat{\overline{Y}}_{regt}\right)}$$

$= 0.482$ m.

Efficiency of the regression estimator compared to the ratio estimator is

$$\frac{M\left(\widehat{\overline{Y}}_{Rt}\right)}{M\left(\widehat{\overline{Y}}_{regt}\right)} \times 100 = 133.307\%.$$

(iii) Optimum proportion of second-phase sample:

The optimum proportion for the ratio estimator is

$$n_{opt}\big/n'_{opt} = \sqrt{\frac{c'}{c}\frac{s_d^2}{s_y^2 - s_d^2}} = \sqrt{\frac{c'}{c}\frac{s_d^2}{s_y^2 - s_d^2}} = \sqrt{\frac{10}{50} \times \frac{2.213}{5.214}} = 0.291$$

The optimum proportion for the regression estimator is

$$n_{opt}\big/n'_{opt} = \sqrt{\frac{c'}{c}\frac{(1 - r^2)}{r^2}} = \sqrt{\frac{10}{50} \times \frac{(1 - 0.911^2)}{0.911^2}} = 0.202$$

10.3 TWO-PHASE SAMPLING FOR STRATIFICATION

It is well known that stratification is an important tool to increase precision of estimators. In stratification, the entire population is divided into a number of disjoint and exhaustive strata so that the each stratum becomes homogeneous with respect to the character under study. If no criterion for formation of strata is known in advance, we may use two-phase sampling for forming the strata. In the first phase, a relatively large sample s' of size n' is selected from the entire population by SRSWOR method, and information on the auxiliary variable (which can be used for stratification) is obtained. Here it is supposed that by noting the value of the auxiliary variable, one can classify the sample s' into a predetermined number of "H" strata. Let s'_h be the set of units of size n'_h falling in the hth stratum. Here we make the assumption that the sample size n' is so large that n'_h's always exceeds 1 for every $h = 1,\ldots, H$; in other words, $Prob\left(n'_h \geq 1\right) = 1$. From each of the selected samples s'_h's, subsamples s_h's of sizes $n_h = \gamma_h n'_h$ are selected independently by SRSWOR method. Here γ_h's are predetermined fractions and n'_h's are assumed to be integers. Hence, for proportional allocation with fixed n, we get $\gamma_h = \gamma = n/n'$ and $n_h = nn'_h/n'$. For example, in an agricultural survey, one can select a large sample s' to acquire information about the size of the farm x (auxiliary variable) only, and on the basis of the observed x values, the farms can be classified into small, medium, large, or very large strata.

10.3.1 Estimation of Mean and Variance

Let y_{hi} and x_{hi} be the value of the study (y) and auxiliary variable (x) for the ith unit of the hth stratum $i = 1,\ldots, N_h$, $h = 1,\ldots, H$. Here the strata sizes N_h's and weights $W_h = N_h/N$'s are not known. Let $\overline{Y}_h = \sum_{i=1}^{N_h} y_{hi}/N_h$ be the population mean for the hth stratum and $\overline{Y} = \sum_{h=1}^{H} W_h \overline{Y}_h$ be the mean for the entire population.

Theorem 10.3.1

Let $w_h = n'_h/n'$, then

(i) $E(w_h) = W_h$, (ii) $V(w_h) = g\, W_h(1 - W_h)/n'$, and

(iii) $Cov(w_h, w_{h'}) = -gW_h W_{h'}/n'$ for $h \neq h'$

where $g = (N - n')/(N - 1)$.

Proof

For a fixed $\sum_{k=1}^{H} n'_k = n'$, $\mathbf{n}' = \left(n'_1, \ldots, n'_k, \ldots, n'_H\right)$ follows the generalized hypergeometric distribution

$$Prob\left(n'_1 = t_1, \ldots, n'_k = t_k, \ldots, n'_H = t_H \,\middle|\, \sum_{k=1}^{H} n'_k = n'\right)$$
$$= \frac{\binom{N_1}{t_1} \cdots \binom{N_k}{t_k} \cdots \binom{N_H}{t_H}}{\binom{N}{n'}}; t_k = 0, 1, \ldots, \min(n', N_k)$$

For this generalized hypergeometric distribution,

$E\left(n'_k\right) = n' W_k$, $V\left(n'_k\right) = n' g W_k (1 - W_k)$ and $Cov\left(n'_k, n'_{k'}\right) = -n' g W_k W_{k'}$

Theorem follows from the aforementioned results.

Finally using Rao (1973), we have the following theorem.

Theorem 10.3.2

(i) The estimator $\widehat{\overline{Y}}_{stt} = \sum_{h=1}^{H} w_h \overline{y}(s_h)$ is unbiased $\overline{Y}$

where $\overline{y}(s_h) = \sum_{j \in s_h} y_{hj} / n_h$

(ii) The variance of $\widehat{\overline{Y}}_{stt}$ is

$$V\left(\widehat{\overline{Y}}_{stt}\right) = \frac{1}{n'} \sum_{h=1}^{H} W_h \left\{ \left(\frac{1}{\gamma_h} - 1\right) + g\left(1 - \frac{1}{N_h}\right) \right\} S_{hy}^2$$
$$+ \frac{g}{n'} \sum_{h=1}^{H} W_h \left(\overline{Y}_h - \overline{Y}\right)^2$$
$$= \left(\frac{1}{n'} - \frac{1}{N}\right) S_y^2 + \sum_{h=1}^{H} \frac{W_h}{n'} \left(\frac{1}{\gamma_h} - 1\right) S_{hy}^2$$

where $(N - 1) S_y^2 = \sum_{h=1}^{H} \sum_{i=1}^{N_h} \left(y_{hi} - \overline{Y}\right)^2$ and $(N_h - 1) S_{hy}^2 = \sum_{i=1}^{N_h} \left(y_{hi} - \overline{Y}_h\right)^2$

(iii) A nonnegative unbiased estimator of $V\left(\widehat{\overline{Y}}_{stt}\right)$ is

$$\widehat{V}\left(\widehat{\overline{Y}}_{stt}\right)=\frac{1}{Nn'}\left[\frac{N-1}{n'-1}\sum_{h=1}^{H}n'_h\left(\frac{1}{\gamma_h}-1\right)s^2_{hy}\right.$$

$$\left.+\frac{N-n'}{n'-1}\left(\sum_{h=1}^{H}\frac{1}{\gamma_h}\sum_{i\in s_h}y^2_{hi}-n'\widehat{\overline{Y}}^2_{stt}\right)\right]$$

$$=\frac{N-1}{N}\sum_{h=1}^{H}\left(\frac{n'_h-1}{n'-1}-\frac{n_h-1}{N-1}\right)\frac{w_hs^2_{hy}}{n_h}+\frac{N-n'}{N(n'-1)}\sum_{h=1}^{H}w_h\left(\overline{y}(s_h)-\widehat{\overline{Y}}_{stt}\right)^2$$

where $g=\dfrac{N-n'}{N-1}$ and $(n_h-1)s^2_{hy}=\sum\limits_{i\in s_h}\left\{y_{hi}-\overline{y}(s_h)\right\}^2$.

Proof

Let $\mathbf{n}'=\left(n'_1,\ldots,n'_h,\ldots,n'_H\right)$. Then,

(i)
$$E\left(\widehat{\overline{Y}}_{stt}\right)=E\left[E\left(\sum_{h=1}^{H}w_h\,\overline{y}(s_h)|\mathbf{n}'\right)\right]$$

$$=E\left[\sum_{h=1}^{H}w_h\,E(\overline{y}(s_h)|\mathbf{n}')\right]$$

$$=E\left[\sum_{h=1}^{H}w_h\,\overline{Y}_h\right]$$

$$=\sum_{h=1}^{H}\{E(w_h)\}\,\overline{Y}_h$$

$$=\sum_{h=1}^{H}W_h\,\overline{Y}_h\ \text{(using Theorem 10.3.1)}$$

$$=\overline{Y}$$

(ii)
$$V\left(\widehat{\overline{Y}}_{stt}\right)=E\left[V\left(\sum_{h=1}^{H}w_h\overline{y}(s_h)|\mathbf{n}'\right)\right]+V\left[E\left(\sum_{h=1}^{H}w_h\overline{y}(s_h)|\mathbf{n}'\right)\right]\tag{10.3.1}$$

Now,

$$E\left[V\left(\sum_{h=1}^{H} w_h\overline{y}(s_h)|\mathbf{n}'\right)\right] = E\left[\sum_{h=1}^{H} w_h^2\ V\{\overline{y}(s_h)|\mathbf{n}'\}\right.$$

$$\left.+\sum_{h\neq}^{H}\sum_{h'=1}^{H} w_h w_{h'}\, Cov\{\overline{y}(s_h),\overline{y}(s_{h'})|\mathbf{n}'\}\right]$$

$$= E\left[\sum_{h=1}^{H} w_h^2\left(\frac{1}{n_h}-\frac{1}{N_h}\right)S_{hy}^2\right] \tag{10.3.2}$$

because $V\{\overline{y}(s_h|\mathbf{n}')\} = \left(\frac{1}{n_h}-\frac{1}{N_h}\right)S_{hy}^2$ and $Cov\{\overline{y}(s_h),\overline{y}(s_{h'})|\mathbf{n}'\} = 0$ as the samples s_h and $s_{h'}$ are selected independently for $h \neq h'$.

Writing $n_h = \gamma_h n_h'$ in Eq. (10.3.2) we get

$$E\left[V\left(\sum_{h=1}^{H} w_h\ \overline{y}(s_h)|\mathbf{n}'\right)\right] = \sum_{h=1}^{H}\left(\frac{E(w_h)}{n'\gamma_h}-\frac{E\left(w_h^2\right)}{N_h}\right)S_{hy}^2$$

$$= \sum_{h=1}^{H} W_h\left\{\left(\frac{1}{n'\gamma_h}-\frac{1}{N}\right)-\frac{g}{n'}\left(\frac{1-W_h}{N_h}\right)\right\}S_{hy}^2 \tag{10.3.3}$$

(using Theorem 10.3.1.)

$$= \frac{1}{n'}\sum_{h=1}^{H} W_h\left\{\left(\frac{1}{\gamma_h}-1\right)+g\left(1-\frac{1}{N_h}\right)\right\}S_{hy}^2$$

Furthermore,

$$V\left[E\left(\sum_{h=1}^{H} w_h\ \overline{y}(s_h)|\mathbf{n}'\right)\right] = V\left(\sum_{h=1}^{H} w_h\overline{Y}_h\right)$$

$$= \frac{g}{n'}\left(\sum_{h=1}^{H} W_h(1-W_h)\overline{Y}_h^2 - \sum_{h\neq}^{H}\sum_{h'}^{H} W_h W_{h'}\overline{Y}_h\overline{Y}_{h'}\right)$$

$$= \frac{g}{n'}\sum_{h=1}^{H} W_h\left(\overline{Y}_h-\overline{Y}\right)^2 \tag{10.3.4}$$

Substituting Eqs. (10.3.3) and (10.3.4) in Eq. (10.3.1) and noting $(N-1)S_y^2 = \sum_{h=1}^{H}(N_h-1)S_{hy}^2 + \sum_{h=1}^{H} N_h\left(\overline{Y}_h-\overline{Y}\right)^2$ we can verify (ii) of the theorem.

(iii) Now, noting $\frac{1}{N-1}\left[N\sum_{h=1}^{H}\frac{w_h}{n_h}\sum_{i\in s_h}y_{hi}^2 - N\widehat{\overline{Y}}_{stt}^2\left\{-\widehat{V}\left(\widehat{\overline{Y}}_{stt}\right)\right\}\right]$ and $\sum_{h=1}^{H}\frac{w_h}{n'}\left(\frac{1}{\gamma_h}-1\right)s_{hy}^2$ are unbiased estimators of S_y^2 and $\sum_{h=1}^{H}\frac{W_h}{n'}\left(\frac{1}{\gamma_h}-1\right)S_h^2$, respectively, we find an unbiased estimator of $V\left(\widehat{\overline{Y}}_{stt}\right)$ from the equation

$$\widehat{V}\left(\widehat{\overline{Y}}_{stt}\right) = \left(\frac{1}{n'}-\frac{1}{N}\right)\frac{1}{N-1}\left[N\sum_{h=1}^{H}\frac{w_h}{n_h}\sum_{i\in s_h}y_{hi}^2 - N\left\{\widehat{\overline{Y}}_{stt}^2 - \widehat{V}\left(\widehat{\overline{Y}}_{stt}\right)\right\}\right] + \sum_{h=1}^{H}\frac{w_h}{n'}\left(\frac{1}{\gamma_h}-1\right)s_{hy}^2 \tag{10.3.5}$$

Eq. (10.3.5) yields

$$\text{i.e., } \widehat{V}\left(\widehat{\overline{Y}}_{stt}\right) = \frac{1}{Nn'}\left[\frac{N-1}{n'-1}\sum_{h=1}^{H}n'_h\left(\frac{1}{\gamma_h}-1\right)s_{hy}^2 + \frac{N-n'}{n'-1}\left(\sum_{h=1}^{H}\frac{1}{\gamma_h}\sum_{i\in s_h}y_{hi}^2 - n'\widehat{\overline{Y}}_{stt}^2\right)\right] \tag{10.3.6}$$

Furthermore, writing $\sum_{i\in s_h}y_{hi}^2 = (n_h-1)s_{hy}^2 + n_h\{\overline{y}(s_h)\}^2$ in Eq. (10.3.6) we find

$$\widehat{V}\left(\widehat{\overline{Y}}_{stt}\right) = \frac{N-1}{N}\sum_{h=1}^{H}\left(\frac{n'_h-1}{n'-1}-\frac{n_h-1}{N-1}\right)\frac{w_h s_{hy}^2}{n_h} + \frac{N-n'}{N(n'-1)}\sum_{h=1}^{H}w_h\left(\overline{y}(s_h)-\widehat{\overline{Y}}_{stt}\right)^2$$

10.3.2 Proportional Allocation

In case the samples are selected from each of the strata by proportional allocation, keeping the total sample size for the second phase fixed as n, then γ_h becomes equal to $\gamma = n/n'$. Furthermore, if the stratum sizes are large for each of the strata, we derive the following result straight from the Theorem 10.3.2.

Theorem 10.3.3

For proportional allocation with large strata sizes

(i) $$V\left(\overline{Y}_{stt}\right) \cong \sum_{h=1}^{H} W_h \left\{ \frac{(1-f_2)}{n} + \frac{(1-f_1)}{n'} \right\} S_{hy}^2 + \frac{(1-f_1)}{n'} \sum_{h=1}^{H} W_h \left(\overline{Y}_h - \overline{Y}\right)^2$$

and

(ii) An unbiased estimator of $V\left(\widehat{\overline{Y}}_{stt}\right)$ is

$$\widehat{V}\left(\widehat{\overline{Y}}\right) = \frac{(1-f_1)}{n'} \sum_{h=1}^{H} w_h \left(\overline{y}(s_h) - \widehat{\overline{Y}}_{stt}\right)^2 + \frac{1}{n} \sum_{h=1}^{H} (1-f) w_h s_{hy}^2$$

where $f = n/N$, $f_1 = n'/N$ and $f_2 = n/n'$.

10.3.3 Estimation of Proportion

To estimate the proportion of persons who possess some attribute A, such as being a smoker or wealthy, we define $y_{hi} = 1$ if the ith unit of the hth stratum belongs to A and $y_{hi} = 0$ otherwise. Then $\overline{Y}$ equals to $\pi = \sum_{h=1}^{H} W_h \pi_h$, the proportion of persons belonging to the group A, and π_h becomes the proportion in the hth stratum. On substituting $\overline{y}(s_h) = \sum_{j \in s_h} y_{hj}/n_h = \widehat{\pi}_h$ sample proportion for the hth stratum, $S_{hy}^2 = \frac{N_h \pi_h (1-\pi_h)}{N_h - 1}$ and $s_{hy}^2 = \frac{n_h \widehat{\pi}_h (1-\widehat{\pi}_h)}{n_h - 1}$ in Theorem 10.3.2, we get the following.

Theorem 10.3.4

(i) An unbiased estimator of π is $\widehat{\pi} = \sum_{h=1}^{H} w_h \widehat{\pi}_h$

(ii) The variance of $\widehat{\pi}$ is

$$V(\widehat{\pi}) = \frac{1}{n'} \sum_{h=1}^{H} W_h \left\{ \left(\frac{1}{\gamma_h} - 1 \right) + g\left(1 - \frac{1}{N_h}\right) \right\} \frac{N_h \pi_h (1-\pi_h)}{N_h - 1} + \frac{g}{n'} \sum_{h=1}^{H} W_h (\pi_h - \pi)^2$$

(iii) An unbiased estimator of $V(\widehat{\pi})$ is

$$\widehat{V}(\widehat{\pi}) = \frac{N-1}{N} \sum_{h=1}^{H} \left(\frac{n_h' - 1}{n' - 1} - \frac{n_h - 1}{N - 1} \right) \frac{w_h \widehat{\pi}_h (1-\widehat{\pi}_h)}{n_h - 1} + \frac{N - n'}{N(n'-1)} \sum_{h=1}^{H} w_h (\widehat{\pi}_h - \widehat{\pi})^2$$

10.3.4 Optimum Allocation of Sample Sizes

Following Rao (1973), we consider the simple cost function

$$C = n'c' + \sum_{h=1}^{H} n_h c_h$$

where c' = cost of surveying a unit in the first phase and c_h = cost of surveying a unit of the hth stratum in the second phase. Clearly, c_h's are expected to be much larger than c''s. Because C is a random variable, we determine the optimum values of γ_h's, which minimize the variance $V\left(\widehat{\overline{Y}}_{stt}\right)$, keeping the expected cost $E(C) = n'c' + n'\sum_{h=1}^{H}\gamma_h W_h c_h$ to a certain level C^*.

$$\text{i.e.,}\quad C^* = n'\left(c' + \sum_{h=1}^{H}\gamma_h W_h c_h\right) \tag{10.3.7}$$

Here we perform minimization in two stages. In the first sage, we minimize

$$V\left(\widehat{\overline{Y}}_{stt}\right) = \left(\frac{1}{n'} - \frac{1}{N}\right)S_y^2 + \sum_{h=1}^{H}\frac{W_h}{n'}\left(\frac{1}{\gamma_h} - 1\right)S_{hy}^2 \tag{10.3.8}$$

subject to Eq. (10.3.8) keeping n' as fixed. For minimization consider

$$\phi = \left(\frac{1}{n'} - \frac{1}{N}\right)S_y^2 + \sum_{h=1}^{H}\frac{W_h}{n'}\left(\frac{1}{\gamma_h} - 1\right)S_{hy}^2 - \lambda\left(C^* - n'c' - n'\sum_{h=1}^{H}\gamma_h W_h c_h\right)$$

where λ is an undetermined Lagrange multiplier.

$\dfrac{\partial\phi}{\partial\gamma_h} = 0$ implies

$$\gamma_h = \frac{1}{n'\sqrt{\lambda}}\frac{S_{hy}}{\sqrt{c_h}} \tag{10.3.9}$$

Eqs. (10.3.7) and (10.3.9) yield

$$\gamma_h = \frac{C^* - n'c'}{n'\left(\sum_{h=1}^{H} W_h S_{hy}\sqrt{c_h}\right)}\frac{S_{hy}}{\sqrt{c_h}} \tag{10.3.10}$$

Finally, putting Eq. (10.3.10) in Eq. (10.3.8) and minimizing it with respect to n', the optimum value of n' comes out as

$$n' = \frac{C^*}{\sqrt{c'}}\left[\sqrt{c'} + \frac{\sum_{h=1}^{H} W_h S_{hy}\sqrt{c_h}}{\sqrt{S_y^2 - \sum_{h=1}^{H} W_h S_{hy}^2}}\right]^{-1} \tag{10.3.11}$$

Substituting Eq. (10.3.11) in Eq. (10.3.10) the optimum value of γ_h is obtained as

$$\gamma_h = \gamma_{h0} = S_{hy}\sqrt{\frac{c'}{c_h}} \Big/ \sqrt{S_y^2 - \sum_{h=1}^{H} W_h S_{hy}^2} \tag{10.3.12}$$

provided $\gamma_h \leq 1$ for all $h = 1,\ldots, H$. It should be noted that $\gamma_h > 0$ because $\left(S_y^2 - \sum_{h=1}^{H} W_h S_{hy}^2\right) > 0$ except in pathological situations. If some γ_h exceed 1, we should set them the value at 1 and determine the rest γ_h's so that all of them are less than or equal to 1.

Now, substituting the optimum values of n' and γ_h from Eqs. (10.3.11) and (10.3.12) in the expression of $V\left(\widehat{\overline{Y}}_{stt}\right)$, we get the minimum value of $V\left(\widehat{\overline{Y}}_{stt}\right)$ as

$$V_{\min}\left(\widehat{\overline{Y}}_{stt}\right) = \frac{1}{C^*}\left[\sum_{h=1}^{H} W_h S_{hy}\sqrt{c_h} + \sqrt{\left(S_y^2 - \sum_{h=1}^{H} W_h S_{hy}^2\right)c'}\right]^2 - \frac{S_y^2}{N} \tag{10.3.13}$$

Example 10.3.1

A pilot sample of 1000 households was selected from 10,000 households in a certain region by SRSWOR method. On the basis of the pilot survey report, the selected households were classified into three categories: The high, middle, and low income groups. From each of the income groups subsamples were selected by SRSWOR method. The sample mean, sample variance of the monthly income, and sample proportion of unemployed persons for the final survey are presented in the following table.

Income group	Pilot survey	Final survey			
	No. of households	No. of households	Sample mean (monthly income in $)	Sample variance (monthly income)	Proportion of unemployed persons
h	n'_h	n_h	$\overline{y}(s_h)$	s^2_{hy}	$\widehat{\pi}_h$
Low	325	100	1575	4000	0.75
Middle	475	75	5750	3500	0.40
High	200	75	8000	2500	0.20

Estimate the average income and proportion of unemployed persons in the region along with their standard error.

Estimation of average income and its standard error

h	$w_h = n'_h/n'$	$w_h\overline{y}(s_h)$	$w_h\left(\overline{y}(s_h) - \widehat{\overline{Y}}_{stt}\right)^2$	$\left(\frac{n'_h-1}{n'-1} - \frac{n_h-1}{N-1}\right)\frac{w_h s^2_{hy}}{n_h}$
1	0.325	511.875	3,471,208.330	4.087
2	0.475	2731.250	390,650.576	10.353
3	0.200	1600.000	1,993,171.953	1.278
Total	1.000	4843.125	5,855,030.859	15.719

$$\text{Estimated average monthly income} = \widehat{\overline{Y}}_{stt} = \sum_{h=1}^{H} w_h\,\overline{y}(s_h) = \$4843.125.$$

$$\text{Estimated variance of } \widehat{\overline{Y}}_{stt} = \widehat{V}\left(\widehat{\overline{Y}}_{stt}\right) \cong \sum_{h=1}^{H}\left(\frac{n'_h - 1}{n' - 1} - \frac{n_h - 1}{N - 1}\right)\frac{w_h s^2_{hy}}{n_h}$$

$$+\frac{N - n'}{N(n' - 1)}\sum_{h=1}^{H} w_h\left(\overline{y}(s_h) - \widehat{\overline{Y}}_{stt}\right)^2$$

$$= 15.719 + \{(10,000 - 1000)/(10,000 \times 999)\} \times 5855030.859$$

$$= \$5290.521$$

$$\text{Estimated standard error of } \widehat{\overline{Y}}_{stt} = \sqrt{\widehat{V}\left(\widehat{\overline{Y}}_{stt}\right)} = \$72.74$$

Estimation of proportion of unemployed persons and its standard error

h	$w_h = n'_h/n'$	$w_h\widehat{\pi}_h$	$w_h(\widehat{\pi}_h - \widehat{\pi})^2$	$\left(\frac{n'_h - 1}{n' - 1} - \frac{n_h - 1}{N - 1}\right)\frac{w_h\widehat{\pi}_h(1 - \widehat{\pi}_h)}{n_h - 1}$
1	0.325	0.244	0.0248	0.00019
2	0.475	0.190	0.0025	0.00072
3	0.200	0.040	0.0149	0.00008
Total	1.000	0.474	0.0423	0.00099

$$\text{Estimated proportion of unemployed persons} = \widehat{\pi} = \sum_{h=1}^{H} w_h\widehat{\pi}_h = 0.4737$$

$$\text{Estimated variance of } \widehat{\pi} = \widehat{V}(\widehat{\pi}) \cong \sum_{h=1}^{H}\left(\frac{n'_h - 1}{n' - 1} - \frac{n_h - 1}{N - 1}\right)\frac{w_h\widehat{\pi}_h(1 - \widehat{\pi}_h)}{n_h - 1} + \frac{N - n'}{N(n' - 1)}\sum_{h=1}^{H} w_h(\widehat{\pi}_h - \widehat{\pi})^2$$

$$= 0.00099 + \{(10000 - 1000)/(10000 \times 999)\} \times 0.0423 = 0.0010$$

$$\text{Estimated standard error of } \widehat{\pi} = \sqrt{\widehat{V}(\widehat{\pi})} = 0.032$$

10.4 TWO-PHASE SAMPLING FOR SELECTION OF SAMPLE

In case the auxiliary information (x) selected in the first phase is positive, it can be used as a measure of size for the selection of sample in the second phase. In this section we describe the use of PPSWR and Rao–Hartley–Cochran (1962, RHC) methods of sampling for the selection of samples in the second phase when the first-phase sample is selected by SRSWOR method.

10.4.1 Probability Proportional to Size With Replacement Sampling

Let us suppose that in the first phase, a sample s' of size n' be selected by SRSWOR method and the values of the auxiliary variable x are obtained. In the second phase, a subsample s of size n is selected from s' by using PPSWR method using x's as measure of size, i.e., the normed size measure for the ith unit is $p'_i = x_i/X'$ for $i \in s'$, where $X' = \sum_{i\in s'} x_i$. The estimator for the population total Y is given by

$$\widehat{Y}_{ppst} = \frac{N}{n'}\left(\frac{1}{n}\sum_{i\in s}\frac{y_i}{p'_i}\right) \tag{10.4.1}$$

Theorem 10.4.1

(i) $\widehat{Y}_{ppst}$ is an unbiased estimator for Y.

(ii) Variance of $\widehat{Y}_{ppst}$ is

$$V\left(\widehat{Y}_{ppst}\right) = N^2\left[\frac{(n'-1)}{N(N-1)n'n}\sum_{i\in U} p_i\left(\frac{y_i}{p_i} - Y\right)^2 + \left(\frac{1}{n'} - \frac{1}{N}\right)S_y^2\right]$$

where $p_i = x_i/X$ and $X = \sum_{i\in U} x_i$.

(iii) An unbiased estimator of $V\left(\widehat{Y}_{ppst}\right)$ is

$$\widehat{V}\left(\widehat{Y}_{ppst}\right) = N^2\left[\frac{N-1}{N(n'-1)n'}\widehat{V}\left(\frac{1}{n}\sum_{i\in s}\frac{y_i}{p'_i}|s'\right) + \frac{N-n'}{N(n'-1)}\left(\frac{1}{n'n}\sum_{i\in s}\frac{y_i^2}{p'_i} - \frac{\widehat{Y}^2_{ppst}}{N^2}\right)\right]$$

where $\widehat{V}\left(\frac{1}{n}\sum_{i\in s}\frac{y_i}{p'_i}|s'\right) = \frac{1}{n(n-1)}\sum_{i\in s}\left(\frac{y_i}{p'_i} - \frac{1}{n}\sum_{i\in s}\frac{y_i}{p'_i}\right)^2.$

Proof

(i)
$$E\left(\widehat{Y}_{ppst}\right) = E\left[\frac{N}{n'}E\left(\frac{1}{n}\sum_{i\in s}\frac{y_i}{p'_i}|s'\right)\right]$$
$$= NE\left(\frac{1}{n'}\sum_{i\in s'} y_i\right)$$
$$= NE(\overline{y}(s')) = Y$$

(ii)
$$V\left(\widehat{Y}_{ppst}\right) = E\left[\frac{N^2}{n'^2}V\left(\frac{1}{n}\sum_{i\in s}\frac{y_i}{p'_i}|s'\right)\right] + V\left[\frac{N}{n'}E\left(\frac{1}{n}\sum_{i\in s}\frac{y_i}{p'_i}|s'\right)\right]$$
$$= E\left[\frac{N^2}{n'^2}\left\{\frac{1}{2n}\sum_{i\neq}\sum_{j\in s'} p'_i p'_j\left(\frac{y_i}{p'_i} - \frac{y_i}{p'_i}\right)^2\right\}\right] + V(N\,\overline{y}(s'))$$
$$= \frac{N^2}{n'^2}\frac{1}{2n}\sum_{i\neq}\sum_{j\in U} p_i p_j\left(\frac{y_i}{p_i} - \frac{y_i}{p_i}\right)^2\pi^0_{ij} + N^2\left(\frac{1}{n'} - \frac{1}{N}\right)S_y^2$$

[where π^0_{ij} = inclusion probability of ith and jth unit $(i \neq j)$ in $s' = n'(n'-1)/\{N(N-1)\}$]

$$= N^2\left[\frac{(n'-1)}{n'nN(N-1)}\sum_{i\in U} p_j\left(\frac{y_i}{p_i} - Y\right)^2 + \left(\frac{1}{n'} - \frac{1}{N}\right)S_y^2\right]$$

(iii) Unbiased estimators of $V\left(\frac{1}{n}\sum_{i\in s}\frac{y_i}{p_i'}|s'\right)$ and S_y^2 are given, respectively, by

$$\widehat{V}\left(\frac{1}{n}\sum_{i\in s}\frac{y_i}{p_i'}|s'\right)=\frac{1}{n(n-1)}\sum_{i\in s}\left(\frac{y_i}{p_i'}-\frac{1}{n}\sum_{i\in s}\frac{y_i}{p_i'}\right)^2$$

and

$$\widehat{S}_y^2=\frac{1}{(N-1)}\left\{\frac{N}{n'n}\sum_{i\in s}\frac{y_i^2}{p_i'}-\frac{\widehat{Y}_{ppst}^2-\widehat{V}\left(\widehat{Y}_{ppst}\right)}{N}\right\}$$

Then an unbiased estimator of $V\left(\widehat{Y}_{ppst}\right)$ is obtained from the following equation

$$\begin{aligned}\widehat{V}\left(\widehat{Y}_{ppst}\right)&=N^2\left[\frac{1}{n'^2}\widehat{V}\left(\frac{1}{n}\sum_{i\in s}\frac{y_i}{p_i'}|s'\right)+\left(\frac{1}{n'}-\frac{1}{N}\right)\widehat{S}_y^2\right]\\&=N^2\left[\frac{1}{n'^2}\widehat{V}\left(\frac{1}{n}\sum_{i\in s}\frac{y_i}{p_i'}|s'\right)\right.\\&\quad\left.+\left(\frac{1}{n'}-\frac{1}{N}\right)\frac{1}{(N-1)}\left\{\frac{N}{n'}\frac{1}{n}\sum_{i\in s}\frac{y_i^2}{p_i'}-\frac{\widehat{Y}_{ppst}^2-\widehat{V}\left(\widehat{Y}_{ppst}\right)}{N}\right\}\right]\end{aligned}\tag{10.4.2}$$

Finally, Eq. (10.4.2) yields

$$\begin{aligned}\widehat{V}\left(\widehat{Y}_{ppst}\right)=N^2&\left[\frac{N-1}{N(n'-1)n'}\widehat{V}\left(\frac{1}{n}\sum_{i\in s}\frac{y_i}{p_i'}|s'\right)\right.\\&\left.+\frac{N-n'}{N(n'-1)}\left(\frac{1}{n'n}\sum_{i\in s}\frac{y_i^2}{p_i'}-\frac{\widehat{Y}_{ppst}^2}{N^2}\right)\right]\end{aligned}$$

10.4.2 Rao–Hartley–Cochran Sampling

Avadhani and Sukhatme (1970) considered two-phase sampling, where in the first phase, a sample s' of size n' is selected by SRSWOR method and in the second phase, a subsample s of size n is selected from s' by using the RHC method (see Section 5.6) using normed size measure $p_i'=x_i/X'$ for the ith unit, $i\in s'$. Let P_i' be the sum of the p' values of the group containing the ith unit in selecting the subsample s by RHC scheme from s'. The proposed estimator for Y is

$$\widehat{Y}_{rhct} = \frac{N}{n'} \sum_{j \in s} \frac{y_j}{p'_j} P'_i \tag{10.4.3}$$

Theorem 10.4.2

(i) $E\left(\widehat{Y}_{rhct}\right) = Y$

(ii) $V\left(\widehat{Y}_{rhct}\right) = N^2 \left[\frac{(n'-n)}{n'nN(N-1)} \sum_{i \in U} p_i \left(\frac{y_i}{p_i} - Y\right)^2 + \left(\frac{1}{n'} - \frac{1}{N}\right) S_y^2\right]$

(iii) $$\widehat{V}\left(\widehat{Y}_{rhct}\right) = N^2 \left[\frac{N-1}{n'N(n'-1)} \widehat{V}\left(\sum_{j \in s} \frac{y_j}{p'_j} P'_j | s'\right) + \frac{N-n'}{N(n'-1)} \left(\frac{1}{n'} \sum_{i \in s} \frac{y_i^2}{p'_i} P'_i - \frac{\widehat{Y}^2_{rhct}}{N^2}\right)\right]$$

where $\widehat{V}\left(\sum_{j \in s} \frac{y_j}{p'_j} P'_j | s'\right) = \frac{n'-n}{n'(n-1)} \sum_{j \in s} P'_j \left(\frac{y_j}{p'_j} - \sum_{j \in s} \frac{y_j}{p'_j} P'_j\right)^2$.

Proof

(i) $$E\left(\widehat{Y}_{rhct}\right) = E\left[\frac{N}{n'} E\left(\sum_{j \in s} \frac{y_j}{p'_j} P'_i | s'\right)\right]$$

$$= E\left(\frac{N}{n'} \sum_{j \in s'} y_j\right)$$

$$= Y$$

(ii) $$V\left(\widehat{Y}_{rhct}\right) = E\left[\frac{N^2}{n'^2} V\left(\sum_{j \in s} \frac{y_j}{p'_j} P'_i | s'\right)\right] + V\left[\frac{N}{n'} E\left(\sum_{j \in s} \frac{y_j}{p'_j} P'_i | s'\right)\right]$$

$$= \frac{N^2}{n'^2} \frac{n'-n}{n(n'-1)} \frac{1}{2} E \sum_{i \neq} \sum_{j \in s'} p_i p_j \left(\frac{y_i}{p_i} - \frac{y_j}{p_j}\right)^2 + N^2 V\left(\frac{1}{n'} \sum_{j \in s'} y_j\right)$$

$$= N^2 \left[\frac{(n'-n)}{n'nN(N-1)} \sum_{i \in U} p_i \left(\frac{y_i}{p_i} - Y\right)^2 + \left(\frac{1}{n'} - \frac{1}{N}\right) S_y^2\right]$$

(iii) Now, noting $\widehat{V}\left(\sum_{j\in s}\frac{y_j}{p'_j}P'_j|s'\right)=\frac{n'-n}{n'(n-1)}\sum_{j\in s}P'_j\left(\frac{y_j}{p'_j}-\sum_{j\in s}\frac{y_j}{p'_j}P'_j\right)^2$

and

$\widehat{S}^2_{rhct}=\frac{1}{N-1}\left\{\frac{N}{n'}\sum_{i\in s}\frac{y_i^2}{p'_i}P'_i-\frac{\widehat{Y}^2_{rhct}-\widehat{V}\left(\widehat{Y}_{rhct}\right)}{N}\right\}$ are unbiased estimators of $V\left(\sum_{j\in s}\frac{y_j}{p'_j}P'_j|s'\right)$ and S_y^2, respectively, we find $\widehat{V}\left(\widehat{Y}_{rhct}\right)$ as an unbiased estimator of $V\left(\widehat{Y}_{rhct}\right)$ from the following equation:

$$\widehat{V}\left(\widehat{Y}_{rhct}\right)=\frac{N^2}{n'^2}\widehat{V}\left(\sum_{j\in s}\frac{y_j}{p'_j}P'_j|s'\right)+N^2\left(\frac{1}{n'}-\frac{1}{N}\right)\widehat{S}^2_{rhct}$$

i.e., $$\widehat{V}\left(\widehat{Y}_{rhct}\right)=N^2\left[\frac{1}{n'^2}\widehat{V}\left(\sum_{j\in s}\frac{y_j}{p'_j}P'_j|s'\right)+\left(\frac{1}{n'}-\frac{1}{N}\right)\frac{1}{N-1}\left\{\frac{N}{n'}\sum_{i\in s}\frac{y_i^2}{p'_i}P'_j-\frac{\widehat{Y}^2_{rhct}-\widehat{V}\left(\widehat{Y}_{rhct}\right)}{N}\right\}\right]$$

i.e., $$\widehat{V}\left(\widehat{Y}_{rhct}\right)=N^2\left[\frac{N-1}{n'N(n'-1)}\widehat{V}\left(\sum_{j\in s}\frac{y_j}{p'_j}P'_j|s'\right)+\frac{N-n'}{N(n'-1)}\left(\frac{1}{n'}\sum_{i\in s}\frac{y_i^2}{p'_i}P'_i-\frac{\widehat{Y}^2_{rhct}}{N^2}\right)\right]$$

10.5 TWO-PHASE SAMPLING FOR STRATIFICATION AND SELECTION OF SAMPLE

Arnab (1991) considered a two-phase sampling procedure, where in the first phase, a PPSWR sample s' of size n' is selected from the entire population U with known size measures z_i's (>0 for $i=1,\ldots,N$), and information of the auxiliary variable x_i's ($i\in s'$) is obtained. By using the information of the x_i's gathered in the initial sample s', the n' sampled units are assigned to a predetermined number of L strata. Let $\widetilde{s}_h$ be the sample of size n'_h corresponding to the hth stratum U_h ($h=1,\ldots,L$). In the second phase, subsamples s_h's of sizes $n_h=\gamma_h n'_h$ (assuming integer and $0\leq\gamma_h\leq 1$) are selected independently from each of the hth stratum by PPSWR method with normed size measure $q_{hi}=(x_{hi}/z_{hi})/\sum_{i\in s'_h}x_{hi}/z_{hi}$ for the ith unit $\left(i\in\widetilde{s}_h\right)$, where z_{hi} and x_{hi} are the values of the variables z and x for the ith unit of the hth stratum, respectively. Here it is assumed that n' is so large that $Prob\left(n'_h \geq 1\right) = 1$. The proposed estimator for the population total Y is

$$\widehat{Y}_A = \sum_h w_h \widehat{Y}_{Ah} \tag{10.5.1}$$

where $$\widehat{Y}_{Ah} = \frac{1}{n_h}\sum\nolimits_{j\in s_h} \frac{y_{hj} - \lambda_h x_{hj}}{\left(n'_h p_{hj}\right) q_{hj}} + \lambda_h \frac{1}{n'_h}\sum\nolimits_{j\in \tilde{s}_h} \frac{x_{hj}}{p_{hj}}, \qquad p_{hj} = z_{hj}/Z,$$

$Z = \sum\limits_{i\in U} z_i$, λ_h is a known constant, and $\sum_{j\in s}$ denotes sum over the units in s with repetition.

Theorem 10.5.1

(i) $E\left(\widehat{Y}_A\right) = Y$

(ii) $$V\left(\widehat{Y}_A\right) = \frac{1}{n'}\left[\sum_h \left(1 - \frac{1}{n' P_h}\right) \frac{\sigma_h^2(d_{\lambda h}|x)}{P_h \gamma_h} + \sigma^2(y|z)\right]$$

where $w_h = n'_h/n'$, $P_h = Z_h/Z$, $Z_h = \sum\limits_{i\in U_h} z_{hi}$, $\sigma_h^2(d_{\lambda h}|x) = \sum\limits_{j\in U_h} \frac{d_{\lambda hj}^2}{x_{hi}} X_h - D_{\lambda h}^2$,

$X_h = \sum\limits_{i\in U_h} x_{hi}$, $\quad d_{\lambda hj} = y_{hj} - \lambda_h x_{hj}$, $\quad D_{\lambda h} = \sum\limits_{j\in U_h} d_{\lambda hj} = Y_h - \lambda_h X_h$, and

$$\sigma^2(y|z) = \sum_{h=1}^{H}\sum_{j\in U_h} \frac{y_{hj}^2}{p_{hj}} - Y^2.$$

Proof

(i) $$\begin{aligned} E\left(\widehat{Y}_A\right) &= E\sum_h w_h E\left(\widehat{Y}_{Ah}\middle|s'_h\right)\\ &= E\sum_h w_h \sum_{j\in \widetilde{s}_h} \frac{y_{hj}}{\left(n'_h p_{hj}\right)}\\ &= E\sum_h \frac{w_h}{P_h} E\left(\sum_{j\in \widetilde{s}_h} \frac{y_{hj} P_h}{\left(n'_h p_{hj}\right)} \middle| \mathbf{w_h}\right)\\ &\qquad (\text{where } \mathbf{w_h} = (\mathrm{w}_1, \ldots, \mathrm{w}_L))\\ &= E\sum_h \frac{w_h}{P_h} Y_h\\ &= Y\\ &\qquad (\text{noting } E(w_h) = P_h) \end{aligned}$$

(ii) $V\left(\widehat{Y}_A\right) = E\left[\sum_h w_h^2 V\left(\widehat{Y}_{Ah}|\widetilde{s}_h\right)\right] + V\left[\sum_h w_h E\left(\widehat{Y}_{Ah}|\widetilde{s}_h\right)\right]$

Now,

$$V\left\{\sum_h w_h E\left(\widehat{Y}_{Ah}|\widetilde{s}_h\right)\right\} = V\left\{\sum_h w_h \sum_{j\in \widetilde{s}_h} \frac{y_{hj}}{\left(n'_h p_{hj}\right)}\right\}$$

$$= E\left\{\sum_h \frac{w_h^2}{P_h^2} V\left(\frac{1}{n'_h}\sum_{j\in \widetilde{s}_h}\frac{y_{hj}}{p'_{hj}}|\mathbf{w_h}\right)\right\}$$

$$+ V\left\{\sum_h \frac{w_h}{P_h} E\left(\frac{1}{n'_h}\sum_{j\in \widetilde{s}_h}\frac{y_{hj}}{p'_{hj}}|\mathbf{w_h}\right)\right\}$$

$\left(\text{where } p'_{hi} = p_{hi}/P_h\right)$

$$= E\left[\sum_h \frac{w_h^2}{P_h^2}\frac{1}{n'_h}\left(\sum_{j\in U_h}\frac{y_{hj}^2}{p'_{hj}} - Y_h^2\right) + V\left(\sum_h \frac{w_h}{P_h} Y_h\right)\right]$$

$$= \frac{1}{n'}\left[\sum_h \frac{1}{P_h}\left(\sum_{j\in U_h}\frac{y_{hj}^2}{p'_{hj}} - Y_h^2\right) + \left(\sum_h \frac{Y_h^2}{P_h} - Y^2\right)\right]$$

$$= \frac{1}{n'}\left(\sum_{h=1}^{H}\sum_{j\in U_h}\frac{y_{hj}^2}{p_{hj}} - Y^2\right)$$

$$= \frac{\sigma^2(y|z)}{n'} \tag{10.5.2}$$

and

$$E\left\{\sum_h w_h^2 V\left(\widehat{Y}_{Ah}|\widetilde{s}_h\right)\right\} = E\left\{\sum_h w_h^2 V\left(\sum_{j\in s_h} \frac{y_{hj} - \lambda_h x_{hj}}{\left(n'_h p_{hj}\right) n_h q_{hj}} |\widetilde{s}_h\right)\right\}$$

$$= E\sum_h \frac{w_h^2}{n_h}\left\{\sum_{j\in \widetilde{s}_h} \frac{d_{\lambda hj}^2}{\left(n'_h p_{hj}\right)^2 q_{hj}} - \left(\sum_{j\in \widetilde{s}_h} \frac{d_{\lambda hj}}{n'_h p_{hj}}\right)^2\right\}$$

$$= E\sum_h \frac{w_h^2}{n_h} E\left\{\sum_{j\in \widetilde{s}_h} \frac{d_{\lambda hj}^2}{\left(n'_h p_{hj}\right)^2} \frac{p_{hj}}{q'_{hj}} \sum_{j\in \widetilde{s}_h} \frac{q'_{hj}}{p_{hj}} - \left(\sum_{j\in \widetilde{s}_h} \frac{d_{\lambda hj}}{n'_h p_{hj}}\right)^2 |\mathbf{w_h}\right\}$$

$$\left(\text{where } q'_{hj} = x_{hj}/X_h\right)$$

$$= E\sum_h \frac{w_h^2}{n_h} E\left\{\sum_{j\in \widetilde{s}_h} \frac{d_{\lambda hj}^2}{\left(n'_h p_{hj}\right)^2} + \sum_{j\in \widetilde{s}_h} \sum_{k(\neq j)\in \widetilde{s}_h} \frac{d_{\lambda hj}^2}{\left(n'_h p_{hj}\right)^2} \frac{p_{hj}}{q'_{hj}} \frac{q'_{hk}}{p_{hk}}\right.$$

$$\left. - \left(\sum_{j\in \widetilde{s}_h} \frac{d_{\lambda hj}}{n'_h p_{hj}}\right)^2 |\mathbf{w_h}\right\}$$

$$= E\sum_h \frac{w_h^2}{n_h}\left[\frac{1}{\left(n'_h\right)^2}\left\{\frac{n'_h}{P_h}\sum_{j\in U_h} \frac{d_{\lambda hj}^2}{p_{hj}} + \frac{n'_h\left(n'_h - 1\right)}{P_h^2}\left(\sum_{j\in U_h} \frac{d_{\lambda hj}^2}{q'_{hj}}\right)\right\}\right.$$

$$\left. - \frac{1}{P_h^2}\left\{\frac{1}{n'_h}\left(\sum_{j\in U_h} \frac{d_{\lambda hj}^2}{p'_{hj}} - D_{\lambda h}^2\right) + D_{\lambda h}^2\right\}\right]$$

$$\left(\text{noting } \sum_{j\in U_h} q'_{hj} = 1\right)$$

$$= E\sum_h \frac{w_h^2}{n_h} \frac{n'_h - 1}{n'_h P_h^2}\left(\sum_{j\in U_h} \frac{d_{\lambda hj}^2}{x_{hj}} X_h - D_{\lambda h}^2\right)$$

$$\left(\text{noting } p'_{hj} = \frac{z_{hj}}{Z_h} = \frac{p_{hj}}{P_h} \text{ and } q'_{hj} = \frac{x_{hj}}{X_h}\right)$$

Finally, writing $n_h = \gamma_h n'_h$, we get

$$E\sum_h w_h^2 V\left(\widehat{Y}_{Ah}|\widetilde{s}_h\right) = \sum_h \left(1 - \frac{1}{n'P_h}\right)\frac{\sigma_h^2(d_{\lambda h}|x)}{n'\gamma_h P_h} \tag{10.5.3}$$

The proof (ii) follows from Eqs. (10.5.2) and (10.5.3).

10.6 EXERCISES

10.6.1 Let an initial sample s' of size n' be selected from a finite population of size N by SRSWOR method and information on the auxiliary variable x is collected. Then an independent sample s of size n is selected by SRSWOR method from the entire population, and information on the auxiliary (x) and study variable (y) is obtained. Let $t = \overline{y}(s) - k\{\overline{x}(s) - \overline{x}(s')\}$.

Show that

(i) $E(t) = \overline{Y}$,

(ii) $V(t) = \left(\frac{1}{n} - \frac{1}{N}\right)\left(S_y^2 + k^2 S_x^2 - 2kS_{xy}\right) + k^2\left(\frac{1}{n'} - \frac{1}{N}\right)S_x^2$

(iii) $\widehat{V}(t) = \left(\frac{1}{n} - \frac{1}{N}\right)\left(s_y^2 + k^2 s_x^2 - 2ks_{xy}\right) + k^2\left(\frac{1}{n'} - \frac{1}{N}\right)s_x^2$ is an unbiased estimator for $V(t)$

(iv) Find the value of k for which $V(t)$ attains a minimum (Raj, 1968).

10.6.2 Initially, a sample s' of size n' is selected from a finite population of size N by PPSWR method using p_i as a normed size measure for the ith unit, and information on the auxiliary variable x is collected. Then an independent sample s of size n is selected by PPSWR method using the same normed size measure for the ith unit, and information on the auxiliary (x) and study variable (y) is obtained.

Let $t = \frac{1}{n}\sum_{i\in s}\frac{y_i}{p_i} - k\left(\frac{1}{n}\sum_{i\in s}\frac{x_i}{p_i} - \frac{1}{n'}\sum_{i\in s'}\frac{x_i}{p_i}\right)$ and $\sum_{i\in s}$ is the sum over units in s with repetition. Show that

(i) $E(t) = \overline{Y}$

(ii) $V(t) = \frac{\sigma^2(d_k|p)}{n} + k^2\frac{\sigma^2(x|p)}{n'}$, where $\sigma^2(d_k|p) = \sum_{i\in U}\frac{d_{ki}^2}{p_i} - D_k^2$,

$\sigma^2(x|p) = \sum_{i\in U}\frac{x_i^2}{p_i} - X^2$, $d_{ki} = y_i - kx_i$, and $D_k = \sum_{i\in U} d_{ki}$

(iii) $\widehat{V}(t) = \frac{\widehat{\sigma}^2(d_k|p)}{n} + k^2\frac{\widehat{\sigma}^2(x|p)}{n'}$ is an unbiased estimator for $V(t)$, where

$$\widehat{\sigma}^2(d_k|p) = \frac{\sum_{i\in s}\left(\frac{d_{ki}}{p_i} - \frac{1}{n}\sum_{i\in s}\frac{d_{ki}}{p_i}\right)^2}{(n-1)} \text{ and } \widehat{\sigma}^2(x|p) = \frac{\sum_{i\in s'}\left(\frac{x_i}{p_i} - \frac{1}{n'}\sum_{i\in s}\frac{x_i}{p_i}\right)^2}{(n'-1)}$$

(iv) Find the value of k for which $V(t)$ attains the minimum (Raj, 1968).

10.6.3 In the first phase an initial sample s' of size n' is selected by SRSWOR method, and information on the auxiliary variable x is collected. At the second phase, a subsample s of size n is selected from s' by Lahiri–Midzuno–Sen (1951, 1952, 1953) sampling procedure using x as a measure of size.

Show that $T = N\dfrac{\sum_{i\in s} y_i}{\sum_{i\in s} x_i}\left(\sum_{i\in s'} x_i/n'\right)$ is an unbiased estimator for the population total Y. Derive the variance of T and find an unbiased estimator of $V(T)$ (Raj, 1968).

10.6.4 In the first phase a sample s' of size n' is selected by RHC sampling scheme using a normed size measure p_i attached to the ith unit. At the second phase a subsample s of size n is selected from s' by SRSWOR method. Let $T = \frac{n'}{n}\sum_{i\in s}\frac{y_i}{p_i}P_i - k\left(\frac{n'}{n}\sum_{i\in s}\frac{x_i}{p_i}P_i - \sum_{i\in s'}\frac{x_i}{p_i}P_i\right)$, where P_i is the sum of p_j values for the group (which was formed for the selection of the sample) containing the ith unit.

(a) Verify the following results:

(i) $E(T) = Y$

(ii) $V(T) = \left(\frac{1}{n} - \frac{1}{n'}\right)\frac{1}{(N-1)}\left[n'\overline{V}(d_k) + (N-n')\sigma^2(d_k|p)\right] + \frac{N-n'}{n'(N-1)}\sigma^2(y|p)$

where $d_{ki} = y_i - kx_i$, $\sigma^2(y|p) = \sum_{i\in U}\frac{y_i^2}{p_i} - Y^2$, $\sigma^2(d_k|p) = \sum_{i\in U}\frac{d_{ki}^2}{p_i} - D_k^2$, $D_k = \sum_{i\in U} d_{ki}$, and $\overline{V}(d_k) = N\sum_{i\in U} d_{ki}^2 - D_k^2$.

(b) Find an unbiased estimator of $\widehat{V}(T)$ and the optimum value of value of k that minimizes $V(T)$ (Arnab, 1979).

10.6.5 Let π_{1i} and π_{1ij} be the first-order and second-order inclusion probabilities for the first-phase sample s_1, and $\pi_{2i|s_1}$ and $\pi_{2ij|s_1}$ be the conditional first-order and second-order inclusion probabilities for the second-phase sample s_2 given s_1. Show that

(i) $\widehat{Y}_{DE} = \sum_{i \in s_2} \frac{y_i}{\pi_i^*}$ with $\pi_i^* = \pi_{1i}\pi_{2i|s_1}$ is an unbiased estimator of the total Y.

(ii) The variance of $\widehat{Y}_{DE}$ can be estimated unbiasedly by using any of the following estimators:

$$\text{(a)}\quad \widehat{V}_1 = \sum_{i \in s_2} \frac{\Delta_{1ij}}{\pi_{ij}^*} \frac{y_i}{\pi_{1i}} \frac{y_j}{\pi_{1j}} + \sum_{i \in s_2} \sum_{j \in s_2} \frac{\Delta_{2ij|s_1}}{\pi_{ij|s_1}} \frac{y_i}{\pi_i^*} \frac{y_j}{\pi_j^*}$$

and

$$\text{(b)}\quad \widehat{V}_2 = \sum_{i \in s_2} \sum_{\substack{j \in s_2 \\ i<j}} \frac{\Delta_{1ij}}{\pi_{ij}^*} \left(\frac{y_i}{\pi_{1i}} - \frac{y_j}{\pi_{1j}} \right)^2 + \sum_{i \in s_2} \sum_{\substack{j \in s_2 \\ i<j}} \frac{\Delta_{2ij|s_1}}{\pi_{ij|s_1}} \left(\frac{y_i}{\pi_i^*} - \frac{y_j}{\pi_j^*} \right)^2$$

where $\pi_{ij}^* = \pi_{1ij}\pi_{2ij|s_1}$, $\Delta_{1ij} = \pi_{1i}\pi_{ij} - \pi_{1ij}$, and $\Delta_{2ij|s_1} = \pi_{2i|s_1}\pi_{2j|s_1} - \pi_{2ij|s_1}$ (Haziza et al., 2011).

10.6.6 To estimate the average price of a two-bedroom flat in a locality, a sample of 75 two-bedroom flats was selected by SRSWOR from the locality of 790 flats and eye estimates of the price of the selected flats were obtained. In phase 2, a random sample of 15 flats was selected from the selected 75 flats by SRSWOR method and the prices of the flats were determined by an appropriate evaluation agency. The findings are given in the following table:

Serial number of flats	Price of the flat by eye estimation (in 000 \$) (x)	Price of the flat by valuation agency (in 000 \$) (y)
1	450	550
2	565	600
3	700	650
4	850	900
5	650	750
6	700	820
7	850	970
8	550	700
9	600	650
10	750	700
11	900	850
12	850	950
13	840	900
14	700	850
15	760	800

Estimate the average price of the flat along with its standard error. If the cost evaluation of a flat by an agency is five times that of the eye estimation, determine the proportion of the second-phase sample that should be selected from the first-phase sample.

10.6.7 To estimate the average daily wages of a factory of 800 workers, an initial sample of 125 workers was selected at random by SRSWOR method. The workers were then stratified according to their skill. Finally, samples of workers were selected from each of the strata by SRSWOR method, and information regarding the daily wages was obtained. The data are given below.

(i) Give an estimate of the average daily wages of the factory workers.

(ii) Estimate the standard error of your estimator.

Workers type	No. of workers in the initial sample	No. of workers in the final sample	Daily wages (in $)
Highly skilled	10	5	625, 750, 1025, 1500, 1000
Skilled	60	10	320, 350, 375, 200, 300, 325, 325, 475, 350, 300
Semiskilled	40	8	200, 175, 150, 240, 250, 180, 120, 170
Unskilled	20	8	150, 175, 100, 140, 200, 160, 100, 100

CHAPTER 11

Repetitive Sampling

11.1 INTRODUCTION

In a repetitive or successive sampling, we survey the same population on different occasions for the purpose of estimating the population parameters for the most recent occasion, for changes over previous occasions, estimating parametric functions over last few occasions, among others. For example, several government agencies collect unemployment data on a monthly basis to estimate the rate of unemployment for the most recent month, change over previous month, average of unemployment rate for the last one year, etc. In repetitive sampling, one has to effectively utilize the accumulated data procured in the course of the survey along with other auxiliary information that may also incidentally be available. The objectives of repetitive sampling include (i) estimating the mean or proportion for the most recent occasion, (ii) difference between the means of two consecutive occasions, or (iii) average of means of all the occasions. For measuring changes over two occasions, we generally observe the same sample over two consecutive periods, while fresh samples for each of the occasions are selected for measuring the mean of all the occasions. For estimating the mean or total of the most recent occasion, a portion of the sample selected on the earlier occasions, technically called a matched sample, is retained and it is supplemented by a fresh sample for the most recent occasion.

From the records we find that the pioneering work on sampling over two occasions was done by Jessen (1942) to estimate mean on the most recent occasion. Yates (1949) extended Jessen (1942) study to include cases where sampling is repeated $h(\geq 2)$ occasions. He, however, worked under the restriction of a constant "total sample size" and "matching sampling fraction" over the occasions and postulated the correlation coefficient of the values for the units over k occasions apart to be of the form ρ^k. Some of his restrictions were later relaxed by Patterson (1950) and Tikkiwal (1951). Ecler (1955) stretched the ideas further to develop the theory of rotation sampling. However, in these investigations, the challenge was to estimate a finite population mean, and the optimalities of the suggested sampling strategies were established under the postulation of an infinite population

Survey Sampling Theory and Applications
ISBN 978-0-12-811848-1
http://dx.doi.org/10.1016/B978-0-12-811848-1.00011-X

setup. Kuldoroff (1963) and Rao and Graham (1964) developed a truly finite population theory for estimating the population mean over various occasions. Raj (1965a,b), Ghangurde and Rao (1969), Avadhan and Sukhatme (1970), Chotai (1974), and Chaudhuri and Arnab (1979a), and Arnab (1991, 1998a), among others, contributed theories involving varying sampling schemes over two, as well as more than two, occasions. Besides, Singh (1968), Singh and Kathuria (1969), and Arnab (1980) are among those who developed multistage sampling techniques for two or more occasions. Duncan and Kalton (1987) and Binder and Hidroglou (1988) cited the application of time series analysis in successive sampling for estimating population characteristics at regular intervals of time.

11.2 ESTIMATION OF MEAN FOR THE MOST RECENT OCCASION

Consider a finite population $U = (u_1,\ldots, u_i,\ldots, u_N)$ of N identifiable units, which is supposed to be sampled over various occasions to estimate the population mean for the most recent occasion. Here, we assume that the units of the population are unchanged over various occasions, i.e., the sampling frames remain the same: no new units are added or deleted from the population. Let y_{hi} be the value of the variate under study y for the ith unit of the population on the hth occasion, $Y_h = \sum_{i\in U} y_{hi}$ be the total and $\overline{Y}_h = Y_h/N$ be its mean.

11.2.1 Sampling on Two Occasions

11.2.1.1 Sampling Scheme

On the first occasion a sample s_1 of size n_1 is selected from the population U with probability $p(s_1)$ by some suitable sampling scheme. On the second occasion a subsample (which is technically called the "matched sample") s_{2m} of suitable size $m(\leq n_1)$ is selected from s_1 with probability $p(s_{2m}|s_1)$ by some appropriate sampling scheme, and an unmatched sample s_{2u} of size $u(=n_2 - m)$ is selected with probability $p(s_{2u}|s_1)$ either from the entire population U or from $U - s_1$, the set of units not selected in the first occasions. The sample size m is chosen in an optimal manner viz. considering either the cost of the survey or the efficiency of the estimators. The sample in the second occasion will be denoted by $s_2 = s_{2m} \cup s_{2u}$.

11.2.1.2 General Method of Estimation

Here, we construct two unbiased estimators of $\overline{Y}_1$, the population mean of the first occasion. One of the estimators is based on the initial sample s_1 and the other is based on the matched sample s_{2m}. They are respectively as follows:

$$\widehat{\overline{Y}}_1 = \sum_{i \in s_1} b(s_1, i) y_{1i} \text{ and } t_{1m} = \sum_{i \in s_{2m}} b(s_{2m}, i) y_{1i} \tag{11.2.1}$$

The estimator t_{1m} is such that $E(t_{1m}|s_1) = \widehat{\overline{Y}}_1$, i.e., $b(s_{2m}, i)$'s satisfy $\sum_{s_{2m} \supset i} b(s_{2m}, i) p(s_{2m}|s_1) = b(s_1, i) \forall i \in s_1$. The unbiasedness conditions of $\widehat{\overline{Y}}_1$ yield $\sum_{s_1 \supset i} b(s_1, i) p(s_1) = 1/N \quad \forall \ i \in U$.

Similarly, from the data collected on the second occasion through the matched sample s_{2m} and the unmatched sample s_{2u}, two unbiased estimators of the mean $\overline{Y}_2$ of the second occasion are constructed as follows:

$$t_{2m} = \sum_{i \in s_{2m}} b(s_{2m}, i) y_{2i} \text{ and } \widehat{\overline{Y}}_{2u} = \sum_{i \in s_{2u}} b(s_{2u}, i) y_{2i}$$

The constants $b(s_{2u}, i)$'s satisfy the unbiasedness condition $\sum_{s_{2u} \supset i} b(s_{2u}, i) p(s_{2u}|s_1) = 1/N \quad \forall \ i \in U$. From the estimators $\widehat{\overline{Y}}_1$, t_{1m}, t_{2m}, and $\widehat{\overline{Y}}_{2u}$, a composite estimator of $\overline{Y}_2$ is obtained as

$$\widehat{\overline{Y}}_{2c} = \phi t_{2m} + \psi_1 t_{1m} + \psi_2 \widehat{\overline{Y}}_1 + \varphi \widehat{\overline{Y}}_{2u} \tag{11.2.2}$$

The weights ϕ, ψ_1, ψ_2, and φ are chosen to make $\widehat{\overline{Y}}_{2c}$ unbiased for $\overline{Y}_2$. The condition $E\left(\widehat{\overline{Y}}_{2c}\right) = \overline{Y}_2$ yields

$$\varphi + \phi = 1 \text{ and } \psi_1 + \psi_2 = 0 \tag{11.2.3}$$

On substituting Eq. (11.2.3) in Eq. (11.2.2) we get

$$\widehat{\overline{Y}}_{2c} = \phi \widehat{\overline{Y}}_{2m} + (1 - \phi) \widehat{\overline{Y}}_{2u} \tag{11.2.4}$$

where $\widehat{\overline{Y}}_{2m} = t_{2m} - \beta\left(t_{1m} - \widehat{\overline{Y}}_1\right)$ and $\beta = -\psi_1/\varphi$.

The variance of $\widehat{\overline{Y}}_{2m}$ is

$$V\left(\widehat{\overline{Y}}_{2m}\right) = E\left\{V\left(\widehat{\overline{Y}}_{2m}\middle|s_1\right)\right\} + V\left\{E\left(\widehat{\overline{Y}}_{2m}\middle|s_1\right)\right\}$$

$$= E\{V(t_{2m}|s_1)\} + \beta^2 E\{V(t_{1m}|s_1)\} - 2\beta E\{Cov(t_{2m}, t_{1m}|s_1)\} + V\left\{\sum_{i \in s_1} b(s_1, i) y_{2i}\right\}$$

The optimum value of β that minimizes the variance $V\left(\widehat{\overline{Y}}_{2m}\right)$ is

$$\beta_0 = E\{Cov(t_{1m}, t_{2m}|s_1)\}/E\{V(t_{1m}|s_1)\} \tag{11.2.5}$$

The estimator $\widehat{\overline{Y}}_{2m}$ with $\beta = \beta_0$ will be denoted by

$$\widehat{\overline{Y}}^{0}_{2m} = t_{2m} - \beta_0\left(t_{1m} - \widehat{\overline{Y}}_1\right) \tag{11.2.6}$$

By replacing $\widehat{\overline{Y}}_{2m}$ with $\widehat{\overline{Y}}^{0}_{2m}$ in Eq. (11.2.4), we derive the composite estimator for $\overline{Y}_2$ as

$$\widehat{\overline{Y}}^{0}_{2c} = \phi\widehat{\overline{Y}}^{0}_{2m} + (1-\phi)\widehat{\overline{Y}}_{2u} \tag{11.2.7}$$

Minimizing the variance of $\widehat{\overline{Y}}^{0}_{2c}$ with respect to ϕ, the optimum value of ϕ is obtained as

$$\phi_0 = \left(1/V'_m + 1/V'_u\right)^{-1}\Big/V'_m \tag{11.2.8}$$

where $V'_m = V\left(\widehat{\overline{Y}}^{0}_{2m}\right) - V_{m,u}$, $V'_u = V\left(\widehat{\overline{Y}}_{2u}\right) - V_{m,u}$ and $V_{m,u} = Cov\left(\widehat{\overline{Y}}^{0}_{2m}, \widehat{\overline{Y}}_{2u}\right)$.

Substituting $\phi = \phi_0$ in Eq. (11.2.7), the optimum estimator of $\overline{Y}_2$ is obtained as

$$\widehat{\overline{Y}}_2(opt) = \phi_0\widehat{\overline{Y}}^{0}_{2m} + (1-\phi_0)\widehat{\overline{Y}}_{2u} \tag{11.2.9}$$

The expression for variance of $\widehat{\overline{Y}}_2(opt)$ is obtained as

$$V_{opt} = \left(1/V'_m + 1/V'_u\right)^{-1} + V_{m,u} \tag{11.2.10}$$

In many situations the expression of V_{opt} is obtained as an explicit function of m and u. In this situation we can find the optimum value of m (i.e., $u = n_2 - m$) by minimizing V_{opt} while keeping total sample size for the second occasion fixed or by keeping total cost of the survey to a certain level using a suitable cost function. The optimum values of m and u so obtained will be denoted by m_0 and u_0, respectively. Finally, replacing the optimum value of m_0 and u_0 in the expressions of $\widehat{\overline{Y}}_2(opt)$ and V_{opt}, the final estimator of $\overline{Y}_2$ and its variance are obtained and they are denoted, respectively, by

$$\widehat{\overline{Y}}_2(\min) = \widehat{\overline{Y}}_2(opt)\Big|_{m=m_0, u=u_o} \tag{11.2.11}$$

and

$$V_{\min} = V\left\{\widehat{\overline{Y}}_2(\min)\right\} = V_{opt}\big|_{m=m_0, u=u_o} \tag{11.2.12}$$

The optimum sampling strategy for sampling over two occasions is a combination of the sampling designs used for the selection of samples over two occasions and the estimator $\widehat{\overline{Y}}_2(\min)$ that yields the minimum value of $V_{\min}$.

In the next section we will present a few sampling strategies over two occasions that are commonly used in practice for estimating $\overline{Y}_2$, the mean of the current (second) occasion. The expressions for $V_{\min}$ and the optimum fraction of the matched sample $\lambda_0 = m_0/n_2$ are also derived.

Remark 11.2.1

In most situations the estimators $\widehat{\overline{Y}}^0_{2m}$ and $\widehat{\overline{Y}}_2(opt)$ cannot be used in practice because they involve unknown parameters. In practice, we replace unknown parameters with their suitable estimates. For example, we take

$$\widehat{\overline{Y}}^0_{2m} = t_{2m} - \widehat{\beta}_0\left(t_{1m} - \widehat{\overline{Y}}_1\right)$$

and

$$\widehat{\overline{Y}}_2(opt) = \widehat{\phi}_0\widehat{\overline{Y}}^0_{2m} + \left(1 - \widehat{\phi}_0\right)\widehat{\overline{Y}}_{2u}$$

where $\widehat{\beta}_0 = \widehat{C}ov(t_{1m}, t_{2m}|s_1)/\widehat{V}(t_{1m}|s_1)$; $\widehat{\phi}_0 = \left(1/\widehat{V}'_m + 1/\widehat{V}'_u\right)^{-1}/\widehat{V}'_m$;

and $\widehat{C}ov(t_{1m}, t_{2m}|s_1)$, $\widehat{V}(t_{1m}|s_1)$, $\widehat{V}'_m$, and $\widehat{V}'_u$ are unbiased estimators of $Cov(t_{1m}, t_{2m}|s_1)$, $V(t_{1m}|s_1)$, V'_m, and V'_u, respectively.

11.2.1.3 Simple Random Sampling Without Replacement

On the first occasion, a sample s_1 of size n_1 is selected from the population U by simple random sampling without replacement (SRSWOR) method. On the second occasion, a matched sample s_{2m} of size $m(\leq n_1)$ is selected from s_1 by SRSWOR method. The unmatched sample s_{2u} of size $u(=n_2 - m \geq 0)$ is selected from $U - s_1$, the set of $N - n_1$ units that were not selected on the first occasion.

The proposed unbiased estimator for $\overline{Y}_2$ based on the matched sample s_{2m} and the initial sample s_1 is given by

$$\widehat{\overline{Y}}_{2m} = \overline{y}_{2m} - \beta\left(\overline{y}_{1m} - \overline{y}_{1n_1}\right) \tag{11.2.13}$$

where $\overline{y}_{2m}=\sum_{i\in s_{2m}} y_{2i}/m=$ sample mean of y for the second occasion based on s_{2m}, $\overline{y}_{1m}=\sum_{i\in s_{2m}} y_{1i}/m=$ sample mean of y for the first occasion based on s_{2m}, $\overline{y}_{1n_1}=\sum_{i\in s_1} y_{1i}/n_1=$ sample mean of y for the first occasion based on s_1, and β is a suitably chosen constant.

Theorem 11.2.1

(i) $E\left(\widehat{\overline{Y}}_{2m}\right)=\overline{Y}_2$ and (ii) $V\left(\widehat{\overline{Y}}_{2m}\right)=\left(\frac{1}{m}-\frac{1}{n_1}\right)S_{d\beta}^2+\left(\frac{1}{n_1}-\frac{1}{N}\right)S_{2y}^2$

where $S_{d\beta}^2=\sum_{i\in U}\left(d_{i\beta}-\overline{D}_\beta\right)^2/(N-1)$, $d_{i\beta}=y_{2i}-\beta y_{1i}$, $\overline{D}_\beta=\overline{Y}_2-\beta\overline{Y}_1$,

and $S_{2y}^2=\sum_{i\in U}\left(y_{2i}-\overline{Y}_2\right)^2/(N-1)$

Proof

(i) $E\left(\widehat{\overline{Y}}_{2m}\right)=E\left\{E\left(\widehat{\overline{Y}}_{2m}\middle|s_1\right)\right\}=E\left(\overline{y}_{2n_1}\right)=\overline{Y}_2$

$\left(\text{where } \overline{y}_{2n_1}=\sum_{i\in s_1} y_{2i}/n_1\right)$

(ii) $V\left(\widehat{\overline{Y}}_{2m}\right)=E\left\{V\left(\widehat{\overline{Y}}_{2m}\middle|s_1\right)\right\}+V\left\{E\left(\widehat{\overline{Y}}_{2m}\middle|s_1\right)\right\}$

$=\left(\frac{1}{m}-\frac{1}{n_1}\right)E\left(s_{d\beta}^2\right)+V\left(\overline{y}_{2n}\right)$

$\left(\text{where } s_{d\beta}^2=\sum_{i\in s_1}\left(d_{i\beta}-\overline{d}_\beta\right)^2\Big/(n_1-1) \text{ and } \overline{d}_\beta=\sum_{i\in s_1} d_{i\beta}/n_1\right)$

$=\left(\frac{1}{m}-\frac{1}{n_1}\right)S_{d\beta}^2+\left(\frac{1}{n_2}-\frac{1}{N}\right)S_{2y}^2$

Now, minimizing $V\left(\widehat{\overline{Y}}_{2m}\right)$ with respect to β, the optimum value of β comes out as

$$\beta=\beta_0=S_{12y}\Big/S_{1y}^2 \tag{11.2.14}$$

where $S_{12y}=\sum_{i\in U}\left(y_{1i}-\overline{Y}_1\right)\left(y_{2i}-\overline{Y}_2\right)/(N-1)$ and $S_{1y}^2=\sum_{i\in U}\left(y_{1i}-\overline{Y}_1\right)^2/(N-1)$.

Placing $\beta = \beta_0$ in Eq. (11.2.13), the optimum estimator $\widehat{\overline{Y}}_{2m}$ and its variances come out as follows:

$$\widehat{\overline{Y}}^0_{2m} = \overline{y}_{2m} - \beta_0\left(\overline{y}_{1m} - \overline{y}_{1n_1}\right) \tag{11.2.15}$$

and

$$V\left(\widehat{\overline{Y}}^0_{2m}\right) = \left[\left(\frac{1}{m} - \frac{1}{n_1}\right)\rho^2 + \left(\frac{1}{n_1} - \frac{1}{N}\right)\right] S^2_{2y} \tag{11.2.16}$$

where $\rho = S_{12y}/(S_{1y}S_{2y})$ = correlation coefficient between y_1 and y_2.

An unbiased estimator for $\overline{Y}_2$ based on the unmatched sample s_{2u} is given by

$$\widehat{\overline{Y}}_{2u} = \overline{y}_{2u} = \sum_{i \in s_{2u}} y_{2i}/u \tag{11.2.17}$$

where $u = n_2 - m$.

The composite estimator for $\overline{Y}_2$ is given by

$$\begin{aligned}\widehat{\overline{Y}}_{2c} &= \phi \widehat{\overline{Y}}^0_{2m} + (1-\phi)\widehat{\overline{Y}}_{2u} \\ &= \phi\left\{\overline{y}_{2m} - \beta_0\left(\overline{y}_{1m} - \overline{y}_{1n_1}\right)\right\} + (1-\phi)\overline{y}_{2u}\end{aligned} \tag{11.2.18}$$

Theorem 11.2.2

The optimum value ϕ that minimizes $V\left(\widehat{\overline{Y}}_{2c}\right)$ and the corresponding value of $V\left(\widehat{\overline{Y}}_{2c}\right)$ are, respectively,

$$\phi_0 = \left\{\left(\frac{1}{m} - \frac{1}{n_1}\right)(1-\rho^2) + \frac{1}{n_1}\right\}^{-1}\left[\left\{\left(\frac{1}{m} - \frac{1}{n_1}\right)(1-\rho^2) + \frac{1}{n_1}\right\}^{-1} + u\right]^{-1}$$

and

$$V_{opt} = \left[\left\{\left(\frac{1}{m} - \frac{1}{n_1}\right)(1-\rho^2) + \frac{1}{n_1}\right\}^{-1} + u\right]^{-1} S^2_{2y} - \frac{S^2_{2y}}{N}$$

Proof

$$V\left(\widehat{\overline{Y}}_{2m}^{0}\right) = E\left[V\left\{\overline{y}_{2m} - \beta_0\left(\overline{y}_{1m} - \overline{y}_{1n_1}\right)\Big|s_1\right\}\right] + V\left[E\left\{\overline{y}_{2m} - \beta_0\left(\overline{y}_{1m} - \overline{y}_{1n_1}\right)\Big|s_1\right\}\right] = \left(\frac{1}{m} - \frac{1}{n_1}\right)(1-\rho^2)S_{2y}^2 + \left(\frac{1}{n_1} - \frac{1}{N}\right)S_{2y}^2 \quad (11.2.19)$$

$$V(\overline{y}_{2u}) = E\left\{V\left(\overline{y}_{2u}|s_1\right)\right\} + V\left\{E\left(\overline{y}_{2u}|s_1\right)\right\} = \left(\frac{1}{u} - \frac{1}{N-u}\right)E\left[\frac{1}{N-u-1}\sum_{i\in U-s_1}\left(y_{2i} - \sum_{i\in U-s_1} y_{2i}/(N-u)\right)^2\right] + V\left(\sum_{i\in U-s_1} y_{2i}/(N-u)\right) = \left(\frac{1}{u} - \frac{1}{N}\right)S_{2y}^2 \quad (11.2.20)$$

and

$$Cov\left(\widehat{\overline{Y}}_{2m}^{0}, \widehat{\overline{Y}}_{2u}\right) = Cov\left(\overline{y}_{2m}, \overline{y}_{2u}\right) = E\left\{Cov\left(\overline{y}_{2m}, \overline{y}_{2u}|s_1\right)\right\} + Cov\left\{E\left(\overline{y}_{2m}|s_1\right), E\left(\overline{y}_{2u}|s_1\right)\right\} = Cov\left(\sum_{i\in s_1} y_{2i}/n_1, \sum_{i\in U-s_1} y_{2i}/(N-n_1)\right) = -n_1^2 V\left(\overline{y}_{2n_1}\right)\Big/\left\{n_1(N-n_1)\right\} \left(\text{where } \overline{y}_{2n_1} = \sum_{i\in s_1} y_{2i}/n_1\right) = -S_{2y}^2\Big/N = V_{m,u} \quad (11.2.21)$$

Writing $V'_m = V\left(\widehat{\overline{Y}}^0_{2m}\right) - V_{m,u} = \left[\left(\frac{1}{m} - \frac{1}{n_1}\right)(1-\rho^2) + \frac{1}{n_1}\right] S^2_{2y}$, $V'_u = V\left(\widehat{\overline{Y}}_{2u}\right) - V_{m,u} = S^2_{2y}/u$ and using Eqs. (11.2.8) and (11.2.10), we get

$$\phi_0 = \left(1/V'_m + 1/V'_u\right)^{-1}/V'_m$$

$$= \left\{\left(\frac{1}{m} - \frac{1}{n_1}\right)(1-\rho^2) + \frac{1}{n_1}\right\}^{-1}\left[\left\{\left(\frac{1}{m} - \frac{1}{n_1}\right)(1-\rho^2) + \frac{1}{n_1}\right\}^{-1} + u\right]^{-1} \tag{11.2.22}$$

and

$$V_{opt} = \left(1/V'_m + 1/V'_u\right)^{-1} + V_{m,u}$$

$$= \left[\left\{\left(\frac{1}{m} - \frac{1}{n_1}\right)(1-\rho^2) + \frac{1}{n_1}\right\}^{-1} + u\right]^{-1} S^2_{2y} - S^2_{2y}/N \tag{11.2.23}$$

11.2.1.3.1 Optimum Allocation of the Matched Sample

Theorem 11.2.3

For fixed n_1 and n_2, the optimum value of the proportion of the matched sample $\lambda = m/n_2$ and the corresponding value of V_{opt} are given, respectively, by

$$\lambda_0 = \gamma\sqrt{1-\rho^2}\left(1+\sqrt{1-\rho^2}\right)^{-1} \quad \text{and}$$

$$V_{\min} = \left[\frac{1}{n_2}\left\{\gamma\frac{2}{1+\sqrt{1-\rho^2}} + (1-\gamma)\right\}^{-1} - \frac{1}{N}\right] S^2_{2y}$$

where $\gamma = n_1/n_2$.

Proof

Putting $u = n_2 - m$ in the expression of V_{opt} in Eq. (11.2.23) and then differentiating it with respect to m and equating the result to zero, we get

$$\frac{(1-\rho^2)}{m^2}\left[\left(\frac{1}{m}-\frac{1}{n_1}\right)(1-\rho^2)+\frac{1}{n_1}\right]^{-2}=1 \tag{11.2.24}$$

Eq. (11.2.24) yields the optimum value of λ as

$\lambda_0 = \gamma\sqrt{1-\rho^2}\left(1+\sqrt{1-\rho^2}\right)^{-1}$.

Finally, placing $\lambda = \lambda_0$ in the expression of V_{opt}, we obtain

$$V_{\min} = \left[\frac{1}{n_2}\left\{\gamma\frac{2}{1+\sqrt{1-\rho^2}}+(1-\gamma)\right\}^{-1}-\frac{1}{N}\right]S_{2y}^2$$

Remark 11.2.2

The estimators $\widehat{\overline{Y}}_{2m}^{0}$, $\widehat{\overline{Y}}_2(opt)$, and the optimum matching fraction λ_0 cannot be used in practice because they involve unknown parameters S_{1y}^2, S_{2y}^2, β_0, and ρ. So, in all practical purposes we replace the parameters by their respective estimates $s_{1y}^2 = \sum_{i\in s_{2m}}\left(y_{1i}-\overline{y}_{1m}\right)^2/(m-1)$, $s_{2y}^2 = \sum_{i\in s_{2m}\cup s_{2u}}\left(y_{2i}-\overline{y}_2\right)^2/(n_2-1)$, $\widehat{\beta}_0 = s_{12y}/s_{1y}^2$, and $\widehat{\rho} = s_{12y}/(s_{1y}\,s_{2y})$, where $s_{12y} = \sum_{i\in s_{2m}}\left(y_{1i}-\overline{y}_{1m}\right)\left(y_{2i}-\overline{y}_{2m}\right)/(m-1)$ and $\overline{y}_2 = \sum_{i\in s_{2m}\cup s_{2u}} y_{2i}/n_2$.

Remark 11.2.3

If no information about the first occasion is used for estimating the second occasion, i.e., in the second occasion an unmatched sample s_2 of size n_2 is selected, then we would use $\overline{y}(s_2) = \sum_{i\in s_2} y_{2i}/n_2$ as an estimator of the population mean $\overline{Y}_2$. The efficiency of the optimum estimator based on the partially matched sample compared to the totally unmatched sample of size n_2 is given by

$$E = \frac{V\{\overline{y}(s_2)\}}{V_{\min}} = (1-f_2)\left[\left\{\gamma\frac{2}{1+\sqrt{1-\rho^2}}+(1-\gamma)\right\}^{-1}-f_2\right]^{-1} \tag{11.2.25}$$

where $f_2 = n_2/N$.

The efficiency $E \geq 1$ as $\rho^2 \leq 1$ and it increases with ρ^2. Hence partial matching produces a gain in efficiency unless $\rho = 0$. The efficiency E attains a maximum value $1 + \dfrac{\gamma}{1-(1+\gamma)f_2}$ when $|\rho| = 1$.

Remark 11.2.4

The optimum proportion of matched sample λ_0 decreases as $|\rho|$ increases. λ_0 attains a maximum $\gamma/2$ when $|\rho| = 0$ and the minimum value of λ_0 is zero when is $|\rho| = 1$.

Remark 11.2.5

The minimum variance $V_{\min}$ decreases with increase of $|\rho|$ and it attains a minimum value $\left(\dfrac{1}{n_2(1+\gamma)} - \dfrac{1}{N}\right)S_{2y}^2$ when the correlation ρ is either $+1$ or -1, and in this case, $\lambda_0 = 0$. But when $\lambda_0 = 0$, no matched sample is used and in this case $\widehat{\overline{Y}}_{2c} = \overline{y}(s_2)$ and $V_{\min} = (1-f_2)S_{2y}^2/n_2$. In fact, the expression of $V_{\min}$ was derived assuming m, $u > 1$. So, when $|\rho|$ is very high we should take the matched sample m to be as small as possible with a minimum of 2 to estimate a parameter like β_0. Similarly, when ρ is very small, we should take m to be as large as possible and a small unmatched sample with a minimum of 2.

Remark 11.2.6

In case $n_1 = n_2 = n$, the expression for the minimum variance is obtained as

$$V_{\min} = \left(\frac{1+\sqrt{1-\rho^2}}{2n} - \frac{1}{N}\right)S_{2y}^2 \tag{11.2.26}$$

Remark 11.2.7

If $n_1 = n_2 = n$ and the unmatched sample s_{2u} is selected from the entire population, then s_1 and s_{2u} become independent and we have $V_{m,u} = 0$. In this situation

$$V_{\min} = \frac{1+\sqrt{1-\rho^2}}{2n}S_{2y}^2 \tag{11.2.27}$$

Remark 11.2.8

Because the magnitude of Eq. (11.2.27) is larger than that of Eq. (11.2.26), we conclude that the selection of the unmatched sample from those units that are not selected in the first occasion produces a gain in efficiency of the combined estimator.

11.2.1.4 Probability Proportional to Size With Replacement Sampling

Raj (1965a,b) proposed the following sampling scheme over two occasions: On the first occasion a sample s_1 of size n is selected from the population U using probability proportional to size with replacement (PPSWR) sampling with known measure of size $z_i(>0)$ for the ith unit, $i = 1,\ldots, N$. On the second occasion a matched subsample s_{2m} of size $m(\leq n)$ is selected from s_1 by SRSWOR method, treating all the selected units in s_1 as distinct. The unmatched sample s_{2u} of size $u(=n-m)$ is selected independently from the entire population by PPSWR method using z_i as the size of measure for the ith unit.

Chaudhuri and Arnab (1979a) proposed the following estimator for Y_2, which is more efficient than the Raj (1965a,b) estimator:

$$\widehat{\widehat{Y}}_{2c} = \phi \widehat{Y}_{2m} + (1-\phi)\widehat{Y}_{2u} \tag{11.2.28}$$

where $\widehat{Y}_{2m} = \frac{1}{m}\sum_{i\in s_{2m}} \frac{y_{2i}}{p_i} - \beta\left(\frac{1}{m}\sum_{i\in s_{2m}} \frac{y_{1i}}{p_i} - \frac{1}{n}\sum_{i\in s_1}\frac{y_{1i}}{p_i}\right)$, $\widehat{Y}_{2u} = \frac{1}{u}\sum_{i\in s_{2u}}\frac{y_{2i}}{p_i}$, $p_i = \frac{z_i}{Z}$, $Z = \sum_{i\in U} z_i$, $\sum_{i\in s_t}$ denotes the sum over the units in s_t with repetition, and β is a constant chosen to minimize $V\left(\widehat{Y}_{2m}\right)$.

Theorem 11.2.4

(i) $E\left(\widehat{Y}_{2m}\right) = Y_2$

(ii) $V\left(\widehat{Y}_{2m}\right) = \left(\frac{1}{m} - \frac{1}{n}\right)\left(V_{2|z} - 2\beta\delta\sqrt{V_{1|z}\cdot V_{2|z}} + \beta^2 V_{1|z}\right) + \frac{1}{n}V_{2|z}$

(iii) The optimum value of β that minimizes $V\left(\widehat{Y}_{2m}\right)$ is $\beta_0 = \delta\sqrt{V_{2|z}/V_{1|z}}$

(iv) $V\left(\widehat{Y}^0_{2m}\right) = \left(\frac{1}{m} - \frac{1}{n}\right)\left(1-\delta^2\right)V_{2|z} + \frac{1}{n}V_{2|z}$

where $\widehat{Y}^0_{2m} = \frac{1}{m}\sum_{i\in s_{2m}}\frac{y_{2i}}{p_i} - \beta_0\left(\frac{1}{m}\sum_{i\in s_{2m}}\frac{y_{1i}}{p_i} - \frac{1}{n}\sum_{i\in s_1}\frac{y_{1i}}{p_i}\right)$,

$V_{1|z} = \sum_{i\in U} p_i\left(\frac{y_{1i}}{p_i} - Y_1\right)^2$, $\quad V_{2|z} = \sum_{i\in U} p_i\left(\frac{y_{2i}}{p_i} - Y_2\right)^2$, $\quad$ and

$\delta = \sum_{i\in U} p_i\left(\frac{y_{1i}}{p_i} - Y_1\right)\left(\frac{y_{2i}}{p_i} - Y_2\right)\Big/\sqrt{\left(V_{1|z}\cdot V_{2|z}\right)}$.

Proof

(i) $E\left(\widehat{Y}_{2m}\right) = E\left\{E\left(\widehat{Y}_{2m}|s_1\right)\right\} = E\left(\frac{1}{n}\sum_{i\in s_1}\frac{y_{2i}}{p_i}\right) = Y_2$

(ii) $V\left(\widehat{Y}_{2m}\right) = E\left\{V\left(\widehat{Y}_{2m}|s_1\right)\right\} + V\left\{E\left(\widehat{Y}_{2m}|s_1\right)\right\}$

$$= \left(\frac{1}{m}-\frac{1}{n}\right)\frac{1}{n-1}E\left[\sum_{i\in s_1}\left\{\frac{y_{2i}-\beta y_{1i}}{p_i}-\frac{1}{n}\sum_{i\in s_1}\frac{y_{2i}-\beta y_{1i}}{p_i}\right\}^2\right]$$

$$+ V\left(\frac{1}{n}\sum_{i\in s_1}\frac{y_{2i}}{p_i}\right)$$

$$= \left(\frac{1}{m}-\frac{1}{n}\right)\sum_{i\in U}\left\{\frac{y_{2i}-\beta y_{1i}}{p_i}-\left(Y_2-\beta Y_1\right)\right\}^2 + V\left(\frac{1}{n}\sum_{i\in s_1}\frac{y_{2i}}{p_i}\right)$$

$$= \left(\frac{1}{m}-\frac{1}{n}\right)\left(V_{2|z}-2\beta\delta\sqrt{V_{1|z}\cdot V_{2|z}}+\beta^2V_{1|z}\right)+\frac{1}{n}V_{2|z}$$

(iii) Differentiating $V\left(\widehat{Y}_{2m}\right)$ with respect to β and equating it to zero we get the optimum value of $\beta = \beta_0 = \delta\sqrt{V_{2|z}/V_{1|z}}$

(iv) Putting $\beta = \beta_0$ in the expressions of $\widehat{Y}_{2m}$ and $V\left(\widehat{Y}_{2m}\right)$, we get

$$\widehat{Y}^0_{2m} = \frac{1}{m}\sum_{i\in s_{2m}}\frac{y_{2i}}{p_i} - \beta_0\left(\frac{1}{m}\sum_{i\in s_{2m}}\frac{y_{1i}}{p_i}-\frac{1}{n}\sum_{i\in s_1}\frac{y_{1i}}{p_i}\right)$$

and

$$V\left(\widehat{Y}^0_{2m}\right) = \left(\frac{1}{m}-\frac{1}{n}\right)\left(1-\delta^2\right)V_{2|z}+\frac{1}{n}V_{2|z}$$

Theorem 11.2.5

The optimum proportion of the matched sample $\lambda = m/n$ that minimizes the variance of $\widehat{\widehat{Y}}_{2c}$ and the expression of minimum variance of $\widehat{\widehat{Y}}_{2c}$ are, respectively,

(i) $\lambda_0 = \dfrac{\sqrt{1-\delta^2}}{1+\sqrt{1-\delta^2}}$ and (ii) $V_{\min} = \dfrac{1+\sqrt{1-\delta^2}}{2n}V_{2|z}$

Proof

Putting $\widehat{Y}_{2m} = \widehat{Y}^{0}_{2m}$ in the expression of $\widehat{\widehat{Y}}_{2c}$ and then minimizing the variance of $\widehat{Y}^{0}_{2c} = \phi \widehat{Y}^{0}_{2m} + (1-\phi)\widehat{Y}_{2u}$ with respect to ϕ, the optimum value of ϕ is obtained as

$$\phi = \phi_0 = \frac{1/V\left(\widehat{Y}^{0}_{2m}\right)}{1/V\left(\widehat{Y}^{0}_{2m}\right) + 1/V\left(\widehat{Y}^{0}_{2u}\right)}$$

Now, noting $V\left(\widehat{Y}_{2u}\right) = V_{2|z}/u$, the optimum value of $V\left(\widehat{Y}^{0}_{2c}\right)$ with $\phi = \phi_0$ is obtained as

$$V_{opt} = \left[\left\{\left(\frac{1}{m} - \frac{1}{n}\right)\left(1-\delta^2\right) + \frac{1}{n}\right\}^{-1} + u\right]^{-1} V_{2|z} \qquad (11.2.29)$$

Minimizing V_{opt} with respect to m while keeping $m + u = n$ as fixed, the optimum proportion of the matched sample is obtained as

$$\lambda_0 = \sqrt{1-\delta^2}\left(1 + \sqrt{1-\delta^2}\right)^{-1}$$

Finally, substituting $m = n\lambda_0$ in Eq. (11.2.29), the expression for the minimum variance of $\widehat{\widehat{Y}}_{2c}$ comes out as

$$V_{\min} = \frac{1 + \sqrt{1-\delta^2}}{2n} V_{2|z}$$

Remark 11.2.9

Here, we note that the optimum proportion of matched sample λ_0 always lies between zero and one. In case information on the first occasion is not used and an unmatched sample s_2 of size n is selected by PPSWR method, one would use an unbiased estimator of the total Y_2 as $\widehat{Y}_{2n} = \frac{1}{n}\sum_{i\in S_2} \frac{y_{2i}}{p_i}$.

Let $\widehat{Y}^{0}_{2c}$ be $\widehat{Y}_{2c}$ with the optimum values β, m, and ϕ. Then the relative efficiency of $\widehat{Y}^{0}_{2c}$ with respect to $\widehat{Y}_{2n}$ is given by

$$E = V\left(\widehat{Y}_{2n}\right)/V_{\min} = \frac{2}{1+\sqrt{1-\delta^2}}$$

The estimator $\widehat{Y}^{0}_{2c}$ remains more efficient than $\widehat{Y}_{2n}$ for all values of δ. The efficiency attains a maximum 2 when $\delta = \pm 1$ and a minimum 1 for $\delta = 0$.

11.2.1.5 Simple Random Sampling With Replacement

Let the initial sample s_1 of size n be selected by simple random sampling with replacement (SRSWR) method and the matched sample s_{2m} of size $m(\leq n)$ be selected from s_1 by SRSWOR method, treating all the units in s_1 as distinct, the unmatched sample s_{2u} of size u is selected from the entire population by SRSWR method. Because PPSWR reduces to SRSWR when $p_i = 1/N$, we derive Theorems 11.2.6 and 11.2.7 by substituting $p_i = 1/N$ in Theorems 11.2.4 and 11.2.5 as follows:

Theorem 11.2.6

(i) $\widehat{Y}_{2m} = N\left[\overline{y}_{2m} - \beta\left(\overline{y}_{1m} - \overline{y}_{1n}\right)\right]$ is unbiased for Y_2

(ii) $V\left(\widehat{Y}_{2m}\right) = N^2\left[\left(\frac{1}{m} - \frac{1}{n}\right)\left(\sigma_{2y}^2 - 2\beta\rho\sigma_{1y}\sigma_{2y} + \beta^2\sigma_{1y}^2\right) + \frac{1}{n}\sigma_{2y}^2\right]$

(iii) The optimum value of β that minimizes $V\left(\widehat{Y}_{2m}\right)$ is $\beta_0 = \rho\sigma_{2y}/\sigma_{1y}$

(iv) The variance of $\widehat{Y}^{0}_{2m}$ is $V\left(\widehat{Y}^{0}_{2m}\right) = N^2\left[\left(\frac{1}{m} - \frac{1}{n}\right)(1 - \rho^2) + \frac{1}{n}\right]\sigma_{2y}^2$

where $\overline{y}_{1n} = \sum_{i\in s_1} y_{1i}/n, \quad \overline{y}_{1m} = \sum_{i\in s_{2m}} y_{1i}/m, \quad \overline{y}_{2m} = \sum_{i\in s_{2m}} y_{2i}/m,$

$\sigma_{1y}^2 = \sum_{i\in U}\left(y_{1i} - \overline{Y}_1\right)^2/N, \ \sigma_{2y}^2 = \sum_{i\in U}\left(y_{2i} - \overline{Y}_2\right)^2/N,$

$\widehat{Y}^{0}_{2m} = N\left\{\overline{y}_{2m} - \beta_0\left(\overline{y}_{1m} - \overline{y}_{1n}\right)\right\}$, and ρ is the correlation coefficient between x and y.

Theorem 11.2.7

(i) $\widehat{Y}_{2c} = \phi\widehat{Y}^{0}_{2m} + (1 - \phi)\widehat{Y}_{2u}$ is unbiased for Y_2

(ii) The optimum proportion of the matched sample $\lambda_0 = \dfrac{\sqrt{1-\rho^2}}{1+\sqrt{1-\rho^2}}$

(iii) The minimum variance of $\widehat{Y}_{2c}$ is

$$V_{\min} = N^2\left(\frac{1+\sqrt{1-\rho^2}}{2n}\right)\sigma_{2y}^2$$

where $\widehat{Y}_{2u} = N\overline{y}_{2u}$ and $\overline{y}_{2u} = \sum_{i\in s_{2u}} y_{2i}/u$.

11.2.1.6 Sampling Over Two Occasions: Stratifying the Initial Sample

Here, we suppose that the population should be stratified into homogeneous strata according to certain criteria, but it cannot be employed as we have no information regarding which unit falls in which particular stratum. For instance, in household surveys, households can be stratified effectively according to various income groups viz. high, upper middle, lower middle, and low, but the income levels of the households may not be known before conducting the survey. Hence for estimation of population characteristics and selection of sample in the second occasion, we may use y_{1i}'s for $i \in s_1$, as a stratification variable.

Let, on the first occasion, a PPSWR sample s_1 of size n be selected from a population U using known size measure z_i's. Using the observed values of y_{1i}'s for $i \in s_1$, the selected n sampled units are assigned to one of the L strata $H_1,\ldots, H_l,\ldots, H_L$. Let y_{1j}^l, y_{2j}^l, and z_j^l be the value of the jth unit of the lth stratum for the variables y_1, y_2 and z, respectively. Typically a random number $n_l\left(0 \leq n_l \leq n, \sum_{l=1}^{L} n_l = n\right)$ of these n units will constitute a sample from the lth strum s_1^l (say). On the second occasion, independent subsamples s_{2m}^l's of sizes m_l's $= m\, n_l/n$ (assumed integer) are chosen from s_1^l $(l = 1,\ldots, L)$ by SRSWOR method. Here, n is assumed to be so large that $Prob(n_l \geq 1) = 1$. Finally, an unmatched sample s_{2u} of size $u = (n - m)$ is selected by PPSWR method from the entire population U using z_i's as a measure of size.

The proposed composite estimator for Y_2 is

$$\widehat{Y}_{2c}(st) = \phi \widehat{Y}_{2m}(st) + (1 - \phi)\widehat{Y}_{2u} \tag{11.2.30}$$

where $\widehat{Y}_{2m}(st) = \sum_{l=1}^{L} w_l \widehat{Y}_{2m}^l$, $\widehat{Y}_{2m}^l = \frac{1}{m_l}\sum_{j\in s_{2m}^l}\frac{d_{\beta j}^l}{p_j^l} + \beta^l \frac{1}{n_l}\sum_{j\in s_1^l}\frac{y_{1j}^l}{p_j^l}$,

$\widehat{Y}_{2u} = \frac{1}{u}\sum_{i\in S_{2u}}\frac{y_{2i}}{p_i}$, $w_l = n_l/n$, $d_{\beta j}^l = y_{2j}^l - \beta^l y_{1j}^l$, $p_j^l = z_j^l \big/ Z$, $p_i = z_i/Z$, $Z = \sum_{j\in U} z_j$, and β^l is a suitably chosen constant.

Theorem 11.2.8

(i) $E\left\{\widehat{Y}_{2m}(st)\right\} = Y_2$

(ii) $V\left\{\widehat{Y}_{2m}(st)\right\} = \left(\frac{1}{m} - \frac{1}{n}\right)\sum_{l=1}^{L}\frac{V_{d|z}^l}{P^l} + \frac{V_{y_2|z}}{n}$

(iii) The optimum value of β^l that minimizes $V\left\{\widehat{Y}_{2m}^l\right\}$ is

$$\beta^{l0} = \delta^l = \left(\sum_{j\in H_l} P_l \frac{y_{1j}y_{2j}}{p_j^l} - Y_1^l Y_2^l\right)\Big/\left(V_{y_1|z}^l V_{y_2|z}^l\right)^{1/2}$$

where $V_{d|z}^l = \sum_{j\in H_l} P_l \frac{\left(d_{\beta j}^l\right)^2}{p_j^l} - \left(D_\beta^l\right)^2$, $D_\beta^l = \sum_{j\in H_l} d_{\beta j}^l$,

$V_{y_1|z}^l = \sum_{j\in H_l} P_l \frac{\left(y_{1j}^l\right)^2}{p_j^l} - \left(Y_1^l\right)^2$, $V_{y_2|z}^l = \sum_{j\in H_l} P_l \frac{\left(y_{2j}^l\right)^2}{p_j^l} - \left(Y_2^l\right)^2$, and

$P_l = \sum_{j\in H_l} p_j^l$.

Proof

Let $E_1(V_1)$ be the expectation (variance) over the random vector $\mathbf{n}=(n_1,\ldots,n_l)$ while $E_2(V_2)$ and $E_3(V_3)$ denote the conditional expectations (variances) over s_1^l for fixed $\mathbf{n}$ and over s_{2m}^l for fixed s_1^l and $\mathbf{n}$, respectively.

$$\text{(i)}\quad \begin{aligned} E\left\{\widehat{Y}_{2m}(st)\right\} &= E_1\left[\sum_{l=1}^{L} w_l E_2\left\{E_3\left(\widehat{Y}^{l}_{2m}\middle| s_1^l,\mathbf{n}\right)\middle|\mathbf{n}\right\}\right]\\ &= E_1\left[\sum_{l=1}^{L} w_l E_2\left(\frac{1}{n_l}\sum_{j\in S_1^l}\frac{y_{2j}^l}{p_j^l}\middle|\mathbf{n}\right)\right]\\ &= E_1\left(\sum_{l=1}^{L} w_l\frac{Y_2^l}{P_l}\right)\\ &= \sum_{l=1}^{L} Y_2^l\\ &= Y_2 \end{aligned} \tag{11.2.31}$$

(noting $Y_2^l=\sum_{j\in H_l} y_{2j}^l$ and $E(w_l)=P_l$)

$$\text{(ii)}\quad \begin{aligned} V\left\{\widehat{Y}_{2m}(st)\right\} &= E_1\left[V_2\left\{E_3\sum_{l=1}^{L} w_l\left(\widehat{Y}^{l}_{2m}\right)\middle| s_1^l,\mathbf{n}\right\}\right]\\ &\quad+E_1\left[E_2\left\{V_3\sum_{l=1}^{L} w_l\left(\widehat{Y}^{l}_{2m}\right)\middle| s_1^l,\mathbf{n}\right\}\right]\\ &\quad+V_1\left[E_2\left\{\sum_{l=1}^{L} w_l E_3\left(\widehat{Y}^{l}_{2m}\right)\middle| s_1^l,\mathbf{n}\right\}\right] \end{aligned} \tag{11.2.32}$$

Now,

$$E_1\left[V_2\left\{E_3\sum_{l=1}^{L}w_l\left(\widehat{Y}^l_{2m}\right)\bigg|s_1^l,\mathbf{n}\right\}\right]=E_1\left[V_2\left\{\sum_{l=1}^{L}w_l\left(\frac{1}{n_l}\sum\nolimits_{j\in s_1^l}\frac{y_{2j}}{p_j^l}\right)\bigg|\mathbf{n}\right\}\right]$$

$$=E_1\left(\sum_{l=1}^{L}w_l^2\frac{V^l_{y_2|z}}{n^lP_l^2}\right)$$

$$=\frac{1}{n}E_1\left(\sum_{l=1}^{L}w_l\frac{V^l_{y_2|z}}{P_l^2}\right)$$

$$=\frac{1}{n}\sum_{l=1}^{L}\frac{V^l_{y_2|z}}{P_l} \tag{11.2.33}$$

and

$$E_1\left[E_2\left\{V_3\sum_{l=1}^{L}w_l\left(\widehat{Y}^l_{2m}\right)\bigg|s_1^l,\mathbf{n}\right\}\right]=E_1\left[E_2\left\{\sum_{l=1}^{L}w_l^2\left(\frac{1}{m_l}-\frac{1}{n_l}\right)\frac{1}{n_l-1}\right.\right.$$

$$\left.\left.\sum_{j\in s_1^l}\left(\frac{d^l_{\beta j}}{p_j^l}-\overline{d}^l_\beta\right)^2\bigg|\mathbf{n}\right\}\right]$$

$$\left(\text{where }\ \overline{d}^l_\beta=\sum_{j\in s_1^l}\frac{d^l_{\beta j}}{p_j^l}\bigg/n_l\right)$$

Now, noting $m_l=mn_l/n$, we get

$$E_1\left[E_2\left\{V_3\sum_{l=1}^{L}w_l\left(\widehat{Y}^l_{2m}\right)\bigg|s_1^l,\mathbf{n}\right\}\right]=\left(\frac{1}{m}-\frac{1}{n}\right)E_1\left(\sum_{l=1}^{L}w_l\frac{V^l_{d|z}}{P_l^2}\right)$$

$$=\left(\frac{1}{m}-\frac{1}{n}\right)\sum_{l=1}^{L}\frac{V^l_{d|z}}{P_l}. \tag{11.2.34}$$

Furthermore,

$$V_1\left[E_2\left\{\sum_{l=1}^{L} w_l E_3\left(\widehat{Y}^{l}_{2m}\right)\Big| s^{l}_{1},\mathbf{n}\right\}\right]=V_1\left[E_2\left\{\sum_{l=1}^{L} w_l\left(\frac{1}{n_l}\sum_{j\in s^l_1}\frac{y_{2j}}{p^l_j}\right)\Big|\mathbf{n}\right\}\right]$$

$$=V_1\left(\sum_{l=1}^{L} w_l\frac{Y^l_2}{P_l}\right)$$

$$=\frac{1}{n}\left(\sum_{l=1}^{L}\frac{1-P_l}{P_l}\left(Y^l_2\right)^2-\sum_{l\neq}^{L}\sum_{k=1}^{L}Y^l_2Y^k_2\right) \tag{11.2.35}$$

(noting $V(w_l)=P_l(1-P_l)$ and $Cov(w_l,w_{l'})=-P_lP_{l'}$)
Substituting Eqs. (11.2.33)–(11.2.35) in Eq. (11.2.32), we get

$$V\left\{\widehat{Y}_{2m}(st)\right\}=\frac{1}{n}\sum_{l=1}^{L}\frac{V^l_{y_2|z}}{P_l}+\left(\frac{1}{m}-\frac{1}{n}\right)\sum_{l=1}^{L}\frac{V^l_{d|z}}{P_l}$$
$$+\frac{1}{n}\left(\sum_{l=1}^{L}\frac{1-P_l}{P_l}\left(Y^l_2\right)^2-\sum_{l\neq}^{L}\sum_{k=1}^{L}Y^l_2Y^k_2\right)$$

Furthermore, noting $\dfrac{V^l_{y_2|z}}{P_l}=\sum_{j\in H_l}\dfrac{\left(y^l_{2j}\right)^2}{p^l_j}-\dfrac{\left(Y^l_2\right)^2}{P_l}$, we get

$$V\left\{\widehat{Y}_{2m}(st)\right\}=\left(\frac{1}{m}-\frac{1}{n}\right)\sum_{l=1}^{L}\frac{V^l_{d|z}}{P_l}+\frac{1}{n}V_{y_2|z}$$

(iii) Minimizing $V\left\{\widehat{Y}_{2m}(st)\right\}$ with respect to β^l, we arrive at the optimum value of β^l as

$$\beta^{l0}=\delta^l=\left(\sum_{j\in H_l}P^l\frac{y_{1j}y_{2j}}{p^l_j}-Y^l_1Y^l_2\right)\Big/\left(V^l_{y_2|z}V^l_{y_1|z}\right)^{1/2}$$

Theorem 11.2.9

The optimum proportion of the matched sample and the expression for the minimum variance of $\widehat{Y}_{2c}(st)$ are given, respectively, by

$$opt\ \lambda=\lambda_0=\frac{\sqrt{A}}{1+\sqrt{A}}\quad\text{and}\quad V_{\min}\left\{\widehat{Y}_{2c}(st)\right\}=\frac{1+\sqrt{A}}{2\,n}V_{y_2|z}$$

where $A=\sum_{l=1}^{L}\dfrac{\left(1-\delta^2_l\right)V^l_{y_2|z}}{V_{y_2|z}P_l}$.

Proof

The estimator $\widehat{Y}_{2m}(st)$ with $\beta^l = \beta^{l0}$ and its variance are given, respectively, by

$$\widehat{Y}^{0}_{2m}(st) = \sum_{l=1}^{L} w_l \left(\frac{1}{m_l} \sum\nolimits_{j \in s^l_{2m}} \frac{y^l_{2j} - \delta^l y^l_{1j}}{p^l_j} + \delta^l \frac{1}{n_l} \sum\nolimits_{j \in s^l_1} \frac{y^l_{1j}}{p^l_j} \right)$$

and

$$V\left(\widehat{Y}^{0}_{2m}\right) = \frac{n-m}{mn} \sum_{l=1}^{L} \frac{\left(1-\delta^2_l\right) V^l_{y_2|z}}{P_l} + \frac{V_{y_2|z}}{n}$$

$$= \frac{1}{n}\left(\frac{n-m}{m} A + 1\right) V_{y_2|z}$$

Substituting $\widehat{Y}_{2m}(st) = \widehat{Y}^{0}_{2m}(st)$ in Eq. (11.2.30) and then minimizing the variance $\widehat{Y}_{2c}(st)$ with respect to ϕ, the optimum variance $\widehat{Y}_{2c}(st)$ comes out as

$$V_{opt}\left\{\widehat{Y}_{2c}(st)\right\} = \left[n \left\{ \frac{n-m}{m} A + 1 \right\}^{-1} + u \right]^{-1} V_{y_2|z} \qquad (11.2.36)$$

Now, minimizing $V_{opt}\left\{\widehat{Y}_{2c}(st)\right\}$ with respect to m and keeping n as fixed, the optimum proportion of matched sample $\lambda = m/n$ and the corresponding value of $V_{opt}\left\{\widehat{Y}_{2c}(st)\right\}$ are obtained, respectively, as

$$\lambda = \lambda_0 = \frac{\sqrt{A}}{1+\sqrt{A}}$$

$$\text{and } V_{\min}\left\{\widehat{Y}_{2c}(st)\right\} = V_{opt}\left\{\widehat{Y}_{2c}(st) \middle| \lambda = \lambda_0\right\}$$

$$= \frac{1}{n}\left(\frac{1+\sqrt{A}}{2}\right) V_{y_2|z}$$

Remark 11.2.10

The relative efficiency of $\widehat{Y}_{2c}(st)$ compared to the Hansen–Hurwitz estimator based on a PPSWR sample of size n selected on the second occasion without using information on the first occasion is

$$E(st) = V\left(\widehat{Y}_{2n}\right) / V_{\min}\left\{\widehat{Y}_{2c}(st)\right\} = \frac{2}{1+\sqrt{A}}$$

The efficiency increases with the value of $|\delta_l|$ and it reaches the maximum 2 when $|\delta_l| = 1\ \forall\ l = 1,\ldots,\ L$. Arnab (1991) showed that $V_{\min}\{\widehat{Y}_{2c}(st)\}$ is more efficient than the strategy proposed by Raj (1965a,b) and Chaudhuri and Arnab (1979a).

11.2.2 Sampling More Than Two Occasions

Consider a finite population $U = (u_1,\ldots,\ u_i,\ldots,\ u_N)$ of N identifiable units that is supposed to be surveyed up to $h(\geq 2)$ occasions. Let y_{ki} be the value of the ith unit of study variable y for the kth occasion, $i = 1,\ldots,\ N$; $k = 1,\ldots,\ h$. Let $Y_k = \sum_{i \in U} y_{ki}$ be the population total of y for the kth occasion.

11.2.2.1 Probability Proportional to Size With Replacement Sampling

Tripathi and Srivastava (1979) considered the following sampling scheme over k occasions. On the first occasion, a sample s_1 of size n is selected by PPSWR method using $p_i(>0)$ as normed size measure for the ith unit. On the second occasion a subsample s_{m_2} of size $m_2(\leq n)$ is selected from s_1 by SRSWOR method, treating all the n units in s_1 as distinct. Clearly, a unit can appear more than once in s_{m_2} if this unit had appeared more than once in s_1. The unmatched sample s_{u_2} of size $u_2(=n - m_2)$ is selected from the entire population by PPSWR method using normed size measures p_i's. The selected sample on the second occasion is denoted by $s_2 = s_{m_2} \cup s_{u_2}$. In general, at the $k(\geq 2)$th occasion, a matched sample s_{m_k} of size $m_k(\leq n)$ is selected by SRSWOR method from the sample s_{k-1}, selected at the $k - 1$th occasion. It is supplemented by an unmatched sample s_{u_k} of size $u_k(=n - m_k)$ selected by PPSWR method from the entire population U using p_i as normed size measures for the ith unit, $i = 1,\ldots,\ N$. The sample selected on the kth occasion is denoted by $s_k = s_{m_k} \cup s_{u_k}$ for $k = 2,\ldots,\ h$.

Theorem 11.2.10

The probability of selection of the jth ($j = 1,\ldots,\ N$) unit at the rth ($r = 1,\ldots,\ n$) draw for selecting the sample s_k is p_j where draws are labeled from 1 to m_k for the selection of the matched sample s_{m_k} and m_{k+1} through n for the selection of unmatched sample s_{u_k}.

Proof

Because the sample s_1 and the unmatched sample s_{u_k}'s are selected by PPSWR method directly from the entire population U, the probability of selection of the jth unit at any draw in s_1 and the also the probability of selection of the jth unit at the m_{k+1}th draw to nth draw in s_{u_k} are also equal to p_j.

For $r = 1,\ldots, m_k$ we note:

Prob (selection of *j*th unit at *r*th draw in $s_{k+1}|j$th unit appeared q times in s_k)

$$= \sum_{x=0}^{r-1} Prob\ (j\text{th unit selected } x \text{ times in the first } (r-1) \text{ draws in } s_{k+1}|j\text{th unit appeared } q \text{ times in } s_k) \times \frac{q-x}{n-(r-1)}$$

$$= \sum_{x=0}^{r-1} \frac{\binom{q}{x}\binom{n-q}{r-1-x}}{\binom{n}{r-1}} \frac{q-x}{n-(r-1)}$$

$$= q/n \left(\text{Since } \frac{\sum_{x=0}^{r-1}\binom{q}{x}\binom{n-q}{r-1-x}}{\binom{n}{r-1}} = 1 \right.$$

$$\left. \text{and } \frac{\sum_{x=0}^{r-1} x \binom{q}{x}\binom{n-q}{r-1-x}}{\binom{n}{r-1}} = \frac{(r-1)q}{n} \right)$$

Hence the unconditional probability of selection of the *j*th unit at *r*th ($r \leq m_2$) draw in s_2

$$= \sum_{q=0}^{n} Prob\ (\text{selection of } j\text{th unit at } r\text{th draw in } s_2|j\text{th unit appears } q \text{ times in } s_1) \times Prob\ (j\text{th unit appears } q \text{ times in } s_1)$$

$$= \sum_{q=0}^{n} \frac{q}{n} \left\{ \binom{n}{q} p_j^q (1-p_j)^{n-q} \right\}$$

$$= p_j$$

Hence we can show that the probability of selecting the jth unit at the rth draw in s_k is p_j by the method of induction.

11.2.2.1.1 Estimator for the Total on hth Occasion

The proposed estimator for $Y_h(\geq 2)$ is

$$\widehat{Y}_h = \phi_h \widehat{Y}_{hm} + (1-\phi_h)\widehat{Y}_{hu} \tag{11.2.37}$$

where $\widehat{Y}_{hm} = \frac{1}{m_h}\sum_{i\in s_{m_h}} \frac{y_{hi}}{p_i} - \beta_h \left\{ \frac{1}{m_h}\sum_{i\in s_{m_h}} \frac{y_{h-1i}}{p_i} - \widehat{Y}_{h-1}(\min) \right\}$,

$\widehat{Y}_{hu} = \frac{1}{u_h}\sum_{i\in s_{u_h}} \frac{y_{hi}}{p_i}$, β_h is a constant chosen to minimize $V\left(\widehat{Y}_{hm}\right)$, and $\widehat{Y}_{h-1}(\min)$ is the estimator $\widehat{Y}_{h-1}$ with the minimum variance (obtained by choosing optimum values of ϕ_{h-1}, m_{h-1} and β_{h-1}).

Here, we make the following assumptions:

$$V_{k|z} = \sum_{i\in U} p_i \left(\frac{y_{ki}}{p_i} - Y_k\right)^2 = V_0 \text{ and}$$

$$\delta_{k,k'} = \sum_{i\in U} p_i \left(\frac{y_{k+1,i}}{p_i} - Y_{k+1}\right)\left(\frac{y_{k'i}}{p_i} - Y_{k'}\right) \Big/ V_o \tag{11.2.38}$$

$$= \delta \quad \text{for } k \neq k';\, k, k' = 1, \ldots, h$$

The optimum value of β_h for $h=2$ was obtained in the Theorem 11.2.4 as $\beta_2 = \delta$. Hence for the sake of simplicity we assume $\beta_k = \delta$ for $k = 1, \ldots, h$ and we write

$$\widehat{Y}^0_{hm} = \frac{1}{m_h}\sum_{i\in s_{m_h}} \frac{y_{hi}}{p_i} - \delta\left\{ \frac{1}{m_h}\sum_{i\in s_{m_h}} \frac{y_{h-1i}}{p_i} - \widehat{Y}_{h-1}(\min) \right\} \tag{11.2.39}$$

Theorem 11.2.11

(i) $E\left(\widehat{Y}^0_{hm}\right) = Y_h$

(ii) $V\left(\widehat{Y}^0_{hm}\right) = \left(1-\delta^2\right)\frac{V_0}{m_h} + \delta^2 V\left\{\widehat{Y}_{h-1}(\min)\right\}$

Proof

(i) From Theorem 11.2.10, we note that s_{m_2} and s_{u_2} are PPSWR samples from U and hence

$$E\left(\frac{1}{m_h}\sum_{i\in s_{m_h}}\frac{y_{ki}}{p_i}\right)=E\left(\frac{1}{u_h}\sum_{i\in s_{u_h}}\frac{y_{ki}}{p_i}\right)=Y_k \quad \text{for } k=1,\ldots,h.$$

For $h=2$, the theorem follows from Theorem 11.2.4. Let us suppose that the theorem holds for $h=k-1$.

Then we have

$$E\left(\widehat{Y}^{0}_{km}\right)=E\left(\frac{1}{m_k}\sum_{i\in s_{m_k}}\frac{y_{ki}}{p_i}\right)-\delta\left[E\left(\frac{1}{m_k}\sum_{i\in s_{m_k}}\frac{y_{k-1i}}{p_i}\right)-E\left\{\widehat{Y}_{k-1}(\min)\right\}\right]$$
$$=Y_k-\delta\left(Y_{k-1}-Y_{k-1}\right)=Y_k \tag{11.2.40}$$

and

$$V\left(\widehat{Y}^{0}_{hm}\right)=V\left(\frac{1}{m_h}\sum_{i\in s_{m_h}}\frac{y_{hi}}{p_i}\right)+\delta^2V\left\{\frac{1}{m_h}\sum_{i\in s_{m_h}}\frac{y_{h-1i}}{p_i}-\widehat{Y}_{h-1}(\min)\right\}$$
$$-2\delta Cov\left[\left(\frac{1}{m_h}\sum_{i\in s_{m_h}}\frac{y_{hi}}{p_i}\right),\left(\frac{1}{m_h}\sum_{i\in s_{m_h}}\frac{y_{h-1i}}{p_i}-\widehat{Y}_{h-1}(\min)\right)\right] \tag{11.2.41}$$

Now, using the assumptions (Eq. 11.2.38) we get

$$V\left(\frac{1}{m_h}\sum_{i\in s_{m_h}}\frac{y_{hi}}{p_i}\right)=\frac{V_o}{m_h}, \tag{11.2.42}$$

$$V\left\{\frac{1}{m_h}\sum_{i\in s_{m_h}}\frac{y_{h-1i}}{p_i}-\widehat{Y}_{h-1}(\min)\right\}=\frac{V_o}{m_h}+V\left\{\widehat{Y}_{h-1}(\min)\right\}$$
$$-2Cov\left\{\frac{1}{m_h}\sum_{i\in s_{m_h}}\frac{y_{h-1i}}{p_i},\widehat{Y}_{h-1}(\min)\right\} \tag{11.2.43}$$

and

$$Cov\left[\frac{1}{m_h}\sum_{i\in s_{m_h}}\frac{y_{hi}}{p_i},\left\{\frac{1}{m_h}\sum_{i\in s_{m_h}}\frac{y_{h-1i}}{p_i}-\widehat{Y}_{h-1}(\min)\right\}\right]$$

$$=\delta\frac{V_0}{m_h}-\delta Cov\left\{\frac{1}{m_h}\sum_{i\in s_{m_h}}\frac{y_{h-1,i}}{p_i},\widehat{Y}_{h-1}(\min)\right\} \tag{11.2.44}$$

Putting Eqs. (11.2.42)–(11.2.44) in Eq. (11.2.41), we get

$$V\left(\widehat{Y}^{0}_{hm}\right)=\frac{V_o}{m_h}+\delta^2\left[\frac{V_o}{m_h}+V\left\{\widehat{Y}_{h-1}(\min)\right\}\right]-2\delta^2\frac{V_0}{m_h}$$

$$=\left(1-\delta^2\right)\frac{V_0}{m_h}+\delta^2 V\left\{\widehat{Y}_{h-1}(\min)\right\}$$

Hence the theorem is proved by the method of induction.

Theorem 11.2.12

The optimum proportion of matched sample $\lambda_h=\dfrac{m_h}{n}$ and minimum variance of $\widehat{Y}_h$ are, respectively, given by

$$\lambda_{h0}=\frac{\sqrt{1-\delta^2}}{1+\sqrt{1-\delta^2}}\frac{1}{g_{h-1}}\quad\text{and}$$

$$V_{\min}\left(\widehat{Y}_h\right)=V_{\min}(h)=\frac{\left(1+\sqrt{1-\delta^2}\right)g_{h-1}}{\left(1+\sqrt{1-\delta^2}\right)g_{h-1}+\left(1-\sqrt{1-\delta^2}\right)}\frac{V_0}{n}$$

where $g_h=\dfrac{V_{\min}(h)}{V_0/n}$ for $h\geq 2$ and $g_1=1$.

Proof

Minimizing the variance of $\widehat{Y}_h\left[=\phi_h\widehat{Y}^{0}_{hm}+(1-\phi_h)\widehat{Y}_{hu}\right]$ with respect to ϕ_h, the optimum value of ϕ_h is obtained as

$$\phi_{ho}=\frac{1}{V\left(\widehat{Y}^{0}_{hm}\right)}\left(\frac{1}{V\left(\widehat{Y}^{0}_{hm}\right)}+\frac{1}{V\left(\widehat{Y}_{hu}\right)}\right)^{-1}$$

Using Theorem 11.2.11, the optimum value of the variance of $\widehat{Y}_h$ with $\phi_h = \phi_{h0}$ is obtained as

$$V_{opt}\left(\widehat{Y}_h\right) = \left(\frac{1}{V\left(\widehat{Y}^0_{hm}\right)} + \frac{1}{V\left(\widehat{Y}_{hu}\right)}\right)^{-1}$$
$$= \left[\left\{\frac{\left(1-\delta^2\right)}{m_h} + \delta^2\frac{V_{\min}(h-1)}{V_0}\right\}^{-1} + u_h\right]^{-1} V_0 \tag{11.2.45}$$

Now, minimizing $V_{opt}\left(\widehat{Y}^0_{hm}\right)$ with respect to m_h, we get the optimum value of λ_h as

$$\lambda_{ho} = \frac{\sqrt{1-\delta^2}}{1+\sqrt{1-\delta^2}}\frac{1}{g_{h-1}}$$

Substituting $m_h = n\ \lambda_{ho}$ in Eq. (11.2.45), we get the minimum variance of $\widehat{Y}_h$ as

$$V_{\min}(h) = \frac{\left(1+\sqrt{1-\delta^2}\right)g_{h-1}}{\left(1+\sqrt{1-\delta^2}\right)g_{h-1} + \left(1-\sqrt{1-\delta^2}\right)}\frac{V_0}{n}$$

Remark 11.2.11

The estimator $\widehat{Y}^0_{hm}$ cannot be used in practice because it involves an unknown parameter δ. So, in practice, we replace δ with its suitable estimate or some known value based on previous surveys.

When h is sufficiently large, we find $g = \underset{h\to\infty}{Lim}\, g_h = \frac{2\sqrt{1-\delta^2}}{1+\sqrt{1-\delta^2}}$, $\lambda = \underset{h\to\infty}{Lim}\, \lambda_{h0} = \frac{1}{2}$, and

$$V_{\min} = \underset{h\to\infty}{Lim}\, V\left\{\widehat{Y}_h(\min)\right\} = \frac{2\sqrt{1-\delta^2}}{1+\sqrt{1-\delta^2}}\frac{V_0}{n}.$$

Remark 11.2.12

The relative efficiency of $\widehat{Y}_h$ based on the data on previous $h-1$ occasions with respect to complete unmatched sample based on the data on hth occasion only is

$$E_h = \frac{V_0/n}{V\left\{\widehat{Y}_h(\min)\right\}} = 1 + \frac{1-\sqrt{1-\delta^2}}{\left(1+\sqrt{1-\delta^2}\right)g_{h-1}}$$

The efficiency $E_h \geq 1$ as $g_h \geq 0$.

11.2.2.2 Simple Random Sampling With Replacement

On the first occasion, a sample s_1 of size n is selected by SRSWR method. On the second occasion, a subsample s_{m_2} of size $m_2(\leq n)$ is selected from s_1 by SRSWOR method treating all the units in s_1 as distinct. The unmatched sample s_{u_2} of size $u_2=(n-m_2)$ is selected from the entire population by SRSWR method. The selected sample on the second occasion is denoted by $s_2 = s_{m_2} \cup s_{u_2}$. In general, at the $k(\geq 2)$th occasion a matched sample s_{m_k} of size $m_k(\leq n)$ is selected by SRSWOR method from the sample s_{k-1}, selected in the $k-1$th occasion and it is supplemented by an unmatched sample s_{u_k} of size $u_k(=n-m_k)$ selected by SRSWR method from the entire population U.

Because PPSWR sampling scheme reduces to SRSWR sampling when $p_i = 1/N$ for $i = 1,\ldots, N$, the assumption (Eq. 11.2.38) reduces to

$$\sigma_{ky}^2 = \frac{1}{N}\sum_{i\in U}\left(y_{ki} - \overline{Y}_k\right)^2 = \sigma_y^2 \text{ and}$$

$$\rho_{k,k'} = \frac{\sum_{i\in U}\left(y_{k,i} - \overline{Y}_k\right)\left(y_{k',i} - \overline{Y}_{k'}\right)}{\left(\sigma_{ky}\sigma_{k'y}\right)} \tag{11.2.46}$$

$$= \rho \quad \text{for } k \neq k';\, k, k' = 1, \ldots, h$$

11.2.2.2.1 Estimator for the Total on the *h*th Occasion

Substituting $p_i = 1/N$ for PPSWR sampling scheme given in Section 11.2.2.1, we get $\delta = \rho$, $\widehat{Y}_{hm}^{0} = N\left[\overline{y}_h(s_{hm}) - \rho\left\{\overline{y}_{h-1}(s_{hm}) - \widehat{\overline{Y}}_{h-1}(\min)\right\}\right]$, $\widehat{Y}_{hu} = N\overline{y}(s_{uh})$, and

$$\widehat{Y}_h = \phi_h \widehat{Y}_{hm}^{0} + (1-\phi_h)\widehat{Y}_{hu} \tag{11.2.47}$$

where

$$\overline{y}(s_{uh}) = \sum\nolimits_{i\in s_{hu}} y_{hi}/u_h, \overline{y}_h(s_{hm}) = \sum\nolimits_{i\in s_{hm}} y_{hi}/m_h, \overline{y}_{h-1}(s_{hm}) = \sum\nolimits_{i\in s_{hm}} y_{h-1,i}/m_h.$$

We have the following results, which follow immediately from Theorems 11.2.11 and 11.2.12 by substituting $p_i = 1/N$ for $i = 1,\ldots, N$.

Theorem 11.2.13

(i) $E\left(\widehat{Y}_{hm}^{0}\right) = Y_h$

(ii) $V\left(\widehat{Y}_{hm}^{0}\right) = N^2\left[\left(1-\rho^2\right)\dfrac{\sigma_y^2}{m_h} + \rho^2\dfrac{a_{h-1}}{n}\right]$

(iii) $V_{opt}\left(\widehat{Y}_h\right) = N^2\left[\left\{\dfrac{(1-\rho^2)}{m_h} + \rho^2\dfrac{a_{h-1}}{n}\right\}^{-1} + u_h\right]^{-1}\sigma_y^2$

(iv) $\lambda_{ho} = \dfrac{\sqrt{1-\rho^2}}{1+\sqrt{1-\rho^2}}\dfrac{1}{a_{h-1}}$

(v) $V\left\{\widehat{Y}_h(\min)\right\} = V_{\min}(h) = N^2\dfrac{\left(1+\sqrt{1-\rho^2}\right)a_{h-1}}{\left(1+\sqrt{1-\rho^2}\right)a_{h-1} + \left(1-\sqrt{1-\rho^2}\right)}\dfrac{\sigma_y^2}{n}$

where $a_h = \dfrac{V_{\min}(h)}{N^2\sigma_y^2/n}$ for $h \geq 2$ and $a_1 = 1$.

Remark 11.2.13

For large h, $a = \underset{h\to\propto}{Lim}\, a_h = \dfrac{2\sqrt{1-\rho^2}}{1+\sqrt{1-\rho^2}}$, $\lambda = \underset{h\to\propto}{Lim}\, \lambda_h = 1/2$, and

$$V_{\min} = \underset{h\to\propto}{Lim}\, V\left\{\widehat{Y}_h(\min)\right\} = N^2\frac{2\sqrt{1-\rho^2}}{1+\sqrt{1-\rho^2}}\frac{\sigma_y^2}{n}$$

11.3 ESTIMATION OF CHANGE OVER TWO OCCASIONS

For estimating the difference $D\left(=\overline{Y}_2 - \overline{Y}_1\right)$ between the two means of the two occasions, we consider the following estimator:

$$\widehat{D} = \phi t_{2m} + \psi_1 t_{1m} + \psi_2 \widehat{\overline{Y}}_1 + \varphi \widehat{\overline{Y}}_{2u} \qquad (11.3.1)$$

where t_{2m}, t_{1m}, $\widehat{\overline{Y}}_1$, $\widehat{\overline{Y}}_{2u}$ are the same as in Section 11.2.1.2 and ϕ, ψ_1, ψ_2, and φ are suitably chosen to make $\widehat{D}$ unbiased for D. The unbiasedness condition $E\left(\widehat{D}\right) = D$ yields

$$\begin{aligned} D &= (\varphi+\phi)\overline{Y}_2 + (\psi_1+\psi_2)\overline{Y}_1 \\ \text{i.e.,}\quad \phi &= 1-\varphi \text{ and } \psi_1+\psi_2 = -1 \end{aligned} \qquad (11.3.2)$$

Substituting Eq. (11.3.2) in Eq. (11.3.1), we get

$$\begin{aligned} \widehat{D} &= \phi\left(\widehat{\overline{Y}}_{2m} - \widehat{\overline{Y}}_1\right) + (1-\phi)\left(\widehat{\overline{Y}}_{2u} - \widehat{\overline{Y}}_1\right) \\ &= \phi\widehat{D}_1 + (1-\phi)\widehat{D}_2 \end{aligned} \qquad (11.3.3)$$

where $\widehat{D}_1 = \widehat{\overline{Y}}_{2m} - \widehat{\overline{Y}}_1$, $\widehat{D}_2 = \widehat{\overline{Y}}_{2u} - \widehat{\overline{Y}}_1$, $\widehat{\overline{Y}}_{2m} = t_{2m} - \beta\left(t_{1m} - \widehat{\overline{Y}}_1\right)$

with $\beta = -\psi_1/\phi$.

The optimum value of β that minimizes the variance of $\widehat{\overline{Y}}_{2m}$ is $\beta = \beta_o = E\{Cov(t_{1m}, t_{2m}|s_1)\}/E\{V(t_{1m}|s_1)\}$.

The estimator $\widehat{\overline{Y}}_{2m}$ with $\beta = \beta_o$ is

$$\widehat{\overline{Y}}{}^{0}_{2m} = t_{2m} - \beta_0\left(t_{1m} - \widehat{\overline{Y}}_1\right) \qquad (11.3.4)$$

Substitution of $\widehat{\overline{Y}}_{2m} = \widehat{\overline{Y}}^{0}_{2m}$ in Eq. (11.3.4) yields

$$\widehat{D} = \phi \widehat{D}_{10} + (1-\phi)\widehat{D}_2 \tag{11.3.5}$$

where $\widehat{D}_{10} = \widehat{\overline{Y}}^{0}_{2m} - \widehat{\overline{Y}}_1$.

The optimum values of ϕ, $\widehat{D}$, and the proportion of matched sample λ can be derived following the same procedure described in Section 11.2.1.2.

11.3.1 Simple Random Sampling Without Replacement

Let s_1, s_{2m}, and s_{2u} be selected by the SRSWOR method as in Section 11.2.1.3. In this situation

$$\widehat{\overline{Y}}^{0}_{2m} = t_{2m} - \beta_0\left(t_{1m} - \widehat{\overline{Y}}_1\right),\ \widehat{\overline{Y}}_1 = \overline{y}_{1n},\ t_{2m} = \overline{y}_{2m},\ \widehat{\overline{Y}}_{2u} = \overline{y}_{2u} \text{ and}$$

$$\beta_0 = S_{12y}\Big/S^2_{1y}$$

The estimator $\widehat{D}$ reduces to

$$\widehat{D} = \phi \widehat{D}_{10} + (1-\phi)\widehat{D}_2 \tag{11.3.6}$$

with $\widehat{D}_{10} = \left\{\overline{y}_{2m} - \beta_0\left(\overline{y}_{1m} - \overline{y}_{1n}\right)\right\} - \overline{y}_{1n}$ and $\widehat{D}_2 = \overline{y}_{2u} - \overline{y}_{1n}$.

The variances of $\widehat{D}_1, \widehat{D}_2$ and covariance of $\widehat{D}_1$ and $\widehat{D}_2$ are given as follows:

$$V\left(\widehat{D}_{10}\right) = V\left\{\overline{y}_{2m} - \beta_0\left(\overline{y}_{1m} - \overline{y}_{1n}\right)\right\} + V\left(\overline{y}_{1n}\right) - 2Cov\left[\left\{\overline{y}_{2m} - \beta_0\left(\overline{y}_{1m} - \overline{y}_{1n}\right)\right\}, \overline{y}_{1n}\right]$$

$$= \left(\frac{1}{m} - \frac{1}{n}\right)\left(1-\rho^2\right)S^2_{2y} + \left(\frac{1}{n} - \frac{1}{N}\right)S^2_{2y} + \left(\frac{1}{n} - \frac{1}{N}\right)S^2_{1y} - 2Cov\left(\overline{y}_{2n}, \overline{y}_{1n}\right)$$

$$= \left(\frac{1}{m} - \frac{1}{n}\right)\left(1-\rho^2\right)S^2_{2y} + \left(\frac{1}{n} - \frac{1}{N}\right)S^2_{2y} + \left(\frac{1}{n} - \frac{1}{N}\right)S^2_{1y} - 2\left(\frac{1}{n} - \frac{1}{N}\right)\rho S_{1y}S_{2y}$$

$$V\left(\widehat{D}_2\right) = \left(\frac{1}{n} - \frac{1}{N}\right)S^2_{1y} + \left(\frac{1}{u} - \frac{1}{N}\right)S^2_{2y} + \frac{2}{N}\rho S_{1y}S_{2y}$$

$$Cov\left(\widehat{D}_{10}, \widehat{D}_2\right) = Cov\left(\overline{y}_{2n} - \overline{y}_{1n}, \overline{y}_{2u}\right) - Cov\left(\overline{y}_{2n} - \overline{y}_{1n}, \overline{y}_{1n}\right)$$

$$= -\frac{1}{N}\left(S^2_{2y} - \rho S_{1y}S_{2y}\right) - \left(\frac{1}{n} - \frac{1}{N}\right)\left(\rho S_{1y}S_{2y} - S^2_{1y}\right)$$

$$= Q_{12}(\text{say})$$

The optimum value of ϕ that minimizes the variance of $\widehat{D}$ for given values of m and $u(=n-m)$ is

$$\phi_0 = (1/Q_1 + 1/Q_2)^{-1}/Q_1 \tag{11.3.7}$$

where $\quad Q_1 = V\left(\widehat{D}_{10}\right) - Q_{12} = \left(\frac{1}{m} - \frac{1}{n}\right)(1-\rho^2)S_{2y}^2 + \frac{1}{n}\left(S_{2y}^2 - S_{12y}\right)$

and $Q_2 = V\left(\widehat{D}_2\right) - Q_{12} = \frac{1}{u}S_{2y}^2 + \frac{1}{n}S_{12y}$.

The variance of $\widehat{D}$ with $\phi = \phi_0$ is given by

$$V_{opt}\left(\widehat{D}\right) = (1/Q_1 + 1/Q_2)^{-1} + Q_{12}. \tag{11.3.8}$$

In particular, if $S_{1y}^2 = S_{2y}^2 = S_y^2$, we have

$$\begin{aligned} Q_1 &= \left(\frac{1-\mu\rho^2}{\lambda} - \rho\right)\frac{S_y^2}{n},\ Q_2 = \left(\frac{1}{\mu} + \rho\right)\frac{S_y^2}{n} \text{ and} \\ Q_{12} &= (1-2f)(1-\rho)S_y^2\Big/n \end{aligned} \tag{11.3.9}$$

where $f = n/N$ and $\mu = 1 - \lambda$.

After a little simplification we arrive at the following expression for $V_{opt}\left(\widehat{D}\right)$ as obtained by Särndal et al. (1992).

$$V_{opt}\left(\widehat{D}\right) = 2(1-\rho)\left(\frac{1}{1-\mu\rho} - f\right)\frac{S_y^2}{n} \tag{11.3.10}$$

From the expression (Eq. 11.3.10), we note that $V_{opt}\left(\widehat{D}\right)$ is an increasing function of μ when ρ is positive. So, our strategy will be to take a 100% matched sample on the second occasion to estimate the difference D if $\rho > 0$. In this case the minimum variance of $\widehat{D}$ is

$$V_{\min}\left(\widehat{D}\right) = 2(1-\rho)(1-f)\frac{S_y^2}{n} \tag{11.3.11}$$

11.4 ESTIMATION OF MEAN OF MEANS

For the estimation of the mean of means of two occasions $M = \left(\overline{Y}_1 + \overline{Y}_2\right)/2$, we may consider the following linear combinations as an unbiased estimator for M

$$\widehat{M} = \frac{1}{2}\left(\phi t_{2m} + \psi_1 t_{1m} + \psi_2 \widehat{\overline{Y}}_1 + \varphi \widehat{\overline{Y}}_{2u}\right) \tag{11.4.1}$$

The unbiasedness condition $E(\widehat{M}) = M$ implies $\varphi + \phi = \psi_1 + \psi_2 = 1$ and hence we can write

$$\widehat{M} = \frac{1}{2}\left(\phi\widehat{M}_1 + (1-\phi)\widehat{M}_2\right) \tag{11.4.2}$$

where $\widehat{M}_1 = \widehat{\overline{Y}}_{2m} + \widehat{\overline{Y}}_1$, $\widehat{\overline{Y}}_{2m} = t_{2m} - \beta\left(t_{1m} - \widehat{\overline{Y}}_1\right)$, $\beta = -\psi_1/\phi$, and $\widehat{M}_2 = \widehat{\overline{Y}}_{2u} + \widehat{\overline{Y}}_1$.

11.4.1 Simple Random Sampling Without Replacement

For SRSWOR sampling, $\widehat{M}_1 = \left\{\overline{y}_{2m} - \beta_0\left(\overline{y}_{1m} - \overline{y}_{1n}\right)\right\} + \overline{y}_{1n}$, $\beta_0 = S_{12y}/S_{1y}^2$, and $\widehat{M}_2 = \overline{y}_{2u} + \overline{y}_{1n}$. The variances of $\widehat{M}_1, \widehat{M}_2$ and covariance of $\widehat{M}_1$ and $\widehat{M}_2$ are given as follows:

$$V\left(\widehat{M}_1\right) = \left[\left(\frac{1}{m} - \frac{1}{n}\right)(1-\rho^2) + \left(\frac{1}{n} - \frac{1}{N}\right)\right]S_{2y}^2 + \left(\frac{1}{n} - \frac{1}{N}\right)S_{1y}^2 + 2\left(\frac{1}{n} - \frac{1}{N}\right)\rho S_{1y}S_{2y}$$

$$V\left(\widehat{M}_2\right) = \left[\left(\frac{1}{n} - \frac{1}{N}\right)S_{1y}^2 + \left(\frac{1}{u} - \frac{1}{N}\right)S_{2y}^2 - \frac{2}{N}\rho S_{1y}S_{2y}\right]$$

$$Cov\left(\widehat{M}_1, \widehat{M}_2\right) = -\frac{1}{N}\left(S_{2y}^2 + \rho S_{1y}S_{2y}\right) + \left(\frac{1}{n} - \frac{1}{N}\right)\left(\rho S_{1y}S_{2y} + S_{1y}^2\right) = \widetilde{Q}_{12}$$

In particular, if $S_{1y}^2 = S_{2y}^2 = S_y^2$, we have

$$\widetilde{Q}_1 = V\left(\widehat{M}_1\right) - Cov\left(\widehat{M}_1, \widehat{M}_2\right)$$

$$= \left[\left(\frac{1}{m} - \frac{1}{n}\right)(1-\rho^2) + \frac{1}{n}(1+\rho)\right]S_y^2$$

$$= \frac{(1+\rho)}{n\lambda}(1-\rho\mu)S_y^2$$

$$\widetilde{Q}_2 = V\left(\widehat{M}_2\right) - Cov\left(\widehat{M}_1, \widehat{M}_2\right) = \left(\frac{1}{u} - \frac{1}{n}\right)\rho S_y^2$$

$$= \left(\frac{1-\mu\rho}{n\mu}\right)S_y^2$$

and

$$\widetilde{Q}_{12} = Cov\left(\widehat{M}_1, \widehat{M}_2\right) = \frac{(1-2f)}{n}(1+\rho)S_y^2$$

The variance of $\widehat{M}$ with the optimum value of ϕ is given by

$$\begin{aligned} V_{opt}\left(\widehat{M}\right) &= \frac{1}{4}\left[\left(1/\widetilde{Q}_1 + 1/\widetilde{Q}_2\right)^{-1} + \widetilde{Q}_{12}\right] \\ &= \frac{(1+\rho)}{2n}\left(\frac{1}{1+\rho\mu} - f\right)S_y^2 \end{aligned} \tag{11.4.3}$$

From the expression (Eq. 11.4.3), we note that $V_{opt}\left(\widehat{M}\right)$ is a deceasing function of μ when ρ is positive. Hence for positive ρ the optimum choice of the matched sample should be zero percent, i.e., an unmatched sample should be the same percent for estimation of M. In this case the minimum variance of $\widehat{M}$ is

$$V_{\min}\left(\widehat{M}\right) = \{1 - f(1+\rho)\}\frac{S_y^2}{2n} \tag{11.4.4}$$

Example 11.4.1

In the year 2008, a sample s_1 of 20 automobile dealers was selected from 50 dealers of a locality by the SRSWOR method. In 2009, a subsample s_{2m} of 8 dealers was selected from s_1 by the SRSWOR method and an unmatched sample of 12 dealers was selected from 30 dealers who were not selected in 2008 by SRSWOR method again. The sales of automobiles are given in the following table.

Dealers	Sales of automobiles		Dealers	Sales of automobiles	
	2008	2009		2008	2009
1	400	700	17	430	
2	350	300	18	225	
3	500	600	19	450	
4	625	725	20	700	
5	700	800	21		800
6	325	400	22		400
7	425	400	23		850

Dealers	Sales of automobiles		Dealers	Sales of automobiles	
	2008	2009		2008	2009
8	500	625	24		900
9	650		25		325
10	750		26		400
11	225		27		600
12	650		28		780
13	450		29		690
14	600		30		900
15	500		31		750
16	375		32		650

(i) Estimate the total sales of automobiles in the year 2009 along with its standard error. Estimate the optimum proportion of the matched sample for the second occasion if the total sample size is kept fixed at 20 for both of the occasions.
(ii) Estimate the change in the average sales of automobiles between these two periods with its standard error.
(iii) Estimate the average sales over two periods and its standard error.

Here, $N = 50$, $n = 20$, $m = 8$, $u = 12$; $\overline{y}_{1n} = 491.5$, $\overline{y}_{1m} = 478.125$, $\overline{y}_{2m} = 568.750$, $\overline{y}_{2u} = 670.417$; $s_{1m}^2 = \sum_{i \in s_{2m}} \left(y_{1i} - \overline{y}_{1m}\right)^2/(m-1)$ $= 17,220.982$, $s_{2m}^2 = \sum_{i \in s_{2m}} \left(y_{2i} - \overline{y}_{2m}\right)^2/(m-1) = 32,633.929$, $s_{2u}^2 = \sum_{i \in s_{2u}} \left(y_{2i} - \overline{y}_{2u}\right)^2/(u-1) = 40,374.811$, $s_{12m} = \dfrac{1}{m-1}\sum_{i \in s_{2m}} \left(y_{1i} - \overline{y}_{1m}\right)\left(y_{2i} - \overline{y}_{2m}\right) = 19,308.035$, $\widehat{\rho} = s_{12m}/(s_{1m}s_{2m}) = 0.8144$, $\widehat{\beta}_0 = \widehat{\rho}s_{2m}/s_{1m} = 1.1212$, $s_{1y}^2 = \sum_{i \in s_1} \left(y_{1i} - \overline{y}_1\right)^2/(n-1) = 24,063.421$, $s_{2y}^2 = \sum_{i \in s_2} \left(y_{2i} - \overline{y}_2\right)^2/(n-1)$ $= 38,009.145$, $s_2 = s_{2m} \cup s_{2u}$, and $\overline{y}_2 = \sum_{i \in s_2} y_{2i}/n = 629.75$.

(i) Estimated total sales in 2009 based on the matched sample is

$$\begin{aligned}\widehat{Y}_{2m}^{0} &= N\left\{\overline{y}_{2m} - \widehat{\beta}_0\left(\overline{y}_{1m} - \overline{y}_1\right)\right\}\\ &= 50\{568.750 - 1.1212(478.125 - 491.5)\}\\ &= 29,187.297\end{aligned}$$

Estimated total sales in 2009 based on unmatched sample

$$\widehat{Y}_{2u} = N\overline{y}_{2u} = 50 \times 670.416 = 33,520.8$$

Composite estimator for the total sales in 2009 is

$$\widehat{Y}_2 = \widehat{\phi}_0 \widehat{Y}^0_{2m} + \left(1 - \widehat{\phi}_0\right) \widehat{Y}_{2u}$$

where

$$\widehat{\phi}_0 = \frac{1/\hat{V}'_m}{1/\hat{V}'_m + 1/\hat{V}'_u} = \left\{\left(\frac{1}{m} - \frac{1}{n}\right)\left(1 - \widehat{\rho}^2\right) + \frac{1}{n}\right\}^{-1}$$

$$\left[\left\{\left(\frac{1}{m} - \frac{1}{n}\right)\left(1 - \widehat{\rho}^2\right) + \frac{1}{n}\right\}^{-1} + u\right]^{-1}$$

(using Eq. 11.2.22)
= 0.526
Hence estimated sales for the year 2009 is

$$\widehat{Y}_2 = 0.526 \times 29,187.297 + (1 - 0.526) \times 33,520.833$$
$$= 31,243.592$$

Estimated variance of $\widehat{Y}_2$ is

$$\widehat{V}\left(\widehat{Y}_2\right) = N^2\left[\left[\left\{\left(\frac{1}{m} - \frac{1}{n}\right)\left(1 - \widehat{\rho}^2\right) + \frac{1}{n}\right\}^{-1} + u\right]^{-1} - 1/N\right] s^2_{2y}$$

where $s^2_{2y} = \sum_{i \in s_2} \left(y_{2i} - \overline{y}_2\right)^2/(n-1) = 38,009.145$ is an unbiased estimate of S^2_{2y} based on the sample s_2.

Hence an estimated standard error of $\widehat{Y}_2$ is

$$Se\left\{\widehat{V}\left(\widehat{Y}_2\right)\right\} = \sqrt{\widehat{V}\left(\widehat{Y}_2\right)} = \sqrt{1,856,963.99} = 1362.704$$

Estimated optimum proportion of the matched sample is

$$\widehat{\lambda}_0 = \frac{\sqrt{1 - \widehat{\rho}^2}}{1 + \sqrt{1 - \widehat{\rho}^2}} = \frac{\sqrt{1 - 0.8144^2}}{1 + \sqrt{1 - 0.8144^2}} = 0.3672$$

(ii) Estimated average change of sales between the period 2008 and 2009 is

$$\widehat{D} = \widehat{\phi}_0 \widehat{D}_1 + \left(1 - \widehat{\phi}_0\right) \widehat{D}_2$$

Here,

$$\widehat{D}_1 = \left\{\overline{y}_{2m} - \widehat{\beta}_0\left(\overline{y}_{2m} - \overline{y}_{1n}\right)\right\} - \overline{y}_{1n} = \widehat{Y}^0_{2m}\Big/50 - \overline{y}_{1n}$$
$$= 583.746 - 491.5 = 92.246 \quad \text{and}$$
$$\widehat{D}_2 = \overline{y}_{2u} - \overline{y}_{1n} = 670.417 - 491.5 = 178.917;$$

$$\widehat{Q}_1 = \left(\frac{1}{m} - \frac{1}{n}\right)\left(1 - \widehat{\rho}^2\right)s^2_{2m} + \frac{1}{n}\left(s^2_{2y} - s_{12m}\right)$$
$$= (1/8 - 1/20)(1 - 0.8144^2) \times 32,633.929 + (38,009.145$$
$$-19,308.035)/20 = 1759.273,$$

$$\widehat{Q}_2 = \frac{1}{u}s^2_{2y} + \frac{1}{n}s_{12m} = 38,009.145/12 + 19,308.035/20 = 4132.831,$$

$$\widehat{Q}_{12} = \left(\frac{1}{n} - \frac{1}{N}\right)\left(s^2_{1m} - S_{12m}\right) - \frac{1}{N}\left(s^2_{2y} - s_{12m}\right) = (1/20 - 1/50)$$
$$(17,220.982 - 19,308.035) - (38,009.145 - 19,308.035)/50$$
$$= -436.633,$$
$$\widehat{\phi}_0 = \frac{1/\hat{Q}_1}{1/\hat{Q}_1 + 1/\hat{Q}_2} = 0.701$$

Hence $\widehat{D} = 0.701 \times 92.246 + 0.299 \times 178.917 = 118.161.$

Estimated standard error of $\widehat{D} = \sqrt{\widehat{V}(\widehat{D})}$

$$= \sqrt{(1/Q_1 + 1/Q_2)^{-1} + Q_{12}} = 28.237.$$

(iii) The estimated average sale of automobiles for the year 2008 and 2009 is

$$\widehat{M} = \frac{1}{2}\left\{\widehat{\phi}_0\widehat{M}_1 + \left(1 - \widehat{\phi}_0\right)\widehat{M}_2\right\}$$

where $\widehat{M}_1 = \widehat{Y}^0_{2m}\Big/50 + \overline{y}_{1n} = 4583.746 + 91.5 = 1075.246;$

$$\widehat{M}_2 = \overline{y}_{2u} + \overline{y}_{1n} = 670.417 + 491.5 = 1161.917;$$
$$\widehat{V}\left(\widehat{M}_1\right) = \left[\left(\frac{1}{m} - \frac{1}{n}\right)\left(1 - \widehat{\rho}^2\right) + \left(\frac{1}{n} - \frac{1}{N}\right)\right]s^2_{2y} + \left(\frac{1}{n} - \frac{1}{N}\right)$$
$$\times s^2_{1y} + 2\left(\frac{1}{n} - \frac{1}{N}\right)s_{12m}$$
$$= \left[(1/8 - 1/20)\left(1 - 0.8144^2\right)\right.$$
$$\left. + (1/20 - 1/50)\right] \times 38,009.145 + (1/20 - 1/50)$$
$$\times 24,063.421 + 2(1/20 - 1/50) \times 19,308.035$$
$$= 3980.635;$$

$$\widehat{V}\left(\widehat{M}_2\right) = \left[\left(\frac{1}{n} - \frac{1}{N}\right)s_{1y}^2 + \left(\frac{1}{u} - \frac{1}{N}\right)s_{2y}^2 - \frac{2}{N}s_{12m}\right]$$
$$= (1/20 - 1/50) \times 24,063.421 + (1/12 - 1/50)$$
$$\times 38,009.145 - 2 \times 19,308.035/50 = 2356.827;$$

$$\widehat{\widehat{Q}}_{12} = \widehat{Cov}\left(\widehat{M}_1, \widehat{M}_2\right) = -\frac{1}{N}\left(s_{2y}^2 + s_{12m}\right) + \left(\frac{1}{n} - \frac{1}{N}\right)\left(s_{12m} + s_{1y}^2\right)$$
$$= -(38,009.145 + 19,308.035)/50 + (1/20 - 1/50)$$
$$(24,063.421 + 19,308.035)$$
$$= 154.800;$$

$$\widehat{\widehat{Q}}_1 = \widehat{V}\left(\widehat{M}_1\right) - \widehat{\widehat{Q}}_{12} = 3825.516, \quad \widehat{\widehat{Q}}_2 = \widehat{V}\left(\widehat{M}_2\right) - \widehat{\widehat{Q}}_{12} = 2202.027,$$

$$\widehat{\phi}_0 = 1/\widehat{\widehat{Q}}_1/(1/\widehat{\widehat{Q}}_1 + 1/\widehat{\widehat{Q}}_2) = 0.365.$$

$$\widehat{M} = \frac{1}{2}\left\{\widehat{\phi}_0\widehat{M}_1 + \left(1 - \widehat{\phi}_0\right)\widehat{M}_2\right\} = 565.141$$

Estimated standard error of $\widehat{M}$ is

$$SE\left(\widehat{M}\right) = \sqrt{\widehat{V}_{opt}\left(\widehat{M}\right)} = \sqrt{\frac{1}{4}\left(\frac{1}{1/\widehat{\widehat{Q}}_1 + 1/\widehat{\widehat{Q}}_2} + \widehat{\widehat{Q}}_{12}\right)} = 19.700$$

11.5 EXERCISES

11.5.1 On the first occasion a sample s_1 of size n is selected from a population U by RHC method of sampling with normed size measure p_i for the ith unit. On the second occasion a subsample s_{2m} of size $m(\leq n)$ is chosen out of s_1 by the SRSWOR method and an independent sample s_{2u} is chosen from U by the RHC method again using the previous normed size measures p_i's. Let $\widehat{Y}_{2m} = \frac{n}{m}\sum_{i\in s_{2m}} \frac{y_{2i} - y_{1i}}{p_i}P_i + \sum_{i\in s_1}\frac{y_{1i}}{p_i}P_i$ and $\widehat{Y}_{2u} = \sum_{i\in s_{2u}}\frac{y_{2i}}{p_i}P_i^*$, where P_i and P_i^* denote the sum of the p_j's for the groups containing the ith unit in selecting samples s_{2m} and s_{2u} by RHC sampling schemes, respectively. Show that

(i) $\widehat{Y}_2 = \phi\widehat{Y}_{2m} + (1-\phi)\widehat{Y}_{2u}$ is an unbiased estimator for the population total Y_2 for any given value of ϕ,

(ii) the optimum proportion of unmatched sample $\mu(=1-\lambda)$ that minimizes the variance of $\widehat{Y}_2$ with the optimum value of ϕ is $\mu_0 = \left[1 + \sqrt{2(1-\delta)(1+\gamma f)}\right]^{-1}$, and

(iii) the minimum variance of $\widehat{Y}_2$ is $V_{\min}\left(\widehat{Y}_2\right) = \frac{N}{2(N-1)n}\left[(1-f) + \sqrt{2(1-\delta)(1+\gamma f)}\right] V_0$

where $f = n/N$, $V_0 = \sum_{i \in U} p_i \left(\frac{y_{1i}}{p_i} - Y_1\right)^2 = \sum_{i \in U} p_i \left(\frac{y_{2i}}{p_i} - Y_2\right)^2$,

$\delta = \sum_{i \in U} p_i \left(\frac{y_{1i}}{p_i} - Y_1\right)\left(\frac{y_{2i}}{p_i} - Y_2\right) / V_0$,

$$\gamma = \frac{(1-\rho)\overline{V}}{(1-\delta)V_0} - 1, \quad \overline{V} = N \sum_{i \in U} (y_{1i} - Y_1)^2 = N \sum_{i \in U} (y_{2i} - Y_2)^2,$$

and $\rho = N \sum_{i \in U} (y_{1i} - Y_1)(y_{2i} - Y_2) / \overline{V}$ (Ghangurde and Rao, 1969; Chotai, 1974).

11.5.2 Let the initial sample s_1 and the unmatched sample s_{2u} be selected as in Exercise 11.5.1, but the matched sample s_{2m} is selected by RHC sampling replacing p_i by P_i defined in Exercise 11.5.1. Show that

(i) $\widehat{Y}_2 = \phi \widehat{Y}'_{2m} + (1-\phi)\widehat{Y}_{2u}$ is an unbiased estimator Y_2,

(ii) the optimum proportion of an unmatched sample with the optimum value of ϕ is $\mu_0 = \left[1 + \sqrt{2(1-\delta)}\right]^{-1}$, and

(iii) the minimum variance of $\widehat{Y}_2$ is $V_{\min}\left(\widehat{Y}_2\right) = \frac{N}{2(N-1)n}\left[(1-f) + \sqrt{2(1-\delta)}\ \right] V_0$

where $f = n/N$, $\widehat{Y}'_{2m} = \sum_{i \in s_{2m}} \frac{y_{2i} - y_{1i}}{p_i} \Delta_i + \sum_{i \in s_1} \frac{y_{1i}}{p_i} P_i$, Δ_i be the sum of P_j-values from the group containing ith unit in selection of s_{2m} from s_1, and $\widehat{Y}_{2u}$, δ, and V_0 are as in Exercise 11.5.1 (Chotai, 1974).

11.5.3 On the first occasion, a sample s_1 of size n is selected from a population U by SRSWOR. On the second occasion, a matched sample s_{2m} of size m is selected from s_1 by the RHC sampling using normed size measure $p_i^* = y_{1i} / \sum_{i \in s_1} y_{1i}$ for the unit $i \in s_1$ and an unmatched sample s_{2u} of size $u = n - m$ is selected from $U - s_1$. Let $\widehat{\overline{Y}}_{2m} = \frac{1}{n} \sum_{i \in s_{2m}} \frac{y_{2i}}{p_i} P_i$, $p_i = y_{1i}/Y_1$, $\widehat{\overline{Y}}_{2u} = \frac{1}{u} \sum_{i \in s_{2u}} y_{2i}$, and $\widehat{\overline{Y}}_2 = \phi \widehat{\overline{Y}}_{2m} + (1-\phi)\widehat{\overline{Y}}_{2u}$, where P_i is the sum of p_j-values of the group containing the ith unit in selecting sample s_{2m} by the RHC procedure. Show that

(i) $E\left(\widehat{\overline{Y}}_{2m}\right) = \overline{Y}_2$,

(ii) $V\left(\widehat{\overline{Y}}_{2m}\right) = \left(\frac{1}{n} - \frac{1}{N}\right) S_{2y}^2 + \left(\frac{1}{m} - \frac{1}{n}\right) \frac{1}{N(N-1)}$

$\sum_{i \in U} \left(y_{2i} - R^* y_{1i}\right)^2,$

(iii) $Cov\left(\widehat{\overline{Y}}_{2m}, \widehat{\overline{Y}}_{2u}\right) = -S_{2y}^2/N,$

(iv) the optimum proportion of the unmatched sample which minimizes the variance of $\widehat{\overline{Y}}_2 = \phi \widehat{\overline{Y}}_{2m} + (1-\phi)\widehat{\overline{Y}}_{2u}$ with the optimum value of ϕ is $\mu_0 = 1/\left(1 + \sqrt{\delta^*}\right)$, and

(v) the minimum variance of $\widehat{\overline{Y}}_2$ is $V_{\min}\left(\widehat{\overline{Y}}_2\right) = \left(\frac{1+\sqrt{\delta^*}}{2n} - \frac{1}{N}\right) S_{2y}^2$

where $R^* = Y_2/Y_1$ and $\delta^* = \dfrac{\sum_{i \in U} \left(y_{2i} - R^* y_{1i}\right)^2 / p_i}{N \sum_{i \in U} \left(y_{2i} - \overline{Y}_2\right)^2}$ (Avadhan and Sukhatme, 1970).

11.5.4 On the first occasion, a sample s_1 of size n is selected from a population U by the SRSWOR method. On the second occasion, a subsample s_{2m} of size m is selected from s_1 by the SRSWOR method and an unmatched sample s_{2u} of size $u(=n-m)$ is selected from the entire population U by SRSWOR. Let $s_{2u} = s_{21} \cup s_{22}$ where $s_{21} = s_{2m} \cap s_{2u}$ denote the units of s_{2u} that come from the match sample and s_{22} be the remaining units of s_{2u}. Let $t = \phi \overline{y}_{2u} + (1-\phi)\overline{y}'_{2m}$ and $t^* = \phi \dfrac{m_2 \overline{y}_{2m} + l_2 \overline{y}_{22}}{m_2 + l_2} + (1-\phi)\,\overline{y}'_{2m}$ where $\overline{y}_{2u} = \sum_{i \in s_{2u}} y_{2i}/u$, $\overline{y}_{2m} = \sum_{i \in s_{2m}} y_{2i}/m$, $\overline{y}'_{2m} = \overline{y}_{2m} - b\left(\overline{y}_{1m} - \overline{y}_1\right)$, $\overline{y}_1 = \sum_{i \in s_1} y_{1i}/n$, $\overline{y}_{22} = \sum_{i \in s_{22}} y_{2i}/l_2$, m_2 is the size of s_{21} and $l_2 = u - m_2$, and b is chosen to minimize the variance of $\overline{y}'_{2m}$. Show that

(i) $E(t) = E\left(t^*\right)$ and (ii) $V(t) - V\left(t^*\right) = \dfrac{m-1}{N-1}\left(\dfrac{1}{n-m} - \dfrac{1}{N}\right) \geq 0$ (Pathak and Rao, 1967).

11.5.5 Consider the Raj (1965a,b) sampling scheme where on the first occasion, a sample s_1 of size n is selected from a population U by the PPSWR method using the normed size measure p_i attached to the ith unit. On the second occasion, a matched sample s_{2m}

of size $m(<n)$ is selected from s_1 by the SRSWOR method treating all the units of s_1 are distinct and an unmatched sample s_{2u} of size $u = n - m$ units from the entire population by the PPSWR method using p_i as normed size measure for the ith unit. Let $t_0 = \phi t_m + (1-\phi)t_u$ and $t_1 = \phi t_m + (1-\phi)t_u^*$, where $t_m = \frac{1}{m}\sum_{i\in s_{2m}} \frac{y_{2i} - y_{1i}}{p_i} + \frac{1}{n}\sum_{i\in s_1} \frac{y_{1i}}{p_i}$, $t_u = \frac{1}{u}\sum_{i\in s_{2u}} \frac{y_{2i}}{p_i}$,

$t_u^* = \frac{1}{u}\left(\sum_{i\in s_{22}} \frac{y_{2i}}{p_i} + m_2 \sum_{i\in s_{2m}} y_{2i} / \sum_{i\in p_{2m}} p_i\right)$, s_{22} is the sample units of s_{2u} that do not belong to s_{2m}, $\sum_{i\in s}$ = sum over the units in s including repetition, and $\sum_{i\in s}$ = sum over the distinct units in s. Show that (i) $E(t_0) = E(t_1) = \overline{Y}_2 = \sum_{i\in U} y_{2i}/N$ and (ii) $V(t_1) \le E(t_0)$ (Singh, 1972).

11.5.6 Consider the sampling scheme over two occasions in Exercise 11.5.2. If we modify the estimator $\widehat{Y}'_{2m}$ by $\widehat{Y}^0_{2m} = \sum_{i\in s_{2m}} \frac{y_{2i} - ky_{1i}}{p_i}\Delta_i + k\sum_{i\in s_1} \frac{y_{1i}}{p_i}P_i$ where k is a suitably chosen constant. Find the optimum value of k and the minimum variance of $\widehat{Y}_2 = \phi\widehat{Y}'_{2m} + (1-\phi)\widehat{Y}_{2u}$, where $\widehat{Y}_{2u}$ as in Exercise 11.5.2 (Chaudhuri and Arnab, 1979a).

11.5.7 On the first occasion, a sample s_1 of size n is selected from a population U of size n by the SRSWOR method. On the second occasion, a matched-sample s_{2m} of size $m(<n)$ is selected from s_1 by the SRSWOR method and an unmatched sample s_{2u} of size u is selected from $U - s_1$ by SRSWOR method. Suppose we want to estimate the population mean $\overline{Y}_1$. Find the optimum values of α_1, α_2, α_3, and α_4 for which the estimator

$$\widehat{\overline{Y}}_1 = \frac{\alpha_1}{n}\sum_{i\in s_1} y_{1i} + \frac{\alpha_2}{m}\sum_{i\in s_{2m}} y_{1i} + \frac{\alpha_3}{m}\sum_{i\in s_{2m}} y_{2i} + \frac{\alpha_4}{u}\sum_{i\in s_{2u}} y_{2i}$$

will be unbiased for $\overline{Y}_1$ and has a minimum variance for a given value of m. Find also the optimum value of m that minimizes the variance of $\widehat{\overline{Y}}_1$ and also obtain the expression for the minimum variance of $\widehat{\overline{Y}}_1$.

11.5.8 To estimate the average sales of a certain product in a region, an independent sample of size n was selected from each establishment by SRSWR method and information of the sales for the current and previous month were collected each month. Let $\overline{y}_h$ and $\overline{x}_{h-1}$ be the sample means of sales on hth and $h-1$ occasion based on the sample selected on the hth occasion. Let the proposed estimator of the mean sales on the hth occasion based on the data collected in the current and previous occasion be of the form $\widehat{\overline{Y}}_h = \overline{y}_h - \alpha_h \overline{x}_{h-1} + \alpha_h \widehat{\overline{Y}}_{h-1}$, where α_h is a suitably chosen constant and $\alpha_1 = 0$. Assuming the variance of sales of each of the occasions and correlation of sales between any two consecutive occasions remain the same as σ^2 and ρ, respectively, show that the minimum variance of $\widehat{\overline{Y}}_h$ is $V_{\min}\left(\widehat{\overline{Y}}_h\right) = (1-\rho\alpha_h)\sigma^2/n$, where α_h is obtained from $\alpha_h = \rho/(2-\rho\alpha_{h-1})$ (Ecler, 1955; Woodroff, 1959).

11.5.9 In 2005, a sample of 30 clinics was selected from a total of 75 clinics in a certain locality by SRSWOR method and the number of HIV/AIDS affected persons treated in the clinics was collected. In 2008, a subsample of 10 clinics was selected from the selected 30 clinics in 2005; in addition, a fresh sample of 20 clinics was selected from the 45 clinics that were not selected in 2005. The data are given below.

	No. of HIV/AIDS persons treated			No. of HIV/AIDS persons treated	
Clinics	**2005**	**2008**	**Clinics**	**2005**	**2008**
1	700	500	26	970	
2	850	800	27	1200	
3	1200	900	28	1870	
4	650	700	29	980	
5	950	800	30	750	
6	750	600	31		550
7	1500	1200	32		750
8	2500	1750	33		1010
9	600	400	34		850
10	300	420	35		600
11	750		36		325
12	700		37		750
13	875		38		895
14	900		39		750
15	750		40		650

	No. of HIV/AIDS persons treated			No. of HIV/AIDS persons treated	
Clinics	2005	2008	Clinics	2005	2008
16	870		41		900
17	1200		42		875
18	280		43		750
19	250		44		480
20	870		45		750
21	600		46		560
22	825		47		420
23	750		48		375
24	890		49		175
25	475		50		600

Based on the data collected in 2005 and 2008 estimate the following:

(i) Total number of HIV/AIDS affected persons treated in the locality in the year 2008 and its standard error.

(ii) Total number of HIV/AIDS affected persons treated in the locality in the year 2005 and its standard error.

(iii) Change in number of HIV/AIDS affected persons treated between the year 2005 and 2008 along with its standard error.

(iv) Total number of HIV/AIDS affected persons treated in the period between 2005 and 2008 altogether and its standard error.

CHAPTER 12

Cluster Sampling

12.1 INTRODUCTION

Suppose we want to select a sample of 2000 matric students from all the matric students in South Africa to estimate the average cost of education. In this case if we use the simple random sampling with replacement (SRSWR), simple random sampling without replacement (SRSWOR), or systematic sampling procedures for selecting the sample, we will first require a sampling frame that comprises a list of all the matric students in South Africa. In most situations such a list is not readily available. We can then look for a list of all the schools in South Africa that offer a matric program. Such a list is much easier to obtain and we may select a sample of schools, instead of selecting students directly, and then survey all the matric students of the selected schools. This type of sampling procedure is called cluster sampling. Technically, a group of units is called a cluster and each unit in a cluster is called an ultimate unit (from which statistical information is to be collected). In cluster sampling, entire population is first divided into a number of mutually exclusive and exhaustive clusters. We then select a sample of clusters by some suitable method of sampling such as SRSWR, SRSWOR, or probability proportional to size with replacement sampling (PPSWR) and survey all the ultimate units belonging to the sampled clusters.

Cluster sampling may be used when the sampling frame of the ultimate units is not readily available or expensive to construct. This method of sampling is much cheaper, easier, and operationally convenient because units within the same cluster are generally located close to each other and hence traveling costs are much less than that of a sampling procedure, which selects units directly from the entire population.

As per efficiency (sampling variance) is concerned, cluster sampling is generally less efficient than a sampling scheme, which selects units directly. Furthermore, when the cluster sizes are unknown or unequal, we cannot select a desired number of ultimate units.

Survey Sampling Theory and Applications
ISBN 978-0-12-811848-1
http://dx.doi.org/10.1016/B978-0-12-811848-1.00012-1

12.2 ESTIMATION OF POPULATION TOTAL AND VARIANCE

Consider a finite population that is divided into N mutually exclusive and exhaustive clusters. Let M_i be the size of the ith cluster, the number of ultimate units that belong to the ith cluster, $i = 1,\ldots, N$. The total number of ultimate units in the population is $M = \sum_{i=1}^{N} M_i$. Let y_{ij} be the value of the character under study for the jth ultimate unit of the ith cluster $j = 1,\ldots, M_i$; $i = 1,\ldots, N$, $Y_i = \sum_{j=1}^{M_i} y_{ij} =$ ith cluster total, and $Y = \sum_{i=1}^{N} Y_i = \sum_{i=1}^{N}\sum_{j=1}^{M_i} y_{ij} =$ population total. Suppose a sample s of size n clusters is selected with probability $p(s)$ by using some suitable sampling design and all the ultimate units that belong to the selected clusters are surveyed. For example, if the kth cluster is selected in the sample s, i.e., if $k \in s$, then values y_{kj}, $j = 1,\ldots, M_k$ and hence the total $Y_k = \sum_{j=1}^{M_k} y_{kj}$ is obtained from the survey.

An unbiased estimator of the population total Y is given by

$$\widehat{Y}_{cs} = \sum_{i \in s} b_{si} Y_i \tag{12.2.1}$$

where b_{si}'s are known constants that satisfy the unbiasedness condition.

$$\sum_{s \supset i} b_{si} p(s) = 1 \tag{12.2.2}$$

Theorem 12.2.1

(i) The variance of $\widehat{Y}_{cs}$ is $V\left(\widehat{Y}_{cs}\right) = \sum_{i=1}^{N} \alpha_i Y_i^2 + \sum_{i \neq}^{N}\sum_{j=1}^{N} \alpha_{ij} Y_i Y_j$

where $\alpha_i = \sum_{s \supset i} b_{si}^2 p(s) - 1$, $\alpha_{ij} = \sum_{s \supset i} b_{si} b_{sj} p(s) - 1$

(ii) An unbiased estimator of the variance of $\widehat{Y}_c$ is

$$\widehat{V}\left(\widehat{Y}_c\right) = \sum_{i \in s} c_{si} Y_i^2 + \sum_{i \neq}\sum_{j \in s} c_{sij} Y_i Y_j$$

where c_{si} and c_{sij} are suitably chosen constants that satisfy $\sum_{s \supset i,j} c_{si} p(s) = \alpha_i$ and $\sum_{s \supset i,j} c_{sij} p(s) = \alpha_{ij}$.

Proof

(i) $V(\widehat{Y}_{cs}) = E(\widehat{Y}_{cs})^2 - Y^2$

$$= \sum_{s}\left(\sum_{i\in s} b_{si}^2 Y_i^2 + \sum_{i\neq}\sum_{j\in s} b_{sij} Y_i Y_j\right)p(s) - Y^2$$

$$= \sum_{i=1}^{N} Y_i^2\left(\sum_{s\supset i} b_{si}^2 p(s) - 1\right)$$

$$+ \sum_{i\neq}^{N}\sum_{j=1}^{N} Y_i Y_j \sum_{s\supset}\sum_{i,j}\left(b_{sij}p(s) - 1\right)$$

$$= \sum_{i=1}^{N} \alpha_i Y_i^2 + \sum_{i\neq}^{N}\sum_{j=1}^{N} \alpha_{ij} Y_i Y_j$$

(ii) $E\left[\widehat{V}(\widehat{Y}_{cs})\right] = E\left(\sum_{i\in s} c_{si} Y_i^2\right) + E\left(\sum_{i\neq}\sum_{j\in s} c_{sij} Y_i Y_j\right)$

$$= \sum_{i=1}^{N} Y_i^2\left(\sum_{s\supset i} c_{si} p(s)\right) + \sum_{i\neq}^{N}\sum_{j=1}^{N} Y_i Y_j\left(\sum_{s\supset}\sum_{i,j} c_{sij} p(s)\right)$$

$$= V(\widehat{Y}_{cs})$$

Remark 12.2.1

We can choose c_{si} and c_{sij} in various ways. The obvious choices are $c_{si} = 1/\pi_i$ and $c_{sij} = 1/\pi_{ij}$, where π_i and $\pi_{ij}(>0)$ are the inclusion probabilities of the ith and ith and jth $(i \neq j)$ clusters, which are assumed to be positive.

12.2.1 Arbitrary Sampling Design

Let a sample s be selected by using a fixed effective size sampling design of size n and let π_i and π_{ij} be the inclusion probabilities of ith and ith and jth units, respectively. Then the following results are obtained from Theorem 12.2.1.

Theorem 12.2.2

(i) $\widehat{Y}_{cs}(ht) = \sum_{i\in s} \frac{Y_i}{\pi_i}$ is an unbiased estimator for the population total Y

(ii) $V\left[\widehat{Y}_{cs}(ht)\right] = \frac{1}{2}\sum_{i\neq}\sum_{j\in U} \Delta_{ij}\left(\frac{Y_i}{\pi_i} - \frac{Y_j}{\pi_j}\right)^2$ and

(iii) $\widehat{V}\left[\widehat{Y}_{cs}(ht)\right] = \frac{1}{2}\sum_{i\neq}\sum_{j\in s} \frac{\Delta_{ij}}{\pi_{ij}}\left(\frac{Y_i}{\pi_i} - \frac{Y_j}{\pi_j}\right)^2$

where $\Delta_{ij} = \pi_i\pi_j - \pi_{ij}$.

12.2.2 Simple Random Sampling Without Replacement

For an SRSWOR sampling, $\pi_i = n/N$ and $\pi_{ij} = n(n-1)/\{N(N-1)\}$. Substituting $\pi_i = n/N$ and $\pi_{ij} = n(n-1)/\{N(N-1)\}$ in Theorem 12.2.2, we obtain

Theorem 12.2.3

(i) $\widehat{Y}_{cs}(wor) = N\widehat{\overline{Y}}_{cs}$ is an unbiased estimator for the population total Y

(ii) $V\left[\widehat{Y}_{cs}(wor)\right] = N^2\left(\dfrac{1}{n} - \dfrac{1}{N}\right)S_Y^2$ and

(iii) $\widehat{V}\left[\widehat{Y}_{cs}(wor)\right] = N^2\left(\dfrac{1}{n} - \dfrac{1}{N}\right)s_Y^2$

where

$$\widehat{\overline{Y}}_{cs} = \sum_{i\in s} Y_i/n,\ S_Y^2 = \sum_{i=1}^{N}\left(Y_i - \overline{Y}_{cs}\right)^2/(N-1),\ \overline{Y}_{cs} = Y/N,$$

$$s_Y^2 = \sum_{i\in s}\left(Y_i - \widehat{\overline{Y}}_{cs}\right)^2\Big/(n-1).$$

Now writing $\overline{M} = \sum_{i=1}^{N} M_i/N =$ average cluster size and $\overline{\overline{Y}} = Y/\left(N\overline{M}\right) = \overline{Y}_{cs}/\overline{M} =$ mean per unit, we get

$$\begin{aligned}
(N-1)S_Y^2 &= \sum_{i=1}^{N}\left(Y_i - \overline{M}\,\overline{\overline{Y}}\right)^2 \\
&= \sum_{i=1}^{N}\left(\sum_{j=1}^{M_i}\left(y_{ij} - \overline{\overline{Y}}\right) + \left(M_i - \overline{M}\right)\overline{\overline{Y}}\right)^2 \\
&= \sum_{i=1}^{N}\left[\sum_{j=1}^{M_i}\left(y_{ij} - \overline{\overline{Y}}\right)^2 + \sum_{j\neq}^{M_i}\sum_{k=1}^{M_i}\left(y_{ij} - \overline{\overline{Y}}\right)\left(y_{ik} - \overline{\overline{Y}}\right)\right. \\
&\qquad \left. + \overline{\overline{Y}}^2\left(M_i - \overline{M}\right)^2 + 2\overline{\overline{Y}}\ M_i\left(M_i - \overline{M}\right)\left(\overline{Y}_i - \overline{\overline{Y}}\right)\right] \\
&= \left(N\overline{M} - 1\right)S_y^2\left[1 + \rho_c\frac{\sum_{i=1}^{N} M_i(M_i - 1)}{N\overline{M}}\right] + Q
\end{aligned} \tag{12.2.3}$$

where

$$S_y^2 = \sum_{i=1}^{N}\sum_{j=1}^{M_i}\left(y_{ij} - \overline{\overline{Y}}\right)^2 \Big/ \left(N\overline{M} - 1\right),$$

$$Q = \overline{\overline{Y}}^2 \sum_{i=1}^{N}\left(M_i - \overline{M}\right)^2 + 2\overline{\overline{Y}}\sum_{i\in U} M_i\left(M_i - \overline{M}\right)\left(\overline{Y}_i - \overline{\overline{Y}}\right)$$

and

$$\rho_c = \frac{\sum_{i=1}^{N}\sum_{j\neq}^{M_i}\sum_{k=1}^{M_i}\left(y_{ij} - \overline{\overline{Y}}\right)\left(y_{ik} - \overline{\overline{Y}}\right) \Big/ \sum_{i=1}^{N} M_i(M_i - 1)}{\sum_{i=1}^{N}\sum_{j=1}^{M_i}\left(y_{ij} - \overline{\overline{Y}}\right)^2 \Big/ \sum_{i=1}^{N} M_i}$$

= intracluster correlation coefficient between the elements within clusters.

Now substituting Eq. (12.2.3) in the expression of $V\left(\widehat{Y}_{cs}\right)$ of the Theorem 12.2.3, an alternative expression of $V\left(\widehat{Y}_{cs}\right)$ comes out as

$$V(\widehat{Y}_{cs}(wor)) = N^2\left(\frac{1}{n} - \frac{1}{N}\right)\frac{1}{N-1} \times \left[\left(N\overline{M} - 1\right)S_y^2\left(1 + \rho_c\frac{\sum_{i=1}^{N} M_i(M_i - 1)}{N\overline{M}}\right) + Q\right] \tag{12.2.4}$$

In the case that clusters are of equal size, i.e., $M_i = \overline{M}$, Q becomes equal to zero and Eq. (12.2.4) reduces to

$$V\left(\widehat{Y}_{cs}(wor)\right) = N^2\left(\frac{1}{n} - \frac{1}{N}\right)\frac{\left(N\overline{M} - 1\right)}{N-1}S_y^2\left[1 + \left(\overline{M} - 1\right)\rho_c\right] \tag{12.2.5}$$

12.3 EFFICIENCY OF CLUSTER SAMPLING

Instead of using cluster sampling, if one selects an SRSWOR sample of size $n\overline{M}$ from a population of size $N\overline{M}$, then the variance of the population total $\widehat{Y} = N\ \overline{y}_s$ ($\overline{y}_s$ = sample mean) is obtained as

$$V\left(\widehat{Y}\right) = \left(N\overline{M}\right)^2\left(\frac{1}{n\overline{M}} - \frac{1}{N\overline{M}}\right)S_y^2 \tag{12.3.1}$$

where $(NM - 1)S_y^2 = \sum_{i=1}^{N}\sum_{j=1}^{M_i}\left(y_{ij} - \overline{\overline{Y}}\right)^2$.

The efficiency of cluster sampling compared with SRSWOR sampling based on the same sample size $n\overline{M}$ is given by

$$E_{cs} = \frac{V\left(\widehat{Y}\right)}{V\left[\widehat{Y}_{cs}(wor)\right]} \tag{12.3.2}$$

In case $M_i = \overline{M}$ for $i = 1, \ldots, N$, and N is very large, the efficiency E_{cs} reduces to

$$E_{cs} = \frac{1}{1 + \left(\overline{M} - 1\right)\rho_c} \tag{12.3.3}$$

From the expression (Eq. 12.3.3), we conclude that the cluster sampling under SRSWOR will be more, equally, or less efficient than the unistage sampling (direct sampling) based on an SRSWOR of the same sample size if the intracluster correlation coefficient ρ_c is negative, zero, or positive, respectively. Clusters are generally formed by taking neighboring units that are expected to produce positive intraclass correlation; hence cluster sampling is generally less efficient than SRSWOR sampling of the same sample size. But as per cost of the survey is concerned, cluster sampling is much cheaper for surveying neighboring units. However, if ρ_c is negative, cluster sampling is favorable in terms of both cost and efficiency.

Because ρ_c is generally positive, the efficiency of cluster sampling E_{cs} decreases with the increase in the cluster size $\overline{M}$. Because the number of clusters decreases with the increase in the size of the clusters, the efficiency of cluster sampling increases with the number of clusters. The cost of the survey, however, rises with the increase in the total number of clusters.

12.3.1 Optimum Choice of Cluster Size

As stated previously, if the total sample size is fixed, the efficiency of the cluster sampling increases with the number of clusters, but the cost of the survey also increases with the number of clusters. Because efficiency and cost are moving in opposite directions, a balance between the cost of the survey and efficiency should be established by choosing the cluster size $\overline{M}$ optimally. In determination of the optimum cluster size, let us assume $M_i = M = \overline{M}$ for $i = 1, \ldots, N$ and the cost of the survey is of the form

$$C = c_0 + c_1 Mn + c_2\sqrt{n} \tag{12.3.4}$$

where c_0 is the fixed overhead cost, c_1 is the cost of surveying a unit, and $c_2\sqrt{n}$ is the total cost of traveling from one cluster to the other, which is assumed to be approximately proportional to the square root of the number of clusters.

Here we determine the optimum values of M and n, which minimize the variance of $\hat{\overline{\overline{Y}}}_{cs}(wor) = \frac{\hat{Y}_{cs}(wor)}{MN}$, the estimator of the mean per unit keeping the total cost of the survey C fixed to a certain level C_0. For this minimization problem consider

$$\phi = V\left[\hat{\overline{\overline{Y}}}_{cs}(wor)\right] + \lambda\left(C^* - c_1 Mn - c_2\sqrt{n}\right) \tag{12.3.5}$$

where $C^* = C_0 - c_0$ and λ is a Lagrange multiplier.

Now

$$\begin{aligned} V\left[\hat{\overline{\overline{Y}}}_{cs}(wor)\right] &= \frac{1}{M^2}\left(\frac{1}{n} - \frac{1}{N}\right)\frac{1}{N-1}\sum_{i=1}^{N}\left(Y_i - \overline{Y}\right)^2 \\ &= \left(\frac{1}{n} - \frac{1}{N}\right)\frac{1}{N-1}\sum_{i=1}^{N}\left(\overline{Y}_i - \overline{\overline{Y}}\right)^2 \end{aligned} \tag{12.3.6}$$

(noting $M_i = M$ for $i = 1,\ldots, N$).

Because the expression (Eq. 12.3.6) is not an explicit function of M, the minimization of ϕ in Eq. (12.3.5) with respect to M is not possible. However, Smith (1938), Mahalanobis (1940, 1942), Jessen (1942), Hendricks (1944), and Sukhatme et al. (1984), among others, modeled within-cluster mean square $S_w^2 = \frac{\sum_{i=1}^{N}\sum_{j=1}^{M}\left(y_{ij} - \overline{Y}_i\right)^2}{N(M-1)}$ as an explicit function of M. Empirical investigations based on agricultural surveys reveal that S_w^2 can be modeled as

$$S_w^2 = \alpha\ M^g$$

where $\alpha(>0)$ and $g(\geq 0)$ are independent of M.

Now writing $M\sum_{i=1}^{N}\left(\overline{Y}_i - \overline{\overline{Y}}\right)^2 = (NM-1)S_y^2 - N(M-1)S_w^2 = (NM-1)S_y^2 - N(M-1)\alpha M^g$ and neglecting finite population correction term we get

$$\phi = \frac{1}{n}\left[S_y^2 - (M-1)\alpha M^{g-1}\right] + \lambda\left(C_0 - c_0 - c_1 Mn - c_2\sqrt{n}\right) \tag{12.3.7}$$

To find the optimum values of m and n, we first find the optimum value of n from the equation $C^* = c_1 Mn + c_2\sqrt{n}$ keeping M fixed. The optimum value of n is obtained as

$$n = \left(\frac{-c_2 + \sqrt{c_2^2 + 4c_1 C^* M}}{2c_1 M}\right)^2 \tag{12.3.8}$$

Differentiating ϕ with respect to m and n and then equating them to zero we get

$$\frac{\partial \phi}{\partial M} = -\frac{\alpha M^{g-2}}{n}[gM - (g-1)] - \lambda c_1 n = 0 \quad (12.3.9)$$

and

$$\frac{\partial \phi}{\partial n} = -\frac{1}{n^2}\left[S_y^2 - (M-1)\alpha M^{g-1}\right] + \lambda\left(-c_1 M - \frac{c_2}{2\sqrt{n}}\right) = 0 \quad (12.3.10)$$

Eliminating λ from Eqs. (12.3.9) and (12.3.10), we have

$$\frac{\alpha M^{g-1}\{gM-(g-1)\}}{S_y^2-(M-1)\alpha M^{g-1}} = \left(1+\frac{c_2}{2c_1 M\sqrt{n}}\right)^{-1} \quad (12.3.11)$$

Now substituting $\sqrt{n}$ from Eq. (12.3.8) in Eq. (12.3.11), we get

$$\frac{\alpha M^{g-1}\{gM-(g-1)\}}{S_y^2-(M-1)\alpha M^{g-1}} = 1 - \frac{c_2}{\sqrt{c_2^2+4c_1 C^* M}} \quad (12.3.12)$$

Because Eq. (12.3.12) is independent of n, we can obtain M numerically by an iterative procedure. The optimum value of n is obtained from Eq. (12.3.8) by substituting the optimum value of M obtained from Eq. (12.3.12).

12.4 PROBABILITY PROPORTIONAL TO SIZE WITH REPLACEMENT SAMPLING

Let a sample s of n clusters be selected by PPSWR method of sampling using normed size measure $p_i\left(>0, \sum_{i=1}^{N} p_i = 1\right)$ attached to the ith unit. The Hansen and Hurwitz (1943) estimator for the population total Y is

$$\widehat{Y}_{cs}(hh) = \frac{1}{n}\sum_{r=1}^{n}\frac{Y(r)}{p(r)} \quad (12.4.1)$$

where $Y(r)$ is the cluster total of the unit selected at the rth draw with probability $p(r)$, i.e., if the rth draw produces the jth cluster then $Y(r) = Y_j$ and $p(r) = p_j$.

Theorem 12.4.1

(i) $E\left[\widehat{Y}_{cs}(hh)\right] = Y$

(ii) $V\left[\widehat{Y}_{cs}(hh)\right] = \frac{1}{n}\sum_{i=1}^{N} p_i\left(\frac{Y_i}{p_i} - Y\right)^2$

(iii) $\widehat{V}\left[\widehat{Y}_{cs}(hh)\right] = \frac{1}{n(n-1)}\sum_{r=1}^{n}\left(\frac{Y(r)}{p(r)} - \widehat{Y}_{cs}(hh)\right)^2$

Proof

This theorem follows directly from Theorem 5.2.2 and hence it is omitted.

12.4.1 Simple Random Sampling With Replacement

Suppose a sample of size n is selected by SRSWR method, then an unbiased estimator of the population total Y is given by

$$\widehat{Y}_{cs}(wr) = \frac{N}{n}\sum_{r=1}^{n} Y(r) \tag{12.4.2}$$

where $Y(r)$ is the cluster selected at the rth draw. We then have the following theorem by substituting $p_i = 1/N$ for $i = 1,\ldots, N$ in Theorem 12.4.1.

Theorem 12.4.2

(i) $E\left[\widehat{Y}_{cs}(wr)\right] = Y$

(ii) $V\left[\widehat{Y}_{cs}(wr)\right] = N^2\sigma_Y^2/n$

(iii) $\widehat{V}\left[\widehat{Y}_{cs}(wr)\right] = N^2\widehat{\sigma}_Y^2/n$

where $\sigma_Y^2 = \frac{1}{N}\sum_{i=1}^{N}\left(Y_i - \overline{Y}\right)^2$, $\widehat{\sigma}_Y^2 = \frac{1}{(n-1)}\sum_{r=1}^{n}\left[Y(r) - \widehat{Y}_{cs}(wr)\right]^2$,

and $\overline{Y} = Y/N$.

12.5 ESTIMATION OF MEAN PER UNIT

The mean per unit is defined by $\overline{\overline{Y}} = Y/\left(N\overline{M}\right)$. Hence if $\overline{M}$ is known, $\overline{\overline{Y}}$ may be estimated unbiasedly by

$$\widehat{\overline{\overline{Y}}}_{cs} = \widehat{Y}_c/\left(N\overline{M}\right) \tag{12.5.1}$$

In practice $\overline{M}$ is unknown. Hence $\overline{\overline{Y}}$ may be estimated through a ratio estimator as follows:

$$\widehat{\overline{\overline{Y}}}_{Rc} = \widehat{Y}_{cs}/\widehat{M}_{cs} \tag{12.5.2}$$

where $\widehat{Y}_{cs} = \sum_{i\in s} b_{si}Y_i$ and $\widehat{M}_{cs} = \sum_{i\in s} b_{si}M_i$.

The ratio estimator $\hat{\overline{\overline{Y}}}_{Rc}$ is biased for $\overline{\overline{Y}}$. However, the amount of bias is negligible for a large sample. Approximate expressions of the bias and the mean square error of $\hat{\overline{\overline{Y}}}_{Rc}$ are obtained from Eqs. (8.2.11) and (8.2.16), respectively, as follows:

$$B\left(\hat{\overline{\overline{Y}}}_{Rc}\right) = -\frac{Cov\left(\widehat{M}_{cs}, \widehat{G}_{cs}\right)}{\left(N\overline{M}\right)^2} \tag{12.5.3}$$

[where $\widehat{G}_{cs} = \sum_{i \in s} b_{si} G_i$ and $G_i = M_i\left(\overline{Y}_i - \overline{\overline{Y}}\right)$] and

$$\begin{aligned} M\left(\hat{\overline{\overline{Y}}}_{Rc}\right) &= \frac{1}{\left(N\overline{M}\right)^2} V\left(\sum_{i \in s} b_{si} G_i\right) \\ &= \frac{1}{\left(N\overline{M}\right)^2}\left(\sum_{i \in U} \alpha_i G_i^2 + \sum_{i \neq}\sum_{j \in U} \alpha_{ij} G_i G_j\right) \end{aligned} \tag{12.5.4}$$

where $\alpha_i = \sum_{s \supset i} b_{si}^2 p(s) - 1$, $\alpha_{ij} = \sum_{s \supset i,j} b_{si} b_{sj} p(s) - 1$, and $p(s)$ is the probability of selection of the sample s.

Using Eqs. (8.2.13) and (8.2.19), approximate unbiased estimators of $B\left(\hat{\overline{\overline{Y}}}_{Rc}\right)$ and $M\left(\hat{\overline{\overline{Y}}}_{Rc}\right)$ are obtained, respectively, as

$$\widehat{B}\left(\hat{\overline{\overline{Y}}}_{Rc}\right) = -\frac{\widehat{Cov}\left(\widehat{M}_{cs}, \widehat{G}_{cs}\right)}{\left(\widehat{M}_{cs}\right)^2}$$

and

$$\widehat{M}\left(\hat{\overline{\overline{Y}}}_{Rc}\right) = \frac{1}{\left(\widehat{M}_{cs}\right)^2}\left(\sum_{i \in s} c_{si} \widehat{G}_i^2 + \sum_{i \neq}\sum_{j \in s} c_{sij} \widehat{G}_i \widehat{G}_j\right) \tag{12.5.5}$$

where $\widehat{G}_i = Y_i - \hat{\overline{\overline{Y}}}_{cs} M_i$, $c_{si} = \alpha_i/\pi_i$, $c_{sij} = \alpha_{ij}/\pi_{ij}$, $\pi_i =$ inclusion probability for the ith unit, and $\pi_{ij} =$ inclusion probability for the ith and $j(\neq i)$th unit.

12.5.1 Examples

12.5.1.1 Arbitrary Sampling Design

Consider a fixed effective size n sampling design with $\pi_i > 0$. For $b_{si} = 1/\pi_i$, we have

$$\widehat{\overline{\overline{Y}}}_{Rc}(ht) = \left(\sum_{i\in s} Y_i/\pi_i\right) \Big/ \left(\sum_{i\in s} M_i/\pi_i\right),$$

$$B\left(\widehat{\overline{\overline{Y}}}_{Rc}\right) = -\frac{1}{\left(N\overline{M}\right)^2}\frac{1}{2}\sum_{i\neq}^{N}\sum_{j=1}^{N}(\pi_i\pi_j - \pi_{ij})\left(\frac{M_i}{\pi_i} - \frac{M_j}{\pi_j}\right)\left(\frac{G_i}{\pi_i} - \frac{G_j}{\pi_j}\right)$$

$$M\left(\widehat{\overline{\overline{Y}}}_{Rc}\right) = \frac{1}{\left(N\overline{M}\right)^2}\frac{1}{2}\sum_{i\neq}\sum_{j\in U}(\pi_i\pi_j - \pi_{ij})\left(\frac{G_i}{\pi_i} - \frac{G_j}{\pi_j}\right)^2,$$

$$\widehat{B}\left(\widehat{\overline{\overline{Y}}}_{Rc}\right) = -\frac{1}{\widehat{M}_{cs}^2}\frac{1}{2}\sum_{i\neq}\sum_{j\in s}\frac{(\pi_i\pi_j - \pi_{ij})}{\pi_{ij}}\left(\frac{M_i}{\pi_i} - \frac{M_j}{\pi_j}\right)\left(\frac{\widehat{G}_i(ht)}{\pi_i} - \frac{\widehat{G}_j(ht)}{\pi_j}\right)$$

and

$$\widehat{M}\left(\widehat{\overline{\overline{Y}}}_{Rc}\right) = \frac{1}{\left(\widehat{M}_{cs}\right)^2}\frac{1}{2}\sum_{i\neq}\sum_{j\in s}\frac{(\pi_i\pi_j - \pi_{ij})}{\pi_{ij}}\left(\frac{\widehat{G}_i(ht)}{\pi_i} - \frac{\widehat{G}_j(ht)}{\pi_j}\right)^2$$

where $\widehat{M}_{cs} = \sum_{i\in s} M_i/\pi_i$ and $\widehat{G}_i(ht) = Y_i - \widehat{\overline{\overline{Y}}}_{Rc}(ht)M_i$.

12.5.1.2 Simple Random Sampling Without Replacement

For SRSWOR sampling $b_{si} = N/n$ and hence

$$\widehat{\overline{\overline{Y}}}_{Rc}(wor) = \sum_{i\in s} Y_i \Big/ \sum_{i\in s} M_i,$$

$$B\left[\widehat{\overline{\overline{Y}}}_{Rc}(wor)\right] = -\frac{(1-f)}{n\overline{M}^2}S_{GM}$$

$$M\left[\widehat{\overline{\overline{Y}}}_{Rc}(wor)\right] = \frac{(1-f)}{n\overline{M}^2}S_G^2$$

$$\widehat{B}\left[\widehat{\overline{\overline{Y}}}_{Rc}(wor)\right] = -\frac{(1-f)}{n\overline{M}_s^2}s_{\hat{G}M}$$

and

$$\widehat{M}\left[\widehat{\overline{\overline{Y}}}_{Rc}(wor)\right]=\frac{(1-f)}{n\widehat{\overline{M}}_s^2}s_{\widehat{G}}^2$$

where $f=n/N$, $\widehat{\overline{M}}_s=\sum_{i\in s}M_i/n$, $S_G^2=\sum_{i\in U}\left(G_i-\overline{G}\right)^2/(N-1)$,

$\overline{G}=\sum_{i\in U}G_i/N=0$, $S_{GM}=\sum_{i=1}^{N}G_i\left(M_i-\overline{M}\right)/(N-1)$,

$s_{\widehat{G}}^2=\sum_{i\in s}\left(\widehat{G}_{iwor}-\widehat{\overline{G}}_s\right)^2\Big/(n-1)$, $\widehat{G}_{iwor}=Y_i-\widehat{\overline{\overline{Y}}}_{Rc}(wor)M_i$, and

$\widehat{\overline{G}}_s=\sum_{i\in s}\widehat{G}_{iwor}\Big/n$.

12.6 EXERCISES

12.6.1 A sample s of n clusters is selected from a population of N clusters of unequal sizes by SRSWOR method. Show that the estimator $t=\frac{1}{n}\sum_{i\in s}\overline{Y}_i$ is biased estimator for the mean per unit $\overline{\overline{Y}}$. Find the expressions of (i) bias, (ii) mean square error, and (iii) their unbiased estimators

12.6.2 Let s be a sample of size n clusters is selected from a population of N clusters by SRSWOR method and all the clusters are of equal size M. Furthermore, let P_i and P be the proportions of units that possess a certain attribute in the ith cluster and entire population, respectively. Show that (i) $\widehat{P}=\sum_{i\in s}\widehat{P}_i/n$ is an unbiased estimator for P and (ii) $V\left(\widehat{P}\right)=\frac{(1-f)}{n}\sum_{i=1}^{N}(P_i-P)^2/(N-1)$.

12.6.3 Let NM units of a population be grouped at random into N clusters each of size M and let s be a sample of n clusters selected from N clusters by SRSWOR method. Show that (i) $t=\frac{1}{n}\sum_{i\in s}\overline{Y}_i$ is an unbiased estimator of the mean per unit $\overline{\overline{Y}}$ and (ii) $V(t)=(1-f)S_y^2/(nM)$, where $S_y^2=\sum_{i=1}^{N}\sum_{j=1}^{M}\left(y_{ij}-\overline{\overline{Y}}\right)^2/(N-1)$ and $f=n/N$.

12.6.4 Let a finite population of M_0 units be divided into N clusters so that the ith cluster contains M_i units $\left(\sum_{i=1}^{N}M_i=M_0\right)$. A sample of m

units is selected by SRSWOR method from the M_0 units and let Y_i be the total of the study variable y that belongs to the ith cluster ($Y_i = 0$ if none of the units fall into the ith cluster). Let a sample s of n clusters be selected from the N clusters by SRSWOR method. Show that $t = \frac{N}{nm} \sum_{i \in s} Y_i$ is an unbiased estimator of the population mean $\overline{\overline{Y}}$ and obtain the variance of t (Ghosh, 1963).

12.6.5 A sample of 10 households is selected from a locality of 50 households by SRSWOR method. The household size, number of earning members, and the total monthly income of the selected households are given in the following table.

Serial no. of households	Household size	Number of earning members	Total income in $
1	5	2	7000
2	3	1	5000
3	4	2	8000
4	2	1	4000
5	4	1	5000
6	1	1	3000
7	4	2	7000
8	5	1	3000
9	4	2	7000
10	2	2	6000

Estimate the average household income and proportion of earning members per household of the locality along with their standard errors.

12.6.6 The following table gives the yearly production of apples in 10 orchards selected at random from 60 orchards in a certain block.

Serial no. of orchards	No. of trees	Yield of orchards in Kg
1	6	200, 150, 720, 420, 400, 200
2	3	250, 200, 200
3	5	320, 470, 125, 375, 120
4	2	250, 400
5	4	175, 125, 195, 370
6	1	300
7	4	180, 120, 320, 170
8	5	250, 175, 300, 420, 180
9	4	370, 480, 200, 150
10	2	400, 250

(i) Estimate the total production of apples of the block and also the average production of apples per orchard. Give unbiased estimates of the variances of the proposed estimators.
(ii) Find a 95% confidence interval of the average production of apple per tree.

12.6.7 To estimate availability of doctors in a locality, 5 clinics were selected from 100 clinics by SRSWOR method. The results are given in the following table.

Serial no. of clinic	No. of doctors	No. of patients treated per day
1	3	20, 15, 40
2	3	25, 20, 20
3	3	20, 27, 12
4	3	25, 40, 20
5	4	15, 25, 20

(i) Estimate the intracluster correlation coefficients between doctors within clinics.
(ii) Estimate the average number of patients treated per clinic when the number of doctors in each of the clinic is fixed at 3. Estimate the efficiency of cluster sampling over SRSWOR sampling for estimating the average number of patients treated by a doctor.

CHAPTER 13

Multistage Sampling

13.1 INTRODUCTION

In cluster sampling, the entire population is divided into a number (N) of mutually exclusive and exhaustive groups called clusters. Each cluster consists of several ultimate units. A sample of $n(<N)$ clusters is chosen by some suitable sampling scheme, and all the ultimate units of the selected clusters are surveyed. In two-stage sampling, "clusters" are termed as "first-stage units (fsu's)" or "primary sampling units." The ultimate units in a cluster are called "second-stage units (ssu's)." In two-stage sampling, a sample s of n fsu's is selected by a sampling procedure. Then from each of the selected fsu's, a sample of ssu's of suitable sizes are selected by a suitable sampling procedure. Clearly, two-stage sampling reduces to a cluster sampling if all of the sampled fsu's are completely enumerated. For example, suppose we want to select 5000 households from KwaZulu-Natal (KZN) to determine their average income. In this situation we can divide KZN into a number of municipalities, which may be regarded as fsu's. Then we may select a sample of fsu's, and from each of the selected fsu's (municipalities) a suitable number of ssu's (households) may be selected by a suitable sampling scheme. If a sample of ultimate units is selected by more than two stages, then the sampling procedure is called multistage sampling. For example, we may divide South Africa into different provinces, then each province is divided again into municipalities, and municipalities may be divided into households. So, in three-stage sampling we first select a sample of provinces, then from each of the selected province, select a sample of municipalities, and finally from the selected municipalities, households may be selected. The multistage sampling is a compromise between a cluster sampling and a unistage sampling (units are directly selected from the population). Hence it is much cheaper and more convenient to draw a sample in a two-stage sampling than a unistage sampling procedure, but more expensive than a cluster sampling. With regard to efficiency, multistage sampling is generally more efficient than

Survey Sampling Theory and Applications
ISBN 978-0-12-811848-1
http://dx.doi.org/10.1016/B978-0-12-811848-1.00013-3

cluster sampling but less efficient than unistage sampling. Multistage sampling is widely used in most large-scale surveys because the sampling frame of the ultimate units is not generally available in practice. For the selection of ssu's, only the sampling frames of the selected fsu's are required. Similarly, for the selection of the third-stage units (tsu's), only the sampling frames of the selected ssu's are required.

13.2 TWO-STAGE SAMPLING SCHEME

Let us consider a population U consisting of N fsu's and the ith fsu consists of M_i ssu's and the total number of ssu's in the population $M_0 = \sum_{i \in U} M_i$. Let y_{ij} be the value of the character y, under study for the jth ssu of the ith fsu, $j = 1,\ldots, M_i$; $i = 1,\ldots, N$ and $Y_i = \sum_{j=1}^{M_i} y_{ij}$ = total of the y-values of the ssu's that belong to the ith fsu. $\overline{Y}_i = Y_i/M_i$ = mean of the ssu's that belong to the ith fsu. The population total for all the ssu's of the population is $Y = \sum_{i=1}^{N}\sum_{j=1}^{M_i} y_{ij} = \sum_{i=1}^{N} Y_i$ and the population mean per unit is $\overline{\overline{Y}} = Y/M_0$.

Let a sample s of size n fsu's be selected from the N fsu's of the population U with probability $p(s)$ by some suitable sampling design. Let π_i and π_{ij} be the inclusion probabilities for the ith and i and jth $(i \neq j)$ fsu's. If the ith fsu is selected in the sample s, i.e., if $i \in s$, then from each of the ith $(i \in s)$ fsu, a subsample s_i of predetermined size m_i ssu's is selected independently by following a suitable sampling design.

13.3 ESTIMATION OF THE POPULATION TOTAL AND VARIANCE

Let $\widehat{Y}_i$ be an unbiased estimator for the total Y_i based on the sample s_i $(i \in s)$. Then we may consider an unbiased estimator for the population total Y as

$$\widehat{Y}_{ts} = \sum_{i \in s} b_{si}\widehat{Y}_i \qquad (13.3.1)$$

The coefficients b_{si}'s are known and satisfy the unbiasedness condition $\sum_{s \supset i} b_{si}p(s) = 1$.

Theorem 13.3.1

(i) $E(\widehat{Y}_{ts}) = Y$

(ii) $$V(\widehat{Y}_{ts}) = \sum_{i\in U} \alpha_i Y_i^2 + \sum_{i\neq}\sum_{j\in U} \alpha_{ij} Y_i Y_j + \sum_{i\in U} \beta_i \sigma_i^2 \qquad (13.3.2)$$

where $\alpha_i = \sum_{s\supset i} b_{si}^2 p(s) - 1, \quad \alpha_{ij} = \sum_{s\supset i,j} b_{si} b_{sj} p(s) - 1, \quad \beta_i = \alpha_i + 1,$ and $\sigma_i^2 = V(\widehat{Y}_i|s)$

Proof

Let E_1 and V_1 be the expectation and variance operators for the selection of the sample s of the fsu's, and let $E_2(\cdot|s)$ and $V_2(\cdot|s)$ be the conditional expectation and variance for a given s. Then,

(i) $$E(\widehat{Y}_{ts}) = E_1\left[\sum_{i\in s} b_{si} E_2(\widehat{Y}_i|s)\right]$$
$$= E_1\left(\sum_{i\in s} b_{si} Y_i\right)$$
$$= Y$$

(ii) $$V(\widehat{Y}_{ts}) = V_1\left[E_2\left(\sum_{i\in s} b_{si}\widehat{Y}_i|s\right)\right] + E_1\left[V_2\left(\sum_{i\in s} b_{si}\widehat{Y}_i|s\right)\right]$$
$$= V_1\left[\sum_{i\in s} b_{si} E_2(\widehat{Y}_i|s)\right] + E_1\left[\sum_{i\in s} b_{si}^2 V_2(\widehat{Y}_i|s)\right]$$
$$= V_1\left(\sum_{i\in s} b_{si} Y_i\right) + E_1\left(\sum_{i\in s} b_{si}^2 \sigma_i^2\right)$$

Now

$$V_1\left(\sum_{i\in s} b_{si} Y_i\right) = \sum_{i\in U} Y_i^2 \sum_{s\supset i} b_{si}^2 p(s) + \sum_{i\neq}\sum_{j\in U} Y_i Y_j \sum_{s\supset i,j} b_{si} b_{sj} p(s) - Y^2$$
$$= \sum_{i\in U} \alpha_i Y_i^2 + \sum_{i\neq}\sum_{j\in U} \alpha_{ij} Y_i Y_j$$

and

$$E_1\left(\sum_{i\in s} b_{si}^2\sigma_i^2\right) = \sum_{i\in U}\sigma_i^2\sum_{s\supset i} b_{si}^2 p(s) = \sum_{i\in U}\beta_i\sigma_i^2 \text{ Q.E.D.}$$

Theorem 13.3.2

Let $Q_s(Y) = \sum_{i\in s} c_{si}Y_i^2 + \sum_{i\neq}\sum_{j\in s} c_{sij}Y_iY_j$ be an unbiased estimator of $V_1\left(\sum_{i\in s} b_{si}Y_i\right)$, satisfying $E_1(Q_s(Y)) = V_1\left(\sum_{i\in s} b_{si}Y_i\right)$, then an unbiased estimator of $V\left(\widehat{Y}_{ts}\right)$ is

$$\widehat{V}\left(\widehat{Y}_{ts}\right) = Q_s\left(\widehat{Y}\right) + \widehat{\Phi}_s$$

where $Q_s\left(\widehat{Y}\right) = \sum_{i\in s} c_{si}\widehat{Y}_i^2 + \sum_{i\neq}\sum_{j\in s} c_{sij}\widehat{Y}_i\widehat{Y}_j$, $\widehat{\Phi}_s$ is an unbiased estimator of $\sum_{i\in U}\sigma_i^2$; c_{si} and c_{sij} are suitably chosen constants.

Proof

Since $Q_s(Y)$ is an unbiased estimator of $V_1\left(\sum_{i\in s} b_{si}Y_i\right)$, c_{si} and c_{sij} must satisfy

$$\sum_{s\supset i} c_{si}p(s) = \alpha_i \text{ and } \sum_{s\supset i,j} c_{sij}p(s) = \alpha_{ij} \tag{13.3.3}$$

Now noting

$$\begin{aligned}
E\left[Q_s\left(\widehat{Y}\right)\right] &= E_1\left[E_2\left(\sum_{i\in s} c_{si}\widehat{Y}_i^2 + \sum_{i\neq}\sum_{j\in s} c_{sij}\widehat{Y}_i\widehat{Y}_j\right)\right] \\
&= E_1\left[\sum_{i\in s} c_{si}E_2\left(\widehat{Y}_i^2\middle|s\right) + \sum_{i\neq}\sum_{j\in s} c_{sij}E_2\left(\widehat{Y}_i\middle|s\right)E_2\left(\widehat{Y}_j\middle|s\right)\right] \\
&= E_1\left[\sum_{i\in s} c_{si}\left(Y_i^2+\sigma_i^2\right) + \sum_{i\neq}\sum_{j\in s} c_{sij}Y_iY_j\right] \\
&= E_1[Q_s(Y)] + \sum_{i\in U}\alpha_i\sigma_i^2 \\
&= V_1\left(\sum_{i\in s} b_{si}Y_i\right) + \sum_{i\in U}\alpha_i\sigma_i^2
\end{aligned}$$

we get

$$
\begin{aligned}
E\left[\widehat{V}\left(\widehat{Y}_{ts}\right)\right] &= E\left[Q_s\left(\widehat{Y}\right)\right] + E\left(\widehat{\Phi}_s\right) \\
&= V_1\left(\sum_{i\in s} b_{si} Y_i\right) + \sum_{i\in U}\alpha_i\sigma_i^2 + \sum_{i\in U}\sigma_i^2 \\
&= \sum_{i\in U}\alpha_i Y_i^2 + \sum_{i\neq}\sum_{j\in U}\alpha_{ij} Y_i Y_j + \sum_{i\in U}\beta_i\sigma_i^2 \\
&= V\left(\widehat{Y}\right) \text{ Q.E.D.}
\end{aligned}
$$

The appropriate choices of c_{si}, c_{sij}, and $\widehat{\Phi}_s$ depend on the sampling design under considerations, but obvious choices are $c_{si} = \alpha_i/\pi_i$, $c_{sij} = \alpha_{ij}/\pi_{ij}$, and $\widehat{\Phi}_s = \sum_{i\in s}\widehat{\sigma}_i^2/\pi_i$ or $\sum_{i\in s} b_{si}\widehat{\sigma}_i^2$ where $\widehat{\sigma}_i^2$ is the conditionally unbiased estimator of σ_i^2, which satisfies $E_2\left(\widehat{\sigma}_i^2\right) = \sigma_i^2$. Detailed discussions of various choices of c_{si}, c_{sij}, and $\widehat{\Phi}_s$ have been given by Chaudhuri and Arnab (1982b).

13.3.1 First-Stage Arbitrary Sampling Designs and Second-Stage Simple Random Sampling Without Replacement

Let the sample s be selected by an arbitrary fixed effective size (n) sampling design with positive inclusion probabilities π_i and π_{ij} for the ith, and ith and $j(\neq i)$ units, respectively. Each of the selected fsu's in s be subsampled independently by simple random sampling without replacement (SRSWOR) method. Let

$$\widehat{Y}_i = M_i\overline{y}(s_i) \tag{13.3.4}$$

with $\overline{y}(s_i) = \sum_{j\in s_i} y_{ij}/m_i$.

Then,

$$\sigma_i^2 = V_2\left(\widehat{Y}_i\middle|s\right) = M_i^2(1-f_i)S_{yi}^2/m_i \tag{13.3.5}$$

where $f_i = m_i/M_i$, $S_{yi}^2 = \sum_{j=1}^{M_i}\left(y_{ij} - \overline{Y}_i\right)^2\Big/(M_i - 1)$ and $\overline{Y}_i = Y_i/M_i$.

Now using the Theorems 13.3.1 and 13.3.2, we note that

$$\widehat{Y}_{ts}(ht) = \sum_{i\in s}\widehat{Y}_i/\pi_i = \sum_{i\in s} M_i\overline{y}(s_i)/\pi_i \tag{13.3.6}$$

is an unbiased estimator of the population total Y with variance

$$V\left[\widehat{Y}_{ts}(ht)\right] = V_1\left(\sum_{i\in s} Y_i/\pi_i\right) + E_1\left(\sum_{i\in s} \sigma_i^2/\pi_i^2\right)$$
$$= \frac{1}{2}\sum_{i\neq}\sum_{j\in U}\Delta_{ij}\left(\frac{Y_i}{\pi_i} - \frac{Y_j}{\pi_j}\right)^2 + \sum_{i\in U}\frac{M_i^2(1-f_i)S_{yi}^2}{m_i\pi_i} \quad (13.3.7)$$

where $\Delta_{ij} = \pi_i\pi_j - \pi_{ij}$, $f_i = m_i/M_i$ and $S_{yi}^2 = \sum_{j=1}^{M_i}\left(y_{ij} - \overline{Y}_i\right)^2$.

Further noting that $Q_s(Y) = \frac{1}{2}\sum_{i\neq}\sum_{j\in s}\frac{\Delta_{ij}}{\pi_{ij}}\left(\frac{Y_i}{\pi_i} - \frac{Y_j}{\pi_j}\right)^2$ and $\widehat{\sigma}_i^2 = M_i^2(1-f_i)s_{yi}^2\Big/m_i$ are unbiased estimators for $V_1\left(\sum_{i\in s} Y_i/\pi_i\right) = \frac{1}{2}\sum_{i\neq}\sum_{j\in U}\Delta_{ij}\left(\frac{Y_i}{\pi_i} - \frac{Y_j}{\pi_j}\right)^2$ and σ_i^2, respectively, we find an unbiased estimator of $V\left(\widehat{Y}_{ms}(ht)\right)$ as

$$\widehat{V}\left(\widehat{Y}_{ts}(ht)\right) = \frac{1}{2}\sum_{i\neq}\sum_{j\in s}\frac{\Delta_{ij}}{\pi_{ij}}\left(\frac{\widehat{Y}_i}{\pi_i} - \frac{\widehat{Y}_j}{\pi_j}\right)^2 + \sum_{i\in s}\frac{M_i^2(1-f_i)s_{yi}^2}{m_i\pi_i} \quad (13.3.8)$$

where $s_{yi}^2 = \sum_{j\in s_i}\left\{y_{ij} - \overline{y}(s_i)\right\}^2/(m_i - 1)$ and $f_i = m_i/M_i$.

13.3.2 Simple Random Sampling Without Replacement Both the Stages

For an SRSWOR sampling, $\pi_i = n/N$, $\pi_{ij} = n(n-1)/\{N(N-1)\}$, and Eqs. (13.3.6)–(13.3.8), reduce to

$$\widehat{Y}_{ts}(wor) = \frac{N}{n}\sum_{i\in s} M_i\overline{y}(s_i) \quad (13.3.9)$$

$$V\left(\widehat{Y}_{ts}(wor)\right) = N^2(1-f)\frac{S_b^2}{n} + \frac{N}{n}\sum_{i\in U}\frac{M_i^2(1-f_i)S_{yi}^2}{m_i} \quad (13.3.10)$$

and

$$\widehat{V}\left(\widehat{Y}_{ts}(wor)\right) = N^2(1-f)\frac{s_b^2}{n} + \frac{N}{n}\sum_{i\in s}\frac{M_i^2(1-f_i)s_{yi}^2}{m_i} \quad (13.3.11)$$

where $S_b^2 = \sum_{i\in U}\left(Y_i - \overline{Y}\right)^2/(N-1)$, $s_b^2 = \sum_{i\in s}\left(\widehat{Y}_i - \overline{\widehat{Y}}_s\right)^2\Big/(n-1)$, $f = n/N$, $\overline{Y} = \sum_{i\in U} Y_i/N$ and $\overline{\widehat{Y}}_s = \sum_{i\in s}\widehat{Y}_i/n = \sum_{i\in s} M_i\overline{y}(s_i)/n$.

13.3.3 First-Stage Rao–Hartley–Cochran and Second-Stage Simple Random Sampling Without Replacement

Here we suppose that the sample s of n fsu's is selected by Rao–Hartley–Cochran (1962) method of sampling using normed size measure p_i attached to the ith unit and P_i is the sum of p_j's for the group containing the ith unit that was formed in selecting the sample s (the detailed selection procedure has been described in Section 5.6). The ssu's are selected independently from the selected fsu's by SRSWOR method. For this sampling scheme, an unbiased estimator for the total Y is given by

$$\widehat{Y}_{ts}(rhc) = \sum_{i\in s}\frac{\widehat{Y}_i}{p_i}P_i \tag{13.3.12}$$

where $\widehat{Y}_i$ is given in Eq. (13.3.4).

The variance of $\widehat{Y}$ is given by

$$V\left(\widehat{Y}_{ts}(rhc)\right) = V_1\left(\sum_{i\in s}\frac{Y_i}{p_i}P_i\right) + E_1\left(\sum_{i\in s}\frac{\sigma_i^2}{p_i^2}P_i^2\right) \tag{13.3.13}$$

Now using Eq. (5.6.7) and Arnab (2004a,b), we get

$$V_1\left(\sum_{i\in s}\frac{Y_i}{p_i}P_i\right) = \frac{N-n}{n(N-1)}\sum_{i\in U}p_i\left(\frac{Y_i}{p_i}-Y\right)^2 \tag{13.3.14}$$

and

$$E_1\left(\sum_{i\in s}\frac{\sigma_i^2}{p_i^2}P_i^2\right) = \frac{N(n-1)}{n(N-1)}\sum_{i\in U}\sigma_i^2 + \frac{N-n}{n(N-1)}\sum_{i\in U}\frac{\sigma_i^2}{p_i} \tag{13.3.15}$$

On substituting Eqs. (13.3.14) and (13.3.15) in Eq. (13.3.13), we get

$$V\left(\widehat{Y}_{ts}(rhc)\right) = \frac{N-n}{n(N-1)}\sum_{i\in U}p_i\left(\frac{Y_i}{p_i}-Y\right)^2 + \frac{N(n-1)}{n(N-1)}\sum_{i\in U}\sigma_i^2 + \frac{N-n}{n(N-1)}\sum_{i\in U}\frac{\sigma_i^2}{p_i} \tag{13.3.16}$$

Noting $Q_s(Y) = \frac{N-n}{N(n-1)} \sum_{i \in s} P_i \left(\frac{Y_i}{p_i} - \sum_{i \in s} \frac{Y_i}{p_i} P_i \right)^2$ is an unbiased estimator of $V_1\left(\sum_{i \in s} \frac{Y_i}{p_i} P_i \right)$ and $\widehat{\Phi}_s = \sum_{i \in s} \frac{\widehat{\sigma}_i^2}{p_i} P_i$ is an unbiased estimator Φ with $\widehat{\sigma}_i^2 = M_i^2(1-f_i)s_{yi}^2 \big/ m_i$, we get an unbiased estimator of $V\left(\widehat{Y}\right)$ as

$$\widehat{V}\left(\widehat{Y}_{ts}(rhc)\right) = Q_s\left(\widehat{Y}\right) + \widehat{\Phi}_s$$

$$= \frac{N-n}{N(n-1)} \sum_{i \in s} P_i \left(\frac{\widehat{Y}_i}{p_i} - \widehat{Y} \right)^2 + \sum_{i \in s} \frac{M_i^2(1-f_i)s_{yi}^2}{m_i p_i} P_i \tag{13.3.17}$$

13.4 FIRST-STAGE UNITS ARE SELECTED BY PPSWR SAMPLING SCHEME

Consider the situation when the sample s of n fsu's are selected by probability proportional to size with replacement (PPSWR) method using the normed size measure p_i attached to the ith unit. If the ith fsu is selected $n_i(s)$ times in the sample, then $n_i(s)$-independent subsamples each of size m_i are selected from the ith fsu using the same sampling design. In this situation, we consider the following Hansen and Hurwitz (1943) type estimator for the total Y as follows:

$$\widehat{Y}_{ts}(hh) = \frac{1}{n} \sum_{r=1}^{n} \frac{\widehat{Y}(r)}{p(r)} \tag{13.4.1}$$

Here $\widehat{Y}(r)$ is an unbiased estimator of the fsu total that was selected at the rth draw with probability $p(r)$, i.e., if the jth fsu is selected at the rth draw, then $\widehat{Y}(r) = \widehat{Y}_j$ and $p(r) = p_j$. In this case

$$E_2\left(\frac{\widehat{Y}(r)}{p(r)} \right) = \frac{Y(r)}{p(r)} \quad \text{and} \quad V_2\left(\frac{\widehat{Y}(r)}{p(r)} \right) = \frac{\sigma^2(r)}{\{p(r)\}^2} \tag{13.4.2}$$

where $Y(r)$ = fsu total of the fsu that is selected at the rth draw and $\sigma^2(r)$ variance of $\widehat{Y}(r)$. Hence

$$E\left(\frac{\widehat{Y}(r)}{p(r)} \right) = E_1\left[E_2\left(\frac{\widehat{Y}(r)}{p(r)} \right) \right] = E_1\left(\frac{Y(r)}{p(r)} \right) = \sum_{i \in U} \frac{Y_i}{p_i} p_i = Y \tag{13.4.3}$$

and

$$V\left(\frac{\widehat{Y}(r)}{p(r)}\right) = E_1\left[V_2\left(\frac{\widehat{Y}(r)}{p(r)}\right)\right] + V_1\left[E_2\left(\frac{\widehat{Y}(r)}{p(r)}\right)\right]$$

$$= E_1\left(\frac{\sigma^2(r)}{p^2(r)}\right) + V_1\left(\frac{Y(r)}{p(r)}\right) \tag{13.4.4}$$

$$= \sum_{i\in U}\frac{\sigma_i^2}{p_i} + \sum_{i\in U} p_i\left(\frac{Y_i}{p_i} - Y\right)^2 = W^2(\text{say})$$

Since $\frac{\widehat{Y}(r)}{p(r)}$'s are independently and identically distributed with common mean Y and common variance W^2, we get the following theorem.

Theorem 13.4.1

(i) $E\left[\widehat{Y}_{ts}(hh)\right] = E\left(\frac{1}{n}\sum_{r=1}^{n}\frac{\widehat{Y}(r)}{p(r)}\right) = Y$

(ii) $V\left[\widehat{Y}_{ts}(hh)\right] = \frac{W^2}{n} = \frac{1}{n}\left[\sum_{i\in U}\frac{\sigma_i^2}{p_i} + \sum_{i\in U} p_i\left(\frac{Y_i}{p_i} - Y\right)^2\right]$

(iii) An unbiased estimator of $V\left[\widehat{Y}_{ts}(hh)\right]$ is

$$\widehat{V}\left[\widehat{Y}_{ts}(hh)\right] = \frac{1}{n(n-1)}\sum_{r=1}^{n}\left(\frac{\widehat{Y}(r)}{p(r)} - \widehat{Y}\right)^2$$

13.4.1 Simple Random Sampling With Replacement

In particular, if the fsu's are selected by simple random sampling with replacement (SRSWR) sampling scheme and if the selected fsu's are subsampled independently by SRSWOR method, we then get $\widehat{Y}(r) = M(r)\overline{y}(r)$, where $M(r)$ and $\overline{y}(r)$ are the total number of ssu's and the sample mean of the ssu's of the fsu that was selected at the rth draw, respectively. In this case, $E_2\left\{\widehat{Y}(r)\right\} = Y(r)$ and $V_2\left\{\widehat{Y}(r)\right\} = \sigma^2(r)$. Thus if the rth draw produces ith unit, we have $E_2\left\{\widehat{Y}(r)\right\} = Y_i$ and $V_2\left\{\widehat{Y}(r)\right\} = \sigma^2(i) = \frac{M_i^2(1-f_i)S_{yi}^2}{m_i}$ with probability $1/N$. Hence, under SRSWR sampling we have the following theorem.

Theorem 13.4.2

Under SRSWR sampling,

(i) $\widehat{Y}_{ts}(wr)=\frac{N}{n}\sum_{r=1}^{n}M(r)\overline{y}(r)$ is an unbiased estimator for Y

(ii) $V\left[\widehat{Y}_{ts}(wr)\right]=\dfrac{N^2}{n}\left[\sum_{i\in U}\dfrac{M_i^2(1-f_i)S_{yi}^2}{m_i}+(N-1)S_b^2\right]$

(iii) $\widehat{V}\left[\widehat{Y}_{ts}(wr)\right]=\dfrac{N^2}{n(n-1)}\sum_{r=1}^{n}\left(M(r)\overline{y}(r)-\frac{1}{n}\sum_{r=1}^{n}M(r)\overline{y}(r)\right)^2$ is an unbiased estimator of $V\left[\widehat{Y}_{ts}(wr)\right]$.

Remark 13.4.1

So far, we have considered σ_i^2 as independent of s, the sample of fsu's. Rao (1975) considered a situation where σ_i^2 may depend on s; i.e., $V_2\left(\widehat{Y}_i\right)=\sigma_{si}^2$. In this situation, following Theorems 13.4.1 and 13.4.2, the variance of $\widehat{Y}_{ts}$ and its unbiased estimator become,

$$V\left(\widehat{Y}_{ts}\right)=\sum_{i\in U}\alpha_i Y_i^2+\sum_{i\neq}\sum_{j\in U}\alpha_{ij}Y_iY_j+\sum_{s}\left(\sum_{i\in s}b_{si}^2\sigma_{si}^2\right)p(s) \quad (13.4.5)$$

and

$$\widehat{V}\left(\widehat{Y}_{ts}\right)=\sum_{i\in s}c_{si}\widehat{Y}_i^2+\sum_{i\neq}\sum_{j\in s}c_{sij}\widehat{Y}_i\widehat{Y}_j+\sum_{i\in s}\left(b_{si}^2-c_{si}\right)\widehat{\sigma}_{si}^2 \quad (13.4.6)$$

respectively, where $\widehat{\sigma}_{si}^2$ is an unbiased estimator of σ_{si}^2 that satisfies $E_2\left(\widehat{\sigma}_{si}^2\right)=\sigma_{si}^2$.

13.4.2 Raj Estimator for Multi-Stage Sampling

Suppose that from a population of N fsu's, a sample s of n fsu's is selected by the PPSWR method using p_i as the normed size measure for the ith unit. If the ith fsu appears in s, $n_i(s)$ times, a subsample of size $n_i(s)\ m_i$ ssu's is selected from the ith fsu by SRSWOR method, assuming $m_i n\leq M_i$. We then have the following theorem according to Raj (1968).

Theorem 13.4.3

(i) $\widehat{Y}^*_{ts}(hh) = \frac{1}{n}\sum_{i=1}^{N}\frac{n_i(s)}{p_i}\widehat{Y}_i$ is an unbiased estimator for Y

(ii) $$V\left[\widehat{Y}^*_{ts}(hh)\right] = \frac{1}{n}\left[\sum_{i\in U} p_i\left(\frac{Y_i}{p_i} - Y\right)^2 + \sum_{i\in U}\left\{\frac{1}{p_i}\left(\frac{1}{m_i} - \frac{1}{M_i}\right) - \frac{n-1}{M_i}\right\}M_i^2 S_{yi}^2\right]$$

(iii) $$\widehat{V}\left[\widehat{Y}^*_{ts}(hh)\right] = \frac{\sum_i n_i(s)\left(\frac{\widehat{Y}_i}{p_i} - \widehat{Y}^*_{ts}(hh)\right)^2}{n(n-1)} + \sum_i \frac{n_i(s)(n_i(s)-1)\widehat{\sigma}_{si}^2}{n(n-1)p_i^2}$$

where $$\widehat{\sigma}_{si}^2 = M_i^2\left(\frac{1}{n_i(s)m_i} - \frac{1}{M_i}\right)\frac{\sum_{j\in s_i}\left\{y_{ij} - \overline{y}(s_i)\right\}^2}{n_i(s)m_i - 1}$$

Proof

(i) $$E\left[\widehat{Y}^*_{ts}(hh)\right] = \frac{1}{n}E_1\left[\sum_{i=1}^{N}\frac{n_i(s)}{p_i}\left\{E_2\left(\widehat{Y}_i\right)\right\}\right]$$
$$= \frac{1}{n}E_1\sum_{i=1}^{N}\frac{n_i(s)}{p_i}Y_i$$
$$= Y$$

(ii) $$V\left[\widehat{Y}^*_{ts}(hh)\right] = V_1\left(\frac{1}{n}\sum_{i=1}^{N}\frac{n_i(s)}{p_i}Y_i\right) + E_1\left(\frac{1}{n^2}\sum_{i=1}^{N}\left(\frac{n_i(s)}{p_i}\right)^2\sigma_i^2(s)\right)$$

where $$\sigma_i^2(s) = M_i^2\left(\frac{1}{n_i(s)m_i} - \frac{1}{M_i}\right)S_{yi}^2$$

Now noting $V\left(\frac{n_i(s)}{np_i}\right) = \frac{(1-p_i)}{np_i}$ and $Cov\left(\frac{n_i(s)}{np_i}, \frac{n_j(s)}{np_j}\right) = -\frac{1}{n}$, we obtain

$$V\left[\widehat{Y}^*_{ts}(hh)\right] = \frac{1}{n}\left[\sum_{i\in U} p_i\left(\frac{Y_i}{p_i} - Y\right)^2 + \sum_{i\in U}\left\{\frac{1}{p_i}\left(\frac{1}{m_i} - \frac{1}{M_i}\right) - \frac{n-1}{M_i}\right\}M_i^2 S_{yi}^2\right]$$

(iii) $n(n-1)E\left[\widehat{V}\left\{\widehat{Y}^*_{ts}(hh)\right\}\right]$

$$= E_1\sum_i \frac{n_i(s)\left\{V_2\left(\widehat{Y}_i|s\right)+Y_i^2\right\}}{p_i^2} - n\left[V\left\{\widehat{Y}^*_{ts}(hh)\right\}+Y^2\right]$$

$$+ E_1\sum_i \frac{n_i(s)\{n_i(s)-1\}E\left(\widehat{\sigma}^2_{si}|s\right)}{p_i^2}$$

$$= E_1\sum_i \frac{n_i(s)Y_i^2}{p_i^2} - n\left(V\left\{\widehat{Y}^*_{ts}(hh)\right\}+Y^2\right) + E_1\sum_i \frac{\{n_i(s)\}^2\sigma^2_{si}}{p_i^2}$$

$$= n\sum_i p_i\left(\frac{Y_i}{p_i}-Y\right)^2 - nV\left\{\widehat{Y}^*_{ts}(hh)\right\} + E_1\sum_i \frac{\{n_i(s)\}^2}{p_i^2}\left(\frac{1}{m_i n_i(s)}-\frac{1}{M_i}\right)S^2_{yi}$$

$$= n\sum_i p_i\left(\frac{Y_i}{p_i}-Y\right)^2 - nV\left\{\widehat{Y}^*_{ts}(hh)\right\} + n\sum_i \frac{1}{p_i}\left(\frac{1}{m_i}-\frac{1+(n-1)p_i}{M_i}\right)M_i^2 S^2_{yi}$$

$$= n(n-1)V\left[\widehat{Y}^*_{ts}(hh)\right]$$

i.e., $E\left[\widehat{V}\left\{\widehat{Y}^*_{ts}(hh)\right\}\right] = V\left[\widehat{Y}^*_{ts}(hh)\right]$

Remark 13.4.2

It is important to note from Theorems 13.3.2 and 13.4.1 that if the fsu's are selected by without replacement sampling schemes or if the estimator of the total $\widehat{Y}_{ts}$ is based on distinct units for a with replacement sampling scheme, then the unbiased estimators for the variance of $\widehat{Y}_{ts}$ involve $\widehat{\sigma}^2_i$, the unbiased estimator of the variance of $\widehat{Y}_i$. But if the fsu's are selected by PPSWR (or SRSWR), then the variance of $\widehat{Y}_{ts}$ can be estimated unbiasedly without estimating σ_i^2, $i\in s$. Thus if the sample of fsu's s is selected by a without replacement sampling scheme and the ssu's are selected by systematic sampling scheme, the variance of $\widehat{Y}_{ts}$ cannot be estimated unbiasedly because an unbiased estimator of σ_i^2 based on the systematic sampling scheme is not available. On the other hand, if the fsu's are selected using either the PPSWR or SRSWR methods and the ssu's are selected by systematic sampling, then the variance of $\widehat{Y}_{ts}$ can be estimated unbiasedly because unbiased estimators of σ_i^2 are not required for estimating the variance of $\widehat{Y}_{ts}$.

13.5 MODIFICATION OF VARIANCE ESTIMATORS

Srinath and Hidiroglou (1980), and Arnab (1988), proposed the following modifications of variance estimators when fsu's are selected by without replacement sampling procedure.

13.5.1 Srinath and Hidiroglou Modification

From each of the selected sample $s_i (i \in s)$, a subsample $\widetilde{s}_i$ of ssu's of suitable size $\widetilde{m}_i$ (to be determined) is selected independently by some suitable sampling scheme, and an unbiased estimator $\widehat{\widehat{Y}}_i$ of Y_i is obtained from $\widetilde{s}_i$, sacrificing observations within s_i but outside $\widetilde{s}_i$. Then,

$$V\left(\widehat{Y}_{ts}\right) = \sum_{i \in U} \alpha_i Y_i^2 + \sum_{i \neq} \sum_{j \in U} \alpha_{ij} Y_i Y_j + \sum_{i \in U} \beta_i \sigma_i^2$$

given in Eq. (13.3.2) can be estimated unbiasedly as a homogeneous quadratic estimator of $\widehat{\widehat{Y}}_i$ in the following form

$$\widetilde{V}\left(\widehat{Y}_{ts}\right) = \sum_{i \in s} c_{si} \widehat{\widehat{Y}}_i^2 + \sum_{i \neq} \sum_{j \in s} c_{sij} \widehat{\widehat{Y}}_i \widehat{\widehat{Y}}_j \tag{13.5.1}$$

provided, $\widetilde{m}_i$'s satisfy

$$(\alpha_i - 1)\widetilde{\sigma}_i^2 = \alpha_i \sigma_i^2 \quad \text{for } i = 1, \ldots, N \tag{13.5.2}$$

where $\widetilde{\sigma}_i^2 = V_2\left(\widehat{\widehat{Y}}_i\right)$ and c_{si} and c_{sij} meet the unbiasedness condition given in Eq. (13.3.3). For $b_{si} = 1/\pi_i$, we get $\alpha_i = 1/\pi_i$ and $\alpha_{ij} = \pi_{ij}/(\pi_i \pi_j) - 1$. Now for the choices of $c_{si} = (\alpha_i - 1)/\pi_i$ and $c_{sij} = (\alpha_{ij} - 1)/\pi_{ij}$, we illustrate how one obtains a $\widetilde{V}\left(\widehat{Y}_{ts}\right)$ in the following examples.

Example 13.5.1

Let s_i be an SRSWOR of size m_i, $\overline{y}(s_i)$ and $\overline{y}(\widetilde{s}_i)$ be the sample means of y_{ij} values based on the samples s_i and $\widetilde{s}_i$, respectively, $\widehat{Y}_i = M_i \overline{y}(s_i)$ and $\widehat{\widehat{Y}}_i = M_i \overline{y}(\widetilde{s}_i)$. Then $\sigma_i^2 = M_i^2(1/m_i - 1/M_i)S_{yi}^2$ and $\widetilde{\sigma}_i^2 = M_i^2(1/\widetilde{m}_i - 1/M_i)S_{yi}^2$. The condition (Eq. 13.5.2) yields

$$\widetilde{m}_i = m_i(1 - \pi_i)/(1 - m_i \pi_i / M_i) \tag{13.5.3}$$

Example 13.5.2

Let s_i of size m_i be chosen by PPSWR method with p_{ij} as normed size measures for the jth ssu of the ith fsu, and let $\widetilde{s}_i$ be an SRSWOR sub-sample of size $\widetilde{m}_i$. Let $\widehat{Y}_i = \sum_{j=1}^{m_i} y_{ij}/(m_i p_{ij})$ and $\widehat{\widetilde{Y}}_i = \sum_{j=1}^{\widetilde{m}_i} y_{ij}/(\widetilde{m}_i p_{ij})$. Then, $\sigma_i^2 = \sum_{j=1}^{M_i} p_{ij}(y_{ij}/p_{ij} - Y_i)^2/m_i$ and $\widetilde{\sigma}_i^2 = m_i\sigma_i^2/\widetilde{m}_i$. The condition (Eq. 13.5.2) yields

$$\widetilde{m}_i = m_i(1 - \pi_i) \tag{13.5.4}$$

13.5.2 Arnab Modification

In order to avoid loss of information, Arnab (1988) divided the sample s_i at random into r groups so that the kth group contains $m_i(k)$ ssu's, $\sum_{k=1}^{r} m_i(k) = m_i$. Let $\widehat{Y}_i(k)$ be an unbiased estimator of Y_i based on the kth group; then we can choose $\widehat{\widetilde{Y}}_i = \sum_{k=1}^{r} w_i(k)\widehat{Y}_i(k)$ as a weighted average of $\widehat{Y}_i(k)$ with suitable weights $w_i(k)$'s $\left(0 \le w_i(k) \le 1, \sum_{k=1}^{r} w_i(k) = 1\right)$. The weights $w_i(k)$ and $m_i(k)$ are determined from the condition (Eq. 13.5.2).

Example 13.5.3

Let s_i be an SRSWOR with $\widehat{Y}_i = M_i\overline{y}(s_i)$ and $\widehat{Y}_i(k) = M_i \sum_{j=1}^{m_i(k)} y_{ij}/m_i(k)$. In this case

$$\sigma_i^2 = V(\widehat{Y}_i) = V\{M_i\overline{y}(s_i)\} = M_i^2\left(\frac{1}{m_i} - \frac{1}{M_i}\right)S_{yi}^2$$

and

$$\begin{aligned}
\widetilde{\sigma}_i^2 = V(\widehat{\widetilde{Y}}_i) &= V\left(\sum_{k=1}^{r} w_i(k)\widehat{Y}_i(k)\right) \\
&= E\left[V\left(\sum_{k=1}^{r} w_i(k)\widehat{Y}_i(k)|s_i\right)\right] + V\left[E\left(\sum_{k=1}^{r} w_i(k)\widehat{Y}_i(k)|s_i\right)\right] \\
&= E\left(\sum_{k=1}^{r} w_i^2(k)V\{\widehat{Y}_i(k)|s_i\}\right. \\
&\quad \left. + \sum_{k\neq}^{r}\sum_{k'=1}^{r} w_i(k)w_i(k')Cov\{\widehat{Y}_i(k), \widehat{Y}_i(k')|s_i\}\right) + V(\widehat{Y}_i)
\end{aligned}$$

Now noting $V\{\widehat{Y}_i(k)|s_i\} = M_i^2\left(\frac{1}{m_i(k)} - \frac{1}{m_i}\right)s_{yi}^2$ and

$Cov\{\widehat{Y}_i(k), \widehat{Y}_i(k')|s_i\} = -\frac{s_{yi}^2}{m_i}$ where $s_{yi}^2 = \frac{1}{m_i - 1}\sum_{j\in s_i}\{y_{ij} - \overline{y}(s_i)\}^2$, we find

$$\widetilde{\sigma}_i^2 = M_i^2 S_{yi}^2\left(\sum_{k=1}^{r}\frac{w_i^2(k)}{m_i(k)} - \frac{1}{M_i}\right).$$

The requirement (Eq. 13.5.2) yields

$$(\alpha_i - 1)\left(\sum_{k=1}^{r}\frac{w_i^2(k)}{m_i(k)} - \frac{1}{M_i}\right) = \alpha_i\left(\frac{1}{m_i} - \frac{1}{M_i}\right) \tag{13.5.5}$$

The solution of $w_i(k)$'s in Eq. (13.5.5) may be obtained by a trial and error method. In particular, for $k = 2$, an easy solution is $w_i(1) = w_i(2) = 1/2$, $m_i(1) = m_i\left(1 + \sqrt{\delta_i}\right)/2$, and $m_i(2) = m_i\left(1 - \sqrt{\delta_i}\right)/2$ with $\delta_i = 1 - (\alpha_i - 1)/(\alpha_i - m_i/M_i)$.

13.6 MORE THAN TWO-STAGE SAMPLING

Here we will define operators E_L and V_L as overall expectation and variance later than the first stage of sampling, respectively. The selected sample of fsu's will be denoted by s as in earlier sections. If the ith fsu is included in the sample s, the sub-sample selected at the second and subsequent stages of sampling from the final (ultimate) stage from the ith fsu will be denoted by s_{iL}. Let $\widehat{Y}_{iL}$ be an unbiased estimator of Y_i based on s_{iL}, then an unbiased estimator of the total Y will be denoted as

$$\widehat{Y}_L = \sum_{i\in s} b_{si}\widehat{Y}_{iL} \tag{13.6.1}$$

where b_{si}'s satisfy the unbiasedness condition $\sum_{s\supset i} b_{si}p(s) = 1$.

Now

$$\begin{aligned}E\left(\widehat{Y}_L\right) &= E_1\left[\sum_{i\in s} b_{si}E_L\left(\widehat{Y}_{iL}\right)\right]\\ &= E_1\left(\sum_{i\in s} b_{si}Y_i\right)\\ &= Y\end{aligned}$$

The expression for the variance of $\widehat{Y}_L$ and its unbiased estimator may be obtained by using Theorems 13.3.1 and 13.3.2 by writing σ^2_{iL} in place of σ^2_i and they are given as follows:

$$V\left(\widehat{Y}_L\right) = \sum_{i\in U} \alpha_i Y_i^2 + \sum_{i\neq}\sum_{j\in U} \alpha_{ij} Y_i Y_j + \sum_{i\in U} \beta_i \sigma^2_{iL} \tag{13.6.2}$$

and

$$\widehat{V}\left(\widehat{Y}\right) = Q_s\left(\widehat{Y}_L\right) + \widehat{\Phi}_{sL} \tag{13.6.3}$$

where $Q_s\left(\widehat{Y}_L\right) = \sum_{i\in s} c_{si}\widehat{Y}^2_{iL} + \sum_{i\neq}\sum_{j\in s} c_{sij}\widehat{Y}_{iL}\widehat{Y}_{jL}$ and $\widehat{\Phi}_{sL}$ are unbiased estimators of $Q\left(\widehat{Y}_L\right) = \sum_{i\in U} \alpha_i Y_i^2 + \sum_{i\neq}\sum_{j\in U} \alpha_{ij} Y_i Y_j$ and $\sum_{i\in U} \sigma^2_{iL}$, respectively; c_{si} and c_{sij} must satisfy $\sum_{s\supset i} c_{si} p(s) = \alpha_i$ and $\sum_{s\supset i,j} c_{sij} p(s) = \alpha_{ij}$ as in Theorem 13.3.2.

13.6.1 Three-Stage Sampling

Suppose a population is divided into N fsu's. The ith fsu consists of M_i ssu's and the jth ssu of the ith fsu consists of M_{ij} tsu's. The total number of tsu's is $M = \sum_{i=1}^{N}\sum_{j=1}^{M_i} M_{ij}$. Let a sample s of n fsu's be selected by SRSWOR method and if the ith fsu is selected in s, then a sub-sample s_i of size m_i ssu's is selected by SRSWOR method from the M_i ssu's of the ith fsu. Finally, if the jth ssu of the ith fsu is selected in the sample s_i, then a sub-sample s_{ij} of m_{ij} tsu's is selected by SRSWOR method from M_{ij} tsu's that belong to the ith fsu and jth ssu. Let y_{ijk} be the value of the study variable for the kth tsu of the jth ssu of ith fsu. Let $Y_{ij} = \sum_{k=1}^{M_{ij}} y_{ijk}$, $Y_i = \sum_{j=1}^{M_i} Y_i$, $Y = \sum_{i=1}^{N} Y_i$, and $\overline{y}(s_{ij}) = \sum_{k\in s_{ij}} y_{ijk}/m_{ij}$. Then an unbiased estimator of the total Y is given by

$$\widehat{Y}_T = \frac{N}{n}\sum_{i\in s}\widehat{Y}_{iL} \tag{13.6.4}$$

where $\widehat{Y}_{iL} = \dfrac{M_i}{m_i}\sum_{j\in s_{ij}}\widehat{Y}_{ij}$ and $\widehat{Y}_{ij} = \dfrac{M_{ij}}{m_{ij}}\overline{y}(s_{ij})$.

Clearly $\widehat{Y}_{ij}$ is an unbiased estimator for Y_{ij}, and $\widehat{Y}_{iL}$ is an unbiased estimator of Y_i. Eqs. (13.6.2) and (13.6.3) yield the variance of $\widehat{Y}_{iL}$ and its unbiased estimator, respectively, as follows:

$$\begin{aligned} V\left(\widehat{Y}_L\right) &= V_1\left(\frac{N}{n}\sum_{i\in s} Y_i\right) + \frac{N^2}{n^2}\left(\sum_{i\in s}\sigma^2_{iL}\right) \\ &= N^2(1-f)\frac{S_b^2}{n} + \frac{N}{n}\sum_{i\in U}\sigma^2_{iL} \end{aligned} \tag{13.6.5}$$

and

$$\widehat{V}\left(\widehat{Y}_L\right) = N^2(1-f)\frac{s_b^2}{n} + \frac{N}{n}\sum_{i\in s}\widehat{\sigma}_{iL}^2 \tag{13.6.6}$$

where $S_b^2 = \frac{1}{N-1}\sum_{i=1}^{N}\left(Y_i - \overline{Y}\right)^2$, $s_b^2 = \frac{1}{n-1}\sum_{i\in s}\left(\widehat{Y}_i - \widehat{\overline{Y}}_s\right)^2$,
$\overline{Y} = Y/N$, $\widehat{\overline{Y}}_s = \sum_{i\in s}\widehat{Y}_i/n$, and $f = n/N$.

The expressions of σ_{iL}^2 and $\widehat{\sigma}_{iL}^2$ are obtained by using Eqs. (13.3.10) and (13.3.11) as

$$\sigma_{iL}^2 = V\left(\widehat{Y}_{iL}\right) = M_i^2(1-f_i)\frac{S_{bi}^2}{m_i} + \frac{M_i}{m_i}\sum_{j=1}^{M_i}\frac{M_{ij}^2(1-f_{ij})S_{yij}^2}{m_{ij}}.$$

and

$$\widehat{\sigma}_{iL}^2 = M_i^2(1-f_i)\frac{s_{bi}^2}{m_i} + \frac{M_i}{m_i}\sum_{j\in s_i}\frac{M_{ij}^2(1-f_{ij})s_{yij}^2}{m_{ij}}$$

where $S_{bi}^2 = \sum_{j=1}^{M_i}\left(Y_{ij} - \overline{Y}_i\right)^2/(M_i - 1)$, $s_{bi}^2 = \sum_{j\in s_i}\left(\widehat{Y}_{ij} - \widehat{\overline{Y}}_{si}\right)^2\Big/(m_i - 1)$,

$f_i = m_i/M_i$, $f_{ij} = m_{ij}/M_{ij}$, $\overline{Y}_i = \sum_{j=1}^{M_i}Y_{ij}/M_i$, $\widehat{\overline{Y}}_{si} = \sum_{j\in s_i}\widehat{Y}_{ij}/m_i$,

$$S_{yij}^2 = \sum_{k=1}^{M_{ij}}\left(y_{ijk} - \overline{Y}_{ij}\right)^2\Big/(M_{ij} - 1),$$

$$s_{yij}^2 = \sum_{k=1}^{m_{ij}}\left\{y_{ijk} - \overline{y}(s_{ij})\right\}^2\Big/(m_{ij} - 1), \text{ and } \overline{y}(s_{ij}) = \sum_{k\in s_{ij}}y_{ijk}/m_{ij}.$$

13.7 ESTIMATION OF MEAN PER UNIT

The mean per unit is defined by $\overline{\overline{Y}} = Y/M_0$, where M_0 is the total number of ultimate units and Y is the population total based on M_0 ultimate units. For a two-stage sampling, $Y = \sum_{i=1}^{N}\sum_{j=1}^{M_i}y_{ij}$ and $M_0 = \sum_{i=1}^{N}M_i$. In case M_0 is known, an unbiased estimate for the mean per unit is given by $\widehat{\overline{\overline{Y}}}_{ts} = \widehat{Y}_{ts}/M_0$. But in most cases, M_0 is unknown and an estimate of $\widehat{\overline{\overline{Y}}}$ is obtained as

$$\widehat{\overline{\overline{Y}}}_{ts}(r) = \widehat{Y}_{ts}\Big/\widehat{M}_0 \tag{13.7.1}$$

where $\widehat{Y}_{ts}$ and $\widehat{M}_0$ are an unbiased estimators of Y and M_0, respectively. For two-stage sampling,

$$\widehat{\overline{\overline{Y}}}_{ts}(r) = \sum_{i \in s} b_{si} \widehat{Y}_i \Big/ \sum_{i \in s} b_{si} M_i$$

Since $\widehat{\overline{\overline{Y}}}_{ts}(r)$ is a ratio estimator, it is biased for $\overline{\overline{Y}}$. The approximate expressions for the bias and mean-square error (MSE) are readily obtained from Theorem 8.2.2 and Eq. (8.2.16) and they are given as follows:

$$\text{Bias of } \widehat{\overline{\overline{Y}}}_{ts}(r) = B\left[\widehat{\overline{\overline{Y}}}_{ts}(r)\right] \cong -\frac{Cov\left[\widehat{Y}_{ts}, \widehat{M}_0\right]}{M_0} \tag{13.7.2}$$

$$\text{Mean square error of } \widehat{\overline{\overline{Y}}}_{ts}(r) = M\left[\widehat{\overline{\overline{Y}}}_{ts}(r)\right] \cong \frac{V\left(\sum_{i \in s} b_{si} \widehat{D}_i\right)}{M_0^2} \tag{13.7.3}$$

where $\widehat{D}_i = \widehat{Y}_i - \overline{\overline{Y}} M_i$

$$\text{Now } V\left(\sum_{i \in s} b_{si} \widehat{D}_i\right) = E\left(\sum_{i \in s} b_{si}^2 V\left(\widehat{D}_i \middle| s_i\right)\right) + V\left(\sum_{i \in s} b_{si} D_i\right)$$

$$= E\left(\sum_{i \in s} b_{si}^2 \sigma_i^2(d)\right) + V\left(\sum_{i \in s} b_{si} D_i\right)$$

$$= \sum_{i \in U} \alpha_i D_i^2 + \sum_{i \neq} \sum_{j \in U} \alpha_{ij} D_i D_j + \sum_{i \in U} \beta_i \sigma_i^2(d)$$

where $D_i = Y_i - \overline{\overline{Y}} M_i$, $\sigma_i^2(d) = V\left(\widehat{D}_i \middle| s_i\right)$; α_i, α_{ij} and β_i are as in Theorem 13.3.1.

Hence

$$M\left[\widehat{\overline{\overline{Y}}}_{ts}(r)\right] \cong \frac{1}{M_0^2}\left[\sum_{i \in U} \alpha_i D_i^2 + \sum_{i \neq} \sum_{j \in U} \alpha_{ij} D_i D_j + \sum_{i \in U} \beta_i \sigma_i^2(d)\right] \tag{13.7.4}$$

An approximate unbiased estimator of $M\left[\widehat{\overline{\overline{Y}}}_{ts}(r)\right]$ is

$$\widehat{M}\left[\widehat{\overline{\overline{Y}}}_{ts}(r)\right] \cong \frac{1}{\widehat{M}_0^2}\left[\sum_{i \in s} \alpha_i \widehat{\widehat{D}}_i^2 \Big/ \pi_i + \sum_{i \neq} \sum_{j \in s} \alpha_{ij} \widehat{\widehat{D}}_i \widehat{\widehat{D}}_j + \sum_{i \in s} \widehat{\sigma}_i^2(d) / \pi_i\right] \tag{13.7.5}$$

where $\widehat{\widehat{D}}_i = \widehat{Y}_i - \widehat{\overline{\overline{Y}}}_{ts}(r) M_i$, and $\widehat{\sigma}_i^2(d)$ is an approximate unbiased estimator of $\sigma_i^2(d)$.

13.7.1 Simple Random Sampling Without Replacement

If s and s_i's are selected by SRSWOR method, we have $b_{si} = N/n$, $\widehat{\overline{\overline{Y}}}_{ts}(wor) = \sum_{i\in s} M_i\overline{y}(s_i)/\sum_{i\in s} M_i$, and the expressions (Eqs. 13.7.4 and 13.7.5) reduce to

$$M\left[\widehat{\overline{\overline{Y}}}_{ts}(wor)\right] \cong \frac{1}{\overline{M}^2}\left[(1-f)S_D^2/n + \frac{1}{nN}\sum_{i=1}^{N} M_i^2(1-f_i)S_{di}^2/m_i\right] \quad (13.7.6)$$

$$\widehat{M}\left[\widehat{\overline{\overline{Y}}}_{ts}(wor)\right] \cong \frac{1}{\widehat{\overline{M}}^2}\left[(1-f)s_{\widehat{D}}^2/n + \frac{1}{nN}\sum_{i\in s} M_i^2(1-f_i)s_{\widehat{d}i}^2\Big/m_i\right] \quad (13.7.7)$$

where $\overline{M} = \dfrac{M_0}{N}$, $S_D^2 = \dfrac{\sum_{i=1}^{N}\left(D_i - \overline{D}\right)^2}{N-1}$, $\overline{D} = \dfrac{\sum_{i=1}^{N} D_i}{N}$,

$$S_{di}^2 = \frac{1}{M_i - 1}\sum_{j=1}^{M_i}\left(D_{ij} - \overline{D}_i\right)^2, D_{ij} = y_{ij} - \overline{\overline{Y}}M_{ij}, \overline{D}_i = \frac{D_i}{M_i},$$

$$s_{\widehat{D}}^2 = \frac{\sum_{i\in s}\left(\widehat{\overline{D}}_i - \widehat{\overline{D}}\right)^2}{n-1}, \widehat{D}_i = \widehat{Y}_i - \widehat{\overline{\overline{Y}}}_{ts}(wor)M_i,$$

$$\widehat{\overline{D}} = \frac{1}{n}\sum_{i\in s}\widehat{\overline{D}}_i, \; s_{\widehat{d}i}^2 = \frac{1}{m_i - 1}\sum_{j\in s_i}\left(\widehat{\overline{D}}_{ij} - \widehat{\overline{D}}_i\right)^2, \widehat{D}_{ij} = y_{ij} - \widehat{\overline{\overline{Y}}}_{ts}(wor)M_{ij},$$

and $\widehat{\overline{D}}_i = \frac{1}{m_i}\sum_{j\in s_i}\widehat{D}_{ij}$.

13.8 OPTIMUM ALLOCATION

In the two-stage sampling considered in the earlier section, we assumed that if the ith fsu is selected in the sample s, a sub-sample s_i of size m_i would be selected from it by some suitable sampling procedure. The size m_i was assumed to be pre-assigned. In this section we will determine the optimal values of m_i's and n that minimize: (i) the cost of the survey, keeping the precision of the estimator at a certain level and (ii) the variance of the estimator, keeping the cost fixed to a certain level. Here we consider the following simple cost function considered by Cochran (1977).

$$C = c_0 + c_1 n + c_2\sum_{i\in s} m_i + c_3\sum_{i\in s} M_i \quad (13.8.1)$$

where c_0 is the overhead cost of the survey, which is fixed for the survey, independent of n and m_i's, c_1 is the cost per fsu, which is assumed to be fixed, c_2 is the cost for surveying an ssu, and c_3 is the cost of listing per ssu. Since the cost C is a random variable that depends on the fsu's selected in s, we minimisze (i) the variance of an estimator, keeping the expected cost $\overline{C} = E(C) = c_0 + c_1 n + c_2 \sum_{i \in U} m_i \pi_i + c_3 \sum_{i \in U} M_i \pi_i$ to a certain level C^* (say) or (ii) the expected cost $\overline{C}$, keeping the variance of the estimator fixed to a certain level V^* (say). Obviously the solutions of (i) and (ii) exist if the variance of the estimator can be expressed as an explicit function of n and m_i's. Consider the case of two-stage sampling where both the fsu's and ssu' are selected by SRSWOR method. In this situation we obtain from Eq. (13.3.10)

$$V\left[\widehat{Y}_{ts}(wor)\right] = N^2(1-f)\frac{S_b^2}{n} + \frac{N}{n}\sum_{i \in U}\frac{M_i^2(1-f_i)S_{yi}^2}{m_i}$$

Now noting $\pi_i = n/N$, we get for the SRSWOR sampling

$$\overline{C} = c_0 + n(\overline{c}_1 + c_2\overline{m}) \tag{13.8.2}$$

where $\overline{c}_1 = c_1 + c_3 \sum_{i \in U} M_i/N$, $\overline{m} = \frac{1}{N}\sum_{i \in U} m_i$.

13.8.1 Fixed Expected Cost

Let us write

$$V\left[\widehat{Y}_{ts}(wor)\right] = \frac{1}{n}\left(A_0 + \sum_{i \in U}\frac{A_i}{m_i}\right) - B \tag{13.8.3}$$

with $A_0 = N^2 S_b^2 - N\sum_{i \in U} M_i S_{yi}^2$, $A_i = NM_i S_{yi}^2$, and $B = NS_b^2$.

Consider

$$\phi = V\left[\widehat{Y}_{ts}(wor)\right] - \lambda\left(\overline{C} - c_0 - \overline{c}_1 n - c_2 n\overline{m}\right) \tag{13.8.4}$$

with λ as an undetermined Lagrange's multiplier.

Now differentiating ϕ with respect to m_i and n and equating them to zero yield

$$m_i = \frac{\sqrt{NA_i}}{n\sqrt{\lambda\, c_2}} \tag{13.8.5}$$

and

$$\lambda n^2 = \frac{A_0 + \sum_{i \in U}\frac{A_i}{m_i}}{\overline{c}_1 + \frac{c_2}{N}\sum_{i \in U} m_i} \tag{13.8.6}$$

Eq. (13.8.5) implies

$$\lambda n^2 = \frac{N \sum_{i \in U} \frac{A_i}{m_i}}{c_2 \sum_{i \in U} m_i} \tag{13.8.7}$$

Eqs. (13.8.6) and (13.8.7) yield

$$\sum_{i \in U} \frac{A_i}{m_i} = \frac{c_2 A_0}{\bar{c}_1 N} \sum_{i \in U} m_i \tag{13.8.8}$$

and

$$\lambda = \frac{A_0}{n^2 \bar{c}_1} \tag{13.8.9}$$

On substituting λ in Eq. (13.8.5) the optimum value of m_i is obtained as

$$opt\, m_i = m_{i0} = \sqrt{\frac{N\bar{c}_1}{c_2 A_0} A_i} \tag{13.8.10}$$

Finally, using the relation $C^* = \overline{C} = c_0 + n\left(\bar{c}_1 + c_2 \frac{1}{N} \sum_{i \in U} m_{i0}\right)$ and substituting m_{i0} from Eq. (13.8.10), the optimum value of n comes out as

$$opt(n) = n_0 = \frac{C^* - c_0}{\bar{c}_1 + \left(\sum_{i \in U} \sqrt{A_i}\right) \sqrt{\frac{\bar{c}_1 c_2}{N A_0}}} \tag{13.8.11}$$

13.8.2 Fixed Variance

Here we find the optimum values of m and n, which minimize the expected cost $\overline{C} = c_0 + n\left(\bar{c}_1 + c_2 \frac{1}{N} \sum_{i \in U} m_i\right)$, keeping the variance to a certain level $\widehat{V}\left[\widehat{Y}_{ts}(wor)\right] = V^* = \frac{1}{n}\left(A_0 + \sum_{i \in U} \frac{A_i}{m_i}\right) - B$. Following Section 13.8.1, the optimum values of m and n are obtained, respectively, as follows:

$$opt(m_i) = m_{i0} = \sqrt{\frac{N\bar{c}_1}{c_2 A_0} A_i} \tag{13.8.12}$$

$$opt(n) = n_0 = \left(\frac{A_0}{V^* + B}\right)\left[1 + \left(\sum_{i \in U}\sqrt{A_i}\right)\sqrt{\frac{c_2}{N\bar{c}_1 A_0}}\right] \quad (13.8.13)$$

13.9 SELF -WEIGHTING DESIGN

A linear homogeneous unbiased estimator of the population total Y based on a sample s is defined as

$$\widehat{Y} = \sum_{i \in s} b_{si} y_i$$

where the coefficient b_{si} is free from y_i but may depend on the unit i and the selected sample s. b_{si}, the coefficient of y_i, is called the multiplier or weight associated with the unit i. An estimator with a constant weight for every unit of the selected sample is termed as a "self-weighting estimator." Hence a self-weighting estimator with the constant weight w is of the form $\widehat{Y} = w\sum_{i \in s} y_i$. A sampling design that yields unbiased estimators for the population mean or total with a constant weight will be called a "self-weighting design." The estimator $\widehat{\overline{Y}} = \frac{1}{n}\sum_{i \in s} y_i$ of the population mean $\overline{Y} = \sum_{i=1}^{N} y_i/N$ based on an SRSWOR or SRSWR sampling provides equal weight $1/n$ for each of the units, and hence $\widehat{\overline{Y}}$ is a self-weighting estimator and SRSWR and SRSWOR are self-weighting designs. The Horvitz–Thompson estimator (1952) $\widehat{Y}_{ht} = \sum_{i \in s} \frac{y_i}{\pi_i}$ is not a self-weighting estimator unless all π_i's are equal. In large-scale surveys we estimate a large number of parameters based on various characteristics. Calculations become much simpler, quicker, and less erroneous if a self -weighing estimator is used. The self-weighting estimator may be obtained by choosing the appropriate sampling design. In multistage sampling design, the estimator $\widehat{Y}_{ts}(ht) = \sum_{i \in s} \widehat{Y}_i/\pi_i = \sum_{i \in s} \frac{M_i}{\pi_i}\frac{1}{m_i}\sum_{j \in s_i} y_{ij}$ given in Eq. (13.3.6) becomes self-weighting if $\pi_i \propto M_i$, and $m_i = m$ is a constant for every $i = 1,\ldots, N$. In stratified sampling (see Chapter 7) the estimator $\widehat{\overline{Y}}_{st} = \frac{1}{N}\sum_{i=1}^{N}\frac{N_i}{n_i}\sum_{j \in s_i} y_{ij}$ becomes a self-weighting estimator under the proportional allocation, where $n_i \propto N_i$.

Example 13.9.1

A sample of 10 orchards from 60 orchards in a certain village is selected by SRSWOR method. From each of the selected orchards, samples of apple plants are selected again by SRSWOR method. The following table gives the production of apples from each of the selected sampled plants.

Selected orchards	Number of plants	Production of apples (in kg) for the sampled plants					
1	50	50	75	125	75	100	120
2	25	35	80	40	70		
3	70	60	100	30	50	75	120
4	20	50	40	40	40		
5	30	50	60	65	70		
6	45	45	50	70	60		
7	50	60	90	80	75	40	80
8	30	55	75	70	70		
9	25	125	100	150	80		
10	25	70	80	75	70		

Estimate the average production of apples per plant when the average number of plants per orchard is (i) known to be 40 and (ii) unknown. Estimate the standard errors of the estimators used in "i" and "ii."

Solution

Here we first prepare the following table.

Selected orchards	# plants	Sampled plants	Sample mean yield	Sample variance	$\widehat{Y}_i$	$\widehat{\sigma}_i^2$	$\widehat{\widehat{D}}_i$
	M_i	m_i	$\overline{y}_i$	s_{yi}^2	$M_i\overline{y}_i$	$M_i^2\left(\frac{1}{m_i}-\frac{1}{M_i}\right)s_{yi}^2$	$\widehat{Y}_i - \widehat{\widehat{\overline{Y}}}_{st}(wor)M_i$
1	50	6	90.83	854.17	4,541.67	313,194.44	961.1486487
2	25	4	56.25	489.58	1,406.25	64,257.81	−384.009009
3	70	6	72.50	1097.50	5,075	819,466.67	62.2747748
4	20	4	42.50	25.00	850.00	200.000	−582.207207
5	30	4	61.25	72.92	1,837.50	14,218.75	−310.810811
6	45	4	56.25	122.92	2,531.25	56,695.31	−691.216216
7	50	6	70.83	324.17	3,541.67	118,861.11	−38.8513513
8	30	4	67.50	75.00	2,025.00	14,625.00	−123.310811
9	25	4	113.75	922.92	2,843.75	121,132.81	1053.490991
10	25	4	73.75	22.92	1,843.75	3,007.8125	53.490991
Total	370				26,495.83	1,527,459.72	

Estimated total production of apples in the village is

$$\widehat{Y}_{ts}(wor) = \frac{N}{n}\sum_{i\in s}\widehat{Y}_i = \frac{60}{10}\times 26,495.83 = 158,974.98\ \text{kg}$$

(i) Here the average number of plants per orchard is $\overline{M} = 40$. Hence the average production of apples per plant is $\widehat{\overline{\overline{Y}}}{}^{*}_{ts}(wor) = \widehat{Y}_{ts}(wor)/\left(N\overline{M}\right) = 158{,}974.98/(60 \times 40) = 66.24$ kg.

An unbiased estimate of variance of $\widehat{\overline{\overline{Y}}}$ is

$$\widehat{V}\left[\widehat{\overline{\overline{Y}}}{}^{*}_{ts}(wor)\right] = \frac{1}{\overline{M}^2}\left[s_b^2 + \frac{1}{Nn}\sum_{i\in s}\widehat{\sigma}_i^2\right]$$

$$= \frac{1}{\overline{M}^2}\left[\left(\frac{1}{n}-\frac{1}{N}\right)\frac{1}{n-1}\sum_{i\in s}\left(\widehat{Y}_i - \widehat{\overline{Y}}\right)^2 + \frac{1}{Nn}\sum_{i\in s}\widehat{\sigma}_i^2\right]$$

$$= \frac{1}{40^2}\left[\left(\frac{1}{10}-\frac{1}{60}\right)\times 1,865,941.165 + \frac{1}{10\times 60}\times 1,527,459.72\right]$$

$$= 98.78$$

Hence an estimated standard error of $\widehat{\overline{\overline{Y}}}{}^{*}_{ts}(wor)$ is $se\left[\widehat{\overline{\overline{Y}}}{}^{*}_{ts}(wor)\right] = \sqrt{\widehat{V}\left[\widehat{\overline{\overline{Y}}}{}^{*}_{ts}(wor)\right]} = 9.94\ \text{kg}$

(ii) In case the average number of plants per orchard $\overline{M}$ is unknown, an estimate of average production of apples per plant is $\widehat{\overline{\overline{Y}}}_{ts}(wor) = \widehat{Y}_{ts}(wor)\Big/\left(N\widehat{\overline{M}}\right) = \sum_{i\in s}\widehat{Y}_i\Big/\sum_{i\in s}M_i = 26,495.83/370 = 71.61\ \text{kg}.$

Following Eq. (13.7.7), an approximate estimate of the MSE of $\widehat{\overline{\overline{Y}}}_{ts}(wor)$ is given by

$$\widehat{M}\left[\widehat{\overline{\overline{Y}}}_{ts}(wor)\right] \cong \frac{1}{\widehat{\overline{M}}^2}\left[\left(\frac{1}{n}-\frac{1}{N}\right)s_{\hat{D}}^2 + \frac{1}{nN}\sum_{i\in s}\widehat{\sigma}_i^2\right]$$

$$= \frac{1}{(370/10)^2}\left[\left(\frac{1}{n}-\frac{1}{N}\right)\frac{1}{n-1}\sum_{i\in s}\left(\widehat{\hat{D}}_i - \widehat{\overline{\hat{D}}}\right)^2 + \frac{1}{nN}\sum_{i\in s}\widehat{\sigma}_i^2\right]$$

$$= \frac{1}{37^2}\left[\left(\frac{1}{10}-\frac{1}{60}\right)\times 346,435.086 + \frac{1}{10\times 60}\times 1,527,459.72\right]$$

$$= 22.95$$

Estimated standard error of $\widehat{\overline{\overline{Y}}}_{ts}(wor)$ is $se\left[\widehat{\overline{\overline{Y}}}_{ts}(wor)\right] = \sqrt{\widehat{M}\left[\widehat{\overline{\overline{Y}}}_{ts}(wor)\right]}$

$= 4.79$ kg

Example 13.9.2

To estimate the total production of paddy in a certain district, a sample of 10 villages was selected from 50 villages in the district by PPSWR method using the area under cultivation of paddy as measure of size variable. From each of the selected villages, a sample of 20 fields was selected by systematic sampling procedure. The data are given below.

Sampled villages	Area under cultivation (in acre)	Total number of fields	Sample mean production per field (000 kg)
1	12.5	100	30
2	10.0	250	20
3	12.0	160	25
4	10.0	300	40
5	8.0	150	30
6	7.5	85	25
7	12.5	100	50
8	7.5	150	30
9	7.5	100	45
10	12.5	105	50

Estimate the total production of paddy and also estimate the standard error of the estimator assuming total area under cultivation in the district is 750 acres.

Solution

Here total number of villages in the district $N = 50$ and number of sampled villages $n = 10$.

The estimated total yield of the sampled villages and normed size measures associated with the villages are given as follows:

Sampled villages (r)	Area under paddy $X(r)$	$M(r)$	Average yield $\bar{y}(r)$	$p(r)$	$\widehat{Y}(r) = M(r)\,\bar{y}(r)$	$\widehat{Y}(r)/p(r)$
1	12.5	100	30	0.016667	3000	180,000
2	10	250	20	0.013333	5000	375,000
3	12	160	25	0.016000	4000	250,000
4	10	300	40	0.013333	12,000	900,000
5	8	150	30	0.010667	4500	421,875
6	7.5	85	25	0.010000	2125	212,500
7	12.5	100	50	0.016667	5000	300,000
8	7.5	150	30	0.010000	4500	450,000
9	7.5	100	45	0.010000	4500	450,000
10	12.5	105	50	0.016667	5250	315,000

$$\text{Estimated the total yield of paddy} = \widehat{Y}_{ts}(hh) = \frac{1}{n}\sum_{r=1}^{n} \widehat{Y}(r)/p(r) \times 1000 \text{ kg}$$

$$= 385,437.5 \times 1000 \text{ kg}$$

$$= 385,437,500 \text{ kg}$$

Estimated standard error of $\widehat{Y}_{ts}(hh)$ is

$$Se\left[\widehat{Y}_{ts}(hh)\right] = \sqrt{\frac{1}{n(n-1)}\sum_{r=1}^{n}\left(\widehat{Y}(r)/p(r) - \frac{1}{n}\sum_{r=1}^{n}\widehat{Y}(r)/p(r)\right)^2} \times 1000$$

$$= \sqrt{4191823351} \times 1000 \text{ kg} = 64,744,292.03 \text{ kg}$$

13.10 EXERCISES

13.10.1 Consider a two-stage sampling procedure where a sample s of n fsu's is selected by the PPSWR method using normed size measure p_i attached to the ith unit, $i = 1, \ldots, N$; $\sum_{i=1}^{N} p_i = 1$, and let

$n_i(s)$ be the number of times the ith fsu appeared in the samples. The selected fsu's are sub-sampled in the following manner.

(a) A single sub-sample of size m_i is selected from the ith fsu, and let $T_1 = \frac{1}{n}\sum_{i \in s}\frac{n_i(s)\widehat{Y}_i}{p_i}$ be an estimator of the population total Y, where $\widehat{Y}_i = M_i\overline{y}_i$ and $\overline{y}_i$ is the sample mean based on the m_i ssu's selected from the ith fsu.

(b) $n_i(s)$-independent sub-samples, each of size m_i, are selected from the ith fsu and let $T_2 = \frac{1}{n}\sum_{r=1}^{n}\frac{\widehat{Y}(r)}{p(r)}$, where $\widehat{Y}(r) = M_j\overline{y}_j(r)$ if the rth draw produces the jth fsu, and $\overline{y}_j(r)$ is the sample mean of the ssu's selected from the jth fsu.

(c) $m_i n_i(s)$ $(\le M_i)$ ssu's are selected from the ith fsu and let $T_2 = \frac{1}{n}\sum_{i \in s}\frac{n_i(s)\widehat{Y}_i^*}{p_i}$, where $\widehat{Y}_i^* = M_i\overline{y}_i^*$ and $\overline{y}_i^*$ = sample mean based on $m_i n_i(s)$ ssu's of the ith fsu.

Show that (i) $E(T_1) = E(T_2) = E(T_3) = Y$ and (ii) $V(T_1) \ge V(T_2) \ge V(T_3)$. Find unbiased estimators of $V(T_1)$, $V(T_2)$, and $V(T_3)$ (Rao, 1961).

13.10.2 Consider a two-stage sampling scheme where a sample of n fsu's is selected by the PPSWOR method using normed size measure p_i attached to the ith unit, $i = 1,\ldots, N$. Let s be an unordered sample obtained from the selected fsu's. If the ith fsu is included in s, a sub-sample s_i of size m_i ssu's is selected from the ith fsu by SRSWOR method. Let $T = \frac{\sum_{i \in s} M_i\overline{y}_i p(s|i)}{P(s)}$, where $\overline{y}_i$ is the sample mean of ssu's selected from the ith fsu, $p(s|i)$ = conditional probability of selection of s given that the ith unit is selected at the first draw, and $p(s)$ is the probability of selection of the unordered sample s. Show that T is an unbiased estimator of Y. Derive $V(T)$, the variance of T and an unbiased estimator of $V(T)$.

13.10.3 Let a sample s of n fsu 's be selected from a population by some suitable sampling schemes with π_i and π_{ij} as inclusion probabilities for the i th and ith and jth $(i \ne j)$ units, respectively. If ith fsu is included in s, a sub-sample s_i of size m_i ssu's is selected by systematic sampling scheme and let t_i be the sample mean.

(a) Show that $T = \sum_{i \in s} M_i t_i / \pi_i$ is an unbiased estimator of the population total Y. Derive the expression for $V(T)$. Explain why $V(T)$ cannot be estimated unbiasedly.

(b) Instead of selecting a single systematic sample s_i, two independent systematic samples $s_i(1)$ and $s_i(2)$ each of sizes $m_i/2$ (assumed to be integer) are selected by a systematic sampling scheme. Let $t_i(1)$ and $t_i(2)$ be the sample mean of ssu's based on samples $s_i(1)$ and $s_i(2)$, respectively. Show that

(i) $T = \sum_{i \in s} \frac{M_i \bar{t}_i}{\pi_i}$ with $\bar{t}_i = \frac{t_i(1) + t_i(2)}{2}$ is an unbiased estimator of Y.

(ii) $V(T) = \frac{1}{2} \sum_{i \neq}^{N} \sum_{j=1}^{N} (\pi_i \pi_j - \pi_{ij}) \left(\frac{Y_i}{\pi_i} - \frac{Y_j}{\pi_j} \right)^2 + \sum_{i=1}^{N} \frac{M_i^2 V(\bar{t}_i | s)}{\pi_i}$

and

(iii) $\widehat{V}(T) = \frac{1}{2} \sum_{i \neq} \sum_{j \in s} \left(\frac{\pi_i \pi_j - \pi_{ij}}{\pi_{ij}} \right) \left(\frac{M_i t_i}{\pi_i} - \frac{M_j t_j}{\pi_j} \right)^2 + \frac{1}{4} \sum_{i \in s} \frac{M_i^2 \{\bar{t}_i(1) - \bar{t}_i(2)\}^2}{\pi_i}$ is an unbiased estimator of $V(T)$.

13.10.4 From a population of N fsu's, a sample of n fsu's is selected by the Lahiri–Midzuno–Sen (1951, 1952, 1953) sampling scheme with p_i as normed size measure for the ith unit. If the ith fsu is selected in the sample, a sub-sample of m_i ssu's is selected by SRSWOR method. Suggest an unbiased estimator of the population total. Derive its variance and an unbiased estimator of the variance.

13.10.5 From a population of N enumeration areas, a sample of n enumerations areas are selected by SRSWOR method. If the ith enumeration area is selected in the sample, a sample s_i of m_i households is selected from the M_i households of the ith enumeration area by SRSWOR method. Let P_i and $\widehat{P}_i$ denote the proportion of households that possess a certain attribute (viz. owning a house) in the entire ith enumeration area and sampled households, respectively. Prove that

(i) $\widehat{P} = \frac{1}{n\overline{M}} \sum_{i \in s} M_i \widehat{P}_i$ is an unbiased estimator of P, the population proportion of houseldholds owning a house where

$\overline{M} = \sum_{i=1}^{N} M_i / N$ is known.

(ii) $V\left(\widehat{P}\right)=\frac{1}{\overline{M}^2}\left[\left(\frac{1}{n}-\frac{1}{N}\right)\frac{1}{N-1}\sum_{i=1}^{N}\left(M_iP_i-\overline{M}P\right)^2\right.$

$\left.+\frac{1}{nN}\sum_{i=1}^{N}M_i^2\left(\frac{1}{m_i}-\frac{1}{M_i}\right)\frac{M_iP_i(1-P_i)}{M_i-1}\right]$

and

$$\widehat{V}\left(\widehat{P}\right)=\frac{1}{\overline{M}^2}\left[\left(\frac{1}{n}-\frac{1}{N}\right)\frac{1}{n-1}\sum_{i\in s}\left(M_i\widehat{P}_i-\overline{M}\widehat{P}\right)^2\right.$$
$$\left.+\frac{1}{nN}\sum_{i=1}^{N}M_i^2\left(\frac{1}{m_i}-\frac{1}{M_i}\right)\frac{m_i\widehat{P}_i\left(1-\widehat{P}_i\right)}{m_i-1}\right]$$

13.10.6 Consider the sampling scheme in Exercise 13.10.5, where $\overline{M}$ is unknown. Show that the estimator $\widehat{P}^* = \sum_{i\in s} M_i\widehat{P}_i \big/ \sum_{i\in s} M_i$ is a biased estimator of P. Derive approximate expressions for the bias and MSE of $\widehat{P}^*$.

13.10.7 From a population of N fsu's, a sample s of n ssu's is selected by SRSWOR method. All the ssu's of the selected fsu's are mixed up to from $M_0 = \sum_{i\in s} M_i$ ssu's. From the M_0 ssu's, a sample of m_0 ssu's are selected by SRSWOR method. Suggest an unbiased estimator of the population total. Derive its variance and unbiased estimator of this variance.

13.10.8 To estimate the average monthly rent of two-bedroom apartments in a certain city, the city was divided into 50 blocks. A sample of five blocks is selected by SRSWOR method, and from each of the blocks, five apartments were selected by SRSWOR method. The data are given below.

Sampled blocks	Number of two-bedroom apartments	Monthly rent of the sampled apartments (in \$)
1	150	3500, 2800, 3500, 4500, 2800
2	150	5000, 3250, 3500, 3000, 3500
3	175	5000, 4500, 4000, 4500, 2800
4	200	4250, 1800, 3500, 4500, 5000
5	150	3500, 2500, 2800, 3500, 4000

Estimate the average rent of the apartment and its standard error when (i) the total number of apartments in the city is known as 875 and (ii) total number of apartment is unknown.

13.10.9 To estimate the total number of HIV/AIDS patients treated in a certain district, a stratified two-stage sampling design was used. Four blocks were selected from each of the strata by PPSWR method using number of people in the block as the measure of size. From each of the selected blocks, five clinics were selected by a systematic sampling procedure. The data obtained are as follows.

Stratum	Total population (in 000)	Sampled blocks	Population (in 000)	Total number of patients treated
1	150	1	5.20	75, 20, 15, 20, 10
		2	8.75	50, 30, 25, 20, 20
		3	12.80	75, 50, 25, 30, 10
		4	10.25	75, 20, 15, 20, 10
2	175	1	8.75	80, 20, 35, 15, 15
		2	12.80	25, 10, 10, 18, 12
		3	15.20	45, 40, 10, 20, 20
		4	16.25	25, 10, 15, 10, 10

Estimate the total number of patients treated and obtain an estimate of its standard error.

13.10.10 In a locality, a sample of six enumeration areas is selected from 50 enumeration areas by SRSWOR method. From each of the selected enumeration areas, a sample of five households is selected by SRSWOR method. The data on the number of HIV infected people in the selected households are given below.

Enumeration area	Total number of households	Sampled households					
1	40	Size	3	4	2	4	5
		#HIV infected	1	2	0	0	1
2	30	Size	4	4	2	1	1
		#HIV infected	0	0	2	0	0
3	20	Size	5	6	4	5	3
		#HIV infected	2	0	2	1	2
4	50	Size	4	5	3	2	1
		#HIV infected	3	2	0	1	0
5	80	Size	5	4	2	4	3
		#HIV infected	2	2	1	2	0
6	75	Size	3	3	4	2	1
		#HIV infected	1	2	2	1	0

Estimate the proportion of HIV infected persons in the locality and its standard error when the total number of people M in the locality (i) is known as 7500 and (ii) is unknown.

CHAPTER 14

Variance/Mean Square Estimation

14.1 INTRODUCTION

From survey data we very often estimate population parameters for various characteristics such as population mean, proportion, coefficient of variation, ratio of two characteristics, correlation coefficient, and so on. The variance or mean square error gives us an idea about the magnitude of the sampling error. The variance estimator is used to determine confidence interval of a parameter and testing of hypothesis related to it. It is also used to determine the optimal sample size for a survey design. The expression of variance of a linear estimator can be determined explicitly. Conditions of unbiasedness of variance estimations can also be derived easily. Although the variance is a nonnegative quantity, the unbiased estimators may not be so. If the variance estimator becomes negative, it cannot be used for interval estimation and testing of hypotheses. In this chapter we will study the method of determining variance of linear estimators along with their nonnegativity properties. The necessary condition for nonnegativity of the variance estimators has been established. However, no necessary and sufficient condition of existence of nonnegative variance estimators is available in general.

14.2 LINEAR UNBIASED ESTIMATORS

Let a sample s of size n be selected from a finite population $U=(U_1,\ldots,U_N)$ of N units with probability $p(s)$. Consider a linear homogeneous unbiased estimator for the population total Y

$$t(s,y)=\sum_{i\in s} b_{si}y_i \tag{14.2.1}$$

where b_{si}'s are constants satisfying unbiasedness condition.

$$\sum_{s\supset i} b_{si}p(s)=1 \ \ \forall i=1,\ldots,N \tag{14.2.2}$$

Survey Sampling Theory and Applications
ISBN 978-0-12-811848-1
http://dx.doi.org/10.1016/B978-0-12-811848-1.00014-5

The variance of $t(s,y)$ is given by

$$\begin{aligned} V(t) &= V(s,y) \\ &= \sum_{i\in U} \alpha_{ii} y_i^2 + \sum_{i\neq}\sum_{j\in U} \alpha_{ij} y_i y_j \end{aligned} \tag{14.2.3}$$

where

$$\alpha_{ii} = \sum_{s\supset i} b_{si}^2 p(s) - 1 \text{ and } \alpha_{ij} = \sum_{s\supset ij} b_{si} b_{sj} p(s) - 1 \tag{14.2.4}$$

14.2.1 Conditions of Unbiased Estimation of Variance

Theorem 14.2.1

For a nonzero α_{ij}, a necessary condition of existence of an unbiased estimation of $V(t)$ is π_{ij}, the inclusion probability for the ith and jth units $(i \neq j)$ should be positive.

Proof

Let $\widehat{V}(s,y)$ be an estimator of $V(t)$ based on the selected sample s. If $\pi_{ij} = 0$, then the units i and j cannot occur together in a sample with positive probability and the estimator $\widehat{V}(s,y)$ will be free from the values of y_i and y_j. Hence the average $E\left[\widehat{V}(s,y)\right] = \sum \widehat{V}(s,y)p(s)$ does not involve y_i and y_j. Hence we cannot find an estimator $\widehat{V}(s,y)$ which is unbiased for $V(s,y)$ because α_{ij}, the coefficient of $y_i y_j$ in $V(s,y)$ is not zero.

Theorem 14.2.2

For $\alpha_{ij} \neq 0\ \forall i \neq j$, a necessary and sufficient condition for a homogeneous quadratic estimator $\widehat{V}(s,y) = \sum_{i\in s} c_{ii}(s) y_i^2 + \sum_{i\neq}\sum_{j\in s} c_{ij}(s) y_i y_j$ to be unbiased for $V(t)$ is $\pi_{ij} > 0$ where $c_{ii}(s)$ and $c_{ij}(s)$ are known constants.

Proof

$\widehat{V}(s,y)$ is unbiased for $V(t)$ if and only if

$$E\left[\widehat{V}(s,y)\right] = V(s,y)\ \forall y \in R^N \tag{14.2.5}$$

The condition (14.2.5) holds if and only if the following conditions are satisfied

$$\sum_{s \supset i} c_{ii}(s)p(s) = \sum_{s} c_{ii}(s)I_{si}p(s) = \alpha_i \quad \forall i = 1, \ldots, N \tag{14.2.6}$$

and

$$\sum_{s \supset ij} c_{ij}(s)p(s) = \sum_{p} c_{ij}(s)p(s)I_{si}I_{sj} = \alpha_{ij} \quad \forall i \neq j = 1, \ldots, N \tag{14.2.7}$$

where $I_{si} = 1$ for $i \in s$ and $I_{si} = 0$ for $i \notin s$.

Because $\alpha_{ij} \neq 0$, Eq. (14.2.5) holds if and only if $I_{si}I_{sj} = 1$ for at least one pair i, j; $i \neq j$. In this situation π_{ij} becomes positive for $\forall i \neq j$.

Furthermore, if $\pi_{ij} > 0$, we can choose $c_{ii}(s)$ and $c_{ij}(s)$ in various ways so that conditions (14.2.6) and (14.2.7) are satisfied. The obvious choices are $c_{ii}(s) = \alpha_{ii}/\pi_i$ and $c_{ij}(s) = \alpha_{ij}/\pi_{ij}$, where π_i is the inclusion probability of the ith unit.

Remark 14.2.1

We can find several unbiased estimators of $V(t)$. For example, the following estimators $\widehat{V}_1(s, y)$ and $\widehat{V}_2(s, y)$ are both unbiased for $V(t)$.

$$\widehat{V}_1(s, y) = \sum_{i \in s} \frac{\alpha_{ii}}{\pi_i} y_i^2 + \sum_{i \neq} \sum_{j \in s} \frac{\alpha_{ij}}{\pi_{ij}} y_i y_j \tag{14.2.8}$$

$$\widehat{V}_2(s, y) = \frac{\sum_{i \in s} \frac{\alpha_{ii}}{\pi_i} M_i(s) y_i^2 + \sum_{i \neq} \sum_{j \in s} \frac{\alpha_{ij}}{\pi_{ij}} M_{ij}(s) y_i y_j}{p(s)} \tag{14.2.9}$$

where $M_i = \sum_s I_{si}$ = number of samples that contain the ith unit and $M_{ij} = \sum_s I_{si}I_{sj}$ = number of samples that contain both the ith and jth units.

Remark 14.2.2

Consider a linear systematic sampling with $N = 12$ and $n = 4$. Here the three possible samples are $s_1 = (1, 4, 7, 10)$, $s_2 = (2, 5, 8, 11)$, and $s_3 = (3, 6, 9, 12)$ each of which has a selection probability 1/3. Here the variance of the sample mean cannot be estimated unbiasedly because the second-order inclusion probabilities for some of the units, e.g., π_{12}, π_{15}, and π_{23} are exactly equal to zero.

14.3 NONNEGATIVE VARIANCE/MEAN SQUARE ESTIMATION

The variance or mean square error of an estimator is a nonnegative quantity, but its unbiased estimators are not always nonnegative. In this section we will establish necessary conditions of existence of nonnegative unbiased estimators of variance/mean square error of an estimator.

Let $T(s,y) = \sum_i c_{si}y_i$ be a linear estimator (not necessary unbiased) of the total Y, where c_{si}'s are constants free of y_i's and $c_{si} = 0$ for $i \notin s$. The mean square error of $T(s, y)$ is given by

$$\begin{aligned} M &= E[T(s,y) - Y]^2 \\ &= \left[\sum_{i=1}^{N}(c_{si} - 1)y_i\right]^2 \\ &= \sum_{i \in U} \beta_{ii}y_i^2 + \sum_{i \neq} \sum_{j \in U} \beta_{ij}y_i y_j \end{aligned} \quad (14.3.1)$$

where

$$\beta_{ii} = E(c_{si} - 1)^2 = \sum_{s \supset i} c_{si}^2 p(s) - 2\sum_{s \supset i} c_{sj}p(s) + 1 \quad \text{and}$$

$$\beta_{ij} = E(c_{si} - 1)(c_{sj} - 1) = \sum_{s \supset ij} c_{si}c_{sj}p(s) - \sum_{s \supset i} c_{si}p(s) - \sum_{s \supset j} c_{sj}p(s) + 1$$

(14.3.2)

We now present the following theorem proposed by Rao and Vijayan (1977).

Theorem 14.3.1

For a fixed sample of size n design, if M, the mean square error of $T(s, y)$, becomes zero when $z_i = y_i/w_i = k$, a constant, and $w_i(\neq 0)$'s are known for $i = 1, \ldots, N$, then

(i) The mean square error M can be written as

$$M = -\frac{1}{2}\sum_{i \neq} \sum_{j \in U} \beta_{ij}w_i w_j (z_i - z_j)^2$$

(ii) A nonnegative quadratic unbiased estimator of M is necessarily of the form

$$\widehat{M}(s,z) = -\frac{1}{2}\sum_{i \neq} \sum_{j \in s} \beta_{ij}(s) w_i\, w_j (z_i - z_j)^2 \quad (14.3.3)$$

where $\beta_{ij}(s)$'s are constants independent of y_i's, equal zero if $i, j \notin s$, and satisfy the unbiasedness condition

$$E\{\beta_{ij}(s)\} = \sum_{s \supset i,j} \beta_{ij}(s)p(s) = \beta_{ij} \text{ for } \forall i \neq j \in U$$

Proof

(i) The mean square error M given in Eq. (14.3.1) can be written as

$$M = \mathbf{y}' \mathbf{B} \mathbf{y} \tag{14.3.4}$$

where $\mathbf{y}' = (y_1, \ldots, y_N)$ and $\mathbf{B} = (\beta_{ij})$ is an $N \times N$ matrix with ith and jth element β_{ij}.

Because $M \geq 0$ for $\mathbf{y} \in R^N$, we find that the matrix $\mathbf{B}$ is a positive semidefinite and hence we can write

$$\mathbf{B} = \mathbf{H}'\mathbf{H} \tag{14.3.5}$$

Substituting $\mathbf{y} = k\mathbf{w} = k(w_1, \ldots, w_N)'$ in Eq. (14.3.4) and using the condition stated in the theorem we get

$$M = k^2\mathbf{w}'\mathbf{B}\mathbf{w} = 0$$
$$\text{i.e., } (\mathbf{H}\mathbf{w})'(\mathbf{H}\mathbf{w}) = 0$$
$$\text{i.e., } \quad \mathbf{H}\mathbf{w} = 0$$
$$\text{(where } \mathbf{0} \text{ is a null column vector)}$$
$$\text{i.e., } \quad \mathbf{B}\mathbf{w} = 0$$
$$\text{i.e., } \quad \beta_{ii}w_i^2 + w_i \sum_{j(\neq i) \in U} \beta_{ij}w_j = 0 \tag{14.3.6}$$

Now

$$\begin{aligned} M &= \sum_{i \in U} \beta_{ii}y_i^2 + \sum_{i \neq} \sum_{j \in U} \beta_{ij}y_iy_j \\ &= \sum_{i \in U} \beta_{ii}w_i^2z_i^2 + \sum_{i \neq} \sum_{j \in U} \beta_{ij}w_iw_jz_iz_j \end{aligned} \tag{14.3.7}$$

$$\text{(writing } z_i = y_i/w_i)$$

Substituting Eq. (14.3.6) in Eq. (14.3.7) yields

$$M = -\sum_{i \neq} \sum_{j \in U} \beta_{ij}w_iw_j\left(z_i^2 - z_iz_j\right)$$

$$= -\frac{1}{2} \sum_{i \neq} \sum_{j \in U} \beta_{ij}w_iw_j(z_i - z_j)^2.$$

(ii) Because $\widehat{M}(s)$ is a nonnegative homogeneous quadratic unbiased estimator for M, we can write $\widehat{M}(s)$ as

$$\widehat{M}(s,y)=\sum_{i\in s}\beta_{ii}(s)y_i^2+\sum_{i\neq}\sum_{j\in s}\beta_{ij}(s)y_iy_j \tag{14.3.8}$$

where $\beta_{ii}(s)$ and $\beta_{ij}(s)$ are known constants independent of y_i's and satisfy the unbiasedness condition

$$\sum_s \widehat{M}(s,y)p(s)=M \tag{14.3.9}$$

Putting $y_i=kw_i\ \forall i=1,\ldots,N$, we find (from the condition of the theorem)

$$\sum_s \widehat{M}(s,w)p(s)=M=0 \tag{14.3.10}$$

Furthermore, the condition $\widehat{M}(s,w)\geq 0\ \ \forall s$ with $p(s)>0$ yields

$$\widehat{M}(s,w)=0\ \forall s \text{ with } p(s)>0$$

$$\text{i.e.,}\quad \sum_{i\in s}\beta_{ii}(s)w_i^2+\sum_{i\neq}\sum_{j\in s}\beta_{ij}(s)w_iw_j=0\ \forall s \text{ with } p(s)>0 \tag{14.3.11}$$

Because $\widehat{M}(s,y)$ is a nonnegative quadratic form and $\widehat{M}(s,y)=0$ for $y_i=kw_i$ for $i\in s$, Eq. (14.3.6) yields

$$\beta_{ii}(s)w_i^2+w_i\sum_{j(\neq i)\in s}\beta_{ij}(s)w_j=0 \tag{14.3.12}$$

$$\text{i.e.,}\quad \beta_{ii}(s)w_i^2z_i^2=-w_iz_i^2\sum_{j(\neq i)\in s}\beta_{ij}(s)w_j$$

$$\text{i.e.,}\quad \sum_{i\in s}\beta_{ii}(s)w_i^2z_i^2=-\sum_{i\in s}w_iz_i^2\sum_{j(\neq i)\in s}\beta_{ij}(s)w_j$$

$$=-\frac{1}{2}\sum_{i\neq}\sum_{j\in s}\beta_{ij}(s)w_iw_j\left(z_i^2+z_j^2\right) \tag{14.3.13}$$

Substituting Eq. (14.3.13) in Eq. (14.3.8), we get

$$\widehat{M}(s,y)=\widehat{M}(s,z)=-\frac{1}{2}\sum_{i\neq}\sum_{j\in s}w_iw_j\beta_{ij}(s)(z_i-z_j)^2.$$

Remark 14.3.1

The constants $\beta_{ij}(s)$'s that satisfy $E\{\beta_{ij}(s)\} = \beta_{ij}$ can be chosen in various ways. The obvious choice is $\beta_{ij}(s) = \beta_{ij}/\pi_{ij}$, provided π_{ij}'s are easy to compute. Alternatively, we may choose $\beta_{ij}(s) = k_{ij}(s) = \beta_{ij}/\{M_{ij}p(s)\}$, where M_{ij} is the number of samples that contain both the ith and jth units. If all $\binom{N}{n}$ possible samples of size n have positive probabilities, then $M_{ij} = M_2 = \binom{N-2}{n-2}$. A more general choice given by Rao (1979) is $\beta_{ij}(s) = \beta_{ij}d_{ij}(s)/E\{d_{ij}(s)\}$, where $d_{ij}(s) = 0$ if the sample s does not contain both the units i and j and $E\{d_{ij}(s)\}$ is easy to compute.

Remark 14.3.2

For a fixed sample size $n = 2$ design, Lanke (1974a,b) relaxed the restriction of the class of quadratic estimators and has shown that any nonnegative unbiased estimator (not necessarily quadratic) of M is necessarily of the form (14.3.3). For $n > 2$, Vijayan (1975) proved that a nonnegative unbiased polynomial estimator of M is necessarily of the form (14.3.3).

14.3.1 Examples

14.3.1.1 Horvitz–Thompson Estimator

Consider the Horvitz–Thompson estimator $\widehat{Y}_{ht} = \sum_{i\in s} \frac{y_i}{\pi_i}$ based on a fixed sample size n design. The estimator $\widehat{Y}_{ht}$ is unbiased and becomes a constant (its variance becomes zero) when $y_i = k\pi_i$ for $i = 1,\ldots, N$ and k is a constant. Here $c_{si} = 1/\pi_i$ and $\beta_{ij} = \sum_{s\supset i,j} \frac{p(s)}{\pi_i\pi_j} - 1 = \frac{\pi_{ij}}{\pi_i\pi_j} - 1$. Hence, the variance of $\widehat{Y}_{ht}$ can be written as

$$V\left(\widehat{Y}_{ht}\right) = M = -\frac{1}{2}\sum_{i\neq}\sum_{j\in U}\beta_{ij}\pi_i\pi_j\left(\frac{y_i}{\pi_i} - \frac{y_j}{\pi_j}\right)^2$$

A nonnegative unbiased variance estimator is necessarily of the form

$$\widehat{V}\left(\widehat{Y}_{ht}\right) = -\frac{1}{2}\sum_{i\neq}\sum_{j\in s}\beta_{ij}(s)\pi_i\pi_j\left(\frac{y_i}{\pi_i} - \frac{y_j}{\pi_j}\right)^2$$

with $\sum_{s\supset i,j} \beta_{ij}(s)p(s) = \frac{\pi_{ij} - \pi_i\pi_j}{\pi_i\pi_j}$.

If we choose $\beta_{ij}(s)=\delta_{ij}$ independent of s, then the constraint $\delta_{ij}\sum_{s\supset i,j}p(s)=\dfrac{\pi_{ij}-\pi_i\pi_j}{\pi_i\pi_j}$ yields $\delta_{ij}=\dfrac{\pi_{ij}-\pi_i\pi_j}{\pi_i\pi_j\pi_{ij}}$. The variance estimator $\widehat{V}\left(\widehat{Y}_{ht}\right)$ reduces to Yates-Grundy's (1953) estimator viz.

$$\widehat{V}\left(\widehat{Y}_{ht}\right)=\widehat{V}_{YG}=\frac{1}{2}\sum_{i\neq}\sum_{j\in s}\frac{\pi_i\pi_j-\pi_{ij}}{\pi_{ij}}\left(\frac{y_i}{\pi_i}-\frac{y_j}{\pi_j}\right)^2 \tag{14.3.14}$$

The estimator $\widehat{V}_{YG}$ is nonnegative for those samples for which $\pi_i\pi_j-\pi_{ij}\geq 0\ \ \forall i,\ j\in s$. However, for a sampling design with $\pi_i\pi_j-\pi_{ij}\geq 0\ \forall i,j\in U,\ \widehat{V}_{YG}\geq 0$.

For $n=2$, the probability of selecting the sample s, containing the units $i,\ j\ (i\neq j)$, is $p(s)=\pi_{ij}$. Hence the condition $\sum_{s\supset i,j}\beta_{ij}(s)p(s)=\dfrac{\pi_{ij}-\pi_i\pi_j}{\pi_i\pi_j}$ reduces to $\beta_{ij}(s)\pi_{ij}=\dfrac{\pi_{ij}-\pi_i\pi_j}{\pi_i\pi_j}$, i.e., $\beta_{ij}(s)=\dfrac{\pi_{ij}-\pi_i\pi_j}{\pi_{ij}\pi_i\pi_j}$. Hence for $n=2$, the nonnegative variance estimator of $V\left(\widehat{Y}_{ht}\right)$ should be necessarily of the form

$$\widehat{V}\left(\widehat{Y}_{ht}\right)=\widehat{V}_{YG}=\frac{\pi_i\pi_j-\pi_{ij}}{\pi_{ij}}\left(\frac{y_i}{\pi_i}-\frac{y_j}{\pi_j}\right)^2$$

which is the Yates-Grundy (1953) estimator. Hence for $n=2$, $\widehat{V}_{YG}=\dfrac{\pi_i\pi_j-\pi_{ij}}{\pi_{ij}}\left(\dfrac{y_i}{\pi_i}-\dfrac{y_j}{\pi_j}\right)^2$ becomes the unique nonnegative estimator of $V\left(\widehat{Y}_{ht}\right)$ in the class of quadratic unbiased estimators if and only if $\pi_i\pi_j-\pi_{ij}\geq 0\ \forall i,j\in U$.

Following Remark 14.3.2, we note that for $n=2$, any nonnegative unbiased estimator (not necessarily quadratic) necessarily becomes the Yates-Grundy (1953) estimator.

The choice of $\beta_{ij}(s)=\dfrac{\pi_i\pi_j-\pi_{ij}}{\pi_i\pi_jM_{ij}p(s)}$ yields an alternative unbiased estimator of $V\left(\widehat{Y}_{ht}\right)$ as

$$\widehat{V}^*\left(\widehat{Y}_{ht}\right)=\frac{1}{2p(s)}\sum_{i\neq}\sum_{j\in s}\frac{\pi_i\pi_j-\pi_{ij}}{M_{ij}}\left(\frac{y_i}{\pi_i}-\frac{y_j}{\pi_j}\right)^2 \tag{14.3.15}$$

where M_{ij}'s are defined in Remark 14.3.1.

14.3.1.2 Hansen–Hurwitz Estimator

The Hansen–Hurwitz estimator based on PPSWR sampling design is given by

$$\widehat{Y}_{hh}=\frac{1}{n}\sum_{i\in s}n_i(s)\frac{y_i}{p_i}$$

where $n_i(s)$ is the number of times the ith unit appears in the sample s and $p_i(>0)$ is the normed size measure for the ith unit.

The estimator $\widehat{Y}_{hh}$ is unbiased for Y. The estimator $\widehat{Y}_{hh}$ becomes a constant if the y_i's are proportional to p_i's and in this case the variance of $\widehat{Y}_{hh}$ is zero. Here $c_{si} = \dfrac{n_i(s)}{np_i}$ and $\beta_{ij} = \dfrac{1}{n^2 p_i p_j} \sum_{s \supset i,j} n_i(s) n_j(s) p(s) - 1 = \dfrac{E[n_i(s)n_j(s)]}{n^2 p_i p_j} - 1 = -\dfrac{1}{n}$. Hence the variance of $\widehat{Y}_{hh}$ can be written as

$$V_{hh} = \frac{1}{2n} \sum_{i \neq} \sum_{j \in U} p_i p_j \left(\frac{y_i}{p_i} - \frac{y_j}{p_j} \right)^2$$

If we choose $\beta_{ij}(s) = \beta_{ij} \dfrac{n_i(s)n_j(s)}{E\{n_i(s)n_j(s)\}} = -\dfrac{n_i(s)n_j(s)}{n^2(n-1)p_i p_j}$, we get an unbiased estimator of V_{hh} as

$$\begin{aligned} \widehat{V}_{hh}(1) &= \frac{1}{2n^2(n-1)} \sum_{i \neq} \sum_{j \in s} n_i(s) n_j(s) \left(\frac{y_i}{p_i} - \frac{y_j}{p_j} \right)^2 \\ &= \frac{1}{n(n-1)} \sum_{i \in s} n_i(s) \left(\frac{y_i}{p_i} - \widehat{Y}_{hh} \right)^2 \end{aligned} \tag{14.3.16}$$

On the other hand, if we choose $\beta_{ij}(s) = -\beta_{ij}/\pi_{ij}$, with $\pi_{ij} = 1 - (1 - p_i)^n - (1 - p_j)^n + (1 - p_i - p_j)^n$, we get an alternative unbiased estimator of V_{hh} as

$$\widehat{V}_{hh}(2) = \frac{1}{2} \sum_{i \neq} \sum_{j \in s} \frac{p_i p_j}{\pi_{ij}} \left(\frac{y_i}{p_i} - \frac{y_j}{p_j} \right)^2 \tag{14.3.17}$$

14.3.1.3 Murthy's Estimator

The Murthy's (1957) estimator for Y based on PPSWOR sampling scheme described in Section 5.3.2.1 is given by $\widehat{Y}_{mur} = \sum_{i \in s} \dfrac{p(s|i)}{p(s)} y_i$, where p_i is the probability of selection of the unit at the first draw and $p(s|i)$ is the conditional probability of the selection of the sample s given that the ith unit is selected at the first draw. Here $c_{si} = \dfrac{p(s|i)}{p(s)}$, $\sum_{s \supset i} c_{si} p(s) = 1$, and $\beta_{ij} = \omega_{ij} - 1$ with $\omega_{ij} = \sum_{s \supset i,j} \dfrac{p(s|i)p(s|j)}{p(s)}$. The estimator $\widehat{Y}_{mur}$ is unbiased for Y and if

$y_i = kp_i$, then $\widehat{Y}_{mur} = \frac{k}{p(s)} \sum_{i \in s} p_i p(s|i) = k =$ constant, because $\sum_{i \in s} p_i p(s|i) = p(s)$. Hence the variance of $\widehat{Y}_{mur}$ can be written as

$$V_{mur} = V\left(\widehat{Y}_{mur}\right) = \frac{1}{2} \sum_{i \neq} \sum_{j \in U} p_i p_j (1 - \omega_{ij}) \left(\frac{y_i}{p_i} - \frac{y_j}{p_j}\right)^2 \tag{14.3.18}$$

Clearly an unbiased estimator of V_{mur} is given by

$$\widehat{V}_{mur}(1) = \frac{1}{2} \sum_{i \neq} \sum_{j \in s} \frac{p_i p_j}{\pi_{ij}} (1 - \omega_{ij}) \left(\frac{y_i}{p_i} - \frac{y_j}{p_j}\right)^2 \tag{14.3.19}$$

The estimator (14.3.19) is found to be nonnegative (Rao, 1979).

Following Rao (1979), if we choose $\beta_{ij}(s) = c_{si} c_{sj} - \frac{p(s|i,j)}{p(s)} = \frac{p(s|i)p(s|j)}{\{p(s)\}^2} - \frac{p(s|i,j)}{p(s)}$ so that $\sum_{s \supset i,j} \beta_{ij}(s) p(s) = \beta_{ij}$, we get an alternative unbiased estimator of V_{mur} as

$$\widehat{V}_{mur}(2) = \frac{1}{2} \sum_{i \neq} \sum_{j \in s} \frac{p_i p_j}{\{p(s)\}^2} \{p(s)p(s|i,j) - p(s|i)p(s|j)\} \left(\frac{y_i}{p_i} - \frac{y_j}{p_j}\right)^2 \tag{14.3.20}$$

The estimator (14.3.20) is nonnegative as observed by Pathak and Shukla (1966).

For $n = 2$, $\widehat{V}_{\text{mur}}(1)$ reduces to a unique nonnegative unbiased variance estimator proposed by Pathak and Shukla (1966) as

$$\widehat{V}_{mur}(1) = \frac{1}{2} \sum_{i} \sum_{j \in s} \frac{(1 - p_i)(1 - p_j)(1 - p_i - p_j)}{(1 - p_i - p_j)^2} \left(\frac{y_i}{p_i} - \frac{y_j}{p_j}\right)^2 \tag{14.3.21}$$

14.3.1.4 Unbiased Ratio Estimator

For Lahiri−Midzuno−Sen (1951, 1952, 1953) sampling scheme, the probability of selecting a sample of size n is $p(s) = \frac{x_s}{M_1 X}$, where $x_s = \sum_{i \in s} x_i$, $X = \sum_{i \in U} x_i$, $M_1 = \binom{N-1}{n-1}$, and $x_i(>0)$ is the size measure for the ith units (see Section 5.5). An unbiased estimator for the population total Y is given by

$$\widehat{Y}_{lms} = \frac{y_s}{x_s} X \text{ with } y_s = \sum_{i \in s} y_i$$

If $y_i = kx_i$ then the estimator $\widehat{Y}_{lms}$ becomes equal to kX, which is a constant and hence the variance of $\widehat{Y}_{lms}$ becomes zero. Here $c_{si} = X/x_s$ and

$$\beta_{ij} = X^2 \sum_{s \supset i,j} \frac{1}{x_s^2} p(s) - 1 = \frac{X}{M_1} \sum_{s \supset i,j} \frac{1}{x_s} - 1 \tag{14.3.22}$$

So the variance of $\widehat{Y}_{lms}$ can be written as

$$V_{lms} = V\left(\widehat{Y}_{lms}\right) = -\frac{1}{2} \sum_{i \neq} \sum_{j \in U} \beta_{ij} x_i x_j \left(\frac{y_i}{x_i} - \frac{y_j}{x_j}\right)^2 \tag{14.3.23}$$

From Eq. (14.3.23), we immediately write an unbiased estimator of V_{lms} as

$$\widehat{V}_{lms}(1) = -\frac{1}{2} \sum_{i \neq} \sum_{j \in s} \frac{\beta_{ij} x_i x_j}{\pi_{ij}} \left(\frac{y_i}{x_i} - \frac{y_j}{x_j}\right)^2 \tag{14.3.24}$$

where β_{ij} is given in Eq. (14.3.22).

Furthermore, writing $\beta_{ij} = \sum_{s \supset i,j} \left(\frac{1}{x_s} \frac{X}{M_1} - \frac{1}{M_2}\right)$ with $M_2 = \binom{N-2}{n-2}$, we can choose

$$\beta_{ij}(s) = \frac{1}{p(s)} \left(\frac{1}{x_s} \frac{X}{M_1} - \frac{1}{M_2}\right) = \frac{X}{x_s} \left(\frac{X}{x_s} - \frac{N-1}{n-1}\right) \tag{14.3.25}$$

so that $\sum_{s \supset i,j} \beta_{ij}(s) p(s) = \beta_{ij}$. The choice of $\beta_{ij}(s)$ in Eq. (14.3.25) yields an alternative unbiased estimator of V_{lms} as

$$\widehat{V}_{lms}(2) = \frac{1}{2} \frac{X}{x_s} \left(\frac{N-1}{n-1} - \frac{X}{x_s}\right) \sum_{i \neq} \sum_{j \in s} x_i x_j \left(\frac{y_i}{x_i} - \frac{y_j}{x_j}\right)^2 \tag{14.3.26}$$

Rao and Vijayan (1977) proved that none of the estimators $\widehat{V}_{lms}(1)$ and $\widehat{V}_{lms}(2)$ are always nonnegative. No nonnegative variance estimators of V_{lms} are as yet available.

14.3.1.5 Ordinary Ratio Estimator

The ratio estimator for the population total Y based on a fixed sample s of size n under simple random sampling without replacement (SRSWOR) sampling is

$$\widehat{Y}_R = \frac{y_s}{x_s} X$$

where $y_s = \sum_{i \in s} y_i$ and $x_s = \sum_{i \in s} x_i$.

Here $c_{si} = (X/x_s)I_{si}$ and $\widehat{Y}_R$ becomes a constant if y_i is proportional to x_i. Hence we can write the mean square error of $\widehat{Y}_R$ as

$$M(\widehat{Y}_R) = -\frac{1}{2}\sum_{i\neq}\sum_{j\in U}\beta_{ij}x_ix_j\left(\frac{y_i}{x_i}-\frac{y_j}{x_j}\right)^2 \tag{14.3.27}$$

where

$$\beta_{ij} = E(c_{si}c_{sj}) - E(c_{si}) - E(c_{sj}) + 1$$
$$= X^2E\left(\frac{I_{si}I_{sj}}{x_s^2}\right) - XE\left(\frac{I_{si}}{x_s}\right) - XE\left(\frac{I_{sj}}{x_s}\right) + 1$$

For SRSWOR $p(s) = 1\Big/\binom{N}{n}$ and hence we can write

$$\beta_{ij} = \binom{N}{n}^{-1}\left[X^2\sum_{s\supset i,j}\frac{1}{x_s^2} - X\sum_{s\supset i}\frac{1}{x_s} - X\sum_{s\supset j}\frac{1}{x_s} + \binom{N}{n}\right]$$

An unbiased estimator of $M(\widehat{Y}_R)$ is given by

$$\widehat{M}(\widehat{Y}_R) = -\frac{1}{2}\sum_{i\neq}\sum_{j\in s}\frac{\beta_{ij}x_ix_j}{\pi_{ij}}\left(\frac{y_i}{x_i}-\frac{y_j}{x_j}\right)^2$$

$$\left(\text{for SRSWOR } \pi_{ij} = \frac{N(N-1)}{n(n-1)}\right)$$

$$= -\frac{1}{2}\frac{N(N-1)}{n(n-1)}\sum_{i\neq}\sum_{j\in s}\beta_{ij}x_ix_j\left(\frac{y_i}{x_i}-\frac{y_j}{x_j}\right)^2 \tag{14.3.28}$$

Remark 14.3.3

It is worth noting that the expressions of mean square error and its unbiased estimators given in Eqs. (14.3.27) and (14.3.28) are exact. The expressions of mean square error and its unbiased estimators given in Eqs. (8.4.7) and (8.4.8) are approximate and are valid when the sample size n is large.

14.3.1.6 Hartley—Ross Estimator

The Hartley—Ross (1954) estimator for the population total Y is given by

$$\widehat{Y}_{hr} = X\overline{r} + \frac{(N-1)n}{n-1}(\overline{y} - \overline{r}\,\overline{x}) \tag{14.3.29}$$

where $\bar{r}=\frac{1}{n}\sum_{i\in s}\frac{y_i}{x_i}$, $\bar{y}=\frac{1}{n}\sum_{i\in s}y_i$, $\bar{x}=\frac{1}{n}\sum_{i\in s}x_i$, and s is a sample of size n selected by SRSWOR method.

We can write

$$\widehat{Y}_{hr}=\sum_{i\in s}c_{si}y_i$$

where $c_{si}=\frac{1}{x_i}\left(\frac{X}{n}-\frac{N-1}{n-1}\bar{x}\right)+\frac{N-1}{n-1}$.

Because $\widehat{Y}_{hr}$ becomes constant when $y_i \propto x_i$, the variance of $\widehat{Y}_{hr}$ can be written as

$$V\left(\widehat{Y}_{hr}\right)=-\frac{1}{2}\sum_{i\neq}\sum_{j\in U}\beta_{ij}x_ix_j\left(\frac{y_i}{x_i}-\frac{y_j}{x_j}\right)^2 \tag{14.3.30}$$

where $\beta_{ij}=E(c_{si}-1)(c_{sj}-1)$.

The expression of β_{ij} is very complex and was obtained by Robson (1957).

An unbiased estimator of $V\left(\widehat{Y}_{hr}\right)$ is

$$\widehat{V}\left(\widehat{Y}_{hr}\right)=-\frac{1}{2}\frac{N(N-1)}{n(n-1)}\sum_{i\neq}\sum_{j\in s}\beta_{ij}x_ix_j\left(\frac{y_i}{x_i}-\frac{y_j}{x_j}\right)^2 \tag{14.3.31}$$

However, nonnegativity of the estimator $\widehat{V}\left(\widehat{Y}_{hr}\right)$ is not easy to check.

14.4 EXERCISES

14.4.1 Let $t(s)=\sum_{i\in s}b_{si}y_i$ be an unbiased estimator of the population total Y based on a sample s of size n selected with probability $p(s)$ and $b_{si}=0$ for $i\notin s$. Show that the following estimators are unbiased for the variance of $t(s)$.

(i) $\widehat{V}_1=\{t(s)\}^2-\left(\sum_{i\in s}b_{si}y_i^2+\sum_{i\neq}\sum_{j\in s}\frac{b_{si}b_{sj}}{\delta_{ij}}y_iy_j\right)$

(ii) $\widehat{V}_2=\{t(s)\}^2-\frac{1}{p(s)}\left(\sum_{i\in s}\frac{y_i^2}{M_i}+\sum_{i\neq}\sum_{j\in s}\frac{y_iy_j}{M_{ij}}\right)$

(iii) $\widehat{V}_3=\sum_{i\in s}b_{si}(b_{si}-1)y_i^2+\sum_{i\neq}\sum_{j\in s}b_{si}b_{sj}\left(1-\frac{1}{\delta_{ij}}\right)y_iy_j$

where $\delta_{ij}=E(b_{si}b_{sj})$, M_i and M_{ij} are, respectively, the total number of samples containing ith, and ith and $j(\neq i)$th units, and all are positive.

14.4.2 Show that the following are unbiased estimators of $V\left(\widehat{Y}_{ht}\right)$ based on a fixed effective size n sampling design.

(i) $\widehat{V}_1 = \frac{1}{2}\sum_{i\neq}\sum_{j\in s}\left(\frac{\pi_i\pi_j - \pi_{ij}}{\pi_{ij}}\right)(z_i - z_j)^2$

(ii) $\widehat{V}_2 = \frac{1}{2}\sum_{i\neq}\sum_{j\in s}\left(\frac{\pi_i\pi_j}{\pi_{ij}} - \frac{\pi_{ij}}{M_{ij}p(s)}\right)(z_i - z_j)^2$

(iii) $\widehat{V}_3 = \frac{1}{2p(s)}\sum_{i\neq}\sum_{j\in s}\left(\frac{\pi_i\pi_j - \pi_{ij}}{M_{ij}}\right)(z_i - z_j)^2$

(iv) $\widehat{V}_4 = \frac{1}{p(s)}\left[\sum_{i\in s}\left(\frac{1-\pi_i}{M_i\pi_i}\right)y_i^2 + \sum_{i\neq}\sum_{j\in s}\left(\frac{\pi_{ij} - \pi_i\pi_j}{M_{ij}\pi_i\pi_j}\right)y_iy_j\right]$

where $z_i = y_i/\pi_i$ and M_i and M_{ij} are as in Exercise 14.4.1.

14.4.3 Let s be a fixed sample of size n distinct units with selection probability $p(s)$ and α and β are fixed constants in [0,1]. Show that

$$\widehat{V}_{ht} = \frac{1}{2}\sum_{i\neq}\sum_{j\in s}\left(\frac{y_i}{\pi_i} - \frac{y_j}{\pi_j}\right)^2\left[\pi_i\pi_j\left\{\frac{\alpha}{p(s)M_{ij}} + \frac{1-\alpha}{\pi_{ij}}\right\} - \pi_{ij}\left\{\frac{\beta}{p(s)M_{ij}} + \frac{1-\beta}{\pi_{ij}}\right\}\right]$$

is an unbiased estimator of the variance of $\widehat{Y}_{ht} = \sum_{i\in s}\frac{y_i}{\pi_i}$.

Furthermore, for the choices $\alpha = 0, \beta = 1$ and $\alpha = 1, \beta = 0$, show that sufficient condition for nonnegativity of $\widehat{V}_{ht}$ is $p(s) > \frac{\pi_{ij}^2}{\pi_i\pi_jM_{ij}}$ $\forall s \supset i, j$ and $p(s) < (\pi_i\pi_j)/M_{ij}$ $\forall s \supset i, j$, respectively, where M_{ij} is defined in Exercise 14.4.1 (Chaudhuri, 1981).

14.4.4 Show that the variance of the unbiased ratio estimator $\widehat{Y}_{lms} = \frac{y_s}{x_s}X$ for the population total Y based on an LMS can be estimated unbiasedly using the formula

$$\widehat{V}_{lms} = \frac{1}{2}\sum_{i\neq}\sum_{j\in s}p_ip_j\left(\frac{y_i}{p_i} - \frac{y_j}{p_j}\right)^2\left[\left\{\frac{\alpha}{\pi_{ij}} + \frac{(1-\alpha)}{M_{ij}p(s)}\right\} - \left\{\frac{\beta}{M_{ij}p(s)} + \frac{1-\beta}{\pi_{ij}}\right\}\frac{1}{M_i^2}\sum_{s\supset i,j}\frac{1}{p(s)}\right]$$

where M_i, M_{ij}, and α, β are as in Exercises 14.4.1 and 14.4.3, respectively (Chaudhuri, 1981).

14.4.5 Let s be a sample of size n selected with probability $p(s)$ by LMS sampling using x_i as a measure of size for the ith unit ($i = 1,\ldots, N$) of a population U of size N. Show that

(i) $t = \dfrac{y_s}{x_s}$ is an unbiased estimator for the population ratio $R = Y/X$, where $y_s = \sum_{i\in s} y_i$ and $x_s = \sum_{i\in s} x_i$,

(ii) The variance of t can be written as follows:

(a) $V_1(t) = \sum_{i\in U} \lambda_i y_i^2 + \sum\sum_{i\neq j\in U} \lambda_{ij} y_i y_j$

(b) $V_2(t) = \sum_{i\in U} (nT_i - N) r_i^2 + \sum\sum_{i< j\in U} (1 - T_{ij})(r_i - r_j)^2$

(c) $V_3(t) = \frac{1}{2} \sum\sum_{i\neq j\in U} Q_{ij}$

(d) $V_4(t) = \frac{1}{2} \sum\sum_{i\neq j\in U} R_{ij}$

where

$$\lambda_i = \frac{1}{MX}\sum_{s\supset i}\frac{1}{x_s} - \frac{1}{X^2},\ \lambda_{ij} = \frac{1}{MX}\sum_{s\supset i,j}\frac{1}{x_s} - \frac{1}{X^2},$$

$$M = \binom{N-1}{n-1},\ r_i = y_i/x_i,\ T_i = 1 + \lambda_i X^2,\ X = \sum_{i\in U} x_i,$$

$$T_{ij} = 1 + \lambda_{ij} X^2,\ Q_{ij} = \frac{T_i - 1}{N-1} r_i^2 + 2(T_{ij} - 1) r_i r_j + \frac{T_j - 1}{N-1} r_j^2 \text{ and}$$

$$R_{ij} = \left(\frac{T_{ij}}{n-1} - \frac{1}{N-1}\right) r_i^2 + 2(T_{ij} - 1) r_i r_j + \left(\frac{T_{ij}}{n-1} - \frac{1}{N-1}\right) r_j^2.$$

(iii) Show that $\widehat{V}_j(t)$ is unbiased for $V_j(t)$ for $j = 1, 2, 3, 4$, where

$$\widehat{V}_1(t) = \sum_{i\in U} \lambda_i y_i^2/\pi_i + \sum_{i\neq}\sum_{j\in U} \lambda_{ij} y_i y_j/\pi_{ij},$$

$$\widehat{V}_2(t) = \sum_{i\in s} (nT_i - N) r_i^2/\pi_i + \sum_{i<}\sum_{j\in s} (1 - T_{ij})(r_i - r_j)^2/\pi_{ij},$$

$$\widehat{V}_3(t) = \frac{1}{2}\sum_{i\neq}\sum_{j\in s} Q_{ij}/\pi_{ij} \text{ and } \widehat{V}_4(t) = \frac{1}{2}\sum_{i\neq}\sum_{j\in s} R_{ij}/\pi_{ij}$$

(π_i is inclusion probability for the ith unit and π_{ij} is the inclusion probability for the ith and $j(\neq i)$th units)

(iv) In case $y_i \geq 0$ for $i = 1,\ldots, N$, a sufficient condition of nonnegativity of $\widehat{V}_1(t)$ is $\lambda_{ij} \geq 0$ (Rao, 1972).
(v) Sufficient condition for nonnegativity of $\widehat{V}_2(t)$ is $T_i \geq N/n$ and $T_{ij} \leq 1$ for $i \neq j = 1,\ldots, N$ (Chaudhuri, 1976).
(vi) Sufficient condition for nonnegativity of $\widehat{V}_3(t)$ is $Q_{ij} \geq 0$ for $i \neq j = 1,\ldots, N$ (Rao, T. J., 1977a,b).
(vii) Sufficient condition for nonnegativity of $\widehat{V}_4(t)$ is $T_{ij} \geq \dfrac{n-1}{N-1}$ and $(T_{ij} - 1)^2 - \left(\dfrac{T_{ij}}{n-1} - \dfrac{1}{N-1}\right)^2 \leq 0$ for $i \neq j = 1,\ldots, N$ (Rao, T. J., 1977a,b).
(viii) Show that the sufficient conditions listed in (iv) to (vii) can never hold for $i \neq j = 1,\ldots, N$ except under trivial situation when $p_i = x_i/X = 1/N\ \forall i = 1, \ldots, N$ (Chaudhuri and Arnab, 1981).

CHAPTER 15

Nonsampling Errors

15.1 INTRODUCTION

Complete enumeration and sample surveys are conducted to estimate unknown population parameters. None of these procedures provide exact results. In sample surveys, we observe a part of the population and draw inferences on the basis of the observed data; hence an error is committed as the entire population is not observed. This type of error is known as the sampling error. Clearly, the sampling error is absent in a complete enumeration or census procedure. However, when we collect information from a unit in a sample survey or complete enumeration, the information regarding the value of the characteristic under study is not free from error. When we estimate population parameters from survey data, all the errors are accumulated and incorrect values of the parameters are obtained. The combination of all the kinds of errors, other than the sampling error, for which one cannot obtain the true value of the parameter by conducting a survey is known as the nonsampling error. The nonsampling error is present in both the sample survey and complete enumeration, while the sampling error is present in sample survey only. The sampling error is determined by the magnitude of mean square error, but no general method of determining the nonsampling error is available. The sampling error decreases with increase in sample size, while the nonsampling error increases with the increase in sample size. The nonsampling error arises at any stage of conducting a survey such as planning, data collection, processing the data, and tabulation as well report writing.

15.2 SOURCES OF NONSAMPLING ERRORS

The sources of nonsampling errors are numerous. So, it is very difficult to give an exhaustive list. Some of the sources are listed here. Good details are given by Murthy (1977).

1. **Inappropriate sampling frame**: In conducting a survey, by either sample survey or complete enumeration, one should completely specify the coverage of the survey and prepare a list of all the units of the target

Survey Sampling Theory and Applications
ISBN 978-0-12-811848-1
http://dx.doi.org/10.1016/B978-0-12-811848-1.00015-7

population, which is technically known as the sampling frame. The list should be accurate with no duplication or omission of units. In a large-scale household survey, sometimes, some of the households on the list either do not exist or are not included in the list. Hence errors arise from the inaccuracy of the sampling frame.

2. **Response error/measurement error**: Sometimes because of error in the instrument, investigators report untrue values. In a crop cutting experiment, enumerators often report eye estimations of the crop yield, which is highly subjective and erroneous. This type of error is called measurement error. Response errors constitute wrong values obtained from the respondents. This error can be caused when the wrong method of interview is employed, for example, when the interviewers influence the respondents' answers. Respondents have a tendency to report ages as even or as multiples of five. Some items in the questionnaire may be difficult to answer because of inappropriate definition, failing memory, or even lack of knowledge of the respondents. For example, respondents are often unable to recall accurately the total amount spent on clothing in the last year. Some questions are of sensitive nature, such as income and HIV infection status, where informants may report untrue values. Response errors occur when the design of questionnaire is inappropriate. An overly long questionnaire may result in the respondent becoming tired and answering false or inaccurately to finish the survey faster. Language barrier poses a threat in the situation when the interviewer and the respondent communicate via different languages. Here, the interviewer may enter a value different from the true value. Investigators may not find the right persons for interview. For example, investigators may collect data from housewives or children for household surveys, which may be inappropriate and may cause errors.
3. **Error in data processing**: Errors can also be committed while entering data into the computer, at the scrutiny stage, as well when inappropriate formulae are used. These are all sources of nonsampling errors.
4. **Nonresponse error**: Nonresponse means failure to gather information on all or some of the items in a schedule or questionnaire. They mainly occur for two reasons: noncoverage and nonresponse. Noncoverage occurs in situations where some of the target population has no chance of being selected into the sample; this is mainly due to an incomplete sampling frame. In a questionnaire, we collect information on several items. For example, in household surveys, information on household size and

composition, age, education level, types of dwelling, income, and consumptions of different items is collected, among others. If all the items of information of a questionnaire or schedule are missing, we call it unit nonresponse, and if information on some of the items is obtained and on the rest is missing, then we call this item nonresponse. Nonresponse occurs for various reasons. Unit nonresponse may occur if the informant is not at home, is busy for other business, refuses to answer (termed hardcore), enumerators fail to make prior appointment or to locate the unit, etc. Item response may occur mainly if the respondent does not know the answer, feels the item of information is sensitive (such as status of HIV/AIDS, income), finds it difficult or does not know the answer, or the enumerator forgets to write down the answer correctly.

15.3 CONTROLLING OF NONSAMPLING ERRORS

It is difficult to have 100% control of nonsampling errors; however, we can take various steps to control them. The sampling frame should be accurate. Every unit of the population should be easily identifiable. Definitions of each of the items of investigations such as households and family should be provided unambiguously. Investigators should be provided with proper training, explaining the meaning and possible answers to each question they might ask. There should be proper supervision of enumerators to ensure collection of quality data. The questionnaire should be designed properly. The order of questioning should follow a logical sequence with easy and nonsensitive questions coming first to make the respondent comfortable. Related questions should be grouped together. The length of the questionnaire should be as short as possible. Questionnaires should have a method of consistency check whenever possible. Before conducting a survey, a pilot survey should be undertaken to test the questionnaire, time taken to administer it, cost of the proposed survey, administration of the survey, etc. The analysis of the questionnaire will help in checking the accuracy of the tables and other anticipated results. After conducting the survey, the filled questionnaires should be scrutinized properly. At least 5% of the collected data should be recollected again by different investigators to check the quality of data collected.

15.4 TREATMENT OF NONRESPONSE ERROR

In almost all large-scale surveys, nonresponse is inevitable. Nonresponse creates bias in the estimates. The amount of bias often increases with the rate of nonresponse. In household income surveys, response from the high income group is considerably lower than the other groups and hence underrepresented in the sample. These types of data therefore produce bias in the estimation of the mean income. Researchers very often take a larger sample size than the required number, taking into account that some of them will not respond [vide Lohr, 1999]. This practice generally produces erroneous results because some of the groups of the units are then underrepresented in the sample. Several methods of handling nonresponse problems in sample surveys are available in the literature. The method of poststratification, response modeling, imputation, and randomized response techniques are popular. Good details are available in Rubin (1987).

15.4.1 Poststratification

15.4.1.1 Hansen–Hurwitz Method

Let a sample s of size n be selected by simple random sampling without replacement (SRSWOR) method. Suppose n_1 units responded and remaining $n_2(=n-n_1)$ did not respond at the first attempt. The set of responded and nonresponded units will be denoted, respectively, by s_1 and s_2. A simple random subsample $\overset{*}{s}_2$ of size $m=\nu n_2$ (assuming integer) is selected from s_2 where ν is a known fraction. Responses from all the units of $\overset{*}{s}_2$ are obtained by using more intensive method, which is obviously expensive. Hansen and Hurwitz (1946) assumed that the population U under study of size N can be divided into two strata according to the nature of respondents. The first stratum H_1 consisting of N_1 (unknown) units that respond at the first attempt and the remaining set of units $N_2(=N-N_1)$ that do not respond at the first attempt comprise the stratum H_2. It is assumed that the members who belong to H_1 always respond at the first attempt while members of H_2 never respond at the first attempt, but they will always respond if a more persuasive method is employed.

Let $\overline{y}(s_1)$ and $\overline{y}\left(\overset{*}{s}_2\right)$ be the sample means of the variable under study y for the samples s_1 and $\overset{*}{s}_2$, respectively. Let

$$\overline{y}_w = w_1\overline{y}(s_1) + w_2\overline{y}\left(\overset{*}{s}_2\right) \tag{15.4.1}$$

with $w_1 = n_1/n$ and $w_2 = n_2/n$.

Let $\overline{Y}$, S_y^2, S_{1y}^2, and S_{2y}^2 denote the population mean, population variance, population variances of response, and nonresponse strata, respectively.

Substituting $n' = n$, $\gamma_1 = 1$, $\gamma_2 = \nu$, $n_2 = m$ in Theorem 10.3.2, we obtain the following result.

Theorem 15.4.1

(i) $\overline{y}_w$ is unbiased for $\overline{Y}$

(ii) The variance of $\overline{y}_w$ is

$$V(\overline{y}_w) = \left(\frac{1}{n} - \frac{1}{N}\right)S_y^2 + \frac{W_2}{n}\left(\frac{1}{\nu} - 1\right)S_{2y}^2$$

(iii) An unbiased estimator of $V(\overline{y}_w)$ is

$$\begin{aligned}\widehat{V}(\overline{y}_w) &= \frac{(N-n)(n_1-1)}{N(n-1)} w_1 \frac{\widehat{S}_{1y}^2}{n_1}\\ &+ \frac{(N-1)(n_2-1)-(n-1)(m-1)}{N(n-1)} w_2 \frac{\widehat{S}_{2y}^2}{m}\\ &+ \frac{N-n}{N(n-1)}\left[w_1\left\{\overline{y}(s_1) - \overline{y}_w\right\}^2 + w_2\left\{\overline{y}(s_2^*) - \overline{y}_w\right\}^2\right]\end{aligned}$$

where $W_2 = N_2/N$, $\widehat{S}_{1y}^2$ and $\widehat{S}_{2y}^2$ denote the sample variance of response and nonresponse stratum, respectively.

Remark 15.4.1

The bias of $\overline{y}(s_1)$ (sample mean of the response stratum) for estimating $\overline{Y}$ is given by

$$\begin{aligned}B[\overline{y}(s_1)] &= E[\overline{y}(s_1)] - \overline{Y}\\ &= \overline{Y}_1 - \overline{Y}\\ &= W_2\left(\overline{Y}_1 - \overline{Y}_2\right)\end{aligned} \tag{15.4.2}$$

where $\overline{Y}_1$ and $\overline{Y}_2$ denote the population mean of response and nonresponse stratum, respectively. The bias in Eq. (15.4.2) is negligible if at least one of the quantities W_2 and $\left(\overline{Y}_1 - \overline{Y}_2\right)$ is negligible. From Theorem 15.4.1, we note that estimator $\overline{y}_w$ is less efficient than the sample mean based on all the n observed, and the loss of efficiency is negligible if W_2 or S_{2y}^2 is small.

15.4.1.1.1 Optimum Value of ν and *n*

Let us consider the cost function of the form

$$C = c_0 n + c_1 n_1 + c_2 m \tag{15.4.3}$$

where c_0 is the fixed per sampled unit, c_1 is the cost per unit to acquire information from the respondent stratum, and c_2 is the cost per unit of acquiring information from the nonresponse stratum. Here, the cost C is a random variable and the expected cost for a given n under SRSWOR is

$$C^* = E(C) = n(c_0 + W_1 c_1 + W_2 \nu c_2) \tag{15.4.4}$$

The optimum values of n and ν may be obtained either by minimizing the expected cost C^* keeping $V(\overline{y}_w)$ to a certain level V_0 or by minimizing $V(\overline{y}_w)$ keeping C^* to a certain level C_0^*.

For minimization consider

$$\left(V(\overline{y}_w)+\frac{S_y^2}{N}\right)C^*=\left(\frac{S_y^2-W_2S_{2y}^2}{n}+\frac{1}{\nu}\frac{W_2}{n}S_{2y}^2\right)\{n(c_0+W_1c_1)+nW_2\nu c_2\} \tag{15.4.5}$$

Using Cauchy's inequality, we find Eq. (15.4.5) attain a minimum if

$$\frac{S_y^2-W_2S_{2y}^2}{(c_0+W_1c_1)}=\frac{S_{2y}^2}{\nu^2c_2}$$

$$\text{i.e.,}\quad Opt\ \nu=\nu_0=\sqrt{\frac{(c_0+c_1W_1)S_{2y}^2}{\left(S_y^2-W_2S_{2y}^2\right)c_2}} \tag{15.4.6}$$

Because ν_0 is independent of n, the optimum value of n that minimizes the cost C^* for a given variance $V(\overline{y}_w)=V_0$ is obtained from the equation

$$V_0=\left(\frac{1}{n}-\frac{1}{N}\right)S_y^2+\frac{W_2}{n}\left(\frac{1}{\nu_0}-1\right)S_{2y}^2$$

$$\text{i.e.,}\quad Opt\ n=n_0=\frac{S_y^2+(1/\nu_0-1)W_2S_{2y}^2}{V_0+S_y^2/N} \tag{15.4.7}$$

Similarly, the optimum n that minimizes $V(\overline{y}_w)$ keeping cost $C=C_0^*$ can be formulated as

$$n_0=\frac{C_0^*}{c_0+W_1c_1+W_2\nu_0c_2}.$$

In case $\nu_0\geq 1$, all the nonrespondents should be interviewed. If $S_{2y}^2=S_y^2$, $\nu_0=\sqrt{\frac{(c_0+c_1W_1)}{c_2W_1}}$ may exceed 1 unless c_2 is much larger than c_1.

The Hansen and Hurwitz's (1946) technique was generalized further by Dalenius (1955), El-Badry (1956), Kish and Hess (1958), and Srinath (1971), among others. Särndal et al. (1992) poststratified the initial stratum by more than two, using information related to the interviewers' background and skills as well as the characteristics of respondents such as sex and dwelling conditions.

15.4.2 Use of Response Probabilities

The formation of a response sample, comprising the set of units from which all the responses are obtained, is totally unknown. Hence probabilistic models are proposed to describe the unknown response distributions. Politz and Simmons (1949) computed the response probability of a respondent as the proportion of time staying at home. The response probability may be directly related to the study variable and hence to the auxiliary variable, which is highly related to the study variable. In household income surveys, people of high income may respond with low probability and hence may be underrepresented in the sample. So, if tax return is considered as an auxiliary variable, then the response probability of an individual may be inversely proportional to the amount of the tax return. Let us define an indicator variable I_i attached to the ith unit such that

$$I_i = \begin{cases} 1 & \text{if } i\text{th unit responds} \\ 0 & \text{otherwise} \end{cases} \tag{15.4.8}$$

In response modeling, we assume

$$P\{I_i = 1\} = \theta_i \tag{15.4.9}$$

The response probability θ_i may depend on a set of known p auxiliary information $\mathbf{x}_i = (x_{1i}, \ldots, x_{ji}, \ldots, x_{pi})$ or the value of the study variable y_i. We classify missing observations into following categories.

15.4.2.1 Classification of Response Probabilities

Missing completely at random (MCAR): If the probability θ_i is independent of the study variable y, then the missing data are called MCAR. Here, missingness does not depend on the values of the data set, observed or unobserved. The missing data are just random subset of the data. For example, if someone lost a schedule, then it may be replaced by a schedule taking at random from the set of filled schedules.

Missing at random (MAR): If the response probability θ_i depends on the observed values of y, but not on the missing values of y, then the missing data are called MAR. Here, missingness does not depend on the unobserved values of the data set but does depend on the observed. MAR is also known as ignorable nonresponse where a model can be used to determine the response mechanism. In household surveys, if consumption on food is missing but household size is not missing, then we estimate missing consumption on food by fitting a linear regression of the consumption of food on the household size based on the observed data where both the consumption of food and household sizes are available.

Not missing at random (NMAR): If θ_i depends on the missing variable y_i, then the nonresponse is called NMAR. Here, missingness depends on the observed values of the data set and also the missing value. NMAR is known as nonignorable nonresponse. For example, respondents with high income less likely to report income or respondents infected with HIV/AIDS respond less likely to report their HIV/AIDS status.

15.4.3 Politz and Simmons Method

Politz and Simmons (1949, 1950) provided a method of finding response probabilities of the sampled households selected by simple random sampling with replacement (SRSWR) method from a population U of N households. The interviewer made a call only once to each of the selected households during the interviewing hours. The time of call may be regarded as random within the interviewing hours. If the respondent is available, then information on the variable of interest y is collected along with the information on how many days in the preceeding 5 consecutive days he/she was at home at this particular point of time. If the respondent is not available, no information from this household is collected. Here, it is assumed that if the respondent is available, then the respondent responds with probability 1. The probability of the ith sampled person being at home θ_i is estimated by $\widehat{\theta}_i = \dfrac{t_i + 1}{6}$, where t_i is the number of times the ith respondent reported to be at home in the last 5 consecutive days. The probability t_i is equal to t, given by

$$P(t_i = t) = \binom{5}{t}\theta_i^t(1-\theta_i)^{5-t}; \quad t = 0, \ldots, 5 \tag{15.4.10}$$

Let $E_h(\cdot|i)$ denote the conditional expectation given that ith respondent was at home.

Then

$$\begin{aligned} E_h\left(\frac{1}{\widehat{\theta}_i}\middle| i\right) &= \sum_{t=0}^{5}\frac{6}{t+1}\binom{5}{t}\theta_i^t(1-\theta_i)^{5-t} \\ &= \sum_{t=0}^{5}\binom{6}{t+1}\theta_i^t(1-\theta_i)^{6-(t+1)} \\ &= \frac{1}{\theta_i}\sum_{x=1}^{6}\binom{6}{x}\theta_i^x(1-\theta_i)^{6-x} \\ &= \frac{1}{\theta_i}\left(1-q_i^6\right) \end{aligned} \tag{15.4.11}$$

(where $q_i = 1 - \theta_i$)

and

$$E_h\left(\frac{1}{\widehat{\theta}_i^2}\middle| i\right)=\sum_{t=0}^{5}\left(\frac{6}{t+1}\right)^2\binom{5}{t}\theta_i^t(1-\theta_i)^{5-t}$$

$$=\sum_{t=0}^{5}\frac{6}{t+1}\binom{6}{t+1}\theta_i^t(1-\theta_i)^{6-(t+1)} \tag{15.4.12}$$

$$=\frac{6}{\theta_i}\sum_{x=1}^{6}\frac{1}{x}\binom{6}{x}\theta_i^x(1-\theta_i)^{6-x}$$

$$=\frac{6}{\theta_i}\delta_i$$

with $\delta_i=\sum_{x=1}^{6}\frac{1}{x}\binom{6}{x}\theta_i^x(1-\theta_i)^{6-x}$.

Politz and Simmons proposed the following biased estimator of the population mean $\overline{Y}$ based on an SRSWR sample of size n:

$$\widehat{\overline{Y}}_{ps}=\frac{1}{n}\sum_{j=1}^{n}\frac{y_{(j)}}{\widehat{\theta}_{(j)}}I(j) \tag{15.4.13}$$

where $I_{(j)}$ is an indicator variable taking the value 1 if the unit selected at the jth draw was at home with probability $\theta_{(j)}$ and 0 otherwise, i.e., if the jth draw produces the ith unit, we get $y_{(j)}=y_i$ and $\widehat{\theta}_{(j)}=\widehat{\theta}_i$.

Theorem 15.4.2

(i) $E\left(\widehat{\overline{Y}}_{ps}\right)=\overline{Y}-\frac{1}{N}\sum_{i\in U}y_iq_i^6$

(ii) $V\left(\widehat{\overline{Y}}_{ps}\right)=\frac{\sigma_{ps}^2}{n}$

(iii) $\widehat{V}\left(\widehat{\overline{Y}}_{ps}\right)=\frac{1}{n(n-1)}\sum_{i\in s}\left(\frac{y_i}{\widehat{\theta}_i}-\widehat{\overline{Y}}_{ps}\right)^2$

where $q_i=1-\theta_i$, $\sigma_{ps}^2=\frac{6}{N}\sum_{i\in U}\delta_iy_i^2-\left(\frac{1}{N}\sum_{i\in U}y_i\left(1-q_i^6\right)\right)^2$, and

$\delta_i=\sum_{x=1}^{6}\frac{1}{x}\binom{6}{x}\theta_i^x(1-\theta_i)^{6-x}$.

Proof

Let $E_h\{\cdot|(j)\}$, E_r, and E_p denote the conditional expectation given that unit selected at the jth draw was at home, unconditional expectation with respect to the person selected at the jth draw was at home and sampling design. Then noting $E_r(I_{(j)}) = \theta_{(j)}$, we get

$$E\left(\frac{y_{(j)}}{\widehat{\theta}_{(j)}} I_{(j)}\right) = E_p\left[y_{(j)}\left\{E_h \frac{1}{\widehat{\theta}_{(j)}}\bigg|(j)\right\}E_r\{I_{(j)}\}\right]$$
$$= E_p\left[y_{(j)}\theta_{(j)}E_h\left\{\frac{1}{\widehat{\theta}_{(j)}}\right\}\right]$$
$$= E_p\left(y_{(j)}\left(1 - q_{(j)}^6\right)\right) \tag{15.4.14}$$
$$= \frac{1}{N}\sum_{i\in U} y_i\left(1 - q_i^6\right)$$
$$= \overline{Y} - \frac{1}{N}\sum_{i\in U} y_i q_i^6$$

$$E\left(\frac{y_{(j)}}{\widehat{\theta}_{(j)}} I_{(j)}\right)^2 = E_p E_h\left(\left(\frac{y_{(j)}}{\widehat{\theta}_{(j)}}\right)^2\bigg|(j)\right)E_r\left(I_{(j)}\right)$$
$$= 6E_p\left[y_{(j)}^2\delta_{(j)}\right] \tag{15.4.15}$$
$$= \frac{6}{N}\sum_{i\in U} y_i^2\delta_i$$

and

$$V\left(\frac{y_{(j)}}{\widehat{\theta}_{(j)}} I_{(j)}\right) = \frac{6}{N}\sum_{j\in U} y_i^2\delta_i - \left(\overline{Y} - \frac{1}{N}\sum_{i\in U} y_i q_i^6\right)^2 \tag{15.4.16}$$
$$= \sigma_{ps}^2$$

Finally, noting $\frac{y_{(j)}}{\widehat{\theta}_{(j)}} I_{(j)}$'s for $j = 1,\ldots, n$ are independently and identically distributed, we verify the theorem.

Remark 15.4.2

The bias of $\widehat{\overline{Y}}_{ps}$ is $-\frac{1}{N}\sum_{i\in U} y_i q_i^6$ and it is certainly negligible if the response probabilities θ_i's are high for every i.

15.4.4 Imputation

Imputation is used for item nonresponse. Here we assign one or more values to a missing item to reduce the bias and control variance due to nonresponse. If we assign a single value, it is called "single imputation." In "multiple imputations" we assign more than one value to a missing item. The data with the imputed values (single or multiple) are analyzed using a standard statistical package assuming imputed values are of true value. Single imputation analysis is very simple because we get a complete data set and assume no observation is missing. Conversely, in multiple imputations we generate more than one data set because more than one value is obtained from the missing units. From each of the data sets, a separate estimate of the parameter of interest is obtained. Finally, a single estimate and its standard error are obtained by combining all these separate estimates. The process of estimation based on multiple imputations is certainly more complex than single imputation. Statistical agencies generally find difficulty in using multiple imputations because of operational challenges in maintaining multiple data sets, especially in large-scale surveys. Several methods of imputations are available in the literature. A few of the popular methods are described as follows.

Deductive imputation: Here, the missing value is imputed through a consistency check—establishing relationship with other available items. For example, suppose the HIV status is missing but information is available that nobody in the household is suffering from HIV infection. So, the missing item of the status of HIV infection is imputed as negative. Deductive imputations can be seen as almost accurate.

Substitution: In this case, if the information of a unit is not available, then it is replaced by a unit nearest to it. In a household survey, if no information is available from the sampled household, then it is substituted by the next door neighbor. Sometimes a unit is substituted randomly from those units that are not selected in the original sample.

Cold deck imputation: The missing data are replaced using records from a recent past survey. Naturally, cold deck imputation cannot provide accurate information, as it is not based on the current survey.

Hot deck imputation: The sampled units are divided into classes using prior information. For example, in household surveys, structure and location of the house may be used to classify the household into the higher, middle, or lower income groups. In hot deck imputation a particular unit in the class is chosen whose entries are complete. Such a responding unit is called the "donor." The information of the donor is substituted for the missing items. If the donor is selected at random, the imputation is called "random hot deck imputation." On the other hand, if the selected donor closest to the recipient (the unit with the missing item), is selected on the basis of a suitable distance function, the imputation is known as "nearest hot deck imputation." In particular, if the information on consumption of food in a middle-class household for an adult Asian male is missing, then the donor should match with the economy level (middle class) age (adult), and race (Asian) as far as possible.

Mean imputation: The mean of the study variable of all possible donors are substituted for each of the missing items.

Ratio and regression imputations: Here, we assume that the auxiliary information is available for all the sampled units (both respondent and nonrespondent units). The value of the study variable y is available for the respondent units but missing for the nonrespondents. In this situation a suitable regression of the study variable on the auxiliary variables is obtained by using the study and auxiliary variables of the respondent units. In regression imputation, the missing observations are predicted using a fitted regression. In case both the study and auxiliary variables are quantitative, a multiple regression may be fitted. Logistic regression may be used if the study variable is binary or qualitative. For a single auxiliary information, the regression imputation reduces to "ratio imputation" if the regression line passes through the origin. The regression imputation may be classified into two categories: deterministic and random imputation. In deterministic regression imputation, the imputed value of y_i is obtained by fitting a regression model

$$y_i = f(\mathbf{x}_i, \boldsymbol{\beta}) + \sigma v_i^{1/2} \epsilon_i \qquad (15.4.17)$$

where $f(\cdot,\cdot)$ is a given function, $\mathbf{x_i} = (x_{1i},\ldots, x_{qi})'$ is a q vector auxiliary variable whose values are available for the response sample s_r and the nonresponse sample $s_m(=s - s_r)$, $\boldsymbol{\beta}$ is a q vector regression coefficient, σ is an unknown parameter, v_i is a known constant, and ϵ_i is an independent identically distributed random variable with mean zero and variance unity. In deterministic imputation the missing observation y_i is imputed as

$$y_i^* = f(x_i, \widehat{\boldsymbol{\beta}}); \quad i \in s_m \qquad (15.4.18)$$

where $f(\mathbf{x}_i, \widehat{\boldsymbol{\beta}})$ is the fitted regression obtained from the response sample s_r, $\widehat{\boldsymbol{\beta}}$ is an estimate of $\boldsymbol{\beta}$. So, in deterministic imputation, the imputation value will be unaffected if the sample s_r is unchanged.

Ratio imputation is a special case of deterministic imputation that uses a single auxiliary variable x_i with $f(\mathbf{x}_i, \boldsymbol{\beta}) = \beta x_i$ and $v_i = 1$. In this case $\widehat{\beta} = \frac{\sum_{i \in s_r} y_i}{\sum_{i \in s_r} x_i}$ and the imputed value of y_i is

$$y_i^* = \frac{\sum_{i \in s_r} y_i}{\sum_{i \in s_r} x_i} x_i \quad \text{for } i \in s_m \tag{15.4.19}$$

The deterministic imputation reduces to the mean imputation if $f(\mathbf{x}, \boldsymbol{\beta}) = \beta$ and $x_i = 1$. In this case

$$y_i^* = \sum_{i \in s_r} y_i / r = \overline{y}_r \quad \text{for } i \in s_m \tag{15.4.20}$$

The random imputation is the modification of the deterministic imputation where a random noise $\widehat{\epsilon}_i$ is added to the imputed y_i^* and it is given by

$$\widehat{y}_i = f(x_i, \widehat{\boldsymbol{\beta}}) + \widehat{\sigma} v_i^{1/2} \widehat{\epsilon}_i; \quad i \in s_m \tag{15.4.21}$$

where $\widehat{\sigma}$ is an estimate of σ obtained from the model (Eq. 15.4.17) and $\widehat{\epsilon}_i$ is a random sample either from the distribution of ϵ_i or from the standardized residuals obtained from the fitted regression (Eq. 15.4.17).

15.4.4.1 Problems of Imputation

The main demerit of imputation is that it underestimates variance if the response rate is not high enough. To illustrate this point, the following example of Chen et al. (2000) is quite useful.

Let a sample s of size n be selected from a population of size N by SRSWOR method and let s_r be the response sample of size r and the nonresponse sample $s - s_r$. Using mean imputation the missing value of jth unit y_j is imputed as

$$\overline{y}_r = \sum_{i \in s_r} y_i / r \quad \text{for } j \in s_m \tag{15.4.22}$$

The imputed estimator of the population mean and the imputed sample variance are given, respectively, by

$$\overline{y}_I = \frac{1}{n} \left\{ \sum_{i \in s_r} y_i + (n - r) \overline{y}_r \right\} \tag{15.4.23}$$

and

$$s_{yI}^2 = \frac{1}{n-1}\sum_{i\in s}\left(y_{i0} - \overline{y}_I\right)^2$$
$$= \frac{(r-1)}{(n-1)}s_{yr}^2 \tag{15.4.24}$$

where $s_{yr}^2 = \frac{1}{r-1}\sum_{i\in s_r}\left(y_{i0} - \overline{y}_r\right)^2$ and $y_{i0} = \begin{cases} y_i & \text{for } i \in s_r \\ \overline{y}_r = \sum_{i\in s_r} y_i/r & \text{for } i \in s_m. \end{cases}$

Under uniform response

$$E\left(s_{yI}^2\middle| r\right) = \frac{r-1}{n-1}S_y^2$$
$$\cong \frac{r}{n}S_y^2 \tag{15.4.25}$$

where S_y^2 = population variance.

If r is not large, Eq. (15.4.25) is much less than S_y^2. Thus the sample variance underestimates the population variance under mean imputation in the case that the response rate is not high. To overcome the underestimation of variance, Rao and Shao (1992), Shao and Sitter (1996), and Chen et al. (2000), among others, proposed alternative methods of imputation where the variance estimators do not underestimate the variance.

Example 15.4.1

The exam marks and CA (continuous assessment) for a sample of 7(=n) students of a class of 50(=N) students that are selected at random are given in the following table.

Student	CA (x)	Exam marks (y)
1	50	60
2	40	30
3	70	65
4	60	70
5	80	90
6	55	?
7	90	?

The exam marks of the sixth and seventh students are missing. Let the average marks of CA for the entire class is known as $\overline{X} = 70$.

Mean Imputation: Here, the sample mean of y based on the response sample is $\overline{y}_r = 63$. So the imputed values of y_6 and y_7 are $y_6^* = 63$ and

$\overset{*}{y}_7 = 63$. Hence the estimated population mean of the exam marks is $\widehat{\overline{Y}} = \overline{y}_r = 63$. The estimated standard error is

$$\sqrt{\left(\frac{1}{n}-\frac{1}{N}\right)s_{yI}^2} = \sqrt{\left(\frac{1}{n}-\frac{1}{N}\right)\frac{1}{n-1}\sum_{i\in s}\left(y_{i0}-\overline{y}_I\right)^2}$$
$$= \sqrt{\left(\frac{1}{7}-\frac{1}{50}\right)} \times 313.33 = 6.20$$

Ratio imputation: Here, $\overset{*}{y}_i = \dfrac{\sum_{i\in s_r} y_i}{\sum_{i\in s_r} x_i} x_i = 1.05x_i$.

So, the imputed values $\overset{*}{y}_6 = 1.05 \times 55 = 57.75$ and $\overset{*}{y}_7 = 1.05 \times 90 = 94.5$.

Regression imputation: Estimated linear regression based on the response sample s_r is $y = -12 + 1.25x$. So, the imputed values are $\overset{*}{y}_6 = -12 + 1.25 \times 55 = 56.75$ and $\overset{*}{y}_7 = -12 + 1.25 \times 90 = 100.5$.

Random imputation: Here, we impute the values of y_6 and y_7 from the response sample s_r by SRSWR method. Let the SRSW sample be (5, 3). Then the imputed values of y_6 and y_7 are, respectively, $\overset{*}{y}_6 = y_5 = 90$ and $\overset{*}{y}_7 = y_3 = 65$.

Nearest hot deck imputation: Here, $x_1 = 50$ and $x_4 = 60$ are nearest to $x_6 = 55$. So, we may impute y_6 either by y_1 or by y_4, i.e., $\overset{*}{y}_6 = y_1 = 60$ or $\overset{*}{y}_6 = y_4 = 70$. Similarly, $\overset{*}{y}_7 = y_5 = 90$.

Random Regression imputation:

Method 1: The residuals are $\widehat{\in}_1 = 9.5$, $\widehat{\in}_2 = -8$, $\widehat{\in}_3 = -10.5$, $\widehat{\in}_4 = 7.0$, and $\widehat{\in}_5 = 2.0$. We choose two residuals at random with replacement. Suppose the selected residuals are $\widehat{\in}_2$ and $\widehat{\in}_1$. The imputed values of y_6 and y_7 are $\overset{*}{y}_6 = 56.75 + \widehat{\in}_2 = 48.75$ and $\overset{*}{y}_7 = 100.5 + \widehat{\in}_1 = 110.0$.

Method 2: The estimated value of σ^2 is $\widehat{\sigma}^2 = \sum_{i\in s_r} \widehat{\in}_i^2/3 = 105.83$. We select two random samples from a uniform distribution (0, 1). Let these be 0.47 and 0.76. The normal deviates corresponding to these sample are $z_1 = -0.08$ and $z_2 = 0.71$. The imputed values $\overset{*}{y}_6 = 56.75 + 10.288 \times (-0.08) = 55.93$ and $\overset{*}{y}_7 = 100.5 + 10.288 \times 0.71 = 107.80$.

15.4.5 Multiple Imputation

In single imputation, the imputed value is treated as the true value, ignoring the fact that the no imputation method can provide the exact value. Single imputation does not reflect the uncertainty about the prediction of the missing values. Multiple imputation was proposed by Rubin (1987). In this method D

imputed values for each of the missing observation is generated and hence we get D complete data set. From each of the complete data set an estimate of the parameter of interest θ is obtained by using a standard technique, assuming no nonresponse is present. This process results in valid statistical inferences that properly reflect the uncertainty due to missing values.

Let $\widehat{\theta}_d$ and $\widehat{\phi}_d$ be an estimator of θ and the estimator of the variance of $\widehat{\theta}_d$ based on the dth complete data set. The combined estimator of θ under multiple imputations is given by

$$\widehat{\theta}_D = \frac{1}{D}\sum_{d=1}^{D}\widehat{\theta}_d \tag{15.4.26}$$

The proposed variance estimator of $\widehat{\theta}_D$ is given by

$$\widehat{\Phi}_D = \overline{\phi}_D + \left(1 + \frac{1}{D}\right)B_D \tag{15.4.27}$$

where $\overline{\phi}_D = \frac{1}{D}\sum_{d=1}^{D}\widehat{\phi}_d$ and $B_D = \frac{1}{D-1}\sum_{d=1}^{D}\left(\widehat{\theta}_d - \overline{\theta}_D\right)^2$. The first component $\overline{\phi}_D$ of Eq. (15.4.27) is the average within-imputation variance and the second component is the product of between-variance component B_D and correction factor $\left(1 + \frac{1}{D}\right)$.

For large sample size the statistic

$$\frac{\widehat{\theta} - \theta}{\sqrt{\widehat{\Phi}_D}} \tag{15.4.28}$$

follows t distribution with ν degrees of freedom where

$$\nu = (D-1)\left(1 + \frac{1}{D+1}\frac{\overline{\phi}_D}{B_D}\right)^2 \tag{15.4.29}$$

For smaller sample size an improved variance formula has been proposed by Barnard and Rubin (1999).

Multiple imputation is more advantageous than the single imputation because it uses several complete data sets and provides both the within-imputation and between-imputation variability. Multiple imputation facilitates simple formula for variance estimation and interval estimation of the parameter of interest. Its only disadvantage is that it is more tedious to analyze the data than the single imputation. But modern computing techniques such as computation do not pose any problem.

15.4.6 Bayesian Imputation

Without loss of generality, let us assume that the response sample s_r comprises the first r units and let $\mathbf{Y}_r = (y_1,\ldots, y_r)'$ and $\mathbf{X}_r = (x_1,\ldots, x_r)'$.

The nonresponse sample $s_m = s - s_r$ comprises the next $n - r$ units and $\mathbf{Y}_{n-r} = (y_{r+1}, \ldots, y_n)'$ and $\mathbf{x}_{n-r} = (x_{r+1}, \ldots, x_n)'$. The Bayesian single or multiple imputations are obtained from the conditional distribution of $\mathbf{Y}_{n-r}$ given the sample data. Here, we will consider the methods of imputations proposed by Wang et al. (1992) and Schenker and Welsh (1988).

15.4.6.1 Wang et al. Method

In this method a linear regression between the study y and the auxiliary variable x is assumed as follows:

$$y_i = \beta_0 + \beta_1 x_i + \in_i, \quad i = 1, \ldots, n \tag{15.4.30}$$

where $\boldsymbol{\beta} = (\beta_0, \beta_1)'$ is the unknown model parameter, x_i's are known values of an auxiliary variable x, and $\in_i$'s are independent normal variable with mean zero and variance σ^2. Here, we assume that the response mechanism cannot be regarded as MAR and the response probability of the ith unit is

$$g(R|y_i) = 1 - e^{-\alpha y_i} \tag{15.4.31}$$

with α as a known constant.

The imputations are obtained from the conditional distribution of $\mathbf{Y}$ corresponding to the nonresponding units given the sample data. For simplicity, Wang et al. (1992) assumed that σ^2 is known and the prior distribution of $\boldsymbol{\beta}$ is noninformative. This implies that f_2, the posterior distribution of $\boldsymbol{\beta}$ given data is proportional to the likelihood of $\boldsymbol{\beta}$.

The imputations are selections, $y_{r+1}, \ldots, y_n$ from the conditional distribution

$$\begin{aligned} & f(y_{r+1}, \ldots, y_n | y_1, \ldots, y_r; x_1, \ldots, x_n; z_1, \ldots, z_n; \sigma^2) \\ & = \int \cdots \int f_1(y_{r+1}, \ldots, y_n | \boldsymbol{\beta}, y_1, \ldots, y_r; x_1, \ldots, x_n; z_1, \ldots, z_n; \sigma^2) \\ & \quad f_2(\boldsymbol{\beta} | y_1, \ldots, y_r; x_1, \ldots, x_n; z_1, \ldots, z_n; \sigma^2) \partial \boldsymbol{\beta} \end{aligned} \tag{15.4.32}$$

where $z_1 = z_2 = \cdots = z_r = 1$ and $z_{r+1} = z_{r+2} = \cdots = z_n = 0$ are indicators for respondents and nonrespondents [vide Wang et al., 1992 and Govindrajulu, 1999]. Rubin (1987) has shown that the distribution of $f_2(\boldsymbol{\beta}|\cdot)$ given in Eq. (15.4.32) is proportional to

$$K\left[\exp\left\{-\alpha\left(\sum_{i=r+1}^{n}(\beta_0 + \beta_1 x_i)\right) - \frac{1}{2\sigma^2}\left(\sum_{i=1}^{r}(y_i - \beta_0 - \beta_1 x_i)^2\right)\right\}\right] \tag{15.4.33}$$

where

$$K = \prod_{i=r+1}^{n}\left[1 - \Phi\left\{-\frac{\beta_0 + \beta_1 x_i}{\sigma} + \alpha\sigma\right\}\right] \times \left[1 - \Phi\left\{-\frac{\beta_0 + \beta_1 x_i}{\sigma}\right\}\right]$$

and Φ is the cumulative distribution function of standard normal variable. For sufficiently large $\dfrac{\beta_0 + \beta_1 x_i}{\sigma}$, the value of K is approximately equal to 1, and in this case $f_2(\boldsymbol{\beta}|y_1,\ldots, y_r; x_1,\ldots, x_n; z_1,\ldots, z_n; \sigma^2)$ can be approximated as a bivariate normal distribution with mean vector $\boldsymbol{\mu}$ and dispersion matrix $\boldsymbol{\Sigma}$ (vide Rubin, 1987) where

$\boldsymbol{\mu}' = -\left(\mathbf{c} - 2\widehat{\boldsymbol{\beta}}_r'\mathbf{A}\right)\mathbf{A}^{-1}/2, \mathbf{c} = 2\alpha\mathbf{1}'\mathbf{X}_r^*$, $\mathbf{1}$ is a column vector of $n - r$ elements each of which is 1,

$$(\mathbf{X}_r^*)' = \begin{pmatrix} 1 & \cdots & 1 \\ x_{r+1} & \cdots & x_n \end{pmatrix},\ \widehat{\boldsymbol{\beta}}_r = (\mathbf{X}_r'\mathbf{X}_r)^{-1}\mathbf{X}_r'\mathbf{Y}_r,\ \ \mathbf{X}_r' = \begin{pmatrix} 1 & \cdots & 1 \\ x_1 & \cdots & x_r \end{pmatrix},$$

$\boldsymbol{\Sigma} = \mathbf{A}^{-1}$, and $\mathbf{A} = (\mathbf{X}_r'\mathbf{X}_r)/\sigma^2$.

To find the imputed values of $y_{r+j}(1)$ for $j = 1,\ldots, n - r$, we follow the following steps:

Step 1: Draw a sample $\widehat{\boldsymbol{\beta}}^* = \left(\widehat{\beta}_0^*, \widehat{\beta}_1^*\right)'$ from the distribution f_2, which is $N(\boldsymbol{\mu}, \boldsymbol{\Sigma})$.

Step 2: Choose a sample $\widehat{y}_{r+j}(1)$ from the population $y_{r+j}\big| x_{r+j}, \widehat{\boldsymbol{\beta}}^*, \sigma^2$, which is $N\left(\widehat{\beta}_0^* + \widehat{\beta}_1^* x_{r+j}, \sigma^2\right)$.

Step 3: Select a sample $\widehat{u}_{r+j}(1)$ from a uniform distribution $U(0, 1)$.

Step 4: Treat $\widehat{y}_{r+j}(1)$ as an imputed value of y_{r+j} if $\widehat{u}_{r+j}(1) \le e^{-\alpha\widehat{y}_{r+j}(1)}$. Otherwise, i.e., if $\widehat{u}_{r+j}(1) > e^{-\alpha\widehat{y}_{r+j}(1)}$, repeat Steps 1–4 until an imputed value of y_{r+j} is obtained.

Using the aforementioned method for $j = r,\ldots, n - r$, the first set of imputed values $\widehat{y}_{r+j}(1), \ldots, \widehat{y}_n(1)$ of $y_{r+j},\ldots, y_n$ are obtained.

To find the second set of imputed values $\widehat{y}_{r+j}(2), \ldots, \widehat{y}_n(2)$, go to

Step 5: Using $\widehat{y}_{r+1}(1), \ldots, \widehat{y}_n(1)$ and the observed data $y_1,\ldots, y_r$, estimate β_0, β_1, and σ^2 using the following formula

$$\widehat{\beta}_0(2) = \overline{y}_n(2) - \widehat{\beta}_1(2)\overline{x}_n, \widehat{\beta}_1(2) = \frac{\sum_{i=1}^{n} y_i^+(2)(x_i - \overline{x}_n)}{\sum_{i=1}^{n}(x_i - \overline{x}_n)^2} \quad \text{and}$$

$$\widehat{\sigma}^2(2) = \frac{\sum_{i=1}^{n}\left\{y_i^+(2) - \widehat{y}_i^+(2)\right\}^2}{(n-2)}$$

where $\quad y_i^+(2) = \begin{cases} y_i & \text{for } i = 1, \ldots, r \\ \widehat{y}_i(1) & \text{for } i = r+1, \ldots, n \end{cases}, \qquad \overline{y}_n(2) = \dfrac{\sum_{i=1}^{n} y_i^+(2)}{n},$

$\overline{x}_n = \dfrac{\sum_{i=1}^{n} x_i}{n}$, and $\widehat{y}_i^+(2) = \widehat{\beta}_0(2) + \widehat{\beta}_1(2)x_i$.

Step 6: Choose a sample $\widehat{y}_{r+j}(2)$ from $N\left(\widehat{\beta}_0(2)+\widehat{\beta}_1(2)x_{r+j}, \widehat{\sigma}^2(2)\right)$.

Step 7: Select a sample $\widehat{u}_{r+j}(2)$ from a uniform distribution $U(0, 1)$.

Step 8: Treat $\widehat{y}_{r+j}(2)$ as an imputed value of y_{r+j} if $\widehat{u}_{r+j}(2) \leq e^{-\alpha\widehat{y}_{r+j}(2)}$. Otherwise, i.e., if $\widehat{u}_{r+j}(2) > e^{-\alpha\widehat{y}_{r+j}(2)}$, repeat Steps 5–8 until an imputed value of y_{r+j} is obtained. Using the aforementioned method the second set of imputed values $\widehat{y}_{r+j}(2), \ldots, \widehat{y}_n(2)$ of $y_{r+j}, \ldots, y_n$ are obtained.

Repeating Steps 5–8, $D-2$ times, the remaining $D-2$ sets of imputed values are obtained using $\widehat{y}_{r+1}(d), \ldots, \widehat{y}_n(d)$ and the observed data $y_1, \ldots, y_r$ to compute $\widehat{y}_{r+1}(d+1), \ldots, \widehat{y}_n(d+1)$.

15.4.6.2 Schenker and Welsh Method

Schenker and Welsh (1988) and Kim (2004) used the following multiple regression model:

$$\begin{aligned} y_i &= \beta_1 + \beta_2 x_{1i} + \cdots + \beta_p x_{pi} + \in_i \\ &= \mathbf{x}_i'\boldsymbol{\beta} + \in_i \end{aligned} \tag{15.4.34}$$

where the $p+1$-dimension vector $\mathbf{x}_i' = (1, x_{1i}, \ldots, x_{pi})$, $i = 1, \ldots, n$, is known; $\boldsymbol{\beta} = (\beta_1, \beta_2, \ldots, \beta_p)'$ is unknown, $\in_i \overset{i.i.d}{\sim} N(0, \sigma^2)$ and σ^2 are unknown.

Let us assume that $\boldsymbol{\beta}$ and σ^2 are independent with constant prior for $(\boldsymbol{\beta}, \log\sigma^2)$, so that the marginal posterior distribution of σ^2 is $\widehat{\sigma}_r^2(r-p-1)\big/\chi^2_{r-p-1}$ and the conditional posterior distribution of $\boldsymbol{\beta}$ given σ^2 is $N\left(\widehat{\boldsymbol{\beta}}, \sigma^2\psi_r^{-1}\right)$, where $\widehat{\sigma}_r^2 = \mathbf{Y}_r'\left[\mathbf{I} - \mathbf{X}_r\left(\mathbf{X}_r'\mathbf{X}_r\right)^{-1}\mathbf{X}_r'\right]\mathbf{Y}_r/(r-p-1)$, $\widehat{\boldsymbol{\beta}}_r = \left(\mathbf{X}_r'\mathbf{X}_r\right)^{-1}\mathbf{X}_r'\mathbf{Y}_r$, χ^2_{r-p-1} is a Chi-square distribution with $r-p-1$ degrees of freedom, and $\psi_r = \left(\mathbf{X}_r'\mathbf{X}_r\right)$.

For multiple imputation of $\mathbf{Y}_{n-r}' = (y_{r+1}, \ldots, y_n)$, the following steps are followed:

Step 1: For the kth repetition, $k(=1, \ldots, D)$, draw a sample of σ^2 as

$$\sigma^2_{(k)} = \widehat{\sigma}_r^2(r-p-1)/g_k$$

where g_k is a random sample from χ^2_{r-p-1}.

Step 2: Draw a sample $\boldsymbol{\beta}_{(k)}$ from $N\left(\widehat{\boldsymbol{\beta}}_r, \left(\mathbf{X}_r'\mathbf{X}_r\right)^{-1}\sigma^2_{(k)}\right)$.

Step 3: For each missing unit $j(=r+1, \ldots, n)$, select a random sample $\in_{j(k)}$ from $N\left(0, \sigma^2_{(k)}\right)$. Then the imputed value of jth unit at kth imputation is

$$\widehat{y}_{j(k)} = \mathbf{x}_j'\boldsymbol{\beta}_{(k)} + \in_{j(k)}$$

15.4.7 Subsampling Method

Let a sample s, of size n, be selected from the population $U = (1,\ldots, N)$ with probability $p(s)$ according to a sampling design p with π_i and π_{ij} as inclusion probabilities for the ith, and ith and jth $(i \neq j)$ units of U. Let $s_r(\subset s)$ be the set of respondent units of size $r(\leq n)$ for which responses y_i's are obtained and the complement $s_m = s - s_r$ (of size $n - r$) be the set of nonresponse units. The formation of the response sample s_r is totally unknown and no exact modeling of s_r is possible. Here, we therefore postulate a probabilistic model that describes the response distribution. Let θ_i be the probability that ith unit $(i = 1,\ldots, N)$ responds if it is selected in the sample s. Hence the response sample s_r is regarded as a subsample of the original sample s according to some unknown sampling design.

15.4.7.1 Arnab and Singh Method

In this method, the inclusion probability of the ith unit in the response sample s_r given s is $\pi_{i|s} = \theta_i$ (response probability of the ith unit). Similarly, the inclusion probability of the ith and jth units $(i \neq j)$ in the sample s_r is $\pi_{ij|s} = \theta_{ij} = \theta_i\theta_j$, assuming responses are independent. The total number of the respondent r, in the sample s_r, is a random variable. Hence, we consider s_r as a subsample from the original sample s selected according to the Poisson sampling design (vide Section 5.4.2.4).

In case θ_i's are known for $i \in s_r$, following Arnab and Singh (2006), we propose a Horvitz–Thompson type estimator for the total Y as follows:

$$\widehat{Y}_{ht}(nr) = \sum_{i\in s_r} \frac{y_i}{\pi_i\theta_i} = \sum_{i\in s} \frac{z_i}{\theta_i} I_{s_r i} \tag{15.4.35}$$

where $z_i = \dfrac{y_i}{\pi_i}$, $I_{s_r i} = 1$ if $i \in s_r$, and $I_{s_r i} = 0$ if $i \in s_m$.

Here, we note that

$$E(I_{s_r i}|s) = \text{The probability that the selected } i\text{th unit will } \textit{respond} = \theta_i$$

$$E(I_{s_r i} I_{s_r j}|s) = \text{The probability that the selected } i\text{th unit and } j\text{th unit } (i \neq j)$$

will respond

$$= \theta_i\theta_j.$$

The expectations, variance, and unbiased estimators of the variance of $\widehat{Y}_{ht}(nr)$ are given in the following theorem.

Theorem 15.4.3

(i) $E\left\{\widehat{Y}_{ht}(nr)\right\} = Y$

(ii) $V\left\{\widehat{Y}_{ht}(nr)\right\} = \sum_i \dfrac{y_i^2}{\pi_i}\left(\dfrac{1}{\theta_i} - 1\right) + \dfrac{1}{2}\sum_{i\neq}\sum_j (\pi_i\pi_j - \pi_{ij})\left(\dfrac{y_i}{\pi_i} - \dfrac{y_j}{\pi_j}\right)^2$

(iii) Two unbiased estimators of $V\{\widehat{Y}_{ht}(nr)\}$ are given by

$$\widehat{V}(1) = \widehat{V}_s(1) + \sum_{i\in s_r} \frac{y_i^2}{\pi_i^2\theta_i}\left(\frac{1}{\theta_i} - 1\right)$$

and

$$\widehat{V}(2) = \widehat{V}_s(2) + \sum_{i\in s_r} \frac{y_i^2}{\pi_i^2\theta_i}\left(\frac{1}{\theta_i} - 1\right)$$

where

$$\widehat{V}_s(1) = \frac{1}{2}\sum_{i\neq j}\sum_{\in s_r} \frac{(\pi_i\pi_j - \pi_{ij})}{\pi_{ij}\theta_i\theta_j}\left(\frac{y_i}{\pi_i} - \frac{y_j}{\pi_j}\right)^2 \text{ and}$$

$$\widehat{V}_s(2) = \sum_{i\in s_r} \frac{y_i^2}{\pi_i\theta_i}\left(\frac{1}{\pi_i} - 1\right) + \sum_{i\neq j}\sum_{\in s_r}\left(\frac{\pi_{ij}}{\pi_i\pi_j} - 1\right)\frac{y_i}{\pi_{ij}\theta_i}\frac{y_j}{\theta_j}$$

Proof

Let $E_p(E_r)$ and $V_p(V_r)$ denote expectations and variances with respect to the sampling design (response mechanism), respectively. Then we have

(i) $E\{\widehat{Y}_{ht}(nr)\} = E_p\left\{\sum_{i\in s} \frac{z_i}{\theta_i} E_r(I_{s_r i})\right\} = E_p\left(\sum_{i\in s} z_i\right) = Y$

(ii) $$V\{\widehat{Y}_{ht}(nr)\} = E_p\left\{\sum_{i\in s} \frac{z_i^2}{\theta_i^2} V_r(I_{s_r i})\right\} + V_p\left(\sum_{i\in s} \frac{z_i}{\theta_i} E_r(I_{s_r i})\right)$$

$$= E_p\left\{\sum_{i\in s} \frac{z_i^2}{\theta_i}(1-\theta_i)\right\} + V_p\left(\sum_{i\in s} z_i\right)$$

$$= \sum_{i\in U} \frac{y_i^2}{\pi_i}\left(\frac{1}{\theta_i} - 1\right) + \frac{1}{2}\sum_{i\neq}\sum_{j\in U}(\pi_i\pi_j - \pi_{ij})\left(\frac{y_i}{\pi_i} - \frac{y_j}{\pi_j}\right)^2$$

(iii) Writing

$$V_p\left(\sum_{i\in s} z_i\right) = V_p\left(\sum_{i\in s} \frac{y_i}{\pi_i}\right) = \frac{1}{2}\sum_{i\neq j}\sum_{\in U}(\pi_i\pi_j - \pi_{ij})\left(\frac{y_i}{\pi_i} - \frac{y_j}{\pi_j}\right)^2$$

$$= \sum_{i\in U} y_i^2\left(\frac{1}{\pi_i} - 1\right) + \sum_{i\neq j}\sum_{\in U}\left(\frac{\pi_{ij}}{\pi_i\pi_j} - 1\right) y_i y_j$$

we get two unbiased estimators of $V_p\left(\sum_{i\in s} z_i\right)$ as follows:

$$\widehat{V}_s(1) = \frac{1}{2}\sum_{i\neq j}\sum_{\in s_r}\frac{(\pi_i\pi_j - \pi_{ij})}{\pi_{ij}\theta_i\theta_j}\left(\frac{y_i}{\pi_i} - \frac{y_j}{\pi_j}\right)^2 \tag{15.4.36}$$

and

$$\widehat{V}_s(2) = \sum_{i\in s_r}\frac{y_i^2}{\pi_i\theta_i}\left(\frac{1}{\pi_i} - 1\right) + \sum_{i\neq j}\sum_{\in s_r}\left(\frac{\pi_{ij}}{\pi_i\pi_j} - 1\right)\frac{y_i}{\pi_{ij}\theta_i}\frac{y_j}{\theta_j} \tag{15.4.37}$$

Further, an unbiased estimator of $\sum_{i\in U}\frac{y_i^2}{\pi_i}\left(\frac{1}{\theta_i} - 1\right)$ is $\sum_{i\in s_r}\frac{y_i^2}{\pi_i^2\theta_i}\left(\frac{1}{\theta_i} - 1\right)$

(15.4.38)

From Eqs. (15.4.36)–(15.4.38), we can verify the result.

15.4.7.1.1 Simple Random Sampling Without Replacement

For an SRSWOR sampling design $\pi_i = n/N$ and $\pi_{ij} = n(n-1)/\{N(N-1)\}$, hence we derive.

Theorem 15.4.4

(i) $\widehat{Y}_{srs}(nr) = \frac{N}{n}\sum_{i\in s_r}\frac{y_i}{\theta_i}$ is an unbiased estimator of Y

(ii) $V\left[\widehat{Y}_{srs}(nr)\right] = \frac{N}{n}\sum_i y_i^2\left(\frac{1}{\theta_i} - 1\right) + N\frac{N-n}{n}S_y^2$

(iii) Two unbiased estimators of $V\left[\widehat{Y}_{nr}(srs)\right]$ are given by

$$\widehat{V}_{srs}(1) = \widehat{V}^*(1) + \frac{N^2}{n^2}\sum_{i\in s_r}\frac{y_i^2}{\theta_i}\left(\frac{1}{\theta_i} - 1\right) \text{ and}$$

$$\widehat{V}_{srs}(2) = \widehat{V}^*(2) + \frac{N^2}{n^2}\sum_{i\in s_r}\frac{y_i^2}{\theta_i}\left(\frac{1}{\theta_i} - 1\right)$$

where

$$\widehat{V}^*(1) = \frac{N(N-n)}{n^2(n-1)}\frac{1}{2}\sum_{i\neq j}\sum_{\in s_r}\frac{(y_i - y_j)^2}{\theta_i\theta_j} \text{ and}$$

$$\widehat{V}^*(2) = N\left(\frac{N-n}{n^2}\right)\left(\sum_{i\in s_r}\frac{y_i^2}{\theta_i} - \frac{1}{n-1}\sum_{i\neq j}\sum_{\in s_r}\frac{y_i}{\theta_i}\frac{y_j}{\theta_j}\right)$$

Corollary 15.4.1

In case all the response probabilities are equal, i.e., $\theta_i = \theta$ for all $i = 1,\dots, N$, then $\widehat{Y}_{nr}(srs) = \dfrac{Nr}{n\theta}\overline{y}(s_r)$ becomes an unbiased estimator of Y with variance

$$V\left[\widehat{Y}_{srs}(nr)\right] = \frac{N}{n}\left(\frac{1}{\theta} - 1\right)\sum_i y_i^2 + N\frac{N-n}{n}S_y^2$$

where $\overline{y}(s_r) = \sum_{i\in s_r} y_i/r$.

Furthermore, if the response probability θ is estimated by $\widehat{\theta} = r/n$, then $\widehat{Y}_{srs}(nr) = N\,\overline{y}(s_r)$.

15.4.7.2 Singh and Singh Method

Singh and Singh (1979) considered the response sample s_r as an SRSWOR subsample of size r from s and proposed an unbiased estimator for the population total as

$$\widehat{Y}_{ss}(nr) = \frac{n}{r}\sum_{i\in s_r}\frac{y_i}{\pi_i} \tag{15.4.39}$$

The expectation, variance, and unbiased estimator of the variance of $\widehat{Y}_{ss}(nr)$ are given in the following theorem.

Theorem 15.4.5

(i) $E\left[\widehat{Y}_{ss}(nr)\right] = Y$

(ii) $V\left[\widehat{Y}_{ss}(nr)\right] = V_{ht} + \left[\sum_{i\in U}\frac{y_i^2}{\pi_i} - \frac{1}{n-1}\sum_{i\neq}\sum_{j\in U}\frac{\pi_{ij}}{\pi_i\pi_j}y_iy_j\right]E_r\left(\frac{n-r}{r}\right)$

(iii) $\widehat{V}\left[\widehat{Y}_{ss}(nr)\right] = \frac{n}{r}\sum_{i\in s_r}\frac{\alpha_i y_i^2}{\pi_i} + \frac{n(n-1)}{r(r-1)}\sum_{i\neq}\sum_{j\in s_r}\frac{\alpha_{ij}y_iy_j}{\pi_{ij}}$

where E_r denotes expectation with respect to the response mechanism; $\alpha_i = \dfrac{n}{r\pi_i} - 1$; $\alpha_{ij} = \dfrac{n(r-1)\pi_{ij}}{r(n-1)\pi_i\pi_j} - 1$; and

$$V_{ht} = \frac{1}{2}\sum_{i\neq}\sum_{j\in U}(\pi_i\pi_j - \pi_i\pi_j)\left(\frac{y_i}{\pi_i} - \frac{y_j}{\pi_j}\right)^2.$$

Proof

(i) $E\left[\widehat{Y}_{ss}(nr)\right] = E_rE_pE_p\left(\frac{n}{r}\sum_{i\in s_r}\frac{y_i}{\pi_i}\Big|s\right) = E_rE_p\left(\sum_{i\in s}\frac{y_i}{\pi_i}\right) = E_r(Y) = Y$

(ii) $V\left[\widehat{Y}_{ss}(nr)\right] = E_r\left[V_p\left\{\widehat{Y}_{ss}(nr)|r\right\}\right] + V_r\left[E_p\left\{\widehat{Y}_{ss}(nr)|r\right\}\right]$

Now, noting

$$E_r\left(V_p\{\widehat{Y}_{ss}(nr)|r\}\right) = E_r\left(E_p\left[V_p\{\widehat{Y}_{ss}(nr)|s_r, r\}\right] + V_p\left[E_p\{\widehat{Y}_{ss}(nr)|s_r, r\}\right]\right)$$

$$= E_r\left(n\frac{n-r}{r(n-1)}E_p\left\{\sum_{i\in s}\left(\frac{y_i}{\pi_i}\right)^2 - \left(\sum_{i\in s}\frac{y_i}{\pi_i}\right)^2 \bigg/ n \bigg| r\right\}\right) + V_p\left(\sum_{i\in s}\frac{y_i}{\pi_i}\right)$$

$$= E_r\left[n\frac{n-r}{r(n-1)}\left\{\sum_{i\in U}\frac{y_i^2}{\pi_i} - \left(\sum_{i\in U}\frac{y_i^2}{\pi_i} + \sum_{I\neq}\sum_{j\in U}\frac{y_i}{\pi_i}\frac{y_j}{\pi_j}\pi_{ij}\right)\bigg/ n\right\}\right] + V_{ht}$$

$$= V_{ht} + \left(\sum_{i\in U}\frac{y_i^2}{\pi_i} - \frac{1}{n-1}\sum_{I\neq}\sum_{j\in U}\frac{y_i}{\pi_i}\frac{y_j}{\pi_j}\pi_{ij}\right)E_r\left(\frac{n-r}{r}\right)$$

and

$$V_r\left[E\{\widehat{Y}_{ss}(nr)|r\}\right] = V_r(Y) = 0,$$

we can verify the part (ii) of the theorem.

(iii) $$E\left[\widehat{V}\{\widehat{Y}_{ss}(nr)\}\right] = E_r\left[\sum_{i\in U}\alpha_i y_i^2 + \sum_{i\neq}\sum_{j\in s_r}\alpha_{ij}y_i y_j\right] = V\{\widehat{Y}_{ss}(nr)\} \qquad (15.4.40)$$

Corollary 15.4.2

For an SRSWOR sampling design, $\pi_i = n/N$, $\pi_{ij} = n(n-1)/\{N(N-1)\}$, and $\widehat{Y}_{ss}(nr) = \frac{N}{r}\sum_{i\in s_r} y_i = N\,\overline{y}_r$, where $\overline{y}_r = \sum_{i\in s_r} y_i/r$. Theorem 15.4.5 therefore reduces to

Theorem 15.4.6

(i) $E\left[N\,\overline{y}_r\right] = Y$

(ii) $V\left[N\,\overline{y}_r\right] = N^2\left[E_r\left(\frac{1}{r}\right) - \frac{1}{N}\right]S_y^2$

(iii) $\widehat{V}\left[\widehat{Y}_{ss}(nr)\right] = N^2\left(\frac{1}{r} - \frac{1}{N}\right)s_{yr}^2$

where

$$s_{yr}^2 = \frac{1}{r-1}\sum_{i\in s_r}\left(y_i - \overline{y}_r\right)^2$$

15.5 MEASUREMENT ERROR

So far, we assumed that if a unit is selected in the sample, the information about the value of the variable under study based on the unit is free from error. But in reality this is far from true. The error may occur because of the use of faulty measuring instrument. For example, if a baumanometer is wrong, one may get incorrect readings of blood pressure. Correct measurement may not be possible to obtain for some of the items such as intelligent quotient (IQ), ability of teaching, and attitude toward work, among others. Faulty responses may also be obtained from the respondents, enumerators, or both because of ambiguous questionnaires, intentionally reporting untrue value from carelessness, or the sensitive nature of questions. Untrue responses can be reported by the respondents if they are influenced by enumerators.

Analysis of measurement errors is reported by various authors. Important works among them include Mahalanobis (1946), Deming (1953), Raj (1968), Sukhatme et al. (1984), and Särndal et al. (1992).

Let y_i be the true value of the ith unit of the study variable y and x_i be the response obtained by an investigator. The response or measurement error for the ith unit is $x_i - y_i$. In this study we will assume the response x_i is a random variable and we will build a model on responses. Because responses from a unit depend on the investigator, enumerator, or measuring instrument, we assume that the response x_i obtained by the ith enumerator based on a sample follows the model

$$x_i = y_i + b_i + \in_{is} \tag{15.5.1}$$

where b_i is the error of measurement associated with the ith unit; $\in_{is}$ is the random error component with $E_m(\in_{is}) = 0$, $V_m(\in_{is}) = \sigma_{is}^2$, and $C_m(\in_{is}, \in_{js}) = \rho_s\sigma_{is}\sigma_{js}$ for $i \neq j$; E_m, V_m, and C_m denote, respectively, expectation, variance and covariance with respect to the model. In the aforementioned model it is assumed that the distribution of $\in_{is}$ depends on the sample s, i.e., environment under which the data were collected. Here, we consider more simple measurement error model

$$x_i = y_i + b_i + \in_i \tag{15.5.2}$$

with $E_m(\in_i) = 0$, $V_m(\in_i) = \sigma_i^2$, and $C_m(\in_i, \in_j) = \sigma_{ij}$ for $i \neq j$.

15.5.1 Measurement Bias and Variance

Let a sample s of size n be selected using a sampling design p and let π_i and π_{ij} be the inclusion probabilities for the ith and ith and jth ($i \neq j$) units, respectively. The Horvitz–Thompson type estimator for the total Y is given by

$$\widehat{Y}_{me}(ht) = \sum_{i \in s} \frac{x_i}{\pi_i} \tag{15.5.3}$$

The expected value of $\widehat{Y}_{me}(ht)$ is

$$\begin{aligned} E\left[\widehat{Y}_{me}(ht)\right] &= E_p E_m\left[\widehat{Y}_{me}(ht)\right] \\ &= E_p \sum_{i \in s} E_m\left(\frac{x_i}{\pi_i}\right) \\ &= E_p\left(\sum_{j \in s} \frac{y_i + b_i}{\pi_i}\right) \\ &= Y + B \end{aligned} \tag{15.5.4}$$

where $B = \sum_{i=1}^{M} b_i$

The quantity

$$B = E\left(\widehat{Y}_{me}(ht)\right) - Y \tag{15.5.5}$$

is known as measurement bias.

The mean square error of $\widehat{Y}_{me}(ht)$ is given by

$$\begin{aligned} M\left(\widehat{Y}_{me}(ht)\right) &= E_p E_m\left[\widehat{Y}_{me}(ht) - Y\right]^2 \\ &= E_p E_m\left[\widehat{Y}_{me}(ht) - E_p E_m\left\{\widehat{Y}_{me}(ht)\right\} + E_p E_m\left\{\widehat{Y}_{me}(ht)\right\} - Y\right]^2 \\ &= E_p E_m\left[\widehat{Y}_{me}(ht) - E_p E_m\left\{\widehat{Y}_{me}(ht)\right\}\right]^2 + B^2 \\ &= V_{pm}\left(\widehat{Y}_{me}(ht)\right) + B^2 \end{aligned} \tag{15.5.6}$$

The quantity $V_{pm}\left(\widehat{Y}_{me}(ht)\right)$ is known as the total variance. The total variance $V_{pm}\left(\widehat{Y}_{me}(ht)\right)$ is decomposed as follows:

$$\begin{aligned} V_{pm}\left(\widehat{Y}_{me}(ht)\right) &= E_p\left[V_m\left(\widehat{Y}_{me}(ht)\right)\right] + V_p\left[E_m\left(\widehat{Y}_{me}(ht)\right)\right] \\ &= E_P\left(\sum_{i \in s} \frac{\sigma_i^2}{\pi_i^2} + \sum_{i \neq} \sum_{j \in s} \frac{\sigma_{ij}}{\pi_i \pi_j}\right) + V_p\left(\sum_{j \in s} \frac{y_i + b_i}{\pi_i}\right) \\ &= \left(\sum_{i \in U} \frac{\sigma_i^2}{\pi_i} + \sum_{i \neq} \sum_{j \in U} \frac{\sigma_{ij}}{\pi_i \pi_j} \pi_{ij}\right) \\ &\quad + \frac{1}{2} \sum_{i \neq} \sum_{j \in U} \left(\pi_i \pi_j - \pi_{ij}\right)\left(\frac{y_i + b_i}{\pi_i} - \frac{y_j + b_j}{\pi_j}\right)^2 \end{aligned} \tag{15.5.7}$$

The first component $\left(\sum_{i\in U}\frac{\sigma_i^2}{\pi_i}+\sum_{i\neq}\sum_{j\in U}\frac{\sigma_{ij}}{\pi_i\pi_j}\pi_{ij}\right)$ is known as measurement variance (MV) and the second component $\frac{1}{2}\sum_{i\neq}\sum_{j\in U}(\pi_i\pi_j-\pi_{ij})\left(\frac{y_i+b_i}{\pi_i}-\frac{y_j+b_j}{\pi_j}\right)^2$ is known as sampling variance.

The summarization of the discussions yield.

Theorem 15.5.1

For the measurement model (Eq. 15.5.2)
(i) The mean square error of $\widehat{Y}_{me}(ht)$ is

$$M\left(\widehat{Y}_{me}(ht)\right)=V_{pm}\left(\widehat{Y}_{me}(ht)\right)+B^2$$

where $V_{pm}\left(\widehat{Y}_{me}(ht)\right)$ is the total variance and B is the measurement bias.

(ii) $V_{pm}\left(\widehat{Y}_{me}(ht)\right)=\text{MV}+\text{SV}$

where $\text{MV}=\left(\sum_{i\in U}\frac{\sigma_i^2}{\pi_i}+\sum_{i\neq}\sum_{j\in U}\frac{\sigma_{ij}}{\pi_i\pi_j}\pi_{ij}\right)$ and

$$\text{SV}=\text{sampling variance}=\frac{1}{2}\sum_{i\neq}\sum_{j\in U}(\pi_i\pi_j-\pi_{ij})\left(\frac{y_i+b_i}{\pi_i}-\frac{y_j+b_j}{\pi_j}\right)^2.$$

Thus the presence of measurement errors increases the variance of the Horvitz–Thompson estimator by

$$\left(\sum_{i\in U}\frac{\sigma_i^2}{\pi_i}+\sum_{i\neq}\sum_{j\in U}\frac{\sigma_{ij}}{\pi_i\pi_j}\pi_{ij}\right)+V_{ht}(b,b)-2V_{ht}(b,y)$$

where $V_{ht}(u,v)=\frac{1}{2}\sum_{i\neq}\sum_{j\in U}(\pi_i\pi_j-\pi_{ij})\left(\frac{u_i}{\pi_i}-\frac{u_j}{\pi_j}\right)\left(\frac{v_i}{\pi_i}\frac{v_j}{\pi_j}\right).$

15.5.1.1 Simple Random Sampling Without Replacement

For SRSWOR $\pi_i=\frac{n}{N}$ and $\pi_{ij}=\frac{n(n-1)}{N(N-1)}$, and Theorem 15.5.1 reduces to

Theorem 15.5.2

For the measurement model (Eq. 15.5.2), the mean square error of $\widehat{Y}_{me}=N\,\overline{x}(s)$ is

$$M[N\,\overline{x}(s)] = N^2\left[\frac{1}{Nn}\left(\sum_{i\in U}\sigma_i^2 + \frac{n-1}{N-1}\sum_{i\neq}\sum_{j\in U}\sigma_{ij}\right)\right.$$
$$\left. + \frac{(1-f)}{n}\left(S_y^2 + S_B^2 + 2S_{by}\right)\right] + B^2$$

where $S_y^2 = \frac{1}{N-1}\sum_{i\in U}\left(y_i - \overline{Y}\right)^2, S_B^2 = \frac{1}{N-1}\sum_{i\in U}\left(b_i - \overline{B}\right)^2,$

$S_{yb} = \frac{1}{N-1}\sum_{i\in U}\left(b_i - \overline{B}\right)\left(y_i - \overline{Y}\right)$, and $\overline{B} = \sum_{i\in U} b_i/N.$

If we assume that the measurement errors are independent, i.e., $\sigma_{ij} = 0$, and measurement bias $b_i = b$ is constant then

$$M\{\overline{x}(s)\} = \left[\overline{B}^2 + \frac{\overline{\sigma}^2}{n} + \frac{(1-f)}{n}S_y^2\right] \tag{15.5.8}$$

where $\overline{\sigma}^2 = \sum_{i\in U}\sigma_i^2/N =$ the average MV.

15.5.2 Interpenetrating Subsamples

The concept of interpenetrating subsamples (IPNS) was introduced by Mahalanobis (1946). In this technique two or more subsamples are selected using the same sampling designs. The subsamples may or may not be independent. Such subsamples are called IPNS. From each of the subsamples an estimate of the population parameter is obtained. This method helps to collect improved quality of data and provides estimate of error of the population parameter irrespective of the complexity of the sampling design used. Each of the subsamples is surveyed by a batch of enumerators and processed by different group of people at tabulation and subsequent stages. If the results and tables show significant discrepancies from one subsample to another, then one have reasons to suspect something is wrong somewhere and the entire work may be checked. Thus the method reduces nonsampling errors substantially. In this section we will consider the role of investigator effects in measurement error using the following model:

$$E_m(z_{ij}) = x_{ij} = y_j + b_i,\ V_m(z_{ij}) = \sigma_{ij}^2,\ C_m(z_{ij}, z_{ik}) = \sigma_{ik} \text{ and } C_m(z_{ij}, z_{i'k})$$
$$= 0 \quad \text{for } i \neq i' \tag{15.5.9}$$

where z_{ij} be the response obtained by ith investigator from the jth unit. Let $E_m(V_m)$, $E_p(V_p)$, and $E_I(V_I)$, respectively, be the expectation (variance) with respect to measurement model, sampling design, and investigator. The operator C_m denotes covariance with respect to the model m.

Here we will consider the following sampling schemes.

Scheme A:

In this scheme k independent samples $s_1,\ldots, s_k$ each of sizes r are selected from a population of N units by SRSWOR method and allotted at random to $k(=n/r)$ (assuming integer) different investigators selected at random from a pool of M investigators. Raj (1968) considered this sampling scheme A when M is very large.

Scheme B:

Here a sample s of size $n(=kr)$ units are selected at random from the population by SRSWOR method and the units are divided at random into k groups $s_1,\ldots, s_k$. Each group is allotted to an investigator as in scheme A.

Let $t_i = \frac{1}{r}\sum_{j\in s_i} z_{ij}$ be an estimator of the population mean $\overline{Y}$ based on the data collected by the ith investigator and

$$\bar{t} = \frac{1}{k}\sum_{i=1}^{k} t_i \tag{15.5.10}$$

be the pooled estimator of $\overline{Y}$ based on the entire sample $s = s_1 \cup \cdots \cup s_k$, then we have the following results.

Theorem 15.5.3

Under the scheme A and model (Eq. 15.5.9), we have

(i) $E(\bar{t}) = \overline{Y} + \overline{B}$

(ii) $V(\bar{t}) = \left(1 - \frac{k}{M}\right)\frac{S_B^2}{k} + \left(1 - \frac{r}{N}\right)\frac{S_y^2}{n} + \frac{A}{n}$

and

(iii) An unbiased estimator of $V(\bar{t})$ is

$$\widehat{V}(\bar{t}) = \frac{1}{k(k-1)}\sum_{i=1}^{k}(t_i - \bar{t})^2 \text{ if } M \text{ is large,}$$

where $\overline{B} = \frac{1}{M}\sum_{i=1}^{B} b_i$, $S_B^2 = \frac{1}{M-1}\sum_{i=1}^{B}\left(b_i - \overline{B}\right)^2$ and

$$A = \frac{1}{NM}\sum_{i=1}^{M}\left(\sum_{j\in U}\sigma_{ij}^2 + \frac{r-1}{(N-1)}\sum_{j\neq}\sum_{l\in U}\sigma_{ij,il}\right).$$

Proof

(i) $$E(\bar{t}) = E_I\left[\frac{1}{k}\sum_{i=1}^{k} E_p\{E_m(t_i)\}\right]$$

$$= E_I\left[\frac{1}{k}\sum_{i=1}^{k} E_p\left(\frac{1}{r}\sum_{j\in s_i} x_{ij}\right)\right]$$

$$= E_I\left(\frac{1}{k}\sum_{i=1}^{k} \overline{X}_i\right)$$

$$= \frac{1}{M}\sum_{i=1}^{M} \overline{X}_i \left(\text{writing } \overline{X}_i = X_i/N, X_i = \sum_{j\in U} x_{ij}\right)$$

$$= \overline{Y} + \overline{B}$$

(ii) Let $E_{pm}(V_{pm})$ denote the expectation (variance) jointly with the sampling design p and model m. Then

$$V(\bar{t}) = V_I\left[E_{pm}(\bar{t})\right] + E_I\left[V_{pm}(\bar{t})\right]$$

$$= V_I\left(\frac{1}{k}\sum_{i=1}^{k} \overline{X}_i\right) + E_I\left[\frac{1}{k^2}\sum_{i=1}^{k}\left\{V_p\left(\frac{1}{r}\sum_{j\in s_i} x_{ij}\right)\right.\right.$$

$$\left.\left. + E_p\frac{1}{r^2}\left(\sum_{j\in s_i}\sigma_{ij}^2 + \sum_{j\neq}\sum_{l\in s_i}\sigma_{ij,il}\right)\right\}\right]$$

$$= \left(\frac{1}{k} - \frac{1}{M}\right)S_B^2 + E_I\left[\frac{1}{k^2}\sum_{i=1}^{k}\left\{\left(\frac{1}{r} - \frac{1}{N}\right)S_y^2\right.\right.$$

$$\left.\left. + \frac{1}{r}\left(\frac{1}{N}\sum_{j\in U}\sigma_{ij}^2 + \frac{r-1}{N(N-1)}\sum_{j\neq}\sum_{l\in U}\sigma_{ij,il}\right)\right\}\right]$$

$$= \left(\frac{1}{k} - \frac{1}{M}\right)S_B^2 + \frac{1}{k}\left[\left(\frac{1}{r} - \frac{1}{N}\right)S_y^2\right.$$

$$\left. + \frac{1}{rNM}\sum_{i=1}^{M}\left\{\sum_{j\in U}\sigma_{ij}^2 + \frac{r-1}{(N-1)}\sum_{j\neq}\sum_{l\in U}\sigma_{ij,il}\right\}\right]$$

$$= \left(1 - \frac{k}{M}\right)\frac{S_B^2}{k} + \left(1 - \frac{r}{N}\right)\frac{S_y^2}{n} + \frac{A}{n}$$

$$\text{(iii)}\ (k-1)E\left[\widehat{V}(\bar{t})\right] = E\left(\frac{1}{k}\sum_{i=1}^{k} t_i^2 - \bar{t}^2\right)$$

$$= E_I\left[E_p\left\{\frac{1}{k}E_m\sum_{i=1}^{k}\left(\frac{1}{r}\sum_{j\in s_r} z_{ij}\right)^2\right\}\right] - \left[V(\bar{t}) + \{E(\bar{t})\}^2\right]$$

$$= E_I\left[E_p\frac{1}{k}\sum_{i=1}^{k}\left\{\frac{1}{r^2}\left(\sum_{j\in s_r}\sigma_{ij}^2 + \sum_{j\neq}\sum_{k\in s_r}\sigma_{ij,ik}\right) + \left(\frac{1}{r}\sum_{j\in s_r} x_{ij}\right)^2\right\}\right]$$

$$-\left[V(\bar{t}) + \overline{X}^2\right]$$

$$\left(\text{writing } \overline{X} = X/(MN), X = \sum_{i=1}^{M} X_i\right)$$

$$= E_I\left[\frac{1}{k}\sum_{i=1}^{k}\left\{\frac{1}{r}\left(\frac{1}{N}\sum_{j\in U}\sigma_{ij}^2 + \frac{r-1}{N(N-1)}\sum_{j\neq}\sum_{k\in U}\sigma_{ij,ik}\right)\right.\right.$$

$$\left.\left. + \left(\frac{1}{r} - \frac{1}{N}\right)S_y^2 + \overline{X}_i^2\right\}\right] - \left[V(\bar{t}) + \overline{X}^2\right]$$

$$= \frac{1}{rMN}\sum_{i=1}^{M}\left(\sum_{j\in U}\sigma_{ij}^2 + \frac{r-1}{(N-1)}\sum_{j\neq}\sum_{k\in U}\sigma_{ij,ik}\right)$$

$$+\left(\frac{1}{r} - \frac{1}{N}\right)S_y^2 + \frac{1}{M}\sum_{i=1}^{M}\overline{X}_i^2 - \left[V(\bar{t}) + \overline{X}^2\right]$$

$$= \frac{A}{r} + \left(1 - \frac{r}{N}\right)\frac{S_y^2}{r} + \frac{M-1}{M}S_B^2 - V(\bar{t})$$

$$= (k-1)\left[V(\bar{t}) + \frac{S_b^2}{M}\right]$$

Thus $E\left[\widehat{V}(\bar{t})\right] \cong V(\bar{t})$ if M is large.

Theorem 15.5.4

Under the scheme B and model (Eq. 15.5.9)

(i) $E(\bar{t}) = \overline{Y} + \overline{B}$

(ii) $V(\bar{t}) = \left(1 - \frac{k}{M}\right)\frac{S_B^2}{k} + \left(1 - \frac{n}{N}\right)\frac{S_y^2}{n} + \frac{A}{n}$

and

(iii) An unbiased estimator of $V(\bar{t})$ is

$$\widehat{V}(\bar{t}) = \frac{1}{k(k-1)} \sum_{i=1}^{k} (t_i - \bar{t})^2$$

if M and N are large.

Proof

$$\text{(i) } E(\bar{t}) = E_I\left[\frac{1}{k}\sum_{i=1}^{k} E_p\{E_m(t_i)\}\right]$$

$$= E_I\left[\frac{1}{k}\sum_{i=1}^{k} E_p\left(\frac{1}{r}\sum_{j\in s_i} x_{ij}\right)\right]$$

$$= E_I\left(\frac{1}{k}\sum_{i=1}^{k} \overline{X}_i\right)$$

$$= \overline{Y} + \overline{B}$$

(ii)

$$V(\bar{t}) = V_I\left[E_{pm}(\bar{t})\right] + E_I\left[V_{pm}(\bar{t})\right]$$

$$= V_I\left(\frac{1}{k}\sum_{i=1}^{k}\overline{X}_i\right) + \frac{1}{k^2}E_I\left[V_p\left\{\sum_{i=1}^{k}\left(\frac{1}{r}\sum_{j\in s_i} x_{ij}\right)\right\}\right.$$

$$\left. + E_p\sum_{i=1}^{k}\frac{1}{r^2}\left(\sum_{j\in s_i}\sigma_{ij}^2 + \sum_{j\neq}\sum_{l\in s_i}\sigma_{ij,il}\right)\right]$$

$$= \left(\frac{1}{k} - \frac{1}{M}\right)S_B^2 + \frac{1}{k^2}E_I\left[\left\{\sum_{i=1}^{k} V_p\left(\frac{1}{r}\sum_{j\in s_i} x_{ij}\right)\right.\right.$$

$$\left. + \sum_{i\neq}^{k}\sum_{l=1}^{k} Cov\left(\frac{1}{r}\sum_{j\in s_i} x_{ij}, \frac{1}{r}\sum_{j\in s_l} x_{lj}\right)\right\}$$

$$\left. + \frac{1}{rN}\sum_{i=1}^{k}\left(\sum_{j\in s_i}\sigma_{ij}^2 + \sum_{j\neq}\sum_{l\in s_i}\sigma_{ij,il}\right)\right]$$

$$= \left(1 - \frac{k}{M}\right)\frac{S_B^2}{k} + \left(1 - \frac{n}{N}\right)\frac{S_y^2}{n} + \frac{A}{n}$$

(noting $Cov\left(\frac{1}{r}\sum_{j\in s_i} x_{ij}, \frac{1}{r}\sum_{j\in s_k} x_{ik}\right) = -S_y^2/N$ and $n = kr$)

$$\text{(iii)}\quad (k-1)E\left[\widehat{V}(\bar{t})\right] = E\left(\frac{1}{k}\sum_{i=1}^{k} t_i^2 - \bar{t}^2\right)$$

$$= E_I E_p\left[\frac{1}{k}\sum_{i=1}^{k} E_m\left(t_i^2\right)\right] - \left(V(\bar{t}) + \overline{X}^2\right)$$

$$= E_I E_p\left[\frac{1}{k}\sum_{i=1}^{k} V_m(t_i) + \left(\frac{1}{r}\sum_{j\in s_i} x_{ij}\right)^2\right] - \left(V(\bar{t}) + \overline{X}^2\right)$$

$$= E_I E_p\left[\frac{1}{k}\sum_{i=1}^{k}\frac{1}{r^2}\left(\sum_{j\in s_i}\sigma_{ij}^2 + \sum_{j\neq}\sum_{l\in s_i}\sigma_{ij,il}\right)\right.$$

$$\left. + \left(\frac{1}{r}\sum_{j\in s_i} x_{ij}\right)^2\right] - \left(V(\bar{t}) + \overline{X}^2\right)$$

$$= \frac{1}{rMN}\sum_{i=1}^{M}\left(\sum_{j\in U}\sigma_{ij}^2 + \frac{r-1}{(N-1)}\sum_{j\neq}\sum_{k\in U}\sigma_{ij,ik}\right)$$

$$+ \left(\frac{1}{r} - \frac{1}{N}\right)S_y^2 + \frac{1}{M}\sum_{i=1}^{M}\overline{X}_i^2 - \left(V(\bar{t}) + \overline{X}^2\right)$$

$$= \frac{A}{r} + \left(1 - \frac{r}{N}\right)\frac{S_y^2}{r} + \frac{M-1}{M}S_b^2 - V(\bar{t})$$

$$= (k-1)\left(V(\bar{t}) + \frac{S_B^2}{M} + \frac{S_y^2}{N}\right)$$

Hence $E\left[\widehat{V}(\bar{t})\right] = V(\bar{t})$ if M and N are large.

Remark 15.5.1

The variance $V(\bar{t})$ based on the sampling scheme B is much lower than that based on the sampling scheme A.

Remark 15.5.2

If we suppose that there are only k investigators, i.e., $k = M$ (Särndal et al., 1992) and each of them is allotted one sample at random for collection of data, then for both the sampling schemes A and B, $V(\bar{t})$ become free of the investigator effect S_B^2 and they are, respectively, given as $\left(1 - \frac{r}{N}\right)\frac{S_y^2}{n} + \frac{A}{n}$ and $\left(1 - \frac{n}{N}\right)\frac{S_y^2}{n} + \frac{A}{n}$. In this situation $\widehat{V}(\bar{t})$ becomes unbiased for $V(\bar{t})$ if

S_B^2 is very small for the sampling scheme A, but for the sampling scheme B, the unbiasedness condition of $\widehat{V}(\bar{t})$ requires N should be large in addition to small value of S_B^2.

Remark 15.5.3

If $\sigma_{ij}^2 = \sigma_i^2$, $\sigma_{ij,ik} = \rho_i\sigma_i^2$, $\overline{\sigma}^2 = \sum_{i=1}^{M} \sigma_i^2/M$, and $\overline{\rho} = \sum_{i=1}^{M} \rho_i\sigma_i^2 / \sum_{i=1}^{M} \sigma_i^2$, then $V(\bar{t})$ reduces to $\left(1 - \frac{k}{M}\right)\frac{S_B^2}{k} + \left(1 - \frac{r}{N}\right)\frac{S_y^2}{n} + \frac{(1+(r-1)\overline{\rho})\overline{\sigma}^2}{n}$ and $\left(1 - \frac{k}{M}\right)\frac{S_b^2}{k} + (1-f)\frac{S_y^2}{n} + \frac{(1+(r-1)\overline{\rho})\overline{\sigma}^2}{n}$ for sampling schemes A and B, respectively.

15.6 EXERCISES

15.6.1 Describe the sources of nonsampling errors and method of controlling these errors.

15.6.2 What is nonresponse? Describe a few methods of controlling nonresponses in a survey.

15.6.3 A sample of 30 households was selected from 100 households in a locality. 20 of them responded and the remaining 10 did not respond. The mean income of the responded households was \$8500. A sample of 4 households was selected from the 10 nonresponded by SRSWOR and the sample mean income was \$12,000. Estimate the mean income of the locality and its standard error. State clearly the assumption made for the estimation.

15.6.4 A sample of 10 households was interviewed in a certain region by telephone. The number of days the respondents were at home for the last 5 consecutive days at the time of the interview was recorded along with their weekly expenditure on fuel. The data collected are given below.

Households	1	2	3	4	5	6	7	8	9	10
No. of days at home	3	2	4	4	5	0	0	1	4	5
Weekly expenditure on fuel (in Rs)	325	400	125	150	215	320	420	370	120	150

Using Politz and Simmons technique, give an estimate of the average weekly consumption of fuel and an unbiased estimator of the variance of the estimator used.

15.6.5 The following table gives the production and area under wheat cultivation of 15 farms selected at random from 150 farms of a block.

Farms	Area (in acre)	Production of wheat (000 kg)
1	150	120
2	160	150
3	70	56
4	120	?
5	80	60
6	100	75
7	150	?
8	65	?
9	120	100
10	80	70
11	90	100
12	150	110
13	60	?
14	100	80
15	80	60

?, indicates missing observation.

Estimate the average production of wheat of the block by (i) mean, (ii) ratio, and (iii) regression methods of imputations. Estimate the standard errors of the estimators used.

15.6.6 A sample of 10 doctors is selected by SRSWOR method from 60 doctors in a hospital and number of patients who consulted on a certain day is given below:

Doctor	1	2	3	4	5	6	7	8	9	10
No. of patients consulted	30	25	20	25	?	10	20	?	?	30

?, indicates missing observation.

Estimate the average number of patients consulted per day by a doctor and its standard error assuming (i) nonresponse is at random,

(ii) probability of obtaining response from each of the doctors is 0.80, and (iii) response sample is an SRSWOR subsample from the selected sample.

15.6.7 A sample of s of n households is selected by some probability sampling with inclusion probability π_i for the ith unit. Selected households were interviewed once from Monday to Friday. If a respondent was found at home, he/she was asked about the expenditure on food (y) for that week along with the information regarding how many days he/she was home for the 4 preceding working days. Let k_i be the number of days the respondent was at home during the preceding 4 working days and s' be the set of respondents. Show that $t = \dfrac{\sum_{i \in s'} \dfrac{5y_i}{(k_i+1)\pi_i}}{\sum_{i \in s'} \dfrac{5}{(k_i+1)\pi_i}}$ is an unbiased estimator for the mean weekly consumption on food. Derive an approximate mean square error of t.

15.6.8 Let a sample s of size n be selected from a population of N units by SRSWOR method. Let the observed response y_j be obtained from the jth unit related with the true response μ_j through the model $y_i = \mu_i + a$ with a as a constant bias. Show that the sample mean $\overline{y}$, ratio estimator $\overline{y}_R = \dfrac{\overline{y}}{\overline{x}}\overline{X}$, and regression estimator $\overline{y}_{lr} = \overline{y} - b(\overline{x} - \overline{X})$ have a constant bias while the estimated error variances of $\overline{y}$, ratio, and regression estimators are free from the measurement bias.

CHAPTER 16

Randomized Response Techniques

16.1 INTRODUCTION

In socioeconomic and biometric research, we very often gather information relating to highly sensitive issues such as induced abortion, drug addiction, HIV infection status, duration of suffering from AIDS, sexual behavior, incidence of domestic violence, and tax evasion, for example. In these situations employing the direct method of interview (asking questions directly to the respondents), the respondents often provide untrue responses or even refuse to respond because of social stigma and/or fear. Under such circumstances the randomized response (RR) techniques may be used to collect more reliable data, protect respondents' confidentiality, and avoid unacceptable rate of nonresponse. The RR technique was originated by Warner (1965).

Warner's (1965) technique was modified by Horvitz et al. (1967), Greenberg et al. (1969), Raghavrao (1971), Franklin (1989), Arnab (1990, 1996), Kuk (1990), Mangat and Singh (1990), Arcos et al. (2015), Rueda et al. (2015), among other researchers, to increase cooperation from respondents and efficiency. Applications of the RR technique in real-life situations were reported by Greenberg et al. (1969): Illegitimacy of offspring; Abernathy et al. (1970): Incidence of induced abortions; Goodstadt and Gruson (1975): Drug uses; Folsom (1973): Drinking and driving; Van der Heijden et al. (1998): Social security fraud, and Arnab and Mothupi (2015): Sexual habits of University students. Details have been given by Chaudhuri and Mukherjee (1988), Singh (2003), and Chaudhuri (2011). In this chapter, some of the RR techniques that are used for qualitative and quantitative characteristics have been described. Methods of estimation of population characteristics for various sampling designs, measures of protection of privacy, and optimality of RR strategies under various superpopulation models have also been described.

Survey Sampling Theory and Applications
ISBN 978-0-12-811848-1
http://dx.doi.org/10.1016/B978-0-12-811848-1.00016-9

16.2 RANDOMIZED RESPONSE TECHNIQUES FOR QUALITATIVE CHARACTERISTICS

16.2.1 Warner's Technique: the Pioneering Method

The RR technique was introduced by Warner (1965) for estimating π_A, the proportion of units in a population possessing a certain stigmatized character A such as HIV infection status. In Warner's method, a sample s of size n is selected from a population U of size N by simple random sampling with replacement (SRSWR) procedure. Because the character A is sensitive, the respondent (unit) need not supply information directly about whether he/she possesses the character A (member of the group A), instead he/she is asked to provide an RR based on an RR device as follows:

The respondent has to select a card at random from a pack of well-scaffolded cards. The pack consists of two types of cards with known proportions, which are identical in appearance. Card type 1, with proportion $P_1(\neq 1/2)$ contains the question "Do you belong to group A?" whereas card type 2 with proportion $1-P_1$ contains the question "Do you belong to group $\overline{A}$?" where $\overline{A}$ is the complement of group A. The respondent will supply a truthful answer "Yes" or "No" for the question mentioned in the selected card. The experiment is performed in the absence of the interviewer and hence the privacy of the respondent is maintained because the interviewer will not know which of the two questions the respondent has answered. We will call this RR model as R_w.

16.2.1.1 Estimation of Proportion

Because the sample s is selected by SRSWR method, a unit may be selected more than once in s. In particular, if a respondent is selected k (≥ 1) times, then he/she has to perform the randomized device k times independently and supplies k RRs. Let λ_{wA} be the proportion of "Yes" answers obtained from the respondents selected in the sample. Then we have the following theorem.

Theorem 16.2.1

(i) $\widehat{\pi}_{Aw} = \dfrac{\lambda_{wA}-(1-P_1)}{2P_1-1}$ is an unbiased estimator of π_A

(ii) $Var(\widehat{\pi}_{Aw}) = \dfrac{\pi_A(1-\pi_A)}{n} + \dfrac{P_1(1-P_1)}{n(2P_1-1)^2}$

(iii) An unbiased estimator of $Var(\widehat{\pi}_{Aw})$ is

$$\widehat{V}ar(\widehat{\pi}_{Aw}) = \frac{\lambda_{wA}(1-\lambda_{wA})}{(n-1)(2P_1-1)^2}$$

Proof

Let θ_w be the probability of getting a "Yes" response from a person selected at rth ($r = 1,\ldots, n$) draw. Then,

θ_w = Conditional probability of getting a "Yes" response from a respondent given that the person belongs to group $A\times$ unconditional probability of the selected person belonging to group $A+$ conditional probability of getting "Yes" response given that the person belongs to group $\overline{A}\times$ unconditional probability of the selected person belonging to group $\overline{A}$

$$= P_1\pi_A + (1 - P_1)(1 - \pi_A) \qquad (16.2.1)$$

Because probability of getting a "Yes" answer at any draw is θ_w and the RRs are independent, the proportion λ_{wA} follows the binomial distribution with parameters and θ_w. So we get

(i) $E(\lambda_{wA}) = \theta_w = \pi_A(2P_1 - 1) + (1 - P_1)$ and hence $E(\widehat{\pi}_A) = \pi_A$

(ii) $$Var(\widehat{\pi}_{Aw}) = \frac{Var(\lambda_{wA})}{(2P_1 - 1)^2}$$

$$= \frac{\theta_w(1 - \theta_w)}{n(2P_1 - 1)^2}$$

$$= \frac{\pi_A(1 - \pi_A)}{n} + \frac{P_1(1 - P_1)}{n(2P_1 - 1)^2}$$

(iii) Since $\widehat{V}ar(\lambda_{wA}) = \dfrac{\lambda_{wA}(1 - \lambda_{wA})}{n - 1}$ is an unbiased estimator of $Var(\lambda_{wA})$, we find that

$$\widehat{V}ar(\widehat{\pi}_{Aw}) = \frac{\widehat{V}ar(\lambda_{wA})}{(2P_1 - 1)^2}$$

$$= \frac{\lambda_{WA}(1 - \lambda_{WA})}{(n - 1)(2P_1 - 1)^2}$$

16.2.1.2 Comparison With Direct Response Surveys

Suppose that the selected individual were asked directly whether he/she belongs to the sensitive group or not and $\widehat{\pi}_A$ is the proportion of "Yes" answers in the sample. In this situation we would get $E(\widehat{\pi}_A) = \pi_A$ and $Var(\widehat{\pi}_A) = \dfrac{\pi_A(1 - \pi_A)}{n}$. Clearly, $Var(\widehat{\pi}_{Aw})$ exceeds $Var(\widehat{\pi}_A)$ and the loss of efficiency of using Warner's RR technique used for the protection of confidentiality of the respondents is

$$1 - Var(\widehat{\pi}_A)/Var(\widehat{\pi}_{Aw}) = \left[1 + \frac{\pi_A(1-\pi_A)}{P_1(1-P_1)}(2P_1-1)^2\right]^{-1} \quad (16.2.2)$$

16.2.1.3 Maximum Likelihood Estimation of Proportion

Let m be the total number of "Yes" answers and $n - m$ be the total number of "No" answers obtained from an RR survey data. The likelihood function is given by

$$L = \binom{n}{m} \theta_w^m \ (1-\theta_w)^{n-m} \quad (16.2.3)$$

Maximizing L with respect to θ_w, Warner observed that the maximum likelihood estimator (MLE) of θ_w is $\widehat{\theta}_w = m/n = \lambda_{wA}$ and consequently the MLE of π_A is

$$\widehat{\pi}_{Aw} = \left\{\widehat{\theta}_w - (1-P_1)\right\}\Big/(2P_1 - 1) \quad \text{with } P_1 \neq 1/2 \quad (16.2.4)$$

Singh (1976, 1978) pointed out that the proposed estimators $\widehat{\theta}_w$, as well as λ_{wA}, are not the MLEs of θ_w and π_A, respectively, because they can take values outside the parametric space [0,1]. For example, $\widehat{\pi}_{Aw} = -0.25$, when $P_1 = 0.7$ and $\lambda_{wA} = 0.2$.

In case $P_1 > 1/2$, then θ_w must lie in the interval $[1 - P_1, P_1]$ as $\pi_A \in [0,1]$.

Hence for $P_1 > 1/2$, the true MLE π_A is

$$\widetilde{\pi}_A = \begin{cases} \widehat{\pi}_{Aw} & \text{if } 1 - P_1 < \lambda_{wA} < P_1 \\ 1 & \text{if } \lambda_{wA} \geq P_1 \\ 0 & \text{if } \lambda_{wA} \leq 1 - P_1 \end{cases} \quad (16.2.5)$$

Similarly for $P_1 < 1/2$, the MLE of π_A is

$$\widetilde{\pi}_A = \begin{cases} \widehat{\pi}_{Aw} & \text{if } P_1 < \lambda_{wA} < 1 - P_1 \\ 1 & \text{if } \lambda_{wA} \leq P_1 \\ 0 & \text{if } \lambda_{wA} \geq 1 - P_1 \end{cases} \quad (16.2.6)$$

Singh (1978) reported that the estimator $\widehat{\pi}_{Aw}$ is inadmissible as the mean square errors of $\widehat{\pi}_{Aw}$ are larger than those of $\widetilde{\pi}_A$. For further information, readers are referred to Devore (1977) and Chaudhuri and Mukherjee (1988).

16.2.2 Greenberg et al.: Unrelated Question Method

Greenberg et al. (1969) modified Warner's method by incorporating a sensitive question (character y) along with a nonsensitive question (character x). In this method, each of the respondents selected in the sample has to pick a card at random from a pack containing two types of identical-looking cards with known proportions as in Warner's model. The type 1 cards bear the sensitive question "Do you belong to the sensitive group A?" with proportion $P_2(\neq 0)$ whereas card type 2 (with proportion $1 - P_2$) bears a question of unrelated or nonsensitive characteristic x such as "Are you an African?". We will call this RR technique as R_g. Greenberg et al. anticipated that this method may receive greater cooperation from respondents as it boosts the degree of privacy.

16.2.2.1 Estimation of Proportion

(i) π_x is known.

Let a sample of n units be selected by SRSWR method. Each of the respondents in the sample has to perform an RR independently. The probability of getting a "Yes" answer for this R_g technique is $\theta_g = P_2\pi_A + (1 - P_2)\pi_x$, where π_A and π_x are the proportions of persons in the population possessing the sensitive character y and nonsensitive character x, respectively. Let λ_g be the proportion of "Yes" answers in the sample and assume that π_x is known. Then we arrive at the following theorem.

Theorem 16.2.2

(i) $\widehat{\pi}_{Ag} = \dfrac{\lambda_g - (1 - P_2)\ \pi_x}{P_2}$ is an unbiased estimator of π_A

(ii) $$Var\left(\widehat{\pi}_{Ag}\right) = \frac{\theta_g(1 - \theta_g)}{nP_2^2}$$

where $\theta_g = P_2\pi_A + (1 - P_2)\pi_x$

(iii) An unbiased estimator of $Var\left(\widehat{\pi}_{Ag}\right)$ is

$$\widehat{Var}\left(\widehat{\pi}_{Ag}\right) = \frac{\lambda_g(1 - \lambda_g)}{(n - 1)P_2^2}$$

Proof

(i) Here we note that $E(\lambda_g) = \theta_g = P_2\pi_A + (1 - P_2)\pi_x$ giving $E\left(\widehat{\pi}_g\right) = \pi_A$
The remaining parts (ii) and (iii) follow from Theorem 16.2.1.

(ii) π_x is unknown.

In case π_x is unknown, two independent SRSWR samples s_1 and s_2 of sizes n_1 and $n_2(=n-n_1)$, respectively, are selected from the population. Each of the respondents in sample s_1 performs a randomized device R_g with $P_2 = P_{21}$ whereas the individuals in sample s_2 perform R_g with $P_2 = P_{22}$ ($\neq P_{21}$). Let $\lambda_g(1)$ and $\lambda_g(2)$ be the proportion of "Yes" answers for the samples s_1 and s_2, respectively.

Theorem 16.2.3

(i) $\widehat{\pi}^*_{Ag} = \dfrac{(1-P_{22})\lambda_g(1) - (1-P_{21})\lambda_g(2)}{P_{21}-P_{22}}$ is an unbiased estimator of π_A

(ii) $$Var\left(\widehat{\pi}^*_{Ag}\right) = \frac{1}{(P_{21}-P_{22})^2}\left[\frac{(1-P_{22})^2\theta_{g1}(1-\theta_{g1})}{n_1} + \frac{(1-P_{21})^2\theta_{g2}(1-\theta_{g2})}{n_2}\right]$$

where $\theta_{gi} = P_{2i}\pi_A + (1-P_{2i})\pi_x$ for $i = 1, 2$.

(iii) An unbiased estimator of $Var\left(\widehat{\pi}^*_{Ag}\right)$ is

$$\widehat{V}ar\left(\widehat{\pi}^*_{Ag}\right) = \frac{1}{(P_{21}-P_{22})^2}\left[\frac{(1-P_{22})^2\lambda_g(1)(1-\lambda_g(1))}{n_1-1} + \frac{(1-P_{21})^2\lambda_g(2)(1-\lambda_g(2))}{n_2-1}\right]$$

Proof

The theorem can be verified easily by noting $E[\lambda_g(i)] = \theta_{gi}$, $Var[\lambda_g(i)] = \dfrac{\theta_{gi}(1-\theta_{gi})}{n_i}$, and an unbiased estimator of $Var[\lambda_g(i)]$ is $\widehat{V}ar[\lambda_g(i)] = \dfrac{\lambda_g(i)(1-\lambda_g(i))}{n_i-1}$ for $i = 1, 2$.

The expression $Var\left(\widehat{\pi}^*_{Ag}\right)$ becomes small if the difference $P_{21} - P_{22}$ is large, hence for the optimal choices of P_{2i}'s, one should choose one of the P_{2i}'s to be as large as possible and the other as small as possible. The optimum value of n_i ($i = 1,2$) obtained by minimizing $Var\left(\widehat{\pi}_{g1}\right)$ with respect to n_i, keeping $n(=n_1+n_2)$ fixed is given by

$$opt(n_i) = n\frac{\sqrt{(1-P_{22})^2\theta_{g1}(1-\theta_{g1})}}{\sqrt{(1-P_{22})^2\theta_{g1}(1-\theta_{g1})} + \sqrt{(1-P_{21})^2\theta_{g2}(1-\theta_{g2})}}$$

The minimum variance of $Var\left(\widehat{\pi}^*_{Ag}\right)$ with $n_i = opt(n_i)$ is given by

$$V_{\min}\left(\widehat{\pi}^*_{Ag}\right) = \frac{\left[\sqrt{(1-P_{22})^2\theta_{g1}(1-\theta_{g1})} + \sqrt{(1-P_{21})^2\theta_{g2}(1-\theta_{g2})}\right]^2}{n(P_{21}-P_{22})^2}$$

16.2.3 Kuk's Model

In Kuk's (1990) randomized device, two boxes each containing black and white cards with known proportions $P_{31}, 1-P_{31}$ and $P_{32}, 1-P_{32} (P_{31} \neq P_{32})$ are constructed. Respondents belonging to group A should choose box 1 whereas others (those belonging to $\overline{A}$) should choose box 2 and draw c cards at random and with replacement. The respondent will then report the number of black cards drawn, as his/her RR (z). We will denote Kuk's RR technique by R_k.

Let a sample of size n be selected from the population by SRSWR and λ_{kr} be the proportion of black cards chosen by the respondent selected at the rth draw, $r = 1, \ldots, n$. Then

$$\begin{aligned} E(\lambda_{kr}) &= \text{Probability}(\text{respondent} \in A) \times E(\lambda_{kr}|\text{respondent} \in A) \\ &\quad + \text{Probability}(\text{respondent} \notin A) \times E_R(\lambda_{kr}|\text{respondent} \notin A) \\ &= \pi_A P_{31} + (1-\pi_A)P_{32} \\ &= \theta_k \end{aligned} \tag{16.2.7}$$

$$\begin{aligned} V(\lambda_{kr}) &= E\left(\lambda_{kr}^2\right) - \theta_k^2 \\ &= \text{Probability}(\text{respondent} \in A) \times E\left(\lambda_{kr}^2|\text{respondent} \in A\right) \\ &\quad + \text{Probability}(\text{respondent} \notin A) \times E\left(\lambda_{kr}^2|\text{the respondent} \notin A\right) - \theta_k^2 \\ &= \pi_A\left(\frac{P_{31}(1-P_{31})}{c} + P_{31}^2\right) + (1-\pi_A)\left(\frac{P_{32}(1-P_{32})}{c} + P_{32}^2\right) - \theta_k^2 \\ &= \frac{\pi_A P_{31} + (1-\pi_A)P_{32}}{c} + \left(1-\frac{1}{c}\right)\left\{\pi_A P_{31}^2 + (1-\pi_A)P_{32}^2\right\} - \theta_k^2 \\ &= \frac{\theta_k(1-\theta_k)}{c} + \left(1-\frac{1}{c}\right)\pi_A(1-\pi_A)(P_{31}-P_{32})^2 \\ &= \Phi_k \end{aligned} \tag{16.2.8}$$

Since λ_{kr} are iid random variables with mean θ_k and variance Φ_k, we have the following result.

Theorem 16.2.4

(i) $\widehat{\pi}_{Ak} = \dfrac{\overline{\lambda}_k - P_{32}}{P_{31} - P_{32}}$ is an unbiased estimator of π_A

where $\overline{\lambda}_k = \dfrac{1}{n}\sum_{r=1}^{n} \lambda_{kr}$ and $P_{31} \neq P_{32}$.

(ii) $$V(\widehat{\pi}_{Ak}) = \frac{\Phi_k}{n(P_{31} - P_{32})^2}$$

$$= \frac{1}{n}\left[\frac{\theta_k(1-\theta_k)}{c(P_{31} - P_{32})^2} + \pi_A(1-\pi_A)\left(1 - \frac{1}{c}\right)\right]$$

(iii) $\widehat{V}(\widehat{\pi}_{Ak}) = \dfrac{1}{n(n-1)(P_{31} - P_{32})^2}\sum_{r=1}^{n}\left(\lambda_{kr} - \overline{\lambda}_k\right)^2$ is an unbiased estimator of $V(\widehat{\pi}_{wk})$

The variance $V(\widehat{\pi}_{wk})$ decreases as c increases. Hence we can increase the efficiency of $\widehat{\pi}_{Ak}$ by increasing c. Kook's RR technique R_k reduces to Warner's technique R_w when $P_{31} = P_{32} = P$ and $c = 1$.

16.2.4 Mangat and Singh Model

Mangat and Singh (1990) proposed the two-stage RR model where in stage 1, the sampled respondent is to pick a card at random from a pack containing two types of cards with proportions T and $1 - T$ with written statements "disclose your membership to A or $\overline{A}$" and "go to randomized device R_w," respectively. If a respondent selects the card written "disclose your membership to A or $\overline{A}$", then the respondent has to reveal his/her membership to A or $\overline{A}$ truthfully, i.e., has to report "Yes" if he/she belongs to A and "No" otherwise. On the other hand if the respondent picks a card written "go to randomized device R_w", then the respondent performs Warner RR device. In this case the respondent is to pick a card from a pack containing two types of cards written "I belong to group A" with proportion P_4 and "I am a member of $\overline{A}$" with proportion $1 - P_4$, respectively, and answer "Yes" or "No." The confidentiality of the respondent is maintained because the entire experiment was performed in absence of the investigator. We call this RR technique as R_{ms}. Mangat and Singh (1990) used SRSWR method for the selection of the sample. Clearly the two-stage model reduces to Warner model if $T = 0$. For this R_{ms} model the probability of obtaining answer "Yes" from a respondent selected by SRSWR sampling procedure is

$$\theta_{ms} = \pi_A\{T + (1-T)(2P_4 - 1)\} + (1 - P_4)(1 - T) \qquad (16.2.9)$$

Let λ_{ms} be the proportion of "Yes" answers obtained from the sample of n respondents. Noting $E(\lambda_{ms}) = \theta_{ms}$, $Var(\lambda_{ms}) = \dfrac{\theta_{ms}(1-\theta_{ms})}{n}$, and $\widehat{V}ar(\lambda_{ms}) = \dfrac{\lambda_{ms}(1-\lambda_{ms})}{n-1}$, an unbiased estimator of $Var(\lambda_{ms})$, we have the following theorem.

Theorem 16.2.5

(i) $\widehat{\pi}_{Ams} = \dfrac{\lambda_{ms} - (1-P_4)(1-T)}{D}$ is an unbiased estimator of π_A

where $D = T + (1-T)(2P_4 - 1)$ and $T \neq \dfrac{1-2P_4}{2(1-P_4)}$

(ii) $V(\widehat{\pi}_{Ams}) = \dfrac{\theta_{ms}(1-\theta_{ms})}{nD^2}$

$$= \frac{1}{n}\left[\pi_A(1-\pi_A) + \frac{(1-T)(1-P_4)\{1-(1-T)(1-P_4)\}}{D^2}\right]$$

(iii) $\widehat{V}(\widehat{\pi}_{Ams}) = \dfrac{\lambda_{ms}(1-\lambda_{ms})}{(n-1)D^2}$ is an unbiased estimator of $V(\widehat{\pi}_{Ams})$

16.3 EXTENSION TO MORE THAN ONE CATEGORIES

So far, we have classified the population into two categories viz. sensitive and nonsensitive. Now, we consider situations where the population is divided into more than two mutually exclusive and exhaustive categories. For example, married women may be classified into three categories, two of which are sensitive viz. having exactly one, or more than one sexual partner other than their husbands and the nonsensitive category comprises of those who have no other sexual partner other than their husbands. To develop the theory, let us suppose that the population is classified into k mutually exclusive and exhaustive categories among which at most $k-1$ of them are designated as sensitive categories. Let the proportion of persons belonging to the jth category be π_j with $\sum \pi_j = 1$. Here we will consider the method of estimating of π_j for $j = 1,\ldots, k$.

16.3.1 Liu and Chow's Technique

In Liu and Chow's (1976) RR technique, two different colors (red and green, for example) of balls but identical in shape are placed inside a flask with a transparent narrow neck, through which only one ball can pass. Green balls are marked with the numbers 1, 2,..., k. The proportions of red balls and green balls marked with the number j are $p(>0)$ and $p_j(>0)$, where

$j = 1, \ldots, k$, respectively. Clearly, $\sum p_j = 1 - p$. Respondents are asked to close the flask, shake it thoroughly, and then turn it upside down and are told to concentrate on the bottommost ball. The respondent is required to report his/her category truthfully if the bottommost ball is red, otherwise (if it is green) he/she should report the number written on the green ball. The whole experiment is to be performed in the absence of the investigator. So the respondent is to report only one number between 1 and k. The confidentiality is maintained because the investigator will know only one number between 1 and k but will not know whether that number represents respondent's true category or the number on the bottommost green ball.

16.3.1.1 Estimation of Proportions

Let a sample of size n be selected from the population by SRSWR procedure and n_j be the number of times the RR "j" was obtained $\left(\sum n_j = n\right)$. Under this RR technique the probability of getting a response "j" is

$$\lambda_j = \text{Probability that the bottom} - \text{most ball is red and the respondent belongs to the } j\text{th group} + \text{Probability that the bottom} - \text{most ball is green and marked with } j$$

$$= p\pi_j + p_j \quad \text{for } j = 1, \ldots, k$$

Since n_j follows multinomial distribution, $\widehat{\lambda}_j = \dfrac{n_j}{n}$ becomes an unbiased estimator of λ_j and we have the following theorem.

Theorem 16.3.1

(i) An unbiased estimator of π_j is $\widehat{\pi}_j = \left(\widehat{\lambda}_j - p_j\right)/p$

(ii) Variance of $\widehat{\pi}_j$ is $V\left(\widehat{\pi}_j\right) = \lambda_j(1 - \lambda_j)/(np^2)$

(iii) An unbiased estimator of $V\left(\widehat{\pi}_j\right)$ is $\widehat{V}\left(\widehat{\pi}_j\right) = \widehat{\lambda}_j\left(1 - \widehat{\lambda}_j\right)/\{(n-1)p^2\}$

Proof

Since n_j follows multinomial distribution with parameter λ_j, we have

(i) $$E\left(\widehat{\pi}_j\right) = E\left(\frac{\hat{\lambda}_j - p_j}{p}\right) = \frac{(\lambda_j - p_j)}{p} = \pi_j;$$

(ii) $$V\left(\widehat{\pi}_j\right) = \frac{V\left(\hat{\lambda}_j\right)}{p^2} = \frac{\lambda_j(1 - \lambda_j)}{np^2} \quad \text{and}$$

(iii) $$E\left\{\widehat{V}\left(\widehat{\pi}_j\right)\right\} = \frac{E\left(\hat{\lambda}_j\right) - \left\{V\left(\hat{\lambda}_j\right) + \lambda_j^2\right\}}{(n-1)p^2}$$

$$= V\left(\widehat{\pi}_j\right)$$

Remark 16.3.1
The estimator $\widehat{\pi}_j$ takes a negative value if $\widehat{\lambda}_j < p_j$.

16.4 RANDOMIZED RESPONSE TECHNIQUES FOR QUANTITATIVE CHARACTERISTICS

Let y be a sensitive and quantitative characteristic, such as expenditure on the prohibitive drug marijuana, and y_i be the value of the character y of the ith unit of the population and $\mathbf{y} = (y_1, \ldots, y_i, \ldots, y_N)$ be the unknown population vector.

16.4.1 Eriksson's Technique

Here we assume that y_i can take any value in the known interval (a,b). In Eriksson's (1973) RR technique, M values $Q_1(=a)$, $Q_2, \ldots, Q_M$ $(=b)$ are chosen in the interval (a,b). The vector $\mathbf{Q} = (Q_1, \ldots, Q_M)$ covers the range (a,b) and the value of M depends on the length of the interval. The respondent is supposed to report either the true value y_i with probability c or the Q_j value with probability $q_j\left(q_j > 0, \sum_j q_j = 1 - c\right)$. For example, in collecting information regarding expenditures on illegal drugs, a randomized device can be framed as follows: Take a jar containing capsules that appear identical. Each of the capsules contains exactly one ticket. Capsules of c proportions contain tickets written "Disclose your true expenditure" and the remaining capsules contain tickets marked \$0(=Q_1), \$5(=Q_2),\$10(=Q_3),..., \$2000(=Q_{401}) with proportions $q_1, q_2, \ldots, q_{401}$, respectively. Here it is assumed that the minimum expenditure on drugs is zero and the maximum \$2000, and $M = 401$. The respondent is to shake the jar well and select a capsule at random. After opening the capsule, if a respondent finds the ticket "Disclose your true expenditure" then he/she has to disclose the true expenditure otherwise he/she will report the amount written in the ticket. The confidentiality of the respondent is preserved since the experiment is performed in the absence of the interviewer, and the interviewer will receive some amount as a response and the interviewer will not be able to identify whether this figure is really respondent's real expenditure or the value written on the ticket.

Let z_i be the response obtained from the ith respondent and let E_R, V_R, and C_R, respectively, denote the expectation, variance, and covariance operators with respect to the RR technique. Now, noting RRs are independent, we obtain

$$E_R(z_i) = c\ y_i + \sum_j q_j Q_j,\quad E_R\left(z_i^2\right) = c\ y_i^2 + \sum q_j Q_j^2,\ \text{and}$$

$$C_R(z_i, z_j) = 0 \ \text{for}\ i \neq j \tag{16.4.1}$$

Denoting

$$r_i = \left(z_i - \sum_j q_j Q_j\right) \Big/ c \tag{16.4.2}$$

as the revised RR for the *i*th unit, we get

$$E_R(r_i) = y_i, \ \ V_R(r_i) = V_R(z_i)/c^2 = \alpha y_i^2 + \beta y_i + \gamma = \sigma_i^2, \ \text{and} \ C_R(r_i, r_j) = 0 \ \text{for} \ i \neq j \tag{16.4.3}$$

where $\alpha = \dfrac{1-c}{c}$, $\beta = -2\sum_j q_j Q_j / c$, and

$$\gamma = \left[\sum_j q_j Q_j^2 - \left(\sum_j q_j Q_j\right)^2\right] \Big/ c^2.$$

An unbiased estimator of σ_i^2 is

$$\widehat{\sigma}_i^2 = \frac{\alpha r_i^2 + \beta r_i + \gamma}{1+\alpha}$$

16.4.2 Arnab's Model

Arnab (1990) proposed a more general RR model as follows:

$$\text{Model } R: E_R(r_i) = y_i, V_R(r_i) = \phi_i = \phi(y_i), \ \text{and} \ C_R(r_i, r_j) = 0 \ \text{for} \ i \neq j \tag{16.4.4}$$

where $\phi_i = \phi_i(y_i)$ is a function of y_i only.

Here we assume that a nonnegative unbiased estimator ϕ_i is available and it will be denoted by $\widehat{\phi}_i$. Most of the RR models for qualitative or quantitative characteristics satisfy the Model R and hence may be considered as a special case of the Model R given in Eq. (16.4.4).

Let us suppose that y is a qualitative variable and $y_i = 1$ if *i*th unit belongs to the sensitive group A and that $y_i = 0$ otherwise. In this case $\pi_A = \sum_{i \in U} y_i / N =$ proportion of individuals in the population who belong to the sensitive group A.

For a qualitative characteristic y, an unbiased estimator of $V_R(r_i) = \phi_i$ is

$$\widehat{\phi}_i = r_i(r_i - 1) \tag{16.4.5}$$

since $E_R\left(\widehat{\phi}_i\right) = E\left(r_i^2\right) - E_R(r_i) = \phi_i + y_i^2 - y_i = \phi_i$ as $y_i^2 = y_i$.

Warner's Model Let z_i be the RR obtained from the ith respondent. Here $z_i = 1$ if the ith respondent answers "Yes" and $z_i = 0$ if answers "No."

In this case we have $E_R(z_i) = y_iP_1 + (1 - y_i)(1 - P_1) = E_R\left(z_i^2\right)$ and $V_R(z_i) = P_1(1 - P_1)$.

Now writing $r_i = \dfrac{z_i - (1 - P_1)}{(2P_1 - 1)}$ and noting P_1 is a known constant, we find

$$E_R(r_i) = y_i,\quad V_R(r_i) = \phi_i = \frac{P_1(1 - P_1)}{(2P_1 - 1)^2} = \widehat{\phi}_i = \phi,\quad \text{a known constant.} \tag{16.4.6}$$

Greenberg et al. Model With Known π_x In this case $y_i = 1$ if the ith individual possesses the sensitive characteristic y and $y_i = 0$ otherwise. Similarly, $x_i = 1$ or 0 according as to whether or not the ith unit possesses the nonsensitive characteristic x. Here also, RR $z_i = 1$ if answer is "Yes" and $z_i = 0$ if answer is "No." Thus we have $E_R(z_i) = y_iP_2 + x_i(1 - P_2) = E_R\left(z_i^2\right)$ and $V_R(z_i) = P_2(1 - P_2)(y_i - x_i)^2$. The revised RR $r_i = \dfrac{z_i - x_i(1 - P_2)}{P_2}$ yields

$$E_R(r_i) = y_i \text{ and } V_R(r_i) = \frac{(1 - P_2)(y_i - x_i)^2}{P_2} = \phi_i(y_i) \text{ and } \widehat{\phi}_i = r_i(r_i - 1) \tag{16.4.7}$$

Here we note that $\phi_i(y_i)$ is unknown since it involves y_i.

Kuk's Model Let z_i be the number of black balls drawn by the ith respondent using Kuk's model R_k described in Section 16.2.3. In this case $E_R(z_i) = c[P_{31}y_i + P_{32}(1 - y_i)] = c[(P_{31} - P_{32})y_i + P_{32}]$ and $V_R(z_i) = c[P_{31}(1 - P_{31})y_i + P_{32}(1 - P_{32})(1 - y_i)]$. The revised RR $r_i = \dfrac{z_i - cP_{32}}{c(P_{31} - P_{32})}$ yields

$$E_R(r_i) = y_i,\ V_R(r_i) = \frac{P_{31}(1 - P_{31})y_i + P_{32}(1 - P_{32})(1 - y_i)}{c(P_{31} - P_{32})^2}$$

$$= \phi_i(y_i) \text{ and } \widehat{\phi}_i = r_i(r_i - 1) \tag{16.4.8}$$

Mangat and Singh (1990) Model Let $z_i = 1(0)$, if the RR response is "Yes" ("No") and $y_i = 1(0)$ if $i \in A(\overline{A})$. Then $E_R(z_i) = y_i T + (1 - T) \times \{y_i P_4 + (1 - y_i)(1 - P_4)\} = E_R(z_i^2)$ and $r_i = \dfrac{z_i - (1 - P_4)(1 - T)}{2T(1 - P_4) + (2P_4 - 1)}$ yield

$$E_R(r_i) = y_i \text{ and } V_R(r_i) = \frac{(1 - P_4)(1 - T)\{1 - (1 - P_4)(1 - T)\}}{\{2T(1 - P_4) + (2P_4 - 1)\}^2} = \phi_i$$
$$= \widehat{\phi}_i = \phi, \text{ a known constant} \tag{16.4.9}$$

16.4.3 Christofides's Model

In Christofides (2003) randomized device, a respondent is asked to pick a card at random from a box containing M different kinds of cards marked 1, 2,…, M with proportions $p_1, p_2, \ldots, p_M$ $\left(\sum_{i=1}^{M} p_i = 1\right)$. If the respondent picks a card with number "x" ($x = 1, \ldots, M$) then he/she supplies the RR $z = M + 1 - x$ provided he/she belongs to the sensitive group "A," otherwise if he/she belongs to the nonsensitive group $\overline{A}$ then respondent reports $z = x$ as his/her RR. Hence the RR obtained from the ith respondent can be expressed as

$$z_i = (M + 1 - x)y_i + x(1 - y_i) = (M + 1)y_i + x(1 - 2y_i)$$

Now writing $E_R(x) = \sum_{k=1}^{M} k p_k = \mu$ and $V_R(x) = \sum_{k=1}^{M} k^2 p_k - \mu^2 = w^2$, we find

$$E_R(z_i) = (M + 1 - 2\mu)y_i + \mu \text{ and } V_R(z_i) = (1 - 2y_i)^2 w^2$$
$$= w^2 \text{ (since } y_i = 0 \text{ or } 1)$$

Let $r_i = \dfrac{z_i - \mu}{M + 1 - 2\mu}$, then we have

$$E_R(r_i) = y_i, \quad V_R(r_i) = \left(\frac{w}{M + 1 - 2\mu}\right)^2 = \phi_i = \widehat{\phi}_i \text{ (known constant)} \tag{16.4.10}$$

16.4.4 Eichhorn and Hayre's Model

In Eichhorn and Hayre's (1983) RR technique, the respondents selected in the sample are advised to draw a random sample from some preassigned distribution such as normal, uniform, chi-square, and so on. The mean θ and variance γ^2 of the distribution are assumed to be known. If the ith respondent is included in the sample and selects a random sample Q_i then he/she is asked to report an RR $z_i = y_i Q_i/\theta$ where y_i is the true value of the sensitive characteristic y. In this situation $r_i = z_i = y_i Q_i/\theta$ and follows the model

$$E_R(r_i) = y_i,\ V_R(r_i) = y_i^2\gamma^2/\theta^2 = \phi_i \text{ and } \widehat{\phi}_i = \gamma^2 r_i^2/\left(\gamma^2+\theta^2\right) \quad (16.4.11)$$

16.4.5 Franklin's Randomized Response Technique

In Franklin's (1989) RR technique, a sample s of n units (respondents) is selected by the SRSWR method. Each of the selected respondents in s has to perform $k(\geq 1)$-independent RR trials. The ith respondent at the trial $j(=1,\ldots, k)$ has to draw a random sample from the density g_{ij} if he/she belongs to the sensitive group A, or if he/she belongs to $\overline{A}$, selects a random sample from the density h_{ij}. Confidentiality of the respondent is maintained because the interviewer will know only the random sample drawn but not the population from which it was selected. The random sample is selected by using some suitable randomized device such as a spinner or a random number table. In developing this theory, Franklin assumed $g_{ij} = g_j$ and $h_{ij} = h_j$ for every $i \in U$. He further assumed that the densities g_j and h_j are normal with known means μ_{1j} and μ_{2j} and known variances σ_{1j}^2 and σ_{2j}^2, respectively. In fact, Franklin made the randomized device much more interesting by using a portable electronic machine. If a respondent pushes a button on this machine, he/she gets two six-digit numbers labeled "Yes" and "No," respectively. If the respondent belongs to the group A ($\overline{A}$), he/she will supply a six-digit number labeled "Yes" ("No"). The first, second, and third two digits will correspond to $3(=k)$ independent samples from g_j and the remaining fourth, fifth, and sixth two digits represent random samples from h_j. Let z_{ij} be the RR obtained from the ith respondent at the jth trial and $y_i = 1(0)$ if $i \in A\left(\in\overline{A}\right)$. Then,

$$\begin{aligned}
E_R(z_{ij}) &= y_i E_R(z_{ij}|i\in A) + (1-y_i)E_R\left(z_{ij}\middle|i\in\overline{A}\right)\\
&= y_i\mu_{1j} + (1-y_i)\mu_{2j}\\
E_R\left(z_{ij}^2\right) &= y_i E_R\left(z_{ij}^2|i\in A\right) + (1-y_i)E_R\left(z_{ij}^2\middle|i\in\overline{A}\right)\\
&= y_i\left(\sigma_{1j}^2+\mu_{1j}^2\right) + (1-y_i)\left(\sigma_{2j}^2+\mu_{2j}^2\right)
\end{aligned}$$

Writing $r_{ij} = (z_{ij}-\mu_{2j})/(\mu_{1j}-\mu_{2j})$, we obtain the following RR model

$$E_R(r_{ij}) = y_i,\ V_R(r_{ij}) = \phi_{ij} = \frac{y_i\sigma_{1j}^2 + (1-y_i)\sigma_{2j}^2}{\left(\mu_{1j}-\mu_{2j}\right)^2},\quad C_R(r_{ij}, r_{i'j'}) = 0$$

$$\text{for } (i,j)\neq(i',j') \text{ and } \widehat{\phi}_{ij} = r_{ij}(r_{ij}-1) \quad (16.4.12)$$

16.4.6 Chaudhuri's Randomized Response

In Chaudhuri's (1987) RR technique, the respondent labeled i is asked to choose independently a pair of numbers $a_i(j)$, $b_i(k)$ at random out of two sets of numbers $a_i(m)$, $b_i(r)$, $m = 1,\ldots, A_i$; $r = 1,\ldots, B_i$, given to him and to report $z_i = a_i(j)y_i + b_i(k)$ as his RR. Denoting $r_i = \left(z_i - \overline{b}_i\right)/\overline{a}_i$ with $\overline{a}_i = \sum_j a_i(j)/A_i$, $\overline{b}_i = \sum_k b_i(k)/B_i$ we get,

$$E_R(r_i) = y_i, \quad V_R(r_i) = \phi_i = \alpha_i y_i^2 + \beta_i, \quad C_R(r_i, r_j) = 0 \text{ for } i \neq j \text{ and } \widehat{\phi}_i = \left(\alpha_i r_i^2 + \beta_i\right)/(1 + \alpha_i) \tag{16.4.13}$$

where $\alpha_i = S_i^2(a)/\overline{a}_i^2, \beta_i = S_i^2(b)/\overline{a}_i^2$, $S_i^2(a) = \sum_j \{a_i(j) - \overline{a}_i\}^2/A_i$ and $S_i^2(b) = \sum_j \left\{b_i(j) - \overline{b}_i\right\}^2 /B_i$.

16.5 GENERAL METHOD OF ESTIMATION

Most surveys in practice are based on complex sampling designs and information regarding more than one character is collected at a time. Some of them are of a confidential nature whereas others are not. Furthermore, the sensitive characters need not be only of a qualitative nature, it may also be quantitative. To cope with this situation, a general method of estimating the population total has been proposed in this section following the methods proposed by Arnab (1994). Expressions for the variances of the proposed estimators and unbiased estimator of the variances have been derived. Here we suppose that a sample s of size n is selected by some arbitrary sampling design p. The inclusion probabilities for the ith unit π_i, and ith and jth units π_{ij} are assumed to be positive. Let y_i be the value of the sensitive character y for the ith unit. The value y_i cannot be directly obtained from the respondent. Hence an RR r_i is obtained by applying some suitable randomized device. The RR r_is are assumed to follow the model given in Eq. (16.4.4) viz.

$$E_R(r_i) = y_i, V_R(r_i) = \phi_i \text{ and } C_R(r_i, r_j) = 0 \text{ for } i \neq j$$

It is further assumed that a nonnegative unbiased estimator of ϕ_i is available and will be denoted by $\widehat{\phi}_i$.

16.5.1 Estimation of Total and Variance

Consider a direct method of survey, where information of y_i's is directly obtained from the respondents. In this case we propose a linear homogeneous unbiased estimator for the total Y as

$$t(s, y) = \sum_{i \in s} b_{si} y_i \qquad (16.5.1)$$

where b_{si}'s are constants free from y_i's and r_i's and satisfy the unbiasedness condition

$$\sum_{s \supset i} b_{si} p(s) = 1 \;\text{ for }\; \forall \;\; i = 1, \ldots, N \qquad (16.5.2)$$

Since y_i's are not directly obtained from the respondent, we replace y_i by its estimate r_i in $t(s,y)$ and obtain the following estimator for RR survey

$$t(s, r) = \sum_{i \in s} b_{si} r_i \qquad (16.5.3)$$

Let $E_p(E_R)$, $V_p(V_R)$, and $C_p(C_R)$, respectively, denote the expectation, variance, and covariance operators with respect to the sampling design p (RR model). Here we note that the commutativity of the operators E_p and E_R holds in the sense $E_pE_R = E_RE_p$ (see Arnab, 1990) for any non-informative sampling design where $p(s)$ does not involve r_i's, $i \in s$.

Theorem 16.5.1

(i) $E[t(s, r)] = Y$

(ii) $V[t(s, r)] = V_p[t(s, y)] + \sum_i \alpha_{ii} \phi_i$

where

$$V_p[t(s, y)] = \sum_{i \in U} (\alpha_{ii} - 1) y_i^2 + \sum_{i \neq} \sum_{j \in U} (\alpha_{ij} - 1) y_i y_j,$$

$$\alpha_{ii} = \sum_{s \supset i} b_{si}^2 p(s) \;\text{ and }\; \alpha_{ij} = \sum_{s \supset i,j} b_{si} b_{sj} p(s)$$

Proof

(i) $E[t(s, r)] = E_p[E_R\{t(s, r)\}]$

$$= E_p\left(\sum_{i \in s} b_{si} y_i\right)$$

$$= Y$$

(ii) $$V[t(s,r)] = V_p[E_R\{t(s,r)\}] + E_p\left[V_R\left(\sum_{i\in s} b_{si}r_i\right)\right]$$
$$= V_p[t(s,y)] + E_p\left(\sum_{i\in s} b_{si}^2\,\phi_i\right)$$

Now noting

$$V_p[t(s,y)] = E_p\left(\sum_{i\in s} b_{si}^2\,y_i + \sum_{i\neq}\sum_{j\in s} b_{si}b_{sj}y_iy_j - Y^2\right)$$
$$= \sum_{i\in U} y_i^2\left(\sum_{s\supset i} b_{si}^2 p(s) - 1\right) + \sum_{i\neq}\sum_{j\in U} y_iy_j\left(\sum_{s\supset i,j} b_{si}b_{sj}p(s) - 1\right)$$
$$= \sum_{i\in U}(\alpha_{ii} - 1)y_i^2 + \sum_{i\neq}\sum_{j\in U}(\alpha_{ij} - 1)y_iy_j$$

and

$$E_p\left(\sum_{i\in s} b_{si}^2\phi_i\right) = \sum_{i\in U}\phi_i\sum_{s\supset i} b_{si}^2 p(s)$$
$$= \sum_{i\in U}\alpha_{ii}\phi_i$$

we can verify the theorem.

Arnab (1994) proposed the method of unbiased estimation of $V[t(s,r)]$ as follows.

Theorem 16.5.2

Let $\widehat{V}[t(s,y)]$ be a homogeneous quadratic unbiased estimator of $V[t(s,y)]$. Then an unbiased estimator of $V[t(s,r)]$ is

$$\widehat{V}^*[t(s,r)] = \widehat{V}[t(s,r)] + \sum_{i\in s} b_{si}\ \widehat{\phi}_i$$

where $\widehat{V}[t(s,r)]$ is obtained by writing r_i in place of y_i in $\widehat{V}[t(s,y)]$.

Proof

A homogenous quadratic unbiased estimator $\widehat{V}[t(s,y)]$ is of the form

$$\widehat{V}[t(s,y)] = \sum_{i\in s} c_{ii}(s)y_i^2 + \sum_{i\neq}\sum_{j\in s} c_{ij}(s)y_iy_j \qquad (16.5.4)$$

where the constants $c_{ii}(s)$ and $c_{ij}(s)$ are free from y_i and r_i s and chosen to make $\widehat{V}[t(s,y)]$ unbiased for $V[t(s,y)]$.

Now $E_p\left[\widehat{V}\{t(s,y)\}\right] = V[t(s,y)]$ for all possible values of $\mathbf{y} = (y_1,\ldots, y_N)$ implies

$$\sum_{s\supset i} c_{ii}(s)p(s) = \alpha_{ii} - 1 \text{ and } \sum_{s\supset i,j} c_{ij}(s)p(s) = \alpha_{ij} - 1 \qquad (16.5.5)$$

Furthermore, writing r_i in place of y_i in $\widehat{V}[t(s,y)]$, we get

$$\widehat{V}[t(s,r)] = \sum_{i\in s} c_{ii}(s)\, r_i^2 + \sum_{i\neq}\sum_{j\in s} c_{ij}(s) r_i r_j$$

and

$$\begin{aligned} E\left[\widehat{V}\{t(s,r)\}\right] &= E_p\left[\sum_{i\in s} c_{ii}(s)E_R\left(r_i^2\right) + \sum_{i\neq}\sum_{j\in s} c_{ij}(s)E_R(r_i r_j)\right] \\ &= E_p\left[\widehat{V}\{t(s,y)\}\right] + E_p\left(\sum_{i\in s} c_{ii}(s)\,\phi_i\right) \\ &= E_p\left(\widehat{V}\{t(s,y)\}\right) + \sum_{i\in U}(\alpha_{ii} - 1)\phi_i \\ &= V[t(s,r)] - \sum_{i\in U}\phi_i \end{aligned}$$

Now noting $E\left(\sum_{i\in s} b_{si}\ \widehat{\phi}_i\right) = E_p\left[\sum_{i\in s} b_{si}E_R\left(\widehat{\phi}_i\right)\right] = E_p\left(\sum_{i\in s} b_{si}\ \phi_i\right) = \sum_{i\in U}\phi_i$, we find

$$E\left[\widehat{V}^*\{t(s,r)\}\right] = V[t(s,r)]$$

16.5.1.1 Horvitz–Thomson Estimator

For $b_{si} = 1/\pi_i$, we have $\alpha_{ii} = 1/\pi_i$ and $\alpha_{ij} = \pi_{ij}/(\pi_i\pi_j)$. In this case the estimator (16.5.3) reduces to the Horvitz–Thomson estimator $\widehat{Y}_{ht}(rr) = \sum_{i\in s} r_i/\pi_i$. Hence substituting $\alpha_{ii} = 1/\pi_i$ and $\alpha_{ij} = \pi_{ij}/(\pi_i\pi_j)$ in Theorems 16.5.1 and 16.5.2, we get the following.

Theorem 16.5.3

For a fixed effective size sampling design

(i) $\widehat{Y}_{ht}(rr) = \sum_{i\in s} r_i/\pi_i$ is an unbiased estimator of Y.

(ii) $V\left[\widehat{Y}_{ht}(rr)\right]=\frac{1}{2}\sum_{i\neq}\sum_{j\in U}(\pi_i\pi_j-\pi_{ij})\left(\frac{y_i}{\pi_i}-\frac{y_j}{\pi_j}\right)^2+\sum_{i\in U}\frac{\phi_i}{\pi_i}$,

and

(iii) $\widehat{V}\left[\widehat{Y}_{ht}(rr)\right]=\frac{1}{2}\sum_{i\neq}\sum_{j\in s}\frac{(\pi_i\pi_j-\pi_{ij})}{\pi_{ij}}\left(\frac{r_i}{\pi_i}-\frac{r_j}{\pi_j}\right)^2+\sum_{i\in s}\frac{\hat{\phi}_i}{\pi_i}$

is an unbiased estimator of $V\left[\widehat{Y}_{ht}(rr)\right]$.

16.5.1.2 Simple Random Sampling Without Replacement

For an simple random sampling without replacement (SRSWOR) $\pi_i = n/N$ and $\pi_{ij} = n(n-1)/\{N(N-1)\}$. Substituting these values of π_i and π_{ij} in Theorem 16.5.3 we get the following theorem.

Theorem 16.5.4

For an SRSWOR sampling design

(i) $\bar{r}(s)=\sum_{i\in s} r_i/n$, the sample mean is an unbiased estimator of the population mean $\overline{Y}$,

(ii) $V[\bar{r}(s)]=\left[(1-f)S_y^2+\overline{\phi}\right]/n$,

and

(iii) $\widehat{V}[\bar{r}(s)]=\left[(1-f)s_r^2+n\overline{\widehat{\phi}}_s/N\right]/n$

where $S_y^2=\sum_{i\in U}\left(y_i-\overline{Y}\right)^2/(N-1)$, $s_r^2=\sum_{i\in s}(r_i-\bar{r}_s)^2/(n-1)$

$f=n/N$, $\overline{\phi}=\sum_{i\in U}\phi_i/N$ and $\overline{\widehat{\phi}}_s=\sum_{i\in s}\widehat{\phi}_i/n$.

Corollary 16.5.1

Let y be a qualitative variable so that $y_i = 1$ if the ith unit possesses an attribute A, and $y_i = 0$ otherwise. Furthermore, if Warner's RR model R_w is used and $z_i = 1(0)$ when "Yes" ("No") response is obtained from the ith respondent we get

$$r_i=\frac{z_i-(1-P_1)}{(2P_1-1)},\quad V_R(r_i)=\phi_i=\frac{P_1(1-P_1)}{(2P_1-1)^2}=\phi=\widehat{\phi}_i \text{ and}$$

$$S_y^2=N\pi_A(1-\pi_A)/(N-1)$$

where $\pi_A=\sum_{i\in U} y_i/N=$ population proportion.

An unbiased estimator for π_A, based on SRSWOR under R_w, is given by

$$\bar{r}(s) = \sum_{i \in s} r_i/n = \frac{\lambda_{WA} - (1 - P_1)}{(2P_1 - 1)} = \widehat{\pi}_A \tag{16.5.6}$$

where λ_{WA} is the proportion of "yes" answers in n RRs.

The expressions of variance of $\widehat{\pi}_A$ and its unbiased estimator are respectively given by

$$V(\widehat{\pi}_A) = \frac{(1-f)N}{n(N-1)}\pi_A(1-\pi_A) + \frac{P_1(1-P_1)}{n(2P_1-1)^2} \tag{16.5.7}$$

and

$$\widehat{V}(\widehat{\pi}_A) = \frac{(1-f)}{n-1}\frac{\lambda_{wA}(1-\lambda_{wA})}{(2P_1-1)^2} + \frac{P_1(1-P_1)}{N(2P_1-1)^2} \tag{16.5.8}$$

Furthermore, writing $\widehat{\pi}_A = \dfrac{\lambda_{wA} - (1-P_1)}{2P_1 - 1}$ in Eq. (16.5.8), we get an alternative expression of $\widehat{V}(\widehat{\pi}_A)$ as

$$\widehat{V}(\widehat{\pi}_A) = \frac{1-f}{n-1}\widehat{\pi}_A(1-\widehat{\pi}_A) + \frac{N-1}{(n-1)N}\frac{P_1(1-P_1)}{(2P_1-1)^2} \tag{16.5.9}$$

16.5.1.3 Rao–Hartley–Cochran Sampling

Let a sample s of size n be selected by Rao–Hartley–Cochran (RHC, 1962) method of sampling assuming N/n is an integer. Unbiased estimators for the population total under direct and RR survey are, respectively, given by

$$t(s,y) = \widehat{Y}_{rhc} = \sum_{i \in s}\frac{y_i}{p_i}P_i \text{ and } t(s,r) = \widehat{Y}_{rhc}(rr) = \sum_{i \in s}\frac{r_i}{p_i}P_i \tag{16.5.10}$$

where p_i is the normed size measure for the ith unit and P_i is the sum of the p_j values for the group containing the ith unit (see Section 5.6 for details).

Theorem 16.5.5

(i) $\widehat{Y}_{rhc}(rr) = \sum_{i \in s}\frac{r_i}{p_i}P_i$ is unbiased estimator of Y,

(ii) $V\left[\widehat{Y}_{rhc}(rr)\right] = \dfrac{N-n}{n(N-1)}\displaystyle\sum_{i \in U} p_i\left(\frac{y_i}{p_i} - Y\right)^2 + \sum_{i \in U} a_i\phi_i$

and

(iii) $\widehat{V}\left[\widehat{Y}_{rhc}(rr)\right] = \dfrac{N-n}{N(n-1)}\sum_{i\in s} P_i\left(\dfrac{r_i}{p_i} - \widehat{Y}_{rhc}(rr)\right)^2 + \sum_{i\in s}\dfrac{\hat{\phi}_i}{p_i}P_i$

where $a_i = \dfrac{1}{n(N-1)}\left[N(n-1) + \dfrac{N-n}{p_i}\right]$

Proof

(i) $E\left[\widehat{Y}_{rhc}(rr)\right] = E_p\sum_{i\in s}\dfrac{E_R(r_i)}{p_i}P_i$

$$= E_p\left(\sum_{i\in s}\frac{y_i}{p_i}P_i\right)$$

$$= Y \text{ (see Theorem 5.6.1.)}$$

(ii) $V\left[\widehat{Y}_{rhc}(rr)\right] = V_p\left(\widehat{Y}_{rhc}\right) + E_p\left(\sum_{i\in s}\dfrac{\phi_i}{p_i^2}P_i^2\right)$

Now

$$V_p\left(\widehat{Y}_{rhc}\right) = \frac{N-n}{n(N-1)}\sum_{i\in U} p_i\left(\frac{y_i}{p_i} - Y\right)^2 \text{ (Theorem 5.6.1)} \qquad (16.5.11)$$

and

$$E_p\left(\sum_{i\in s}\frac{\phi_i}{p_i^2}P_i^2\right) = nE_p\left(\frac{\phi_i}{p_i^2}P_i^2\right)$$

$$= nE_{G_i}\left(\sum_{j\in G_i}\frac{\phi_i}{p_j}\right)\left(\sum_{j\in G_i}p_j\right)$$

(where G_i is the ith group $(i = 1, \ldots, n)$, see Section 5.6)

$$= nE_{G_i}\left(\sum_{j\in G_i}\phi_j + \sum_{j\neq}\sum_{k\in G_i}\phi_j p_k/p_j\right)$$

$$= n\left(\frac{N/n}{N}\sum_{j\in U}\phi_j + \frac{(N/n)(N/n-1)}{N(N-1)}\sum_{j\neq}\sum_{k\in U}\phi_j p_k/p_j\right)$$

$$= \sum_{j \in U} \phi_j \left\{ 1 + \frac{(N/n - 1)}{(N-1)} (1 - p_j)/p_j \right\}$$

$$= \sum_{j \in U} \phi_j \left[\left(1 - \frac{(N-n)}{n(N-1)} \right) + \frac{(N-n)}{n(N-1)p_j} \right]$$

$$= \sum_{j \in U} \phi_j a_j \tag{16.5.12}$$

The second part of the theorem follows from Eqs. (16.5.11) and (16.5.12).

(iii) Noting $\widehat{V}\left(\widehat{Y}_{rhc}\right) = \dfrac{N-n}{N(n-1)} \sum_{i \in s} P_i \left(\dfrac{y_i}{p_i} - \widehat{Y}_{rhc} \right)^2$ and $\sum_{i \in s} \dfrac{\hat{\phi}_i}{p_i} P_i$ are unbiased estimators of $V\left(\widehat{Y}_{rhc}\right)$ and $\sum_{i \in U} \phi_i$ respectively, and using Theorem 16.5.2, we find that $\widehat{V}\left[\widehat{Y}_{rhc}(rr)\right]$ is an unbiased estimator of $V\left[\widehat{Y}_{rhc}(rr)\right]$.

16.5.1.4 Probability Proportional to Aggregate Size Sampling

In Lahiri-Midzuno-Sen (1951, 1952, 1953) probability proportional to aggregate size (PPAS) sampling scheme (see Section 5.5), the probability of selecting a samples s of size n is $p(s) = x_s/(M_1 X)$ where $x_s = \sum_{i \in s} x_i$, $X = \sum_{i \in U} x_i$, $M_1 = \begin{pmatrix} N-1 \\ n-1 \end{pmatrix}$ and $x_i(>0)$ is the measure of size for the ith unit. For this sampling scheme

$$t(s, y) = \widehat{Y}_{lms} = \frac{y_s}{x_s} X \text{ and } t(s, r) = \widehat{Y}_{lms}(rr) = \frac{r_s}{x_s} X \tag{16.5.13}$$

where $y_s = \sum_{i \in s} y_i$ and $r_s = \sum_{i \in s} r_i$.

Now

$$V\left[\widehat{Y}_{lms}(rr)\right] = V_p\left(\frac{y_s}{x_s} X \right) + E_p\left\{ \sum_{i \in s} \phi_i \left(\frac{X}{x_s} \right)^2 \right\}$$
$$= -\frac{1}{2} \sum_{i \neq} \sum_{j \in U} \beta_{ij} x_i x_j \left(\frac{y_i}{x_i} - \frac{y_j}{x_j} \right)^2 + \sum_{i \in U} \beta_i \phi_i \tag{16.5.14}$$

where $\beta_{ij} = X \sum_{s \supset i,j} 1/(M_1 x_s) - 1$ and $\beta_i = X \sum_{s \supset i} 1/(M_1 x_s)$(for detail, see Section 14.3.1.4).

Since $\widehat{V}\left(\widehat{Y}_{lms}\right) = B_s \frac{1}{2} \sum_{i \neq} \sum_{j \in s} x_i x_j \left(\frac{y_i}{x_i} - \frac{y_j}{x_j}\right)^2$ and $\frac{\hat{\phi}_s}{x_s} X$ are unbiased estimators of $V\left(\widehat{Y}_{lms}\right)$ and $\sum_{i \in U} \phi_i$, respectively, we obtain an unbiased estimator for $V\left[\widehat{Y}_{lms}(rr)\right]$ as

$$\widehat{V}\left[\widehat{Y}_{lms}(rr)\right] = B_s \frac{1}{2} \sum_{i \neq} \sum_{j \in s} x_i x_j \left(\frac{r_i}{x_i} - \frac{r_j}{x_j}\right)^2 + X\widehat{\phi}_s / x_s \qquad (16.5.15)$$

where $\widehat{\phi}_s = \sum_{i \in s} \widehat{\phi}_i$ and $B_s = \left(\frac{N-1}{n-1} - \frac{X}{x_s}\right) \frac{X}{x_s}$

16.5.1.5 Probability Proportional to Size With Replacement Sampling

Let a sample s of size n be selected by probability proportional to size with replacement (PPSWR) sampling scheme with the normed size measure p_i attached to the ith unit. Here we will consider the following cases:

Case I: If the ith respondent is selected in the sample $n_i(s)$ times, he/she has to perform randomized devices $n_i(s)$ times independently. Let $r(k)$, $y(k)$ and $p(k)$ be the RR, value of the variable y and the selection probability of the unit selected at the kth ($k = 1,\ldots, n$) draw. Then, $E_R\{r(k)\} = y(k) = y_j$, $V_R\{r(k)\} = \phi(k) = \phi_j$, and $p(k) = p_j$ if the rth draw produces the jth unit with probability $p_j (j = 1,\ldots, N)$. In this case we get

$$E\left(\frac{r(k)}{p(k)}\right) = E_p\left[E_R\left(\frac{r(k)}{p(k)}\right)\right] = E_p\left(\frac{y(k)}{p(k)}\right) = \sum_{j=1}^{N} \frac{y_j}{p_j} p_j = Y \qquad (16.5.16)$$

$$\begin{aligned} V\left(\frac{r(k)}{p(k)}\right) &= E_p\left[V_R\left(\frac{r(k)}{p(k)}\right)\right] + V_p\left[E_R\left(\frac{r(k)}{p(k)}\right)\right] \\ &= E_p\left[\frac{\phi(k)}{\{p(k)\}^2}\right] + V_p\left(\frac{y(k)}{p(k)}\right) \qquad (16.5.17) \\ &= \sum_{i \in U} \frac{\phi_i}{p_i} + \left(\sum_{i \in U} \frac{y_i^2}{p_i} - Y^2\right) \end{aligned}$$

Since $\frac{r(k)}{p(k)}$'s are independently and identically distributed random variables, we find from Arnab (1990) as follows:

Theorem 16.5.6

(i) $\widehat{Y}_{hh}\left(rr\right) = \frac{1}{n}\sum_{r=1}^{n}\frac{r(k)}{p(k)}$ is an unbiased estimator of Y.

(ii) $V\left[\widehat{Y}_{hh}(rr)\right] = \left[\left(\sum_{i\in U}\frac{y_i^2}{p_i} - Y^2\right) + \sum_{i\in U}\frac{\phi_i}{p_i}\right]\Big/ n$

and

(iii) $\widehat{V}\left[\widehat{Y}_{hh}(rr)\right] = \frac{1}{n(n-1)}\sum_{k=1}^{n}\left\{\frac{r(k)}{p(k)} - \widehat{Y}_{hh}(rr)\right\}^2$ is an unbiased estimator of $V\left[\widehat{Y}_{hh}(rr)\right]$

Case II: Here we suppose that each of the units in the sample produces only one RR even if it is selected more than once in the sample. More specifically, suppose the ith unit is selected $n_i(s)$ times and we receive only one revised RR r_i from it. Let $\widehat{Y}^*_{hh}(rr) = \frac{1}{n}\sum_{i\in U} n_i(s)\frac{r_i}{p_i}$, then we have the following from Arnab (1990).

Theorem 16.5.7

(i) $E\left[\widehat{Y}^*_{hh}(rr)\right] = Y$

(ii) $V\left[\widehat{Y}^*_{hh}(rr)\right] = \left[\left(\sum_{i=1}^{N}\frac{y_i^2}{p_i} - Y^2\right) + \sum_{i=1}^{N}\{1+(n-1)p_i\}\frac{\phi_i}{p_i}\right]\Big/ n$

and

(iii) $\widehat{V}\left[\widehat{Y}^*_{hh}(rr)\right] = \frac{1}{n(n-1)}\sum_{i\in U} n_i(s)\left\{\frac{r_i}{p_i} - \widehat{Y}^*_{hh}(rr)\right\}^2 + \frac{1}{n}\sum_{i\in U} n_i(s)\frac{\hat{\phi}_i}{p_i}$

Proof

The theorem can be proved by noting $E(n_i(s)) = np_i$, $V(n_i(s)) = np_i(1-p_i)$ and $Cov(n_i(s), n_j(s)) = -np_ip_j$ for $i \neq j$.

Remark 16.5.1

The estimator $\widehat{Y}_{hh}(rr)$ is more efficient than $\widehat{Y}^*_{hh}(rr)$ since $V\left[\widehat{Y}_{hh}(rr)\right] \leq \widehat{V}\left[\widehat{Y}^*_{hh}(rr)\right]$. This is because the estimator $\widehat{Y}_{hh}(rr)$ is based on $n_i(s)(\geq 1)$ RRs from the ith ($i \in s$) unit while $\widehat{Y}^*_{hh}(rr)$ is based on a single RR from the ith unit even it is selected $n_i(s)$ times.

16.5.1.6 Simple Random Sampling With Replacement

Substituting $p_i = 1/N$ in Theorems 16.5.6 and 16.5.7, we derive the following results.

Theorem 16.5.8

For an SRSWR sampling

(i) $\bar{r} = \frac{1}{n}\sum_{k=1}^{n} r(k)$ is an unbiased estimator for the population mean $\overline{Y}$.

(ii) $V(\bar{r}) = \left(\sigma_y^2 + \overline{\phi}\right)/n$

and

(iii) $\widehat{V}(\bar{r}) = \frac{1}{n(n-1)}\sum_{k=1}^{n}\left(r(k) - \bar{r}\right)^2$

where $\overline{\phi} = \sum_{i\in U}\phi_i/N$ and $\sigma_y^2 = \sum_{i\in U}\left(y_i - \overline{Y}\right)^2/N$

Theorem 16.5.9

For an SRSWR sampling where a single RR response is obtained from the ith respondent even he/she is selected more than once.

(i) $\bar{r}^* = \frac{1}{n}\sum_{i\in U} n_i(s) r_i$ is an unbiased estimator of $\overline{Y}$.

(ii) $V(\bar{r}^*) = \left[\sigma_y^2 + \left(1 + \frac{n-1}{N}\right)\overline{\phi}\right]/n$

and

(iii) $\widehat{V}[\bar{r}^*] = \left[\frac{1}{(n-1)}\sum_{i\in U} n_i(s)(r_i - \bar{r}^*)^2 + \frac{1}{N}\sum_{i\in U} n_i(s)\widehat{\phi}_i\right]/n$

The Table 16.5.1 below shows the estimators for the population total Y, its variances, and unbiased estimators of the variances under various sampling designs.

Table 16.5.1 Unbiased estimators, variances and unbiased estimators of variances for the population total for various sampling design

Sampling design	Unbiased estimator $\widehat{Y}$	Variance $V(\widehat{Y})$	Unbiased estimator of variance $\widehat{V}(\widehat{Y})$
Arbitrary	$t(s,r)=\sum_{i\in s} b_{si}r_i$	$\sum_{i\in U}(\alpha_{ii}-1)y_i^2+\sum_{i\neq}\sum_{j\in U}(\alpha_{ij}-1)y_iy_j + \sum_{i\in U}\alpha_{ii}\phi_i$	$\sum_{i\in s} c_{ii}(s)r_i^2+\sum_{i\neq}\sum_{j\in s} c_{ij}(s)r_ir_j + \sum_{i\in s} b_{si}\widehat{\phi}_i$
Fixed effective sample size	$\widehat{Y}_{ht}(rr)=\sum_{i\in s}\frac{r_i}{\pi_i}$	$\frac{1}{2}\sum_{i\neq}\sum_{j\in U}(\pi_i\pi_j-\pi_{ij})\left(\frac{y_i}{\pi_i}-\frac{y_j}{\pi_j}\right)^2 + \sum_{i\in U}\frac{\phi_i}{\pi_i}$	$\frac{1}{2}\sum_{i\neq}\sum_{j\in s}\frac{(\pi_i\pi_j-\pi_{ij})}{\pi_{ij}}\left(\frac{r_i}{\pi_i}-\frac{r_j}{\pi_j}\right)^2 + \sum_{i\in s}\frac{\widehat{\phi}_i}{\pi_i}$
Rao–Hartley–Cochran	$\widehat{Y}_{rhc}(rr)=\sum_{i\in s}\frac{r_i}{p_i}P_i$	$\frac{N-n}{n(N-1)}\sum_{i\in U}p_i\left(\frac{y_i}{p_i}-Y\right)^2 + \sum_{i\in U}a_i\phi_i$	$\frac{N-n}{N(n-1)}\sum_{i\in s}P_i\left\{\frac{r_i}{p_i}-\widehat{Y}_{rhc}(rr)\right\}^2 + \sum_{i\in s}\frac{\widehat{\phi}_i}{p_i}P_i$
PPAS	$\widehat{Y}_{lms}(rr)=\frac{r_s}{x_s}X$	$-\frac{1}{2}\sum_{i\neq}\sum_{j\in U}\beta_{ij}x_ix_j\left(\frac{y_i}{x_i}-\frac{y_j}{x_j}\right)^2 + \sum_{i\in U}\beta_i\phi_i$	$B_s\frac{1}{2}\sum_{i\neq}\sum_{j\in s}x_ix_j\left(\frac{r_i}{x_i}-\frac{r_j}{x_j}\right)^2 + \frac{X}{x_s}\sum_{i\in s}\widehat{\phi}_i$

Continued

Table 16.5.1 Unbiased estimators, variances and unbiased estimators of variances for the population total for various sampling design—cont'd

Sampling design	Unbiased estimator $\widehat{Y}$	Variance $V(\widehat{Y})$	Unbiased estimator of variance $\widehat{V}(\widehat{Y})$
SRSWOR	$\widehat{Y}_{wor}(rr) = N\frac{1}{n}\sum_{i\in s} r_i$	$N^2\left\{(1-f)S_y^2 + \frac{1}{N}\sum_{i\in U}\phi_i\right\}\Big/ n$	$N^2\left\{(1-f)s_r^2 + \frac{1}{N}\sum_{i\in s}\widehat{\phi}_i\right\}\Big/ n$
PPSWR	$\widehat{Y}_{hh}(rr) = \frac{1}{n}\sum_{k=1}^{n}\frac{r(k)}{p(k)}$	$\frac{1}{n}\left[\sum_{i\in U} p_i\left(\frac{y_i}{p_i} - Y\right)^2 + \sum_{i\in U}\frac{\phi_i}{p_i}\right]$	$\frac{1}{n(n-1)}\sum_{k=1}^{n}\left\{\frac{r(k)}{p(k)} - \widehat{Y}_{hh}(rr)\right\}^2$
PPSWR	$\widehat{Y}^{*}_{hh}(rr) = \frac{1}{n}\sum_{i\in U} n_i(s)\frac{r_i}{p_i}$	$\sum_{i\in U} p_i\left(\frac{y_i}{p_i} - Y\right)^2\Big/ n$ $+\sum_{i\in U}\{1+(n-1)p_i\}\frac{\phi_i}{p_i}\Big/ n$	$\frac{1}{n(n-1)}\sum_{i\in U} n_i(s)\left\{\frac{r_i}{p_i} - \widehat{Y}^{*}_{hh}(rr)\right\}^2$ $+\frac{1}{n}\sum_{i\in U} n_i(s)\frac{\widehat{\phi}_i}{p_i}$
SRSWR	$\widehat{Y}_{wr}(rr) = N\frac{1}{n}\sum_{k=1}^{n} r(k)$	$N^2\left[\sigma_y^2 + \frac{1}{N}\sum_{i\in U}\phi_i\right]\Big/ n$	$\frac{N^2}{n(n-1)}\sum_{r=1}^{n}\left\{r(k) - \frac{1}{n}\sum_{j=1}^{n} r(j)\right\}^2$
SRSWR	$\widehat{Y}^{*}_{wr}(rr) = \frac{N}{n}\sum_{i\in U} n_i(s) r_i$	$\frac{N^2}{n}\left[\sigma_y^2 + \left\{1 + \frac{(n-1)}{N}\right\}\frac{1}{N}\sum_{i\in U}\phi_i\right]$	$N^2\left[\frac{1}{n(n-1)}\sum_{i\in U} n_i(s)\left\{r_i - \frac{1}{n}\sum_{i\in U} n_i(s) r_i\right\}^2 + \frac{1}{Nn}\sum_{i\in U} n_i(s)\widehat{\phi}_i\right]$

16.6 OPTIONAL RANDOMIZED RESPONSE TECHNIQUES

In optional RR technique (ORT) it is assumed that the aspects of inquiry are felt to be sensitive by most of the respondents but some are more willing to answer directly. In ORT, respondents are given an option either to supply RR using a specified randomized device or to respond directly according to the extent to which the respondent feels the question is sensitive or not. Most of the methods developed for ORT are limited to SRSWR sampling only. A few of the ORT techniques are available for complex surveys. ORT can be classified into two categories: Full ORT (FORT) and Partial ORT (PORT). The ORT is more efficient than compulsory RRT (CRT) because the probability of obtaining true responses in ORT is much higher than that in the CRT (Vide Arnab, 2004a).

FORT: Here respondents are given an option either to supply RR using a specified randomized device or to respond directly according to whether the respondent feels that the question is sensitive or not. In this method it is assumed that respondents who feel that the character under investigation is confidential, belong to certain group G and produce RR with probability 1, whereas the respondents who feel the character is not confidential, belong to the complementary group $\overline{G}$ and supply direct response with probability 1. FORT was proposed by Chaudhuri and Mukherjee (1988), Arnab (2004a), Chaudhuri and Saha (2005), Huang (2008), among others.

PORT: Here it is assumed that the respondents may supply direct or RR with certain probability depending on their judgment of sensitivity (mood) at the particular time of answering the question. In other words, the respondent may sometimes supply RRs and at other times direct responses during the period of the survey. Most of the researchers developed various methods of PORT, e.g., Mangat and Singh (1994), Gupta (2001), Gupta et al. (2002), Pal (2008), among others.

16.6.1 Full Optional Randomized Response Technique

Arnab (2004a) proposed a theory of estimation of population characteristics for FORT, which is applicable for both qualitative and quantitative characteristics. Let $s_G = s \cap G$ be the set of respondents selected in the sample s that belong to group G and $s_{\overline{G}} = s \cap \overline{G} = s - s_G$. In this method, respondents belonging to group s_G provide RRs using some suitable randomized device whereas respondents belonging to group $s_{\overline{G}}$ provide direct responses. Here we assume that the respondents do not disclose their membership to group G or $\overline{G}$.

16.6.1.1 Estimation of Population Total

Let r_i be the revised RR obtained from the ith respondent if he/she is included in the sample s_G and let y_i be the direct response if the ith unit belongs to $S_{\overline{G}}$. Here we suppose that the r_i's follow Model (16.4.4). Let us define

$$\widetilde{r}_i = \delta_i y_i + (1 - \delta_i) r_i \tag{16.6.1}$$

where $\delta_i = 1$ if $i \in \overline{G}$ and $\delta_i = 0$ if $i \in G$.

Clearly, $\widetilde{r}_i$ follows the following model

$$\begin{aligned}\text{Model } R_o\text{: } & E_R(\widetilde{r}_i) = y_i,\ V_R(\widetilde{r}_i) = (1 - \delta_i) V_R(r_i) \\ & = (1 - \delta_i)\phi_i \text{ and } C_R(\widetilde{r}_i, \widetilde{r}_j) = 0 \text{ for } i \neq j\end{aligned} \tag{16.6.2}$$

Replacing r_i by $\widetilde{r}_i$ in Eq. (16.5.3), we find a linear unbiased estimator for the population total Y under ORR technique as

$$\widetilde{t}(s, \widetilde{r}) = \sum_{i \in s} b_{si} \widetilde{r}_i \tag{16.6.3}$$

Theorem 16.6.1

(i) $\widetilde{t}(s, \widetilde{r})$ is an unbiased estimator for Y

(ii) $V\left[\widetilde{t}(s, \widetilde{r})\right] = \sum_{i \in U} (\alpha_i - 1) y_i^2 + \sum\sum_{i \neq j \in U} (\alpha_{ij} - 1) y_i y_j + \sum_{i \in G} \alpha_i \phi_i$

(iii) An unbiased estimator of $V\left[\widetilde{t}(s, \widetilde{r})\right]$ is

$$\widehat{V}\left[\widetilde{t}(s, \widetilde{r})\right] = \sum_{i \in s} c_{ii}(s) \widetilde{r}_i^2 + \sum_{i \neq} \sum_{j \in s} c_{ij}(s) \widetilde{r}_i \widetilde{r}_j + \sum_{i \in s} (1 - \delta_i) b_{si} \widehat{\phi}_i$$

where $\alpha_i = \sum_{s \supset i} b_{si}^2 p(s)$, $\alpha_{ij} = \sum_{s \supset i} b_{si} b_{sj} p(s)$, $\sum_{s \supset i} c_{ii}(s) p(s) = \alpha_i - 1$ and $\sum_{s \supset i} c_{ij}(s) p(s) = \alpha_{ij} - 1$

Proof

(i) $E\left[\widetilde{t}(s, \widetilde{r})\right] = E_p\left[\sum_{i \in s} b_{si} E_R(\widetilde{r}_i)\right] = \sum_i y_i \sum_{s \supset i} b_{si} p(s) = Y$

(using the unbiasedness condition $\sum_{s \supset i} b_{si} p(s) = 1$)

$$\text{(ii) } V\left[\tilde{t}(s,\tilde{r})\right] = V_p\left[E_R\{\tilde{t}(s,\tilde{r})\}\right] + E_p\left[V_R\{\tilde{t}(s,\tilde{r})\}\right]$$

$$= V_p\left(\sum_{i \subset s} b_{si} y_i\right) + E_p\left(\sum_{i \in s} b_{si}^2 (1-\delta_i)\phi_i\right)$$

$$= \sum_{i \in U} (\alpha_i - 1) y_i^2 + \sum_{i \neq} \sum_{j \in U} (\alpha_{ij} - 1) y_i y_j + \sum_{i \in U} \alpha_i (1-\delta_i)\phi_i$$

$$= \sum_{i \in U} (\alpha_i - 1) y_i^2 + \sum_{i \neq} \sum_{j \in U} (\alpha_{ij} - 1) y_i y_j + \sum_{i \in G} \alpha_i \phi_i$$

$$\text{(iii) } E\left[\widehat{V}\{\tilde{t}(s,\tilde{r})\}\right] = E_p\left[\sum_{i \in s} c_{ii}(s)\left\{y_i^2 + (1-\delta_i)\phi_i\right\} + \sum_{i \neq} \sum_{j \in s} c_{ij}(s) y_i y_j + \sum_{i \in s} (1-\delta_i) b_{si} \phi_i\right]$$

$$= \sum_{i \in U} (\alpha_i - 1)\left\{y_i^2 + (1-\delta_i)\phi_i\right\} + \sum_{i \neq} \sum_{j \in U} (\alpha_{ij} - 1) y_i y_j + \sum_{i \in U} (1-\delta_i)\phi_i$$

$$= V\left[\tilde{t}(s,\tilde{r})\right]$$

Remark 16.6.1

The variance of $\tilde{t}(s,\tilde{r})$ is smaller than that of $t(s,r)$ by an amount of $\sum_{i \in \overline{G}} \alpha_i \phi_i$.

Hence ORR technique is more efficient than the compulsory RR technique.

16.6.1.2 Horvitz–Thompson Estimator Based on a Fixed Sample Size Design

Substituting $b_{si} = \dfrac{1}{\pi_i}$ in Eq. (16.6.3), the Horvitz–Thompson estimator for the population total Y is obtained as

$$t(s,\tilde{r}) = t_{ht} = \sum_{i \in s} \frac{\tilde{r}_i}{\pi_i} \tag{16.6.4}$$

The expression of variance and its unbiased estimators of t_{ht} are as follows:

$$V(t_{ht}) = \frac{1}{2}\sum_{i\neq}\sum_{j\in U}(\pi_i\pi_j - \pi_{ij})\left(\frac{y_i}{\pi_i} - \frac{y_j}{\pi_j}\right)^2 + \sum_{i\in G}\frac{\phi_i}{\pi_i} \tag{16.6.5}$$

and

$$\widehat{V}(t_{ht}) = \frac{1}{2}\sum_{i\neq}\sum_{j\in s}\frac{(\pi_i\pi_j - \pi_{ij})}{\pi_{ij}}\left(\frac{\widetilde{r}_i}{\pi_i} - \frac{\widetilde{r}_j}{\pi_j}\right)^2 + \sum_{i\in s_G}\frac{\widehat{\phi}_i}{\pi_i} \tag{16.6.6}$$

16.6.1.3 Simple Random Sampling Without Replacement

For SRSWOR $\pi_i = n/N$ and $\pi_{ij} = \dfrac{n(n-1)}{N(N-1)}$, we get

$$t(s,\widetilde{r}) = t_{wor} = N\,\overset{\approx}{r}_s \tag{16.6.7}$$

$$V(t_{wor}) = \frac{N(N-n)}{n}S_y^2 + \frac{N}{n}\sum_{i\in G}\phi_i \tag{16.6.8}$$

and an unbiased estimator of $V(t_{wor})$ is

$$\widehat{V}(t_{wor}) = \frac{N(N-n)}{n}s_{\widetilde{r}}^2 + \frac{N}{n}\sum_{i\in s_G}\widehat{\phi}_i \tag{16.6.9}$$

where $(n-1)s_{\widetilde{r}}^2 = \sum_{i\in s}(\widetilde{r}_i - \overset{\approx}{r}_s)^2$ and $\overset{\approx}{r}_s = \sum_{i\in s}\dfrac{\widetilde{r}_i}{n}$.

16.6.1.4 Rao–Hartley–Cochran Sampling

For the RHC sampling scheme given in Section 5.6, the expression for the unbiased estimator for the population total Y is

$$t(s,\widetilde{r}) = t_{rhc} = \sum_{i\in s}\frac{\widetilde{r}_i}{p_i}P_i \tag{16.6.10}$$

The expressions for the variance and its unbiased estimators are given by Arnab (2004a) as follows:

$$V(t_{rhc}) = \frac{N-n}{n(N-1)}\left(\sum_{i\in U}\frac{y_i^2}{p_i} - Y^2\right) + \sum_{j\in G}a_j\phi_j \tag{16.6.11}$$

and

$$\widehat{V}(t_{rhc}) = \frac{N-n}{N(n-1)}\sum_{i\in s}\left(\frac{\widetilde{r}_i}{p_i} - t_{rhc}\right)^2 P_i + \sum_{j\in s_G}P_j\frac{\widehat{\phi}_j}{p_j} \tag{16.6.12}$$

where $a_j = \dfrac{1}{n(N-1)}\left(N(n-1) + \dfrac{N-n}{p_j}\right)$.

16.6.1.5 Probability Proportional to Size With Replacement Sampling

For PPSWR sampling, the estimator for the total Y is

$$t(s,\widetilde{r}) = t_{pps} = \frac{1}{n}\sum_{k=1}^{n}\frac{\widetilde{r}(k)}{p(k)} \tag{16.6.13}$$

where $\widetilde{r}(k) = \widetilde{r}_i, p(k) = p_i$ if the kth draw produces ith unit, $i = 1,\ldots, N$.

In this case the variance of $t(s,\widetilde{r})$ and its unbiased estimators are obtained as follows:

$$V(t_{pps}) = \frac{1}{n}\left[\sum_{i\in U} p_i\left(\frac{y_i}{p_i} - Y\right)^2 + \sum_{i\in G}\frac{\phi_i}{p_i}\right] \tag{16.6.14}$$

and

$$\widehat{V}(t_{pps}) = \frac{1}{n(n-1)}\sum_{k=1}^{n}\left(\frac{\widetilde{r}(k)}{p(k)} - t_{pps}\right)^2 \tag{16.6.15}$$

16.6.1.6 Simple Random Sampling With Replacement

In case sample s is selected by SRSWR, an unbiased estimator of Y, its variance and an unbiased estimator of variance are obtained by substituting $p_i = 1/N$ in Eqs. (16.6.13)–(16.6.15) as follows:

$$t(s,\widetilde{r}) = t_{wr} = N\,\overset{\approx}{r} \tag{16.6.16}$$

$$V(t_{wr}) = \frac{N^2}{n}\left[\sigma_y^2 + \frac{1}{N}\sum_{i\in G}\phi_i\right] \tag{16.6.17}$$

and

$$\widehat{V}(t_{wr}) = \frac{N^2}{n(n-1)}\sum_{k=1}^{n}\left(r(k) - \overset{\approx}{r}\right)^2 \tag{16.6.18}$$

where $\sigma_y^2 = \frac{1}{N}\left(y_i - \overline{Y}\right)^2$ and $\overset{\approx}{r} = \frac{1}{N}\sum_{k=1}^{n} r(k)$.

16.6.2 Partial Optional Randomized Response Technique

PORT was originally proposed by Mangat and Singh (1994) for qualitative variables, and it was extended by several authors including Singh and Joarder (1997), Gupta et al. (2002, 2006, 2010, 2013), Gupta and Shabbir (2004), Huang (2004, 2008), Pal (2008), Chaudhuri and Dihidar (2009), among others. Most of the researchers proposed PORT for SWRWR sampling whereas a few of them viz. Pal (2008), Chaudhuri and Dihidar (2009) for complex survey designs. Here we will describe PORT proposed by Gupta et al. (2002) only.

16.6.2.1 Gupta et al. Model

A sample s of size n units is selected from a population by SRSWR method. Each of the selected respondents was asked to choose one of the following two options:
(a) Report the true value of y or (b) report RR yx, where x is a random sample from a population with known mean $\mu_x = 1$ and known variance γ^2. Here also the interviewer will not know whether the respondents supplied the true response or RR. Let us denote the response obtained from the ith respondent as

$$z_i = x_i^k y_i$$

where $k = 1$, if the response is scrambled and $k = 0$, otherwise.
Gupta et al. (2002) derived the following results:

Theorem 16.6.2

(i) $\widehat{\mu}_y = \frac{1}{n}\sum_{i=1}^{n} z_i$ is an unbiased estimator of the μ_y

(ii) $V\left(\widehat{\mu}_y\right) = \frac{1}{n}\left[\sigma_y^2 + \gamma^2 W_G^2\left(\sigma_y^2 + \mu_x^2\right)\right]$

where W_G is the probability that a person will report the scramble response, which is generally unknown. W_G is called the degree of sensitivity of the attribute A.

Noting

$$W_G \cong \frac{E(\log z) - E(\log y)}{E(\log x)},$$

an approximate estimator of W_G was obtained by Gupta et al. (2002) as

$$\widehat{W}_G = \frac{\frac{1}{n}\sum_{i=1}^{n}\log z_i - \log\left(\frac{1}{n}\sum_{i=1}^{n} z_i\right)}{\delta}$$

where $\delta = E(\log x)$.

16.7 MEASURE OF PROTECTION OF PRIVACY

To measure the success of an RR technique, a statistician's objective is to obtain efficient estimators of the parameters of interest whereas the interviewee's objective is to protect his/her privacy. It is seen that the efficiency of an estimator and maintenance of privacy are general goes in opposite directions. Hence one should compare the efficiencies of different RR techniques by keeping the degree of confidentiality fixed to a certain level. In this section, we will present a few measures of the degree of confidentiality proposed by Lanke (1975a,b, 1976), Leysieffer and Warner (1976) and Anderson (1975a,b,c), among others. More details have been given by Chaudhuri and Mukherjee (1988), Singh (2003) and Hong and Yan (2012).

16.7.1 Qualitative Characteristic With "Yes—No" Response

Suppose the units of a population are classified in two categories A (possessing a sensitive characteristic) and $\overline{A}$ (complement of A) with unknown proportions π_A and $1-\pi_A$, respectively. Each unit provides an RR response "R" either "Yes = Y" or "No = N" by using a suitable randomized device R^*. The conditional probabilities of obtaining an RR "R" from a unit, which belong to the group A and $\overline{A}$ are $P(R|A)$ and $P\left(R|\overline{A}\right)$, respectively. These quantities are at the disposal of the investigator and are called design probabilities. The posterior probability of classifying an individual in group $A\left(\overline{A}\right)$ when he/she reports R is $P(A|R)\left(P\left(\overline{A}|R\right)\right)$. The probabilities $P(A|R)$ and $P\left(\overline{A}|R\right)$ are called revealing probabilities.

16.7.1.1 *Leysieffer and Warner's Measure*

According to Leysieffer and Warner (1976), the response "R" is said to be jeopardizing with respect to A if the posterior probability of classification increases given that a response "R," i.e.,

$$P(A|R) > \pi_A = P(A) \tag{16.7.1}$$

In this case the respondent feels exposed rather than protected. Similarly the response "R" is jeopardized with respect to $\overline{A}$ if

$$P\left(\overline{A}|R\right) > 1-\pi_A = P\left(\overline{A}\right) \tag{16.7.2}$$

Since $P(A|R) > \pi_A$ implies $P\left(\overline{A}|R\right) < 1-\pi_A$, we find that whenever R is jeopardized with respect to A, R is not jeopardized with respect to $\overline{A}$ and vice versa.

Leysieffer and Warner (1976) provided us with an index of measure of jeopardy, which is known as the jeopardy function. Jeopardy of R with respect to A is

$$g(R,A) = P(R|A)/P(R|\overline{A}) \tag{16.7.3}$$

Similarly, jeopardy of R with respect to $\overline{A}$ is

$$g(R,\overline{A}) = P(R|\overline{A})/P(R|A) = 1/g(R,A) \tag{16.7.4}$$

Now using Bayes theorem we find

$$\begin{aligned} P(A|R) &= \frac{P(R|A)\pi_A}{P(R|A)\pi_A + P(R|\overline{A})(1-\pi_A)} \quad \text{and} \\ P(\overline{A}|R) &= \frac{P(R|\overline{A})(1-\pi_A)}{P(R|A)\pi_A + P(R|\overline{A})(1-\pi_A)} \end{aligned} \tag{16.7.5}$$

Eq. (16.7.5) yields

$$g(R,A) = \frac{P(A|R)(1-\pi_A)}{P(\overline{A}|R)\pi_A} \quad \text{and} \quad g(R,\overline{A}) = \frac{P(\overline{A}|R)\pi_A}{P(A|R)(1-\pi_A)} \tag{16.7.6}$$

Hence the response R is jeopardized with respect to A if $g(R,A) > 1$ because in this case $P(A|R) > \pi_A$. Similarly R is jeopardized with respect to $\overline{A}$ if $g(R,A) < 1$. For $g(R,A) = 1$, R is nonjeopardizing with respect to A or $\overline{A}$

Let a sample of size n be selected by SRSWR method and let each of the selected units provide an RR "R" independently using Warner's (1965) method. It is assumed that if a respondent is selected $t(\geq 1)$ times in the sample he/she supplies t RRs independently. Then the probability of getting response $R = Y$(Yes) is

$$\theta = P(Y|A)\pi_A + P(Y|\overline{A})(1-\pi_A) \tag{16.7.7}$$

Let m be the total number of "Y" answers and $\widehat{\lambda} = m/n$, then

$$\widehat{\pi}_A = \frac{\widehat{\lambda} - P(Y|\overline{A})}{P(Y|A) - P(Y|\overline{A})} \tag{16.7.8}$$

is an unbiased estimator of π_A provided $P(R = Y|A) \neq P(R = Y|\overline{A})$, i.e., $g(R,A) \neq 1$.

The variance of $\widehat{\pi}_A$ is

$$V(\widehat{\pi}_A) = \frac{\pi_A(1-\pi_A)}{n} + \frac{P(Y|A)\{1-P(Y|A)\}\pi_A + P(Y|\overline{A})\{1-P(Y|\overline{A})\}(1-\pi_A)}{n\left[P(Y|A)-P(Y|\overline{A})\right]^2}$$

$$= \frac{\pi_A(1-\pi_A)}{n} + \frac{\pi_A g(Y,A) + (1-\pi_A)g(N,\overline{A})}{n\{g(Y,A)-1\}\{g(N,\overline{A})-1\}} \quad (16.7.9)$$

Without loss of generality, let us suppose that $P(Y|A) > P(Y|\overline{A})$. Then $g(Y,A) > 1$ and $g(N,\overline{A}) > 1$, i.e., "Y" answer and "N" answers, are jeopardizing with respect to A and $\overline{A}$, respectively. It can be easily checked that $V(\widehat{\pi}_A)$ is a decreasing function of $g(Y,A)$ and $g(N,\overline{A})$. Hence to minimize $V(\widehat{\pi}_A)$, one should choose $g(Y,A)$ and $g(N,\overline{A})$ to be as large as possible. But for the purpose of protection of confidentiality, we cannot choose $g(Y,A)$ and $g(N,\overline{A})$ more than certain level. Hence as compromise, one should minimize $V(\widehat{\pi}_A)$ keeping the jeopardy $g(Y,A)$ and $g(N,\overline{A})$ to the maximum acceptable levels, say k_1 and k_2, respectively, which still allow cooperation. Thus one should choose the design parameters of the RR model R^*, which minimize the $V(\widehat{\pi}_A)$ subject to the constraint $1 < g(Y|A) < k_1$ and $1 < g(N|\overline{A}) < k_2$. In particular if we choose $g(Y,A) = g(N,\overline{A}) = k$, the design parameters for the optimum RR model R^* should be set so that

$$P(Y|A) = k/(k+1) \text{ and } P(\overline{Y}|A) = 1/(k+1) \quad (16.7.10)$$

Example 16.7.1 Comparison Between Warner and Mangat & Singh Model
For Warner's model R_w described in Section 16.2.1,

$$P(Y|A) = P_1, \; P(Y|\overline{A}) = 1-P_1, P(N|\overline{A}) = P_1 \text{ and}$$
$$P(N|A) = 1-P_1 \text{ implies}$$

$$g(Y,A) = g_w(Y,A) = \frac{P(Y|A)}{P(Y|\overline{A})} = \frac{P_1}{1-P_1}$$

$$\text{and } g(N,\overline{A}) = g_w(N,\overline{A}) = \frac{P(N|\overline{A})}{P(N|A)} = \frac{P_1}{1-P_1} \quad (16.7.11)$$

here we choose the design parameter P_1 such that maximum allowable values of $g_w(Y,A)$ and $g_w(N,\overline{A})$ attains to certain levels k_1 and k_2, respectively. Since $g_w(Y,A)=g_w(N,\overline{A})$, we cannot choose k_1 value different from k_2. Hence we should choose P_1 such that $k_1=k_2=k$. So, our optimum choice of P_1 for a given value of $k(>1)$ is

$$P_1=\frac{k}{1+k} \tag{16.7.12}$$

The variance of $\widehat{\pi}_{Aw}$ under Warner model R_w with the optimal value $P_1=\dfrac{k}{1+k}$ comes out as

$$V_{opt}(\widehat{\pi}_{Aw})=\frac{1}{n}\left(\pi_A(1-\pi_A)+\frac{k}{(k-1)^2}\right) \tag{16.7.13}$$

For Mangat and Singh (1990) model R_{ms} described in Section 16.2.4.

$P(Y|A)=T+(1-T)P_4,\ P(Y|\overline{A})=(1-T)(1-P_4)$,
$P(N|\overline{A})=T+(1-T)P_4,\ P(N|A)=(1-T)(1-P_4)$ implies

$$g(Y,A)=g_{ms}(Y,A)=\frac{P(Y|A)}{P(Y|\overline{A})}=\frac{T+(1-T)(1-P_4)}{(1-T)(1-P_4)} \text{ and}$$

$$g(N,\overline{A})=g_{ms}(N,\overline{A})=\frac{P(N|\overline{A})}{P(N|A)}=\frac{T+(1-T)P_4}{(1-T)(1-P_4)} \tag{16.7.14}$$

Here also $g_{ms}(Y,A)=g_{ms}(N,\overline{A})=\dfrac{T+(1-T)P_4}{(1-T)(1-P_4)}$, so we choose the maximum allowable values of $g_{ms}(Y,A)$ and $g_{ms}(N,\overline{A})$ are each equal to

$$k=\frac{T+(1-T)P_4}{(1-T)(1-P_4)} \tag{16.7.15}$$

Eq. (16.7.15) yields the optimum value of P_4 for a given value of T as

$$P_4=\frac{(1-T)k-T}{(1-T)(1+k)} \tag{16.7.16}$$

The expression for the variance of $\widehat{\pi}_{ms}$ under RR model R_{ms} with the optimum value of P_4 in Eq. (16.7.16) yields

$$V_{opt}(\widehat{\pi}_{Ams})=\frac{1}{n}\left(\pi_A(1-\pi_A)+\frac{k}{(k-1)^2}\right) \tag{16.7.17}$$

Eqs. (16.7.13) and (16.7.17) lead the following result obtained by Singh (2003).

Theorem 16.7.1

The Warner's RR technique R_w and Mangat–Singh RR technique R_{ms} are equally efficient under the same level of privacy protection measure suggested by Leysieffer and Warner (1976).

Example 16.7.2 Comparison Between Warner and Unrelated Model

For the Greenberg et al. (1969) unrelated model R_g with known π_x we have
$P(Y|A) = P_2 + (1-P_2)\pi_x$, $P(Y|\overline{A}) = (1-P_2)\pi_x$,
$P(N|\overline{A}) = P_2 + (1-P_2)(1-\pi_x)$ and $P(N|A) = (1-P_2)(1-\pi_x)$ yields

$$g(Y,A) = g_g(Y,A) = \frac{P_2 + (1-P_2)\pi_x}{(1-P_2)\pi_x} \text{ and }$$
$$g(N,\overline{A}) = g_g(N,\overline{A}) = \frac{P_2 + (1-P_2)(1-\pi_x)}{(1-P_2)(1-\pi_x)} \tag{16.7.18}$$

Now setting k_1 and k_2 as the maximum allowable values of $g_g(Y,A)$ and $g_g(N,\overline{A})$, the optimum choices of P_2 and π_x come out as

$$P_2 = \frac{(k_1-1)(k_2-1)}{k_1k_2-1} \text{ and } \pi_x = \frac{k_2-1}{k_1+k_2-2} \tag{16.7.19}$$

In case $\overline{A}$ is nonsensitive, we may set $k_2 - \infty$. In this case the optimum values of P_2 and π_x come out, respectively, as

$$P_2 = \frac{(k_1-1)}{k_1} \text{ and } \pi_x = 1 \tag{16.7.20}$$

Substituting Eq. (16.7.20) in the expression of $Var(\widehat{\pi}_{Ag})$ in Theorem 16.2.2, the optimum value of $Var(\widehat{\pi}_G)$ for a given value of $k_1 = k$ comes out as

$$V_{opt}(\widehat{\pi}_{Ag}) = \frac{1}{n}\left(\pi_A(1-\pi_A) + \frac{(1-\pi_A)}{k-1}\right) \tag{16.7.21}$$

Now

$$V_{opt}(\widehat{\pi}_{ms}) - V_{opt}(\widehat{\pi}_{Ag}) = \frac{1}{n(k-1)}\left(\frac{1}{k-1} + \pi_A\right) \tag{16.7.22}$$

The expression (16.7.23) is positive since $k > 1$. Thus we have the following theorem.

Theorem 16.7.2

The Greenberg et al. unrelated question model R_2 with $\pi_x = 1$ is more efficient than the Warner model R_1 with the same degree of privacy protection measure suggested by Leysieffer and Warner (1976).

Remark 16.7.1

For Greenberg et al. RR technique with $\pi_x = 1$, each respondent has to draw a card from a pack of cards containing two types of card. The type 1 card bears sensitive question "Do you belong to the sensitive group A?" with proportion P_2 and the card type 2 bears the statement "Are you an African." Since $\pi_x = 1$, all the respondents are African. So, the RR "No" will come only from the respondents of the nonsensitive group $\overline{A}$. Hence if $\pi_x = 1$, respondents belong to the nonsensitive group $\overline{A}$ will be identified with probability 1.

16.7.1.2 Lanke's Measure

Lanke (1976) pointed out that the respondent possessing the attribute A (such as being HIV +ve) may feel embarrassed to disclose his/her membership to group A whereas membership of $\overline{A}$ (HIV −ve) may not be embarrassing. Hence, larger the conditional probability of becoming a member of A given a certain answer, the greater the embarrassment caused by providing that answer. Let $P(A|Y)$ $(P(A|N))$ be the conditional probability of a respondent belonging to group A given that he/she provided "Y" ("N") answer. Lanke's (1976) measure of protection based on the RR technique R^* is defined as

$$\mathcal{L}(R^*) = Max[P(A|Y), P(A|N)] \tag{16.7.23}$$

The smaller value of $\mathcal{L}(R^*)$ is more privacy protected.

Thus an RR technique R_1 is more protective than that of R_2 if $\mathcal{L}(R_1) < \mathcal{L}(R_2)$. R_1 and R_2 are equivalent if $\mathcal{L}(R_1) = \mathcal{L}(R_2)$.

Example 16.7.3 Comparison Between Warner and Mangat & Singh Model

From Eq. (16.7.5), we find that the values of $P(A|Y)$ and $P(A|N)$ for Warner's RR technique are, respectively,

$$P_w(A|Y) = \frac{\pi_A P_1}{\pi_A P_1 + (1-\pi_A)(1-P_1)} \quad \text{and}$$

$$P_w(A|N) = \frac{\pi_A(1-P_1)}{\pi_A(1-P_1) + (1-\pi_A)P_1}$$

In this case Lanke's measure is given by

$$\mathcal{L}_w = Max[P_w(A|Y), P_w(A|N)]$$

$$= \begin{cases} P_w(A|Y) = \dfrac{\pi_A P_1}{\pi_A P_1 + (1-\pi_A)(1-P_1)} & \text{if} \quad P_1 > 1/2 \\ \\ P_w(A|Y) = \dfrac{\pi_A(1-P_1)}{\pi_A(1-P_1) + (1-\pi_A)P_1} & \text{if} \quad P_1 < 1/2 \end{cases} \tag{16.7.24}$$

Similarly for Mangat–Singh RR model, Lanke's measure is

$$\mathcal{L}_{ms} = \begin{cases} P_{ms}(A|Y) = \dfrac{\pi_A P^*}{\pi_A P^* + (1-\pi_A)(1-P^*)} & \text{if} \quad P^* > 1/2 \\ P_{ms}(A|N) = \dfrac{\pi_A(1-P^*)}{\pi_A(1-P^*) + (1-\pi_A)P^*} & \text{if} \quad P^* < 1/2 \end{cases}$$

where $P^* = T + (1-T)P_4$.

Consider the situation (i) $P_1 > 1/2$ and $P^* > 1/2$. In this case $\mathcal{L}_w = \mathcal{L}_{ms}$ yields $P_1 = P^*$, which in turn gives $Var(\widehat{\pi}_w) = Var(\widehat{\pi}_{ms})$. Similarly, for each of the other three situations viz. (ii) $P_1 > 1/2$ and $P^* < 1/2$, (iii) $P_1 < 1/2$ and $P^* > 1/2$ and (iv) $P_1 < 1/2$ and $P^* < 1/2$, we find $\mathcal{L}_w = \mathcal{L}_{ms}$, which implies $Var(\widehat{\pi}_w) = Var(\widehat{\pi}_{ms})$. Hence we have the following theorem obtained by Singh (2003).

Theorem 16.7.3

Under Lanke's measure, Warner's model and Mangat–Singh's model are equally efficient for maintaining the same level of privacy protection.

16.7.1.3 Anderson's Measure

Anderson (1975a) defined $P(A|R)$ and $P(\overline{A}|R)$ as two "risk of suspicion" corresponding to response R and suggested to restrict them such that

$$P(A|R) \le \xi_2 < 1 \text{ and } P(\overline{A}|R) \le 1 - \xi_1 < 1 \tag{16.7.25}$$

Also since $P(\overline{A}|R) = 1 - P(A|R)$, Eq. (16.7.25) implies

$$\xi_1 \le P(A|R) \le \xi_2 \tag{16.7.26}$$

Since $P(A|R)$ depends on π_A, Eq. (16.7.26) gives

$$g(Y,A) \le \frac{1-\pi_A}{\pi_A}\frac{\xi_2}{1-\xi_2} \text{ and } g(N,\overline{A}) \le \frac{\pi_A}{1-\pi_A}\frac{1-\xi_1}{\xi_1} \tag{16.7.27}$$

Thus, Anderson's criteria for protecting confidentiality is to set upper bounds for $g(Y,A)$ and $g(N,\overline{A})$ and then minimize $Var(\widehat{\pi}_A)$ subject to this restrictions. For further details, readers are referred to Chaudhuri and Mukherjee (1988).

Flinger et al. (1977) provided with a measure of jeopardy as

$$\mathcal{F} = \frac{1 - \max\{P(A|Y), P(A|N)\}}{1 - \pi_A} = \frac{1 - \mathcal{L}}{1 - \pi_A} \tag{16.7.28}$$

Nayak (2007) pointed out that the respondents' protection increases as posterior probabilities $P(A|Y)$ and $P(A|N)$ decreases. Hence the RR device R_1 is better than R_2 for estimating π_A if

$$\begin{aligned} &P_{R_1}(A|Y) \leq P_{R_2}(A|Y), \quad P_{R_1}(A|N) \leq P_{R_2}(A|N) \text{ and} \\ &V_{R_1}(\widehat{\pi}_A) \leq V_{R_2}(\widehat{\pi}_A) \end{aligned} \tag{16.7.29}$$

where P_{Ri} and V_{Ri} denote, respectively, probability and variance with respect to the design R_i; $i = 1.2$.

16.7.2 Quantitative Characteristics

Anderson (1977) proposed a measure of privacy protection, which can be used for quantitative characteristics also. Suppose that the embarrassing characteristic y follows an unknown distribution function $F_y(t)$. Since y is a sensitive characteristic, it cannot be obtained directly from the respondent. So, an RR "R" is obtained from the respondent. The distribution of R depends on the unknown y-value of the respondent. Let the probability density of R for a given $y = t$ be $h_R(r|t)$. The density $h_R(r|t)$ is called the response density. The density $h_R(r|t)$ is generated by the RR device proposed by a statistician. The unconditional density of R

$$g_R(r) = \int h_R(r|t)dF_y(t) = E[h_R(r|t)] \tag{16.7.30}$$

is a mixture of response densities with $F_y(t)$ as mixing distribution.

The conditional density of t given $R = r$

$$f_y(t|r) = h_R(r|t)f_y(t)/g_R(r), \quad r \in \Omega_r \tag{16.7.31}$$

where $f_y(t)$ is the marginal probability density of y and Ω_r is the set of possible RR values. The density $f_y(t|r)$ is known as revealing density. It depends both on response distributions and the unknown distribution of y.

After obtaining an RR "R = r" from a respondent, the revealing density $f_y(t|r)$ provides information about y, while the discrepancy between

$f_y(t|r)$ and $f_y(t)$ gives the amount of invasion of privacy caused by the RR "R = r". If the density $f_y(t|r)$ has a high concentration about the true value y the maintenance of privacy is very small, and alternatively if the spread of $f_y(t|r)$ is high, privacy is well maintained. Anderson (1977) proposed a measure of privacy protection associated with the response $R = r$ as $V(y|R = r)$ and the overall measure as

$$\tau = E\{V(y|R)\} \tag{16.7.32}$$

Alternatively, one can use the relative measures $V(y|R = r)/V(y)$ and $E\{V(y|R)\}/V(y)$.

Example 16.7.4

Consider an RR technique where a respondent provides an RR $R = y + X$ with y as the true value of the study variable y. Suppose that of y is $N(\mu_y, \sigma_y)$, normal with unknown mean μ_y and variance σ_y^2, and X is $N(\mu_x, \sigma_x)$. Assuming X and y are independent, we find $h_R(r|y)$, the condition distribution of R, given y is $N(y + \mu_x, \sigma_x)$. The marginal distribution of R, $g_R(r)$, is normal with mean $\mu = \mu_x + \mu_y$ and variance $\sigma^2 = \sigma_x^2 + \sigma_y^2$. The conditional distribution of y given $R = r$ is

$$\text{i.e. } f_y(y|r) = h_R(r|y) f_y(y)/g_R(r)$$

$$= \frac{1}{\sqrt{2\pi}\,\widetilde{\sigma}} e^{-\frac{1}{\widetilde{\sigma}^2}(y - \widetilde{\mu})^2} \tag{16.7.33}$$

i.e., $f_y(y|r)$ is $N(\widetilde{\mu}, \widetilde{\sigma})$ where $\widetilde{\mu} = \left(\frac{r - \mu_x}{\sigma_x^2} + \frac{\mu_y}{\sigma_y^2}\right) \Big/ \left(\frac{1}{\sigma_x^2} + \frac{1}{\sigma_y^2}\right)$ and

$\widetilde{\sigma}^2 = \left(\frac{1}{\sigma_x^2} + \frac{1}{\sigma_y^2}\right)^{-1}$.

Hence,

$$\tau = E[V(y|R)] = \widetilde{\sigma}^2 = \left(1/\sigma_x^2 + 1\big/\sigma_y^2\right)^{-1} \tag{16.7.34}$$

The larger σ_x^2 implies larger spread of the revealing distribution, i.e., the respondents are more protected.

16.8 OPTIMALITY UNDER SUPERPOPULATION MODEL

In this section, we will present few optimal strategies based on RR model. The results of this section were mainly derived by Arnab (1990, 1995a,b, 1998a, 2004a) after modification of the results stated in Chapter 6. Here we will assume that the population vector $\mathbf{y} = (y_1,\ldots, y_N)$ is a random variable that follows a superpopulation model ξ, which was described in Chapter 6. Let E_ξ, V_ξ and C_ξ denote respectively the expectation, variance, and covariance with respect to the model ξ. As stated earlier, the values of y_i's are not available directly from the respondents. If the ith unit (respondent) is included in the sample$s$, a revised RR r_i is obtained from it by using some randomized device described in Sections 16.2 and 16.4. The responses r_i's are independent random variables satisfying the following RR model defined in Eq. (16.4.4) viz.

$$E_R(r_i) = y_i,\ V_R(r_i) = \phi_i = \phi_i(y_i) \text{ and } C_R(r_i, r_j) = 0 \text{ for } i \neq j$$

We have defined in Chapter 6 that the class C_{pu} consists of the p-unbiased estimators $t = t(s,y)$ based on y_i's for $i \in s$ satisfying the unbiasedness condition $E_p(t) = Y\ \forall \mathbf{y} \in R_N$. Replacing y_i by r_i in t, we define the pR unbiased (or simply unbiased) estimator based on the RR model as $t_r = t(s,r)$, which satisfies the unbiasedness condition

$$E_pE_R(t_r) = E_RE_p(t_r) = Y\ \ \forall\ \ \mathbf{y} \in R_N \tag{16.8.1}$$

The class of unbiased estimators for RR model will be denoted by C_{pur}. Similarly, C_{plr}, the class of linear unbiased estimators based on RR survey data consists of the estimators of the form

$$t_{lr} = b_s + \sum_{i \in s} b_{si} r_i \tag{16.8.2}$$

The estimator t_{lr} satisfies the unbiasedness condition

$$E_pE_R(t_{lr}) = E_RE_p(t_{lr}) = Y\ \ \forall\ \ \mathbf{y} \in R_N \tag{16.8.3}$$

The unbiasedness condition (16.8.3) yields

$$E_p(b_s) = \sum_s b_s p(s) = 0 \text{ and } \sum_{s \supset i} b_{si} p(s) = 1\ \ \forall i \in U \tag{16.8.4}$$

16.8.1 Product Measure Model

Model $M1$: y_i's are independently distributed with known $E_\xi(y_i) = \mu_i$ and unknown variance $V_\xi(y_i) = \sigma_i^2$. Since r_i's are independently distributed we have under model $M1$

$$E_{\xi R}(r_i) = E_\xi\{E_R(r_i)\} = \mu_i \text{ and}$$
$$V_{\xi R}(r_i) = E_\xi\{V_R(r_i)\} + V_\xi\{E_R(r_i)\} = E_\xi(\phi_i) + \sigma_i^2 \tag{16.8.5}$$

where $E_{\xi R}$ and $V_{\xi R}$ denote operators of overall expectation and variance for the combination of the superpopulation model $M1$ and RR model R.

Using Theorem 6.3.2, we get the following theorem.

Theorem 16.8.1

Under the model $M1$ and a noninformative sampling design p with inclusion probability $\pi_i > 0 \ \forall \ i = 1,\ldots, N$

$$E_\xi E_R E_p(t_r - Y)^2 = E_{\xi R} V_p(t) \geq \sum_{i\in U}\{E_\xi(\phi_i) + \sigma_i^2\}\left(\frac{1}{\pi_i} - 1\right)$$
$$= E_{\xi R} V_p(t_{0r}) \quad \forall\, t_r \in C_{pur}$$

where $t_{0r} = \sum_{i\in s}\frac{r_i - \mu_i}{\pi_i} + \sum_{i\in U}\mu_i$.

The estimator t_{0r} is analogous to the generalized difference estimator t_0 defined in Eq. 6.3.9. But t_{0r} cannot be used in practice because μ_i's are generally unknown. Consider a special case of the model $M1$ with $\mu_i = \beta\, x_i$, where β is an unknown constant but $x_i(>0)$ is a known value of an auxiliary variable x for the ith unit $i = 1,\ldots, N$. Let $p(\pi,x)$ be a fixed effective sample of size n design ($FED(n)$) with $\pi_i = np_i$, $p_i = x_i/X$, and $X = \sum_{i\in U} x_i$. For such a $p(\pi,x)$ design, t_{0r} reduces to $\widehat{Y}_{ht}(r) = \frac{1}{n}\sum_{i\in s}\frac{r_i}{p_i}$ and we get the following result similar to Theorem 6.3.4.

Theorem 16.8.2

Under the model $M1$ with $\mu_i = \beta\, x_i$,

$$E_{\xi R} V_{p(\pi,x)}(t_r) \geq \sum_{i\in U}\{E_\xi(\phi_i) + \sigma_i^2\}\left(\frac{X}{n\, x_i} - 1\right) = E_{\xi R} V_{p(\pi,x)}\left(\widehat{Y}_{ht}(r)\right)$$

$\forall\, t_r \in C_{pur}$.

Furthermore, for the model $M1$ with $\mu_i = \beta\, x_i$, $\sigma_i^2 = \sigma^2 x_i^2$, $E_\xi(\phi_i) = \lambda\, x_i^2$ and λ as a constant, Theorem 16.8.2 yields

$$E_{\xi R} V_p(t_r) \geq (\sigma^2 + \lambda) \sum_{i \in U} x_i^2 \left(\frac{1}{\pi_i} - 1 \right) \quad \forall\, t_r \in C_{pur} \tag{16.8.6}$$

Minimizing the right hand side of Eq. (16.8.6) with respect to π_i while keeping $\sum_{i \in U} \pi_i = n$ as fixed, we find the optimum value of $\pi_i = n\, x_i/X$ and the estimator t_{0r} reduces to $\widehat{Y}_{ht}(r) = \frac{1}{n} \sum_{i \in s} \frac{r_i}{p_i}$. Denoting the class of estimators with fixed effective size n sampling design $\mathcal{P}_n$, we derive the following result parallel to Theorem 6.3.5.

Theorem 16.8.3

Under the model $M1$ with $\mu_i = \beta\, x_i$, $\sigma_i^2 = \sigma^2 x_i^2$ and $E_\xi(\phi_i) = \lambda x_i^2$

$$\begin{aligned} E_{\xi R} V_p(t_r) &\geq (\sigma^2 + \lambda) \left(\frac{X^2}{n} - \sum_{i \in U} x_i^2 \right) \\ &= E_{\xi R} V_{p(\pi, x)} \left(\widehat{Y}_{ht}(r) \right) \quad \forall \;\; t \in C_{pur}, p \in \mathcal{P}_n \end{aligned}$$

From the theorem above, we note that the strategy $h_{0r} = \left(p_{(\pi,x)}, \widehat{Y}_{ht}(r) \right)$ is the optimum in the class of strategies $H = (p, t_r)$, $p \in \mathcal{P}_n$, $t_r \in C_{pur}$ under an RR technique with $E_\xi(\phi_i)$ proportional to x_i^2. The construction of such an optimum RR technique was provided by Arnab (1998a), and it is given in Section 16.8.3. Furthermore, if $x_i = 1 \;\; \forall \;\; i = 1, .., N$, $p_{(\pi,x)}$ reduces to p_0, where $\pi_i = \pi_0 = n/N$. In this case $\widehat{Y}_{ht}(r)$ reduces to $N\, \bar{r}_s$, where $\bar{r}_s = \sum_{i \in s} r_i/n$. The design p_0 includes SRSWOR. In this case Theorem 16.8.3 reduces to the following theorem.

Theorem 16.8.4

Under the model $M1$ with $\mu_i = \beta$, $\sigma_i^2 = \sigma^2$ and $E_\xi(\phi_i) = \lambda$

$$E_{\xi R} V_p(t_r) \geq (\sigma^2 + \lambda) N \left(\frac{N}{n} - 1 \right) = N^2 E_{\xi R} \left(V_{p_0}(\bar{r}_s) \right) \quad \forall\, p \in \mathcal{P}_n, t_r \in C_{pur}$$

16.8.2 Equicorrelation Model

Consider the model

$$\begin{aligned} M2: E_\xi(y_i) = \beta x_i, \;\; V_\xi(y_i) = \sigma^2 x_i^2 \;\; \text{and} \;\; C_\xi(y_i, y_j) \\ = \rho x_i x_j \;\; \text{with} \;\; -1/(N-1) \leq \rho \leq 1 \end{aligned} \tag{16.8.7}$$

Under the RR Model (16.4.4) we get

$$E_{\xi R}(r_i) = \beta x_i,\ V_{\xi R}(y_i) = \sigma^2 x_i^2 + \phi_i \text{ and } C_{\xi R}(r_i, r_j) = E_\xi\{C_R(r_i, r_j)\} + C_\xi\{E_R(r_i), E_R(r_j)\} = \rho x_i, x_j \tag{16.8.8}$$

For an estimator t_{lr} that belongs to the class C_{plr}, we have under model $M2$ given in Eq. (16.8.8)

$$\begin{aligned} E_{\xi R} V_p(t_{lr}) &= E_\xi E_R E_p (t_{lr} - Y)^2 \\ &= E_\xi E_p E_R (t_{lr} - Y)^2 \\ &= E_\xi E_p [\{E_R(t_{lr} - Y)\}^2 + V_R(t_{lr} - Y)] \\ &= E_\xi V_p(t_l) + \sum_{i \in U} \alpha_i E_\xi(\phi_i) \end{aligned} \tag{16.8.9}$$

$$\left(\text{where } t_l = b_s + \sum_{i \in s} b_{si} y_i \text{ and } \alpha_i = \sum_{s \supset i} b_{si}^2 p(s)\right)$$

Using Theorem 6.3.9, we note that

$$E_\xi V_p(t_l) \geq (1 - \rho)\sigma^2 \left(\frac{X^2}{n} - \sum_{i \in U} x_i^2\right) = E_\xi V_{p(\pi,x)}\left(\widehat{Y}_{ht}(r)\right) \text{ for } p \in \mathscr{P}_n \tag{16.8.10}$$

The second part of Eq. (16.8.9) becomes

$$\sum_{i \in U} \alpha_i E_\xi(\phi_i) \geq \sum_{i \in U} E_\xi(\phi_i)/\pi_i \tag{16.8.11}$$

$$\begin{aligned} \text{Since } \alpha_i &\geq \sum_{s \supset i} b_{si} p(s)^2 \Big/ \sum_{s \supset i} b_{si} p(s) \\ &= 1/\pi_i [\text{using unbiasedness condition (16.8.4)}]. \end{aligned} \tag{16.8.12}$$

The right hand side of Eq. (16.8.12) is minimized for a fixed sample size design with $\sum_i \pi_i = n$ when

$$\pi_i = n\sqrt{E_\xi(\phi_i)} \Big/ \sum_{i \in U} \sqrt{E_\xi(\phi_i)} \tag{16.8.13}$$

Now if we choose an RR model for which $E_\xi(\phi_i) = \lambda\, x_i^2$ with λ as a constant, then π_i becomes equal to $n\, x_i/X$. In this situation we arrive at the following theorem.

Theorem 16.8.5
Under model M_2 with $E_\xi(\phi_i) = \lambda\, x_i^2$,

$$E_{\xi R} V_p(t_{lr}) \geq [\lambda + (1-\rho)\sigma^2]\frac{X^2}{n} - (1-\rho)\sigma^2 \sum_{i \in U} x_i^2$$
$$= E_{\xi R} V_{p(\pi,x)}\left\{\widehat{Y}_{ht}(r)\right\} \text{ for } p \in \mathcal{P}_n,\ t_{lr} \in C_{plr}$$

From Eq. (16.8.13) and Theorem 16.8.5, we note that the optimal estimators for the population total can be obtained if we can construct an RR model for which $E_\xi(\phi_i) = \lambda\, x_i^2$. We will call an RR model for which $E_\xi(\phi_i) = \lambda\, x_i^2$ is the optimal RR technique.

16.8.3 Construction of an Optimal Randomized Response Technique

Arnab (1998a) proposed modifications of Eriksson's (1973) and Chaudhuri's (1987) RR techniques so that $E_\xi(\phi_i)$ becomes proportional to x_i^2. For the model M_2 with $E_\xi(y_i) = \beta^* x_i$, we may choose constants $k_1, \ldots, k_L$, which anticipate the possible range of β^* so that $k_1 x_i, \ldots, k_L x_i$ in turn cover the range of y_i. The modification of the Eriksson RR technique is given as follows:

The respondent labeled i is to report either the true y_i with probability c or $Q_j(i) = k_j x_i$ with probability q_j for $j = 1, \ldots, L\left(q_j > 0, \sum q_j = 1 - c\right)$. Denoting z_i as the RR obtained from the ith respondent and $r_i = \left(z_i - x_i \sum_j k_j q_j\right)/c$, we find for the modified RR technique

$E_\xi V_R(r_i) = \lambda x_i^2$, where $\lambda = \left[c(1-c)\left(\sigma^2 + \beta^{*2}\right) - 2c\beta^*\left(\sum_j k_j q_j\right) + \left(\sum_j k_j^2 q_j\right) - \left(\sum_j k_j q_j\right)^2\right]/c^2$.

Similarly we can modify Chaudhuri's (1987) RR technique by taking $A_i = A$, $B_i = B$, $a_i(j) = a(j)$, and $b_i(k) = b(k)x_i$ for $i = 1, \ldots, N$; $j = 1, \ldots, A$, and $k = 1, \ldots, B$. This modification yields $r_i = \dfrac{z_i - \overline{b}\, x_i}{\overline{a}}$ and $E_\xi(\phi_i) = \left[\left(\sigma^2 + \beta^{*2}\right)s_a^2 + s_b^2\right]x_i^2/\overline{a}^2$, where $s_a^2 = \sum_j \{a(j) - \overline{a}\}^2/A$,

$s_b^2 = \sum_j \left\{b(j) - \overline{b}\right\}^2/B$, $\overline{a} = \sum_j a(j)/A$ and $\overline{b} = \sum_j b(j)/B$.

Remark 16.8.1
For the model $M2$ with $x_i = 1\ \forall i \in U$, $\widehat{Y}_{ht}(r)$ reduces to $N\ \overline{r}_s$ and $p_{(\pi,x)}$ reduces to p_0. Hence $N\ \overline{r}_s$ based on SRSWOR provides the optimal

strategy in the class of strategies $\mathscr{K} = (p, t)$, $p \in \mathscr{P}_n$, $t \in C_{lpr}$ when $E_\xi(\phi_i) = \phi_0$ is a constant. For the modified Eriksson's (1973) and Chaudhuri's (1987) RR techniques mentioned above $E_\xi(\phi_i) = \phi_0$ when $x_i = 1 \ \forall i \in U$.

Remark 16.8.2

Under a random permutation model, the probability of realizing the vector $y = (y_1 = Y_{i_1}, \ldots y_N = Y_{i_N}) = 1/N!$ for every permutation $(i_1, \ldots, i_N)$ of $(1, \ldots, N)$, where $Y_1, \ldots, Y_N$ are fixed numbers. In this model $E_\xi(y_i) = \overline{Y} = \sum_{i \in U} y_i / N$, $V_\xi(y_i) = \sum_{i \in U} (y_i - \overline{Y})^2 / N$ and $C_\xi(y_i, y_j) = -1/(N-1)$ for $i \neq j$. Hence for the random permutation model, $\overline{r}_s$ based on an SRSWOR is the optimum strategy for estimating the population mean $\overline{Y}$ under the RR model with $E_\xi(\phi_i)$ as a constant.

16.9 EXERCISES

16.9.1 An RR survey was conducted to find out the proportion of HIV+ students in a university. A sample of 500 students was selected from 15,500 students by SRSWOR method and each student selected was asked to answer "Yes" or "No" to one of the questions "Are you HIV+ ?" and "I passed the Matriculation examination with a C grade" with a probability 0.7 and 0.3, respectively. Among the students selected in the sample, 324 students answer "Yes." Estimate the proportion of HIV+ students in the university and obtain 95% confidence interval of the proportion when the proportion of students received grade C in Matriculation examination is 20%.

16.9.2 Let a sample s of size n be selected from a finite population by varying probability sampling design with inclusion probability π_i for the ith unit. From each of the selected respondents, RRs were obtained by using Kuk's device described in Section 16.2.3. Show that $\widehat{\pi}_A = \dfrac{1}{Nc(P_{31} - P_{32})} \sum_{i \in s} \dfrac{z_i - cP_{32}}{\pi_i}$ is an unbiased estimator of the population proportion π_A. Derive the variance of $\widehat{\pi}_A$ and suggest an unbiased estimator of its variance.

16.9.3 A sample of 15 industrial workers was selected at random from 60 workers by SRSWOR method. Each of the workers was asked to select a ticket at random from a jar and multiply the number written on the ticket by its actual expenditure on gambling as his/her RR. The responses have been given in the following table.

Workers	1	2	3	4	5	6	7	8	9	10
Randomized responses	35	95	100	125	250	375	125	100	75	60

Assuming that the number on a ticket follows Gamma distribution with mean 20 and variance 5, obtain an unbiased estimator of the average expenditure on gambling. Also, estimate the standard error of the estimator used.

16.9.4 Consider Greenberg's RR technique where a respondent answers the question "Are you HIV +ve?" with probability $p = 0.2$ and the question "Are you a black African" with probability $1 - p = 0.8$. A sample of 150 students is selected from 7500 students of a university by SRSWOR method. The proportion of "Yes" answer was 0.65. Estimate the proportion of HIV+ students and 90% confidence interval of the proportion when it is known that 60% of students come from the black African community.

16.9.5 Consider Greenberg et al. (1969)'s RR model described in Section 16.2.2 where π_x is unknown. Suppose two independent samples of sizes n_1 and n_2 are selected by SRSWOR method. Show that (i) $\widehat{\pi}_G(1) = \dfrac{(1 - P_{22})\lambda_g(1) - (1 - P_{21})\lambda_g(2)}{P_{21} - P_{22}}$ is an unbiased estimator of π_A, (ii) $Var(\widehat{\pi}_{G1}) = \dfrac{1}{(P_{21} - P_{22})^2}\left[\dfrac{(1 - P_{22})^2\theta_{g1}(1 - \theta_{g1})}{n_1} + \dfrac{(1 - P_{21})^2\theta_{g2}(1 - \theta_{g2})}{n_2}\right] - Q_1 - Q_2$ where

$$Q_1 = \frac{(1 - P_{22})^2(n_1 - 1)\left\{P_{21}^2\pi_y(1 - \pi_y) + (1 - P_{21})^2\pi_x(1 - \pi_x)\right\}}{n_1(N - 1)(P_{21} - P_{22})^2} \text{ and}$$

$$Q_2 = \frac{(1 - P_{21})^2(n_2 - 1)\left\{P_{22}^2\pi_y(1 - \pi_y) + (1 - P_{22})^2\pi_x(1 - \pi_x)\right\}}{n_2(N - 1)(P_{21} - P_{22})^2}, \text{ and}$$

(iii) find an unbiased estimator of $Var(\widehat{\pi}_{G1})$ (Kim, 1978).

16.9.6 Consider the RR technique R, proposed by Greenberg et al. (1969) where a respondent answers whether or not he/she possesses the sensitive character x with probability p and nonsensitive character y with probability $1 - p$. Suppose a sample s of size n is selected from a finite population of size N by SRSWOR method.

The sample s is portioned at random into two subsamples s_1 and s_2 of sizes n_1 and $n_2(=n-n_1)$, respectively. Respondents belonging to s_1 are asked to perform the randomized device R whereas respondents belonging to the subsample s_2 are directly asked whether or not they possess character y. Let $\widehat{\theta}_1$ and $\widehat{\pi}_y$ be the proportion of "Yes" answers in s_1 and s_2, respectively, and π_x be the proportion of respondents in the population that possesses the sensitive character x. Show that (i) $\widehat{\pi}_x = \left\{\widehat{\theta}_1 - (1-p)\widehat{\pi}_y\right\}/p$ is an unbiased estimator of π_x and (ii) the minimum variance of $\widehat{\pi}_x$ with the optimum choice of n_1 and n_2 is $V_{\min}(\widehat{\pi}_x) = (\frac{\sqrt{A}+\sqrt{B}}{n} - C)/p^2$, where

$$A = \theta(1-\theta) + \left[p^2\pi_x(1-\pi_x) + (1-p)^2\ \pi_y(1-\pi_y)\right]/(N-1),$$

$$B = \frac{N(1-p)^2\pi_y(1-\pi_y)}{N-1},\ C = \frac{p^2\pi_x(1-\pi_x)}{N-1}, \text{ and}$$

$\theta = p\pi_x + (1-p)\pi_y$ (Arnab, 2006).

16.9.7 Consider an RR device where a person was asked to say "Yes" if he/she belongs to a certain sensitive group A. If the person does not belong to the group A, then he/she was asked to provide an RR using Warner's technique where he/she needs to select a card at random from a pack containing two different types of cards, which are identical in shape. The type I card with known proportion p bears the statement "I belong to the sensitive group A" whereas the type II card with proportion $1-p$ bares the statement "I do not belong to the group A." The respondent needs to answer "Yes" or "No". Let a sample s of size n be selected from a population by SRSWR method and $\widehat{\lambda}$ be the proportion of "Yes" answers obtained from the respondents based on the above sampling procedure. Show that

(i) $\widehat{\pi} = \left\{\widehat{\lambda} - (1-p)\right\}/p$ is an unbiased estimator of π, the proportion of persons belongs to the group A in the population.

(ii) $V(\widehat{\pi}) = (1-\pi)\{\pi + (1-p)/p\}/n$

(iii) The MLE of π is $\widehat{\pi}_M = \begin{cases} 1-p \text{ if } \widehat{\lambda} \le 1-p \\ \widehat{\lambda} \text{ if } 1-p < \widehat{\lambda} \le 1 \end{cases}$

(iv) Let us further assume that π has a prior beta distribution

$$f(\pi|\alpha,\beta)=\frac{\Gamma(\alpha+\beta)}{\Gamma\alpha)\Gamma\beta}\,\pi^{\alpha-1}(1-\pi)^{\beta-1},\ \ 0<\pi<1$$

with known parameters α and β. Show that

(a) the posterior distribution of π, given $\widehat{\lambda}$, is

$$g(\pi|T=t,\alpha,\beta)=\frac{\sum_{j=0}^{t}\binom{t}{j}\,d^{t-j}\pi^{\alpha+j-1}(1-\pi)^{n+\beta-t-1}}{\sum_{j=0}^{t}\binom{t}{j}\,d^{t-j}B(\alpha+j,n+\beta-t)}$$

where $T=$ total number of "Yes" answers.

(b) The Bayes estimator under squared error loss is

$$\widehat{\pi}_B=\frac{\sum_{j=0}^{t}\binom{t}{j}\,d^{t-j}B(\alpha+j+1,n+\beta-t)}{\sum_{j=0}^{t}\binom{t}{j}\,d^{t-j}B(\alpha+j,n+\beta-t)}$$

(Kim et al., 2006)

16.9.8 Let a sample s of size n be selected by SRSWR method. The ith respondent in the sample s reports the true value y_i if he/she feels that the characteristic y is not confidential. Otherwise, if the respondent feels that the character y is confidential, he/she reports $z_i=s_iy_i/\theta$ as an RR where s_i is a random sample from a gamma population with known mean θ and variance γ^2. Let $r_i=y_i$ in case that the ith respondent reports the true value and $r_i=z_i$ if he/she reports a scrambled response. Show that (i) $\overline{r}=\sum_{i=1}^{n}r_i/n$ is an unbiased estimator of the population mean μ_y and (ii) $V(\overline{r})=\left[\sigma_y^2+W\gamma^2\left(\sigma_y^2+\mu_y^2\right)\right]\Big/n$, where σ_y is the population variance and W is the probability of reporting a scrambled response (Gupta et al., 2002).

16.9.9 For the RR model $E_R(r_i)=y_i$, $E_R(r_i)=\phi_i$ and $C_R(r_i,r_j)=0$ for $i\neq j$ prove the following results: (i) Under PPSWR sampling, the Hansen–Hurwitz estimator $\widehat{Y}_{hh}(r)=\frac{1}{n}\sum_{i\in s}\frac{r_i}{p_i}$ is admissible in the class of linear unbiased estimators of the population total.

(ii) Under SRSWR sampling, the sample mean $\overline{r}=\sum_{i\in s}r_i/n$ based on all the units (including repetition) is admissible in the class of linear unbiased estimators of the population mean (Arnab, 1995a).

CHAPTER 17

Domain and Small Area Estimation

17.1 INTRODUCTION

Large-scale surveys cover wide geographical areas, and information on various items is included in the scope of the survey. In most situations, estimates from different sections of populations are required. For example, the "Botswana Aids Impact Survey II" (BAIS II) was conducted in 2003 to cover the whole Botswana to collect data on exposure to HIV infections as well as socioeconomic, demographic, and behavioral patterns, among others. The HIV infection status for the country as a whole was not the only important item required but information on different sections of populations (e.g., districts or counties) was also important for the management of HIV infections. A subpopulation of a population is called a domain. Here, district or counties may be considered as a domain (large domain) covering large geographical areas. In case sampling frames of different domains are available, one can select samples by employing a stratified sampling procedure and treating a domain as a stratum. But in reality, sampling frames of the domains under study would not be available at the planning stage of the survey, e.g., domain comprising HIV infection rates among business executives, middle class families, or child-headed households. Furthermore, if the number of domains increases, which is the case for a multicharacter survey (covering information on several characteristics), one cannot select samples from each of the domains directly. If the domain is large and a relatively large sample size is obtained, one may get reliable estimates of the parameters of interest. But reliable estimates of HIV infection rates of (for example) immigrants of different nationalities in different districts may not be obtained directly from the sample because the sample size belonging to such a subpopulation is not reasonably large enough. In this case, design-based estimates become very unstable. We call a section of population, whose representation in a sample is small or absent, a small area or small domain. We will consider the methods of estimation from domains and small areas separately. Theories of domain and small area estimation have been considered by Purcell and Kish (1979),

Survey Sampling Theory and Applications
ISBN 978-0-12-811848-1
http://dx.doi.org/10.1016/B978-0-12-811848-1.00017-0

Gonzalez (1973), Platek and Singh (1986), Platek et al. (1987), and Brackstone (1987), among others. Details have been given by Ghosh and Rao (1994), Chaudhuri and Stenger (1992), Rao (2003), and Särndal et al. (1992).

17.2 DOMAIN ESTIMATION

Consider a finite population U of N identifiable units from which a sample s of size n is selected with probability $p(s)$ such that inclusion probabilities for the ith unit π_i is positive for $i = 1,\ldots, N$. Let the population U be partitioned into mutually exclusive and exhaustive domains $U_1,\ldots, U_d,\ldots, U_D$; and y_i be the value of the ith unit of the variable of interest y and $Y_d = \sum_{i\in U_d} y_i$ be the dth domain total, $d = 1,\ldots, D$. Let

$$\widehat{Y} = \sum_{i\in s} b_{si}y_i \tag{17.2.1}$$

be an unbiased estimator of the population total $Y = \sum_{i\in U} y_i$, where b_{si}'s are constants that satisfy the unbiasedness condition $\sum_{s\supset i} b_{si}p(s) = 1$.

Let $s_d = s\cap U_d$ be the part of the sample s that belong to U_d of size n_d, which is not only positive but also reasonably large for every $d = 1,\ldots, D$. Then we find an unbiased estimator of Y_d as

$$\widehat{Y}_d = \sum_{i\in s} b_{si}y_iI_{di} = \sum_{i\in s_d} b_{si}y_i \tag{17.2.2}$$

where

$$I_{di} = 1 \quad \text{if } i\in U_d \text{ and } I_{di} = 0 \quad \text{if } i \notin U_d \tag{17.2.3}$$

The variance of $\widehat{Y}_d$ is given by

$$\begin{aligned} V\left(\widehat{Y}_d\right) &= \sum_{i\in U} \alpha_i I_{di}y_i^2 + \sum_{i\neq}\sum_{j\in U} \alpha_{ij}I_{di}I_{dj}y_iy_j \\ &= \sum_{i\in U_d} \alpha_i y_i^2 + \sum_{i\neq}\sum_{j\in U_d} \alpha_{ij}y_iy_j \end{aligned} \tag{17.2.4}$$

where $\alpha_i = \sum_{s\supset i} b_{si}^2p(s) - 1$ and $\alpha_{ij} = \sum_{s\supset i,j} b_{si}b_{sj}p(s) - 1$.

An unbiased estimator of $Var\left(\widehat{Y}_d\right)$ is given by

$$\begin{aligned} \widehat{V}\left(\widehat{Y}_d\right) &= \sum_{i\in s} c_{si}I_{di}y_i^2 + \sum_{i\neq}\sum_{j\in s} c_{sij}I_{di}I_{dj}y_iy_j \\ &= \sum_{i\in s_d} c_{si}\, y_i^2 + \sum_{i\neq}\sum_{j\in s_d} c_{sij}\, y_iy_j \end{aligned} \tag{17.2.5}$$

where $\sum_{s\supset i} c_{si}p(s) = \alpha_i$ and $\sum_{s\supset i,j} c_{sij}p(s) = \alpha_{ij}$.

The population mean $\overline{Y}_d = Y_d/N_d$ of the domain U_d of known size N_d may be estimated by

$$\widehat{\overline{Y}}_d = \widehat{Y}_d/N_d$$

If N_d is unknown, the following ratio estimator may be used.

$$\widehat{\overline{Y}}_{dR} = \frac{\sum_{i\in s} b_{si} y_i I_{di}}{\sum_{i\in s} b_{si} I_{di}} = \frac{\sum_{i\in s_d} b_{si} y_i}{\sum_{i\in s_d} b_{si}} \quad (17.2.6)$$

Approximate expressions of mean-square error (MSE) of $\widehat{\overline{Y}}_{Rd}$ and its unbiased estimators are obtained by using Theorem 8.2.2 as follows:

$$M\left(\widehat{\overline{Y}}_{dR}\right) \cong \sum_{i\in U_d} \alpha_i z_i^2 + \sum_{i\neq}\sum_{j\in U_d} \alpha_{ij} z_i z_j \quad (17.2.7)$$

$$\widehat{M}\left(\widehat{\overline{Y}}_{dR}\right) \cong \sum_{i\in s_d} \frac{\alpha_i}{\pi_i} \widehat{z}_i^2 + \sum_{i\neq}\sum_{j\in s_d} \frac{\alpha_{ij}}{\pi_{ij}} \widehat{z}_i \widehat{z}_j \quad (17.2.8)$$

where $z_i = y_i - \overline{Y}_d$ and $\widehat{z}_i = y_i - \widehat{\overline{Y}}_{dR}$.

17.2.1 Horvitz–Thomson Estimator

In particular, if $b_{si} = 1/\pi_i$, then the expressions (Eqs. 17.2.2, 17.2.4, and 17.2.5) become

$$\widehat{Y}_d = \widehat{Y}_d(ht) = \sum_{i\in s_d} y_i/\pi_i \quad (17.2.9)$$

$$V\left[\widehat{Y}_d(ht)\right] = \sum_{i\in U_d}\left(\frac{1}{\pi_i} - 1\right) y_i^2 + \sum_{i\neq}\sum_{j\in U_d}\left(\frac{\pi_{ij}}{\pi_i\pi_j} - 1\right) y_i y_j, \quad (17.2.10)$$

and

$$\widehat{V}\left[\widehat{Y}_d(ht)\right] = \sum_{i\in s_d}\left(\frac{1}{\pi_i} - 1\right)\frac{y_i^2}{\pi_i} + \sum_{i\neq}\sum_{j\in s_d}\left(\frac{\pi_{ij}}{\pi_i\pi_j} - 1\right)\frac{y_i y_j}{\pi_{ij}}, \quad (17.2.11)$$

respectively.

In case the domain total N_d is known, one can modify the conventional estimator $\widehat{Y}_d(ht)$ as

$$\widehat{Y}_{dR}(ht) = \frac{\widehat{Y}_d(ht)}{\widehat{N}_d} N_d \quad (17.2.12)$$

where $\widehat{N}_d = \sum_{i\in s_d} 1/\pi_i$.

Approximate expressions for the MSE of $\widehat{Y}_{dR}(ht)$ and its unbiased estimators are obtained using Eqs. (8.4.2) and (8.4.4) as follows:

$$M\left[\widehat{Y}_{dR}(ht)\right] \cong V\left(\sum_{i\in s_d}\frac{z_i}{\pi_i}\right)$$

$$= V\left[\widehat{Y}_d(ht)\right] \tag{17.2.13}$$

$$= \sum_{i\in U_d}\left(\frac{1}{\pi_i}-1\right)z_i^2 + \sum_{i\neq}\sum_{j\in U_d}\left(\frac{\pi_{ij}}{\pi_i\pi_j}-1\right)z_i z_j$$

and

$$\widehat{M}\left[\widehat{Y}_{dR}(ht)\right] \cong \sum_{i\in s_d}\left(\frac{1}{\pi_i}-1\right)\frac{\widehat{z}_i^2}{\pi_i} + \sum_{i\neq}\sum_{j\in s_d}\left(\frac{\pi_{ij}}{\pi_i\pi_j}-1\right)\frac{\widehat{z}_i\widehat{z}_j}{\pi_{ij}} \tag{17.2.14}$$

where $z_i = y_i - \overline{Y}_d$, $\overline{Y}_d = Y_d/N_d$, $\widehat{z}_i = y_i - \widehat{\overline{Y}}_d$ and $\widehat{\overline{Y}}_d = \widehat{Y}_d(ht)/\sum_{i\in s_d} 1/\pi_i$.

For estimation of the population mean $\overline{Y}_d = Y_d/N_d$, we may use an unbiased estimator

$$\widehat{\overline{Y}}_d(ht) = \widehat{Y}_d(ht)/N_d \tag{17.2.15}$$

when N_d is known. However, a better estimator

$$\widehat{\overline{Y}}_d(ht) = \widehat{Y}_d(ht)/\sum_{i\in s} 1/\pi_i \tag{17.2.16}$$

may be used if strictly unbiasedness condition is not required.

17.3 SMALL AREA ESTIMATION

The terms "small area" or "small domain" are commonly used to denote small geographic areas or small subpopulations of a population. Sample sizes that belong to the small areas are generally very small because the overall sample size in a survey is determined with consideration of the accuracy of the estimates of the parameters for the entire population. This is why direct estimators based on the small areas very often yield a large variance. Small area estimation is given considerable importance because of the recent increasing demand for reliable estimates from government and public sector enterprises to determine, for example, birth rate, death rate, school enrollment, and revenue from different municipalities for counties. Various methods of small area estimation are available in the literature. Some of the important methods include symptomatic accounting technique (SAT), direct method, synthetic method, composite method, and various methods based on statistical modeling. Detailed discussions are available in Brackstone (1987), Ghosh and Rao (1994), and Lehtonen et al. (2003).

17.3.1 Symptomatic Accounting Technique

Demographers have proposed various methods of small area estimation, which are broadly known as SATs which use logical relationship in conjunction with statistical relationship based on previous data. SAT utilizes current data gathered from administrative records as well as related data from the latest census. This method was developed for estimation of birth rate, death rate, population projections, types of dwelling, and school enrollment based on diverse registration data, which are known as "symptomatic" variables. A detailed review has been given by Purcell and Kish (1980) and Ghosh and Rao (1994). SAT includes vital rates (VRs) method (Bogue, 1950), composite method (Bogue and Duncan, 1959), census component (CC) method, housing unit (HU) method (Smith and Lewis, 1980), ratio correlation, and difference correlation method, among others.

17.3.1.1 Vital Rates Method

The VRs technique was given by Bogue (1950). Let b_t and d_t be the estimated total number of births and deaths for a smaller area of interest at time t. The estimated total numbers of births and deaths at the larger area containing the smaller area are, respectively, B_t and D_t. Suppose from the last census period $t = 0$, the crude birth (death) rate for the smaller and larger areas are $g_0(h_0)$ and $G_0(H_0)$, respectively, while $G_t(H_t)$ denotes a reliable estimate of birth rate (death rate) at time t for the larger area. The birth rate g_t and death rate h_t for the period t are obtained assuming $g_t/g_0 = G_t/G_0$ and $h_t/h_0 = H_t/H_0$, respectively, and are given as follows:

$$g_t = g_0(G_t/G_0)$$
$$h_t = h_0(H_t/H_0)$$

Let P_t be the total population at time t for the small area of interest. Then $g_t \cong b_t/P_t$ and $h_t \cong d_t/P_t$. Hence P_t can be estimated by

$$\widehat{P}_t = \frac{1}{2}\left(\frac{b_t}{g_t} + \frac{d_t}{h_t}\right) \tag{17.3.1}$$

17.3.1.2 Composite Method

The composite method is an improvement on the VR method. The method was proposed by Bogue and Duncan (1959). In this method, local area population is divided into distinct age subgroup. From each of these subgroups, population estimates are obtained separately by VR method and then summing these estimates across the groups, the "composite" estimate of the current population is obtained. The composite method uses the group-specific birth and death counts for the local area as well as current population of each group for the larger area containing the local area.

17.3.1.3 Census Component Method

In the CC method the population at the local area at a particular time is estimated by using the formula

$$\widehat{P}_t = P_0 + b_t - d_t + m_t \tag{17.3.2}$$

where P_0 is the population in the local area at the last census period $t = 0$ and m_t is the net migration during the period 0 and t.

17.3.1.4 Housing Unit Method

Let $\widehat{U}_t$ and $\widehat{Q}_t$ be the estimated total number of occupied housing units (HUs) and group quarters at the local area at time t. In the HU method, the estimated population total at a time t is

$$\widehat{P}_t = \widehat{U}_t \overline{x}_t + \widehat{Q}_t \overline{x}_t^* \tag{17.3.3}$$

where $\overline{x}_t$ and $\overline{x}_t^*$ denote the estimated average number of persons per occupying HU and group quarters, respectively.

17.3.1.5 Ratio Correlation Method

Suppose we want to estimate the population count p_{ti} for the ith small area at the time t based on the recent two census population count data for the period 0 and 1 along with q symptomatic variables $x_1, \ldots, x_q$ for the periods 0, 1, and t. Let p_{ki} be the population count and x_{kij} be the value of the jth symptomatic variable for the ith area at the time k; $k = 0, 1, t$; $j = 1, \ldots, q$; $i = 1, \ldots, A$. In the ratio correlation method, the change in the population counts between the periods 0 and 1 of the ith area is measured as

$$r_i = \frac{p_{1i}/P_1}{p_{0i}/P_0} \text{ with } P_k = \sum_{i \in A} p_{ki}, k = 0, 1, t$$

Similarly, the change in the ratio of jth symptomatic variable is measured as

$$z_{ij} = \frac{x_{1ij}/X_{1j}}{x_{0ij}/X_{0j}} \text{ with } X_{kj} = \sum_{i \in A} x_{kij}, k = 0, 1, t$$

Consider a multiple regression of r_i on $z_{i1}, \ldots, z_{iq}$ as

$$r_i = \beta_0 + \beta_1 z_{i1} + \cdots + \beta_j z_{ij} + \cdots + \beta_q z_{iq} + \in_i$$

where $\in_i$'s are usual independent identically distributed (iid) error components with mean zero and variance σ^2. Let the fitted multiple regression based on ordinary least squares (OLSs) method be

$$\widehat{r}_i = \widehat{\beta}_0 + \widehat{\beta}_1 z_{i1} + \cdots + \widehat{\beta}_j z_{ij} + \cdots + \widehat{\beta}_q z_{iq} \quad (17.3.4)$$

with $\widehat{\beta}_j$ as the least squares estimator of β_j.

Let $r_i^* = \dfrac{p_{ti}/P_t}{p_{1i}/P_1}$ and $z_{ij}^* = \dfrac{x_{tij}/X_{tj}}{x_{1ij}/X_{1j}}$.

Then r_i^* can be estimated from the given values of $z_{i1}^*, \ldots, z_{iq}^*$ using the regression (Eq. 17.3.4) above, as

$$\widehat{r}_i^* = \widehat{\beta}_0 + \widehat{\beta}_1 z_{i1}^* + \cdots + \widehat{\beta}_j z_{ij}^* + \cdots + \widehat{\beta}_q z_{iq}^* \quad (17.3.5)$$

Now noting $\widehat{r}_i^* = \dfrac{\hat{p}_{ti}/\hat{P}_t}{p_{1i}/P_1}$, we find an estimated value of the population count for the period t as

$$\widehat{p}_{ti} = \frac{p_{1i}}{P_1} \widehat{r}_i^* \widehat{P}_t \quad (17.3.6)$$

where $\widehat{P}_t$ is the estimated reliable population count at time t assumed to be available for the larger area.

17.3.1.6 Difference Correlation Method

In the difference correlation method, the ratio of changes in population count and the jth symptomatic variable between the periods 0 and 1 are measured, respectively, by the differences

$$\widetilde{r}_i = p_{1i}/P_1 - p_{0i}/P_0 \text{ and } \widetilde{z}_{ij} = x_{1ij}/X_{1j} - x_{0ij}/X_{0j}$$

Denoting the multiple regression of $\widetilde{r}_i$ on $\widetilde{z}_{i1}, \ldots, \widetilde{z}_{iq}$ as

$$\widehat{\widetilde{r}}_i = \widehat{\widetilde{\beta}}_0 + \widehat{\widetilde{\beta}}_1 \widetilde{z}_{i1} + \cdots + \widehat{\widetilde{\beta}}_j \widetilde{z}_{ij} + \cdots + \widehat{\beta}_q \widetilde{z}_{iq},$$

$\widetilde{r}_i^* = p_{ti}/P_t - p_{1i}/P_1$ and $\widetilde{z}_{ij}^* = x_{tij}/X_{tj} - x_{1ij}/X_{1j}$, we estimate $\widetilde{r}_i^*$ for given values $\widetilde{z}_{i1}^*, \ldots, \widetilde{z}_{iq}^*$ as

$$\widehat{\widetilde{r}}_i^* = \widehat{\widetilde{\beta}}_0 + \widehat{\widetilde{\beta}}_1 z_{i1}^* + \cdots + \widehat{\widetilde{\beta}}_j z_{ij}^* + \cdots + \widehat{\widetilde{\beta}}_q z_{iq}^*$$

Finally, noting $\widetilde{r}_i^* = p_{ti}/P_t - p_{1i}/P_1$, the population count p_{ti} is estimated by

$$\widehat{\widetilde{p}}_{ti} = \left(\widehat{\widetilde{r}}_i^* + p_{1i}/P_1\right)\widehat{P}_t \tag{17.3.7}$$

17.3.2 Direct Estimation

Suppose that we are interested in A small areas $\widetilde{U}_a$, $a = 1, \ldots, A$ with $\widetilde{U}_1 \cup \ldots \cup \widetilde{U}_A = \widetilde{U}$. Let the size of $\widetilde{U}_a$ and the part of the sample s that has intersection with the small area $\widetilde{U}_a$ be $\widetilde{N}_a$ and s_a, respectively. The direct estimators for the population total $\widetilde{Y}_a = \sum_{i \in \widetilde{U}_a} y_i$ and mean $\overline{\widetilde{Y}}_a = \sum_{i \in \widetilde{U}_a} y_i/\widetilde{N}_a$ are estimated by

$$\widehat{\widetilde{Y}}_a = \sum_{i \in s_a} y_i/\pi_i \text{ and } \overline{\widetilde{Y}}_a = \frac{\sum_{i \in s_a} y_i/\pi_i}{\sum_{i \in s_a} 1/\pi_i} \tag{17.3.8}$$

The estimators given in Eq. (17.3.8) are generally unstable because they are often based on the small sample size n_a. So we consider the following alternative methods of estimation.

17.3.3 Synthetic Estimation

Synthetic estimation was proposed by Gonzalez (1973). This method is based on the assumption that the small areas have the same characteristics as larger area. Let $\mathcal{D}'$ be the set of larger domains that have the same characteristics as the smaller area $\widetilde{U}_a$ $\left(\widetilde{U}_a \subset \mathcal{D}'\right)$. Furthermore, let us assume that the auxiliary information x is available for the domains $\mathcal{D}'$ and the smaller area $\widetilde{U}_a$. Suppose $s_{ad} = s_a \cap U_d$ is the intersection of the small area sample s_a with the large domain $U_d(U_d \subset \mathcal{D}')$ with $s_a = \bigcup_{d \in \mathcal{D}'} s_{ad}$, and t_d is a reliable estimator for the domain total Y_d based on a larger sample $s_d = s \cap U_d$, then the synthetic estimator for the small area total $\widetilde{Y}_a$ is obtained as

$$\widehat{\widetilde{Y}}_a^S = \sum_{d \in \mathcal{D}'} \frac{X_{ad}}{X_d} t_d \tag{17.3.9}$$

where $X_{ad} = \sum_{i \in \widetilde{U}_a \cap U_d} x_i$ and $X_d = \sum_{i \in U_d} x_i$ are known totals and x_i be the value of the auxiliary variable for the ith unit.

In particular, if t_d is a ratio estimator of the form

$$t_d = \frac{\widehat{Y}_d}{\widehat{X}_d} X_d$$

with $\widehat{Y}_d = \sum_{i \in s_d} y_i/\pi_i$ and $\widehat{X}_d = \sum_{i \in s_d} x_i/\pi_i$, $\widehat{\widetilde{Y}}_a^S$ reduces to

$$\widehat{\widetilde{Y}}_a^S = \sum_{d \in \mathcal{D}'} \frac{\widehat{Y}_d}{\widehat{X}_d} X_{ad} \tag{17.3.10}$$

Corollary 17.3.1

For simple random sampling without replacement (SRSWOR) sampling $\pi_i = n/N$, $\widehat{X}_d = N\frac{n_d}{n}\overline{x}_{sd}$, $\widehat{Y}_d = N\frac{n_d}{n}\overline{y}_{sd}$, $\overline{x}_{sd} = \frac{1}{n_d}\sum_{i \in s_d} x_i$, and $\overline{y}_{sd} = \frac{1}{n_d}\sum_{i \in s_d} y_i$, we have

$$\widehat{\widetilde{Y}}_a^S = \sum_{d \in \mathcal{D}'} N_{ad} \frac{\overline{y}_{sd}}{\overline{x}_{sd}} \overline{X}_{ad} \tag{17.3.11}$$

where $\overline{X}_{ad} = X_{ad}/N_{ad}$ and N_{ad} = size of the subpopulation $U_d \cap \widetilde{U}_a$.

Case 1: If only one domain is considered, i.e., $U_d = U$, $s_d = s$, $N_{ad} = \widetilde{N}_a$ and $\overline{X}_{ad} = \overline{\widetilde{X}}_a = \sum_{i \in \widetilde{U}_a} x_i/\widetilde{N}_a$, the estimator (Eq. 17.3.11) reduces to

$$\widehat{\widetilde{Y}}_a^S = N_a \frac{\overline{y}_s}{\overline{x}_s} \overline{\widetilde{X}}_a \tag{17.3.12}$$

Furthermore, if no auxiliary information is available and we take $x_i = 1$ for very i, Eq. (17.3.12) reduces to

$$\widehat{\widetilde{Y}}_a^S = N_a \overline{y}_s \tag{17.3.13}$$

Let $y_i = 1$ if, the ith individual belongs to certain group viz. HIV+ and $y_i = 0$ otherwise, then $\overline{y}_s = \widehat{\pi}_s$ = proportion of HIV+ persons in the sample s. The estimator of the total number of HIV+ persons in the small area "a" is

$$\widehat{\widetilde{Y}}_a^S = N_a \frac{n_+}{n}$$

where n_+ = total number of HIV+ persons in the samples of size n.

17.3.4 Composite Estimation

The synthetic estimator has a potential bias whereas the direct estimator has large sampling variation. A composite estimator is obtained by taking the weighted average of the synthetic and direct estimator as

$$\hat{\tilde{Y}}_a^S = \phi_a \hat{\tilde{Y}}_a + (1-\phi_a)\,\hat{\tilde{Y}}_a^S \tag{17.3.14}$$

Here $\hat{\tilde{Y}}_a$ is the direct unbiased estimator of the total $\tilde{Y}_a$ and ϕ_a $(0 \le \phi_a \le 1)$ is a suitably chosen weight. The optimum value of ϕ_a that minimizes the MSE of $\hat{\tilde{Y}}_a^C$, assuming $Cov\left(\hat{\tilde{Y}}_a^S, \hat{\tilde{Y}}_a\right) \cong 0$, is given by

$$Opt(\phi_a) = \phi_{a0} = \mathrm{MSE}\left(\hat{\tilde{Y}}_a^S\right) \Big/ \left[\mathrm{MSE}\left(\hat{\tilde{Y}}_a^S\right) + Var\left(\hat{\tilde{Y}}_a\right)\right] \tag{17.3.15}$$

The MSE of $\hat{\tilde{Y}}_a^C$ with $\phi_a = \phi_{a0}$ is given by

$$\mathrm{MSE}\left(\hat{\tilde{Y}}_a^C\right) = \left[\frac{1}{\mathrm{MSE}\left(\hat{\tilde{Y}}_a^S\right)} + \frac{1}{Var\left(\hat{\tilde{Y}}_a\right)}\right]^{-1} \tag{17.3.16}$$

The optimum weight ϕ_{a0} cannot be used in practice because it involves unknown parameters $\mathrm{MSE}\left(\hat{\tilde{Y}}_a^S\right)$ and $Var\left(\hat{\tilde{Y}}_a\right)$. One can use an estimate of the weight ϕ_{a0} as

$$\hat{\phi}_{a0} = \hat{M}\left(\hat{\tilde{Y}}_a^S\right) \Big/ \left[\hat{M}\left(\hat{\tilde{Y}}_a^S\right) + \hat{V}\left(\hat{\tilde{Y}}_a\right)\right] \tag{17.3.17}$$

where $\hat{M}\left(\hat{\tilde{Y}}_a^S\right)$ = an estimator of $\mathrm{MSE}\left(\hat{\tilde{Y}}_a^S\right)$ and $\hat{V}\left(\hat{\tilde{Y}}_a\right)$ = an estimator of $V\left(\hat{\tilde{Y}}_a\right)$.

Purcell and Kish (1979) used a common weight ϕ for estimating $\tilde{Y}_a$ for every $a = 1, \ldots, A$. The optimum value ϕ was obtained by minimizing $\frac{1}{A}\sum_{a=1}^{A} \mathrm{MSE}\left(\hat{\tilde{Y}}_a^C\right)$ with respect to ϕ and it can be written as:

$$\phi_0 = 1 - \sum_{i=1}^{A} Var\left(\hat{\tilde{Y}}_a\right) \Big/ \sum_{i=1}^{A} Var\left(\hat{\tilde{Y}}_a^S - \hat{\tilde{Y}}_a\right)$$

An estimator of ϕ_0 was chosen as

$$\widehat{\phi}_0 = 1 - \sum_{i=1}^{A} \widehat{V}\left(\widehat{\widehat{Y}}_a\right) \Big/ \sum_{i=1}^{A} \left(\widehat{\widehat{Y}}_a^S - \widehat{\widehat{Y}}_a\right)^2 \quad (17.3.18)$$

Obviously, the common weight cannot provide an efficient estimator if the variation of $Var\left(\widehat{\widehat{Y}}_a\right)$ values is very large.

In case ϕ_{a0} depends on N_a, the population size of the area a, Drew et al. (1982) recommended the following weight:

$$\widehat{\phi}_a(1) = \begin{cases} 1 & \text{if } \widehat{N}_a \geq \delta N_a \\ \widehat{N}_a / \delta N_a, & \text{otherwise} \end{cases} \quad (17.3.19)$$

where $\widehat{N}_a$ is an unbiased estimator of N_a, and δ is a subjectivity factor to be determined by a statistician through his personal experience.

A similar recommendation was suggested by Särndal and Hidiroglou (1989) as

$$\widehat{\phi}_a(2) = \begin{cases} 1 & \text{if } \widehat{N}_a \geq \delta N_a \\ \left(\widehat{N}_a / N_a\right)^{k-1}, & \text{otherwise} \end{cases} \quad (17.3.20)$$

with a subjective factor k.

Clearly, $\widehat{\phi}_a(1) = \widehat{\phi}_a(2)$ if $\delta = 1$ and $k = 2$.

17.3.5 Borrowing Strength From Related Areas

Here we assume that the variables of interest y follow the following superpopulation model,

$$y_i = \beta_0 + \beta_1 x_{1i} + \cdots + \beta_p x_{qi} + \varepsilon_i \quad (17.3.21)$$

where $x_{1i}, \ldots, x_{qi}$ are known auxiliary variables, $\beta_0, \beta_1, \ldots, \beta_q$ are unknown model parameters, ε_i's are iid random components with model expectation $E_m(\varepsilon_i) = 0$ and model variance $V_m(\varepsilon_i) = \sigma^2$.

Reliable estimates of the model parameters $\beta_0, \beta_1, \ldots, \beta_q$ cannot be obtained from the small sample s_a. This problem can be overcome if we suppose that the model (Eq. 17.3.21) holds not only for the area U_a under study but also for a larger area $\Im(\subset U)$, which contains the area U_a and the reliable estimators for the model parameters $\beta_0, \beta_1, \ldots, \beta_q$ based on the relatively larger sample obtained $s_{\Im} = s \cap \Im$ by applying least squares method. The technique of using information from related areas other than the small area to get reliable estimators of the model parameters is known as borrowing strength (or information). This technique was recommended by Ghosh and Rao (1994) and Pfeffermann (2002), among others. The vector of model parameter $\boldsymbol{\beta}=\left(\beta_0,\beta_1,\ldots,\beta_q\right)'$ may be estimated by using generalized least squares (GLSs) method (vide Lehtonen et al., 2003) as

$$\widehat{\boldsymbol{\beta}}=\left(\widehat{\beta}_0,\widehat{\beta}_1,\ldots,\widehat{\beta}_q\right)'=\left(\sum_{i\in s_{\mathfrak{Z}}}\mathbf{x}_i\mathbf{x}_i'/(c_i\pi_i)\right)^{-1}\left(\sum_{i\in s_{\mathfrak{Z}}}\mathbf{x}_i y_i/(c_i\pi_i)\right) \tag{17.3.22}$$

where $\mathbf{x}_i'=(1,x_{1i},\ldots,x_{qi})$ and c_i's are known positive weights.

Lehtonen et al. (2003) proposed the following types of estimators for $\widetilde{Y}_a$.

17.3.5.1 Synthetic Estimator

$$\widehat{\widetilde{Y}}_a^{*S}=\sum_{i\in U_a}\widehat{y}_i \tag{17.3.23}$$

where $\widehat{y}_i=\widehat{\beta}_0+\widehat{\beta}_1x_{1i}+\ldots+\widehat{\beta}_qx_{qi}$ and $\widehat{\beta}_0,\widehat{\beta}_1,\ldots,\widehat{\beta}_q$ are obtained from Eq. (17.3.22).

17.3.5.2 Generalized Regression Estimator

$$\begin{aligned}\widehat{\widetilde{Y}}_a^{*G}&=\sum_{i\in U_a}\widehat{y}_i+\sum_{i\in s_a}\left(y_i-\widehat{y}_i\right)/\pi_i\\&=\widehat{\widetilde{Y}}_a^{*S}+\sum_{i\in s_a}\left(y_i-\widehat{y}_i\right)/\pi_i\end{aligned} \tag{17.3.24}$$

17.3.5.3 Composite Estimator

$$\begin{aligned}\widehat{\widetilde{Y}}_a^{*C}&=\phi^*\widehat{\widetilde{Y}}_a^{*S}+\left(1-\phi^*\right)\widehat{\widetilde{Y}}_a^{*G}\\&=\phi^*\widehat{\widetilde{Y}}_a^{*S}+\left(1-\phi^*\right)\left(\widehat{\widetilde{Y}}_a^{*S}+\sum_{i\in s}\frac{y_i-\widehat{y}_i}{\pi_i}\right)\\&=\widehat{\widetilde{Y}}_a^{*S}+\left(1-\phi^*\right)\sum_{i\in s}\frac{y_i-\widehat{y}_i}{\pi_i}\end{aligned} \tag{17.3.25}$$

where $\phi^*(0\le\phi^*\le 1)$ is a suitably chosen weight.

Example 17.3.1

The following table gives the estimated total number of births and deaths for the three cities Durban, Pietermaritzburg, and Richards Bay of the state KwaZulu-Natal (KZN) based on a household survey conducted in 2010 along with the birth and death rates for the last census year 2000. The estimated birth rate and death rate of KZN for the year 2010 were obtained 2.15% and 1.9%, respectively.

	2010 survey report		2000 census report	
	Births (b_t)	Deaths (d_t)	Percentage birth rate ($g_0 \times 100$)	Percentage death rate ($h_0 \times 100$)
Durban	2500	2000	2.25	1.75
Pietermaritzburg	3000	2500	2.00	1.25
Richards Bay	1000	6000	1.80	1.50
KZN	10000	7500	2.35	1.80
	(B_t)	(D_t)	(G_0)	(H_0)

The estimated birth rate $g_t = g_0(G_t/G_0)$, death rate $h_t = h_0(H_t/H_0)$, and the estimated number of population $P_t = (b_t/g_t + d_t/h_t)/2$ for the three cities for the year 2010 are obtained by VR method as follows:

	Percentage birth rate ($g_t \times 100$)	Percentage death rate ($h_t \times 100$)	Total population (P_t)
Durban	2.058	1.847	114,858
Pietermaritzburg	1.830	1.319	176,713
Richards Bay	1.647	1.583	219,835

Example 17.3.2

A sample of 25 agricultural farms was selected from a list of 280 farms of a certain district by SRSWOR method and information of the production of a certain crop (y) and area of the farms (x) was obtained. The farms were classified into three categories large (1), medium (2) and small (3). The total number of farms and their average sizes were obtained from the last census. The data are given below:

Farm type	Production of crop (kg)	Farm size (acre)	Total number of farms	Mean area of farm (acre)
1	100	20	50	28.5
	180	27		
	150	25		
	175	28		
	160	30		
2	100	15	100	14.5
	87	12		
	60	10		
	75	12		
	65	14		
	70	14		
	80	15		
	85	17		
3	50	8	130	7.5
	60	8		
	60	7		
	50	6		
	45	5		
	45	5		
	50	8		
	40	7		
	30	5		
	20	6		
	20	5		
	30	6		

Let y_{ij} be the production of crop for the jth farm of the ith category, $j = 1, \ldots, n_i$; $i = 1, 2, 3$, and $n_1 = 5$, $n_2 = 8$ and $n_3 = 12$. The total numbers of large, medium, and small farms in the population and their average sizes are, respectively, $N_1 = 50$, $N_2 = 100$, $N_3 = 130$ and $\overline{X}_1 = 28.5$, $\overline{X}_2 = 14.5$, and $\overline{X}_3 = 7.5$ acres.

Direct Estimates

The direct estimates (sample mean) for mean production of three types of farms are $\overline{y}_1 = 153.00, \overline{y}_2 = 77.750$, and $\overline{y}_3 = 41.667$, respectively.

Synthetic and Composite Estimates

Case I: Sample Mean as an Estimator of Overall Mean

Synthetic Estimates

The estimated overall mean $(\overline{Y})$ production is $\overline{y} = 75.48$.

The mean production of crop $(\overline{Y}_i)$ for the ith type of farm can be estimated by using synthetic method and have been given in the following table:

Farm type	Number of farms N_i	Average farm size $\overline{X}_i$	Estimated mean production per farm $\widehat{\overline{Y}}_i^S = \frac{N_i \overline{X}_i}{N\overline{X}} \overline{y}$
Large	50	28.5	27.937
Medium	100	14.5	28.427
Small	130	7.5	19.115
Overall	280	13.75	

Composite Estimators

The composite estimator for $\overline{Y}_i$ is given by

$$\widehat{\overline{Y}}_i^C = \phi_i \overline{y}_i + (1 - \phi_i) \widehat{\overline{Y}}_i^S$$

Now using Drew et al. (1982) with $\delta = 1$, we find

$$\phi_i = \begin{cases} 1 & \text{if } \widehat{N}_i > N_i \\ \dfrac{\widehat{N}_i}{N_i} & \text{otherwise} \end{cases}$$

where $\widehat{N}_i = n_i \times N/n$. Since $\widehat{N}_1 = 56$, $\widehat{N}_2 = 89.5$ and $\widehat{N}_3 = 134.4$, we find $\phi_1 = 1$, $\phi_2 = 0.896$, and $\phi_3 = 1$. Hence we obtain the composite estimators as $\widehat{\overline{Y}}_1^C = \overline{y}_1 = 153.0$, $\widehat{\overline{Y}}_2^C = 0.896 \times \overline{y}_2 + (1 - 0.896) \times \widehat{\overline{Y}}_2^S = 72.602$, and $\widehat{\overline{Y}}_3^C = \overline{y}_3 = 41.667$.

Case II: Overall Mean Is Estimated by Ratio Estimator

The ratio estimator for the overall mean $\overline{Y}$ is given by

$$\widehat{\overline{Y}}_R = \frac{\overline{y}}{\overline{x}}\overline{X} = (75.458/12.6) \times 13.75 = 82.369.$$

The synthetic and composite estimators for the ith type farm is obtained by using the formula $\widehat{\overline{Y}}_i^S(R) = \frac{N_i\overline{X}_i}{N\overline{X}}\widehat{\overline{Y}}_R$ and $\widehat{\overline{Y}}_i^C(R) = \phi_i\overline{y}_i + (1-\phi_i)\widehat{\overline{Y}}_i^S(R)$, respectively. The estimates are given as follows:

Farm type	Number of farms N_i	Average farm size $\overline{X}_i$	Estimated mean production per farm		
			Synthetic $\widehat{\overline{Y}}_i^S(R)$	Weight ϕ_i	Composite $\widehat{\overline{Y}}_i^C(R)$
Large	50	28.50	30.487	1.000	153.00
Medium	100	14.50	31.022	0.896	72.890
Small	130	7.50	20.859	1.000	41.667
Overall	280	13.75			

Case III: Borrowing of Strength

Here we fit a linear regression $y_{ij} = \beta_0 + \beta_1 x_{ij} + \in_{ij}$ over 25 observations.

We find estimates of β_0 and β_1 are $\widehat{\beta}_0 = \overline{y} - \widehat{\beta}_1\overline{x} = 4.316$ and $\widehat{\beta}_1 = \sum_{i\in s}(y_i - \overline{y})(x_i - \overline{x})/\sum_{i\in s}(x_i - \overline{x})^2 = 5.648$, respectively, where s denotes the sample of 25 observations.

Synthetic Estimators

Synthetic estimates for the mean production of large, medium, and small farms are $\widehat{\overline{Y}}_1^{*S} = \widehat{\beta}_0 + \widehat{\beta}_1\overline{X}_1 = 165.281$, $\widehat{\overline{Y}}_2^{*S} = \widehat{\beta}_0 + \widehat{\beta}_1\overline{X}_2 = 86.211$, and $\widehat{\overline{Y}}_3^{*S} = \widehat{\beta}_0 + \widehat{\beta}_1\overline{X}_3 = 46.675$, respectively.

Generalized Regression Estimator

Let s_1, s_2, and s_3 be the sample corresponding to large, medium, and small farms with respective sizes $n_1 = 5$, $n_2 = 8$, and $n_3 = 12$. The generalized regression estimators for the mean production of large, medium, and small farms are obtained as follows:

$$\widehat{\overline{Y}}_1^{*G} = \frac{N}{nN_1}\sum_{i\in s_1}(y_i - \widehat{y}_i) + \widehat{\overline{Y}}_1^{*S} = \frac{280}{50\times 25}\times 9.189 + 165.281 = 167.340$$

$$\widehat{\overline{Y}}_2^{*G} = \frac{N}{nN_2}\sum_{i\in s_2}(y_i - \widehat{y}_i) + \widehat{\overline{Y}}_2^{*S} = \frac{280}{100\times 25}\times (-28.152) + 86.257 = 83.057$$

$$\widehat{\overline{Y}}_3^{*G} = \frac{N}{nN_3}\sum_{i\in s_3}(y_i - \widehat{y}_i) + \widehat{\overline{Y}}_3^{*S} = \frac{280}{130\times 25}\times 18.963 + 46.721 = 48.309$$

17.3.6 Use of Models

Indirect estimators such as synthetic and composite estimators are based on implicit or explicit models that connect small areas through supplementary data. In this section we will consider the general mixed effect model proposed by Henderson (1975). The random effects account for the area variation that cannot be explained by auxiliary variables. The general mixed effect model can be partitioned into area-level model, a unit-level model, and a hybrid. Area-level models relate small area direct estimators to area-specific auxiliary data. This model is useful if unit-level auxiliary information is not available. In unit-level model the study variable of each unit is related to a set of concomitant variables. Hybrid models are combination of unit-level and area-level models.

17.3.6.1 General Linear Mixed Model

Henderson (1975) considered the following general linear mixed model, where y_{ij}, the value of the study variable y of the jth unit of the ith area is related to q auxiliary variables through the following model

$$y_{ij} = x_{ij1}\beta_1 + \cdots + x_{ijt}\beta_t + \cdots + x_{ijq}\beta_q + z_{i1}v_1 + \cdots + z_{iA}v_A + \in_{ij};$$
$$j = 1, \ldots, N_i;\quad i = 1, \ldots, A \qquad (17.3.26)$$

where x_{ijt} is the value of the auxiliary variable x_t of jth unit of the ith small area is assumed to be known, z_{ij} is a known positive constant, and $v_1, \ldots, v_A$ are area-specific random effects that are assumed to be independently and identically distributed with

$$E_m(v_i) = 0 \text{ and } V_m(v_i) = \sigma_v^2;\quad i = 1, \ldots, A \qquad (17.3.27)$$

where E_m and V_m are expectation and variance operators with respect to the model (Eq. 17.3.26). The error components $\in_{ij}$'s are independently distributed with

$$E_m(\in_{ij}) = 0 \text{ and } V_m(\in_{ij}) = \sigma_e^2 \tag{17.3.28}$$

We further assume that $\in_{ij}$'s are independent of v_i's. Let a sample s of n units be selected from the population by some suitable sampling design, and let n_i be the number of units that falls in the ith small area $i = 1,\ldots, A$. Here we assume that the model (Eq. 17.3.26) is valid for the sampled data also, i.e.,

$$y_{ij} = x_{ij1}\beta_1 + \cdots + x_{ijt}\beta_t + \cdots + x_{ijq}\beta_q + z_{i1}v_1 + \cdots + z_{iA}v_A + \in_{ij};$$
$$j = 1, \ldots, n_i; \quad i = 1, \ldots, A \tag{17.3.29}$$

Battese et al. (1988) used the model (Eq. 17.3.29) to estimate county crop areas using satellite information as auxiliary variables, whereas Rao and Choudhry (1995) used this model to estimate total wages and salaries for Nova Scotia province using gross business as an auxiliary variable. Further applications of the model (Eq. 17.3.29) with real live data were provided by Kleffe and Rao (1992), Datta and Ghosh (1991), and Ghosh and Lahiri (1998), among others.

In matrix notation, Eq. (17.3.29) can be written as

$$\mathbf{y} = \mathbf{X\beta} + \mathbf{Zv} + \mathbf{e} \tag{17.3.30}$$

where

$$\mathbf{y} = \begin{pmatrix} y_{11} \\ \cdot \\ y_{1n_1} \\ \cdot \\ y_{A1} \\ \cdot \\ y_{An_A} \end{pmatrix}^{n\times 1}, \mathbf{X} = \begin{pmatrix} x_{111} & \cdot & x_{11q} \\ & \cdot & \\ x_{1n_11} & \cdot & x_{1n_1q} \\ & \cdot & \\ x_{A11} & \cdot & x_{A1q} \\ & \cdot & \\ x_{An_A1} & \cdot & x_{An_Aq} \end{pmatrix}^{n\times q}, \mathbf{Z} = \begin{pmatrix} z_{11} & \cdot & z_{1A} \\ & \cdot & \\ z_{11} & \cdot & z_{1A} \\ & \cdot & \\ z_{A1} & \cdot & z_{AA} \\ & \cdot & \\ z_{A1} & \cdot & z_{AA} \end{pmatrix}^{n\times A},$$

$$\mathbf{e} = \begin{pmatrix} \in_{11} \\ \cdot \\ \in_{1n_1} \\ \cdot \\ \in_{A1} \\ \cdot \\ \in_{An_A} \end{pmatrix}^{n\times 1}, \boldsymbol{\beta} = \begin{pmatrix} \beta_1 \\ \cdot \\ \beta_q \end{pmatrix}^{q\times 1}, n = \sum n_i \text{ and } \mathbf{v}' = (v_1, \quad \cdot \quad, v_A). \tag{17.3.31}$$

Matrices **X** and **Z** are known and are of rank q and A, respectively; $\beta_1,\ldots, \beta_q$ are unknown regression coefficients (fixed effect) and **v** is an unknown vector of random effects and **e** is an unknown random error vector. The vectors **e** and **v** are distributed independently with mean **0** and variance–covariance matrices $\mathbf{R} = \sigma_e^2\mathbf{I}_n$ and $\mathbf{G} = \sigma_v^2\mathbf{I}_A$, respectively. **R** and **G** involve parameter vector $\boldsymbol{\tau} = \left(\sigma_v^2, \sigma_e^2\right)$, which is called the variance component vector. For the time being, we will assume that the vector $\boldsymbol{\tau}$ is known.

Here we are interested in the method of estimating (or predicting) a linear combination

$$\mu = \mathbf{l}'\boldsymbol{\beta} + \mathbf{m}'\mathbf{v} \tag{17.3.32}$$

where $\mathbf{l}'$ and $\mathbf{m}'$ are vectors of known constants.

We concentrate on the class C_l of linear unbiased estimators of μ, which consists of the estimators of the form

$$\widehat{\mu} = \mathbf{u}'\mathbf{y} + b \tag{17.3.33}$$

satisfying $E_m(\widehat{\mu}) = \mu$, where $\mathbf{u}'$ and b's are known constants. For a known parameter vector $\boldsymbol{\tau}$, we define the best linear unbiased prediction (BLUP) estimator of μ as one which minimizes $V_m(\widehat{\mu}) = E_m(\widehat{\mu} - \mu)^2$ for $\widehat{\mu} \in C_l$. Following Henderson (1975), we note that the BLUP estimator for μ is

$$\widehat{\mu}_h(\boldsymbol{\tau}) = \mathbf{l}'\widehat{\boldsymbol{\beta}} + \mathbf{m}'\mathbf{G}\mathbf{Z}'\mathbf{V}^{-1}\left(\mathbf{y} - \mathbf{X}\widehat{\boldsymbol{\beta}}\right) \tag{17.3.34}$$

where $\widehat{\boldsymbol{\beta}} = (\mathbf{X}'\mathbf{V}^{-1}\mathbf{X})^{-1}(\mathbf{X}'\mathbf{V}^{-1}\mathbf{y})$ is the GLS estimator of $\boldsymbol{\beta}$ and $\mathbf{V} = \mathbf{R} + \mathbf{Z}\mathbf{G}\mathbf{Z}'$ is the variance–covariance matrix of **y**. The MSE of $\widehat{\mu}_h(\boldsymbol{\tau})$ was obtained by Henderson (1975) is as follows:

$$\text{MSE}\ [\widehat{\mu}_h(\boldsymbol{\tau})] = (\mathbf{l}', \mathbf{m}')\begin{pmatrix} \mathbf{C}_{11} & \mathbf{C}_{12} \\ \mathbf{C}_{12}' & \mathbf{C}_{22} \end{pmatrix}\begin{pmatrix} \mathbf{l} \\ \mathbf{m} \end{pmatrix} \tag{17.3.35}$$

where

$$\begin{pmatrix} \mathbf{C}_{11} & \mathbf{C}_{12} \\ \mathbf{C}_{12}' & \mathbf{C}_{22} \end{pmatrix} = \begin{pmatrix} \mathbf{X}'\mathbf{R}^{-1}\mathbf{X} & \mathbf{X}'\mathbf{R}^{-1}\mathbf{Z} \\ \mathbf{Z}'\mathbf{R}^{-1}\mathbf{X} & \mathbf{Z}'\mathbf{R}^{-1}\mathbf{Z} + \mathbf{G}^{-1} \end{pmatrix}^{-1}.$$

17.3.6.2 Nested Error Regression Model

Let y_{ij} be the value of the study variable for the unit j of the ith area, $j = 1,\ldots, N_i$, $i = 1,\ldots, A$. Assume for each unit $\mathbf{x_{ij}} = (x_{ij1}, \ldots, x_{ijq})'$, a

q-vector auxiliary information available. A unit-level model relates y_{ij} to the auxiliary variable through the following nested error regression model

$$y_{ij} = x_{ij1}\beta_1 + \cdots + x_{ijk}\beta_k + \cdots + x_{ijq}\beta_q + v_i + \in_{ij};\; j = 1, \ldots, N_i;\\ i = 1, \ldots, A \qquad (17.3.36)$$

where $\boldsymbol{\beta} = \left(\beta_1, \ldots, \beta_q\right)'$ is a q-vector regression parameter, v_i is the ith specific random effect, and $\in_{ij}$ is the error component. Battese et al. (1988) considered the model (Eq. 17.3.36) for estimating the mean acreage under crop for 36 counties in Iowa using satellite and survey data.

The population mean of the ith small area is given by

$$\overline{Y}_i = \sum_{j=1}^{N_i} y_{ij}/N_i = \overline{X}_{i\cdot 1}\beta_1 + \cdots + \overline{X}_{i\cdot k}\beta_k + \cdots + \overline{X}_{i\cdot q}\beta_q + v_i + \overline{E}_i$$

where $\overline{X}_{i\cdot k} = \sum_{j=1}^{N_i} x_{ijk}/N_i$ and $\overline{E}_i = \sum_{j=1}^{N_i} \in_{ij}/N_i$.

For large N_i we can write

$$\begin{aligned}\overline{Y}_i &= \overline{X}_{i\cdot 1}\beta_1 + \cdots + \overline{X}_{i\cdot j}\beta_j + \cdots + \overline{X}_{i\cdot q}\beta_q + v_i\\ &= \overline{\mathbf{X}}_i'\boldsymbol{\beta} + v_i \\ &= \mu_i(\text{say})\end{aligned} \qquad (17.3.37)$$

where $\overline{\mathbf{X}}_i' = \left(\overline{X}_{i\cdot 1}, \ldots, \overline{X}_{i\cdot q}\right)'$.

Suppose our objective is to estimate the mean μ_i from known $\overline{\mathbf{X}}_i'$ on the basis of the sampled data satisfying

$$y_{ij} = x_{ij1}\beta_1 + \cdots + x_{ijk}\beta_k + \cdots + x_{ijq}\beta_q + v_i + \in_{ij};\; j = 1, \ldots, n_i;\\ i = 1, \ldots, A$$

The BLUP estimator of μ_i is obtained from Eq. (17.3.34) by substituting $\mathbf{l}' = \overline{\mathbf{X}}_i' = \left(\overline{X}_{i\cdot 1}, \ldots, \overline{X}_{i\cdot q}\right)$, $\mathbf{m}' = (0, \ldots, 0, 1, 0, \ldots, 0)$ with 1 in the ith position and it is given by

$$\widehat{\mu}_h^{(i)}(\boldsymbol{\tau}) = \mathbf{l}'\widehat{\boldsymbol{\beta}} + \mathbf{m}'\mathbf{G}\mathbf{Z}'\mathbf{V}^{-1}\left(\mathbf{y} - \mathbf{X}\widehat{\boldsymbol{\beta}}\right) \qquad (17.3.38)$$

where $\widehat{\boldsymbol{\beta}} = \left(\mathbf{X}'\mathbf{V}^{-1}\mathbf{X}\right)^{-1}\left(\mathbf{X}'\mathbf{V}^{-1}\mathbf{y}\right)$, $\mathbf{V} = Diag(\mathbf{V}_1, \ldots, \mathbf{V_i}, \ldots, \mathbf{V_A})$, $\mathbf{V}_i = \sigma_e^2\mathbf{I}_i + \sigma_v^2\mathbf{E}_{i,i}$, $\mathbf{I}_i$ is a unit matrix of order n_i, $\mathbf{E}_{i,i}$ is a $n_i \times n_i$ matrix with each element is 1 and

$$\mathbf{Z}' = \begin{pmatrix} 1 & \cdot & 1 & \cdot & 0 & \cdot & 0 \\ \cdot & \cdot & \cdot & \cdot & \cdot & \cdot & \cdot \\ 0 & \cdot & 0 & \cdot & 1 & \cdot & 1 \end{pmatrix}^{A \times n}$$

Furthermore, noting $\mathbf{V}_i^{-1} = \frac{1}{\sigma_e^2}\left(\mathbf{I}_i - \gamma_i \mathbf{E}_{i,i}/n_i\right)$ with $\gamma_i = n_i\sigma_v^2/\ \left(n_i\sigma_v^2 + \sigma_e^2\right)$, the BLUP estimator $\mu_i = \overline{Y}_i = \overline{\mathbf{X}}_i'\boldsymbol{\beta} + v_i$ is obtained by Prasad and Rao (1990) as

$$\begin{aligned}\widehat{\mu}_h^{(i)}(\boldsymbol{\tau}) &= \overline{X}_i'\widehat{\boldsymbol{\beta}} + \gamma_i\left(\overline{y}_i - \overline{\mathbf{x}}_{\mathbf{i}\cdot}'\widehat{\boldsymbol{\beta}}\right) \\ &= \gamma_i\left\{\overline{y}_i - \left(\overline{\mathbf{x}}_{i\cdot}' - \overline{\mathbf{X}}_i'\right)\widehat{\boldsymbol{\beta}}\right\} + (1-\gamma_i)\overline{\mathbf{X}}_i'\widehat{\boldsymbol{\beta}}\end{aligned} \tag{17.3.39}$$

where $\overline{y}_i = \sum_{j=1}^{n_i} y_{ij}/n_i$, $\overline{\mathbf{x}}_{i\cdot}' = \left(\overline{x}_{i\cdot 1}, \ldots, \overline{x}_{i\cdot q}\right)$ and $\overline{x}_{i\cdot k} = \sum_j x_{ijk}/n_i$.

The estimator (Eq. 17.3.39) is a weighted average of the sample regression estimator $\overline{y}_i - \left(\overline{\mathbf{x}}_{i\cdot}' - \overline{\mathbf{X}}_i'\right)\widehat{\boldsymbol{\beta}}$ and the synthetic estimator $\overline{\mathbf{X}}_i'\widehat{\boldsymbol{\beta}}$ of $\overline{Y}_i$ with weights γ_i and $1-\gamma_i$, respectively. For areas i with no samples, $\widehat{\mu}_h^{(i)}(\boldsymbol{\tau}) = \overline{X}_i'\widehat{\boldsymbol{\beta}}$. The coefficient γ_i is a "shrinkage factor" providing a compromise between the large variance of the regression predictor $\overline{y}_i - \left(\overline{\mathbf{x}}_{i\cdot}' - \overline{\mathbf{X}}_i'\right)\widehat{\boldsymbol{\beta}}$ and the bias of the synthetic estimator $\overline{X}_i'\widehat{\boldsymbol{\beta}}$. The estimator $\widehat{\mu}_h^{(i)}(\boldsymbol{\tau})$ reduces to the regression estimator $\overline{y}_i - \left(\overline{\mathbf{x}}_{i\cdot}' - \overline{\mathbf{X}}_i'\right)\widehat{\boldsymbol{\beta}}$ if the sample size n_i is very large or the model variance σ_v^2 is sufficiently large compared to σ_e^2. Conversely, if the model variance σ_v^2 is small compared to $\sigma_v^2 + \sigma_e^2/n_i$, then $\gamma_i \cong 0$ and the estimator $\widehat{\mu}_h^{(i)}(\boldsymbol{\tau})$ reduces to the synthetic estimator $\overline{\mathbf{X}}_{\mathbf{i}}'\widehat{\boldsymbol{\beta}}$.

The MSE of the BLUP estimator $\widehat{\mu}_h^{(i)}(\boldsymbol{\tau})$ was given by Prasad and Rao (1990) as

$$\begin{aligned}\text{MSE}\left[\widehat{\mu}_h^{(i)}(\boldsymbol{\tau})\right] &= E\left[\widehat{\mu}_h^{(i)}(\boldsymbol{\tau}) - \mu_i\right]^2 \\ &= (1-\gamma_i)\sigma_v^2 + \left(\overline{\mathbf{X}}_i - \gamma_i\overline{\mathbf{x}}_{i\cdot}'\right)'\left(\mathbf{X}'\mathbf{V}^{-1}\mathbf{X}\right)^{-1}\left(\overline{\mathbf{X}}_i - \gamma_i\overline{\mathbf{x}}_{i\cdot}'\right)\end{aligned} \tag{17.3.40}$$

17.3.6.3 Area-Level Model

Area-level random effect model is used when the auxiliary information is available at the area level. Here, direct survey estimates are models to area-specific auxiliary data. In an area-level model, we are often interested in estimating $\theta_i = g\left(\overline{Y}_i\right)$, a function of ith area population mean $\overline{Y}_i$ using the following linear model

$$\theta_i = x_{i1}\widetilde{\beta}_1 + \cdots + x_{ij}\widetilde{\beta}_j + \cdots + x_{iq}\widetilde{\beta}_q + z_i v_i;\quad i = 1, \ldots, A \tag{17.3.41}$$

where x_{ij}'s are known values of the jth auxiliary variable x_j for the ith area, z_i's are known positive constants, v_i's are area-specific random effects assumed to be independently and identically distributed with

$$E_m(v_i) = 0 \text{ and } V_m(v_i) = \widetilde{\sigma}_v^2; \quad i = 1, \ldots, A \tag{17.3.42}$$

Numerous applications of the area-level model are available in the literature. Fay and Herriot (1979) used $\theta_i = \overline{Y}_i$, the true per capita income in "local government unit" i, while National Research Council, USA (2000) used $\theta_i = \log Y_i$ with Y_i as the poverty count for the ith area. Ericksen and Kadane (1985) took $\theta_i = (Y_i - C_i)/Y_i$, where Y_i and C_i denote respectively the true and census count for the ith area.

To make inference about θ_i, we assume

$$\widehat{\theta}_i = g\left(\widehat{\overline{Y}}_i\right) = \theta_i + \widetilde{\in}_i \tag{17.3.43}$$

where $\widehat{\overline{Y}}_i$ is a direct estimator based on the selected sample s using a suitable sampling design p and $\widetilde{\in}_i$'s are sampling errors distributed independently with

$$E_m(\widetilde{\in}_i|\theta_i) = 0 \text{ and } V_m(\widetilde{\in}_i|\theta_i) = \widetilde{\sigma}_i^2 \tag{17.3.44}$$

Here the sampling variance $\widetilde{\sigma}_i^2$ is assumed to be known for all areas $i = 1, \ldots, A$.

Eqs. (17.3.43) and (17.3.44) yield

$$\begin{aligned}\widehat{\theta}_i &= x_{i1}\widetilde{\beta}_1 + \cdots + x_{ij}\widetilde{\beta}_j + \cdots + x_{iq}\widetilde{\beta}_q + z_i v_i + \widetilde{\in}_i; \quad i = 1, \ldots, A \\ &= \mathbf{x}_i'\widetilde{\boldsymbol{\beta}} + z_i v_i + \widetilde{\in}_i\end{aligned} \tag{17.3.45}$$

where $\mathbf{x}_i' = (x_{i1}, \ldots, x_{ij}, \ldots, x_{iq})$, $\widetilde{\boldsymbol{\beta}} = \left(\widetilde{\beta}_1, \ldots, \widetilde{\beta}_j, \ldots, \widetilde{\beta}_q\right)'$, and z_i is a known positive constant.

Note that the model (Eq. 17.3.45) involves both the design-induced random variable $\widetilde{\in}_i$ as well the model-based random variable v_i. Now using Henderson's (1975) result given in Eq. (17.3.34), we find the BLUP estimator θ_i when the parametric function $\widetilde{\boldsymbol{\sigma}} = \left(\widetilde{\sigma}_v^2, \widetilde{\boldsymbol{\sigma}}_e^2\right)$ with $\widetilde{\boldsymbol{\sigma}}_e^2 = \left(\widetilde{\sigma}_1^2, \ldots, \widetilde{\sigma}_A^2\right)$ is known was obtained by Ghosh and Rao (1994) as

$$\widehat{\theta}_i^h(\widetilde{\boldsymbol{\tau}}) = \mathbf{x}_i'\widehat{\widetilde{\boldsymbol{\beta}}} + \widetilde{\gamma}_i\left(\widehat{\theta}_i - \mathbf{x}_i'\widehat{\widetilde{\boldsymbol{\beta}}}\right) \tag{17.3.46}$$

where $\widetilde{\gamma}_i = z_i^2\widetilde{\sigma}_v^2/\left(z_i^2\widetilde{\sigma}_v^2 + \widetilde{\sigma}_i^2\right)$ and $\widehat{\widetilde{\boldsymbol{\beta}}} = \left(\sum_{i=1}^{A} \frac{\mathbf{x}_i\hat{\theta}_i}{\widetilde{\sigma}_i^2 + z_i^2\widetilde{\sigma}_v^2}\right) \Big/ \left(\sum_{i=1}^{A} \frac{\mathbf{x}_i\mathbf{x}_i'}{\widetilde{\sigma}_i^2 + z_i^2\widetilde{\sigma}_v^2}\right)$.

The expression (Eq. 17.3.46) can be written as a weighted average of the direct estimator $\widehat{\theta}_i$ and synthetic estimator $\mathbf{x}_i'\widehat{\widehat{\boldsymbol{\beta}}}$ with weights $\widetilde{\gamma}_i$ and $1-\widetilde{\gamma}_i$ as follows:

$$\widehat{\theta}_i^h(\widetilde{\boldsymbol{\tau}}) = \widetilde{\gamma}_i\widehat{\theta}_i + (1-\widetilde{\gamma}_i)\mathbf{x}_i'\widehat{\widehat{\boldsymbol{\beta}}} \tag{17.3.47}$$

The design bias and MSE of $\widehat{\theta}_i^h(\widetilde{\boldsymbol{\tau}})$ are given by Ghosh and Rao (1994) and Rao (2003) as follows:

$$\begin{aligned} B\left[\widehat{\theta}_i^h(\widetilde{\boldsymbol{\tau}})\right] &= E_p\left[\widehat{\theta}_i^h(\widetilde{\boldsymbol{\tau}}) - \theta_i\right] \\ &\cong (1-\widetilde{\gamma}_i)\left(\mathbf{x}_i'\widehat{\widehat{\boldsymbol{\beta}}}^* - \theta_i\right) \end{aligned} \tag{17.3.48}$$

where $\widehat{\widehat{\boldsymbol{\beta}}}^*$ is the conditional expectation of $\widehat{\widehat{\boldsymbol{\beta}}}$ given $\boldsymbol{\theta} = (\theta_1, \ldots, \theta_q)$

$$\begin{aligned} \text{MSE}\left[\widehat{\theta}_i^h(\widetilde{\boldsymbol{\tau}})\right] &= E_p\left[\widehat{\theta}_i^h(\widetilde{\boldsymbol{\tau}}) - \theta_i\right]^2 \\ &= \widetilde{\gamma}_i\widetilde{\sigma}_i^2 + (1-\widetilde{\gamma}_i)^2\mathbf{x}_i'\left(\sum_{i=1}^{A}\frac{\mathbf{x}_i\mathbf{x}_i'}{\widetilde{\sigma}_i^2 + z_i^2\widetilde{\sigma}_v^2}\right)^{-1}\mathbf{x}_i \end{aligned} \tag{17.3.49}$$

In practice, the variances $\widetilde{\sigma}_i^2$ and $\widetilde{\sigma}_v^2$ are unknown and they are replaced by sample estimates.

17.3.6.4 Fay–Herriot Model

In the Fay and Herriot (1979) model, parameter of interest $\theta_i = \overline{Y}_i = \mu_i$, the population mean of the ith small area is related to the auxiliary information $\mathbf{x}_i' = (x_{i1}, \ldots, x_{ij}, \ldots, x_{iq})$ through the following model

$$\theta_i = x_{i1}\widetilde{\beta}_1 + \cdots + x_{ij}\widetilde{\beta}_j + \cdots + x_{iq}\widetilde{\beta}_p + v_i \quad \text{for } i = 1, \ldots, A \tag{17.3.50}$$

The sample mean of the ith area $\overline{y}_i\left(=\widehat{\theta}_i\right)$ is again related to $\overline{Y}_i(=\theta_i)$ as follows:

$$\overline{y}_i = \theta_i + \widetilde{\epsilon}_i = \mathbf{x}_i'\widetilde{\boldsymbol{\beta}} + v_i + \widetilde{\epsilon}_i \tag{17.3.51}$$

It is assumed that v_i's and $\widetilde{\epsilon}_i$'s are independently distributed with $E_m(v_i) = 0$, $V_m(v_i) = \widetilde{\sigma}_v^2$, and $E_m(\widetilde{\epsilon}_i|\theta_i) = 0$ and $V_m(\widetilde{\epsilon}_i|\theta_i) = \widetilde{\sigma}_i^2$. The BLUP estimator for θ_i is given by

$$\begin{aligned} \widehat{\theta}_i^{fh}(\widetilde{\boldsymbol{\tau}}) &= \widehat{\mu}_i^{fh}(\widetilde{\boldsymbol{\tau}}) \\ &= w_i\overline{y}_i + (1-w_i)\mathbf{x}_i'\widehat{\widehat{\boldsymbol{\beta}}} \end{aligned} \tag{17.3.52}$$

where $\widehat{\widetilde{\boldsymbol{\beta}}} = \left(\mathbf{X}'\widetilde{\mathbf{V}}^{-1}\mathbf{X}\right)^{-1}\left(\mathbf{X}'\widetilde{\mathbf{V}}^{-1}\overline{\mathbf{y}}\right) = \left(\sum_{i=1}^{A}\frac{\mathbf{x}_i\overline{y}_i}{\widetilde{\sigma}_i^2+\widetilde{\sigma}_v^2}\right)\Big/\left(\sum_{i=1}^{A}\frac{\mathbf{x}_i\mathbf{x}_i'}{\widetilde{\sigma}_i^2+\widetilde{\sigma}_v^2}\right)$,
$\overline{y}' = \left(\overline{y}_1,\ldots,\overline{y}_A\right)$, $\mathbf{V} = diag\left(\widetilde{\sigma}_1^2+\widetilde{\sigma}_v^2,\ldots,\widetilde{\sigma}_A^2+\widetilde{\sigma}_v^2\right)$ and

$$w_i = \widetilde{\sigma}_v^2\Big/\left(\widetilde{\sigma}_v^2+\widetilde{\sigma}_i^2\right) \tag{17.3.53}$$

Furthermore, $\widehat{\theta}_i^{fh}(\widetilde{\boldsymbol{\tau}})$ reduces to the synthetic estimator $\mathbf{x}_i\widehat{\widetilde{\boldsymbol{\beta}}}$ if $\widetilde{\sigma}_i^2$ is relatively large compared to $\widetilde{\sigma}_v^2$. Alternatively if $\widetilde{\sigma}_v^2$ is large compared to $\widetilde{\sigma}_i^2$, the estimator $\widehat{\mu}_i^{fh}$ becomes the sample mean $\overline{y}_i$.

17.3.7 Empirical Best Linear Unbiased Prediction, Empirical Bayes, and Hierarchical Bayes Methods

17.3.7.1 Empirical Best Linear Unbiased Prediction

The BLUP estimator $\widehat{\mu}_h(\boldsymbol{\tau})$ and $\widehat{\theta}_i^h(\widetilde{\boldsymbol{\tau}})$ given in Eqs. (17.3.34) and (17.3.46) involve model parameters $\boldsymbol{\tau}$ and $\widetilde{\boldsymbol{\tau}}$, respectively. The parameters $\boldsymbol{\tau}$ and $\widetilde{\boldsymbol{\tau}}$ are unknown in most practical situations. Hence, to use the BLUP estimator in practice, the parameters $\boldsymbol{\tau}$ and $\widetilde{\boldsymbol{\tau}}$ are estimated through a selected sample. Various methods of estimating the variance component parameters are available in literature. Popular among them are the method of moments, maximum likelihood (ML), and restricted ML (RML) methods (Rao, 2003). All these methods yield consistent estimators under general regularity conditions. Replacing the parameters $\boldsymbol{\tau}$ and $\widetilde{\boldsymbol{\tau}}$ by their suitable estimates $\widetilde{\boldsymbol{\tau}}$ and $\widehat{\widetilde{\boldsymbol{\tau}}}$, in BLUP estimators, empirical best linear unbiased prediction (EBLUP) estimators are obtained. For details, readers are referred to Rao (2003). A few examples have been given below.

17.3.7.1.1 Onefold Nested Error Regression Model

For the nested error regression model (Eq. 17.3.36), the EBLUP estimator of $\overline{Y}_i = \overline{\mathbf{X}}_i'\boldsymbol{\beta} + v_i$ is obtained by replacing $\boldsymbol{\tau} = \left(\sigma_v^2,\sigma_e^2\right)$ by its unbiased estimator $\widehat{\boldsymbol{\tau}} = \left(\widehat{\sigma}_v^2,\widehat{\sigma}_e^2\right)$ in the expression of $\widehat{\mu}_h^{(i)}(\boldsymbol{\tau})$ of Eq. (17.3.38) and it will be denoted by

$$\widehat{\mu}_i^{\mathrm{BLUP}} = \widehat{\mu}_i(\widetilde{\boldsymbol{\tau}}) \tag{17.3.54}$$

The unbiased estimators of $\widehat{\sigma}_e^2$ and $\widehat{\sigma}_v^2$ obtained by Prasad and Rao (1990) are

$$\widehat{\sigma}_e^2 = \sum_i\sum_j \widehat{e}_{ij}^2\Big/(n-A-q) \text{ and } \widehat{\sigma}_v^2 = \frac{1}{n^*}\left[\sum_i\sum_j \widehat{u}_{ij}^2 - (n-q)\widehat{\sigma}_e^2\right] \tag{17.3.55}$$

where $n^* = n - \text{tr}\left[(\mathbf{X}'\mathbf{X})^{-1}\sum_{i=1}^{A} n_i^2 \overline{\mathbf{x}}_i \overline{\mathbf{x}}_i'\right]$, $\overline{\mathbf{x}}_i' = \left(\overline{x}_{i\cdot 1}, \ldots, \overline{x}_{i\cdot q}\right)$ and $\left\{\widehat{e}_{ij}\right\}$'s are the residuals from the OLS regression of $y_{ij} - \overline{y}_i$ on $\left(x_{ij1} - \overline{x}_{i\cdot 1}, \ldots, x_{ijq} - \overline{x}_{i\cdot q}\right)$ and $\left\{\widehat{u}_{ij}\right\}$'s are the residuals from the OLS regression of y_{ij} on $(x_{ij1}, \ldots, x_{ijq})$. The estimator $\widehat{\sigma}_v^2$ in Eq. (17.3.55) may turnout to be negative but the probability of the estimator taking a negative value becomes zero if $A \to \infty$. Hence we take the value of $\widehat{\sigma}_v^2$ as zero if it is negative.

17.3.7.1.2 Fay—Herriot Model

For Fay—Herriot model (Eq. 17.3.50), an unbiased estimator of $\widetilde{\sigma}_v^2$ assuming $\widetilde{\sigma}_i^2$ is known, is given by

$$\widehat{\widetilde{\sigma}}_v^2 = \left[\sum_{i=1}^{A} \widehat{u}_i^2 - \sum_{i=1}^{A} \widetilde{\sigma}_i^2\left(1 - \mathbf{x}_i'(\mathbf{X}'\mathbf{X})^{-1}\mathbf{x}_i\right)\right] \Big/ (A - q) \qquad (17.3.56)$$

where $\widehat{u}_i = \overline{y}_i - \overline{\mathbf{x}}_i'\widehat{\widehat{\boldsymbol{\beta}}}$ and $\widehat{\widehat{\boldsymbol{\beta}}} = (\mathbf{X}'\mathbf{X})^{-1}\mathbf{X}'\mathbf{y}$.

Here again, $\widehat{\widetilde{\sigma}}_v^2$ can take a negative value, and we set it equal to zero when it is negative. The EBLUP estimator for θ_i is obtained by placing $\widetilde{\sigma}_v^2 = \widehat{\widetilde{\sigma}}_v^2$ in the expression of Eq. (17.3.52) and it will be denoted by

$$\begin{aligned} \widehat{\theta}_i^{\text{BLUP}} &= \widehat{\theta}_i^{fh}\left(\widehat{\boldsymbol{\tau}}\right) \\ &= \widehat{w}_i\overline{y}_i + (1 - \widehat{w}_i)\overline{\mathbf{x}}_i'\widehat{\widehat{\boldsymbol{\beta}}} \end{aligned} \qquad (17.3.57)$$

where

$$\widehat{w}_i = \widehat{\widetilde{\sigma}}_v^2 \Big/ \left(\widehat{\widetilde{\sigma}}_v^2 + \widetilde{\sigma}_i^2\right)$$

17.3.7.2 Empirical Bayes Approach

Here we assume that the parameter of interest θ has some prior distribution $\xi(\theta|\boldsymbol{\lambda})$ with unknown parameter $\boldsymbol{\lambda}$. The posterior distribution of θ, given the data, is first obtained assuming that the parameter $\boldsymbol{\lambda}$ is known. The parameter $\boldsymbol{\lambda}$ is estimated through the marginal distribution of the data. Inference of θ is obtained from the estimated posterior distribution $\xi\left(\theta|\widehat{\boldsymbol{\lambda}}\right)$. Good details have been given by Morris (1983) and Ghosh and Rao (1994).

Let us assume that $\widehat{\theta}_i$'s given in Eq. (17.3.45) are independently normally distributed with unknown mean θ_i but known variance $\widetilde{\sigma}_i^2$. Let us further assume that the distribution of θ_i is normal with mean $\boldsymbol{\lambda} = \mathbf{x}_i'\widetilde{\boldsymbol{\beta}}$ and variance $z_i^2\widetilde{\sigma}_v^2$, where $\widetilde{\boldsymbol{\beta}}$ and $\widetilde{\sigma}_v^2$ are unknown model parameters. The posterior distribution of θ_i, given $\widehat{\theta}_i$, $\widetilde{\boldsymbol{\beta}}$, and $\widetilde{\sigma}_v^2$, has the mean $\widehat{\theta}_i^B = \widetilde{\gamma}_i\widehat{\theta}_i + (1-\widetilde{\gamma}_i)\mathbf{x}_i'\widetilde{\boldsymbol{\beta}}$ and variance $\widetilde{\gamma}_i\widetilde{\sigma}_i^2$, where $\widetilde{\gamma}_i$ is given in Eq. (17.3.46). It can be shown that $\widehat{\theta}_i^B$ is the Bayes estimator of $\widehat{\theta}_i$ under squared error loss. This Bayes estimator $\widehat{\theta}_i^B$ involves unknown parameters $\widetilde{\boldsymbol{\beta}}$ and $\widetilde{\sigma}_v^2$. These unknown parameters can be estimated from the marginal distribution of $\widehat{\theta}_i$ which is normal with mean $\mathbf{x}_i'\widetilde{\boldsymbol{\beta}}$ and variance $z_i^2\widetilde{\sigma}_v^2 + \widetilde{\sigma}_i^2$. Let $\widehat{\widetilde{\boldsymbol{\beta}}}^{(ml)}$ and $\widehat{\widetilde{\sigma}}_v^{2(ml)}$ be the ML or RML estimates of $\widetilde{\boldsymbol{\beta}}$ and $\widetilde{\sigma}_v^2$, respectively, then the empirical Bayes estimator of θ_i is given by

$$\widehat{\theta}_i^{\text{EB}} = \widehat{\widetilde{\gamma}}_i^{(ml)}\widehat{\theta}_i + \left(1 - \widehat{\widetilde{\gamma}}_i^{(ml)}\right)\mathbf{x}_i'\widehat{\widetilde{\boldsymbol{\beta}}}^{(ml)} \tag{17.3.58}$$

with

$$\widehat{\widetilde{\gamma}}_i^{(ml)} = z_i^2\widehat{\widetilde{\sigma}}_v^{2(ml)} \Big/ \left(z_i^2\widehat{\widetilde{\sigma}}_v^{2(ml)} + \widetilde{\sigma}_i^2\right)$$

Further details have been given by Rao (2003).

17.3.7.3 Hierarchical Bayes Approach

In the hierarchical Bayes (HB) approach, we assume that the parameter of interest has some prior distribution with unknown parameters. The unknown parameters have again some prior distribution with unknown parameters and so on. At the ultimate stage, it is assumed that all the parameters of the prior distribution of the ultimate stage are known. The inference is based on the posterior distribution of the parameter of interest. In particular, under squared error loss, the parameter is estimated by the posterior mean, and its performance is measured by its posterior variance.

Consider the model (Eq. 17.3.45). Let us assume that the distribution of $\widehat{\theta}_i$ given $\theta_i, \widetilde{\boldsymbol{\beta}}, \widetilde{\sigma}_v^2$ are independent normal with mean θ_i and variance $\widetilde{\sigma}_i^2$. Furthermore, the distribution of θ_i given $\widetilde{\boldsymbol{\beta}}, \widetilde{\sigma}_v^2$ are independent normal with mean $\overline{\mathbf{x}}_i'\widetilde{\boldsymbol{\beta}}$ and variance $z_i^2\widetilde{\sigma}_v^2$. Let us assume that the prior distribution of $\widetilde{\boldsymbol{\beta}}$ is uniform, and $\widetilde{\sigma}_v^2$ and $\widetilde{\sigma}_i^2$ are known. Then the posterior distribution of θ_i given $\widehat{\theta}_i$ and $\widetilde{\boldsymbol{\beta}}$ is normal with mean

$$E\left(\theta_i|\widehat{\theta}_i, \widetilde{\sigma}_v^2, \widetilde{\sigma}_i^2\right) = \widetilde{\gamma}_i\widehat{\theta}_i + (1-\widetilde{\gamma}_i)\mathbf{x}_i'\widehat{\widetilde{\boldsymbol{\beta}}} = \widehat{\theta}_i^{HB} \quad (17.3.59)$$

and variance $\widetilde{\gamma}_i\widetilde{\sigma}_i^2$, where $\widetilde{\gamma}_i$ is given in Eq. (17.3.46). Hence the HB estimator for θ_i is $\widehat{\theta}_i^{HB} = \widehat{\theta}_i^{h}(\widetilde{\boldsymbol{\tau}})$, the BLUP estimator given in Eq. (17.3.47).

To compare the performances of the estimators of the parameters of interest generated by different methods of estimation and estimation of confidence intervals based on these estimators, one needs to compute the MSEs of the estimators. Generally, exact expressions of MSEs are hard to obtain. However, Morris (1983), Kass and Steffey (1989), Prasad and Rao (1990), Singh et al. (1998), Butar and Lahiri (2001), Datta et al. (2002), among others, have provided approximate expressions of MSE under various small-area models. Ghosh and Lahiri (1987), Lahiri (1990), Ghosh (1992), Ghosh and Rao (1994), Chottopadhyaya et al. (1999), Prasad and Rao (1990), Kleffe and Rao (1992), Singh et al. (1994), Stukel and Rao (1999), Jiang et al. (2002), among others, presented simulation studies based on live data to compare efficiencies and biases of various small area estimators under different models.

17.4 EXERCISES

17.4.1 The total number of admission and the total number of dropout of secondary students of three countries of a certain district in the year 2008 are given below. The admission and dropout rates as recorded by the last census of schools in 2001 were 2.5% and 1.8%, respectively, whereas the estimated admission and dropout rates for the district in 2008 were 1.8 and 1.6, respectively. Use VR method and estimate the total number of students for the year 2008.

Counties	Admission 2008	Dropout 2008
1	525	480
2	625	550
3	300	250

17.4.2 A sample of 20 factories was selected at random from 250 factories of a certain district. The factories were classified into small, medium, and large according to the number of workers. The monthly output (in 000$) and number of workers in the sampled farms along with the total number of factories and their mean number of workers are given in the following table.

Factories	Monthly output (000$) y	Number of workers x	Total number of factories	Mean number of workers
Large	200	100	50	140
	180	150		
	175	125		
	280	120		
Medium	100	75	80	45
	120	70		
	90	50		
	75	40		
	65	40		
	70	30		
	80	30		
	60	40		
Small	18	10	120	15
	25	15		
	20	15		
	40	20		
	30	15		
	25	15		
	35	20		
	30	15		

Estimate the average monthly output of the different types of factories by synthetic and composite method taking direct estimator as (i) sample mean and (ii) ratio estimator. Suppose that a linear model $y_i = \beta x_i + \in_i$ is appropriate, estimate the mean output of the different types of factories by (i) synthetic and (ii) generalized regression method.

17.4.3 A sample of 35 agricultural farms was selected from a list of 1200 farms of a certain district at random. The farms were classified into three groups large (L), medium (M), and small (S) according to their size as well as irrigation facilities. The yield of wheat (y), farm size (x), and irrigation facilities are recorded from the sampled farms and is given in the following table.

Yield of wheat and farm sizes of the selected farms.

Irrigation facility	Farm	Yield of wheat (00 kg) y	Farm size (acre) x
Yes	L	100	20
		180	27
		150	25
	M	100	15
		87	12
		60	10
		75	12
		65	14
		70	14
		80	15
		85	17
	S	50	8
		60	8
		60	7
		50	6
		45	5
		45	5
		50	8
		40	7
		45	5
No	L	50	25
		70	30
		85	40
	M	30	15
		25	12
		30	10
		35	12
		45	15
	S	20	5
		15	5
		12	6
		15	6
		12	5
		15	5
		20	6

Estimate the average yield of wheat of the different types of farm by (i) direct, (ii) synthetic, and (iii) composite method (a) without assuming a model and (b) fitting a linear regression of yield of a farm on the area of the farm.

CHAPTER 18

Variance Estimation: Complex Survey Designs

18.1 INTRODUCTION

A sampling design other than simple random sampling is known as a complex sampling design. Most real-life surveys are complex surveys, and for such surveys we often need to estimate nonlinear parametric functions such as the population ratio of the total of two characteristics, population coefficient of variation, population regression coefficient, and population correlation coefficient, among others. Variance estimation is essential for estimating the precision of the survey estimates, calculation of confidence intervals, determination of optimum sample sizes, and testing of hypotheses, among others. In particular, finding the optimum sample size is the key factor in the determination of the cost of a survey and the subsequent precision of estimates. In most situations unbiased estimators of the nonlinear parametric functions are not available. We get neither the exact expressions of the variance/mean square of the nonlinear estimators nor the exact unbiased estimators of the variance/mean square errors. In this section we will discuss a few popular methods of approximate variance estimation of nonlinear estimators that can be used for complex survey designs. The methods include (i) linearization method (LR), (ii) random group method (RG), (iii) jackknife method (JK), (iv) balanced repeated replication (BRR), (v) bootstrap method (BT), and (vi) generalized variance functions (GVF). By and large, the proposed estimators are approximately unbiased and consistent. So, in our present discussion we will use the terms variance and mean square error in the same sense.

18.2 LINEARIZATION METHOD

Let $\theta_1,\ldots,\theta_k$ be k parametric functions of a finite population parameter $\mathbf{y}=(y_1,\ldots,y_N)$ and $\widehat{\theta}_1(s)=\widehat{\theta}_1,\ldots,\widehat{\theta}_k(s)=\widehat{\theta}_k$ be the estimators of $\theta_1,\ldots,\theta_k$ based on a sample s of size n selected with probability p(s) according to some suitable sampling design. In most situations, $\widehat{\theta}_j$'s are either unbiased or

Survey Sampling Theory and Applications
ISBN 978-0-12-811848-1
http://dx.doi.org/10.1016/B978-0-12-811848-1.00018-2

consistent estimators of θ_j's. Suppose we want to estimate a parametric function $f(\boldsymbol{\theta}) = f(\theta_1,\ldots,\theta_k)$. In this situation we generally choose $T(s) = T = f\left(\widehat{\theta}_1, \ldots, \widehat{\theta}_k\right)$ as an estimator of $f(\boldsymbol{\theta})$. In case T is a differentiable function of $\widehat{\theta}_j$'s, for $j = 1,\ldots,k$, we can expand $T = f\left(\widehat{\theta}_1, \ldots, \widehat{\theta}_k\right)$ around $\theta_1,\ldots,\theta_k$ by Taylor's theorem as follows:

$$T = f\left(\widehat{\theta}_1, \ldots, \widehat{\theta}_k\right)\Big|_{\hat{\theta}_j = E\left(\hat{\theta}_j\right)} + \sum_{j=1}^{k}\left(\widehat{\theta}_j - \theta_j\right)\frac{\partial f\left(\widehat{\theta}_1, \ldots, \widehat{\theta}_k\right)}{\partial \widehat{\theta}_j}\Bigg|_{\hat{\theta}_j = E\left(\hat{\theta}_j\right)} + R_2 \tag{18.2.1}$$

where R_2 is the error or remainder term.

Letting $\widehat{\theta}_j$ be an unbiased or consistent estimator for θ_j for $j = 1,\ldots,k$, and assuming R_2 is small, at least for large n, we approximate

$$T \cong f(\theta_1, \ldots, \theta_k) + \sum_{j=1}^{k}\omega_j\left(\widehat{\theta}_j - \widehat{\theta}_j\right) \tag{18.2.2}$$

where $\omega_j = \dfrac{\partial f\left(\hat{\theta}_1, \ldots, \hat{\theta}_k\right)}{\partial \hat{\theta}_j}\Bigg|_{\hat{\theta}_j = \theta_j}$.

From expression (18.2.2), we get the following:

Theorem 18.2.1

(i) $E(T) \cong f(\theta_1,\ldots,\theta_k)$

(ii) Variance of $T = V(T) \cong \sum_{j=1}^{k}\omega_j^2 V\left(\widehat{\theta}_j\right) + \sum_{i\neq}^{k}\sum_{j=1}^{k}\omega_j\omega_k Cov\left(\widehat{\theta}_j, \widehat{\theta}_k\right)$

$= \boldsymbol{\omega}'\boldsymbol{\Lambda}\boldsymbol{\omega}$

(iii) An approximate unbiased estimator of $V(T)$ is

$$\widehat{V}(T) \cong \sum_{j=1}^{k}\widehat{\omega}_j^2\widehat{V}\left(\widehat{\theta}_j\right) + \sum_{i\neq}^{k}\sum_{j=1}^{k}\widehat{\omega}_j\widehat{\omega}_k C\widehat{o}v\left(\widehat{\theta}_j, \widehat{\theta}_k\right)$$
$$= \widehat{\boldsymbol{\omega}}'\widehat{\boldsymbol{\Lambda}}\widehat{\omega}$$

where $\boldsymbol{\omega}' = (\omega_1, \ldots, \omega_k)$, $\boldsymbol{\Lambda}$ = variance covariance matrix of $\widehat{\theta}_j$'s, $\widehat{\boldsymbol{\omega}}' = (\widehat{\omega}_1, \ldots, \widehat{\omega}_k)$ = unbiased or approximately unbiased estimator of $\boldsymbol{\omega}'$, $\widehat{V}\left(\widehat{\theta}_j\right)$, $C\widehat{ov}\left(\widehat{\theta}_j, \widehat{\theta}_k\right)$ are unbiased estimators of $V\left(\widehat{\theta}_j\right)$ and $Cov\left(\widehat{\theta}_j, \widehat{\theta}_k\right)$, respectively, and $\widehat{\boldsymbol{\Lambda}}$ is an unbiased estimator of $\boldsymbol{\Lambda}$.

In case $\widehat{\theta}_j = \sum_{i \in s} b_{si}\xi_i(j)$ is a linear homogeneous unbiased estimator of $\theta_j = \sum_{i \in U} \xi_i(j)$ for $j = 1,\ldots,k$ and b_{si} are constants free from $\xi_i(j)$'s satisfy $\sum_{s \supset i} b_{si} = 1$ and $b_{si} = 0$ for $i \notin s$, we can write following Woodruff (1971)

$$\begin{aligned} V(T) &\cong V\left(\sum_{j=1}^{k} \omega_j \widehat{\theta}_j\right) \\ &= V\left[\sum_{i \in s}\left(\sum_{j=1}^{k} \omega_j b_{si}\right)\xi_i(j)\right] \\ &= V\left(\sum_{i \in s} b_{si} h_i\right) \quad \text{with } h_i = \sum_{j=1}^{k} \omega_j \xi_i(j) \\ &= \sum_{i \in U} \beta_{ii} h_i^2 + \sum_{i \neq} \sum_{j \in U} \beta_{ij} h_i h_j \end{aligned} \tag{18.2.3}$$

where $\beta_{ii} = \sum_{s \supset i} b_{si}^2 p(s) - 1$ and $\beta_{ij} = \sum_{s \supset i,j} b_{si} b_{sj} p(s) - 1$.

An approximate unbiased estimator of $V(T)$ is

$$\widehat{V}_{LR}(T) = \sum_{i \in s} \frac{\beta_{ii}}{\pi_i} \widehat{h}_i^2 + \sum_{i \neq} \sum_{j \in s} \frac{\beta_{ij}}{\pi_{ij}} \widehat{h}_i \widehat{h}_j \tag{18.2.4}$$

where $\widehat{h}_i$ is a suitable unbiased or consistent estimator of h_i.

The performance of $\widehat{V}_{LR}(T)$ obviously depends on the validity of the expansion of the Taylor series (18.2.2). The inclusion of second- and higher-order terms of the expansion of the Taylor series will certainly increase the performance of the variance estimator but at the same time yield a more complex variance formula. Wolter (1985) pointed out that the LR method provides efficient variance estimators in complex surveys when the sample size is fairly large. However, the method may provide unreliable estimators if the Taylor series is not convergent or the population is highly skewed. The LR technique fails for the parameter that cannot be expressed as a simple function of the population total, e.g., population median.

18.2.1 Ratio Estimator

The ratio estimator for the population ratio $R = Y/X$ is given by

$$\widehat{R} = \frac{\widehat{Y}}{\widehat{X}}$$

where $\widehat{Y} = \sum_{i\in s} b_{si}y_i$ and $\widehat{X} = \sum_{i\in s} b_{si}x_i$ are unbiased estimators of $Y = \sum_{i=1}^{N} y_i$ and $X = \sum_{i=1}^{N} x_i$, respectively, based on a sample s with probability $p(s)$. The coefficients b_{si}'s satisfy $\sum_{s\supset i} b_{si}p(s) = 1$.

Taking $y_i = \xi_i(1)$, $x_i = \xi_i(2)$, $\theta_1 = Y = \sum_{i\in U} \xi_i(1)$, $\theta_2 = X = \sum_{i\in U} \xi_i(2)$, $\widehat{\theta}_1 = \widehat{Y}$, $\widehat{\theta}_2 = \widehat{X}$, and $f(\theta_1, \theta_2) = \dfrac{\theta_1}{\theta_2} = R$, we can write

$$\widehat{R} = T = f\left(\widehat{\theta}_1, \widehat{\theta}_2\right) = \frac{\widehat{\theta}_1}{\widehat{\theta}_2} = f(\theta_1, \theta_2) + \omega_1\left(\widehat{\theta}_1 - \theta_1\right) + \omega_2\left(\widehat{\theta}_2 - \theta_2\right) \tag{18.2.5}$$

where

$$\omega_1 = \left.\frac{\partial T}{\partial \widehat{\theta}_1}\right|_{\hat{\theta}_1=\theta_1,\hat{\theta}_2=\theta_2} = \frac{1}{\theta_2} \text{ and } \omega_2 = \left.\frac{\partial T}{\partial \widehat{\theta}_2}\right|_{\hat{\theta}_1=\theta_1,\hat{\theta}_2=\theta_2} = -\left.\frac{\hat{\theta}_1}{\hat{\theta}_2^2}\right|_{\hat{\theta}_1=\theta_1,\hat{\theta}_2=\theta_2} = -\frac{\theta_1}{\theta_2^2}.$$

Writing $h_i = \sum_{j=1}^{2} \omega_j \xi_i(j) = \dfrac{1}{X}(y_i - Rx_i)$

we get $\widehat{R} \cong R + \dfrac{1}{X}\sum_{i\in s} b_{si}(y_i - Rx_i)$

and

$$V\left(\widehat{R}\right) \cong \frac{1}{X^2}\left[\sum_{i\in U} \beta_{ii}(y_i - Rx_i)^2 + \sum_{i\neq}\sum_{j\in U} \beta_{ij}(y_i - Rx_i)(y_j - Rx_j)\right] \tag{18.2.6}$$

For a fixed sample size n design with $b_{si} = 1/\pi_i$, the expression (18.2.6) reduces to

$$V\left(\widehat{R}\right) \cong \frac{1}{2}\frac{1}{X^2}\sum_{i\neq}\sum_{j\in U}(\pi_i\pi_j - \pi_{ij})\left(\frac{d_i}{\pi_i} - \frac{d_j}{\pi_j}\right)^2 \tag{18.2.7}$$

where $d_i = y_i - Rx_i$.

An approximate unbiased estimator of $V\left(\widehat{R}\right)$ is

$$\widehat{V}_{LR}\left(\widehat{R}\right) \cong \frac{1}{2}\frac{1}{X^2}\sum_{i\neq}\sum_{j\in s}\frac{\pi_i\pi_j - \pi_{ij}}{\pi_{ij}}\left(\frac{\widehat{d}_i}{\pi_i} - \frac{\widehat{d}_j}{\pi_j}\right)^2 \tag{18.2.8}$$

with $\widehat{d}_i = y_i - \widehat{R}x_i$.

For a simple random sampling without replacement (SRSWOR) sampling design with $b_{si} = N/n$. Hence the expressions (18.2.7) and (18.2.8) become, respectively,

$$V\left(\widehat{R}\right) \cong (1-f)\frac{1}{n\overline{X}^2}S_d^2 \text{ and } \widehat{V}_{LR}\left(\widehat{R}\right) \cong (1-f)\frac{1}{n\overline{X}^2}s_{\hat{d}}^2$$

where $f = n/N$, $(N-1)S_d^2 = \sum_{i\in U}\left(d_i - \overline{D}\right)^2$, $\overline{D} = \sum_{i\in U} d_i/N$, $(n-1)s_{\hat{d}}^2 = \sum_{i\in s}\left(\widehat{d}_i - \overline{\widehat{d}}\right)^2$, and $\overline{\widehat{d}} = \sum_{i\in s}\widehat{d}_i\Big/n$.

18.2.2 Coefficient of Variation

The population coefficient of variation of a finite population is defined as

$$\theta = \frac{\sqrt{\sum_{i\in U}\left(y_i - \overline{Y}\right)^2/N}}{\overline{Y}} = \frac{\sqrt{N\sum_{i\in U} y_i^2 - \left(\sum_{i\in U} y_i\right)^2}}{\sum_{i\in U} y_i} = \frac{\sqrt{\theta_0\theta_2 - \theta_1^2}}{\theta_1}$$
$$= f(\theta_0, \theta_1, \theta_2)$$

where $\theta_j = \sum_{i\in U} y_i^j$ for $j = 0,\ 1,\ 2$.

Let us choose $\widehat{\theta} = f\left(\widehat{\theta}_0, \widehat{\theta}_1, \widehat{\theta}_2\right)$ with $\widehat{\theta}_j = \sum_{i\in s} y_i^j\Big/\pi_i$. The variance of $\widehat{\theta}$ becomes

$$V\left(\widehat{\theta}\right) = \boldsymbol{\omega}'\boldsymbol{\Lambda}\boldsymbol{\omega}$$

where $\boldsymbol{\omega}' = (\omega_0, \omega_1, \omega_2)$, $\omega_0 = \left.\frac{\partial\hat{\theta}}{\partial\hat{\theta}_0}\right|_{\hat{\theta}_0=\theta_0,\hat{\theta}_1=\theta_1,\hat{\theta}_2=\theta_2} = \frac{1}{2\theta}\frac{\theta_2}{\theta_1^2}$,
$\omega_1 = \left.\frac{\partial\hat{\theta}}{\partial\hat{\theta}_1}\right|_{\hat{\theta}_0=\theta_0,\hat{\theta}_1=\theta_1,\hat{\theta}_2=\theta_2} = -\frac{\theta_0}{\theta}\frac{\theta_2}{\theta_1^3}$, $\omega_2 = \left.\frac{\partial\hat{\theta}}{\partial\hat{\theta}_2}\right|_{\hat{\theta}_0=\theta_0,\hat{\theta}_1=\theta_1,\hat{\theta}_2=\theta_2} = \frac{1}{2\theta}\frac{\theta_0}{\theta_1^2}$, and
$\boldsymbol{\Lambda} = (V_{jk})$ with

$$V_{jk} = Cov\left(\widehat{\theta}_j, \widehat{\theta}_k\right) = \sum_{r\in U}\left(\frac{1}{\pi_r} - 1\right)y_r^{j+k} + \sum_{r\neq}\sum_{t\in U}\left(\frac{\pi_{rt}}{\pi_r\pi_t} - 1\right)y_r^j y_t^k.$$

An approximate unbiased estimator of $V\left(\widehat{\theta}\right)$ is

$$\widehat{V}_{LR}\left(\widehat{\theta}\right) = \widehat{\boldsymbol{\omega}}'\widehat{\boldsymbol{\Lambda}}\widehat{\omega} \quad (18.2.9)$$

where $\widehat{\boldsymbol{\omega}}' = (\widehat{\omega}_0, \widehat{\omega}_1, \widehat{\omega}_2)$, $\widehat{\omega}_0 = \frac{1}{2\hat{\theta}}\frac{\hat{\theta}_2}{\hat{\theta}_1^2}$, $\widehat{\omega}_1 = -\frac{\hat{\theta}_0}{\hat{\theta}}\frac{\hat{\theta}_2}{\hat{\theta}_1^3}$, $\widehat{\omega}_2 = \frac{1}{2\hat{\theta}}\frac{\hat{\theta}_0}{\hat{\theta}_1^2}$,

$$\widehat{\Lambda} = \left(\widehat{V}_{jk}\right), \text{ and } \widehat{V}_{jk} = \sum_{r\in s}\left(\frac{1}{\pi_r} - 1\right)\frac{y_r^{j+k}}{\pi_r} + \sum_{r\neq}\sum_{t\in s}\left(\frac{\pi_{rt}}{\pi_r\pi_t} - 1\right)\frac{y_r^j y_t^k}{\pi_{rt}}.$$

18.3 RANDOM GROUP METHOD

The pioneering work of the RG method was due to Mahalanobis (1946) who proposed the interpenetrating network of subsampling procedure (IPNS). The main objective of IPNS was to maintain the quality of the data collected by different investigators by comparing the responses obtained from those units that are common in two or more samples. In this method k independent samples of the same size are selected from the population using the same sampling design. For $k = 2$, the method is called half sampling. Let $\widehat{\theta}_{(i)}$ be an unbiased estimator of the parameter θ obtained from the ith sample $i = 1,\ldots,k$ and $\widehat{\overline{\theta}} = \sum_{i=1}^{k}\widehat{\theta}_{(i)}\Big/k$ be a pooled estimator of θ, then we can easily verify the following theorem.

Theorem 18.3.1

(i) $E\left(\widehat{\overline{\theta}}\right) = \theta$

(ii) $V\left(\widehat{\overline{\theta}}\right) = \sigma^2\Big/k$

(iii) $\widehat{V}\left(\widehat{\overline{\theta}}\right) = \frac{1}{k(k-1)}\sum_{i=1}^{k}\left(\widehat{\theta}_{(i)} - \widehat{\overline{\theta}}\right)^2$

where $\widehat{V}\left(\widehat{\overline{\theta}}\right)$ is an unbiased estimator for $V\left(\widehat{\overline{\theta}}\right)$ and $\sigma^2 = V\left(\widehat{\theta}_{(i)}\right)$ for $i = 1,\ldots,k$.

This theorem may be generalized by selecting different samples independently by using different sampling procedures so that the $E\left(\widehat{\theta}_{(i)}\right)$'s are equal to θ, while the variances $V\left(\widehat{\theta}_{(i)}\right) = \sigma_i^2$ may vary. In this situation we also get the following theorem, which is similar to the previous one.

Theorem 18.3.2

(i) $E\left(\widehat{\overline{\theta}}\right) = \theta$

(ii) $V\left(\widehat{\overline{\theta}}\right) = \sum_{i=1}^{k}\sigma_i^2/k^2$

(iii) $\widehat{V}\left(\widehat{\overline{\theta}}\right) = \frac{1}{k(k-1)}\sum_{i=1}^{k}\left(\widehat{\theta}_{(i)} - \widehat{\overline{\theta}}\right)^2$

Thus if the samples are selected independently by any sampling procedure, an unbiased estimator of the variance of the estimator of the parameter θ may be obtained easily by applying the aforementioned theorems. But in practice, the sample is selected once and unbiased estimators of the parameters of interest (such as population ratio, regression coefficient, coefficient of variance, etc.) cannot be obtained easily. In this situation we divide the original sample s of size n at random into k disjoint groups. The ith RG s_i is a random subsample of size $m = n/k$ (assuming m is an integer) selected from the original sample s, by SRSWOR method. Let $\widehat{\theta}$ be an estimator of θ based on the original sample s, $\widehat{\theta}_{(i)}$ estimator of θ based on the sample s_i, and $\widehat{\overline{\theta}} = \sum_{i=1}^{k} \widehat{\theta}_{(i)} \Big/ k$ be a combined estimator of θ. Then the variance of $\widehat{\theta}$ can be estimated by any of the following formulae.

$$\widehat{V}_{RG1}\left(\widehat{\theta}\right) = \frac{1}{k(k-1)} \sum_{i=1}^{k} \left(\widehat{\theta}_{(i)} - \widehat{\overline{\theta}}\right)^2 \tag{18.3.1}$$

$$\widehat{V}_{RG2}\left(\widehat{\theta}\right) = \frac{1}{k(k-1)} \sum_{i=1}^{k} \left(\widehat{\theta}_{(i)} - \widehat{\theta}\right)^2 \tag{18.3.2}$$

None of the aforementioned estimators are unbiased for $V\left(\widehat{\theta}\right)$. The estimator $\widehat{V}_{RG2}\left(\widehat{\theta}\right)$ is conservative in the sense that it has an upward bias. If $\widehat{\theta}$ and $\widehat{\theta}_{(i)}$ are linear unbiased estimators of θ, then $\widehat{\overline{\theta}} = \widehat{\theta}$. But, if $\widehat{\theta}$ and $\widehat{\theta}_{(i)}$ are nonlinear estimators of θ such as the population ratio $R = Y/X$, then $\widehat{\theta} = \widehat{Y} \Big/ \widehat{X}$ is not, in general, equal to $\widehat{\overline{\theta}} = \frac{1}{k} \sum_{i=1}^{K} \widehat{Y}_i / \widehat{X}_i$, where $\widehat{Y}, \widehat{X}$ and $\widehat{Y}_i, \widehat{X}_i$ are unbiased estimators for totals of Y and X, respectively, based on the samples s and s_i. As per the magnitude of $\widehat{V}_{RG1}\left(\widehat{\theta}\right)$ and $\widehat{V}_{RG2}\left(\widehat{\theta}\right)$, we note that $\widehat{V}_{RG1}\left(\widehat{\theta}\right) \leq \widehat{V}_{RG2}\left(\widehat{\theta}\right)$ because $\widehat{V}_{RG2}\left(\widehat{\theta}\right) = \frac{1}{k(k-1)} \sum_{i=1}^{k} \left(\widehat{\theta}_{(i)} - \widehat{\theta}\right)^2 = \widehat{V}_{RG1}\left(\widehat{\theta}\right) + \frac{\left(\widehat{\overline{\theta}} - \widehat{\theta}\right)^2}{k-1}$. However, for large sample size $\widehat{V}_{RG1}\left(\widehat{\theta}\right)$ and $\widehat{V}_{RG2}\left(\widehat{\theta}\right)$ are approximately equal.

18.3.1 Simple Random Sampling With Replacement

Let $\widehat{\theta} = \overline{y}(s)$ be the sample mean based on a sample s of size n ($=mk$) selected by the simple random sampling with replacement (SRSWR) method. The sample s is divided into k groups at random and $\widehat{\theta}_{(i)} = \overline{y}(s_i)$ is the sample mean of the ith group s_i, $i = 1,\ldots,k$. Here, $\widehat{\theta}$, $\widehat{\theta}_{(i)}$, and the combined estimator $\widehat{\overline{\theta}} = \sum_{i=1}^{k} \widehat{\theta}_{(i)} \Big/ k$ are all unbiased estimators for the population mean $\overline{Y} = \sum_{i\in U} y_i/N$. Furthermore, $\widehat{\theta} = \overline{y}(s) = \widehat{\overline{\theta}} = \sum_{i=1}^{k} \overline{y}(s_i)/k$, $\widehat{V}_{RG1}\left(\widehat{\theta}\right) = \widehat{V}_{RG2}\left(\widehat{\theta}\right) = \frac{1}{k(k-1)} \sum_{i=1}^{k} \{\overline{y}(s_i) - \overline{y}(s)\}^2$, and

$$E\left\{\widehat{V}_{RG1}\left(\widehat{\theta}\right)\right\} = \frac{1}{k(k-1)} E\left(\sum_{i=1}^{k} \widehat{\theta}_{(i)}^2 - k\widehat{\theta}^2\right) = \frac{V\left(\widehat{\theta}_{RG1}\right) - V\left(\widehat{\theta}\right)}{(k-1)}$$

$$= \frac{\sigma^2}{n} = V(\overline{y}_s)$$

where $\sigma^2 = \sum_{i\in U} (y_i - \overline{Y})^2/N$. Hence $\widehat{V}_{RG1}\left(\widehat{\theta}\right)$ (i.e., $\widehat{V}_{RG2}\left(\widehat{\theta}\right)$) is unbiased for $\frac{\sigma^2}{n} = V\left(\widehat{\theta}\right)$. It is important to note that $\widehat{V}_{RG1}\left(\widehat{\theta}\right)$ is different from the traditional unbiased estimator $\widehat{V}(\overline{y}_s) = s_y^2 \big/ n$, where $s_y^2 = \sum_{i\in s} \{y_i - \overline{y}(s)\}^2/(n-1)$.

18.3.2 Simple Random Sampling Without Replacement

Let a sample s of size n be selected by SRSWOR method and be divided at random into k groups of m units each ($n = mk$). The estimator for $\overline{Y}$ based on the sample s and s_i are the sample means $\widehat{\theta} = \overline{y}(s)$ and $\widehat{\theta}_{(i)} = \overline{y}(s_i)$, respectively. In this case the combined estimator of θ is $\widehat{\theta} = \overline{y}(s) = \widehat{\overline{\theta}} = \sum_{i=1}^{k} \widehat{\theta}_{(i)} \Big/ k = \sum_{i=1}^{k} \overline{y}(s_i)/k$. Each of the estimated variances $\widehat{V}_{RG1}\left(\widehat{\overline{\theta}}\right)$ and $\widehat{V}_{RG2}\left(\widehat{\theta}\right)$ of $\widehat{\theta}$ is equal to $\frac{1}{k(k-1)} \sum_{i=1}^{k} \{\overline{y}(s_i) - \overline{y}(s)\}^2$ and

$$E\left\{\widehat{V}_{RG1}\left(\widehat{\theta}\right)\right\} = E\left\{\widehat{V}_{RG2}\left(\widehat{\theta}\right)\right\} = \frac{V\left(\widehat{\theta}_{(i)}\right) - V\left(\widehat{\theta}\right)}{(k-1)}$$

$$= \frac{1}{k-1}\left[\left(\frac{1}{m} - \frac{1}{N}\right) - \left(\frac{1}{n} - \frac{1}{N}\right)\right] S_y^2$$

$$= \frac{S_y^2}{n}$$

The estimator $\widehat{V}_{RG1}\left(\widehat{\theta}\right)\left(=\widehat{V}_{RG2}\left(\widehat{\theta}\right)\right)$ is therefore biased for $V\left(\widehat{\theta}\right)$ and the amount of bias is $E\left\{\widehat{V}_{RG1}\left(\widehat{\theta}\right)-V\left(\widehat{\theta}\right)\right\}=S_y^2\big/N$. Thus the estimator $\widehat{V}_{RG1}\left(\widehat{\theta}\right)$ (i.e., $\widehat{V}_{RG2}\left(\widehat{\theta}\right)$) overestimates the variance of $V\left(\widehat{\theta}\right)$. The amount of bias of $\widehat{V}_{RG1}\left(\widehat{\theta}\right)$ is obviously negligible if N is large.

18.3.3 Varying Probability Sampling

Let a sample s of size n be selected by the varying probability sampling scheme with positive inclusion probability for the ith unit π_i for $i=1,\ldots,N$. The Horvitz–Thompson estimator for the population total Y based on the sample s is $\widehat{\theta}=\sum_{i\in s} y_i/\pi_i$. The sample s is divided at random into k groups each of size m. The estimator for Y based on the ith group is $\widehat{\theta}_{(i)}=\dfrac{n}{m}\sum_{i\in s_i} y_i/\pi_i$ for $i=1,\ldots,k$ and the combined estimator is

$$\widehat{\widehat{\theta}}=\sum_{i=1}^{k}\widehat{\theta}_{(i)}\Big/k=\sum_{i=1}^{k}\sum_{i\in s_i} y_i/\pi_i=\sum_{i\in s} y_i/\pi_i=\widehat{\theta}.$$

Here $\widehat{V}_{RG1}\left(\widehat{\theta}\right)=\widehat{V}_{RG2}\left(\widehat{\theta}\right)=\dfrac{1}{k(k-1)}\displaystyle\sum_{i=1}^{k}\left(\frac{n}{m}\sum_{i\in s_i} y_i/\pi_i-\sum_{i\in s} y_i/\pi_i\right)^2$ and

$$\begin{aligned}
E\left[\widehat{V}_{RG1}\left(\widehat{\theta}\right)\right]&=\frac{1}{(k-1)}\left[V\left(\frac{n}{m}\sum_{i\in s_i} y_i/\pi_i\right)-V\left(\sum_{i\in s} y_i/\pi_i\right)\right]\\
&=\frac{1}{(k-1)}\left[E\left\{V\left(\frac{n}{m}\sum_{i\in s_i} y_i/\pi_i|s\right)\right\}\right.\\
&\qquad\left.+V\left\{E\left(\frac{n}{m}\sum_{i\in s_i} y_i/\pi_i|s\right)\right\}-V\left(\sum_{i\in s} y_i/\pi_i\right)\right]\\
&=\frac{1}{(k-1)}E\left[V\left(\frac{n}{m}\sum_{i\in s_i} y_i/\pi_i|s\right)\right]\\
&=\frac{1}{(k-1)}E\left[n^2\left(\frac{1}{m}-\frac{1}{n}\right)\frac{1}{n-1}\left\{\sum_{i\in s} y_i^2/\pi_i^2-\frac{1}{n}\left(\sum_{i\in s} y_i/\pi_i\right)^2\right\}\right]\\
&=\frac{n}{n-1}\left[\sum_{i\in U} y_i^2/\pi_i-\frac{Y^2}{n}-\frac{1}{n}V\left(\sum_{i\in s} y_i/\pi_i\right)\right]
\end{aligned}$$

The amount of bias of the estimator $\widehat{V}_{RG1}\left(\widehat{\theta}\right)$ for estimation of $V\left(\widehat{\theta}\right)$ is given by

$$
\begin{aligned}
B\left[\widehat{V}_{RG1}\left(\widehat{\overline{\theta}}\right)\right] &= E\left[\widehat{V}_{RG1}\left(\widehat{\overline{\theta}}\right)\right] - V\left(\widehat{\theta}\right)\\
&= \frac{n}{n-1}\left[\sum_{i\in U} y_i^2/\pi_i - \frac{Y^2}{n} - V\left(\sum_{i\in s} y_i/\pi_i\right)\right]
\end{aligned}
\tag{18.3.3}
$$

For an inclusion probability proportional to size (IPPS or πps) sampling design $\pi_i = np_i$, the expression of bias (18.3.3) reduces to

$$
B\left[\widehat{V}_{RG1}\left(\widehat{\overline{\theta}}\right)\right] = \frac{n}{n-1}\left[\frac{1}{n}\sum_{i\in U} p_i\left(\frac{y_i}{p_i} - Y\right)^2 - V\left(\sum_{i\in s}\frac{y_i}{\pi_i}\right)\right] \tag{18.3.4}
$$

The expression $\frac{1}{n}\sum_{i\in U} p_i\left(\frac{y_i}{p_i} - Y\right)^2$ is the variance of the Hansen–Hurwitz estimator based on a sample of size n selected by probability proportional to size with replacement sampling (PPSWR) sampling scheme, which is expected to be higher than the variance of the Horvitz–Thompson estimator based on a πps design of the same sample size n. Hence for a πps sampling design $\widehat{V}_{RG1}\left(\widehat{\overline{\theta}}\right)$, i.e., $\widehat{V}_{RG2}\left(\widehat{\theta}\right)$ over-estimates the variance $V\left(\widehat{\theta}\right)$ in general.

18.3.4 Multistage Sampling

In the BAIS II survey (Botswana HIV/AIDS Impact Survey II) each district was divided into a number of enumeration areas. At first, a sample of s of n enumeration areas of a district is selected from a total of N enumeration areas by the IPPS (or πps) sampling scheme using Goodman and Kish (1950) sampling taking number of households x_i as measure of size for the ith enumeration area. Here, $\pi_i = np_i$, $\left(p_i = x_i/X, X = \sum_{i\in U} x_i\right)$ is the inclusion probability of the ith enumeration area. If the ith enumeration area is included in the sample, a subsample s_i of m_i households is selected from the M_i households by a systematic sampling procedure. Let y_{ij} be the value of the variate of interest for the jth household of the ith enumeration area, then the population total is $Y = \sum_{i \in U} Y_i$, where $Y_i = \sum_{j=1}^{M_i} y_{ij}$. An unbiased estimator for the population total Y is

$$\widehat{\theta} = \sum_{i \in s} \frac{\widehat{Y}_i}{\pi_i} \tag{18.3.5}$$

where $\widehat{Y}_i = M_i \overline{y}(s_i)$ and $\overline{y}(s_i) = \sum_{j \in s_i} y_{ij}/m_i$. The variance of $\widehat{Y}$ is

$$\begin{aligned} V\left(\widehat{\theta}\right) &= V\left[\left(E\sum_{i \in s}\frac{\widehat{Y}_i}{\pi_i}\bigg|s\right)\right] + E\left[V\left(\sum_{i \in s}\frac{\widehat{Y}_i}{\pi_i}\bigg|s\right)\right] \\ &= V\left(\sum_{i \in s}\frac{Y_i}{\pi_i}\right) + E\left(\sum_{i \in s}\frac{\sigma_i^2}{\pi_i^2}\right) \end{aligned} \tag{18.3.6}$$

(where σ_i^2 is the variance of $\widehat{Y}_i$ given s).

$$= \frac{1}{2}\sum_{i \neq}\sum_{j \in U}(\pi_i\pi_j - \pi_{ij})\left(\frac{Y_i}{\pi_i} - \frac{Y_j}{\pi_j}\right)^2 + \sum_{i \in U}\frac{\sigma_i^2}{\pi_i}$$

The variance $V\left(\widehat{\theta}\right)$ cannot be estimated unbiasedly because an unbiased estimator of the variance σ_i^2, based on a single systematic sample, is not available. However, if we divide the sample into n groups, taking only one enumeration unit in a group and $\widehat{\theta}_{(i)} = \frac{\hat{Y}_i}{p_i}$ be an estimator of the total Y based on the ith enumeration area, then the combined estimator for θ is

$$\widehat{\widehat{\theta}} = \frac{1}{n}\sum_{i \in s}\frac{\widehat{Y}_i}{p_i} = \widehat{\theta} \tag{18.3.7}$$

The RG variance estimator of $V\left(\widehat{\theta}\right)$ is given by

$$\widehat{V}_{RG1}\left(\widehat{\theta}\right) = \widehat{V}_{RG2}\left(\widehat{\theta}\right) = \frac{1}{n(n-1)}\sum_{i \in s}\left(\frac{\widehat{Y}_i}{p_i} - \widehat{\theta}\right)^2 \tag{18.3.8}$$

$$\begin{aligned}
\text{The bias of } \widehat{V}_{RG1}\left(\widehat{\theta}\right) &= E\left[\widehat{V}_{RG1}\left(\widehat{\theta}\right)\right] - V\left(\widehat{\theta}\right)\\
&= \frac{1}{n(n-1)} E\left[\sum_{i\in s}\left(\frac{\widehat{Y}_i}{p_i}\right)^2 - n\widehat{\theta}^2\right] - V\left(\widehat{\theta}\right)\\
&= \frac{1}{n(n-1)}\left[E\left(\sum_{i\in s}\frac{Y_i^2+\sigma_i^2}{p_i^2}\right) - n\left(V\left(\widehat{\theta}\right)+Y^2\right)\right] - V\left(\widehat{\theta}\right)\\
&= \frac{1}{n(n-1)}\left[\sum_{i\in U}\frac{Y_i^2+\sigma_i^2}{p_i^2}\pi_i - n\left(V\left(\widehat{\theta}\right)+Y^2\right)\right] - V\left(\widehat{\theta}\right)\\
&= \frac{1}{(n-1)}\left[\left(\sum_{i\in U}\frac{Y_i^2}{p_i} - Y^2\right) + \sum_{i\in U}\frac{\sigma_i^2}{p_i} - V\left(\widehat{\theta}\right)\right]\\
&\qquad (\text{since } \pi_i = np_i)\\
&= \frac{n}{(n-1)}\left[\frac{1}{n}\left(\sum_{i\in U}\frac{Y_i^2}{p_i} - Y^2\right) - V\left(\sum_{i\in s}\frac{Y_i}{\pi_i}\right)\right]
\end{aligned} \tag{18.3.9}$$

Because $V\left(\sum_{i\in s}\frac{Y_i}{\pi_i}\right)$, the variance of the Horvitz–Thompson estimator based on a πps sampling scheme, is expected to be smaller than $\frac{1}{n}\left(\sum_{i\in U}\frac{Y_i^2}{p_i} - Y^2\right)$, the variance of the Hansen–Hurwitz estimator based on the PPSWR sampling scheme of the same size, the RG estimator $\widehat{V}_{RG1}\left(\widehat{\theta}\right)\left(\widehat{V}_{RG2}\left(\widehat{\theta}\right)\right)$ is expected to overestimate the variance.

18.3.5 Numerical Example

A sample s of size 15 is selected from a population of 50 households by the SRSWOR method. The medical expenditures (y) and family sizes (x) of households are given in the following Table 18.3.1. Our objective is to estimate the average medical consumption per household given that the average household size of the population is 3.5.

Table 18.3.1 Medical Expenditures and family size of the sampled households

Households	**1**	**2**	**3**	**4**	**5**	**6**	**7**	**8**
y	1500	6000	4500	4000	8000	6800	9750	8800
x	1	2	2	2	4	2	3	4
Households	**9**	**10**	**11**	**12**	**13**	**14**	**15**	
y	7620	7500	4500	5000	6500	7500	4500	
x	3	3	4	3	3	4	2	

Suppose that the sample s is divided at random into three groups each of size 5 as follows; $s_1 = (1,2,3,4,5)$, $s_2 = (6,7,8,9,10)$, and $s_3 = (11,12,13,14,15)$.

(i) **Estimates without using auxiliary information**

The estimator of the population mean medical expenditure per household $= \theta = \overline{Y}$ is $\widehat{\theta} = \overline{y}(s) = \sum_{i\in s} y_i/15 = 6164.667$ and an unbiased estimate of the variance $V\left(\widehat{\theta}\right)$ is $\widehat{V}\left(\widehat{\theta}\right) = \left(\frac{1}{n} - \frac{1}{N}\right)s_y^2 = 2,18,367.244$ (taking $n = 15$, $N = 50$, and $s_y^2 = \frac{1}{n-1}\sum_{i\in s}\{y_i - \overline{y}(s)\}^2 = 46,79,298.095$). To use RG method, the three group estimates are computed as $\widehat{\theta}_{(1)} = \overline{y}(s_1) = 4800$, $\widehat{\theta}_{(2)} = \overline{y}(s_2) = 8094$, and $\widehat{\theta}_{(3)} = \overline{y}(s_3) = 5600$. The RG estimator for θ is $\widehat{\overline{\theta}} = \frac{4800 + 8094 + 5600}{3} = 6164.667 = \widehat{\theta}$. The estimator of $V\left(\widehat{\theta}\right)$ by RG method is $\widehat{V}_{RG1}\left(\widehat{\theta}\right) = \widehat{V}_{RG2}\left(\widehat{\theta}\right) = \frac{1}{3\times 2}\sum_{i=1}^{3}\left(\widehat{\theta}_{(i)} - \widehat{\overline{\theta}}\right)^2$ $= 9,83,918.111$, which is much larger than $\widehat{V}\left(\widehat{\theta}\right) = 2,18,367.2$. This is because the estimated amount of bias of $\widehat{V}_{RG1}\left(\widehat{\theta}\right) = S_y^2\big/N = 93,585.961$ is large.

(ii) **Estimates using auxiliary information**

The ratio estimator for the average expenditure $\theta = \overline{Y}$ is $\widehat{\theta} = \frac{\overline{y}(s)}{\overline{x}(s)}\overline{X} = \frac{6164.67}{2.8} \times 3.5 = 7705.833$. The conventional estimate of variance of $\widehat{\theta}$ based on LR method is $\widehat{V}\left(\widehat{\theta}\right) = (1/n - 1/N)s_d^2 =$ $(1/15 - 1/60) \times 31,55,850 = 1,47,273$. Ratio estimates based on the three groups are $\widehat{\theta}_{(1)} = 4800 \times 3.5\big/2.2 = 7636.36$, $\widehat{\theta}_{(2)} = 8094 \times 3.5\big/3$ $= 9443.000$, and $\widehat{\theta}_{(3)} = 5600 \times 3.5\big/3.2 = 6125.000$. The RG estimate of θ is $\widehat{\overline{\theta}} = 7734.788$, which is different from $\widehat{\theta} = 7705.83$. The two estimates of the variance of $\widehat{\theta}$ are $\widehat{V}_{RG1}\left(\widehat{\theta}\right) = \frac{1}{3\times 2}\sum_{i=1}^{3}\left(\widehat{\theta}_{(i)} - \widehat{\overline{\theta}}\right)^2 = 9,19,848.433$ and $\widehat{V}_{RG2}\left(\widehat{\theta}\right) = \frac{1}{3\times 2}\sum_{i=1}^{3}\left(\widehat{\theta}_{(i)} - \widehat{\theta}\right)^2 = 9,20,268.016$.

18.4 JACKKNIFE METHOD

The most widely used method of estimation of variance in a complex survey design is the Jackknife (JK) method. The JK method was introduced by

Quenouille (1949) for reduction of bias of an estimator of a serial correlation coefficient. Quenouille (1956) extended this technique for bias reduction, in general, under infinite population setup. Tuckey (1958) used the JK technique for estimation of variance under an infinite population setup while Durbin (1959) proposed the JK method for estimation of variance in finite population sampling. Good details of the JK method are given by Gray and Schucany (1972), Miller (1974), Efron (1982), and Wolter (1985), among others.

18.4.1 Jackknife Method for an Infinite Population

Let a random sample s of size n be selected from an infinite population with distribution function $\mathcal{F}$ and let $\widehat{\theta}(s) = \widehat{\theta}$ be an estimator for the population parameter θ based on the full sample s. The sample s is partitioned into k disjoined groups, $s_1,\ldots,s_k$ each of size $m(=n/k$, assuming an integer). Let $\widehat{\theta}_{(-j)}$ be an estimator of θ, which is the same functional form as $\widehat{\theta}$, but based on the reduced sample $s - s_j$ of size $n - m$, obtained by deleting the sample s_j from s. Let us define the pseudovalue computed from $s - s_j$ as

$$\widehat{\theta}^{(j)} = k\widehat{\theta} - (k-1)\widehat{\theta}_{(-j)} \tag{18.4.1}$$

The Jackknife (JK) estimator for θ is defined as the average of the pseudovalues $\widehat{\theta}^{(j)}$, and is denoted by

$$\begin{aligned}\widehat{\theta}_J &= \frac{1}{k}\sum_{j=1}^{k}\widehat{\theta}^{(j)} \\ &= k\widehat{\theta} - (k-1)\ \widehat{\theta}_{\bullet}\end{aligned} \tag{18.4.2}$$

where

$$\widehat{\theta}_{\bullet} = \frac{1}{k}\sum_{j=1}^{k}\widehat{\theta}_{(-j)} \tag{18.4.3}$$

The JK variance estimators of the variances of both the estimators of $\widehat{\theta}$ and $\widehat{\theta}_J$ are given by

$$\begin{aligned}\widehat{V}_J(1) &= \frac{1}{k(k-1)}\sum_{j=1}^{k}\left(\widehat{\theta}^{(j)} - \widehat{\theta}_J\right)^2 \\ &= \frac{(k-1)}{k}\sum_{j=1}^{k}\left(\widehat{\theta}_{(-j)} - \widehat{\theta}_{\bullet}\right)^2\end{aligned} \tag{18.4.4}$$

An alternative JK variance estimator is given by

$$\widehat{V}_J(2) = \frac{1}{k(k-1)} \sum_{j=1}^{k} \left(\widehat{\theta}^{(j)} - \widehat{\theta}\right)^2 \tag{18.4.5}$$

The estimator $\widehat{V}_J(2)$ is conservative in the sense $\widehat{V}_J(2) \geq \widehat{V}_J(1)$.

Theorem 18.4.1

If $B\left(\widehat{\theta}\right)$, the bias of $\widehat{\theta}$ is of order $1/n$ and can be expressed as

$$B\left(\widehat{\theta}\right) = \frac{b_1(\theta)}{n} + \frac{b_2(\theta)}{n^2} + \cdots$$

then the bias of $\widehat{\theta}_J$ is of order $1/n^2$ and is given by

$$B\left(\widehat{\theta}_J\right) = \frac{c_2(\theta)}{n^2} + \frac{c_3(\theta)}{n^3} + \cdots$$

where the constants $b_1(\theta)$, $b_2(\theta)$, $b_3(\theta)$,... and $c_2(\theta)$, $c_3(\theta)$,... are free from n but may depend on θ.

Proof

Given $E\left(\widehat{\theta}\right) = \theta + B\left(\widehat{\theta}\right) = \theta + \dfrac{b_1(\theta)}{n} + \dfrac{b_2(\theta)}{n^2} + \cdots$

Hence $E\left(\widehat{\theta}_{(-j)}\right) = \theta + \dfrac{b_1(\theta)}{m(k-1)} + \dfrac{b_2(\theta)}{\{m(k-1)\}^2} + \cdots$

and

$$\begin{aligned} E\left(\widehat{\theta}^{(j)}\right) &= kE\left(\widehat{\theta}\right) - (k-1)E\left(\widehat{\theta}_{(-j)}\right) \\ &= k\left(\theta + \frac{b_1(\theta)}{n} + \frac{b_2(\theta)}{n^2} + \cdots\right) \\ &\quad - (k-1)\left(\theta + \frac{b_1(\theta)}{m(k-1)} + \frac{b_2(\theta)}{\{m(k-1)\}^2} + \cdots\right). \end{aligned}$$

The bias of $\widehat{\theta}_J$ is

$$\begin{aligned} B\left(\widehat{\theta}_J\right) &= E\left(\widehat{\theta}_J\right) - \theta \\ &= \frac{1}{k}\sum_{j=1}^{k} E\left(\widehat{\theta}^{(j)}\right) - \theta \\ &= \frac{c_2(\theta)}{n^2} + \cdots \end{aligned}$$

where $c_2(\theta) = -kb_2(\theta)/(k-1)$.

Remark 18.4.1

As for the optimum number of the group k that maximizes efficiency of the JK estimator, no general rule is available. However, studies by Rao and Webster (1966), Chakrabarty and Rao (1968), and Rao and Rao (1971) reveal that for ratio estimation, the bias and variance of the JK estimator decreases with an increase in k when the sample size n is small or moderate. More detailed discussions are given by Wolter (1985).

Example 18.4.1.1

Let $X_1, X_2,\ldots,X_n$ be a random sample from a population with mean μ and variance σ^2. Consider a biased estimator for $\theta = \sigma^2$ as

$$\widehat{\theta} = \frac{1}{n}\sum_{i=1}^{n}\left(X_i - \overline{X}\right)^2 = \frac{1}{2n^2}\sum_{i=1}^{n}\sum_{j=1}^{n}(X_i - X_j)^2.$$

The bias of $\widehat{\theta}$ is $B\left(\widehat{\theta}\right) = -\sigma^2/n$. Let $m = 1$ (i.e., $k = n$), then

$$\begin{aligned}
\widehat{\theta}_{(-j)} &= \frac{1}{n-1}\sum_{\substack{i=1\\ i\neq j}}^{n}\left(X_i - \frac{n\overline{X} - X_j}{n-1}\right)^2\\
&= \frac{1}{n-1}\sum_{\substack{i=1\\ i\neq j}}^{n}\left(X_i - \overline{X} - \frac{\overline{X} - X_j}{n-1}\right)^2\\
&= \frac{1}{n-1}\left[\sum_{i=1}^{n}\left(X_i - \overline{X} - \frac{\overline{X} - X_j}{n-1}\right)^2 - \left(\frac{n}{n-1}\right)^2\left(X_j - \overline{X}\right)^2\right]\\
&= \frac{1}{n-1}\left[\sum_{i=1}^{n}\left(X_i - \overline{X}\right)^2 + \frac{n}{(n-1)^2}\left(X_j - \overline{X}\right)^2 - \left(\frac{n}{n-1}\right)^2\left(X_j - \overline{X}\right)^2\right]\\
&= \frac{1}{n-1}\left[n\widehat{\theta} - \frac{n}{n-1}\left(X_j - \overline{X}\right)^2\right],
\end{aligned}$$

$$\begin{aligned}
\widehat{\theta}_{\bullet} &= \frac{1}{n}\sum_{j=1}^{n}\widehat{\theta}_{(-j)} = \frac{1}{n-1}\left[n\widehat{\theta} - \frac{1}{n-1}\sum_{j=1}^{n}\left(X_j - \overline{X}\right)^2\right]\\
&= \frac{n(n-2)}{(n-1)^2}\widehat{\theta}
\end{aligned}$$

and

$$\widehat{\theta}_J = n\widehat{\theta} - (n-1)\widehat{\theta}.$$
$$= \frac{n}{n-1}\widehat{\theta}$$
$$= \frac{1}{n-1}\sum_{i=1}^{n}(X_i - \overline{X})^2$$

Clearly, $\widehat{\theta}_J$ is an unbiased estimator of θ.

Now,

$$\sum_{j=1}^{n}\left(\widehat{\theta}_{(-j)} - \widehat{\theta}.\right)^2 = \frac{n^2}{(n-1)^4}\sum_{j=1}^{n}\left((X_j - \overline{X})^2 - \sum_{j=1}^{n}(X_j - \overline{X})^2/n\right)^2$$
$$= \frac{n^2}{(n-1)^4}\left[\sum_{j=1}^{n}(X_j - \overline{X})^4 - \left\{\sum_{j=1}^{n}(X_j - \overline{X})^2\right\}^2 \Big/ n\right]$$
$$= \frac{n^3}{(n-1)^4}\left(m_4 - m_2^2\right)$$

where $m_4 = \frac{1}{n}\sum_{i=1}^{n}\left(X_i - \overline{X}\right)^4$ and $m_2 = \frac{1}{n}\sum_{i=1}^{n}\left(X_i - \overline{X}\right)^2$.

The JK variance estimator of $\widehat{\theta}_J$ is

$$\widehat{V}_J(1) = \frac{n-1}{n}\sum_{j=1}^{n}\left(\widehat{\theta}_{(-j)} - \widehat{\theta}.\right)^2$$
$$= \frac{n^2}{(n-1)^3}\left(m_4 - m_2^2\right)$$

Example 18.4.1.2

Let a random sample $X_1,\ldots,X_n$ of size n be selected from a uniform distribution with density $f(x,\theta) = 1/\theta;\ 0 < x < \theta$. The nth order statistic $X_{(n)}$ is the maximum likelihood estimator of θ. The estimator $\widehat{\theta} = X_{(n)}$ is a biased estimator of θ. Noting $E\{X_{(n)}\} = n\theta/(n+1)$, we find the bias of $\widehat{\theta}\left(= X_{(n)}\right)$ as $B\left(\widehat{\theta}\right) = E\left\{X_{(n)}\right\} - \theta = -\theta/(n+1)$, which is of order n^{-1}. Now,

$$\widehat{\theta}_{(-j)} = \begin{cases} X_{(n)} & \text{for } j = 1, \ldots, n-1 \\ X_{(n-1)} & \text{for } j = n \end{cases}$$

where $X_{(n-1)}$ is the $(n-1)$th order statistic.

Furthermore, noting $\widehat{\theta}_{\bullet} = \frac{1}{n}\sum_{j=1}^{n}\widehat{\theta}_{(-j)} = \frac{(n-1)X_{(n)} + X_{(n-1)}}{n}$, we find the JK estimator of θ is

$$\begin{aligned}\widehat{\theta}_J &= n\widehat{\theta} - (n-1)\widehat{\theta}_{\bullet} \\ &= X_{(n)} + \frac{n-1}{n}\left(X_{(n)} - X_{(n-1)}\right)\end{aligned}$$

Noting $E\{X_{(n-1)}\} = (n-1)\theta/(n+1)$, the bias of θ_J is obtained as

$$B\left(\widehat{\theta}_J\right) = E\left(\widehat{\theta}_J\right) - \theta = -\frac{\theta}{n(n+1)} \text{ is of order } n^{-2}.$$

The JK variance estimator of $\widehat{\theta}_J$ is

$$\begin{aligned}\widehat{V}_J(1) &= \frac{n-1}{n}\sum_{j=1}^{n}\left(\widehat{\theta}_{(-j)} - \widehat{\theta}_{\bullet}\right)^2 \\ &= \frac{n-1}{n}\left[(n-1)\left\{X_{(n)} - \frac{(n-1)X_{(n)} + X_{(n-1)}}{n}\right\}^2 \right. \\ &\quad \left. + \left\{X_{(n-1)} - \frac{(n-1)X_{(n)} + X_{(n-1)}}{n}\right\}^2\right] \\ &= \frac{(n-1)^2}{n^2}\left(X_{(n)} - X_{(n-1)}\right)^2 \\ &\cong \left(X_{(n)} - X_{(n-1)}\right)^2 \text{ for large } n.\end{aligned}$$

18.4.1.1 Higher-Order Jackknife Estimator

Let $\widehat{\theta}_{J(-j)}$ be the estimator of θ based on the sample size $n-1$ obtained from $\widehat{\theta}_J$ by deleting jth observation. The second-order JK estimator of θ was proposed by Quenouille (1956) as

$$\begin{aligned}\widehat{\theta}_J^{(2)} &= \frac{n^2\widehat{\theta}_J - \frac{(n-1)^2}{n}\sum_{j=1}^{n}\widehat{\theta}_{J(-j)}}{n^2 - (n-1)^2} \\ &= \frac{1}{(2n-1)}\left[n^3\widehat{\theta} - (n-1)\{2n(n-1)+1\}\widehat{\theta}_{\bullet} \right. \\ &\quad \left. + (n-1)^2(n-2)\left\{\frac{1}{n(n-1)}\sum_{i\neq}^{n}\sum_{j=1}^{n}\widehat{\theta}_{-i,j}\right\}\right] \qquad (18.4.6)\end{aligned}$$

where $\widehat{\theta}_{-i,j}$ is the same functional as $\widehat{\theta}$ obtained from the sample size $n-2$ after deleting the ith and $j(\neq i)$th observations from the original sample.

The second-order JK estimator $\widehat{\theta}_J^{(2)}$ eliminates the order $1/n^2$ term from the bias. Similarly, the higher-order JK estimators are defined to eliminate third-, fourth-, and higher-order bias terms.

18.4.1.2 Generalized Jackknife Estimator

Schucany et al. (1971) generalized the biased reduction procedure further. Consider two biased estimators $\widehat{\theta}_1$ and $\widehat{\theta}_2$ of θ with

$$E\left(\widehat{\theta}_1\right) = \theta + f_1(n)a(\theta),$$
$$E\left(\widehat{\theta}_2\right) = \theta + f_2(n)a(\theta)$$

and

$$\begin{vmatrix} 1 & 1 \\ f_1(n) & f_2(n) \end{vmatrix} \neq 0.$$

The generalized JK estimator (defined by Wolter, 1985) as

$$GJ\left(\widehat{\theta}_1, \widehat{\theta}_2\right) = \frac{\begin{vmatrix} \widehat{\theta}_1 & \widehat{\theta}_2 \\ f_1(n) & f_2(n) \end{vmatrix}}{\begin{vmatrix} 1 & 1 \\ f_1(n) & f_2(n) \end{vmatrix}} \tag{18.4.7}$$

is strictly unbiased for θ.

If we take $\widehat{\theta}_1 = \widehat{\theta}$, $\widehat{\theta}_2 = \frac{1}{n}\sum_{j=1}^{n}\widehat{\theta}_{(-j)} = \widehat{\theta}_\bullet$, $f_1(n) = 1/n$, and $f_2(n) = 1/(n-1)$, the $GJ\left(\widehat{\theta}_1, \widehat{\theta}_2\right)$ reduces to $\widehat{\theta}_J$.

More general, consider $p+1$ estimators of θ based on a sample size n such that

$$B\left(\widehat{\theta}_j\right) = E\left(\widehat{\theta}_j\right) - \theta = f_{1j}(n)b_1(\theta) + \cdots + f_{kj}(n)b_k(\theta) + \cdots$$

for $j = 1, 2, \ldots, p+1$

$$\text{with} \quad \begin{vmatrix} 1 & \cdot & \cdot & 1 \\ f_{11}(n) & \cdot & \cdot & f_{1p+1}(n) \\ \cdot & \cdot & \cdot & \cdot \\ \cdot & & & \cdot \\ f_{p1}(n) & \cdot & \cdot & f_{pp+1}(n) \end{vmatrix} \neq 0. \tag{18.4.8}$$

Schucany et al. (1971) showed that the generalized JK estimator

$$GJ\left(\widehat{\theta}_1, \ldots, \widehat{\theta}_{p+1}\right) = \frac{\begin{vmatrix} \widehat{\theta}_1 & \cdot & \cdot & \widehat{\theta}_{p+1} \\ f_{11}(n) & \cdot & \cdot & f_{1p+1}(n) \\ \cdot & \cdot & \cdot & \cdot \\ \cdot & \cdot & \cdot & \cdot \\ f_{p1}(n) & \cdot & \cdot & f_{pp+1}(n) \end{vmatrix}}{\begin{vmatrix} 1 & \cdot & \cdot & 1 \\ f_{11}(n) & \cdot & \cdot & f_{1p+1}(n) \\ \cdot & \cdot & \cdot & \cdot \\ \cdot & \cdot & \cdot & \cdot \\ f_{p1}(n) & \cdot & \cdot & f_{pp+1}(n) \end{vmatrix}} \quad (18.4.9)$$

eliminates the first p terms of the bias of Eq. (18.4.8).

18.4.2 Jackknife Method for a Finite Population

Here a sample s of size n is selected from a finite population using some sampling design. The sample s is divided at random into k groups each of size m assuming $n/m = k$ is an integer. The formulae $\widehat{\theta}_{(-j)}$, $\widehat{\theta}^{(j)}$, $\widehat{\theta}_J$, $\widehat{\theta}_\bullet$, and the JK variance estimators $\widehat{V}_J(1)$ and $\widehat{V}_J(2)$ used for infinite population in Section 18.4.1 remain the same for the finite population.

18.4.2.1 Probability Proportional to Size With Replacement Sampling

Let a sample s of size n be selected from a finite population U of N units by PPSWR method using $p_i\left(>0, \sum_{i\in U} p_i = 1\right)$ as the normed size measure for the ith unit. The Hansen–Hurwitz estimator for the population total Y is

$$\widehat{\theta} = \widehat{Y}_{hh} = \frac{1}{n}\sum\nolimits_{i\in s} \frac{y_i}{p_i} \quad (18.4.10)$$

where $\sum_{i\in s}$ denotes the sum over units in s with repetition.

The expressions for the variance of $\widehat{\theta}$ and its unbiased estimators are, respectively, given by $V\left(\widehat{\theta}\right) = \frac{1}{n}\sum_{i\in U} p_i\left(\frac{y_i}{p_i} - Y\right)^2$ and $\widehat{V}\left(\widehat{\theta}\right) = \frac{1}{n(n-1)}\sum_{i\in s}\left(\frac{y_i}{p_i} - \widehat{Y}_{hh}\right)^2$.

Let the sample s be divided at random into k groups each of size m. The estimator for the total Y based on the sample after deleting the jth group (i.e., based on $s - s_j$) is given by $\widehat{\theta}_{(-j)} = \dfrac{1}{n-m}\sum_{i\in s-s_j}\dfrac{y_i}{p_i}$. The pseudovalue computed from $s - s_j$ is

$$\widehat{\theta}^{(j)} = k\widehat{\theta} - \frac{k-1}{n-m}\sum_{i\in s-s_j}\frac{y_i}{p_i}$$

Hence the JK estimator for Y is given by

$$\begin{aligned}\widehat{\theta}_J &= \frac{1}{k}\sum_{j=1}^{k}\widehat{\theta}^{(j)} = k\widehat{\theta} - \frac{k-1}{k(n-m)}\sum_{j=1}^{k}\sum_{i\in s-s_j}\frac{y_i}{p_i}\\ &= k\widehat{\theta} - \frac{k-1}{k(n-m)}\sum_{j=1}^{k}\left(\sum_{i\in s}\frac{y_i}{p_i} - \sum_{i\in s_j}\frac{y_i}{p_i}\right) = \widehat{\theta} = \widehat{\theta}.\end{aligned} \quad (18.4.11)$$

The JK variance estimator is given by

$$\begin{aligned}\widehat{V}_J(1) = \widehat{V}_J(2) &= \frac{k-1}{k}\sum_{j=1}^{k}\left(\widehat{\theta}_{(-j)} - \widehat{\theta}.\right)^2\\ &= \frac{k-1}{k}\sum_{j=1}^{k}\left\{\frac{1}{n-m}\sum_{i\in s-s_j}\left(\frac{y_i}{p_i} - \widehat{Y}_{hh}\right)\right\}^2\end{aligned} \quad (18.4.12)$$

Clearly, the JK variance estimator $\widehat{V}_J(1)$ (i.e., $\widehat{V}_J(2)$) is not equal to $\widehat{V}\left(\widehat{\theta}\right)$ in general. However, for $m = 1$, i.e., $k = n$, we get

$$\begin{aligned}\widehat{V}_J(1) = \widehat{V}_J(2) &= \frac{1}{n(n-1)}\sum_{i\in s}\left(\frac{y_i}{p_i} - \widehat{Y}_{hh}\right)^2\\ &= \widehat{V}\left(\widehat{\theta}\right)\end{aligned} \quad (18.4.13)$$

18.4.2.1.1 Bias of Jackknife Variance Estimator

$$\begin{aligned}E\{\widehat{V}_J(1)\} &= \frac{k-1}{k}\left\{\sum_{j=1}^{k}E\left(\widehat{\theta}_{(-j)}\right)^2 - k\left(\widehat{Y}_{hh}\right)^2\right\}\\ &= \frac{k-1}{k}\left\{k\left(Y^2 + \frac{\sigma_0^2}{n-m}\right) - k\left(Y^2 + \frac{\sigma_0^2}{n}\right)\right\}\\ &\qquad \left(\text{where } \sigma_0^2 = \sum_{i\in U}p_i\left(\frac{y_i}{p_i} - Y\right)^2\right)\\ &= \frac{1}{n}\sum_{i\in U}p_i\left(\frac{y_i}{p_i} - Y\right)^2 = V\left(\widehat{\theta}\right)\end{aligned}$$

Thus for PPSWR sampling the JK estimator is unbiased for $V\left(\widehat{\theta}\right)$.

18.4.2.2 Simple Random Sampling With Replacement

PPSWR reduces to simple random sampling (SRSWR) if $p_i = 1/N$ for $\forall\ i \in U$. Hence the JK estimator $\widehat{\theta}_J$ for Y is equal to the full sample estimator $N\,\overline{y}(s)$.

18.4.2.3 Inclusion Probability Proportional to Size or πps Sampling Design

Suppose a sample s of size n is selected from a population with $\pi_i = np_i$ as the inclusion probability for the ith unit with $p_i\left(>0, \sum_{i\in U} p_i = 1\right)$ as the normed size measure for the ith unit. The Horvitz–Thompson estimator for the total Y based on the full sample s is given by

$$\widehat{\theta} = \widehat{Y}_{ht} = \sum_{i\in s}\frac{y_i}{\pi_i} = \frac{1}{n}\sum_{i\in s}\frac{y_i}{p_i} \tag{18.4.14}$$

The variance of $\widehat{\theta}$ and its unbiased estimator are, respectively, given by

$$V\left(\widehat{\theta}\right) = \frac{1}{2}\sum_{i\neq}\sum_{j\in U}\left(\pi_i\pi_j - \pi_{ij}\right)\left(\frac{y_i}{\pi_i} - \frac{y_j}{\pi_j}\right)^2 \tag{18.4.15}$$

and

$$\widehat{V}\left(\widehat{\theta}\right) = \frac{1}{2}\sum_{i\neq}\sum_{j\in s}\frac{\pi_i\pi_j - \pi_{ij}}{\pi_{ij}}\left(\frac{y_i}{\pi_i} - \frac{y_j}{\pi_j}\right)^2 \tag{18.4.16}$$

The sample s is partitioned at random in to k groups each of size m ($n = mk$). The Horvitz–Thompson estimator for the total Y based on the sample $s - s_j$ is given by

$$\widehat{\theta}_{(-j)} = \frac{1}{n-m}\sum_{i\in s-s_j}\frac{y_i}{p_i}$$

The ith pseudovalue is

$$\begin{aligned}\widehat{\theta}^{(j)} &= k\sum_{i\in s}\frac{y_i}{np_i} - (k-1)\sum_{i\in s-s_j}\frac{y_i}{(n-m)p_i}\\ &= k\sum_{i\in s}\frac{y_i}{np_i} - k\left(\sum_{i\in s}\frac{y_i}{np_i} - \sum_{i\in s_j}\frac{y_i}{np_i}\right)\\ &= k\sum_{i\in s_j}\frac{y_i}{np_i}\end{aligned}$$

The JK estimator of θ is the same as the original estimator $\widehat{Y}_{hte}$ because

$$\begin{aligned}\widehat{\theta}_J &= \frac{1}{k}\sum_{j=1}^{k}\widehat{\theta}^{(j)}\\ &= \sum_{i\in s}\frac{y_i}{np_i}\\ &= \widehat{\theta} = \widehat{Y}_{ht} = \widehat{\theta}.\end{aligned} \tag{18.4.17}$$

The JK variance estimator is given by

$$\widehat{V}_J(1) = \widehat{V}_J(2) = \frac{k-1}{k}\sum_{j=1}^{k}\left(\widehat{\theta}_{(-j)} - \widehat{\theta}.\right)^2 \tag{18.4.18}$$

The estimators $\widehat{V}_J(1)\left(\text{ i.e., } \widehat{V}_J(2)\right)$ are not equal to $\widehat{V}\left(\widehat{\theta}\right)$ in general.

18.4.2.3.1 Bias of Jackknife Variance Estimator

The bias of the JK estimator is given by

$$B\left[\widehat{V}_J(1)\right] = E\left\{\widehat{V}_J(1)\right\} - V\left(\widehat{Y}_{ht}\right) \tag{18.4.19}$$

Now,

$$\begin{aligned}E\left[\widehat{V}_J(1)\right] &= \frac{k-1}{k}E\left[\sum_{j=1}^{k}\left(\widehat{\theta}_{(-j)}\right)^2 - k\left(\widehat{Y}_{ht}\right)^2\right]\\ &= \frac{k-1}{k}\left[E\left\{\sum_{j=1}^{k}V\left(\widehat{\theta}_{(-j)}\middle|s\right)\right\}\right]\\ &= (k-1)\left(\frac{1}{n-m} - \frac{1}{n}\right)\frac{1}{n-1}E\left(\sum_{i\in s}\frac{y_i^2}{p_i^2} - n\widehat{Y}_{ht}^2\right)\\ &= \frac{1}{n-1}\left[\sum_{i\in U}p_i\left(\frac{y_i}{p_i} - Y\right)^2 - V\left(\widehat{Y}_{ht}\right)\right]\end{aligned} \tag{18.4.20}$$

Substituting Eq. (18.4.20) in Eq. (18.4.19) we get

$$B\left[\widehat{V}_J(1)\right] = \frac{n}{n-1}\left[\frac{1}{n}\sum_{i\in U}p_i\left(\frac{y_i}{p_i} - Y\right)^2 - V\left(\widehat{Y}_{ht}\right)\right] \tag{18.4.21}$$

From expression (18.4.21), we note that the JK estimator overestimates or underestimates the variance if the variance of the Hansen–Hurwitz estimator based on PPSWR sampling is more or less than the variance of Horvitz–Thompson estimator based on a πps sampling scheme of the same sample size. In practice Eq. (18.4.21) is expected to be positive because the Horvitz–Thompson estimator is expected to be more efficient than the Hansen–Hurwitz estimator for estimating total Y.

18.4.2.4 Simple Random Sampling Without Replacement

On substituting $\pi_i = n/N$ in Eq. (18.4.17), we find that the JK estimators for SRSWOR sampling scheme are equal to full sample estimate $N\overline{y}(s)$ and that the JK variance estimator overestimates the variance as variance of the sample mean based on SRSWR is larger than that of the sample mean based on SRSWOR.

18.4.2.5 Regression Estimator

Let a sample s of size n be selected by the SRSWOR method. Let y_i and x_i be the value of the study (y) and auxiliary variable (x) for the ith unit and $\overline{X} = \sum\limits_{i\in U} x_i/N$ be the population mean of x, which is assumed to be known.
The regression estimator for the population mean $\theta = \overline{Y}$ based on the full sample s is given by

$$\widehat{\theta} = \widehat{Y}_{reg}(s) = \overline{y}(s) - \widehat{\beta}(s)\left(\overline{x}(s) - \overline{X}\right) \tag{18.4.22}$$

where $\overline{x}(s)$, $\overline{y}(s)$, and $\widehat{\beta}(s) = \dfrac{\sum\limits_{i\in s} y_i x_i - n\overline{x}(s)\overline{y}(s)}{\sum\limits_{i\in s} x_i^2 - n\{\overline{x}(s)\}^2}$ are, respectively, the sample mean of x, sample mean of y, and the sample regression coefficient of y on x.

The regression estimator after deleting the group s_j is given by

$$\widehat{\theta}_{(-j)} = \overline{y}_{(-j)} - \widehat{\beta}_{(-j)}\left(\overline{x}_{(-j)} - \overline{X}\right)$$

where $\overline{x}_{(-j)}$, $\overline{y}_{(-j)}$, and $\widehat{\beta}_{(-j)} = \dfrac{\sum\limits_{i\in s-s_j}\left(y_i - \overline{y}_{(-j)}\right)\left(x_i - \overline{x}_{(-j)}\right)}{\sum\limits_{i\in s-s_j}\left(x_i - \overline{x}_{(-j)}\right)^2}$ are, respectively, the sample mean of x, sample mean of y, and sample regression coefficient obtained from $s - s_j$. Noting

$$\widehat{\theta}^{(j)} = k\left\{\overline{y}(s) - \widehat{\beta}(s)\left(\overline{x}(s) - \overline{X}\right)\right\} - (k-1)\left(\overline{y}_{(-j)} - \widehat{\beta}_{(-j)}\left(\overline{x}_{(-j)} - \overline{X}\right)\right) \tag{18.4.23}$$

we find the JK estimator of θ as

$$\widehat{\theta}_J = \frac{1}{k}\sum_{j=1}^{k}\widehat{\theta}^{(j)} \tag{18.4.24}$$

The estimator $\widehat{\theta}_J$ is quite different from $\widehat{\theta}$. The JK variance estimators

$$\widehat{V}_J(1) = \frac{1}{k(k-1)}\sum_{j=1}^{k}\left(\widehat{\theta}^{(j)} - \widehat{\theta}_J\right)^2 \text{ and}$$
$$\widehat{V}_J(2) = \frac{1}{k(k-1)}\sum_{j=1}^{k}\left(\widehat{\theta}^{(j)} - \widehat{\theta}\right)^2 \tag{18.4.25}$$

are also different.

18.4.2.6 Numerical Example

The following data give the number of patients treated and the cost of treatment per day at 15 clinics selected at random from 50 clinics by the SRSWOR method (Table 18.4.1).

Let Y and X be the total cost per day and the total number of patients in 50 clinics, respectively, and we want to estimate $R = \frac{Y}{X}$ = cost per patient based on the selected sample s of 15 (=n) clinics. Here estimated cost per patient θ, based on the full sample, is $\widehat{\theta} = \widehat{R} = \frac{\overline{y}(s)}{\overline{x}(s)} = 90.361$, where $\overline{y}(s) = 5000$ and $\overline{x}(s) = 55.333$ are the sample means of x and y, respectively. The estimated approximate variance of $\widehat{R}$ based on the full sample of size 15 is $\widehat{V}\left(\widehat{\theta}\right) = \widehat{V}\left(\widehat{R}\right) = \frac{1}{\overline{x}_s^2}\left(\frac{1}{15} - \frac{1}{50}\right)\frac{1}{14}\sum_{i\in s}\widehat{d}_i^2 = 24.390$, where $\widehat{d}_i = y_i - \widehat{R}x_i$. Let us divide the sample s into 15 (=k) groups each group consisting of just one (=m) unit. Writing $\widehat{\theta}_{(-j)} = \widehat{R}_{(-j)} = \frac{n\overline{y}(s) - y_j}{nx(s) - x_j}$ and $\widehat{\theta}^{(j)} = k\widehat{\theta} - (k-1)\widehat{\theta}_{(-j)}$, we arrive at the JK estimator for θ as $\widehat{\theta}_J = \frac{1}{k}\sum_{j=1}^{k}\widehat{\theta}^{(j)} = 89.491$.

Two alternative JK variance estimators are, respectively, given by $\widehat{V}_J(1) = \frac{1}{k(k-1)}\sum_{j=1}^{k}\left(\widehat{\theta}^{(j)} - \theta_J\right)^2 = 40.526$ and $\frac{1}{k(k-1)}\sum_{j=1}^{k}\left(\widehat{\theta}^{(j)} - \widehat{\theta}\right)^2 = 40.579$.

Here we note that the full sample estimate and JK estimate of the population ratio R are very close to each other and hence the two JK variance estimators are almost equal but the JK variance estimates are much larger than the full sample estimate. Numerical computations are shown in Table 18.4.2.

Table 18.4.1 Number of patients and cost of treatment of selected clinics

Clinic	1	2	3	4	5	6	7	8	9	10	11	12	13	14	15
No. of patients (x)	100	40	50	70	50	60	45	150	60	15	20	40	30	45	55
Cost in rands (y)	8,000	4,000	5,000	5,000	6,000	6,000	4,000	10,000	5,000	2,000	3,000	5,000	3,000	4,000	5,000

Table 18.4.2 Compution of Jackknife variance estimation

j	x	y	$n\overline{y}(s) - y_j$	$n\overline{x}(s) - x_j$	$\widehat{R}_{(-j)}$	$\widehat{\theta}^{(j)}$	$\widehat{\theta}^{(j)} - \widehat{\theta}$	$\left(\widehat{\theta}^{(j)} - \widehat{\theta}\right)^2$	$\widehat{d}_j = y_j - \widehat{R}x_j$	$\widehat{d}_j^2$
1	100	8,000	67,000	730	91.781	70.488	−19.873	394.936	−1,036.145	1,073,596.461
2	40	4,000	71,000	790	89.873	97.20000	6.839	46.772	385.542	148,642.634
3	50	5,000	70,000	780	89.744	99.006	8.645	74.736	481.928	232,254.597
4	70	5,000	70,000	760	92.105	65.952	−24.409	595.799	−1,325.301	1,756,422.741
5	50	6,000	69,000	780	88.462	116.954	26.593	707.188	1481.928	2,196,110.597
6	60	6,000	69,000	770	89.61	100.882	10.521	110.691	578.313	334,445.926
7	45	4,000	71,000	785	90.446	89.178	−1.183	1.399	−66.265	4,391.050
8	150	10,000	65,000	680	95.588	17.1900	−73.171	5,353.995	−3,554.217	12,632,458.480
9	60	5,000	70,000	770	90.909	82.696	−7.665	58.752	−421.687	177,819.926
10	15	2,000	73,000	818	89.571	101.428	11.067	122.478	644.578	4,18,480.798
11	20	3,000	72,000	810	88.889	110.976	20.618	424.978	1,192.771	1,422,702.658
12	40	5,000	70,000	790	88.608	114.91	24.549	602.653	1,385.542	1,919,726.634
13	30	3,000	72,000	800	90.000	95.422	5.061	25.614	289.187	83,611.771
14	45	4,000	71,000	785	90.446	89.178	−1.183	1.399	−66.265	4,391.050
15	55	5,000	70,000	775	90.323	90.900	0.539	0.291	30.120	907.214
Total	830	75,000				1,342.360		8,523.681		22,402,962.540

18.5 BALANCED REPEATED REPLICATION METHOD

Let a population of N units be stratified into L strata and N_h be the size of the $h(=1,\ldots,L)$th stratum. Let y_{hi} be the value of the variable of interest y for u_{hi}, the ith unit of the hth stratum, $W_h = N_h/N$, $\overline{Y}_h = \sum_{i=1}^{N_h} y_{hi}/N_h$, and $\overline{Y} = \sum_{h=1}^{L} W_h\overline{Y}_h$. Suppose from each of the strata a sample $\Im_h$ of size n_h units is selected by some suitable sampling scheme. Let $\widehat{\theta}_{st}$ be an estimator of the population parameter of interest θ based on the full sample $s = \bigcup_{h=1}^{L} \Im_h$ of size $n = \sum_h n_h$. In this section we shall consider the various methods of estimation of variance of $\widehat{\theta}_{st}$ using the BRR method. The BRR method was introduced by McCarthy (1969).

18.5.1 Stratified Sampling With $n_h = 2$

Suppose a sample $\Im_h = (u_{h1}, u_{h2})$ of size $n_h = 2$ is selected from the hth stratum by the SRSWOR method, $h = 1,\ldots,L$. Let y_{h1} and y_{h2} be the value of the study variable y for the units u_{h1} and u_{h2}, respectively. In this case, an unbiased estimator for the population mean $\overline{Y}$ is $\widehat{\overline{Y}}_{st} = \sum_{h=1}^{L} W_h\overline{y}_h$, where $\overline{y}_h = (y_{h1} + y_{h2})/2$. If the finite population correction term $f_h = n_h/N = 2/N$ is neglected for each of the strata, then the variance of $\widehat{\overline{Y}}_{st}$ reduces to

$$V\left(\widehat{\overline{Y}}_{st}\right) = \frac{1}{2}\sum_{h=1}^{L} W_h^2\, S_h^2 \tag{18.5.1}$$

where $S_h^2 = \sum_{j=1}^{N_h} \left(y_{hj} - \overline{Y}_h\right)^2 \Big/ (N_h - 1)$.

An unbiased estimator of $V\left(\widehat{\overline{Y}}_{st}\right)$ is given by

$$\begin{aligned}\widehat{V}\left(\widehat{\overline{Y}}_{st}\right) &= \frac{1}{2}\sum_{h=1}^{L} W_h^2 s_h^2 \\ &= \frac{1}{4}\sum_{h=1}^{L} W_h^2 d_h^2\end{aligned} \tag{18.5.2}$$

where $s_h^2 = \sum_{i=1}^{n_h} \left(y_{hi} - \overline{y}_h\right)^2/(n_h - 1) = (y_{h1} - y_{h2})^2/2$, $n_h = 2$, and $d_h = y_{h1} - y_{h2}$.

Let us divide the selected sample $\Im = (\Im_1, \ldots, \Im_L)$ into two half-samples $\alpha_1 = (u_{11},\ldots,u_{h1},\ldots,u_{L1})$ and $\alpha_2 = (u_{12},\ldots,u_{h2},\ldots,u_{L2})$. From each of the half-samples α_1 and α_2, we can construct estimators for the mean $\overline{Y}$ as

$$t_{\alpha_1} = \sum_{h=1}^{L} W_h y_{h1} \text{ and } t_{\alpha_2} = \sum_{h=1}^{L} W_h y_{h2} \tag{18.5.3}$$

From the estimators t_{α_1} and t_{α_2}, a combined estimator of $\overline{Y}$ is obtained as

$$\bar{t} = \frac{t_{\alpha_1} + t_{\alpha_2}}{2} = \widehat{\overline{Y}}_{st} \tag{18.5.4}$$

Assuming that the half-samples are independent, we find an unbiased estimator of $V\left(\widehat{\overline{Y}}_{st}\right)$ using the RG method as

$$\begin{aligned} \widehat{V}_{RG}(\bar{t}) &= \frac{1}{2}\sum_{j=1}^{2}\left(t_{\alpha_j} - \bar{t}\right)^2 \\ &= \frac{1}{4}(t_{\alpha_1} - t_{\alpha_2})^2 \end{aligned} \tag{18.5.5}$$

Clearly, $\widehat{V}_{RG}(\bar{t})$ may be different from $V\left(\widehat{\overline{Y}}_{st}\right)$ given in Eq. (18.5.2).

In general, we can construct 2^L half-samples by taking one unit from each of the stratum. Let us select one half-sample α (say) by choosing one unit at random from each of the stratum and construct an unbiased estimator of $\overline{Y}$ as follows:

$$t_\alpha = \sum_{h=1}^{L} W_h\{\phi_{h\alpha} y_{h1} + (1 - \phi_{h\alpha}) y_{h2}\} \tag{18.5.6}$$

where $\phi_{h\alpha} = 1$ if the unit u_{h1} belongs to the half-sample α and $\phi_{h\alpha} = 0$ otherwise. Let $\widehat{V}(t_\alpha) = \left(t_\alpha - \widehat{\overline{Y}}_{st}\right)^2$. Then, we have the following theorem:

Theorem 18.5.1

(i) $E(t_\alpha) = \overline{Y}$ and (ii) $E\left[\widehat{V}(t_\alpha)\right] = V\left(\widehat{\overline{Y}}_{st}\right)$

Proof

(i) $E(t_\alpha) = E[E(t_\alpha|\Im)]$

$$= E\left[\frac{1}{2^L}\sum_{\alpha=1}^{2^L} t(\alpha)\right]$$

$$= E\left[\frac{1}{2^L}\sum_{h=1}^{L} W_h\left\{y_{h1}\sum_{\alpha=1}^{2^L}\phi_{h\alpha} + y_{h2}\sum_{\alpha=1}^{2^L}(1 - \phi_{h\alpha})\right\}\right]$$

Now noting $\sum_{\alpha=1}^{2^L} \phi_{h\alpha}$ = number of times the unit u_{h1} is repeated in all the 2^L half-samples $= 2^{L-1}$, we find

$$E(t_\alpha) = E\left\{\frac{1}{2}\sum_{h=1}^{L} W_h(y_{h1} + y_{h2})\right\}$$
$$= E\left(\widehat{\overline{Y}}_{st}\right)$$
$$= \overline{Y}$$

(ii)
$$E\left[\widehat{V}(t_\alpha)\right] = E\left[E\left(\widehat{V}(t_\alpha)|\Im\right)\right]$$
$$= E\left[\frac{1}{2^L}\sum_{\alpha=1}^{2^L}\left(t_\alpha - \widehat{\overline{Y}}_{st}\right)^2\right] \tag{18.5.7}$$

Now,

$$\frac{1}{2^L}\sum_{\alpha=1}^{2^L}\left(t_\alpha - \widehat{\overline{Y}}_{st}\right)^2 = \frac{1}{2^L}\sum_{\alpha=1}^{2^L}\left[\sum_{h=1}^{L} W_h\{\phi_{h\alpha}y_{h1} + (1-\phi_{h\alpha})y_{h2}\} - \frac{1}{2}(y_{h1} + y_{h2})\right]^2$$
$$= \frac{1}{2^L}\sum_{\alpha=1}^{2^L}\left[\sum_{h=1}^{L} W_h\left\{y_{h1}\left(\phi_{h\alpha} - \frac{1}{2}\right) + \left(\frac{1}{2} - \phi_{h\alpha}\right)y_{h2}\right\}\right]^2$$
$$= \frac{1}{2^L}\frac{1}{4}\sum_{\alpha=1}^{2^L}\left(\sum_{h=1}^{L} W_h d_h \psi_{h\alpha}\right)^2 \tag{18.5.8}$$

where

$$\psi_{h\alpha} = 2\phi_{h\alpha} - 1 \tag{18.5.9}$$

From expression (18.5.9) we note that $\psi_{h\alpha} = 1$ if the unit u_{h1} is selected in the half-sample α and $\psi_{h\alpha} = -1$ otherwise. Furthermore,

$$\frac{1}{2^L}\sum_{\alpha=1}^{2^L}\left(\sum_{h=1}^{L} W_h d_h \psi_{h\alpha}\right)^2 = \frac{1}{2^L}\sum_{\alpha=1}^{2^L}\left(\sum_{h=1}^{L} W_h^2 d_h^2 + \sum_{h\neq}^{L}\sum_{h'=1}^{L} W_h W_{h'} d_h d_{h'} \psi_{h\alpha}\psi_{h'\alpha}\right)$$
$$= \sum_{h=1}^{L} W_h^2 d_h^2$$
$$= \widehat{V}\left(\widehat{\overline{Y}}_{st}\right) \tag{18.5.10}$$

because $\sum_{\alpha=1}^{2^L} \psi_{h\alpha}\psi_{h'\alpha} = 0$ for $h \neq h'$.

Finally, from Eqs. (18.5.7), (18.5.8), and (18.5.10), we find

$$\begin{aligned} E\left[\widehat{V}(t_\alpha)\right] &= E\left(\frac{1}{4}\sum_{h=1}^{L} W_h^2 d_h^2\right) \\ &= E\left[\widehat{V}\left(\widehat{\overline{Y}}_{st}\right)\right] \\ &= V\left(\widehat{\overline{Y}}_{st}\right) \end{aligned}$$

Hence $\widehat{V}(t_\alpha) = \frac{1}{2^L}\sum_{\alpha=1}^{2^L}\left(t_\alpha - \widehat{\overline{Y}}_{st}\right)^2$ is an unbiased estimator of $V\left(\widehat{\overline{Y}}_{st}\right)$. The estimator $\widehat{V}$ cannot be used in practice because one cannot compute all possible 2^L half-samples as 2^L is a huge number, e.g., $L = 6$ produces $2^L = 64$ half-samples. Furthermore, if we choose any random subset of $k(\leq 2^L)$ half-samples and construct an estimator of $V\left(\widehat{\overline{Y}}_{st}\right)$ as

$$\widehat{V} = \frac{1}{k}\sum_{\alpha=1}^{k}\left(t_\alpha - \widehat{\overline{Y}}_{st}\right)^2$$

then $\widehat{V}$ will be less efficient than $\widehat{V}\left(\widehat{\overline{Y}}_{st}\right)$ because

$$\begin{aligned} V\left(\widehat{V}(t)\right) &= V\left[E\left(\widehat{V}\middle|\Im\right)\right] + E\left[\widehat{V}\left(\widehat{V}\middle|\Im\right)\right] \\ &= V\left[\widehat{V}\left(\widehat{\overline{Y}}_{st}\right)\right] + E\left[V\left(\widehat{V}\middle|\Im\right)\right] \\ &\geq V\left[\widehat{V}\left(\widehat{\overline{Y}}_{st}\right)\right] \end{aligned} \tag{18.5.11}$$

To overcome this difficulty, we choose a specific subset Ψ of k half-samples from 2^L half subsamples for which

$$\widehat{V}_{BRR} = \frac{1}{k}\sum_{\alpha\in\Psi}\left(t_\alpha - \widehat{\overline{Y}}_{st}\right)^2 = \widehat{V}\left(\widehat{\overline{Y}}_{st}\right) \tag{18.5.12}$$

Now noting from Eq. (18.5.8)

$$\begin{aligned} \widehat{V}_{BRR} &= \frac{1}{4k}\sum_{\alpha\in\Psi}\left(\sum_{h=1}^{L} W_h d_h \psi_{h\alpha}\right)^2 \\ &= \frac{1}{4}\left[\sum_{h=1}^{L} W_h^2 d_h^2 + \frac{1}{k}\sum_{h\neq}^{L}\sum_{h'=1}^{L} W_h W_{h'} d_h d_{h'} \sum_{\alpha\in\Psi}\psi_{h\alpha}\psi_{h'\alpha}\right] \end{aligned}$$

we find that the condition (18.5.12) holds if we choose a subset Ψ of k half-samples that satisfies

$$\sum_{\alpha \in \Psi} \psi_{h\alpha}\psi_{h'\alpha} = 0 \tag{18.5.13}$$

A set of half-samples satisfying property (18.5.13) is called balanced half-samples. The method of variance estimation based on a balanced set Ψ of half-samples is known as the Balanced Repeated Replication (BRR) method. The balanced half-samples can be constructed by using Hadamard matrices. A Hadamard matrix is a square matrix of order multiple of 4 with elements $+1$ and -1. The columns of the Hadamard matrices are orthogonal to each other. Details are given by Plackett and Burman (1946). An example of 8×8 Hadamard matrix is given as follows.

Hadamard matrix of order 8

$$\begin{matrix} +1 & +1 & +1 & +1 & +1 & +1 & +1 & +1 \\ +1 & -1 & +1 & -1 & +1 & -1 & -1 & +1 \\ +1 & -1 & -1 & +1 & +1 & -1 & +1 & -1 \\ +1 & +1 & -1 & -1 & +1 & +1 & -1 & -1 \\ +1 & +1 & +1 & +1 & -1 & -1 & -1 & -1 \\ +1 & -1 & +1 & -1 & -1 & +1 & +1 & -1 \\ +1 & -1 & -1 & +1 & -1 & +1 & -1 & +1 \\ +1 & +1 & -1 & -1 & -1 & -1 & +1 & +1 \end{matrix} \tag{18.5.14}$$

We take the rows of a Hadamard matrix as half-samples and columns as strata. The entry $+1$ of the stratum h indicates that the unit u_{h1} is included in the half-sample and -1 indicates the inclusion of u_{h2} in the half-sample. Thus an 8×8 Hadamard matrix can be used for a maximum number of strata $L=8$. If we want to consider any smaller number of strata, say 5, we take just any of the five columns of the Hadamard matrix. Thus, the rows of the Hadamard matrix (18.5.14) form the half-samples

$$\alpha_1 = (u_{11}, u_{21}, u_{31}, u_{41}, u_{51}, u_{61}, u_{71}, u_{81}),\ \alpha_2 = (u_{11}, u_{22}, u_{31}, u_{42}, u_{51}, u_{62}, u_{72}, u_{81}),$$
$$\alpha_3 = (u_{11}, u_{22}, u_{32}, u_{41}, u_{51}, u_{62}, u_{71}, u_{82}),\ \alpha_4 = (u_{11}, u_{21}, u_{32}, u_{42}, u_{51}, u_{61}, u_{72}, u_{82}),$$
$$\alpha_5 = (u_{11}, u_{21}, u_{31}, u_{41}, u_{52}, u_{62}, u_{72}, u_{82}),\ \alpha_6 = (u_{11}, u_{22}, u_{31}, u_{42}, u_{52}, u_{61}, u_{71}, u_{82}),$$
$$\alpha_7 = (u_{11}, u_{22}, u_{32}, u_{41}, u_{52}, u_{61}, u_{72}, u_{81}),\ \alpha_8 = (u_{11}, u_{21}, u_{32}, u_{42}, u_{52}, u_{62}, u_{71}, u_{81}).$$

For estimation of variance we choose a set of balanced half-samples consisting of k half-samples from the totality of 2^L half-samples so that

$\widehat{V}_{BRR}$ becomes exactly unbiased for $V\left(\widehat{\overline{Y}}_{st}\right)$. The number k should be as small as possible. A minimal set of balanced samples can be readily obtained from a Hadamard matrix of order $k \times k$ by choosing any L columns excluding the column of all +1's, where $L+1 \leq k \leq L+4$ (Rao and Shao, 1999).

An estimator for $\overline{Y}$ based on k half-samples is given by

$$\begin{aligned}
t_{BRR} &= \frac{1}{k}\sum_{\alpha\in\Psi} t(\alpha) \\
&= \frac{1}{k}\sum_{h=1}^{L} W_h\left\{y_{h1}\sum_{\alpha\in\Psi}\left(\frac{1}{2}+\frac{\psi_{h\alpha}}{2}\right)+y_{h2}\sum_{\alpha\in\Psi}\left(\frac{1}{2}-\frac{\psi_{h\alpha}}{2}\right)\right\} \\
&= \widehat{\overline{Y}}_{st}+\frac{1}{2k}\sum_{h=1}^{L} W_h(y_{h1}-y_{h2})\sum_{\alpha\in\Psi}\psi_{h\alpha}
\end{aligned}$$

Thus $t_{BRR} = \widehat{\overline{Y}}_{st}$ if

$$\sum_{\alpha\in\Psi}\psi_{h\alpha} = 0 \tag{18.5.15}$$

A set of balanced half-samples satisfying the properties of Eqs. (18.5.13) and (18.5.15) is known as full orthogonal balanced half-samples. For full orthogonal balanced half-samples, k should be a multiple of 4 and $k > L$. For example if $L = 8$, one needs minimum 12 replicates to achieve full orthogonal balance. Thus for reduction of computation labor, we should choose k as a smallest multiple of 4 but greater than L. If $k = L$, then only the balance is achieved but the full orthogonal balance is not. In case k is less than L, neither the balance nor the full orthogonal balance is achieved. Because Hadamard matrices are not unique, full orthogonal balance may be achieved with alternative sets of half-samples.

18.5.2 Methods of Variance Estimation

Let $\widehat{\theta}_{st}$ be an estimator (not necessarily linear) of a population parameter θ based on a stratified sample of 2 units per stratum using any sampling design. Here we choose a set of k balanced half-samples. Let $\widehat{\theta}_{\alpha}$ be an estimator of θ based on the αth half-sample, $\alpha = 1,\ldots,k$ and $\overline{\theta} = \sum_{\alpha=1}^{k}\widehat{\theta}_{\alpha}\Big/k$. The estimators $\widehat{\theta}_{\alpha}$ should necessarily be the same functional form of the estimator

$\widehat{\theta}_{st}$. Then the variance or mean square estimator of $\widehat{\theta}_{st}$ can be estimated using any of the following formula.

$$\widehat{V}_{BRR}(1) = \frac{1}{k}\sum_{\alpha=1}^{k}\left(\widehat{\theta}_{\alpha} - \widehat{\theta}_{st}\right)^2 \tag{18.5.16}$$

$$\widehat{V}_{BRR}(2) = \frac{1}{k}\sum_{\alpha=1}^{k}\left(\widehat{\theta}_{\alpha} - \overline{\theta}\right)^2 \tag{18.5.17}$$

Let $\widehat{\theta}^*_{\alpha}$ be an estimator of θ based on the complement of the αth half-sample. We can then also get the following alternative estimators

$$\widehat{V}_{BRR}(3) = \frac{1}{k}\sum_{\alpha=1}^{k}\left(\widehat{\theta}^*_{\alpha} - \widehat{\theta}_{st}\right)^2 \tag{18.5.18}$$

$$\widehat{V}_{BRR}(4) = \frac{1}{k}\sum_{\alpha=1}^{k}\left(\widehat{\theta}^*_{\alpha} - \overline{\theta}^*\right)^2 \tag{18.5.19}$$

$$\widehat{V}_{BRR}(5) = \frac{1}{4k}\sum_{\alpha=1}^{k}\left(\widehat{\theta}_{\alpha} - \widehat{\theta}^*_{\alpha}\right)^2 \tag{18.5.20}$$

where $\overline{\theta}^* = \sum_{\alpha=1}^{k}\widehat{\theta}^*_{\alpha}\Big/k$.

The estimators of the variance of a linear estimator described previously based on a set of balanced half-samples do not provide a more improved estimator than the conventional variance estimator of the stratified sampling. However, the BRR method can be gainfully used for nonlinear estimators where elegant and unbiased variance estimators are not available. For very large L it is difficult to find balanced replicates and in this situation one may use partially balanced replicates by dividing L into a number of groups. Details are given by Wolter (1985).

18.5.3 Applications

18.5.3.1 Population Ratio

Consider a stratified sampling where two units are selected at random from each of the stratum. The conventional combined stratified ratio estimator of the population ratio $R = Y/X$ is $\widehat{R}_{com} = \widehat{Y}_{st}\big/\widehat{X}_{st}$, where $\widehat{Y}_{st} = \sum_{h=1}^{L} N_h\overline{y}_h$, $\widehat{X}_{st} = \sum_{h=1}^{L} N_h\overline{x}_h$. An estimator R based on αth half-sample is $\widehat{R}_{\alpha(com)} = \widehat{Y}_{\alpha}\big/\widehat{X}_{\alpha}$, where $\widehat{Y}_{\alpha} = \sum_{h=1}^{L} N_h\ \{\phi_{h\alpha}y_{h1} + (1-\phi_{h\alpha})y_{h2}\}$ and

$\widehat{X}_\alpha = \sum_{h=1}^{L} N_h\{\phi_{h\alpha}x_{h1} + (1-\phi_{h\alpha})x_{h2}\}$. The estimator for R based on a set of k balanced half-samples is given by

$$\widehat{R}_{BRR} = \frac{1}{k}\sum_{\alpha=1}^{k} \widehat{Y}_\alpha / \widehat{X}_\alpha$$

An estimator for the variance of $\widehat{R}_{com}$ based on k balanced half-samples is

$$\widehat{V}_{BRR} = \frac{1}{k}\sum_{\alpha=1}^{k}\left(\widehat{R}_{\alpha(com)} - \widehat{R}_{com}\right)^2$$

Clearly, $\widehat{V}_{BRR}$ is quite different from the conventional variance estimator

$$\widehat{V}_{st}\left(\widehat{R}_{com}\right) = \frac{1}{4\widehat{X}_{st}^2}\sum_{h=1}^{L} N_h^2\left\{(y_{h1}-y_{h2}) - \widehat{R}_{com}(x_{h1}-x_{h2})\right\}^2.$$

18.5.3.2 Inclusion Probability Proportional to Size Sampling Scheme

Suppose from each of the stratum two units are selected with IPPS sampling design. Let the inclusion probability for the ith unit of the hth stratum be $\pi_{hi} = 2p_{hi}$. Here the conventional estimator for the total Y and its variance are, respectively, given by

$$\widehat{Y}_{st} = \frac{1}{2}\sum_{h=1}^{L}\left(\frac{y_{h1}}{p_{h1}} + \frac{y_{h2}}{p_{h2}}\right)$$

and

$$V\left(\widehat{Y}_{st}\right) = \frac{1}{4}\sum_{h=1}^{L}\sum_{i\neq}^{N_h}\sum_{j=1}^{N_h}\frac{(p_{hi}-p_{hj})^2}{\pi_{hij}}\left(\frac{y_{hi}}{p_{hi}} - \frac{y_{hj}}{p_{hj}}\right)^2$$

The estimator based on the αth half-sample is

$$t_\alpha = \sum_{h=1}^{L}\left\{\phi_{h\alpha}\frac{y_{h1}}{p_{h1}} + (1-\phi_{h\alpha})\frac{y_{h2}}{p_{h2}}\right\} \quad \text{for } \alpha = 1, \ldots, k$$

The variance estimator of $\widehat{Y}_{st}$ based on BRR is

$$\widehat{V}_{BRR} = \frac{1}{k}\sum_{\alpha=1}^{k}\left(t_\alpha - \widehat{Y}_{st}\right)^2$$

18.5.4 Numerical Example

The following table relates to the number of patients (y) treated per day and the number of doctors (x) at the clinics in a certain city. The clinics are classified into five different zones (strata). From each of the zones two clinics are selected by the SRSWR method (Table 18.5.1).

Table 18.5.1 Number of patients treated and number of doctors

Zone	**1**		**2**		**3**		**4**		**5**	
Number of clinics	15		25		20		30		10	
Sampled clinics	1	2	1	2	1	2	1	2	1	2
y	30	20	30	25	40	15	40	50	20	20
x	2	1	3	2	2	1	2	3	2	1

18.5.4.1 Population Mean

Here the parameter of interest is $\theta = \sum W_h \overline{Y}_h = \overline{Y} =$ average number of patients treated per clinic per day. An unbiased estimator of θ is $\widehat{\theta} = \overline{y}_{st} = \sum W_h \overline{y}_h = 0.15 \times 25 + 2.5 \times 27.5 + 0.2 \times 27.5 + 0.3 \times 45 + 0.1 \times 20 = 31.625$. The estimated variance of $\overline{y}_{st}$ is

$$\widehat{V}\left(\overline{y}_{st}\right) = \sum W_h^2 (y_{h1} - y_{h2})^2 / 4 = 9.453$$

Let us now consider the BRR method with $L = 5$ and $k = 8$. For this method we delete the first and the last two columns of the Hadamard matrix (18.5.14) to achieve orthogonal balance. The Hadamard matrix (after deletion of columns) and the units belonging to the replicates are given in Tables 18.5.2 and 18.5.3.

Table 18.5.2 BRR method of selection of sample

	Stratum *h*					**Units belong to the replicates — Stratum *h***				
Replicates (α)	**1**	**2**	**3**	**4**	**5**	**1**	**2**	**3**	**4**	**5**
1	+1	+1	+1	+1	+1	u_{11}	u_{21}	u_{31}	u_{41}	u_{51}
2	−1	+1	−1	+1	−1	u_{12}	u_{21}	u_{32}	u_{41}	u_{52}
3	−1	−1	+1	+1	−1	u_{12}	u_{22}	u_{31}	u_{41}	u_{52}
4	+1	−1	−1	+1	+1	u_{11}	u_{22}	u_{32}	u_{41}	u_{51}
5	+1	+1	+1	−1	−1	u_{11}	u_{21}	u_{31}	u_{42}	u_{52}
6	−1	+1	−1	−1	+1	u_{12}	u_{21}	u_{32}	u_{42}	u_{51}
7	−1	−1	+1	−1	+1	u_{12}	u_{22}	u_{31}	u_{42}	u_{51}
8	+1	−1	−1	−1	−1	u_{11}	u_{22}	u_{32}	u_{42}	u_{52}

Table 18.5.3 BRR esimates

(α)	$\overline{y}_{st}(\alpha)$	$\overline{x}_{st}(\alpha)$	$\widehat{R}_{sep}(\alpha)$	$\widehat{R}_{com}(\alpha)$
1	34.00	2.25	15.750	15.1111
2	27.50	1.80	16.500	15.2777
3	31.25	1.75	18.125	17.8571
4	27.75	1.80	15.375	15.4166
5	37.00	2.45	15.750	15.1020
6	30.50	2.20	14.500	13.8636
7	34.25	2.15	16.125	15.9302
8	30.75	2.00	15.375	15.3750
Mean	31.625	2.05	15.9375	15.4917

$\overline{y}_{BRR} = \sum_{\alpha=1}^{8} \overline{y}_{st}(\alpha)/8 = 31.625 = \overline{y}_{st}$. An estimated variance of $\overline{y}_{st}$ by the BRR method is $\widehat{V}(\overline{y}_{BRR}) = \sum_{\alpha=1}^{k} \{\overline{y}_{st}(\alpha) - \overline{y}_{st}\}^2/k = 9.453 = \widehat{V}(\overline{y}_{st})$. Because the replicates are not only balanced but also orthogonally balanced, we find for linear estimates $\overline{y}_{BRR} = \overline{y}_{st}$ and $\widehat{V}(\overline{y}_{BRR}) = \widehat{V}(\overline{y}_{st})$.

18.5.4.2 Population Ratio

Here the parameter of interest is θ = average number of patients treated by a doctor per day $= R = Y/X$. The combined and separate ratio estimators of R are, respectively, given by $\widehat{R}_{com} = \overline{y}_{st}/\overline{x}_{st} = 31.625/2.05 = 15.426$ and $\widehat{R}_{sep} = \sum W_h \widehat{R}_h = 15.65$, where $\widehat{R}_h = \overline{y}_h/\overline{x}_h$.

The estimated approximate variances or mean square errors of $\widehat{R}_{com}$ and $\widehat{R}_{sep}$ are, respectively,

$$\widehat{V}(\widehat{R}_{com}) = \frac{1}{4(\overline{x}_{st})^2}\sum W_h^2\left[(y_{h1} - \widehat{R}_{com}x_{h1}) - (y_{h2} - \widehat{R}_{com}x_{h2})\right]^2 = 0.9609$$

$$\widehat{V}(\widehat{R}_{sep}) = \frac{1}{4(\overline{x}_{st})^2}\sum W_h^2\left[(y_{h1} - \widehat{R}_{sep}x_{h1}) - (y_{h2} - \widehat{R}_{sep}x_{h2})\right]^2 = 0.9890$$

The combined ratio estimates based on the balanced sample is given by $\widehat{R}_{BRR(com)} = \sum_{\alpha=1}^{8} \widehat{R}_{com}(\alpha)/8 = 15.4917$, which is not equal to $\widehat{R}_{com} = 15.426$. The variance of $\widehat{R}_{com}$ can be estimated using any of the following formulae

$$\widehat{V}_{BRR(com)}(1) = \frac{1}{8}\sum_{\alpha=1}^{8}\left(\widehat{R}_{com}(\alpha) - \widehat{R}_{com}\right)^2 = 1.1041$$

$$\widehat{V}_{BRR(com)}(2) = \frac{1}{8}\sum_{\alpha=1}^{8}\left(\widehat{R}_{com}(\alpha) - \widehat{\overline{R}}_{com}\right)^2 = 1.099$$

The separate ratio estimates based on the balanced sample is given by $\widehat{R}_{BRR}(sep) = \sum_{\alpha=1}^{8} \widehat{R}_{sep}(\alpha)/8 = 15.9375$, which is not equal to $\widehat{R}_{sep} = 15.65$. The variance of $\widehat{R}_{sep}$ can be estimated by using any of the following formulae

$$\widehat{V}_{BRR(sep)}(1) = \frac{1}{8}\sum_{\alpha=1}^{8}\left(\widehat{R}_{sep}(\alpha) - \widehat{R}_{sep}\right)^2 = 1.071$$

$$\widehat{V}_{BRR(sep)}(2) = \frac{1}{8}\sum_{\alpha=1}^{8}\left(\widehat{R}_{sep}(\alpha) - \widehat{\overline{R}}_{sep}\right)^2 = 0.998$$

Thus we see for the nonlinear statistics that the BRR estimates are quite different from the conventional estimates based on the full sample.

18.5.4.3 Correlation Coefficient

The sample correlation coefficient between x and y is given by

$$r = \frac{\sum_h N_h(y_{h1}x_{h1} + y_{h2}x_{h2}) - 2N\overline{y}_{st}\overline{x}_{st}}{\sqrt{\left(\sum_h N_h\left(y_{h1}^2 + y_{h2}^2\right) - 2N\overline{y}_{st}^2\right)}\sqrt{\left(\sum_h N_h\left(x_{h1}^2 + x_{h2}^2\right) - 2N\overline{x}_{st}^2\right)}}$$
$$= 0.717$$

The correction coefficient r_α and $\overset{*}{r_\alpha}$ based on the αth half-sample and its complementary sample are given in Table 18.5.4.

Table 18.5.4 BRR method of variance estimation

α	r_α	$\overset{*}{r_\alpha}$	$\frac{\left(r_\alpha - \overset{*}{r_\alpha}\right)^2}{4}$
1	−0.3481	0.9543	0.42406
2	0.6356	0.8405	0.01049
3	0.7276	0.8012	0.00135
4	0.6866	0.7003	0.00004
5	0.5726	0.6960	0.00380
6	0.8141	0.5547	0.01682
7	0.8480	0.5824	0.01763
8	0.9604	0.2414	0.12924
Mean	0.6121	0.6713	0.0754

The estimates of variance of r obtained by using different formulae are as follows: $\widehat{V}_{BRR}(1) = \frac{1}{k}\sum_{\alpha=1}^{k}(r_\alpha - r)^2 = 0.1562$, $\widehat{V}_{BRR}(2) = \frac{1}{k}\sum_{\alpha=1}^{k}(r_\alpha - \bar{r})^2 = 0.1451$, $\widehat{V}_{BRR}(3) = \frac{1}{k}\sum_{\alpha=1}^{k}\left(r_\alpha^* - r\right)^2 = 0.0438$, $\widehat{V}_{BRR}(4) = \frac{1}{k}\sum_{\alpha=1}^{k}\left(r_\alpha^* - \bar{r}^*\right)^2 = 0.0416$, $\widehat{V}_{BRR}(5) = \frac{1}{4k}\sum_{\alpha=1}^{k}\left(r_\alpha - r_\alpha^*\right)^2 = 0.0754$, $\widehat{V}_{BRR}(6) = \left(\widehat{V}_{BRR}(1) + \widehat{V}_{BRR}(3)\right)/2 = 0.1000$, and $\widehat{V}_{BRR}(7) = \left(\widehat{V}_{BRR}(2) + \widehat{V}_{BRR}(4)\right)/2 = 0.0933$, where $\bar{r} = \frac{1}{k}\sum_{\alpha=1}^{k} r_\alpha = 0.6121$, $\bar{r}^* = \frac{1}{k}\sum_{\alpha=1}^{k} r_\alpha^* = 0.6713$, and $k = 8$.

18.5.5 Stratum Size $n_h \geq 2$

18.5.5.1 Grouped Balanced Half-Sample Method

Let $\overline{Y}_h$ and $\bar{y}_h$ be the population mean and sample mean of the hth stratum, respectively. The parameter of interest is $\theta = g(\overline{\mathbf{Y}})$, a function of $\overline{\mathbf{Y}} = \left(\overline{Y}_1, \ldots, \overline{Y}_L\right)$ the vector of the strata means. Let a full sample estimate of θ is $\widehat{\theta} = g(\bar{\mathbf{y}})$, where $\bar{\mathbf{y}} = \left(\bar{y}_1, \ldots, \bar{y}_L\right)$. In the grouped balanced half-sample (GBHS) method, the selected sample $n_h(>2)$ units from the stratum h are divided at random into two groups of sizes $m_{h1} = [n_h/2]$ and $m_{h2} = n_h - [n_h/2]$, respectively. A set of k balanced half-samples of groups are selected. Let $\bar{y}_{h1}$ and $\bar{y}_{h2}$ denote the sample means of the first and second groups, respectively. The estimator of the population mean θ based on the αth half-sample is denoted by

$$\widehat{\theta}_\alpha = g(\bar{\mathbf{y}}'_\alpha)$$

where $\bar{\mathbf{y}}'_\alpha = (\bar{\mathbf{y}}'_{1\alpha}, \ldots, \bar{\mathbf{y}}'_{h\alpha}, \ldots, \bar{\mathbf{y}}'_{L\alpha})$, $\bar{\mathbf{y}}'_{h\alpha} = \phi_{h\alpha}\bar{y}_{h1} + (1 - \phi_{h\alpha})\bar{y}_{h2}$, $\phi_{h\alpha} = 1$ if the group 1 of the stratum h is selected in the half-sample and zero otherwise.

A GBHS variance estimator of $\widehat{\theta}$ is given by

$$\widehat{V}_{GB}\left(\widehat{\theta}\right) = \frac{1}{k}\sum_{\alpha=1}^{k}\left(\widehat{\theta}_\alpha - \widehat{\theta}\right)^2 \tag{18.5.21}$$

The variance estimator $\widehat{V}_{GB}\left(\widehat{\theta}\right)$ in general is quite different from the usual unbiased variance estimator $\widehat{V}_{st} = \sum_{h=1}^{L} W_h^2 s_h^2 / n_h$, where s_h^2 is the hth stratum sample variance. However, it should be noted that for the linear function $\theta = \sum_{h=1}^{L} W_h \overline{Y}_h = \overline{Y}$, $\widehat{\theta} = \sum_{h=1}^{L} W_h \overline{y}_h = \widehat{\overline{Y}}_{st}$ with $\overline{y}_h$ as the hth stratum sample mean, the GBHS variance estimator (18.5.21) reduces to

$$\widehat{V}\left(\widehat{\overline{Y}}_{st}\right) = \frac{1}{4}\sum_{h=1}^{L} W_h^2 \left(\overline{y}_{h1} - \overline{y}_{h2}\right)^2 = \widehat{V}_{st}$$

Rao and Shao (1996) have shown that the GBHS method leads to asymptotically incorrect inferences as strata sample sizes $n_h \rightarrow \infty$ with L fixed. To overcome this difficulty, they proposed an alternative method known as repeatedly grouped balanced half-sample (RGBHS) method where the random grouping is repeated T times independently and then taking average of the GBHS variance estimators $\widehat{V}^{t}_{GB}\left(\widehat{\theta}\right)$, obtained from the tth grouping. The resulting RGBHS variance estimator

$$\widehat{V}_{RGB}\left(\widehat{\theta}\right) = \frac{1}{T}\sum_{t=1}^{T} \widehat{V}^{t}_{GB}\left(\widehat{\theta}\right) \tag{18.5.22}$$

possesses asymptotic validity. Rao and Shao (1996) proposed modification of $\widehat{V}_{GB}\left(\widehat{\theta}\right)$ by replacing W_h by $W_h' = W_h\sqrt{\lambda_h}$ with $\lambda_h = 1 - n_h/N_h$ in the calculation of $\widehat{\theta}_\alpha$ and $\widehat{\theta}$.

18.5.5.2 Subdivision of Strata

This method is suitable with a small number of strata and relatively large sample sizes within strata. Let us assume that the stratum size $n_h = 2m_h$ with m_h is an integer for all h. The hth stratum is subdivided into m_h artificial strata each of size 2 so that the total number of strata is $H = \sum_h m_h$. A balanced set of k half-samples is obtained by using a Hadamard matrix $k \times k$ $\left(\sum_h m_h \leq k \leq \sum_h m_h + 3\right)$. The proposed variance estimator of $\widehat{\theta}$ is

$$\widehat{V}_{SG}\left(\widehat{\theta}\right) = \frac{1}{k}\sum_{\alpha=1}^{k} \left(\widehat{\theta}_{\alpha^*} - \widehat{\theta}\right)^2 \tag{18.5.23}$$

where $\widehat{\theta}_{\alpha^*} = g(\overline{\mathbf{y}}'_{\alpha^*})$, $\overline{\mathbf{y}}'_{\alpha^*} = (\overline{\mathbf{y}}'_{1\alpha^*}, \ldots, \overline{\mathbf{y}}'_{h\alpha^*}, \ldots, \overline{\mathbf{y}}'_{H\alpha^*})$, $\overline{\mathbf{y}}'_{h\alpha^*} = \frac{1}{m_h}\sum_{i=1}^{m_h}\{\phi_{hi\alpha} y_{hi1} + (1 - \phi_{hi\alpha}) y_{hi2}\}$, $\phi_{hi\alpha} = 1$ if the first unit of the artificial strata with y-value y_{hi1} belongs to the half-sample α and $\phi_{hi\alpha} = 0$ if the second unit of the artificial strata with y-value y_{hi2} belongs to the half-sample α.

For the linear estimator $\widehat{\theta} = \sum_h W_h \overline{y}_h = \widehat{\overline{Y}}_{st}$, the variance estimator (18.5.23) reduces to

$$\widehat{V}_{SG}\left(\widehat{\theta}\right) = \sum_{h=1}^{L} \frac{W_h^2}{m_h^2} \sum_{i=1}^{m_h} \frac{\left(y_{hi1} - y_{hi2}\right)^2}{4}$$

The variance estimator $\widehat{V}_{SG}\left(\widehat{\theta}\right)$ does not agree with the usual unbiased variance estimator $\widehat{V}_{st} = \sum_{h=1}^{L} W_h^2 s_h^2 / n_h$.

18.5.6 Stratified Multistage Sampling

Consider a stratified multistage sampling design where the population comprises of large number of L strata. The hth strata consists of N_h clusters. The ith cluster of the hth stratum comprises with M_{hi} ultimate units. A sample of n_h clusters is selected from the hth stratum with probability proportional to sizes without replacement (PPSWOR) method and each of the selected clusters are subsampled independently using the PPSWOR method again. Let y_{hik} be the value of the study variable y associated with (h,i,k), the kth ultimate unit of the ith cluster of the hth stratum. Let w_{hik} be the survey weights attached to (h,i,k) if it is included in the selected sample s. Consider the class of unbiased estimators of the population total $Y = \sum_{h=1}^{L} \sum_{i=1}^{N_h} \sum_{k=1}^{M_{hi}} y_{ijk}$ of the form

$$\widehat{Y} = \sum_{(h,i,k)\in s} w_{hik} y_{hik} \tag{18.5.24}$$

where w_{hik} is the weight associated with (h,i,k).

Suppose we are interested in estimating parametric function of the form $\theta = g(\mathbf{A})$ where $\mathbf{A}$ is a vector of population totals. For example, $\mathbf{A} = (A_1, A_2, A_3, A_4, A_5)$ with $A_1 = \sum_{h=1}^{L} \sum_{i=1}^{N_h} \sum_{k=1}^{M_{hi}} x_{hik}$, $A_2 = \sum_{h=1}^{L} \sum_{i=1}^{N_h} \sum_{k=1}^{M_{hi}} y_{hik}$, $A_3 = \sum_{h=1}^{L} \sum_{i=1}^{N_h} \sum_{k=1}^{M_{hi}} x_{hik}^2$, $A_4 = \sum_{h=1}^{L} \sum_{i=1}^{N_h} \sum_{k=1}^{M_{hi}} y_{hik}^2$, and $A_5 = \sum_{h=1}^{L} \sum_{i=1}^{N_h} \sum_{k=1}^{M_{hi}} x_{hik} y_{hik}$, we can express θ as a population ratio, variance, correlation coefficient, and coefficient of variation of two characters x and y, among others. Let $\widehat{\mathbf{A}} = \left(\widehat{A}_1, \ldots, \widehat{A}_5\right)$ be a consistent or unbiased estimator of $\mathbf{A}$, where $\widehat{A}_1 = \sum_{ijk\in s} w_{ijk} x_{ijk}$, $\widehat{A}_2 = \sum_{ijk\in s} w_{ijk} y_{ijk}$, $\widehat{A}_3 = \sum_{ijk\in s} w_{ijk} x_{ijk}^2$, $\widehat{A}_4 = \sum_{ijk\in s} w_{ijk} y_{ijk}^2$, $\widehat{A}_5 = \sum_{ijk\in s} w_{ijk} x_{ijk} y_{ijk}$, and w_{ijk} are suitably chosen weight. Consider the estimator $\widehat{\theta}^* = g(\widehat{\mathbf{A}})$ of θ based on the full sample s as $\widehat{\theta}^* = g^*(\widehat{\mathbf{A}})$. For variance estimation, we select two units ($n_h = 2$) from each of the stratum $h(=1,\ldots, L)$ and construct k balanced half-samples using Hadamard matrix as described in Section 18.5.1. The estimator $\widehat{\theta}^*_\alpha$ is obtained from the αth half-sample by using the same formula $\widehat{\theta}^*$ with weight w_{hik} changed to w^α_{hik}, which is equal to $2w_{hik}$ or 0 according to whether or not (h,i) cluster is selected in the αth half-sample. The variance estimator of $\widehat{\theta}^*$ is given by

$$\widehat{V}\left(\widehat{\theta}^*\right) = \frac{1}{k}\sum_{\alpha=1}^{k}\left(\widehat{\theta}^*_\alpha - \widehat{\theta}^*\right)^2 \tag{18.5.25}$$

For $n_h > 2$, variance estimator can be obtained by constructing k balanced half-samples of clusters following Sections 18.5.5.1 or 18.5.5.2 and adjusting weight appropriately.

18.5.7 Fay's Method

Consider stratified sampling where $n_h = 2$ units are selected from each of the strata by the SRSWOR method. In this case the conventional estimator for the population mean $\overline{Y}$ is $\overline{Y}_{st} = \sum_{h=1}^{L} W_h \frac{(y_{h1} + y_{h2})}{2}$. The weight associated with each of the units of the hth stratum is $W_h' = W_h/2$. The estimator for the population mean based on the half-sample α defined in Eq. (18.5.6) is

$$t_\alpha = \sum_{h=1}^{L} 2W_h'\{\phi_{h\alpha} y_{h1} + (1 - \phi_{h\alpha}) y_{h2}\}$$

Here, a weight $2W_h'$ is attached to each of the units selected in the half-sample α and a zero weight is attached to the units not selected in the half-sample α. In Fay's (1989) adjustment of weights, the selected unit in the half-sample is given a less weight $W_h'(1+\in)$ with $0 < \in \leq 1$ if it is selected in the half-sample α while a positive weight $W_h'(1-\in)$ is assigned to the units not selected in the half-sample. So, Fay's estimator based on the half-sample α is

$$t^F_\alpha = \sum_{h=1}^{L} \frac{W_h}{2}[\{1 + \in(2\phi_{h\alpha} - 1)\}y_{h1} + \{1 - \in(2\phi_{h\alpha} - 1)\}y_{h2}] \tag{18.5.26}$$

where $\phi_{h\alpha} = 1$ if the unit u_{h1} is selected in the half-sample α and $\phi_{h\alpha} = 0$ if u_{h2} is selected in the half-sample.

The Fay's adjusted variance estimator of $\widehat{\overline{Y}}_{st}$ based on a set k balanced half-samples is given by

$$\widehat{V}^F_{BRR} = \frac{1}{k\in^2}\sum_{\alpha=1}^{k}\left(t^F_\alpha - \widehat{\overline{Y}}_{st}\right)^2 \tag{18.5.27}$$

Theorem 18.5.2

For a set of k balanced half-samples

$$\widehat{V}^F_{BRR} = \frac{1}{4}\sum_{h=1}^{L} W_h^2 (y_{h1} - y_{h2})^2 = \widehat{V}\left(\overline{Y}_{st}\right)$$

Proof

$$\widehat{V}^F_{BRR} = \frac{1}{k\epsilon^2}\sum_{\alpha=1}^{k}\left[\sum_{h=1}^{L}\frac{W_h}{2}\{(1+\epsilon(2\phi_{h\alpha}-1))y_{h1} + (1-\epsilon(2\phi_{h\alpha}-1))y_{h2} - (y_{h1}+y_{h2})\}\right]^2$$

$$= \frac{1}{4k}\sum_{\alpha=1}^{k}\left[\sum_{h=1}^{L} W_h\psi_{h\alpha}(y_{h1}-y_{h2})\right]^2 \quad \text{where } \psi_{h\alpha} = 2\phi_{h\alpha} - 1$$

$$= \frac{1}{4k}\sum_{\alpha=1}^{k}\left[\sum_{h=1}^{L} W_h^2\psi_{h\alpha}^2(y_{h1}-y_{h2})^2 + \sum_{h\neq}^{L}\sum_{h'}^{L} W_h\psi_{h\alpha}W_{h'}\psi_{h'\alpha}(y_{h1}-y_{h2})(y_{h'1}-y_{h'2})\right]$$

$$= \frac{1}{4k}\left[\sum_{h=1}^{L} W_h^2(y_{h1}-y_{h2})^2\sum_{\alpha=1}^{k}\psi_{h\alpha}^2 + \sum_{h\neq}^{L}\sum_{h'}^{L} W_hW_{h'}(y_{h1}-y_{h2})\right.$$
$$\left.(y_{h'1}-y_{h'2})\sum_{\alpha=1}^{k}\psi_{h\alpha}\psi_{h'\alpha}\right]$$

Now, noting $\psi_{h\alpha} = 1$ if u_{h1} belongs to the half-sample α and $\psi_{h\alpha} = -1$ if u_{h1} does not belong to the half-sample α, and $\sum_{\alpha=1}^{k}\psi_{h\alpha}\psi_{h'\alpha} = 0$ (vide Eq. 18.5.13) for the set of k balanced half-samples, we find $\widehat{V}^F_{BRR} = \widehat{V}\left(\overline{Y}_{st}\right)$.

For the multistage sampling design described in Section 18.5.6, the Fay-adjusted variance estimator based on a set of k balanced half-samples is given by

$$\widehat{V}^F\left(\widehat{\theta}^*\right) = \frac{1}{k\epsilon^2}\sum_{\alpha=1}^{k}\left(\widehat{\theta}^{*F}_{\alpha} - \widehat{\theta}^*\right)^2$$

where $\widehat{\theta}^{*F}_{\alpha}$ is computed using the formula for $\widehat{\theta}^*$ but with w_{hik} changed to $w^F_{hik} = w_{hik}(1+\epsilon)$ if the (hik) is included in the half-sample α otherwise $w^F_{hik} = w_{hik}(1-\epsilon)$. When $\widehat{\theta}^*$ is linear, $\widehat{V}^F\left(\widehat{\theta}^*\right)$ reduces to the standard variance estimator (Judkins, 1990).

18.6 BOOTSTRAP METHOD

18.6.1 Bootstrap for Infinite Population

The Bootstrap (BT) method was introduced by Efron (1979). This method can be used for the estimation of variance of an estimator and determination of confidence interval of the parameter of interest. Let $X_1,\ldots, X_n$ be a random sample from a population with distribution function F_θ indexed by a parameter θ. Let $\widehat{\theta} = \widehat{\theta}(X_1, \ldots, X_n)$ be an estimator of the parameter θ. The empirical distribution of F_n of F is obtained by assigning a mass $1/n$ to each of the observations $X_1, \ldots, X_n$. From the empirical distribution F_n, a sample $X_1^*, \ldots, X_n^*$ of size n is selected with replacement. Let $\widehat{\theta}^* = \widehat{\theta}^*\left(X_1^*, \ldots, X_n^*\right)$ be an estimator of θ based on the BT sample $X_1^*, \ldots, X_n^*$. The selection of the BT sample is then repeated independently a large number of times B (at least 1000 times). Let $\widehat{\theta}_*^b$ be the value of $\widehat{\theta}^*$ based on the BT sample $b(=1,\ldots, B)$. The variance of this distribution of $\widehat{\theta}^*$ is considered as an estimator of the variance of $\widehat{\theta}$ and it is given by

$$\widehat{V}_B\left(\widehat{\theta}\right) = \frac{1}{B}\sum_{b=1}^{B}\left(\widehat{\theta}_b^* - \overline{\theta}^*\right)^2 \tag{18.6.1}$$

where $\overline{\theta}^* = \frac{1}{B}\sum_{b=1}^{B}\widehat{\theta}_b^*$.

In the Eq. (18.6.1), $\overline{\theta}^*$ may be replaced by $\widehat{\theta}$.

18.6.1.1 Bootstrap Confidence Interval

To determine the confidence interval of θ, we may use (i) percentile method and (ii) BT t-method.

18.6.1.1.1 Percentile Method

In the percentile method, we arrange the values of $\widehat{\theta}_b^*$ in the ascending order of magnitude. Let $\widehat{\theta}_{b,U}^*$ and $\widehat{\theta}_{b,L}^*$ be the upper and lower $(\alpha/2)100$ percent points of the distribution of $\widehat{\theta}_b^*$, respectively. Then 100 $(1 - \alpha)$% BT confidence interval of θ is given by $\left(\widehat{\theta}_{b,L}^*, \widehat{\theta}_{b,U}^*\right)$.

18.6.1.1.2 Bootstrap t-Method

In BT t-method we compute $t_b = \left(\widehat{\theta}_b^* - \widehat{\theta}\right)\Big/\sqrt{\widehat{V}_B\left(\widehat{\theta}\right)}$ for $b = 1,\ldots,B$ and assume distribution of t_b approximately the same as $t = \left(\widehat{\theta} - \theta\right)\Big/\sqrt{\widehat{V}_B\left(\widehat{\theta}\right)}$. The $100(1 - \alpha)$% confidence interval of θ is given by $\left(\widehat{\theta} - t_U^*\sqrt{\widehat{V}_B\left(\widehat{\theta}\right)}, \widehat{\theta} - t_L^*\sqrt{\widehat{V}_B\left(\widehat{\theta}\right)}\right)$, where t_L^* and t_U^* are lower and upper $100\alpha/2$ percent points of the BT histogram generated by the values of t_b. Although the BT t-method involves tedious computation, it is better in terms of coverage probability.

18.6.2 Bootstrap for Finite Population

Application of BT technique in finite sampling, especially in complex surveys, is not straightforward as demonstrated by Rao and Wu (1988),

Gross (1980), Bickel and Freedman (1984), and Sitter (1992a,b), among others. However, few applications are presented here.

18.6.2.1 Bootstrap for Simple Random Sampling With Replacement

Suppose a sample s of n units is selected from a finite population of N units by the SRSWR method and let $\mathbf{y}(n) = (y_1,\ldots, y_i,\ldots,y_n)$ be the observed values of the character y. Our objective is to estimate a population parameter θ, such as population mean, median, coefficient of variation, etc. Suppose that $\widehat{\theta} = \widehat{\theta}(\mathbf{y}(n))$ is an estimator of θ. We select a with-replacement random sample of size n from s assuming all the elements in s are distinct. Let us denote the selected sample by s^* and let the observed y-values be denoted by $\mathbf{y}^*(n) = \left(y_1^*,\ldots,y_i^*,\ldots,y_n^*\right)$. Let $\widehat{\theta}_{s^*} = \widehat{\theta}\left(\mathbf{y}^*(n)\right)$ be an estimator of θ based on the sample s^*. The selection of sample s^* is then repeated independently a large number of B times. The BT variance of $\widehat{\theta}$ may be calculated by using any of the following formulae

$$\widehat{V}_B(1) = \sum_{b=1}^{B}\left(\widehat{\theta}_{s^*}^{b} - \overline{\widehat{\theta}}\right)^2 \Big/ B \text{ or } \widehat{V}_B(2) = \sum_{b=1}^{B}\left(\widehat{\theta}_{s^*}^{b} - \widehat{\theta}\right)^2 \Big/ B \quad (18.6.2)$$

where $\widehat{\theta}_{s^*}^{b}$ is the value of $\widehat{\theta}_{s^*}$ based on bth BT sample $b = 1,\ldots,B$ and $\overline{\widehat{\theta}} = \sum_{b=1}^{B}\widehat{\theta}_{s^*}^{b}\Big/B.$

The BT confidence interval of θ may be computed using any of the formulae

$$\left(\widehat{\theta}_L^*, \widehat{\theta}_U^*\right) \text{ and } \left(\widehat{\theta} - t_U^*\sqrt{\widehat{V}_B(\bullet)}, \widehat{\theta} - t_L^*\sqrt{\widehat{V}_B(\bullet)}\right) \quad (18.6.3)$$

where $\widehat{\theta}_L^*$ and $\widehat{\theta}_U^*$ are the lower and upper $(\alpha/2)100$ percent points of the values of $\widehat{\theta}_{s_*}^{b}$; t_L^* and t_U^* denote lower and upper $(\alpha/2)100$ points of BT t distribution; $\widehat{V}_B(\bullet)$ may be taken as $\widehat{V}_B(1)$ or $\widehat{V}_B(2)$.

Example 18.6.1

Let the daily wages (in US$) of random sample s of five factory workers be as follows:

$$20, 30, 50, 80, \text{ and } 40$$

Suppose we want to estimate the population coefficient of variation (θ) and confidence intervals of θ by BT method, we proceed as follows:

First, we compute the coefficient of variation based on the full sample s as $\widehat{\theta} = 0.5232$. Then we select $B = 60$ BT samples each of size 5 from the sample s by the SRSWR method and compute the sample coefficient of variation $\widehat{\theta}_{s^*}^{b}$ as follows:

b	bootstrap sample s^*					$\widehat{\theta}_{s^*}^{b}$	t_b	b	bootstrap sample s^*					$\widehat{\theta}_{s^*}^{b}$	t_b	b	bootstrap sample s^*					$\widehat{\theta}_{s^*}^{b}$	t_b
1	30	80	30	80	40	0.498	−0.195	21	30	20	20	50	80	0.637	0.879	41	40	50	80	20	40	0.476	−0.365
2	30	40	50	50	20	0.343	−1.393	22	50	20	40	30	80	0.523	0.000	42	40	50	50	40	20	0.306	−1.679
3	40	80	50	50	30	0.374	−1.153	23	80	20	40	50	50	0.452	−0.550	43	40	20	80	30	30	0.586	0.485
4	20	40	50	80	30	0.523	−0.002	24	50	30	40	50	80	0.374	−1.150	44	20	30	80	30	20	0.697	1.343
5	40	20	80	80	30	0.566	0.331	25	50	20	30	50	40	0.343	−1.390	45	30	50	80	30	30	0.498	−0.195
6	40	40	50	40	50	0.124	−3.085	26	80	30	20	80	20	0.681	1.220	46	20	30	40	30	20	0.299	−1.733
7	50	20	80	50	30	0.500	−0.179	27	80	40	40	50	40	0.346	−1.370	47	20	80	40	20	50	0.593	0.539
8	50	40	80	80	50	0.312	−1.632	28	40	40	20	80	20	0.612	0.686	48	20	50	40	50	50	0.310	−1.648
9	80	30	80	40	50	0.411	−0.867	29	20	50	30	80	30	0.568	0.346	49	30	40	20	50	80	0.523	−0.002
10	40	30	40	40	50	0.177	−2.675	30	20	80	20	80	50	0.600	0.594	50	50	80	20	30	40	0.523	−0.002
11	80	50	80	20	80	0.433	−0.697	31	30	40	30	40	80	0.471	−0.400	51	30	40	20	50	30	0.335	−1.454
12	80	80	40	20	80	0.471	−0.403	32	30	80	20	80	30	0.614	0.702	52	30	40	20	30	80	0.586	0.485
13	80	30	20	20	40	0.655	1.019	33	50	80	30	40	50	0.374	−1.150	53	30	20	20	80	40	0.655	1.019
14	40	50	50	20	20	0.421	−0.790	34	80	40	30	20	80	0.566	0.331	54	50	20	30	80	40	0.523	−0.002
15	20	50	30	20	30	0.408	−0.890	35	80	20	30	30	50	0.568	0.346	55	80	30	30	40	20	0.586	0.485
16	30	80	50	30	50	0.427	−0.743	36	50	40	50	20	40	0.306	−1.680	56	50	20	50	80	20	0.570	0.362
17	30	40	80	20	30	0.586	0.485	37	80	30	30	80	40	0.498	−0.200	57	50	30	30	40	30	0.248	−2.127
18	20	40	80	30	80	0.566	0.331	38	80	40	50	40	20	0.476	−0.370	58	50	30	30	40	20	0.335	−1.454
19	30	20	80	30	50	0.568	0.346	39	80	20	40	20	40	0.612	0.686	59	80	30	50	40	30	0.451	−0.558
20	30	40	40	30	50	0.220	−2.343	40	40	30	50	80	20	0.523	0.000	60	40	50	50	20	20	0.421	−0.790

BT variance of the estimator of the sample coefficient of variation $\widehat{\theta}$(=sample standard deviation/sample mean) is

$$\widehat{V}_B(1) = \sum_{b=1}^{B}\left(\widehat{\theta}_{s^*}^{b} - \widehat{\overline{\theta}}\right)^2 \Big/ B = 0.0167$$

where $\widehat{\overline{\theta}} = \frac{1}{60}\sum_{b=1}^{60}\widehat{\theta}_{s^*}^{b} = 0.4708.$

To determine the confidence interval for θ by percentile method, we arrange $\widehat{\theta}_{s^*}^{b}$ in the ascending order of magnitude as follows:

0.124	0.177	0.220	0.248	0.299	0.306	0.306	0.310	0.312	0.335	0.335	0.343
0.343	0.346	0.374	0.374	0.374	0.408	0.411	0.421	0.421	0.427	0.433	0.451
0.452	0.471	0.471	0.476	0.476	0.498	0.498	0.498	0.500	0.523	0.523	0.523
0.523	0.523	0.523	0.566	0.566	0.566	0.568	0.568	0.568	0.570	0.586	0.586
0.586	0.586	0.593	0.600	0.612	0.612	0.614	0.637	0.655	0.655	0.681	0.697

Here, lower and upper 5% points of the $\widehat{\theta}_{s^*}^{b}$ values are 0.220 and 0.655, respectively. Hence 90% confidence interval for θ obtained by the percentile method is (0.220, 0.655).

For BT t-method, we arrange t_b values in the ascending order as follows:

−3.085	−2.675	−2.343	−2.127	−1.733	−1.679	−1.679	−1.648	−1.632	−1.454	−1.454	−1.393
−1.393	−1.369	−1.153	−1.153	−1.153	−0.890	−0.867	−0.790	−0.79	−0.743	−0.697	−0.558
−0.550	−0.403	−0.403	−0.365	−0.365	−0.195	−0.195	−0.195	−0.179	−0.002	−0.002	−0.002
−0.002	−0.002	−0.002	0.331	0.331	0.331	0.346	0.346	0.346	0.362	0.485	0.485
0.485	0.485	0.539	0.594	0.686	0.686	0.702	0.879	1.019	1.019	1.22	1.343

The lower and upper 5% percentile points of the BT t-distribution are −2.343 and 1.019, respectively. Hence 90% confidence interval for θ is

$$\left(\widehat{\theta} - (1.019)\sqrt{\widehat{V}_B\left(\widehat{\theta}\right)} - (-0.243)\sqrt{\widehat{V}_B\left(\widehat{\theta}\right)}\right) = (0.391, 0492).$$

18.6.2.2 Rescaling Bootstrap

Rao and Wu (1988) proposed various rescaling BT procedures for estimation of variance for stratified, multistage, and the Rao–Hartley–Cochran (1962) method of sampling designs. They all reproduce to usual variance estimators in the linear case. Consider the case where a sample s of size n is selected by the SRSWR sampling design. Then the variance of the sample mean $\overline{y}_s = \sum_{i \in s} y_i / n$ is unbiasedly estimated by $\widehat{V}(\overline{y}_s) = s_y^2 / n$, where $s_y^2 = \dfrac{\sum_{i \in s} (y_i - \overline{y}_s)^2}{n-1}$ and $\sum_{i \in s}$ denote the sum over units in s with repetition. In BT method we select B independent samples each of size n from s by SRSWR method treating all the units in s as distinct. Then the BT estimator of the variance of $\overline{y}_s$ is given by

$$\widehat{V}_B(1) = \sum_{b=1}^{B} \left(\overline{y}_b^* - \overline{y}^*\right)^2 / B \tag{18.6.4}$$

where $\overline{y}_b^*$ is the sample mean of the bth BT sample and $\overline{y}^* = \sum_{b=1}^{B} \overline{y}_b^* / B$. Let E_* denote the conditional expectation with respect to BT sampling given s, then $E_*\left(\widehat{V}_B(1)\right) = \dfrac{(n-1)}{n} \dfrac{s_y^2}{n}$, which is not equal to $\widehat{V}(\overline{y}_s) = \dfrac{s_y^2}{n}$, particularly if n is small. Rao and Wu (1988) proposed a rescaling BT method where the BT variance estimator for the variance of $\overline{y}_s$ is exactly equal to $\widehat{V}(\overline{y}_s)$. The proposed rescaling method is as follows:

Step 1: Draw BT sample (with SRSWR method) $s^* = \left(y_1^*, \ldots, y_i^*, \ldots, y_n^*\right)$ and calculate

$$\widetilde{y}_i = \overline{y}_s + \{n/(n-1)\}^{1/2} \left(y_i^* - \overline{y}_s\right) \tag{18.6.5}$$

and

$$\widetilde{\theta}^* = \widetilde{y}_{s^*} = \sum_{i \in s^*} \widetilde{y}_i / n \tag{18.6.6}$$

Step 2: Repeat step 1 independently B times and compute the corresponding estimates $\widetilde{y}_{s^*}^1, \ldots, \widetilde{y}_{s^*}^B$.

Step 3: The BT variance for $\overline{y}_s$ based on the rescaling technique is given by

$$\widehat{V}_{BRS}(1) = \sum_{b=1}^{B} \left(\widetilde{y}_{s^*}^b - \widetilde{y}^*\right)^2 / B \text{ or } \widehat{V}_{BRS}(2) = \sum_{b=1}^{B} \left(\widetilde{y}_{s^*}^b - \overline{y}_s\right)^2 / B \tag{18.6.7}$$

where $\widetilde{y}^* = \sum_{b=1}^{B} \widetilde{y}_{s^*}^b / B$.

We note that $E_*\left(\widehat{V}_{BRS}(1)\right) = E_*\left(\widehat{V}_{BRS}(2)\right) = \widehat{V}(\overline{y}_s) = s_y^2 / n$.

18.6.2.3 Bootstrap Without Replacement Method

Bootstrap without replacement (BWO) was proposed by Gross (1980) for variance estimation for the SRSWOR sampling design. Suppose a sample $s=(y_1,\ldots,y_n)$ of size n is selected from a finite population of size N by SRSWOR method and $N/n=k$ is an integer. Let the estimator of the parameter of interest θ, based on the sample s be $\widehat{\theta}=\widehat{\theta}(s)$. The variance of $\widehat{\theta}$ is computed by the BWO procedure is as follows:

Step 1: Generate a pseudopopulation by replication of the sample s, k times. The pseudopopulation is denoted by

$$U^*=\left(\overbrace{y_1,\ldots,y_n}^{1},\ldots,\overbrace{y_1,\ldots,y_n}^{k}\right).$$

Step 2: Draw an SRSWOR sample $s^*=\left(y_1^*,\ldots,y_n^*\right)$ of size n from the population U^* of size $N^*=nk$ assuming all elements in U^* are distinct. Let $\widehat{\theta}^*$ be the value of $\widehat{\theta}(s^*)$ based on the sample s^*.

Step 3: Repeat step 2 a large number times B and compute $\widehat{\theta}_1^*,\ldots,\widehat{\theta}_B^*$. The BT variance estimator of $\widehat{\theta}$ is given by

$$\widehat{V}_{BWO}\left(\widehat{\theta}\right)=\frac{1}{B}\sum_{b=1}^{B}\left(\widehat{\theta}_b^*-\overline{\theta}^*\right)^2 \tag{18.6.8}$$

where $\overline{\theta}^*=\frac{1}{B}\sum_{b=1}^{B}\widehat{\theta}_b^*$ ($\overline{\theta}^*$ may be replaced by $\widehat{\theta}$).

Sitter (1992a) observed that the $\widehat{V}_{BWO}\left(\widehat{\theta}\right)$ is quite different from $\widehat{V}\left(\overline{y}_s\right)=(1-f)s_y^2\big/n$, where $\widehat{\theta}=\sum_{i\in s}y_i/n$ and $f=n/N$.

18.6.2.4 Mirror-Match Bootstrap

Sitter (1992b) proposed the mirror-match BT method, which is applicable to the stratified random sampling, two-stage cluster sampling, and Rao–Hartley–Cochran (1962) sampling designs. All the variance estimates based on this method reduce to the standard ones in the linear case. For simplicity, let us consider the SRSWOR sampling scheme where a sample s of size n is selected from a finite population and let $d=(y_1,\ldots,y_n)$ be the observed values of the study variable y. Let $\widehat{\theta}=\theta(d)$ be an estimator for the parameter of interest θ. The mirror-match BT method is described as follows:

Step 1: Select a sample s' of size $n'(<n)$ from s by SRSWOR method. Let corresponding observations be $d=\left(\widetilde{y}_1,\ldots,\widetilde{y}_{n'}\right)$.

Step 2: Repeat step 1, $k = n(1-f^*)/\{n'(1-f)\}$ (assuming integer) times independently to generate y-values $\widetilde{\mathbf{y}}^* = (\widetilde{y}_1, \ldots, \widetilde{y}_{n^*})$ where $f^* = n'/n$ and $n^* = n(1-f^*)/(1-f)$. Let $\widehat{\theta}^m = \widehat{\theta}^m\left(\widetilde{\mathbf{y}}^*\right)$ be an estimator of θ based on $\widetilde{\mathbf{y}}^*$.

Step 3: Repeat steps 1 and 2, a large number of times B, and compute $\widehat{\theta}^m_1, \ldots, \widehat{\theta}^m_B$, where $\widehat{\theta}^m_j$ is the value of $\widehat{\theta}^m$ based on the jth iteration. The mirror-match estimator for the variance of $\widehat{\theta}$ is given by

$$\widehat{V}_{BM}\left(\widehat{\theta}\right) = \frac{1}{B}\sum_{b=1}^{B}\left(\widehat{\theta}^m_b - \overline{\theta}^m\right)^2 \text{ with } \overline{\theta}^m = \frac{1}{B}\sum_{b=1}^{B}\widehat{\theta}^m_b \tag{18.6.9}$$

($\overline{\theta}^m$ may be replaced by $\widehat{\theta}$).

In case $\widehat{\theta} = \overline{y}_s =$ the sample mean,

$E_*\left[\widehat{V}_{BM}\left(\widehat{\theta}\right)\right] = \widehat{V}(\overline{y}_s) = (1-f)s_y^2\big/n.$

18.6.2.5 Bootstrap for Varying Probability Sampling Without Replacement

Särndal et al. (1992) proposed the following method of BT for varying probability sampling schemes. Suppose a sample s of size n is selected from a finite population U of size N using a varying probability sampling scheme. Let π_i be the inclusion probability of the ith unit and $\widehat{\theta} = \widehat{Y}_{ht} = \sum_{i \in s} y_i/\pi_i$ be the Horvitz–Thompson estimator of the total $\theta = \sum_{i \in U} y_i$. We generate a population U^* by repeating units from the selected sample s. If a unit u_j is included in the sample s, then the unit u_j will appear $f_j = 1/\pi_j$ times in the population U^* assuming f_j is an integer for $j \in s$ so that the total number of units in U^* is $\widehat{N} = \sum_{j \in s} 1/\pi_j$. Now, from the population U^*, a sample s^* of size n is selected by PPSWR method using normed size measure $p_j = \pi_j/n$ for the unit $u_j \in s$. Clearly, $\sum_{k \in U^*} p_k = 1$. From the sample s^*, an estimator for θ is obtained as $\widehat{\theta}^* = \sum_{j \in s^*}\frac{y_j}{np_j}$. In the proposed BT method, we repeat the selection of sample s^* a large number of B times. Then the BT estimator of the variance of $\widehat{\theta}$ is

$$\widehat{V}_{BVP}\left(\widehat{\theta}\right) = \frac{1}{B}\sum_{b=1}^{B}\left(\widehat{\theta}^*_b - \overline{\theta}^*\right)^2 \tag{18.6.10}$$

where $\widehat{\theta}^*_b$ is the value of $\widehat{\theta}^*$ based on the bth BT sample and $\overline{\theta}^* = \frac{1}{B}\sum_{b=1}^{B}\widehat{\theta}^*_b$.

18.7 GENERALIZED VARIANCE FUNCTIONS

In many large-scale surveys such as US Current Population Surveys (CPS), National Health Improvement Surveys (USA), Medical Expenditure Panel Surveys (USA), Botswana Aids Impact Surveys (BAIS), hundreds or even thousands of estimates of population characteristics are calculated. In such a situation direct computation of standard errors of each of the estimates may not be possible even with modern computers because of cost and time. Even if the cost is affordable, timely presentation of hundreds or thousands of estimates with their standard errors is quite impractical because of limitations of tabular presentations. However, such estimates are of interest to users. The GVFs provide variances for groups of statistics rather than individual variances of each of the statistic.

18.7.1 Generalized Variance Function Model

A GVF is a simple mathematical relationship between the variance or relative variance of a statistic and its expectation. Let $\widehat{\theta}$ be an unbiased estimator of a parameter θ of interest and $V\left(\widehat{\theta}\right)$ be the variance of $\widehat{\theta}$. Then the relative variance is defined by

$$\Delta_\theta^2 = \frac{V\left(\widehat{\theta}\right)}{\theta^2} \tag{18.7.1}$$

The GVF may be modeled as follows:

$$\Delta_\theta^2 = \alpha + \frac{\beta}{\theta} \tag{18.7.2}$$

$$\Delta_\theta^2 = \alpha + \frac{\beta}{\theta} + \frac{\gamma}{\theta^2} \tag{18.7.3}$$

$$\Delta_\theta^2 = \frac{1}{\alpha + \beta\theta + \gamma\theta^2} \tag{18.7.4}$$

$$log\left(\Delta_\theta^2\right) = \alpha - \beta \log \theta \tag{18.7.5}$$

The model parameters α, β, and γ are unknown and these are estimated from the survey data. The simplest and most popular model is (18.7.2) where relative variance Δ_θ^2 is a decreasing function of θ when β is positive. The US Census Bureau used this model for CPS survey.

18.7.2 Justification of Generalized Variance Function Model

There is little theoretical justification of the aforementioned models available in the literature. However, some justifications have been provided by Wolter (1985) and Kish (1965). These are described as follows.

Let a sample s of n clusters be selected from a population of N clusters each of size M by the SRSWOR method. Then an unbiased estimator of the population total $Y = \theta$ is $\widehat{\theta} = \frac{N}{n}\sum_{i\in s} Y_i$, where $Y_i = \sum_{j=1}^{M} y_{ij}$ and y_{ij} is the value of the characteristic y for the jth unit of the ith cluster. The variance of $\widehat{\theta}$ is obtained from Eq. (12.2.5) as

$$V\left(\widehat{\theta}\right) = N^2\left(\frac{1}{n} - \frac{1}{N}\right)\frac{NM-1}{N-1}S_y^2\{1 + (M-1)\rho_c\}$$

where ρ is the intracluster correlation coefficient and S_y^2 is the population variance of y.

For Large M and N, $V\left(\widehat{\theta}\right)$ reduces to

$$V\left(\widehat{\theta}\right) = \frac{N^2M}{n}S_y^2\{1 + (M-1)\rho\}$$

The relative variance is given by

$$\Delta_{\widehat{\theta}}^2 = \frac{N^2M}{nY^2}S_y^2\{1 + (M-1)\rho\}$$

Let y_{ij} be binary that takes value 1 if jth unit of the ith cluster possesses a certain attribute and y_{ij} is 0 otherwise, and $\pi = Y/(NM)$ is the proportion of persons possessing the attribute, then

$$\Delta_{\widehat{\theta}}^2 = \frac{1}{nM}\frac{1-\pi}{\pi}\{1 + (M-1)\rho\}$$

$$= \alpha + \frac{\beta}{\theta}$$

where $\alpha = -\frac{1 + (M-1)\rho}{nM}$, $\beta = \frac{N\{1 + (M-1)\rho\}}{n}$, and $\theta = Y = NM\pi$.

Let $\theta(=\pi)$ be the population proportion and $\widehat{\pi}$ be the sample proportion based on an SRSWR sample of size n. Then the design effect (*Deff*) of fixed sample size n design (p) with respect to the estimator $\widehat{\theta}$ is defined by Kish (1965) as

$$Deff = \frac{V_p\left(\widehat{\theta}\right)}{V(\widehat{\pi})} = \frac{V_p\left(\widehat{\theta}\right)}{\pi(1-\pi)/n}$$

In this case the relative variance is

$$\Delta_\pi^2 = \frac{V_p\left(\widehat{\theta}\right)}{\pi^2}$$
$$= \frac{Deff(1-\pi)}{n\pi}$$

If the design effect for a class of statistics happens to be independent of the parameter π, then the relative variance Δ_π^2 can be written as

$$\Delta_\pi^2 = \alpha + \frac{\beta}{\pi}$$

with $\alpha = -Deff/n$ and $\beta = Deff/n$.

18.7.3 Generalized Variance Function Method for Variance Estimation

First, we need to group the survey estimates so that they follow a common model. The group may comprise estimates related to the same demographic or economic characteristics, similar geographical areas, the same race ethnicity, the same intraclass correlation ρ, or similar *Deffs*. The items to be included in a group may also be based on the past experience.

After forming a group, some estimates $\widehat{\theta}_1, \ldots, \widehat{\theta}_k$ belonging to a group are chosen and their variances $V\left(\widehat{\theta}_1\right), \ldots, V\left(\widehat{\theta}_k\right)$ are estimated using a direct method such as LR, RG, BRR, JK, or BT method. Then estimates of GVFs $\left(\Delta_\theta^2\right)$ are computed using the formula $\widehat{\Delta}_{\theta_j}^2 = \widehat{V}\left(\widehat{\theta}_j\right)/\widehat{\theta}_j^2$.

To determine appropriate model, scatter plots $\widehat{\Delta}_{\theta_j}^2$ against $\widehat{\theta}_j$ may be useful. One may remove statistics from the group that do not seem to follow the same model.

After selection of the appropriate model, the model parameters may be determined by the ordinary least square (OLS) procedure. However, the OLS procedure may not be totally satisfactory because some of the estimates with low value may have high value of the estimated relative variance. This difficulty can be overcome by using the weighted least square method with weights inversely proportional to the variance of the estimates.

Finally, the estimated GVF of $\widehat{\theta}_j$ based on the model is obtained from the estimated model, e.g., for the model (18.7.2) GVF of $\widehat{\theta}_j$ is

$$\widehat{\Delta}_{\theta_j}^2 = \widehat{\alpha} + \frac{\widehat{\beta}}{\widehat{\theta}_j} \tag{18.7.6}$$

where $\widehat{\alpha}$ and $\widehat{\beta}$ are the estimates of the model parameters α and β.

The variance of $\widehat{\theta}_j$ based on the GVF method is given by

$$\widehat{V}_{gvf}\left(\widehat{\theta}_j\right) = \widehat{\Delta}^2_{\hat{\theta}_j} \times \widehat{\theta}^2_j \tag{18.7.7}$$

18.7.4 Applicability Generalized Variance Function Model

There is a limited theoretical justification of the GVF models discussed earlier. Furthermore, grouping of estimates following the same model is not an easy task. However, the GVF method can be used when sufficient information is not available for direct computation of variance. GVF saves time and cost for production of reports. For further information readers are referred to Wolter (1985).

18.8 COMPARISON BETWEEN THE VARIANCE ESTIMATORS

For most complex survey designs, unbiased variance estimators of statistics that are expressible as linear functions of the observations can be derived. For nonlinear statistics and functional this is generally not the case. Various methods of obtaining variance estimation have been proposed in the literature. The merits of the proposed variance estimators can be judged with respect to the criteria of bias, mean square error, setting the confidence interval of the estimators, and applicability. It is very difficult to make theoretical comparisons among estimators. Little has been done to compare relative performances of the proposed estimators. Limited empirical studies reveal that none of the proposed variance estimators is the best in all situations.

The biases of LR, RG, JK, BRR, and BT are equal to a first-order of approximation, but regarding the mean square errors of the variance estimators, no definite conclusion can be made. Kreweski and Rao (1981) proved that the pivotal quantity $z = \dfrac{\hat{\theta} - \theta}{\sqrt{\hat{V}\left(\hat{\theta}\right)}}$ asymptotically follows standardized normal distribution for all the proposed methods (LR, RG, JK, BRR and BT) where $\widehat{\theta}$ and $\widehat{V}\left(\widehat{\theta}\right)$, respectively, denote estimator of θ and $V\left(\widehat{\theta}\right)$. So, all the variance estimators can be used for determination of confidence interval as well as testing hypothesis regarding the parameter θ when the sample size is large, but little is known about the distribution z when the sample size is small.

The RG method is easy to apply and works for any sampling design. The LR method needs theoretical derivation of variance and hence no standard programming can be used. This can make it cumbersome to implement. JK method is easy to apply and routine programming is available. BRR method is useful only for stratified sampling and can be complex for large stratum sample size. BT method is easy to apply with replacement sampling scheme but not readily applicable for varying probability without replacement sampling scheme. Although the GFS method does not possess adequate theoretical justification, it is useful where the variances of a large number of estimators are required to be computed or when sufficient information is not available for direct computation. Further details are given by Rao and Shao (1996, 1999), Rao and Wu (1988), Shao et al. (1998), Sitter (1992a,b), Saigo et al. (2001), and Wolter (1985).

18.9 EXERCISES

18.9.1 What is meant by a complex sampling design? Describe different methods of variance estimation in complex survey designs.

18.9.2 Explain the linearization method of variance estimation. Describe merits and demerits of the linearization method.

18.9.3 Let a sample s of size n be selected by a varying probability sampling scheme with inclusion probability π_i attached to the ith unit. Determine the expressions of bias and approximate mean square error of $t = \dfrac{\sum_{i\in s} y_i/\pi_i}{\sum_{i\in s} 1/\pi_i}$ as an estimator of the population mean $\overline{Y}$.

18.9.4 The population correlation coefficient between x and y is defined by

$$\rho = \frac{\sum_{i=1}^{N} x_i y_i - \frac{1}{N}\left(\sum_{i=1}^{N} x_i\right)\left(\sum_{i=1}^{N} y_i\right)}{\sqrt{\sum_{i=1}^{N} x_i^2 - \frac{1}{N}\left(\sum_{i=1}^{N} x_i\right)^2}\sqrt{\sum_{i=1}^{N} y_i^2 - \frac{1}{N}\left(\sum_{i=1}^{N} y_i\right)^2}}$$

Using LR method, derive approximate expressions of bias and mean square error of r when a sample of size n selected by the SRSWR method.

18.9.5 Express the population regression coefficient
$$\beta = \frac{\sum_{i=1}^{N} x_i y_i - \frac{1}{N}\left(\sum_{i=1}^{N} x_i\right)\left(\sum_{i=1}^{N} y_i\right)}{\sqrt{\sum_{i=1}^{N} x_i^2 - \frac{1}{N}\left(\sum_{i=1}^{N} x_i\right)^2}}$$
as a function of $\theta_0 = N$,

$\theta_1 = \frac{1}{N}\sum_{i=1}^{N} x_i, \theta_2 = \frac{1}{N}\sum_{i=1}^{N} x_i^2, \theta_3 = \frac{1}{N}\sum_{i=1}^{N} y_i$, and

$\theta_4 = \frac{1}{N}\sum_{i=1}^{N} x_i y_i$. Estimate the bias and mean square error of the estimate of β by LR technique based on a sample selected by varying probability sampling scheme with inclusion probabilities π_i and π_{ij} for the ith and ith and jth ($i \neq j$)th units.

18.9.6 The following table gives the number of cows and daily production of milk of 10 farms selected by SRSWOR method from 125 farms in a certain agricultural block.

Farm	1	2	3	4	5	6	7	8	9	10
No of cows	45	25	50	30	45	10	30	45	35	40
Production of milk in (L)	180	40	100	70	80	30	45	75	50	80

Estimate the average daily production of milk per cow. Obtain the standard error of the estimator used by using (i) linearization and (ii) jackknife method.

18.9.7 The following data give the monthly income distribution of 1000 families selected at random from 10,000 families in Durban. Estimate the income inequality using (i) sample coefficient of variation and (ii) sample variance as measure of inequality. Estimate the standard errors of your estimates by using (i) jackknife and (ii) random group method.

Income (in \$)	0 −1,000	1,001 −2,500	2,501 −5,000	5,001 −8,000	8,001 −18,000	18,001 −30,000	30,001 −50,000	50,001 −80,000
No. of families	100	250	300	180	100	75	20	5

18.9.8 The following table gives the IQ of 10 students selected at random from a school. Estimate the median IQ of the students and its standard error by bootstrap and jackknife method. Obtain 90% confidence interval of the population median.

Students	1	2	3	4	5	6	7	8	9	10
IQ	55	75	80	80	95	80	90	75	65	60

18.9.9 Samples each of size 2 are selected by SRSWOR from each of the six strata and information of the expenditure on food (x) and total income (y) are gathered.

Strata 1		Strata 2		Strata 3		Strata 4		Strata 5		Strata 6	
Size: 20		Size: 30		Size: 10		Size: 25		Size: 18		Size: 20	
x	y	x	y	x	y	x	y	x	y	x	y
1,000	2,500	1,250	3,000	3,000	5,000	3,500	6,000	4,000	8,000	4,000	9,500
1,500	2,000	2,500	4,000	2,500	5,500	3,000	7,500	5,000	8,500	6,000	10,000

(i) Estimate the proportion of income expended on food. (ii) Estimate the standard error of the proportion by (a) BRR, (b) linearization, and (c) random group method.

18.9.10 The following data give the distribution of daily wages of 1000 factory workers.

Daily wages (in $)	0–50	51–100	101–200	201–250
No. of workers	18	200	300	250
Daily wages (in $)	251–300	301–500	501–600	601–800
No. of workers	100	30	25	20

Estimate the median income and coefficient of variation of the income. Estimate the variances of the median income and coefficient of variance by (i) jackknife, (ii) random group, and (iii) bootstrap methods.

CHAPTER 19

Complex Surveys: Categorical Data Analysis

19.1 INTRODUCTION

Chi-square test statistics are extensively used in goodness of fit, tests of homogeneity, and independency. For example, in the Botswana Aids Impact Survey-II, importance was given to determine if the new HIV/AIDS management program yielded any improvement on the HIV/AIDS prevalence rate. In "Household Income and Expenditure Surveys," one may be interested in finding out if there is an association between income and level of education, or to test whether the income distribution remains unchanged over the last census period or not. Researchers often answer these questions by computing chi-square test statistics using SPSS, SAS, or other statistical packages for data analysis. The chi-square test statistics are valid if the sample size is large and the sample is selected by simple random sampling with replacement (SRSWR) method. But in practice complex survey designs involving stratification, clustering, and unequal probability sampling are used for collection of data. The effect of stratification, clustering, and unequal methods of selection of sample invalidates chi-square tests. So, the standard statistical packages report wrong p-values. Rao and Scott (1981) reported that the chi-square tests based on complex survey designs produced actual level of type I error as high as 40% when the target type 1 error is at the 5% level. In this chapter, we will consider the effect of unequal probability sampling, stratification, and clustering on the usual chi-square tests for goodness of fit, independency, and homogeneity. We also propose some modification of the existing chi-square tests, which yield reasonably valid tests.

19.2 PEARSONIAN CHI-SQUARE TEST FOR GOODNESS OF FIT

Suppose that a population $U = (U_1, \ldots, U_i, \ldots, U_N)$ of N units is divided into k mutually exclusive and exhaustive classes $\Omega_1, \ldots, \Omega_i, \ldots, \Omega_k$ according

Survey Sampling Theory and Applications
ISBN 978-0-12-811848-1
http://dx.doi.org/10.1016/B978-0-12-811848-1.00019-4

to some criteria. Let N_i be the number of units that belong to the ith class and the respective proportion be $p_i = N_i/N \left(0 \le p_i \le 1, \sum_{i=1}^{k} p_i = 1\right)$. The values of p_i's are not known and we are interested to test the following hypothesis of goodness of fit:

$$H_0: p_i = p_{i0} \quad \text{for } i = 1, \ldots, k \tag{19.2.1}$$

against the alternative H_1:$p_i \ne p_{i0}$ for at least one $i(=1,\ldots, k)$ where p_{i0}'s are known nonnegative constants with $\sum_{i=1}^{k} p_{i0} = 1$.

Suppose a sample s of size n units is selected by SRSWR method and n_i is the number of units falling in the ith class. Then under H_0, the Pearson chi-square test statistic

$$X_P^2 = n \sum_{i=1}^{k} \frac{\left(\widetilde{p}_i - p_{i0}\right)^2}{p_{i0}} \tag{19.2.2}$$

follows χ_{k-1}^2, chi-square distribution with $k - 1$ degrees of freedom (df) with $\widetilde{p}_i = n_i/n$. We reject the hypothesis H_0 at 100 α% level of significance if the computed value of X_P^2 exceeds $\chi_{\alpha,k-1}^2$, the upper 100 α% point of chi-square distribution with $k - 1$ df. The test is valid if the sample size n is large and the expected cell frequency ($n_{io} = np_{i0}$) for each cell exceeds 5.

The likelihood-ratio chi-squared statistic

$$G^2 = 2 \sum_{i=1}^{k} n_i \log(n_i/n_{i0}) \tag{19.2.3}$$

may be used as an alternative to Pearsonian chi-square statistic X_P^2. Under the null hypothesis H_0, G^2 follows a chi-square distribution with $k - 1$ df when the sample size n is large. Convergence of chi-squared is quicker than G^2 and the approximation of G^2 to chi-square is poor if $n/k < 5$ (Agresti, 2002).

19.3 GOODNESS OF FIT FOR A GENERAL SAMPLING DESIGN

Suppose a sample s of size n is selected from the population $U(=\Omega_1 \cup \ldots \cup \Omega_k)$ by some sampling design with probability $p(s)$. Let $\pi_i(>0)$ and $\pi_{ij}(i \ne j)$ be the inclusion probabilities of the ith, ith and jth units of the population. Let s_i be the set of units that belong to the ith class

Ω_i and y_j be the value of a certain character y for the jth ($j = 1,\ldots, N$) unit. Consider an unbiased estimator for the population total Y as

$$\widehat{Y} = \sum_{j\in s} b_{sj} y_j \tag{19.3.1}$$

where b_{sj}'s are constants that satisfy the unbiasedness condition

$$\sum_{s\supset j} b_{sj} p(s) = 1 \quad \text{for } j = 1, \ldots, N$$

Using Särndal et al. (1992) notation, an unbiased estimator of $Y_i\left(= \sum_{j\in\Omega_i} y_i\right)$, the total of the y-values of the units belonging to Ω_i is given by

$$\widehat{Y}_i = \sum_{j\in s} b_{sj} y_j I_j(i) \tag{19.3.2}$$

where $I_j(i) = \begin{cases} 1 & \text{if the } j\text{th unit belong to } \Omega_i \\ 0 & \text{otherwise} \end{cases}$.

Substituting $y_j = 1$ for $j = 1,\ldots, N$ in Eq. (19.3.2) we find an unbiased estimator of $p_i = N_i/N$ as

$$\widehat{p}_i = \frac{\sum_{j\in s} b_{sj} I_j(i)}{N} \tag{19.3.3}$$

The variance of $\widehat{p}_i$ and covariance of $\widehat{p}_i$ and $\widehat{p}_{i'}$ are, respectively, given by

$$\begin{aligned} Var(\widehat{p}_i) &= \sum_{j\in U} I_j(i)\alpha_{jj} + \sum_{j\neq}\sum_{k\in U} I_j(i) I_k(i) \alpha_{jk} \\ &= \sum_{j\in\Omega_i} \alpha_{jj} + \sum_{j\neq}\sum_{k\in\Omega_i} \alpha_{jk} \\ &= V_{ii}/n \end{aligned} \tag{19.3.4}$$

and

$$\begin{aligned} Cov(\widehat{p}_i, \widehat{p}_{i'}) &= \sum_{j\in\Omega_i}\sum_{k\in\Omega_{i'}} \alpha_{jk} \\ &= V_{ii'}/n \end{aligned} \tag{19.3.5}$$

where $\alpha_{jj} = \left(\sum_{s\supset j} b_{sj}^2 p(s) - 1\right) \Big/ N^2$ and $\alpha_{jk} = \left(\sum_{s\supset j,k} b_{sj} b_{sk} p(s) - 1\right) \Big/ N^2$.

Unbiased estimators of V_{ii} and $V_{ii'}$ are given by

$$\widehat{V}_{ii}/n = \left(\sum_{j\in s_i}\frac{\alpha_{jj}}{\pi_j} + \sum_{j\neq}\sum_{k\in s_i}\frac{\alpha_{jk}}{\pi_{jk}}\right) \text{ and } \widehat{V}_{ii'}/n = \left(\sum_{j\neq s_i}\sum_{k\in s_{i'}}\frac{\alpha_{jk}}{\pi_{jk}}\right). \tag{19.3.6}$$

Example 19.3.1

In case $b_{sj} = 1/\pi_j$, we get $\widehat{p}_i = \dfrac{\sum_{j\in s} I_j(i)/\pi_j}{N}$,

$$V_{ii}/n = \left[\sum_{j\in\Omega_i}\left(\frac{1}{\pi_j}-1\right) + \sum_{j\neq}\sum_{k\in\Omega_i}\left(\frac{\pi_{jk}}{\pi_j\pi_k}-1\right)\right]\Big/N^2,$$

$$V_{ii'}/n = \sum_{j\in\Omega_i}\sum_{j'\in\Omega_{i'}}\left(\frac{\pi_{jj'}}{\pi_j\pi_{j'}}-1\right)\Big/N^2,$$

$$\widehat{V}_{ii}/n = \left[\sum_{j\in s_i}\frac{1}{\pi_j}\left(\frac{1}{\pi_j}-1\right) + \sum_{j\neq}\sum_{k\in s_i}\frac{1}{\pi_{jk}}\left(\frac{\pi_{jk}}{\pi_j\pi_k}-1\right)\right]\Big/N^2, \text{ and}$$

$$\widehat{V}_{ii'}/n = \sum_{j\in s_i}\sum_{k\in s_{i'}}\frac{1}{\pi_{jk}}\left(\frac{\pi_{jk}}{\pi_j\pi_k}-1\right)\Big/N^2.$$

Example 19.3.2

For simple random sampling without replacement (SRSWOR), $\pi_j = n/N$ and $\pi_{jk} = n(n-1)/\{N(N-1)\}$. In this case, we have from Example 19.3.1

$$\widehat{p}_i = \sum_{j\in s} I_j(i)/n = n_i/n = \widetilde{p}_i,\ V_{ii} = \frac{N-n}{(N-1)}p_i(1-p_i),\ V_{ii'} = -\frac{N-n}{(N-1)}p_ip_{i'},$$

$$\widehat{V}_{ii}/n = \frac{N-n}{N(n-1)}\widetilde{p}_i\left(1-\widetilde{p}_i\right) \text{ and } \widehat{V}_{ii'}/n = -\frac{N-n}{N(n-1)}\widetilde{p}_i\widetilde{p}_{i'}.$$

Example 19.3.3

If we choose $\widehat{p}_i = \dfrac{\sum_{j\in s} I_j(i)/\pi_j}{\sum_{j\in s} 1/\pi_j}$ and $\widehat{p}_{i'} = \dfrac{\sum_{j\in s} I_j(i')/\pi_j}{\sum_{j\in s} 1/\pi_j}$, as ratio estimators for p_i and $p_{i'}$, respectively, then the approximate consistent estimators of the variance of $\widehat{p}_i$ and covariance of $\widehat{p}_i$ and $\widehat{p}_{i'}$ are obtained from the Theorem 8.2.2.

19.3.1 Wald Statistic for Goodness of Fit

Let $\widehat{p}_i \geq 0$ for $i = 1,\ldots, k-1$, $\widehat{p}_k = 1 - \sum_{i=1}^{k-1} \widehat{p}_i \geq 0$, $\widehat{\mathbf{p}} = (\widehat{p}_1, \ldots, \widehat{p}_{k-1})'$, $\mathbf{p} = (p_1, \ldots, p_{k-1})'$, $\mathbf{p}_0 = (p_{10}, \ldots, p_{k-1,0})'$, $\mathbf{V}/n = (V_{ij})/n =$ variance–covariance matrix of $\widehat{\mathbf{p}}$ of rank $k-1$, and $\widehat{\mathbf{V}}/n = (\widehat{V}_{ij})/n$ be unbiased or consistent estimator of $\mathbf{V}/n$ of rank $k-1$. The Wald statistic for goodness of fit is defined by

$$
\begin{aligned}
X_W^2 &= n(\widehat{\mathbf{p}} - \mathbf{p}_0)'\mathbf{V}^{-1}(\widehat{\mathbf{p}} - \mathbf{p}_0) \\
&\cong n(\widehat{\mathbf{p}} - \mathbf{p}_0)'\widehat{\mathbf{V}}^{-1}(\widehat{\mathbf{p}} - \mathbf{p}_0)
\end{aligned}
\tag{19.3.7}
$$

For a large n, X_W^2 is distributed as a chi-square distribution with $k-1$ df (χ^2_{k-1}) when H_0 is true. To use X_W^2 in practice, one needs a consistent estimator $\widehat{\mathbf{V}}$ of $\mathbf{V}$ in addition to the requirement of a large sample size. In case an unbiased estimator of $\mathbf{V}$ is not available because of complexity of the survey design, one may estimate variance by using methods such as linearization (LR), jackknife, balanced repeated replication (BRR), and bootstrap (BT) discussed in Chapter 18. The main drawback of the Wald Statistic is that if the number of classes is large then the sample sizes for some of the classes may become too small to provide consistent estimators for variances of the respective classes (see Fay, 1985). Furthermore, survey reports very often give estimates with their standard error but rarely publish estimates of the covariances.

19.3.1.1 Simple Random Sampling With Replacement

For SRSWR sampling, $\widehat{p}_i = \widetilde{p}_i = n_i/n$ where $n_i =$ number of unit following the ith class and $\widehat{\mathbf{p}}$ follows multinomial distribution with variance–covariance matrix $\mathbf{V}/n = (V_{ij})/n$, where $V_{ii} = p_i(1-p_i)$ and $V_{ij} = -p_ip_j$. Writing $\mathbf{V}^{-1}\big|_{\mathbf{p}=\mathbf{p}_0} = \mathbf{P}_0^{-1} = \left(\delta_0^{ij}\right)$ with $\delta_0^{ii} = \left(1/p_{i,0} + 1/p_{k,0}\right)$ and $\delta_0^{ij} = 1/p_{k,0}$, we find under H_0,

$$
\begin{aligned}
X_W^2 &= n(\widehat{\mathbf{p}} - \mathbf{p}_0)'\mathbf{P}_0^{-1}(\widehat{\mathbf{p}} - \mathbf{p}_0) \\
&= n\sum_{i=1}^{k} \frac{(\widetilde{p}_i - p_{i0})^2}{p_{i0}} \\
&= X_P^2
\end{aligned}
\tag{19.3.8}
$$

Alternatively, writing $\widehat{V}_{ii} = \widetilde{p}_i(1-\widetilde{p}_i)$ and $\widehat{V}_{ij} = -\widetilde{p}_i\widetilde{p}_j$, we find

$$\begin{aligned} X_W^2 &= n(\widehat{\mathbf{p}} - \mathbf{p}_0)'\widehat{\mathbf{V}}^{-1}(\widehat{\mathbf{p}} - \mathbf{p}_0) \\ &= n\sum_{i=1}^{k}\frac{(\widetilde{p}_i - p_{i0})^2}{\widetilde{p}_i} \\ &= X_N^2 \end{aligned} \tag{19.3.9}$$

The statistic X_N^2 is known as Neyman statistic.

Replacing $\widetilde{p}_i$ by a consistent estimator $\widehat{p}_i$, the expression (Eq. 19.3.9) becomes

$$X_M^2 = n\sum_{i=1}^{k}\frac{(\widehat{p}_i - p_{i0})^2}{\widehat{p}_i} \tag{19.3.10}$$

The statistic X_M^2 is known as modified chi-square statistic. Bhapkar (1966) established equivalency of the modified chi-square and Wald statistic for linear hypothesis.

19.3.2 Generalized Pearsonian Chi-Square Statistic

The estimator $\widetilde{p}_i = n_i/n$ is not a consistent estimator of p_i unless the sampling design is self-weighing. So, replacing $\widetilde{p}_i$ by $\widehat{p}_i$, an unbiased or consistent estimator of p_i in the expression of X_P^2 given in Eq. (19.2.2), we derive the generalized Pearsonian chi-square statistic for a complex survey design as follows:

$$X_G^2 = n\sum_{i=1}^{k}\frac{(\widehat{p}_i - p_{i0})^2}{p_{i0}} \tag{19.3.11}$$

Rao and Scott (1981) provided the asymptotic distribution of X_G^2, which is stated in the following theorem without derivation.

Theorem 19.3.1

Let $\lambda_{10},\ldots,\lambda_{k-1,0}$ be the eigenvalues of $\mathbf{D}_0 = \mathbf{P}_0^{-1}\mathbf{V}_0$, where $\mathbf{V}_0$ and $\mathbf{P}_0$ are the values of $\mathbf{V}$, and $\mathbf{P} = diag(\mathbf{p}) - \mathbf{pp}'$ for $\mathbf{p} = \mathbf{p}_0$, then under the null hypothesis H_0:$\mathbf{p} = \mathbf{p}_0$, the asymptotic distribution of X_G^2 is

$$X_G^2 \sim \sum_{i=1}^{k-1}\lambda_{i0}Z_i^2 \tag{19.3.12}$$

where $Z_1, \ldots, Z_{k-1}$ are independent normal variates with mean zero and variance unity.

Hence X_G^2 is asymptotically distributed as a weighted sum of independent χ_1^2 (chi-square with 1 df) variables.

19.3.2.1 Design Effect

The design effect (deff) of a sampling design d with respect to an estimator T is defined as $V_d(T)/V_{srs}(t)$, where $V_d(T)$ is the variance of T with respect to the design d and $V_{srs}(t)$ is the variance of a comparable estimator based on an SRSWR sampling design.

Let $\lambda_1, \ldots, \lambda_{k-1}$ be the eigenvalues of $\mathbf{D} = \mathbf{P}^{-1}\mathbf{V}$, then following Rao and Scott (1981), we find for any arbitrary vector $\mathbf{c}' = (c_1, \ldots, c_{k-1})$,

$$\lambda_{\max} = \sup_c \frac{\mathbf{c}'\mathbf{V}\mathbf{c}}{\mathbf{c}'\mathbf{P}\mathbf{c}} = \sup_c \frac{V_d\left(\sum_{i=1}^{k-1} c_i\widehat{p}_i\right)}{V_{srs}\left(\sum_{i=1}^{k-1} c_i n_i/n\right)} \quad (19.3.13)$$

and

$$\lambda_{\min} = \inf_c \frac{\mathbf{c}'\mathbf{V}\mathbf{c}}{\mathbf{c}'\mathbf{P}\mathbf{c}} = \inf_c \frac{V_d\left(\sum_{i=1}^{k-1} c_i\widehat{p}_i\right)}{V_{srs}\left(\sum_{i=1}^{k-1} c_i n_i/n\right)} \quad (19.3.14)$$

Thus $\lambda_{\max}(\lambda_{\min})$ is the largest (smallest) deffs over all possible linear combination of the $\widehat{p}_i$'s. Rao and Scott (1981) termed λ_i's as generalized deffs.

In case λ_i's or their consistent estimates $\widehat{\lambda}_i$ are known, one can attain good approximations for the percentage points of the asymptotic distribution of X_G^2 using Solomon and Stephens (1977). But knowledge of λ_i's (or $\widehat{\lambda}_i$'s) require the knowledge of $\mathbf{V}$ or $\widehat{\mathbf{V}}$ and if $\mathbf{V}$ or $\widehat{\mathbf{V}}$ is known, one could use the Wald Statistic.

19.3.3 Modifications to X_G^2

19.3.3.1 Use of Maximum or Minimum Eigenvalues

From the expression (Eq. 19.3.12) we note

$$\frac{X_G^2}{\lambda_{\max,0}} \le \sum_{i=1}^{k-1} Z_i^2 = \chi_{k-1}^2 \quad (19.3.15)$$

where χ^2_{k-1} follows chi-square distribution with $k-1$ df and $\lambda_{\max,0} = \max\{\lambda_{i0}\}$.

If $\lambda_{\max,0}$ or its reliable estimate is known and one treats $\frac{X_G^2}{\lambda_{\max,0}}$ as a chi-square variable with $k-1$ df for testing the hypothesis $H_0{:}\mathbf{p} = \mathbf{p}_0$, then the test would result in a conservative test, which produces a lower significance level. In other words, suppose we set a level of significance α and reject H_0 if $\frac{X_G^2}{\lambda_{\max,0}} > \chi^2_{\alpha,k-1}$, where $\chi^2_{\alpha,k-1}$ is the upper 100 α% point of χ^2_{k-1}, then the true level of significance (type I error probability) of $\frac{X_G^2}{\lambda_{\max,0}}$ is lower than α. Similarly, treating $\frac{X_G^2}{\lambda_{\min,0}}$ with $\lambda_{\min,0} = \min\{\lambda_{i0}\}$ as a chi-square variable with $k-1$ df will produce a test with much higher significance level than the desired significance level.

19.3.3.2 Rao–Scott First-Order Corrections

Consider the test statistic

$$X_A^2 = X_G^2/\overline{\lambda}_0 \tag{19.3.16}$$

where

$$\overline{\lambda}_0 = \sum_{i=1}^{k-1} \lambda_{i0}/(k-1) \tag{19.3.17}$$

Noting

$$E\left(X_A^2\right) = E\left(X_G^2\right)/\overline{\lambda}_0 = \sum_{i=}^{k-1} \lambda_{i0} E\left(Z_i^2\right)/\overline{\lambda}_0 = (k-1)$$

one may treat $\widehat{X}_A^2 = X_G^2 \Big/ \widehat{\overline{\lambda}}_0$ as χ^2_{k-1} with $\widehat{\overline{\lambda}}_0$ as an estimated value of $\overline{\lambda}_0$ and reject H_0 if $\widehat{X}_A^2 > \chi^2_{\alpha;k-1}$. In this case, the desired level of significance may be achieved approximately provided $\widehat{\lambda}_{i0}$'s do not vary much among themselves. One advantage of using $\widehat{\overline{\lambda}}_0$ is that it requires only estimation of V_{ii}'s but does not require estimation of V_{ij}'s because

$$\widehat{\overline{\lambda}}_0 = \frac{1}{(k-1)} \sum_{i=1}^{k} \frac{\widehat{V}_{ii}}{p_{i0}} = \frac{1}{(k-1)} \sum_{i=1}^{k} \frac{\widehat{p}_i\left(1-\widehat{p}_i\right)\widehat{d}_i}{p_{i0}}$$

with $\widehat{d}_i = \widehat{V}_{ii}/\left[\widehat{p}_i\left(1-\widehat{p}_i\right)\right]$ = estimated deff of the ith cell proportion.

19.3.3.3 Rao–Scott Second-Order Corrections

Consider

$$X_B^2 = \frac{X_A^2}{(1+a^2)} = \frac{X_G^2}{\overline{\lambda}_0(1+a^2)} \tag{19.3.18}$$

where a^2 is a constant, which makes

$$E\left(X_B^2\right) = E\left(\chi_\nu^2\right) = \nu \tag{19.3.19}$$

and

$$\begin{aligned} V\left(X_B^2\right) &= \frac{2}{\left(1+a^2\right)^2}\left[(k-1) + \sum_{i=1}^{k-1} \frac{\left(\lambda_{i0} - \overline{\lambda}_0\right)^2}{\overline{\lambda}_0^2}\right] \\ &= V\left(\chi_\nu^2\right) = 2\nu \end{aligned} \tag{19.3.20}$$

Eqs. (19.3.19) and (19.3.20) yield

$$\nu = \frac{(k-1)}{1+a^2} \text{ and } a^2 = \frac{\sum_{i=1}^{k-1}\left(\lambda_{i0} - \overline{\lambda}_0\right)^2}{(k-1)\overline{\lambda}_0^2} \tag{19.3.21}$$

Thus under H_0, the statistic $\widehat{X}_B^2 = \dfrac{\widehat{X}_A^2}{\left(1+\widehat{a}^2\right)}$ with $\widehat{a}^2 = \dfrac{\sum_{i=1}^{k-1}\left(\widehat{\lambda}_{i0} - \widehat{\overline{\lambda}}_0\right)^2}{(k-1)\widehat{\overline{\lambda}}_0^2}$ is a good approximation of χ_ν^2 in the sense that it has first two order moments exactly equal to χ_ν^2. Thus the test statistic $\widehat{X}_B^2$ provides the desired level of significance. The main disadvantage of using $\widehat{X}_B^2$ is that we need to estimate a^2, which in turn requires estimation of $\sum_{i=1}^{k-1} \lambda_{i0}^2 = \sum_{i=1}^{k}\sum_{j=1}^{k} V_{ij}^2 \Big/ \left(p_{i0}p_{j0}\right)$, i.e., estimates of covariances V_{ij}/n's are needed in the calculation of $\widehat{X}_B^2$.

19.3.3.4 Fellegi Correction

In case estimates of V_{ij}'s are either not available or reliable, the following mean deff adjustment was proposed by Fellegi (1980)

$$X_F^2 = X_G^2 \Big/ \widehat{\overline{d}} \tag{19.3.22}$$

where $\widehat{\overline{d}} = \sum_{i=1}^{k} \widehat{d}_i \Big/ k$ and $\widehat{d}_i = \widehat{V}_{ii} / \left[\widehat{p}_i\left(1-\widehat{p}_i\right)\right]$.

19.3.4 Simple Random Sampling Without Replacement

For SRSWOR sampling, $\mathbf{V} = (1 - n/N)\mathbf{P}$ and hence $\mathbf{D}_0 = \mathbf{P}_0^{-1}\mathbf{V}_0 = (1 - n/N)\mathbf{I}_{\mathbf{k}-1}$. In this case $\lambda_{10} = \ldots = \lambda_{k-10} = 1 - n/N$ and $X_G^2/(1 - n/N)$ is asymptotically distributed as χ_{k-1}^2. Thus treating X_G^2 as χ_{k-1}^2, one achieves the desired level of significance provided the sampling fraction n/N is negligible. But if n/N is not negligible, the test statistic X_G^2 will provide a lower significance level.

19.3.5 Stratified Sampling

Consider a population stratified into L strata. Let W_h and p_{hi}, respectively, denote the population proportion of units from the stratum h and proportion of elements from stratum h that belong to the ith category. Let a sample s_h of size m_h be selected from the hth stratum by SRSWR and m_{hi} be the number of units that belong to ith category. Then $p_i = \sum_h W_h p_{hi}$, the population proportion of units that belong to the ith category can be estimated by

$$\widehat{p}_i = \sum_h W_h \widehat{p}_{hi} \quad \text{where } \widehat{p}_{hi} = m_{hi}/m_h \tag{19.3.23}$$

For proportional allocation $m_h = nW_h$ with $n = \sum_{h=1}^{L} m_h$. In this case $\widehat{p}_i$ reduces to n_i/n, where n_i is the total number of units in the sample $s(=s_1 \cup \ldots \cup s_h)$ that fall in the ith category. The variance of $\widehat{p}_i$ and the covariance of $\widehat{p}_i$ and $\widehat{p}_j$ under proportional allocation are, respectively, given by

$$Var\left(\widehat{p}_i\right) = \frac{1}{n}\sum_{h=1}^{L} W_h p_{hi}(1 - p_{hi}) = \frac{p_i}{n} - \frac{1}{n}\sum_{h=1}^{L} W_h p_{hi}^2 \text{ and}$$

$$Cov\left(\widehat{p}_i, \widehat{p}_j\right) = -\frac{1}{n}\sum_{h=1}^{L} W_h p_{hi} p_{hj}.$$

The variance–covariance matrix of $\widehat{\mathbf{p}} = \left(\widehat{p}_1, \ldots, \widehat{p}_{k-1}\right)'$ is $\mathbf{V}(st)/n$ with

$$\begin{aligned}\mathbf{V}(st) &= diag(\mathbf{p}) - \sum_{h=1}^{L} W_h \mathbf{p}_h \mathbf{p}_h' \\ &= \mathbf{P} - \sum_{h=1}^{L} W_h (\mathbf{p}_h - \mathbf{p})(\mathbf{p}_h - \mathbf{p})'\end{aligned} \tag{19.3.24}$$

where $\mathbf{p}_h = (p_{h,1}, \ldots, p_{h,k-1})'$ and $\mathbf{P}$ is as in Theorem 19.3.1.

Noting $0 \le \mathbf{c}'\mathbf{V}(st)\mathbf{c}/\mathbf{c}'\mathbf{P}\mathbf{c} \le 1$, we find that all the eigenvalues of $\mathbf{D}_0 = \mathbf{P}^{-1}\mathbf{V}(st)\big|_{\mathbf{p}=\mathbf{p}_0}$ are less than 1. Hence under $\mathbf{H}_0$,

$$0 \le X_G^2 \le \sum_{i=1}^{k-1} Z_i^2 \approx \chi_{k-1}^2 \tag{19.3.25}$$

Thus for stratified sampling, Pearsonian chi-square X_G^2 in Eq. (19.3.11) always becomes an asymptotically conservative test. Rao and Scott (1981) showed that for two strata ($L = 2$)

$$X_G^2 \cong \chi_{k-2}^2 + (1-\delta_0)\chi_1^2 \tag{19.3.26}$$

where $\delta_0 = W_1 W_2 \sum_{i=1}^{k} (p_{1i} - p_{2i})^2 / p_{i0}$ is the minimum eigenvalue of $\mathbf{D}_0$.

Hence, X_G^2 can be asymptotically well approximated by χ_{k-1}^2 unless k is small.

19.3.6 Two-Stage Sampling

Consider a two-stage sampling where a population consists of R first-stage units (fsu's) and hth fsu consists of M_h second-stage units (ssu's). A sample s of r fsu's is selected from R fsu's by probability proportional to size with replacement sampling using $W_h = M_h/M$ $\left(M = \sum_{h=1}^{R} M_h\right)$ as a normed size measure for the hth fsu. If the hth fsu is selected in the sample s, a subsample s_h of m ssu's is selected from it by SRSWR method. So, the total number of the selected ssu's in the sample s is $n = rm$. Let us denote $y_{hj}(i)$ as 1 if the jth ssu of the hth fsu belong to the ith category and $y_{hj}(i)$ is 0 otherwise. Then the proportion of units that belong to the ith category in s_h is $\widehat{p}_{hi} = m_{hi}/m = \overline{y}_h(i) = \sum_{j \in s_h} y_{hj}(i)/m$ where m_{hi} is the total number of ssu's that belong to the ith category and $\sum_{j \in s_h}$ denotes the sum over fsu's in the sample s_h including repetition. The proportion of units belonging to the ith category in the population is $p_i = \sum_{h=1}^{R} W_h \overline{Y}_h(i) = \sum_{h=1}^{R} W_h p_{hi}$, where $\overline{Y}_h(i) = \sum_{j=1}^{M_h} y_{hj}(i)/M_h = p_{hi}$.

Theorem 19.3.2

(i) $\widehat{p}_i = \frac{1}{r}\sum_{h\in s}\overline{y}_h(i) = \frac{1}{r}\sum_{h\in s}\widehat{p}_{hi}$ is unbiased for p_i

(ii) $V\left(\widehat{p}_i\right) = \frac{1}{n}\left[p_i(1-p_i) + (m-1)\sum_{h=1}^{R} W_h(p_{hi}-p_i)^2\right]$

(iii) $Cov\left(\widehat{p}_i, \widehat{p}_j\right) = \frac{1}{n}\left[-p_ip_j + (m-1)\sum_{h=1}^{R} W_h\{p_h(i)-p_i\}\{p_h(j)-p_j\}\right]$

Proof

(i) $$E\left(\widehat{p}_i\right) = E\left[\frac{1}{Mr}\sum_{h\in s}\frac{1}{W_h}M_h E\{\overline{y}_h(i)|s\}\right]$$

$$= \frac{1}{M}\sum_{h=1}^{R} M_h \overline{Y}_h(i) = p_i.$$

(ii) $$V\left(\widehat{p}_i\right) = E\left[V\left(\widehat{p}_i|s\right)\right] + V\left[E\left(\widehat{p}_i|s\right)\right]$$

$$= E\left[\frac{1}{M^2r^2}\sum_{h\in s}\frac{1}{W_h^2}M_h^2\frac{p_{hi}(1-p_{hi})}{m}\right] + V\left(\frac{1}{Mr}\sum_{h\in s}\frac{1}{W_h}M_h p_{hi}\right)$$

$$= \frac{1}{n}\left[\sum_{h=1}^{R} W_h p_{hi}(1-p_{hi}) + m\left(\sum_{h=1}^{R} W_h p_{hi}^2 - p_i^2\right)\right]$$

$$= \frac{1}{n}\left[p_i(1-p_i) + (m-1)\sum_{h=1}^{R} W_h(p_{hi}-p_i)^2\right]$$

$$\text{(iii) } Cov\left(\widehat{p}_i, \widehat{p}_j\right) = Cov\left[E\left(\widehat{p}_i, \widehat{p}_j \middle| s\right)\right] + E\left[Cov\left(\widehat{p}_i, \widehat{p}_j \middle| s\right)\right]$$

$$= \frac{1}{M^2}\left[Cov\left\{\frac{1}{r}\sum_{h\in s}\frac{1}{W_h}M_h\overline{Y}_h(i), \frac{1}{r}\sum_{h\in s}\frac{1}{W_h}M_h\overline{Y}_h(j)\right\}\right]$$

$$+ \frac{1}{M^2}\left[E\left\{Cov\left(\frac{1}{r}\sum_{h\in s}\frac{1}{W_h}M_h\overline{y}_h(i), \frac{1}{r}\sum_{h\in s}\frac{1}{W_h}M_h\overline{y}_h(j) \middle| s\right)\right\}\right]$$

$$= \frac{1}{M^2}\left[\frac{1}{r^2}\sum_{h\in s}Cov\left(\frac{M_h}{W_h}\overline{Y}_h(i), \frac{M_h}{W_h}\overline{Y}_h(j)\right)\right.$$

$$\left. - E\left\{\frac{1}{r^2 m}\sum_{h\in s}\left(\frac{M_h}{W_h}\right)^2\overline{Y}_h(i)\overline{Y}_h(j)\right\}\right]$$

$$= \frac{1}{M^2}\frac{1}{r}\left[\left\{E\left(\frac{M_h}{W_h}p_{hi}\frac{M_h}{W_h}p_{hj}\right) - M^2 p_i p_j\right\} - \frac{1}{m}\sum_{l=1}^{R}\frac{M_h^2}{W_h}p_{hi}p_{hj}\right]$$

$$= \frac{1}{r}\left[\sum_{h=1}^{R}W_h p_{hi}p_{hj} - p_i p_j - \frac{1}{m}\sum_{l=1}^{R}W_h p_{hi}p_{hj}\right]$$

$$= \frac{1}{n}\left[-p_i p_j + (m-1)\sum_{h=1}^{R}W_h(p_{hi} - p_i)(p_{hj} - p_j)\right]$$

For large r, the distribution of $\widehat{\mathbf{p}} = \left(\widehat{p}_1, \ldots, \widehat{p}_{k-1}\right)'$ is approximately normal with mean $\mathbf{p} = (p_1, \ldots, p_{k-1})'$ and variance−covariance matrix $\mathbf{V}(t)/n$, where

$$\mathbf{V}(t) = \mathbf{P} + (m-1)\mathbf{A}, \tag{19.3.27}$$

$\mathbf{A} = \sum_{h=1}^{R} W_h(\mathbf{p_h} - \mathbf{p})(\mathbf{p_h} - \mathbf{p})'$, $\mathbf{p_h} = \left(p_{h1}, \ldots, p_{h,k-1}\right)'$, and $\mathbf{P} = diag(\mathbf{p}) - \mathbf{pp}'$.

Hence under $\mathbf{H}_0$: $\mathbf{p} = \mathbf{p}_0$,

$$X_G^2 = n\sum_{i=1}^{k-1}\frac{\left(\widehat{p}_i - p_{i0}\right)^2}{p_{i0}}$$

$$= \sum_{i=1}^{k-1}\{1 + (m-1)\rho_{i0}\}Z_i^2 \tag{19.3.28}$$

where $\rho_{10},\ldots,\ \rho_{k-1,0}$ are the eigenvalues of $\mathbf{P}^{-1}\mathbf{A}$ for $\mathbf{p} = \mathbf{p}_0$. Noting $0 \le c'\mathbf{A}c/c'\mathbf{P}c \le 1$, we find that $\rho_1,\ldots,\ \rho_{k-1}$, the eigenvalues of $\mathbf{A}$, are all positive and less than unity. Hence we can write

$$\mathbf{X}_G^2 \le [1+(m-1)\rho_{\max 0}]\sum_{i=1}^{k-1} Z_i^2 \text{ with } \rho_{\max,0} = \max\{\rho_{10},\ldots,\rho_{k-1,0}\}$$

$$\text{i.e., } \frac{X_G^2}{m} \le \sum_{i=1}^{k-1} Z_i^2 \tag{19.3.29}$$

Thus $\frac{X_G^2}{m}$ provides asymptotically a conservative test. Rao and Scott (1981) called ρ_i's a generalized measure of homogeneity similar to intraclass correlation ρ. However, one can use the Wald statistic if $\widehat{\mathbf{p}}_h$ are available because $\widehat{\mathbf{V}}(t) = m\sum_{h=1}^{r}(\widehat{\mathbf{p}}_h - \widehat{\mathbf{p}})(\widehat{\mathbf{p}}_h - \widehat{\mathbf{p}})'/(r-1)$, the estimator of $\mathbf{V}(t)$ is easy to compute. Thomas and Rao (1987) showed that the Wald test provides poor control of type1 error if the degree of freedom f for estimating $\widehat{\mathbf{V}}$ is not more than the degrees of freedom for the hypothesis. So, the application of Wald statistics is limited because the degrees of freedom $\widehat{\mathbf{V}}$ of most survey designs are at most moderate. To overcome the instability problem of $\widehat{\mathbf{V}}$, the following F-corrected Wald statistics are proposed by Rao and Thomas (1988).

$$F_{1w} = \frac{f-k+2}{f(k-1)}X_W^2 \text{ and } F_{2w} = X_W^2/(k-1) \tag{19.3.30}$$

where f = number of sampled clusters − number of strata.

The statistics F_{1w} and F_{2w} are treated as F distribution with degrees of freedom $k-1,\ f-k+2$ and $k-1,\ f$, respectively. Here we note that for $k=2$ both the statistics F_{1w} and F_{2w} reproduce the original Wald statistic. For further details readers are referred to Lehtonen and Pahkinen (2004).

19.3.7 Residual Analysis

In case the hypothesis of goodness of fit $H_0{:}\,\mathbf{p}=\mathbf{p}_0$ is rejected, it is important to check which of the p_i's differ significantly from p_{i0}. In this situation we test the following k hypotheses separately. The hypothesis

$$H_{0i}{:}\ p_i = p_{i0} \text{ against alternative } H_{1i}{:}\ p_i \ne p_{i0} \quad \text{for } i=1,\ldots,k$$

should be rejected at $\alpha \times 100\%$ level of significance if $|\widehat{e}_i| > z_{\alpha/2}$, where $\widehat{e}_i = \frac{\hat{p}_i - p_{i0}}{\sqrt{\widehat{V}(\hat{p}_i)}} = \frac{(\hat{p}_i - p_{i0})}{\sqrt{\widehat{d}_i \hat{p}_i (1-\hat{p}_i)/n}} = \frac{e_i}{\sqrt{\widehat{d}_i}}$, $e_i = \frac{(\hat{p}_i - p_{i0})}{\sqrt{\hat{p}_i(1-\hat{p}_i)/n}} =$ standardized residual under SRSWR sampling, and z_α is the upper $\alpha \times 100\%$ point of the standardized normal distribution.

Example 19.3.4

Consider an artificial example where a population of 300 enumeration areas (EAs), which was stratified into 15 strata. From each of the strata two EAs were selected using a suitable inclusion probability proportional to size sampling scheme using the number of people in the EA as a measure of size variable. From each of the selected EAs a sample of 50 individuals was selected by SRSWOR method. The selected sample of 1500 individuals was classified into six categories according to their income level. The estimated and hypothesized proportions for the six categories and estimated variance—covariance matrix $\left(\widehat{V}/n\right)$ were computed as follows:

Income level	Estimated proportion ($\hat{p}_i$)	Hypothesized proportion (p_{i0})
A	0.120	0.120
B	0.205	0.190
C	0.210	0.230
D	0.214	0.250
E	0.126	0.120
F	0.125	0.090

$$\widehat{V}/n = 10^{-5} \times \begin{pmatrix} 20.875 & 1.024 & -0.624 & -5.070 & -3.258 & -12.947 \\ 1.024 & 10.258 & 3.646 & 0.177 & -4.816 & -10.289 \\ -0.624 & 3.646 & 20.745 & -6.317 & -4.586 & -12.864 \\ -5.070 & 0.177 & -6.317 & 30.785 & -5.718 & -13.857 \\ -3.258 & -4.816 & -4.586 & -5.718 & 20.650 & -2.272 \\ -12.947 & -10.289 & -12.864 & -13.857 & -2.272 & 52.229 \end{pmatrix}$$

Here, the generalized Pearsonian chi-square: $X_G^2 = n\sum_{i=1}^{k} \frac{(\hat{p}_i - p_{i0})^2}{p_{i0}} = 1500 \sum_{i=1}^{6} \frac{(\hat{p}_i - p_{i0})^2}{p_{i0}} = 33.028$ with 5 $(=k-1)$ df and has a p-value 0.0000.

Modified chi-square:

$$X_M^2 = n\sum_{i=1}^{k}\frac{\left(\hat{p}_i - p_{i0}\right)^2}{\hat{p}_i} = 1500\sum_{i=1}^{6}\frac{\left(\hat{p}_i - p_{i0}\right)^2}{\hat{p}_i} = 28.716 \text{ with 5 df and}$$

has a p-value 0.0000.

Rao–Scott first-order correction:

The estimated deffs for the six categories that are obtained using the formula $\widehat{d}_i = \widehat{V}_{ii}/\left[\widehat{p}_i(1-\widehat{p})\right]$ are as follows:

$\widehat{d}_1 = 2.965, \widehat{d}_2 = 0.944, \widehat{d}_3 = 1.876, \widehat{d}_4 = 2.745, \widehat{d}_5 = 2.813$, and $\widehat{d}_6 = 7.163$.

The estimated value of

$$\overline{\lambda}_0 = \sum_{i=1}^{6}\widehat{V}_{ii}/(5p_{i0}) = \sum_{i=1}^{6}\widehat{p}_i\left(1-\widehat{p}_i\right)\ \widehat{d}_i\Big/(5p_{i0}) = 3.581.$$

$\widehat{X}_A^2 = X_G^2\Big/\widehat{\overline{\lambda}}_0 = 30.028/3.581 = 9.222$ with 5 df and has a p-value 0.1005.

Rao–Scott second-order correction:

$$1+\widehat{a}^2 = \frac{1}{\left(5\widehat{\overline{\lambda}}^2\right)}\sum_{i=1}^{6}\sum_{j=1}^{6}\frac{\widehat{V}_{ij}^2}{p_{i0}p_{j0}} = 1.794 \text{ and } \nu = 5/(1+\widehat{a}^2) = 2.787.$$

$\widehat{X}_B^2 = \widehat{X}_A^2\Big/(1+\widehat{a}^2) = 9.222/1.794 = 5.140$ with 2.787 df and has a p-value 0.1618.

Fellegi correction:

$\widehat{X}_F^2 = \widehat{X}_G^2\Big/\widehat{\overline{d}} = 33.028/3.084 = 10.708$ with 5 df and has a p-value 0.0575.

Wald statistic:

$X_W^2 = \left(\widehat{\mathbf{p}} - \mathbf{p}_0\right)'\left(\frac{\mathbf{V}}{n}\right)^{-1}\left(\widehat{\mathbf{p}} - \mathbf{p}_0\right) = 13.121$ with 5 df and ha a p-value 0.0223.

F-corrections for Wald statistic:

Noting $f = m - h = 50 - 15 = 35$, we get

$F_1 = \dfrac{f-k+2}{f(k-1)}X_W^2 = 1.771 \times 13.121 = 2.324$ with 5 and 31 df and has a p-value 0.0665.

$F_2 = X_W^2/(k-1) = 13.121/5 = 2.624$ with 5 and 35 df and has a p-value 0.0407.

Residual analysis:

The standardized residuals are $e_1 = 0$, $e_2 = 1.481$, $e_3 = -1.388$, $e_4 = -2.051$, $e_5 = 0.417$, and $e_6 = 1.531$.

Here we note that the values of statistics X_G^2 and X_M^2 are highly significant and provide strong evidence of rejecting the hypothesis H_0 while the statistics $\widehat{X}_A^2$, $\widehat{X}_B^2$, F_1, and F_2 provide strong evidence of accepting the hypothesis. The Wald statistic X_W^2 favors acceptance of hypothesis H_0. The residual analysis reveals that the population proportions for the income groups do not deviate significantly from the respective hypothesized proportions.

19.4 TEST OF INDEPENDENCE

Consider a finite population of N identifiable units, which has been classified into "a" levels of factor A and "b" levels of the factor B. Let the proportion of individuals belonging to the ith level of factor A and jth level of the factor B in the population be p_{ij}, $i = 1,\ldots, a; j = 1,\ldots, b$. Here we are interested in testing the independence of the two factors A and B, i.e., we want to test

$$\mathbf{H}_0: \theta_{ij} = p_{ij} - p_{i\cdot}p_{\cdot j} = 0 \quad \text{for } i = 1, \ldots, a;\ j = 1, \ldots, b \qquad (19.4.1)$$

against alternative $\mathbf{H}_1: \theta_{ij} \neq 0$ for at least one of the combination i, j where $p_{i\cdot} = \sum_{j=1}^{b} p_{ij}$ and $p_{\cdot j} = \sum_{i=1}^{a} p_{ij}$.

Let a sample s of size n be selected from the population with probability $p(s)$ using a complex survey design and $\widehat{p}_{ij}$ be an unbiased or consistent estimator of p_{ij} based on the sample s.

19.4.1 Wald Statistic

For large n, under $\mathbf{H}_0$ the Wald statistic

$$X_w^2 = n\widehat{\boldsymbol{\theta}}'\,\widehat{\mathbf{V}}_0^{-1}\,\widehat{\boldsymbol{\theta}} \qquad (19.4.2)$$

follows a χ_t^2 (chi-square distribution with t df) with $t = (a-1)\ (b-1)$ where $\widehat{\boldsymbol{\theta}}' = \left(\widehat{\theta}_{11}, \widehat{\theta}_{12}, \ldots, \widehat{\theta}_{a-1,b-1}\right)$ and $\widehat{\mathbf{V}}_\theta/n$ is an estimated variance–covariance matrix of $\widehat{\boldsymbol{\theta}} = V\left(\widehat{\boldsymbol{\theta}}\right)\Big/n = \mathbf{V}_\theta/n$. The estimator $\widehat{\mathbf{V}}_\theta$ can be obtained by applying any of the methods of variance estimation, e.g. LR, RG, BRR, or BT described in Chapter 18. The estimators of covariances become unstable, if the cell frequencies are small. Hence in practice, the performance of the Wald statistic for a large contingency table is poor (Lohr, 1999) because of small cell frequencies.

19.4.2 Bonferroni Test

To avoid calculations of covariances, one may perform the following $(a-1)(b-1)$ separate tests

$$H_{011}: \theta_{11} = 0;\ H_{012}: \theta_{12} = 0; \ldots, H_{0(a-1)(b-1)}: \theta_{a-1,b-1} = 0$$

The hypothesis $\mathbf{H}_0$ stated in Eq. (19.4.1) is rejected at α 100% level of significance if any of the test H_{0ij} is rejected at $\frac{\alpha}{2t} \times 100\%$ level of significance, i.e., if any one

$$\frac{\left|\widehat{\theta}_{ij}\right|}{\sqrt{\widehat{V}\left(\widehat{\theta}_{ij}\right)}} > t_{\frac{\alpha}{2t},\, q} \tag{19.4.3}$$

where $t_{\frac{\alpha}{2t},\, q}$ is the upper $(\alpha/2t)$ 100% point of t distribution with q df and q is the degree of freedom carried by the estimator of the variance. More details are given by Lohr (1999) and Thomas (1989).

19.4.3 Modified Chi-Square

The modified chi-square statistic for complex survey design for testing $\mathbf{H}_0$ stated in Eq. (19.4.1) is given by

$$X_M^2 = n \sum_{i=1}^{a-1} \sum_{j=1}^{b-1} \frac{\left(\widehat{p}_{ij} - \widehat{p}_{i\cdot}\widehat{p}_{\cdot j}\right)^2}{\widehat{p}_{i\cdot}\widehat{p}_{\cdot j}} \tag{19.4.4}$$

Now, writing $\widehat{\boldsymbol{\theta}} = \left(\widehat{\theta}_{11}, \widehat{\theta}_{12}, \ldots, \widehat{\theta}_{a-1,b-1}\right)'$, $\mathbf{p} = \left(p_{11}, p_{12}, \ldots, p_{a-1,b-1}\right)'$, $\widehat{\mathbf{p}}_r = \left(\widehat{p}_{1\cdot}, \ldots, \widehat{p}_{a-1\cdot}\right)'$, $\widehat{\mathbf{p}}_c = \left(\widehat{p}_{\cdot 1}, \ldots, \widehat{p}_{\cdot c-1}\right)'$, $\widehat{\mathbf{P}}_r = diag\left(\widehat{\mathbf{p}}_r\right) - \widehat{\mathbf{p}}_r\widehat{\mathbf{p}}_r'$, and $\widehat{\mathbf{P}}_c = diag\left(\widehat{\mathbf{p}}_c\right) - \widehat{\mathbf{p}}_c\widehat{\mathbf{p}}_c'$, the expression (Eq. 19.4.4) can be written as

$$X_M^2 = n\widehat{\boldsymbol{\theta}}'\left(\widehat{\mathbf{P}}_r^{-1} \otimes \widehat{\mathbf{P}}_c^{-1}\right)\widehat{\boldsymbol{\theta}} \tag{19.4.5}$$

where $\otimes$ denotes the kronecker product.

Rao and Scott (1981) showed that under $\mathbf{H}_0$, X_M^2 is asymptotically distributed as

$$X_M^2 \approx \sum_{i=1}^{t} \delta_{i0} Z_i^2 \tag{19.4.6}$$

where Z_i's are iid $N(0, 1)$ and δ_{i0}'s are eigenvalues of $\mathbf{D}_{\boldsymbol{\theta}} = \mathbf{P}_r^{-1} \otimes \mathbf{P}_c^{-1}\mathbf{V}_{\boldsymbol{\theta}}$ under H_0. Let $\widehat{\delta}_{i0}$ be an estimator of δ_{i0}, $\widehat{\delta}_{\max,0} = \max\left\{\widehat{\delta}_{i0}\right\}$, and

$\widehat{\overline{\delta}}_0 = \sum_{i=1}^{t} \widehat{\delta}_{i0} \Big/ t$. Then, $X_M^2 \Big/ \widehat{\delta}_{\max,0}$ provides a conservative test, while $X_M^2 \Big/ \widehat{\overline{\delta}}_0 \approx \sum_{i=1}^{t} \left(\widehat{\delta}_{i0} \Big/ \widehat{\overline{\delta}}_0 \right) Z_i^2$ proves a better test.

19.5 TESTS OF HOMOGENEITY

Suppose we have g populations each of which is classified into k categories. Let the number of units belonging to the ith category of the jth population be N_{ij}, $i = 1,\ldots, k$; $j = 1,\ldots, g$ and their proportion be $p_{ij} = N_{ij}/N_i$, where $N_j = \sum_{i=1}^{k} N_{ij}$ is the size of the jth population and $\sum_{i=1}^{k} p_{ij} = 1$ for $j = 1,\ldots, g$. The problem of homogeneity consists of testing the hypothesis

$$H_0: p_{i1} = p_{i2} = \cdots = p_{ig} \quad \text{for } i = 1,\ldots,k-1 \tag{19.5.1}$$

against alternative H_1:H_0 is false.

In matrix notation, H_0 can be written as

$$\mathbf{H}_0: \mathbf{p}_1 = \mathbf{p}_2 = \cdots = \mathbf{p_g} \text{ with } \mathbf{p_j} = \left(p_{1j}, p_{2j}, \ldots, p_{k-1,j}\right)'$$

From each of the populations, samples are selected independently using some complex sampling design. Let s_j be a sample of size n_j selected from the jth population and let the number of units falling in the ith class be n_{ij}. For simplicity, let us consider $g = 2$ and consider the null hypothesis

$$\mathbf{H}_0: \mathbf{p}_1 = \mathbf{p}_2 \text{ against } \mathbf{H}_1: \mathbf{p}_1 \neq \mathbf{p}_2 \tag{19.5.2}$$

As usual, let $\widehat{p}_{ij}$ be an unbiased or a consistent estimator of p_{ij} and $\widehat{\mathbf{p}}_j = \left(\widehat{p}_{1j}, \widehat{p}_{2j}, \ldots, \widehat{p}_{k-1,j}\right)'$.

19.5.1 Wald Statistic

Let $\widehat{V}_i/n_i$ be a consistent estimator of the variance−covariance matrix of $\widehat{\mathbf{p}}_\mathbf{i}$ for $i = 1, 2$. Under the null hypothesis $\mathbf{H}_0$, the Wald statistic

$$X_{WH}^2 = \left(\widehat{\mathbf{p}}_1 - \widehat{\mathbf{p}}_2\right)' \left(\frac{\widehat{\mathbf{V}}_1}{n_1} + \frac{\widehat{\mathbf{V}}_2}{n_2} \right)^{-1} \left(\widehat{\mathbf{p}}_1 - \widehat{\mathbf{p}}_2\right) \tag{19.5.3}$$

follows χ_{k-1}^2, chi-square distribution with $k-1$ df when sample sizes n_1 and n_2 are so large that each of $\widehat{\mathbf{p}}_1$ and $\widehat{\mathbf{p}}_2$ are distributed independently as $k-1$ normal variables. The statistic X_{WH}^2 has limited application because of the problem of reliable estimation of covariances, especially when the number of classes is large.

19.5.2 Modified Chi-Square Statistics

The Pearsonian chi-square statistic for testing $\mathbf{H}_0$ is given by

$$X^2_{PH} = n\sum_{i=1}^{2}\sum_{j=1}^{k}\frac{\left(n_{ij} - n_{i\cdot}n_{\cdot j}/n\right)^2}{n_{i\cdot}n_{\cdot j}} \tag{19.5.4}$$

with $n_{i\cdot} = \sum_{j=1}^{k} n_{ij}$ and $n_{\cdot j} = \sum_{i=1}^{2} n_{ij}$.

The statistic X^2_{PH} asymptotically follows χ^2_{k-1}, if the samples s_1 and s_2 are selected by SRSWR method. For a complex survey design, however, one should use the following modified chi-square statistic given by Scott and Rao (1981):

$$\begin{aligned} X^2_{MPH} &= \sum_{i=1}^{2}\sum_{j=1}^{k}\frac{n_i\left(\widehat{p}_{ij} - \widehat{p}_{+j}\right)^2}{\widehat{p}_{+j}} \\ &= \frac{n_1 n_2}{n_1 + n_2}\left(\widehat{\mathbf{p}}_1 - \widehat{\mathbf{p}}_2\right)'\widehat{\mathbf{P}}^{-1}\left(\widehat{\mathbf{p}}_1 - \widehat{\mathbf{p}}_2\right) \end{aligned} \tag{19.5.5}$$

where $\widehat{p}_{+j} = \dfrac{n_1\hat{p}_{1j} + n_2\hat{p}_{2j}}{n_1 + n_2}$, $\widehat{\mathbf{P}} = diag\left(\widehat{\mathbf{p}}_+\right) - \widehat{\mathbf{p}}_+\widehat{\mathbf{p}}'_+$, and $\widehat{\mathbf{p}}_+ = \left(\widehat{p}_{+1}, \ldots, \widehat{p}_{+k-1}\right)'$.

So, the statistic X^2_{MPH} is a particular case of Wald statistic X^2_{WH} given in Eq. (19.5.3) when $\widehat{\mathbf{V}}_1 = \widehat{\mathbf{V}}_2 = \widehat{\mathbf{P}}$. The statistic X^2_{MPH} does not asymptotically follow chi-square distribution. Scott and Rao (1981) showed that under the null hypothesis $\mathbf{H}_0$: $\mathbf{p}_1 = \mathbf{p}_2 = \mathbf{p}$,

$$X^2_{MPH} = \sum_{j=1}^{k-1}\lambda_j Z_j^2 \tag{19.5.6}$$

where Z_i's are iid $N(0,\ 1)$ and $\lambda_1,\ldots,\ \lambda_{k-1}$ are eigenvalues of $\mathbf{D} = (n_2\mathbf{D}_1 + n_1\mathbf{D}_2)/(n_1 + n_2)$ with $\mathbf{D_i} = \mathbf{P}^{-1}\mathbf{V_i}$ and $\mathbf{P} = diag\ \mathbf{p} - \mathbf{pp}'$. The application of the statistics X^2_{MPH} has already been discussed in Section 19.3.3. On the basis of extensive empirical studies based on two-stage sampling with varying probabilities, Scott and Rao (1981) concluded that treating X^2_{MPH} as an ordinary chi-square results in severe distortion of significance level.

19.6 CHI-SQUARE TEST BASED ON SUPERPOPULATION MODEL

19.6.1 Altham's Model

Cohen (1976) proposed chi-square statistic based on a cluster sampling with a fixed cluster of size 2, using a superpopulation model. Altham (1976) extended the theory to any fixed cluster size M. Rao and Scott (1981) extended this further to cover a general two-stage sampling. The Rao and Scott (1981) procedure is described as follows: Consider a two-stage sampling of R fsu's and the hth fsu consists of M_h ssu's. A sample s of r fsu's is selected from R fsu's by some suitable sampling scheme. If the hth fsu is selected in the sample s, a subsample s_h of size m_h ssu's is selected from the hth fsu by using a suitable sampling scheme. The total number of ssu's in the sample $s = (s_1 \cup \ldots \cup s_r)$ is $\sum_{h=1}^{r} m_h = n$. Let $y_{hj}(i) = 1$ if the jth ssu of the hth fsu belong to the ith category and $y_{hj}(i) = 0$ otherwise; $h = 1,\ldots, R$; $i = 1,\ldots, k$. Hence the total number of ssu's that belong to the ith category in the entire sample s is $n_i = \sum_{h\in s}\sum_{j\in s_h} y_{hj}(i)$ and $n = \sum_{i=1}^{k} n_i$. Altham (1976) considered the following superpopulation model:

(i) random variables $y_{hj}(i)$ for different clusters (fsu's) are independent,
(ii) $E_m\{y_{hj}(i)\} = \pi_i$ and (iii) $C_m\{y_{hj}(i), y_{hj'}(i')\} = \sigma_{ii'}$ for $j \neq j'$

(19.6.1)

where E_m and C_m denote expectation and covariance operators with respect to the model.

From the model (Eq. 19.6.1), we have for a given s,

(i) the model expectation of $n_i = E_m(n_i) = \sum_{h\in s}\sum_{j\in s_h} E_m\{y_{hj}(i)\} = n\pi_i$,

(19.6.2)

(ii) the model variance of

$$n_i = V_m(n_i) = \left[\sum_{h\in s} V_m\left\{\sum_{j\in s_h} y_{hj}(i)\right\} + \sum_{h\neq}\sum_{h'\in s} C_m\left\{\sum_{j\in s_h} y_{hj}(i), \sum_{j\in s_{h'}} y_{h'j}(i)\right\}\right]$$
$$= \pi_i(1-\pi_i)n + \sigma_{ii}\left\{\sum_{h\in s} m_h(m_h-1)\right\}$$
$$= n\pi_i(1-\pi_i) + \sigma_{ii}\left(\sum_{h\in s} m_h^2 - n\right)$$
$$\text{(noting } V_m\{y_{hj}(i)\} = \pi_i(1-\pi_i))$$

(19.6.3)

and
(iii) the model covariance of n_i and $n_{i'}$ is

$$
\begin{aligned}
C_m(n_i, n_{i'}) &= \sum_{h\in s} C_m\left(\sum_{j\in s_h} y_{hj}(i), \sum_{j\in s_h} y_{hj}(i')\right) \\
&= \sum_{h\in s}\left[\sum_{j\in s_h} C_m\{y_{hj}(i), y_{hj}(i')\}\right. \\
&\quad \left. + \sum_{j\neq}\sum_{j'\in s_h} C_m\{y_{hj}(i), y_{hj'}(i')\}\right] \qquad (19.6.4)\\
&= -n\pi_i\pi_{i'} + \sigma_{ii'}\left(\sum_{h\in s} m_h^2 - n\right) \\
&\qquad \left(\text{noting } C_m\{y_{hj}(i), y_{hj}(i')\} = -\pi_i\pi_{i'}\right)
\end{aligned}
$$

Let $\widehat{\boldsymbol{\pi}} = (\widehat{\pi}_1, \ldots, \widehat{\pi}_i, \ldots, \widehat{\pi}_{k-1})'$ with $\widehat{\pi}_i = n_i/n$. Then from Eqs. (19.6.2)–(19.6.4) we note that under the model (Eq. 19.6.1) and a given s, the expectation and variance–covariance matrix of $\widehat{\boldsymbol{\pi}}$ are, respectively,

$$E_m(\widehat{\boldsymbol{\pi}}) = \boldsymbol{\pi} = (\pi_1, \ldots, \pi_i, \ldots, \pi_{k-1})' \text{ and } V_m(\widehat{\boldsymbol{\pi}}) = \mathbf{Q}_{ms}/n \qquad (19.6.5)$$

where $\mathbf{Q}_{ms} = \{\boldsymbol{\Delta}(\boldsymbol{\pi}) + (m_{0s} - 1)\boldsymbol{\Phi}\}$, $\boldsymbol{\Delta}(\boldsymbol{\pi}) = diag(\boldsymbol{\pi}) - \boldsymbol{\pi}\boldsymbol{\pi}'$, $\boldsymbol{\Phi} = (\sigma_{ii'})$, and $m_{0s} = \sum_{h\in s} m_h^2/n$.

For large r, $\widehat{\boldsymbol{\pi}}$ follows the $k-1$ variate normal distribution with mean $\boldsymbol{\pi}$ and variance $\mathbf{Q}_{ms}/n$. Hence for testing the hypothesis $\mathbf{H}_0$: $\boldsymbol{\pi} = \boldsymbol{\pi}_0$ against alternative $\mathbf{H}_1$: $\boldsymbol{\pi} \neq \boldsymbol{\pi}_0$, the test statistic

$$X^2(\pi) = n(\widehat{\boldsymbol{\pi}} - \boldsymbol{\pi}_0)'\mathbf{Q}_{\mathbf{ms}}^{-1}(\widehat{\boldsymbol{\pi}} - \boldsymbol{\pi}_0) \qquad (19.6.6)$$

follows a chi-square distribution with $k-1$ df for large r when the null hypothesis $\mathbf{H}_0$ is true. Let $\overline{\lambda}_1, \ldots, \overline{\lambda}_{k-1}$ be the eigenvalues of $(\boldsymbol{\Delta}(\boldsymbol{\pi}))^{-1}\boldsymbol{\Phi}$, then we can write, following Rao and Scott (1981),

$$X^2(\pi) \cong \sum_{i=1}^{k-1}\{1 + (m_{s0} - 1)\overline{\lambda}_i\}Z_i^2 \qquad (19.6.7)$$

Noting that $\boldsymbol{\Delta}(\boldsymbol{\pi}) - \boldsymbol{\Phi}$ is a nonnegative definite (Rao and Scott, 1981), so that $0 \leq \overline{\lambda}_i \leq 1$ for $i = 1,\ldots, k-1$, we find $X^2(\pi) \leq m_{0s}\sum_{i=1}^{k-1} Z_i^2 = m_{0s}\chi_{k-1}^2$. Hence treating $X^2(\pi)/m_{0s}$ as χ_{k-1}^2 under $\mathbf{H}_0$, we arrive at a conservative test. In case all m_h's are equal to m, $X^2(\pi)/m_{0s}$ reduces to $X^2(\pi)/m$.

19.6.1.1 A Simpler Model

Consider a special case of the model (Eq. 19.6.1) where

$$C_m\{y_{hj}(i), y_{hj'}(i')\} = \sigma_{ii'} = \begin{cases} \overline{\rho}\pi_i(1-\pi_i) & \text{for } i = i' \\ -\overline{\rho}\pi_i\pi_j & \text{for } i \neq i' \end{cases}$$

In this situation $\boldsymbol{\Phi} = \overline{\rho}\Delta(\boldsymbol{\pi})$ and the asymptotic distribution of $X^2(\pi)/\{1+(m_{s0}-1)\overline{\rho}\}$ under $\mathbf{H}_0$ is an exact chi-square distribution with $k-1$ df.

19.6.2 Brier Model

Brier (1978) considered two-stage sampling where the sample sizes for the ssu's m_h's are selected independently. Let m_{hi} be the number of sampled units from the hth fsu that belong to the ith category with $\sum_{i=1}^{k} m_{hi} = n_h$. Brier (1978) assumed that the vector $\mathbf{m_h} = (m_{h1}, \ldots, m_{hk-1})'$ follows a multinomial distribution with parameter $\mathbf{p_h} = (p_{h1}, \ldots, p_{h,k})'$, $0 < p_{hj} < 1, \sum_{j=1}^{k} p_{hj} = 1$. The parameter vector $\mathbf{p_h}$ is assumed to be distributed as the Dirichlet distribution with probability density function

$$f(\mathbf{p_h}|\boldsymbol{\pi}, \nu) = \frac{\Gamma\nu}{\prod_{i=1}^{k}\Gamma(\nu\pi_i)}\prod_{i=1}^{k} p_{hi}^{\nu\pi_i - 1}; \quad \nu, \pi_i > 0, \quad \sum_{i=1}^{k}\pi_i = 1 \qquad (19.6.8)$$

Rao and Scott (1981) showed that for $m_h = m$, $E_m(\mathbf{m}_h) = m\boldsymbol{\pi}$, and $E_m(\mathbf{m}_h - m\boldsymbol{\pi})(\mathbf{m}_h - m\boldsymbol{\pi})' = \frac{m(\nu+m)}{\nu+1}\Delta(\boldsymbol{\pi})$, where $\Delta(\boldsymbol{\pi})$ is given in Eq. (19.6.5). Further noting that the mean and variance of $\mathbf{n} = (n_1, \ldots, n_{k-1})'$ as $E(\mathbf{n}) = n\boldsymbol{\pi}$ and $V(\mathbf{n}) = n[1+(m-1)\overline{\rho}]\Delta(\boldsymbol{\pi})$ with $\overline{\rho} = 1/(\nu+1)$, we find that under H_0, the asymptotic distribution of the modified statistic $\frac{(\nu+1)}{\nu+m}X^2(\pi)$ is chi-square with $k-1$ df. Rao and Scott (1981) also extended Brier's results to multistage sampling when the ssu's of unequal sizes are selected by SRSWOR method.

19.7 CONCLUDING REMARKS

Categorical data analysis is used extensively in analyzing survey data. Practitioners use standard statistical packages such as SPSS, BMDP, SAS, etc. to compute chi-square test statistics for goodness of fit, tests of independence, and homogeneity. The software packages give erroneous results if the data are collected through a complex survey design, as software computations are based on SRSWR sampling. Several methods of analyzing categorical data for complex survey designs are available in the literature. The use of Wald statistics and modifications of standard chi-square statistics are popular. The Wald statistic has limitations in that it requires estimation of covariances that are not generally available from the published report. Furthermore, if the number of cells is large but the cell frequencies are not large enough, the estimates of variance and covariances are unstable. The first-order correction proposed by Rao and Scott (1981) is quite effective because it needs only estimation of variances, while second-order correction needs not only estimation variances but also covariances. Testing of significance of goodness of fit of the log-linear model based on a complex survey data can be performed following the procedure outlined by Rao and Scott (1981). Hidiroglou and Rao (1987) provided with practical application of chi-square test for goodness of fit, homogeneity, and independency for the Community Health Survey data collected though complex survey designs. For further information, readers are referred to the works of Rao and Scott (1987), Rao and Thomas (1988), Roberts et al. (1987), and Fay (1985), among others.

19.8 EXERCISES

19.8.1 In a survey conducted in 2010, the district of Gaborone was stratified into 20 strata each containing 20 EAs. From each of the EAs, a sample of two EAs was selected by probability proportional to size without replacement (PPSWOR) sampling scheme taking the number of people in the EA as measure of size variable. From each of the selected EAs 30 individuals were selected by SRSWOR method. The sampled 1200 individuals were classified into three age groups. The following table gives estimated proportions, last 2001 census proportion and the estimated variance—covariance matrix of the estimated proportions for the survey. Test if there is any change in the distribution of age between the periods 2010 and 2001 using (i) Pearsonian chi-square, (ii) modified chi-square, (iii) Wald statistic, (iv) Rao—Scott first- and second-order corrections, (v) Fellegi, and (vi) F-corrected to Wald statistic.

Age	Estimated proportion $\hat{p}_i$	Census proportion
Below 20	0.525	0.410
21—60	0.355	0.490
61 and above	0.120	0.100

Variance—covariance matrix

$$= \widehat{\mathbf{V}}/1200 = 10^{-5}\begin{pmatrix} 12.500 & -3.384 & -9.115 \\ & 8.950 & -5.565 \\ & & 14.680 \end{pmatrix}$$

19.8.2 Suppose a sample of 1800 individuals is selected from the district Francistown using a sampling design similar to that in Exercise 19.8.1. The age distribution and variance covariance matrix of the estimated cell proportions are given as follows:

Age	Estimated proportion $\hat{p}_i$
Below 20	0.525
21–60	0.355
61 and above	0.120

Variance—covariance matrix

$$= \widehat{\mathbf{V}}/1800 = 10^{-5}\begin{pmatrix} 15.750 & -3.384 & -12.365 \\ & 25.500 & -5.566 \\ & & 17.931 \end{pmatrix}$$

Test if there is any significant difference in age distribution between the district Gaborone (given in Exercise 19.8.1) and Francistown using (i) modified chi-square and (ii) Wald statistic.

19.8.3 A sample of 1500 households was selected by a complex survey design and classified according to income and education (head of the house hold). The estimated cell probabilities are given as follows:

	Income group			
Education	**Poor**	**Middle**	**High**	**Total**
Nil	0.15	0.10	0.05	0.30
Primary	0.10	0.08	0.02	0.20
Secondary	0.10	0.15	0.05	0.30
Tertiary	0.02	0.08	0.10	0.20
Total	0.37	0.41	0.22	

Fit an appropriate log-linear model on the estimated cell probabilities and use the model test if income depends on the level of education.

19.8.4 A sample of 6000 factory employees was selected using a complex survey design and classified according to gender and income (monthly salary). The estimated cell probabilities and deff's (in brackets) are given in the following table:

Gender	Income level				
	Below $1000	$1001 to $2500	$2501 to $5000	$5000 and above	Total
Male	0.215	0.15	0.08	0.012	0.457
	(1.82)	(0.75)	(2.75)	(4.32)	(1.854)
Female	0.325	0.175	0.038	0.005	0.543
	(1.75)	(1.95)	(1.78)	(5.86)	(2.10)
Total	0.540	0.325	0.118	0.017	
	(1.85)	(2.85)	(3.05)	(2.15)	

Test if there is any discrimination of salaries between male and female workers using modified chi-square and Bonferroni test.

19.8.5 The sample of 600 employees is divided at random into three groups. For each group the estimated cell proportions are given in the following table:

Gender	Income level				
	Below $1000	$1001 to $2500	$2501 to $5000	$5000 and above	Total
Group 1					
Male	0.265	0.12	0.12	0.005	0.51
Female	0.215	0.135	0.137	0.003	0.49
Total	0.48	0.255	0.257	0.008	
Group 2					
Male	0.218	0.21	0.15	0.01	0.588
Female	0.155	0.125	0.127	0.005	0.412
Total	0.373	0.335	0.277	0.015	
Group 3					
Male	0.258	0.222	0.12	0.02	0.62
Female	0.125	0.095	0.15	0.01	0.38
Total	0.383	0.317	0.27	0.03	

Test if the distribution of the hypothesis of dependency of income on gender using (i) modified chi-square, (ii) Wald statistic, and (iii) Bonferroni test.

19.8.6 The HIV status (estimated cell probabilities with the deff's in brackets) of two districts based on large-scale surveys is given in the following table:

Gender	**HIV status**		
	Positive	**Negative**	**Total**
District 1			
Male	0.10 (1.85)	0.36 (0.98)	0.46 (1.98)
Female	0.16 (2.15)	0.38 (1.75)	0.54 (2.08)
Total	0.26 (2.12)	0.74 (1.95)	
District 2			
Male	0.06 (1.85)	0.40 (1.75)	0.46 (1.08)
Female	0.11 (3.25)	0.43 (1.65)	0.54 (2.15)
Total	0.27 (2.02)	0.83 (1.75)	

Test if there is any significant difference in the HIV infection pattern between the two districts.

CHAPTER 20

Complex Survey Design: Regression Analysis

20.1 INTRODUCTION

In regression analysis we describe the relationship between a response (dependent) variable and a number of explanatory (independent) variables. We also predict the future value of the dependent variable using the established relationship. The relationships are explained through variances, simple and multiple correlations, regression coefficients, fitting regression of one variable on others, and so on. For example, in BAIS II survey one may be interested in finding the relationship between HIV infection rates and age, sex, economic and social conditions, etc. The classical method of regression analysis is based on the assumption that the data are collected through simple random sampling with replacement method and software packages such as SAS and SPSS are based on this assumption. But in reality, large-scale surveys are generally based on complex survey designs involving stratification, clustering, and unequal probability of selection of samples. Therefore the assumption of independency of observations is rarely valid; hence the output of the standard software packages are erroneous or misleading when the underlying sampling design is ignored. In general, assumption of independency of observations underestimates the variance and hence underestimates confidence intervals also. Kish and Frankel (1974) recommended that "Standard errors should be computed in accordance with the complexity of the sampling designs; neglect of that complexity is a common source of serious mistakes." In this chapter we will consider regression analysis from complex survey designs under design-based, model-based, and model/design-based approaches. In regression analysis, we first indentify the appropriate model and parameters for inference and then we need to consider the type of inference that is required viz. point or interval estimation, or testing of hypotheses (see Nathan, 1988).

Survey Sampling Theory and Applications
ISBN 978-0-12-811848-1
http://dx.doi.org/10.1016/B978-0-12-811848-1.00020-0

20.2 DESIGN-BASED APPROACH

In a design-based approach, we consider a finite population $U = (U_1,\ldots, U_i,\ldots, U_N)$ of N units. Associated with the unit U_i, we have y_i, a study variable, and $\mathbf{x}_i = (x_{0i}, x_{1i},\ldots, x_{ji},\ldots, x_{pi})$, a vector of p auxiliary (independent) variables $x_1,\ldots, x_p$ with $x_{0i} = 1$. Here our parameters of interest (see Nathan, 1988) $B_0, B_1,\ldots, B_j,\ldots, B_p$ are such that

$$\sum_{i=1}^{N} \left(y_i - B_0 - B_1 x_{1i} - \cdots - B_j x_{ji} - \cdots B_p x_{pi}\right)^2 \tag{20.2.1}$$

attains a minimum, i.e., parameter of interest $\mathbf{B} = (B_0, B_1, \ldots, B_p)'$ is the ordinary least square (OLS) estimate if the whole population is observed. Here the regression coefficient $\mathbf{B}$ is a descriptive parameter because its value can be determined without any error if the entire population is surveyed and if no response and measurement error is present. Similarly, the finite population mean, variance, and correlation coefficient are also descriptive parameters. In contrast, the model parameters such as mean and variance of a normal distribution can never be calculated exactly.

Let $\mathbf{X} = \begin{pmatrix} \mathbf{x}_1 \\ \cdot \\ \mathbf{x}_i \\ \cdot \\ \mathbf{x}_N \end{pmatrix}$ be of full rank and $\mathbf{y} = \begin{pmatrix} y_1 \\ \cdot \\ y_i \\ \cdot \\ y_N \end{pmatrix}$, then the solution of Eq. (20.2.1) is

$$\mathbf{B} = (\mathbf{X}'\mathbf{X})^{-1}\mathbf{X}'\mathbf{y} \tag{20.2.2}$$

Denoting $T_{yj} = \sum_{i\in U} y_i x_{ji}$ and $T_{jk} = \sum_{i\in U} x_{ji} x_{ki}$ for $j,\ k = 0,\ldots,\ p$, Eq. (20.2.2) can be written as

$$\mathbf{B} = \mathbf{T}_{xx}^{-1}\mathbf{T}_{xy} \tag{20.2.3}$$

where $\mathbf{T}_{xx} = (\mathbf{X}'\mathbf{X})$ is a $(p+1) \times (p+1)$ matrix with the jth and kth element as T_{jk} and $\mathbf{T}_{xy} = \mathbf{X}'\mathbf{y}$ is a $(p+1) \times 1$ column matrix with jth element T_{yj}.

To estimate $\mathbf{B}$, we select a sample s of size n from the population U with probability $p(s)$ by using a suitable sampling design p. Let π_i (>0) and π_{ij} (>0) denote the inclusion probabilities for the ith, and ith and jth units. Let the quantities T_{yj} and T_{jk} be estimated unbiasedly using the estimators

$$\widehat{T}_{yj} = \sum_{i\in s} b_{si} y_i x_{ji} \text{ and } \widehat{T}_{jk} = \sum_{i\in s} b_{si} x_{ji} x_{ki} \tag{20.2.4}$$

where b_{si}'s are suitably chosen constants satisfying the unbiasedness condition $\sum_{s\supset i} b_{si} p(s) = 1$ for $i = 1,\ldots, N$.

Without loss of generality, let us suppose that the sample s consists of first n units with labels $1,2,\ldots, n$. Then we can denote

$$\widehat{T}_{wyj} = (x_{j1}, x_{j2}, \ldots, x_{jn})\mathbf{W}_s\mathbf{y}_s \text{ for } j = 0, 1, \ldots, p$$

where $\mathbf{y}_s = (y_1, \ldots, y_n)'$ and $\mathbf{W}_s = \mathbf{diag}(b_{s1},\ldots, b_{sn})$.

In case $\mathbf{X}_s = \begin{pmatrix} \mathbf{x}_1 \\ \cdot \\ \cdot \\ \mathbf{x}_n \end{pmatrix}$ is a full rank, we can set an estimator of $\mathbf{B}$ as

$$\widehat{\mathbf{B}}_w = \widehat{\mathbf{T}}_{wxx}^{-1} \widehat{\mathbf{T}}_{wxy} \tag{20.2.5}$$

where $\widehat{\mathbf{T}}_{wxx} = \mathbf{X}_s'\mathbf{W}_s\mathbf{X}_s$ is a $(p + 1) \times (p + 1)$ matrix with jth and kth elements as $\widehat{T}_{wjk}$ and $\widehat{\mathbf{T}}_{wxy} = \mathbf{X}_s'\mathbf{W}_s\mathbf{y}_s$ is a $(p + 1) \times 1$ column matrix with the jth element $\widehat{T}_{wyj}$.

In particular, if we take $b_{si} = 1/\pi_i$, then Eq. (20.2.5) reduces to

$$\widehat{\mathbf{B}}_\pi = \widehat{\mathbf{T}}_\pi^{-1} \widehat{\mathbf{T}}_{\pi y} \tag{20.2.6}$$

where $\widehat{\mathbf{T}}_\pi = \mathbf{X}_s'\mathbf{W}_\pi\mathbf{X}_s$, $\mathbf{W}_\pi = \mathbf{diag}(1/\pi_1,\ldots,1/\pi_n)$ and $\mathbf{T}_{\pi y}$ is a $(p + 1) \times 1$ column matrix with jth element $\widehat{T}_{\pi yj} = \sum_{i\in s} y_i x_{ji}/\pi_i$.

Obviously, the estimators $\widehat{\mathbf{B}}_w$ and $\widehat{\mathbf{B}}_\pi$ are not a design unbiased in general but they are design consistent for $\mathbf{B}$.

Example 20.2.1

In case of a single auxiliary variable ($p = 1$), we find

$$B_0 = \overline{Y} - B_1\overline{X}_1 = T_{y0}/T_{00} - B_1 T_{11}/T_{00} \text{ and}$$

$$B_1 = \frac{\sum_{i\in U} y_i x_{1i} - \left(\sum_{i\in U} x_{1i}\right)\left(\sum_{i\in U} y_i\right)\Big/N}{\sum_{i\in U} x_{1i}^2 - \left(\sum_{i\in U} x_{1i}\right)^2\Big/N} = \frac{T_{y1} - T_{01}T_{y0}/T_{00}}{T_{11} - T_{01}^2/T_{00}} \tag{20.2.7}$$

where $T_{y0} = \sum_{i\in U} y_i$, $T_{y1} = \sum_{i\in U} y_i x_{1i}$, $T_{00} = N$, $T_{01} = \sum_{i\in U} x_{1i}$, and $T_{11} = \sum_{i\in U} x_{1i}^2$.

Now, writing $\widehat{T}_{w00} = \sum_{i\in s} b_{si}$, $\widehat{T}_{w01} = \sum_{i\in s} b_{si}x_{1i}$, $\widehat{T}_{w11} = \sum_{i\in s} b_{si}x_{1i}^2$, $\widehat{T}_{wy0} = \sum_{i\in s} b_{si}y_i$, and $\widehat{T}_{wy1} = \sum_{i\in s} b_{si}x_{1i}y_i$, the estimators of B_0 and B_1 are obtained as

$$\widehat{B}_{w0} = \widehat{T}_{wy0}\Big/\widehat{T}_{w00} - \widehat{B}_{w1}\widehat{T}_{w01}\Big/\widehat{T}_{w00} = \frac{\sum_{i\in s} b_{si}y_i}{\sum_{i\in s} b_{si}} - \widehat{B}_{w1}\frac{\sum_{i\in s} b_{si}x_{1i}}{\sum_{i\in s} b_{si}} \quad (20.2.8)$$

and

$$\widehat{B}_{w1} = \frac{\widehat{T}_{wy1} - \widehat{T}_{w01}\widehat{T}_{wy0}\Big/\widehat{T}_{w00}}{\widehat{T}_{w11} - \widehat{T}^2_{w01}\Big/\widehat{T}_{w00}}$$

$$= \frac{\sum_{i\in s} b_{si}y_i x_{1i} - \left(\sum_{i\in s} b_{si}x_{1i}\right)\left(\sum_{i\in s} b_{si}y_i\right)\Big/\left(\sum_{i\in s} b_{si}\right)}{\sum_{i\in s} b_{si}x_{1i}^2 - \left(\sum_{i\in s} b_{si}x_{1i}\right)^2\Big/\left(\sum_{i\in s} b_{si}\right)} \quad (20.2.9)$$

In case, $b_{si} = 1/\pi_i$, the estimators (20.2.8) and (20.2.9) reduce to

$$\widehat{B}_{\pi 0} = \left(\sum_{i\in s}\frac{y_i}{\pi_i}\right)\Big/\left(\sum_{i\in s}\frac{1}{\pi_i}\right) - \widehat{B}_{\pi 1}\left(\sum_{i\in s}\frac{x_{1i}}{\pi_i}\right)\Big/\left(\sum_{i\in s}\frac{1}{\pi_i}\right) \text{ and}$$

$$\widehat{B}_{\pi 1} = \frac{\sum_{i\in s}\frac{y_i x_{1i}}{\pi_i} - \left(\sum_{i\in s}\frac{y_i}{\pi_i}\right)\left(\sum_{i\in s}\frac{x_{1i}}{\pi_i}\right)\Big/\sum_{i\in s}\frac{1}{\pi_i}}{\sum_{i\in s}\frac{x_{1i}^2}{\pi_i} - \left(\sum_{i\in s}\frac{x_{1i}}{\pi_i}\right)^2\Big/\sum_{i\in s}\frac{1}{\pi_i}} \quad (20.2.10)$$

For SRSWOR sampling $\pi_i = n/N$ and we get

$$\widehat{B}_0 = \widehat{B}_{\pi 0} = \overline{y}_s - \widehat{B}_{\pi 1}\,\overline{x}_s \text{ and } \widehat{B}_1 = \widehat{B}_{\pi 1}$$

$$= \frac{\sum_{i\in s} y_i x_{1i} - \left(\sum_{i\in s} y_i\right)\left(\sum_{i\in s} x_{1i}\right)\Big/n}{\sum_{i\in s} x_{1i}^2 - \left(\sum_{i\in s} x_{1i}\right)^2\Big/n} \quad (20.2.11)$$

20.2.1 Estimation of Variance

The variance of $\widehat{\mathbf{B}}_w$ can be estimated by using standard variance estimation techniques for complex survey designs such as the random group, balanced repeated replication, jackknife, and bootstrap methods. Shah et al. (1977) and Binder (1983) derived an approximate expression of variance and its estimator by using Taylor series expansion as follows.

Let us write

$$\begin{aligned}\mathbf{W}(\mathbf{B}) &= \left(W_1(\mathbf{B}), \ldots, W_{p+1}(\mathbf{B})\right)' \\ &= \sum_{i=1}^{N}(y_i - \mathbf{x}_i\mathbf{B})\mathbf{x}_i' = \mathbf{0}\end{aligned} \tag{20.2.12}$$

Then an unbiased estimator of $\mathbf{W}(\mathbf{B})$ when $\mathbf{B}$ is known is

$$\widehat{\mathbf{W}}(\mathbf{B}) = \sum_{i\in s} b_{si}(y_i - \mathbf{x}_i\mathbf{B})\mathbf{x}_i'. \tag{20.2.13}$$

Now, writing $\widehat{e}_i = y_i - \mathbf{x}_i\widehat{\mathbf{B}}_w$, we find an approximate estimator of $\widehat{\mathbf{W}}(\mathbf{B})$ as

$$\widehat{\mathbf{W}}\left(\widehat{\mathbf{B}}_w\right) = \sum_{i\in s} b_{si}\widehat{e}_i\mathbf{x}_i' \tag{20.2.14}$$

Taylor series expansion of $\widehat{\mathbf{W}}\left(\widehat{\mathbf{B}}_w\right)$ around the point $\widehat{\mathbf{B}}_w = \mathbf{B}$ yields

$$\widehat{\mathbf{W}}\left(\widehat{\mathbf{B}}_w\right) \simeq \widehat{\mathbf{W}}(\mathbf{B}) + \frac{\partial\widehat{\mathbf{W}}(\mathbf{B})}{\partial\mathbf{B}}\left(\widehat{\mathbf{B}}_w - \mathbf{B}\right) \tag{20.2.15}$$

where $\dfrac{\partial\hat{\mathbf{W}}\left(\hat{\mathbf{B}}_w\right)}{\partial\mathbf{B}}$ is a $(p+1)\times(p+1)$ matrix whose i, jth element is the partial derivatives $\dfrac{\partial\hat{W}_i\left(\hat{\mathbf{B}}_w\right)}{\partial B_j}$.

Now, writing $\widehat{\mathbf{W}}\left(\widehat{\mathbf{B}}_w\right) = 0$, we obtain

$$\widehat{\mathbf{W}}(\mathbf{B}) \cong -\frac{\partial\widehat{\mathbf{W}}(\mathbf{B})}{\partial\mathbf{B}}\left(\widehat{\mathbf{B}}_w - \mathbf{B}\right)$$

Taking variances of both sides, and taking limit (following Binder, 1983), we get

$$V\left[\widehat{\mathbf{W}}(\mathbf{B})\right] \cong \left(\frac{\partial\mathbf{W}(\mathbf{B})}{\partial\mathbf{B}}\right)V\left(\widehat{\mathbf{B}}_w\right)\left(\frac{\partial\mathbf{W}(\mathbf{B})}{\partial\mathbf{B}}\right)' \tag{20.2.16}$$

Now, assuming $\frac{\partial \mathbf{W}(\mathbf{B})}{\partial \mathbf{B}}$ is of full rank, we find

$$V\left(\widehat{\mathbf{B}}_w\right) \simeq \left(\frac{\partial \mathbf{W}(\mathbf{B})}{\partial \mathbf{B}}\right)^{-1} V\left(\widehat{\mathbf{W}}(\mathbf{B})\right)\left(\frac{\partial \mathbf{W}(\mathbf{B})}{\partial \mathbf{B}}\right)^{\prime -1} \tag{20.2.17}$$

Now, noting $\frac{\partial \mathbf{W}(\mathbf{B})}{\partial \mathbf{B}} = -\mathbf{X}'\mathbf{X}$ and $\widehat{\mathbf{W}}(\mathbf{B}) = \sum_{i \in s} b_{si} e_i \mathbf{x}_i'$ with $e_i = y_i - \mathbf{x}_i \mathbf{B}_w$, we get

$$V\left(\widehat{\mathbf{B}}_w\right) \simeq (\mathbf{X}'\mathbf{X})^{-1} V\left(\sum_{i \in s} b_{si} e_i \mathbf{x}_i'\right)(\mathbf{X}'\mathbf{X})^{-1} \tag{20.2.18}$$

Finally, we get an approximate expression of $V\left(\widehat{\mathbf{B}}_w\right)$ as

$$\widehat{V}\left(\widehat{\mathbf{B}}_w\right) = \widehat{\mathbf{T}}_{wxx}^{-1}\, \widehat{\boldsymbol{\Sigma}}_{ex}\, \widehat{\mathbf{T}}_{wxx}^{-1} \tag{20.2.19}$$

where $\widehat{\mathbf{T}}_{wxx}$ given in Eq. (20.2.5) is an unbiased estimator of $\mathbf{X}'\mathbf{X}$ and $\widehat{\boldsymbol{\Sigma}}_{ex}$ is an unbiased estimator of $V\left(\sum_{i \in s} b_{si} e_i \mathbf{x}_i'\right)$.

In particular, if $b_{si} = 1/\pi_i$, the variance of $\widehat{\mathbf{B}}_\pi$ becomes

$$V\left(\widehat{\mathbf{B}}_\pi\right) = (\mathbf{X}'\mathbf{X})^{-1}\mathbf{Q}(\mathbf{X}'\mathbf{X})$$

where $\mathbf{Q}$ is a $(p+1) \times (p+1)$ with j, kth element

$$Q_{jk} = \frac{1}{2}\sum_{i \neq}\sum_{t \in U}(\pi_i\pi_t - \pi_{it})\left(\frac{e_i x_{ji}}{\pi_i} - \frac{e_t x_{jt}}{\pi_t}\right)\left(\frac{e_i x_{ki}}{\pi_i} - \frac{e_t x_{kt}}{\pi_t}\right).$$

A consistent estimator of $V\left(\widehat{\mathbf{B}}_\pi\right)$ is

$$\widehat{V}\left(\widehat{\mathbf{B}}_\pi\right) = (\mathbf{X}'\mathbf{X})^{-1}\widehat{\mathbf{Q}}(\mathbf{X}'\mathbf{X})$$

where j, k element of $\widehat{\mathbf{Q}}$ is

$$\widehat{Q}_{jk} = \frac{1}{2}\sum_{i \neq}\sum_{t \in s}\frac{(\pi_i\pi_t - \pi_{it})}{\pi_{it}}\left(\frac{\hat{e}_i x_{ji}}{\pi_i} - \frac{\hat{e}_t x_{jt}}{\pi_t}\right)\left(\frac{\hat{e}_i x_{ki}}{\pi_i} - \frac{\hat{e}_t x_{kt}}{\pi_t}\right)$$

where $\widehat{e}_i = y_i - \mathbf{x}_i \widehat{\mathbf{B}}_\pi$.

A $100 \times (1-\alpha)\%$ confidence interval for B_j can be worked out using the formula

$$\widehat{B}_{wj} \pm t_{\alpha/2, n-p-1}\sqrt{\widehat{V}\left(\widehat{B}_{wj}\right)} \text{ for } j = 0, 1, .., p \tag{20.2.20}$$

where $t_{\alpha/2, n-p-1}$ is the upper $\alpha/2 \times 100$ point of t-distribution with $n-p-1$ df.

20.2.2 Logistic Regression

In linear regression, the dependent variable y is generally taken as a continuous variable while in logistic regression the dependent variable y is binary taking values 1 if it possesses some attribute and 0 otherwise. For example, the HIV status (y) of person may be denoted as 1 if the person is HIV positive and 0 otherwise. We generally establish relationship between HIV status with age, race, occupation, economic condition, gender, etc. through a logistic regression model. The logistic regression with a vector of independent variable $\mathbf{x}' = (x_0, x_1, \ldots, x_p)$, $x_0 = 1$ and a parameter $\boldsymbol{\beta}' = (\beta_0, \beta_1, \ldots, \beta_p)$ is defined as

$$Prob(y = 1) = p(\mathbf{x}) = \frac{e^{\beta_0 + x_1\beta_1 + \cdots + x_p\beta_p}}{1 + e^{\beta_0 + x_1\beta_1 + \cdots + x_p\beta_p}} = \frac{e^{\mathbf{x}'\boldsymbol{\beta}}}{1 + e^{\mathbf{x}'\boldsymbol{\beta}}} \quad (20.2.21)$$

where $\mathbf{x}' = (1, x_1, \ldots, x_p)$ and $\boldsymbol{\beta}' = (\beta_0, \beta_1, \ldots, \beta_p)$.

The probability y will take value zero is $Prob(y = 0) = 1 - p(\mathbf{x}) = \frac{1}{1 + e^{\mathbf{x}'\boldsymbol{\beta}}}$. The odd ratio $p(\mathbf{x})/\{1 - p(\mathbf{x})\}$ is called $\log\text{it}\{p(\mathbf{x})\}$ and it is linear in $\mathbf{x}$, i.e., $\log\text{it}\{p(\mathbf{x})\} = \mathbf{x}'\boldsymbol{\beta}$. If the entire population of N units is surveyed, we would get the likelihood function assuming y_i's are independent as

$$L(\boldsymbol{\beta}) = \prod_{i=1}^{N} p_i^{y_i}(1 - p_i)^{1 - y_i} \quad (20.2.22)$$

where $p_i = Prob(y_i = 1) = p(\mathbf{x}_i)$; y_i and $\mathbf{x}_i = (x_{i0}, x_{i1}, \ldots, x_{ip})'$ are the values of y and $\mathbf{x}$ associated with the ith unit.

The likelihood equation $\frac{\partial \log L(\boldsymbol{\beta})}{\partial \boldsymbol{\beta}} = 0$ yields

$$\sum_{i=1}^{N}\left(y_i - \frac{e^{\mathbf{x}_i'\boldsymbol{\beta}}}{1 + e^{\mathbf{x}_i'\boldsymbol{\beta}}}\right)x_{ij} = 0 \quad \text{for} \quad j = 0, 1, \ldots p \quad (20.2.23)$$

The finite population parameter $\mathbf{B} = (B_0, B_1, \ldots, B_p)'$ is defined as the solution for $\mathbf{B}$ of the equation

$$\sum_{i=1}^{N}\left(y_i - \frac{e^{\mathbf{x}_i'\mathbf{B}}}{1 + e^{\mathbf{x}_i'\mathbf{B}}}\right)x_{ij} = 0 \quad \text{for} \quad j = 0, 1, \ldots p \quad (20.2.24)$$

No explicit expression of $\mathbf{B}$ can be obtained from Eq. (20.2.24). A design-based estimate of $\mathbf{B}$ is obtained as a solution of $\widehat{\mathbf{B}}_w$ of the equation

$$\sum_{i \in s} b_{si}\left(y_i - \frac{e^{\mathbf{x}_i'\hat{\mathbf{B}}_w}}{1 + e^{\mathbf{x}_i'\hat{\mathbf{B}}_w}}\right)x_{ij} = 0 \quad \text{for} \quad j = 0, 1, \ldots p \quad (20.2.25)$$

where b_{si}'s are appropriately chosen weights. The solution of $\widehat{\mathbf{B}}_w$ may be obtained by Newton–Raphson iterative procedure. The variance of $\widehat{\mathbf{B}}_w$ can be obtained following the method described in Section 20.2.1. However, $\widehat{\boldsymbol{\beta}}$, the estimator of the model parameter $\boldsymbol{\beta}$ of the model (20.2.21) is obtained from the equation

$$\sum_{i\in s}\left(y_i - \frac{e^{\mathbf{x}_i'\widehat{\boldsymbol{\beta}}}}{1+e^{\mathbf{x}_i'\widehat{\boldsymbol{\beta}}}}\right)x_{ij} = 0 \quad \text{for} \quad j = 0, 1, \ldots p \tag{20.2.26}$$

20.3 MODEL-BASED APPROACH

In the model-based approach, the finite population is a realization of a random vector $\mathbf{Y} = (Y_1,\ldots, Y_N)$ under a superpopulation model ξ (see Chapter 6). Under this approach, we assume that the study variable y_i is related to the vector of auxiliary variables $\mathbf{x}_i$ through the following superpopulation model:

$$y_i = \boldsymbol{\beta}\mathbf{x}_i + \in_i \text{ for } i = 1, \ldots, N \tag{20.3.1}$$

where $\boldsymbol{\beta} = (\beta_0, \beta_1, \ldots, \beta_p)'$; $E_\xi(\in_i) = 0$, $V_\xi(\in_i) = \sigma_i^2$ and $C_\xi(\in_i, \in_j) = 0$ for $i \neq j$; and E_ξ, V_ξ and C_ξ denote expectation, variance, and covariance with respect to the model ξ.

In case the entire population is surveyed and $\boldsymbol{\Sigma} = \text{diag}(\sigma_1^2, \ldots, \sigma_N^2)$ is known, one would get the GLS (generalized least square) estimator

$$\mathbf{B}^* = (\mathbf{X}'\boldsymbol{\Sigma}^{-1}\mathbf{X})^{-1}\mathbf{X}'\boldsymbol{\Sigma}^{-1}\mathbf{y} \tag{20.3.2}$$

which is the parameter of the present interest.

In practice, surveying the entire population is not possible; hence a suitable estimator for $\mathbf{B}^*$ based on a sample s is the GLS estimator of $\boldsymbol{\beta}$, which is given by

$$\widehat{\boldsymbol{\beta}}_{gls} = (\mathbf{X}_s'\boldsymbol{\Sigma}_s^{-1}\mathbf{X}_s)^{-1}\mathbf{X}_s'\boldsymbol{\Sigma}_s^{-1}\mathbf{y}_s \tag{20.3.3}$$

where $\boldsymbol{\Sigma}_s$ denotes an $n \times n$ submatrix of $\boldsymbol{\Sigma}$ associated with the selected sample s, $\mathbf{X}_s$, and y_s are as defined in Section 20.2.

Obviously, $\widehat{\boldsymbol{\beta}}_{gls}$ is useful if $\boldsymbol{\Sigma}_s$ is known. In case $\sigma_i^2 = \sigma^2$ for $i = 1,\ldots, N$

$$\text{i.e.,} \quad \boldsymbol{\Sigma} = \sigma^2\ \mathbf{I}_N \tag{20.3.4}$$

the best linear unbiased estimator (BLUE) of $\boldsymbol{\beta}$ is the OLS estimator

$$\widehat{\boldsymbol{\beta}}_{ols} = (\mathbf{X}_s'\mathbf{X}_s)^{-1}\mathbf{X}_s'\mathrm{y}_s \tag{20.3.5}$$

The estimator $\widehat{\mathbf{B}}_{\pi} = \widehat{\boldsymbol{\beta}}_{gls}$ if $\pi_i \propto \sigma_i^2$ and $\widehat{\boldsymbol{\beta}}_{gls} = \widehat{\boldsymbol{\beta}}_{ols}$ if $\sigma_i^2 = \sigma^2$ for $i = 1,\ldots, N$.

20.3.1 Performances of the Proposed Estimators

All the estimators $\widehat{\mathbf{B}}_w$, $\widehat{\mathbf{B}}_{\pi}$, $\widehat{\boldsymbol{\beta}}_{gls}$, and $\widehat{\boldsymbol{\beta}}_{ols}$ are model unbiased for $\boldsymbol{\beta}$ in the sense that

$$E_{\xi}\left(\widehat{\mathbf{B}}_w\right) = E_{\xi}\left(\widehat{\mathbf{B}}_{\pi}\right) = E_{\xi}\left(\widehat{\boldsymbol{\beta}}_{gls}\right) = E_{\xi}\left(\widehat{\boldsymbol{\beta}}_{ols}\right) = \boldsymbol{\beta} \tag{20.3.6}$$

The model variances of $\widehat{\boldsymbol{\beta}}_{gls}$ and $\widehat{\boldsymbol{\beta}}_{ols}$ under the models (20.3.1) and (20.3.4) are, respectively,

$$V_{\xi}\left(\widehat{\boldsymbol{\beta}}_{gls}\right) = \left(\mathbf{X}'_s\boldsymbol{\Sigma}_s^{-1}\mathbf{X}_s\right)^{-1} \text{ and } V_{\xi}\left(\widehat{\boldsymbol{\beta}}_{ols}\right) = \sigma^2(\mathbf{X}'_s\mathbf{X}_s)^{-1} \tag{20.3.7}$$

The estimator $\widehat{\boldsymbol{\beta}}_{gls} = \left(\widehat{\beta}_{0,gls}, \widehat{\beta}_{1,gls}, \ldots, \widehat{\beta}_{p,gls}\right)'$ is the optimum in the class of linear model unbiased estimators for $\boldsymbol{\beta} = (\beta_0, \beta_1, \ldots, \beta_p)'$ under the model (20.3.1) in the sense $V_{\xi}\left(\widehat{\beta}_{j,gls}\right) \leq V_{\xi}\left(\widehat{\beta}_j^*\right)$ for every $j = 0,1,\ldots, p$, where $\widehat{\beta}_j^*$ is any other linear model unbiased estimator of β_j. Similarly, $\widehat{\boldsymbol{\beta}}_{ols}$ is optimal under model (20.3.4). However, no definite conclusion can be reached if the estimators are restricted to the class of model/design unbiased estimators, where the estimators $\widehat{\beta}_j^*$ satisfy $E_p E_{\xi}\left(\widehat{\beta}_j^*\right) = \beta_j$ and E_p is the expectation with respect to the sampling design p.

20.3.2 Variance Estimation

In case $\boldsymbol{\Sigma} = \sigma^2\boldsymbol{\Sigma}^*$, where $\boldsymbol{\Sigma}^*$ is known, unbiased estimators of $V_{\xi}\left(\widehat{\boldsymbol{\beta}}_{gls}\right)$ and $V_{\xi}\left(\widehat{\boldsymbol{\beta}}_{ols}\right)$ under the model (20.3.1) and (20.3.4) are given, respectively, by

$$\widehat{V}_{\xi}\left(\widehat{\boldsymbol{\beta}}_{gls}\right) = \widehat{\sigma}_1^2\left(\mathbf{X}'_s\boldsymbol{\Sigma}_s^{*-1}\mathbf{X}_s\right)^{-1} \tag{20.3.8}$$

and

$$\widehat{V}_{\xi}\left(\widehat{\boldsymbol{\beta}}_{ols}\right) = \widehat{\sigma}_2^2(\mathbf{X}'_s\mathbf{X}_s)^{-1} \tag{20.3.9}$$

where

$$\widehat{\sigma}_1^2 = \frac{1}{n-p-1}\left(\mathbf{y}_s - \mathbf{X}_s\widehat{\boldsymbol{\beta}}_{gls}\right)'\left(\mathbf{y}_s - \mathbf{X}_s\widehat{\boldsymbol{\beta}}_{gls}\right) \text{ and}$$

$$\widehat{\sigma}_2^2 = \frac{1}{n-p-1}\left(\mathbf{y}_s - \mathbf{X}_s\widehat{\boldsymbol{\beta}}_{ols}\right)'\left(\mathbf{y}_s - \mathbf{X}_s\widehat{\boldsymbol{\beta}}_{ols}\right) \tag{20.3.10}$$

The model variances of $\widehat{\mathbf{B}}_w$ and $\widehat{\mathbf{B}}_\pi$ under model (20.3.1) with $\boldsymbol{\Sigma} = \boldsymbol{\Sigma}^*$ are given by

$$V_m\left(\widehat{\mathbf{B}}_w\right) = \sigma^2\left(\mathbf{X}_s'\mathbf{W}_s\mathbf{X}_s\right)^{-1}\left(\mathbf{X}_s'\mathbf{W}_s\boldsymbol{\Sigma}_s^*\mathbf{W}_s\mathbf{X}_s\right)\left(\mathbf{X}_s'\mathbf{W}_s\mathbf{X}_s\right)^{-1} \tag{20.3.11}$$

and

$$V_m\left(\widehat{\mathbf{B}}_\pi\right) = \sigma^2\left(\mathbf{X}_s'\mathbf{W}_\pi\mathbf{X}_s\right)^{-1}\left(\mathbf{X}_s'\mathbf{W}_\pi\boldsymbol{\Sigma}_s^*\mathbf{W}_\pi\mathbf{X}_s\right)\left(\mathbf{X}_s'\mathbf{W}_\pi\mathbf{X}_s\right)^{-1} \tag{20.3.12}$$

The values of $\widehat{\boldsymbol{\beta}}_{ols}$, $\widehat{\boldsymbol{\beta}}_{gls}$, and $\widehat{\boldsymbol{B}}_\pi$ and estimated variances of $\widehat{\boldsymbol{\beta}}_{ols}$ and $\widehat{\boldsymbol{\beta}}_{gls}$ may be available by using standard software packages (e.g., BMDP), but the variance formulas (20.3.11) and (20.3.12) cannot be obtained by standard statistical packages unless the design is self-weighting ($\mathbf{W}_s \propto \mathbf{I}_n$) and $\boldsymbol{\Sigma}_s^* = \mathbf{I}_n$, where $\mathbf{I}_n$ is an identity matrix of order n.

20.3.3 Multistage Sampling

Consider a finite population consisting of N first-stage units (fsu's) and the ith fsu consists of M_i second-stage units (ssu's), $i = 1,\ldots, N$. Let a sample s of n fsu's be selected from the population by some suitable sampling design. If the ith fsu is selected in s, a subsample s_i of m_i ssu's is selected from the M_i ssu's of the ith fsu by some sampling procedure. Let y_{ij} be the value of the study variable y for the jth ssu of the ith fsu, $j = 1,\ldots, M_i$; $i = 1,\ldots, N$. Let us suppose that the study variable y is related to the explanatory variables $x_1,\ldots, x_p$ through the following superpopulation model

$$y_{ij} = \beta_0 x_{ij0} + \beta_1 x_{ij1} + \cdots + \beta_p x_{ijp} + \in_{ij} \tag{20.3.13}$$

where $x_{ij0} = 1$, $\beta_0, \beta_1,\ldots, \beta_p$ are unknown model parameters, $\in_{ij}'s$ are error components with

$$E_\xi(\in_{ij}) = 0,\ V_\xi(\in_{ij}) = \sigma^2,\ C_\xi(\in_{ij}, \in_{ik}) = \rho\sigma^2,\ C_\xi(\in_{ij}, \in_{i'k}) = 0 \text{ for } i \neq i';$$
$$j = 1, \ldots, M_i;\ i, i' = 1, \ldots, N$$

The model (20.3.13) indicates that any two units belonging to the same fsu are correlated with a common intracluster correlation ρ, while ssu's belonging to different fsu's are uncorrelated. This type of model (20.3.13) was considered by Fuller (1975), Campbell (1977) and Holt and Scott (1981). For simplicity, let us assume that the selected fsu's in the sample s consist of first n units (i.e., fsu's labeled 1,2,..., n) and the subsample s_1 comprises of first m_1 ssu's of the fsu labeled 1, s_2 comprises of the first m_2 ssu's from the fsu labeled 2, and so on. The model for the selected sample can be written as

$$\mathbf{Y} = \mathbf{X}\boldsymbol{\beta} + \boldsymbol{\varepsilon} \tag{20.3.14}$$

where
$$\mathbf{Y} = \begin{pmatrix} y_{11} \\ \cdot \\ y_{1m_1} \\ y_{21} \\ \cdot \\ y_{2m_2} \\ \cdot \\ y_{n1} \\ \cdot \\ y_{nm_n} \end{pmatrix}, \quad \mathbf{X} = \begin{pmatrix} 1 & x_{111} & x_{112} & \cdot & x_{11p} \\ \cdot & \cdot & \cdot & \cdot & \cdot \\ 1 & x_{1m_11} & x_{1m_12} & \cdot & x_{1m_1p} \\ 1 & x_{211} & x_{212} & \cdot & x_{21p} \\ \cdot & \cdot & \cdot & \cdot & \cdot \\ 1 & x_{2m_21} & x_{2m_22} & \cdot & x_{2m_2p} \\ \cdot & \cdot & \cdot & \cdot & \cdot \\ 1 & x_{n11} & x_{n12} & \cdot & x_{n1p} \\ & & \cdot & & \\ 1 & x_{nm_n1} & x_{nm_n2} & \cdot & x_{nm_np} \end{pmatrix},$$

$$\boldsymbol{\beta} = \begin{pmatrix} \beta_0 \\ \beta_1 \\ \cdot \\ \beta_p \end{pmatrix} \text{ and. } \boldsymbol{\varepsilon} = \begin{pmatrix} \epsilon_{11} \\ \cdot \\ \epsilon_{1m_1} \\ \epsilon_{21} \\ \cdot \\ \epsilon_{2m_2} \\ \cdot \\ \epsilon_{n1} \\ \cdot \\ \epsilon_{nm_n} \end{pmatrix}$$

The error component $\boldsymbol{\varepsilon}$ has $E_\xi(\boldsymbol{\varepsilon}) = 0$ and $V_\xi(\boldsymbol{\varepsilon}) = \sigma^2\mathbf{G}$, where $\mathbf{G} = Diag(\mathbf{G}_1, \ldots, \mathbf{G_n})$ with $\mathbf{G}_i = [(1-\rho)\mathbf{I}_{mi} + \rho\mathbf{E}_{mi,mi}]$, $\mathbf{I}_{mi}$ = unit matrix of order m_i, and $\mathbf{E}_{mi,mi}$ is a $m_i \times m_i$ matrix of each element 1. In case ρ is known and $\mathbf{X}$ is of full rank, the BLUE of $\boldsymbol{\beta}$ is the GLS estimator

$$\widehat{\boldsymbol{\beta}}_{gls} = \left(\mathbf{X}'\mathbf{G}^{-1}\mathbf{X}\right)^{-1}\mathbf{X}'\mathbf{G}^{-1}\mathbf{Y} \tag{20.3.15}$$

The OLS estimator of $\boldsymbol{\beta}$ is

$$\widehat{\boldsymbol{\beta}}_{ols} = \left(\mathbf{X}'\mathbf{X}\right)^{-1}\mathbf{X}'\mathbf{Y} \tag{20.3.16}$$

Both the estimators $\widehat{\boldsymbol{\beta}}_{gls}$ and $\widehat{\boldsymbol{\beta}}_{ols}$ are model unbiased for $\boldsymbol{\beta}$ in the sense $E_\xi\left(\widehat{\boldsymbol{\beta}}_{gls}\right) = E_\xi\left(\widehat{\boldsymbol{\beta}}_{ols}\right) = \boldsymbol{\beta}$. The estimator $\widehat{\boldsymbol{\beta}}_{gls}$ is the optimal in the sense $V_\xi\left(\widehat{\beta}_{j,gls}\right) \le V_\xi\left(\widehat{\beta}_{j,ols}\right)$ for $j = 0, \ldots, p$, where $\widehat{\beta}'_{gls} = \left(\widehat{\beta}_{0,gls}, \widehat{\beta}_{1,gls}, \ldots, \widehat{\beta}_{p,gls}\right)$ and $\widehat{\boldsymbol{\beta}}'_{ols} = \left(\widehat{\beta}_{0,ols}, \widehat{\beta}_{1,ols}, \ldots, \widehat{\beta}_{p,ols}\right)$.

In application of the GLS procedure the correlation coefficient ρ must be known. If it is not known, one may use iterative procedure following Fuller and Battese (1973) or use a computer package such as SUPERCARP (Hidiroglou et al., 1980). Furthermore, to compute $\widehat{\boldsymbol{\beta}}_{gls}$, one needs to know which observation comes from which cluster, and this information is rarely available, especially from a secondary data. The main advantage of $\widehat{\boldsymbol{\beta}}_{ols}$ is that it is easy to compute and does not require such information. For small value of ρ, the loss of efficiency of using $\widehat{\boldsymbol{\beta}}_{ols}$ is not substantial (Scott and Holt, 1982). The variances of $\widehat{\boldsymbol{\beta}}_{gls}$ and $\widehat{\boldsymbol{\beta}}_{ols}$ under the model (20.3.13) are given, respectively, by

$$V_\xi\left(\widehat{\boldsymbol{\beta}}_{gls}\right) = \sigma^2\left(\mathbf{X}'\mathbf{G}^{-1}\mathbf{X}\right)^{-1} \tag{20.3.17}$$

and

$$\begin{aligned} V_\xi\left(\widehat{\boldsymbol{\beta}}_{ols}\right) &= \sigma^2\left(\mathbf{X}'\mathbf{X}\right)^{-1}\left(\mathbf{X}'\mathbf{G}\mathbf{X}\right)\left(\mathbf{X}'\mathbf{X}\right)^{-1} \\ &= \sigma^2\left(\mathbf{X}'\mathbf{X}\right)^{-1}\mathbf{D} \end{aligned} \tag{20.3.18}$$

where $\mathbf{D} = \left(\mathbf{X}'\mathbf{G}\mathbf{X}\right)\left(\mathbf{X}'\mathbf{X}\right)^{-1}$.

If we ignore the effect of clustering, i.e., treating the correlation $\rho = 0$ and accept the $V_\xi\left(\widehat{\boldsymbol{\beta}}_{ols}\right) = \sigma^2\left(\mathbf{X}'\mathbf{X}\right)^{-1}$, then we will be committing a serious mistake of evaluating $V_\xi\left(\widehat{\boldsymbol{\beta}}_{ols}\right)$ while ignoring the factor $\mathbf{D}$. Kish and Frankel (1974) termed the factor **D** as a design effect although **D** does not depend on a sampling design. Hence, Scott and Holt (1982) termed **D** as a model "misspecification" effect. The effect of ignoring the factor **D** has a serious effect in estimating confidence interval and hypothesis testing based on $\widehat{\boldsymbol{\beta}}_{ols}$. The usual estimator of error variance σ^2 based on OLS (ignoring **D**) is

$$\widehat{\sigma}^2_{ols} = \mathbf{Y}'(\mathbf{1} - \mathbf{P}_0)\mathbf{Y}/(n - p - 1) \tag{20.3.19}$$

where $\mathbf{P}_0 = \mathbf{X}(\mathbf{X}'\mathbf{X})^{-1}\mathbf{X}'$.

Scott and Holt (1982) showed that under the model (20.3.13)

$$E_\xi\left(\widehat{\sigma}^2_{ols}\right) = \sigma^2\frac{1 - \text{trace}(\mathbf{D})}{n - p - 1} = \sigma^2\frac{1 - (p + 1)\overline{d}}{n - p - 1} \tag{20.3.20}$$

where $\overline{d}$ is the average diagonal element of **D**.

It is seen in most practical situations $\overline{d}$ exceeds 1, hence $\widehat{\sigma}^2_{ols}$ underestimates σ^2, but the effect is negligible if the sample size is reasonably large.

20.3.4 Separate Regression for Each First-Stage Unit

Konijn (1962), Porter (1973), and Pfeffermann and Nathan (1981) considered a separate regression model for each of the first-stage units. In this model, the dependent variable y for the ith fsu is related to a single independent variable x through the model

$$y_{ij} = \beta^{(i)}x_{ij} + \in_{ij} \text{ for } i = 1, \ldots, N; j = 1, \ldots, M_i \tag{20.3.21}$$

Let $\mathbf{Y}_i = \begin{pmatrix} y_{i1} \\ \cdot \\ y_{im_i} \end{pmatrix}$ and $\mathbf{X}_i = \begin{pmatrix} x_{i1} \\ \cdot \\ x_{m_i1} \end{pmatrix}$ denote, respectively, the vector of dependent and independent variables for the sampled m_i ssu's that belong to the ith selected fsu in s, $E_\xi(\in_{ij}) = 0$, $V_\xi(\in_{ij}) = \sigma_i^2$ and $C_\xi(\in_{ij}, \in_{i'j'}) = 0$ for $(i,j) \neq (i'j')$.

Konijn (1962) selected a sample s with inclusion probability $\pi_i = \sum_{s \supset i} p(s)$ for the ith fsu and proposed an unbiased estimator of the average regression coefficient $\overline{\beta} = \dfrac{\sum_{i=1}^{N} M_i\beta^{(i)}}{M_0}$ as

$$\widehat{\overline{\beta}} = \frac{\sum_{i \in s} M_i\widehat{\beta}^{(i)}_{ols}}{M_0} \tag{20.3.22}$$

where $M_0 = \sum_{i=1}^{N} M_i$ and $\widehat{\beta}_{ols}^{(i)} = \left(\mathbf{X}_i'\mathbf{X}_i\right)^{-1}\mathbf{X}_i'\mathbf{Y}_i$, the OLS estimator of $\beta^{(i)}$. The estimator $\widehat{\overline{\beta}}$ is a model design unbiased estimator of $\overline{\beta}$ because it satisfies $E_pE_m\left(\widehat{\overline{\beta}}\right) = \overline{\beta}$. Porter (1973) considered estimating $\overline{\beta}^* = \sum_{i=1}^{N} \beta^{(i)} \Big/ N$. Pfeffermann and Nathan (1981) proposed an alternative approach where the regression parameters $\beta^{(i)}$'s have a joint distribution $m_\beta(\theta)$. Although this model involves unknown parameters θ, it generally improves inference by using relationship between the variables $\beta^{(i)\prime}s(i = 1,\ldots, N)$. The model is very popular in econometrics, and good details are given by Maddala (1977), Fay and Herriot (1979), Rubin (1976), etc. In the proposed Pfeffermann and Nathan's (1981) model

$$\beta^{(i)} = \beta + \nu_i \tag{20.3.23}$$

$$E_m(\nu_i) = 0,\ V_m(\nu_i) = \delta^2 \text{ and } C_m(\nu_i, \nu_j) = 0 \text{ for } i \neq j;\ i,j = 1, \ldots, N$$

Following Haitovsky (1973), Pfeffermann and Nathan (1981) derived the extended least square estimators of the individual coefficients $\beta^{(i)}$ assuming δ^2 and σ_i^2 are known, as

$$\widehat{\widehat{\beta}}^{(i)} = \begin{cases} \lambda_i\widehat{\beta}^{(i)} + (1-\lambda_i)\widehat{\beta} & \text{for } i \in s \\ \widehat{\beta} & \text{otherwise} \end{cases} \tag{20.3.24}$$

where $\widehat{\beta}^{(i)} = \sum_{j\in s_i} x_{ij}y_{ij} \Big/ \sum_{j\in s_i} x_{ij}^2$, $\widehat{\beta} = \sum_{i\in s} \lambda_i\widehat{\beta}^{(i)} \Big/ \sum_{i\in s} \lambda_i$, and

$$\lambda_i = \left(1 + \frac{\sigma_i^2}{\delta^2 \sum_{j\in s_i} x_{ij}^2}\right)^{-1}.$$

The estimator $\widehat{\widehat{\beta}}^{(i)}$ cannot be used in practice because the parameters δ^2 and σ_i^2 are generally unknown. Pfeffermann and Nathan (1981) proposed estimators for σ_i^2 $(i \in s)$ as

$$\widehat{\sigma}_i^2 = \sum_{j\in s_i} \left(y_{ij} - \widehat{\beta}^{(i)}x_{ij}\right)^2 \Big/ (m_i - 1) \tag{20.3.25}$$

Denoting $\widehat{\lambda}_i = \left(1 + \dfrac{\hat{\sigma}_i^2}{\delta^2 \sum_{j\in s_i} x_{ij}^2}\right)^{-1}$, Pfeffermann and Nathan (1981) proposed the estimator δ^2 as the largest solution of the equation

$$\frac{1}{(n-1)}\sum_{i\in s}\widehat{\lambda}_i\left(\widehat{\beta}^{(i)}-\widehat{\overline{\beta}}^{(\hat{\lambda})}\right)^2=\delta^2 \tag{20.3.26}$$

where $\widehat{\overline{\beta}}^{(\hat{\lambda})}=\sum_{i\in s}\widehat{\beta}^{(i)}\widehat{\lambda}_i\Big/\sum_{i\in s}\widehat{\lambda}_i$.

Pfeffermann and Nathan (1981) studied in detail the existence and uniqueness of a positive solution and proximity of that solution to the true variance. Pfeffermann and Smith (1985) extended the results for p independent variables and studied the properties of the regression estimators and their application in large-scale real-life survey data.

20.4 CONCLUDING REMARKS

In descriptive inference the population parameter is supposed to be a known function of finite population values such as the population mean, variance, and correlation coefficient. The descriptive inference may be model dependent or design based. Analytic inference is model based. Kish and Frankel (1974) considered regression analysis in a descriptive approach. Jönrup and Rennermalm (1976) and Shah et al. (1977) followed a similar approach. On the basis of empirical studies, Kish and Frankel (1974) showed that if clustering of the population is ignored and OLS is used for regression analysis, then the estimates of the variances of the regression coefficients underestimate the variance. The underestimation of variances has serious effects on interval estimation and testing of hypotheses of parameters. The standard software packages provide wrong estimates of standard errors. However, Konjin (1962), Fuller (1975), and Pfeffermann and Smith (1985) favor analytic approach where the finite population is regarded as sample from an infinite population.

20.5 EXERCISES

20.5.1 Let a sample of 10 households be selected from a locality of 50 households by PPSWR method using household size (z) as a measure of size variable. The following table gives the monthly household expenditure on food (y), monthly household income (x), and household size (z). Fit a finite population regression y on x with intercept assuming mean household size of the population is 3.5. Find 95% confidence intervals of the regression coefficients.

House holds	1	2	3	4	5	6	7	8	9	10
z	4	5	2	6	4	2	3	3	4	1
y(in \$)	2000	2500	1000	4500	3500	2000	3000	2500	2800	1500
x (in \$)	5000	4000	2000	5600	4500	2500	6000	4000	3200	5000

20.5.2 Consider the Exercise 20.5.1., and assume that y is related to x through the regression model $y_i = \beta_0 + \beta_1 x_i + \in_i$; $E_m(\in_i) = 0$, $V_m(\in_i) = \sigma^2 x_i$, and $C_m(\in_i, \in_j) = 0$ for $i \neq j$. Estimate regression coefficients by OLS and GLS methods.

20.5.3 Let a sample of size 10 be selected from a finite population of 50 using a complex survey design. The following table gives the values of the study variable y, auxiliary variables x_1 and x_2, and inclusion probabilities π of the selected units in the sample. Estimate the finite population regression of y on x_1 and x_2, and also estimate the multiple correlation of y on x_1 and x_2.

Units	1	2	3	4	5	6	7	8	9	10
y	20	25	32	60	40	20	35	40	40	20
x_1	15	10	25	40	30	10	25	15	25	10
x_2	25	10	20	60	20	25	30	20	25	15
π	0.20	0.10	0.05	0.15	0.15	0.20	0.10	0.25	0.15	0.10

20.5.4 Let a sample of n first-stage units be selected from a population of N first-stage units. From each of the selected first-stage units, a subsample of m second-stage units is selected by some suitable sampling procedure. Let y_{ij} and x_{ijk} be the value of the study variable y and auxiliary variable x_k for the jth second-stage units of the ith first-stage units, $k = 1,2,\ldots, p$. Assume y_{ij} is related to x_{ij} through the superpopulation model:

$$y_{ij} = \beta_0 x_{ij0} + \beta_1 x_{ij1} + \ldots + \beta_p x_{ijp} + \in_{ij}$$

where $x_{ij0} = 1$, $\beta_0, \beta_1, \ldots, \beta_p$ are unknown model parameters, $\in'_{ij}s$ are error components with

$$E_\xi(\in_{ij}) = 0,\ V_\xi(\in_{ij}) = \sigma^2,\ C_\xi(\in_{ij}, \in_{ik}) = \rho\sigma^2,$$
$$C_\xi(\in_{ij}, \in_{i'k}) = 0 \text{ for } i \neq i'; j = 1, \ldots, M_i; i, i' = 1, \ldots, N$$

Let $\widehat{\boldsymbol{\beta}}_G$ and $\widehat{\boldsymbol{\beta}}_0$ be the GLS and OLS of $\boldsymbol{\beta} = \left(\beta_0, \ldots, \beta_p\right)'$, respectively. Show that (i) $\widehat{\boldsymbol{\beta}}_0$ is unbiased estimator of $\boldsymbol{\beta}$,

(ii) $\dfrac{4\lambda_1\lambda_n}{(\lambda_1 + \lambda_n)^2} \leq E = \dfrac{\operatorname{Var}\left(c'\hat{\beta}_G\right)}{\operatorname{Var}\left(c'\hat{\beta}_0\right)} \leq 1$, where $\mathbf{c}$ is an arbitrary coefficient vector and $\lambda_1 \geq \lambda_2 \geq \cdots \geq \lambda_n$ are eigenvalues of

$$\mathbf{V} = \begin{bmatrix} 1 & \rho & \ldots & \rho \\ \rho & 1 & \ldots & \rho \\ \cdot & \cdot & \ldots & \cdot \\ \rho & \rho & \ldots & 1 \end{bmatrix}$$

, and (iii) the upper bound of the loss of efficiency of using $\widehat{\boldsymbol{\beta}}_0$ in place of $\widehat{\boldsymbol{\beta}}_G$ is

$$1 - E = \left(1 + \frac{4(1-\rho)[1+(m-1)\rho]}{m^2\rho^2}\right)^{-1}$$ (Scott and Holt, 1982).

CHAPTER 21

Ranked Set Sampling

21.1 INTRODUCTION

Ranked set sampling (RSS) was introduced by McIntyre (1952) to estimate the mean pasture and forage yield. The RSS is used when precise measurement of the variable of interest is difficult or expensive but the variable can be ranked easily without measuring the actual variable by an inexpensive method such as visual perception, judgment, and auxiliary information. For example, in estimating the mean height of trees in a forest, the heights of a small sample of two or three trees standing nearby can be ranked easily by visual inspection without measuring them. In estimating the number of bacterial cells per unit volume, we can rearrange two or three test tubes easily in the order of concentration using optical instruments without measuring exact values. In RSS, instead of selecting a single sample of size m, we select m sets of samples each of size m. In each of the sets, all the elements are ranked but only one is measured. Finally, an average of the m measured units is taken as an estimate of the population mean. The sample mean based on RSS is unbiased for the population mean regardless of the errors of ranking. The RSS mean is at least as precise as the sample mean of the simple random sampling with replacement (SRSWR) sampling scheme of the same size. Stokes (1980a,b, 1988) showed that RSS provides precise estimators for cumulative distribution function (cdf), population variance, and correlation coefficient.

21.2 RANKED SET SAMPLING BY SIMPLE RANDOM SAMPLING WITH REPLACEMENT METHOD

First, we choose a small number m (set size) such that one can easily rank m elements of the population with sufficient accuracy. Then the selection procedure of RSS is as follows: Select a sample of m^2 units from a population U by the SRSWR method. Allocate these m^2 units at random into m sets each of size m. Rank all the units in a set with respect to the values of the variable of interest y from 1 (minimum) to m (maximum) by a very inexpensive method such as eye inspection. No actual measurement is done

Survey Sampling Theory and Applications
ISBN 978-0-12-811848-1
http://dx.doi.org/10.1016/B978-0-12-811848-1.00021-2

at this stage. After the ranking has been completed, the unit holding rank 1 of the set 1, unit holding rank 2 of the set 2,..., and finally the unit holding rank m of the set m are measured accurately by using a suitable instrument. This completes a cycle of the sampling. The process is repeated for r cycles to obtain the desired sample of size $n = mr$ units. Thus in RSS, a total of m^2r units have been drawn from the population, but only mr of them are measured and the rest $mr(m - 1)$ are discarded. These measured mr observations are called a "ranked set sample." Because the ordering of a large number of observations is difficult, increase in sample size $n(=mr)$ is done by increasing the number of cycles r. Let $y_{i1|k},\ldots, y_{ij|k},\ldots, y_{im|k}$ be the value of the variable of interest y of the ith set of elements of the kth cycle, $i = 1,\ldots, m$; $k = 1,\ldots, r$. Furthermore, let $y_{i(j)|k}$ be the smallest jth observation (order statistic) of $y_{i1|k},\ldots, y_{ij|k},\ldots, y_{im|k}$. From the ith set the ith order statistic is measured, i.e., the value of $y_{i(i)|k}$ is obtained. This can be represented as follows.

Set	Cycle *k* after rearrangement of *y* values					Observed *y*
1	$y_{1(1)\|k}$	$\cdots$	$y_{1(i)\|k}$	$\cdots$	$y_{1(m)\|k}$	$y_{1(1)\|k}$
2	$y_{2(1)\|k}$	$\cdots$	$y_{2(i)\|k}$	$\cdots$	$y_{2(m)\|k}$	$y_{2(2)\|k}$
$\cdots$	$\cdots$	$\cdots$	$\cdots$	$\cdots$	$\cdots$	$\cdots$
i	$y_{i(1)\|k}$	$\cdots$	$y_{i(i)\|k}$	$\cdots$	$y_{i(m)\|k}$	$y_{i(i)\|k}$
$\cdots$	$\cdots$	$\cdots$	$\cdots$	$\cdots$	$\cdots$	$\cdots$
m	$y_{m(1)\|k}$	$\cdots$	$y_{m(i)\|k}$	$\cdots$	$y_{m(m)\|k}$	$y_{m(m)\|k}$

Thus we have the following data:

$$d = \left\{ \overbrace{y_{1(1)|1} \quad y_{2(2)|1} \quad \cdots \quad y_{m(m)|1}}^{\text{cycle } 1};\ldots; \overbrace{y_{1(1)|k} \quad y_{2(2)|k} \quad \cdots \quad y_{m(m)|k}}^{\text{cycle } k};\ldots; \overbrace{y_{1(1)|r} \quad y_{2(2)|r} \quad \cdots \quad y_{m(m)|r}}^{\text{cycle } r} \right\}$$

21.2.1 A Fundamental Equality

Let $y_{i1|k},\ldots, y_{ij|k},\ldots, y_{im|k}$ be a random sample from a population with cdf $F(y)$ and probability density function (pdf) $f(y)$. Let the mean and variance of y be μ and σ^2, respectively. Then we have the following equalities:

$$\sum_{j=1}^{m} y_{ij|k} = \sum_{j=1}^{m} y_{i(j)|k} \tag{21.2.1}$$

and

$$\sum_{j=1}^{m} \left(y_{ij|k} - \mu\right)^2 = \sum_{j=1}^{m} \left(y_{i(j)|k} - \mu\right)^2 \tag{21.2.2}$$

Let $\mu_{(j)|m} = E(y_{i(j)|k})$ be the mean of the jth order statistic of a random sample of size m of the cycle k, which depends on m but is independent of the set i and the cycle k.

Eq. (21.2.1) yields

$$E\left(\frac{1}{m}\sum_{j=1}^{m} y_{ij|k}\right) = E\left(\frac{1}{m}\sum_{j=1}^{m} y_{i(j)|k}\right)$$

$$\text{i.e.,} \quad \mu = \frac{1}{m}\sum_{j=1}^{m} \mu_{(j)|m} \tag{21.2.3}$$

Similarly, Eq. (21.2.2) yields

$$\sum_{j=1}^{m} E\left(y_{ij|k} - \mu\right)^2 = \sum_{j=1}^{m} E\left(y_{i(j)|k} - \mu\right)^2$$

$$\text{i.e.,} \quad m\sigma^2 = \sum_{j=1}^{m} \left\{\sigma^2_{(j)|m} + \left(\mu_{(j)|m} - \mu\right)^2\right\}$$

where $\sigma^2_{(j)|m}$ = variance of $y_{i(j)|m}$.

$$\text{i.e.,} \quad \sigma^2 = \frac{1}{m}\sum_{j=1}^{m} \sigma^2_{(j)|m} + \frac{1}{m}\sum_{i=1}^{m} \left(\mu_{(j)|m} - \mu\right)^2 \tag{21.2.4}$$

21.2.2 Estimation of the Mean

Let $\overline{y}_{[m]|k} = \frac{1}{m}\sum_{i=1}^{m} y_{i(i)|k}$ = arithmetic mean of the m quantified values of the variable y for the cycle k and

$$\widehat{\mu}_{rss} = \frac{1}{r}\sum_{k=1}^{r} \overline{y}_{[m]|k} = \frac{1}{n}\sum_{k=1}^{r}\sum_{i=1}^{m} y_{i(i)|k} \tag{21.2.5}$$

is the mean of $n(=mr)$ quantified variables based on all the r cycles. The following theorem shows that the estimator $\widehat{\mu}_{rss}$ is unbiased for μ and possesses a lower variance than $\widehat{\mu}_{srs}$, the sample mean based on an SRSWR sample of the same size n.

Theorem 21.2.1

(i) $E(\widehat{\mu}_{rss}) = \mu$

(ii) $V(\widehat{\mu}_{rss}) = \dfrac{\sigma^2_{[m]}}{n} = \dfrac{1}{n}\left[\sigma^2 - \dfrac{1}{m}\sum_{i=1}^{m}\left(\mu_{(j)|m} - \mu\right)^2\right] \leq \sigma^2/n$

(iii) An unbiased estimator of the variance of $V(\widehat{\mu}_{rss})$ is

$$\widehat{V}(\widehat{\mu}_{rss}) = \frac{1}{r(r-1)}\sum_{k=1}^{r}\left(\overline{y}_{[m]|k} - \widehat{\mu}_{rss}\right)^2.$$

where $\sigma^2_{[m]} = \dfrac{1}{m}\sum_{j=1}^{m}\sigma^2_{(j)|m}$.

Proof

(i) $E(\widehat{\mu}_{rss}) = \dfrac{1}{r}E\left(\sum_{k=1}^{r}\overline{y}_{[m]|k}\right)$

Eq. (21.2.1) yields

$$E\left(\overline{y}_{[m]|k}\right) = \frac{1}{m}\sum_{i=1}^{m}E\left(y_{i(i)|k}\right) = \frac{1}{m}\sum_{i=1}^{m}\mu_{(i)|m} = \mu$$

Hence, $E(\widehat{\mu}_{rss}) = \mu$.

(ii) Because $\overline{y}_{[m]|1}, \ldots, \overline{y}_{[m]|r}$ are iid random variables, we have

$$V(\widehat{\mu}_{rss}) = \frac{1}{r^2}\sum_{k-1}^{r}V\left(\overline{y}_{[m]|k}\right) = \frac{V\left(\overline{y}_{[m]|k}\right)}{r} \tag{21.2.6}$$

Now

$$V\left(\overline{y}_{[m]|k}\right) = \frac{1}{m^2}\left[\sum_{i=1}^{m}V\left(y_{i(i)|k}\right)\right] = \frac{1}{m^2}\left[\sum_{i=1}^{m}\sigma^2_{(i)|m)}\right] = \frac{1}{m}\left[\sigma^2 - \frac{1}{m}\sum_{i=1}^{m}\left(\mu_{(i)|m} - \mu\right)^2\right]$$

(using Eq. 21.2.4).

Hence $$V(\widehat{\mu}_{rss}) = \frac{1}{n}\left[\sigma^2 - \frac{1}{m}\sum_{i=1}^{m}\left(\mu_{(i)|m} - \mu\right)^2\right] \leq \sigma^2/n \tag{21.2.7}$$

(iii) The result follows from the fact that $\overline{y}_{[m]|k}$ are distributed independently.

21.2.3 Precision of the Ranked Set Sampling

The relative precision of $\widehat{\mu}_{rss}$ compared with $\widehat{\mu}_{srs}$, sample mean of an SRSWR sample of size n, is

$$RP = \frac{V(\widehat{\mu}_{srs})}{V(\widehat{\mu}_{rss})} = \frac{\sigma^2}{\sigma^2_{[m]}} \tag{21.2.8}$$

Example 21.2.1: Uniform Distribution

Let $X_1,\ldots, X_m$ be a random sample from a uniform distribution over (0,1). In this case $E(X_i) = \mu = 1/2$ and $\sigma^2 = V(X_i) = 1/12$. The distribution of the jth order statistic $X_{(j,m)}$ has the density

$$f_{(j,m)} = \begin{cases} \dfrac{\Gamma(m+1)}{\Gamma j\,\Gamma(m-j+1)} x^{j-1}(1-x)^{m-j} & \text{for } 0 < x < 1 \\ 0 & \text{elsewhere} \end{cases} \tag{21.2.9}$$

The variance of $X_{(j,m)} = \sigma^2_{(j)|m} = \dfrac{(m+1-j)j}{(m+1)^2(m+2)}$ and $\dfrac{1}{m^2}\sum_{j=1}^{m}\sigma^2_{(j)|m} = \dfrac{1}{6m(m+1)}$. Hence the relative efficiency of $\widehat{\mu}_{rss}$ with respect to $\widehat{\mu}_{srs}$ is

$$RP = \frac{V(\widehat{\mu}_{srs})}{V(\widehat{\mu}_{rss})} = \frac{1/(12mr)}{1/\{6m(m+1)r\}} = \frac{m+1}{2}$$

Example 21.2.2: Exponential Distribution

Following Takahasi and Wakimoto (1968), we find for the exponential distribution $\sigma^2_{(j)|m} = \sigma^2 \sum_{k=1}^{m} \dfrac{1}{(m-k+1)^2}$ and $\sigma^2_{[m]} = \dfrac{\sigma^2}{m}\sum_{j=1}^{m}\dfrac{1}{j}$. Hence the relative precision is

$$RP = 1\Big/\sum_{j=1}^{m}\frac{1}{j}$$

The exact expressions of the relative precisions of the most of distributions including gamma, normal, Weibull, and double exponential distribution are not simple. Takahasi and Wakimoto (1968) derived the relative precisions for specific values of m viz. 2, 3, 4, etc. McIntyre (1952) computed the values of RP for several populations and concluded that the upper bound of RP is $(m + 1)/2$. Takahasi and Wakimoto (1968) proved that for the class of continuous distribution, the maximum value of RP is $(m + 1)/2$ and the upper bound is attained by the uniform distribution.

However, the lower bound, in general, cannot be improved on zero. McIntyre (1952) reported that RP is very small for symmetric distributions.

21.2.4 Optimum Value of *m*

In a ranked set sampling we select samples in the form of r cycles, each consisting of m units. We need to find the optimum number of cycles r and the number of elements per cycles m, which minimize the variance of $\widehat{\mu}_{rss}$, keeping the total sample size $n(=mr)$ fixed. Let $\widehat{\mu}_{rss}(\widetilde{m},\widetilde{r})$ denote the mean of the ranked set sample of $\widetilde{r}$ cycles with $\widetilde{m}$ units so that $n=\widetilde{m}\widetilde{r}$. Then from the Theorem 21.2.1, we find the difference

$$\begin{aligned}
V\{\widehat{\mu}_{rss}(m,r)\}-V\{\widehat{\mu}_{rss}(\widetilde{m},\widetilde{r})\} &= \frac{1}{r^2}\sum_{k=1}^{r}V\left(\overline{y}_{[m]|k}\right)-\frac{1}{\widetilde{r}^2}\sum_{k=1}^{\widetilde{r}}V\left(\overline{y}_{[\widetilde{m}]|k}\right)\\
&= \frac{1}{rm^2}\left(\sum_{i=1}^{m}\sigma^2_{(i)|m}\right)-\frac{1}{\widetilde{r}\widetilde{m}^2}\left(\sum_{i=1}^{\widetilde{m}}\sigma^2_{(i)|\widetilde{m}}\right)\\
&= \frac{1}{n}\left(\sigma^2_{[m]}-\sigma^2_{[\widetilde{m}]}\right)
\end{aligned} \tag{21.2.10}$$

where $\sigma^2_{[m]}=\frac{1}{m}\sum_{i=1}^{m}\sigma^2_{(i,m)}$ and $\sigma^2_{[\widetilde{m}]}=\frac{1}{\widetilde{m}}\sum_{i=1}^{\widetilde{m}}\sigma^2_{(i,\widetilde{m})}$.

First we will prove the following result due to Takahasi and Wakimoto (1968).

Theorem 21.2.2

$$\sigma^2_{[m+1]}<\sigma^2_{[m]}$$

Proof

Let $X_1,\ldots,X_m$ be a random sample from a population with cdf $F(x)$ and density function $f(x)$, then the density function of jth order statistic $X_{(j,m)}$ is given by

$$f_{(j,m)}(x)=\frac{\Gamma(m+1)}{\Gamma j\,\Gamma(m-j+1)}F^{j-1}(x)\{1-F(x)\}^{m-j}f(x) \quad \text{for } j=1,\ldots,m \tag{21.2.11}$$

Similarly, the density function for the tth order statistic $X_{(t,m+1)}$, based on a random sample $(m+1)$ observation, is

$$f_{(t,m+1)}(x)=\frac{\Gamma(m+2)}{\Gamma t\,\Gamma(m-t+2)}F^{t-1}(x)\{1-F(x)\}^{m+1-t}f(x) \quad \text{for}$$

$$t=1,\ldots,m+1 \tag{21.2.12}$$

From Eqs. (21.2.11) and (21.2.12), we have

$$f_{(j,m)}(x) = \frac{m+1-j}{m+1} f_{(j,m+1)}(x) + \frac{j}{m+1} f_{(j+1,m+1)}(x) \tag{21.2.13}$$

Eq. (21.2.13) yields

$$\begin{aligned} \mu_{(j)|m} &= E\{X_{(j,m)}\} \\ &= \frac{m+1-j}{m+1} E\{X_{(j,m+1)}\} + \frac{j}{m+1} E\{X_{(j+1,m+1)}\} \\ &= \frac{m+1-j}{m+1} \mu_{(j)|m+1} + \frac{j}{m+1} \mu_{(j+1)|m+1} \end{aligned} \tag{21.2.14}$$

Similarly, the variance of $X_{(j,m)}$ is

$$\begin{aligned} \sigma^2_{(j,m)} &= E\left\{X^2_{(j,m)}\right\} - \mu^2_{(j,m)} \\ &= \frac{m+1-j}{m+1}\sigma^2_{(j)|m+1} + \frac{j}{m+1}\sigma^2_{(j+1)|m+1} + \left\{\frac{m+1-j}{m+1}\mu^2_{(j)|m+1}\right. \\ &\quad \left. + \frac{j}{m+1}\mu^2_{(j+1)|m+1} - \left(\frac{m+1-j}{m+1}\mu_{(j)|m+1} + \frac{j}{m+1}\mu_{(j+1)|m+1}\right)^2\right\} \end{aligned} \tag{21.2.15}$$

Eq. (21.2.15) yields

$$\begin{aligned} \sum_{j=1}^{m}\sigma^2_{(j)|m} &= \frac{1}{(m+1)}\left\{\sum_{j=1}^{m+1}(m+1-j)\sigma^2_{(j)|m+1)} + \sum_{j=1}^{m+1}(j-1)\sigma^2_{(j)|m+1}\right\} \\ &\quad + \frac{1}{m+1}\left\{\sum_{j=1}^{m}(m+1-j)\mu^2_{(j)|m+1} + \sum_{j=1}^{m} j\mu^2_{(j+1)|m+1)}\right. \\ &\quad \left. - \frac{1}{(m+1)}\sum_{j=1}^{m}\left((m+1-j)\mu_{(j)|m+1} + j\mu_{(j+1)|m+1)}\right)^2\right\} \\ &= \frac{m}{(m+1)}\sum_{j=1}^{m+1}\sigma^2_{(j)|m+1} \\ &\quad + \frac{1}{(m+1)^2}\sum_{j=1}^{m} j(m+1-j)\left(\mu_{(j+1)|m+1} - \mu_{(j)|m+1}\right)^2 \end{aligned}$$

Thus,

$$\sigma^2_{[m]} - \sigma^2_{[m+1]} = \frac{1}{m(m+1)^2} \sum_{j=1}^{m} j(m+1-j)\left(\mu_{(j+1)|m+1} - \mu_{(j)|m+1}\right)^2 > 0$$

This completes the proof of the theorem.

From the Theorem 21.2.2, we arrive at

Theorem 21.2.3

$V\{\widehat{\mu}_{rss}(m,r)\} - V\{\widehat{\mu}_{rss}(s,t)\} > 0$ for $s > m$ with $n = rm = st$

From Theorem 21.2.3, we note that efficiency $\widehat{\mu}_{rss}(m,r)$ increases with m when the total sample size n is fixed and maximum efficiency is achieved when $m = n$ and $r = 1$. However, a large $m(= n)$ is impractical because the cost of ordering a large number of elements will be expensive. In most practical cases m is taken as 3 or 4.

21.2.5 Optimum Allocation

In our description of RSS in Section 21.2, each of the elements holding rank 1, rank 2,..., and rank m were measured only once in each of the r cycles. Instead of measuring each of the elements holding rank 1, rank 2 ... and rank m an equal number of times (r), we may measure the element holding rank j, $r_j(\geq 1)$ times with $\sum_{j=1}^{m} r_j = n$. In this situation an unbiased estimator of the population mean μ is

$$\widehat{\mu} = \frac{1}{m} \sum_{j=1}^{m} \overline{y}_{(j)|m} \tag{21.2.16}$$

where $\overline{y}_{(j)|m} = \frac{1}{r_j} \sum_{k=1}^{r_j} y_{j(j)|k}$ and $y_{j(j)|k}$ represent the actual value of the jth order statistic of a random sample of m elements in the kth cycle. In case $\sigma_{(j)|m}$ is known, the optimum value of r_j is obtained using Neyman's optimum allocation as

$$r_{j0} = n \frac{\sigma_{(j)|m}}{\sum_{j=1}^{m} \sigma_{(j)|m}} \tag{21.2.17}$$

where $\sigma^2_{(j)|m} = V\left(y_{j(j)|k}\right)$.

The variance of $\widehat{\mu}$ with the optimum value of $r_j = r_{j0}$ is

$$V_{opt}(\widehat{\mu}) = \frac{\left(\sum_{j=1}^{m} \sigma_{(j)|m}\right)^2}{m^2 n} \tag{21.2.18}$$

21.2.5.1 Right-Tail Allocation Model

In practice, the implementation of the optimum allocation is not possible because $\sigma_{(j)|m}$ values are generally unknown. In this case it is of interest to study "near" optimal allocation where full knowledge of the value $\sigma_{(j)|m}$ is not required. It is well known for a positively skewed population that the variance of the order statistic increases with rank order: $\sigma^2_{(1)|m} \le \sigma^2_{(2)|m} \le \dots \le \sigma^2_{(m)|m}$. For highly skewed distributions, the variance of the highest order statistic $\sigma^2_{(m)|m}$ is very high compared with the variances of the rest of the order statistics. Keeping this in mind, Kaur et al. (1997) proposed the following t-allocation

$$r' = r_1 = r_2 = \dots = r_{m-1} = r_m/t \qquad (21.2.19)$$

For $n = (m - 1 + t)r'$, the variance of $\widehat{\mu}$ under t-allocation is

$$V(\widehat{\mu}_t) = \frac{1}{m^2}\sum_{j=1}^{m}\frac{\sigma^2_{(j)|m}}{r_j} = \frac{1}{m^2}\left(\sum_{j=1}^{m-1}\frac{\sigma^2_{(j)|m}}{r'} + \frac{\sigma^2_{(m)|m}}{tr'}\right) = \frac{1}{m^2 r'}\left(d' + \sigma^2_{(m)|m}/t\right) \qquad (21.2.20)$$

where $d' = \sum_{j=1}^{m-1} \sigma^2_{(j)|m}$.

The relative precision of $\widehat{\mu}_t$ compared with $\widehat{\mu}_{srs}$ is

$$RP_t = \frac{m^2 r' \sigma^2/n}{d' + \sigma^2_{(m)|m}/t} = \frac{m^2\sigma^2}{(m-1+t)\left(d' + \sigma^2_{(m)|m}/t\right)} \qquad (21.2.21)$$

Under equal allocation $r_1 = r_2 = \dots = r_m = r$, $\widehat{\mu}_t$ reduces to $\widehat{\mu}_e = \widehat{\mu} = \frac{1}{mr}\sum_{j=1}^{m}\sum_{k=1}^{r} y_{j(j)|k}$ and the relative precision of $\widehat{\mu}_t$ compared with equal allocation is

$$RP_{e,t} = \frac{m^2 r'\left(\frac{1}{m^2 r}\sum_{j=1}^{m}\sigma^2_{(j)|m}\right)}{\left(d' + \sigma^2_{(m)|m}/t\right)} = \frac{m^2}{(m-1+t)}\frac{\left(\frac{1}{m}\sum_{j=1}^{m}\sigma^2_{(j)|m}\right)}{\left(d' + \sigma^2_{(m)|m}/t\right)} \qquad (21.2.22)$$

On maximizing Eq. (21.2.22) the optimum value of t is obtained as

$$t_{opt} = \sqrt{\frac{\sigma^2_{(m)|m}(m-1)}{d'}} \tag{21.2.23}$$

It can be shown that RP_t is a monotonically increasing function of t for $t = (1, t_{opt})$ and decreasing for $t = (t_{opt}, \infty)$. On the basis of empirical studies based on of 225 positively skewed distributions belonging to different parametric populations including inverse Gaussian, reciprocal gamma, lognormal, Pareto and beta (type II) distribution, Kaur et al. (1997) provided guidelines for selection of the appropriate value of t. For details readers are referred to Kaur et al. (1997).

21.2.6 Judgment Ranking

In judgment ranking, each of the selected samples is ranked by a visual or judgment process, which may include the use of concomitant variable. Hence sometimes ranking may be imperfect. Let $y_{i\langle j\rangle|k}$ be the smallest jth "judgment order statistic" in the ith set of the cycle k. In case the judgment ranking is perfect, $y_{i\langle j\rangle|k}$ becomes equal to $y_{i(j)|k}$, otherwise if the judgment process is imperfect, we find $y_{i\langle j\rangle|k} \neq y_{i(j)|k}$. Here the selected data are $d^* = (y_{i\langle i\rangle|k};\ i = 1, \ldots, m,\ k = 1, \ldots, r)$.

21.2.6.1 Moment of the Judgment Order Statistic

For any constants c and p, the following relations hold

$$\sum_{i=1}^{m}\left(y_{ij|k} - c\right)^p = \sum_{i=1}^{m}\left(y_{i\langle j\rangle|k} - c\right)^p \quad \text{for } j = 1, \ldots, m;\ k = 1, \ldots, r \tag{21.2.24}$$

From Eq. (21.2.24) it follows that

$$m\,E\left(y_{ij|k} - c\right)^p = \sum_{i=1}^{m} E\left(y_{i\langle j\rangle|k} - c\right)^p$$

Substituting $p = 1$, $p = 2$, and $c = \mu$, we get

$$\mu = \frac{1}{m}\sum_{j=1}^{m}\mu_{\langle j\rangle|m} \quad \text{and } \sigma^2 = \frac{1}{m}\sum_{j=1}^{m}\sigma^2_{\langle j\rangle|m} + \frac{1}{m}\sum_{j=1}^{m}\tau^2_{\langle j\rangle|m} \tag{21.2.25}$$

where $\mu_{\langle j\rangle|m} = E\left(y_{i\langle j\rangle|k}\right)$, $\sigma^2_{\langle j\rangle|m} = V\left(y_{i\langle j\rangle|k}\right)$, and $\tau_{\langle j\rangle|m} = \mu_{\langle j\rangle|m} - \mu$.

Because (i) $\overline{y}_{\langle m\rangle|k} = \frac{1}{m}\sum_{i=1}^{m} y_{i\langle i\rangle|k}$ are independent random variables with mean $E\left(\overline{y}_{\langle m\rangle|k}\right) = \mu$ and variance $V\left(\overline{y}_{\langle m\rangle|k}\right) = \frac{1}{m}\sum_{i=1}^{m}\sigma^2_{\langle j\rangle|m}$, we find the following theorem analogous to Theorem 21.2.1.

Theorem 21.2.4

(i) $\widehat{\mu}_{\langle rss\rangle} = \frac{1}{r}\sum_{k=1}^{r}\overline{y}_{\langle m\rangle|k}$ is an unbiased estimator for μ.

(ii) The variance of $\widehat{\mu}_{\langle rss\rangle}$ is

$$V\left(\widehat{\mu}_{\langle rss\rangle}\right) = \frac{1}{n}\left[\sigma^2 - \frac{1}{m}\sum_{i=1}^{m}\left(\mu_{\langle j\rangle|m} - \mu\right)^2\right]$$

(iii) An unbiased estimator of the variance of $V\left(\widehat{\mu}_{\langle rss\rangle}\right)$ is

$$\widehat{V}\left(\widehat{\mu}_{\langle rss\rangle}\right) = \frac{1}{r(r-1)}\sum_{k=1}^{r}\left(\overline{y}_{\langle m\rangle|k} - \widehat{\mu}_{\langle rss\rangle}\right)^2$$

Consider the extreme situation where making a judgment is impossible. So, random ranking is assigned to all the elements of this set. In this case $y_{i\langle j\rangle|k}$ is a random sample from the original population and $\widehat{\mu}_{\langle rss\rangle}$ remains unbiased as it becomes equal to $\widehat{\mu}_{srs}$. In practice, ranking ability is expected to be between perfect and extreme random ranking. Actually, the error in ranking has little effect on the precision of the estimator. Dell and Clutter (1972) conducted simulation studies on the effect of errors in ranking of various populations including rectangular, exponential, and normal populations. They showed that the relative precision of $\widehat{\mu}_{\langle srs\rangle}$ with respect to $\widehat{\mu}_{srs}$ ranges between 1.05 and 3.00.

21.2.7 Estimation of Population Variance

It is well known that for simple random sampling, the sample variance is an unbiased estimator of the population variance σ^2. Stokes (1980a) showed that for RSS the sample variance

$$\widehat{\sigma}^2_{\langle rss\rangle} = \frac{1}{mr-1}\sum_{k=1}^{r}\sum_{j=1}^{m}\left(y_{j,\langle j\rangle|k} - \widehat{\mu}_{\langle rss\rangle}\right)^2$$

is not an unbiased estimator of σ^2.

Theorem 21.2.5
$\widehat{\sigma}^2_{\langle rss\rangle}$ is a biased estimator of σ^2 and the amount of bias of $\widehat{\sigma}^2_{\langle rss\rangle}$ is

$$\frac{1}{m(mr-1)}\sum_{j=1}^{m}\left(\mu_{\langle j\rangle|m)}-\mu\right)^2.$$

Proof

$$\begin{aligned}
E\left(\widehat{\sigma}^2_{\langle rss\rangle}\right) &= \frac{1}{mr-1}\left[\sum_{k=1}^{r}\sum_{j=1}^{m}E\left(y^2_{j,\langle j\rangle|k}\right)-mrE\left(\widehat{\mu}_{\langle rss\rangle}\right)^2\right]\\
&= \frac{1}{mr-1}\left[\sum_{k=1}^{r}\sum_{j=1}^{m}\left\{\sigma^2_{\langle j\rangle|m)}+\mu^2_{\langle j\rangle|m}\right\}-mr\left\{V\left(\widehat{\mu}_{\langle rss\rangle}\right)+\mu^2\right\}\right]\\
&= \frac{1}{mr-1}\left[r\sum_{j=1}^{m}\sigma^2_{\langle j\rangle|m)}-\frac{1}{m}\sum_{j=1}^{m}\sigma^2_{\langle j\rangle|m)}+r\sum_{j=1}^{m}\left(\mu_{\langle j\rangle|m}-\mu\right)^2\right]\\
&= \frac{1}{mr-1}\left[\frac{(mr-1)}{m}\sum_{j=1}^{m}\sigma^2_{\langle j\rangle|m)}+r\sum_{j=1}^{m}\left(\mu_{\langle j\rangle|m}-\mu\right)^2\right]
\end{aligned} \quad (21.2.26)$$

Now using Eq. (21.2.25) we find

$$E\left(\widehat{\sigma}^2_{\langle rss\rangle}\right)=\sigma^2+\frac{1}{m(mr-1)}\sum_{j=1}^{m}\left(\mu_{\langle j\rangle|m)}-\mu\right)^2$$

21.2.7.1 Efficiency of $\widehat{\sigma}^2_{\langle rss\rangle}$

The estimator $\widehat{\sigma}^2_{\langle rss\rangle}$ is asymptotically unbiased for σ^2 if either r or m becomes large. Eq. (21.2.25) shows that the upper bound of the bias of $\widehat{\sigma}^2_{\langle rss\rangle}$ is $\sigma^2/(mr-1)$. Stokes (1980a) reported that RSS may not always provide an efficient estimator for σ^2. The variance of $\widehat{\sigma}^2_{\langle rss\rangle}$ was given by Stokes (1980a) as

$$\begin{aligned}
V\left(\widehat{\sigma}^2_{\langle rss\rangle}\right) &= \frac{r}{(mr-1)^2}\left[\left(\frac{mr-1}{rm}\right)^2\sum_{j=1}^{m}\mu_{\langle 4j\rangle|m}+4\sum_{j=1}^{m}\left(\mu_{\langle j\rangle|m}-\mu\right)^2\sigma^2_{\langle j\rangle|m)}\right.\\
&+4\left(\frac{mr-1}{mr}\right)\sum_{j=1}^{m}\left(\mu_{\langle j\rangle|m}-\mu\right)\mu_{\langle 3j\rangle|m}+\frac{2r}{(mr)^2}\sum_{j\neq}^{m}\sum_{k=1}^{m}\sigma^2_{\langle j\rangle|m}\sigma^2_{\langle k\rangle|m}\\
&\left.+\frac{2(r-1)-(mr-1)^2}{(mr)^2}\sum_{j=1}^{m}\sigma^4_{\langle j\rangle|m}\right]
\end{aligned} \quad (21.2.27)$$

where $\mu_{\langle pj\rangle|m)}=E(y_{j\langle j\rangle|m}-\mu_{\langle j\rangle|m})^p$.

In case the ranking is totally imperfect in the sense that it is random, $\mu_{\langle pj\rangle|m} = E(y-\mu)^p = \mu_p$ and Eq. (21.2.27) yields

$$V\left(\widehat{\sigma}^2_{\langle rss\rangle}\right) = \frac{\mu_4}{mr} - \frac{(mr-3)}{mr(mr-1)}\sigma^4$$
$$= V(s^2)$$

where s^2 is the sample variance based on rm observations.

Stokes (1980a) proved that there exists N^* such that the relative precision of $\widehat{\sigma}^2_{\langle rss\rangle}$,

$$RP\left(\widehat{\sigma}^2_{\langle rss\rangle}\right) = \frac{V(s^2)}{V\left(\widehat{\sigma}^2_{\langle rss\rangle}\right)} \geq 1 \text{ if } mr > N^* \tag{21.2.28}$$

Equality in Eq. (21.2.28) holds if and only if $E(y_{j\langle j\rangle|m} - \mu)^2 = E(y-\mu)^2$ for all $j = 1,\ldots, m$. The increased precision can be realized by increasing either m or the number of cycles r. Because a substantial increase in m is impractical, the asymptotic results hold for large r.

21.2.8 Use of Concomitant Variables

Stokes (1977) considered the situation where the study variable y cannot be ordered easily but a related variable x that is observable and can be easily ordered accurately. For example, y be the internal characteristic of a patient or a laboratory animal, which is expensive, painful, or inconvenient to measure. In this case, RSS can be employed if judgment ordering can be accomplished on the y's by ordering some external characteristic x, which is correlated to y. So, in this case we have data of judgment ordering

$$d_J = \left\{ \overbrace{y_{1\langle 1\rangle|1} \quad y_{2\langle 2\rangle|1} \quad \cdots \quad y_{m\langle m\rangle|1}}^{\text{cycle } 1}; \ldots; \overbrace{y_{1\langle 1\rangle|k} \quad y_{2\langle 2\rangle|k} \quad \cdots \quad y_{m\langle m\rangle|k}}^{\text{cycle } k}; \ldots; \right.$$
$$\left. \overbrace{y_{1\langle 1\rangle|r} \quad y_{2\langle 2\rangle|r} \quad \cdots \quad y_{m\langle m\rangle|r}}^{\text{cycle } r} \right\} \tag{21.2.29}$$

where $y_{i\langle j\rangle|k}$ be the smallest jth "judgment order statistic" of y corresponding to the smallest jth "order statistic" $x_{i(j)|k}$ of x in the ith set of the cycle k.

Let $\mu_{x(i)|m} = E(x_{i(i)|k})$, $\mu_{y(i)|m} = E(y_{i(i)|k})$ be the mean of the ith order statistic of x and y in a random sample of size m. The population means and variances of x and y will be denoted by μ_x, μ_y and σ_x^2, σ_y^2, respectively. The proposed estimator for the population mean μ_y is

$$\widehat{\mu}_{y\langle rss\rangle} = \frac{1}{r}\sum_{k=1}^{r} \overline{y}_{\langle m\rangle|k} \tag{21.2.30}$$

where $\overline{y}_{\langle m\rangle|k} = \frac{1}{m}\sum_{i=1}^{m} y_{i\langle i\rangle|k}$.

Stokes (1977) assumed that the regression of y on x is linear, i.e.,

$$y = \mu_y + \rho\frac{\sigma_y}{\sigma_x}(x - \mu_x) + \in \tag{21.2.31}$$

where x and $\in$ are independent, $E(\in|x) = 0$, $V(\in|x) = \sigma_y^2(1-\rho^2)$, and ρ is the correlation coefficient between x and y.

Then we have from Eq. (21.2.31)

$$y_{i\langle i\rangle|k} = \mu_y + \rho\frac{\sigma_y}{\sigma_x}\left(x_{i(i)|k} - \mu_x\right) + \in_{i(i)|k}$$

Stokes further assumed that $\dfrac{y-\mu_y}{\sigma_y}$ and $\dfrac{x-\mu_x}{\sigma_x}$ have same distribution that holds for bivariate normal and bivariate Pareto distributions. Stokes (1977) derived the following theorem:

Theorem 21.2.6

(i) $E\left(\widehat{\mu}_{y\langle rss\rangle}\right) = \mu_y$

(ii) $V\left(\widehat{\mu}_{y\langle rss\rangle}\right) = \dfrac{1}{mr}\left[\sigma_y^2 - \dfrac{\rho^2}{m}\sum_{i=1}^{m}\tau_{y(i)|m}^2\right]$

where $\tau_{y(i)|m} = E(y_{(i)|m}) - \mu_y$.

Proof

(i)
$$E\left(\overline{y}_{\langle m\rangle|k}\right) = \frac{1}{m}\sum_{i=1}^{m} E\left(y_{i\langle i\rangle|k}\middle|x_{i(i)|k}\right)$$

$$= \mu_y + \rho\frac{\sigma_y}{\sigma_x}\frac{1}{m}\sum_{i=1}^{m} E\left(x_{i(i)|k} - \mu_x\right)$$

$$= \mu_y + \rho\frac{\sigma_y}{\sigma_x}\frac{1}{m}\sum_{i=1}^{m}\left(\mu_{x(i)|m} - \mu_x\right)$$

$$= \mu_y$$

(ii) $$V\left(\widehat{\mu}_{y\langle rss\rangle}\right) = \frac{V\left(\overline{y}_{\langle m\rangle|k}\right)}{r}$$

$$= \frac{1}{m^2 r}\sum_{i=1}^{m}\left[E\left\{V\left(y_{i\langle i\rangle|k}\middle|x_{i(i)|k}\right)\right\} + V\left\{E\left(y_{i\langle i\rangle|k}\middle|x_{i(i)|k}\right)\right\}\right]$$

$$= \frac{1}{m^2 r}\sum_{i=1}^{m}\left[\sigma_y^2(1-\rho^2) + \rho^2\frac{\sigma_y^2}{\sigma_x^2}V\left(\mu_{x(i)|m} - \mu_x\right)\right]$$

(21.2.32)

Now using the assumption that $\dfrac{y-\mu}{\sigma_y}$ and $\dfrac{x-\mu_x}{\sigma_x}$ have same distribution, we find

$$\sum_{i=1}^{m}V\left(\mu_{x(i)|m} - \mu_x\right) = \frac{\sigma_x^2}{\sigma_y^2}\sum_{i=1}^{m}V\left(\mu_{y(i)|m} - \mu_y\right)$$

$$= \frac{\sigma_x^2}{\sigma_y^2}\left(m\sigma_y^2 - \sum_{i=1}^{m}\tau_{y(i)|m}^2\right) \quad \text{(using Eq. 21.2.4)}$$

(21.2.33)

Part (ii) of theorem follows from Eqs. (21.2.32) and (21.2.33).

So, the relative efficiency of the ranking with the concomitant variable compared with the perfect ranking is

$$\frac{V\left(\overline{y}_{(rss)}\right)}{V\left(\widehat{\mu}_{y\langle rss\rangle}\right)} = \frac{1 - \dfrac{1}{m}\displaystyle\sum_{i=1}^{m}\tau_{y(i)|m}^2}{1 - \dfrac{\rho^2}{m}\displaystyle\sum_{i=1}^{m}\tau_{y(i)|m}^2}$$

Likewise, the efficiency of $\widehat{\mu}_{y\langle rss\rangle}$ with $\widehat{\mu}_{srs}$ is

$$\frac{V\left(\overline{y}_{srs}\right)}{V\left(\widehat{\mu}_{y\langle rss\rangle}\right)} = \frac{1}{1 - \dfrac{\rho^2}{m}\displaystyle\sum_{i=1}^{m}\tau_{y(i)|m}^2}.$$

Stokes (1977) used concomitant variable x for ranking y only. Yu and Lam (1997) used the variable x for estimation of μ_y. They proposed the following difference estimator when μ_x, the population mean of the concomitant variable x, is known.

$$\widehat{\mu}_y^{(d)} = \widehat{\mu}_{y\langle rss\rangle} - B\left(\widehat{\mu}_{x(rss)} - \widehat{\mu}_x\right)$$

where $\widehat{\mu}_{x(rss)} = \dfrac{1}{mr}\displaystyle\sum_{k=1}^{r}\sum_{i=1}^{m}x_{i(i)|k}$ and B is a constant to be determined.

The estimator $\widehat{\mu}_y^{(d)}$ is unbiased for μ, as $\widehat{\mu}_{\langle rss\rangle}$ and $\widehat{\mu}_{x(rss)}$ are unbiased for μ_y and μ_x, respectively. The optimum value of B is obtained by minimizing

$$\begin{aligned} V\left(\widehat{\mu}_y^{(d)}\right) &= V\left(\widehat{\mu}_{y\langle rss\rangle}\right) + B^2 V\left(\widehat{\mu}_{x(rss)}\right) - 2BCov\left(\widehat{\mu}_{y\langle rss\rangle}, \widehat{\mu}_{x(rss)}\right) \\ &= V\left(\widehat{\mu}_{y\langle rss\rangle}\right) + B^2 V\left(\widehat{\mu}_{x(rss)}\right) - 2B\rho\frac{\sigma_y}{\sigma_x} V\left(\widehat{\mu}_{x(rss)}\right) \end{aligned} \tag{21.2.34}$$

where

$$\begin{aligned} V\left(\widehat{\mu}_{y\langle rss\rangle}\right) &= \frac{1}{mr}\left[\sigma_y^2 - \frac{\rho^2}{m}\sum_{i=1}^{m}\tau_{y(i)|m}^2\right] \quad \text{(using Theorem 21.2.6)} \\ &= \frac{\sigma_y^2}{mr}\left[1 - \frac{\rho^2}{m}\sum_{i=1}^{m}\left(\frac{\tau_{x(i)|m}}{\sigma_x}\right)^2\right], \end{aligned} \tag{21.2.35}$$

$$V\left(\widehat{\mu}_{x(rss)}\right) = \frac{1}{mr}\sigma_x^2\left[1 - \frac{1}{m}\sum_{i=1}^{m}\left(\frac{\tau_{x(i)|m}}{\sigma_x}\right)^2\right] \quad \text{(using Theorem 21.2.1)} \tag{21.2.36}$$

and $\tau_{x(i)|m} = \mu_{x(i)|m} - \mu_x$.

Eq. (21.2.34) yields the optimum value of B as

$$B^* = \rho\frac{\sigma_y}{\sigma_x}$$

A natural estimator of B^* is

$$\widehat{B} = \frac{\sum_{k=1}^{r}\sum_{i=1}^{m}\left(x_{i(i)|k} - \widehat{\mu}_{x(rss)}\right)\left(y_{i\langle i\rangle|k} - \widehat{\mu}_{y\langle rss\rangle}\right)}{\sum_{k=1}^{r}\sum_{i=1}^{m}\left(x_{i(i)|k} - \widehat{\mu}_{x(rss)}\right)^2}$$

Hence, RSS regression estimator of μ_y is defined as

$$\widehat{\mu}_{reg} = \widehat{\mu}_{y\langle rss\rangle} - \widehat{B}\left(\widehat{\mu}_{x(rss)} - \mu_x\right) \tag{21.2.37}$$

Theorem 21.2.7

Under the model (21.2.31),

(i) $E\left(\widehat{\mu}_{reg}\right) = \mu_y$,

(ii) $V\left(\widehat{\mu}_{reg}\right) = \frac{\sigma_y^2}{mr}(1-\rho^2)\left[1 + E\left(\frac{\overline{z}_{rss}^2}{s_z^2}\right)\right]$

where $\overline{z}_{rss} = \frac{1}{mr}\sum_{k=1}^{r}\sum_{i=1}^{m} z_{i(i)|k}$, $s_z^2 = \frac{1}{mr}\sum_{k=1}^{r}\sum_{i=1}^{m}\left(z_{i(i)|k} - \overline{z}_{rss}\right)^2$, and $z_{i(i)|k} = \frac{x_{i(i)|k} - \mu_x}{\sigma_x}$.

Proof

(i) $$E\left(\widehat{\mu}_{reg}\right) = E\left\{\widehat{\mu}_{y\langle rss\rangle} - \widehat{B}\left(\widehat{\mu}_{x(rss)} - \mu_x\right)\right\}$$

$$= \mu_y - E\left[\left(\widehat{\mu}_{x(rss)} - \mu_x\right)\left\{E\left(\widehat{B}\middle|x\right)\right\}\right]$$

$$= \mu_y - BE\left(\widehat{\mu}_{x(rss)} - \mu_x\right)$$

$$= \mu_y$$

(ii) $$V\left(\widehat{\mu}_{reg}\right) = E\left(\widehat{\mu}_{reg} - \mu_y\right)^2$$

$$= E\left\{-\left(\widehat{B} - B\right)\left(\widehat{\mu}_{x(rss)} - \mu_x\right) + \overline{\in}_{rss}\right\}^2$$

$$\left(\text{where } \overline{\in}_{rss} = \frac{1}{rm}\sum_{k=1}^{r}\sum_{i=1}^{m} \in_{i(i)|k}\right)$$

$$= E\left[\left(\widehat{\mu}_{x(rss)} - \mu_x\right)^2 E\left\{\left(\widehat{B} - B\right)^2\middle|x\right\} + E\left(\overline{\in}_{rss}\right)^2\right.$$

$$\left. - 2E\left\{\left(\widehat{B} - B\right)\left(\widehat{\mu}_{x(rss)} - \mu_x\right)\overline{\in}_{rss}\middle|x\right\}\right]$$

$$= E\left[\left(\widehat{\mu}_{x(rss)} - \mu_x\right)^2 \frac{E\left\{\left(\sum_{k=1}^{r}\sum_{i=1}^{m}\left(x_{i(i)|k} - \widehat{\mu}_{x(rss)}\right)\in_{i(i)|r}\right)^2\middle|x\right\}}{\left(\sum_{k=1}^{r}\sum_{i=1}^{m}\left(x_{i(i)|k} - \widehat{\mu}_{x(rss)}\right)^2\right)^2} + E\left(\overline{\in}_{rss}\right)^2\right]$$

$$= \sigma_y^2(1-\rho^2)E\left[\frac{\left(\widehat{\mu}_{x(rss)} - \mu_x\right)^2}{\sum_{k=1}^{r}\sum_{i=1}^{m}\left(x_{i(i)|k} - \widehat{\mu}_{x(rss)}\right)^2} + \frac{1}{mr}\right]$$

$$= \frac{\sigma_y^2}{mr}(1-\rho^2)\left[1 + E\left\{\frac{\left(\widehat{\mu}_{x(rss)} - \mu_x\right)^2}{\frac{1}{mr}\sum_{k=1}^{r}\sum_{i=1}^{m}\left(x_{i(i)|k} - \widehat{\mu}_{x(rss)}\right)^2}\right\}\right]$$

$$= \frac{\sigma_y^2}{mr}(1-\rho^2)\left[1 + E\left(\frac{\overline{z}_{rss}^2}{s_z^2}\right)\right]$$

21.2.8.1 Relative Precision of $\widehat{\mu}_{reg}$

The relative precision of the RSS regression estimator with respect to the RSS naïve estimator $\widehat{\mu}_{(rss)}$ based on the model (21.2.31) is

$$RP\left(\widehat{\mu}_{reg}, \widehat{\mu}_{(rss)}\right) = \frac{1 - \rho^2 E\left(\frac{1}{m}\sum_{i=1}^{n} z^2_{(i)r}\right)}{(1-\rho^2)\left[1 + E\left(\frac{\overline{z}^2_{rss}}{s^2_z}\right)\right]}$$

Because $\widehat{\mu}_{(rss)}$ does not use any information on the concomitant variable x, a fairer comparison of the RSS regression estimator $\widehat{\mu}_{reg}$ is to compare with the naïve regression estimator based on SRSWR sampling defined by

$$\widehat{\mu}^*_{reg} = \overline{y}_{srs} - \widehat{\beta}(\overline{x}_{srs} - \mu_x)$$

where $\overline{x}_{srs}, \overline{y}_{srs}$, and $\widehat{\beta}$ denote the sample means of x,y and the sample regression coefficient based on an SRSWR sample of size $n(= mr)$. For the bivariate normal distribution $\widehat{\mu}^*_{reg}$ is unbiased for μ_y and the variance of $\widehat{\mu}^*_{reg}$ is $V\left(\widehat{\mu}^*_{reg}\right) = \frac{\sigma^2_y}{mr}(1-\rho^2)\left(1 + \frac{1}{mr-3}\right)$ (see Sukhatme & Sukhatme, 1970).

Hence the relative efficiency of $\widehat{\mu}_{reg}$ with respect to $\widehat{\mu}^*_{reg}$ is defined as follows:

$$RP\left(\widehat{\mu}_{reg}, \widehat{\mu}^*_{reg}\right) = \frac{V\left(\widehat{\mu}^*_{reg}\right)}{V\left(\widehat{\mu}_{reg}\right)} = \frac{1 + \frac{1}{mr-3}}{1 + E\left(\frac{\overline{z}^2_{rss}}{s^2_z}\right)}$$

21.3 SIMPLE RANDOM SAMPLING WITHOUT REPLACEMENT

RSS for SRSWOR sampling was proposed by Patil et al. (1995). The method is described as follows. Consider a finite population $U = (U_1, \ldots, U_i, \ldots, U_N)$ of N identifiable units and let y_i be the value of the variable under study, y, for the ith unit. Let us assume $y_1 < y_2 < \cdots < y_N$. From the population U, a sample s of $n = m^2r$ units is selected by the SRSWOR method. The sample s is partitioned at random into mr sets each containing m distinct units. All the units in a set are ranked by some inexpensive procedure. The lowest rank holder is quantified in each of the first r sets:

$$y_{(1,m)1}, \ldots, y_{(1,m)i}, \ldots, y_{(1,m)r}$$

The second ranked unit is quantified for each of the next r sets to yield:

$$y_{(2,m)1}, \ldots, y_{(2,m)i}, \ldots, y_{(2,m)r}$$

The process is continued until the highest ranked unit is quantified for each of the last r sets:

$$y_{(m,m)1}, \ldots, y_{(m,m)i}, \ldots, y_{(m,m)r}$$

The ranked set estimator for the finite population mean $\overline{Y} = \sum_{i=1}^{N} y_i/N$ is given by

$$\widehat{\overline{Y}}_{rss} = \frac{1}{rm}\sum_{\alpha=1}^{r}\sum_{i=1}^{m} y_{(i,m)\alpha} \tag{21.3.1}$$

Let us consider a particular set s_α, consisting of m units. The set s_α may be considered as an SRSWOR sample of size m from the population U. Let us define the event

$$\{i \rightarrow t\} \tag{21.3.2}$$

to denote that the ith ranked unit in the set s_α is the tth ranked unit in the population U. We also define the probability of the event as

$$A_i^t = \Pr\{i \rightarrow t\} \tag{21.3.3}$$

Theorem 21.3.1

(i) $A_i^t = \dfrac{\binom{t-1}{i-1}\binom{N-t}{m-i}}{\binom{N}{m}}$

(ii) The inclusion probability of the unit t in a set is m/N.

Proof

(i) The event $\{i \rightarrow t\}$ will hold if exactly $i-1$ units are selected from the smallest $t-1$ members of the population U and $m-i$ must be selected from the $N-t$ units of the population holding rank greater than t. So, probability of the event is A_i^t.

(ii) The inclusion probability of the unit t in a set is

$$\sum_{i=1}^{m} A_i^t = \sum_{i=1}^{m} \frac{\binom{t-1}{i-1}\binom{N-t}{m-i}}{\binom{N}{m}}$$

$$= m/N$$

Let us define the event that the ith ranked unit from the set s_α has rank t in the population and jth ranked unit from the set $s_\beta(\beta \neq \alpha)$ has rank l in the population as

$$\{i \rightarrow t, j \rightarrow l\}$$

The probability of the event $\{i \rightarrow t, j \rightarrow l\}$ is denoted by

$$B_{ij}^{t,l} = \Pr\{i \rightarrow t, j \rightarrow l\}$$

Let $\mathbf{B}_{ij}$ be the $N \times N$ matrix with $B_{ij}^{t,l}$ as its (t,l) component. Clearly, $\mathbf{B}_{ij} = \mathbf{B}_{ji}^T$ as $B_{ij}^{t,l} = B_{ji}^{l,t}$. Patil et al. (1995) derived the following theorem.

Theorem 21.3.2

(i)

$$B_{ij}^{t,l} = \begin{cases} \displaystyle\sum_{\lambda=0}^{m-i} \frac{\binom{t-1}{i-1}\binom{l-t-1}{\lambda}\binom{N-l}{m-i-\lambda}\binom{l-1-i-\lambda}{j-1}\binom{N-l-m+i+\lambda}{m-j}}{\binom{N}{m,m}} & \text{for } t < l \\ 0 & \text{for t} = l \\ B_{ji}^{l,t} & \text{for t} > l \end{cases}$$

(ii) The probability of inclusion of tth unit in a set α and lth in a set β is

$$\frac{m^2}{N(N-1)}(1 - \delta_{tl})$$

where $\binom{N}{m, m} = \dfrac{N!}{m!m!(N-2m)!}$, λ is the number of units in set α that lies between y_t and y_l, $\delta_{tl} = 1$ if $t = l$, and $\delta_{tl} = 0$ if $t \neq l$.

Proof

(i) The proof follows from the similar argument in the Theorem 21.3.1. In fact λ must satisfy the following restrictions:

$$0 \le \lambda \le l - t - 1;\ 2m - i - j + l - N \le \lambda \le m - i;\ \lambda \le l - i - j$$

(ii) Probability of inclusion of the tth unit in set α and lth unit in set the β is

$$\sum_{i=1}^{m}\sum_{j=1}^{m} B_{ij}^{tl} = \frac{m^2}{N(N-1)} \text{ for } l > t \tag{21.3.4}$$

and for $t = l$, $\sum_{i=1}^{m}\sum_{j=1}^{m} B_{ij}^{tt} = 0$ because $B_{ij}^{tt} = 0$ for $t = 1,\ldots, N$.

Theorem 21.3.3

(i) $\mu_{(i,m)} = E\left[y_{(i,m)\alpha}\right] = \mathbf{A}_i^T\mathbf{y}$

(ii) $\sigma^2_{(i,m)} = V\left[y_{(i,m)\alpha}\right] = \mathbf{A}_i^T\mathbf{y}^{(2)} - \left(\mathbf{A}_i^T\mathbf{y}\right)^2$

(iii) $\sigma_{(ij,m)} = Cov\left[y_{(i,m)\alpha}, y_{(j,m)\beta}\right] = \mathbf{y}^T\left(\mathbf{B}_{ij} - \mathbf{A}_i\mathbf{A}_i^T\right)\mathbf{y}$

where $\mathbf{A}_i^T = \left(A_i^1, \ldots, A_i^t, \ldots, A_i^N\right)$ and $\mathbf{y}^{(2)} = \left(y_1^2, \ldots, y_i^2, \ldots, y_N^2\right)^T$.

Proof

(i)

$$\begin{aligned}
\mu_{(i,m)} &= E\left[y_{(i,m)\alpha}\right] = \sum_{t=1}^{N} y_t \Pr\left(y_{(i,m)\alpha} = y_t\right) \\
&= \sum_{t=1}^{N} y_t \Pr(\{i \to t\}) \\
&= \sum_{t=1}^{N} y_t A_i^t \\
&= \mathbf{A}_i^T\mathbf{y}
\end{aligned} \tag{21.3.5}$$

(ii) $\sigma^2_{(i,m)} = V\left[y_{(i,m)\alpha}\right] = E\left[y_{(i,m)\alpha}\right]^2 - \mu^2_{(i,m)}$

$$\begin{aligned}
&= \sum_{t=1}^{N} y_t^2 \Pr\left(y_{(i,m)j} = y_t\right) - \mu^2_{(i,m)} \\
&= \sum_{t=1}^{N} y_t^2 A_i^t - \mu^2_{(i,m)} \\
&= \mathbf{A}_i^T\mathbf{y}^{(2)} - \left(\mathbf{A}_i^T\mathbf{y}\right)^2
\end{aligned}$$

(iii)
$$\begin{aligned}\sigma_{(ij,m)} &= Cov\left[y_{(i,m)\alpha}, y_{(j,m)\beta}\right]\\ &= E\left[y_{(i,m)\alpha}y_{(j,m)\beta}\right] - \mu_{(i,m)}\mu_{(j,m)}\\ &= \sum_{t=1}^{N}\sum_{l=1}^{N} y_t y_l B_{ij}^{tl} - \mu_{(i,m)}\mu_{(j,m)}\\ &= \mathbf{y}^T\mathbf{B}_{ij}\mathbf{y} - \left(\mathbf{A}_i^T\mathbf{y}\right)\left(\mathbf{A}_j^T\mathbf{y}\right)\\ &= \mathbf{y}^T\left(\mathbf{B}_{ij} - \mathbf{A}_i\mathbf{A}_i^T\right)\mathbf{y}\end{aligned}$$

Theorem 21.3.4

(i) $\widehat{\overline{Y}}_{rss} = \frac{1}{rm}\sum_{\alpha=1}^{r}\sum_{i=1}^{m} y_{(i,m)\alpha}$ is unbiased for $\overline{Y}$

(ii) $V\left(\widehat{\overline{Y}}_{rss}\right) = \left\{\frac{1}{rm}\ \frac{N-1-mr}{N-1}\sigma_y^2 - \frac{1}{m}(\mathbf{y}-\boldsymbol{\mu})^T\left(\sum_{i=1}^{m}\mathbf{B}_{ii}\right)(\mathbf{y}-\boldsymbol{\mu})\right\}$

where $\sigma_y^2 = \sum_{i=1}^{N}\left(y_i - \overline{Y}\right)^2/N$, $\boldsymbol{\mu}^T = (\mu,\ldots\mu)$ and $\mu = \overline{Y}$.

Proof

(i)
$$\begin{aligned}E\left(\widehat{\overline{Y}}_{rss}\right) &= \frac{1}{rm}\sum_{\alpha=1}^{r}\sum_{i=1}^{m} E\left(y_{(i,m)\alpha}\right)\\ &= \frac{1}{rm}\sum_{\alpha=1}^{r}\sum_{i=1}^{m}\mathbf{A}_i^T\mathbf{y}\\ &= \frac{1}{rm}\sum_{\alpha=1}^{r}\sum_{i=1}^{m}\sum_{t=1}^{N} A_i^t y_t\\ &= \frac{1}{m}\sum_{t=1}^{N}\left(\sum_{i=1}^{m} A_i^t\right) y_t\\ &= \frac{1}{N}\sum_{t=1}^{N} y_t \quad \text{(using Theorem 21.3.1)}\\ &= \overline{Y} = \mu\end{aligned}$$

(ii)
$$V\left(\widehat{\overline{Y}}_{rss}\right) = \frac{1}{(rm)^2} V\left(\sum_{\alpha=1}^{r} \sum_{i=1}^{m} y_{(i,m)\alpha}\right)$$

$$= \frac{1}{(rm)^2} \left\{ \sum_{\alpha=1}^{r} V\left(\sum_{i=1}^{m} y_{(i,m)\alpha}\right) + \sum_{\alpha \neq}^{r} \sum_{\alpha'=1}^{r} Cov\left(\sum_{i=1}^{m} y_{(i,m)\alpha}, \sum_{i=1}^{m} y_{(i,m)\alpha'}\right)\right\}$$

Now, writing $V\left(y_{(i,m)\alpha}\right) = \sigma^2_{(i,m)}$, $Cov\left(y_{(i,m)\alpha}, y_{(j,m)\alpha}\right) = \sigma^2_{(ij,m)}$ for $i \neq j$, $Cov\left(y_{(i,m)\alpha}, y_{(i,m)\alpha'}\right) = \sigma^2_{(ii,m)}$ for $\alpha \neq \alpha'$, and $Cov\left(y_{(i,m)\alpha}, y_{(j,m)\alpha'}\right) = \sigma^2_{(ij,m)}$ for $\alpha \neq \alpha'$, we have

$$V\left(\widehat{\overline{Y}}_{rss}\right) = \frac{1}{(rm)^2} \left\{ \sum_{\alpha=1}^{r} \left(\sum_{i=1}^{m} \sigma^2_{(i,m)} + \sum_{i \neq}^{m} \sum_{j=1}^{m} \sigma_{(ij,m)} \right) + \sum_{\alpha \neq}^{r} \sum_{\alpha'=1}^{r} \left(\sum_{i=1}^{m} \sigma_{(ii,m)} + \sum_{i \neq}^{m} \sum_{j=1}^{m} \sigma_{(ij,m)} \right) \right\}$$

$$= \frac{1}{(rm)^2} \left\{ r\left(\sum_{i=1}^{m} \sigma^2_{(i,m)} + \sum_{i \neq}^{m} \sum_{j=1}^{m} \sigma_{(ij,m)} \right) + r(r-1)\left(\sum_{i=1}^{m} \sigma_{(ii,m)} + \sum_{i \neq}^{m} \sum_{j=1}^{m} \sigma_{(ij,m)} \right) \right\}$$

$$= \frac{1}{rm^2} \left(\sum_{i=1}^{m} \sigma^2_{(i,m)} + r \sum_{i=1}^{m} \sum_{j=1}^{m} \sigma_{(ij,m)} - \sum_{i=1}^{m} \sigma_{(ii,m)} \right) \qquad (21.3.6)$$

Now,

$$\begin{aligned}\sum_{i=1}^{m}\sigma^2_{(i,m)} &= \sum_{i=1}^{m}E\left\{y_{(im)r}-\mu_{(i,m)}\right\}^2\\ &= \sum_{i=1}^{m}E\left(y_{(im)r}-\mu\right)^2-\sum_{i=1}^{m}\left(\mu_{(i,m)}-\mu\right)^2\\ &= \sum_{i=1}^{m}\sum_{t=1}^{N}A_i^t(y_t-\mu)^2-\sum_{i=1}^{m}\left(\mu_{(i,m)}-\mu\right)^2\\ &= \sum_{t=1}^{N}(y_t-\mu)^2\sum_{i=1}^{m}A_i^t-\sum_{i=1}^{m}\left(\mu_{(i,m)}-\mu\right)^2\\ &= m\sigma_y^2-\sum_{i=1}^{m}\left(\mu_{(i,m)}-\mu\right)^2 \qquad (21.3.7)\end{aligned}$$

(using Theorem 21.3.1)

Furthermore,

$$\begin{aligned}\sum_{i=1}^{m}\sum_{j=1}^{m}\sigma_{(ij,m)} &= \sum_{i=1}^{m}\sum_{j=1}^{m}\left\{E\left(y_{(i,m)\alpha}y_{(j,m)\beta}\right)-\mu_{(i,m)}\mu_{(j,m)}\right\}\\ &= \sum_{i=1}^{m}\sum_{j=1}^{m}\left(\sum_{t=1}^{N}\sum_{l=1}^{N}y_ty_lB_{ij}^{tl}-\mu_{(i,m)}\mu_{(j,m)}\right)\\ &= \sum_{t\neq}^{N}\sum_{l=1}^{N}y_ty_l\left(\sum_{i=1}^{m}\sum_{j=1}^{m}B_{ij}^{tl}\right)-m^2\mu^2 \quad \left(\text{since } B_{ij}^{tt}=0\right)\\ &= \frac{m^2}{N(N-1)}\sum_{t\neq}^{N}\sum_{l=1}^{N}y_ty_l-m^2\mu^2 \quad (\text{using Eq. 21.3.5})\\ &= -\frac{m^2}{(N-1)}\sigma_y^2 \qquad (21.3.8)\end{aligned}$$

Substituting Eqs. (21.3.7) and (21.3.8) in Eq. (21.3.6), we find

$$V\left(\widehat{\overline{Y}}_{rss}\right)=\frac{r}{(rm)^2}\left(\frac{m(N-1-mr)}{N-1}\sigma_y^2-\sum_{i=1}^{m}\left(\mu_{(i,m)}-\mu\right)^2-\sum_{i=1}^{m}\sigma_{(ii,m)}\right) \tag{21.3.9}$$

We note that $\sum_{i=1}^{m}\sigma_{(ii,m)}$ remains unchanged if the population is centered, i.e., if y_i is replaced by $y_i-\mu$ and $\mu_{(i,m)}$ is replaced by $\mu_{(i,m)}-\mu$. Thus

$$\begin{aligned}\sigma_{(ii,m)}&=\mathbf{y}^T\mathbf{B}_{ii}\mathbf{y}-\mu_{(i,m)}^2\\&=(\mathbf{y}-\boldsymbol{\mu})^T\mathbf{B}_{ii}(\mathbf{y}-\boldsymbol{\mu})-\left(\mu_{(i,m)}-\mu\right)^2\end{aligned} \tag{21.3.10}$$

Substituting Eq. (21.3.10) in Eq. (21.3.9), we obtain

$$V\left(\widehat{\overline{Y}}_{rss}\right)=\frac{1}{rm}\left(\frac{N-1-mr}{N-1}\sigma_y^2-\frac{1}{m}(\mathbf{y}-\boldsymbol{\mu})^T\left(\sum_{i=1}^{m}\mathbf{B}_{ii}\right)(\mathbf{y}-\boldsymbol{\mu})\right) \tag{21.3.11}$$

21.3.1 Relative Precision

The variance of the sample mean $\overline{y}_{wor}$ based on a sample of size mr selected by the SRSWOR method is

$$V\left(\overline{y}_{wor}\right)=\frac{N-mr}{mr(N-1)}\sigma_y^2$$

So, the relative precision of RSS compared with the sample mean of an SRSWOR sample based on the same sample size mr is

$$\begin{aligned}RP_{wor}&=\frac{V\left(\overline{y}_{wor}\right)}{V\left(\widehat{\overline{Y}}_{rss}\right)}\\&=\frac{1}{1-\dfrac{N-m}{N-mr}\tau}\end{aligned} \tag{21.3.12}$$

where

$$\tau=\frac{1}{N-m}+\frac{N-1}{N-m}\frac{\gamma}{\sigma_y^2}\quad\text{and}\quad\gamma=\frac{1}{m}(\mathbf{y}-\boldsymbol{\mu})^T\left(\sum_{i=1}^{m}\mathbf{B}_{ii}\right)(\mathbf{y}-\boldsymbol{\mu}).$$

The expression for the relative savings is

$$RS_{wor} = 1 - 1/RP_{wor}$$
$$= \frac{1 - f/r}{1 - f}\tau \tag{21.3.13}$$

where $f = mr/N = n/N$.

The expression (21.3.12) indicates that the relative precision depends on the replication factors r, m, γ, and σ_y^2. The relative savings is a monotonic increasing function of r for a given value of m.

21.4 SIZE-BIASED PROBABILITY OF SELECTION

Cox (1969) proposed the harmonic mean as an estimator of the mean of a random variable y, if the sample is selected from a population with probability proportional to y. For example, y may be lake size if we are using point-intercept sampling for lakes in a region (Muttlak and McDonald, 1990), fiber length when using point-intercept sampling for fibers in yarn (Cox, 1969), or particle width if we are using line-intercept sampling to estimate the number of particles in a region (McDonald, 1980). Let $f(y)$ be the pdf of y. If each unit has a probability of selection proportional to its size y, then Cox (1969) showed that the pdf of the observed sizes is

$$g(y) = \frac{yf(y)}{\mu}, \quad y > 0 \tag{21.4.1}$$

where $g(y)$ is called the weighted (or length-biased) pdf and μ is the mean of the unweighted density $f(y)$.

Cox (1969) derived the following results:

$$\text{(i) } E_g\left(\frac{1}{y}\right) = \frac{1}{\mu} \text{ (ii) } E_g(\widetilde{\mu}) = \frac{1}{\mu} \text{ and (iii) } V_g(\widetilde{\mu}) = \frac{1}{n}\left(\frac{\mu\mu_{-1} - 1}{\mu^2}\right) \tag{21.4.2}$$

where E_g denotes expectation with respect to the weighted density $g(y)$, $\widetilde{\mu} = \frac{1}{n}\sum_{i=1}^{n}\frac{1}{y_i}$ and $\mu_{-1} = E\left(\frac{1}{y}\right)$.

Cox (1969) has shown that for large n, the harmonic mean, $\widehat{\mu}_h = \left(\frac{1}{\widetilde{\mu}}\right)$, is asymptotically normal with mean μ and variance

$$\frac{\mu^2(\mu\,\mu_{-1} - 1)}{n} \tag{21.4.3}$$

provided $V_g(\widetilde{\mu})$ is finite.

Muttlak and McDonald (1990) applied size-biased probability selection in RSS. In this method, initially m^2 units are selected with size-biased probability of selection with density function $f(y)$. The selected m^2 units are partitioned at random into m sets each of size m. Hence, the pdf of each of the selected m^2 observed y values is $g(y)$, given in Eq. (21.4.1). All the units in a set are ranked by some inexpensive procedure. The lowest rank holder is quantified for the first set, the second ranked holder is quantified for the next set, and the process is continued until the highest rank unit is quantified for the last mth set. The cycle is repeated r times. Let $y_{i(i)|k}$ be the value of the variate of interest y based on ith ranked unit of the ith set of the kth cycle, $i = 1,\ldots, m$; $k = 1,\ldots, r$.

Consider

$$\widetilde{\mu}_k = \frac{1}{m}\sum_{i=1}^{m}\frac{1}{y_{i(i)|k}} \tag{21.4.4}$$

as an estimator of $1/\mu$.

We can write

$$g(y) = \frac{1}{m}\sum_{i=1}^{m} g_{m,i}(y) \tag{21.4.5}$$

where $g_{m,i}(y) = \dfrac{\Gamma(m+1)}{\Gamma(m)\Gamma(m-i+1))}[G(y)]^{i-1}[1-G(y)]^{m-i}g(y)$ and $G(y) = \int_{-\infty}^{y} g(y)dy$.

Eq. (21.4.5) yields

$$\begin{aligned} E(g(y)) &= \frac{1}{m}\sum_{i=1}^{m} E\left(\frac{1}{y_{i(i)|k}}\right) \\ \text{i.e., } \frac{1}{\mu} &= \frac{1}{m}\sum_{i=1}^{m}\mu_{-(i)|m} \end{aligned} \tag{21.4.6}$$

where $E\left(\dfrac{1}{y_{i(i)|k}}\right) = \mu_{-(i)|m}$.

Hence from Theorem 21.2.1, we get

$$E(\widetilde{\mu}_k) = \frac{1}{m}\sum_{i=1}^{m}\frac{1}{\mu_{-(i)|m}} = \frac{1}{\mu} \text{ and } V(\widetilde{\mu}_k) = \frac{1}{m^2}\sum_{i=1}^{m}\sigma^2_{-(i)|m} \le V\left(\frac{1}{y}\right)\Big/ m \tag{21.4.7}$$

where $\sigma^2_{-(i)|m} = V\left(\dfrac{1}{y_{i(i)|k}}\right)$.

Eq. (21.4.7) leads to the following theorem:

Theorem 21.4.1

(i) $\widetilde{\mu}_{\bullet} = \frac{1}{r}\sum_{k=1}^{r}\widetilde{\mu}_k = \frac{1}{mr}\sum_{k=1}^{r}\sum_{i=1}^{m}\frac{1}{y_{i(i)|k}}$ is an unbiased estimator of $\frac{1}{\mu}$

(ii) $V(\widetilde{\mu}_{\bullet}) \le V(\widetilde{\mu}_{mr})$

where $\widetilde{\mu}_{mr} = \frac{1}{mr}\sum_{i=1}^{mr}\frac{1}{y_i}$ and $y_1, \ldots, y_{mr}$ is a size-biased sample from a population with density of $f(y)$.

Following Cox (1969), the harmonic mean $1/\widetilde{\mu}_{\bullet}$ based on the RSS is approximately unbiased for the population mean μ. In addition, one would expect that the variance of $1/\widetilde{\mu}_{\bullet}$ is smaller than that of $1/\widetilde{\mu}_{mr}$ because the variance of $\widetilde{\mu}_{\bullet}$ is smaller than that of $\widetilde{\mu}_{mr}$. Muttlak and McDonald (1990) supported correctness of this inequality by using computer simulations.

21.5 CONCLUDING REMARKS

RSS is useful for the situations where quantification of an element is difficult but elements can be ranked fairly by eye or some other method with negligible cost. RSS provides an unbiased and efficient estimate for the population mean; even the ranking is imperfect. RSS does not provide unbiased estimates of the population variance. Unlike estimation of the mean, RSS does not provide appreciable gain in efficiency for estimation of variance. However, RSS is found to be most beneficial when estimates of both mean and variances are needed (Stokes, 1980a). RSS for estimating the population mean under SRSWOR sampling was proposed by Patil et al. (1995). However, the expression of variance was not simple as it involves a quantity $\gamma = (\mathbf{y} - \boldsymbol{\mu})^T\boldsymbol{\Gamma}(\mathbf{y} - \boldsymbol{\mu})/m$, where $\boldsymbol{\Gamma} = \left(\sum_{i=1}^{m}\mathbf{B}_{ii}\right)$ is an $N \times N$ matrix whose entries are a function of the population size N and the set size m but independent of the study variable y. A computer program for computation of $\boldsymbol{\Gamma}$ was provided by Patil et al. (1995). Several extensions to the theories and applications of RSS have been proposed; that is, RSS with imperfect ranking, RSS by ranking with a concomitant variable, and RSS with multivariate samples and random selection method were considered by Stokes (1977, 1980a), Patil et al. (1993), and Li et al. (1999), among others. Application of RSS in Mann–Whitney Wilcoxon test was provided by Bonn and Wolfe (1992, 1994) while an RSS version of the sign test was provided by Koti and Babu (1996). RSS for density estimation was investigated by Chen (1999), and RSS for parametric families was considered by Shen (1994) and Sinha et al. (1996).

21.6 EXERCISES

21.6.1 Let m sets of samples each of size m be selected independently from a population by SRSWR method. A sample $s_1,\ldots, s_r$ of $r(<m)$ sets is selected at random from the selected m sets by SRSWOR method. The units of the selected samples $s_1,\ldots, s_r$ are ordered with respect to the value of the study variable y. Let a subset of integers $k_1,\ldots, k_r$ be selected from the set of integers $1,\ldots, n$ by SRSWOR method. Let $y_{(i,k_i)}$ be the k_ith ordered statistic from the set s_i, $i = 1,\ldots, r$.

(a) Show that

(i) $\widehat{\mu}_{mrss} = \frac{1}{r}\sum_{i=1}^{r} y_{(i,k_i)}$ is an unbiased estimator of μ, the population mean of y.

(ii) $$Var(\widehat{\mu}_{mrss}) = \frac{Q}{mr}\,\frac{m-r}{m-1} + \frac{\sum_{j=1}^{m}\sigma^2_{j:m}}{mr}$$

where $Q = \sum_{j=1}^{m}\left(\mu_{j:m} - \mu\right)^2$, $\mu_{j:m} = E(y_{(i,j)})$, and $\sigma^2_{j:m} = Var\left(y_{(i,j)}\right)$.

(b) Let $\overline{y}_m$ be the sample mean of size m selected by SRSWR method. Show that a necessary condition for $Var(\widehat{\mu}_{mrss}) < Var\left(\overline{y}_m\right)$ is $r > m^2/(2m-1)$ (Li et al., 1999).

21.6.2 Consider a finite population $U = \{1,\ldots, N\}$ of N units. Let y_i and x_i be the values of the study and auxiliary variables y and x, respectively. From the population U, m independent samples $s_1,\ldots, s_m$ each of size m are selected by SRSWOR method. The selected units in $s_j(j = 1,\ldots, m)$ are ranked with respect to the x values. Let $y_{j\langle j\rangle}$ be the judgment jth order statistic corresponding to $x_{j(j)}$, jth order statistic of x for the sample s_j, $j = 1,\ldots, m$. Thus, we get the following ranked set sample $rss = (y_{1\langle 1\rangle},\ldots, y_{j\langle j\rangle},\ldots, y_{m\langle m\rangle})$. The selection of ranked set sample is repeated r times (cycles). The ranked set sample based on the kth cycle is denoted by $S = \bigcup_{k=1}^{r} S_k$, where $S_k = (y_{1\langle 1\rangle|k},\ldots, y_{j\langle j\rangle|k},\ldots, y_{m\langle m\rangle|k})$, $k = 1,\ldots, r$. Let S^* be the set of distinct units in S.

(a) Show that the inclusion probabilities for the ith unit, i and $t(\neq i)$ units are, respectively,

$$\pi_i = 1 - \prod_{j=1}^{m}(1-\alpha_{ij})^r \text{ and}$$

$$\pi_{it} = 1 - \prod_{j=1}^{m}(1-\alpha_{ij})^r - \prod_{j=1}^{m}(1-\alpha_{tj})^r + \prod_{j=1}^{m}(1-\alpha_{ij}-\alpha_{tj})^r$$

where $\alpha_{ij} = \binom{i-1}{j-1}\binom{N-i}{m-j}\Big/\binom{N}{m}$

(b) Let δ_i be the number of times ith unit appears in S. Show that $E(\delta_i) = mr/N$.

(c) Let $T_{yk} = \frac{1}{m}\sum_{j=1}^{m} y_{j\langle j\rangle|k}$ and $T_y = \frac{1}{r}\sum_{k=1}^{r} T_{yk}$. Show that

(i) $E(T_y) = \overline{Y}$

(ii) The variance of T_y is

$$V(T_y) = \frac{1}{mr}\sigma_y^2 - \frac{1}{rm^2}\sum_{j=1}^{m}\left(\mu_{y\langle j\rangle} - \overline{Y}\right)^2$$

and

(iii) An unbiased estimator of $V(T_y)$ is

$$\widehat{V}(T_y) = \frac{1}{r(r-1)}\sum_{j=1}^{r}(T_{yr} - T_y)^2$$

where $\overline{Y} = \sum_{i=1}^{N} y_i/N$, $\sigma_y^2 = \sum_{i\in U}\left(y_i - \overline{Y}\right)^2/N$,

and $\mu_{y\langle j\rangle} = E(y_{j\langle j\rangle|k})$ (Jozani and Johnson, 2011).

21.6.3 Let $y_{i(i)|k}$, $i = 1,\ldots, m$; $k = 1,\ldots, r$, be a ranked set sample of size mr selected from a population with cdf F with r cycles each of which consists of m sets of size m each. The empirical distribution function (edf) based on the ranked set sample is denoted by $F^*(t) = \frac{1}{mr}\sum_{k=1}^{r}\sum_{i=1}^{m}\Delta\left(y_{i(i)|k} - t\right)$, where $\Delta(u) = 1$ if $u \leq 0$ and $\Delta(u) = 0$ if $u > 0$. Show that

(i) $F^*(t)$ is an unbiased estimator of $F(t)$

(ii) $Var(F^*(t)) = \frac{1}{rm^2}\sum_{j=1}^{m} F_{(j)}\left(1 - F_{(j)}\right)$, where $F_{(j)}$ is the distribution function of $y_{i(i)|k}$ and

(iii) $Var(F^*(t)) \leq Var\left(\widehat{F}(t)\right)$, where $\widehat{F}(t)$ is the edf of a random sample of size mr (Stokes and Sager, 1988).

21.6.4 Let a ranked set sample of size $n(=mr)$ be selected from a population with r cycles each of which consists of m sets of size m each. Let $y_{i\langle i\rangle|k}$ $(x_{i(i)|k})$ denote ith judgment ordering (order statistic) in the ith set of the cycle r for the study (auxiliary) variable $y(x)$.

Consider the ratio estimator $\widehat{\mu}_{rRSS} = \dfrac{\overline{y}_{\langle n\rangle}}{\overline{x}_{[n]}}\mu_x$ of the population mean μ_y of the study variable y, where $\overline{y}_{\langle n\rangle} = \dfrac{1}{rm}\sum_{k=1}^{r}\sum_{i=1}^{m} y_{i\langle i\rangle|k}$, $\overline{x}_{[n]} = \dfrac{1}{rm}\sum_{k=1}^{r}\sum_{i=1}^{m} x_{i(i)|k}$, and μ_x is the population mean of the auxiliary variable x, which is assumed to be known. Show that the bias and MSE of $\widehat{\mu}_{rRSS}$ are as follows:

$$B(\widehat{\mu}_{rRSS}) \cong \mu_y\left[\frac{1}{mr}\left(C_x^2 - C_{yx}\right) - \left(W_{x(i)}^2 - W_{yx(i)}\right)\right],$$

$$M(\widehat{\mu}_{rRSS}) \cong \frac{1}{mr}\left(\sigma_y^2 - 2R\sigma_{yx} + R^2\sigma_x^2\right) - \frac{1}{m^2 r}\left(\sum_{i=1}^{m}\tau_{y\langle i\rangle}^2 - 2R\sum_{i=1}^{m}\tau_{yx(i)}^2 + R^2\sum_{i=1}^{m}\tau_{x(i)}^2\right)$$

where $C_x(C_y) =$ population CV of $x(y)$, $C_{yx} = \rho C_x C_y$, $\rho =$ correlation coefficient between y and x, $W_{x(i)}^2 = \dfrac{1}{m^2 r\mu_x^2}\sum_{i=1}^{m}\tau_{x(i)}^2$, $W_{yx(i)} = \dfrac{1}{m^2 r\mu_x\mu_y}\sum_{i=1}^{m}\tau_{yx(i)}$, $\tau_{x(i)} = E(x_{i(i)|k}) - \mu_x$, $\tau_{y\langle i\rangle} = E(y_{i\langle i\rangle|k}) - \mu_y$, $\tau_{yx(i)} = E(y_{i\langle i\rangle|k} - \mu_y)(x_{i(i)|k} - \mu_x)$, $\sigma_y^2\left(\sigma_x^2\right) =$ population variance of $y(x)$, and $R = \mu_y/\mu_x$.

(Kadilar et al., 2009).

CHAPTER 22

Estimating Functions

22.1 INTRODUCTION

In the classical theory of estimation, we construct an estimator $t(\mathbf{y})$ of a population parameter θ based on a sample observation $\mathbf{y} = (y_1, \ldots, y_n)$. The estimator $t(\mathbf{y})$ is a function of $\mathbf{y}$ only (independent of θ) and expected to satisfy certain optimal criteria such as unbiasedness, sufficiency, and efficiency. The estimator $t(\mathbf{y})$ is constructed using some standard methods of estimation such as maximum likelihood (ML), least squares (LS), methods of moments, minimum chi-square, among others. These methods are ad hoc because no optimal criteria were used in developing them. However, in developing the minimum variance unbiased estimator by Rao–Blackwell and Lehmann–Scheffe approach, the criteria of sufficient statistic were used. The method of LS, ML, method of moments, and minimum chi-square have a similarity in which the estimator $t(\mathbf{y})$ is a solution of θ from the equation of the form $g(\mathbf{y},\theta) = 0$. The method of LS and ML are extensively used in statistical applications. Each method has some advantages and limitations. The method of estimating function (EF) was introduced by Godambe (1960a,b) and Durbin (1960). The EF method is more general since it includes the LS, ML, methods of moments and minimum chi-square methods of estimation. The EF has more strengths and fewer weaknesses than the other methods of estimation (Godambe and Kale, 1991). The EF method has wide applications in the areas of biostatistics, stochastic process, survey sampling, among others. In this chapter we will consider the application of EF only in the area of survey sampling. The confidence intervals in survey sampling will be determined using the framework of EFs.

22.2 ESTIMATING FUNCTION AND ESTIMATING EQUATIONS

Let $\mathbf{y} = (y_1, y_2, \ldots, y_n)$ be a random sample from a population with density function $f(\mathbf{y},\theta)$ with respect to the measure μ, where θ is an unknown parameter that belongs to a known parametric space Ω_θ. The methods of

Survey Sampling Theory and Applications
ISBN 978-0-12-811848-1
http://dx.doi.org/10.1016/B978-0-12-811848-1.00022-4

LS, ML, and minimum chi-square yield equation $g(\mathbf{y},\theta)=0$, and by solving the equation we get an estimate of θ. If θ is a k-dimension vector, then we get k-independent equations $g_i(\mathbf{y},\theta)=0$ for $i=1,\ldots,k$. A function $g(\mathbf{y},\theta)$ of observation $\mathbf{y}$ and θ is called an estimating function if an estimate of θ can be derived from the equation $g(\mathbf{y},\theta)=0$. The equation $g(\mathbf{y},\theta)=0$ is called an estimating equation.

Example 22.2.1 (Godame and Kale, 1991)

Let $y_1,\ldots,y_n$ be independent random variables with

$$E(y_i)=\theta \text{ and } Var(y_i)=\sigma^2 \text{ for } i=1,\ldots,n \tag{22.2.1}$$

The LS estimator of θ is obtained by minimizing

$$S^2=\sum_{i=1}^{n}(y_i-\theta)^2 \tag{22.2.2}$$

with respect to θ.

Now $\dfrac{\partial S^2}{\partial\theta}=0$ gives $(\overline{y}-\theta)=0$. If we choose the EF $g(\mathbf{y},\theta)=g(y_1,\ldots,y_n;\theta)=(\overline{y}-\theta)$, then the estimating equation $g(\mathbf{y},\theta)=0$ and the LS method provide the same solution of θ as $\overline{y}$. Furthermore, if we assume the distribution of y is normal, then the estimator of θ based on LS, ML, and EF method will be same as $\overline{y}$.

Now consider the class of estimation function G of the form

$$g(\mathbf{y},\theta)=\sum_{i=1}^{n}(y_i-\theta)b_i \tag{22.2.3}$$

where b_i's are known constants such that $\sum_{i=1}^{n}b_i\neq 0$.

The estimating equation $g(\mathbf{y},\theta)=0$ yields the solution of θ as

$$\widehat{\theta}_g=\sum_{i=1}^{n}b_iy_i\Big/\sum_{i=1}^{n}b_i \tag{22.2.4}$$

The variance of $\widehat{\theta}_g$ is

$$V\left(\widehat{\theta}_g\right)=\sigma^2\frac{\sum b_i^2}{\left(\sum b_i\right)^2} \tag{22.2.5}$$

Furthermore, if we impose the restriction $\sum_{i=1}^{n}b_i=c$, a constant, then the variance of $g(\mathbf{y},\theta)$ is minimized when $b_i=c/n$ for $i=1,\ldots,n$. Under this situation, both EF and Gauss–Markov approach lead the $\overline{y}$ as the solution of θ.

An estimation function $g \in G$ is unbiased if $E(g) = 0$.

The EF $g' = kg$ with k as a constant belongs to the class G. Although, the estimating equations $g = 0$ and $g' = 0$ yield the same unbiased estimate $\widehat{\theta}_g$, the variance of the EF g' is $Var(g') = k^2 Var(g)$, which can be made arbitrary small by appropriate choice of the constant k. Hence for the sake of comparability the concept of standardized estimation function has been introduced. A standardized estimation function is defined as

$$g_s = \frac{g}{E\left(\frac{\partial g}{\partial \theta}\right)} \tag{22.2.6}$$

Consider a larger class of EFs G_1 consisting of EFs of the form

$$g_1(y, \theta) = \sum b_i(\theta)(y_i - \theta) \tag{22.2.7}$$

where $b_i(\theta)$ is a differentiable function of θ.

The EF $g_1(\mathbf{y},\theta)$ is unbiased, but some solutions of the EF $g_1(\mathbf{y},\theta) = 0$ may not yield unbiased estimators of θ. The standardized EF derived from $g_1(\mathbf{y},\theta)$ is given by

$$g_{1s} = \frac{g_1}{E\left(\frac{\partial g_1}{\partial \theta}\right)} = \frac{g_1}{-\sum b_i(\theta)} \tag{22.2.8}$$

The variance of g_{1s} is

$$Var(g_{1s}) = \sigma^2 \frac{\sum b_i^2(\theta)}{\left(\sum b_i(\theta)\right)^2} \tag{22.2.9}$$

The variance $Var(g_{1s})$ is minimized when $b_i(\theta) = b(\theta) \neq 0$ for all $\theta \in \Omega_\theta$. Thus the optimum EF that minimizes the variance of $Var(g_{1s})$ in the class G_1 is

$$g_{10} = b(\theta) \sum (y_i - \theta) \tag{22.2.10}$$

The estimating Eq. (22.2.10) yields the optimal estimator of θ as $\overline{y} = \sum y_i/n$.

22.2.1 Optimal Properties of Estimating Functions

The following regularity conditions on $f(y,\theta)$ and $g(\mathbf{y},\theta)$ were imposed by Godambe (1960a,b) for derivation of the optimal properties of the estimation functions.

Regularity conditions on $f(y,\theta)$ (conditions A):

(i) Ω_θ is an open interval of the real line.

(ii) For almost all y, $\dfrac{\partial \log f(y,\theta)}{\partial\theta}$ and $\dfrac{\partial^2 \log f(y,\theta)}{\partial\theta^2}$ exist for $\forall\theta\in\Omega_\theta$.

(iii) $\int f(y,\theta)d\mu$ and $\displaystyle\int \frac{\partial \log f(y,\theta)}{\partial\theta} f(y,\theta)d\mu$ are differentiable under the sign of integration.

(iv) $E\left(\dfrac{\partial \log f(y,\theta)}{\partial\theta}\Big|\theta\right)^2 > 0\,\forall\theta\in\Omega_\theta$.

Regularity conditions on $g(\mathbf{y},\theta)$ (conditions B):

(i) $E[g(\mathbf{y},\theta)] = 0\ \forall\theta\in\Omega_\theta$

(ii) For almost all $\mathbf{y}$, $\dfrac{\partial g(\mathbf{y},\theta)}{\partial\theta}$ exists $\forall\theta\in\Omega_\theta$.

(iii) $\int g(\mathbf{y},\theta) f(\mathbf{y},\theta)\partial\mu$ is differentiable under the sign of integration.

(iv) $E\left[\dfrac{\partial g(\mathbf{y},\theta)}{\partial\theta}\right]^2 \geq 0\ \forall\theta\in\Omega_\theta$

The condition (i) of **B** is known as an unbiasedness condition of an EF. If $L(\mathbf{y},\theta)$, the likelihood function of θ is differentiable in θ, then $g(\mathbf{y},\theta) = \dfrac{\partial L(\mathbf{y},\theta)}{\partial\theta}$ is an EF. An EF satisfying all the regularity conditions **B** is known as a regular EF.

Let $\widehat{\theta}_g$ be a solution of $g(\mathbf{y},\theta) = 0$, then using Taylor expansion, we can write

$$g(\mathbf{y},\theta) = g\left(\mathbf{y},\widehat{\theta}_g\right) + \left(\theta - \widehat{\theta}_g\right)\frac{\partial g(\mathbf{y},\theta^*)}{\partial\theta} \tag{22.2.11}$$

where $\theta^*\in\left(\widehat{\theta}_g,\theta\right)$.

Now noting $g\left(\mathbf{y},\widehat{\theta}_g\right) = 0$, we find

$$\theta - \widehat{\theta}_g = -g(\mathbf{y},\theta)/g'(\mathbf{y},\theta^*) \text{ assuming } g'(\mathbf{y},\theta^*) \neq 0 \tag{22.2.12}$$

A good EF should have the difference $\theta - \widehat{\theta}_g$ small. Hence it is desirable that the estimator $\widehat{\theta}_g$ should have

$$E\left(\theta - \widehat{\theta}_g\right)^2 = E(g^2)\Big/\left[E\left(\frac{\partial g}{\partial\theta}\right)\right]^2 \tag{22.2.13}$$

as small as possible. So the criterion

$$\lambda_g(\theta) = \frac{E(g^2)}{\left[E\left(\dfrac{\partial g}{\partial\theta}\right)\right]^2}$$

can be used as a measure of efficiency of an EF g.

Let G be the class of EF satisfying conditions (i) to (v) of B. Godambe (1960a,b) defined the EF $g_0 = g_0(\mathbf{y},\theta)$ as optimum EF in the class G if

$$\frac{E\left(g_0^2\right)}{\left[E\left(\dfrac{\partial g_0}{\partial \theta}\right)\right]^2} \le \frac{E\left(g^2\right)}{\left[E\left(\dfrac{\partial g}{\partial \theta}\right)\right]^2} \quad \forall \theta \in \Omega_\theta \tag{22.2.14}$$

The equation $g_0(\mathbf{y},\theta) = 0$ is called an optimal estimating equation and the corresponding solution of θ will be called the optimal estimate.

Let $g^s = g/E\left(\dfrac{\partial g}{\partial \theta}\right)$ be the standardized unbiased EF derived form $g(\in G)$. Then the Eq. (21.1.14) above can be expressed as

$$Var\left(g_0^s\right) \le Var\left(g^s\right) \ \forall g \in G \tag{22.2.15}$$

Under certain regularity conditions, Kale (1962) proved the following inequality analogous to the Rao–Cramer inequality

$$Var(g) \ge \frac{\left\{E\left(\dfrac{\partial g}{\partial \theta}\right)\right\}^2}{I(\theta)} \quad \forall g \in G \tag{22.2.16}$$

where $I(\theta) = E\left[\dfrac{\partial \log f(y,\theta)}{\partial \theta}\right]^2$ is the Fisher information.

The equality in Eq. (22.2.16) is attained when

$$g = g^* = \frac{\partial \log f}{\partial \theta}$$

$g^* = \dfrac{\partial \log f}{\partial \theta}$ is known as score function of the ML.

In particular if $g = t(\mathbf{y}) - \nu(\theta)$, where $t(\mathbf{y})$ is an unbiased estimator of $\nu(\theta)$, then the inequality (22.2.16) reduces to Rao–Cramer inequality

$$Var(t(\mathbf{y})) \ge \frac{\{\nu'(\theta)\}^2}{I(\theta)} \tag{22.2.17}$$

22.3 ESTIMATING FUNCTION FROM SUPERPOPULATION MODEL

Consider a finite population $U = (U_1,\ldots, U_i,\ldots, U_N)$ of N identifiable units, and let y_i be the value of the character under study y associated with the ith unit of the population U. Here we assume that the vector $\mathbf{y} = (y_1,\ldots, y_i,\ldots, y_N)$ is generated from a distribution ξ, which is known to be a member of a class C. The class C is called a superpopulation model. A superpopulation parameter θ is a real-valued function defined on C. Here we will define $g(\mathbf{y},\theta)$ as an unbiased EF for θ if

$$E_\xi\{g(y,\theta(\xi)\} = 0 \ \forall \xi \in C \tag{22.3.1}$$

where E_ξ denotes expectation under ξ.

The solution of the estimating equation $g(\mathbf{y},\theta) = 0$ has two interpretations:

(i) If the vector $\mathbf{y}$ is known, i.e., the population is completely enumerated, then $\theta_N(\mathbf{y})$, the solution of Eq. (22.3.1) is an estimate of the superpopulation parameter θ.

(ii) If the vector $\mathbf{y}$ is partially known, then $\theta_N(\mathbf{y})$ is the parameter of the survey population.

Consider a superpopulation model $C = \{\xi\}$ under which $y_1,\ldots, y_N$ are independent.

An unbiased EF $g^*(y,\theta) \in G$ is said to be optimal in class G and superpopulation $\xi \in C$ under the regularity conditions B stated in Section 22.2.1, if

$$\frac{E_\xi\left(g^{*2}\right)}{\left\{E_\xi\left(\frac{\partial g^*}{\partial\theta}\right)\Big|_{\theta=\theta(\xi)}\right\}^2} \le \frac{E_\xi\left(g^2\right)}{\left\{E_\xi\left(\frac{\partial g}{\partial\theta}\right)\Big|_{\theta=\theta(\xi)}\right\}^2} \tag{22.3.2}$$

for every $g \in G$ and every $\xi \in C$.

22.3.1 Optimal and Linearly Optimal Estimating Functions

Consider the superpopulation C under which $y_1,\ldots, y_N$ are independent. We define the EF $g_l(\mathbf{y},\theta)$ as linear in $\phi_1,\ldots, \phi_N$ if

$$g_l(\mathbf{y},\theta) = \sum_{i=1}^{N}\phi_i(y_i,\theta)a_i(\theta) \tag{22.3.3}$$

where $a_i(\theta)$ are real differentiable functions of θ and free from y_i's, and ϕ_i are functions satisfying $E_\xi\{\phi_i(y_i,\theta)\} = 0$ for $i = 1,\ldots, N$. An example, one may choose $\phi_i = y_i - \theta$, where $E_\xi(y_i) = \theta$ for $i = 1,\ldots, N$. We will denote the class of linear EF defined in Eq. (22.3.3) as G_L.

The EF $g_l^*(\mathbf{y},\theta)$ is said to be linearly optimal if it belongs to G_L and satisfies Eq. (22.3.2). The equation $g_l^*(y,\theta) = 0$ is called an optimal estimating equation, and the corresponding solution of θ will be called the optimal estimate of θ.

Theorem 22.3.1

For the superpopulation model $C = \{\xi\}$, where $y_1,\ldots, y_N$ are independently distributed, $E_\xi\{\phi_i(y_i,\theta)\} = 0$ for $i = 1,\ldots, N$ and $\xi \in C$, the EF $g_{l0}(\mathbf{y},\theta) = \sum_{i=1}^{N} \phi_i(y_i,\theta)$ is linearly optimal if

$$E_\xi\left(\frac{\partial\phi_i(y_i,\theta)}{\partial\theta}\Big|_{\theta=\theta(\xi)}\right) = k\{(\theta(\xi)\}E_\xi\{\phi_i^2\} \quad \text{for } \forall i = 1,\ldots,N \quad (22.3.4)$$

where $k\{(\theta(\xi)\}$ is a function of $\theta(\xi)$ only.

Proof

$$E_\xi\left(\frac{g_l}{E_\xi\left(\frac{\partial g_l}{\partial\theta}\right)\Big|_{\theta=\theta(\xi)}} - \frac{g_{l0}}{E_\xi\left(\frac{\partial g_{l0}}{\partial\theta}\right)\Big|_{\theta=\theta(\xi)}}\right)^2$$

$$= \frac{E_\xi(g_l)^2}{\left\{E_\xi\left(\frac{\partial g_l}{\partial\theta}\right)\Big|_{\theta=\theta(\xi)}\right\}^2} + \frac{E_\xi(g_{l0})^2}{\left\{E_\xi\left(\frac{\partial g_{l0}}{\partial\theta}\right)\Big|_{\theta=\theta(\xi)}\right\}^2} \quad (22.3.5)$$

$$- 2\frac{E_\xi(g_l g_{l0})}{\left\{E_\xi\left(\frac{\partial g_l}{\partial\theta}\right)\Big|_{\theta=\theta(\xi)}\right\}\left\{E_\xi\left(\frac{\partial g_{l0}}{\partial\theta}\right)\Big|_{\theta=\theta(\xi)}\right\}}$$

$$\geq 0$$

Now using the condition $E_\xi\left(\frac{\partial\phi_i(y_i,\theta)}{\partial\theta}\Big|_{\theta=\theta(\xi)}\right) = k(\theta(\xi))E_\xi\{\phi_i(y_i,\theta)\}^2$, we find

$$\frac{E_\xi(g_l g_{l0})}{\left\{E_\xi\left(\frac{\partial g_l}{\partial\theta}\right)\Big|_{\theta=\theta(\xi)}\right\}\left\{E_\xi\left(\frac{\partial g_{l0}}{\partial\theta}\right)\Big|_{\theta=\theta(\xi)}\right\}}$$

$$= \frac{\sum_{i=1}^{N} a_i(\theta)E_\xi(\phi_i^2)}{\{k(\theta(\xi))\}^2\left\{\sum_{i=1}^{N} a_i(\theta)E_\xi(\phi_i^2)\right\}\left\{\sum_{i=1}^{N} E_\xi(\phi_i^2)\right\}} \quad (22.3.6)$$

$$= \frac{1}{\{k(\theta(\xi))\}^2\left\{\sum_{i=1}^{N} E_\xi(\phi_i^2)\right\}}$$

and

$$\frac{E_\xi(g_{l0})^2}{\left\{E_\xi\left(\frac{\partial g_{l0}}{\partial\theta}\right)\Big|_{\theta=\theta(\xi)}\right\}^2}=\frac{1}{\{k(\theta(\xi))\}^2\left(\sum_{i=1}^{N}E_\xi\left(\phi_i^2\right)\right)} \tag{22.3.7}$$

Substituting Eqs. (22.3.6) and (22.3.7) in Eq. (22.3.5), we get

$$E_\xi\left(\frac{g_l}{E_\xi\left(\frac{\partial g_l}{\partial\theta}\right)\Big|_{\theta=\theta(\xi)}}\right)^2\geq E_\xi\left(\frac{g_{l0}}{E_\xi\left(\frac{\partial g_{l0}}{\partial\theta}\right)\Big|_{\theta=\theta(\xi)}}\right)^2$$

Corollary 22.3.1

If $y_1,\ldots,\ y_N$ are independently and identically distributed and $\phi_i=\phi$ for $i=1,\ldots,\ N$, then $g_{l0}(\mathbf{y},\theta)=\sum_{i=1}^{N}\phi(y_i,\theta)$ is an optimal EF for θ(Godambe and Thompson, 1978, 1984) if

$$E_\xi\left(\frac{\partial\phi(y_i,\theta)}{\partial\theta}\Big|_{\theta=\theta(\xi)}\right)=k\{\theta(\xi)\}(E_\xi\{\phi(y_i,\theta)\}^2$$

Remark 22.3.1

The different choices of ϕ_i yield different survey parameters. For example, the choice $\phi_i(\mathbf{y}_i,\theta)=y_i-\theta$, the estimating equation $\sum_{i=1}^{N}\phi_i(y_i,\theta)=0$, yields $\theta_N=\overline{Y}=\sum_{i=1}^{N}y_i/N$, the mean of survey population. Similarly, for the choice $\phi_i(y_i,\theta)=(y_i-\theta x_i)$, θ_N becomes $\sum_{i=1}^{N}y_i/\sum_{i=1}^{N}x_i$, the population ratio. Again, if we choose $\phi_i(y_i,\theta)=\varphi_i(y_i,\theta)-p$ with $\varphi_i(y_i,\theta)=1$ for $y_i\leq\theta$ and $\varphi_i(y_i,\theta)=0$ otherwise, then the $\sum_{i=1}^{N}\phi_i(y_i,\theta)=0$ yields θ_N as pth quantile of the survey population.

Example 22.3.1

Suppose that $y_1,\ldots,\ y_N$ are independent with $E_\xi(y_i)=\theta(\xi)x_i$ and $V_\xi(y_i)=\sigma_i^2$. Then condition (22.3.4) of Theorem 22.3.1 holds for $\phi_i=\frac{x_i(y_i\ -\ \theta x_i)}{\sigma_i^2}$. Hence, $g_{l0}(\mathbf{y},\theta)=\sum_{i=1}^{N}\frac{x_i(y_i\ -\ \theta x_i)}{\sigma_i^2}$ is linearly optimal and $\theta_N = \dfrac{\sum_{i=1}^{N} y_i x_i/\sigma_i^2}{\sum_{i=1}^{N} x_i^2/\sigma_i^2}$ is the optimal estimate of θ as well as the parameter of the survey population (Chaudhuri and Stenger, 1992).

Example 22.3.2

In Example 22.3.1 above, if we assume $V_\xi(y_i) = \sigma^2$ (constant), then $\theta_N = \frac{\sum_{i=1}^{N} y_i x_i}{\sum_{i=1}^{N} x_i^2}$ is the optimal estimate of θ as well as the parameter of the survey population (Ghosh, 1991).

Example 22.3.3

In Example 22.3.2 if $x_i = 1$ for $i = 1,\ldots, N$, then $g_{l0}(\mathbf{y}, \theta) = \sum_{i=1}^{N} (y_i - \theta)$ is an optimal EF of θ and $\theta_N = \sum_{i=1}^{N} y_i/N$ becomes the optimal estimate of θ and also the survey parameter (Godambe and Thompson, 1986a,b).

22.4 ESTIMATING FUNCTION FOR A SURVEY POPULATION

Suppose a sample s of size n is selected from a finite population U of size N with probability $p(s)$ using a sampling design p. Let the inclusion probability for the ith unit be $\pi_i = \sum_{s \supset i} p(s) > 0$ for $i = 1,\ldots, N$. Furthermore, let $y_1,\ldots, y_N$ be distributed independently as in Theorem 22.3.1 and θ_N be the solution of the linearly optimal EF $g_{l0}(\mathbf{y}, \theta) = \sum_{i=1}^{N} \phi_i(y_i, \theta)$. Since y_i for $i \notin s$ are not known, we construct an EF $h(d,\theta)$ as a function of the collected data $d = (y_i, i \in s)$. Such a function $h(d,\theta)$ will be called a sample EF. The solution $\widehat{\theta}_s$ of the sample estimating equation $h(d,\theta) = 0$ provides an estimate of the survey parameter θ_N. The function $h(d,\theta)$ is said to be design unbiased of $\sum_{i=1}^{N} \phi_i(y_i, \theta)$ if

$$E_p[h(d, \theta)] = \sum_{s} h(d, \theta) p(s) = \sum_{i=1}^{N} \phi_i(y_i, \theta) \tag{22.4.1}$$

for each population vector $\mathbf{y} = (y_1,\ldots, y_N)$ and θ, where E_p denotes expectation with respect to the sampling design p. The class of unbiased EFs $h(d,\theta)$ that satisfy Eq. (22.4.1) will be denoted by C_u. The function

$$h^*(d, \theta) = \sum_{i \in s} \frac{\phi_i(y_i, \theta)}{\pi_i} \tag{22.4.2}$$

is an unbiased EF of $\sum_{i=1}^{N} \phi_i(y_i, \theta)$.

Godambe and Thompson (1986a,b) defined an EF $h(d,\theta)$ satisfying Eq. (22.4.1) as optimum if it minimizes

$$E_\xi E_p\{h^2(d,\theta)\} \Big/ \left\{E_\xi E_p\left(\frac{\partial h(d,\theta)}{\partial\theta}\right)\right\}^2 \tag{22.4.3}$$

Here we note that

$$E_\xi E_p\left(\frac{\partial h(d,\theta)}{\partial\theta}\right) = E_\xi\left(\frac{\partial}{\partial\theta}\sum_{i=1}^{N}\phi_i\right) \tag{22.4.4}$$

as

$$E_\xi E_p\left(\frac{\partial h(d,\theta)}{\partial\theta}\right) = E_\xi\left\{\sum_s \frac{\partial h(d,\theta)}{\partial\theta}p(s)\right\} = E_\xi\left\{\frac{\partial}{\partial\theta}E_p\left(\sum_s h(d,\theta)\right)\right\} =$$
$$E_\xi\left(\frac{\partial}{\partial\theta}\sum_{i=1}^{N}\phi_i\right)$$

Since $E_\xi\left(\frac{\partial}{\partial\theta}\sum_{i=1}^{N}\phi_i\right)$ is independent of h, our problem is to find an $h(d,\theta)$, which minimizes $E_\xi E_p\{h^2(d,\theta)\}$ subject to $E_p[h(d,\theta)] = \sum_{i=1}^{N}\phi_i(y_i,\theta)$. The following theorem (Godambe and Thompson, 1986a,b) establishes optimality of the EF $h^*(d,\theta)$ given in Eq. (22.4.2).

Theorem 22.4.1

For the superpopulation model $C = \{\xi\}$ where $\phi_i(y_i,\theta)$ are independently distributed with $E_\xi\{\phi_i(y_i,\theta(\xi))\} = 0$ for $i = 1,\ldots, N$, the EF $h^*(d,\theta) = \sum_{i\in s}\frac{\phi_i(y_i,\theta)}{\pi_i}$ is the optimal (minimizes Eq. 22.4.3) in the class of unbiased EFs C_u (satisfying Eq. 22.4.1).

Proof

Let

$$Q(d,\theta) = h(d,\theta) + h^*(d,\theta) \tag{22.4.5}$$

The unbiasedness condition (22.4.1) yields

$$E_p[Q(d,\theta)] = \sum_s Q(d,\theta)p(s) = 0 \tag{22.4.6}$$

Furthermore,

$$E_p[h(d,\theta)]^2 = E_p[h^*(d,\theta)]^2 + E_p[Q(d,\theta)^2 + 2E_p[h^*(d,\theta)Q(d,\theta)] \tag{22.4.7}$$

and

$$E_\xi E_p[h^*(d,\theta)Q(d,\theta)] = E_\xi E_p\left[Q(d,\theta)\sum_{i\in s}\frac{\phi_i(y_i,\theta)}{\pi_i}\right]$$

$$= E_\xi \sum_s \left[Q(d,\theta)\sum_{i\in s}\frac{\phi_i(y_i,\theta)}{\pi_i}\right]p(s)$$

$$= E_\xi \sum_{i=1}^{N}\frac{\phi_i(y_i,\theta)}{\pi_i}\sum_{s\supset i} Q(d,\theta)p(s)$$

$$= -E_\xi \sum_{i=1}^{N}\frac{\phi_i(y_i,\theta)}{\pi_i}\sum_{\bar{s}_i} Q(d,\theta)p(s)$$

(noting $E_p[Q(d,\theta)] = \sum_{s\supset i} Q(d,\theta)p(s) + \sum_{\bar{s}_i} Q(d,\theta)p(s) = 0$, where $\sum_{\bar{s}_i}$ denotes sum over the samples that do not contain i)

Since $\phi_i(y_i,\theta)$ are independent and $E_\xi\{\phi_i(y_i,\theta)\} = 0$, we have

$$E_\xi E_p[h^*(d,\theta)Q(d,\theta)] = -\sum_{i=1}^{N} E_\xi\frac{\phi_i(y_i,\theta)}{\pi_i}\sum_{\bar{s}_i} E_\xi[Q(d,\theta)]p(s) = 0 \tag{22.4.8}$$

Finally, Eqs. (22.4.7) and (22.4.8) yield

$$E_p[h(d,\theta)]^2 \geq E_p[h^*(d,\theta)]^2$$

Hence the theorem.

For a fixed effective sample size n[FES(n)] design, $\sum_{i=1}^{N}\pi_i = n$ and the expression

$$E_\xi E_p[h^*(d,\theta)]^2 = \sum_s\sum_{i\in s}\frac{E_\xi[\phi_i(y_i,\theta)]^2}{\pi_i^2}p(s) = \sum_{i=1}^{N}\frac{E_\xi[\phi_i(y_i,\theta)]^2}{\pi_i}$$

is minimized when $\pi_i \propto \sqrt{E_\xi\phi_i^2(y_i,\theta)}$ and the minimum value of $E_\xi E_p[h^*(d,\theta)]^2$ is

$$\left(\sum_{i=1}^{N}\sqrt{E_\xi[\phi_i(y_i,\theta)]^2}\right)^2 \Big/ n \tag{22.4.9}$$

Thus we have the following as a corollary of Theorem 22.3.1.

Corollary 22.4.1

For the model ξ where $\phi_i(y_i,\theta)$ are independently distributed with $E_\xi\{\phi_i(y_i,\theta)\} = 0$, $i = 1,\ldots, N$, the sampling design p_0 with $\pi_i = \pi_{i0} = n\sqrt{E_\xi\left(\phi_i^2(y_i,\theta)\right)} \Big/ \sum_{i=1}^{N}\sqrt{E_\xi\left(\phi_i^2(y_i,\theta)\right)}$, is the optimal in the class $\mathcal{P}_n$ of FES(n) designs in the sense

$$E_\xi E_p(h(d,\theta))^2 \geq E_\xi E_{p_0}\left(\sum_{i\in s}\frac{\phi_i(y_i,\theta)}{\pi_{i0}}\right)^2$$

$$= \left(\sum_{i=1}^{N}\sqrt{E_\xi[\phi_i(y_i,\theta)]^2}\right)^2 \Big/ n \;\; \forall p \in \mathcal{P}_n$$

Remark 22.4.1

The estimator $\widehat{\theta}_s$ of the survey parameter θ_N obtained from the optimal unbiased EF $h^*(d,\theta) = 0$ is called an optimal estimator. The estimator $\widehat{\theta}_s$ may not be design unbiased for θ_N. The optimality of $h^*(d,\theta)$ is independent of the variance structure $E_\xi\{\phi_i(y,\theta)\}^2$.

Example 22.4.1

Consider the superpopulation model

$$y_i = \theta(\xi)x_i + \epsilon_i \tag{22.4.10}$$

where ϵ_i's are independent with $E_\xi(\epsilon_i) = 0$, $V_\xi(\epsilon_i) = \sigma_i^2$. The constants x_i's and σ_i^2's are positive and known. Since $\phi_i(y_i,\theta) = \dfrac{(y_i - \theta x_i)x_i}{\sigma_i^2}$ satisfies the condition (22.3.4), the linearly optimal EF is

$$g_{l0}(\mathbf{y},\theta) = \sum_{i=1}^{N}\frac{x_i(y_i - \theta x_i)}{\sigma_i^2}$$

as $\phi_i(y_i,\theta) = \dfrac{(y_i - \theta x_i)x_i}{\sigma_i^2}$ satisfies the condition (22.3.4). In this case the survey parameter is

$$\theta_N = \frac{\sum_{i=1}^{N} y_i x_i/\sigma_i^2}{\sum_{i=1}^{N} x_i^2/\sigma_i^2} \tag{22.4.11}$$

The survey estimator of θ_N is obtained from the optimum sample estimating equation $\sum_{i\in s}\dfrac{x_i(y_i - \theta x_i)}{\pi_i\sigma_i^2} = 0$ and it is given by

$$\widehat{\theta}_s = \frac{\sum_{i\in s} y_i x_i/\left(\pi_i\sigma_i^2\right)}{\sum_{i\in s} x_i^2/\left(\pi_i\sigma_i^2\right)} \tag{22.4.12}$$

In this case the inclusion probability for the optimum sampling design p_0 is given by

$$\pi_{i0} = \frac{nx_i/\sigma_i}{\sum_{i=1}^{N} x_i/\sigma_i} \tag{22.4.13}$$

For practical importance, consider the following special cases:

Case 1: $\sigma_i^2 \propto x_i$ yields $\theta_N = \sum_{i=1}^{N} y_i \Big/ \sum_{i=1}^{N} x_i$, $\widehat{\theta}_s = \dfrac{\sum_{i\in s} y_i/\pi_i}{\sum_{i\in s} x_i/\pi_i}$, and $\pi_{i0} = \dfrac{n\sqrt{x_i}}{\sum_{i=1}^{N}\sqrt{x_i}}$.

Case 2: $\sigma_i^2 \propto x_i^2$ yields $\theta_N = \dfrac{1}{N}\sum_{i=1}^{N}\dfrac{y_i}{x_i}$, $\widehat{\theta}_s = \dfrac{\sum_{i\in s} y_i/(\pi_i x_i)}{\sum_{i\in s} 1/\pi_i}$, and $\pi_{i0} = n/N$.

Case 3: $\sigma_i^2 = \sigma^2$ and $x_i = 1$ for $i = 1,\ldots, N$ yield $\theta_N = \overline{Y}$, $\widehat{\theta}_s = \dfrac{\sum_{i\in s} y_i/\pi_i}{\sum_{i\in s} 1/\pi_i}$, and $\pi_{i0} = n/N$. The estimator $\widehat{\theta}_s = \dfrac{\sum_{i\in s} y_i/\pi_i}{\sum_{i\in s} 1/\pi_i}$ was proposed by Brewer (1963a,b) and Hájek (1971).

Example 22.4.2

Consider the model

$$y_i = \theta + \epsilon_i \tag{23.3.14}$$

where ϵ_i's are independent with $E_\xi(\epsilon_i) = 0$ and $V_\xi(\epsilon_i) = \sigma_i^2$.

Substituting $x_1 = \cdots = x_N = 1$, in Example 23.4.1, we get

$$\theta_N = \frac{\sum_{i=1}^{N} y_i/\sigma_i^2}{\sum_{i=1}^{N} 1/\sigma_i^2},\quad \widehat{\theta}_s = \frac{\sum_{i\in s} y_i/\left(\pi_i\sigma_i^2\right)}{\sum_{i\in s} 1/\left(\pi_i\sigma_i^2\right)},\ \text{and}\ \pi_{i0} = \frac{n/\sigma_i}{\sum_{i=1}^{N} 1/\sigma_i}.$$

Example 22.4.3

Consider the multiple regression model given in Mantel (1991), where

(i) y_i's are independent.

(ii) $E_\xi(y_i) = \mathbf{x}_i\boldsymbol{\beta}$, where $\mathbf{x}_i = (x_{i1},\ldots, x_{ip})$, is a known vector of auxiliary information and $\boldsymbol{\beta}' = \left(\beta_1, \ldots, \beta_p\right)$ is a vector of unknown parameters.

(iii) $V_\xi(y_i) = \sigma^2 v_i$, where v_i is known but σ^2 is unknown.

Following Theorem 22.3.1, the linearly optimal estimating equation comes out as

$$g_{l0}(\mathbf{y},\theta)=\sum_{i=1}^{N}(y_i-\mathbf{x}_i\boldsymbol{\beta})\mathbf{x}_i'/v_i=0 \qquad (22.4.15)$$

Eq. (22.4.15) yields the survey parameter as

$$\boldsymbol{\beta}_{\mathbf{N}}=\left(\mathbf{X}_N'\mathbf{V}_N^{-1}\mathbf{X}_N\right)^{-1}\mathbf{X}_N'\mathbf{V}_N^{-1}\mathbf{y}_N \qquad (22.4.16)$$

where $\mathbf{y}_N'=(y_1,\ldots,y_N)$ and $\mathbf{V}_N=diag(v_1,\ldots,v_N)$; $\mathbf{X}_N$ is a matrix of order $N\times p$ with ith row, $\mathbf{x}_i$.

The survey estimator of $\boldsymbol{\beta}_N$ is found from the optimum sample estimating equation

$$h(d,\theta)=\sum_{i\in s}(y_i-\mathbf{x}_i\boldsymbol{\beta})\mathbf{x}_i'/(\pi_i v_i)=0 \qquad (22.4.17)$$

to be

$$\widehat{\boldsymbol{\beta}}_s=\left(\mathbf{X}_s'\boldsymbol{\Pi}_s^{-1}\mathbf{V}_s^{-1}\mathbf{X}_s\right)^{-1}\mathbf{X}_s'\boldsymbol{\Pi}_s^{-1}\mathbf{V}_s^{-1}\mathbf{y}_s \qquad (22.4.18)$$

where $\mathbf{y}_s$ is the vector of y_i for $i\in s$; $\boldsymbol{\Pi}_s$ and $\mathbf{V}_s$ are diagonal matrices with ith diagonal element π_i and v_i, respectively, for $i\in s$; and $\mathbf{X}_s$ is a matrix with ith row $\mathbf{x}_i$ for $i\in s$.

For estimating the finite population mean $\overline{Y}_N=\sum_{i=1}^{N}y_i/N$, we may consider the following generalized regression (*greg*) predictor proposed by Cassel et al. (1977) as

$$\overline{y}_{greg}=\overline{\mathbf{X}}_N'\widehat{\boldsymbol{\beta}}_s+\mathbf{1}_s'\boldsymbol{\Pi}_s^{-1}\widehat{\mathbf{e}}_s/N \qquad (22.4.19)$$

where $\widehat{\mathbf{e}}_s=\mathbf{y}_s-\mathbf{X}_s\widehat{\boldsymbol{\beta}}_s$ and I_s' is a vector of 1s whose length is the size of the sample s.

Mantel (1991) proposed an alternative estimator (*areg*) for $\overline{Y}_N$ as

$$\overline{y}_{areg}=\overline{\mathbf{X}}_N'\widehat{\boldsymbol{\beta}}_s+(c_1/c_2)\mathbf{1}_s'\boldsymbol{\Pi}_s^{-1}\widehat{\mathbf{e}}_s/N \qquad (22.4.20)$$

where $c_1=\mathbf{I}_N'\left(\mathbf{V}_N\mathbf{I}_N-\mathbf{X}_N\left(\mathbf{X}_s'\mathbf{V}_s^{-1}\boldsymbol{\Pi}_s^{-1}\mathbf{X}_s\right)^{-1}\mathbf{X}_s'\boldsymbol{\Pi}_s^{-1}\mathbf{I}_s\right)$,

$c_2=\mathbf{I}_s'\boldsymbol{\Pi}_s^{-1}\left(\mathbf{V}_s\mathbf{I}_s-\mathbf{X}_s\left(\mathbf{X}_s'\mathbf{V}_s^{-1}\boldsymbol{\Pi}_s^{-1}\mathbf{X}_s\right)^{-1}\mathbf{X}_s'\boldsymbol{\Pi}_s^{-1}\mathbf{I}_s\right)$, and $\mathbf{I}_N$ is a column vector of 1s.

The estimator $\overline{y}_{areg}$ is very similar to $\overline{y}_{greg}$ except for an adjusted weight c_1/c_2 in the second part. The estimator $\overline{y}_{areg}$ is more design based whereas $\overline{y}_{areg}$ is more model based. However, both the estimators are design consistent. Further details are given by Mantel (1991).

22.5 INTERVAL ESTIMATION

In survey sampling, the confidence interval of a parameter θ is obtained by inverting the pivotal $\left(\widehat{\theta}-\theta\right)\Big/\sqrt{\widehat{V}\left(\widehat{\theta}\right)}$, where $\widehat{\theta}$ is a suitable estimator of θ and $\widehat{V}\left(\widehat{\theta}\right)$ is an estimated variance of $\widehat{\theta}$. Godambe and Thompson (1999) provided a more direct method of constructing a pivotal quantity to find the confidence interval. This alternative method provides better confidence intervals than the conventional method. In this section we will mainly discuss the methods of determination of confidence intervals cited by Godambe (1991) and Godambe and Thompson (1999).

22.5.1 Confidence Interval for θ

Let $\phi_i(y_i,\theta)$ be independent elementary EF with

$$E_\xi[\phi_i(y_i,\theta)]=0 \text{ and } V_\xi[\phi_i(y_i,\theta)]=\sigma_i^2 \quad \text{for } i\in s \tag{22.5.1}$$

and $T(s,\theta)=\sum_{i\in s} g_i(y,\theta)$ be an unbiased sample EF, which satisfies

$$E_p\{T(s,\theta)\}=\sum_{i=1}^{N}\phi_i(y_i,\theta)$$

The EF $T(s,\theta)$ may or may not be the optimum EF for estimating θ. An approximate $100(1-\alpha)\%$ confidence interval for θ under the super-population model (22.5.1) can be obtained by inverting

$$\left|\frac{\sum_{i\in s} g_i(y_i,\theta)}{\sqrt{\sum_{i\in s} g_i^2(y_i,\theta)}}\right|=z_{1-\alpha/2} \tag{22.5.2}$$

where $z_{1-\alpha/2}$ is the $(1-\alpha/2)$ quantile of the $N(0,1)$ distribution.

22.5.1.1 *Confidence Interval for Survey Parameter θ_N*

Let θ_N be the survey parameter obtained by solving the estimating equation $\sum_{i=1}^{N}\phi_i(y_i,\theta)=0$. The confidence interval of θ_N can be obtained using any of the following methods.

Method 1: Confidence interval of θ_N is obtained by inverting

$$\left|\frac{\sum_{i\in s} g_i(y_i,\theta_N)}{\sqrt{\widehat{V}_\xi\{T(s,\theta_N)\}}}\right|=z_{1-\alpha/2} \tag{22.5.3}$$

where $\widehat{V}_{\xi}\{T(s,\theta)\}$ is a model-unbiased estimator of the model variance $V_{\xi}\{T(s,\theta)\}$, i.e., $E_{\xi}\left[\widehat{V}_{\xi}\{T(s,\theta)\}\right] = V_{\xi}\{T(s,\theta)\}$.

Method 2: By inverting

$$\left|\frac{\sum_{i\in s} g_i(y_i,\theta_N)}{\sqrt{\widehat{V}_p\{T(s,\theta_N)\}}}\right| = z_{1-\alpha/2} \tag{22.5.4}$$

where $\widehat{V}_p\{T(s,\theta_N)\}$ is a design-unbiased estimator of the design variance $V_p\{T(s,\theta_N)\}$, i.e.,

$$E_p\left[\widehat{V}_p\{T(s,\theta)\}\right] = E_p\left\{\sum_{i\in s} g_i(y_i,\theta) - \sum_{i=1}^{N}\phi_i(y_i,\theta)\right\}^2.$$

The confidence intervals derived from the two methods above are expected to be close to one another if the chosen sampling design and superpopulation model are appropriate.

Example 22.5.1 (Godambe and Thompson, 2009)

Suppose that $y_1,\ldots, y_N$ are independent with $E_{\xi}(y_i) = \theta x_i$ and $V_{\xi}(y_i) = \sigma^2$. Let $\phi_i = y_i - \theta x_i$, then the estimating equation $\sum_{i=1}^{N}\phi_i(y_i,\theta) = 0$ gives $\theta = \theta_N = Y/X = R$, the population ratio. Let a sample of size n be selected by the simple random sampling without replacement (SRSWOR) method. Consider a sample estimating equation $T(s,R) = \sum_{i\in s}(y_i - Rx_i)$, which yields $\widehat{R}_s = \sum_{i\in s} y_i\Big/\sum_{i\in s} x_i$ as an estimator of R. The design variance of $T(s,R)$ is $V_p[T(s,R)] = n^2\left(\frac{1}{n} - \frac{1}{N}\right)\frac{1}{N-1}\sum_{i=1}^{N}\left(z_i - \overline{Z}\right)^2$, where $z_i = y_i - Rx_i$. An approximate design-unbiased estimator for $V_p[T(s,R)]$ is $\widehat{V}_p\left[T\left(s,\widehat{R}_s\right)\right] = n^2\left(\frac{1}{n} - \frac{1}{N}\right)\frac{1}{n-1}\sum_{i\in s}\widehat{z}_i^2$, where $\widehat{z}_i = y_i - \widehat{R}_s x_i$. The $100(1-\alpha)\%$ confidence interval of R is obtained from the equation

$$\frac{\sum_{i\in s}(y_i - Rx_i)}{\sqrt{\left(1-\frac{n}{N}\right)\frac{n}{n-1}\sum_{i\in s}\left(y_j - \widehat{R}_s x_j\right)^2}} = \pm\, z_{1-\alpha/2} \tag{22.5.5}$$

Eq. (22.5.5) yields an approximate confidence interval for R as

$$\widehat{R}_s \pm z_{1-\alpha/2}\sqrt{\left(1-\frac{n}{N}\right)\frac{1}{n}\left(s_y^2 - 2\widehat{R}_s s_{xy} + \widehat{R}_s^2 s_x^2\right)}\Big/\overline{x}_s \tag{22.5.6}$$

where $(n-1)s_x^2 = \sum_{i\in s}(x_i - \overline{x}_s)^2$, $(n-1)s_y^2 = \sum_{i\in s}(y_i - \overline{y}_s)^2$ and $(n-1)s_{xy} = \sum_{i\in s}(x_i - \overline{x}_s)(y_i - \overline{y}_s)$.

22.5.1.2 Stratified Sampling

Consider a population U of N units classified into H strata and let N_h be the size of the hth stratum. From the hth stratum, a sample s_h of size n_h is selected by SRSWOR method. Let y_{hi} be the value of the study variable y of the ith unit of the hth stratum $i = 1,\ldots, N_h$; $h = 1,\ldots, H$. Suppose that $y_{hi}'s$ are distributed independently with

$$E_\xi(y_{hj}) = \theta \text{ and } V_\xi(y_{hj}) = \sigma_h^2 \quad \text{for } i = 1,\ldots,N_h; h = 1,\ldots,H \tag{22.5.7}$$

Consider the elementary EF $\phi_{hj}(y_{ij},\theta) = y_{hj} - \theta$, which yields $\theta_N = \overline{Y} = \sum_{h=1}^{H}\sum_{j=1}^{N_h} y_{hj}/N$ as the solution of the estimating equation $\frac{1}{N}\sum_{h=1}^{H}\sum_{j=1}^{N_h}\phi_{hj}(y_{hj},\theta) = 0$. Noting that the inclusion probability for the jth unit of the hth stratum under SRSWOR sampling is n_h/N_h, we find the optimal sample EF is

$$T = \sum_{h=1}^{H}\frac{W_h}{n_h}\sum_{j\in s_h}\phi_{hj}\left(y_{hj},\overline{Y}\right) = \sum_{h=1}^{H}\frac{W_h}{n_h}\sum_{j\in s_h}\left(y_{hj} - \overline{Y}\right) \tag{22.5.8}$$

where $W_h = N_h/N$.

Thus the optimal estimate of $\overline{Y}$ is

$$\widehat{\theta}_s = \sum_{h=1}^{H} W_h\overline{y}_h = \overline{y}_{st}$$

where $\overline{y}_h = \sum_{j\in s_h} y_{hj}/n_h$.

The design variance of T is

$$V_p(T) = \sum_{h=1}^{H} W_h^2(1-f_h)S_{hi}^2/n_h \tag{22.5.9}$$

where $f_h = n_h/N_h$, $S_{hi}^2 = \sum_{j=1}^{N_h}\left(y_{hj} - \overline{Y}_h\right)^2/(N_h - 1)$ and $\overline{Y}_h = \sum_{j=1}^{N_h} y_{hj}/N_h$.

The model $E_\xi(y_{ij}) = \theta$ suggests that the strata means $\overline{Y}_h$ should be approximately equal to $\overline{Y}$ when the strata sizes N_h are large. Hence, we find a design-unbiased estimate of $V_p(T)$ as

$$\widehat{V}_p(T) = \sum_{h=1}^{H} W_h^2 \frac{(1-f_h)N_h}{n_h(N_h-1)} \sum_{j \in s_h} \left(y_{hj} - \overline{Y}\right)^2 \Big/ n_h \qquad (22.5.10)$$

$$= \widehat{V}_0 + Q$$

where $\widehat{V}_0 = \sum_{h=1}^{H} W_h^2(1-f_h)s_h^2/n_h$, $s_h^2 = \sum_{j \in s_h}\left(y_{hj} - \overline{y}_h\right)^2 \Big/ (n_h - 1)$ and Q is of the order $O\left(1/n_h^2\right)$. When n_h's are large for every h, we can neglect Q and assume $T\Big/\sqrt{\widehat{V}_0} \sim N(0,1)$ and find the confidence interval of $\overline{Y}$ by inverting $T\Big/\sqrt{\widehat{V}_0}$. On the other hand if all n_h's are not large, we can find the confidence interval of $\overline{Y}$ by inverting the distribution of $T\Big/\sqrt{\widehat{V}_p(T)}$, which is asymptotically $N(0,1)$.

The model variance $V_\xi(T)$ can by estimated by

$$\widehat{V}_\xi(T) = \sum_{h=1}^{H} \frac{W_h^2}{n_h^2} \sum_{j \in s_h} \left(y_{hj} - \overline{Y}\right)^2 = \widehat{V}_\xi \qquad (22.5.11)$$

Hence the confidence interval of $\overline{Y}$ can be obtained by inverting $T\Big/\sqrt{\widehat{V}_\xi}$, which is asymptotically $N(0,1)$. Consider the superpopulation model

$$E_\xi(y_{hj}) = \theta x_{hj} \text{ and } V_\xi(y_{hj}) = \sigma_h^2 \text{ for } j = 1, \ldots, N_h; h = 1, \ldots, H \qquad (22.5.12)$$

Here the EF

$$\frac{1}{N} \sum_{h=1}^{H} \sum_{j=1}^{N_h} (y_{hj} - \theta x_{hj}) \qquad (22.5.13)$$

yields the population ratio as the survey parameter, which is given by

$$\theta_N = \frac{\sum_{h=1}^{H}\sum_{j=1}^{N_h} y_{hj}}{\sum_{h=1}^{H}\sum_{j=1}^{N_h} x_{hj}} = \frac{\sum_{h=1}^{H} Y_h}{\sum_{h=1}^{H} X_h} = \frac{Y}{Y} = R \qquad (22.5.14)$$

where $Y = \sum_{h=1}^{H} Y_h$, $Y_h = \sum_{j=1}^{N_h} y_{hj}$, $X = \sum_{h=1}^{H} X_h$, and $X_h = \sum_{j=1}^{N_h} x_{hj}$.

The optimal sample EF for estimating R is

$$T=\sum_{h=1}^{H}\frac{W_h}{n_h}\sum_{j=1}^{n_h}(y_{hj}-R\,x_{hj}) \tag{22.5.15}$$

The EF (22.5.15) yields the estimate of R as. $\widehat{R}_s=\sum_{h=1}^{H}W_h\overline{y}_h\Big/\sum_{h=1}^{H}W_h\overline{x}_h$

For the model (22.5.12) with large strata sizes N_h, we may assume that $R=Y_h/X_h$ and neglect the differences Y_h-RX_h. In this case an approximate estimator of the design variance of T is given by

$$\widehat{V}_p(T)=\sum_{h=1}^{H}\frac{W_h^2(1-f_h)}{n_h}\frac{N_h}{N_h-1}\frac{1}{n_h}\sum_{j\in s_h}(y_{hj}-Rx_{hj})^2 \tag{22.5.16}$$

Finally the confidence interval of R is obtained by inversion of the distribution of $T/\sqrt{\widehat{V}_p(T)}$, which is asymptotically $N(0,1)$.

22.5.1.3 Confidence Intervals for Quantiles

Consider the stratified population as described in Section 22.5.1.2 above. Here we define

$$\delta(y,\theta)=\begin{cases}1 & \text{if } \; y\le\theta\\ 0 & \text{otherwise}\end{cases}$$

The pth quantile of the survey population θ_N is a solution of the equation

$$F=\sum_{h=1}^{H}\sum_{j=1}^{N_h}\{\delta(y_{hj},\theta)-p\}=0 \tag{22.5.17}$$

The survey estimate $\widehat{\theta}_s$ of θ_N based on the stratified sample is obtained by equating the sample optimum EF

$$\widehat{F}=\sum_{h=1}^{H}\frac{N_h}{n_h}\sum_{j\in s_h}\{\delta(y_{hj},\theta)-p\} \tag{22.5.18}$$

to zero.

The expectation of $\widehat{F}$ is zero for $\theta=\theta_N$. The estimated variance of $\widehat{F}$ is

$$\widehat{V}\left(\widehat{F}\right)=\sum_{h=1}^{H}N_h^2(1-f_h)\frac{\alpha_h(1-\alpha_h)}{n_h-1} \tag{22.5.19}$$

where $\alpha_h = \frac{1}{n_h}\sum_{j\in s_h}\delta(y_{hj}, \theta_N)$ is the proportion of y_{hj}'s that is less than equal to the pth quantile θ_N in the selected sample s_h of the hth stratum. Since θ_N is unknown, we replace θ_N by $\widehat{\theta}_s$ in Eq. (22.5.19) for computation $\widehat{V}(\widehat{F})$. Hence the confidence interval of θ_N is obtain by inversion of the distribution

$$\widehat{F}_{\theta=\theta_N}\Big/\widehat{V}(\widehat{F})_{\theta=\hat{\theta}_s} \tag{22.5.20}$$

which is asymptotically $N(0,1)$.

The confidence interval for the $p-th$ quantile was derived by Woodruff (1952) while Godambe and Thompson (1999) considered Eq. (22.5.20) from the angle of the EF. A lucid presentation was provided by Godambe (1991). The main feature of the derivation of this confidence interval is that it is not obtained from the traditional method of inverting the distribution of the estimator $\widehat{\theta}_s$ but it was derived from the distribution of $\widehat{F}$.

22.6 NONRESPONSE

Let a sample s of size n be selected from a finite population U with probability $p(s)$ using a sampling design p and let $\pi_i(>0)$ be the inclusion probability of the ith unit. Suppose $r(\geq 0)$ of the selected n units responded and the remaining $n-r$ did not response. Let the set of response sample be denoted by $s'(\subset s)$. Here the data in hand are $d(s, s') = \{(s, s') : (i, y_i), i\in s'\}$ with y_i as the value of the study variable y for the ith unit. Let us assume that the response probability of the unit i is $q_i(>0)$ and it is known for $i = 1,\ldots, N$.

Consider the superpopulation model where $y_1,\ldots, y_N$ are independent with

$$E_\xi(y_i) = \theta x_i, \theta\in\Omega_\theta \text{ for } i = 1,\ldots, N. \tag{22.6.1}$$

Suppose we are interested in estimating the survey parameter $\theta_N = \sum_{i=1}^{N}\alpha_i y_i\Big/\sum_{i=1}^{N}\alpha_i x_i$ from the estimating equation

$$g = \sum_{i=1}^{N}(y_i-\theta x_i)\alpha_i = 0 \tag{22.6.2}$$

where x_i and α_i are known constants.

We can estimate θ_N using the following two approaches:

Approach 1: Let $\mathcal{H}^{(1)}$ be the class of unbiased estimating functions (UEFs), which comprises of EFs h_1 based on the data $d(s, s')$ satisfying

$$E_pE_R(h_1) = \sum_{i=1}^{N}(y_i-\theta x_i)\alpha_i \quad \forall \mathbf{y} = (y_1,\ldots, y_N) \text{ and } \theta\in\Omega_\theta \tag{22.6.3}$$

where E_R denotes expectation over the response mechanism R.

The optimum EF in the class $\mathcal{H}^{(1)}$ is one which $E_\xi E_p E_R\left(h_1^2\right)$ minimizes for $h_1 \in H^1$. Since the inclusion probability for the ith unit in s' is $\pi_i q_i$, the optimal EF is obtained from Theorem 22.3.1 as

$$\widehat{h}_1^* = \sum_{i\in s'} \frac{(y_i - \theta x_i)\alpha_i}{\pi_i q_i} \tag{22.6.4}$$

Approach 2: In case 100% response was available, i.e., $r = n$, then the optimum UEF based on the data $d(s) = \{s:(i,y_i), i \in s\}$ for estimating θ_N would be

$$h^*(s) = \sum_{i\in s} \frac{(y_i - \theta x_i)\alpha_i}{\pi_i} \tag{22.6.5}$$

Since $h^*(s)$ is unknown, we consider the class $\mathcal{H}^{(2)}$ UEFs $h_2(s, s')$ satisfying

$$E_p[h_2(s, s') - h^*(s)|s] = 0 \quad \forall \mathbf{y} \text{ and } \theta \in \Omega_\theta \tag{22.6.6}$$

The optimum UEFs is defined by $\widehat{h}_2^*$ that satisfies

$$E_\xi E_p\left(\widehat{h}_2^*\right)^2 \le E_\xi E_p\left(h_2^2\right) \quad \forall \mathrm{h}_2 \in \mathcal{H}^{(2)} \text{ and } \theta \in \Omega_\theta$$

The optimal EF in the class $\mathcal{H}^{(2)}$ is obtained from Theorem 22.4.1 as

$$\widehat{h}_2^* = \sum_{i\in s'} \frac{(y_i - \theta x_i)\alpha_i}{\pi_i q_i} \tag{22.6.7}$$

Thus the two approaches lead to the same optimal EF $\widehat{h}_1^* = \widehat{h}_2^* = \sum_{i\in s'} \frac{(y_i - \theta x_i)\alpha_i}{\pi_i q_i} = \widehat{h}^*$.

The following theorems of Godambe and Thompson (1986a,b, 1987) summarize the discussions above.

Theorem 22.6.1

For the superpopulation model (Eq. (22.6.1)) the EF $\widehat{h}^* = \sum_{i\in s'} \frac{(y_i - \theta x_i)\alpha_i}{\pi_i q_i}$ is optimal in the class of unbiased EFs $\mathcal{H}^{(1)}$.

Theorem 22.6.2

Let $\mathcal{H}^{(2)}$ be a subclass of $\mathcal{H}^{(1)}$ for which $h_2(s, s')$ depends only on s'. Then the EF $\widehat{h}^*$ given in Eq. (22.6.7) is also optimum in the subclass $\mathcal{H}^{(2)}$.

Example 22.6.1

Let us assume $E_\xi(y_i - \theta x_i)^2 = \sigma^2 v(x_i)$, where $v(x_i)$ is a known function of x_i and σ^2 is positive but unknown. In this case

$$E_\xi E_p\left(\widehat{h}*\right)^2 = \sigma^2 \sum_{i=1}^{N} \frac{\alpha_i^2 v(x_i)}{\pi_i q_i} \tag{22.6.8}$$

The optimum values of the π_i's, which minimize (25.5.8) are obtained under the following two constraints.

In case $\sum_{i=1}^{N} \pi_i = n =$ expected sample size $= E_p|s|$ is kept fixed, the optimum value of π_i becomes

$$\pi_{i0} = n \frac{\alpha_i \sqrt{v(x_i)/q_i}}{\sum_{i=1}^{N} \alpha_i \sqrt{v(x_i)/q_i}} \tag{22.6.9}$$

On the other hand in case $\sum_{i=1}^{N} \pi_i q_i = n^* =$ expected sample size of s' is kept fixed, the optimum value of π_i is obtained as

$$\pi_{i0}^* = n^* \frac{\left(\alpha_i \sqrt{v(x_i)}\right)/q_i}{\sum_{i=1}^{N} \alpha_i \sqrt{v(x_i)}} \tag{22.6.10}$$

22.7 CONCLUDING REMARKS

The concept of the EF was developed by Godambe (1960a,b) and Durbin (1960). It unifies the principal methods in the theory of estimation, namely, the method of LS, ML, minimum chi-square, and the method of minimum variance unbiased estimation. The EFs provide suitable solutions when the least square theory and ML methods fail to give reasonable solutions. Godambe and Thompson (1986a,b) provided a unified approach of finding optimal EFs in presence of a superpopulation model. They also provided methods of determination of confidence intervals through EFs. The theory of estimation functions has been applied successfully in various fields of estimation problems such as survey sampling, biostatistics, econometrics, time series, and stochastic processes. Interested readers are referred to Kalbfleisch and Lawless (1988), Godambe (1985, 1991), Godambe and Thompson (1986a,b), Heyde and Lin (1991), Vijayan (1991) and Baswa (2000), among others.

22.8 EXERCISES

22.8.1 Find the optimal EF for θ in the class of linear EFs when $y_1,\ldots,y_N$ are independent and follow the following superpopulation models. In each case, obtain the survey parameter θ_N and its sample estimate $\widehat{\theta}_s$. (i) $y_i=\theta\log x_i+\epsilon_i$, (ii) $y_i=\theta/x_i+\epsilon_i$, and (iii) $y_i=\theta x_i+\epsilon_i$.

$$E_\xi(\epsilon_i)=0,\ V_\xi(\epsilon_i)=\sigma_i^2,$$

22.8.2 Let $y_1,\ldots,y_N$ be independent with $E_\xi(y_i)=\theta x_i$, $V_\xi(y_i)=\sigma^2x_i^g, g\geq 0$. Find θ_N, the optimal estimator of the superpopulation parameter θ and also optimal sample estimator of θ_N when $g=0$, 1 and 2.

22.8.3 Consider the model $y_i=\theta x_i+\epsilon_i$ with $\epsilon_i>0$, ϵ_i's are independent with $E_m(\epsilon_i)=0$ and $V_m(\epsilon_i)=\sigma_i^2$. Find the survey parameter θ_N and sample estimate $\widehat{\theta}_s$ of θ_N. Find also the optimum sampling design when (i) $\sigma_i^2\propto x_i$ and (ii) $\sigma_i^2\propto x_i^2$.

22.8.4 Consider the superpopulation model $y_i=\alpha_i(\theta)+\epsilon_i$, $i=1,\ldots,N$, ϵ_i's are independent with mean zero and variance $V(\epsilon_i)=\sigma^2x_i$

and x_i's are known positive constants. Show that the EF $g^*(\mathbf{y},\theta)=$

$a(\theta)\sum_{i=1}^{N}\left(\frac{\partial\alpha_i(\theta)}{\partial\theta}\right)\left(\frac{y_i-\alpha_i(\theta)}{x_i}\right)$ is optimal in the class of unbiased

EFs of the form $g(\mathbf{y},\theta)=\sum_{i=1}^{N}a_i(\theta)(y_i-\alpha_i(\theta))$ (Vijayan, 1991).

22.8.5 Let $y_1,\ldots,y_N$ be independent with $E_\xi(y_i)=\theta x_i$, $V_\xi(y_i)=\sigma^2x_i^g, g\geq 0$. Find θ_N, the optimal estimator of the superpopulation parameter θ and also optimal sample estimator of θ_N when $g=0$, 1, and 2.

CHAPTER 23

Estimation of Distribution Functions and Quantiles

23.1 INTRODUCTION

In earlier sections, we have considered various methods of estimating the finite population totals, means, and ratios of two variables. Most of these methods, when extended to the estimation of distribution function, yield unsatisfactory results because these estimators may not satisfy the basic properties of the distribution functions. Estimation of distribution functions from the survey data is often a part of the objectives of a survey. In particular, it may be of interest to planners who would want to know the proportion of people living below the poverty line, unemployed, or money spent on education. Furthermore, it is well known that the median is considered to be a more appropriate measure of location than the mean of a skewed distribution such as income. It is also important to estimate the income inequality through the Gini coefficient, which is a function of a distribution function. In this section, different methods of estimation of distribution function have been considered. Estimation of quantiles and medians will also be obtained from the inversion of the estimates of distribution functions.

23.2 ESTIMATION OF DISTRIBUTION FUNCTIONS

The distribution function of a random variable X is defined as $F(x) = P(X \leq x)$. The distribution function $F(x)$ has the following properties:

$$\text{(i) } F(x) \geq 0, \text{ (ii) nondecreasing, (iii) } F(-\infty) = 0, \text{ (iv) } F(+\infty) = 1, \text{ and (v) } F(x) \text{ is right continuous} \tag{23.2.1}$$

For a finite population of N units, we define the distribution function $F(t)$ of the study variable y, which is the proportion of the units of the population U whose values are less than or equal to t, i.e.,

$$F(t) = \frac{1}{N}\sum_{i=1}^{N} I(y_i \leq t) \tag{23.2.2}$$

Survey Sampling Theory and Applications
ISBN 978-0-12-811848-1
http://dx.doi.org/10.1016/B978-0-12-811848-1.00023-6

where $I(y_i \le t) = \begin{cases} 1 & \text{if } y_i \le t \\ 0 & \text{if } y_i > t \end{cases}$ and y_i is the value of the study variable y for the ith unit of the population; $i = 1,\ldots, N$.

23.2.1 Design-Based Estimation

Let a sample s of size n be selected from a population U using a sampling design p with $\pi_i(>0)$ and $\pi_{ij}(>0)$ as the inclusion probabilities for the ith, and ith and jth unit ($i \ne j$), respectively. In case the population size N is known, an unbiased estimator of $F(t)$ is given by

$$\widehat{F}(t) = \frac{1}{N}\sum_{i\in s}\frac{I(y_i \le t)}{\pi_i} \tag{23.2.3}$$

For simple random sampling without replacement (SRSWOR) sampling $\pi_i = n/N$ and $\widehat{F}(t)$ reduces to the sample empirical distribution function

$$s(t) = \frac{r(t)}{n} \tag{23.2.4}$$

where $r(t)$ is number of y_i's in the sample less than or equal to t.

The estimator $s(t)$ is admissible and also minimax under certain loss function (Dorfman, 2009). KuK (1988) proposed following alternative estimators based on complementary proportion

$$\widehat{F}_c(t) = 1 - \frac{1}{N}\sum_{i\in s}\frac{1 - I(y_i \le t)}{\pi_i} \tag{23.2.5}$$

and

$$\widehat{F}_w(t) = w\widehat{F}(t) + (1-w)\widehat{F}_c(t) \tag{23.2.6}$$

where w is a suitably chosen weight.

It should be noted that none of the estimators $\widehat{F}(t)$, $\widehat{F}_c(t)$, and $\widehat{F}_w(t)$ mentioned above possess all the properties of the distribution function stated in Eq. (23.2.1). In fact $\widehat{F}(t)$ may exceed unity and $\widehat{F}(\infty) = \frac{1}{N}\sum_{i\in s}\frac{1}{\pi_i}$ may not equal to 1. Details on discussions of the performances of the estimators have been given by Dorfman (2009). To overcome this difficulty, the following alternative Hájek type estimator has been proposed.

$$\widehat{F}_h(t) = \sum_{i\in s}\frac{I(y_i \le t)}{\pi_i}\Big/\sum_{i\in s}\frac{1}{\pi_i} \tag{23.2.7}$$

Though the estimator (23.2.7) is not design unbiased for $F(t)$, it is design consistent and approximately unbiased. It satisfies all the properties of the distribution function stated in Eq. (23.2.1).

23.2.2 Design-Based Estimators Using Auxiliary Information

Rao et al. (1990) considered design-based ratio and difference estimators of the population distribution function $F(t)$ when the values of the auxiliary variable x_i's are known for $i = 1,\ldots, N$. The conventional estimator for the population ratio $R = Y/X$ is

$$\widehat{R} = \left(\sum_{i\in s}\frac{y_i}{\pi_i}\right)\Big/\left(\sum_{i\in s}\frac{x_i}{\pi_i}\right) \tag{23.2.8}$$

where $X = \sum_{i=1}^{N} x_i$ and $Y = \sum_{i=1}^{N} y_i$.

Treating $\widehat{y}_i = \widehat{R}x_i$ as an estimator of y_i, Rao et al. (1990) proposed the following ratio estimator of $F(t)$ as

$$\widehat{F}_R(t) = \frac{1}{N}\widehat{\lambda} \times \left(\sum_{i\in U} I(\widehat{y}_i \le t)\right) \tag{23.2.9}$$

where

$$\widehat{y}_i = \widehat{R}x_i \quad \text{and} \quad \widehat{\lambda} = \left(\frac{\sum_{i\in s}\frac{I(y_i \le t)}{\pi_i}}{\sum_{i\in s}\frac{I(\widehat{y}_i \le t)}{\pi_i}}\right) \tag{23.2.10}$$

The proposed ratio estimator $\widehat{F}_R(t)$ is design consistent and exactly equal to the population distribution function $F(t)$ if y_i is proportional to x_i for every $i \in U$. The proposed estimator is expected to gain in efficiency compared to the conventional estimator $\widehat{F}(t)$ if y_i is approximately proportional to x_i. But the estimator $\widehat{F}_R(t)$ suffers from a drawback that it may not possess the desirable properties of the distribution function given in Eq. (23.2.1): it may have a value outside the interval [0, 1]. For a large sample size n, an approximate expression of variance of $\widehat{F}_R(t)$ is given by Rao et al. (1990) as

$$\begin{aligned} Var\left[\widehat{F}_R(t)\right] &\cong V\left(\frac{1}{N}\sum_{i\in s}\frac{Q_i(t)}{\pi_i}\right) \\ &= \frac{1}{2N^2}\sum_{i\neq}\sum_{j\in U}(\pi_i\pi_j - \pi_{ij})\left(\frac{Q_i}{\pi_i} - \frac{Q_j}{\pi_j}\right)^2 \end{aligned} \tag{23.2.11}$$

where

$$Q_i = I(y_i \le t) - \lambda I(Rx_i \le t) \text{ and } \lambda = \frac{\sum_{i \in U} I(y_i \le t)}{\sum_{i \in U} I(Rx_i \le t)}. \qquad (23.2.12)$$

An approximate unbiased estimator of $V_p\left[\widehat{F}_R(t)\right]$ is

$$\widehat{V}\left[\widehat{F}_R(t)\right] = \frac{1}{2N^2} \sum_{i \ne} \sum_{j \in s} \frac{(\pi_i \pi_j - \pi_{ij})}{\pi_{ij}} \left(\frac{\widehat{Q}_i}{\pi_i} - \frac{\widehat{Q}_j}{\pi_j} \right)^2 \qquad (23.2.13)$$

where $\widehat{Q}_i = I(y_i \le t) - \widehat{\lambda} I(\widehat{y}_i \le t)$ and $\widehat{\lambda}$ is given in Eq. (23.2.10).

Rao et al. (1990) also proposed a difference estimator of $F(t)$ as

$$\widehat{F}_D(t) = \frac{1}{N}\left\{ \left(\sum_{i \in s} \frac{I(y_i \le t)}{\pi_i} \right) - \left(\sum_{i \in s} \frac{I(\widehat{y}_i \le t)}{\pi_i} - \sum_{i \in U} I(\widehat{y}_i \le t) \right) \right\} \qquad (23.2.14)$$

An approximate expression of the variance of $\widehat{F}_D(t)$ and its unbiased estimators are respectively given by

$$\begin{aligned} Var\left[\widehat{F}_D(t)\right] &\cong V_p\left(\frac{1}{N} \sum_{i \in s} \frac{G_i(t)}{\pi_i} \right) \\ &= \frac{1}{2N^2} \sum_{i \ne} \sum_{j \in U} (\pi_i \pi_j - \pi_{ij}) \left(\frac{G_i(t)}{\pi_i} - \frac{G_j(t)}{\pi_j} \right)^2 \end{aligned} \qquad (23.2.15)$$

and

$$\widehat{V}\left[\widehat{F}_D(t)\right] = \frac{1}{2N^2} \sum_{i \ne} \sum_{j \in s} \frac{(\pi_i \pi_j - \pi_{ij})}{\pi_{ij}} \left(\frac{\widehat{G}_i(t)}{\pi_i} - \frac{\widehat{G}_j(t)}{\pi_j} \right)^2 \qquad (23.2.16)$$

where $G_i(t) = I(y_i \le t) - I(Rx_i \le t)$ and $\widehat{G}_i(t) = I(y_i \le t) - I\left(\widehat{R}x_i \le t\right)$.

23.2.3 Model-Based Estimators

Let us suppose that the study variable y is related to the auxiliary variable through the superpopulation model

$$y_i = \mu(x_i) + \sqrt{v(x_i)}\, \epsilon_i \qquad (23.2.17)$$

where $\mu(x_i)$ is a function of x_i but involves unknown model parameters, $\nu(x_i)$ is known and positive, and $\in_i$'s are independently and identically distributed random variables with $E_m(\in_i) = 0$ and $V_m(\in_i) = \sigma^2(>0)$. Here E_m and V_m denote expectation and variance operators, respectively, with respect to the model (23.2.17). Chambers and Dunstan (1986) and Rao et al. (1990) used the model (23.2.17) with $\mu(x_i) = \beta x_i$, where β is an unknown parameter.

Let $\widehat{\mu}(x_i)$ be a model-unbiased estimator of $\mu(x_i)$, i.e., $E_m[\widehat{\mu}(x_i)] = \mu(x_i)$. Then $\widehat{r}_i(t) = \dfrac{t - \hat{\mu}(x_i)}{\sqrt{\nu(x_i)}}$ becomes model-unbiased estimator of $r_i(t) = \dfrac{t - \mu(x_i)}{\sqrt{\nu(x_i)}}$.

Let $A_i(t)$ be the model expectation of $I(y_i \leq t)$ in the sense that

$$A_i(t) = E_m\{I(y_i \leq t)\} = P(y_i \leq t) = P(\in_i \leq r_i(t)) \tag{23.2.18}$$

An approximate model-unbiased estimator of $A_i(t)$ is given by

$$\widehat{A}_i(t) = \frac{1}{n}\sum_{j\in s} I\left\{\widehat{r}_j(y_j) \leq \widehat{r}_i(t)\right\} \tag{23.2.19}$$

where $\widehat{r}_j(y_j) = \dfrac{y_j - \hat{\mu}(x_j)}{\sqrt{\nu(x_j)}}$.

Under the model (23.2.17), Johnson (2003) proposed the following model-based estimator of $F(t)$ as

$$\widehat{F}_J(t) = \frac{1}{N}\left(\sum_{i\in s} I(y_i \leq t) + \sum_{i\in U-s} \widehat{A}_i(t)\right) \tag{23.2.20}$$

where $U - s$ denotes the set of nonsampled units.

Chambers and Dunstan (1986) considered the estimator (23.2.20) when $\mu(x_i) = \beta x_i$ and the model parameter β was estimated by the weighted least squares method as

$$\widehat{\beta} = \left(\sum_{i\in s}\frac{y_i x_i}{\nu(x_i)}\right) \Bigg/ \left(\sum_{i\in s}\frac{x_i^2}{\nu(x_i)}\right) \tag{23.2.21}$$

Substituting $\widehat{\mu}(x_i) = \widehat{\beta}x_i$ in the expression (23.2.20), Chambers and Dunstan (1986) estimator is obtained as

$$\widehat{F}_{cd}(t) = \frac{1}{N}\left[\sum_{i\in s} I(y_i \leq t) + \sum_{i\in U-s}\left\{\frac{1}{n}\sum_{j\in s} I\left(\frac{y_j - \widehat{\beta}x_j}{\sqrt{\nu(x_j)}} \leq \frac{t - \widehat{\beta}x_i}{\sqrt{\nu(x_i)}}\right)\right\}\right] \tag{23.2.22}$$

The estimator $\widehat{F}_{cd}$ satisfies all the properties of a distribution function. Although the estimator $\widehat{F}_{cd}(t)$ is independent of the sampling design, it is asymptotically model unbiased and has the property that $\widehat{F}_{cd}(t)$ equals to $F(t)$ when $y_i \propto x_i \ \forall i \in U$. Since the estimator $\widehat{F}_{cd}$ depends on the assumed model, it is highly efficient if the model (23.2.17) holds but if the model is incorrect, the estimator becomes biased and can perform much worse even than the naive estimator $s(t)$.

23.2.4 Model-Assisted Estimators

Rao et al. (1990) proposed the following model assisted difference estimators:

$$\widehat{F}^*(t) = \frac{1}{N}\left[\sum_{i\in s}\frac{I(y_i \le t)}{\pi_i} - \left(\sum_{i\in s}\frac{B_i(t)}{\pi_i} - \sum_{i\in U} B_i(t)\right)\right] \tag{23.2.23}$$

with

$$B_i(t) = \frac{1}{N}\sum_{j\in U} I\left(\frac{y_j - \beta x_j}{\sqrt{v(x_j)}} \le \frac{t - \beta x_i}{\sqrt{v(x_i)}}\right) \tag{23.2.24}$$

The estimator $\widehat{F}^*(t)$ does not possess all the properties of distribution function. It is both design unbiased and asymptotically model unbiased for $F(t)$. It is calibrated in the sense that $\widehat{F}^*(t)$ reduces $F(t)$ if y_i is exactly proportional to x_i for $\forall i \in U$. The main demerit of this estimator is that it cannot be used in practice as $B_i(t)$'s are unknown. To overcome this difficulty, Rao et al. (1990) proposed the following alternative estimator:

$$\widehat{F}_{rkm}(t) = \frac{1}{N}\left[\sum_{i\in s}\frac{I(y_i \le t)}{\pi_i} - \left(\sum_{i\in s}\frac{\widehat{B}_{ic}(t)}{\pi_i} - \sum_{i\in U} \widehat{B}_i(t)\right)\right] \tag{23.2.25}$$

where

$$\widehat{B}_i(t) = \left(\sum_{j\in s}\frac{1}{\pi_j}\right)^{-1}\left[\sum_{j\in s}\frac{1}{\pi_j} I\left(\frac{y_j - \widehat{\beta} x_j}{\sqrt{v(x_j)}} \le \frac{t - \widehat{\beta} x_i}{\sqrt{v(x_i)}}\right)\right] \text{ and}$$

$$\widehat{B}_{ic}(t) = \left(\sum_{j\in s}\frac{\pi_i}{\pi_{ij}}\right)^{-1}\left[\sum_{j\in s}\frac{\pi_i}{\pi_{ij}} I\left(\frac{y_j - \widehat{\beta} x_j}{\sqrt{v(x_j)}} \le \frac{t - \widehat{\beta} x_i}{\sqrt{v(x_i)}}\right)\right] \tag{23.2.26}$$

The estimator $\widehat{B}_i(t)$ is asymptotically design unbiased for $B_i(t)$ whereas $\widehat{B}_{ic}(t)$ is asymptotically and conditionally design unbiased for $B_i(t)$ given

$i \in s$. The estimator $\widehat{F}_{rkm}(t)$ is more complex than $\widehat{F}^*(t)$ because it involves second-order inclusion probabilities otherwise it shares the same properties of the estimator $\widehat{F}^*(t)$. Under model misspecification, $\widehat{F}_{rkm}(t)$ performs better than $\widehat{F}_{cd}$.

The asymptotic design variance of $\widehat{F}_{rkm}(t)$ and an estimator of the variance was provided by Rao et al. (1990) as follows:

$$Var\left\{\widehat{F}_{rkm}(t)\right\} = \frac{1}{N^2}\frac{1}{2}\sum_{i\neq}\sum_{j\in U}(\pi_i\pi_j - \pi_{ij}) \left(\frac{I(y_i \le t) - B_i(t)}{\pi_i} - \frac{I(y_j \le t) - B_j(t)}{\pi_j}\right)^2 \tag{23.2.27}$$

$$\widehat{V}\left\{\widehat{F}_{rkm}(t)\right\} = \frac{1}{N^2}\frac{1}{2}\sum_{i\neq}\sum_{j\in s}\frac{\pi_i\pi_j - \pi_{ij}}{\pi_{ij}} \left(\frac{I(y_i \le t) - \widehat{B}^*_{ic}(j)}{\pi_i} - \frac{I(y_j \le t) - \widehat{B}^*_{jc}(i)}{\pi_j}\right)^2 \tag{23.2.28}$$

where $\widehat{B}^*_{ic}(j) = \sum_{k\in s}\frac{\pi_{ij}}{\pi_{ijk}} I\left(\frac{y_k - \hat{\beta}x_k}{\sqrt{v(x_k)}} \le \frac{t - \hat{\beta}x_i}{\sqrt{v(x_i)}}\right) \Big/ \left(\sum_{k\in s}\frac{\pi_{ij}}{\pi_{ijk}}\right)$ and $\pi_{ijk}(i \neq j \neq k)$ is the inclusion probability for the ith, jth, and kth unit in the sample.

Dorfman (2009) proposed a modified estimator of $\widehat{F}_{rkm}(t)$ as follows:

$$\begin{aligned}\widehat{F}_{dm}(t) &= \frac{1}{N}\left[\sum_{i\in s} I(y_i \le t) + \sum_{i\in U-s}\left\{\frac{1}{n}\sum_{j\in s} I\left(\frac{y_i - \widehat{\beta}x_j}{\sqrt{v(x_j)}} \le \frac{t - \widehat{\beta}x_i}{\sqrt{v(x_i)}}\right)\right\} + \sum_{i\in s}\left(\frac{1}{\pi_i} - 1\right)R_i\right] \\ &= \widehat{F}_{cd}(t) + \frac{1}{N}\sum_{i\in s}\left(\frac{1}{\pi_i} - 1\right)R_i\end{aligned} \tag{23.2.29}$$

where $R_i = I(y_i \le t) - \frac{1}{n}\sum_{j\in s} I\left(\frac{y_j - \hat{\beta}x_j}{\sqrt{v(x_j)}} \le \frac{t - \hat{\beta}x_i}{\sqrt{v(x_i)}}\right)$.

Both the estimator $\widehat{F}_{dm}(t)$ and $\widehat{F}_{rkm}(t)$ are equally efficient (Dorfman, 2009). $\widehat{F}_{dm}(t)$ has additional advantage as it does not require computation of second-order inclusion probabilities. Wang and Dorfman (1996) proposed an alternative estimator $\widehat{F}_{wdr}(t) = w\widehat{F}_{dm}(t) + (1-w)\widehat{F}_{rkm}(t)$, which is the weighted average of $\widehat{F}_{dm}(t)$ and $\widehat{F}_{rkm}(t)$. Simulation studies showed that the estimator $\widehat{F}_{wdr}(t)$ performs better than $\widehat{F}_{dm}(t)$ and $\widehat{F}_{rkm}(t)$. Mak and Kuk (1993) proposed a modification of $\widehat{F}_{cd}(t)$ estimator as follows:

$$\widehat{F}_{mk}(t) = \frac{1}{N}\left[\sum_{i\in s} I(y_i \le t) + \sum_{i\in U-s}\left\{\frac{1}{n}\sum_{j\in s}\Phi\left(\frac{t-\widehat{\beta}x_i}{\widehat{\sigma}\sqrt{v(x_i)}}\right)\right\}\right]$$

where $\Phi\left(\frac{t-\hat{\beta}x_i}{\hat{\sigma}\sqrt{v(x_i)}}\right)$ is the standard normal distribution function and $\widehat{\sigma}^2$ is the weighted list square estimate of σ^2 obtained from the model (23.2.17). The main advantage of using $\widehat{F}_{mk}(t)$ over $\widehat{F}_{cd}(t)$ is ease of computation but their relative performances are not known.

23.2.5 Nonparametric Regression Method

For the model-based and model-assisted estimators, we used parametric regression where a single smooth function $\mu(x)$ was estimated over the entire possible range of the auxiliary variable x. In a nonparametric regression we approximate $\mu(x)$ locally by placing more weight on y_i's corresponding to x_i, which are close to x. In particular, suppose we wish to approximate $\mu(x)$ by a polynomial of degrees p viz.

$$\mu_{lp}(x) = \beta_0 + \beta_1(x_i - x) + \cdots + \beta_p(x_i - x)^p \tag{23.2.30}$$

then the regression coefficients β_0, $\beta_1,\ldots,$ β_p may be estimated by using weighted least square method using kernel weight

$$K\left(\frac{x_i - x}{b}\right)$$

attached to the ith observation. The kernel function $K(u)$ is a symmetric positive function of u, which decreases as $|u|$ increases. The parameter $b(>0)$ is the smoothing parameter, which is referred to bandwidth. Thus the normal equations are obtained from the selected sample by minimizing

$$\sum_{i\in s}\left(y_i - \beta_0 - \beta_1(x_i - x) - \cdots - \beta_p(x_i - x)\right)^2\left\{\frac{1}{\pi_i}K\left(\frac{x_i - x}{b}\right)\right\}$$

with respect to $\beta_0, \beta_1, \ldots, \beta_p$. Here we assume, without loss of generality, that the selected sample s contains first n units of the population. The estimates of $\boldsymbol{\beta}' = (\beta_0, \beta_1, \ldots, \beta_p)$ are given by

$$\widehat{\boldsymbol{\beta}} = \left(\mathbf{X}_x'\mathbf{W}_x\mathbf{X}_x\right)^{-1}\mathbf{X}_x'\mathbf{W}_x\mathbf{y}_s \tag{23.2.31}$$

where

$$\mathbf{X}_x = \begin{pmatrix} 1 & x_1 - x & \ldots & (x_1 - x)^p \\ \cdot & \cdot & \cdot & \cdot \\ \cdot & \cdot & \cdot & \cdot \\ 1 & x_n - x & \cdot & (x_n - x)^p \end{pmatrix},$$

$$\mathbf{W}_x = diag\left(\frac{1}{\pi_1}K\left(\frac{x_1 - x}{b}\right)\cdots\frac{1}{\pi_n}K\left(\frac{x_n - x}{b}\right)\right) \text{and } \mathbf{y}_s' = (y_1, \ldots, y_n).$$

The predicted value of y at the point $x = x_i$ is the intercept. So we write

$$\widehat{y}_i = \widehat{\mu}_{lp}(x_i) = \widehat{\beta}_0 = \mathbf{e}_1'\left(\mathbf{X}_{x_i}'\mathbf{W}_{x_i}\mathbf{X}_{x_i}\right)^{-1}\mathbf{X}_{x_i}'\mathbf{W}_{x_i}\mathbf{y} \tag{23.2.32}$$

where $\mathbf{e}_1$ is a $(p+1)$ column vector with 1 as the first element and 0 elsewhere.

23.2.5.1 Nandaraya–Watson Estimator

For $p = 0$, Eq. (23.2.32) reduces to Nandaraya–Watson (1964) estimator:

$$\widehat{\mu}_{lp}(x_i) = \frac{\sum_{i \in s} K\left(\frac{x_i - x}{b}\right)\frac{y_i}{\pi_i}}{\sum_{i \in s} K\left(\frac{x_i - x}{b}\right)\frac{1}{\pi_i}} \tag{23.2.33}$$

The data analyst must choose p and b suitably. For further details, readers are referred to Ruppert et al. (2003).

23.2.5.2 Breidt and Opsomer Estimator

The local polynomial estimator for the population total is given by

$$\widehat{Y}_{bo} = \sum_{i \in s}\frac{y_i - \widehat{\mu}_{lp}(x_i)}{\pi_i} + \sum_{i \in U}\widehat{\mu}_{lp}(x_i) \tag{23.2.34}$$

Now replacing y_i with $I(y_i \leq t)$ in Eq. (23.2.34), we get the Breidt and Opsomer (2000) estimator of the distribution function as

$$\widehat{F}_{lp}(t) = \frac{1}{N}\sum_{i \in s}\frac{I(y_i \leq t) - \widehat{\mu}_{lp}^*(x_i, t)}{\pi_i} + \frac{1}{N}\sum_{i \in U}\widehat{\mu}_{lp}^*(x_i, t) \tag{23.2.35}$$

where $\widehat{\mu}_{lp}^{*}(x_i, t) = \mathbf{e}_1'\left(\mathbf{X}_{x_i}'\mathbf{W}_{x_i}\mathbf{X}_{x_i}\right)^{-1}\mathbf{X}_{x_i}'\mathbf{W}_{x_i}\mathbf{I}_{st}$ and $\mathbf{I}_{st} = (I(y_1 \le t), \ldots, I(y_i \le t), \ldots, I(y_n \le t))'$.

From Breidt and Opsomer (2000), we see that $\widehat{F}_{lp}(t)$ maintains design consistency and asymptotic design-unbiased properties. The estimated variance of $\widehat{F}_{lp}(t)$ was given by Johnson (2003) as

$$\widehat{V}\left(\widehat{F}_{lp}(t)\right) = \frac{1}{2}\sum_{i\neq}\sum_{j\in s}\frac{\pi_{ij} - \pi_i\pi_j}{\pi_{ij}\pi_i\pi_j}\left(I(y_i \le t) - \widehat{\mu}_{lp}^{*}(x_i, t)\right)\left(I(y_j \le t) - \widehat{\mu}_{lp}^{*}(x_j, t)\right) \tag{23.2.36}$$

23.2.5.3 Kuo Estimator

Kuo (1988) proposed the following estimator for the distribution function $F(t)$:

$$\widehat{F}_{ko}(t) = \frac{1}{N}\left[\sum_{i\in s} I(y_i \le t) + \sum_{j\in U-s}\sum_{i\in s} w_{ij} I(y_i \le t)\right] \tag{23.2.37}$$

where $w_{ij} = \dfrac{K\left(\dfrac{x_j - x_i}{b}\right)}{\sum_{i\in s} K\left(\dfrac{x_j - x_i}{b}\right)}$, $K(z) = e^{-z^2/2}$ is the standard normal density (kernel).

Dorfman and Hall (1993) provided the expressions of asymptotic bias and variance of $\widehat{F}_{ko}(t)$.

23.2.5.4 Kuk Estimator

Kuk (1993) nonparametric regression estimator for $F(t)$ is given by

$$\widehat{F}_{kk}(t) = \frac{1}{N}\sum_{j\in U}\widehat{R}_j(t) \tag{23.2.38}$$

where

$$\widehat{R}_j(t) = \frac{\sum_{i\in s}\dfrac{1}{\pi_i} w\left(\dfrac{x_j - x_i}{b}\right) W\left(\dfrac{t - y_i}{b}\right)}{\sum_{i\in s}\dfrac{1}{\pi_i} w\left(\dfrac{x_j - x_i}{b}\right)},$$

$W(z) = e^z/(1 + e^z)$ is the standard logistic distribution function with density $w(z) = e^z/(1 + e^z)^2$, and b is the band with parameter used to control the amount of smoothing.

It should be noted that both the estimators $\widehat{F}_{ko}(t)$ and $\widehat{F}_{kk}(t)$ meet the properties of the distribution function given in Eq. (23.2.1). For more details, readers are referred to Kuo (1988), Kuk (1993), Dorfman (2009), among others. Dorfman and Hall (1993) considered a design-adjusted version of Kuo (1988) and Rao–Kovar–Mantel (1990) estimators while Silva and Skinner (1995) suggested a poststratified estimator of $F(t)$.

23.2.6 Calibration Method

In Deville and Särndal's (1992) calibration method, the Horvitz–Thompson estimator $\widehat{\overline{Y}}_{ht} = \frac{1}{N}\sum_{i\in s} d_i y_i$ with $d_i = 1/\pi_i$ was calibrated as

$$\widehat{\overline{Y}}_c = \sum_{i\in s} w_i y_i$$

The weights w_i's were chosen to minimize the distance

$$\Phi = \sum_{i\in s} \frac{(w_i - d_i)^2}{d_i q_i} \tag{23.2.39}$$

subject to the calibrating constraints $\sum_{i\in s} w_i x_i = X$. Here x_i's are the values of the auxiliary variable x with known total X and q_i are suitably chosen weights. Details have been given in Section 9.7. Suppose that the study variable y is related to the auxiliary variable through the following superpopulation model

$$y_i = \mu(x_i, \theta) + \sqrt{v(x_i)} \in_i$$

where θ is an unknown model parameter and $\in_i$'s are independently identically distributed with mean zero and variance σ^2.

Let $\widehat{\theta}$ be a suitable estimator of θ obtained from the selected sample using some standard procedure so that $\widehat{\mu}_i = \mu\left(x_i, \widehat{\theta}\right)$ is an estimator for $\mu_i = \mu(x_i, \theta)$. Minimizing Eq. (23.2.39) subject to the calibrating constraints (i) $\sum_{i\in s} w_i = N$ and (ii) $\sum_{i\in s} w_i \widehat{\mu}_i = \sum_{i\in U} \widehat{\mu}_i$, Wu and Sitter (2001) derived the calibrated estimator for the population total Y as

$$\widehat{\overline{Y}}_c = \widehat{\overline{Y}}_{ht} - \widehat{B}\left(\frac{1}{N}\sum_{i\in s}\frac{\widehat{\mu}_i}{\pi_i} - \frac{1}{N}\sum_{i\in U}\widehat{\mu}_i\right) \tag{23.2.40}$$

where $$\widehat{B} = \left(\sum_{i\in s} d_i q_i (\widehat{\mu}_i - \overline{\mu})(y_i - \overline{y})\right)\left(\sum_{i\in s} d_i q_i (\widehat{\mu}_i - \overline{\mu})^2\right)^{-1},$$

$\overline{y} = \sum_{i\in s} d_i q_i y_i / \sum_{i\in s} d_i q_i$, and $\overline{\mu} = \sum_{i\in s} d_i q_i \widehat{\mu}_i / \sum_{i\in s} d_i q_i$.

For estimating distribution function $F(t)$, Wu and Sitter (2001) replaced y_i by $I(y_i \leq t)$, $\widehat{\mu}_i$ by $\widehat{G}_i = \frac{1}{n}\sum_{j\in s} I\left(\frac{y_j - \hat{\mu}_j}{\sqrt{v(x_j)}} \leq \frac{t - \hat{\mu}_i}{\sqrt{v(x_i)}}\right)$ and substituted $q_i = 1$ in the expression (23.2.40). The resultant model calibrated estimator was $F(t)$ obtained as

$$\widehat{F}_{ws}(t) = \widehat{F}_{ht}(t) - \widehat{B}_{ws}\left(\frac{1}{N}\sum_{i\in s}\frac{\widehat{G}_i}{\pi_i} - \frac{1}{N}\sum_{i\in U}\widehat{G}_i\right) \tag{23.2.41}$$

where $\widehat{F}_{ht}(t) = \frac{1}{N}\sum_{i\in s}\frac{I(y_i \leq t)}{\pi_i}$, $\widehat{B}_{us} = \frac{\left(\sum_{i\in s} d_i\left(\hat{G}_i - \overline{G}\right)\left(I(y_i \leq t) - \overline{I}\right)\right)}{\left(\sum_{i\in s} d_i\left(\hat{G}_i - \overline{G}\right)^2\right)}$,

$\overline{G} = \frac{\sum_{i\in s} d_i\hat{G}_i}{\sum_{i\in s} d_i}$, and $\overline{I} = \frac{\sum_{i\in s} d_i I(y_i \leq t)}{\sum_{i\in s} d_i}$.

Wu and Sitter (2001) also proposed pseudoempirical likelihood estimator of $F(t)$ as

$$F^*_{ws}(t) = \sum_{i\in s}\widehat{p}_i I(y_i \leq t) \tag{23.2.42}$$

where the weights $\widehat{p}_i$ are obtained by maximizing pseudoempirical likelihood

$$l(p) = \sum_{i\in s} p_i I(y_i \leq t)$$

subject to (i) $p_i \geq 0$, (ii) $\sum_{i\in s} p_i = 1$, and (iii) $\sum_{i\in s} p_i\widehat{G}_i = \frac{1}{N}\sum_{i\in U}\widehat{G}_i$.

Since $p_i \geq 0$, $F^*_{ws}(t)$ is a genuine distribution function. The calibrated estimator $\widehat{F}_{ws}$ and the pseudocalibrated estimator F^*_{ws} are asymptotically equivalent.

Rueda et al. (2007) considered a vector of auxiliary variables $\mathbf{x}_i = (x_{i1}, \ldots, x_{ij}, \ldots, x_{iq})'$, which are known for $i = 1,\ldots, N$. They defined a pseudovariable $g_i = \widehat{\boldsymbol{\beta}}'\mathbf{x}_i$ for $i = 1,\ldots, N$, where

$\widehat{\boldsymbol{\beta}} = \left(\sum_{i \in s} d_i q_i \mathbf{x}_i \mathbf{x}'_i\right)^{-1} \sum_{i \in s} d_i q_i \mathbf{x}_i y_i$ and q_i are known positive constants unrelated to $d_i = 1/\pi_i$. The proposed calibrated estimator is

$$\widehat{F}_r = \frac{1}{N} \sum_{i \in s} w_i I(y_i \leq t) \tag{23.2.43}$$

where the weights w_i were derived by minimizing the chi-square distance $\Phi_s = \sum_{i \in s} \frac{(w_i - d_i)^2}{d_i q_i}$ subject to the constraints

$$\frac{1}{N} \sum_{i \in s} w_i I(g_i \leq t_j) = F_g(t_j) = \frac{1}{N} \sum_{i \in U} I(g_i \leq t_j),\ j = 1, \ldots, p$$

for suitably chosen p points $t_1 < t_2 < \cdots < t_p$. Rueda et al. (2007), showed by simulation studies based on the actual data, that their estimator $\widehat{F}_r$ performs similar to $\widehat{F}_{cd}$ but better than $\widehat{F}_{rkm}$.

23.2.7 Method of Poststratification

Silva and Skinner (1995) poststratified the initial sample s into H strata $U_1, \ldots, U_h, \ldots, U_H$ where the unit $i \in U_h$ if $x_{(h-1)} < x_i \leq x_{(h)}$ for $h = 1, \ldots, H$, $-\infty = x_{(0)} < x_{(1)} < x_{(2)} < \cdots < x_{(H)} = \infty$. Let s_h be the sample of size n_h from the hth stratum and $n = \sum_{h=1}^{H} n_h$. Silva and Skinner (1995) assumed that N_h, the stratum size of U_h, is known and n is so large that the probability of $n_h = 0$ is zero. Under this assumption, they proposed the following estimator of $F(t)$ as

$$\widehat{F}_{ss}(t) = \sum_{h=1}^{H} W_h \widehat{F}_h^*(t) \tag{23.2.44}$$

where $W_h = \frac{N_h}{N}$ and $\widehat{F}_h^*(t) = \left(\sum_{i \in s_h} \frac{I(y_i \leq t)}{\pi_i}\right) \Big/ \left(\sum_{i \in s_h} \frac{1}{\pi_i}\right)$

It can be easily checked that the estimator $\widehat{F}_{ss}(t)$ possesses all the properties of the distribution function stated in Eq. (23.2.1). Following Rao et al. (1990), Silva and Skinner (1995) derived the approximate expression of variance of $\widehat{F}_{ss}(t)$ as

$$Var\left\{\widehat{F}_{ss}(t)\right\} \cong \frac{1}{N^2} \frac{1}{2} \sum_{i \neq} \sum_{j \in U} (\pi_i \pi_j - \pi_{ij}) \times \left(\frac{I(y_i \leq t) - F^*_{h(i)}(t)}{\pi_i} - \frac{I(y_j \leq t) - F^*_{h(j)}(t)}{\pi_j}\right)^2 \tag{23.2.45}$$

where $h(k)$ is the poststratum to which the unit k belongs and $F_h^*(t) = \frac{1}{N_h} \sum_{i \in U_h} I(y_i \leq t)$ is the population distribution function of the hth stratum $h = 1, \ldots, H$.

An estimator of variance of $\widehat{F}_{ss}(t)$ was presented by Silva and Skinner (1995) as

$$\widehat{Var}\left\{\widehat{F}_{ss}(t)\right\} \cong \frac{1}{N^2}\frac{1}{2}\sum_{i \neq}\sum_{j \in s}\frac{(\pi_i\pi_j - \pi_{ij})}{\pi_{ij}} \left(\frac{I(y_i \leq t) - \widehat{F}^*_{h(i)}}{\pi_i} - \frac{I(y_j \leq t) - \widehat{F}^*_{h(j)}(t)}{\pi_j}\right)^2 \tag{23.2.46}$$

where $\widehat{F}^*_{h(i)}$ is an estimate of $F^*_{h(i)}$.

23.2.8 Empirical Comparison of the Estimators

Rao et al. (1990) did comprehensive studies of the performances of the design-based estimators $\widehat{F}_h(t), \widehat{F}_R(t), \widehat{F}_D(t), \widehat{F}^*(t)$ and the model-based estimator $\widehat{F}_{cd}(t)$ based on a population of sugarcane farms considered by Chambers and Dunstan (1986). The gross value of cane was treated as the study variable y while area under sugarcane was taken as auxiliary variable x. The population was found to obey the model (23.2.17) with $v(x) = \sqrt{x}$. Under simple random sampling, the relative biases of $\widehat{F}_D(t)$ and $\widehat{F}_{rkm}(t)$ are negligible and much less than that of $\widehat{F}_R(t)$. Whence efficiency is concerned, $\widehat{F}_{rkm}(t)$ is more efficient than $\widehat{F}_D(t)$ and $\widehat{F}(t)$. The ratio estimator $\widehat{F}_R(t)$ is less efficient than $\widehat{F}_D(t)$. The model-based estimator $\widehat{F}_{cd}(t)$ was found to be more efficient than the design-based estimators possibly because the population obeys the model. Rao et al. (1990) also considered the Hansen et al. (1983) population with $N = 14{,}000$ units to study the effect of model misspecification and computed relative bias and relative root mean square errors of the estimators $\widehat{F}_h(t), \widehat{F}_R(t), \widehat{F}_D(t), \widehat{F}_{cd}(t)$, and $\widehat{F}_{rkm}(t)$. It was found that design-based estimators have negligible bias, even less than 1%, whereas the model-based estimator $\widehat{F}_{cd}(t)$ has a much larger relative bias of about 20%. The design-based estimator $\widehat{F}_{rkm}(t)$ was found to be much more efficient than the other design-based estimators $\widehat{F}_R(t)$ and $\widehat{F}_D(t)$ whereas the model-based estimator $\widehat{F}_{cd}(t)$ was found least efficient.

Silva and Skinner (1995) conducted simulation studies to compare performances of the estimators $\widehat{F}_h(t)$, $\widehat{F}_{ss}(t)$, $\widehat{F}_{ko}(t)$, $\widehat{F}_{kk}(t)$, $\widehat{F}_{cd}(t)$, and $\widehat{F}_{rkm}(t)$ based on two populations. The first population was the sugarcane farm data mentioned above, which was originally considered by Chambers and Dunstan (1986) and later Rao et al. (1990) and Kuk (1993). The second population comprises of 430 farms with 50 or more beef cattle, which was originally used by Chambers et al. (1993) and then Kuk (1993). Population 1 is a good working model for Eq. (23.2.17) as stated earlier whereas Population 2 is not. As far as relative bias is concerned, for both the populations, $\widehat{F}_{ss}(t)$ was found to have least bias while $\widehat{F}_{rkm}(t)$ occupied second place. $\widehat{F}_{ko}(t)$, $\widehat{F}_{kk}(t)$, and $\widehat{F}_{cd}(t)$ were found to have relatively high absolute relative biases. For Population 1, $\widehat{F}_{cd}(t)$ had least mean squared error (MSE) as expected because the model fits the population. The second and third places were occupied by $\widehat{F}_{kk}(t)$ and $\widehat{F}_{rkm}(t)$, respectively. Surprisingly, for Population 2, $\widehat{F}_{cd}(t)$ also had the smallest MSE followed by $\widehat{F}_{kk}(t)$. The estimators $\widehat{F}_{ss}(t)$ and $\widehat{F}_{ko}(t)$ were found to have moderate MSE. For further details, interested readers are referred to Rao et al. (1990) and Silva and Skinner (1995).

23.3 ESTIMATION OF QUANTILES

The αth quantile $\theta_y(\alpha), 0 < \alpha < 1$ of a finite population vector $\mathbf{y} = (y_1, \ldots, y_N)$ is defined as

$$\theta_y(\alpha) = inf\{t : F_y(t) \geq \alpha\} \tag{23.3.1}$$

where $F_y(t)$ is the distribution function of y. In case $\widehat{F}_y(t)$, an estimator of $F_y(t)$, is a monotonic nondecreasing function of t, the customary estimator of $\theta_y(\alpha)$ is obtained as

$$\widehat{\theta}_y(\alpha) = inf\left\{t : \widehat{F}_y(t) \geq \alpha\right\} \tag{23.3.2}$$

Let $\widehat{F}_x(t)$ be the customary estimator of $F_x(t)$. In case the population αth quantile $\theta_x(\alpha)$ of x is known, the ratio estimator of $\theta_y(\alpha)$ is given by

$$\widehat{\theta}_{ry}(\alpha) = \frac{\widehat{\theta}_y(\alpha)}{\widehat{\theta}_x(\alpha)}\theta_x(\alpha) \tag{23.3.3}$$

Similarly, a difference estimator of $\theta_y(\alpha)$ is given by

$$\widehat{\theta}_{dy}(\alpha) = \widehat{\theta}_y(\alpha) - \widehat{R}\left\{\widehat{\theta}_x(\alpha) - \theta_x(\alpha)\right\} \tag{23.3.4}$$

where $\widehat{R} = \dfrac{\sum_{i\in s} y_i/\pi_i}{\sum_{i\in s} x_i/\pi_i}$ is a consistent estimator of the population ratio $R = Y/X$.

Both the estimators $\widehat{\theta}_{ry}(\alpha)$ and $\widehat{\theta}_{dy}(\alpha)$ reduce to $\theta_y(\alpha)$ if $y_i \propto x_i\ \forall i \in U$. In this case the variances of the estimators become zero. Hence the estimators $\widehat{\theta}_{ry}(\alpha)$ and $\widehat{\theta}_{dy}(\alpha)$ are expected to produce a considerable gain in efficiency over $\widehat{\theta}_y(\alpha)$ if y_i is approximately proportional to x_i. Rao et al. (1990) derived the variances of $\widehat{\theta}_{ry}(\alpha)$ and $\widehat{\theta}_{dy}(\alpha)$ and also their unbiased estimators. These are omitted here because of their complexities.

23.4 ESTIMATION OF MEDIAN

The median of a variable y is obtained by substituting $\alpha = 1/2$ in Eq. (23.3.1). Thus the population median of y and x are, respectively,

$$\begin{aligned} \widetilde{\mu}_y &= \theta_y(1/2) = inf\{t : F_y(t) \geq 1/2\} \text{ and} \\ \widetilde{\mu}_x &= \theta_x(1/2) = inf\{t : F_x(t) \geq 1/2\} \end{aligned} \tag{23.4.1}$$

If the population median $\widetilde{\mu}_x = \theta_x(1/2)$ of the auxiliary variable x is known, then the ratio estimator of the population median $\widetilde{\mu}_y$ is obtained from Eq. (23.3.3) as

$$\widehat{\widetilde{\mu}}_{ry} = \widehat{\theta}_{ry}(1/2) = \frac{\widehat{\widetilde{\mu}}_y}{\widehat{\widetilde{\mu}}_x}\widetilde{\mu}_x \tag{23.4.2}$$

where $\widehat{\widetilde{\mu}}_x$ and $\widehat{\widetilde{\mu}}_y$ are the sample medians of x and y, respectively.

Kuk and Mak (1989) proposed the following modifications of the ratio estimator of $\widehat{\widetilde{\mu}}_{ry}$ based on simple random sampling.

23.4.1 Position Estimator and Stratification Estimator

At first, let us arrange y_i's, $i \in s$ in order of magnitude as $y_{(1)} \leq y_{(2)} \leq \cdots \leq y_{(n)}$. Let i_0 be the number of observations of y_i's less than or equal to the population median $\widetilde{\mu}_y$, i.e., $y_{(i_0)} \leq \widetilde{\mu}_y < y_{(i_0+1)}$. Clearly, i_0 is unknown because $\widetilde{\mu}_y$ is unknown. Let $p = i_0/n$, then $\widetilde{\mu}_y$ is approximately the sample pth quantile $\widehat{\theta}_y(p)$. Suppose $\widehat{p}$ is an estimator of p, then $\widehat{\theta}_y(\widehat{p})$ is an estimator of the median $\widetilde{\mu}_y$. $\widehat{\widetilde{\mu}}_y$, the sample median of y can be viewed as the special estimator $\widehat{\theta}_y(\widehat{p})$ with $\widehat{p} = 1/2$. To estimate $\widehat{p}$, Kuk and Mak (1989) considered the following two-way classified table.

	$x \le \widetilde{\mu}_x$	$x > \widetilde{\mu}_x$	Total
$y \le \widetilde{\mu}_y$	P_{11}	P_{12}	P_{10}
$y > \widetilde{\mu}_y$	P_{21}	P_{22}	P_{20}
Total	P_{01}	P_{02}	1

In the table above, P_{ij} denotes the proportion of units in the population that belongs to the (i, j)th cell. Let n_x be the number of observations in the sample s with x_i's less than or equal to $\widetilde{\mu}_x$. In case P_{ij} are known, an estimate of p is given by

$$\begin{aligned}\widehat{p} &= \frac{1}{n}\left(n_x\frac{P_{11}}{P_{01}} + (n - n_x)\frac{P_{12}}{P_{02}}\right)\\ &\cong \frac{2}{n}\left(n_xP_{11} + (n - n_x)\left(\frac{1}{2} - P_{11}\right)\right)\end{aligned} \tag{23.4.3}$$

where $P_{0j} = P_{1j} + P_{2j} \cong 1/2$ for $j = 1, 2$.

In practice, P_{ij} are generally unknown. So, we estimate P_{ij} by p_{ij}, the proportion of units in the sample that fall in the (i, j)th cell, i.e., p_{11} is the proportion of observation in the class $x \le \widehat{\widetilde{\mu}}_x$(=sample median of x) and $y \le \widehat{\widetilde{\mu}}_y$. Now replacing P_{11} by p_{11} in Eq. (23.4.3), we find an estimate of p as

$$\widehat{p}_1 \cong \frac{2}{n}\left(n_xp_{11} + (n - n_x)\left(\frac{1}{2} - p_{11}\right)\right) \tag{23.4.4}$$

Thus an estimator of $\widetilde{\mu}_y$ is given by

$$\widehat{\widetilde{\mu}}_{py} = \widehat{\theta}_y(\widehat{p}_1) \tag{23.4.5}$$

The estimator $\widehat{\widetilde{\mu}}_{py}$ was termed as the "position estimator" by Kuk and Mak (1989).

For a given value of $y = t$, let $\widetilde{F}_{1y}(t)$ be the proportion of those units in the sample with $x \le \widetilde{\mu}_x$, which have y-values less than equal to t and $\widetilde{F}_{2y}(t)$ be the proportion of those units in the sample with $x > \widetilde{\mu}_x$, which have y-values less than equal to t. Then $F_y(t)$ can be estimated as

$$\begin{aligned}\widetilde{F}_y(t) &= \frac{1}{N}\left[N_x\widetilde{F}_{1y}(t) + (N - N_x)\widetilde{F}_{2y}(t)\right]\\ &\cong \frac{1}{2}\left[\widetilde{F}_{1y}(t) + \widetilde{F}_{2y}(t)\right]\end{aligned} \tag{23.4.6}$$

where N_x is the number of units in the population with $x \le \widetilde{\mu}_x$.

Since $\widetilde{F}_y(t)$ is a distribution function, an estimator of the median $\widetilde{\mu}_y$ is obtained as

$$\widehat{\widetilde{\mu}}_{sy} = inf\left\{y : \widetilde{F}_y(t) \geq \frac{1}{2}\right\} \tag{23.4.7}$$

Kuk and Mak (1989) called the estimator $\widehat{\widetilde{\mu}}_{sy}$ as the "stratification estimator."

23.4.2 Comparison of the Efficiencies

Gross (1980) and Kuk and Mak (1989) derived the asymptotic distributions of the median estimators $\widehat{\widetilde{\mu}}_y$, $\widehat{\widetilde{\mu}}_{ry}$, $\widehat{\widetilde{\mu}}_{py}$, and $\widehat{\widetilde{\mu}}_{sy}$ when $N \to \infty$, $n \to \infty$ and $n/N \to f$, $0 \leq f \leq 1$. They assumed that as $N \to \infty$, the bivariate distribution of (x, y) approaches a continuous distribution with marginal densities $f_x(x)$ and $f_y(y)$ for x and y, respectively. Gross (1980) proved that the sample median $\widehat{\widetilde{\mu}}_y$ is consistent and asymptotically normally distributed with mean $\widetilde{\mu}_y$ and variance

$$Var\left(\widehat{\widetilde{\mu}}_y\right) = \frac{(1-f)}{4n}\left\{f_y\left(\widetilde{\mu}_y\right)\right\}^{-2} \tag{23.4.8}$$

Kuk and Mak (1989) derived the following results:
(i) $\widehat{\widetilde{\mu}}_{ry}$ is asymptotically normal with mean $\widetilde{\mu}_y$ and variance

$$Var\left(\widehat{\widetilde{\mu}}_{ry}\right) = \frac{(1-f)}{n}\left[\frac{1}{4}\left(f_y\left(\widetilde{\mu}_y\right)\right)^{-2} + \frac{1}{4}\widetilde{R}^2\left(f_x(\widetilde{\mu}_x)\right)^{-2} - 2\widetilde{R}\left(\frac{f_y\left(\widetilde{\mu}_y\right)}{f_x(\widetilde{\mu}_x)}\right)^{-1}\left(P_{11} - \frac{1}{4}\right)\right] \tag{23.4.9}$$

where $\widetilde{R} = \widetilde{\mu}_y/\widetilde{\mu}_x$.

(ii) $\widehat{\widetilde{\mu}}_{py}$ and $\widehat{\widetilde{\mu}}_{sy}$ both asymptotically follow the same distribution that is normal with mean $\widetilde{\mu}_y$ and variance

$$Var\left(\widehat{\widetilde{\mu}}_{py}\right) = Var\left(\widehat{\widetilde{\mu}}_{sy}\right) = \frac{2(1-f)P_{11}(1-2P_{11})}{n}\left(f_y\left(\widetilde{\mu}_y\right)\right)^{-2} \tag{23.4.10}$$

where P_{11} is the proportion of units in the population with $x \leq \widetilde{\mu}_x$ and $y \leq \widetilde{\mu}_y$. The probability P_{11} can be regarded as a measure of concordance.

It is important to note that if $P_{11} = 1/2$, then the asymptotic variances $Var\left(\widehat{\widetilde{\mu}}_{py}\right)$ and $Var\left(\widehat{\widetilde{\mu}}_{sy}\right)$ are both equal to zero. The expressions (23.4.8) and (23.4.10) show that the estimators $\widehat{\widetilde{\mu}}_{py}$ and $\widehat{\widetilde{\mu}}_{sy}$ are asymptotically more efficient than the sample median because

$$Var\left(\widehat{\widetilde{\mu}}_{sy}\right) - Var\left(\widehat{\widetilde{\mu}}_{y}\right) = \frac{1}{4}(4P_{11} - 1)^2 \geq 0 \tag{23.4.11}$$

The estimator $\widehat{\widetilde{\mu}}_{ry}$ becomes asymptotically more efficient than the sample median $\widehat{\widetilde{\mu}}_{y}$ if

$$\rho_c > \frac{1}{2}\left(\widetilde{R}\frac{f_Y\left(\widetilde{\mu}_y\right)}{f_X(\widetilde{\mu}_x)}\right) \tag{23.4.12}$$

where $\rho_c = 4(P_{11} - 1/4)$ varies from -1 to 1 as P_{11} increases from 0 to 1/2.

23.4.3 Further Generalization

Let the vector of the auxiliary variable $\mathbf{x} = (x_1, \ldots, x_N)$ be known and the range of the auxiliary variable be partitioned into r mutually exclusive and exhaustive class intervals $(a_0, a_1], (a_1, a_2], \ldots, (a_{r-1}, a_r]$ with $a_0 = 0$, $a_r = \infty$, and $a_j = \theta_x(\alpha_j)$. Let P_{1j} be the proportion of the units in the population with $y \leq \widetilde{\mu}_y$ and x falling in the class $(a_{j-1}, a_j]$ for $j = 1, \ldots, r$ and let p_{1j} be the proportion of units in s with $y \leq \widehat{\widetilde{\mu}}_y$ and x in $\left(\widehat{a}_{j-1}, \widehat{a}_j\right]$ where $\widehat{a}_j = \widehat{\theta}_x(\alpha_j)$. Also, let P_{0j} be the proportion of units in the population with $x \in (a_{j-1}, a_j]$ and p_{oj} be the proportion in the sample with $x \in \left(\widehat{a}_{j-1}, \widehat{a}_j\right]$, whereas n_{xj} is the number of units in the sample with $x \in (a_{j-1}, a_j]$.

Proportions in population

	$(a_0, a_1]$	—	$(a_{j-1}, a_j]$	—	$(a_{r-1}, a_r]$
$y \leq \widetilde{\mu}_y$	P_{11}	—	P_{1j}	—	P_{1r}
$y > \widetilde{\mu}_y$	P_{21}	—	P_{2j}	—	P_{2r}
Total	P_{01}	—	P_{0j}	—	P_{0r}

Proportions in sample

	$(a_0, a_1]$	—	$(a_{j-1}, a_j]$	—	$(a_{r-1}, a_r]$
$y \leq \widehat{\widetilde{\mu}}_y$	p_{11}	—	p_{1j}	—	p_{1r}
$y > \widehat{\widetilde{\mu}}_y$	p_{21}	—	p_{2j}	—	p_{2r}
Total	p_{01}	—	p_{0j}	—	p_{0r}

The estimated proportion of y's in the sample that is less than equal to $\widetilde{\mu}_y$ is given by Kuk and Mak (1989) as

$$\begin{aligned}\widetilde{p}_{g1} &= \frac{1}{n}\sum_{j=1}^{r} n_{xj}\frac{p_{1j}}{p_{0j}}\\ &= \frac{1}{n}\sum_{j=1}^{r} n_{xj}\frac{p_{1j}}{(\alpha_j - \alpha_{j-1})}\end{aligned} \quad (23.4.13)$$

Finally, the estimated populated median is

$$\widehat{\widetilde{\mu}}_{gpy} = \widehat{\theta}_y\left(\widetilde{p}_{g1}\right) \quad (23.4.14)$$

Let $\widehat{F}_{yj}(t)$ be the proportion of the units in the sample with $x \in (a_{j-1}, a_j]$ that have y-values less than or equal to t. Then $F_y(t)$ may be estimated by

$$\widehat{F}_y(t) = \sum_{j=1}^{r}(\alpha_j - \alpha_{j-1})\widehat{F}_{yj}(t)$$

Consequently the estimated median is

$$\widehat{\widetilde{\mu}}_{gsy} = inf\left\{y : \widehat{F}_y(t) \geq 1/2\right\} \quad (23.4.15)$$

Kuk and Mak (1989) showed that the asymptotic variances of $\widehat{\widetilde{\mu}}_{gpy}$ and $\widehat{\widetilde{\mu}}_{gsy}$ are the same and equal to

$$V_{ar}\left(\widehat{\widetilde{\mu}}_{gpy}\right) = V_{ar}\left(\widehat{\widetilde{\mu}}_{gsy}\right) = \frac{1-f}{n}\frac{1}{\left(f_y\left(\widetilde{\mu}_y\right)\right)^2}\left(\frac{1}{2} - \sum_{j=1}^{r}\frac{P_{1j}^2}{(\alpha_j - \alpha_{j-1})}\right) \quad (23.4.16)$$

23.4.4 Empirical Comparison

Kuk and Mak (1989) compared efficiencies of the estimators $\widehat{\widetilde{\mu}}_y$, $\widehat{\widetilde{\mu}}_{ry}$, $\widehat{\widetilde{\mu}}_{py}$, and $\widehat{\widetilde{\mu}}_{sy}$ empirically using four populations named Hospitals, Counties 70, Villages, and Factories. The first two populations were used by Royall and Cumberland (1981), where x and y are well correlated and a linearity relationship holds. For the population Villages (Murthy, 1967), where x (area in 1951) and y (number of households in 1961) are poorly correlated, the probability of concordance P_{11} is also low. For the population Factories (Murthy, 1967), x (number of workers) and y (output) are not linearly related. From the first two populations, 500 independent samples of suitable sizes were selected by SRSWOR method while from each of the remaining other two populations, 1000 independent samples of suitable sizes were selected. Empirical studies reveal that for the populations Hospitals and Counties where x and y are linearly related, all three estimators $\widehat{\widetilde{\mu}}_{ry}$, $\widehat{\widetilde{\mu}}_{py}$, and $\widehat{\widetilde{\mu}}_{sy}$ have much lower mean square errors than that of the sample median $\widetilde{\mu}_y$. For the population Villages where x and y are poorly related, the ratio estimator $\widehat{\widetilde{\mu}}_{ry}$ performed worse than the sample median $\widehat{\widetilde{\mu}}_y$. But the other two estimators $\widehat{\widetilde{\mu}}_{py}$ and $\widehat{\widetilde{\mu}}_{sy}$ fare better than $\widehat{\widetilde{\mu}}_y$. For the nonlinear population Factories, $\widehat{\widetilde{\mu}}_{py}$ and $\widehat{\widetilde{\mu}}_{sy}$ still performed better than $\widehat{\widetilde{\mu}}_y$ where the ratio estimator $\widehat{\widetilde{\mu}}_{ry}$ is again outperformed by the sample median. Thus the efficiencies of the position estimator $\widehat{\widetilde{\mu}}_{py}$ and stratification estimator $\widehat{\widetilde{\mu}}_{sy}$ do not depend on the validity of linearity assumption and hence it is much safer to use, than the ratio estimator $\widehat{\widetilde{\mu}}_{ry}$.

23.5 CONFIDENCE INTERVAL FOR DISTRIBUTION FUNCTION AND QUANTILES

The conventional $(1 - \alpha)100\%$ confidence interval of the distribution function $F(t)$ is

$$\left(\widehat{F}(t) - z_{\alpha/2}\sqrt{\widehat{V}\left[\widehat{F}(t)\right]},\ \widehat{F}(t) + z_{\alpha/2}\sqrt{\widehat{V}\left[\widehat{F}(t)\right]}\right) \qquad (23.5.1)$$

where $z_{1-\alpha/2}$ is the $1 - \alpha/2$ quantile from $N(0, 1)$ and $\widehat{V}\left[\widehat{F}(t)\right]$ is an estimator of $V\left[\widehat{F}(t)\right]$. The validity of the confidence interval obviously depends on the asymptotic normality of the distribution

$$\frac{\widehat{F}(t) - F(t)}{\sqrt{\widehat{V}\left[\widehat{F}(t)\right]}}$$

which is justified when the sample size is large. However, for small to moderate sample size, because of range constraint $0 \leq \widehat{F}(t) \leq 1$, the distribution of $\widehat{F}(t)$ for t at large or small quantiles is usually not symmetric. For small and moderate sample size, the performance of the interval (23.5.1) is very often unsatisfactory; coverage probability is generally lower than the nominal value and two tail probabilities are unbalanced (Wu, 1999).

Chen and Wu (2002) proposed an alternative method. In this method a smooth and monotone function g is chosen so that the distribution of

$\widehat{W} = \widehat{W}(t) = g\left[\widehat{F}(t)\right]$ is better approximated by the normal distributions. Two such popular transformations are logit transformation and complementary log–log transformation: $\widehat{W} = \log\left(\frac{\hat{F}(t)}{1 - \hat{F}(t)}\right)$ and $\log\left(-\log\left\{\widehat{F}(t)\right\}\right)$. Noting that $\widehat{W}$ asymptotical normal with mean $W = W(t) = g[F(t)]$ and variance $V\left(\widehat{W}\right) = \left(g'\{F(t)\}\right)^2 V\left\{\widehat{F}(t)\right\}$, we can find a $(1-\alpha)100\%$ confidence interval of W as

$$\left(\widehat{W} - z_{\alpha/2}\sqrt{\widehat{V}\left(\widehat{W}\right)},\ \widehat{W} + z_{\alpha/2}\sqrt{\widehat{V}\left(\widehat{W}\right)}\right)$$

where $\widehat{V}\left(\widehat{W}\right) = \left(g'\left\{\widehat{F}(t)\right\}\right)^2 \widehat{V}\left\{\widehat{F}(t)\right\}$ is a suitable estimator of $V\left(\widehat{W}\right)$.

Finally, transformed confidence interval for $F(t)$ is obtained as

$$\left(g^{-1}\left\{\widehat{W} - z_{\alpha/2}\sqrt{\widehat{V}\left(\widehat{W}\right)}\right\},\ g^{-1}\left\{\widehat{W} + z_{\alpha/2}\sqrt{\widehat{V}\left(\widehat{W}\right)}\right\}\right) \quad (23.5.2)$$

On the basis of simulation studies, Wu (1999) showed that performance of the intervals (23.5.1) and (23.5.2) is similar when t is in the middle range of quantiles but Eq. (23.5.2) is much better when t is small or large quantiles.

Let $\widehat{\theta}(\alpha)$ be an estimator of the $\alpha(0 < \alpha < 1)$th quantile $\theta(\alpha)$ obtained from the inversion of estimated distribution function $\widehat{F}(t)$, i.e., $\widehat{F}\left\{\widehat{\theta}(\alpha)\right\} = \alpha$. Furthermore, let $L(t)$ and $U(t)$ be the lower and upper $(1-\alpha)100\%$ confidence intervals of $F(t)$ obtained from any of the formulas (23.5.1) and (23.5.2). Then following Woodruff (1952), the confidence interval of the $\alpha(0 < \alpha < 1)$th quantile $\theta(\alpha)$ is obtained as

$$\left(\widehat{F}^{-1}\left[L\left\{\widehat{\theta}(\alpha)\right\}\right],\ \widehat{F}^{-1}\left[U\left\{\widehat{\theta}(\alpha)\right\}\right]\right) \quad (23.5.3)$$

Obviously the confidence interval based on the formula (23.5.2) is expected to perform better than that on Eq. (23.5.1).

Godambe and Thompson (1999) used estimating functions for determination of confidence intervals of distribution functions and quantiles while Chen and Sitter (1999), Wu and Sitter (2001), among others, used empirical likelihood methods. Details have been given in Chapters 22 and 25, respectively.

23.6 CONCLUDING REMARKS

In surveys, estimation of distribution function and quantiles often play an important role. Rao et al. (1990) showed how one can use auxiliary information x to construct customary design-based ratio $\widehat{F}_R(t)$ and difference estimator $\widehat{F}_D(t)$ of the population distribution function $F(t)$, which can improve upon $\widehat{F}(t)$, the conventional estimator of the distribution function $F(t)$. The proposed estimators $\widehat{F}_R(t)$ and $\widehat{F}_D(t)$ are asymptotically design unbiased but not model unbiased. Chambers and Dunstan (1986) proposed a model-based estimator $\widehat{F}_{cd}(t)$, which is efficient if the assumed model is valid, but this estimator can perform poorly under model misspecification. Rao et al. (1990) also proposed model-based estimators $\widehat{F}^*(t)$ and $\widehat{F}_{rkm}(t)$, which are both asymptotically design unbiased and model-unbiased under the model (23.2.17) with $\nu(x_i) = \sqrt{x_i}$. Moreover, $\widehat{F}_{rkm}(t)$ can be modified using multiauxiliary variables. Poststratification estimator $\widehat{F}_{ss}(t)$ was proposed by Silva and Skinner (1995) and it retains the property of asymptotic design unbiasedness. To compute Kuk's (1993) estimator $\widehat{F}_{kk}(t)$, the values of the study variable y need to be scaled since the bandwidth that is used to control the smoothing is the same for both the variables x and y. The estimators $\widehat{F}_R(t)$, $\widehat{F}_D(t)$, $\widehat{F}^*(t)$, $\widehat{F}_{cd}(t)$, and $\widehat{F}_{ss}(t)$ are advantageous because the variance estimators are readily available and easy to compute. Although the estimator of variance of $\widehat{F}_{rkm}(t)$ is readily available, it can be extremely complex for a varying probability sampling scheme because of the involvement of third-order inclusion probabilities π_{ijk} that are difficult to compute. Extensive simulation studies were done by Chambers and Dunstan (1986), Rao et al. (1990), Chambers et al. (1993), Kuk (1993), and Silva and Skinner (1995), among others, to compare performances among the proposed estimators. The simulation studies do not finger out any particular estimator as the best in all situations. However, $\widehat{F}_{rkm}(t)$ seems to perform well in most of the situations irrespective of the failure of the model and so it is safe to be used.

Estimators of the distribution functions can be extended to estimation of population quantiles and medians. Kuk and Mak (1989) proposed the alternative position estimator $\widehat{\widetilde{\mu}}_{py}$ and stratification estimator $\widehat{\widetilde{\mu}}_{sy}$ to estimate the median under SRSWOR sampling. Empirical studies of Kuk and Mak (1989) reveal that the estimators $\widehat{\widetilde{\mu}}_{py}$ and $\widehat{\widetilde{\mu}}_{sy}$ are more efficient than the sample median $\widehat{\widetilde{\mu}}_y$ and the ratio estimator $\widehat{\widetilde{\mu}}_{ry}$ even if the relation between x and y is not linear. The use of empirical likelihood method for estimating finite population distribution function was considered by Owen (1988), Chen and Sitter (1999), Wu and Sitter (2001). The proposed estimators use auxiliary information effectively at the estimation stage and possess attractive properties. Godambe and Thompson (1999) used the method of estimating function whereas Chen and Wu (2002) used calibrated pseudoempirical likelihood methods for determining confidence intervals. Further discussions are given by Johnson (2003), Drofman (2009), among others.

23.7 EXERCISES

23.7.1 Let a sample of size 10 be selected from a population of size 50 by simple random sampling with replacement method.

Selected sample	1	2	3	4	5	6	7	8	9	10
y-values	10	8	6	15	3	8	5	6	3	1

(i) Sketch the empirical distribution function.
(ii) Estimate the 25th, 50th, and 80th percentiles from the graph. Estimate the standard errors of the estimators used. Determine 95% confidence interval of the population median.

23.7.2 A sample of size of size 8 is selected from a finite population of size 50 using Poisson sampling scheme. The following table gives the units selected in the sample, y-values, and inclusion probabilities (π_i) of the selected units.

Selected sample	1	2	3	4	5	6	7	8
y-values	15	10	5	4	6	15	20	15
π_i	0.10	0.15	0.20	0.25	0.15	0.10	0.25	0.10

Sketch the distribution function using the formulae.

(i) $\widehat{F}(t) = \frac{1}{N} \sum_{i \in s} \frac{I(y_i \le t)}{\pi_i}$

(ii) $\widehat{F}_h(t) = \sum_{i \in s} \frac{I(y_i \le t)}{\pi_i} \Big/ \sum_{i \in s} \frac{1}{\pi_i}$

(iii) $\widehat{F}_c(t) = 1 - \frac{1}{N} \sum_{i \in s} \frac{1 - I(y_i \le t)}{\pi_i}$

Discuss the appropriateness of the formulae. From the sketches, obtain estimates of the population median. Compute standard errors of the estimators used.

23.7.3 A sample of 10 plants is selected at random from 40 plants of a garden. The following table gives the height (y) of 10 plants along with the diameter (x) of all the 40 plants in that garden. Fit distribution functions over the data using the following methods:

(i) Ratio method: $\widehat{F}_R(t) = \frac{1}{N}\widehat{\lambda}\sum_{i\in U}\frac{I(\hat{y}_i \le t)}{\pi_i}$

where $\widehat{\lambda} = \left(\dfrac{\sum_{i\in s}\frac{I(y_i \le t)}{\pi_i}}{\sum_{i\in s}\frac{I(\widehat{y}_i \le t)}{\pi_i}}\right)$ and $\widehat{y}_i = \left(\dfrac{\sum_{i\in s} y_i}{\sum_{i\in s} y_i}\right)x_i$

(ii) Difference method:

$$\widehat{F}_D(t) = \frac{1}{N}\left[\sum_{i\in s}\frac{I(y_i \le t)}{\pi_i} - \left\{\sum_{i\in s}\frac{I(\hat{y}_i \le t)}{\pi_i} - \sum_{i\in U} I(\widehat{y}_i \le t)\right\}\right]$$

From the fitted distribution functions, estimate median and 80th percentiles of the heights of the plants. Also estimate standard errors of the estimators used.

Plants	1	2	3	4	5	6	7	8	9	10
Diameter (in cm)	15	20	30	25	30	40	15	12	20	30
Height (in cm)	80	90	120	100	100					

Plants	11	12	13	14	15	16	17	18	19	20
Diameter (in cm)	15	20	30	25	30	40	15	12	20	30

Plants	21	22	23	24	25	26	27	28	29	30
Diameter (in cm)	10	20	18	28	50	60	45	20	25	30

Plants	31	32	33	34	35	36	37	38	39	40
Diameter (in cm)	15	25	20	10	10	30	25	40	60	30

23.7.4 Continuation of Exercise 27.7.3. Assume that y and x are related to the model $y_i = \beta x_i + \in_i$, where $\in_i$'s are independent with $E_m(\in_i) = 0$ and $V_m(\in_i) = \sigma^2 x_i$. Estimate the median height of the plant using the Chamber and Dunstan estimator

$$\widehat{F}_{cd}(t) = \frac{1}{N}\left[\sum_{i\in s} I(y_i \le t) - \sum_{i\in U-s}\left\{\frac{1}{n}\sum_{j\in s} I\left(\frac{\left(y_i - \widehat{\beta}x_j\right)}{\sqrt{v(x_j)}} \le \frac{\left(t - \widehat{\beta}x_j\right)}{\sqrt{v(x_j)}}\right)\right\}\right]$$

where $\widehat{\beta} = \sum_{j\in s}\frac{y_i x_i}{\sqrt{v(x_j)}} \Big/ \sum_{j\in s}\frac{x_i^2}{\sqrt{v(x_j)}}$ and $v(x_j) = \sigma^2 x_j^2$.

23.7.5 The following table relates to the daily wages in (US$) of 40 factory workers selected at random from 120 workers.

120	80	160	200	400	550	300	400	500	150
140	90	150	180	300	250	200	180	150	250
270	380	920	150	200	225	420	100	120	150
80	75	150	200	500	590	250	140	150	200

Estimate the median wage of the factory workers and obtain 90% confidence interval of the median wages.

CHAPTER 24

Controlled Sampling

24.1 INTRODUCTION

In selecting a sample of size N from a finite population of N units by simple random sampling without replacement (SRSWOR) procedure, all the possible $\binom{N}{n}$ samples have equal probability of selection, but all the samples are not equally advantageous for surveying purposes. Sampling units within a sample may be so widespread that the cost of data collection may be very expensive due to travel costs, and at the same time non-sampling errors involving nonresponse and investigator's bias increase because of inadequate supervision of fieldwork. Such samples, which are uneconomical and also create organizational and other difficulties, are termed nonpreferred or undesirable samples by Goodman and Kish (1950). The controlled sampling method, originated by Goodman and Kish (1950), reduces the probability of selection of undesirable samples while retaining properties associated with a probability sampling design. The scope of control sampling may include appropriate distribution of sampling units over different subgroups of the population to obtain reliable estimates from each of the subgroups. Controlled sampling is also used for increasing efficiency of key estimates for a multicharacter survey. Controlled selection can be easily achieved by stratification. For example, if we choose a sample of 6 students from a group of 24 students comprising of 12 male and 12 female, there is a possibility of selecting all 6 male students or all 6 female students. If we wish to control the selection of male or female students, we may stratify the 24 students into two strata comprising of 12 male and 12 female students, and then select 3 students from each of the strata. We thereby control the selection of male and female students to a fixed number of three each. Goodman and Kish (1950) pointed out that the control selection cannot be obtained by applying stratification alone. The use of controlled sampling is not always safe for multistage sampling because unbiased variance estimation may not always be possible. Using data available from a survey of Scottish schools, Waterton (1983) showed that controlled sampling provides more efficient

Survey Sampling Theory and Applications
ISBN 978-0-12-811848-1
http://dx.doi.org/10.1016/B978-0-12-811848-1.00024-8

estimates than multiproportionate stratified sampling. The sample coordination problem is similar to controlled sampling where the overlaps of two or more samples drawn in different occasions are controlled. It is either positive or negative. In positive coordination, the expected overlap is maximized whereas in negative coordination, it is minimized.

Five different approaches of controlled sampling available in the literature include the following: (i) experimental design configurations, (ii) linear programming (LP), (iii) nearest proportional to size design, (iv) nonlinear programming, and (v) coordination of samples overtime. Combinatorial properties of experimental designs in controlled sampling designs were used by Chakrabarti (1963), Avadhani and Sukhatme (1973), Foody and Hedayat (1977), Gupta et al. (1982, 2012), Nigam et al. (1984) etc. The application of linear and nonlinear programming in controlled sampling was used by Rao and Nigam (1990, 1992), Mandal et al. (2010, 2011) and Tiwari et al. (2007), while Gabler (1987a,b) used nearest proportional to size sampling design for selection of controlled sample. Coordination of samples overtime was considered by Keyfitz (1951), Fellegi (1963), Lanke (1974a,b), among others. A detailed review has been given by Arnab (2013). Some of the controlled sampling techniques have been described below.

24.2 PIONEERING METHOD

Consider the example of Goodman and Kish (1950) where a population is stratified into two strata. Strata 1 comprises 6 units A, B, C, D, E, and F whereas strata 2 comprises 5 units a, b, c, d, and e. The units B, C, F of the stratum 1 are identified as coastal units and the rest (A, D and E) are inland units. Similarly, for stratum 2 units a, b, c, and e are inland while the unit d is a coastal unit. The probability of selection assigned to each of the units is given below:

Stratum 1		Stratum 2	
Unit	**Probability**	**Unit**	**Probability**
A	0.10	a	0.15
B	0.15	b	0.30
C	0.10	c	0.10
D	0.20	d	0.20
E	0.25	e	0.25
F	0.20		
P (inland) = 0.55		P (inland) = 0.80	
P (coastal) = 0.45		P (coastal) = 0.20	

It is desirable to select one inland and one coastal unit. The selection of two coastal units is undesirable. Under stratified random sampling selecting one unit from each of the stratum, the probabilities of selection of different combinations of units are as follows:

$$
\begin{aligned}
P\,(\text{one inland, one coastal}) &= P\,(\text{inland from stratum 1}) \\
&\quad \times P\,(\text{coastal from stratum 2}) \\
&\quad + P\,(\text{coastal from stratum 1}) \\
&\quad \times P\,(\text{inland from stratum 2}) \\
&= 0.55 \times 0.20 + 0.45 \times 0.80 = 0.47
\end{aligned}
$$

$$
\begin{aligned}
P\,(\text{two inland}) &= P\,(\text{inland from stratum 1}) \times P\,(\text{inland from stratum 2}) \\
&= 0.55 \times 0.80 = 0.44 \\
P\,(\text{two coastal}) &= P\,(\text{coastal from stratum 1}) \times P\,(\text{coastal from stratum 2}) \\
&= 0.45 \times 0.20 = 0.09
\end{aligned}
$$

Goodman and Kish (1950) in their proposed method rearranged units in the stratum 1 by listing *B*, *C*, *F* first, followed by *A*, *D*, and *E*. Then they rearranged the units in the stratum 2 by shifting *d* to the end, i.e., by placing the unit *e* above *d*.

Stratum 1		Stratum 2	
Unit	**Probability**	**Unit**	**Probability**
B	0.15	*a*	0.15
C	0.10	*b*	0.30
F	0.20	*c*	0.10
A	0.10	*e*	0.25
D	0.20	*d*	0.20
E	0.25		

In this method the selection of units from both the strata was done by drawing a single random number from 1 to 100. If the selected random number is 45 or less, a coastal unit is selected from stratum 1 and an inland unit from stratum 2. If the selected number is between 46 and 80, an inland unit is selected from both the strata. If the selected number is greater than 80, an inland unit is selected from stratum 1 and a coastal unit from stratum 2. In this method the probabilities of different combinations of selection of units are:

P (one inland one coastal unit) = 0.65 and P (two inland units) = 0.35. In this procedure the original assigned probabilities of all the units are rigorously maintained and the probability of selection of desirable samples (one inland and one coastal unit) is made as large as possible within the limitation of probability sampling.

24.3 EXPERIMENTAL DESIGN CONFIGURATIONS

In an experimental design setup, one to one correspondence between a sampling design and a block design are established. The treatment and block of a block design are termed as the unit and sample of a sampling design, respectively. Thus, the total number of treatments v is equal to the total number of units N and the sample size n is equal to the block size k. The total number of blocks of an experimental design b will be treated as the total number of possible samples in a sampling design. In constructing a controlled sampling, properties of various incomplete block designs with minimum number of support (block) sizes are used. Preferred samples are assigned as many blocks as possible by trial and error method while the remaining blocks are associated with the undesirable samples. One block (sample) is then selected at random with a preassigned probability so that the property of a probability sampling is maintained.

24.3.1 Equal Probability Sampling Design

Here, one block (sample) is selected at random from b blocks so that first- and second-order inclusion probabilities for the ith unit and i and jth ($i \neq j$) units become, respectively, equal to

$$\pi_i = \frac{n}{N} \text{ and } \pi_{ij} = \frac{n(n-1)}{N(N-1)} \tag{24.3.1}$$

Avadhani and Sukhatme (1973) used the properties of the balanced incomplete block design (BIBD) in the construction of a controlled sampling design. In this method a BIBD is constructed with parameters (v, b, r, k, λ) assuming it exists, where $v = N =$ population size, $b =$ total number of blocks (samples), $r=$ replication of a treatment $=$ total number of times a unit appears in b samples, which is the same for all units, $k(=n)$ block (sample) size and $\lambda=$ number of times any two treatments (units) appear together in the same block (sample). Let us identify blocks of the BIBD with the preferred samples or maximum possible number of preferred samples and the rest with nonpreferred samples. One block is selected at random from the b blocks. The selected block constitutes the controlled sample. In this controlled sampling, the inclusion probabilities are $\pi_i = r/b$ and $\pi_{ij}=\lambda/b$. Furthermore, from the properties of a BIBD, *viz.*, (i) $bk=vr$ and (ii) $\lambda(v-1)=r(k-1)$ (Raghavrao, 1971), we find $\pi_i=\frac{n}{N}$ and $\pi_{ij}=\frac{n(n-1)}{N(N-1)}$. Thus the number of supports for a controlled sampling b is much less than the number of supports $\binom{N}{n}$ of an SRSWOR sampling design of size n. The number of supports becomes the minimum when $b=v=N$, i.e., if the BIBD is symmetric. Since the number of the preferred samples is identified as much as possible with the block, the probability of selection of preferred samples for controlled sampling is much higher than that of uncontrolled SRSWOR sampling using the same sample size. Where efficiency is concerned, the sample mean of the controlled sampling design is equally precise as the sample mean of SRSWOR sampling because both the designs possess the same first- and second-order inclusion probabilities.

Example 24.3.1

Consider the following example of Avadhani and Sukhatme (1973), which comprises $N=7$ and $n=3$. Suppose that the units are located as follows:

	2		1	
7		5		4
	6		3	

From the point of view of travel and the inconvenience of fieldwork, the following 14 samples are considered as undesirable samples:

$$(1,2,3),(1,2,6),(1,3,6),(1,3,7),(1,4,6),(1,4,7),(1,6,7),$$
$$(2,3,4),(2,3,6),(2,3,7),(2,4,6),(2,4,7),(3,4,7),(4,6,7)$$

Consider the following BIBD with parameters $v=7=b$, $r=k=3$, and $\lambda=1$ with layout.

Block 1: (1, 2, 4); Block 2: (2, 3, 5); Block 3: (3, 4, 6); Block 4: (4, 5, 7); Block 5: (5, 6, 1); Block 6: (6, 7, 2), and Block 7: (7, 1, 3)*; here *denotes nonpreferred sample.

For controlled sampling design, take the above seven blocks as possible samples, each of which has the same selection probability 1/7. In the above seven possible samples, only the Block 7: (7, 1, 3)* is an undesirable sample and the other six are desirable samples. The probability of selection of the undesirable sample under controlled sampling is 1/7, which is much less than probability of selection of an undesirable sample (14/35) from the uncontrolled SRSWOR sampling design of size 3. Since the solution of a BIBD is not unique, one can get different solutions if another BIBD is chosen. For example, Rao and Nigam (1990) have shown the following alternative layout of a BIBD with parameters $v = 7 = b$, $r = k = 3$, and $\lambda = 1$, whence the probability of nonpreferred samples is 3/7.

Block 1: (1, 3, 4); Block 2: (2, 4, 5); Block 3: (3, 5, 6); Block 4: (4, 6, 7)*; Block 5: (5, 7, 1); Block 6: (6, 1, 2)* and Block 7: (7, 2, 3)*.

Thus the choice of an appropriate BIBD requires trial and error practices.

For large N and n, a BIBD of the required type may not exist or even if it does exist, it is difficult to construct. Furthermore, the identification of the undesirable sample at the initial stage may not be possible because of a lack of adequate information. In this situation we may construct controlled sampling using the following method provided by Avadhani and Sukhatme (1973).

Mechanism for controlled sampling:

(i) Divide the population of N units at random into k disjoint groups containing $N_1, N_2, \ldots, N_k$ units with $\sum_{i=1}^{K} N_i = N$.

(ii) Let $n_i = nN_i/N$ be an integer for $i = 1, 2, \ldots, k$. Choose an integer n'_i such that $n_i < n'_i < N_i$ and there exists a BIBD with parameters $(n'_i, b_i, r_i, n_i, \lambda_i)$ for $i = 1, 2, \ldots, k$. Then select a simple random sub-sample of the n'_i units from the N_i units of the ith group and do independently for $i = 1, 2, \ldots, k$.

(iii) Determine the preferred combination of n_i from n'_i units and establish a one to one correspondence between the blocks of BIBD's in (ii) and the preferred combinations. Select one block at random from the BIBD of each of the k groups independently. Then the collection of the selected blocks of the BIBD's will constitute the controlled sample of size n.

Here we can easily verify the following theorem.

Theorem 24.3.1

Let $\overline{y}_i$ be the sample mean based on n_i units selected from the ith group and $\overline{y}_w = \frac{1}{N}\sum_{i=1}^{k} N_i \overline{y}_i$. Then,

(i) $E(\overline{y}_w) = \overline{Y} =$ population mean and

(ii) $V(\overline{y}_w) = \left(\frac{1}{n} - \frac{1}{N}\right) S_y^2$, where S_y^2 is the population variance.

The theorem above indicates that the weighted controlled sample mean $\overline{y}_w$ is unbiased for the population mean and at the same time it is as efficient as the sample mean based on an uncontrolled SRSWOR sample of the same size n. Furthermore, the controlled selection given above reduces the probability of selection of the nonpreferred sample. Wynn (1977) and Foody and Hedayat (1977) used BIBD with repeated blocks for controlled sampling when nontrivial BIBD do not exist.

24.3.2 Unequal Probability Sampling Design

Let s be a sample of size n selected with probability $p(s)$ using a sampling design p. Let S be the support of p, i.e., S is the collection of all possible samples such that $p(s) > 0$ with $\sum_{s\in S} p(s) = 1$. Let us further suppose that p be an IPPS or π ps (inclusion probability proportional to the measure of size) sampling design with the inclusion probability of the ith unit as $\pi_i = np_i$, where $p_i(>0)$ is the normed size measure for the ith unit with $\sum_{i=1}^{N} p_i = 1$.

Controlled IPPS sampling design:

Gupta et al. (1982) proposed the following controlled sampling design:

(i) Select a BIBD with parameters ($v = N$, b, r, $k \geq n$, λ) assuming it exists.

(ii) Select one block s_j from the BIBD mentioned above with probability

$$p(s_j) = \frac{v\left(r\sum_{i\in s_j} p_i - \lambda\right)}{b(r-\lambda)} \text{ for } j = 1,\ldots, b \text{ (assuming } p(s_j) \geq 0)$$

If $k = n$, the selected block constitutes the required sample of size n.

(i) If $k > n$, select a subsample of size n units from k units of the selected block s_j by SRSWOR method.

(ii) Associate the blocks of the BIBD with the maximum possible number with preferred samples and the rest with nonpreferred samples.

Inclusion probability for the ith unit in the above controlled sampling design is $\pi_i = np_i$. Hedayat and Stufken (1989) and Nigam et al. (1984) proposed alternative controlled IPPS sampling scheme, which possesses nonnegative Yates-Grundy (1953) variance estimators. Further details have been given in Section 5.4.2.5.

24.3.3 Balanced Sampling Plan Without Contiguous Units

The first step of selection of a sample is to determine the sampling frame where the units of the populations are labeled by the numbers 1, 2,..., N; N is the total number of units in the population. In general, the units are labeled according to their physical positions. For examples, in household surveys adjacent enumeration areas receive contiguous numbers, e.g., 101 and 102. Similarly, households within the enumeration areas are numbered serially according to their physical positions. In most situations the contiguous units possess similar information especially when ordering is done in time or space. In such situations, samples containing contiguous units are treated as undesirable samples. Hedayat et al. (1988) proposed balanced sampling plan excluding contiguous (BSEC) units, where each sample contains same number (n) of distinct units and no pair of contiguous units appear together in the same sample whereas all other pairs appear equally often in the samples. A sampling design with support BSEC constitutes the desired controlled sampling design.

Example 24.3.2

Hedayat et al. (1988) provided the following example of a BSEC sampling plan with population size $N = 9$ and sample size $n = 3$ where i and $(i + 1)$ mod 9 as contiguous units, i.e., units 9 and 1 are treated as contiguous units.

$$\{1,3,6\},\{1,4,8\},\{1,5,7\},\{2,4,7\},\{2,5,9\},\{2,6,8\},\{3,5,8\},$$
$$\{3,7,9\},\{4,6,9\}$$

If the samples are selected with equal probability, then the first- and second-order inclusion probabilities become

$$\pi_i = n/N = 1/3 \text{ and}$$
$$\pi_{ij} = \begin{cases} 1/9 & \text{if } i \text{ and } j \text{ are noncontiguous} \\ 0 & \text{otherwise} \end{cases} \quad \text{for } i \neq j.$$

Consider a population $U = \{1,..., N\}$ of N units from which a sample s of size n is selected using a sampling design, which assign equal probability to each of the samples of a BSEC sampling plan where units i and $(i + 1)$ mod N are treated as contiguous units. Let $\overline{y}_{BSEC}$ be the sample mean of the variable under study y based on the selected sample s. Then we have the following theorem from Hedayat et al. (1988).

Theorem 24.3.2

(i) Inclusion probabilities for the ith, and ith and jth ($j \neq i$) units are

$$\pi_i = n/N \text{ and } \pi_{ij} = \begin{cases} \dfrac{n(n-1)}{N(N-3)} & \text{if } (i,j) \text{ noncontiguous} \\ 0 & \text{if } (i,j) \text{ contiguous} \end{cases}$$

(ii) $\overline{y}_{\text{BSEC}}$ is an unbiased estimator for the population mean $\overline{Y}$

(iii) $V\left(\overline{y}_{\text{BSEC}}\right) = \dfrac{\sigma_y^2}{n}\left(1 - \dfrac{(1+2\rho_1)(n-1)}{N-3}\right)$

where $\sigma_y^2 = \dfrac{1}{N}\sum_{i \in U}\left(y_i - \overline{Y}\right)^2$ and $\rho_1 = \dfrac{\sum_{i=1}^{N}\left(y_i - \overline{Y}\right)\left(y_{i+1} - \overline{Y}\right)}{N\sigma_y^2}$ is the serial correlation of first order.

(iv) $V\left(\overline{y}_{\text{BSEC}}\right) \leq V\left(\overline{y}_{sr}\right)$ if $\rho_1 \geq -1/(N-1)$

where $\overline{y}_{sr}$ sample mean of SRSWOR sample of size n.

Proof

(i) From construction of the sampling design we have $\pi_i = \alpha$ (constant), $\pi_{ij} = \beta$ (constant) when the units (i, j) noncontiguous and $\pi_{ij} = 0$ when the units are contiguous. The set of contiguous units is $C = \{(1, 2), (2, 3),\ldots, (N-1, N), (N, 1)\}$ and the cardinality of C is N. Now using the consistency conditions of inclusion probabilities $\sum_{i=1}^{N}\pi_i = n$ and $\sum_{i\neq}^{N}\sum_{j=1}^{N}\pi_{ij} = n(n-1)$ given in Section 1.3.3 yield $N\lambda = n$ and $\{N(N-1) - 2N\}\beta = n(n-1)$, i.e., $\pi_i = \alpha = \dfrac{n}{N}$ and $\beta = \pi_{ij} = \dfrac{n(n-1)}{N(N-3)}$ for (i, j) noncontiguous.

(ii)
$$E\left(\overline{y}_{\text{BSEC}}\right) = \frac{1}{n}E\left(\sum_{i \in s} y_i\right)$$

$$= \frac{1}{n}\sum_{i \in U} y_i E(I_{si}) \text{ where } I_{si} = \begin{cases} 1 & \text{if } i \in s \\ 0 & \text{if } i \notin s \end{cases}$$

$$= \frac{1}{n}\sum_{i \in U} y_i \pi_i$$

$$= \overline{Y}$$

$$\text{(iii)}\ V\left(\bar{y}_{\text{BSEC}}\right)=\frac{1}{n^2}E\left(\sum_{i\in s}(y_i-\bar{Y})\right)^2$$

$$=\frac{1}{n^2}E\left(\sum_{i\in U}(y_i-\bar{Y})I_{si}\right)^2$$

$$=\frac{1}{n^2}\left[\sum_{i\in U}(y_i-\bar{Y})^2\pi_i+\sum_{i\neq}\sum_{j\in U}(y_i-\bar{Y})\left(y_j-\bar{Y}\right)\pi_{ij}\right]$$

$$=\frac{1}{n^2}\left[\frac{n}{N}\sum_{i\in U}(y_i-\bar{Y})^2+\frac{n(n-1)}{N(N-3)}\sum_{i\neq}\sum_{j\in U|i,j\text{ noncontiguous}}(y_i-\bar{Y})\left(y_j-\bar{Y}\right)\right]$$

$$=\frac{1}{n^2}\left[\frac{n}{N}\sum_{i\in U}(y_i-\bar{Y})^2+\frac{n(n-1)}{N(N-3)}\sum_{i\neq}\sum_{j\in U}(y_i-\bar{Y})\left(y_j-\bar{Y}\right)\right.$$

$$\left.-\frac{n(n-1)}{N(N-3)}\sum_{i\neq}\sum_{j\in U|i,j\text{ contiguous}}(y_i-\bar{Y})\left(y_j-\bar{Y}\right)\right]$$

$$=\frac{1}{n^2}\left[\frac{n}{N}\left(1-\frac{n-1}{N-3}\right)\sum_{i\in U}(y_i-\bar{Y})^2-\frac{2n(n-1)}{N(N-3)}\sum_{i=1}^{N}(y_i-\bar{Y})(y_{i+1}-\bar{Y})\right]$$

$$=\frac{\sigma_y^2}{n}\left(1-\frac{(n-1)(1+2\rho_1)}{N-3}\right)$$

$$\text{(iv)}\quad V\left(\bar{y}_{\text{BSEC}}\right)-V\left(\bar{y}_{sr}\right)=\frac{\sigma_y^2}{n}\left(1-\frac{(n-1)(1+2\rho_1)}{N-3}-\frac{N-n}{N-1}\right)$$

$$=-\frac{2(n-1)}{(N-1)(N-3)n}\{1+(N-1)\rho_1\}\sigma_y^2$$

$$\leq 0 \text{ if } \rho_1\geq -1/(N-1)$$

From the theorem above, it follows that sample mean based on BSEC is more precise than the sample mean based on SRSWOR sample of the same size if the serial correlation $\rho_1 > -1/(N-1)$. The condition $\rho_1 > -1/(N-1)$ is likely to be realized in practice especially when the contiguous units have high positive correlation. Hedayat et al. (1988) studied the existence and constructions of such BSEC.

The main demerit of the BSEC is that $V(\bar{y}_{BSEC})$ cannot be estimated unbiasedly because the inclusion probabilities of the two contiguous units are zero. Several extensions of BSEC have been proposed. Stufken (1993) proposed a balanced sampling plans excluding adjacent units [BSA(m)] where all pairs of units whose distance are less than or equal to $m(\geq 1)$ are excluded. Clearly, BSA(1) is equivalent to BSEC. Here also the unbiased estimator of the variance of the sample mean is not available. Stufken et al. (1999) introduced polygonal designs, which is a generalization of BIBD to obtain a [BSA(m)]. Mandal et al. (2008) studied existence and constructions of such designs.

Mandal et al. (2009) proposed distance balanced sampling plan (DBSP) whose first- and second-order inclusion probabilities are

$$\pi_i = n/N, \text{ for } i = 1, \ldots, N \text{ and } \pi_{ij} = \frac{n(n-1)}{N} \frac{f_{ij}}{\sum\limits_{j(\neq i)=1}^{N} f_{ij}} \text{ for}$$

$$i \neq j = 1, \ldots, N$$

where f_{ij} is a suitably defined nonnegative distance function between the units i and j. So for the proposed DBSP, the variance of the sample mean can be estimated unbiasedly. More details have been given by Rao and Vijayan (2008), Mandal et al. (2010, 2011), Gupta et al. (2012), among others.

24.4 APPLICATION OF LINEAR PROGRAMMING

The methods of controlled sampling focus on the reduction of support size through applications of experimental designs and increase of preferred samples by trial and error methods. The criterion of minimum support size is not even relevant for controlled sampling design (Rao and Nigam, 1992). Rao and Nigam (1990, 1992) constructed the optimum controlled sampling by applying LP method. The proposed method not only minimizes the probability of selection of undesirable samples but also maintains conditions such as unbiasedness properties, controlling sampling variance to a certain level, and nonnegative variance estimation, which are desirable to sampling designs.

Let S be the collection of all possible samples such that $p(s) > 0$ for $s \in S$, $S_1(\subset S)$ is the collection of all undesirable samples. Our objective is to minimize $\sum\limits_{s \in S_1} p(s)$ subject to (i) IPPS condition: $\pi_i = np_i$, (ii) Nonnegative Yates and Grundy's variance estimation: $\pi_i\pi_j \geq \pi_{ij}$ for

$i \neq j$, and (iii) controlling the magnitude of variance: $c\pi_i\pi_j \leq \pi_{ij} \leq \pi_i\pi_j$ with $c(<1)$, a prespecified constant such as $c = 1/2$. Thus we need to find a solution of $p(s)$ from the following LP problem:

Objective function:

$$\text{Minimize } \phi = \sum_{s \in S_1} p(s)$$

Constraints:

(i) $p(s) \geq 0$ for $s \in S$

(ii) $\sum_{s \in S} p(s) = 1$

(iii) $\sum_{s \supset i} p(s) = np_i$ for $i = 1, \ldots, N$

(iv) $\sum_{s \supset i,j} p(s) \leq n^2 p_i p_j;\ i \neq j = 1, \ldots, N$

(v) $\sum_{s \supset i,j} p(s) \geq cn^2 p_i p_j;\ i \neq j = 1, \ldots, N$

One can obviously choose a more general objective function $\phi = \sum_{s \in S} c(s)p(s)$ with suitable weight $c(s)$. The objective function ϕ reduces to the expected cost of the survey when $c(s)$ is the cost of selecting the sample s.

The solution of the LP, although not unique, can be obtained numerically by using the simplex method. The computer software package for LP is available. However, the proposed LP method becomes impractical if both N and n become large, because in that case the number of variables as well as the number of constraints increases very rapidly. Rao and Nigam (1992) suggested the use of stratified sampling if the undesirable samples can be identified separately in each of the stratum. Lahiri and Mukherjee (2000) suggested an alternative method that reduces the dimensionality of the problem and hence reduces the computing time to a great extent. Mandal et al. (2010, 2011) proposed linear integer programming approach for controlled sampling designs, particularly the balanced sampling plan and DBSPs.

24.5 NEAREST PROPORTIONAL TO SIZE DESIGN

Suppose that a sampling design p with a support S is desirable due to theoretical considerations but the set of samples $S_1(\subset S)$ is considered undesirable because of practical considerations. To eliminate the undesirable samples, we may consider the following sampling design p_0, which assigns zero probability of selection for each of the nonpreferred samples in S_1

$$p_0(s) = \begin{cases} \dfrac{p(s)}{1 - \sum\limits_{s\in S_1} p(s)} & \text{for } s\in S - S_1 \\ 0 & \text{otherwise} \end{cases}$$

where $p_0(s)$ and $p(s)$ denote, respectively, the selection probabilities of the sample s under p_0 and p.

The newly constructed sampling design p_0 may not have the desirable properties as the original sampling design p. So, we need to construct a sampling design $\widetilde{p}$ with support $\widetilde{S}(= S - S_1)$ consisting of preferred samples while at the same time retaining properties of the desirable design p. We can construct such a design using the method of Gabler (1987a,b). In this method the design $\widetilde{p}$ is constructed in such a way that the distance between $\widetilde{p}$ and p_0 becomes as small as possible. Gabler (1987a,b) proposed the following distance measure analogous to the chi-square and used by Cassel and Särandal (1972).

$$D(p_0, \widetilde{p}) = \sum_{s\in S-S_1} \frac{\{\widetilde{p}(s) - p_0(s)\}^2}{p_0(s)}$$

Details on construction of such a design $\widetilde{p}$ have been given in Section 5.4.2.6.

24.6 APPLICATION OF NONLINEAR PROGRAMMING

Tiwari et al. (2007) obtained the solution of $\widetilde{p}$ by applying nonlinear programming method as follows:

$$\text{Minimize } \varphi = \sum_{s\in \widetilde{S}} \frac{\{\widetilde{p}(s) - p_0(s)\}^2}{p_0(s)}; \quad \widetilde{S} = S - S_1$$

subject to the following constraints:

(i) $\widetilde{p}(s) \geq 0 \ \ \forall s\in\widetilde{S}$,

(ii) $\sum\limits_{s\in\widetilde{S}} \widetilde{p}(s) = 1$,

(iii) $\sum\limits_{s\supset i} \widetilde{p}(s) = \widetilde{\pi}_i \ \ \forall i = 1, \ldots, N$,

(iv) $\sum_{s \supset i,j} \widetilde{p}(s) = \widetilde{\pi}_{ij} > 0 \ \ \forall i \neq j = 1, \ldots, N$, and

(v) $\sum_{s \supset i,j} \widetilde{p}(s) \leq \widetilde{\pi}_i \widetilde{\pi}_j \ \ \forall i \neq j = 1, \ldots, N$.

Tiwari et al. (2007) also constructed a controlled sampling design by using the following alternative objective function suggested by Takeuchi et al. (1983).

$$\varphi^* = \sum_{s \in \widetilde{S}} \frac{\{\widetilde{p}(s) - p_0(s)\}^2}{\widetilde{p}(s) + p_0(s)}$$

Tiwari et al. (2007) reported that both objective functions provide similar results on various numerical problems. They also observed that a feasible solution to the quadratic programming may not always exist. In this condition, one should try to get solutions by relaxing some of the constraints. One of the limitations of the proposed quadratic as well as LP methods is that the methods become impractical if $\binom{N}{n}$ is very large since enumeration of all possible samples and formulation of the objective functions with numerous constraints become highly tedious. Tiwari et al. (2007) studied different controlled sampling procedures using numerical data. The empirical findings reveal that the quadratic programming method performs better than the others.

24.7 COORDINATION OF SAMPLES OVERTIME

The sample coordination problem consists of managing the overlap of two or more samples drawn in different occasions. It is either positive or negative. In positive coordination, the expected overlap is maximized whereas in negative coordination it is minimized. This is important because the measure of size associated with the population unit changes overtime because of growth, birth, deaths, and mergers. Such changes in the auxiliary information should be incorporated to increase the efficiency of the estimates of the current occasion. Because of the high cost of obtaining information on the auxiliary and study variables, and of familiarizing new respondents with reporting procedures, it is often desirable to retain as many respondents as possible from the original sample (based on the outdated measure of size) for the new sample (based on the updated measure of size). So the sample coordination problem reduces to controlled selection if the selected sample in the first occasion is treated as the desirable (or undesirable) sample in the second occasion. The pioneering work in this field evolved from Patterson (1950) and Keyfitz (1951). Other contributors include Fellegi (1963, 1966), Kish (1963), Gray and Platek (1963), Raj (1968), Kish and Scott (1971), Brewer et al. (1972), Lanke (1974a,b), Arthanari and Doge (1981), and Matei and Tillé (2005).

Consider a finite population $U=(1,\ldots,N)$ of N identifiable units. Let, at a certain point of time, a sample $s(1)=\{i_1(1),\ldots,i_n(1)\}$ of size n be selected using auxiliary information $\mathbf{x}(1)=(x_1(1),\ldots,x_N(1))$. Later on, $\mathbf{x}(1)$ changes into $\mathbf{x}(2)=(x_1(2),\ldots,x_N(2))$ and it is required to revise the sample accordingly, but on the other hand, one does not want to exchange units unnecessarily. Let $s(2)=\{i_1(2),\ldots,i_n(2)\}$ be a sample of size n selected on the second occasion using $\mathbf{x}(2)=(x_1(2),\ldots,x_N(2))$ as the auxiliary available. Our problem is to maximize $E\{|s(1)\cap s(2)|\}$, the expected number of units common between the samples $s(1)$ and $s(2)$. Keyfitz (1951) gave a procedure applicable for the probability proportional to size with replacement (PPSWR) method for selection of one unit. His procedure is optimal in the sense that it maximizes the probability of the units drawn on the first occasion to be retained for the second occasion. Keyfitz's procedure can be easily extended to the general sample size n. Keyfitz method is given below.

24.7.1 Keyfitz Method

Let $p_i(1)=x_i(1)/X(1)$, $p_i(2)=x_i(2)/X(2)$, $X(1)=\sum_{i\in U}x_i(1)$, and $X(2)=\sum_{i\in U}x_i(2)$. Partition the population U into two groups $U(1)=\{i|p_i(1)>p_i(2)\}$ and $U(2)=\{i|p_i(1)\le p_i(2)\}$. Draw one unit i (say) using normed size measure $p_i(1)$ and take $s(1)=\{i_1(1)=i\}$. If the selected unit $i\in U(2)$, then retain i as the selected sample for $s(2)$, i.e., $s(2)=\{i_1(2)=i\}$. But if $i\in U(1)$, then perform a Bernoulli trial with success probability $p_i(2)/p_i(1)$. If the trial results in a success, retain the unit i for the sample $s(2)$, i.e., $\{i_1(2)=i\}$. On the other hand, if the trial results in failure, retain the jth unit from $U(2)$ with probability proportional to $p_j(2)-p_j(1)$.

Theorem 24.7.1

Keyfitz method selects samples $s(1)=\{i_1(1)=i\}$ and $s(2)=\{i_1(2)=i\}$ with probabilities $p_i(1)$ and $p_i(2)$, respectively.

Proof

The theorem is true for selection of $s(1)$ because in this method the unit is selected by PPSWR method using normed size measure $p_i(1)$ for the ith unit. For selection of sample $s(2)$, there are two scenarios:

If the unit $i \in U(1)$, then probability of selecting the unit i is

$$\text{Prob}\{i_1(2) = i\} = p_i(1) \times \frac{p_i(2)}{p_i(1)} = p_i(2) \qquad (24.7.1)$$

If the unit $i \in U(2)$, then probability of selecting the unit i is

$$\text{Prob}\{i_1(2) = i\} = \text{Prob}\{i_1(1) = i\} + \sum_{j\in U(1)} \text{Prob}\{i_1(1) = j, i_1(2) = i\}$$

$$= p_i(1) + \sum_{j\in U(1)} p_j(1)\left(1 - \frac{p_j(2)}{p_j(1)}\right) \frac{p_i(2) - p_i(1)}{\Delta(2)}$$

$$\left(\text{where } \Delta(2) = \sum_{i\in U(2)} \{p_i(2) - p_i(1)\}\right)$$

Now noting, $\Delta = \sum_{i\in U}\{p_i(2) - p_i(1)\} = \sum_{i\in U(1)} \{p_i(2) - p_i(1)\} + \sum_{i\in U(2)} \{p_i(2) - p_i(1)\} = 0$, we find

$$\text{Prob}\{i_1(2) = i\} = p_i(1) + \{p_i(2) - p_i(1)\} = p_i(2)$$

Theorem 24.7.2

Let $s(1) = \{i_1(1)\}$ and $s(2) = \{i_1(2)\}$ be PPSWR samples of size 1 each. Then the expected number of common units between $s(1)$ and $s(2)$ is

$$E(|s(1) \cap s(2)|) \leq \sum_{i=1}^{N} \min\{p_i(1), p_i(2)\}$$

Proof

$$\begin{aligned} E(|s(1) \cap s(2)|) &= \sum_{i=1}^{N} \text{Prob}\{i_1(1) = i, i_1(2) = i\} \\ &\leq \sum_{i=1}^{N} \min[\text{Prob}\{i_1(1) = i\}, \text{Prob}\{i_1(2) = i\}] \\ &= \sum_{i=1}^{N} \min\{p_i(1), p_i(2)\} \end{aligned}$$

Theorem 24.7.3
For the Keyfitz method

$$\text{Prob}\{i_1(1) = i_1(2)\} = \sum_{i=1}^{N} \min\{p_i(1), p_i(2)\}$$

Proof
If $i \in U(2)$, then $\text{Prob}\{i_1(1) = i_1(2) = i\} = \text{Prob}\{i_1(1) = i\}$

$$= p_i(1)(\leq p_i(2)) = \min(p_i(1), p_i(2)) \quad (24.7.2)$$

If $i \in U(1)$, then $\text{Prob}\{i_1(1) = i_1(2) = i\} = p_i(1) \times \frac{p_i(2)}{p_i(1)} = p_i(2)(\leq p_i(1))$

$$= \min(p_i(1), p_i(2)) \quad (24.7.3)$$

The theorem follows from Eqs. (24.7.2) and (24.7.3).

24.7.2 Probability Proportional to Aggregate Size Sampling Scheme

Lanke (1974a,b) considered the problem of selection of two samples $s(1)$ and $s(2)$, each of size n so that probability of selection of $s(1)$ and $s(2)$ are proportional to the aggregate measure of size (PPAS) $x(1)$ and $x(2)$, respectively, at the same time the expected number of common units between $s(1)$ and $s(2)$ is maximized. The method is described as follows.

24.7.2.1 Lanke Method

Draw a pair of units $\{i_1(1), i_1(2)\}$ by the Keyfitz method using normed size measures $p_1(1),\ldots, p_N(1)$ and $p_1(2),\ldots, p_N(2)$, respectively. Then draw an SRSWOR sample $s_0(1)$ of size $n-1$ from the $U - \{i_1(1)\}$ and take

$$s(1) = \{i_1(1)\} \cup s_0(1)$$

$$s(2) = \begin{cases} s(1) & \text{if} \quad i_1(2) \in s(1) \\ \{i_1(2)\} \cup s_0(1) & \text{if} \quad i_1(2) \notin s(1) \end{cases}$$

The first unit $i_1(1)$ of $s(1)$ is selected with probability $p_i(1)$ and the remaining $n-1$ units are selected from $U - \{i_1(1)\}$. Hence, $s(1)$ is selected by the Lahiri–Mizuno–Sen (1951, 1952, 1953) sampling method and the probability of selection of $s(1)$ is $p\{s(1)\} = \left(\sum_{i \in s(1)} p_i(1)\right) \Big/ M_1$ where

$M_1 = \binom{N-1}{n-1}$. To prove that $p\{s(2)\} = \left(\sum_{i\in s(2)} p_i(2)\right)/M_1$, we need to show that $s_0(2) = s(2) - i_1(2)$ is an SRSWOR sample from $U - \{i_1(2)\}$. If $i_1(1) = i_1(2)$, then $s_0(1) = s_0(2)$. If $i_1(1) \neq i_1(2)$, then

$$s_0(2) = \begin{cases} s_0(1) & \text{if} \quad i_1(2) \notin s_0(1) \\ \{i_1(1)\} \cup s_0(1) - \{i_1(2)\} & \text{if} \quad i_1(2) \in s_0(1) \end{cases}$$

Thus $s_0(2)$ is an SRSWOR sample from $U - \{i_1(1)\}$ where $i_1(2)$, whenever selected in the sample, is replaced by $i_1(1)$. Hence $s_0(2)$ is an SRSWOR sample selected from $U - \{i_1(2)\}$.

Theorem 24.7.4

Let $s(1)$ and $s(2)$ be PPAS samples each of size n, with normed size measures $p_1(1),\ldots, p_N(1)$ and $p_1(2),\ldots, p_N(2)$, respectively, then the expected number of common units between $s(1)$ and $s(2)$ satisfies

$$E\{|s(1)\cap s(2)|\} \leq \frac{N(n-1)}{N-1} + \frac{N-n}{N-1}\sum_{i=1}^{N} \min\{p_i(1), p_i(2)\}$$

Proof

Let $\gamma_i = \text{Prob}\{s(1) \cap s(2) \supset i\}$. Then

$$\begin{aligned} E\{s(1)\cap s(2)\} &= \sum_{i\in U} \gamma_i \\ &\leq \sum_{i\in U} \min[\text{Prob}\{s(1)\supset i\}, \text{Prob}\{s(2)\supset i\}] \qquad (24.7.4) \\ &= \sum_{i\in U} \min[\pi_i(1), \pi_i(2)] \end{aligned}$$

where $\pi_i(1)$ = inclusion probability of the ith unit for selection of sample $s(1)$ according to PPAS sampling design with normed size measure $p_1(1),\ldots, p_N(1)$.

$$= \frac{n-1}{N-1} + \frac{N-n}{N-1} p_i(1) \qquad (24.7.5)$$

Similarly $\pi_i(2)$ = inclusion probability of the ith unit for selection of sample $s(2)$ according to PPAS sampling design with normed size measure $p_1(2),\ldots, p_N(2)$

$$= \frac{n-1}{N-1} + \frac{N-n}{N-1} p_i(2) \qquad (24.7.6)$$

Substituting Eqs. (24.7.5) and (24.7.6) in Eq. (24.7.4), we get

$$E\{|s(1)\cap s(2)|\} \le \frac{N(n-1)}{N-1} + \frac{N-n}{N-1}\sum_{i=1}^{N}\min\{p_i(1), p_i(2)\} \qquad (24.7.7)$$

Theorem 24.7.7

For Lanke sampling design the upper bound (24.7.7) is attained

$$\text{i.e., } E\{|s(1)\cap s(2)|\} = \frac{N(n-1)}{N-1} + \frac{N-n}{N-1}\sum_{i=1}^{N}\min\{p_i(1), p_i(2)\}$$

Proof

$$\begin{aligned}
\text{Prob}\{s(1)=s(2)\} &= \text{Prob}\{i_1(1)=i_1(2)\} + \text{Prob}\{i_1(1) \neq i_1(2)\} \\
&\quad \times \text{Prob}\{s(1)=s(2)|i_1(1) \neq i_1(2)\} \\
&= \left(\sum_{i\in U}\min\{p_i(1), p_i(2)\}\right) \\
&\quad + \left(1-\sum_{i\in U}\min\{p_i(1), p_i(2)\}\right)\frac{n-1}{N-1} \\
&= \frac{n-1}{N-1} + \frac{N-n}{N-1}\sum_{i\in U}\min\{p_i(1), p_i(2)\}
\end{aligned}$$

Since $|s(1)\cap s(2)|$ can take two values $n-1$ and n, we have

$$\begin{aligned}
E\{|s(1)\cap s(2)|\} &= (n-1)\text{Prob}\{s(1) \neq s(2)\} + n\text{Prob}\{s(1)=s(2)\} \\
&= (n-1) + \text{Prob}\{s(1)=s(2)\} \\
&= \frac{N(n-1)}{N-1} + \frac{N-n}{N-1}\sum_{i\in U}\min\{p_i(1), p_i(2)\}
\end{aligned}$$

Remark 24.7.1

Lanke's scheme is not uniquely optimum, i.e., there exists at least one other method for which $E\{|s(1)\cap s(2)|\}$ attains the upper bound (24.7.7). Lanke also generalized this method for drawing $d(>2)$ PPAS samples, each of size n with different sets of measures of size.

24.8 DISCUSSIONS

The main purpose of selection of a sample using an appropriate sampling design is to obtain efficient estimates of parameters of interest. But the selected units within the sample sometimes may be so widespread that the cost of data collection becomes very high because of travel costs, and it may be inconvenient for administrative purposes also. Samples, which are uneconomical and also create organizational and other difficulties, are termed as nonpreferred or undesirable samples. Controlled sampling procedure has been proposed to overcome such difficulties. Broadly, the methods are classified into five categories: (i) use of experimental designs for selection of sample, (ii) LP, (iii) nonlinear programming, (iv) nearest proportional to size design, and (v) coordination of samples overtime. However, none of the proposed methods is optimum in all the situations. The selection of samples using combinatorial properties of experimental designs, especially using balanced sampling, reduces the selection of nonpreferred samples drastically. But, it does not always yield optimum solution. The method of linear and nonlinear programming can produce optimal solution numerically by using suitable objective function and constraints. Both methods have limited applications when the population and sample sizes are both very large. In general, linear and nonlinear programming methods cannot be used to study the properties of the estimators theoretically. In nearest proportional to size sampling design, one selects sample from a sampling design that is closed to the target sampling design. This method is quite advantageous because it is applicable for large sample sizes. It can also be used for various varying probability sampling designs. The main demerit of this method is that it may fail to exist in some situations. Sampling coordination may be successfully achieved for some situations but has limited application especially for varying probability sampling designs. For practical purposes, the situation dictates the controlled sampling procedure to be used.

24.9 EXERCISES

24.9.1 Agricultural farms are stratified into two strata. Stratum 1 comprises seven farms *A, B, C, D, E, F,* and *G* while stratum 2 comprises five farms *a, b, c, d,* and *e*. The farms *A, B, C, D* and *c, d, e* have irrigation facilities and the remaining farms have no such facilities. The normed size measures for the farms have been given in the following table. Select one farm from each of the stratum with the given normed size measure at the same time probability of selection of one irrigated and another nonirrigated farm is maximized.

Stratum 1		Stratum 2	
Units	Normed size measures	Units	Normed size measures
A	0.10	*a*	0.20
B	0.20	*b*	0.25
C	0.15	*c*	0.25
D	0.15	*d*	0.20
E	0.20	*e*	0.10
F	0.10		
G	0.10		

24.9.2 Consider the Example 24.3.1, where we need to select a sample of $n = 3$ units from a population of $N = 7$ units. Use appropriate LP problem to select 3 units so that inclusion probabilities of each of the 7 units becomes exactly 1/7 and at the same time the probability selection of the following sets of nonpreferred samples becomes a minimum.

$$\text{Nonpreferred units}: (1,2,3),(1,2,6),(1,3,6),(1,3,7),(1,4,6),(1,4,7),\\(1,6,7),(2,3,4),(2,3,6),(2,3,7),(2,4,6),(2,4,7),\\(3,4,7),(4,6,7)$$

24.9.3 The first- and second-order inclusion probability matrix of a sample of 3 units from a population of 6 units has been given in the following table. Using an LP method, select a sample of 3 units realizing the following inclusion probabilities (First-order inclusion probabilities have been given in the diagonal).

	Units					
Units	1	2	3	4	5	6
1	0.55	0.25	0.2	0.25	0.2	0.2
2	0.25	0.5	0.2	0.1	0.25	0.2
3	0.2	0.2	0.475	0.25	0.1	0.2
4	0.25	0.1	0.25	0.475	0.1	0.25
5	0.2	0.25	0.1	0.1	0.45	0.25
6	0.2	0.2	0.2	0.25	0.25	0.55

24.9.4 Expenditures on food and school fees of households are approximately proportional to the household size (hh) and number of school going children (z). It is decided to select two samples S_1 and S_2 each of size 3 from the same 10 households with replacement so that probability of selection of a unit in sample S_1 is proportional to the household size while probability of selection of a unit in S_2 is proportional to number of children. Use Keyfitz method, select such samples so that the number of common units in S_1 and S_2 is maximized. Find the expected number of common units for such sampling design.

House hold (hh)	1	2	3	4	5	6	7	8	9	10
hh size	4	3	6	5	5	4	7	6	6	4
Number of children	1	1	2	1	2	2	3	2	2	1

24.9.5 Consider the data of Example 24.9.4. Select two samples each of size 4 with probability proportional to the aggregate measure of sizes of households and number of children, respectively, and at the same time maximize the expected number of common units between them.

24.9.6 The following table gives the cost of travel between six villages and their measure of size.

Villages	1	2	3	4	5	6
1		112	115	124	130	120
2			120	130	115	118
3				135	130	20
4					140	250
5						240
Measure of size	10	15	20	30	25	15

Select a sample of three villages with inclusion probability proportional to their measure of size, and at the same time cost of travel for collecting data is minimized. Find also the minimum cost.

CHAPTER 25

Empirical Likelihood Method in Survey Sampling

25.1 INTRODUCTION

Likelihood is the most important tool for parametric inference whereas empirical likelihood (EL) is a powerful nonparametric approach to statistical inference. EL was first introduced in survey sampling by Hartley and Rao (1968) in the name of scale load approach. The modern concept of EL was introduced by Owen (1988). The application of EL approach in survey sampling was introduced by Chen and Quin (1993), Chen and Sitter (1999), Sitter and Wu (2002), and Rao and Wu (2009), among others. In this chapter, we will present an overview of recent developments in EL methods in estimating finite population characteristics such as population mean, variance, and distribution function. We also introduce the concept of pseudo—empirical likelihood (PEL) for the probability sampling designs and their applications to superpopulation models and raking estimators. The uses of EL in determination of confidence intervals have also been discussed.

25.2 SCALE LOAD APPROACH

Consider a finite population U that consists of N units and y_i be the value of the study variable y for the ith unit of the population. In a fixed population approach the vector $\mathbf{y} = (y_1,\ldots,\ y_i,\ldots,\ y_N)$ is treated as a parameter and the corresponding parameter space is $\Omega_y = (-\infty < y_1 < \infty,\ldots, -\infty < y_i < \infty,\ldots,\ -\infty < y_N < \infty)$. Suppose a sample s is selected from the population U with probability $p(s)$ following a sampling design p. The likelihood function of $\mathbf{y}$ is the conditional probability of obtaining the data $d = (y_i, i \in s)$ for the given $\mathbf{y}$ and it can be written as

$$L(\mathbf{y}) = Prob(d|\mathbf{y}) = \begin{cases} p(s) & \text{for} \quad y_i, i \in s \\ 0 & \text{otherwise} \end{cases} \tag{25.2.1}$$

Survey Sampling Theory and Applications
ISBN 978-0-12-811848-1
http://dx.doi.org/10.1016/B978-0-12-811848-1.00025-X

Godambe (1966) concluded that though this aforementioned likelihood function (25.2.1) is well defined, it is noninformative in the sense that all possible nonobserved values $y_i, i \notin s$ have the same likelihood. To overcome this difficulty, Hartley and Rao (1968) introduced the scale load approach where the likelihood function becomes informative. In this approach, the variable y can take only a finite set of values $\widetilde{y}_t$, $t = 1,\ldots, T$ and N_t is the frequency of $\widetilde{y}_t$ so that $N = \sum_{t=1}^{T} N_t$, the population size and $\overline{Y} = \sum_{t=1}^{T} N_t \widetilde{y}_t / N$, the population mean of $y \cdot \widetilde{y}_t$ is called the scale point and N_t is the corresponding scale load. Let a sample s of size n be selected from the population U using simple random sampling without replacement (SRSWOR) sampling design and n_t be the number of units having y-values equal to $\widetilde{y}_t$. Then the likelihood function of $\mathbf{N} = (N_1,\ldots, N_i,\ldots, N_T)$ is given by

$$L(\mathbf{N}) = \frac{\binom{N_1}{n_1}\cdots\binom{N_t}{n_t}\cdots\binom{N_T}{n_T}}{\binom{N}{n}} \tag{25.2.2}$$

For $N \to \infty$, $N_i/N \to p_i$, and $n/N \to 0$, the likelihood function of $\mathbf{p} = (p_1,\ldots, p_i,\ldots, p_T)$ can be approximated as

$$L(\mathbf{p}) \cong \frac{n!}{n_1!\cdots n_i!\cdots n_T!}\; p_1^{n_1}\cdots p_i^{n_i}\cdots p_T^{n_T} \tag{25.2.3}$$

The maximum likelihood estimate (MLE) of $\mathbf{p}$ is $\widehat{\mathbf{p}} = \left(\widehat{p}_1,\ldots, \widehat{p}_i,\ldots, \widehat{p}_T\right)$ where $\widehat{p}_i = n_i/n$ and the MLE of $\overline{Y} = \sum_{t=1}^{T} p_t \widetilde{y}_t$ is $\overline{y} = \sum_{t=1}^{T} \widehat{p}_t \widetilde{y}_t$, the sample mean.

Consider the situation where the information on the auxiliary variable x, closely related to the study variable y, is available and the scale points of x be $\widetilde{x}_j$ with $j = 1,\ldots, J$. Let the scale load of $\left(\widetilde{y}_t, \widetilde{x}_j\right)$ in the population (sample of size n) be N_{tj} (n_{tj}). In this case we have $\overline{X} = \sum_{t=1}^{T}\sum_{j=1}^{J} \widetilde{x}_j p_{tj}$ and $\overline{Y} = \sum_{t=1}^{T}\sum_{j=1}^{J} \widetilde{y}_j p_{tj}$ where $p_{tj} = N_{tj}/N$ and $\sum_{t=1}^{T}\sum_{j=1}^{J} p_{tj} = 1$. The scale load estimator of $\overline{Y}$ proposed by Hartley and Rao (1968) is

$$\widehat{\overline{Y}}_{HR} = \sum_{t=1}^{T}\sum_{j=1}^{J} \widehat{p}_{tj}\widetilde{y}_j \tag{25.2.4}$$

where $\widehat{p}_{tj}\left(= n_{tj}/n\right)$'s maximize the log likelihood function

$$\sum_{t=1}^{T}\sum_{j-1}^{J} n_{tj} \log(p_{tj})$$

subject to constraints $p_{tj} > 0, \sum_{t=1}^{T}\sum_{j=1}^{J} p_{tj} = 1$ and $\overline{X} = \sum_{t=1}^{T}\sum_{j=1}^{J} \widetilde{x}_j p_{tj}$.

Hartley and Rao (1968) showed that the estimator $\widehat{\overline{Y}}_{HR}$ is asymptotically equivalent to the customary regression estimator of $\overline{Y}$. Hartley and Rao (1968) also considered the probability proportional to size with replacement (PPSWR) sampling scheme with x as a measure of size variable and taking scale points of y_i/x_i as $\widetilde{r}_t$, $t = 1,\ldots, T$. The resulting scale load estimator was reduced to the well-known Hansen–Hurwitz estimator. Rao and Wu (2009) pointed out that scale load approach for unequal probability sampling scheme without replacement may not yield any useful solution.

25.3 EMPIRICAL LIKELIHOOD APPROACH

Let $y_1,\ldots, y_i,\ldots, y_n$ be independently and identically distributed random variables with the common cumulative distribution function (CDF) $F(y_i) = P(y \le y_i)$. The EL (nonparametric) function of the CDF (F) is

$$L(F) = \prod_{i=1}^{n}[F(y_i) - F(y_i-)] \tag{25.3.1}$$

where $F(y_i-) = P(y < y_i)$.

Thus $L(F)$ is the probability of getting exactly the observation $y_1,\ldots, y_i,\ldots, y_n$ from the distribution function F. Thus $L(F) = 0$ if F is continuous. Let $p_i = F(y_i) - F(y_i-)$ be the probability of $y = y_i$ and we denote the EL function as

$$L(F) = L(\mathbf{p}) = \prod_{i=1}^{n} p_i \tag{25.3.2}$$

The empirical CDF (ECDF) of $y_1,\ldots, y_i,\ldots, y_n$ is

$$S_n(y) = \frac{1}{n}\sum_{i=1}^{n} I(y_i \le y) \tag{25.3.3}$$

where

$$I(y_i \le y) = \begin{cases} 1 & \text{if } y_i \le y \\ 0 & \text{otherwise} \end{cases}$$

25.4 EMPIRICAL LIKELIHOOD FOR SIMPLE RANDOM SAMPLING

The application of EL method in survey sampling was introduced by Chen and Quin (1993). The method is described as follows. Let a sample s of size n be selected from a finite population U of N units by simple random sampling with or without replacement. For the case of without replacement the sampling fraction n/N is assumed to be negligible so that the Owen's EL results described in Section 25.3 can be applied directly. Chen and Quin (1993) defined EL function for the finite population as

$$L(\mathbf{p}) = \prod_{i \in s} p_i \tag{25.4.1}$$

where p_i is the probability mass associated with the ith unit.

Maximizing $L(\mathbf{p})$ subject to $p_i > 0$ and $\sum_{i \in s} p_i = 1$, the optimum value of p_i is obtained as $\widehat{p}_i = 1/n$. The maximum empirical likelihood (MEL) estimator of the parameter $\theta = \sum_{i \in U} g(y_i)/N$ is given by

$$\widehat{\theta}_{mel} = \sum_{i \in s} \widehat{p}_i g(y_i) = \sum_{i \in s} g(y_i)/n \tag{25.4.2}$$

where $g(y_i)$ is a known function of y_i.

In particular if $g(y_i) = y_i$, we get MEL estimator of the population mean $\overline{Y} = \sum_{i \in U} y_i/N$ is the sample mean

$$\overline{Y}_{mel} = \sum_{i \in s} y_i/n = \overline{y} \tag{25.4.3}$$

For the choice of $g(y_i) = I(y_i \leq y)$, θ reduces to the population distribution function $F(y) = \sum_{i \in U} I(y_i \leq y)/N$ and the MEL of $F(y)$ is the sample distribution function

$$\widehat{F}_{mel}(y) = \sum_{i \in s} I(y_i \leq y)/n \tag{25.4.4}$$

Consider the situation where the auxiliary information of the finite population can be summarized as

$$E\{w(x)\} = \sum_{i=1}^{N} w(x_i)/N = \{w(x_1) + \cdots + w(x_i) + \cdots + w(x_N)\}/N = 0 \tag{25.4.5}$$

where $w(x)$ is a known function of the auxiliary variable x. In this case, MEL of θ is obtained by maximizing

$$l(\mathbf{p}) = \log\{L(\mathbf{p})\} = \sum_{i\in s} \log(p_i)$$

subject to

$$p_i > 0, \sum_{i\in s} p_i = 1 \text{ and } \sum_{i\in s} p_i w_i(x) = 0 \tag{25.4.6}$$

No explicit expression of the solution of Eq. (25.4.6) is obtained. However, numerical solution of p_i's is obtained from

$$\widehat{p}_i = \frac{1}{n\{1+\lambda w(x_i)\}} \text{ for } i \in s \tag{25.4.7}$$

where λ is a Lagrange multiplier satisfying

$$\sum_{i\in s} \frac{w(x_i)}{1+\lambda w(x_i)} = 0$$

As regards to the existence of solution of p_i, Chen and Quin (1993) pointed out that for large n, the solution is most likely to exist. However, Eq. (25.4.7) fails to have proper solution when the convex hull $\{w(x_i), i \in s\}$ does not contain zero. For $w(x_i) = x_i - \overline{X}$ with $\overline{X} = \sum_{i=1}^{N} x_i/N$, Rao and Wu (2009) pointed out that unique solution of p_i of Eq. (25.4.7) exists if $\overline{X}$ is an inner point of the convex hull $\{x_i, i \in s\}$ and this happen with probability 1 as $n \to \infty$.

Thus MEL estimator of the population mean $\overline{Y}$ and distribution function $F(y)$ are, respectively, given by

$$\widehat{\overline{Y}}_{mel} = \sum_{i\in s} \widehat{p}_i y_i \text{ and } \widehat{F}_{mel}(y) = \sum_{i\in s} \widehat{p}_i I(y_i \leq y) \tag{25.4.8}$$

where $\widehat{p}_i$ satisfies Eq. (25.4.7).

It should be noted that $\widehat{F}_{mel}(y)$ is a proper distribution function in the sense that it satisfies all the properties of a distribution function.

25.5 PSEUDO–EMPIRICAL LIKELIHOOD METHOD

The expression of the EL function for the general unequal probability without replacement sampling design cannot be obtained because the expression for the joint probability function of the sample is not feasible.

Chen and Sitter (1999) proposed pseudo—empirical likelihood (PEL) for the unequal probability sampling schemes. Under this sampling scheme, it is assumed that population vector $\mathbf{y} = (y_1, \ldots, y_i, \ldots, y_N)$ is a sample from a superpopulation. For simplicity let us assume that y_i's are independently and identically distributed random variables with distribution function $F(y)$. Then the empirical likelihood of the entire population $\mathbf{y}$ is $L(\mathbf{p}) = \prod_{i=1}^{N} p_i$ and log empirical likelihood is

$$l(\mathbf{p}) = \sum_{i=1}^{N} \log\ p_i \tag{25.5.1}$$

Suppose a sample s is selected from the population with probability $p(s)$ so that the inclusion probability of the ith unit $\pi_i\left(=\sum_{s\supset i} p(s)\right)$ is positive for every $i = 1, \ldots, N$. An unbiased estimator of $l(\mathbf{p})$ based on the sample s is given by

$$\widehat{l}(\mathbf{p}) = \sum_{i\in s} w_i(s) \log p_i \tag{25.5.2}$$

where $w_i(s)$ are suitably chosen weights satisfying unbiasedness condition $\sum_{s\supset i} w_i(s)p(s) = 1$ for $i = 1, \ldots, N$. The estimator $\widehat{l}(\mathbf{p})$ is termed by Chen and Sitter (1999) as "pseudo—empirical log likelihood function." Maximizing $\widehat{l}(\mathbf{p})$ subject to constraints $p_i > 0$ and $\sum_{i\in s} p_i = 1$, yield the optimum value of p_i as

$$\widehat{p}_i = \frac{w_i(s)}{\sum_{i\in s} w_i(s)} \tag{25.5.3}$$

The maximum PEL (MPEL) for a parametric function $\overline{G} = \sum_{i\in U} g(y_i)/N$ is given by

$$\begin{aligned}\widehat{\overline{G}} &= \sum_{i\in s} \widehat{p}_i g(y_i)\\ &= \frac{\sum_{i\in s} w_i(s) g(y_i)}{\sum_{i\in s} w_i(s)}\end{aligned} \tag{25.5.4}$$

The estimator (25.5.4) is similar to the Hájek (1964) estimator.

25.5.1 MPEL Estimator for the Population Mean

Substituting $g(y_i) = y_i$ in Eq. (25.5.4), we get MPEL estimator for the population mean $\overline{Y}$ as

$$\widehat{\overline{Y}}_{mpel} = \frac{\sum_{i \in s} w_i(s) y_i}{\sum_{i \in s} w_i(s)} \tag{25.5.5}$$

In case $w_i(s) = 1/\pi_i$, $\widehat{\overline{Y}}_{mpel}$ reduces to $\widehat{\overline{Y}}_{mpel} = \dfrac{\sum_{i \in s} y_i/\pi_i}{\sum_{i \in s} 1/\pi_i}$. For SRSWOR sampling $\pi_i = n/N$ and $\widehat{\overline{Y}}_{pmel} = \overline{y} =$ sample mean.

25.5.2 MPEL Estimator for the Population Distribution Function

Substituting $g(y_i) = I(y_i \le t)$ in Eq. (25.5.4) MPEL estimator for the distribution function $F(t)$ is obtained as

$$\widehat{F}_{mpel}(t) = \frac{\sum_{i \in s} w_i(s) I(y_i \le t)}{\sum_{i \in s} w_i(s)} \tag{25.5.6}$$

In particular for $w_i(s) = 1/\pi_i$, the estimator $\widehat{F}_{pmel}(t)$ reduces to $\widehat{F}_{mpel}(t) = \dfrac{\sum_{i \in s} I(y_i \le t)/\pi_i}{\sum_{i \in s} 1/\pi_i}$. For SRSWOR $\widehat{F}_{mpel}(t)$ reduces to empirical distribution function $S(t) = \dfrac{1}{n} \sum_{i \in s} I(y_i \le t)$.

25.5.3 MPEL Estimator Under Linear Constraints

Suppose that the population mean $\overline{X}$ of the auxiliary variable x in is known. In this situation, p_i's for $i \in s$ are obtained by maximizing the pseudo–empirical log likelihood $\hat{l}(\mathbf{p}) = \sum_{i \in s} w_i(s) \log p_i$ subject to the constraints

$$\text{(i) } p_i \ge 0, \text{(ii) } \sum_{i \in s} p_i = 1 \text{ and (iii) } \sum_{i \in s} p_i u(x_i) = 0 \tag{25.5.7}$$

where $u_i = u(x_i) = x_i - \overline{X}$.

The solution of the minimization problem above has no closed form. It is obtained numerically by solving the following equations.

$$\widehat{p}_i = \frac{\widetilde{w}_i(s)}{1+\lambda\, u_i} \tag{25.5.8}$$

where $\widetilde{w}_i(s) = w_i(s)/\sum_{i\in s} w_i(s)$ and λ is the Lagrange multiplier satisfying

$$\sum_{i\in s}\frac{\widetilde{w}_i(s)u_i}{1+\lambda\ u_i} = 0 \tag{25.5.9}$$

Rao and Wu (2009) presented a detailed algorithm for the solution of λ and $\widehat{p}_i$ given in Eq. (25.5.8).

Thus the MPEL estimator for the population distribution function $F(t)$ under constraint $\sum_{i\in U} u(x_i) = 0$ is given by

$$\widehat{F}_{mpel}(t) = \sum_{i\in s}\widehat{p}_i I(y_i \le t) \tag{25.5.10}$$

where $\widehat{p}_i$ is given in Eq. (25.5.8). The estimator $\widehat{F}_{mpel}(t)$ is a proper distribution function since $\widehat{p}_i > 0$ and $\sum_{i\in s}\widehat{p}_i = 1$ and estimators of quintiles can be obtained by direct inversion of $\widehat{F}_{mpel}(t)$.

25.6 ASYMPTOTIC BEHAVIOR OF MPEL ESTIMATOR

Chen and Sitter (1999) derived the asymptotic behavior of the MPEL estimator of the population mean. The theorem is given below without derivation.

Theorem 25.6.1

Under the regularity conditions stated below, $\widehat{\overline{Y}}_{mpel}$, the MPEL estimator of the population mean $\overline{Y}$, when $\overline{X}$, the population mean of the auxiliary variable is known, is asymptotically equivalent to the generalized regression estimator (GREG), i.e.,

$$\widehat{\overline{Y}}_{mpel} = \widehat{\overline{Y}}_{greg} + o_p\left(n^{-1/2}\right)$$

where $\widehat{\overline{Y}}_{greg} = \overline{y}_w - \widehat{B}_{greg}\left(\overline{x}_w - \overline{X}\right)$, $\overline{x}_w = \sum_{i\in s}\widetilde{w}_i(s)\ x_i$, $\overline{y}_w = \sum_{i\in s}\widetilde{w}_i(s)\ y_i$,

$$\widetilde{w}_i(s) = w_i(s)\Big/\sum_{i\in s} w_i(s), \text{ and } \widehat{B}_{greg} = \frac{\sum_{i\in s}\widetilde{w}_i(s)(x_i-\overline{x}_w)y_i}{\sum_{i\in s}\widetilde{w}_i(s)(x_i-\overline{x}_w)^2}.$$

In this case the Lagrange multiplier λ is given by

$$\lambda = \frac{\overline{x}_w - \overline{X}}{\sum_{i\in s} \widetilde{w}_i(s)(x_i - \overline{x}_w)^2} + o_p\left(n^{-1/2}\right)$$

The regularity conditions:

(i) $u^* = \max_{i\in s}|u_i| = o_p\left(n^{1/2}\right)$ and (ii) $\dfrac{\sum_{i\in s} w_i(s)u_i}{\sum_{i\in s} w_i(s)u_i^2} = O_p\left(n^{-1/2}\right)$,

where $u_i = x_i - \overline{X}$.

Remark 25.6.1

Chen and Sitter (1999) showed that many commonly used sampling designs satisfy the regularity conditions above such as PPSWR and Rao-Hartley-Cochran (1962) sampling designs.

25.6.1 GREG Estimator Versus MPEL Estimator

Deville and Särndal (1992) proposed calibrated Horvitz–Thompson estimator $\widehat{\overline{Y}}_{HT} = \frac{1}{N}\sum_{i\in s}\frac{y_i}{\pi_i}$ of the population mean $\overline{Y}$ as

$$\widehat{\overline{Y}}_c = \sum_{i\in s} w_i y_i \tag{25.6.1}$$

Here w_i's are the calibrated design weights that are obtained by minimizing distance between w_i's and $d_i\left(=\frac{1}{N\pi_i}\right)$'s subject to the constraints

$$\sum_{i\in s} w_i x_i = \overline{X} \tag{25.6.2}$$

The commonly used distance measure is the chi-squared distance

$$\phi = \sum_{i\in s}\frac{(w_i - d_i)^2}{d_i q_i} \tag{25.6.3}$$

with q_i's as prespecified weights. Several alternative distance measures were proposed by Deville and Särndal (1992).

The minimization of Eq. (25.6.3) subject to Eq. (25.6.2) yields

$$w_i = d_i - \frac{d_i x_i q_i}{\sum_{i\in s} d_i x_i^2 q_i}\left(\widehat{\overline{X}}_{HT} - \overline{X}\right) \tag{25.6.4}$$

where $\widehat{\overline{X}}_{HT} = \sum_{i\in s} d_i x_i$.

Substituting the values of w_i obtained from Eq. (25.6.4) in Eq. (25.6.1) yields calibrated estimator as

$$\widehat{Y}_c = \widehat{\overline{Y}}_{ht} - \widehat{B}_{ht}\left(\widehat{\overline{X}}_{HT} - \overline{X}\right) \tag{25.6.5}$$

where

$$\widehat{B}_{ht} = \frac{\sum_{i \in s} y_i d_i q_i x_i}{\sum_{i \in s} y_i d_i q_i x_i^2} \tag{25.6.6}$$

The estimator Eq. (25.6.6) is the well-known generalized regression estimator. It is worth to note that the weights w_i's given in Eq. (25.6.4) may take negative values whereas the weights $\widehat{p}_i$'s of the MPEL estimator are always nonnegative. Thus the GREG estimator for a distribution function may not be a genuine distribution function whereas an MPEL estimator for a distribution function is always a genuine distribution function.

25.7 EMPIRICAL LIKELIHOOD FOR STRATIFIED SAMPLING

Suppose that a population of N units is stratified into H strata of sizes $N_1,\ldots,$ N_H and $W_h = N_h/N$. Let y_{hi} and x_{hi} be the values of the study and auxiliary variables for the ith unit of the hth stratum, $i = 1,\ldots, N_h$; $h = 1,\ldots, H$. Suppose a sample s_h of size n_h is selected from the hth stratum by SRSWOR. Let the samples be selected independently from each of the stratum and sampling fractions n_h/N_h are negligible for each $h = 1,\ldots, H$. Assuming $z_{hi} = (x_{hi}, y_{hi})$'s are independently distributed with distribution function $F_h, i = 1,\ldots, n_h, h = 1,\ldots, H$. Following Chen and Quin (1993), the log empirical likelihood for the stratified random sampling is obtained as

$$l(\mathbf{p}_1, \ldots, \mathbf{p}_\mathbf{H}) = \sum_{h=1}^{H} \sum_{i \in s_h} \log(p_{hi}) \tag{25.7.1}$$

where p_{hi} is the probability mass assigned to z_{hi}, $i \in s_h$, and $\mathbf{p_h} = (p_{h1},\ldots, p_{hn_h})'$. In case the population mean $\overline{X}$ of the auxiliary variable x is known, the MEL estimator of the population mean $\overline{Y}$ and distribution function F are, respectively, given by

$$\widehat{\overline{Y}}_{EL}^{st} = \sum_{h=1}^{H} W_h \sum_{i \in s_h} \widehat{p}_{hi} y_{hi} \tag{25.7.2}$$

and

$$\widehat{F}_{EL}^{st}(t) = \sum_{h=1}^{H} W_h \sum_{i \in s_h} \widehat{p}_{hi} I(y_{hi} \le t) \tag{25.7.3}$$

where $\widehat{p}_{hi}$'s maximize $l(\mathbf{p_1},\ldots, \mathbf{p_H})$ subject to,

(i) $p_{hi} \geq 0$, (ii) $\sum_{i\in s_h} p_{hi} = 1$, and (iii) $\sum_{h=1}^{H} W_h \sum_{i\in s_h} p_{hi}x_{hi} = \overline{X}$.

For maximization of $l(\mathbf{p_1},\ldots, \mathbf{p_H})$, we consider the following function ϕ where λ_h and λ are Lagrange multipliers and $n = \sum_h n_h$.

$$\begin{aligned}\phi = &\sum_h \sum_{i\in s_h} \log p_{hi} - \sum_h \lambda_h \left(\sum_{i\in s_h} p_{hi} - 1 \right) \\ &- n\,\lambda \left(\sum_h W_h \sum_{i\in s_h} p_{hi}x_{hi} - \overline{X} \right)\end{aligned} \tag{25.7.4}$$

Maximization of $l(\mathbf{p_1},\ldots, \mathbf{p_H})$ yields

$$p_{hi} = \frac{1}{n_h\left[1 + m_h\,\lambda\left(x_{hi} - \overline{X}_h^*\right)\right]} \tag{25.7.5}$$

where $\overline{X}_h^* = \sum_{i\in s_h} p_{hi}x_{hi}$ and $m_h = nW_h/n_h$.

Zhong and Rao (2000) showed that the solution of the system of Eq. (25.7.3) exists with probability tending to 1 as the sample sizes tend to infinity for each stratum. However, for deep stratification where the number of strata H is large and n_h's are small, the nature of solution is unknown.

25.7.1 Asymptotic Properties

Zhong and Rao (2000) derived the following result relating to asymptotic properties of empirical likelihood estimators by considering a sequence of stratified populations indexed by a label ν. The result is stated below without derivation.

Theorem 25.7.1

Suppose that, as $\nu \to \infty$, N_h, n_h and $N_h - n_h$ tends to infinity, $n_h/n \to k_h(>0)$, both $\sum_h W_h^2 \frac{1}{N_h} \sum_{i=1}^{N_h} |x_{hi}|^3$ and $\sum_h W_h^2 \frac{1}{N_h} \sum_{i=1}^{N_h} |y_{hi}|^3$ have an upper bound independent of ν, and $S_{hxx} = \frac{1}{N_h - 1} \sum_{i=1}^{N_h} \left(x_{hi} - \overline{X}_h\right)^2 \geq S_0 > 0 \;\; \forall h$ and ν. Then

$$\frac{\widehat{\overline{Y}}{}^{st}_{EL} - \overline{Y}}{\sigma_E}$$

asymptotically normally distributed with mean zero and variance unity as $\nu \to \infty$ where

$$\sigma_E^2 = \sum_h W_h^2\left(\frac{1}{n_h} - \frac{1}{N_h}\right)\left(S_{hyy} - 2BS_{hxy} + B^2 S_{hxx}\right),$$

$$B = \left(\sum_h \frac{W_h^2 S_{hxy}}{n_h}\right) \Big/ \left(\sum_h \frac{W_h^2 S_{hxx}}{n_h}\right),$$

$$S_{hyy} = \frac{1}{N_h - 1}\sum_{i=1}^{N_h}\left(y_{hi} - \overline{Y}_h\right)^2 \text{ and } S_{hxy} = \frac{\sum_{i=1}^{N_h}\left(x_{hi} - \overline{X}_h\right)y_{hi}}{N_h - 1}.$$

From the Theorem 25.7.1 above, it follows that $\widehat{Y}_{EL}^{st}$ approximately possesses the same asymptotic variance of the optimum regression estimator

$$\widehat{\overline{Y}}_{lr} = \overline{y}_{st} - \widehat{B}_0\left(\overline{x}_{st} - \overline{X}\right) \quad (25.7.6)$$

where $\overline{x}_{st} = \sum_{h=1}^{H} W_h \frac{1}{n_h}\sum_{i\in s_h} x_{hi}$, $\overline{y}_{st} = \sum_{h=1}^{H} W_h \frac{1}{n_h}\sum_{i\in s_h} y_{hi}$,

$\widehat{B}_0 = \sum_h W_h^2\left(\frac{1}{n_h} - \frac{1}{N_h}\right)s_{hxy} \Big/ \sum_h W_h^2\left(\frac{1}{n_h} - \frac{1}{N_h}\right)s_{hxx}$ with

$$s_{hxy} = \frac{\sum_{i\in s_h}\left(x_{hi} - \overline{x}_h\right)y_{hi}}{n_h - 1}, \text{ and } s_{hxx} = \frac{\sum_{i\in s_h}\left(x_{hi} - \overline{x}_h\right)^2}{n_h - 1}.$$

25.7.1.1 Variance Estimation

Under the conditions of Theorem 25.7.1, it can be shown that a consistent estimator of σ_E^2 (the asymptotic variance of $\widehat{Y}_{EL}^{st}$) is

$$s_E^2 = \sum_h W_h^2\left(\frac{1}{n_h} - \frac{1}{N_h}\right)\frac{1}{n_h - 1}\sum_{i\in s_h}\left\{y_{hi} - \overline{y}_h^* - \widehat{B}\left(x_{hi} - \overline{x}_h^*\right)\right\}^2$$

where

$$\widehat{B} = \sum_h \frac{W_h^2}{n_h}\sum_{i\in s_h}\left(y_{hi} - \overline{y}_h^*\right)\left(x_{hi} - \overline{x}_h^*\right) \Big/ \sum_h \frac{W_h^2}{n_h}\sum_{i\in s_h}\left(x_{hi} - \overline{x}_h^*\right)^2,$$

$\overline{y}_h^* = \sum_{i\in s_h}\widehat{p}_{hi}y_{hi}$, and $\overline{x}_h^* = \sum_{i\in s_h}\widehat{p}_{hi}x_{hi}$.

25.7.1.2 *Jackknife Variance Estimation*

Let $\widehat{\overline{Y}}^{st}_{EL}(h_j)$ be the estimator of $\overline{Y}$ when jth sample observation of the h stratum is deleted, then the Jackknife variance estimator of σ_E^2 is given by

$$s_{JE}^2 = \sum_h (n_h - 1)\left(\frac{1}{n_h} - \frac{1}{N_h}\right)\sum_{j=1}^{n_h}\left(\widehat{\overline{Y}}^{st}_{EL}(h_j) - \widehat{\overline{Y}}_{EL}\right)^2$$

Zhong and Rao (2000) showed that $n\left(s_{JE}^2 - \sigma_E^2\right)$ converges in probability to 0.

25.7.2 Pseudo–empirical Likelihood Estimator

Suppose that samples s_h of size n_h are selected independently from each of the strata by using unequal probability sampling schemes so that the inclusion probability of the ith unit of the hth stratum π_i^h is positive for every $i = 1,\ldots, N_h$, $h = 1,\ldots, H$. Let us further assume that z_{hi}'s independently distributed with distribution function F_h, then the log likelihood function of the entire stratified population becomes

$$l(\mathbf{p}_1, \ldots, \mathbf{p_H}) = \sum_{h=1}^{H}\sum_{i=1}^{N_h}\log(p_{hi})$$

A design-unbiased estimate of the population empirical log likelihood $l(\mathbf{p_1},\ldots, \mathbf{p_H})$ is given by

$$\widehat{l}(\mathbf{p}_1, \ldots, \mathbf{p_H}) = \sum_{h=1}^{H}\sum_{i\in s_h} d_{hi}\log(p_{hi}) \qquad (25.7.7)$$

where $d_{hi} = 1/\pi_i^h$.

In case $\overline{X}$ is known, maximizing $\widehat{l}(\mathbf{p}_1, \ldots, \mathbf{p_H})$ subject to (i) $p_{hi} \geq 0$, (ii) $\sum_{i\in s_h} p_{hi} = 1$, and (iii) $\sum_{h=1}^{H}\sum_{i\in s_h} p_{hi}\left(x_{hi} - \overline{X}\right) = 0$, Chen and Sitter (1999) derived the expression of the MPEL estimator of the population mean $\overline{Y}$ as

$$\widehat{\overline{Y}}^{st}_{MPEL} = \overline{y}_w - \widehat{B}^{st}_{greg}\left(\overline{x}_w - \overline{X}\right) + o_p\left(n^{-1/2}\right) \qquad (25.7.8)$$

where $\widehat{B}^{st}_{greg} = \dfrac{\sum_h\sum_{i\in s_h} w_{hi}(s)(x_{hi} - \overline{x}_w)y_{hi}}{\sum_h\sum_{i\in s_h} w_{hi}(s)(x_{hi} - \overline{x}_w)^2}$, $n = \sum_h n_h$, $\overline{y}_w = \sum_h\sum_{i\in s_h} w_{hi}(s)\, y_{hi}$, $\overline{x}_w = \sum_h\sum_{i\in s_h} w_{hi}(s)\, x_{hi}$, and $w_{hi}(s) = d_{hi}/\sum_h\sum_{i\in s_h} d_{hi}$

For SRSWOR, $d_{hi} = N_h/n_h$ and the above expression (25.7.8) reduces to

$$\widehat{\overline{Y}}_{MPEL} = \overline{y}^{st}_{greg} + o_p\left(n^{-1/2}\right) \tag{25.7.9}$$

where $\overline{y}^{st}_{greg} = \overline{y}_{st} - \dfrac{\sum\limits_h \sum\limits_{i\in s_h} W_h(x_{hi} - \overline{x}_{st})y_{hi}/n_h}{\sum\limits_h \sum\limits_{i\in s_h} W_h(x_{hi} - \overline{x}_{st})^2/n_h}\left(\overline{x}_{st} - \overline{X}\right)$, $\overline{y}_{st} = \sum W_h\overline{y}_h$

and $\overline{x}_{st} = \sum W_h\overline{x}_h$.

25.7.2.1 Multistage Sampling

Let us consider a finite population U, which is stratified into H strata. The hth strata consists of N_h first-stage units (fsu's) and the jth fsu of the hth stratum consists of N_{hj} second-stage units (ssu's). Let y_{hij} and x_{hij} be the values of the study and auxiliary variables for the jth ssu of the ith fsu of the hth stratum. Then the population totals of y and x are $Y = \sum\limits_{h=1}^{H}\sum\limits_{i=1}^{N_h}\sum\limits_{j=1}^{N_{hi}} y_{hij}$ and $X = \sum\limits_{h=1}^{H}\sum\limits_{i=1}^{N_h}\sum\limits_{j=1}^{N_{hi}} x_{hij}$, respectively. From the h (=1,..., H)th stratum, a subsample s_h of n_h fsu's is selected by same suitable sampling design with inclusion probability $\pi^h_i(>0)$ for the ith fsu of the hth stratum. If the ith fsu is selected in s_h, a subsample s_{hi} of n_{hi} units is selected from it by some suitable sampling design with inclusion probability $\pi^{ih}_j(>0)$ for the jth ssu of the ith fsu of the hth stratum. Here the sample sizes n_{hj} are predetermined numbers. Let $y_{hi1},\ldots, y_{hiNi}$ be distributed independently with distribution function F_{hi} and p_{hij} probability mass attached to y_{hij}. The log likelihood function of the entire population is given by

$$L = \sum_{h=1}^{H}\sum_{i=1}^{N_h}\sum_{j=1}^{N_{hi}} \log(p_{hij})$$

An unbiased estimate of the log likelihood L is given by

$$\widehat{L} = \sum_{h=1}^{H}\frac{1}{\pi^h_i}\sum_{i\in s_h}\frac{1}{\pi^{hi}_j}\sum_{j\in s_{hi}}\log(p_{hij}) = \sum_{h=1}^{H}\sum_{i\in s_h}\sum_{j\in s_{hi}} d_{hij}\log(p_{hij})$$

where $d_{hij} = \dfrac{1}{\pi^h_i\pi^{hi}_j}$.

Now maximizing $\widehat{L}$ subject to (i) $p_{hij} > 0$ and (ii) $\sum\limits_{i\in s_h}\sum\limits_{j\in s_{hi}} p_{hij} = 1$, the MPEL estimator for the population mean becomes a ratio estimator

$$\widehat{\overline{Y}}_{mst} = \sum_{h=1}^{H}\sum_{i\in s_h}\sum_{j\in s_{hi}}\widehat{p}_{hij}y_{hij} \tag{25.7.10}$$

where $\widehat{p}_{hij} = d_{hij}\Big/\sum\limits_{h=1}^{H}\sum\limits_{i\in s_h}\sum\limits_{j\in s_{hi}} d_{hij}$.

25.8 MODEL-CALIBRATED PSEUDOEMPIRICAL LIKELIHOOD

Wu and Sitter (2001) introduced the model-calibrated PEL method where the study variable y is related to a vector of auxiliary variable $\mathbf{x}_i' = (1, x_{i1}, \ldots, x_{ip})$ through the following superpopulation model ξ

$$E_\xi(y_i|\mathbf{x}_i) = \mu(\mathbf{x}_i, \boldsymbol{\theta}) = \mu_i \text{ and } V_\xi(y_i|\mathbf{x}_i) = \sigma^2 v_i^2; i = 1, \ldots, N \qquad (25.8.1)$$

where $\boldsymbol{\theta} = (\theta_0, \theta_1, \ldots, \theta_p)'$ and σ^2 are unknown model parameters, $\mu(\mathbf{x}_i, \boldsymbol{\theta})$ is a known function of $\mathbf{x}_i$ and $\boldsymbol{\theta}$, v_i is a known function of $\mathbf{x}_i$, and E_ξ and V_ξ denote the expectation and variance, respectively, with respect to the model ξ. We also assume that $(y_1, \mathbf{x}_1), \ldots, (y_N, \mathbf{x}_N)$ are mutually independent.

Let a sample s of size n be selected from the population with inclusion probability for the ith, and i and jth unit $(i \neq j)$ be π_i and π_{ij}, respectively. In the model-based approach, $(y_i, \mathbf{x}_i), i \in s$ are regarded as a random sample from the superpopulation ξ. Here we consider the design-based approach where $(y_i, \mathbf{x}_i), i \in s$ are not viewed as a random sample from the superpopulation. The model parameter $\boldsymbol{\theta}$ is regarded an estimate of $\boldsymbol{\theta}$ based on the entire population of N units and will be denoted by $\boldsymbol{\theta}_N$. For a linear regression model $\mu_i = \mathbf{x}_i'\boldsymbol{\theta}$, $\boldsymbol{\theta}_N$ is defined as

$$\boldsymbol{\theta}_N = (\mathbf{X}_N'\mathbf{X}_N)^{-1}\mathbf{X}_N'\mathbf{y}_N \qquad (25.8.2)$$

where $\mathbf{X}_N = \begin{pmatrix} \mathbf{x}_1' \\ \cdots \\ \mathbf{x}_N' \end{pmatrix}$, $\mathbf{y}_N = \begin{pmatrix} y_1 \\ \cdots \\ y_N \end{pmatrix}$.

A design-based estimate of $\boldsymbol{\theta}_N$ is

$$\widehat{\boldsymbol{\theta}} = (\mathbf{X}_n'\Pi^{-1}\mathbf{X}_n)^{-1}\mathbf{X}_n'\Pi^{-1}\mathbf{y}_n$$

where $\Pi = diag(\pi_1, \ldots, \pi_n)$, $\mathbf{X}_n = \begin{pmatrix} \mathbf{x}_1' \\ \cdots \\ \mathbf{x}_n' \end{pmatrix}$, $\mathbf{y}_n = \begin{pmatrix} y_1 \\ \cdots \\ y_n \end{pmatrix}$ assuming $s = (1, \ldots, n)$.

25.8.1 Estimation of the Population Mean

Let $\widehat{\mu}_i = \mu(\mathbf{x}_i, \widehat{\boldsymbol{\theta}})$ is the predicted value of y_i based on the model ξ and $\widehat{\boldsymbol{\theta}}$ is design-based estimator of $\boldsymbol{\theta}$. Here we maximize the PEL function

$$\widehat{l}(\mathbf{p}) = \sum_{i \in s} d_i \log p_i \text{ with } d_i = 1/\pi_i \qquad (25.8.3)$$

under constraints (i) $\sum_{i\in s} p_i = 1$ and (ii) $\sum_{i\in s} p_i\widehat{\mu}_i = \frac{1}{N}\sum_{i\in U}\widehat{\mu}_i$ (25.8.4)

It is to be noted that for the linear model $\mu_i = \mathbf{x}_i'\boldsymbol{\theta}$, $\frac{1}{N}\sum_{i\in U}\widehat{\mu}_i$ reduces to $\overline{\mathbf{X}}'\widehat{\boldsymbol{\theta}}$, so only the vector of the population mean $\overline{\mathbf{X}}$ of the auxiliary variables are in need for the constraint (25.8.4) but nonlinear model $\mu_i = \mathbf{x}_i'\boldsymbol{\theta}$, the constraint (25.8.4) requires complete information of $\mathbf{x}_1, \ldots, \mathbf{x}_N$.

The model-calibrated MPEL estimator of the population mean $\overline{Y}$ is given by

$$\widehat{\overline{Y}}_{CMPEL} = \sum_{i\in s}\widehat{p}_i y_i$$

where $\widehat{p}_i$'s maximize $\widehat{l}(\mathbf{p})$ subject to Eq. (25.8.4).

It follows from Eq. (25.5.8) that $\widehat{p}_i = \dfrac{w_i}{1 + \lambda u_i}$ where the Lagrange multiplier λ satisfies

$$\sum_{i\in s}\frac{w_i u_i}{1+\lambda\ u_i} = 0$$

with $w_i = d_i \Big/ \sum_{i\in s} d_i$ and $u_i = \widehat{\mu}_i - \frac{1}{N}\sum_{i\in U}\widehat{\mu}_i$.

25.8.2 Estimation of the Population Distribution Function

The population distribution function is $F_N(t) = \frac{1}{N}\sum_{i\in U} I(y_i \le t)$ and the Horvitz–Thompson estimator of $F_N(t)$ is $\widehat{F}_{ht}(t) = \frac{1}{N}\sum_{i\in s} d_i I(y_i \le t)$.

Under superpopulation model ξ,

$$E_\xi\{I(y_i \le t)\} = P(y_i \le t) = G\left(\frac{t-\mu_i}{v(x_i)}\right)$$

where $G(\cdot)$ is the CDF of the error term $\epsilon_i = \dfrac{y_i - \mu_i}{v(x_i)}$.

For a given value of t, let us define $\widehat{G}_i = \frac{1}{n}\sum_{i\in s} I\left(\widehat{\epsilon}_i \le \frac{t-\hat{\mu}_i}{v(\mathbf{x}_i)}\right)$ with $\widehat{\epsilon}_i = \dfrac{y_i - \hat{\mu}_i}{v(x_i)}$.

The MPEL estimator of $F_N(t)$ is given by

$$\widehat{F}_{MPEL}(t) = \sum_{i\in s}\widehat{p}_i I(y_i \le t) \tag{25.8.5}$$

where $\widehat{p}_i$ maximizes the PEL $\widehat{l}(\mathbf{p})$ given in Eq. (25.8.3) subject to

$$\text{(i) } p_i \geq 0, \text{(ii) } \sum_{i\in s} p_i = 1 \text{ and } \sum_{i\in s} p_i\widehat{G}_i = \frac{1}{N}\sum_{i\in U}\widehat{G}_i$$

The MPEL estimator $\widehat{F}_{MPEL}(t)$ derived in Eq. (25.8.5) always satisfies the properties of distribution function.

25.8.3 Model-Calibrated MPEL Estimation for Population Quadratic Parameters

The population variance and covariances can be expressed as a quadratic function of the form

$$T = \frac{1}{2}\sum_{i\neq}\sum_{j\in s}\phi(y_i, y_j) \tag{25.8.6}$$

where $\phi(y_i,y_j)$ is a symmetric function of y_i and y_j.

The pseudo log empirical likelihood function for the quadratic parameters is defined by

$$\widehat{l}^*(\mathbf{p}) = \frac{1}{2}\sum_{i\neq}\sum_{j\in s} d_{ij}\log(p_{ij})$$

where $d_{ij} = 1/\pi_{ij}$, π_{ij} inclusion probability for the ith and jth$(i \neq j)$ unit (assuming positive) and p_{ij} is the probability mass assigned to the pair of units (i,j).

The model-calibrated MPEL estimator of T is given by

$$\widehat{T}_{CMPEL} = \frac{1}{2N^*}\sum_{i\neq}\sum_{j\in s}\widehat{p}_{ij}\phi(y_i, y_j)$$

where $N^* = N(N-1)$ and $\widehat{p}_{ij}$ maximizes $\widehat{l}^*(\mathbf{p})$ subject to

$$\text{(i) } p_{ij} \geq 0, \text{(ii) } \frac{1}{2}\sum_{i\neq}\sum_{j\in s} p_{ij} = 1 \text{ and (iii) } \frac{1}{2}\sum_{i\neq}\sum_{j\in s} p_{ij}\widehat{u}_{ij} = \frac{1}{2N^*}\sum_{i\neq}\sum_{j\in U}\widehat{u}_{ij} \tag{25.8.7}$$

In the constraint (25.8.7) above, $\widehat{u}_{ij}$ is the estimate of $u_{ij} = E_\xi\{\phi(y_i,y_j)\}$, which is obtained by replacing the model parameters by their suitable design-based estimators. For example, consider the finite population variance

$$S_y^2 = \frac{1}{N-1}\sum_{i \in U}\left(y_i - \overline{Y}\right)^2 = \frac{1}{2N(N-1)}\sum_{i \neq}\sum_{j \in U}(y_i - y_j)^2$$

Here $u_{ij} = E_\xi\left(y_i - y_j\right)^2 = \{\mu(\mathbf{x}_i, \boldsymbol{\theta}) - \mu(\mathbf{x}_j, \boldsymbol{\theta})\}^2 + \sigma^2\{v_i + v_j\}$ and $\widehat{u}_{ij} = \left\{\mu\left(\mathbf{x}_i, \widehat{\boldsymbol{\theta}}\right) - \mu\left(\mathbf{x}_j, \widehat{\boldsymbol{\theta}}\right)\right\}^2 + \widehat{\sigma}^2(v_i + v_j)$, where $\widehat{\boldsymbol{\theta}}$ and $\widehat{\sigma}^2$ are design-based estimators of $\boldsymbol{\theta}$ and σ^2, respectively.

25.9 PSEUDO–EMPIRICAL LIKELIHOOD TO RAKING

Let U be a population of N units, which is cross-classified into $r \times c$ table and let U_{ij} be the set of N_{ij} units falling in the cell (i,j) such that $\sum_{i=1}^{r}\sum_{j=1}^{c} N_{ij} = N$. Let the marginal totals $N_i\cdot = \sum_{j=1}^{r} N_{ij}$ and $N._j = \sum_{i=1}^{c} N_{ij}$ of the cells are known but N_{ij}'s are unknown. To estimate the cell frequencies N_{ij}'s, a sample s of size n is selected from the population U and let $s_{ij} = s \cap U_{ij}$ be the sample of size n_{ij} for the (i,j) the cell. We can estimate N_{ij} by using estimates $\widehat{N}_{ij}(1) = n_{ij}(N/n)$, $\widehat{N}_{ij}(2) = n_{ij}(N_i\bullet/n_i\bullet)$, or $\widehat{N}_{ij}(3) = n_{ij}(N\bullet_j/n\bullet_j)$ where $n_i\bullet = \sum_{j=1}^{c} n_{ij}$ and $n\bullet_j = \sum_{i=1}^{r} n_{ij}$. But none of the estimates $\widehat{N}_{ij}(k)$ satisfies $\sum_{j=1}^{c}\widehat{N}_{ij}(k) = N_i\bullet$ and $\sum_{i=1}^{r}\widehat{N}_{ij}(k) = N\bullet_j$ for $k = 1,2,3$.

The purpose of the raking procedure is to find estimates $\widehat{N}_{ij}$ satisfying constraints

$$\sum_{j=1}^{c}\widehat{N}_{ij} = N_i. \text{ and } \sum_{i=1}^{r}\widehat{N}_{ij} = N._j. \tag{25.9.1}$$

Deming and Steaphan (1940) proposed a raking estimator of N_{ij} as

$$\widehat{N}_{ij}^{ds} = m_{ij}(N/n)$$

where m_{ij}'s are obtained by minimizing

$$\phi = \sum_{i=1}^{r}\sum_{j=1}^{c}(m_{ij} - n_{ij})^2/n_{ij} \tag{25.9.2}$$

subject to constraints

$$\sum_{j=1}^{c} m_{ij} = N_i.\,n/N, i = 1, \ldots, r-1 \text{ and } \sum_{i=1}^{r} m_{ij} = N._j n/N, \quad j = 1, \ldots, c-1 \tag{25.9.3}$$

Deming and Steaphan (1940) used the iterative proportional fitting procedure (IPFP) to solve the minimization problem. Rao and Wu (2009) pointed out that although IPFP satisfies Eq. (25.9.1), the solution does not minimize the least square distance (25.9.2). Rao and Wu (2009) gave the following solutions:

Let p_{ij} be the probability that an observation will fall in (i,j)th cell, then the log likelihood function is given by

$$l_0(\mathbf{p}) = \sum_{i=1}^{r}\sum_{j=1}^{c} n_{ij}\log(p_{ij}) \tag{25.9.4}$$

An empirical likelihood estimate of N_{ij} is given by

$$\widehat{N}_{ij}^{EL} = \widehat{p}_{ij}N \tag{25.9.5}$$

where $\widehat{p}_{ij}$'s are obtained by maximizing $l_0(\mathbf{p})$ subject to (i) $p_{ij} \geq 0$, (ii) $\sum_{i=1}^{r}\sum_{j=1}^{c} p_{ij} = 1$, and (iii) $\sum_{j=1}^{c} p_{ij} = N_{i\cdot}/N$, $i = 1,\ldots, r-1$ and $\sum_{i=1}^{r} p_{ij} = N_{\cdot j}/N$, $j = 1,\ldots, c-1$.

The solution (25.9.5) satisfies the raking conditions in (25.9.1).

25.10 EMPIRICAL LIKELIHOOD RATIO CONFIDENCE INTERVALS

Owen (1988) showed that the empirical likelihood approach provides nonparametric confidence intervals for the parameter of interest θ similar to parametric likelihood ratio confidence intervals. Consider the situation where the parameter of interest θ can be defined as a unique solution of the estimating equation $E[g(y,\theta)] = 0$. For example, $g(y,\theta) = y - \theta$ and $g(y,\theta) = I(y \leq t) - \theta$ yield $\theta = E(y)$, the population mean, and $\theta = E[I(y \leq t)] = F(t)$, the population distribution function, respectively.

25.10.1 Simple Random Sampling

Let a sample of size n is selected from a finite population of size N by simple random sampling with or without replacement. For SRSWOR, we assume that the sampling fraction is negligible so that we can write the likelihood function as $\prod_{i=1}^{n} p_i$. A profile likelihood function is then defined as

$$R(\theta) = \text{Max}\left\{\prod_{i=1}^{n}(np_i) \,\middle|\, \sum_{i=1}^{n} p_i g(y_i,\theta) = 0, p_i > 0, \sum_{i=1}^{n} p_i = 1\right\} \tag{25.10.1}$$

Owen (1988) proved that under mild moment conditions, for the case $\theta = E(y)$ or $F(t)$ that

$$r(\theta) = -2\log\{R(\theta)\} \tag{25.10.2}$$

converges to χ_1^2 (chi-square distribution with one degree of freedom) as $n \to \infty$. Hence $(1-\alpha)$-level confidence interval for θ is obtained as $\{\theta | r(\theta) \leq \chi_1^2(\alpha)\}$ where $\chi_1^2(\alpha)$ is the upper α-quantile of χ_1^2. It should be noted that unlike confidence interval based on normal approximation, empirical likelihood intervals do not require estimation of standard error of estimators and yield more balanced tail error rates.

25.10.2 Complex Sampling Designs

Let a sample of size n be selected from the population using a sampling design p and $\pi_i(>0)$ be the inclusion probability of the ith unit. In this case the pseudo−empirical log likelihood is given by

$$l_{ns}(\mathbf{p}) = \sum_{i\in s} d_i \log p_i \;\text{ with }\; d_i = 1/\pi_i \tag{25.10.3}$$

Maximizing $l_{ns}(\mathbf{p})$ with respect to $\mathbf{p}$ subject to (i) $p_i > 0$ and (ii) $\sum_{i\in s} p_i = 1$ yield

$$\widehat{p}_i = \widetilde{d}_i(s) = d_i \Big/ \sum_{i\in s} d_i$$

Similarly, let $\widehat{p}_i(\theta)$ be the value of p_i that maximizes the log likelihood $l_{ns}(\mathbf{p})$ for a fixed θ subject to

$$\text{(i) } p_i > 0, \text{(ii) } \sum_{i\in s} p_i = 1 \text{ and (iii) } \sum_{i\in s} p_i g(y_i) = \theta$$

where $g(y_i)$ is a function of y_i.

The pseudo−empirical log likelihood function is given by

$$r_{ns}(\theta) = -2\left[l_{ns}\{\widehat{\mathbf{p}}(\theta)\} - l_{ns}(\widehat{\mathbf{p}})\right] \tag{25.10.4}$$

where $\widehat{\mathbf{p}}(\theta)$ and $\widehat{\mathbf{p}}$ are the vector $\mathbf{p}$ with $p_i = \widehat{p}_i(\theta)$ and $p_i = \widehat{p}_i$, respectively.

Consider a simple situation where $g(y_i) = y_i$ and $\theta = \overline{Y}$. The design effect of estimating $\overline{Y}$ using the Hájek (1964) estimator $\widehat{\overline{Y}}_H = \sum_{i\in s} d_i y_i \Big/ \sum_{i\in s} d_i$ is given by

$$deff_H = V_p\left(\widehat{\overline{Y}}_H\right) \Big/ \left(S_y^2 \big/ n\right) \tag{25.10.5}$$

where $V_p\left(\widehat{\overline{Y}}_H\right)$ is the design variance of $\widehat{\overline{Y}}_H$ with respect to the sampling design p, S_y^2 is the population variance, and S_y^2/n is the variance of $\widehat{\overline{Y}}_H$ under SRSWOR sampling design (ignoring the sampling fraction).

Wu and Rao (2006) derived the following theorem (given without derivation) relating to the asymptotic distribution of $r_{ns}(\theta)$.

Theorem 25.10.1

Under the regularity conditions stated below, the adjusted pseudo–empirical log likelihood ratio statistic

$$r_{ns}^{[H]}(\theta) = \{r_{ns}(\theta)\}/deff_H$$

is asymptotically distributed as χ_1^2 when $\theta = \overline{Y}$.

The regularity conditions comprise the following:

C1: The sampling design p and the study variable y satisfy $\max_{i\in s}|y_i| = o_p(n^{1/2})$, where the stochastic order $o_p(\bullet)$ is with respect to the sampling design p.

C2: The sampling design p satisfies $\frac{1}{N}\sum_{i\in s} d_i - 1 = O_p\left(n^{-1/2}\right)$.

C3: The Horvitz–Thompson estimator $\widehat{\overline{Y}}_{HT} = \frac{1}{N}\sum_{i\in s} d_i y_i$ of the population mean $\overline{Y}$ is asymptotically normally distributed.

Using Theorem 25.10.1, we can set a $(1-\alpha)$ level confidence interval for $\theta = \overline{Y}$ as

$$\left\{\overline{Y} \middle| r_{ns}^{[H]}\left(\overline{Y}\right) \le \chi_1^2(\alpha)\right\}.$$

Now let us suppose that the we have a vector of auxiliary variable $\mathbf{x}$ with known population mean $\overline{\mathbf{X}}$. In this case the pseudoempirical log likelihood function for $\overline{Y}$ given in Eq. (25.10.3) should be obtained incorporating the additional constraint $\sum_{i\in s} p_i\mathbf{x}_i = \overline{\mathbf{X}}$ in finding both $\widehat{p}_i$ and $\widehat{p}_i(\theta)$. In this case the pseudo–empirical log likelihood ratio function is adjusted by the design effect associated with the GREG estimator

$$\widehat{\overline{Y}}_G = \widehat{\overline{Y}}_H - \mathbf{B}'\left(\widehat{\overline{\mathbf{X}}}_H - \overline{\mathbf{X}}\right)$$

where $\widehat{\overline{\mathbf{X}}}_H = \sum_{i\in s}\widetilde{d}_i(s)\mathbf{x}_i$, $\widehat{\overline{\mathbf{Y}}}_H = \sum_{i\in s}\widetilde{d}_i(s)y_i$ and $\mathbf{B} = \dfrac{\sum_{i\in U}(\mathbf{x}_i - \overline{\mathbf{x}})\left(y_i - \overline{\mathbf{Y}}\right)}{\sum_{i\in U}(\mathbf{x}_i - \overline{\mathbf{x}})(\mathbf{x}_i - \overline{\mathbf{x}})'}$.

The design effect associated with $\widehat{\overline{Y}}_G$ is given by

$$deff_G = V_p\left(\widehat{\overline{Y}}_G\right)\Big/\left(S_r^2/n\right) \quad (25.10.7)$$

where $V_p\left(\widehat{\overline{Y}}_G\right) = V_p\left(\sum_{i\in s}\widetilde{d}_i(s)r_i\right)$, $r_i = y_i - \overline{Y} - \mathbf{B}'\mathbf{u}_i$, $\mathbf{u}_i = \mathbf{x}_i - \overline{\mathbf{X}}$, S_r^2/n is the variance of $\widehat{\overline{Y}}_G$ under SRSWOR ignoring sampling fraction n/N, and $S_r^2 = \sum_{i=1}^{N} r_i^2/(N-1)$.

Wu and Rao (2006) derived the asymptotic distribution of distribution of the adjusted pseudoempirical log likelihood ratio statistic $r_{ns}^{(G)}(\theta) = \{\widetilde{r}_{ns}(\theta)\}/deff_G$ under the following additional constraints.

C4: $\max_{i\in s}||\mathbf{x}_i|| = o_p(n^{1/2})$

C5: $\widehat{\overline{\mathbf{X}}}_{HT} = \frac{1}{N}\sum_{i\in s} d_i\mathbf{x}_i$ is asymptotically normally distributed.

The result has been given in the following theorem without derivation.

Theorem 25.10.2

Let $\widetilde{\mathbf{p}}$ maximizer of $l_{ns}(\widetilde{\mathbf{p}})$ subject to (i) $p_i > 0$, (ii) $\sum_{i\in s} p_i = 1$, and (iii) $\sum_{i\in s} p_i\mathbf{x}_i = \overline{\mathbf{X}}$; and let $\widetilde{\mathbf{p}}(\theta)$ be obtained by maximizing $l_{ns}(\widetilde{\mathbf{p}})$ under the constraints (i) $p_i > 0$, (ii) $\sum_{i\in s} p_i = 1$, (iii) $\sum_{i\in s} p_i\mathbf{x}_i = \overline{\mathbf{X}}$, and (iv) $\sum_{i\in s} p_i y_i = \theta$ for a fixed θ. Then under the regularity conditions stated C1 to C5, the adjusted pseudoempirical log likelihood ratio statistic

$$r_{ns}^{(G)}(\theta) = \{\widetilde{r}_{ns}(\theta)\}/deff_G$$

is asymptotically distributed as χ_1^2 when $\theta = \overline{Y}$ where $\widetilde{r}_{ns}(\theta) = -2\left[l_{ns}\{\widetilde{\mathbf{p}}(\theta)\} - l_{ns}(\widetilde{\mathbf{p}})\right]$.

In practice the design effects $deff_H$ and $deff_G$ are unknown because of involvement of unknown parameters. The common practice is to replace the unknown parameters by their consistent estimators. Wu and Rao (2006) reported that asymptotic distributions of $r_{ns}^{(H)}(\theta)$ and $r_{ns}^{(G)}(\theta)$ remain unchanged if designs effects $deff_H$ and $deff_G$ are replaced by their consistent estimators. The estimated design effects are

$$\widehat{deff}_H = \widehat{V}_p\left(\widehat{\overline{Y}}_H\right)\Big/\left(\widehat{S}_y^2/n\right) \text{ and } \widehat{deff}_G = \widehat{V}_p\left(\widehat{\overline{Y}}_G\right)\Big/\left(\widehat{S}_y^2/n\right)$$

where $\widehat{V}_p\left(\widehat{\overline{Y}}_H\right) = \frac{1}{\widehat{N}^2}\frac{1}{2}\sum\sum_{i\neq j\in s}\frac{\pi_i\pi_j - \pi_{ij}}{\pi_{ij}}\left(\frac{e_i}{\pi_i} - \frac{e_j}{\pi_j}\right)^2$ with $\widehat{N} = \sum_{i\in s} d_i$

and $e_i = y_i - \widehat{\overline{Y}}_H$; $\widehat{S}_y^2 = \frac{1}{N(N-1)}\frac{1}{2}\sum\sum_{i\neq j\in s}\frac{(y_i - y_j)^2}{\pi_{ij}}$;

$$\widehat{V}_p\left(\widehat{\overline{Y}}_G\right) = \frac{1}{\widehat{N}^2}\frac{1}{2}\sum\sum_{i\neq j\in s}\frac{\pi_i\pi_j - \pi_{ij}}{\pi_{ij}}\left(\frac{\hat{r}_i}{\pi_i} - \frac{\hat{r}_j}{\pi_j}\right)^2$$

with $\hat{r}_i = y_i - \hat{\hat{\overline{Y}}}_H - \hat{\mathbf{B}}'(\mathbf{x}_i - \overline{\mathbf{X}})$, $\hat{B} = \dfrac{\sum_{i\in s}(\mathbf{x}_i - \hat{\overline{\mathbf{X}}}_h)y_i/\pi_i}{\sum_{i\in s}(\mathbf{x}_i - \hat{\overline{\mathbf{X}}}_h)(\mathbf{x}_i - \hat{\overline{\mathbf{X}}}_h)'/\pi_i}$,

and $\hat{S}_r^2 = \dfrac{1}{N(N-1)}\dfrac{1}{2}\sum_{i\neq}\sum_{j\in s}\dfrac{(r_i - r_j)^2}{\pi_{ij}}$.

Using Theorems 25.10.1 and 25.10.2 we can set a $(1-\alpha)$ level PEL confidence interval for $\theta = \overline{Y}$ as $\left\{\overline{Y}\middle|\hat{r}_{ns}^{(H)}(\overline{Y}) \le \chi_1^2(\alpha)\right\}$ for the case of no auxiliary information is available. In case the vector of the population mean $\overline{\mathbf{X}}$ is available, the confidence interval of $\theta = \overline{Y}$ is obtained as $\left\{\overline{Y}\middle|r_{ns}^{(\hat{G})}(\overline{Y}) \le \chi_1^2(\alpha)\right\}$

where $\hat{r}_{ns}^{(H)}(\overline{Y}) = \{r_{ns}(\theta)\}/\widehat{deff}_H$ and $r_{ns}^{(\hat{G})}(\overline{Y}) = \{\tilde{r}_{ns}(\overline{Y})\}/\widehat{deff}_G$.

25.10.3 Stratified Sampling

Consider the stratified sampling described in Section 25.7, where independent samples s_h's of sizes n_h's are selected from the hth stratum with inclusion probability $\pi_i^h = 1/d_{hi}$ (>0) for the ith unit of the hth stratum and $\sum_{h=1}^{H} n_h = n$ be the overall sample size. The PEL function under stratified sampling was given by Wu and Rao (2006) as

$$l_{st}(\hat{\mathbf{p}}_1, \ldots, \hat{\mathbf{p}}_H) = n\sum_{h=1}^{H} W_h \sum_{i\in s_h} \tilde{d}_{hi}\log(p_{hi}) \qquad (25.10.9)$$

where $\tilde{d}_{hi} = d_{hi}/\sum_{i\in s_h} d_{hi}$.

Suppose that the mean vectors $\overline{\mathbf{X}}_h$' s of auxiliary variable of the strata are unknown but the overall population mean $\overline{\mathbf{X}} = \sum_{h=1}^{H} W_h\overline{\mathbf{X}}_h$ is known. In this case the pseudo–empirical log likelihood ratio statistic $r_{st}(\theta)$ is given by Rao and Wu (2009) as

$$r_{st}(\theta) = -2[l_{st}\{\hat{\mathbf{p}}_1(\theta), \ldots, \hat{\mathbf{p}}_H(\theta)\} - l_{st}\{\hat{\mathbf{p}}_1, \ldots, \hat{\mathbf{p}}_H\}] \qquad (25.10.10)$$

where

(a) $\hat{p}_{hi}$ maximize $l_{st}(\hat{\mathbf{p}}_1, \ldots, \hat{\mathbf{p}}_H) = n\sum_{h=1}^{H} W_h \sum_{i\in s_h} \tilde{d}_{hi}\log(p_{hi})$ subject to the set of constraints

$$\text{(i) } p_{hi} > 0,\ \text{(ii) } \sum_{i\in s_h} p_{hi} = 1, h = 1, \ldots, H \text{ and (iii) } \sum_{h=1}^{H} W_h \sum_{i\in s_h} p_{hi}\mathbf{x}_{hi} = \overline{\mathbf{X}} \qquad (25.10.11)$$

and

(b) $\widehat{p}_{hi}(\theta)$ maximize $l_{st}(\widehat{\mathbf{p}}_1, \ldots, \widehat{\mathbf{p}}_H)$ subject to (25.10.11) with an additional constraint $\sum_{h=1}^{H} W_h \sum_{i \in s_h} p_{hi} y_{hi} = \theta$.

Theorem 25.10.3

Under the regularity condition C1 to C5 within each stratum h, the adjusted PEL ratio statistic

$$r_{st}^{[a]}(\theta) = \{r_{st}(\theta)\} / deff_{G(st)}$$

is asymptotically distributed as χ_1^2 when $\theta = \overline{Y}$. The design effect $deff_{G(st)}$ is given by Wu and Rao (2006).

The $(1 - \alpha)$ level confidence interval for the population mean can be constructed from

$$\left\{ \overline{Y} \middle| r_{st}^{[a]}(\overline{Y}) \leq \chi_1^2(\alpha) \right\} \quad (25.10.12)$$

25.10.4 Confidence Interval for Distribution Function

Pseudo–empirical likely ratio confidence intervals of the distribution function $F(t)$ for a given t can be obtained from Theorems 25.10.1–25.10.3 by writing $I(y_i \leq t)$ in place of y_i. The bench mark condition $\sum_{i \in s} p_i \mathbf{x}_i = \overline{\mathbf{X}}$ for making inference on $F(t)$ may not produce efficient result as the correlation between the indicator variable $I(y_i \leq t)$ and $\mathbf{x}_i$ is seldom high. However, if the entire auxiliary variables of the population are known, different constraints may lead more efficient inference on $F(t)$ (Chen and Wu, 2002).

25.11 CONCLUDING REMARKS

Hartley and Rao (1968) introduced the concept of scale load approach, which produces useful results for simple random sampling in absence and presence of auxiliary variables. However, the generalization of scale load approach in unequal probability sampling without replacement does not provide any useful result. The modern concept of EL was introduced by Owen (1988). Chen and Quin (1993) provided the asymptotic variance of EL estimator of the finite population mean and a consistent estimator of the variance of the proposed estimator. Their method was limited to simple random sampling only. Chen and Sitter (1999) brought the concept of PEL approach, which can be used for any complex survey design. For simple random sampling, the PEL reduces to EL approach proposed by Chen and Quin (1993). For estimating the finite population mean or distribution function under complex survey design, the PEL approach yields asymptotically generalized regression estimator when the population means of the vector of auxiliary variables are known. Wu (2004) developed a PEL approach that combines information from two or more independent surveys from the same population with some common variable of interest. The method can be used to handle data from independent samples taken from two or more incomplete frames covering the entire population of interest and to produce efficient estimators. Sitter and Wu (2002) proposed the model-calibrated PEL method for estimating quadratic population parameters, which includes population variance and covariance. Chen and Quin (1993) used EL approach for nonparametric confidence intervals of the population mean and distribution function. Quin and Lawless (1994) used EL and estimating equations for constructing confidence intervals and hypothesis testing problems. Wu and Rao (2006) used PEL method for estimating confidence interval for finite population parameter with or without auxiliary information. They pointed out that PEL ratio confidence intervals are better than those based on normal approximation in terms of coverage probability and length of the confidence intervals. PEL confidence intervals are based on the asymptotic distribution of PEL ratio statistic with the adjustment factor related to design effect. Estimation of design effect involves variance estimation, which may not be easy for complex surveys. Wu and Rao (2009) introduced bootstrap procedures for PEL ratio confidence intervals, which do not require variance estimation and at the same time provide superior intervals in respect of coverage probability for moderate or small sample sizes. Wu (2005) provided computer software for computation of confidence intervals. Wang and Rao (2002) and Quin et al. (2006) used empirical and PEL methods for kernel regression imputation for nonresponse. Rao (2010) introduced Baysian PEL intervals for complex survey designs. Some good reviews have been given by Rao (2006) and Rao and Wu (2009).

25.12 EXERCISES

25.12.1 Let $y_1, \ldots, y_i, \ldots, y_n$ be independent random variables with common CDF F_0 and S_n be their ECDF. Let F be any CDF, then $L(F) < L(S_n)$ for $F \neq S_n$ (Owen, 2001).

25.12.2 Let $y_1,\ldots, y_i,\ldots, y_n$ be a random sample from a discrete distribution with $P(y = y_i) = p_i$, $p_i > 0, \sum_{i=1}^{N} p_i = 1$. Find the MEL estimator for $\mu_k = E(y^k)$.

25.12.3 A sample of n units using PPSWR method with normed size measure q_i attached to the ith unit. Show that the MPEL estimator for the population mean $\overline{Y}_N$ without using any auxiliary information is $\widehat{\overline{Y}}_N = \left(\sum_{i=1}^{N} y_i/p_i\right) \Big/ \left(\sum_{i=1}^{N} 1/p_i\right)$ (Chen and Sitter, 1999).

25.12.4 Suppose a sample of n is selected from a population by Rao–Hartley–Cochran sampling scheme using p_i as normed size measure for the ith unit. Find the MPEL estimator for the population mean when no auxiliary variable is available (Chen and Sitter, 1999).

25.12.5 Let a sample of n is selected by Poisson sampling scheme with inclusion probability π_i for the ith unit.

(i) Show that the MPEL estimator for the population mean is $\left(\sum_{j=1}^{n} y_i/\pi_i\right) \Big/ \left(\sum_{j=1}^{n} 1/\pi_i\right)$ when no auxiliary information is available.

(ii) If the auxiliary information is used as $\left(\sum_{j=1}^{n} p_i x_i\right) = \mu_x$, show that p_i is obtainable as a solution of the equations

$$p_i = \frac{1}{\lambda_1 \pi_i + \lambda_2 (x_i - \mu_x)}, \sum_{i=1}^{n} \frac{1}{\lambda_1 \pi_i + \lambda_2 (x_i - \mu_x)} = 1$$

$$\text{and} \sum_{i=1}^{n} \frac{x_i - \mu_x}{\lambda_1 \pi_i + \lambda_2 (x_i - \mu_x)} = 0 \ \text{(Kim, 2009)}.$$

CHAPTER 26

Sampling Rare and Mobile Populations

26.1 INTRODUCTION

A rare population is defined as a small subset of a population of interest. Smallness, although subjective, is generally treated as one-tenth, one-hundredth, or even less. For example, surveys on physically disabled persons, child labors, ethnic minority groups, households with very high income, persons with rare diseases, etc. focus on rare populations. Mobile populations include migratory birds, visitors to places of historical interest or shopping malls, pavement dwellers, and hospital outpatients. The main objectives of surveying rare and mobile populations are as follows:

(i) To find the number (M) or the prevalence rate $P = M/N$ where N is the size of the total population. For example, P may be the prevalence of the HIV infection of the total population.

(ii) To estimate characteristics y of the rare or mobile population such as mean income of a child labor. Let y_i be the value of y for the ith unit and let $\delta_i = 1$ if the ith unit belongs to the rare population and $\delta_i = 0$ otherwise, then the size of the rare population and the rare population mean of the character y are, respectively,

$$M = \sum_{i=1}^{N} \delta_i \text{ and } \overline{Y} = \sum_{i=1}^{N} y_i\delta_i/M$$

If there is a separate frame for the rare population, then the sample of rare population can be selected very easily by following any standard sampling technique. But in reality such a frame is rarely available. Furthermore, most surveys are multicharacter surveys where information on several characteristics are collected at the same time with some of the characteristics being rare and others not. For example, in a household survey, we collect information on the income of the household as well as if they have been a victim of theft, among others. In this case, being victim of theft is a rare event but inquiring about household income is not. A number of methods have been proposed

Survey Sampling Theory and Applications
ISBN 978-0-12-811848-1
http://dx.doi.org/10.1016/B978-0-12-811848-1.00026-1

for sampling of rare and mobile populations when a separate list is not available. Good details are given by Kish (1965, 1991), Sudman (1976), Kalton and Anderson (1986), Sudman and Kalton (1986), Thompson and Seber (1996), and Kalton (2001). Some of the methods are as follows:

(i) screening, (ii) disproportionate sampling, (iii) multiplicity or network sampling, (iv) multiframe sampling, (v) snowball sampling, (vi) location sampling, (vii) sequential sampling, (viii) adaptive sampling, and (ix) capture—recapture method.

26.2 SCREENING

In the case that members of the rare population can be identified from the information given in the sampling frame, the frame should be cleaned by deleting the nonrare population from the list and the sample should be selected from the cleaned frame. If any nonrare member is selected in the sample, he/she may be eliminated at the time of data collection. However, the sampling frame generally does not contain enough information to identify whether a unit belongs to the rare population or not. In this case a relatively large sample may be selected from the population and then the members of the rare population may be identified at the stage of data collection. The cost of screening for rare population will be high if the degree of rarity is high. In this case the following routes may be used.

26.2.1 Telephonic Interview

If the majority of households (units) have telephone facilities, then the population can be screened by telephone calls, which will incur a small cost. The sampled people can then be interviewed face to face. Obviously, telephone interviewing underrepresents households that cannot be accessible through telephone calls, e.g., households with low income, the deaf, those with limited activity due to illness, single or divorced heads. The telephonic interview is widely used in the United States for various surveys.

26.2.2 Mail Questionnaire

The mail questionnaire method can be used if full addresses of the households are available. The questionnaires should include items of information that can identify whether the household is a member of rare population or not. The mail questionnaire method was found very useful in the United Kingdom. The response was very high at 86% (Kalton and Anderson, 1986) and even more in certain areas.

26.2.3 Cluster Sampling

The cluster sampling method, although less efficient, from the efficiency point of view is useful for reducing the cost of data collection, as the travel cost from one unit to the other is minimum. If the rare population is underrepresented in many clusters and cannot be identified in advance, then the sampling of the rare population becomes unproductive and expensive because of the high screening cost. Sudman (1972, 1978, 1985) and Wakesberg (1978) provided a method of selection of clusters that prevents selection of clusters that do not contain a rare population. In this method clusters are selected with probability proportional to the number of rare population of the cluster.

26.2.3.1 Sudman–Waksberg Method

Suppose a population consists of N primary stage units (psu's) and each of them consists of M subunits. Let X_i be the number of a rare population that belongs to the ith psu, $i = 1,\ldots,N$. The number X_i is unknown and no listing of the subunits is available. Here, we assume that a cluster either contains no member of rare population or contains at least $n(k+1)$ members of rare population. Here, n and $k+1$ denote the desired number of psu's and the number of members of rare population from each of psu's to be selected in the sample, respectively. The values of n and k are determined optimally from the consideration of efficiency or cost of the survey.

In the Sudman–Waksberg method a psu is chosen at random and one unit (subunit) is selected from it randomly. If the selected subunit is a member of the rare population, then the psu is retained in the sample and additional k subunits are selected from this psu by simple random sampling without replacement (SRSWOR) method. If the selected subunit does not contain a member of the rare population, then the psu is not selected. The procedure is continued until a sample of n psu's is selected. The Sudman–Waksberg method selects a sample of n psu's by probability proportional to the measure of size with replacement (PPSSWR) method where the measure of size is the unknown number of the rare population of the psu. This follows from the fact that the probability of selection of ith psu in an effective draw (where a psu is selected)

$$\begin{aligned}&= \text{Probability of the selection of } i\text{th psu in the first draw and it is retained}\\&\quad + \text{Probability that the first draw does not select a psu and it is selected in}\\&\quad \text{the second draw } + \cdots\\&= X_i/(NM) + (1-\pi)X_i/(NM) + (1-\pi)^2 X_i/(NM) + \cdots\end{aligned}$$

$$\left(\text{where } \pi = \frac{1}{N}\sum_{i=1}^{N}\frac{X_i}{M}\right)$$

$$= X_i \Big/ \sum_{i=1}^{N} X_i.$$

26.2.4 Two-Phase Sampling

Identification of the rare population in some situations may be expensive, e.g., a medical diagnosis may be required for the prevalence of a neurological disorder. In this approach, a large sample is selected from the entire population. It is then classified into two or more strata according to the likelihood of rarity of the units by some relatively cheap but imperfect screening procedure. Subsamples from each of the strata are selected with numbers proportional to degree of likelihood of being a member of the rare population. Finally, an expensive method is used to detect whether the selected units are a member of the rare population or not. The two-phase method is useful if the cost of the first-phase method of screening is much less than that of the second phase. Deming (1977) recommended that the two-phase method should be used if the ratio of cost of screening for the second to first phase is at least 6:1.

26.3 DISPROPORTIONATE SAMPLING

Sometimes, rare populations are more heavily concentrated in certain pockets of the populations. In this situation, we treat various pockets as strata and samples are selected with higher sampling fractions for those strata with higher concentrations of the rare population. This procedure is cost-efficient because less screening is needed to identify the rare population in the strata with the higher concentration of the rare population.

Let $\overline{Y}_h$ and S_h^2 denote the population mean and population variance of the variable under study y for the stratum h, respectively, and let c_{sh} and c_{Rh} denote the cost of screening a member of the nonrare population and collecting data for a member of a rare population for the stratum h, respectively. Under the assumptions (i) $\overline{Y}_h = \overline{Y}$, (ii) $S_h^2 = S^2$, (iii) $c_{sh} = c_s$, and (iv) $c_{Rh} = c_R$ for all h, Kalton (2001) derived the optimum sampling fraction for the stratum h under simple random sampling with replacement (SRSWR) sampling as

$$f_h \propto \sqrt{\frac{P_h}{1 + (r - 1)P_h}} \tag{26.3.1}$$

where $P_h = M_h/N_h$ = prevalence of the rare population in the stratum h, M_h = number of the rare population, and N_h is the size of the corresponding stratum h and $r = c_R/c_s$.

The value of r generally exceeds 1. For $r = 1$, the optimum sampling fraction reduces to

$$f_h \propto \sqrt{P_h}$$

The gain in precision with the use of disproportionate sampling with the optimum sampling fraction over the proportionate stratification with $r = 1$ is provided by Kalton (2001) as

$$R = \left(\sum \sqrt{A_h W_h}\right)^2 \tag{26.3.2}$$

where A_h and W_h are the proportion of rare and total population in the stratum h, respectively.

It is observed that the disproportionate stratification produces appreciable gain in efficiency only when the following two conditions are realized: (i) the strata to be oversampled should have high concentration of the rare population and (ii) the strata need to contain a substantial proportion of the rare population. For further details readers are referred to Kalton and Anderson (1986) and Kalton (2001).

26.4 MULTIPLICITY OR NETWORK SAMPLING

Multiplicity or network sampling was originally proposed by Birnbaum and Sirken (1965) and was later developed by Sirken (1970, 1972), Sirken and Levy (1974), and Nathan (1976), among others. Sampling of a rare population requires many contacts to identify sampling units possessing the rare trait. The objective of the multiplicity sampling is to spread the identification of the rare population more broadly over the total population and hence it reduces the number of contacts needed. In a conventional household survey, a number of households are selected in the initial sample, and information of the rare trait is collected from the members of the sampled households if they possess the required characteristic. In multiplicity sampling design, information regarding the membership of the rare trait is collected not only from each of the selected households but also from other households that are linked to that household. The linkage should be clearly defined. For example, the linkage may include any of the family relationship such as children, brothers, parents, or siblings; coworkers; members of the same church or social organization; and adjacent neighbors of the selected households. An individual living in an institution has no chance of selection in a conventional household survey, but he/she may have a chance of selection in the multiplicity sampling. The linkage provides necessary information of the existence of the rare trait. Sirken et al. (1978) and Czaja et al. (1884) used close relatives as linkage for survey of rare illnesses, Nathan (1976) for survey of births and deaths, whereas Brown and Ritchie (1981) for the survey of ethnic minorities.

Multiplicity sampling may be useful in estimating prevalence of the rare population or simply to identify members of the rare population.

The selection probabilities of the sampled members of the rare trait are determined by the defined linkage. For example, in an equal probability selection of households, all members of a household with a rare trait are given equal weight, but in a multiplicity sampling, the weight associated with an individual with a rare trait is inversely proportional to the number of households in which an individual member has a link including the household.

The main advantage of a multiplicity sample is that it needs a smaller sample to yield the required sample size of the rare population. The disadvantage is that the sampling design yields unequal probability of selection of members of the rare population and hence requires adjustment of weights. In addition, in the case of nonresponse caused by failure to trace the identified members of the rare population, the weight adjustments become tedious. Furthermore, the cost tracing and contacting all the individuals in a multiplicity sampling may be considerably high. The main demerit of a multiplicity sampling is that it may jeopardize the privacy of the individuals possessing the rare trait, e.g., an HIV infected person may not wish to be identified.

26.5 MULTIFRAME SAMPLING

A complete list of the rare population may not be available, but different partial (incomplete) lists of the rare population may be available from different sources. The lists are not generally disjoint, but their union may cover the entire rare population under study. For example, a complete list of antiretroviral (ARV) treatment recipients is difficult to obtain, but several lists such as hospital records, and records from government and nongovernment clinics and pharmacies may cover almost all ARV recipients. The lists may not be disjoint because a person may visit different clinics or hospitals.

One can prepare a complete frame by combining the incomplete frames and omitting duplicates. But in practice, removing duplication is not easy and the process may be highly error-prone because the spelling of the name or address of the same person may be different in the two lists. On the contrary, an incomplete list but containing a high proportion of rare individuals may provide a better result and be cheaper to sample. Selection of the ARV treatment recipients from the sampled clinics is easy and less expensive than surveying households because the hospital lists comprise of a high proportion of ARV treatment recipients than the complete household lists. Hartley (1962, 1974) pointed out that sampling from both the incomplete and complete frame may produce better estimate than sampling from the complete frame only.

The methodology of multiframe survey was introduced by Hartley (1962). The theory was extended by Lund (1968), Fuller and Burmeister (1972), Bankier (1986), Kalton and Anderson (1986), Singh and Wu (1996), and Skinner and Rao (1996), among others. In this section we will consider estimation of population characteristics from dual frame survey only. The theory can be extended for multiframe surveys.

26.5.1 Methods of Estimation

Let U_A and U_B be two incomplete frames that cover the population U under consideration, i.e., $U = U_A \cup U_B$. The frames U_A and U_B have overlap $U_{ab} = U_A \cap U_B = U_{ba}$. Let $U_a = U_A \cap U_B^c =$ consisting units of U_A only while $U_b = U_B \cap U_A^c$ consisting of units of U_B only. Let N, N_A, N_B, N_a, N_b, and N_{ab} ($=N_{ba}$) be sizes of U, U_A, U_B, U_a, U_b, and $U_{ab} = (U_{ba})$, respectively. Clearly, $N = N_a + N_b + N_{ab}$, and $N_A = N_a + N_{ab}$ and $N_B = N_b + N_{ab}$. Let y_i be the value of the variable under study for the ith unit of the population U. Let us define $Y = \sum_{i \in U} y_i$, $Y_A = \sum_{i \in U_A} y_i$, $Y_B = \sum_{i \in U_B} y_i$, $Y_a = \sum_{i \in U_a} y_i$, $Y_b = \sum_{i \in U_b} y_i$, and $Y_{ab} = \sum_{i \in U_{ab}} y_i = Y_{ba}$. Then we have

$$Y = Y_a + Y_b + Y_{ab}$$
$$Y_A = Y_a + Y_{ab}$$
$$Y_B = Y_b + Y_{ab}$$

Let samples s_A and s_B of sizes n_A and n_B be selected independently from the frames A and B, respectively, using suitable sampling designs p_A and p_B. Let samples s_a and s_{ab} ($s_A = s_a \cup s_{ab}$) of sizes n_a and $n_{ab}(n_A = n_a + n_{ab})$ fall in the population U_a and U_{ab}, respectively. Similarly, samples s_b and $s_{ba}(s_B = s_b \cup s_{ba})$ of sizes n_b and $n_{ba}(n_B = n_b + n_{ba})$ fall in the population U_b and U_{ab}, respectively. It should be noted that $s_{ab} \neq s_{ba}$ and $n_{ab} \neq n_{ba}$ in general. The estimation problem depends on the knowledge of the size of the population sizes N_A, N_B, and N_{ab}. There are three scenarios (i) all N_A, N_B, and N_{ab} are known, (ii) N_A, and N_B are known but N_{ab} unknown, and (iii) all N_A, N_B, and N_{ab} are unknown.

Let $\widehat{Y}_a$ and $\widehat{Y}_{ab}$ be suitable estimators of the totals Y_a and Y_{ab} obtained from the samples s_a and s_{ab}, respectively. Similarly, $\widehat{Y}_b$ and $\widehat{Y}_{ba}$ denote the estimators of Y_b and $Y_{ba}(=Y_{ab})$ obtained from the samples s_b and s_{ba}, respectively.

26.5.2 Simple Random Sampling Without Replacement

Let us assume that the samples s_A and s_B are selected by SRSWOR method independently. Here we consider the following cases:

Case 1: N_a, N_b, and N_{ab} are known

Hartley (1962) proposed the following result relating to estimation of the population total Y.

Theorem 26.5.1

(i) $\widehat{Y}_H = N_a\overline{y}_a + N_{ab}\{p\overline{y}_{ab} + (1-p)\overline{y}_{ba}\} + N_b\overline{y}_b$ is an unbiased estimator of Y (26.5.1)

where $\overline{y}_a = \sum_{i\in s_a} y_i/n_a, \overline{y}_{ab} = \sum_{i\in s_{ab}} y_i/n_{ab}, \overline{y}_b = \sum_{i\in s_b} y_i/n_b, \overline{y}_{ba} = \sum_{i\in s_{ba}} y_i/n_{ba}$, and $p(0 \leq p \leq 1)$ is a suitably chosen constant.

$$(ii)\ V\left(\widehat{Y}_H\right) \cong \frac{N_A^2}{n_A}\left\{(1-\alpha)S_a^2 + \alpha p^2 S_{ab}^2\right\} + \frac{N_B^2}{n_B}\left\{(1-\beta)S_b^2 + \beta(1-p)^2 S_{ab}^2\right\} \tag{26.5.2}$$

where

$$S_a^2 = \frac{1}{N_a - 1}\sum_{i\in U_a}\left(y_i - \overline{Y}_a\right)^2, S_b^2 = \frac{1}{N_b - 1}\sum_{i\in U_b}\left(y_i - \overline{Y}_b\right)^2,$$

$$S_{ab}^2 = \frac{1}{N_{ab} - 1}\sum_{i\in U_{ab}}\left(y_i - \overline{Y}_{ab}\right)^2, \overline{Y}_a = Y_a/N_a, \overline{Y}_b = Y_b/N_b,$$

$$\overline{Y}_{ab} = Y_{ab}/N_{ab} = \overline{Y}_{ba},\ \alpha = N_{ab}/N_A \text{ and } \beta = N_{ab}/N_B.$$

Proof

(i) $E(\widehat{Y}_H) = E\{E(\widehat{Y}_H | n_a, n_b, n_{ab}, n_{ba})\} = E(Y) = Y$

(ii) $V(\widehat{Y}_H) = E\{V(\widehat{Y}_H | n_a, n_b, n_{ab}, n_{ba})\} + V\{E(\widehat{Y}_H | n_a, n_b, n_{ab}, n_{ba})\}$

Now noting $E(\widehat{Y}_H | n_a, n_b, n_{ab}, n_{ba}) = Y$, we find

$$V(\widehat{Y}_H) = E\left[N_a^2\left(\frac{1}{n_a} - \frac{1}{N_a}\right)S_a^2 + p^2N_{ab}^2\left(\frac{1}{n_{ab}} - \frac{1}{N_{ab}}\right)S_{ab}^2 \right.$$
$$\left. + N_b^2\left(\frac{1}{n_b} - \frac{1}{N_b}\right)S_b^2 + (1-p)^2N_{ab}^2\left(\frac{1}{n_{ba}} - \frac{1}{N_{ab}}\right)S_{ab}^2 \middle| n_a, n_b, n_{ab}, n_{ba}\right]$$

Now neglecting the sampling fractions $f_a = n_a/N_a$, $f_b = n_b/N_b$, $f_{ab} = n_{ab}/N_{ab}$, and $f_{ba} = n_{ba}/N_{ab}$, we get

$$\begin{aligned} V(\widehat{Y}_H) &= E\left[N_a^2S_a^2/n_a + p^2N_{ab}^2S_{ab}^2/n_{ab} \right. \\ &\quad \left. + N_b^2S_b^2/n_b + (1-p)^2N_{ab}^2S_{ab}^2/n_{ba} \middle| n_a, n_b, n_{ab}, n_{ba}\right] \\ &\cong \frac{N_a^2}{E(n_a)}S_a^2 + p^2\frac{N_{ab}^2}{E(n_{ab})}S_{ab}^2 + \frac{N_b^2}{E(n_b)}S_b^2 + (1-p)^2\frac{N_{ab}^2}{E(n_{ba})}S_{ab}^2 \end{aligned} \tag{26.5.3}$$

Now, noting $E(n_a) = n_AN_a/N_A$, $E(n_b) = n_BN_b/N_B$, $E(n_{ab}) = n_AN_{ab}/N_A$, and $E(n_{ba}) = n_BN_{ab}/N_B$, we find

$$\begin{aligned} V(\widehat{Y}_H) &= \frac{N_A^2}{n_A}\{(1-\alpha)S_a^2 + \alpha p^2S_{ab}^2\} + \frac{N_B^2}{n_B}\{(1-\beta)S_b^2 + \beta(1-p)^2S_{ab}^2\} \\ &= \frac{N_A^2}{n_A}\{(1-\alpha)S_a^2 + \alpha p^2S_{ab}^2\} + \frac{N_B^2}{n_B}\{(1-\beta)S_b^2 + \beta(1-p)^2S_{ab}^2\} \end{aligned}$$

Lund (1968) obtained the optimum value of p by minimizing the expression (Eq. 26.5.3) as a function of p as

$$p_0 = \frac{n_{ab}}{n_{ab} + n_{ba}} \tag{26.5.4}$$

Substituting the optimum value of $p = p_0$ in expression (Eq. 26.5.1), Lund (1968) derived an improved estimator of Y as

$$\widehat{Y}_L = N_a\overline{y}_a + N_{ab}\overline{y}_{ab}^{(w)} + N_b\overline{y}_b \tag{26.5.5}$$

where

$$\overline{y}_{ab}^{(w)} = p_0\overline{y}_{ab} + (1-p_0)\overline{y}_{ba} \tag{26.5.6}$$

Ignoring finite population correction terms, Lund (1968) derived the approximate expression of the variance of $\widehat{Y}_L$ as

$$V\left(\widehat{Y}_L\right)=\frac{N_A^2}{n_A}(1-\alpha)S_a^2+\frac{N_AN_B\alpha\beta}{\alpha n_A+\beta n_B}S_{ab}^2+\frac{N_B^2}{n_B}(1-\beta)S_b^2 \qquad (26.5.7)$$

Although the variance of $\widehat{Y}_L$ is less than or equal to the variance of $\widehat{Y}_H$, substituting Eq. (26.5.4) in Eq. (26.5.2) provides the same approximate variance expressions.

Case 2: N_a, N_b, and N_{ab} are unknown

In this case the proportions $w_a=\frac{n_a}{n_A}$, $w_{ab}=\frac{n_{ab}}{n_A}$, $w_b=\frac{n_b}{n_B}$, and $w_{ba}=\frac{n_{ba}}{n_B}$ are unbiased estimators of $\frac{N_a}{N_A}, \frac{N_{ab}}{N_A}, \frac{N_b}{N_B}$, and $\frac{N_{ab}}{N_B}$, respectively. Here, N_{ab} possesses two unbiased estimators: N_Aw_{ab} and N_Bw_{ba}. Now, replacing N_a, N_b, and N_{ab} in Eq. (26.5.1) by their estimates, Hartley obtained the following unbiased estimator for Y.

$$\widehat{Y}_H^*=N_A\left\{w_a\overline{y}_a+pw_{ab}\overline{y}_{ab}\right\}+N_B\left\{w_b\overline{y}_b+(1-p)w_{ba}\overline{y}_{ba}\right\} \qquad (26.5.8)$$

Theorem 26.5.2

(i) $E\left(\widehat{Y}_H^*\right)=Y$

(ii) $$V\left(\widehat{Y}_H^*\right)\cong N_a\left(\frac{1}{f_A}-1\right)S_a^2+N_{ab}\left\{\left(\frac{1}{f_A}-1\right)p^2+(1-p)^2\left(\frac{1}{f_B}-1\right)\right\}S_{ab}^2$$
$$+N_b\left(\frac{1}{f_B}-1\right)S_b^2+\frac{g_AN_aN_{ab}}{n_A}\left(\overline{Y}_a-p\overline{Y}_{ab}\right)^2$$
$$+\frac{g_BN_bN_{ab}}{n_B}\left\{\overline{Y}_b-(1-p)\overline{Y}_{ab}\right\}^2$$

where $f_A=\frac{n_A}{N_A}$, $f_B=\frac{n_B}{N_B}$, $g_A=\frac{N_A-n_A}{N_A-1}$ and $g_B=\frac{N_B-n_B}{N_B-1}$

Proof

(i) $$E\left(\widehat{Y}_H^*\right)=E\left[E\left(\widehat{Y}_H^*\middle|n_a,n_{ab},n_b,n_{ba}\right)\right]$$
$$=E\left[N_A\left\{w_a\overline{Y}_a+pw_{ab}\overline{Y}_{ab}\right\}+N_B\left\{w_b\overline{Y}_b\right.\right.$$
$$\left.\left.+(1-p)w_{ba}\overline{Y}_{ba}\right\}\middle|n_a,n_{ab},n_b,n_{ba}\right]=\overline{Y}$$

$$\text{(ii) } V\left(\widehat{Y}_H^*\right) = V\left\{E\left(\widehat{Y}_H^* \middle| n_a, n_{ab}, n_b, n_{ba}\right)\right\} + E\left\{V\left(\widehat{Y}_H^* \middle| n_a, n_{ab}, n_b, n_{ba}\right)\right\}$$
$$= V\left[N_A\left(w_a\overline{Y}_a + pw_{ab}\overline{Y}_{ab}\right)\right] + V\left[N_B\left\{w_b\overline{Y}_b + (1-p)w_{ba}\overline{Y}_{ba}\right\}\right]$$
$$+ E\left[V\left(\widehat{Y}_H^* \middle| n_a, n_{ab}, n_b, n_{ba}\right)\right] \tag{26.5.9}$$

The first component of Eq. (26.5.9) is

$$V\left[N_A\left(w_a\overline{Y}_a + pw_{ab}\overline{Y}_{ab}\right)\right]$$
$$= N_A^2\left\{\overline{Y}_a^2 V(w_a) + p^2\overline{Y}_{ab}^2 V(w_{ab}) + 2p\overline{Y}_a\overline{Y}_{ab}Cov(w_a, w_{ab})\right\}$$
$$= N_A^2\left(\frac{1}{n_A} - \frac{1}{N_A}\right)\frac{N_A}{N_A - 1}\left[W_a(1 - W_a)\overline{Y}_a^2 + p^2\overline{Y}_{ab}^2 W_{ab}(1 - W_{ab})\right.$$
$$\left. -2p\overline{Y}_a\overline{Y}_{ab}W_aW_{ab}\right]$$

(where $W_a = N_a/N_A$ and $W_{ab} = N_{ab}/N_A$)

$$= N_A^2\left(\frac{1}{n_A} - \frac{1}{N_A}\right)\frac{N_A}{N_A - 1}W_aW_{ab}\left(\overline{Y}_a - p\overline{Y}_{ab}\right)^2$$

(noting $W_a = 1 - W_{ab}$)

$$= \frac{g_AN_aN_{ab}}{n_A}\left(\overline{Y}_a - p\overline{Y}_{ab}\right)^2 \tag{26.5.10}$$

Similarly, the second component of Eq. (26.5.9) is

$$V\left[N_B\left\{w_b\overline{Y}_b + (1-p)w_{ba}\overline{Y}_{ba}\right\}\right]$$
$$= \frac{g_BN_bN_{ab}}{n_B}\left\{\overline{Y}_b - (1-p)\overline{Y}_{ab}\right\}^2. \tag{26.5.11}$$

The third component of Eq. (26.5.9) is

$$E\left\{V\left(\widehat{Y}_H^* \middle| n_a, n_{ab}, n_b, n_{ba}\right)\right\}$$
$$= N_A^2E\left\{w_a^2\left(\frac{1}{n_a} - \frac{1}{N_a}\right)S_a^2 + p^2w_{ab}^2\left(\frac{1}{n_{ab}} - \frac{1}{N_{ab}}\right)S_{ab}^2\right\}$$
$$+ N_B^2E\left\{w_b^2\left(\frac{1}{n_b} - \frac{1}{N_b}\right)S_b^2 + (1-p)^2w_{ba}^2\left(\frac{1}{n_{ba}} - \frac{1}{N_{ab}}\right)S_{ab}^2\right\} \tag{26.5.12}$$

Now, noting

$$E(w_a) = \frac{N_a}{N_A} = W_a,\ V(w_a) = \left(\frac{1}{n_A} - \frac{1}{N_A}\right) \frac{N_A}{N_A - 1} W_a(1 - W_a),$$

we find

$$\begin{aligned}
& N_A^2 E\left\{w_a^2\left(\frac{1}{n_a} - \frac{1}{N_a}\right)\right\} S_a^2 \\
&= N_A^2\left\{\frac{W_a}{n_A} - \frac{W_a^2 + V(w_a)}{N_a}\right\} S_a^2 \\
&= N_A^2\left[\frac{W_a}{n_A} - \frac{W_a^2}{N_a} - \frac{1}{N_a}\left(\frac{1}{n_A} - \frac{1}{N_A}\right) \frac{N_A W_a(1 - W_a)}{N_A - 1}\right] S_a^2 \\
&= N_A^2 W_a\left[\frac{1}{n_A} - \frac{W_a}{N_a} - \frac{1}{N_a}\left(\frac{1}{n_A} - \frac{1}{N_A}\right) \frac{N_A(1 - W_a)}{N_A - 1}\right] S_a^2 \\
&= N_a\left[\frac{N_A}{n_A} - 1 - \frac{1}{N_a}\left(\frac{N_A}{n_A} - 1\right) \frac{N_A(1 - W_a)}{N_A - 1}\right] S_a^2 \\
&= N_a\left(\frac{1}{f_A} - 1\right)\left\{1 - \frac{1}{N_a} \frac{N_A(1 - W_a)}{N_A - 1}\right\} S_a^2 \\
&\cong N_a\left(\frac{1}{f_A} - 1\right)\left\{1 - \frac{(1 - W_a)}{N_a}\right\} S_a^2 \\
&\cong N_a\left(\frac{1}{f_A} - 1\right) S_a^2 \left\{\text{neglecting the term } \left(\frac{1}{N_a} - \frac{1}{N_A}\right)\right\}
\end{aligned} \tag{26.5.13}$$

Similarly,

$$\begin{aligned}
& N_A^2 E\left\{w_{ab}^2\left(\frac{1}{n_{ab}} - \frac{1}{N_{ab}}\right) S_{ab}^2\right\} \cong N_{ab}\left(\frac{1}{f_A} - 1\right) S_{ab}^2, \\
& N_B^2 E\left\{w_b^2\left(\frac{1}{n_b} - \frac{1}{N_b}\right)\right\} S_{ab}^2 \cong N_b\left(\frac{1}{f_B} - 1\right) S_b^2 \text{ and} \\
& N_B^2 E\left\{w_{ba}^2\left(\frac{1}{n_{ba}} - \frac{1}{N_{ab}}\right) S_{ab}^2\right\} \cong N_{ab}\left(\frac{1}{f_B} - 1\right) S_{ab}^2
\end{aligned} \tag{26.5.14}$$

Expressions (Eqs. 26.5.12–26.5.14) yield

$$\begin{aligned}
E\left\{V\left(\widehat{Y}_H^* \middle| n_a, n_{ab}, n_b, n_{ba}\right)\right\} &\cong N_a\left(\frac{1}{f_A} - 1\right) S_a^2 \\
&+ N_{ab}\left\{\left(\frac{1}{f_A} - 1\right) p^2 + \left(\frac{1}{f_B} - 1\right)(1 - p)^2\right\} S_{ab}^2 + N_b\left(\frac{1}{f_B} - 1\right) S_b^2
\end{aligned} \tag{26.5.15}$$

The proof of the theorem follows from Eqs. (26.5.9)−(26.5.11) and (26.5.15).

Corollary 26.5.1

If sampling fraction f_A and f_B are ignored we get

$$V\left(\widehat{Y}_H^*\right) \cong \frac{N_A^2}{n_A}\left\{(1-\alpha)S_a^2 + \alpha p^2 S_{ab}^2\right\} + \frac{N_B^2}{n_B}\left\{(1-\beta)S_b^2 + \beta(1-p)^2 S_{ab}^2\right\}$$
$$+ \frac{N_A^2\alpha(1-\alpha)}{n_A}\left(\overline{Y}_a - p\overline{Y}_{ab}\right)^2 + \frac{N_B^2\beta(1-\beta)}{n_B}\left\{\overline{Y}_b - (1-p)\overline{Y}_{ab}\right\}^2$$

Lund (1968) modified Hartley estimator $\widehat{Y}_H^*$ using the combined mean of the overlapping frame (including repetition of units) $\overline{y}_w = \dfrac{n_{ab}\overline{y}_{ab} + n_{ba}\overline{y}_{ba}}{n_{ab} + n_{ba}}$ as an estimator of $\overline{Y}_{ab}$. Lund estimator is as follows:

$$\widehat{Y}_L^* = N_A w_a \overline{y}_a + \left\{N_A p w_{ab} + N_B(1-p)w_{ba}\right\}\overline{y}_w + N_B w_b \overline{y}_b$$

Using Theorem 26.5.2, we can derive the following:

Theorem 26.5.3

(i) $E\left(\widehat{Y}_L^*\right) = Y$

(ii) $$V\left(\widehat{Y}_L^*\right) = N_a\left(\frac{1}{f_A} - 1\right)S_a^2 + \frac{N_{ab}}{(f_A + f_B)^2}\left\{f_A(1-f_A) + f_B(1-f_B)\right\}S_{ab}^2$$
$$+ N_B\left(\frac{1}{f_B} - 1\right)S_b^2 + \frac{g_A}{n_A}N_{ab}N_a\left(\overline{Y}_a - p\overline{Y}_{ab}\right)^2$$
$$+ \frac{g_B}{n_B}N_{ab}N_b\left\{\overline{Y}_b - (1-p)\overline{Y}_{ab}\right\}^2$$

Lund (1968) derived the optimum value of p as

$$p_{0L} = \frac{(1-\alpha)\overline{y}_a/f_A + (1-\beta)\left(\overline{y}_w - \overline{y}_b\right)/f_B}{[(1-\alpha)/f_A + (1-\beta)/f_B]\overline{y}_w} \quad (26.5.16)$$

Lund proposed the following estimator for p_{0L}

$$\widehat{p}_{0L} = \frac{n_a n_B \overline{y}_a/f_A + n_A n_b\left(\overline{y}_w - \overline{y}_b\right)/f_B}{[n_B n_a/f_A + n_A n_b/f_B]\overline{y}_w} \quad (26.5.17)$$

The expression of the variance $V\left(\widehat{Y}_L^*\right)$ given in Theorem 26.5.3 was provided by Fuller and Burmeister (1972). The estimator $\widehat{Y}_L^*$ possesses a much lower variance than that of Hartley estimator $\widehat{Y}_H^*$.

Fuller and Burmeister (1972) proposed an alternative estimator for Y as

$$\widehat{Y}_{FB} = \left(N_A - \widehat{N}_{ab}\right)\overline{y}_a + \widehat{N}_{ab}\overline{y}_{w^*} + \left(N_B - \widehat{N}_{ab}\right)\overline{y}_b \qquad (26.5.18)$$

where $\overline{y}_{w^*} = w^*\overline{y}_{ab} + (1 - w^*)\overline{y}_{ba}$, $w^* = \dfrac{n_{ab}(1 - f_B)}{n_{ab}(1 - f_B) + n_{ba}(1 - f_A)}$, and $\widehat{N}_{ab}$ is the smallest root of the quadratic,

$$(n_A g_B + n_B g_A)\widehat{N}_{ab}^2 - (n_A N_B g_B + n_B N_A g_A + n_{ab} N_A g_B + n_{ba} N_B g_A)\widehat{N}_{ab} + (n_{ab} g_B + n_{ba} g_A) N_A N_B = 0 \qquad (26.5.19)$$

We state the following theorem (without derivation) that is related to the mean and variance of $\widehat{Y}_{FB}$ derived by Fuller and Burmeister (1972).

Theorem 26.5.4

(i) $E\left(\widehat{Y}_{FB}\right) = Y + o\left(\dfrac{1}{n}\right)$

(ii) $V\left(\widehat{Y}_{FB}\right) = N_a\left(\dfrac{1}{f_A} - 1\right)S_a^2 + \dfrac{(1 - f_A)(1 - f_B)N_{ab}}{(1 - f_B)f_A + (1 - f_A)f_B}S_{ab}^2 + N_b\left(\dfrac{1}{f_B} - 1\right)S_b^2 + \dfrac{N_{ab}N_a N_b g_A g_B}{n_A N_b g_B + n_B N_a g_A}\left(\overline{Y}_{ab} - \overline{Y}_a - \overline{Y}_b\right)^2 + o(1)$

Fuller and Burmeister (1972) pointed out that the bias in the Lund estimator $\widehat{Y}_L^*$ is $o(1)$ while that of $\widehat{Y}_{FB}$ is $o\left(\dfrac{1}{n}\right)$. Furthermore, $\widehat{Y}_{FB}$ is advantageous over $\widehat{Y}_L^*$ in the computational point of view. In cases where the units in samples s_{ab} and s_{ba} are identifiable, the following estimator $\widehat{Y}_{FB}^*$ has smaller variance than $\widehat{Y}_{FB}$.

$$\widehat{Y}_{FB}^* = \left(N_A - \widehat{N}_{ab,m}\right)\overline{y}_a + \widehat{N}_{ab,m}\overline{y}_d + \left(N_B - \widehat{N}_{ab,m}\right)\overline{y}_b \qquad (26.5.20)$$

where $\overline{y}_d = \dfrac{1}{n_d}\sum_{i\in s_d} y_i$, s_d is the set of distinct units in $s_{ab} \cup s_{ba}$, n_d is the number of distinct units in s_d, and $\widehat{N}_{ab,m}$ is the smallest root of the quadratic equation

$$(n_a + n_d + n_b)\widehat{N}_{ab,m}^2 - \{n_a N_B + n_d(N_A + N_B) + n_b N_A - n_a n_b\}\widehat{N}_{ab,m} + n_d N_A N_B = 0 \qquad (26.5.21)$$

Fuller and Burmeister (1972) recommended that $\widehat{N}_{ab,m}$ should be considered as a maximum likelihood estimate (MLE) of N_{ab} with score (Eq. 26.5.21), whereas Skinner (1991) provided reasons for considering $\widehat{Y}^*_{FB}$ as an MLE of Y.

26.5.3 General Sampling Procedures

Let π_{Ai} and π_{Bi} be the inclusion probabilities of selection of ith unit of the population U_A based on sampling design p_A and U_B based on sampling design p_B, respectively. In case N_A is known, Y_a and Y_{ab} can be estimated by

$$\widehat{Y}_a = \sum_{i \in s_a} w_{Ai} y_i \text{ and } \widehat{Y}_{ab} = \sum_{i \in s_{ab}} w_{Ai} y_i \qquad (26.5.22)$$

where $w_{Ai} = N_A \left(\frac{1}{\pi_{Ai}} \Big/ \sum_{i \in s_A} \frac{1}{\pi_{Ai}} \right)$.

Similarly, if N_B is known, estimators for Y_b and Y_{ab} are, respectively,

$$\widehat{Y}_b = \sum_{i \in s_b} w_{Bi} y_i \text{ and } \widehat{Y}_{ba} = \sum_{i \in s_{ba}} w_{Bi} y_i \qquad (26.5.23)$$

where $w_{Bi} = N_B \left(\frac{1}{\pi_{Bi}} \Big/ \sum_{i \in s_B} \frac{1}{\pi_{Bi}} \right)$.

The Hartley (1974) estimator for Y is

$$\widetilde{Y}_H = \widehat{Y}_a + \widehat{Y}_b + \varphi \widehat{Y}_{ab} + (1 - \varphi) \widehat{Y}_{ba} \qquad (26.5.24)$$

where the constant φ is chosen to minimize $V\left(\widetilde{Y}_H\right)$. Clearly, the optimum φ will involve unknown parameters, which should be estimated from the selected sample.

Fuller and Burmeister (1972) proposed the following estimator for Y

$$\widetilde{Y}_{FB} = \widehat{Y}_a + \widehat{Y}_b + \varphi \widehat{Y}_{ab} + (1 - \varphi) \widehat{Y}_{ba} + \phi\left(\widehat{N}_{ab} - \widehat{N}_{ba}\right) \qquad (26.5.25)$$

where $\widehat{N}_{ab} = \sum_{i \in s_{ab}} w_{Ai}$ and $\widehat{N}_{ba} = \sum_{i \in s_{ba}} w_{Bi}$.

Here also φ and ϕ are chosen to minimize the variance of $\widetilde{Y}_{FB}$. The optimum values of φ and ϕ will involve unknown parameters, which need to be estimated from the data.

The pseudo-MLE for Y proposed by Skinner and Rao (1996) is given by

$$\widehat{Y}_{PML} = \left(N_A - \widehat{N}_{ab,PML}\right)\widehat{\mu}_a + \left(N_B - \widehat{N}_{ab,PML}\right)\widehat{\mu}_b + \widehat{N}_{ab,PML}\widehat{\mu}_w \qquad (26.5.26)$$

where

$\widehat{\mu}_a = \widehat{Y}_a \Big/ \widehat{N}_a$, $\widehat{N}_a = \sum_{i \in s_a} w_{Ai}$, $\widehat{\mu}_b = \widehat{Y}_b \Big/ \widehat{N}_b$, $\widehat{N}_b = \sum_{i \in s_b} w_{Bi}$,

$\widehat{\mu}_w = \left(\frac{n_A}{N_A} \widehat{N}_{ab} \widehat{\mu}_{ab} + \frac{n_B}{N_B} \widehat{N}_{ba} \widehat{\mu}_{ba} \right) \Big/ \left(\frac{n_A}{N_A} \widehat{N}_{ab} + \frac{n_B}{N_B} \widehat{N}_{ba} \right)$,

$\widehat{\mu}_{ab} = \widehat{Y}_{ab} \Big/ \widehat{N}_{ab}$, and $\widehat{\mu}_{ba} = \widehat{Y}_{ba} \Big/ \widehat{N}_{ba}$.

Furthermore, $\widehat{N}_{ab,PML}$ is the smallest root of the quadratic equation

$$(n_A + n_B)x^2 - \left(n_A N_B + n_B N_A + n_A \widehat{N}_{ab} + n_B \widehat{N}_{ba} \right) x + n_A \widehat{N}_{ab} N_B + n_B \widehat{N}_{ba} N_A = 0 \tag{26.5.27}$$

Skinner and Rao (1996) showed that the estimators $\widetilde{Y}_H$ and $\widetilde{Y}_{FB}$ are all consistent for Y and are asymptotically normally distributed with mean Y. They also provided expressions of the asymptotic variances of the estimators.

In case N_A, N_B, and N_{ab} are known, we may write $\widehat{N}_{ab} = \widehat{N}_{ba} = \widehat{N}_{ab,PML} = N_{ab}$ in the expression of $\widehat{Y}_{PML}$ given in Eq. (26.5.26) and obtain the following modified estimator

$$\widehat{Y}_{PML} = N_a \widehat{\mu}_a + N_b \widehat{\mu}_b + N_{ab} \widehat{\mu}_w$$

where $\widehat{\mu}_w = w \widehat{\mu}_{ab} + (1 - w) \widehat{\mu}_{ba}$ and $w = \dfrac{(n_A/N_A)}{(n_A/N_A) + (n_B/N_B)}$.

In case none of N_A, N_B, or N_{ab} is known, the estimators $\widetilde{Y}_H$ and $\widetilde{Y}_{FB}$ can be modified by simply writing $w_{Ai} = 1/\pi_{Ai}$, $w_{Bi} = 1/\pi_{Bi}$ in expressions (Eqs. 26.5.24 and 26.5.25), respectively. The estimator $\widehat{Y}_{PML}$ can also be modified by replacing N_A and N_B with their estimates $\sum_{i \in s_A} 1/\pi_{Ai}$ and $\sum_{i \in s_B} 1/\pi_{Bi}$, respectively.

26.5.4 Horvitz–Thompson-Based Estimators

If the units in the overlapping population U_{ab} can be identified, the sample $s = s_A \cup s_B$ can be regarded as a sample from the population U with inclusion probability for the ith unit

$$\pi_i = \begin{cases} \pi_{Ai} & \text{for } i \in s_a \\ \pi_{Bi} & \text{for } i \in s_b \\ \pi_{ABi} & \text{for } i \in s_{AB} \end{cases} \tag{26.5.28}$$

where $\pi_{ABi} = \pi_{Ai} + \pi_{Bi} - \pi_{Ai}\pi_{Bi}$ and $s_{AB} = s_A \cap s_B$.

In this case the Horvitz–Thompson estimator for the population total Y is given by

$$\widehat{Y}_{ht} = \sum_{i \in s_a} \frac{y_i}{\pi_{Ai}} + \sum_{i \in s_b} \frac{y_i}{\pi_{Bi}} + \sum_{i \in s_{AB}} \frac{y_i}{\pi_{ABi}} \qquad (26.5.29)$$

It is very difficult to compare the performance of $\widehat{Y}_{ht}$ with the alternatives $\widetilde{Y}_H$, $\widetilde{Y}_{FB}$, and $\widehat{Y}_{PML}$. However, $\widehat{Y}_{ht}$ is expected to perform better as it is based on distinct units only.

26.5.5 Concluding Remarks

The estimation problem from the dual frame survey depends on the knowledge of the size of the population sizes N_A, N_B, and N_{ab}. Estimation of the population total and the mean were considered by Hartley (1962), Lund (1968), and Fuller and Burmeister (1972) for SRSWOR sampling while Hartley (1974), Fuller and Burmeister (1972), Skinner (1991), Skinner and Rao (1996), and Singh and Wu (1996) studied for the general sampling designs. Fuller and Burmeister considered "pseudo"-MLEs under SRSWR and complex survey designs. The "pseudo"-MLE uses the same survey weights for all the variables, unlike Hartley (1974) and Fuller and Burmeister (1972). Hence "pseudo"-MLEs possess a computational advantage. The asymptotic properties of the proposed estimators are studied by Skinner and Rao (1996). They conducted limited simulation studies relating to the performances of these estimators and recommended the use of "pseudo"-MLEs, as they perform better than the others and bring significant gain in efficiency over the single frame estimators. Multiframe surveys for multistage sampling design were studied by Saxena et al. (1984). Singh and Wu (1996) used multiauxiliary variables for multiframe complex surveys. More comprehensive researches are required to obtain conclusive results relating to the performances of the proposed estimators used for multiframe surveys.

26.6 SNOWBALL SAMPLING

Snowball sampling is used for surveying an extremely rare population where the cost of the survey using one of the methods discussed earlier remains prohibitively high. In snowball sampling, members of the rare population are assumed to be known to each other. This condition is very restrictive, but may hold true for some rare populations such as ethnic minorities and religious groups. Initially, a few members of the rare population are identified and each of them is asked to identify the other members. Then each of these members are contacted and asked to identify other members and so on. After a certain stage, no new member is found, i.e., the list of the rare population is completed. Finally, from the completed list of the rare population a sample is selected using a suitable sampling design. For example, snowball sampling can be used to select doctors in a locality by identifying a few doctors and asking them if they know other doctors and so on. After a certain stage no new identification of doctors could be found.

Some applications of snowball sampling avoid construction of the sampling frame for the rare population. Instead, the snowball sampling process is continued till a sizable number of the rare population is identified and information is collected from these identified individuals. Because this sampling is not a probability sampling, unbiased estimation of population characteristics is not possible.

Snowball sampling was used by Snow et al. (1981) to select a sample of Hispanics in Atlanta, by Welch (1975) for selection of samples of Mexican Americans in Omaha, and Biernacki and Waldrof (1981) for sampling ex—heroin addicts.

26.7 LOCATION SAMPLING

Location sampling is widely used to sample populations that have no fixed abode for both census and surveys: nomads may be sampled at water points when they take their animals for water, and homeless persons may be sampled at soup kitchens when they go for food.

Location sampling is used to sample rare mobile populations at the time of their visits to specific locations such as airports, game parks, churches, shopping malls, and playgrounds. Location sampling fails to cover those who do not visit such places during the survey period. Here, the unit of analysis may be visits or visitors (Kalton, 1991).

Location sampling can readily produce a probability sample of visits, with known probabilities, and hence visits are easily analyzed. Visits may be the appropriate unit of analysis for, say, a survey about satisfaction with visits to a museum. In this case no issues of multiplicity arise.

However, for many surveys, the visitor is the appropriate unit of analysis, for example, in a survey of visitors to soup kitchens over a week to estimate the number of homeless, a survey of nomads visiting watering holes to estimate the size of the nomadic population. In this situation issues of multiplicity arise because a visitor can visit more than once during the survey reference period. To avoid this problem one may treat the first visits during the time period as an eligible visit. Otherwise, multiplicity adjustments will be required. For example, each sampled person is asked whether this visit is the first since the start of the survey or not. If the answer is "yes" the person is selected and if the answer is "no" the person is not selected. In this procedure most visits near the start of the survey will be accepted, as they are the first visits, whereas at the end of the survey most of the visits will not be accepted because these are not the first visits.

Kalton (1991, 2009) recommended a two-stage sampling procedure where psu's are combination of locations (entrances or exits) and time segments when the locations are open (e.g., on week days from 10:00 a.m. to 5:00 p.m. and on holidays from 8:00 a.m. to 4:00 p.m.). The psu's may be selected with probability proportional to size, with careful stratification by location and time. Then systematic sampling may be used to select visitors entering or exiting the location.

Location sampling has been used to sample men, with locations being gay bars, bathhouses, and bookstores (Kalton, 1993; MacKellar et al., 1996), who have sex with men. The Young Men's Survey conducted in seven cities in 1994–98 in 194 public locations is a major survey of this type (Vallerory et al., 2000). Mckenzie and Mistiaen (2009) carried out studies to compare location sampling with area sampling and snowball sampling for sampling on Nikkei (Brazilians of Japanese descents) in Sao Paulo and Panama. The locations were places where Nikkei often visit such as the metro station, ethnic grocery stores, sports clubs, and other locations where family members of Nikkei community congregate. The studies reveal that the location and snowball sampling are unlikely to provide a representative sample.

26.8 SEQUENTIAL SAMPLING

Because the size of the rare population is unknown, it is difficult to determine the sample of the desired size needed for estimation of the rare population characteristics (for example, prevalence) with a specified degree of accuracy. The problem is acute, especially when there are no past survey data of the rare population available. In this case, sample size may be determined sequentially. At the first stage, a preliminary estimate is made on the basis of expert judgment or small pilot survey, and on the basis of this estimate the desired sample size is determined. At the second stage a survey is conducted with half of the required sample size and a revised estimate of the population characteristic is obtained. Using the estimates in the second step, the optimal sample size is determined and the additional required sample is selected at the third step. The sequential sampling procedure generally increases time and cost of the study but is worth the expense to obtain an efficient estimate of the rare population.

26.9 ADAPTIVE SAMPLING

In adaptive sampling, the selection of the sample depends on the values of the characteristic under study y. Adaptive sampling was motivated for sampling rare and mobile populations such as rare contagious diseases, drug use, rare species of animals or birds, and density of animals in a forest (Thompson, 1990). In sampling animals or rare species of trees in a forest, the entire forest may be divided into number of square plots of equal size. A sample of n_1 plots can be selected by some probability sampling such as SRSWOR method. If a sampled unit (plot) satisfies a certain condition C, for example, the number of animals y_i of the plot i exceeds a certain pre-specified number c, then the neighborhood units are added in the sample. Here, neighborhoods are defined in a certain manner such as adjacent four plots viz. north, east, south, and west. If the other units in the neighborhood satisfy the condition C, then their neighborhoods are also included in the sample. This process is continued until a cluster of units is obtained that contains a boundary of units (called edge units) that do not satisfy the condition C. Thus from the selection of initial n_1 plots, we select n_1 clusters of plots. The selected clusters may not be distinct. It should be noted that if no animal is selected in the initial selected plot, the neighboring plots are not sampled. In this case, it becomes a cluster of one unit plot. The details of the selection procedures are given by Thompson and Seber (1996).

The neighborhood relationship can be defined in various ways and neighborhoods need not be contiguous. For example, the neighborhood relation may be brothers, sisters, siblings, or the same ethnic group. This type of sampling is known as adaptive cluster sampling. Here, the total number of sampled units (plots) is a random variable. Thus adaptive sampling includes sequential sampling where sampling is continued until some condition is satisfied.

26.9.1 Unbiased Estimation of Population Mean

Let a region U be divided into a finite number of N square plots and let a sample s of n_1 plots be selected by SRSWOR method. We define a network A_i for the unit i as a cluster generated by the unit i, but its edge units are removed. Thus selection of any unit in A_i will lead the selection of all the units of A_i. If a unit i is the only one unit in a cluster satisfying C, then A_i is itself a network of size 1. We also define a unit that does not satisfy C as a network of size 1. Note that any edge unit is also a network of size 1. Thus all the clusters of size 1 are also a network of size 1. Hence according to the definition any two different networks are disjoint and networks form the partition of the population U. Let $B_1,\ldots,B_K$ be the distinct networks that form the partition of the population U. Following Thompson and Seber (1996) we propose the following unbiased estimators of the population mean.

26.9.1.1 Use of Intersection Probabilities

Here we define the population total as

$$Y = \sum_{i=1}^{N} y_i = \sum_{j=1}^{K} y_j^* \tag{26.9.1}$$

where $y_j^* = \sum_{k\in B_j} y_k$.

Let $I_j(s) = 1$ if $s \cap B_j \neq \phi$, i.e., some of the selected units in s also belong to B_j, $I_j(s) = 0$ if $s \cap B_j = \phi$, and x_j is the number of units in the jth network.

Then, we have

$$\begin{aligned}
\alpha_j &= \Pr ob\{I_j(s) = 1\} \\
&= \text{Probability that the network } B_j \text{ will have non} - \text{null intersectionwith } s \\
&= \text{Probability that at least one of the } x_j \text{ units of the network} \\
&\quad\ B_j \text{ is included in the sample } s \\
&= 1 - \text{ Probability that none of the } x_j \text{ units of the network } B_j \text{ is selected in } s \\
&= 1 - \binom{N - x_j}{n_1} \Big/ \binom{N}{n_1}
\end{aligned} \tag{26.9.2}$$

and

$$\begin{aligned}
\alpha_{jk} &= \Pr ob\{I_j(s) = 1, I_k(s) = 1\} \text{ for } j \neq k \\
&= \text{Probability that both the network } B_j \text{ and } B_k \text{ will have non-} \\
&\quad\ \text{null intersections with } s \\
&= 1 - \left[\binom{N - x_j}{n_1} + \binom{N - x_k}{n_1} - \binom{N - x_j - x_k}{n_1}\right] \Big/ \binom{N}{n_1}
\end{aligned} \tag{26.9.3}$$

Theorem 26.9.1

(i) $\widehat{\overline{Y}}_1 = \frac{1}{N}\sum_{j=1}^{K}\frac{y_j^*}{\alpha_j}I_j(s)$ is an unbiased for $\overline{Y} = \sum_{j=1}^{K} y_j^* \Big/ N$

(ii) $$V\left(\widehat{\overline{Y}}_1\right) = \frac{1}{N^2}\left[\sum_{j=1}^{K} y_j^{*2}\left(\frac{1}{\alpha_j} - 1\right) + \sum_{j\neq}^{K}\sum_{k=1}^{K}\frac{y_j^*}{\alpha_j}\frac{y_k^*}{\alpha_k}\left(\frac{\alpha_{jk}}{\alpha_j\alpha_k} - 1\right)\right]$$

(iii) $$\widehat{V}\left(\widehat{\overline{Y}}_1\right) = \frac{1}{N^2}\left[\sum_{j=1}^{K} y_j^{*2}\left(\frac{1}{\alpha_j} - 1\right)\frac{I_j(s)}{\alpha_j}\right.$$
$$\left. + \sum_{j\neq}^{K}\sum_{k=1}^{K}\frac{y_j^*}{\alpha_j}\frac{y_k^*}{\alpha_k}\left(\frac{\alpha_{jk}}{\alpha_j\alpha_k} - 1\right)\frac{I_j(s)I_k(s)}{\alpha_{jk}}\right]$$

Proof

(i) $$E\left(\widehat{\overline{Y}}_1\right) = \frac{1}{N}\sum_{j=1}^{K}\frac{y_j^*}{\alpha_j}E\{I_j(s)\} = \overline{Y}$$

(ii) $$V\left(\widehat{\overline{Y}}_1\right) = \frac{1}{N^2}E\left\{\sum_{j=1}^{K}\frac{y_j^*}{\alpha_j}I_j(s)\right\}^2 - \overline{Y}^2$$

$$= \frac{1}{N^2}\left[\sum_{j=1}^{K}\frac{y_j^{*2}}{\alpha_j^2}E\{I_j(s)\} + \sum_{j\neq}^{K}\sum_{k=1}^{K}\frac{y_j^*}{\alpha_j}\frac{y_k^*}{\alpha_k}E\{I_j(s)I_k(s)\}\right] - \overline{Y}^2$$

$$= \frac{1}{N^2}\left[\sum_{j=1}^{K} y_j^{*2}\left(\frac{1}{\alpha_j} - 1\right) + \sum_{j\neq}^{K}\sum_{k=1}^{K}\frac{y_j^*}{\alpha_j}\frac{y_k^*}{\alpha_k}\left(\frac{\alpha_{jk}}{\alpha_j\alpha_k} - 1\right)\right]$$

(iii) Noting $E\{I_j(s)\} = \alpha_j$ and $E\{I_j(s)I_k(s)\} = \alpha_{jk}$, we can prove $E\left[\widehat{V}\left(\widehat{\overline{Y}}_1\right)\right] = V\left(\widehat{\overline{Y}}_1\right)$.

26.9.1.2 Use of the Number of Intersections

Let A_i be the network containing the ith unit, $i = 1,\ldots, N$. The networks A_i's need not be distinct. The networks A_1 and A_2 are identical if both of them contain the units 1 and 2. Let $f_i(s)$ be the number of units in the initial sample s that belongs to the network A_i (ignoring the edge units of the clusters). Clearly, $f_i(s) = 0$ if no unit in the initial sample intersects A_i. Hence $f_i(s)$ follows hypergeometric distribution and

$$\nu_i = E[f_i(s)] = n_1\frac{m_i}{N} \tag{26.9.4}$$

where m_i is the number of units in the ith network.

Hence

$$\widehat{\overline{Y}}_2 = \frac{1}{N}\sum_{i=1}^{N} y_i \frac{f_i(s)}{\nu_i}$$

is an unbiased estimator of the population mean $\overline{Y}$. We can write $\widehat{\overline{Y}}_2$ as follows:

$$\begin{aligned}\widehat{\overline{Y}}_2 &= \frac{1}{n_1}\sum_{i=1}^{N} y_i \frac{f_i(s)}{m_i}\\ &= \frac{1}{n_1}\sum_{i\in s} w_i \\ &= \overline{w}_s\end{aligned} \qquad (26.9.5)$$

where $w_i = \frac{1}{m_i}\sum_{j\in A_i} y_j$.

From the expression of $\widehat{\overline{Y}}_2$ given in Eq. (26.9.5), we get the following theorem.

Theorem 26.9.6

(i) $E(\overline{w}_s) = \overline{Y}$

(ii) $V(\overline{w}_s) = \frac{N-n_1}{n_1 N} S_w^2$

(iii) $\widehat{V}(\overline{w}_s) = \frac{N-n_1}{n_1 N} s_w^2$

where $S_w^2 = \frac{1}{N-1}\sum_{i=1}^{N}\left(w_i - \overline{Y}\right)^2$ and $s_w^2 = \frac{1}{n_1-1}\sum_{i\in s}(w_i - w_s)^2$.

Proof

(i) Noting $\overline{w}_s$ is the sample mean of n_1 units selected by SRSWOR method, we have

$$E(\overline{w}_s) = \frac{1}{N}\sum_{i=1}^{N} w_i = \frac{1}{N}\sum_{i=1}^{K}\sum_{i\in B_k} y_i = \overline{Y}$$

Proofs of (ii) and (iii) follow straight from the properties of SRSWOR sampling.

Details of adaptive sampling for unequal sampling schemes are given by Thompson and Seber (1996). But we have omitted them here because of their highly technical nature.

26.10 CAPTURE—RECAPTURE METHOD

The capture—recapture method was developed at least in the 100 years ago in the field of ecology and wildlife studies for estimating the unknown population size (N). The application of this method in epidemiological problems (e.g., incidence of disease) and demography (e.g., population size, survival, recruitment, emigration, and migration) came relatively late to the literature. Other names of the method include capture—mark—recapture, mark—recapture, mark—release—recapture, and band recovery. The basic method involves capturing a sample of animals and then marking and releasing them into the population. A second sample is then selected from the population and the number of marked animals discovered. From the data thus collected, the population size is estimated.

A population that remains unchanged during the period of investigation will be called a closed population. In this case there is neither addition in the population due to birth or immigration nor removal due to migration or death. Here, the only parameter of interest is the population size N. A population is called an open population if it may be subject to change because of processes such as immigration, migration, births, deaths, etc. In the open population interest is given to the population dynamics such as birth and death rates, survival rates, and population change. Capture—recapture method does not correspond to the sampling of a finite population paradigm because the frame does not exist. The capture—recapture method was considered by Peterson (1896) and Lincoln (1930) and great details are provided by Seber (1982).

26.10.1 Closed Population

26.10.1.1 Peterson and Lincoln Method

Let N be the total number of the unknown population (birds or animals). From the population a sample s_1 of n_1 animals are caught, marked (or tagged), and released into the population. After sometime when the animals have resettled with the unmarked animals, another sample s_2 of size n_2 is selected from the population. Let $s_{2m}(\subset s_2)$ of size m_2 be the set of marked animals and $s_{2u}(s_2 - s_{2m})$ be the sample of unmarked animals of size $u_2 = n_2 - m_2$. Now, we make the following assumptions:

(a) Each animal is equally likely to be caught in the two samples, i.e., healthy, unhealthy, and animals once caught have the same chance of being captured again.

(b) Animals do not lose their marks

Under these assumptions, we find that the proportion of marked individuals that are caught (m_2/n_1) should be equal to the proportion of total animals caught (n_2/N), i.e.,

$$\frac{m_2}{n_1} = \frac{n_2}{N}$$
$$\text{i.e.,} \quad \widehat{N} = \widehat{N}_P = \frac{n_1 n_2}{m_2} \tag{26.10.1}$$

Eq. (26.10.1) is called the Peterson (1896) estimate or Lincoln (1930) Index.

26.10.1.2 Hypergeometric Model

Under the assumptions (a) and (b), the conditional probability distribution of m_2 given n_1, n_2, and N is the hypergeometric distribution

$$f(m_2|n_1, n_2, N) = \frac{\binom{n_1}{m_2}\binom{N-n_1}{n_2-m_2}}{\binom{N}{n_2}}; m_2 = 0, 1, \ldots, \min(n_1, n_2) \tag{26.10.2}$$

Here, $E(m_2) = n_2\frac{n_1}{N}$, hence we can choose an estimator of N to be $\widehat{N} = \widehat{N}_P = \frac{n_1 n_2}{m_2}$. Clearly, $\widehat{N}_P$ is a biased estimator. Chapman (1951) showed that $\widehat{N}_P$ is asymptotically normal as $N \to \infty$, but the bias of $\widehat{N}_P$ cannot be negligible for a small sample. For $n_1 + n_2 \geq N$, Chapman proposed the following exactly unbiased estimator for N.

$$\widehat{N}_c = \frac{(n_1+1)(n_2+1)}{(m_2+1)} - 1 \tag{26.10.3}$$

However, for $n_1 + n_2 < N$, the estimator $\widehat{N}_c$ is biased. The expression of bias was obtained by Robson and Regier (1964).

Assuming Poisson approximation to Eq. (26.10.2), Chapman (1951) computed the variance of $\widehat{N}_c$ as

$$V\left(\widehat{N}_c \middle| n_1, n_2\right) = N^2\left(\frac{1}{\mu} + \frac{2}{\mu^2} + \frac{6}{\mu^3}\right) \tag{26.10.4}$$

where $\mu = E(m_2|n_1, n_2) = n_1 n_2/N$.

Seber (1970) proposed an approximate unbiased estimator of $V\left(\widehat{N}_c\right)$ as

$$\widehat{V}\left(\widehat{N}_c\right) = \frac{(n_1+1)(n_2+1)(n_1-m_2)(n_2-m_2)}{(m_2+1)^2(m_2+2)} \tag{26.10.5}$$

26.10.1.3 Bailey's Binomial Model

Using Binomial approximation to Eq. (26.10.2), Bailey (1951) obtained the probability distribution of m_2 as

$$f(m_2|n_1, n_2) = \binom{n_2}{m_2}\left(\frac{n_1}{N}\right)^{m_2}\left(1-\frac{n_1}{N}\right)^{n_2-m_2}; m_2 = 0, 1, \ldots, n_2 \quad (26.10.6)$$

Because $E(m_2) = n_2\frac{n_1}{N}$, the MLE of N comes out as $\widehat{N}_P$. Because $\widehat{N}_P$ is biased, Bailey proposed the following alternative estimator of N as

$$\widehat{N}_b = \frac{n_1(n_2+1)}{(m_2+1)} \quad (26.10.7)$$

The proposed estimator of the variance of $\widehat{N}_b$ is

$$\widehat{V}\left(\widehat{N}_b\right) = \frac{n_1^2(n_2+1)(n_2-m_2)}{(m_2+1)^2(m_2+2)} \quad (26.10.8)$$

26.10.1.4 Ratio Method

Let us attach a variable y_i for the ith member of the population such that $y_i = 1$. Then $Y = \sum_{i=1}^{N} y_i = N =$ population size. Similarly, we can attach a variable x_i to the ith individual such that $x_i = 1$ if the ith individual is marked and 0 if it is unmarked. In this case $X = \sum_{i=1}^{N} x_i = n_1, \overline{y}(s_{2m}) = 1$ and $\overline{x}(s_2) = m_2/n_2$. Hence the ratio estimator of N is given by

$$\widehat{N}_R = \frac{\overline{y}}{\overline{x}}X = \frac{n_2}{m_2}n_1 = \widehat{N}_P \quad (26.10.9)$$

The ratio estimator $\widehat{N}_R$ is biased as usual and an approximate estimator of the variance of $\widehat{N}_R$ was obtained by Lohr (1999) as

$$\widehat{V}\left(\widehat{N}_R\right) \cong \frac{n_1^2 n_2(n_2-m_2)}{m_2^3} \quad (26.10.10)$$

26.10.1.5 Inverse Sampling Methods

26.10.1.5.1 Without Replacement Method

Here, the sample s_1 of size n_1 animals is selected as before. The selection of sample s_2 is continued until m_2 of the tagged n_1 animals are recaptured again. Here, n_2, the sample size of s_2, is a random variable. The condition probability distribution of n_2 given n_1 and m_2 is the negative hypergeometric distribution

$$f(n_2|n_1, m_2, N) = \frac{\binom{n_1}{m_2 - 1}\binom{N - n_1}{n_2 - m_2}}{\binom{N}{n_2 - 1}} \frac{n_1 - m_2 + 1}{N - n_2 + 1}; \quad (26.10.11)$$

$$n_2 = m_2, m_2 + 1, \ldots, N + m_2 - n_1$$

Here the MLE of N is

$$\widehat{N} = \widehat{N}_I = \frac{n_2(n_1 + 1)}{m_2} - 1 \quad (26.10.12)$$

The estimator $\widehat{N}_I$ is unbiased for N with variance

$$V\left(\widehat{N}_I \middle| n_1, m_2\right) = \frac{(N + 1)(N - n_1)(n_1 - m_2 + 1)}{m_2(n_1 + 2)} \cong \frac{N^2}{m_2} \quad (26.10.13)$$

The coefficient of variation of $\widehat{N}_I$ is approximately equal to

$$C\left(\widehat{N}_I\right) = \left(\frac{n_1 - m_2 + 1}{m_2(n_1 + 2)}\right)^{1/2} \quad (26.10.14)$$

Because $C\left(\widehat{N}_I\right)$ is independent of N, one can choose m_2 for prescribed values of $C\left(\widehat{N}_I\right)$ and n_1.

26.10.1.5.2 With Replacement Method

Here, animals in the second sample s_2 are caught one by one and released into the original population. The catch and release procedure is continued until m_2 (a prespecified) marked animals are caught. In this case the probability distribution of n_2 given n_1 and m_2 is the negative binomial distribution

$$f(n_2|n_1, m_2, N) = \binom{n_2 - 1}{m_2 - 1}\left(\frac{n_1}{N}\right)^{m_2}\left(1 - \frac{n_1}{N}\right)^{n_2 - m_2}; n_2 = m_2, m_2 + 1, \ldots \quad (26.10.15)$$

Because $E(n_2) = N\dfrac{m_2}{n_1}$, an unbiased estimator of N is

$$\widehat{N}_{IW} = \widehat{N}_P = n_1 n_2/m_2 \qquad (26.10.16)$$

The variance of $\widehat{N}_{IW}$ and its unbiased estimator are given, respectively, as follows:

$$V\left(\widehat{N}_{IW}\right) = \left(N^2 - Nn_1\right)/m_2 \text{ and } \widehat{V}\left(\widehat{N}_{IW}\right) = \frac{n_2 n_1^2 (n_2 - m_2)}{m_2^2(m_2+1)} \qquad (26.10.17)$$

A serious disadvantage of the use of inverse sampling is that the variance of n_2 becomes large for inappropriate choices of n_1 and m_2. The coefficient of variation of $\widehat{N}_{IW}$ under this sampling scheme is

$$C\left(\widehat{N}_{IW}\right) \cong \frac{1}{\sqrt{m_2}} \qquad (26.10.18)$$

So, one can easily find the value of m_2 for a given value of $C\left(\widehat{N}_{IW}\right)$.

26.10.1.6 Interval Estimation

In determining the confidence interval of N based on capture–recapture method, one should be certain about the normality of the distribution of $\widehat{N}$. For a small sample the distribution of $\widehat{N}$ is skewed, which results in poor coverage of the probability of the confidence interval. Several methods of determining the confidence interval of N have been proposed by Seber (1970). Some of them are reported below as follows.

As $N \to \infty$, $\widehat{N}_c$ is asymptotically normally distributed with mean N and estimated variance $\widehat{V}\left(\widehat{N}_c\right)$. Hence $100(1-\alpha)\%$ confidence interval of N is given by

$$\widehat{N}_c \pm z_{\alpha/2}\sqrt{\widehat{V}\left(\widehat{N}_c\right)} \qquad (26.10.19)$$

where $z_{\alpha/2}$ is the upper $100\alpha/2$ percent point of standard normal distribution.

For $\widehat{p} = m_2/n_2 < 0.1$ and $m_2 > 50$, normal approximation of $\widehat{p}$ is valid and the $100(1-\alpha)\%$ confidence interval of $p = n_1/N$ is

$$\widehat{p} \pm \left\{ z_{\alpha/2}\sqrt{(1-f)\widehat{p}(1-\widehat{p})/(n_2-1)} + 1/(2n_2) \right\} \qquad (26.10.20)$$

The confidence interval of N is obtained from the inversion of Eq. (26.10.20). Here, the unknown sampling fraction $f(=n_2/N)$ can be ignored if its estimate m_2/n_1 is assumed to be less than 0.1.

In case $N > 150$, $n_1 > 50$ and $n_2 > 50$, the distribution of m_2 is asymptotically normal (Seber, 1973). In this case the two largest roots of the equation

$$\frac{\left(m_2 - \frac{n_1 n_2}{N}\right)^2}{n_2 \frac{n_1}{N}\left(1 - \frac{n_1}{N}\right) \frac{N - n_2}{N - 1}} = \left(z_{\alpha/2}\right)^2 \qquad (26.10.21)$$

provide the confidence interval of N.

26.10.1.7 Multiple Marking

Consider the Schnabel census (1938) where a series of samples $s_1, s_2,..., s_t$ of sizes $n_1, n_2,..., n_t$ were selected independently from the entire population. All the captured animals were tagged or marked and returned to the population. Tags are unique, so that the capture history of each of the individual animal can be followed separately. Let

m_i = number of marked animals in the sample s_i,

$u_i = n_i - m_i$ = number of unmarked animals in the sample s_i,

$M_i = u_1 + u_2 + \cdots + u_{i-1}; i = 1, ..., t+1$

= number of marked animals in the population just before the ith sample s_i is taken.

Clearly, $m_1 = M_1 = 0$, $M_2 = u_1 = n_1$, and $M_{t+1}(=r$ say) is the number of marked animals in the population after performing the experiment.

Let $w = x_1 x_2 ... x_t$, with $x_j = 1$ if an animal is caught at the jth sample s_j and $x_j = 0$ if it is not caught at the jth sample. Following Seber (1986) we denote a_w as the number of animals with a capture history, e.g., $t = 2$, a_{10} = number of animals caught in the sample s_1 but not in s_2; a_{01} = number of animals caught in the sample s_2 but not in s_1; a_{11} = number of animals caught in the sample s_1 and also in s_2, and a_{00} = number of animals that were not captured in any of the samples s_1 and $s_2 = N - r$.

Here, we assume that all animals irrespective of capture history will have the same probability of being caught in a particular sample. The caught in the samples $s_1, s_2,..., s_t$ are independent. Let $p_i(=1 - q_i)$ = probability that an animal will be captured in the ith sample s_i. Then for $t = 2$, the probability of the vector $\{a_w\} = \{a_{00}, a_{01}, a_{10}, a_{11}\}$ follows the multinomial distribution

$$\begin{aligned} f\{a_w\} &= \frac{N!}{a_{10}!a_{01}!a_{11}!(N-r)!}(p_1q_2)^{a_{10}}(q_1p_2)^{a_{01}}(p_1p_2)^{a_{11}}(q_1q_2)^{N-r} \\ &\propto \frac{N!}{(N-r)!}p_1^{n_1}p_2^{n_2}q_1^{N-n_1}q_2^{N-n_2} \end{aligned} \tag{26.10.22}$$

For general t, the probability distribution of $\{a_w\}$ is given by

$$f\{a_w\} \propto \frac{N!}{(N-r)!}\prod_{i=1}^{t}\left(p_i^{n_i}q_i^{N-n_i}\right) \tag{26.10.23}$$

Eq. (26.10.23) indicates that n_i's are independent binomial variable with distribution

$$f(n_i) = \binom{N}{n_i}p_i^{n_i}q_i^{N-n_i} \tag{26.10.24}$$

The conditional distribution of $\{a_w\}$ for fixed sample sizes $\{n_i\}$ is given by

$$f(\{a_w\}|\{n_i\}) \propto \frac{N!}{(N-r)!}\prod_{t=1}^{t}\binom{N}{n_i}^{-1} \tag{26.10.25}$$

The MLE of N is the unique root greater than r of the following $(t-1)$ degree polynomial (Seber, 1970)

$$1-\frac{r}{N} = \prod_{i=1}^{t}\left(1-\frac{n_i}{N}\right) \tag{26.10.26}$$

For $t = 2$, MLE of N is $\widehat{N}_P = n_1n_2/m_2$, the Peterson estimate.

Chapman (1952) derived Eq. (26.10.26) by considering the conditional distribution of $\{m_2,\ldots, m_t\}$ given $\{n_1, n_2,\ldots, n_t\}$ as a product of hypergeometric distributions

$$f(m_2,\ldots,m_t|n_1,\ldots,n_t) = \prod_{i=2}^{t}\frac{\binom{M_i}{m_i}\binom{N-M_i}{u_i}}{\binom{N}{n_i}} \propto \frac{N!}{(N-r)!}\prod_{i=2}^{t}\binom{N}{n_i}^{-1} \tag{26.10.27}$$

For further details readers are referred to Seber (1986).

26.10.2 Open Model

In the open model, provision is made for death, birth, immigration, and migration. The models are broadly classified into two categorics: The first one is based on the bird-banding and fish-tagging studies where a number of animals are caught and banded for several (k) periods. Each band or tag carries a unique identification number. Data were recorded of bands or tags collected from dead animals for each of the k periods. The recovery data are analyzed for estimation of the annual survival rate and the annual band recovery rate for each of the k periods. A compressive review is given by Brownie et al. (1985). The second one, known as the Jolly—Seber model, deals with multiple recaptures of alive marked animals. The links between the two models were established by Brownie et al. (1985). The details are given by Seber (1973). Because there is not much difference in the collection and analysis of Brand recovery data and live animal capture data, our present discussions are limited to the Jolly—Seber model only.

26.10.2.1 Jolly—Seber Model

Jolly (1965) and Seber (1965) independently provided the most important stochastic model for capture—recapture sampling in the open population setup. The model allows estimates of survival, capture probability, and population size for each sampling time, and recruitment between sampling times.

Experimental protocol: Here, we capture and recapture animals over $k(>1)$ successive occasions. In each occasion, animals are captured, tagged uniquely, and then released. The capture or sighting history of each individual is recorded. On the first occasion a sample s_1 of n_1 animals is captured, all the animals are tagged, a few d_1 of them died in the capturing process, and the remaining $R_1(=n_1 - d_1)$ animals are released. On the second occasion, a sample s_2 of n_2 animals is captured of which u_2 are untagged and the remaining $n_2 - u_2$ are tagged (recaptured). All the recaptured and newly captured animals are uniquely marked and capture history is recorded. If d_2 of them die during the capture process, the remaining $R_2(=n_2 - d_2)$ are released and treated as cohort of the sample s_2. In general, at the ith occasion, a sample s_i of size n_i is captured of which u_i are untagged, $n_i - m_i$ are tagged, and d_i dies. All the $R_i(=n_i - d_i)$ animals are tagged uniquely, released, and treated as cohort of ith sample s_i. Let m_{ij} be the number of animals recaptured the first time from the released R_i animals in the jth sample, $i = 1,\ldots, k-1; j = i+1,\ldots, k$.

26.10.2.1.1 Summary Data

The summary of the data is presented as follows:

					s_1	s_2	s_3	…	s_{i+1}	…	s_k		
Samples	n_i	u_i	d_i	R_i		m_{ij}						Total r_i	z_{i+1}
s_1	n_1	u_1	d_1	R_1	0	m_{12}	m_{13}	…	m_{1i+1}	…	m_{1k}	r_1	$z_2 = r_1 - m_{12}$
s_2	n_2	u_2	d_2	R_2			m_{23}	…	m_{2i+1}	…	m_{2k}	r_2	$z_3 = r_2 - m_{23}$
…								…	…	…	…		
s_i	n_i	u_i	d_i	R_i					m_{ii+1}	…	m_{ik}	r_i	$z_{i+1} = r_i - m_{i,i+1}$
…													
s_{k-1}	n_{k-1}	u_{k-1}	d_{k-1}	R_{k-1}							m_{k-1k}	r_{k-1}	$z_k = r_{k-1} - m_{k-1,k}$
Total					$m_1 = 0$	m_2	m_3	…	m_{i+1}	…	m_k		

Descriptions of the statistics:

r_i = number of marked animal released in the sample s_i, which were ever captured again = ith row total.

m_i = number of animals captured in the sample s_i = ith column total.

z_i = number of animals released before the ith sample, not caught in the sample s_i but are captured in the subsequent samples $= r_{i-1} - m_{i-1,i}$.

26.10.2.1.2 Assumptions of the Model

(i) Capture probability p_i: Each animal (marked or unmarked) in the population has the same probability p_i of capture at the time of ith sample $(i = 1,\ldots, k)$.

(ii) Survival probability ϕ_i: Each marked animal has the same survival probability ϕ_i for the period of taking ith sample to $i+1$th sample $(i = 1,\ldots, k-1)$.

(iii) Tags are not lost or damaged.

(iv) All samples are instantaneous: each release is made immediately after the selection of sample.

26.10.2.1.3 Estimation of Parameters

The conditional distribution of $m_{ii+1}, m_{ii+2},\ldots, m_{ii+k}, R_i - \sum_{j=i+1}^{k} m_{ij} = R_i - r_i$ given R_i follows multinomial distribution with cell probabilities:

$$\begin{aligned}\pi_{ii+1} &= \{\phi_i p_{i+1}\}, \pi_{ii+2} = \{\phi_i(1-p_{i+1})\}\{\phi_{i+1}p_{i+2}\},\ldots,\\ \pi_{ij} &= [\{\phi_i(1-p_{i+1})\}\{\phi_{i+1}(1-p_{i+2})\}\cdots\{\phi_{j-1}p_j\}],\ldots,\\ \pi_{ik} &= [\{\phi_i(1-p_{i+1})\}\{\phi_{i+1}(1-p_{i+2})\}\cdots\{\phi_{k-1}p_k\}].\end{aligned}$$

Hence the likelihood function of the recaptured conditional to $R_1,\ldots, R_{k-1}$ is

$$L(\mathbf{m}) \propto \prod_{i=1}^{k-1} \lambda_i^{R_i - r_i} \prod_{j=i+1}^{k} \{\pi_{ij}\}^{m_{ij}} \tag{26.10.28}$$

Let U_i be the number of unmarked animals just before the ith sample is taken, then the likelihood function of $u_1,\ldots, u_k$ is

$$L(\mathbf{u}) \propto \prod_{i=1}^{k-1} \binom{U_i}{u_i} p_i^{u_i}(1-p_i)^{U_i - u_i} \tag{26.10.29}$$

Let M_i be the number of marked animal just before the ith sample is taken. $M_1 = 0$ but M_i's for $i = 2, \ldots, k$ are unknown. The M_i's are estimated by Seber (1973) as follows.

Among the marked M_i animals, m_i are caught in the sample s_i and the remaining $M_i - m_i$ are not caught. Out of $M_i - m_i$ marked animals, z_i are subsequently caught. R_i animals are realized in the sample s_i of which r_i are caught. Hence,

$$\frac{z_i}{M_i - m_i} = \frac{r_i}{R_i} \tag{26.10.30}$$

Eq. (26.10.30) yields an estimate of M_i as

$$\widehat{M}_i = m_i + \frac{R_i z_i}{r_i} \tag{26.10.31}$$

Population size N_i:

The number of the marked animal in the sample s_i of size n_i is m_i and M_i is the number of the marked sample in the population of size N_i when the sample s_i was taken. Hence

$$\frac{m_i}{n_i} \cong \frac{M_i}{N_i} \tag{26.10.32}$$

Eq. (26.10.32) leads

$$\widehat{N}_i = \widehat{M}_i \frac{n_i}{m_i} \tag{26.10.33}$$

Capture probability p_i:

Proportion of marked animals captured in ith sample s_i to the total number of animals at that time

$$\text{i.e.,} \quad \widehat{p}_i = \frac{n_i}{\widehat{N}_i} = \frac{m_i}{\widehat{M}_i} \tag{26.10.34}$$

Survival rate ϕ_i:

The number of marked animals in the population immediately after sample s_i is $M_i - m_i + R_i$. The number of the marked animals in the population just before sample s_{i+1} is selected is M_{i+1}. Hence

$$\widehat{\phi}_i = \frac{\widehat{M}_{i+1}}{\widehat{M}_i - m_i + \widehat{R}_i} \tag{26.10.35}$$

Recruitment B_i:

Estimated number of survival between the sampling period i to $i+1$ is $(N_i - n_i + R_i)\phi_i$. Hence recruitment between the sampling period i to $i+1$ is

$$\widehat{B}_i = \widehat{N}_{i+1} - \left(\widehat{N}_i - n_i + R_i\right)\widehat{\phi}_i \tag{26.10.36}$$

Out of the five estimators only $\widehat{\phi}_i$ and $\widehat{p}_i$ are MLEs and the remaining $\widehat{M}_i$, N_i, and $\widehat{B}_i$ are intuitive estimators. These estimators are not unbiased. Seber (1982) proposed approximate unbiased estimators of these five estimators and their variances.

Example 26.10.1

Consider the following artificial data where animals are captured and tagged for 7 consecutive months.

Month (i)	Sample size (n_i)	Released (R_i)	(m_{ij}) 1	2	3	4	5	6	7
1	80	80		15	10	6	2	1	0
2	125	120			30	10	3	0	1
3	150	146				35	18	5	4
4	180	178					35	20	10
5	140	135						30	15
6	100	98							20

Estimates of parameters:

i	n_i	m_i	r_i	R_i	z_i	$\widehat{M}_i$	$\widehat{N}_i$	$\widehat{\phi}_i$	$\widehat{p}_i$	$\widehat{B}_i$
1	80	0	34	80						
2	125	15	44	120	19	66.9	556.8	0.835	0.225	39.3
3	150	40	62	146	14	73.0	273.6	0.425	0.548	252.7
4	180	51	65	178	27	125.0	441.0	0.698	0.484	99.4
5	140	58	45	135	30	148.0	357.2	0.875	0.392	28.5
6	100	56	20	98	15	129.5	231.5	0.576	0.432	
7		50			0					

The Jolly–Seber model assumes that the marked and unmarked animals have equal capture and survival probabilities, which may not hold in practice. This model estimates time-specific apparent survival rates and is restricted to specific localities only. The estimators may be highly biased in the presence of unequal capture and survival probabilities. Several extensions of the Jolly–Seber model are available in the literature. Pollock (1975) allowed behavioral effects in the captured and survivors; Brownie et al. (1986) considered survival and captured parameters unchanged over time; while Cormack (1981) used log-linear models on open population. A comprehensive review of the recent development in the capture–recapture model for the open population was given by Seber (1986), Boswell et al. (1988), Pollock et al. (1990), and Barker (1995).

26.11 EXERCISES

26.11.1 A sample of 800 fish were captured from a lake, then marked uniquely, and realized alive. One month later, a sample of 1000 fish was captured from the same lake and 80 marked fish captured earlier were found. Estimate the total number of fish of the lake and find 95% confidence interval of the number of fishes using (i) Peterson–Lincoln, (ii) Chapman, (iii) Bailey, and (iv) ratio methods.

26.11.2 A sample of 1000 people is selected from a locality and 20 of them were found to be illegal immigrants. One month later another sample of 825 people was selected by the inverse sampling method without replacement from the same locality till 10 of them were found to be illegal immigrants selected in the earlier sample. Estimate the number of illegal immigrant in that locality and compute its standard error.

26.11.3 Samples of animals are captured from a certain game park for 8 consecutive years. All the animals that were captured are uniquely marked and released unless they die in the capturing process. Estimate the total number of animals, capture probability rates, survival rates, and new recruitments for each of the sampling years. The following table gives the capture record:

Year	i		1	2	3	4	5	6	7	8
i	n_i	R_i					m_{ij}			
1	100	98		20	10	8	6	4	0	2
2	125	125			30	20	7	5	4	3
3	160	158				30	25	10	2	0
4	180	175					40	15	10	5
5	150	150						25	10	2
6	130	130							30	15
7	100	100								20

26.11.4 Consider two incomplete frames U_A and U_B, which cover the population $U(=U_A \cup U_B)$. Let us define $U_{ab} = U_A \cap U_B$, $U_a = U_A \cap U_B^c$, and $U_b = U_B \cap U_A^c$. Two independent samples s_A and s_B of sizes n_A and n_B are selected from U_A and U_B by the PPSWR method using normed size measures p_{i1} and p_{j2} of the ith unit of U_A and the jth unit of U_B, respectively. Let $t_a = \frac{1}{n_a}\sum_{s_a}\frac{y_i}{p_{1i}}$, $t_b = \frac{1}{n_b}\sum_{s_b}\frac{y_i}{p_{2i}}$, $t_{ab} = \frac{1}{n_{ab}}\sum_{s_{ab}}\frac{y_i}{p_{1i}}$, and $t_{ba} = \frac{1}{n_{ba}}\sum_{s_{ba}}\frac{y_i}{p_{2i}}$, where $s_a = s_A \cap U_a$, $s_{ab} = s_A \cap U_{ab}$, $s_b = s_B \cap U_b$, $s_{ba} = s_B \cap U_{ab}$; $\sum_{s_t}$ and n_t denote respectively the sum over the units and its size in s_t with repletion, $t = a, b, ab, ba$. Show that $T = w_a t_a + t_w + w_b t_b$ is an unbiased estimator of the population total of $Y = \sum_{i \in U} y_i$, where $t_w = \alpha w_{ab} t_{ab} + (1-\alpha) w_{ba} t_{ba}$, $w_a = \frac{n_a}{n_A}$, $w_b = \frac{n_b}{n_B}$, $w_{ab} = \frac{n_{ab}}{n_A}$, $w_{ba} = \frac{n_{ba}}{n_B}$ and α is a known constant. Find the variance of T and an unbiased estimator of the variance.

26.11.5 Consider a population consisting of N units, which are classified into K overlapping clusters. The ith cluster consists of N_i primary units on unknown size and $\sum_{i=1}^{K} N_i = M \geq N$. A primary unit may be found in more than one cluster. Let y_{ij} be the value of y for the jth unit of the ith cluster and f_{ij} its frequency of occurring in K clusters. Let a sample s of size k clusters be selected by the PPSWR method with selection probability $p_i = N_i/M$ for the ith cluster. If the ith cluster is selected in s, a subsample s_i of n_i primary units are selected from the ith cluster by the SRSWOR method. Define $Z_{ij} = y_{ij}/f_{ij}$, $w_{ij} = 1/f_{ij}$, $\overline{z}_i = \frac{1}{n_i}\sum_{i \in s_i} z_{ij}$, $\overline{Z}_i = \frac{1}{N_i}\sum_{i=1}^{N_i} z_{ij}$, $\overline{w}_i = \frac{1}{n_i}\sum_{j \in s_i} w_{ij}$, and $\overline{W}_i = \frac{1}{N_i}\sum_{i=1}^{N_i} w_{ij}$. Show that the ratio estimator $\widehat{Y}_R = \sum_{i=1}^{k}\overline{z}_i \Big/ \sum_{i=1}^{k}\overline{w}_i$ is a biased estimator of the population mean $\overline{Y}$ and the mean square error of $\widehat{Y}_R$ is

$$MSE\left(\widehat{Y}_R\right) \cong \frac{M}{kN^2}\sum_{i=1}^{K} N_i\left[\left(\overline{Z}_i - \overline{Y}\overline{W}_i\right)^2\left(\frac{1}{n_i} - \frac{1}{N_i}\right)D_i^2\right],$$

where $D_i^2 = \frac{1}{N_i - 1}\sum_{j=1}^{N_i}\left\{\left(z_{ij} - \overline{Z}_i\right) - \overline{Y}\left(w_{ij} - W_i\right)\right\}^2$ (Tracy and Osahan, 1994).

REFERENCES

Abernathy, J. R., Greenberg, B. G., & Horvitz, D. G. (1970). Estimates of induced abortion in urban North Carolina. *Demography, 7*, 19–29.

Aggarwal, O. P. (1959). Bayes and minimax procedures in sampling from finite and infinite populations I. *Annals of Mathematical Statistics, 30*, 206–218.

Aggarwal, O. P. (1966). Bayes and minimax procedures for estimating the arithmetic mean of a population with two-stage sampling. *Annals of Mathematical Statistics, 37*, 1186–1195.

Agresti, A. (2002). *Categorical data analysis*. New Jersey: Wiley.

Altham, P. A. E. (1976). Discrete variable analysis for individuals grouped into families. *Biometrika, 63*, 263–269.

Anderson, H. (1975a). *Efficiency versus protection in RR designs for estimating proportions*. Technical Report, 9. Sweden: University of Lund.

Anderson, H. (1975b). *Efficiency versus protection in a general RR model*. Technical Report, 10. Sweden: University of Lund.

Anderson, H. (1975c). Efficiency versus protection in a general RR model. *Scandinavian Journal of Statistics, 37*, 177–188.

Anderson, H. (1977). Efficiency versus protection in general randomized response model. *Scandinavian Journal of Statistics, 4*, 11–19.

Anderson, P. G., & Thorburn, D. (2005). An optimal calibration distance leading to the optimal regression estimator. *Survey Methodology, 31*, 95–99.

Arcos, A., Rueda, M., & Singh, S. A. (2015). Generalized approach to randomized response for quantitative variables. *Quality & Quantity, 49*, 1239–1256.

Arnab, R. (1979). *Contributions to theories of repetitive survey sampling strategies*. Indian Statistical Institute (Unpublished Ph.D. thesis).

Arnab, R. (1980). Two-stage sampling over two occasions. *Australian Journal of Statistics, 22*, 349–357.

Arnab, R. (1986). Optimal prediction for a finite population total with connected designs and related model based results. *Metrika, 33*, 79–84.

Arnab, R. (1988). Variance estimation in multi-stage sampling. *Australian Journal of Statistics, 30*, 107–110.

Arnab, R. (1990). On commutativity of design and model expectations in randomized response surveys. *Communications in Statistics – Theory and Methods, 19*, 3751–3757.

Arnab, R. (1991). On sampling over two occasions using varying probabilities. *Journal of the Indian Society of Agricultural Statistics, 43*, 282–290.

Arnab, R. (1994). Non-negative variance estimation in randomized response surveys. *Communications in Statistics – Theory and Methods, 23*, 1743–1752.

Arnab, R. (1995a). On admissibility and optimality of sampling strategies in randomized response surveys. *Sankhyā, 57*, 385–390.

Arnab, R. (1995b). Optimal estimation of a finite population total under randomized response surveys. *Statistics, 27*, 175–180.

Arnab, R. (1996). Randomized response trials: A unified approach for qualitative data. *Communications in Statistics – Theory and Methods, 25*(6), 1173–1183.

Arnab, R. (1998a). Randomized response surveys: Optimum estimation of a finite population total. *Statistical Papers, 39*, 405–408.

Arnab, R. (1998b). Sampling on two occasions: Estimation of population total. *Survey Methodology, 24*, 171–184.

Arnab, R. (2004a). Optional randomized response techniques for complex survey designs. *Biometrical Journal, 46*(1), 114–124.
Arnab, R. (2004b). A note on the nearest proportional to size sampling design. *Journal of the Indian Society of Agricultural Statistics, 58*(2), 212–223.
Arnab, R. (2006). Randomized response technique for complex survey designs. *Statistical Papers, 48*, 131–141.
Arnab, R. (2013). Controlled sampling: A review. *Statistics and Applications, 11*(1&2), 127–146.
Arnab, R., & Mothupi, T. (2015). Randomized response techniques: A case study of the risky behaviors' of students of a certain University. *Model Assisted Statistics and Applications, 10*, 421–430.
Arnab, R., & Roy, D. (1990). On use of symmetrical balanced incomplete block design in construction of sampling design realizing preassigned sets of inclusion probabilities first two orders. *Communications in Statistics — Theory and Methods, 19*, 3223–3232.
Arnab, R., & Singh, S. (2006). Estimation of variance from missing data. *Metron, LXIV*(2), 166–177.
Arthanari, T., & Doge, Y. (1981). *Mathematical programming in statistics*. New York: Wiley.
Asok, C. (1974). *Contribution to the theory of unequal probability sampling without replacement*. Ames, Iowa: Iowa State University (Unpublished Ph.D. thesis).
Asok, C., & Sukhatme, B. V. (1976). On Sampford's procedure of unequal probability sampling without replacement. *Journal of the American Statistical Association, 71*, 912–918.
Asok, C., & Sukhatme, B. V. (1978). A note on Midzuno scheme of sampling. In *Paper presented at the 32nd Annual Conference of the Indian Society of Agricultural Statistics, Ludhiana, India*.
Avadhani, M. S., & Sukhatme, B. V. (1970). A comparison of two sampling procedures with applications to successive sampling. *Applied Statistics, 19*, 251–259.
Avadhani, M. S., & Sukhatme, B. V. (1973). Controlled sampling with equal probabilities and without replacement. *International Statistical Review, 41*, 175–182.
Bailey, N. T. J. (1951). On estimating the size of mobile populations from capture-recapture data. *Biometrika, 38*, 293–306.
Bankier, M. D. (1986). Estimators based on several stratified samples with applications to multiple frame surveys. *Journal of the American Statistical Association, 81*, 1074–1079.
Barker, R. J. (1995). *Open population mark-recapture models including sightings*. Palmerston North, New Zealand: Messy University (Ph.D. thesis).
Barnard, J., & Rubin, D. B. (1999). Small-sample degrees of freedom with multiple imputation. *Biometrika, 86*, 949–955.
Basu, D. (1958). On sampling with and without replacement. *Sankhyā, 20*, 287–294.
Basu, D. (1969). Role of sufficiency and likelihood principles in sample survey theory. *Sankhyā, A, 26*, 3–16.
Basu, D. (1971). An essay on the logical foundations of survey sampling, Part 1 (with discussion). In V. P. Godambe, & D. A. Sprott (Eds.), *Foundations of statistical inference* (pp. 203–242). Toronto: Holt, Rinehart and Winston.
Baswa, I. V. (May 2000). Inference for stochastic processes, via estimating equations. In *Symposium for stochastic process*. Athens: University of Georgia.
Battese, G. E., Hartler, R. M., & Fuller, W. A. (1988). An error component model for prediction of county crop areas using survey and satellite data. *Journal of the American Statistical Association, 92*, 999–1005.
Beale, E. M. L. (1962). Some uses of computers in operations research. *Industrielle Organization, 31*, 51–52.

Bellhouse, D. R. (1977). Optimal designs for systematic sampling in two dimensions. *Biometrika, 64*, 605–611.

Bellhouse, D. R. (1981). Spatial surveys in presence of a trend. *Journal of Statistical Planning and Inference, 5*, 365–375.

Bellhouse, D. R. (1988). Systematic sampling. In P. R. Krishnaiah, & C. R. Rao (Eds.), *Handbook of statistics* (Vol. 6, pp. 125–145). Amsterdam: North-Holland.

Bhapkar, V. P. (1966). A note on equivalence of two test criteria of hypotheses in categorical data. *Journal of the American Statistical Association, 61*, 228–235.

Bickel, P. J., & Freedman, D. A. (1984). Asymptotic normality and the bootstrap in stratified sampling. *Annals of Statistics, 12*, 470–482.

Biernacki, P., & Waldrof, D. (1981). Snowball sampling. Problems and techniques of chain referral. *Sociological Methods and Research, 10*(1), 141–163.

Binder, D. A. (1983). On the variance of asymptotically normal estimators from complex surveys. *International Statistical Review, 51*, 279–292.

Binder, D. A., & Hidroglou, M. A. (1988). Sampling over time. In P. R. Krishnaiah, & C. R. Rao (Eds.), *Handbook of statistics* (Vol. 6, pp. 187–211). Amsterdam: North-Holland.

Birnbaum, Z. W., & Sirken, M. G. (1965). *Design of sample surveys to estimate the prevalence of rare diseases: Three unbiased estimates*. National center for health statistics, ser 2, no. 11. Washington, DC, U.S.: Government Printing Office.

Bogue, D. J. (1950). A technique for making extensive postcensal estimates. *Journal of the American Statistical Association, 45*, 149–163.

Bogue, D. J., & Duncan, B. D. (1959). *A composite method of estimating postcensal population of small areas by age, sex and colour*. Vital Statistics-special report, 47, No. 6. Washington, DC: National Office of Vital Statistics.

Bonn, L. L., & Wolfe, D. A. (1992). Nonparametric two-sample procedures for ranked-set samples data. *Journal of the American Statistical Association, 87*, 552–561.

Bonn, L. L., & Wolfe, D. A. (1994). The effect of imperfect judgment ranking on properties of procedures based on the ranked-set samples analog to the Mann-Whitney – Wilcoxon statistic. *Journal of the American Statistical Association, 89*, 168–176.

Boswell, M. T., Brunham, K. P., & Patil, G. P. (1988). Role and use of composite sampling and capture-recapture sampling in ecological studies. In P. R. Krishnaiah, & C. R. Rao (Eds.), *Handbook of statistics* (Vol. 6, pp. 469–488). Amsterdam: North-Holland.

Bowley, A. L. (1926). Measurement of precision attained in sampling. *Bulletin of the International Statistical Institute, 22*, 1–62.

Brackstone, G. J. (1987). Small area data: Policy issues and technical challenges. In R. Platek, J. N. K. Rao, C. E. Särndal, & M. P. Singh (Eds.), *Small area statistics* (pp. 3–20). New York: Wiley.

Breidt, F., & Opsomer, J. D. (2000). Local polynomial regression estimators in survey sampling. *Annals of Statistics, 28*, 1026–1053.

Brewer, K. R. W. (1963a). A model of systematic sampling schemes of unequal probabilities. *Australian Journal of Statistics, 5*, 5–13.

Brewer, K. R. W. (1963b). Ratio estimation and finite populations: Some results deducible from the assumption of an underlying stochastic process. *Australian Journal of Statistics, 5*, 93–105.

Brewer, K. R. W., Early, L., & Joyce, S. (1972). Selecting several samples from a single population. *Australian Journal of Statistics, 14*, 231–239.

Brewer, K. R. W., & Hanif, M. (1983). *Sampling with unequal probabilities*. Lecture notes in statistics. New York: Springer-Verlag.

Brier, S. S. (1978). *Discrete data models with random effects*. Technical report. University of Minnesota, School of Statistics.

Brown, C., & Ritchie, J. (1981). *Focussed enumeration: The development of a method for sampling ethnic minority groups*. London: Policy Studies Institute of Social and Community Planning Research.

Brownie, C., Anderson, D. R., Burnham, K. P., & Robson, D. S. (1985). *Statistical inference from brand recovery data — A Handbook* (2nd ed., p. 156). U.S.: Fish and Wildlife Service Resource Publication.

Brownie, C., Hines, J. E., & Nicholas, J. D. (1986). Constant parameter capture-recapture models. *Biometrics, 42*, 561—574.

Butar, F. B., & Lahiri, P. (2001). *On measures of uncertainty of empirical Bayes small area estimators*. Technical report. Lincoln: Department of Statistics, University of Nebraska.

Calvin, L. D. (1954). Doubly balanced incomplete block designs for experiments in which treatments are correlate. *Biometrics, 10*, 61—83.

Campbell, C. (1977). *BMDP: Biomedical computer programs, P-series*. Berkeley: University of California Press.

Cassel, C. M., & Särandal, C. E. (1972). A model for studying robustness of estimators in informativeness of labels in sampling with varying probabilities. *Journal of the Royal Statistical Society, Series B, 34*, 279—289.

Cassel, C. M., Särndal, C. E., & Wretman, J. H. (1976). Some results on generalized difference estimation and generalized regression estimation for finite populations. *Biometrika, 63*, 615—620.

Cassel, C. M., Särndal, C. E., & Wretman, J. H. (1977). *Foundations of inference in survey sampling*. New York: Wiley.

Chakrabarti, M. C. (1963). On the use of incidence matrices of designs in sampling from finite populations. *Journal of the Indian Statistical Association, 1*, 78—85.

Chakrabarty, R. P. (1968). *Contribution to the theory of ratio type estimators*. Texas A&M University (Ph.D. thesis).

Chakrabarty, R. P., & Rao, J. N. K. (1968). The bias and stability of the jackknife variance estimator in ratio estimation (abstract). *Journal of the American Statistical Association, 63*, 748.

Chambers, R. L., Drofman, A. H., & Wehrly, T. E. (1993). Bias robust estimation in finite populations using nonparametric calibration. *Journal of the American Statistical Association, 88*, 268—277.

Chambers, R. L., & Dunstan, R. (1986). Estimating distribution functions from survey data. *Biometrika, 73*, 597—604.

Chapman, D. G. (1951). *Some properties of hypergeometric distribution with application to zoological censuses. 1* pp. 131—160). University of California Publications in Statistics.

Chapman, D. G. (1952). Inverse, multiple and sequential sample censuses. *Biometrics, 8*, 286—306.

Chaudhuri, A. (1969). Minimax solutions of some problems in sampling from a finite population. *Calcutta Statistical Association Bulletin, 18*, 1—24.

Chaudhuri, A. (1976). A non-negativity criterion for a certain variance estimator. *Metrika, 23*, 201—205.

Chaudhuri, A. (1981). Non-negative unbiased variance estimators. In D. Krewski, R. Platek, & J. N. K. Rao (Eds.), *Current topics in survey sampling* (pp. 317—328). New York: Academic Press.

Chaudhuri, A. (1987). Randomized response surveys of finite population: A unified approach with quantitative data. *Journal of Statistical Planning and Inference, 15*, 157—165.

Chaudhuri, A. (2011). *Randomized response and indirect questioning techniques in surveys*. New York: Chapman & Hall/CRC.

Chaudhuri, A., & Arnab, R. (1978). On the role of sample-size in determining efficiency of Horvitz-Thompson estimators. *Sankhyā, C, 40*, 104—109.

Chaudhuri, A., & Arnab, R. (1979a). On the relative efficiencies of the sampling strategies under a superpopulation model. *Sankhyā, C, 41*, 40–53.
Chaudhuri, A., & Arnab, R. (1979b). On estimating the mean of a finite population on two occasions with varying probabilities. *Australian Journal of Statistics, 21*, 162–165.
Chaudhuri, A., & Arnab, R. (1981). On non-negative variance estimation. *Metrika, 28*, 1–12.
Chaudhuri, A., & Arnab, R. (1982a). On unbiased product type estimators. *Journal of the Indian Society of Agricultural Statistics, 34*, 65–70.
Chaudhuri, A., & Arnab, R. (1982b). On unbiased variance estimation with various multistage sampling strategies. *Sankhyā, B, 44*, 92–101.
Chaudhuri, A., & Dihidar, K. (2009). Estimating means of stigmatizing qualitative and quantitative variables from discretionary responses randomized or direct. *Sankhyā, B, 71*, 123–136.
Chaudhuri, A., & Mukherjee, R. (1988). *Randomized response: Theory and techniques*. New York: Marcel Dekker.
Chaudhuri, A., & Saha, A. (2005). Optional versus compulsory randomized response techniques in complex surveys. *Journal of Statistical Planning and Inference, 135*, 516–527.
Chaudhuri, A., & Stenger, H. (1992). *Survey sampling theory and methods*. New York: Marcel Dekker.
Chen, Z. (1999). Density estimation using ranked-set sampling data. *Environmental and Ecological Statistics, 6*, 135–146.
Chen, J., & Quin, J. (1993). Empirical likelihood estimation for finite population and the effective usages of auxiliary information. *Biometrika, 80*, 107–116.
Chen, J., Rao, J. N. K., & Sitter, R. R. (2000). Efficient random imputation for missing survey data in complex surveys. *Statistica Sinica, 10*, 1153–1159.
Chen, J., & Sitter, R. R. (1999). A pseudo empirical likelihood approach to the effective use of auxiliary information in complex surveys. *Statistica Sinica, 9*, 385–406.
Chen, J., & Wu, C. (2002). Estimation of distribution function and quantiles using the model-calibrated pseudo empirical likelihood method. *Statistica Sinica, 12*, 1223–1239.
Chotai, J. (1974). A note on Rao-Hartley-Cochran method for pps sampling over two occasions. *Sankhyā, C, 36*, 173–180.
Chottopadhyaya, M., Lahiri, P., Laren, M., & Reimnitz, J. (1999). Composite estimation of drug preferences for sub-state areas. *Survey Methodology, 25*, 81–86.
Christofides, T. C. (2003). A generalized randomized response technique. *Metrika, 57*, 195–200.
Cochran, W. G. (1946). Relative accuracy of systematic and stratified random samples for a certain class of population. *Annals of Mathematical Statistics, 17*, 164–177.
Cochran, W. G. (1961). Comparison of methods for determining stratum boundaries. *Bulletin of the International Statistical Institute, 38*, 345–358.
Cochran, W. G. (1977). *Sampling techniques* (3rd ed.). New York: Wiley.
Cohen, J. E. (1976). The distribution of chi-squared statistics under cluster sampling from contingency tables. *Journal of the American Statistical Association, 71*, 665–670.
Cormack, R. M. (1981). Loglinear models for capture-recapture experiments on open populations. In R. W. Horons, & D. Cooke (Eds.), *The mathematical theory of the biological populations II* (pp. 217–235). London: Academic Press.
Cox, D. R. (1969). Some sampling problems in technology. In N. L. Johnson, & J. R. Smith (Eds.), *New developments in survey sampling* (pp. 506–527). New York: Wiley.
Czaja, R., Warnecke, R. B., Eastman, E., Royston, P., Sirken, M., & Tuteur, D. (1984). Locating patients with rare diseases using network sampling: Frequency and quality of reporting. In *Health survey research methods: Proceedings of the fourth conference on health survey research methods* (pp. 311–324). Washington, DC: Department of Health and Human Services.

Dalenius, T. (1953). Multivariate sampling problem. *Skandinavisk Actuarietidskrift, 36*, 92–102.

Dalenius, T. (1955). The problem of not-at-homes. *Statistisk Tidskrift, 4*, 208–211.

Dalenius, T., & Gurney, M. (1951). The problem of optimum stratification II. *Scandinavian Actuarial Journal, 34*, 133–148.

Dalenius, T., & Hodges, J. L. (1959). Minimum variance stratification. *Journal of the American Statistical Association, 54*, 88–101.

Das, A. C. (1951). Systematic sampling. *Bulletin of the International Statistical Institute, 33*, 119–132.

Datta, G. S., Day, B., & Basawa, I. (1999). Empirical best linear unbiased and empirical bayes prediction in multivariate small area estimation. *Journal of Statistical Planning and Inference, 75*, 269–279.

Datta, G. S., & Ghosh, M. (1991). Bayesian prediction in linear models: Applications to small area estimation. *Annals of Statistics, 19*, 1748–1770.

Datta, G. S., Lahiri, P., & Maiti, T. (2002). Empirical Bayes estimation of median income of four-person families by state using time series and cross-sectional data. *Journal of Statistical Planning and Inference, 102*, 83–97.

De Pascal, N. (1961). Unbiased ratio estimators in stratified sampling. *Journal of the American Statistical Association, 56*, 70–87.

Dell, T. R., & Clutter, J. L. (1972). Ranked-set sampling theory with order statistic background. *Biometrics, 28*, 545–555.

Deming, W. E. (1953). On probability mechanism to attain an economic balance between the resulting error of response bias of non-response. *Journal of the American Statistical Association, 48*, 743–772.

Deming, W. E. (1977). An essay on screening, or on two-phase sampling, applied to surveys of a community. *International Statistical Review, 45*, 29–37.

Deming, W. E., & Steaphan, F. F. (1940). On a least squares adjustment of a sample frequency when the expected marginal totals are known. *Annals of Mathematical Statistics, 11*, 427–444.

Deville, J. C., & Särndal, C. E. (1992). Calibration estimation in survey sampling. *Journal of the American Statistical Association, 87*, 376–382.

Devore, J. L. (1977). A note on the RR techniques. *Communications in Statistics – Theory and Methods, 6*, 1525–1529.

Dorfman, A. H. (2009). Inference on distribution functions and quantiles. In D. Pfeffermann, & C. R. Rao (Eds.), *Handbook of statistics, sample surveys: Inference and analysis* (Vol. 29B, pp. 371–395). Amsterdam, North-Holland: Elsevier.

Dorfman, A. H., & Hall, P. (1993). Estimators for the finite population distribution function using nonparametric regression. *Annals of Statistics, 21*, 1452–1475.

Drew, D., Singh, M. P., & Choudhry, G. H. (1982). Evaluation of small area estimation techniques for Canadian Labour Force Survey. *Survey Methodology, 8*, 17–47.

Duncan, G. J., & Kalton, G. (1987). Issues of design and analysis of surveys across time. *International Statistical Review, 55*, 97–117.

Durbin, J. (1959). A note on the application of Quenouille's method of bias reduction to the estimation of ratios. *Biometrika, 46*, 477–480.

Durbin, J. (1960). Estimation of parameters in time series regression models. *Journal of the Royal Statistical Society, Series B, 22*, 139–153.

Durbin, J. (1967). Designs of multi-stage survey for estimation of sampling error. *Applied Statistics, 16*, 152–164.

Ecler, A. R. (1955). Rotation sampling. *Annals of Mathematical Statistics, 26*, 664–685.

Efron, B. (1979). Bootstrap method: Another look of jackknife. *Annals of Statistics, 7*, 1–26.

Efron, B. (1982). *The jackknife, the bootstrap and other resampling plans*. Philadelphia: Society for Industrial and Applied Mathematics.

Eichhorn, B. H., & Hayre, L. S. (1983). Scrambled randomized response methods for obtaining sensitive quantitative data. *Journal of Statistical Planning and Inference, 7*, 306–316.

El-Badry, M. A. (1956). A sampling procedure for mail questionnaires. *Journal of the American Statistical Association, 51*, 209–227.

Ericson, W. A. (1969a). Subjective Bayesian in sampling finite populations (with discussion). *Journal of the Royal Statistical Society, Series B, 31*, 195–233.

Ericson, W. A. (1969b). A note on the posterior mean of a population mean. *Journal of the Royal Statistical Society, Series B, 31*, 332–334.

Ericson, W. A. (1970). On a class uniformly admissible estimators of a finite population total. *Annals of Mathematical Statistics, 41*, 1369–1372.

Eriksson, S. A. (1973). A new model for randomized response. *International Statistical Review, 41*, 40–43.

Ericksen, E. P., & Kadane, J. B. (1985). Estimating the population census year: 1980 and beyond (with discussions). *Journal of the American Statistical Association, 84*, 927–944.

Fay, R. E. (1985). A jackknifed chi-squared test for complex samples. *Journal of the American Statistical Association, 80*, 148–157.

Fay, R. E. (1989). Theory and application of replicate weighting for variance calculations. In *Proceedings of the survey research methods section of the American Statistical Association* (pp. 495–500).

Fay, R. E., & Herriot, R. A. (1979). Estimating of income from small places: An application of James – Stein procedure to census data. *Journal of the American Statistical Association, 74*, 269–277.

Fellegi, I. P. (1963). Sampling with varying probabilities without replacement, rotation and non-rotating samples. *Journal of the American Statistical Association, 58*, 183–201.

Fellegi, I. P. (1966). Changing the probabilities of selection when two units are selected with PPS without replacement. In *Proceedings of the survey research methods section of the American Statistical Association* (pp. 434–442).

Fellegi, I. P. (1980). Approximate goodness of fit based on stratified multi-stage samples. *Journal of the American Statistical Association, 75*, 261–278.

Feller, W. (1957). *An introduction to probability theory and its applications* (Vol. 1). New York: Wiley.

de Finetti, B. (1937). La prévision: Ses lois logiques, ses sources subjectives. *Annales Institute Henri Poincare, 7*, 1–68.

Flinger, M. A., Policello, G. E., & Singh, J. (1977). A comparison of two RR survey methods with consideration for the level of respondent protection. *Communications in Statistics – Theory and Methods, 6*, 1511–1524.

Folsom, S. A. (1973). The two alternative questions randomized response model for human surveys. *Journal of the American Statistical Association, 68*, 525–530.

Foody, W., & Hedayat, A. (1977). On theory and applications of BIBD designs and repeated blocks. *Annals of Statistics, 5*, 932–945.

Franklin, L. A. (1989). A comparison of estimators for randomized response sampling with continuous distribution from dichotomous populations. *Communications in Statistics – Theory and Methods, 18*, 489–505.

Fuller, W. A. (1975). Regression Analysis for sample surveys. *Sankhyä, C, 37*, 117–132.

Fuller, W. A., & Battese, G. (1973). Transformations for estimation of linear models with nested error structure. *Journal of the American Statistical Association, 68*, 626–632.

Fuller, W. A., & Burmeister, L. F. (1972). Estimators for samples selected for two over-sampling frames. In *Proceedings of the social statistics, American Statistical Association* (pp. 245–249).

Gabler, S. (1987). The nearest proportional to size sampling design. *Communications in Statistics – Theory and Methods, 16*, 1117–1131.

Gautschi, W. (1957). Some remarks on systematic sampling. *Annals of Mathematical Statistics, 28*, 385–394.

Ghangurde, P. D., & Rao, J. N. K. (1969). Some results on sampling over two occasions. *Sankhyā, A, 31*, 463–472.

Ghosh, M. (1991). Estimating functions in survey sampling: A review. In V. P. Godambe (Ed.), *Estimating functions* (pp. 201–209). Oxford: Oxford University Press.

Ghosh, M. (1992). Constrained Bayes estimation with applications. *Journal of the American Statistical Association, 87*, 533–540.

Ghosh, M., & Lahiri, P. (1987). Robust empirical Bayes estimation of means from stratified samples. *Journal of the American Statistical Association, 82*, 1153–1162.

Ghosh, M., & Lahiri, P. (1998). Bayes and empirical Bayes analysis in multistage sampling. In S. S. Gupta, & J. O. Berger (Eds.), *Statistical decision theory and related topics IV* (Vol. 1, pp. 195–212). New York: Springer.

Ghosh, M., & Rao, J. N. K. (1994). Small area estimation: An appraisal. *Statistical Science, 9*, 55–93.

Ghosh, S. P. (1963). Post cluster sampling. *Annals of Mathematical Statistics, 34*, 578–597.

Godambe, V. P. (1955). A unified theory of sampling from finite populations. *Journal of the Royal Statistical Society, Series B, 17*, 269–278.

Godambe, V. P. (1960a). An admissible estimate for any sampling design. *Sankhyā, 22*, 285–288.

Godambe, V. P. (1960b). An optimum property of regular maximum likelihood estimation. *Annals of Mathematical Statistics, 31*, 1208–1211.

Godambe, V. P. (1966). A new approach to sampling from finite populations, I and II. *Journal of the Royal Statistical Society, Series B, 28*, 310–328.

Godambe, V. P. (1968). Bayesian sufficiency in sampling. *Annals of the Institute of Statistical Mathematics, 20*, 363–373.

Godambe, V. P. (1969). Admissibility and Bayes estimation in sampling finite population-V. *Annals of Mathematical Statistics, 40*, 672–676.

Godambe, V. P. (1985). The foundations of finite sample estimation in stochastic processes. *Biometrika, 72*, 419–428.

Godambe, V. P. (1991). Confidence interval for quantiles. In V. P. Godambe (Ed.), *Estimating functions* (pp. 211–215). Oxford: Oxford University Press.

Godambe, V. P., & Joshi, V. M. (1965). Admissibility and Bayes estimation in sampling finite population I. *Annals of Mathematical Statistics, 36*, 1707–1722.

Godambe, V. P., & Kale, B. K. (1991). Estimating functions: An overview. In V. P. Godambe (Ed.), *Estimating functions* (pp. 3–20). Oxford: Oxford University Press.

Godambe, V. P., & Thompson, M. E. (1978). Some aspects of the theory of estimating equations. *Journal of Statistical Planning and Inference, 2*, 95–104.

Godambe, V. P., & Thompson, M. E. (1984). Robust estimation through estimating equations. *Biometrika, 71*, 115–125.

Godambe, V. P., & Thompson, M. E. (1986a). Parameters of superpoulation and survey population, their relationship and estimation. *International Statistical Review, 54*, 127–138.

Godambe, V. P., & Thompson, M. E. (1986b). Some optimality results in presence of non-response. *Survey Methodology, 12*, 29–36.

Godambe, V. P., & Thompson, M. E. (1987). Corrigendum. *Survey Methodology, 13*, 123.

Godambe, V. P., & Thompson, M. E. (1999). A new look at confidence intervals in survey sampling. *Survey Methodology, 25*, 161–173.

Godambe, V. P., & Thompson, M. E. (2009). Estimating functions and survey sampling. In D. Pfeffermann, & C. R. Rao (Eds.), *Handbook of statistics, sample surveys: Inference and analysis* (Vol. 29B, pp. 669–687). Amsterdam, North-Holland: Elsevier.

Gonzalez, M. E. (1973). Use and evaluation of synthetic estimators. In *Proceedings of the survey research methods section of the American Statistical Association* (Vol. 82, pp. 1153–1162).

Goodman, L. A. (1960). On the exact variance of product. *Journal of the American Statistical Association, 55*, 708–713.

Goodman, L. A., & Hartley, H. O. (1958). The precision of unbiased ratio-type estimators. *Journal of the American Statistical Association, 53*, 491–508.

Goodman, R., & Kish, L. (1950). Controlled selection—a technique in probability sampling. *Journal of the American Statistical Association, 45*, 350–372.

Goodstadt, M. S., & Gruson, V. (1975). The randomized response technique; a test on drug use. *Journal of the American Statistical Association, 70*, 814–818.

Govindrajulu, Z. (1999). *Elements of sampling theory and method.* NJ: Prentice-Hall.

Gray, G., & Platek, R. (1963). Several methods of re-designing area samples utilizing probabilities proportion to size change significantly. *Journal of the American Statistical Association, 63*, 1280–1297.

Gray, H. L., & Schucany, W. R. (1972). *The generalized jackknife statistics.* New York: Marcel Deckker.

Greenberg, B. D., Abul-Ela, A. L. A., Simmons, W. R., & Horvitz, D. G. (1969). The unrelated question randomized response model: Theoretical framework. *Journal of the American Statistical Association, 64*, 520–539.

Gross, S. T. (1980). Median estimation in sample surveys. In *Proceedings of the survey research methods section of the American Statistical Association* (pp. 181–184).

Gunning, P., & Horgan, J. M. (2004). A simple algorithm for stratifying skewed populations. *Survey Methodology, 30*, 159–166.

Gupta, S. (2001). Qualifying the sensitivity level of binary response personal interview survey questions. *Journal of Combinatorics, Information & System Sciences, 26*(1–4), 101–109.

Gupta, S., Gupta, B., & Singh, S. (2002). Estimation of sensitivity level of personal interview survey question. *Journal of Statistical Planning and Inference, 100*, 239–247.

Gupta, S., Mehta, S., Shabbir, J., & Dass, B. K. (2013). Generalized scrambling in quantitative optional randomized response models. *Communications in Statistics – Theory and Methods, 42*, 4034–4042.

Gupta, S., & Shabbir, J. (2004). Sensitivity estimation for personal interview survey questions. *Statistica, 64*(3), 643–653.

Gupta, S., Shabbir, J., & Sehra, S. (2010). A comparison of multiplicative and additive optional RRT models. *Journal of Statistical Planning and Inference, 140*(10), 2870–2874.

Gupta, S., Thornton, B., Shabbir, J., & Singhal, S. (2006). A comparison of multiplicative an additive optional RRT models. *Journal of Statistical Theory and Applications, 64*, 226–239.

Gupta, V. K., Mandal, B. N., & Prasad, R. (2012). *Combinatorics in sample surveys vis-à-vis controlled selection.* Germany: Lambert Academic Publishing Company.

Gupta, V. K., Nigam, A. K., & Kumar, P. (1982). On a family of sampling scheme with inclusion probability proportional to size. *Biometrika, 69*, 191–196.

Haitovsky, Y. (1973). *Maximum joint probability estimates of the linear hierarchical model.* Unpublished paper. Hebrew University.

Hájek, J. (1959). Optimum strategy and other problems in probability sampling. *Časopis Pro Pěstování Matematiky, 84*, 387–423.

Hájek, J. (1964). Asymptotic theory of rejective sampling with varying probabilities from finite population. *Annals of Mathematical Statistics, 35*, 1491–1523.

Hájek, J. (1971). Discussion of 'an essay on the logical foundations of survey sampling, part one' by D. Basu. In V. P. Godambe, & D. A. Sprott (Eds.), *Foundations of statistical inference* (p. 236). Toronto, ON, Canada: Holt, Rinehart and Winston.

Haldane, J. B. S. (1946). On method of estimating frequencies. *Biometrika, 33*, 222–225.

Hansen, M. H., & Hurwitz, W. N. (1943). On the theory of sampling from finite populations. *Annals of Mathematical Statistics, 14*, 333–362.

Hansen, M. H., & Hurwitz, W. N. (1946). The problems on non-response in sample surveys. *Journal of the American Statistical Association, 41*, 517–529.

Hansen, M. H., Madaw, W. G., & Tepping, B. J. (1983). An evaluation of model-dependent and probability-sampling inference in sample surveys. *Journal of the American Statistical Association, 78*, 776–793.

Hanurav, T. V. (1965). *Optimum sampling strategies and some related problems* (Ph.D. thesis). Indian Statistical Institute.

Hanurav, T. V. (1966). Some aspects of unified sampling theory. *Sankhyā, A, 28*, 175–204.

Hanurav, T. V. (1967). Optimum utilization of auxiliary information: πps sampling of two units from a stratum. *Journal of the Royal Statistical Society, Series B, 29*, 374–391.

Hartley, H. O. (1962). Multiple frame surveys. In *Proceedings of the survey research methods section of the American Statistical Association* (pp. 203–206).

Hartley, H. O. (1974). Multiple frame methodology and selected applications. *Sankhyā, C, 36*, 99–118.

Hartley, H. O., & Rao, J. N. K. (1962). Sampling with unequal probabilities and without replacement. *Annals of Mathematical Statistics, 33*, 350–374.

Hartley, H. O., & Rao, J. N. K. (1968). A new estimation theory for sample surveys. *Biometrika, 55*, 547–557.

Hartley, H. O., & Ross, A. (1954). Unbiased ratio estimators. *Nature, 174*, 270–271.

Haziza, D., Hidiroglou, M. A., & Rao, J. N. K. (2011). Comparison of variance estimators in two-phase sampling: An empirical investigation. *Pakistan Journal of Statistics, 27*, 477–492.

Hedayat, A., & Kageyama, S. (1980). The family of t-designs part I. *Journal of Statistical Planning and Inference, 4*, 173–212.

Hedayat, A. S., Rao, C. R., & Stufken, J. (1988). Sampling plan excluding contiguous units. *Journal of Statistical Planning and Inference, 19*, 159–170.

Hedayat, A. S., & Stufken, J. (1989). The construction of IPPS sampling designs through a method of emptying boxes. *Annals of Statistics, 17*, 1886–1905.

Henderson, C. R. (1975). Best linear unbiased estimation and prediction under a selection model. *Biometrics, 31*, 423–447.

Hendricks, W. A. (1944). The relative efficiencies of group of farms as sampling units. *Journal of the American Statistical Association, 39*, 336–376.

Heyde, C. C., & Lin, Y.-X. (1991). Approximate confidence zones in an estimating function context. In V. P. Godambe (Ed.), *Estimating functions* (pp. 161–168). New York: Oxford University Press.

Hidiroglou, M. A., Fuller, W. A., & Hickman, R. D. (1980). *SUPERCARP* (6th ed.). Ames, IA: Statistical Laboratory, Survey Section, Iowa State University.

Hidiroglou, M. A., & Rao, J. N. K. (1987). Chi-squared tests with categorical data from complex surveys: Part I-simple goodness-of fit, homogeneity and independency in a two-way table with applications to the Canada Health Survey (1978–1979). *Journal of Official Statistics, 3*, 117–132.

Holt, D., & Scott, A. J. (1981). Regression analysis using survey data. *Statistician, 30*, 169–178.

Hong, Z., & Yan, Z. (2012). Measure of privacy in randomized response model. *Quality and Quantity, 46*, 1167–1180.

Horvitz, D. G., Shah, B. V., & Simmons, W. R. (1967). The unrelated question randomized response model. In *Proceedings of the survey research methods section of the American Statistical Association* (pp. 65–72).

Horvitz, D. G., & Thompson, D. J. (1952). A generalization of sampling without replacement from a finite population. *Journal of the American Statistical Association, 47*, 663–685.

Huang, K. C. (2004). A survey technique for estimating the proportion and sensitivity in a dichotomous finite population. *Statistica Neeralandica, 58*, 75–82.

Huang, K. C. (2008). Estimation of sensitive characteristics using optional randomized technique. *Quality and Quantity, 42*, 679–686.

Jessen, R. J. (1942). Statistical investigation of sample survey for obtaining farm facts. *Iowa Agricultural Experiment Station, Research Bulletin, 304*.

Jiang, J., Lahiri, P., & Wan, S.-M. (2002). A unified Jackknife theory. *Annals of Statistics, 30*, 1782–1810.

Johnson, A. (2003). *Estimating distribution functions from survey data using nonparametric regression*. Retrieved from www.stat.colostate.edu/~nsu/starmap/johnsonaa.report.pdf.

Jojani, M. J., & Johnson, B. C. (2011). Design based estimation for ranked set sampling in finite population. *Environmental and Ecological Statistics, 18*, 663–685.

Jolly, G. M. (1965). Explicit estimates from capture-recapture data with both death and immigration-stochastic models. *Biometrika, 52*, 226–247.

Jönrup, H., & Rennermalm, B. (1976). Regression analysis in samples from finite populations. *Scandinavian Journal of Statistics, 3*, 33–37.

Joshi, V. M. (1965a). Admissibility and Bayes estimation in sampling finite population II. *Annals of Mathematical Statistics, 36*, 1723–1729.

Joshi, V. M. (1965b). Admissibility and Bayes estimation in sampling finite population III. *Annals of Mathematical Statistics, 36*, 1730–1742.

Joshi, V. M. (1966). Admissibility and Bayes estimation in sampling finite populations IV. *Annals of Mathematical Statistics, 37*, 1658–1670.

Jowette, H. H. (1952). They accuracy of systematic sampling from conveyor belts. *Applied Statistics, 1*, 50–59.

Judkins, R. (1990). Fay's method for variance estimation. *Journal of Official Statistics, 6*, 223–239.

Kadilar, C., Unyazici, Y., & Cingi, H. (2009). Ratio estimator for the population mean using ranked set sampling. *Statistical Papers, 50*, 301–309.

Kalbfleisch, J. D., & Lawless, J. (1988). Estimation of reliability in field performance studies. *Technometrics, 30*, 365–388.

Kale, B. K. (1962). An extension of the Cramer-Rao inequality for statistical estimation function. *Skandinaviske Akturietidskrift, 45*, 60–89.

Kalton, G. (1991). Sampling flows of mobile human populations. *Survey Methodology, 17*, 183–194.

Kalton, G. (1993). Sampling considerations in research on HIV risk and illness. In D. G. Ostrow, & R. C. Kessler (Eds.), *Methodological issues in AIDS behavioural research*. New York: Plenum Press.

Kalton, G. (2001). Practical methods for sampling rare and elusive populations. In *Proceedings of the American Statistical Association*.

Kalton, G. (2009). Methods for oversampling rare subpopulations in social surveys. *Survey Methodology, 35*, 126–141.

Kalton, G., & Anderson, D. W. (1986). Sampling rare populations. *Journal of the Royal Statistical Society, Series A, 149*, 65–82.

Kass, R. E., & Steffey, D. (1989). An approximate Bayesian inference in conditionally independent hierarchical models (parametric empirical Bayes models). *Journal of the American Statistical Association, 84*, 717–726.

Kaur, A., Patil, G. P., & Taillie, C. (1997). Unequal allocation model for ranked set sampling with skew distribution. *Biometrics, 53*, 123–130.

Kempthorne, O. (1969). Some remark on statistical inference in finite sampling. In N. L. Johnson, & H. Smith, Jr. (Eds.), *New development in survey sampling* (pp. 671–695). New York: Wiley.

Kerkvliet, J. M. (1994). Estimating a logit model with randomized data; the case of cocaine use. *Australian Journal of Statistics, 36*, 9–20.
Keyfitz, N. (1951). Sampling with probabilities proportional to size, adjustment for changes in the probabilities. *Journal of the American Statistical Association, 46*, 105–109.
Kim, J. M. (1978). *Randomized response technique for surveying human populations* (Ph.D. dissertation). Philadelphia, USA: Temple University.
Kim, J. M., Tebbs, J., & An, S. W. (2006). Extensions of Mangat's randomized response model. *Journal of Statistical Planning and Inference, 136*, 1554–1567.
Kim, J. W. (2004). Finite sample properties of multiple imputation estimators. *Annals of Statistics, 32*, 766–783.
Kim, J. W. (2009). Calibration estimation using empirical likelihood in survey sampling. *Statistica Sinica, 19*, 145–157.
Kish, L. (1963). Changing strata and selection probabilities. In *Proceedings of the social statistics section of the American Statistical Association, Washington* (pp. 124–131).
Kish, L. (1965). *Survey sampling*. New York: Wiley.
Kish, L. (1991). Taxonomy of elusive populations. *Journal of Official Statistics, 7*, 339–347.
Kish, L., & Frankel, M. R. (1974). Inference from complex surveys (with discussions). *Journal of the Royal Statistical Society, Series B, 36*, 1–37.
Kish, L., & Hess, I. (1958). On non coverage of sampling dwellings. *Journal of the American Statistical Association, 53*, 509–524.
Kish, L., & Scott, A. (1971). Retaining units after changing strata and probabilities. *Journal of the American Statistical Association, 66*, 461–470.
Kleffe, J., & Rao, J. N. K. (1992). Estimation of mean square error of empirical best linear unbiased predictors under a random error variance linear model. *Journal of Multivariate Analysis, 43*, 1–15.
Konijn, H. (1962). Regression analysis in sample surveys. *Journal of the American Statistical Association, 57*, 590–605.
Koop, J. C. (1971). On splitting systematic sample for variance estimation. *Annals of Mathematical Statistics, 42*, 1084–1087.
Koop, J. C. (1976). *Systematic sampling in two dimensional surfaces and related problems*. Technical Report. NC: Research Triangle Institute.
Koti, K. M., & Babu, G. J. (1996). Sign test for ranked-set sampling. *Communications in Statistics – Theory and Methods, 25*(7), 1617–1630.
Kreweski, D., & Chakrabarti, R. P. (1981). On the stability of the jackknife variance estimator in ratio estimation. *Journal of Statistical Planning and Inference, 5*, 71–79.
Kreweski, D., & Rao, J. N. K. (1981). Inference from stratified samples: Properties of linearization, jackknife and balanced repeated replication methods. *Annals of Statistics, 9*, 1010–1019.
Kuk, A. Y. C. (1988). Estimation of distribution functions and medians under sampling with unequal probabilities. *Biometrika, 75*, 97–103.
Kuk, A. Y. C. (1990). Asking sensitive question indirectly. *Biometrika, 77*, 436–438.
Kuk, A. Y. C. (1993). A kernel method for estimating finite population distribution functions using auxiliary information. *Biometrika, 80*, 385–392.
Kuk, A. Y. C., & Mak, T. K. (1989). Median estimation in presence of auxiliary information. *Journal of the Royal Statistical Society, Series B, 51*, 261–269.
Kuldoroff, G. (1963). Some problems of optimum allocation for sampling on two occasions. *Review of the International Statistical Institute, 31*, 24–57.
Kuo, L. (1988). Classical and prediction approaches to estimating distribution functions from survey data. In *Proceedings of the section on survey research methods, American Statistical Association* (pp. 280–285).
Lahiri, D. B. (1951). A method of sample selection for providing unbiased ratio estimates. *Bulletin of the International Statistical Institute, 33*, 133–140.

Lahiri, D. B. (1954). On the question of bias of systematic sampling. In *Proceedings of world population conference* (Vol. 6, pp. 349–362).
Lahiri, P. (1990). "Adjusted" Bayes and empirical Bayes estimation in finite population sampling. *Sankhyā, B, 52*, 50–60.
Lahiri, P., & Mukherjee, R. (2000). On simplification of the linear programming approach to controlled sampling. *Statistica Sinica, 10*, 1171–1178.
Lanke, J. (1974a). On non-negative variance estimation in survey sampling. *Sankhyā, C, 36*, 33–42.
Lanke, J. (1974b). *Some contribution to the theory of survey sampling*. AV-Centralen I Lund.
Lanke, J. (1975a). On the choice of unrelated questions in Simmons' version of RR. *Journal of the American Statistical Association, 68*, 525–530.
Lanke, J. (1975b). *Some contribution to the theories of survey sampling*. Sweden: University of Lund (Unpublished Ph.D. thesis).
Lanke, J. (1976). On the degree of protection in randomized interviews. *International Statistical Review, 44*, 197–203.
Lehtonen, R., & Pahkinen, E. (2004). *Practical methods for design analysis of complex surveys*. New York: Wiley.
Lehtonen, R., Särandal, C. E., & Veijanen, A. (2003). The effect of model choice in estimation for domains. *Survey Methodology, 29*, 33–44.
Leysieffer, R. W., & Warner, S. L. (1976). Respondent jeopardy and optimal designs in RR models. *Journal of the American Statistical Association, 71*, 649–656.
Li, D., Sinha, B. K., & Perron, F. (1999). Random selection in ranked set sampling and its applications. *Journal of Statistical Planning and Inference, 76*, 185–201.
Lincoln, F. C. (1930). *Calculating waterfowl abundance on the basis of banding returns. 118* pp. 1–4). United States Department of Agriculture Circular.
Liu, T. P., & Chow, L. P. (1976). A new discrete quantitative RR model. *Journal of the American Statistical Association, 64*, 520–539.
Lohr, S. (1999). *Sampling: Design and analysis* (2nd ed.). Pacific Grove, CA: Duxbury Press.
Lund, R. E. (1968). Estimation in multiple frame surveys. In *Proceedings of the social statistics of the American Statistical Association* (pp. 282–288).
MacKellar, D., Valleroy, L., Karon, J., Lemp, G., & Janssen, R. (1996). The Young Men's Survey: Methods for estimating HIV seroprevalence and risk factors among young men who have sex with men. *Public Health Reports, 111*(Suppl. 1), 138–144.
Madaw, W. G. (1949). On the theory of systematic sampling II. *Annals of Mathematical Statistics, 20*, 333–354.
Madaw, W. G., & Madaw, L. H. (1944). On the theory of systematic sampling. *Annals of Mathematical Statistics, 15*, 1–24.
Maddala, G. S. (1977). *Econometrics*. New York: McGraw-Hill.
Mahalanobis, P. C. (1940). A sample survey of acreage under jute in Bengal. *Sankhyā, 4*, 511–530.
Mahalanobis, P. C. (1942). *General report on the sample census of area under jute in Bengal*. Indian Central Jute Committee.
Mahalanobis, P. C. (1946). Recent experiment in statistical sampling in the Indian statistical institute. *Journal of the Royal Statistical Society, Series A, 109*, 325–378.
Mahalanobis, P. C. (1952). Some aspects of the design of sample surveys. *Sankhyā, 12*, 1–7.
Mak, T. K., & Kuk, A. Y. C. (1993). A new method for estimating finite population quantiles using auxiliary information. *The Canadian Journal of Statistics, 21*, 29–38.
Mandal, B. N., Prasad, R., & Gupta, V. K. (2008). Computer aided construction of balanced sampling plans excluding contiguous units. *Statistics and Applications, 3*, 59–85.
Mandal, B. N., Prasad, R., & Gupta, V. K. (2010). Linear programming approach to construct distance balanced sampling plan. *Journal of the Indian Society of Agricultural Statistics, 64*, 303–312.

Mandal, B. N., Prasad, R., & Gupta, V. K. (2011). Construction of polygonal designs using linear integer programming. *Communications in Statistics — Theory and Methods, 40*, 1787—1794.

Mandal, B. N., Prasad, R., Gupta, V. K., & Sud, U. C. (2009). A family of distance balanced sampling plans. *Journal of Statistical Planning and Inference, 139*, 860—874.

Mangat, N. S., & Singh, R. (1990). An alternative randomized response procedure. *Biometrika, 77*, 349—442.

Mangat, N. S., & Singh, S. (1994). An optional randomized response sampling technique. *Journal of the Indian Society of Agricultural Statistics, 32*, 71—75.

Mantel, H. (1991). Making use of a regression model for inference about a finite population mean. In V. P. Godambe (Ed.), *Estimating functions* (pp. 216—221). New York: Oxford University Press.

Matei, A., & Tillé, Y. (2005). Maximal and minimal sampling co-ordination. *Sankhyā, 67*, 590—612.

McCarthy, P. J. (1969). Pseudo-replication: Half samples. *International Statistical Review, 37*, 239—264.

McDonald, L. L. (1980). Line-intercept sampling for attributes other than coverage and density. *Journal of Wildlife Management, 44*, 530—533.

McIntyre, G. A. (1952). A method of unbiased selective sampling, using ranked sets. *Australian Journal of Agricultural Research, 3*, 385—390.

Mckenzie, D. J., & Mistiaen, J. (2009). Surveying migrant households: A comparison of census-based, snowball and intercept point surveys. *Journal of the Royal Statistical Society, Series A, 172*, 339—360.

Midzuno, H. (1952). On the sampling system with probabilities proportionate to sum of sizes. *Annals of the Institute of Statistical Mathematics, 3*, 99—107.

Miller, R. G. (1974). The Jackknife — a review. *Biometrika, 61*, 1—18.

Morris, C. A. (1983). Parametric empirical Bayes inference: Theory and applications. *Journal of the American Statistical Association, 78*, 47—54.

Mukherjee, R., & Sengupta, S. (1989). Optimal estimation of finite population total under a general correlated model. *Biometrika, 76*, 789—794.

Murthy, M. N. (1957). Ordered and unordered estimators in sampling without replacement. *Sankyā, 18*, 379—390.

Murthy, M. N. (1964). Product method of estimation. *Sankhyā, 21*, 381—392.

Murthy, M. N. (1967). *Sampling theory and methods*. Calcutta: Statistical Publishing Society.

Murthy, M. N. (1977). *Sampling theory and methods* (2nd ed.). Calcutta: Statistical Publishing Society.

Murthy, M. N., & Nanjamma, N. S. (1959). Almost unbiased ratio estimates based on interpenetrating sub-sample estimates. *Sankhyā, 21*, 381—392.

Murthy, M. N., & Rao, T. J. (1988). Systematic sampling with illustrative examples. In P. R. Krishnaiah, & C. R. Rao (Eds.), *Handbook of statistics* (Vol. 6, pp. 147—185). Amsterdam: Elsevier Science Publishers.

Muttlak, A. H., & McDonald, L. L. (1990). Ranked based sampling with size-based probability of selection. *Biometrics, 46*, 435—445.

Nandaraya, E. A. (1964). On estimating regression. *Theory of Probability and Its Applications, 9*, 141—142.

Narain, R. D. (1951). On sampling without replacement with varying probabilities. *Journal of the Indian Society of Agricultural Statistics, 3*, 169—174.

Nathan, G. (1976). An empirial study of response and sampling errors for multiplicity estimates with different counting rules. *Journal of the American Statistical Association, 71*, 808—815.

Nathan, G. (1988). Inference based on data from complex survey designs. In P. R. Krishnaiah, & C. R. Rao (Eds.), *Handbook of statistics* (Vol. 6, pp. 247—266). Amsterdam: Elsevier Science.

National Research Council. (2000). Small-area estimation of school-age children in poverty: Evaluation of current methodology. In C. F. Citro, & G. Kalton (Eds.), *Committee on national statistics*. Washington, DC: National Academy Press.

Nayak, T. (2007). On randomized response surveys for estimating a proportion. *Communications in Statistics — Theory and Methods, 23*, 3303—3321.

Neyman, J. (1934). On two different aspects of the representative method. *Journal of the Royal Statistical Society, 97*, 558—606.

Neyman, J. (1938). Contribution to the theory of sampling human populations. *Journal of the American Statistical Association, 33*, 101—116.

Nigam, A. K., Kumar, P., & Gupta, V. K. (1984). Some methods of inclusion probability proportional to size sampling. *Journal of the Royal Statistical Society, Series B, 46*, 546—571.

Olkin, I. (1958). Multivariate ratio estimation for finite population. *Biometrika, 43*, 154—163.

Owen, A. B. (1988). Empirical likelihood ratio confidence intervals for a single functional. *Biometrika, 75*, 237—249.

Owen, A. B. (2001). *Empirical likelihood*. New York: Chapman and Hall.

Pal, S. (2008). Unbiasedly estimating the total of a stigmatizing variable from a complex survey on permitting options for direct or randomized responses. *Statistical Papers, 49*, 157—164.

Patel, H. C., & Dharmadikari, S. K. (1978). Admissibility of Murthy's and Midzuno's estimators within the class of linear unbiased estimators of a finite population total. *Sankhyā, C, 40*, 21—28.

Pathak, P. K. (1961). On the evaluation of moments of distinct units in a sample. *Sankhyā, A, 23*, 415—420.

Pathak, P. K., & Rao, T. J. (1967). In admissibility of customary estimators in sampling over two occasions. *Sankhyā, A, 29*, 49—54.

Pathak, P. K., & Shukla, N. D. (1966). Non-negativity of a variance estimator. *Sankhyā, A, 28*, 41—46.

Patil, G. P., Sinha, A. K., & Taillie, C. (1993). Relative precision of ranked-set sampling: A comparison with the regression estimator. *Environmetrics, 4*, 399—412.

Patil, G. P., Sinha, A. K., & Taillie, C. (1995). Finite population corrections for ranked set sampling. *Annals of the Institute of Statistical Mathematics, 47*, 621—636.

Patterson, H. (1950). Sampling on successive occasions with partial replacement of units. *Journal of the Royal Statistical Society, Series B, 12*, 241—255.

Peterson, C. G. J. (1896). The yearly immigration of young plaice into the Limfjord form the German Sea. *Report of the Danish Biological Station, 6*, 1—48.

Pfeffermann, D. (2002). Small area estimation — new developments and directions. *International Statistical Review, 70*, 125—143.

Pfeffermann, D., & Nathan, G. (1981). Regression analysis of data from a cluster sample. *Journal of the American Statistical Association, 76*, 681—689.

Pfeffermann, D., & Smith, T. M. F. (1985). Regression models for grouped populations in cross-section surveys. *International Statistical Review, 53*, 37—59.

Plackett, R. L., & Burman, J. P. (1946). The design of optimum multifactorial experiments. *Biometrika, 33*, 305—325.

Platek, R., Rao, J. N. K., Särndal, C. E., & Singh, M. P. (1987). *Small area statistics*. New York: Wiley.

Platek, R., & Singh, M. P. (Eds.). (1986). *Small area statistics: Contributed papers*. Laboratory for Research in Statistics and Probability, Carleton University.

Politz, A., & Simmons, W. (1949). An attempt to get the "not-at-homes" into the sample without call-backs. *Journal of the American Statistical Association, 44*, 9—31.

Politz, A., & Simmons, W. (1950). Note on an attempt to get the not-at-homes into the sample without call-backs. *Journal of the American Statistical Association, 45*, 136—137.

Pollock, K. H. (1975). A K-sample tag-recapture model allowing for unequal survival and catchability. *Biometrika, 62*, 577–583.
Pollock, K. H., Nichols, J. D., Brownie, C., & Hines, J. E. (1990). Statistical inference for capture-recapture experiments. *Wildlife Monograph, 107.*
Porter, R. M. (1973). On the use of survey sample weights in the linear model. *Annals of Economic and Social Measurement, 2*, 141–158.
Prasad, N. G. N., & Rao, J. N. K. (1990). The estimation of mean squared errors of small area estimators. *Journal of the American Statistical Association, 85*, 163–171.
Purcell, N. J., & Kish, L. (1979). Estimation for small domain. *Biometrics, 35*, 365–384.
Purcell, N. J., & Kish, L. (1980). Postcensal estimates for local areas (or domains). *International Statistical Review, 48*, 3–18.
Quenouille, M. H. (1949). Problems in plane sampling. *Annals of Mathematical Statistics, 20*, 335–375.
Quenouille, M. H. (1956). Notes on bias in estimation. *Biometrika, 43*, 353–360.
Quin, J., & Lawless, J. (1994). Empirical likelihood and general estimating equations. *Annals of Statistics, 22*, 300–325.
Quin, Y., Rao, J. N. K., & Ren, Q. (2006). *Confidence intervals for parameters of the response variable in a linear model with missing data.* Technical Report. Ottawa, Canada: Laboratory for Research in Statistics and Probability, Carleton University.
Raghavarao, D. (1971). *Constructions and combinatorial problems in design of experiments.* New York: Wiley.
Raghavarao, D. (1978). On estimation problem in Warner's randomized response techniques. *Biometrics, 34*, 87–90.
Raj, D. (1956). Some estimators in sampling with varying probabilities without replacement. *Journal of the American Statistical Association, 51*, 269–284.
Raj, D. (1958). On the relative accuracy of some sampling techniques. *Journal of the American Statistical Association, 53*, 98–101.
Raj, D. (1965a). On a method of using multi-auxiliary in sample surveys. *Journal of the American Statistical Association, 60*, 270–277.
Raj, D. (1965b). Sampling over two occasions with probability proportional to size. *Annals of Mathematical Statistics, 36*, 327–330.
Raj, D. (1968). *Sampling theory*. New York: McGraw-Hill.
Ramachandran, G., & Rao, T. J. (1974). Allocation to strata and relative efficiencies of stratified and unstratified πps sampling schemes. *Journal of the Royal Statistical Society, Series B, 36*, 558–606.
Ramakrishnan, M. K. (1975). Choice of an optimum sampling strategy-I. *Annals of Statistics, 3*, 669–679.
Rao, C. R. (1971). Some aspects of statistical inference in problems of sampling from finite populations. In V. P. Godambe, & D. A. Sprott (Eds.), *Foundations of statistical inference* (pp. 177–202). Toronto: Holt, Rinehart and Winston.
Rao, J. N. K. (1961). On sampling with varying probabilities in sub-sampling designs. *Journal of the Indian Society of Agricultural Statistics, 13*, 211–217.
Rao, J. N. K. (1963). On two systems of unequal probability sampling. *Annals of the Institute of Statistical Mathematics, 15*, 67–72.
Rao, J. N. K. (1965). On two sample schemes of unequal probability sampling without replacement. *Journal of the Indian Society of Agricultural Statistics, 3*, 169–174.
Rao, J. N. K. (1966a). Alternative estimators in PPS sampling for multiple characteristics. *Sankhyā, A, 28*, 47–60.
Rao, J. N. K. (1966b). On the relative efficiency of some estimators in PPS sampling for multiple characteristics. *Sankhyā, A, 28*, 61–70.
Rao, J. N. K. (1969). Ratio and regression estimators. In N. L. Johnson, & H. Smith (Eds.), *New development in survey sampling* (pp. 213–234). New York: Wiley.

Rao, J. N. K. (1973). On double sampling for stratification and analytical surveys. *Biometrika, 60*, 125–133.

Rao, J. N. K. (1975). Unbiased variance estimation for multi-stage designs. *Sankhyā, C, 37*, 133–139.

Rao, J. N. K. (1979). On deriving mean square errors and their non-negative unbiased estimators in finite population sampling. *Journal of Indian Statistical Association, 17*, 125–136.

Rao, J. N. K. (1994). Estimating totals and distribution functions using auxiliary information at the estimation stage. *Journal of Official Statistics, 10*, 153–165.

Rao, J. N. K. (2003). *Small area estimation*. New Jersey: Wiley.

Rao, J. N. K. (2006). Empirical likelihood for sample survey data: An overview. *Austrian Journal of Statistics, 35*, 191–196.

Rao, J. N. K. (2010). Bayesian pseudo empirical likelihood intervals for complex surveys. *Journal of the Royal Statistical Society, Series B, 72*, 533–544.

Rao, J. N. K., & Bellhouse, D. R. (1978). Estimation of finite population mean under generalized random permutation model. *Journal of Statistical Planning and Inference, 2*, 125–141.

Rao, J. N. K., & Choudhry, G. H. (1995). Small area estimation: Overview and empirical study. In B. G. Cox, D. A. Binder, B. N. Chinnappa, A. Christianson, M. J. Colledge, & P. S. Kott (Eds.), *Business survey method* (pp. 527–542). New York: Wiley.

Rao, J. N. K., & Graham, J. B. (1964). Rotation designs for sampling on repeated occasions. *Journal of the American Statistical Association, 59*, 492–509.

Rao, J. N. K., Hartley, H. O., & Cochran, W. G. (1962). On a simple procedure of unequal probability sampling without replacement. *Journal of the Royal Statistical Society, Series B, 24*, 482–491.

Rao, J. N. K., Kovar, J. G., & Mantel, H. J. (1990). On estimating distribution functions and quantiles from survey data using auxiliary information. *Biometrika, 77*, 365–375.

Rao, J. N. K., & Nigam, A. K. (1990). Optimum controlled sampling designs. *Biometrika, 77*, 807–814.

Rao, J. N. K., & Nigam, A. K. (1992). Optimal controlled sampling: A unified approach. *International Statistical Review, 60*, 89–98.

Rao, J. N. K., & Shao, J. (1992). Jackknife variance estimation with survey data under hot deck imputation. *Biometrika, 79*, 811–822.

Rao, J. N. K., & Scott, A. J. (1981). The analysis of categorical data from complex sample surveys: Chi-squared test for goodness of fit and independence in two-way tables. *Journal of the American Statistical Association, 76*, 221–230.

Rao, J. N. K., & Scott, A. J. (1984). On chi-squares tests for multi-way tables with cell proportions estimated from survey data. *Annals of Statistics, 15*, 385–397.

Rao, J. N. K., & Scott, A. J. (1987). On simple adjustments to chi-square tests with sample survey data. *Annals of Statistics, 15*, 385–397.

Rao, J. N. K., & Shao, J. (1996). On balanced half-sample variance estimation in stratified random sampling. *Journal of the American Statistical Association, 91*, 343–348.

Rao, J. N. K., & Shao, J. (1999). Modified balanced repeated replication for complex survey data. *Biometrika, 86*, 403–415.

Rao, J. N. K., & Thomas, D. R. (1988). The analysis of cross-classified categorical data from complex surveys. *Sociological Methodology, 18*, 213–269.

Rao, J. N. K., & Vijayan, K. (1977). On estimating the variance in sampling with probability proportional to aggregate size. *Journal of the American Statistical Association, 72*, 579–584.

Rao, J. N. K., & Vijayan, K. (2008). Application of experimental designs in survey sampling. *Journal of the Indian Society of Agricultural Statistics, 62*, 126–131.

Rao, J. N. K., & Webster, K. (1966). On two methods of bias reduction in estimation of ratios. *Biometrika, 53*, 571–577.

Rao, J. N. K., & Wu, C. (2009). Empirical likelihood methods. In C. R. Rao, & D. Pfeffermann (Eds.), *Handbook of statistics* (Vol. 29B, pp. 189–207). Oxford: Elsevier.

Rao, J. N. K., & Wu, C. F. G. (1988). Resampling inference with complex survey data. *Journal of the American Statistical Association, 72*, 579–584.

Rao, P. S. R. S. (1981). Estimation of the mean square error of the ratio estimator. In D. Krewski, R. Platek, & J. N. K. Rao (Eds.), *Current topics in survey sampling* (pp. 305–315). New York: Academic Press.

Rao, P. S. R. S., & Rao, J. N. K. (1971). Small sample results for ratio estimators. *Biometrika, 58*, 625–630.

Rao, T. J. (1966). On certain unbiased estimators. *Annals of the Institute of Statistical Mathematics, 18*, 117–121.

Rao, T. J. (1967a). On the choice of a strategy for a ratio method of estimation. *Journal of the Royal Statistical Society, Series B, 29*, 392–397.

Rao, T. J. (1967b). *Contribution to the theory of sampling strategies*. Calcutta: I.S.I (Ph.D. thesis).

Rao, T. J. (1968). On the allocation of sample size in stratified sampling. *Annals of the Institute of Statistical Mathematics, 20*, 159–166.

Rao, T. J. (1971). πps sampling designs and Horvitz-Thompson estimator. *Journal of the American Statistical Association, 66*, 872–875.

Rao, T. J. (1972). On the variance of ratio estimator for the Midzuno-Sen sampling scheme. *Metrika, 18*, 209–215.

Rao, T. J. (1977a). Estimating variance of the ratio estimator for the Midzuno-Sen sampling scheme. *Metrika, 24*, 203–208.

Rao, T. J. (1977b). Optimum allocation of sample size and prior distributions: A review. *International Statistical Review, 45*, 173–179.

Rao, T. J. (1983). Horvitz-Thompson strategy vs. stratified random sampling strategy. *Journal of Statistical Planning and Inference, 8*, 43–50.

Roberts, G., Rao, J. N. K., & Kumar, S. (1987). Logistic regression analysis of sample survey data. *Biometrika, 74*, 1–12.

Robson, D. S. (1957). Application of multivariate polykays to the theory of unbiased ratio type estimation. *Journal of the American Statistical Association, 52*, 511–522.

Robson, D. S., & Regier, H. A. (1964). Sample size in Peterson mark-recapture experiments. *Transactions of the American Fisheries Society, 93*, 215–226.

Royall, R. M. (1970). On finite population sampling theory under certain linear regression models. *Biometrika, 57*, 377–387.

Royall, R. M., & Cumberland, W. G. (1981). An empirical study of the ratio estimator and estimator of variance. *Journal of the American Statistical Association, 76*, 66–77.

Royall, R. M., & Herson, J. (1973). Robust estimation in finite population I. *Journal of the American Statistical Association, 68*, 880–889.

Rubin, D. B. (1976). Inference and missing data. *Biometrika, 63*, 581–592.

Rubin, D. B. (1987). *Multiple imputation for nonresponse in surveys*. New York: Wiley.

Rueda, M., Cobo, B., & Arcos, A. (2015). *Package 'RRTCS': Randomized response techniques for complex surveys*. Retrieved from http://cran.r-project.org/web/packages/RRTCS.

Rueda, M., Martinez, S., Martinez, H., & Arcos, A. (2007). Estimation of the distribution function with calibrated methods. *Journal of Statistical Planning and Inference, 137*, 435–448.

Ruppert, D., Wand, M. P., & Carroll, R. J. (2003). *Semiparametric regression*. New York: Cambridge University Press.

Saigo, H., Shao, J., & Sitter, R. (2001). A repeated half-sample bootstrap and balanced repeated replications for randomly imputed data. *Survey Methodology, 27*, 189–196.

Sampford, M. R. (1967). On sampling without replacement with unequal probability selection. *Biometrika, 67*, 639–650.

Särndal, C. E. (1982). Implications of survey design for generalized regression estimation of linear functions. *Journal of Statistical Planning and Inference, 7*, 155–170.
Särndal, C. E., & Hidiroglou, M. A. (1989). Small domain estimation: A conditional analysis. *Journal of the American Statistical Association, 84*, 266–275.
Särndal, C. E., Swensson, B., & Wretman, J. (1992). *Model assisted survey sampling*. New York: Springer-Verlag.
Saxena, B. C., Narian, P., & Srivastava, A. K. (1984). Multiple frame surveys in two stage sampling. *Sankhyā, B*, 75–82.
Schenker, N., & Welsh, A. (1988). Asymptotic results for multiple imputation. *Annals of Statistics, 16*, 1550–1566.
Schnabel, Z. E. (1938). The estimation of total fish population of a lake. *American Mathematical Monthly, 45*, 348–352.
Schucany, W. R., Gray, H. L., & Owen, D. B. (1971). On bias reduction in estimation. *Journal of the American Statistical Association, 66*, 524–533.
Scott, A. J., & Holt, D. (1982). The effect of two-stage sampling on ordinary least squares. *Journal of the American Statistical Association, 77*, 848–854.
Scott, A. J., & Rao, J. N. K. (1981). Chi-squared tests for contingency tables with proportions estimated from survey data. In D. Krewski, & J. N. K. Rao (Eds.), *Current topics in survey sampling*. New York: Academic Press.
Searls, D. T. (1964). The utilization of a known coefficient of variation in estimation procedure. *Journal of the American Statistical Association, 21*, 20–21.
Seber, G. A. F. (1965). A note on multiple-recapture census. *Biometrka, 52*, 249–269.
Seber, G. A. F. (1970). The effects of trap response on tags-recapture estimates. *Biometrics, 26*, 13–22.
Seber, G. A. F. (1973). *The estimation of animal abundance and related parameters*. London: Griffin.
Seber, G. A. F. (1982). *The estimation of animal abundance and related parameters*. New York: Macmillan.
Seber, G. A. F. (1986). A review of estimating animal abundance. *Biometrics, 42*, 267–292.
Sen, A. R. (1953). On estimate of variance in sampling with varying probabilities. *Journal of the Indian Society of Agricultural Statistics, 5*, 119–127.
Sengupta, S. (1980). On admissibility of the generalized Des Raj estimator for PPSWR-sampling of size two. *Calcutta Statistical Association Bulletin, 29*, 35–40.
Seth, G. R. (1966). On estimates of variance of estimates of population total in varying probabilities. *Journal of the Indian Society of Agricultural Statistics, 5*, 119–127.
Sethi, V. K. (1965). On optimum pairing of units. *Sankhyā, B, 27*, 315–320.
Shah, B. V., Holt, M. M., & Folsom, R. E. (1977). Inference about regression models from sample survey data. *Bulletin of the International Statistical Institute, 47*(3), 43–57.
Shao, J., Chen, Y., & Chen, Y. (1998). Balanced repeated replication for multistage survey data under imputation. *Journal of the American Statistical Association, 93*, 819–831.
Shao, J., & Sitter, R. R. (1996). Bootstrap for imputed survey data. *Journal of the American Statistical Association, 91*, 755–765.
Shen, W. (1994). Use of ranked-set sampling for test of a normal mean. *Calcutta Statistical Association Bulletin, 44*, 183–193.
Shukla, N. D. (1976). Almost unbiased product-type estimator. *Metrika, 23*, 127–133.
Silva, P. L. D., & Skinner, C. J. (1995). Estimating distribution function with auxiliary information using poststratification. *Journal of Official Statistics, 11*, 277–294.
Singh, A. C., Mantel, J. H., & Thomas, B. W. (1994). Time series EBLUPs for small areas using survey data. *Survey Methodology, 20*, 33–43.
Singh, A. C., & Mohl, C. A. (1996). Understanding calibration estimators in survey sampling. *Survey Methodology, 22*, 107–115.
Singh, A. C., Stukel, D. M., & Pfeffermann, D. (1998). Bayesian versus frequentist measures of error in small area estimation. *Journal of the Royal Statistical Society, Series B, 60*, 377–396.

Singh, A. C., & Wu, S. (1996). Estimation for multiframe complex surveys by modified regression. In *Proceedings of the survey method section of the American Statistical Association* (pp. 69–77).

Singh, D. (1968). Estimates in successive sampling using a multi-stage design. *Journal of the American Statistical Association, 63*, 99–112.

Singh, D., Jindal, K. K., & Grag, J. N. (1968). On modified systematic sampling. *Bometrika, 55*, 541–546.

Singh, D., & Kathuria, O. P. (1969). On two-stage successive sampling. *Australian Journal of Statistics, 11*, 59–66.

Singh, J. (1976). A note on RR techniques. In *Proceedings of the survey research methods section of the American Statistical Association* (p. 772).

Singh, J. (1978). A note on maximum likelihood estimation from randomized response models. In *Proceedings of the survey research methods section of the American Statistical Association* (pp. 282–283).

Singh, P., & Srivastava, A. K. (1980). Sampling schemes providing unbiased regression estimators. *Biometrika, 67*, 205–209.

Singh, R. (1972). On Pathak and Rao's estimates in pps with replacement over two occasions. *Sankhyā, A, 34*, 301–303.

Singh, S. (2003). *Advanced sampling theory with applications*. Netherlands: Kluwer Academic Publishers.

Singh, S., & Joarder, A. H. (1997). Optional randomized response technique for sensitive quantitative variable. *Metron, 15*, 151–157.

Singh, S., & Singh, R. (1979). On random non-response in unequal probability sampling. *Sankhyā, C, 41*, 127–137.

Sinha, B. K. (1973). On sampling schemes to realise pre-assigned sets of inclusion probabilities of first two orders. *Calcutta Statistical Association Bulletin, 22*, 69–110.

Sinha, B. K., Sinha, B. K., & Purkayastha, S. (1996). On some aspects of ranked set sampling for estimation of normal and exponential parameters. *Statistics and Decisions, 14*, 223–240.

Sirken, M. G. (1970). Household surveys with multiplicity. *Journal of the American Statistical Association, 65*, 257–266.

Sirken, M. G. (1972). Variance components of multiplicity estimators. *Biometrics, 28*, 869–873.

Sirken, M. G., Grabard, B. I., & Mcdaniel, M. J. (1978). National network surveys of diabetes. In *Proceedings of the survey research methods section of the American Statistical Association* (pp. 631–635).

Sirken, M. G., & Levy, P. S. (1974). Multiplicity estimation of proportions based on ratios of random variables. *Journal of the American Statistical Association, 69*, 68–73.

Sitter, R. R. (1992a). A resampling procedure for complex survey data. *Journal of the American Statistical Association, 87*, 755–765.

Sitter, R. R. (1992b). Comparing three bootstrap methods for survey data. *The Canadian Journal of Statistics, 20*, 135–184.

Sitter, R. R., & Wu, C. (2002). Efficient estimation of quadratic finite population functions in the presence of auxiliary information. *Journal of the American Statistical Association, 97*, 535–543.

Skinner, C. J. (1991). On the efficiency of raking ratio estimation for multiple frame surveys. *Journal of the American Statistical Association, 86*, 779–784.

Skinner, C. J., & Rao, J. N. K. (1996). Estimation in dual frame surveys with complex designs. *Journal of the American Statistical Association, 91*, 349–356.

Smith, H. F. (1938). An empirical law describing heterogeneity in the yield of agricultural crops. *Journal of Agricultural Science, 28*, 1–23.

Smith, S. K., & Lewis, B. B. (1980). Some new techniques for applying the housing unit method of local population estimations. *Demography, 17*, 323–340.

Snow, R. E., Hutcheson, J. D., & Prather, J. E. (1981). Using reputational sampling to identify residential clusters of minorities in a large urban region: Hispanics in Atlanta, Georgia. In *Proceedings of the survey research methods section of the American Statistical Association* (pp. 101–106).

Solomon, H., & Stephens, M. A. (1977). Distribution of a sum of weighted chi-square variables. *Journal of the American Statistical Association, 72*, 881–885.

Srinath, K. P. (1971). Multiphase sampling in non-response problems. *Journal of the American Statistical Association, 66*, 583–586.

Srinath, K. P., & Hidiroglou, M. A. (1980). Estimation of variance in multi-stage sampling. *Metrika, 27*, 121–125.

Srivastava, S. K. (1967). An estimator using auxiliary information in sample surveys. *Calcutta Statistical Association Bulletin, 16*, 121–132.

Srivenkataramana, T. (1980). A dual to ratio estimator in sample surveys. *Biometrika, 67*, 199–204.

Stephan, F. F. (1945). The expected value and variance of the reciprocal and other negative powers of a positive Bernoulli variate. *Annals of Mathematical Statistics, 16*, 50–61.

Stokes, S. L. (1977). Ranked set sampling with concomitant variables. *Communications in Statistics – Theory and Methods, 12*, 1207–1211.

Stokes, S. L. (1980a). Estimation of variance using judgment ordered ranked-set samples. *Biometrics, 36*, 35–42.

Stokes, S. L. (1980b). Inference on correlation coefficient in bivariate normal populations from ranked-set sampling. *Journal of the American Statistical Association, 75*, 989–995.

Stokes, S. L. (1988). Characterization of a ranked-set sample with application to estimating distribution functions. *Journal of the American Statistical Association, 83*, 374–381.

Stokes, S. L., & Sager, T. W. (1988). Characterization of ranked-set sample with application to estimating distribution function. *Journal of the American Statistical Association, 83*, 374–381.

Stufken, J. (1993). Combinatorial and statistical aspects of sampling plans to avoid the selection of adjustment units. *Journal of Combinatorics, Information and System Sciences, 18*, 81–92.

Stufken, J., Song, S. Y., See, K., & Driessel, K. R. (1999). Polygonal design: Some existence and non-existence results. *Journal of Statistical Planning and Inference, 77*, 155–166.

Stukel, D. M., & Rao, J. N. K. (1999). Small-area estimation under two-fold nested error regression models. *Journal of Statistical Planning and Inference, 78*, 131–147.

Sudakar, K. (1978). A note on circular systematic sampling design. *Sankhyā, C, 40*, 72.

Sudman, S. (1972). On sampling very rare human populations. *Journal of the American Statistical Association, 67*, 335–339.

Sudman, S. (1976). *Applied sampling*. New York: Academic Press.

Sudman, S. (1978). Optimum cluster designs within a primary unit using combined telephone screening and face-to-face interviewing. *Journal of the American Statistical Association, 73*, 300–304.

Sudman, S. (1985). Efficient screening methods for the sampling of geographically clustered special populations. *Journal of Marketing Research, 22*, 20–29.

Sudman, S., & Kalton, G. (1986). New development in the sampling of special populations. *Annual Review of Sociology, 12*, 401–429.

Sukhatme, P. V. (1944). Moments and product moments of moment statistics for samples of finite and infinite populations. *Sankhyā, 6*, 363–382.

Sukhatme, P. V., & Sukhatme, B. V. (1970). *Sampling theory and applications*. Ames: Iowa State University Press.

Sukhatme, P. V., Sukhatme, B. V., Sukhatme, S., & Asok, C. (1984). *Sampling theory of surveys with applications*. New Delhi: Iowa State University Press and Indian Society of Agricultural Statistics.

Takahasi, K., & Wakimoto, K. (1968). On unbiased estimates of the population mean based on the sample stratified by means of ordering. *Annals of the Institute of Statistical Mathematics, 30*, 814–824.

Takeuchi, K., Yanai, H., & Mukherjee, B. N. (1983). *The foundations of multivariate analysis* (1st ed.). New Delhi: Wiley Eastern Ltd.

Tam, S. M. (1984). Optimal estimation in survey sampling under a regression superpopulation model. *Biometrika, 71*, 645–647.

Thomas, D. R. (1989). Simultaneous confidence intervals for proportions under cluster sampling. *Survey Methodology, 15*, 557–559.

Thomas, D. R., & Rao, J. N. K. (1987). Small sample comparison level and power for simple goodness-of-fit statistics under cluster sampling. *Journal of the American Statistical Association, 82*, 630–636.

Thompson, S. K. (1990). Adaptive cluster sampling. *Journal of the American Statistical Association, 85*, 1050–1059.

Thompson, S. K., & Seber, G. A. F. (1996). *Adaptive sampling*. New York: Wiley.

Tikkiwal, B. D. (1951). *Theory of successive sampling*. New Delhi: ICAR (Unpublished thesis for diploma).

Tin, M. (1965). Comparison of some ratio estimators. *Journal of the American Statistical Association, 60*, 294–307.

Tiwari, N., Nigam, A. K., & Pant, I. (2007). On an optimum controlled nearest proportional to size sampling scheme. *Survey Methodology, 33*, 87–94.

Tracy, D. S., & Osahan, S. S. (1994). Estimating in overlapping clusters with unknown population size. *Survey Methodology, 20*, 53–57.

Tripathi, T. P., & Srivastava, O. P. (1979). Estimation on successive occasions using PPSWR sampling. *Sankhyā, C, 41*, 84–91.

Tuckey, J. W. (1958). Bias and confidence in not quite large samples. *Annals of Statistics, 29*, 614.

Vallerory, L. A., Mackellar, D., & Karon, J. (2000). HIV prevalence and associated risks in young men who have sex with men. *Journal of the American Statistical Association, 284*, 198–204.

Van der Heijden, P. G. M., Van Gils, G., Bouts, J., & Hox, J. J. (1998). A comparison of randomized response, CASAQ, and direct questioning; eliciting sensitive information in the context of social security fraud. *Kwantitatieve Methoden, 59*, 15–34.

Vijayan, K. (1975). On estimating the variance in unequal probability sampling. *Journal of the American Statistical Association, 70*, 713–716.

Vijayan, K. (1991). Estimating function in Survey Sampling: Estimation of superpopulation regression parameters. In V. P. Godambe (Ed.), *Estimating functions* (pp. 223–238). Clarendon Press: Oxford Univ.

Wakesberg, J. (1978). Sampling method for random digit dialing. *Journal of the American Statistical Association, 73*, 40–46.

Wang, Q., & Rao, J. N. K. (2002). Empirical likelihood-based inference in linear models with missing data. *Scandinavian Journal of Statistics, 29*, 563–576.

Wang, R., Sedransk, J., & Jinn, J. H. (1992). Data analysis when there are missing observations. *Journal of the American Statistical Association, 87*, 952–961.

Wang, S., & Dorfman, A. H. (1996). A new estimator for the finite population distribution function. *Biometrika, 83*, 639–652.

Warner, S. L. (1965). Randomize response: A survey technique for eliminating evasive answer bias. *Journal of the American Statistical Association, 60*, 63–69.

Waterton, J. J. (1983). A exercise in controlled selections. *Applied Statistics, 32*, 150–164.

Watson, G. S. (1964). Smooth regression analysis. *Sankhyā, A*, 359–372.

Welch, S. (1975). Sampling by referral in a dispersed population. *Public Opinion Quarterly, 39*, 237–245.

Wolter, K. M. (1984). An investigation of some estimators of variance for systematic sampling. *Journal of the American Statistical Association, 79*, 781–790.

Wolter, K. M. (1985). *Introduction to variance estimation*. New York: Springer-Verlag.

Woodruff, R. S. (1952). Confidence interval for medians and other position measures. *Journal of the American Statistical Association, 47*, 635–636.

Woodruff, R. S. (1959). The use of rotation samples in Census Bureau's monthly surveys. In *Proceedings of the survey research methods section of the American Statistical Association* (pp. 130–138).

Woodruff, R. S. (1971). A simple method of approximation of variance of a complicated estimate. *Journal of the American Statistical Association, 79*, 781–790.

Wu, C. (1999). *The effective use of complete auxiliary information from survey data*. Canada: Simon Fraser University (Unpublished doctoral dissertation).

Wu, C. (2004). Combine information from multiple surveys through empirical likelihood. *The Canadian Journal of Statistics, 34*, 15–26.

Wu, C. (2005). Algorithms and R codes for the pseudo empirical likelihood method in survey sampling. *Survey Methodology, 31*, 239–243.

Wu, C., & Rao, J. N. K. (2006). Pseudo empirical likelihood ratio confidence intervals for complex surveys. *The Canadian Journal of Statistics, 34*, 359–375.

Wu, C., & Rao, J. N. K. (2009). *Bootstrap procedure for pseudo empirical likelihood method in sample surveys*. Working paper series, 2009–2. Department of Statistics and Actuarial Science, University of Waterloo.

Wu, C., & Sitter, R. R. (2001). A model-calibration approach to use complex auxiliary information from survey data. *Journal of the American Statistical Association, 96*, 185–193.

Wu, C. F. J. (1982). Estimation of variance of the ratio estimator. *Biometrika, 69*, 183–189.

Wynn, H. P. (1977). Convex sets of finite population plans. *Annals of Statistics, 5*, 414–418.

Yates, F. (1948). Systematic sampling. *Philosophical Transactions of the Royal Statistical Society, A, 214*, 345–377.

Yates, F. (1949). *Sampling methods for censuses and surveys*. London: Charles Griffins.

Yates, F., & Grundy, P. M. (1953). Selection without replacement from within strata with probability proportional to size. *Journal of the Royal Statistical Society, Series B, 15*, 235–261.

Yu, P. L. H., & Lam, K. (1997). Regression estimator in ranked set sampling. *Biometrics, 53*, 1070–1080.

Zhong, B., & Rao, J. N. K. (2000). Empirical likelihood under stratified random sampling using auxiliary population information. *Biometrika, 87*, 929–938.

AUTHOR INDEX

T

U

V

W

Y

Z

SUBJECT INDEX

'*Note*: Page numbers followed by "f" indicate figures, "t" indicate tables.'

N

O

P

Q

R

S

FSC
www.fsc.org
MIX
Papier | Fördert
gute Waldnutzung
FSC® C083411

Kolibri 360 GmbH
Ferdinand-Jühlke-Straße 7
99095 Erfurt, Deutschland
produktsicherheit@kolibri360.de